QUINAZOLINES

Supplement I

This is the fifty-fifth volume in the series

THE CHEMISTRY OF HETEROCYCLIC COMPOUNDS

THE CHEMISTRY OF HETEROCYCLIC COMPOUNDS

A SERIES OF MONOGRAPHS

EDWARD C. TAYLOR, *Editor*

ARNOLD WEISSBERGER, *Founding Editor*

QUINAZOLINES
Supplement I

D. J. Brown

Research School of Chemistry
Australian National University
Canberra

AN INTERSCIENCE® PUBLICATION

JOHN WILEY & SONS, INC.

NEW YORK · CHICHESTER · BRISBANE · TORONTO · SINGAPORE · WEINHEIM

Sep/Ae
chem

This text is printed on acid-free paper

Library of Congress Cataloging in Publication Data:

Brown, D. J.
Quinazolines. Supplement I / D. J. Brown.
p. cm.—(This is the fifty-fifth volume in the series the Chemistry of heterocyclic compounds)
"An Interscience publication."
Includes bibliographical references and index.
ISBN 0-471-14565-3 (cloth : alk. paper)
1. Quinazoline. I. Title. II. Series: Chemistry of heterocyclic compounds; v. 55.
QD401.F96 pt. 1 Suppl.
547′.593—dc20 96-6182

Printed in the United States of America

10 9 8 7 6 5 4 3 2 1

Dedicated to
the Memory of

Arnold Weissberger
1899–1984

Zichrono livracha

The Chemistry of Heterocyclic Compounds
Introduction to the Series

The chemistry of heterocyclic compounds is one of the most complex and intriguing branches of organic chemistry, of equal interest for its theoretical implications, for the diversity of its synthetic procedures, and for the physiological and industrial significance of heterocycles.

The Chemistry of Heterocyclic Compounds, published since 1950 under the initial editorship of Arnold Weissberger, and later, until Dr. Weissberger's death in 1984, under our joint editorship, has attempted to make the extraordinarily complex and diverse field of heterocyclic chemistry as organized and readily accessible as possible. Each volume has traditionally dealt with syntheses, reactions, properties, structure, physical chemistry, and utility of compounds belonging to a specific ring system or class (e.g., pyridines, thiophenes, pyrimidines, three-membered ring systems). This series has become the basic reference collection for information on heterocyclic compounds.

Many broader aspects of heterocyclic chemistry are recognized as disciplines of general significance that impinge on almost all aspects of modern organic chemistry, medicinal chemistry, and biochemistry, and for this reason we initiated several years ago a parallel series entitled General Heterocyclic Chemistry, which treated such topics as nuclear magnetic resonance, mass spectra, and photochemistry of heterocyclic compounds, the utility of heterocycles in organic synthesis, and the synthesis of heterocycles by means of 1,3-dipolar cycloaddition reactions. These volumes were intended to be of interest to all organic, medicinal, and biochemically-oriented chemists, as well as to those whose particular concern is heterocyclic chemistry. It has, however, become increasingly clear that the above distinction between the two series was unnecessary and somewhat confusing, and we have therefore elected to discontinue *General Heterocyclic Chemistry* and to publish all forthcoming volumes in this general area in *The Chemistry of Heterocyclic Compounds* series.

This series, together with the international community of chemists concerned with heterocyclic chemistry, is indebted once again to the indefatigable efforts of Dr. D. J. Brown. His record of authoring classics in nitrogen heterocyclic chemistry (*The Pyrimidines*, *The Pyrimidines Supplement I*, *The Pyrimidines Supplement II*, *Pteridines*) is now extended by the publication of an exhaustive supplement to Wilf Armarego's initial volume on *Quinazolines*. We extend once again our congratulations and our thanks to Dr. Brown for a further outstanding contribution to the literature of heterocyclic chemistry.

Department of Chemistry
Princeton University
Princeton, New Jersey

EDWARD C. TAYLOR

Preface

Dr. Wilf Armarego's original volume, *Quinazolines*, appeared within this series in 1967. Not only did it represent an excellent summary of quinazoline chemistry to the end of 1965, but it clearly facilitated and stimulated considerable subsequent research in the field. Thus the need for a supplementary volume covering the last 30 years' literature has become pressing. On account of a radical change in his research interests during that period, Dr. Armarego felt disinclined to undertake such an updating task, which consequently fell to the present author, whose interests have remained broadly within this area.

Because of a great expansion in the scope of quinazoline chemistry and for other pragmatic reasons, it has been necessary to inaugurate a massive chapter on primary syntheses and to reorganize completely the content of remaining chapters so that they might better reflect and emphasize current research trends. However, the status of the present volume as a *supplement* has been maintained by sectional cross-references (e.g., *H* 42) to pages of the original volume (*Hauptwerk*) where earlier relevant information may be found. Moreover, in view of the vast increase in the number and types of individual quinazolines described in recent literature, it has been necessary to abandon the myriad classified tables of known quinazolines in favor of a single alphabetical table of simple quinazolines. To facilitate recovery of any earlier data from tables in the original volume, a cross-reference (e.g., *H* 151) has been added (when appropriate) to each individual entry in the new table. The opportunity has been taken to bring the chemical nomenclature into line with current IUPAC recommendations [*Nomenclature of Organic Chemistry, Sections A–F, H* (Eds. J. Rigaudy and S. P. Klesney, Pergamon Press, Oxford, 1970)] with one important exception: in order to keep "quinazoline" as the principal part of each name, those groups that would normally qualify as principal suffixes, but that are not attached directly to the nucleus, are rendered as prefixes. For example, 2-carboxymethyl-4(3*H*)-quinazolinone is used instead of α-(4-oxo-3,4-dihydro-2-quinazolinyl)acetic acid. Secondary or tertiary amino groups are rendered invariably as prefixes. Trivial names, still occasionally used for some naturally occurring oxyquinazolines, are included as appropriate in the table of simple quinazolines and/or in Section 4.8.4. Finally, to avoid repetition and inevitable confusion, literature references are presented as a single list rather than as smaller lists at the end of each chapter.

In preparing this *supplement*, the massive patent literature has been ignored in the belief that useful factual material therein has appeared subsequently in the regular literature. It must be mentioned also that a small but significant proportion of the research papers quoted as references have proved very disappointing in terms of essential detail, thereby reflecting badly on their authors and on the editorial policies of the journals in which they appeared. The popularity of quinazoline research in India and Egypt is both noteworthy and rather puzzling.

Although the original papers on quinazoline chemistry during the last 30 years came from no less than 55 countries, they appeared in only 16 languages (ignoring any subsequent translations): from the following percentages, it is evident that the laudable trend toward publication of research results in a widely understood major language has continued.

English	75.9%
German	11.0%
Russian	5.9%
French	2.3%
Japanese	1.4%
Italian	1.1%
Romanian	0.6%
Polish	0.4%
Ukrainian	0.4%
Spanish	0.2%
Bulgarian	0.2%
Chinese	0.2%
Hungarian	0.1%
Czech	0.1%
Korean	0.1%
Portugese	<0.1%
Slovenian	<0.1%

I am greatly indebted to my former colleagues, Drs W. L. F. Armarego and G. B. Barlin, for invaluable discussions; to successive Deans of the Research School of Chemistry (Professors A. L. G. Beckwith, L. N. Mander, and J. W. White) for the provision of excellent postretirement accommodation and facilities within the School; to Dr. Adam Vincze (Israel Institute for Biological Research) for kind advice; to the branch librarian, Mrs J. Smith, for unfailing cooperation; and to my wife, Jan, for her cheerful forebearance and unstinting assistance during indexing, proofreading, and the like.

Research School of Chemistry   DES J. BROWN
Australian National University
Canberra

Contents

CHAPTER 1

Primary Syntheses

The primary synthesis of quinazolines may be accomplished by cyclization of benzene substrates already bearing appropriate substituents; by treatment of benzene substrates with synthons to provide one or more of the ring atoms required to complete the pyrimidine ring; by analogous processing of preformed pyrimidine substrates; by elaboration from several acyclic synthons; or by rearrangement, ring expansion/contraction, degradation, or modification of appropriate derivatives of other heterocyclic systems. Partially or even fully reduced quinazolines may often be made by rather similar procedures; such cases are usually illustrated toward the end of each subsection. Examples of any pre-1966 syntheses in each category may be found from the cross-references (e.g., *H* 48) to Armarego's parent volume;[2414] some post-1966 material has been reviewed elsewhere in brief.[409,2382,2383]

1.1. FROM A SINGLE BENZENE SUBSTRATE

A remarkable number of quinazoline syntheses have been carried out by pre-forming appropriate benzene derivatives for cyclization to required quinazolines by the formation of one remaining bond on the pyrimidine side of the molecule.

1.1.1. By Formation of the 1,2-Bond (*H* 11,48,394)

Such a process is typified by the facile conversion of 3-hydroxy-6- nitrobenzaldehyde into 3-(diformamido)methyl-4-nitrophenol (1) with subsequent reductive cyclization (Zn/AcOH) to 6-quinazolinol (2) (57%),[2359] representing a classic example of Reidel's synthesis (*H* 48). Rather similar reductive procedures gave 5,6,8-trimethoxy-7-methylquinazoline (48%),[2363] 5,8-dimethoxyquinazoline

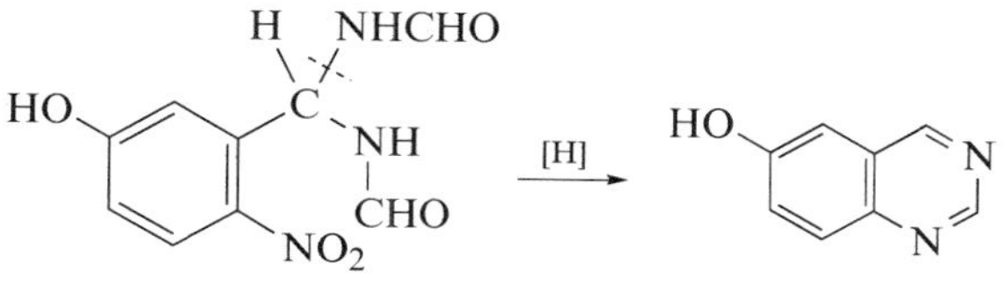

(1)   (2)

1

(41% or 93%),[585,1604] and 5,8-dibutoxyquinazoline (79%).[595] Other reductive processes and several completely different ways to achieve cyclization by 1,2-bond formation are illustrated in the following examples.

Reductive cyclizations

N-Acetyl-2-nitrobenzamide (**3**, R = H) gave 2-methyl-4(3*H*)-quinazolinone 1-oxide or 2-methyl-4(3*H*)-quinazolinone (electrolytic: according to conditions).[2399]

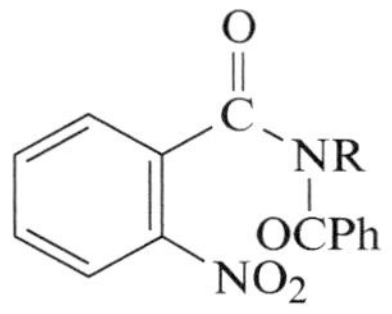

(3)

N-Acetyl-*N*-methyl-2-nitrobenzamide (**3**, R = Me) gave 2,3-dimethyl-4(3*H*)-quinazolinone [CO, $Ru_2(CO)_{12}$, dioxane, 10 atm, 140°, 16 h: 93%].[2406]

N-Ethoxycarbonyl-*N*-methyl-2-nitrobenzamide gave 1-hydroxy-3-methyl-2,4(1*H*,3*H*)-quinazolinedione ($NaBH_4$, NaOH, H_2O, 20°, 90 min: 24%).[425]

5-*o*-Aminophenyl-3-phenyl-1,2,4-oxadiazole (**4**) gave 2-phenyl-4(3*H*)-quinazolinone (H_2, Pd/C, EtOH, 1 atm, 20°: 72%).[991]

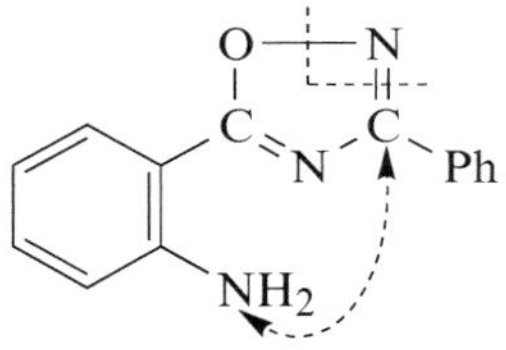

(4)

Also other examples.[1766,2067]

Thermal cyclizations

2-*p*-Fluorobenzylamino-*N*-(α-methylamino-α-methylthiomethylene)benzamide (**5**) gave 1-*p*-fluorobenzyl-2-methylamino-4(3*H*)-quinaolinone [(MeOCH₂CH₂)₂O, trace NaOH, reflux, 2 h: 74%].[906]

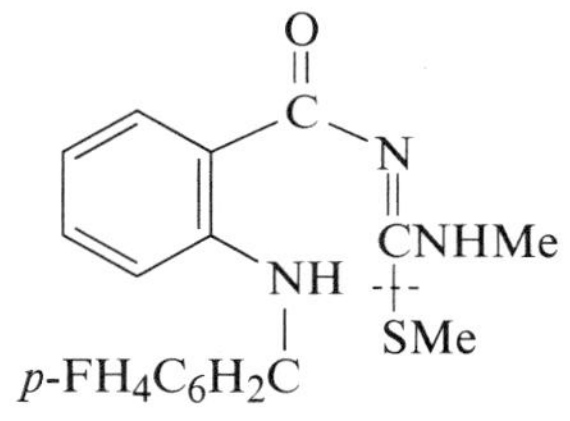

(5)

o-(Semicarbazidocarbonyl)aniline (**6**) gave 3-amino-2,4(1*H*,3*H*)-quinazo-linedione (**7**) (decalin, reflux, 2 h: 50%; involving a rearrangement step).[485]

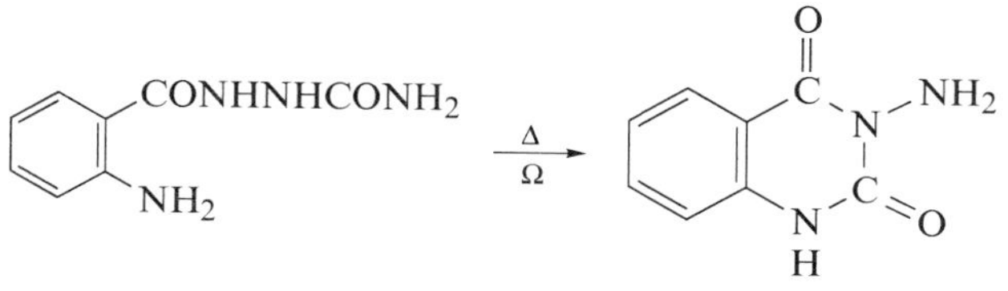

(6)   **(7)**

Also other examples.[2554]

Cyclizations in acid

o-Amino-*N*-benzoylbenzamidine (**8**) gave 2-phenyl-4-quinazolinamine (<0.1M HCl, 20°, 1 min: 95%).[1232]

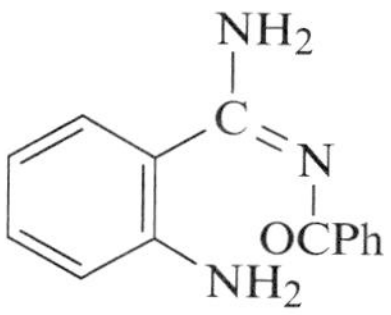

(8)

o-(Methoxycarbonylmethyl)amino-*N*-methyl-*N*-(phenacylcarbonyl)benzam-ide (**9**) gave 1-methoxycarbonylmethyl-3-methyl-2,4(1*H*,3*H*)-quinazo-linedione (10% AcOH/EtOH, reflux, 2 h: 42%).[2270]

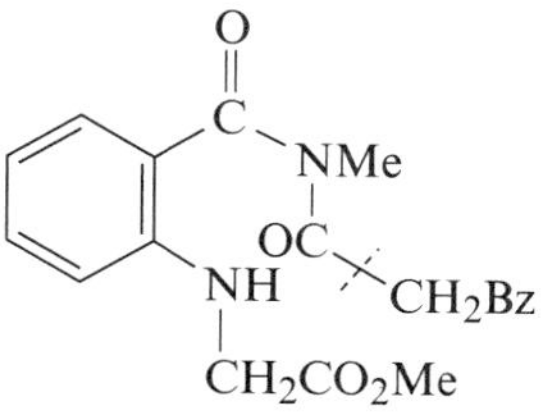

(9)

N'-Acetyl-*o*-amino-*N'*-methylbenzohydrazide (**10**) gave 2-methy-3-methyl-amino-4(3*H*)-quinazolinone (**11**) (10% H_2SO_4, 80°, 30 min: 78%; involves rearrangement).[921]

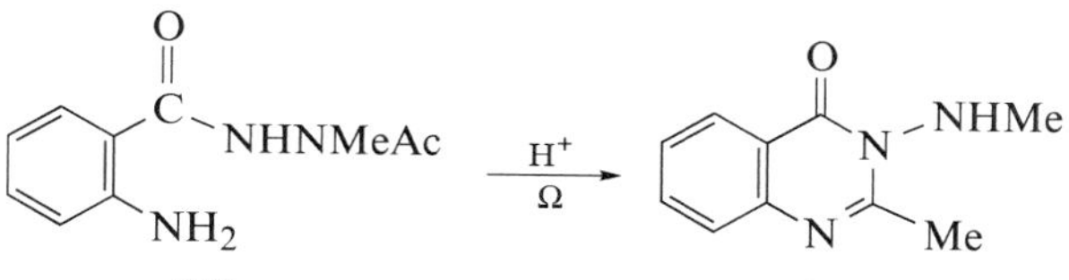

(10)   **(11)**

Also other examples.[773,2400]

Cyclizations in base

N-Cyanomethyl-*o*-nitrobenzamide   (**12**, R = H)   gave   2-ethoxy-4(3*H*)-quinazolinone 1-oxide (**13**, R = H)(EtONa, EtOH, reflux, 1 h: $\sim$ 35%; involves CN replacement) and some 1-hydroxy-2,4(1*H*, 3*H*)-quinazolinedione (**14**, R = H) (from the aqueous mother liquors).[182]

In contrast, *N*-cyanomethyl-*N*-methyl-*o*-nitrobenzamide (**12**, R = Me) gave only 1-hydroxy-3-methyl-2,4(1*H*, 3*H*)-quinazolinedione (**14**, R = Me) (EtONa, EtOH, reflux, 1 h, aqueous workup: 93%; or NaOH/H_2O, reflux, 30 min: $\sim$ 90%).[187]

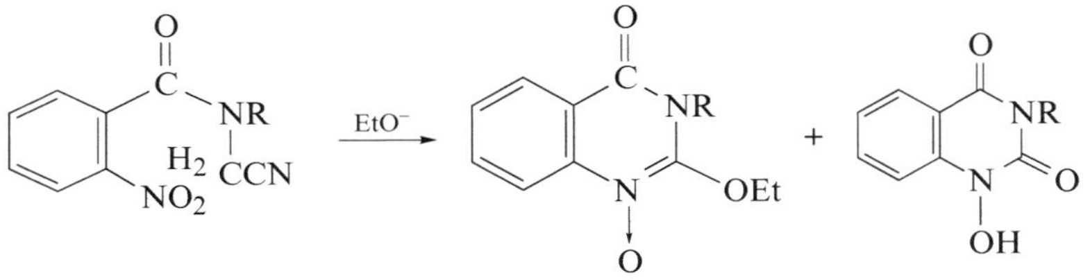

(12)   (13)   (14)

Wittig-assisted cyclizations

N-Acetyl-2-azido-*N*-methylbenzamide (**15**, R = Me) gave 2,3-dimethyl-4(3*H*)-quinazolinone (Ph$_3$P, xylene, 20°, 2 h: > 95%); likewise, 2-azido-*N*-cinnamoyl-*N*-methylbenzamide (**15**, R = CH:CHPh) gave 3-methyl-2-styryl-4(3*H*)-quinazolinone (> 95%).[2079,2131]

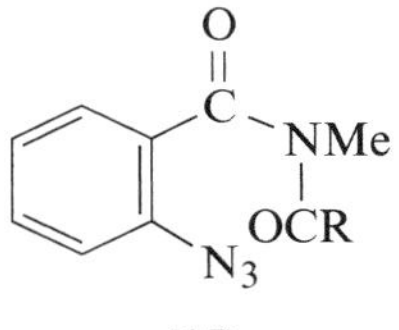

(15)

Cyclizations to hydroquinazolines

2-(Dithiocarboxy)aminomethyl-4-nitroaniline (**16**) gave 6-nitro-3,4-dihydro-2(1*H*)-quinazolinethione (**17**) (0.1M NaOH, 85°, 3 h: $\sim$ 50%).[374]

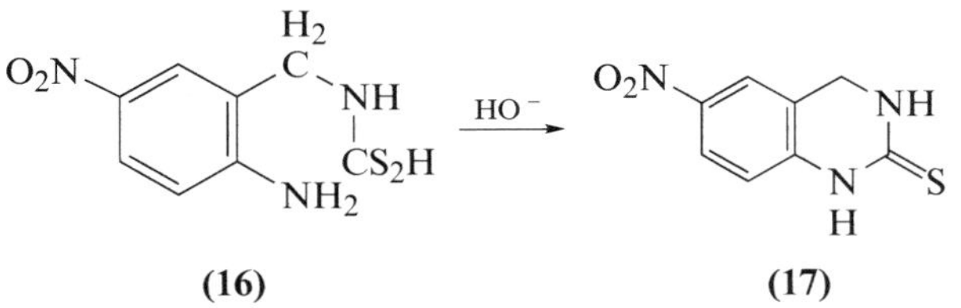

(16)   (17)

Also other examples.[430,1811,1890]

1.1.2. By Formation of the 1,8a-Bond

This unappealing route has been used to make a few aromatic and reduced quinazolines. Thus treatment of N,N'-dimethyl-N-pentafluorobenzoylurea (18) with potassium fluoride in refluxing dimethylformamide for 7 h gave 5,6,7,8-tetrafluoro-1,3-dimethyl-2,4($1H$,$3H$)-quinazolinedione (19, R = F) (49%) but, when sodium hydride was used in place of potassium fluoride, the only product was the 7-dimethylamino derivative (19, R = NMe$_2$) (66%);[1317] prolonged irradiation of 2-benzyl-3-phenyl-1,2,4-oxadiazol-5($2H$)-one (20) gave a little 2-phenylquinazoline;[225] 2-ureidomethylenecyclohexanone (21) in boiling dilute alkali gave 5,6,7,8-tetrahydro-2($1H$)-quinazolinone (22) (92%);[190] and other examples have been described.[423,1692,2046]

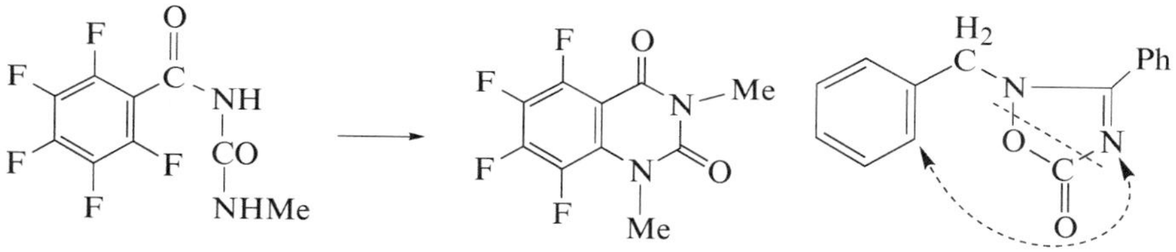

(18)   (19)   (20)

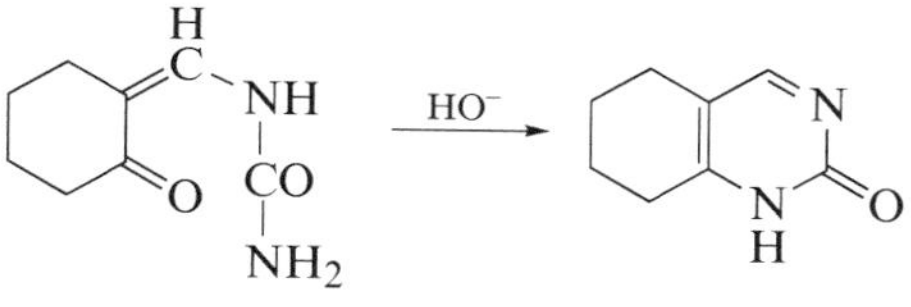

(21)   (22)

1.1.3. By Formation of the 2,3-Bond

This route has been used widely to prepare quinazolines (and a few hydroquinazolines) from a variety of substrate types, as described in the following subsections.

1.1.3.1. From o-Acylaminobenzamides (H 77,101)

The cyclization of o-acylaminobenzamides, usually to 4-quinazolinones, has been done in several ways as indicated in the following examples.

Thermal cyclizations

3-Bromo-6-(N-methyl-o-nitrobenzamido)benzamide (23) gave 6-bromo-1-methyl-2-o-nitrophenyl-4($1H$)-quinazolinone (24) (neat, 255°, 30 min: 59%).[45]

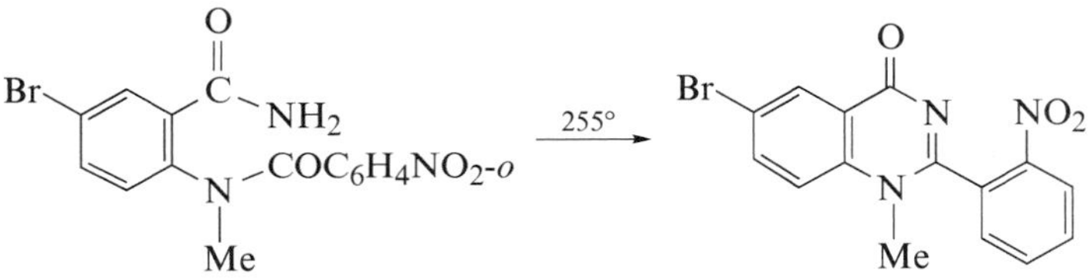

(23)   **(24)**

N-(But-3-enyl)-*o*-(trifluoroacetamido)benzamide (**25**) gave 3-but-3′-enyl-2-trifluoromethyl-4(3*H*)-quinazolinone (neat, 200°, 1 h: 85%).[1727]

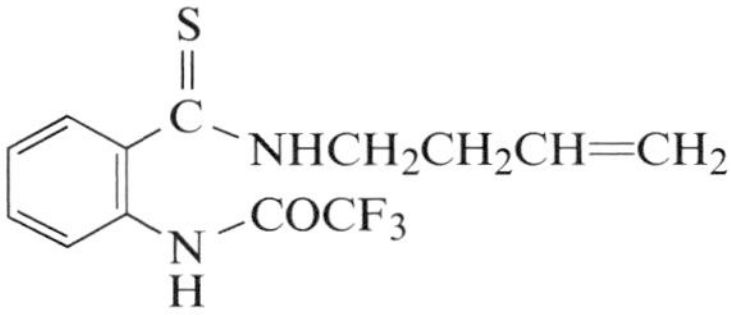

(25)

o-(Pentafluoropropionamido)benzamide gave 2-pentafluoroethyl-4(3*H*)-quinazolinone (Me$_2$NCHO, reflux, 3 h: 76%).[914]

o-Benzamido(thiobenzamide) (**26**) gave 4-ethylthio-2-phenylquinazoline (Et$_3$OBF$_4$, CH$_2$Cl$_2$, reflux, 1 h: 82%; note the *S*-ethylation).[1060]

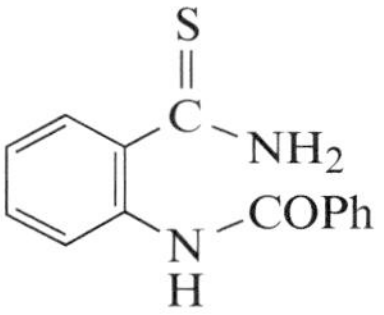

(26)

o-Cinnamamido-*N*-methylbenzamide gave 3-methyl-2-styryl-4(3*H*)-quinazolinone (neat, > 200°, 20 min).[537]

4-Nitro-2-(2,3,3-trimethylbutyramido)benzohydrazide (**27**) gave 3-amino-7-nitro-2-*α*,*β*,*β*-trimethylpropyl-4(3*H*)-quinazolinone (**28**) (EtOH, 165°, sealed, 12 h: 51%).[1954]

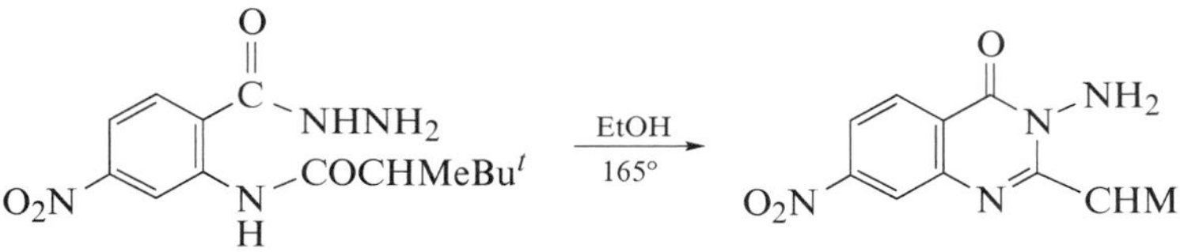

(27)   **(28)**

Also other examples.[519,960,968,983,992,1197,1311,1345,1371,1391,1535,2116,2287]

Cyclizations in acid

1-Acetamido-2-(α-amino-β-benzoylvinyl)benzene (**29**) gave 4-β-hydroxy-styryl-2-methylquinazoline (**30**) (HCl, EtOH–H$_2$O, reflux, 30 min: $\sim$ 75%).[1781]

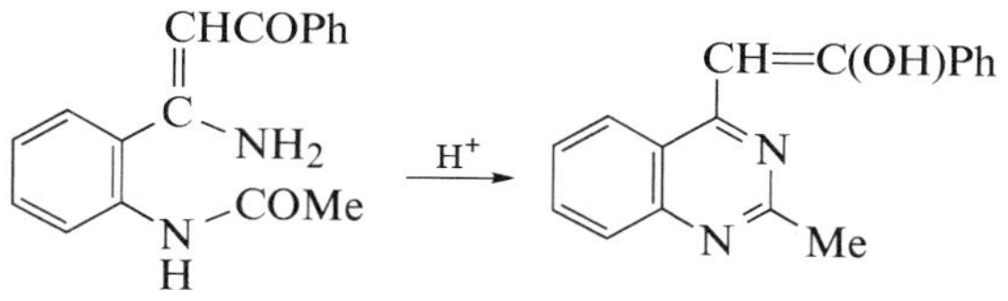

(**29**)   (**30**)

o-Chloroacetamido-N-(o-tolyl)benzamide gave 2-chloromethyl-3-o-tolyl-4(3H)-quinazolinone (AcOH, reflux).[2304]

o-(N-Methylformamido)benzanilide (**31**) gave 3-phenyl-4(3H)-quinazolinone 1-methobromide (**32**) (HBr, EtOH–H$_2$O, 20°, briefly: 84%).[352]

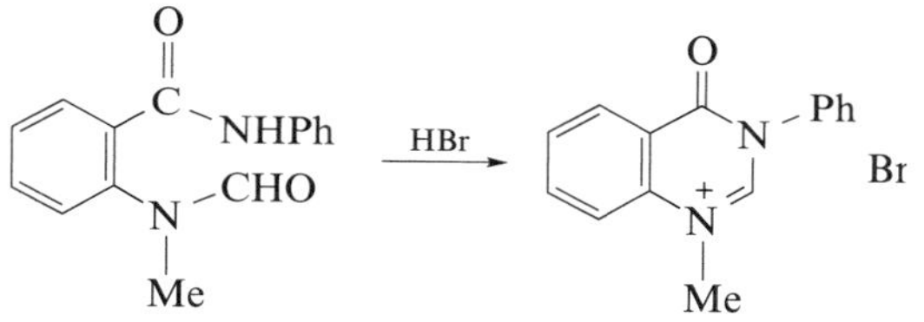

(**31**)   (**32**)

Also other examples.[687,1033,1526]

Cyclizations in base

3,4-Dimethoxy-6-pivalamidobenzamide (**33**) gave 2-t-butyl-6,7-dimethoxy-4(3H)-quinazolinone (**34**) (1M NaOH, 70°, 5 min: 85%).[1354]

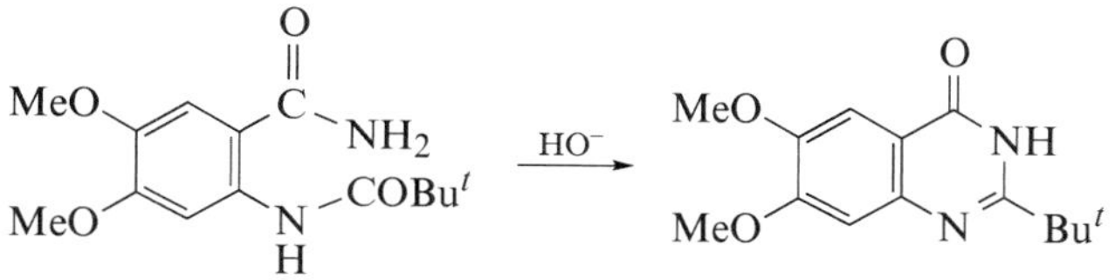

(**33**)   (**34**)

o-(o-Azidobenzamido)benzamide gave 2-(o-azidophenyl)-4(3H)-quinazo-linone (1.2M NaOH, 95°, 1 h: 85%).[1277]

o-Pyruvamidobenzamide (**35**) gave 2-acetyl-4(3H)-quinazolinone (2M NaOH, reflux, 2 h: 60%).[185]

2,5-Dimethoxy-6-(phenoxyacetamido)benzamide gave 5,8-dimethoxy-2-phenoxymethyl-4(3H)-quinazolinone (NaOH, EtOH–H$_2$O, reflux, 24 h: 88%).[1999]

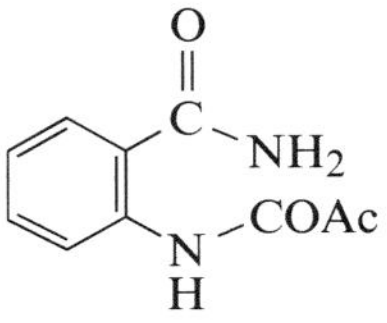

$$
\begin{array}{c}
\overset{\displaystyle O}{\underset{\displaystyle \|}{}} \\
\overset{\displaystyle C}{\diagdown} NH_2 \\
\underset{\displaystyle H}{N} \diagdown COAc
\end{array}
$$

(35)

o-(2-Acetoxypropionamido)benzamide (36) gave 2-α-hydroxyethyl-4(3*H*)-quinazolinone (K$_2$CO$_3$, H$_2$O, 20°, 3 days: 75%; note the deacetylation).[2069]

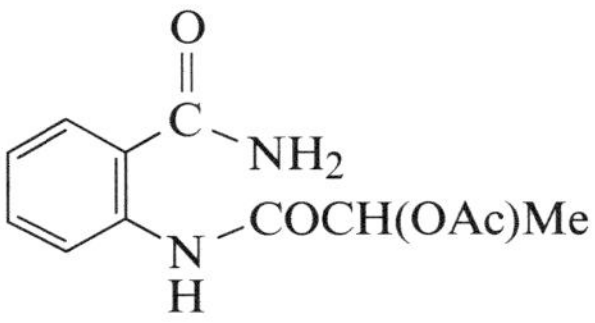

$$
\begin{array}{c}
\overset{\displaystyle O}{\underset{\displaystyle \|}{}} \\
\overset{\displaystyle C}{\diagdown} NH_2 \\
\underset{\displaystyle H}{N} \diagdown COCH(OAc)Me
\end{array}
$$

(36)

2-Acetamido-3,5-dibromo-*N*-methylbenzamide gave 6,8-dibromo-2,3-dimethyl-4(3*H*)-quinazolinone (EtOH, base, 20°, 2 days; EtNH$_2$ or Et$_2$NH: > 95%; Et$_3$N: 30%).[2184]

Also other examples, some illustrating the formation of hydroquinazolines or the further use of organic bases.[17,181,411,482,513,557,584,1346,1639,1981,2081,2098,2430,2505] Mechanistic aspects have been studied.[2555]

Cyclizations with dehydrating reagents

o-(2-Phenylacetamido)benzamide (37) gave 2-benzyl-4(3*H*)-quinazolinone (P$_2$O$_5$, xylene, reflux, 4 h: 54%).[1306]

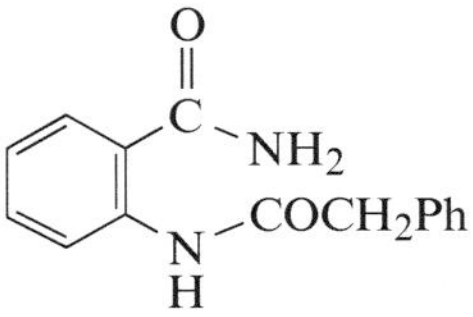

$$
\begin{array}{c}
\overset{\displaystyle O}{\underset{\displaystyle \|}{}} \\
\overset{\displaystyle C}{\diagdown} NH_2 \\
\underset{\displaystyle H}{N} \diagdown COCH_2Ph
\end{array}
$$

(37)

o-(Ethoxalylamino)benzanilide (38) gave ethyl 4-oxo-3-phenyl-3,4-dihydro-2-quinazolinecarboxylate (PCl$_3$, PhMe, reflux, 1 h: 75%).[781]

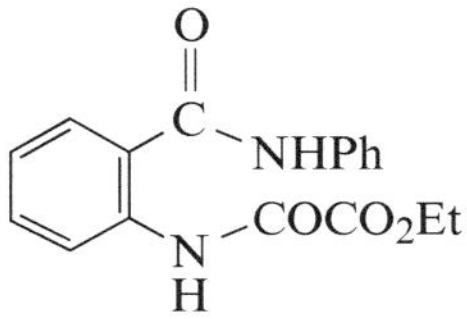

$$
\begin{array}{c}
\overset{\displaystyle O}{\underset{\displaystyle \|}{}} \\
\overset{\displaystyle C}{\diagdown} NHPh \\
\underset{\displaystyle H}{N} \diagdown COCO_2Et
\end{array}
$$

(38)

N-(*m*-Chlorophenyl)-*o*-nicotinamidobenzamide gave 3-*m*-chlorophenyl-2-pyridin-3′-yl-4(3*H*)-quinazolinone (**39**) (PCl$_3$, xylene, reflux, 4 h: 75%).[370]

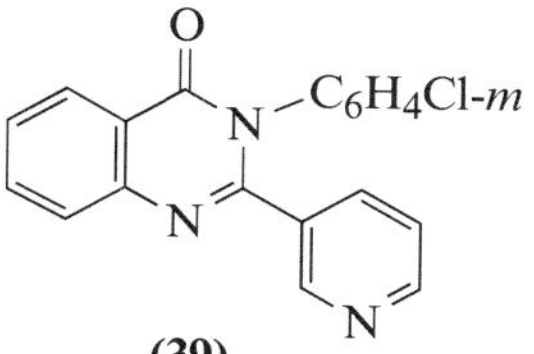

(39)

2-Acetamido-3,5-dibromobenzanilide gave 6,8-dibromo-2-methyl-3-phenyl-4(3*H*)-quinazolinone (POCl$_3$, PhMe, reflux, 6 h: 60%).[1170]

o-[(Methyloxamoyl)amino]benzamide (**40**) gave *N*-methyl-4-oxo-3,4-dihydro-2-quinazolinecarboxamide (Ac$_2$O, reflux, 2 h: 78%).[768]

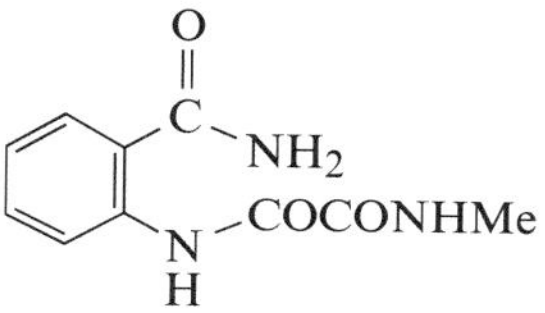

(40)

2-Acetamido-5-acetoxy-4-methoxybenzanilide gave 6-acetoxy-7-methoxy-2-methyl-3-phenyl-4(3*H*)-quinazolinone (Ac$_2$O, reflux, 1 h: 90%).[143]

Also other examples in the foregoing references and elsewhere.[159,301,304, 608,761,937,964,981,1072,1272,1322,1397,1800]

Wittig-assisted cyclizations

o-Acetamido-*N*-methylbenzamide gave 2,4-dimethyl-4(3*H*)-quinazolinone (Ph$_3$PBr$_2$, CH$_2$Cl$_2$, reflux, 3 h: 74%, apparently by a complicated mechanism involving rearrangement).[2267]

1.1.3.2. From *o*-Acylaminobenzamide Oximes

According to conditions, the cyclization of several amide oximes has led to 4-quinazolinone oximes (4-hydroxyaminoquinazolines), to 4-quinazolinamine 3-oxides, or to other products.[60] Thus, *o*-benzamidobenzamide oxime (**41**) in boiling dilute aqueous–ethanolic alkali for 6 h, gave a 67% yield of 4-hydroxyamino-2-phenylquinazoline (**42**), whereas the same substrate in boiling dilute aqueous–alcoholic hydrochloric acid gave 2-phenyl-4-quinazolinamine 3-oxide (**43**) in 75% yield.[1872,2078] Analogs, such as 4-hydroxyaminoquinazoline and 2-methyl-4-quinazolinamine 3-oxide, have been made rather similarly.[60]

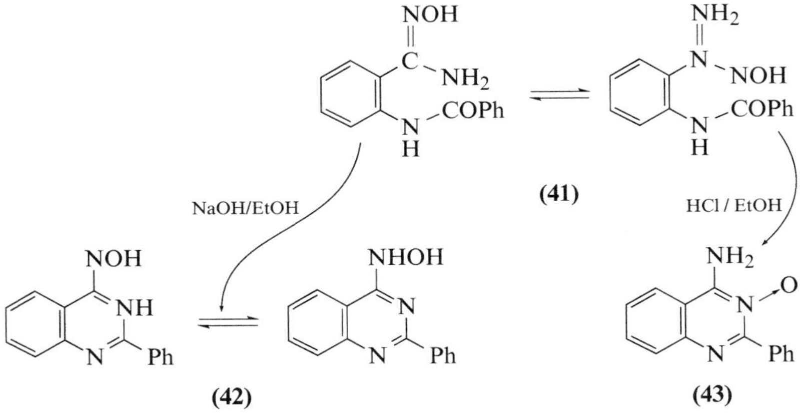

(41)

(42)   (43)

1.1.3.3. From o-(Alkoxycarbonylamino)benzamides

The cyclization of such urethanes has been used occasionally to afford 2,4-quinazolinediones. Thus o-ethoxycarbonylamino-N-(1-ethoxycarbonylpiperidin-4-yl)benzamide (44) gave 3-(1-ethoxycarbonylpiperidin-4-yl)-2,4(1H,3H)-quinazolinedione (45) in 55% yield by refluxing in ethanolic alkali for 3 h;[1706] o-(ethoxycarbonylamino)benzanilide underwent thiation and cyclization to 3-phenyl-4-thioxo-3,4-dihydro-2(1H)-quinazolinone (46) (35%) by boiling with phosphorus pentasulfide in 1,2,4-trimethylbenzene for 30 min;[86] o-ethoxycarbonylamino-N-(methoxycarbonylmethyl)benzamide gave 3-carboxymethyl-2,4(1H,3H)-quinazolinedione (62%) on boiling in aqueous–ethanolic alkali for 1 h;[1838] and other examples, as well as a mechanistic study, have been reported.[314,320,494,2589]

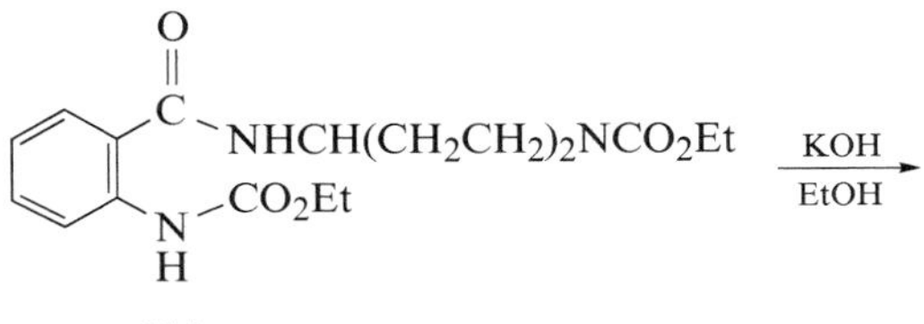

(44)

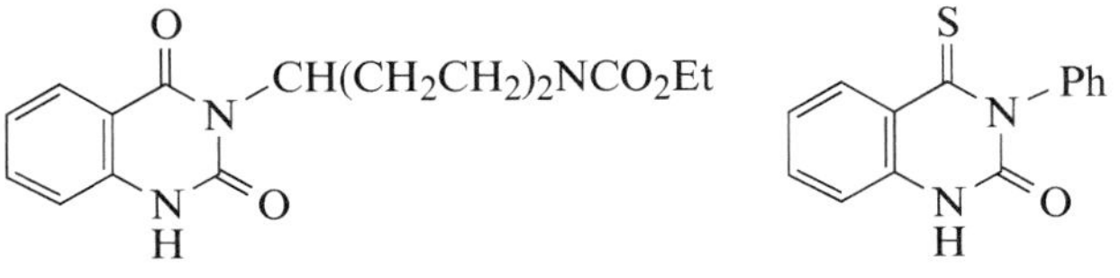

(45)   (46)

1.1.3.4.   From o-Ureidobenzamides

o-Ureidobenzamides (**48**) can undergo cyclization to 2,4-quinazolinediones by formation of the 2,3-bond to give product (**47**) or of the 3,4-bond to give product (**49**): if both terminal amino groups in the substrate (**48**) are unsubstituted (Q = R = H) or similarly monosubstituted (Q = R), no ambiguity arises because the products (**47** and **49**) are the same; if one such amino group is disubstituted (tertiary), again no ambiguity arises because it must be the amine lost during cyclization; but if NHQ ≠ NHR in substrate (**48**), cyclization can give either or both of the products (**47** and **49**). The experimental data on such ambiguous cyclizations suggest that both the conditions of reaction and the nature of Q and/or R will determine the product(s) formed, but any formulation of definitive rules would seem premature.

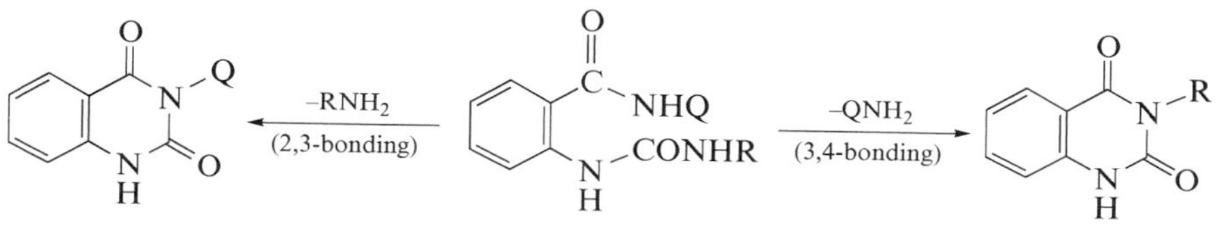

| (47) | (48) | (49) |

The following examples will illustrate the foregoing and other variations in this type of ring closure.

Unambiguous cyclizations

o(N'-Phenylureido)benzanilide (**48**, Q = R = Ph) gave 3-phenyl-2,4(1H,3H)-quinazolinedione (**47**, Q = Ph) (240°, 20 min: 93%).[709]

3-Chloro-4,5-dimethoxy-2-ureidobenzamide gave 8-chloro-6,7-dimethoxy-2,4(1H,3H)-quinazolinedione (2M NaOH, 95°, 30 min: 95%).[1146]

o-(N',N'-Dimethylureido)benzamide gave 2,4(1H,3H)-quinazolinedione (Ac$_2$O. reflux).[1530]

2-(N'-Aminoureido)cyclohexanecarboxamide (**50**) gave 4a,5,6,7,8,8a-hexahydro-2,4(1H,3H)-quinazolinedione (**51**) (I$_2$, H$_2$O, hot: 60%; note the oxidative removal of the N-amino group during cyclization).[1855]

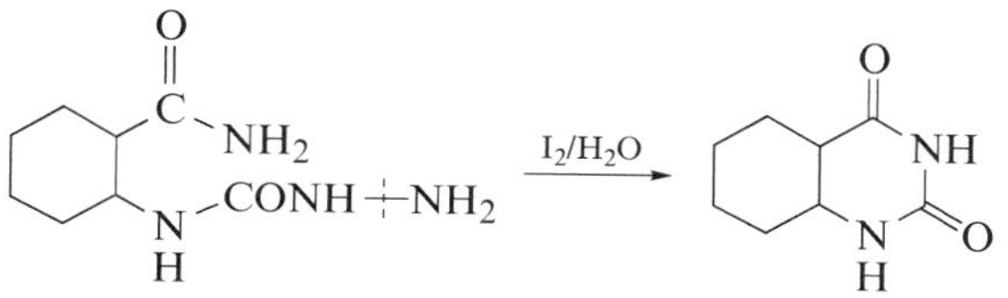

| (50) | (51) |

Also other examples in the foregoing references and elsewhere.[385,1987,1341]

Ambiguous cyclizations

o-(N'-β-Chloroethylureido)benzamide (**48**, Q = H, R = CH$_2$CH$_2$Cl) gave only 3-β-chloroethyl-2,4(1H,3H)-quinazolinedione (**48**, R = CH$_2$CH$_2$Cl) (HCl, EtOH–H$_2$O, reflux, 30 min: 77%).[1931]

3-Bromo-N-methoxycarbonylmethyl-6-ureidobenzamide (**52**, R = Br) gave 6-bromo-3-methoxycarbonylmethyl-2,4(1H,3H)-quinazolinedione (200°, N$_2$, 15 min: 65%), whereas the closely related debromo substrate (**52**, R = H) gave 2,4(1H,3H)-quinazolinedione (KOH, EtOH–H$_2$O, 95°, 2 h: 68%).[1708]

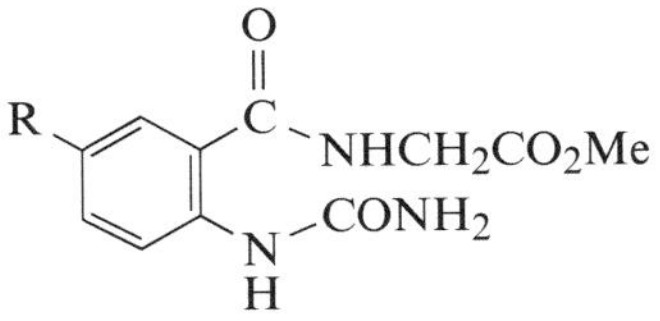

(52)

N-(o-N-Butylcarbamoylphenyl)-o-ureidobenzamide gave only 3-(o-N-butyl-carbamoylphenyl)-2,4(1H,3H)-quinazolinedione (pyridine, reflux, 8 h: 79%).[40]

o-(N'-Ethoxycarbonylureido)-N-methoxybenzamide (**48**, Q = OMe, R = CO$_2$Et) gave only 3-methoxy-2,4(1H,3H)-quinazolinedione (**47**, Q = OMe) (pyridine, reflux: 82%), whereas the closely related ethoxycar-bonylmethylureido substrate (**48**, Q = OMe, R = CH$_2$CO$_2$Et) under the same conditions gave only 3-ethoxycarbonylmethyl-2,4(1H,3H)-quinazolinedione (**49**, R = CH$_2$CO$_2$Et) (55%).[738]

N-Methyl-o-[N'-phenyl(thioureido)]thiobenzamide (**53**) in refluxing ethanol gave a separable mixture of 3-methyl-2,4(1H,3H)-quinazolinedithione (**54**) and 4-anilino-3-methyl-2(3H)-quinazolinethione (**56**), the latter formed by Dimroth rearrangement of its isomer (**55**).[88]

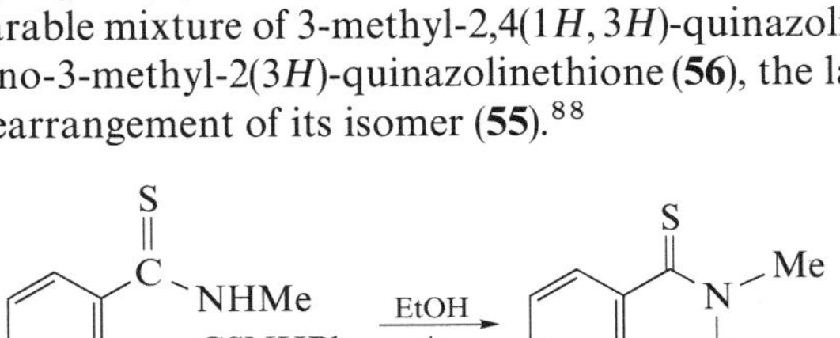

(53)   (54)

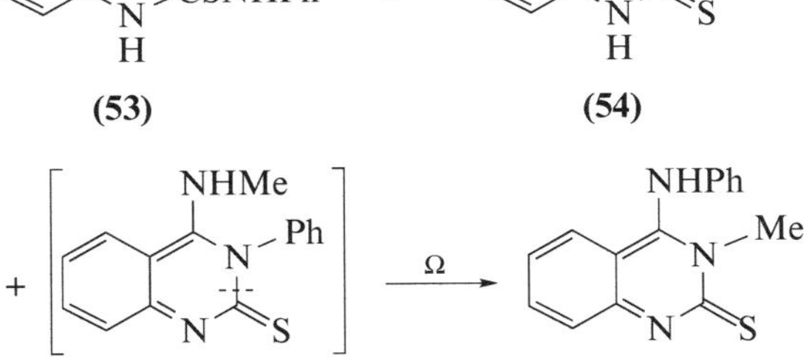

(55)   (56)

o-Thioureidobenzohydrazide (**57**) in a hot ethanolic solution of chloroacetic acid gave a 68% yield of 2,3-diamino-4(3*H*)-quinazolinone (**58**), probably because initial *S*-alkylation made the resulting alkylthio group a highly preferred target for intramolecular aminolysis. [1715]

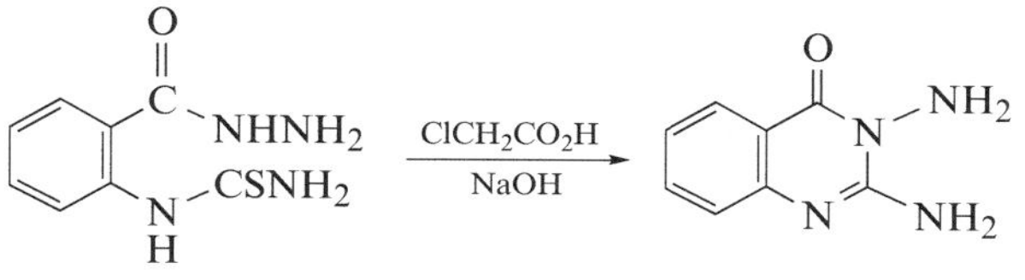

(**57**)   (**58**)

Also other examples. [87,379,843,1774]

1.1.3.5. From *o*-(Benzylideneamino)benzamides (*H* 392)

This type of synthesis must lead to a dihydroquinazolinone unless an oxidizing agent is present. Thus *o*-(*o*-nitrobenzylideneamino)benzamide (**59**) in refluxing alkali gave 2-*o*-nitrophenyl-1,2-dihydro-4(3*H*)-quinazolinone (55%);[638] *o*-(*p*-chlorobenzylideneamino)benzamide treated in a vacuum at 150–180° gave 2-*p*-chlorophenyl-1,2-dihydro-4(3*H*)-quinazolinone, which underwent subsequent oxidation by permanganate in acetone to afford 2-*p*-chlorophenyl-4(3*H*)-quinazolinone;[566] and *N'*-(*p*-methylbenzylidene)-*o*-(*p*-methylbenzylideneamino)benzohydrazide (**60**) underwent both cyclization and oxidation in boiling acetonic permanganate to give 3-(*p*-methylbenzylideneamino)-2-*p*-tolyl-4(3*H*)-quinazolinone (**61**) in 80% yield.[1796]

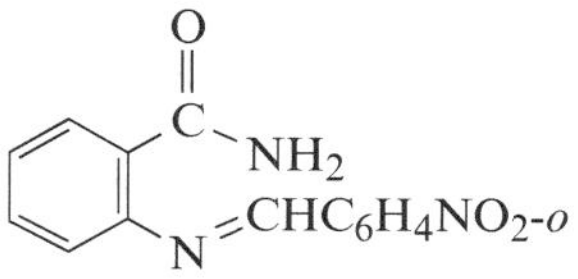

(**59**)

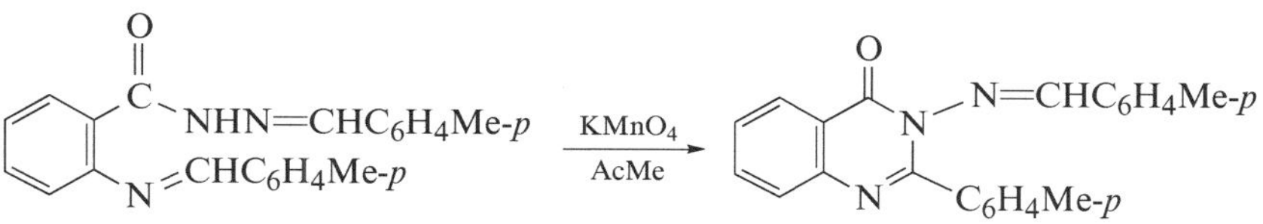

(**60**)   (**61**)

1.1.3.6. From *o*-Acylaminobenzonitriles (*H* 81,238,277)

Most such cyclizations have occurred on treatment with reagents that can initially convert the nitrile into an amide, amine, imidic ester, or the like.

The following examples illustrate reagents, conditions, and products to be expected.

2-Acetamido-4-trifluoromethylbenzonitrile (**62**, $R = CF_3$) gave 2-methyl-7-trifluoromethyl-4(3H)-quinazolinone (HCl gas, EtOH, reflux, 1 h: 83%).[1712]

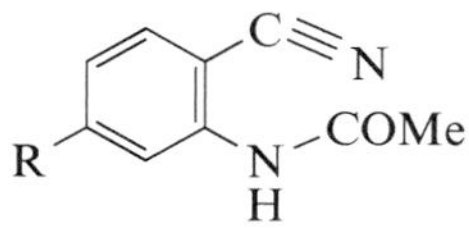

(**62**)

o-Benzamidobenzonitrile (**63**) gave 2-phenyl-4-quinazolinamine (**64**, $R = H$) (NH_4Cl, P_2O_5, $C_6H_{11}NMe_2$, 240°, 2–5 h: 81%) or 2-phenyl-4-propylaminoquinazoline (**64**, R = Pr) ($PrNH_3Cl$, etc. 5 h: 65%).[1300]

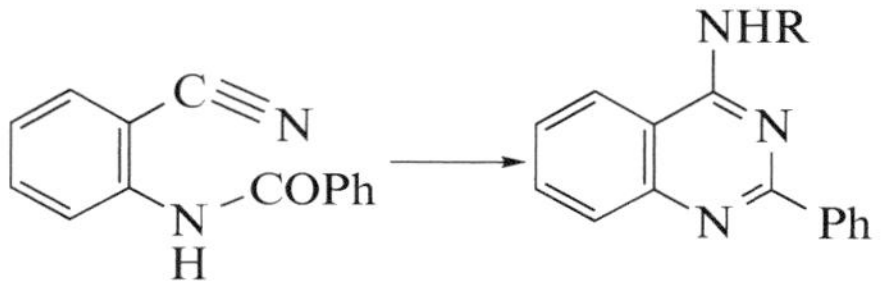

(**63**)   (**64**)

o-Acetamidobenzonitrile (**62**, $R = H$) gave 2-methyl-4(3H)-quinazolinone (**65**, R = H) (5M HCl, AcOEt, stirred, 20°, 10 min: 91%)[1290] or 1,2-dimethyl-4(1H)-quinazolinone (**65**, R = Me) (Me_2SO_4, NaOH, H_2O; then HCl gas, EtOH, reflux, 30 min: 60% as hydrochloride).[244]

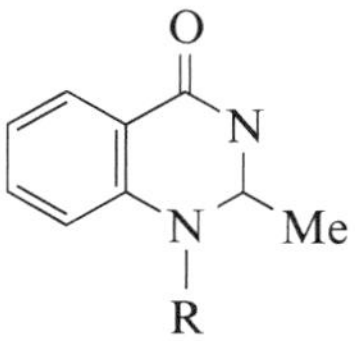

(**65**)

2-Acetamido-5-iodobenzonitrile gave 6-iodo-2-methyl-4(3H)-quinazolinone ($NaBO_3$, H_2O–dioxane, reflux, 24 h: 67%).[2475]

o-(Heptafluorobutyramido)benzonitrile gave 2-heptafluoropropyl-4(3H)-quinazolinone (H_2O_2, NaOH, H_2O–dioxane–AcOEt, reflux, 12 h: 73%).[1287]

o-(C-Chloroformamido)- (**66**, X = O) or o-[C-chloro(thioformamido)]ben-zonitrile (**66**, X = S) gave 4-chloro-2(1H)-quinazolinone (**67**, X = O) or the corresponding thione (**67**, X = S), respectively (HCl gas, Bu_2O, 70°: 87%, 70%; use of HBr gas gave the 4-bromo analog of the first product in 78% yield).[72]

o-(*N*'-Methylureido)benzaldehyde dimethyl acetal (**107**) gave 3-methyl-2(3*H*)-quinazolinone (**108**), isolated as its stable covalent hydrate (**109**) (10M HCl, dioxane, 75°, briefly: 60%); homologs were made similarly.[1027]

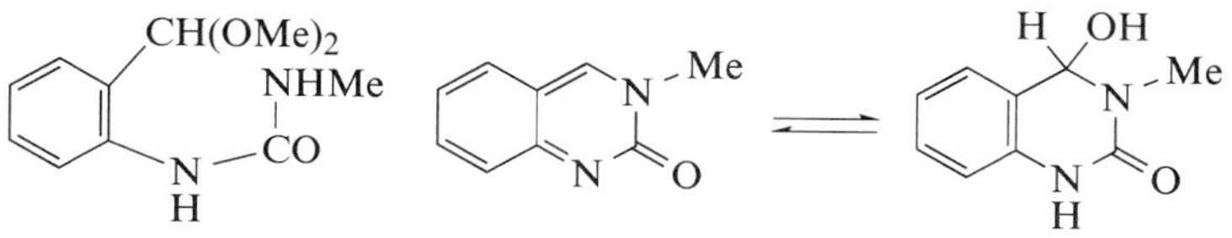

(107)   **(108)**   **(109)**

α,α-Dichloro-2-ureidotoluene (**110**) ($\simeq$ *o*-Ureidobenzaldehyde) gave 2(1*H*)-quinazolinone (3M HCl, reflux, until clear: 95%); several analogs were made similarly.[849]

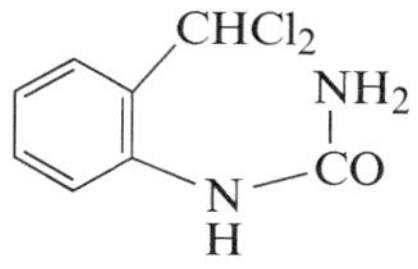

(110)

3-Chloro-6-(*N*-methylureido)benzophenone (**111**) gave 6-chloro-1-methyl-4-phenyl-2(1*H*)-quinazolinone (**112**) (PhMe, reflux, 2.5 h: 90%).[493]

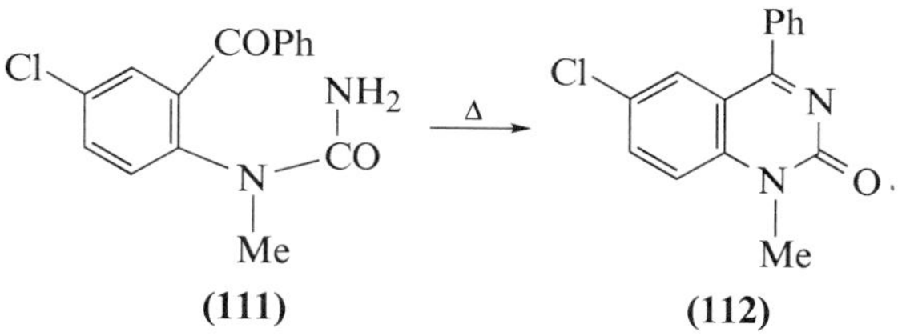

(111)   **(112)**

α-Dicyanomethylene-*o*-ureidobenzyl alcohol (**113**) gave 4-dicyanomethylene-3,4-dihydro-2(1*H*)-quinazolinone, tautomeric with the aromatic 4-dicyanomethyl-2(1*H*)-quinazolinone (**114**) (Ac$_2$O, pyridine, 20°, 12 h: ~ 50%).[1350]

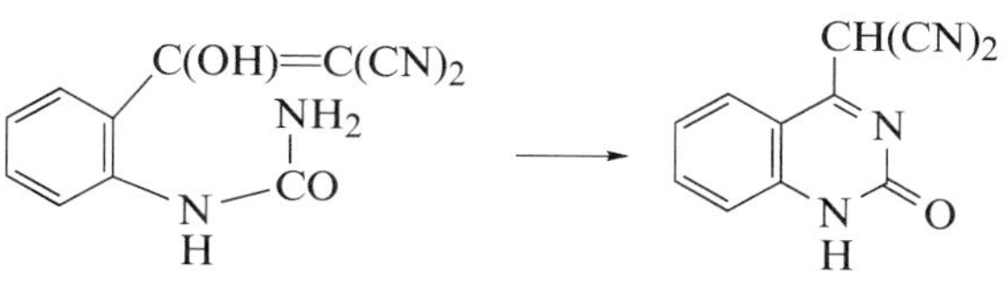

(113)   **(114)**

Also other examples.[130,470,492,540,797]

1.1.4.6.   From *o*-(Aminomethyleneamino)benzoic Acids, Esters, or Amides

Cyclization of such substrates will usually afford 4-quinazolinones bearing an alkyl, aryl, or no substituent at the 2-position. Most possibilities are illustrated by the following examples.

o-[(α-Anilinoethylidene)amino]benzoic acid (**115**) gave 2-methyl-3-phenyl-4(3*H*)-quinazolinone (**116**) (0.5M NaOH, 20°, 30 min: ~ 80%).[931,932]

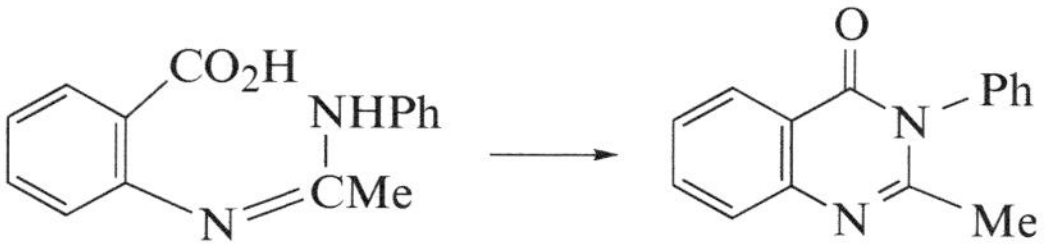

(**115**)   (**116**)

o-[α-(β-Cyanoacetylhydrazino)ethylidene]aminobenzoic acid (**117**) gave 3-cyanoacetamido-2-methyl-4(3*H*)-quinazolinone (**118**) (PhH, TsOH, 95°, 2.5 h: ~ 25%; or CH$_2$N$_2$, Et$_2$O–MeOH, 58%, via the ester?).[109]

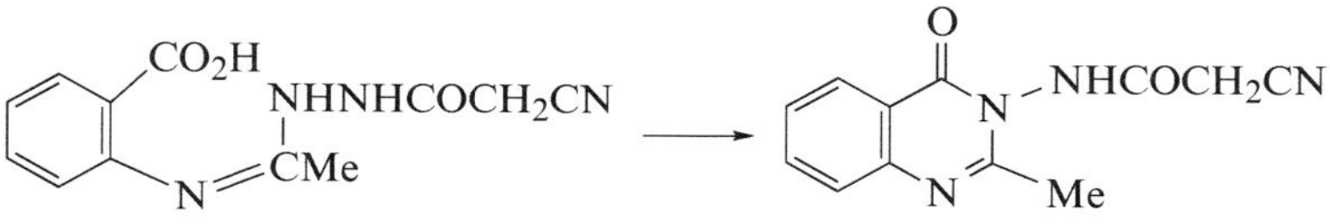

(**117**)   (**118**)

Methyl *o*-[α-(isopropylamino)benzylideneamino]benzoate (**119**) gave 3-iso-propyl-2-phenyl-4(3*H*)-quinazolinone (NaOMe, MeOH, reflux, 2 h: 70%).[412]

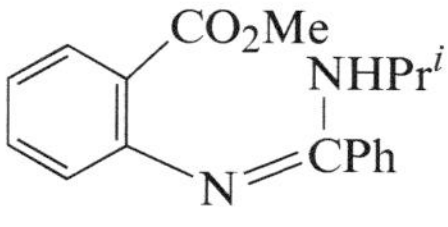

(**119**)

Methyl *o*-(tetrazol-5-ylaminomethyleneamino)benzoate (**120**) gave 3-tetrazol-5′-yl-4(3*H*)-quinazolinone (NH$_4$OH, MeOH, 20°, 1 h: 89% as ammonium salt).[2217]

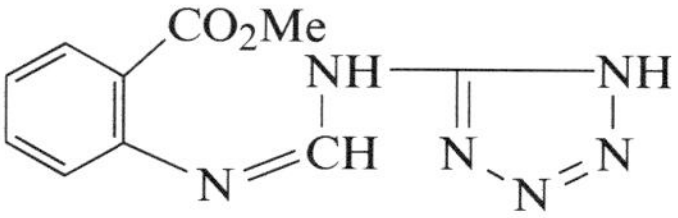

(**120**)

Methyl o-[α-(hydroxyamino)phenacylideneamino]benzoate(121) gave 2-benzoyl-4(1H)-quinazolinone 3-oxide (10M HCl, 70°, 20 min: 50%).[101]

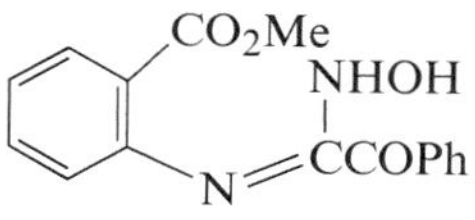

(121)

2-(α-Aminobenzylideneamino)-4,4-dimethyl-6-oxocyclohex-1-enecarbothioanilide (122) proved exceptional in giving 4-anilino-7,7-dimethyl-2-phenyl-5,6,7,8-tetrahydro-5-quinazolinone (dioxane, trace H_3PO_4, reflux: 57%).[287]

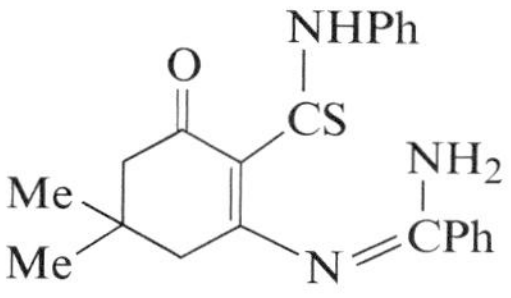

(122)

Also other examples,[318,793,856,1732,2571] including one in which a trifluoromethyl group acted as the equivalent of a carboxyl function in the substrate.[95]

1.1.4.7. From Miscellaneous Substrates

Some other types of substrate have been used to afford quinazolines by 3,4-bond formation: most have required rearrangement or other modification during the ring-closure process, as illustrated in the following examples.

Ethyl 2-(N-cyano-N-methylamino)-1-phenylcyclohexanecarboxylate (123) gave 1-methyl-4a-phenyl-4a,5,6,7,8,8a-hexahydro-2,4(1H,3H)-quinazolinedione (124) (10M HCl, 95°, 3 h: 65%; undoubtedly via the amide).[108]

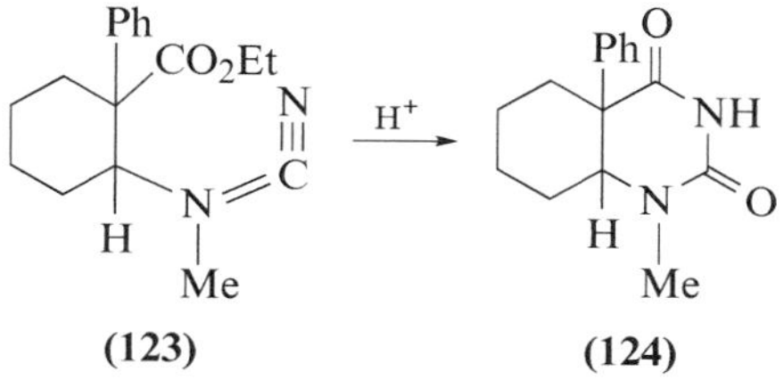

(123)   **(124)**

2-[2-(N-Acetoxy-N-phenoxycarbonylamino)acetamido]-5-chlorobenzophenone (125) gave 6-chloro-4-phenyl-2(1H)-quinazolinone (126) (NH4OH, MeOCH2CH2OMe, 20°, 3 days: ~ 80%; by a postulated multistage mechanism).[859]

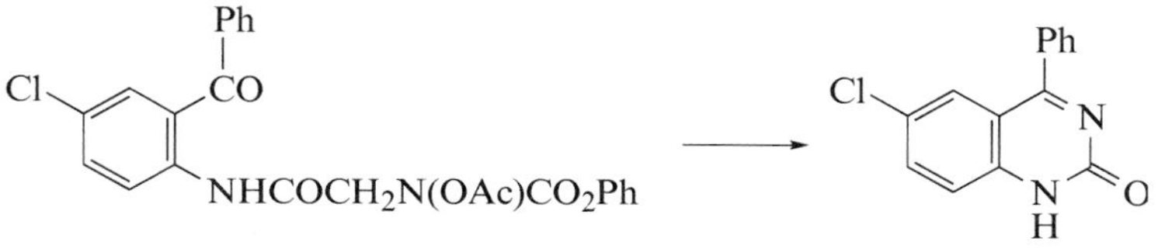

(125)   **(126)**

o-[Methoxycarbonyl(triphenylphosphoranylidene)methylazo]benzaldehyde
(**127**) gave methyl 4-oxo-3,4-dihydro-2-quinazolinecarboxylate (**128**)
(PhMe, reflux, 1 h: 86%; by a rearrangement involving N=N cleavage,
etc.);[1329] rather similarly, 1-bromo-4-(α-chloro-α-methoxycarbonylmethy-
lene)hydrazino-3-oxalobenzene (**129**) gave methyl 6-bromo-4-oxo-3,4-
dihydro-2-quinazolinecarboxylate (NaOAc, H$_2$O–MeOH, 20°, 3 h:
?%).[1945]

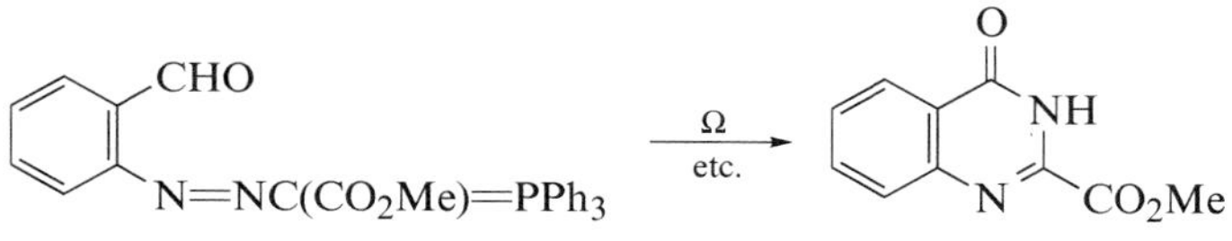

(127)   **(128)**

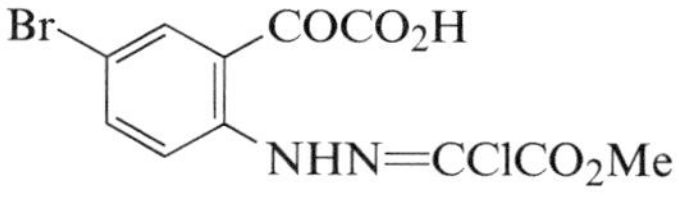

(129)

o-(*N*-Benzylhydrazino)benzonitrile (**130**) gave 2-phenyl-4-quinazolinamine
[NaH, (MeOCH$_2$CH$_2$)$_2$O, reflux, 16 h: 32%].[258]

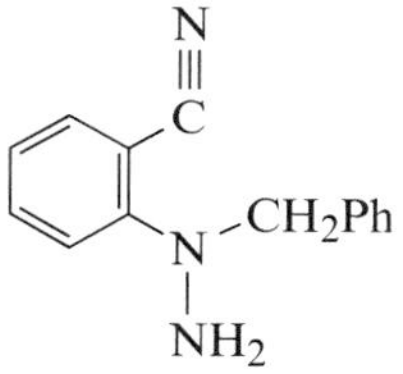

(130)

o-(*N″*-Acetyl-*N‴*-phenylguanidino)benzonitrile gave 2-acetamido-3-phenyl-
4(3*H*)-quinazolinimine (EtOH, reflux, 1 h: 89%; or NaOH, H$_2$O–EtOH,
20°, 5 min: 85%), which easily underwent Dimroth rearrangement and
deacylation to 4-anilino-2-quinazolinamine (NaOH, H$_2$O–EtOH, reflux,
10 min: 89%).[2418]
1-(*o*-Carbamoylphenyl)-3-(*o*-methoxycarbonylphenyl)triazene (**131**) gave
2-(*o*-piperidinoazophenyl)-4(3*H*)-quinazolinone (**132**)   [(CH$_2$)$_5$NH,

EtOH, reflux, 5 h: 70%; suggested mechanism involved an initial 3,4-bond formation).[703]

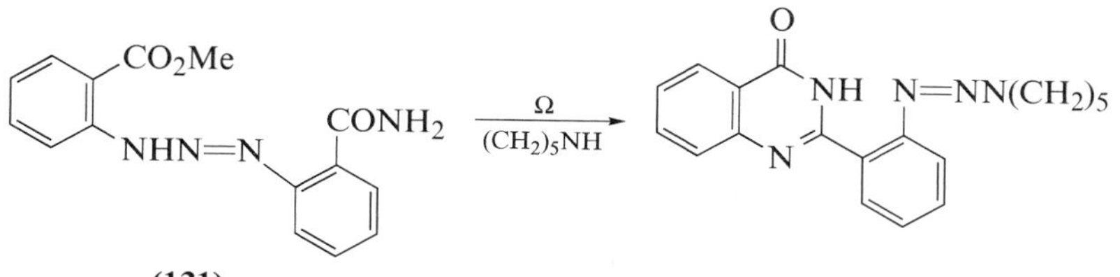

(131)   (132)

1.1.5.   By Formation of the 4,4a-Bond

All examples of this cyclization involve bond formation between C4 of an —N—C—N—C—R chain attached through N to a benzene (or cyclohexane) ring and an *unsubstituted ortho*-position: no customary leaving group is required on the ring, but C4 of the chain must be appropriately activated as part of a microenvironment of double or triple bonds and/or suitable attached groups. The following illustrations are subdivided according to the nature of C4 in the substrate chain.

Cyclizations through an ester (all thermal)

(N'-Ethoxycarbonyl-S-ethylisothioureido)benzene (**133**) gave 2-ethylthio-4($3H$)-quinazolinone (**134**) (175°, 3 h: 36%).[1459]

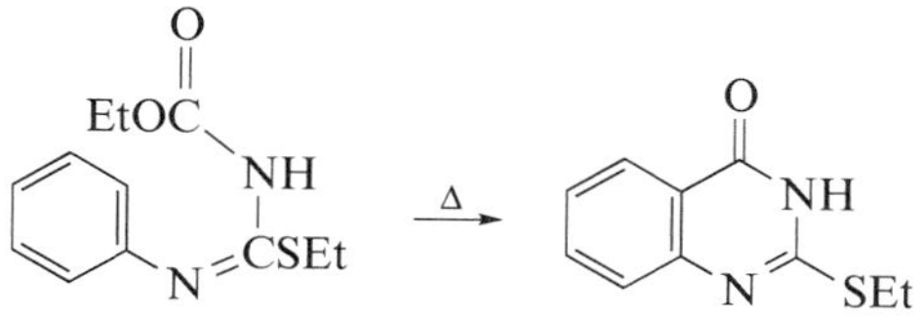

(133)   (134)

(α-Ethoxycarbonylaminobenzylidene)aniline (**135**) gave 2-phenyl-4($3H$)-quinazolinone (quinoline, boil, 4 min: 84%).[1447]

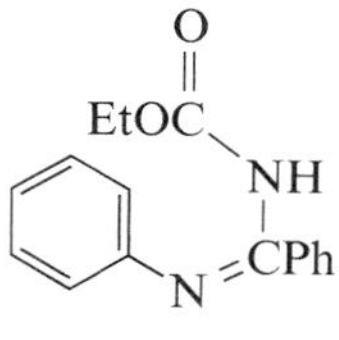

(135)

Also other examples.[453,491,930,2145]

Cyclizations through a thioamide

α-[N'-Phenyl(thioureido)]benzylideneaniline (**136**) gave 4-anilino-2-phenyl-quinazoline (HgO, CH$_2$Cl$_2$, 20°, 2 h: 68%).[728,1545]

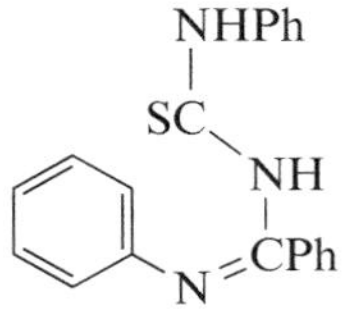

(136)

Also analogs.[1545]

Cyclizations through an N-acyl group

1-(N'-Acylureido)-3,5-dimethoxybenzene (**137**) gave 5,7-dimethoxy-4-methyl-2(1H)-quinazolinone (polyphosphoric acid, 140°, 3 h: 93%).[1592]

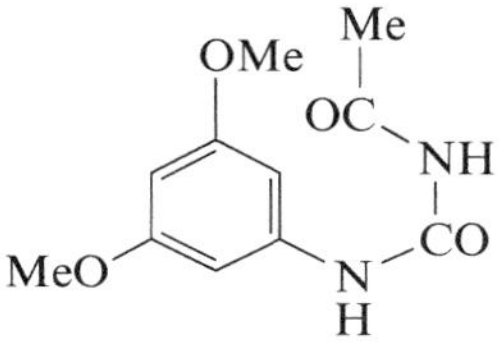

(137)

Also analogous examples.[79,1148,2501]

Cyclizations through a nitrile

Cyanoaminomethyleneaminobenzene (**138**) gave 4-quinazolinamine (AlCl$_3$, PhCl, reflux, 30 min: 50%); also several derivatives likewise).[826]

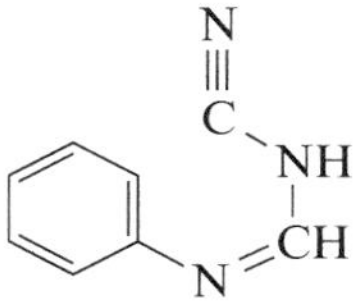

(138)

Cyclizations through an isothiocyanate or isocyanate

α-t-Butyl-α-isothiocyanatomethyleneaminobenzene (**139**, X = S) gave 2-t-butyl-4(3H)-quinazolinethione (Me$_2$NCHO, reflux, 90 min: 87%).[724]

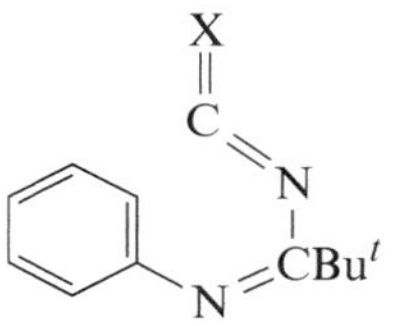

(139)

o-(α-Isothiocyanatobenzylideneamino)toluene (**140**) gave 8-methyl-2-phenyl-4(3*H*)-quinazolinethione (*m*-xylene, reflux, 2 h: 40%).[756]

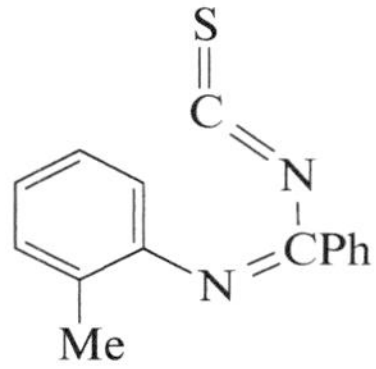

(140)

α-*t*-Butyl-α-isocyanatomethyleneaminobenzene (**139**, X = O; made *in situ*) gave 2-*t*-butyl-4(3*H*)-quinazolinone (CH$_2$Cl$_2$, reflux, 8 h: 85%).[852]
Also other examples.[413,707,835,2044,2050,2480]

Cyclizations through a methylene group

α-[(α-Ethoxycarbonylethylidene)methyleneamino]-*p*-methoxybenzylidene-aniline (**141**) gave 4-α-ethoxycarbonylethyl-2-*p*-methoxyphenylquinazo-line (PhMe, 20°, 24 h: 77%).[1702]

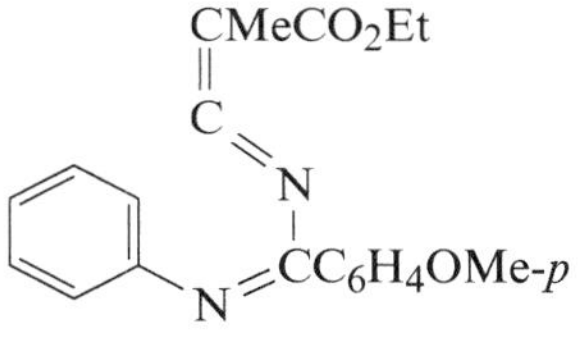

(141)

α-(Dimethylaminomethyleneamino)benzylideneaniline (**142**) gave 2-phenyl-quinazoline (polyphosphoric acid, 140°, 30 min: 72%).[1206]

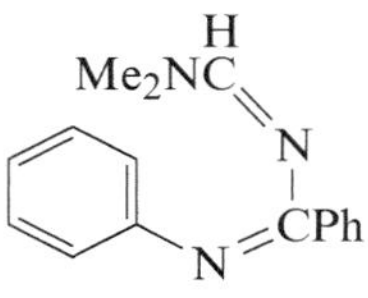

(142)

α-(Methylaminomethyleneamino)benzylideneaniline (**143**) gave 4-methyl-amino-2-phenylquinazoline (CHCl$_3$, reflux, 2 h: 92%).[1666]

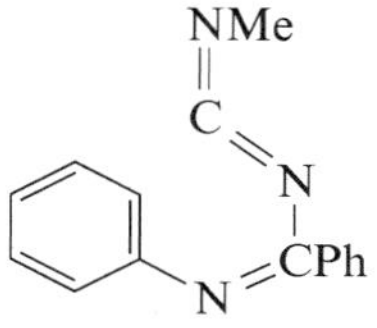

$$
\begin{array}{c}
NMe \\
\parallel \\
C \\
\diagdown N
\end{array}
$$

(**143**)

Also other examples.[850,989,1331]

Cyclizations through a methyl group

(α-Allylaminoethylidene)aminobenzene (**144**) gave 4-ethyl-2-methyl-3,4-dihydroquinazoline (polyphosphoric acid, < 170°, < 7 h: 75%) and thence 4-ethyl-2-methylquinazoline [K$_3$Fe(CN)$_6$, 1.3M KOH, 50°, 1 h: 86%].[421]

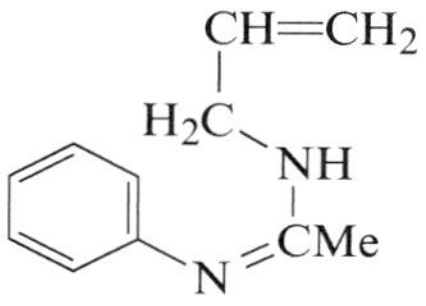

$$
\begin{array}{c}
CH{=}CH_2 \\
H_2C \\
\diagdown NH
\end{array}
$$

(**144**)

N'-(β,β-Dimethoxyethyl)-N,N'-dimethylureidobenzene (**145**) gave 4-chloro-methyl-1,3-dimethyl-3,4-dihydro-2(1H)-quinazolinone (**146**) (6M HCl, re-flux, 10 min: 60%).[348]

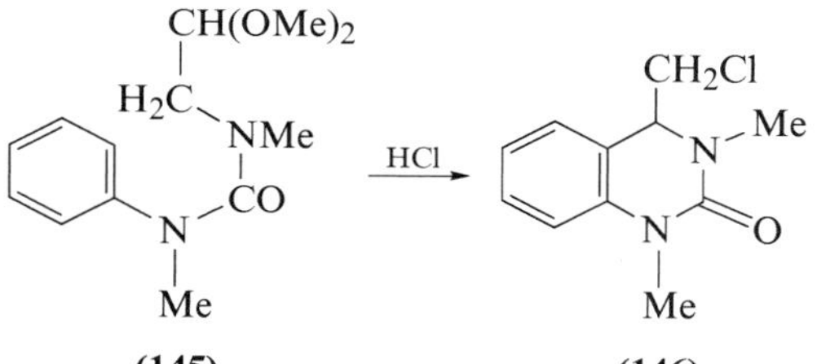

(**145**) →HCl→ (**146**)

α-(β,β,β-Trichloro-α-hydroxyethylamino)benzylideneaniline (**147**) gave a sep-arable mixture of 3-acetyl-2-phenyl-4-trichloromethyl-3,4-dihydroquina-zoline (**148**) and 4-dichloromethyl-2-phenylquinazoline (by loss of AcCl) (Ac$_2$O, reflux, 5 h: 24% and 10%, respectively).[438]

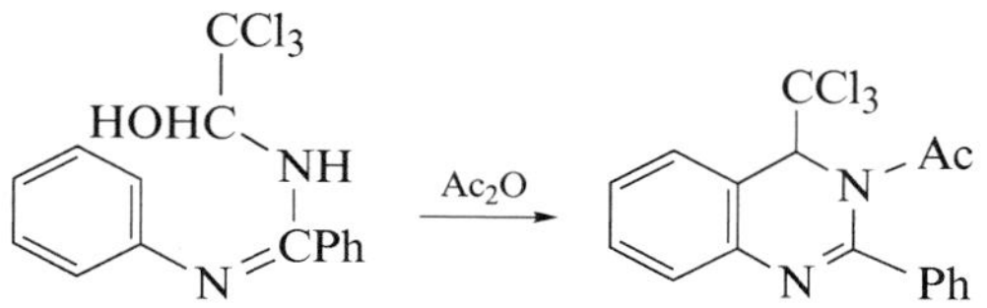

(**147**) →Ac$_2$O→ (**148**)

1.2. FROM A BENZENE SUBSTRATE AND ANCILLARY SYNTHON(S)

The bulk of quinazoline syntheses fall into this broad category, which is subdivided according to the ring atom(s) supplied by the ancillary synthon(s). The benzene substrate may, of course, be replaced by an appropriate cyclohexane derivative to afford a quinazoline reduced in its benzene ring.

1.2.1. Where the Synthon Supplies N1

This rarely used type of synthesis is illustrated by the reaction of N-ethoxycar-bonyl-2-morpholinocyclohex-1-enecarbothioamide (**149**) with ethanolic aniline at 20° to give 1-phenyl-4-thioxo-3,4,5,6,7,8-hexahydro-2(1H)-quinazolinone (**151**) in 89% yield, via the labile intermediate (**150**);[223] several analogs were made similarly.[223]

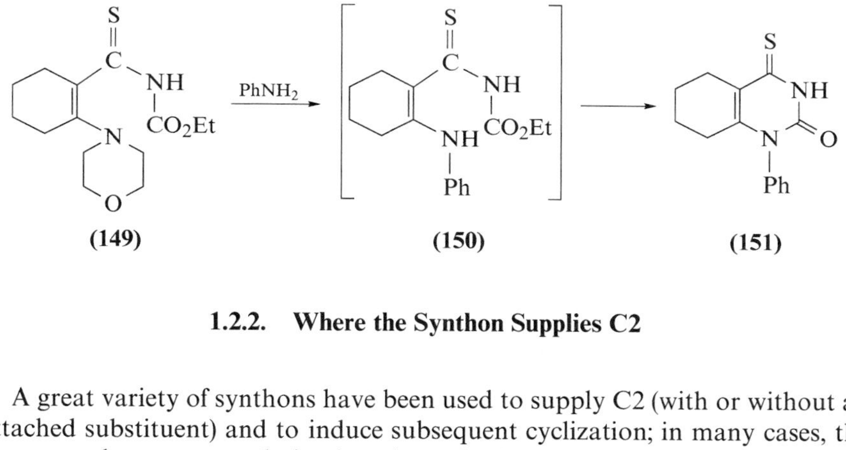

(**149**)   (**150**)   (**151**)

1.2.2. Where the Synthon Supplies C2

A great variety of synthons have been used to supply C2 (with or without an attached substituent) and to induce subsequent cyclization; in many cases, the benzene substrate was a derivative of o-aminobenzamide (anthranilamide).

1.2.2.1. The Use of Carboxylic Acids and Related Synthons

The following examples illustrate the use of carboxylic acids and their derivatives, aldehydes, ketones, alcohols, and some equivalents to provide C2 in quinazoline syntheses.

2,4-Diamino-5-methylbenzamide (**152**) gave 7-amino-6-methyl-4(3H)-quinazolinone (**153**) (HCO_2H, reflux, 8 h: 97%).[1356]

o-Aminobenzohydrazide gave 3-amino-4(3H)-quinazolinone (HCO_2H, reflux, 8 h: 87%).[1657]

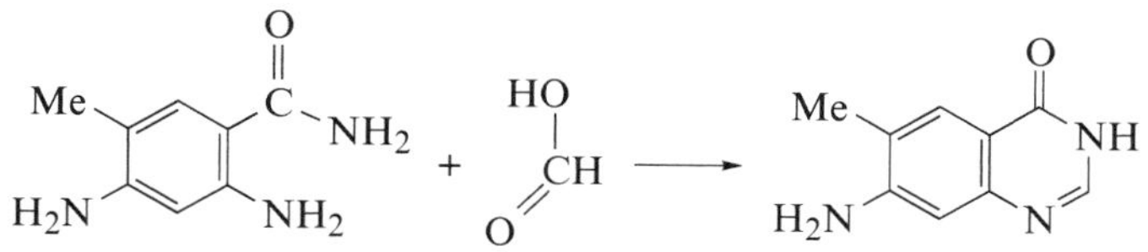

(152)   **(153)**

o-Amino-N-(2-oxopiperidin-3-yl)benzamide (**154**) gave 2-methyl-3-(2-oxopi-
peridin-3-yl)-4(3H)-quinazolinone (AcOH, 170°, sealed, 3 h: 43%).[2047]

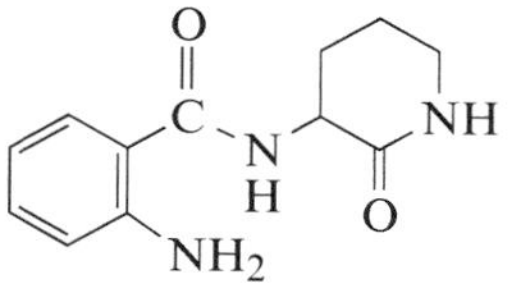

(154)

o-(Aminomethyl)aniline with butyrolactone ($\simeq$ 4-hydroxybutyric acid) gave
2-γ-hydroxybutyl-3,4-dihydroquinazoline (200°, N$_2$, 2 h: 17%).[677]

o-Aminobenzamide with triphenyl(phenylethynyl)phosphonium bromide
(Ph$_3\overset{+}{P}$C≡CPh Br$^-$; $\simeq$ PhCO$_2$H!) gave 2-phenyl-4(3H)-quinazolinone
(MeCN, reflux, 5 days: 45%).[936]

o-Aminobenzamide   with   hexafluoropropene   (CF$_3$CF=CF$_2$;
$\simeq$ CF$_3$CHFCO$_2$H) gave 2-(α,β,β,β-tetrafluoroethyl)-4(3H)-quinazolinone
(Me$_2$NCHO), 80°, sealed, 8 h: $\sim$ 40% after separation of a byproduct);[712]
o-aminobenzylamine rather similarly gave 2-(α,β,β,β-tetrafluoroethyl)-3,4-
dihydroquinazoline (56%).[350]

o-Aminobenzamide with carbon monoxide ($\simeq$ HCO$_2$H), assisted by Se or
S, gave 2,4(1H,3H)-quinazolinedione [CO, Se, Me$_2$NCHO, MeN(CH$_2$)$_4$,
30 atm, 100°, 20 h: 21%;[1794] S for Se, similarly: > 95%);[2100] o-
(aminomethyl)aniline likewise gave 3,4-dihydro-2(1H)-quinazolinone (Se,
etc.: 86%).[1751,1794]

Also   other   examples.[235,328,330,394,448,648,820,1240,17551897,2017,2112,2239,
2324,2564]

Carboxylic esters as synthons

o-Aminobenzamide (**155**) with ethyl glycolate gave 2-hydroxymethyl-4(3H)-
quinazolinone (**156**, R = CH$_2$OH) (no solvent, 180°, 6 h: $\sim$ 55%).[123]

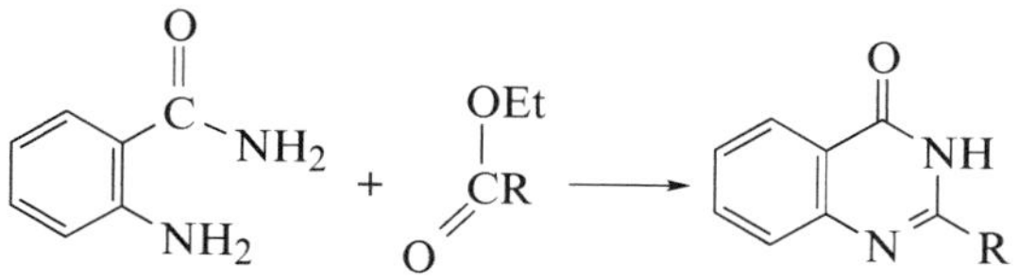

(155)   **(156)**

o-(α-Aminoethyl)aniline (**185**) with acetamidine hydrochloride gave 2,4-dimethyl-3,4-dihydroquinazoline (**186**) (conditions?: 81%).[80]

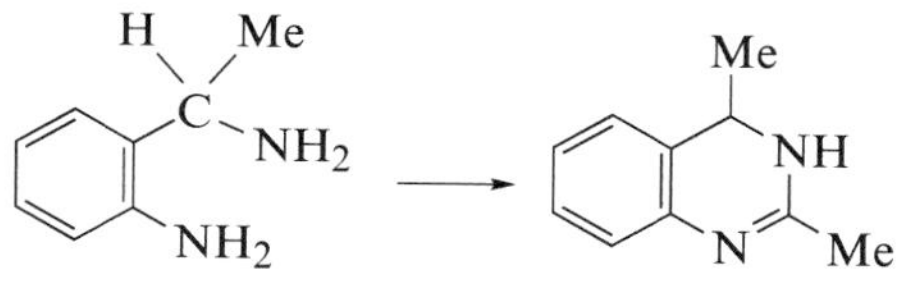

(**185**)   (**186**)

o-Methylaminobenzylamine with formamidine acetate gave 1-methyl-1,4-dihydroquinazoline (**187**) (EtOH, reflux, 16 h: 67%).[732]

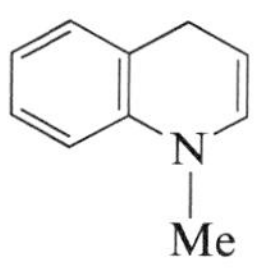

(**187**)

Also other examples in the foregoing references.

Aldehydes as synthons

Note: With the usual substrates, aldehyde synthons necessarily afford dihydroquinazolinones, which may tautomerize[205] and may undergo dehydrogenation, spontaneous or otherwise.

o-Amino-*N*-methylbenzamide (**188**) with formaldehyde gave 3-methyl-1,2-dihydro-4(3*H*)-quinazolinone (**189**, R = H) (EtOH–H$_2$O, NaOH, reflux, 15 min: ∼ 60%); substrate (**188**) with acetaldehyde acetal gave the 2,3-dimethyl homolog (**189**, R = Me) (EtOH, trace TsOH, reflux, 3 h: ∼ 60%).[649]

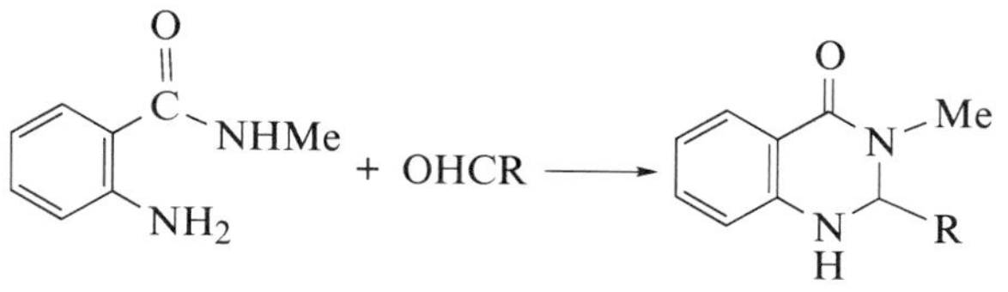

(**188**)   (**189**)

2-Amino-4-chloro-5-sulfamoylbenzamide with benzaldehyde gave a separable mixture of 7-chloro-4-oxo-2-phenyl-3,4-dihydro-6-quinazolinesulfonamide (**190**) and its 1,2,3,4-tetrahydro analog (MeOCH$_2$CH$_2$OH, 115°, 15 h: ∼ 5% and ∼ 70%, respectively).[583, cf. 585]

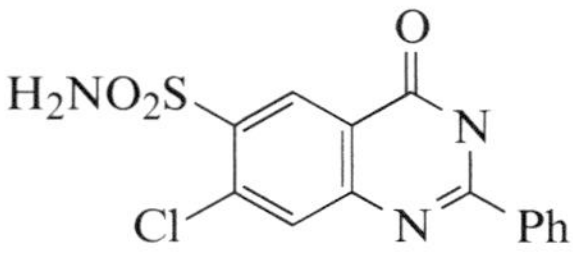

(190)

o-Ethylaminobenzanilide with acetaldehyde acetal gave 1-ethyl-2-methyl-3-phenyl-1,2-dihydro-4(3*H*)-quinazolinone (**191**) (EtOH, trace H_2SO_4, reflux, 4.5 h: 33%; oxidation blocked).[645]

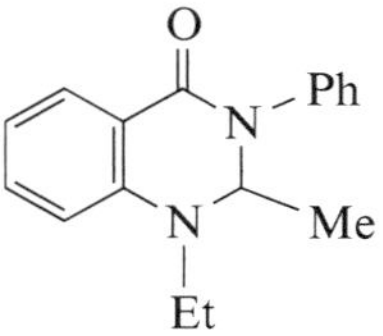

(191)

o-Aminobenzamide oxime with benzaldehyde gave 2-phenyl-1,2-dihydro-4-quinazolinamine 3-oxide (**192**) (EtOH, 20°, 1 h: 88%), which underwent thermal Dimroth rearrangement to 4-hydroxyamino-2-phenyl-1,2-dihydroquinazoline (**193**) (PhMe, 140°, sealed, 4 h: 87%).[1828]

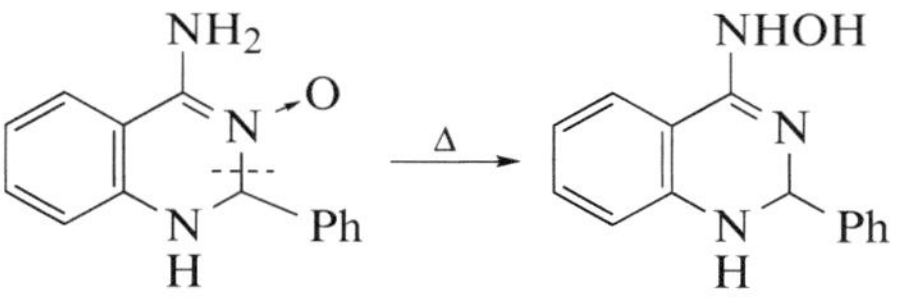

(192)   **(193)**

1-Methylamino-2-methylaminomethylcyclohexane with formaldehyde gave 1,3-dimethyldecahydroquinazoline (**194**) (H_2O, 20°, 12 h: 91% as dipicrate);[193] likewise, 2-aminomethylcyclohexylamine gave decahydroquinazoline (91%).[179]

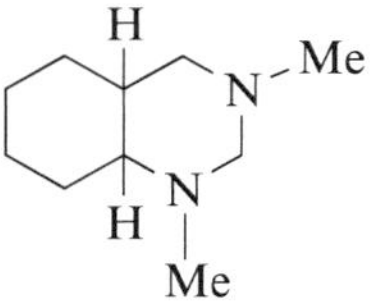

(194)

o-Amino-*N*-pyridin-2′-ylbenzamide with 3-pyridinecarbaldehyde gave 3-pyridin-2′-yl-2-pyridin-3′-yl-1,2-dihydro-4(3*H*)-quinazolinone (EtOH, reflux, 5 h: 94%), which easily underwent oxidation to product (**195**) (KMnO₄, AcMe, reflux, 1 h: 60%).[983]

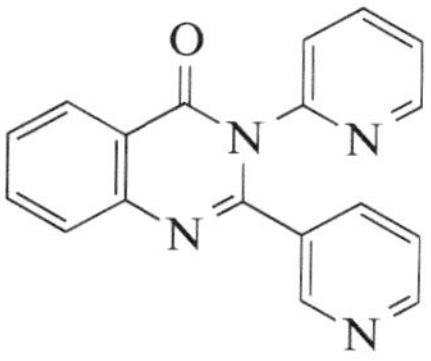

(195)

o-Azidobenzamide with benzaldehyde gave *directly* 2-phenyl-4(3*H*)-quinazolinone (no solvent, 120°, 2 h: 74%).[1050]

o-Anilino-*N'*-methylbenzohydrazide with benzaldehyde gave 3-methylamino-1,2-diphenyl-1,2-dihydro-4(3*H*)-quinazolinone (EtOH, reflux, 48 h: 33%).[858]

o-Aminobenzamide with benzaldehyde (in the presence of NaHSO$_3$) gave the dehydrogenated product, 2-phenyl-4(3*H*)-quinazolinone (NAHSO$_3$, AcNMe$_2$, 150°, 2 h: 95%; dehydrogenation mechanism is unknown).[1211]

N-Methyl-*o*-methylaminobenzamide with ethyl propiolate ($\simeq$ OHCCH$_2$CO$_2$Et) gave 2-ethoxycarbonylmethyl-1,3-dimethyl-1,2-dihydro-4(3*H*)-quinazolinone (**196**) (EtOH, reflux, 18 h, then NaOEt, reflux, 1 h: 63%).[1423]

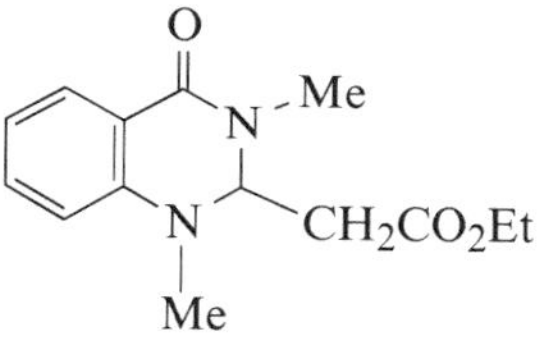

(196)

o-Aminobenzylamine with *N*-(α-chloropropyl)pyridinium chloride ($\simeq$ EtCHO) gave 2-ethyl-1,2,3,4-tetrahydroquinazoline (AcONa, CH$_2$Cl$_2$–H$_2$O, 20°, 1 h: 95%) and thence 2-ethylquinazoline (2,3-dichloro-5,6-dicyanobenzoquinone, PhH, 20°, 1 h: 72%).[2478]

Other aldehyde-equivalent synthons include dimethylmethyleneammonium chloride (Me$_2\overset{+}{\text{N}}$=CH$_2$ Cl$^-$: $\simeq$ CH$_2$O),[2233] 4-methyl-2-phenyl (**197**, R = H), or 2-*p*-methoxyphenyl-4-methyltetrahydrooxazole (**197**, R = OMe) ($\simeq$ benzaldehyde or *p*-anisaldehyde, respectively),[2233] and benzylideneaniline (PhCH=NPh: $\simeq$ PhCHO, with the advantage that an excess acts as a dehydrogenating agent to afford directly the aromatic rather than the dihydroquinazoline).[1082]

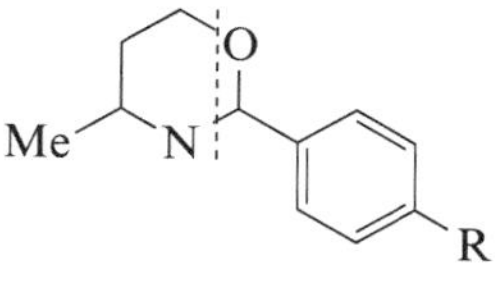

(197)

o-Aminobenzamide with the "unsaturated aldehyde," trifluoroacetylketene diethyl acetal [$F_3CC(=O)CH=C(OEt)_2$], gave directly the aromatic product, 2-(trifluorocetyl)methyl-4(3*H*)-quinazolinone (PhMe, 90°, 3 h: 71%).[2534]

Also many other examples.[18,61,73,102,192,205,230,233,265,273,328,329,344,454,500, 660,662,686,720,772,819,854,901,1256,1275,1406,1470,1534,1590,1630,1634,1665, 1705,1823,1860,1943,1961,1978,2086,2107,2262,2280,2290,2490,2499,2559,2560,2596]

Ketones as synthons

Note: Ketone synthons necessarily give 2,2-disubstituted-1,2-dihydroquinazolines, not prone to simple dehydrogenation.

o-Aminobenzamide (**198**) with an appropriate ketone gave 2-chloromethyl-2-methyl-(**199**, X = CH_2Cl, Y = Me) (ClCH$_2$Ac, PhH, reflux, 4 h: 94%),[256] 2,2-bistrifluoromethyl-(**199**, X = Y = CF_3) [(F$_3$C)$_2$CO, dioxane, reflux, 2 h: 71%],[1391] and 2-benzyl-2-methyl-1,2-dihydro-4(3*H*)-quinazolinone (**199**, X = CH_2Ph, Y = Me) (PhCH$_2$Ac, ion-exchange medium, EtOH, reflux, 8 h: 61%);[817] also 2-methyl-4-oxo-1,2,3,4-tetrahydro-2-quinazolinecarbaldehyde diethyl acetal [**199**, X = Me, Y = $CH(OEt)_2$] [AcCH(OEt)$_2$, PhMe, trace TsOH, reflux, Dean–Stark apparatus, 10 min: 51%],[1561] 2-benzyl-4-oxo-1,2,3,4-tetrahydro-2-quinazolinecarboxylic acid (**199**, X = CH_2Ph, Y = CO_2H) (PhCH$_2$COCO$_2$H, PhH, reflux, Dean–Stark, 3 h: 93%),[948] and 4-oxo-1,2,3,4-tetrahydroquinazoline-2-spiro-4'-(*N*-benzyl-piperidine) [**199**, XY = C(CH$_2$CH$_2$)$_2$NCH$_2$Ph] [PhCH$_2$N(CH$_2$CH$_2$)$_2$CO, TsOH, PhH, reflux, Dean–Stark, 6 h: 88%].[1721]

(**198**)   (**199**)

o-Amino-*N*-(methoxycarbonylmethyl)benzamide with acetone gave 3-methoxycarbonylmethyl-2,2-dimethyl-1,2-dihydro-4(3*H*)-quinazolinone (excess AcMe, HCl, reflux, 6 h: 15%).[1705]

o-Aminobenzophenone *anti*-oxime with ethyl ω-chloroacetoacetate gave 2-chloromethyl-2-ethoxycarbonylmethyl-4-phenyl-1,2-dihydroquinazoline 3-oxide (**200**) (PhH, AcOH, reflux, 1 h: 70%).[2038]

(**200**)

o-Aminobenzamide with benzil gave 2-benzoyl-2-phenyl-1,2-dihydro-4(3*H*)-quinazolinone (**201**) (equimolar, AcOH, ZnCl$_2$, N$_2$, reflux, 3 h: ∼ 50% via ZnCl$_2$ complex),[251] but 2,2'-diphenyl-1,1',2,2'-tetrahydro-2,2'-biquinazoline-4,4' (3*H*, 3'*H*)-dione (**202**) has been isolated (61%) under different (?) conditions;[2007] a similar condensation with benzoin is not simple.[872]

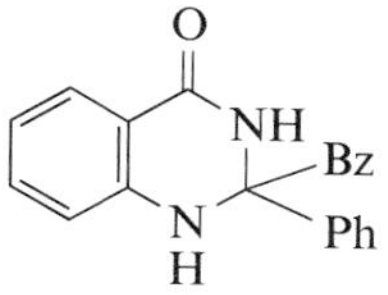

(201) 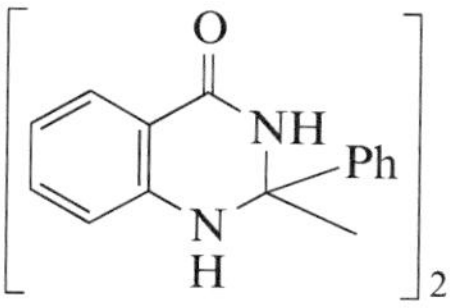　　　　　　**(202)**

2-Amino-5-chlorobenzophenone oxime with dimethyl acetylenedicarboxylate (≃ MeO$_2$CCH$_2$COCO$_2$Me) gave methyl 6-chloro-2-methoxycarbonyl-methyl-4-phenyl-1,2-dihydro-2-quinazolinecarboxylate 3-oxide (**203**) (MeOH, 20°, 1 h: 72%);[490] the same synthon has been used in other such condensations.[480,879,2407]

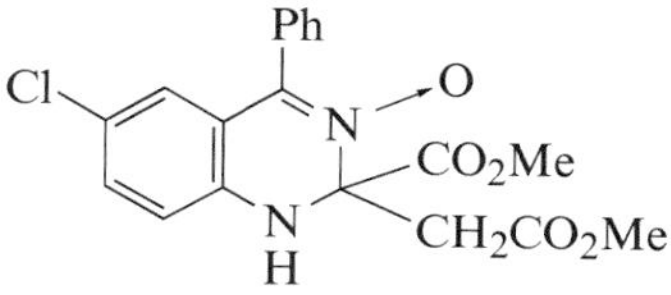

(203)

o-Aminobenzamide with 3-aminocrotonanilide [MeC(NH$_2$)=CHCONHPh, ≃ MeCOCH$_2$CONHPh] gave 2-methyl-2-(*N*-phenylcar-bamoylmethyl)-1,2-dihydro-4(3*H*)-quinazolinone (**204**) (no solvent, 120°, 12 h: 42%).[1370]

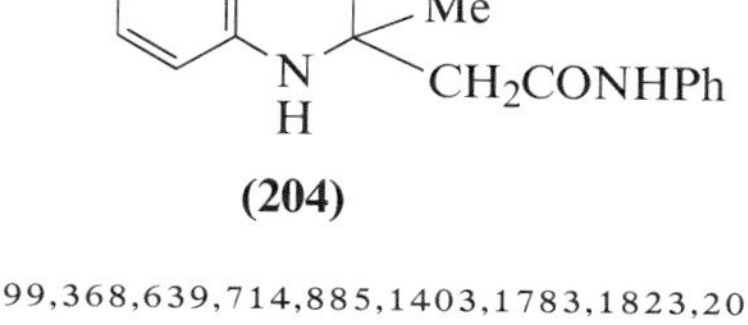

(204)

Also other examples.[299,368,639,714,885,1403,1783,1823,2094,2253,2438,2596]

Alcohols as synthons

Note: Alcohols can condense with the usual substrates only when some oxidative capacity is provided.

o-Aminobenzohydrazide (**205**), on irradiation in ethanol, gave, inter alia, 3-amino-2-methyl-4(3*H*)-quinazolinone (**206**, R = NH$_2$) and 2-methyl-

4(3H)-quinazolinone (**206**, R = H), both arising from a complicated sequence initiated by photochemical oxidation of ethanal to acetaldehyde.[706]

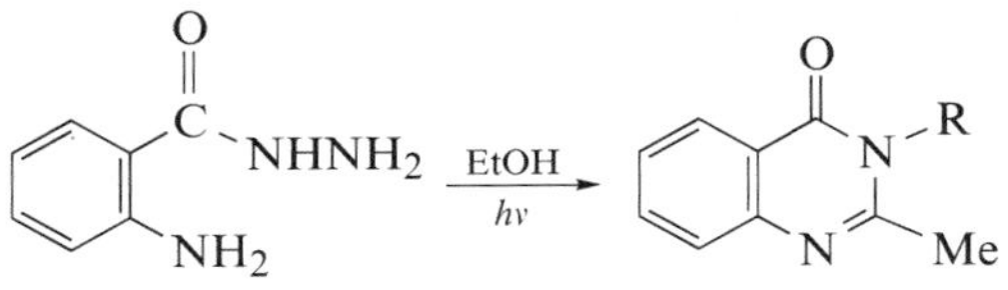

(**205**)   (**206**)

o-Nitrobenzamide (**207**) with benzylamine ($\simeq$ PhCh$_2$OH) gave 2-phenyl-4(3H)-quinazolinone (**208**) (Ph$_2$O, reflux, 1 h: $\sim$ 80%; oxidation of the synthon by the nitro function of the substrate must clearly be involved).[1226]

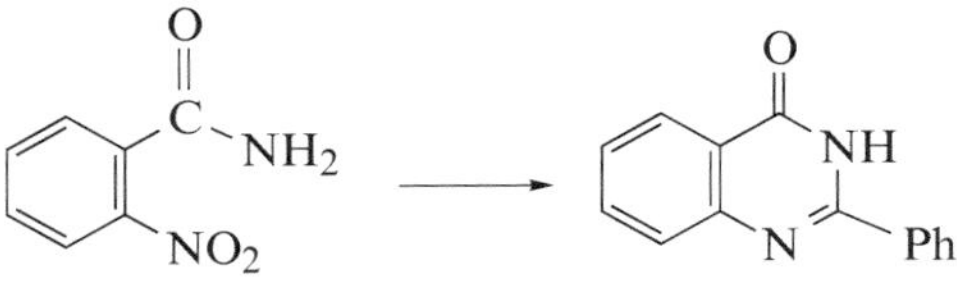

(**207**)   (**208**)

1.2.2.2.   The Use of Carbonic Acid–Derived Synthons (*H* 126)

The following examples illustrate the use of all manner of carbonic acid derivatives to supply C2 in quinazoline syntheses.

Carbon dioxide and carbon disulfide as synthons

Note: Although carbon disulfide reacts easily with the usual substrates (in the presence of a base), carbon dioxide requires an activated substrate for any such reaction.

o-Aminobenzamide   (**209**,   R = H)   gave   2-thioxo-1,2-dihydro-4(3H)-quinazolinone (**210**, R = H) (CS$_2$, KOH, EtOH, reflux, 8 h: 65%).[807]

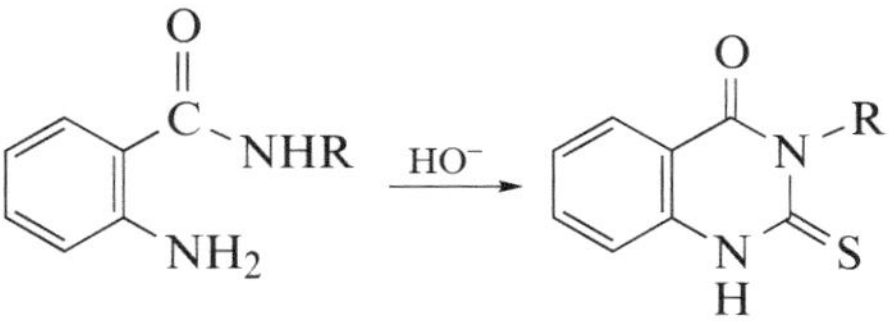

(**209**)   (**210**)

o-Amino-N'-(β-hydroxy-β-methylpropyl)-N'-methylbenzohydrazide   (**209**,
R = NMeCH$_2$CMe$_2$OH)   gave   3-[N-(β-hydroxy-β-methylpropyl)-N-methylamino]-2-thioxo-1,2-dihydro-4(3H)-quinazolinone   (**210**,   R =
NMeCH$_2$CMe$_2$OH) (CS$_2$, NaOH, EtOH–H$_2$O, reflux, 8 h: 94%).[1444]

o-Aminobenzylamine gave 3,4-dihydro-2[1H]-quinazolinethione (**211**) (CS$_2$, KOH, EtOH: 88%).[1573]

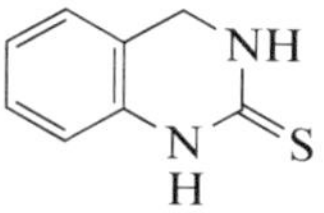

(211)

o-Aminobenzonitrile gave 2,4(1H,3H)-quinazolinedithione (**212**, X = NH, Y = S), via a Dimroth rearrangement of an intermediate 3,1-benzothiazine (**212**, X = S, Y = NH) (CS$_2$, pyridine, reflux, 2 h: 97%);[38] likewise, 2-aminocyclohex-1-ene-1-carbonitrile gave 5,6,7,8-tetrahydro-2,4(1H,3H)-quinazolinedithione (32%).[21]

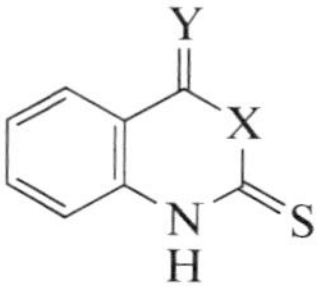

(212)

N-Methyl-o-(triphenylphosphoranylideneamino)benzamide (**213**) gave 3-methyl-2,4(1H,3H)-quinazolinedione (**214**, X = O) (solid CO$_2$, PhMe, 90°, sealed, 15 h: 96%) or 3-methyl-2-thioxo-1,2-dihydro-4(3H)-quinazolinone (**214**, X = S) (CS$_2$, PhMe, 90°, sealed, 10 h: 96%).[1758,2093]

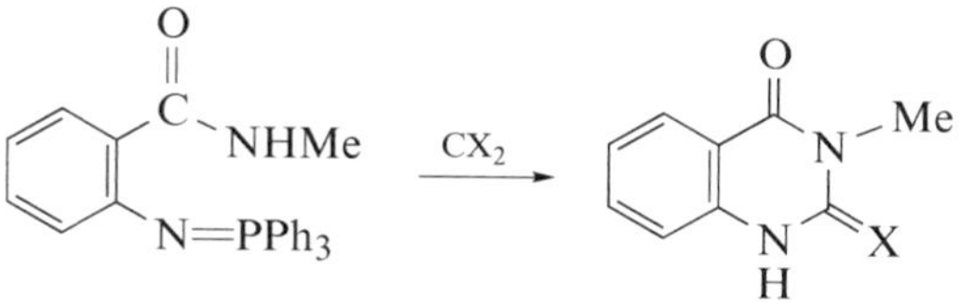

(213)   **(214)**

Also other examples.[125,715,866,1073,1124,1166,1730,2057,2089]

Phosgene and thiophosgene as synthons

o-Amino-N-s-butylbenzamide (**215**) gave 3-s-butyl-2,4(1H,3H)-quinazoline-dione (**216**) (COCl$_2$, dioxane, reflux, 2.5 h: 83%); also many analogs like-wise.[227]

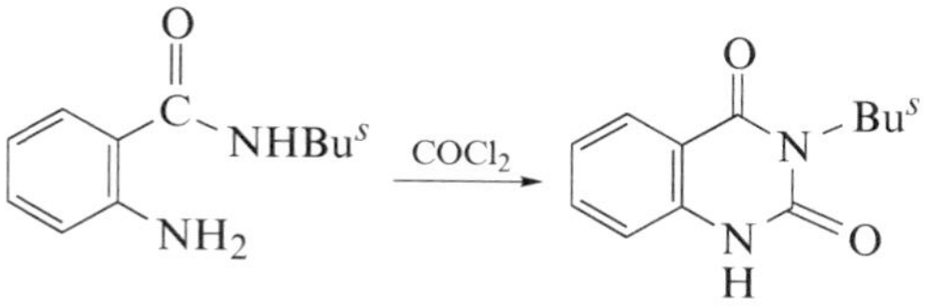

(215)   **(216)**

o-Amino-*N*-(*o*-methoxyphenyl)benzamide gave 3-*o*-methoxyphenyl-2-thioxo-1,2-dihydro-4(3*H*)-quinazolinone (CSCl$_2$, CHCl$_3$, AcOH, H$_2$O, HCl, 20°, 12 h: 84%).[954]

o-(Aminomethyl)aniline gave 3,4-dihydro-2(1*H*)-quinazolinethione (**217**) (CSCl$_2$, NEt$_3$, Et$_2$O, − 78°, 1 h: 58%).[1569,1929]

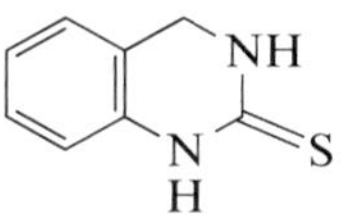

(217)

o-Amino-*N*-phenethylbenzamide gave 3-phenethyl-2,4(1*H*, 3*H*)-quinazolinedione ["triphosgene" (Cl$_3$C–O–CO–O–CCl$_3$), CH$_2$Cl$_2$, reflux, 2 h: 78%].[216]

Also other examples.[89,685,867,1038,1176,1288,1454,1603,1654,1876]

Ethyl chloroformate and the like as synthons

o-Amino(thiobenzamide) gave 4-thioxo-3,4-dihydro-2(1*H*)-quinazolinone (**218**) (EtO$_2$CCl, pyridine, reflux, 3 h: 81%).[405]

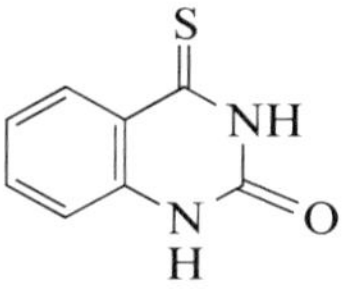

(218)

o-(Cyclopentylaminomethyl)aniline gave 3-cyclopentyl-3,4-dihydro-2(1*H*)-quinazolinone (**219**) (EtO$_2$CCl, pyridine, reflux, 24 h: 75%).[2339]

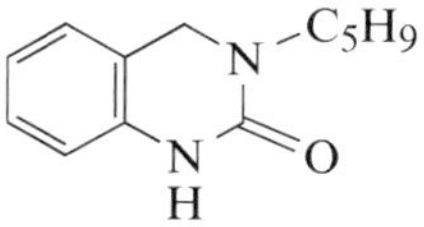

(219)

o-Amino-*N*-(isoquinolin-5-yl)benzamide gave 3-(isoquinolin-5-yl)-2,4(1*H*, 3*H*)-quinazolinedione (EtO$_2$CCl, pyridine, reflux, 20 h: 57%).[2187]

o-(Methoxycarbonylmethyl)aminobenzamide gave 1-methoxycarbonylmethyl-2,4(1*H*, 3*H*)-quinazolinedione (**220**) ["diphosgene" (ClCO$_2$CCl$_3$), dioxane, reflux, 10 h: 34%].[1822]

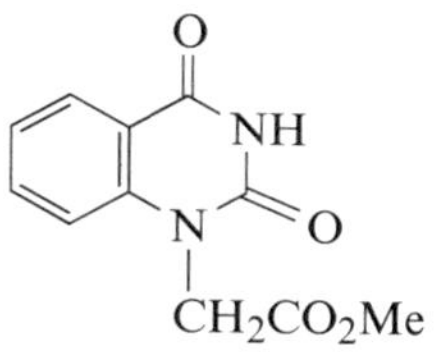

O

NH

N

O

CH$_2$CO$_2$Me

(220)

2-Amino-5-nitrobenzonitrile gave 6-nitro-2,4(1H,3H)-quinazolinedithione (221) (EtOCS$_2$K, BuOH, reflux, 4 h: 80%; mechanism unproved).[77]

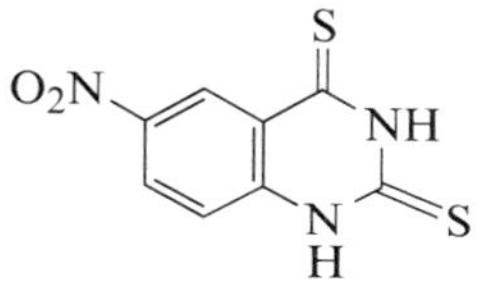

O$_2$N

S

NH

N
H

S

(221)

o-Aminobenzamide gave 2,4(1H,3H)-quinazolinedione (ClCOSCl, dioxane, reflux, 30 min: 52%).[1920]

Also other examples.[116,1073,1243,1437,1450,1596]

Urea and the like as synthons

o-Aminobenzanilide gave 3-phenyl-2-thioxo-1,2-dihydro-4(3H)-quinazolinone (222) [(H$_2$N)$_2$CS, 190°, 1 h: 80%].[357]

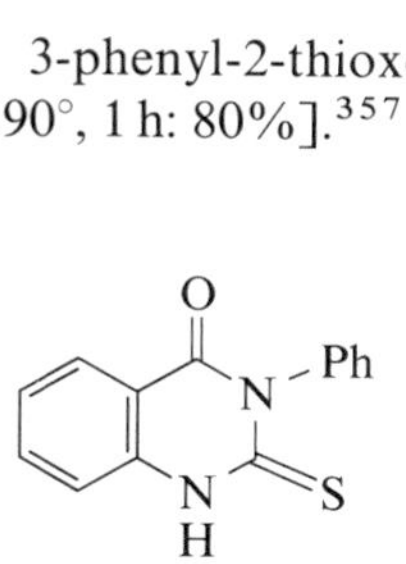

O

N — Ph

N
H

S

(222)

4-Amino-5-benzyloxy-4-methoxybenzamide gave 6-benzyloxy-7-methoxy-2, 4(1H,3H)-quinazolinedione [(H$_2$N)$_2$CO, pyridine, trace HCl, reflux, 24 h: 71%].[785]

o-(Methylaminomethyl)aniline gave 3-methyl-3,4-dihydro-2(1H)-quinazolinone (223) [(H$_2$N)$_2$CO, 195°, 40 min: 77%;[656] (H$_2$N)$_2$CO, H$_2$O, HCl, reflux, 24 h: 58%;[1936] or 1,1′-carbonyldiimidazole (224; ≃ urea), THF, reflux, 19 h: 46%].[2339]

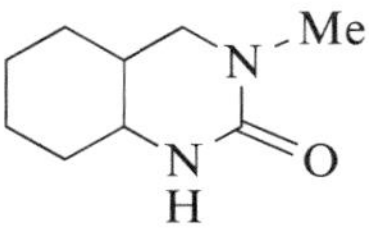

(223) 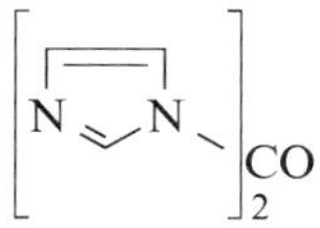   **(224)**

o-(Methylamino)benzylamine (**225**) gave 1-methyl-1,4-dihydro-2-quinazo-linamine (**226**) [MeSC($=\overset{+}{N}H_2$)NH$_2$.H$_2$SO$_4$, EtOCH$_2$CH$_2$OH, reflux, 24 h: 61%; note the elimination of SMe and retention of one NH$_2$];[173] somewhat similarly, *o*-(aminomethyl)aniline gave 2-ethoxycarbonylamino-3,4-dihydroquinazoline [MeSC($=$NCO$_2$Et)NHCO$_2$Et, EtOH, reflux, 12 h: 52%].[1186]

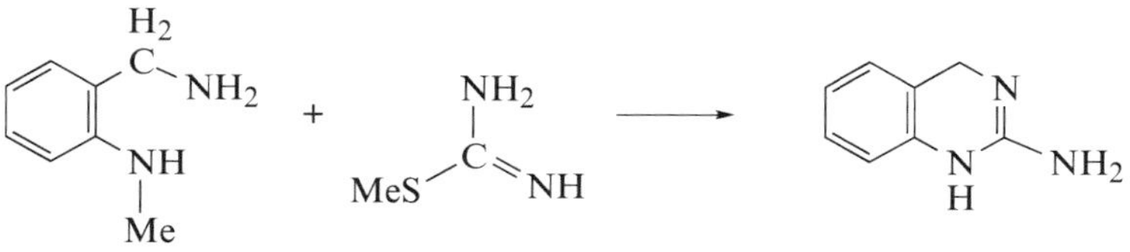

(225)   **(226)**

N-Methyl-*o*-(triphenylphosphoranylideneamino)benzamide (**227**) gave 2-ani-lino-3-methyl-4(3*H*)-quinazolinone (**228**) (PhNCO, PhMe, reflux, 24 h: 97%).[1758,2093]

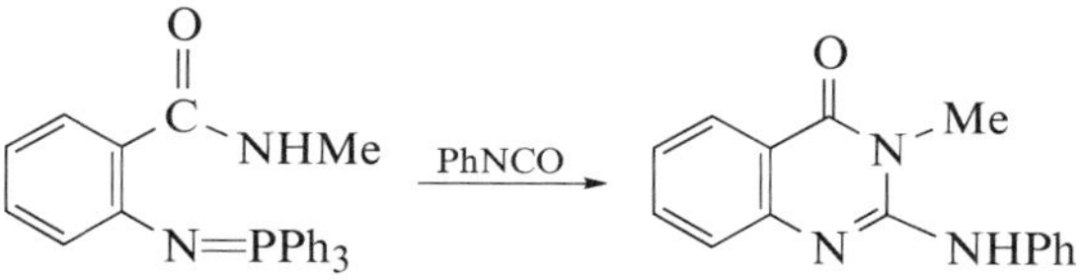

(227)   **(228)**

Also other examples in the foregoing references and elsewhere.[654,1004,1466, 2074,2262,2550]

Cyanogen bromide and chloroformamidine as synthons

o-Aminomethyl-*N*-benzylaniline (**229**) gave 1-benzyl-1,4-dihydro-2-quin-azolinamine (**230**) (BrCN, EtOH, reflux, 5 h: 70%).[400]

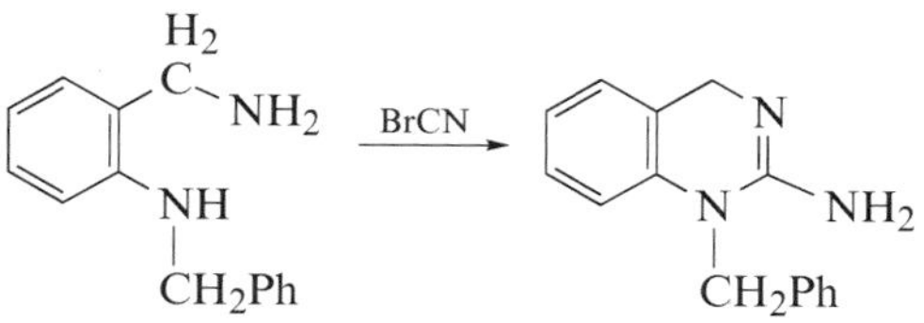

(229)   **(230)**

2-Aminomethyl-3-chloroaniline   gave   5-chloro-3,4-dihydro-2-quinazo-linamine (BrCN, PhMe, 100°, 4 h: 90%).[1711]

2-Amino-4-trifluoromethylbenzamide (**231**) gave 2-amino-7-trifluoromethyl-4(3H)-quinazolinone (**232**) [ClC(=NH)NH$_2$.HCl, (MeOCH$_2$CH$_2$)$_2$O, 130° then reflux, 15 min: 60%].[678]

(**231**)   (**232**)

2-Amino-6-chlorobenzamide   gave   2-amino-5-chloro-4(3H)-quinazolinone [ClC(=NH)NH$_2$.HCl, Me$_2$SO$_2$, 160°, 1 h: > 95%].[2015]

Dialkyl carbonates as synthons

4-Chloro-2-(C-cyclopropyl-C-aminomethyl)aniline (prepared *in situ* from 2-amino-5-chlorobenzonitrile and cyclopropyl magnesium bromide) gave 6-chloro-4-cyclopropyl-2(1H)-quinazolinone (Me$_2$CO$_3$, THF, 50°, 30 min: 79% overall).[2563]

1.2.3.   Where the Synthon Supplies N3

In this widely used synthesis, a primary amine has been employed almost always as the synthon to supply N3; in contrast, many types of benzene substrate have been used, thus affording a convenient basis for subdivision in the following treatment.

1.2.3.1.   With *o*-Acylaminobenzoic Acids as Substrates (*H* 93,98)

The easily obtained *o*-acylaminobenzoic acids have been used extensively as starting materials for the synthesis of 4-quinazolinones and thiones. The following examples indicate typical condensing agents, conditions, and yields.

o-Acetamidobenzoic acid (**233**) gave 2-methyl-3-*o*-tolyl- (**234**, R = C$_6$H$_4$Me-*o*) (*o*-MeC$_6$H$_4$NH$_2$, PhMe, PCl$_3$, reflux, 3 h: 66%),[2223] 3-*p*-methoxycar-bonylphenyl-2-methyl- (**234**, R = C$_6$H$_4$CO$_2$Me-*p*) (*p*-H$_2$NC$_6$H$_4$CO$_2$Me, PhMe, PCl$_3$, reflux, 90 min: 63%),[1580] or 2-methyl-3-phenyl-4(3H)-quinazolinone (**234**, R = Ph) [PhNH$_2$, PhMe, PCl$_3$, reflux, 2 h: ~ 70%;[146] PhNH$_2$, pyridine, P(OPh)$_3$, N$_2$, 100°, 3 h: 84%;[1422] or PhNCS, trace pyridine, 160°, 30 min: 54%][1644] (cf. the Grimmel–Guenther–Morgan synthesis; *H* 93).

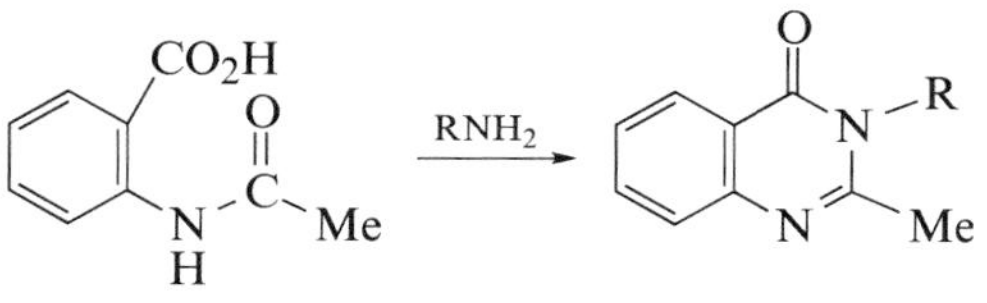

(233)   **(234)**

o-(α-Chloroacetamido)benzoic acid gave 2-chloromethyl-3-*p*-chlorophenyl-4(3*H*)-quinazolinone ($H_2NC_6H_4Cl$-*p*, $POCl_3$, PhMe, reflux, 3 h: 70%).[1806]

o-(α-Phenylacetamido)benzoic acid gave 2-benzyl-3-phenyl-4(3*H*)-quinazolinone (PhNH$_2$, polyphosphoric acid, 190°, 1 h: 40%).[525]

o-Formamidobenzoic acid gave 3-*p*-ethylphenyl-4(3*H*)-quinazolinone ($H_2NC_6H_4Et$-*p*, PhOH, reflux, 4 h: 40%).[669]

o-Propionamidobenzoic acid gave 3-amino-2-ethyl- (**235**, R = H) (H_2NNH_2, < 190°, 45 min: 52%) or 2-ethyl-3-thioureido-4(3*H*)-quinazolinone (**235**, R = $CSNH_2$) (H_2NCSNH_2, < 190°, 45 min: 55%).[1979]

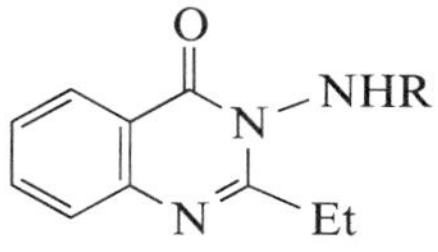

(235)

o-(Cyclohexanecarbonylamino)benzoic acid gave 2-cyclohexyl-4(3*H*)-quinazolinone (H_2NCHO, 175°, 3 h: ∼ 80%; such use of formamide represents one stage in Niementowski's synthesis; see Section 1.2.6.1) or 2-cyclohexyl-4(1*H*)-quinazolinone 3-oxide (**236**) ($HONH_2$.HCl, pyridine, reflux, 6 h: 93%).[487]

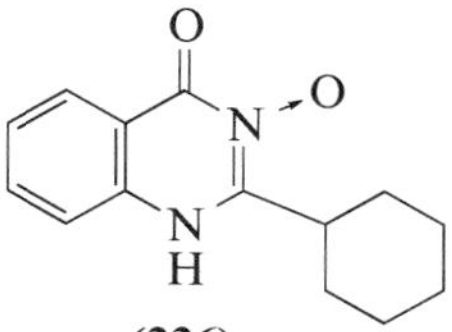

(236)

Also additional examples using PCl$_3$ or POCl$_3$,[135,142,143,149,151,269, 282,323,394,489,548,579,587,617,692,791,884,983,1198,1384,1389,1402,1587,2306, 2308,2372] heat with or without a solvent,[114,331,391,695,1289,1671] polyphosphoric acid,[14,702,1111,1865] the Niementowski conditions (H_2NCHO),[147, 564,1731] or other ways.[358,461, 509,786,1081,1151,1608,1849]

1.2.3.2. With *o*-Acylaminobenzoic Esters as Substrates

Most of the cyclizations of *o*-acylaminobenzoic esters by amines have been done simply by warming without any additional condensing agent. The following examples illustrate conditions for this and for some exceptional procedures.

Methyl o-(α-chloroacetamido)benzoate (237) gave 2-chloromethyl-4(3H)-quinazolinone (238) (NH₃ gas, MeOH, 25°, 3 days: ~80%, allowing for recovered substrate).[2195]

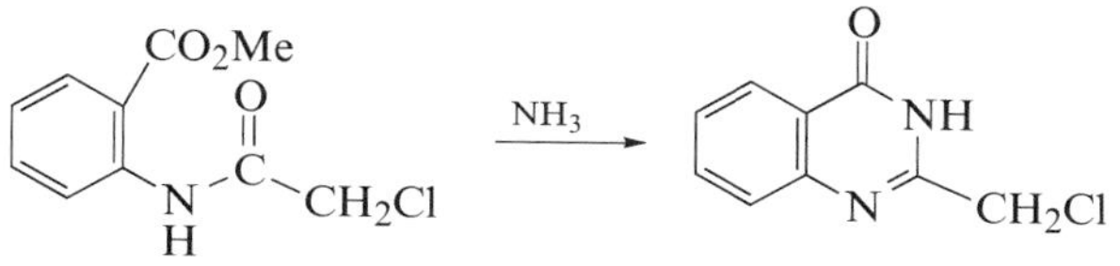

(237)   (238)

Methyl o-acetamidobenzoate gave 3-amino-2-methyl-4(3H)-quinazolinone (239, R = H) (H₂NNH₂.H₂O, EtOH, reflux, 19 h: ~80%).[200, cf. 2128,2317]

Methyl o-(α-diethylaminoacetamido)benzoate gave 3-amino-2-diethylaminomethyl-4(3H)-quinazolinone (239, R = NEt₂) (H₂NNH₂.H₂O, BuOH, reflux, 26 h: 98%).[816]

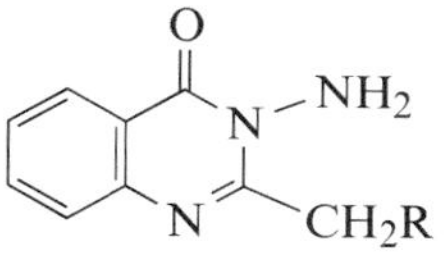

(239)

Methyl o-acetamidobenzoate gave a separable mixture of 3-s-butyl-2-methyl-4(3H)-quinazolinone (240) and 4-s-butylamino-2-methylquinazoline (241) (s-BuNH₂.HCl, P₂O₅, Me₂NC₆H₁₁, 180°, 45 min: 71% and unstated small yield, respectively).[1267]

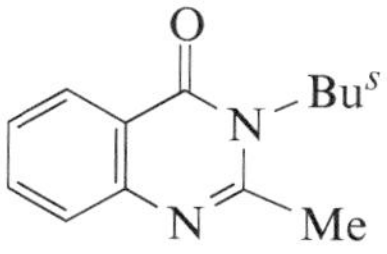

(240) 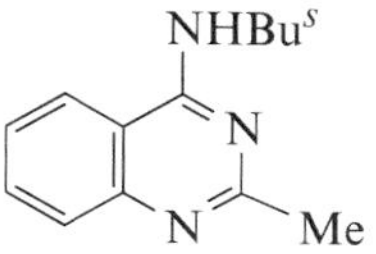   (241)

Ethyl o-(α,α,α-trichloroacetamido)benzoate gave 2,4(1H,3H)-quinazolinedione rather than 2-trichloromethyl-4(3H)-quinazolinone (NH₄OAc, Me₂SO, 95°, 4 h: 78%).[1060]

Methyl o-acetamidobenzoate gave 2,3-dimethyl-4(3H)-quinazolinone [PhOP(O)(NHMe)₂, ~280°, 10 min: 76%].[853]

Also additional examples without a condensing agent,[183,340,851,1308,1351, 1355,1473,1659,1818,1826,1888,1944,1947,1952,1964,2063] using P₂O₅,[1615,2155] or using PCl₃ or POCl₃.[530,1045,1177]

1.2.3.3. With o-Acylaminobenzonitriles as Substrates

This cyclization has been little used. It is typified by the reaction of o-(N-benzylbenzamido)benzonitrile (**242**) with hydroxylamine (MeOH, 95°, sealed, 4 h) to give 1-benzyl-2-phenyl-4(1H)-quinazolinone oxime (**243**) (69%)[2132] and of o-acetamidobenzonitrile with aniline (P_2O_5, $Me_2NC_6H_{11}$.HCl, $PhNH_2$.HCl, limited H_2O, reflux, 200°, 30 min; then substrate added, 180°, 20 min) to give 2-methyl-3-phenyl-4(3H)-quinazolinone (**244**) (43%); when the water was omitted in the latter case, 4-anilino-2-methylquinazoline was obtained.[1782]

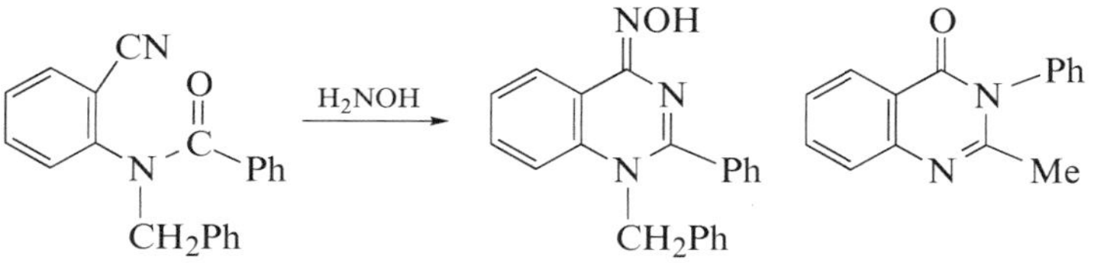

(242)   (243)   (244)

1.2.3.4. With o-Acylaminobenzaldehydes or Related Ketones as Substrates (H 13,39,99,337)

The reaction of ammonia or a primary amine with such a substrate, a process sometimes known as *Bischler's quinazoline synthesis* (H 39), must give initially a 3,4-dihydroquinazoline that can undergo spontaneous dehydrogenation to a quinazoline when N3 bears a proton (from ammonia); however, when N3 bears an alkyl or other group (from an amine), no such oxidation is possible unless the product is acceptable as a quaternary salt. The products arising from such cyclizations and the conditions required are illustrated as follows.

o-(o-Methoxybenzamido)benzaldehyde (**245**) gave 2-o-methoxyphenyl-quinazoline (**247**), via the unisolated hydrate (**246**) (NH_3, EtOH, 160°, sealed, 16 h: ?%).[226]

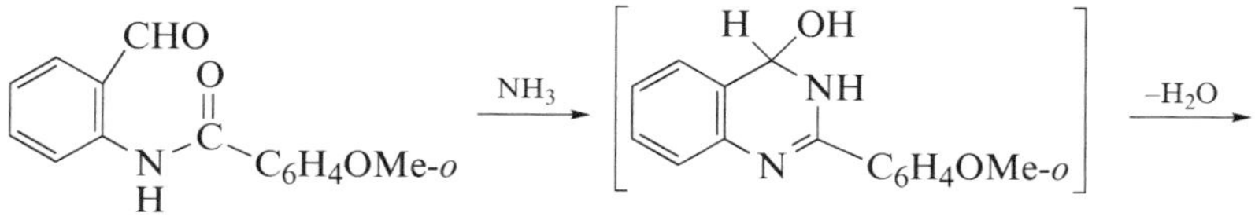

(245)   (246)

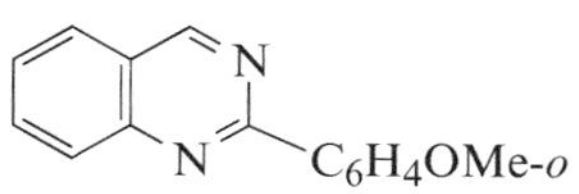

(247)

2-Formamido-4,5-dimethoxypropiophenone gave 4-ethyl-6,7-dimethoxy-quinazoline (HCO$_2$NH$_4$, NH$_3$ gas, 125°, 6 h: 69%; NH$_3$ gas, EtOH, sealed, 150°, 12 h: 90%).[261]

o-Formamidoacetophenone gave 4-methylquinazoline (AcONH$_4$, NH$_3$ gas, 137°, 3 h: 60%).[1819]

N,N'-Bis(2-benzyl-4-chlorobenzoyl)-1,3-propanedicarboxamide gave 1,3-bis(6-chloro-4-phenylquinazolin-2-yl)propane (248) (AcONH$_4$, NH$_3$ gas, 175°, 6 h: 91%).[804]

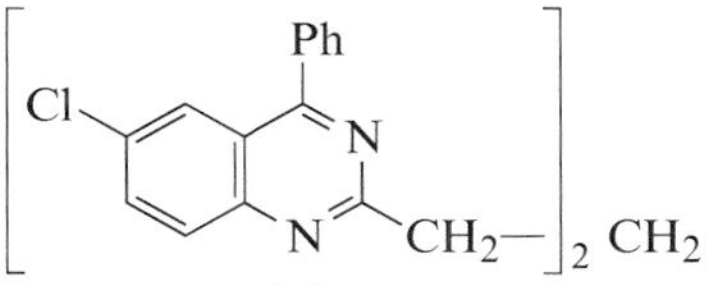

(248)

2-Formamido-5-chlorobenzophenone (249) gave 3-amino-6-chloro-4-phenyl-3,4-dihydro-4-quinazolinol (250) (H$_2$NNH$_2$, EtOH, 20°, 12 h: 92%).[257]

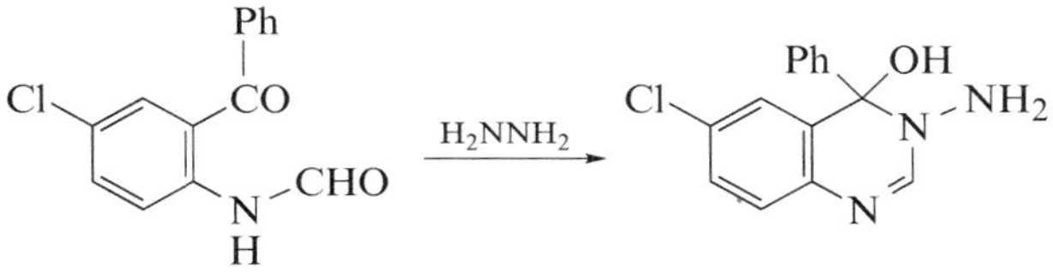

(249)   **(250)**

o-(N-Methylformamido)benzophenone gave 1-methyl-4-phenylquinazolinium iodide (251) (NH$_4$I, EtOH, 150°, sealed, 10 h: 5%; also prepared by dissolution of 1-methyl-4-phenyl-1,4-dihydro-4-quinazolinol in HI/PrOH: 89%).[353]

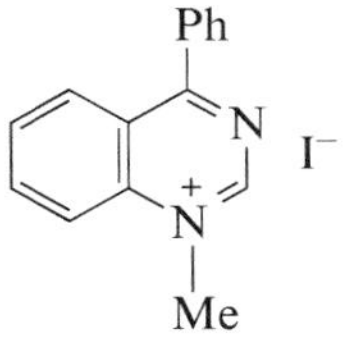

(251)

Also other examples.[33,239,271,658,742,773,836,865,1305,1425,1511,1532,1570,1941,2496]

1.2.3.5. With o-(Alkoxycarbonylamino)benzoic Esters or Related Ketones as Substrates (H 70,122)

These condensations are more useful than their sparse use might suggest. Typical examples follow.

Methyl o-ethoxy(thiocarbonyl)aminobenzoate (**252**) gave 2-ethoxy-4(3H)-quinazolinone (**253**) (NH₃, EtOH, 20°, 4 days: 94%).[1473]

Ethyl 3-bromo-6-ethoxycarbonylaminobenzoate (**254**) gave 6-bromo-3-β-hydroxyethyl-2,4(1H,3H)-quinazolinedione (**255**) [neat HOCH₂CH₂NH₂, 160°, open, 30 min: 91%; compare the course of this reaction with that of the previous example).[1611]

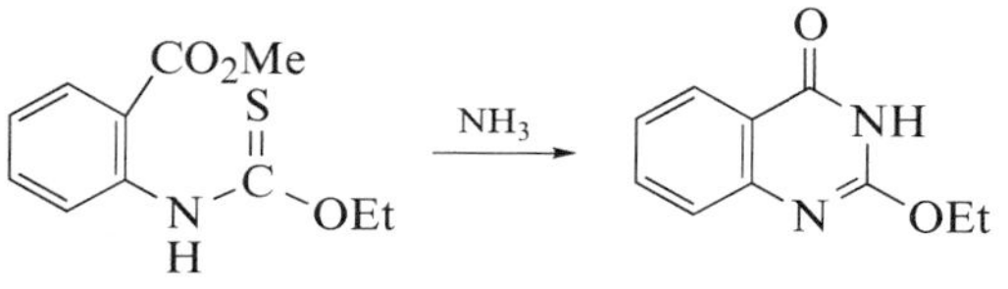

(**252**)   (**253**)

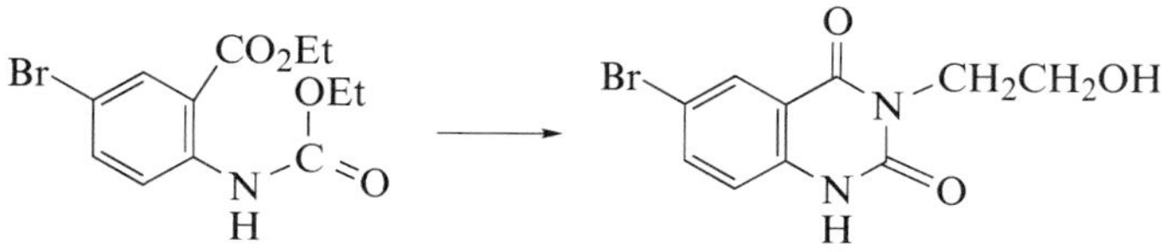

(**254**)   (**255**)

2-Ethoxycarbonylamino-5,6-dimethoxypropiophenone gave 5,6-dimethoxy-4-methyl-2(1H)-quinazolinone (neat AcONH₄, 125°, 2 h: 74% as hydrochloride).[1592]

3-Chloro-6-ethoxycarbonylaminobenzophenone (**256**) gave 3-benzyl-6-chloro-4-hydroxy-4-phenyl-3,4-dihydro-2(1H)-quinazolinone (**257**) (PhCH₂NH₂, xylene, reflux, Dean–Stark, 15 h: ~ 30%), which underwent subsequent dehydrative oxidation to give 3-benzyl-6-chloro-4-phenyl-2(3H)-quinazolinone (**258**) (sublimation, 250°/0.2 mmHg).[552]

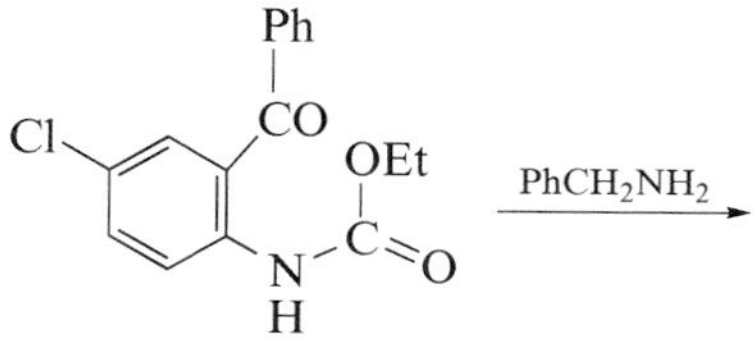

(**256**)

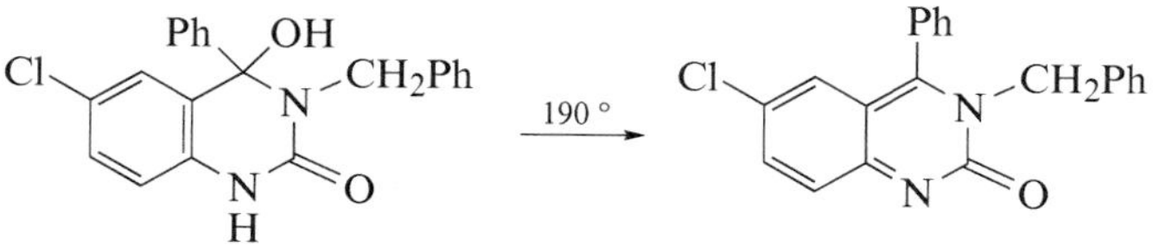

(**257**)   (**258**)

Also other examples.[928,1400,2410]

1.2.3.6. With o-(Substituted Methyleneamino)benzoic Esters or Related Ketones as Substrates

These substrates embrace a variety of substituent types and also some closely related tautomeric or isomeric structures, as indicated in the following examples.

Methyl o-(ethoxymethyleneamino)benzoate (259) gave 3-methylamino-4(3H)-quinazolinone (260) (neat H_2NNHMe, warm, 5 min: 92%).[934]

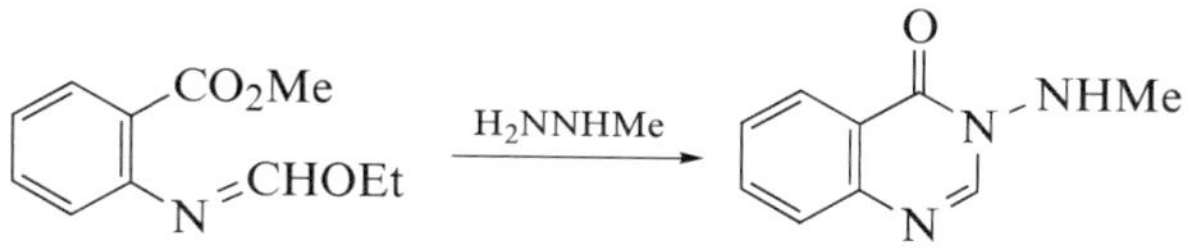

CO2Me

N=CHOEt

H2NNHMe

O

N—NHMe

N

(259)　　　　(260)

Methyl o-(phenyliminomethyleneamino)benzoate (261) gave 2-anilino-4(3H)-quinazolinone (NH_3, THF, 25°, 2 h: 76%).[2242]

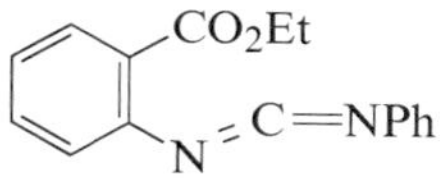

CO2Et

N=C=NPh

(261)

Methyl o-(dimethylaminomethyleneamino)benzoate (262) gave 3-butyl-4(3H)-quinazolinone (BuNH₂ TsOH, dioxane, N₂, reflux, 4 h: 61% after purification).[1970]

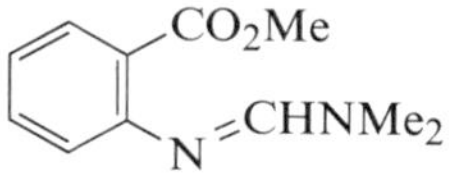

CO2Me

N=CHNMe2

(262)

Methyl o-[mercapto(methylthio)methyleneamino]benzoate (263) gave 3-amino-2-thioxo-1,2-dihydro-4(3H)-quinazolinone ($H_2NNH_2 \cdot H_2O$, 20°, 30 min: 43%).[1736]

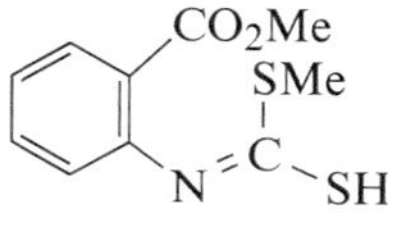

CO2Me

SMe

N=C—SH

(263)

Methyl o-(N'-cyano-O-phenylisoureido)benzoate, formulated as (264), gave 2-cyanoamino-4(3H)-quinazolinone (NH_3, PrOH, 20°, 1 h: 84%).[2000]

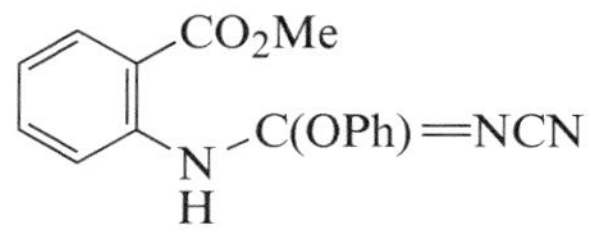

(264)

Methyl o-(β,β-diethoxycarbonylvinylamino)benzoate (**265**) gave 3-amino-4(3H)-quinazolinone ($H_2NNH_2.H_2O$, EtOH, 20°, 60 h: 82%).[614]

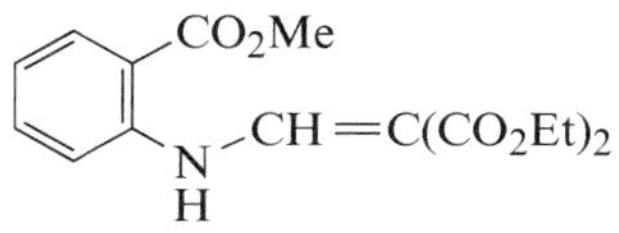

(265)

3-Chloro-6-(α-ethoxyethylideneamino)benzophenone (**266**) gave 3-[α-(benzyloxycarbonylamino)acetamido]-6-chloro-2-methyl-4-phenyl-3,4-dihydro-4-quinazolinol (**267**) ($PhCH_2O_2CNHCH_2CONHNH_2$, AcOH, EtOH, 20°, 6 h: 60%).[1560]

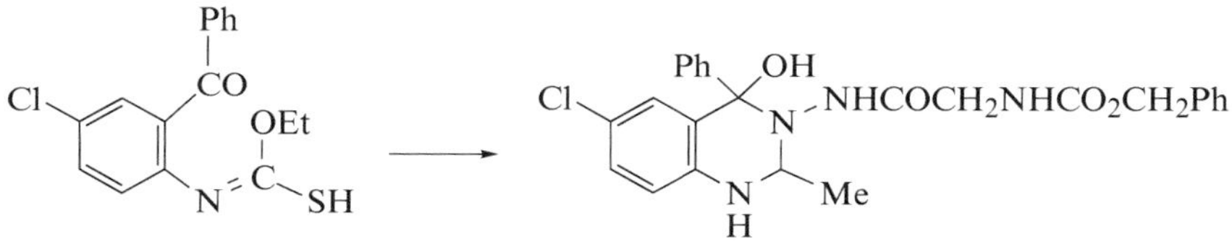

(266)   **(267)**

3-Chloro-6-(β,β-diethoxycarbonylvinylamino)benzophenone (**268**) gave 6-chloro-4-phenylquinazoline (NH_3, MeOH, 20°, 60 h: 50%).[2415]

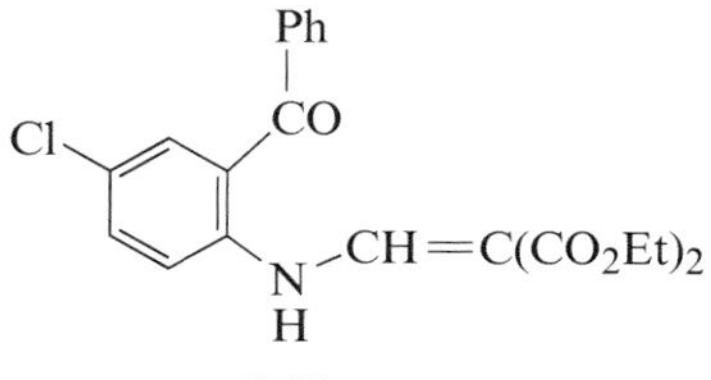

(268)

Also other examples in the foregoing references and elsewhere.[369,690,927, 1994,2562]

1.2.3.7.   With o-(Substituted Methyleneamino)benzonitriles as Substrates

These cyclizations have been used to produce 4-quinazolinamines or corresponding (fixed) imines, as illustrated in the following examples.

o-(Phenyliminomethyleneamino)benzonitrile   (**269**)   gave   2-anilino-4-quinazolinamine (NH$_3$, THF, 25°, 2 h; then MeOH, reflux, 18 h: 53%).[2242]

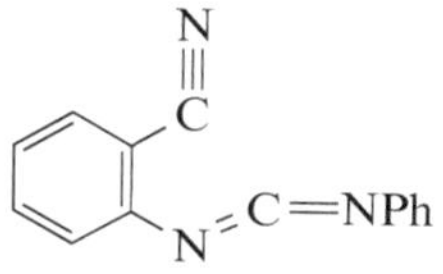

(269)

o-(Ethoxymethyleneamino)benzonitrile (**270**) gave 3-methyl- (**271**, R = Me) (MeNH$_2$, EtOH, 20°, 15 min: 97%)[1000] or 3-benzamido-4(3*H*)-quinazolinimine (**271**, R = NHBz) (BzHNNH$_2$, EtOH, reflux, 1 h: 67%).[2345]

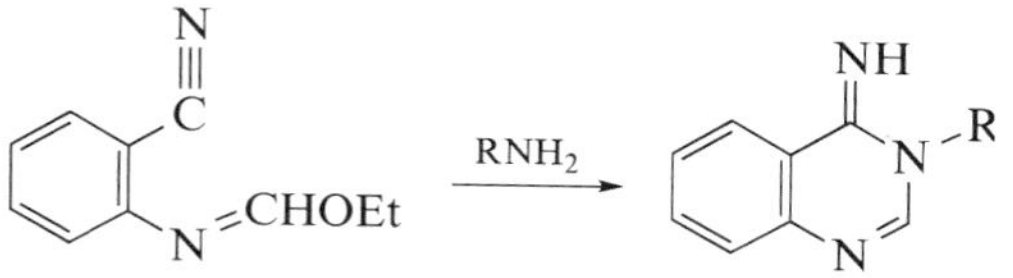

(270)   **(271)**

o-(Dimethylaminomethyleneamino)benzonitrile   (**272**)   gave   4-quinazolinamine (AcONH$_4$, EtOH–H$_2$O, reflux, 2 h: 71%) or its 3-oxide (H$_2$NOH.HCl, EtOH, reflux, 20 min: 91%).[507, cf. 916]

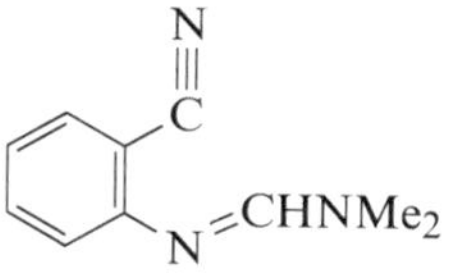

(272)

2-[(4-Formylpiperazin-1-yl)(methylthio)methylene]amino-4,5-dimethoxy-benzonitrile (**273**) gave 2-(4-formylpiperazin-1-yl)-6,7-dimethoxy-4-quinazolinamine (NH$_4$Cl, H$_2$NCHO, 80°, 1 h: 60%).[2288]

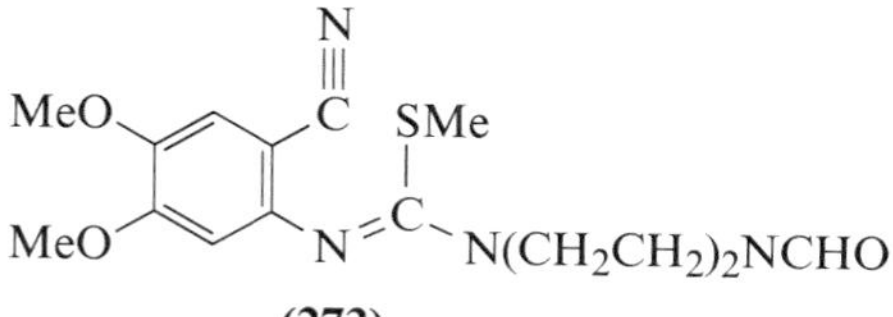

(273)

o-(Benzylideneamino)benzonitrile   (**274**)   gave   2-phenyl-1,2-dihydro-4-quinazolinamine 3-oxide (**275**) (H$_2$NOH, EtOH, 25°, 3 days: 88%).[1828]

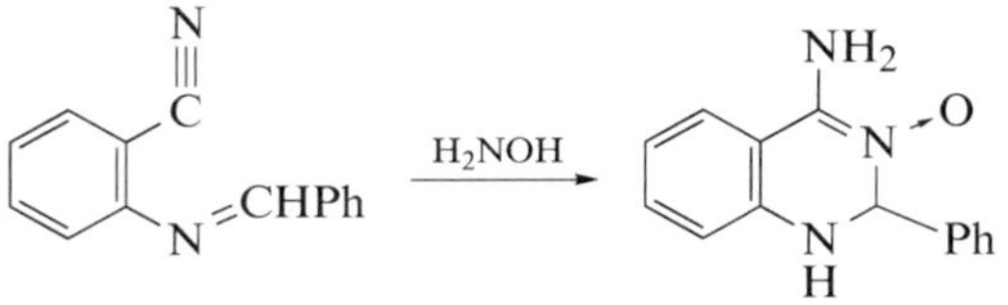

(274)   (275)

Also other examples in the foregoing references and elsewhere.[422,916,1576, 1805,2238]

1.2.3.8.   With *o*-Isocyanatobenzoyl Chlorides as Substrates

o-Isocyanatobenzoyl chloride is easily obtained[857] and has been used to make several 3-substituted-2,4-quinazolinediones. Thus this substrate (**276**) with 2,3- or 2,4-dimethylaniline gave 3-(2,3-dimethylphenyl)- (**277**, R = $C_6H_3Me_2$-2,3) (85%)[1436] or 3-(2,4-dimethylphenyl)-2,4(1*H*,3*H*)-quinazolinedione (**277**, R = $C_6H_3Me_2$-2,4) (90%), respectively (arylamine, Bu_3N, PhMe, 0°, 1 h; then 20°, 12 h; then 100°, 7 h); likewise, with other appropriate amines, it gave 3-(benzothiazol-2-yl)-(Et_3N, PhMe, as before: 50%),[1309] 3-(tetrazol-5-yl)-(THF–H_2O, 20°, 30 min: 31%),[1911] and 3-(*o*-benzoylphenyl)-2,4(1*H*,3*H*)-quinazolinedione (**277**, R = C_6H_4Bz-*o*) (PhH, 20°, 30 min: 64%).[1241] An interesting variation was the self-condensation of *o*-isocyanatobenzoyl chloride (**276**) to give 3-*o*-carboxyphenyl-2,4(1*H*,3*H*)-quinazolinedione (**278**) ($AlCl_3$, MeOH–CH_2Cl_2, 55°, 1 h: 7%; note the hydrolysis of the chlorocarbonyl group during aqueous workup) plus several other separable byproducts.[1742]

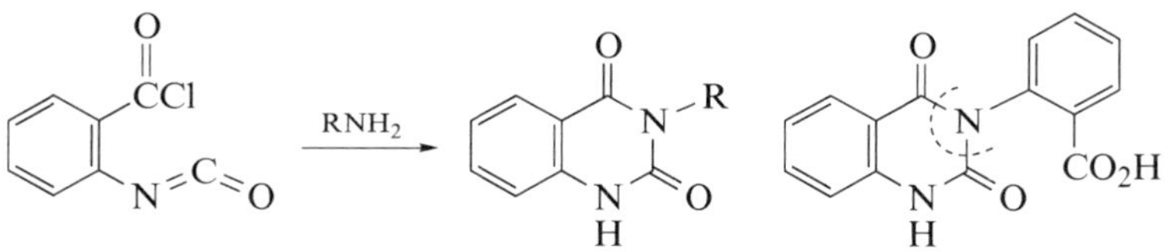

(276)   (277)   (278)

1.2.3.9.   With *o*-Isocyanatobenzoic Esters and Related Ketones or Nitriles as Substrates

These substrates have not been as widely used, as the utility of the following examples might justify.

Methyl *o*-isocyanatobenzoate (**279**) gave 3-dimethylamino-2,4(1*H*,3*H*)-quinazolinedione (**280**) (H_2NNMe_2, PhMe, reflux, 4 days: 72%).[1437]

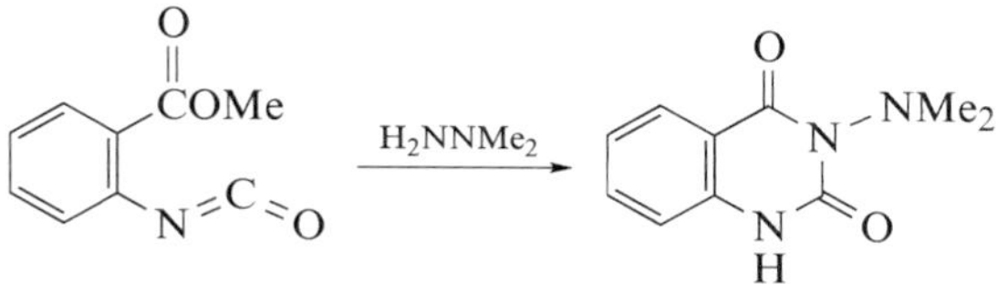

(279)   (280)

Methyl 2-isocyanato-3-nitrobenzoate gave 3-carboxymethyl-8-nitro-2,4($1H,3H$)-quinazolinedione ($H_2NCH_2CO_2H$, dioxane, H_2O, NaOH, 50°, 4 h: 85%).[2435]

Methyl o-isothiocyanatobenzoate gave 3-methoxycarbonylmethyl- (**281**, R = CH_2CO_2Me) ($MeO_2CCH_2NH_2$.HCl, Et_3N, THF, 20°, 20 h: $\sim$93%)[1137] or 3-amino-2-thioxo-1,2-dihydro-4($3H$)-quinazolinone (**281**, R = NH_2) (H_2NNH_2, Et_2O, 20°, 3 h: 88%).[52]

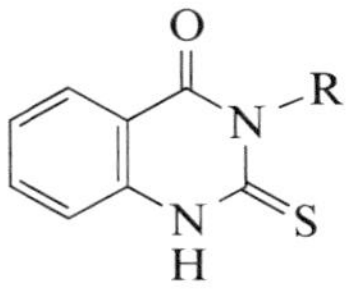

(281)

o-Isothiocyanatoacetophenone gave 3-benzyl-4-hydroxy-4-methyl-3,4-dihydro-2($1H$)-quinazolinethione (**282**) ($PhCH_2NH_2$, conditions?) and thence by dehydration, 3-benzyl-4-methyl-2($3H$)-quinazolinethione (**283**).[2333]

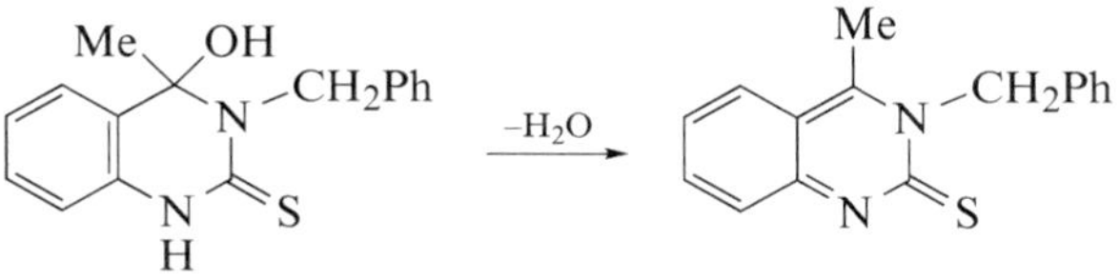

(282)   (283)

o-Isocyanatobenzonitrile (**284**) gave 3,4-diamino-2($3H$)-quinazolinone (**285**) (H_2NNH_2, conditions?).[2004]

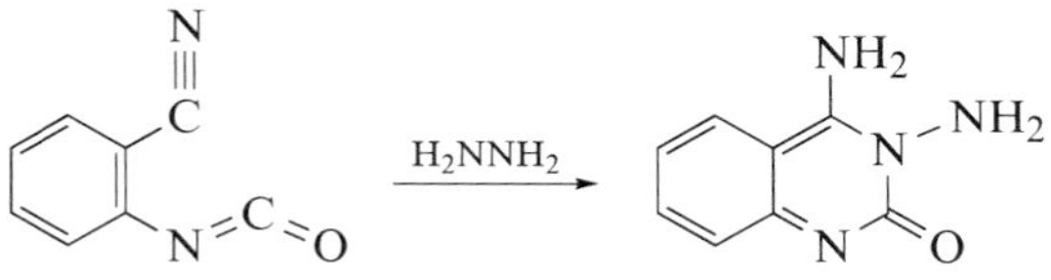

(284)   (285)

o-Isothiocyanatobenzonitrile gave 4-amino-3-ethyl-2($3H$)-quinazolinethione (EtNH$_2$, CH_2Cl_2–light petroleum, 20°, ? h: 95%).[2417]

Also other examples in the foregoing references and elsewhere.[51,54,55,1442,1495,1857,2070,2222]

1.2.3.10. With o-Cyanoamino- or o-Ureidobenzoic Esters and Related Ketones or Nitriles as Substrates

These substrates have significant but largely untapped potential in quinazoline syntheses, as indicated by the paucity of the following examples.

Methyl o-cyanoaminobenzoate (**286**) gave 2,3-diamino-4(3H)-quinazolinone (**287**) (H$_2$NNH$_2$, EtOH, reflux, 1 h: ~ 80%).[1347]

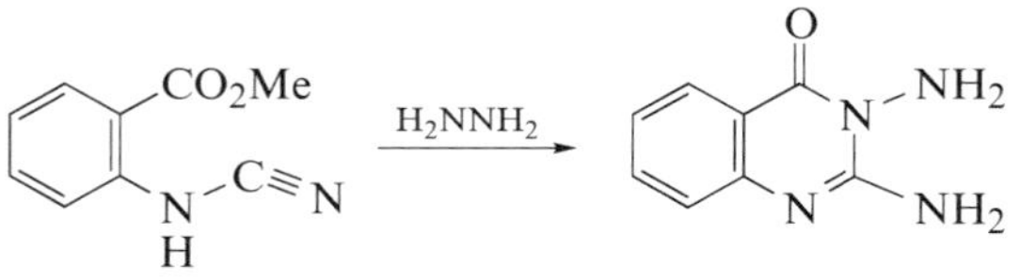

2-Ureidocyclohex-1-enecarbonitrile (**289**) gave 4-amino-3-phenyl-5,6,7,8-tetrahydro-2(3H)-quinazolinone (**288**) (PhNH$_2$, xylene, ~ 140°, short time: 94%),[1851] but under more vigorous conditions this initial product underwent Dimroth rearrangement to afford 4-anilino-5,6,7,8-tetrahydro-2(1H)-quinazolinone (**290**) (PhNH$_2$, decalin, ~ 230°, 2 h: 33%);[1853] substrate (**289**) also gave 4-amino-3-anilino-5,6,7,8-tetrahydro-2(3H)-quinazolinone (**291**) (PhHNNH$_2$, xylene, reflux, 2 h: 53%).[2311]

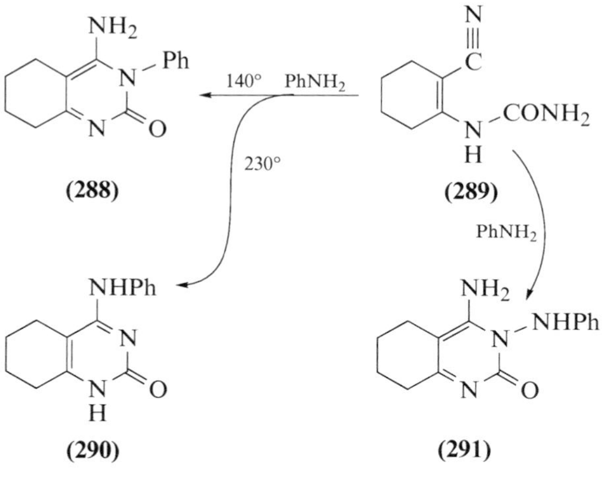

3-Chloro-6-[N'-phenyl(thioureido)]benzophenone (**292**) gave 2-anilino-4-phenylquinazoline 3-oxide (**293**) (H$_2$NOH.HCl, Et$_3$N, reflux, 24 h: 87%).[399]

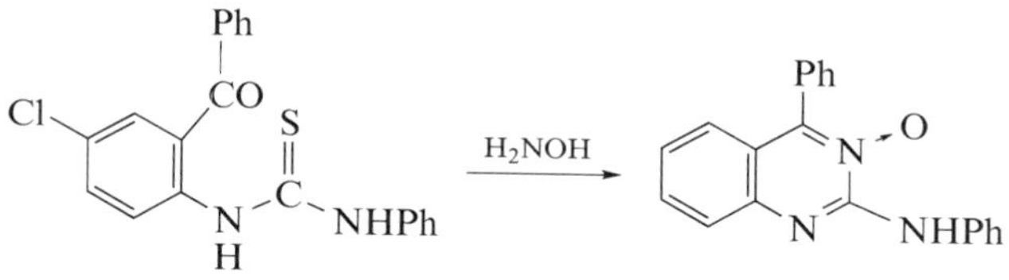

(292)　　　　　　　　**(293)**

Also a few analogous examples in the foregoing references.

1.2.3.11.　With Miscellaneous Substrates

Several other substrates accept N3 from common synthons to afford quinazolines. Thus 3-*p*-chlorobenzamidophthalic anhydride **(294)** with ammonium hydroxide at 150° (sealed, 6 h) gave 2-*p*-chlorophenyl-4-oxo-3,4-dihydro-5-quinazolinecarboxylic acid **(295)** (92%);[672] *N*-benzenesulfonyloxyphthalimide **(296)** with neat aniline at 200° for 3 h gave 3-phenyl-2,4(1*H*, 3*H*)-quinazolinedione **(297)** by a complicated mechanism involving Lossen rearrangement;[1007] and 3-chloro-6-(Chloromethane)sulfonamidobenzophenone **(298)**, sealed with liquid ammonia at 60° for 4 days, afforded a 90% yield of 6-chloro-4-phenyl-1,2-dihydroquinazoline **(300)**, possibly by sulfur extrusion from an heterobicyclic intermediate **(299)**.[753]

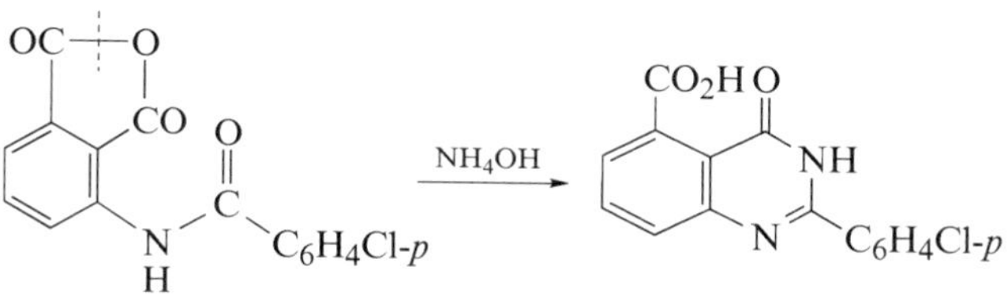

(294)　　　　　　　　　　**(295)**

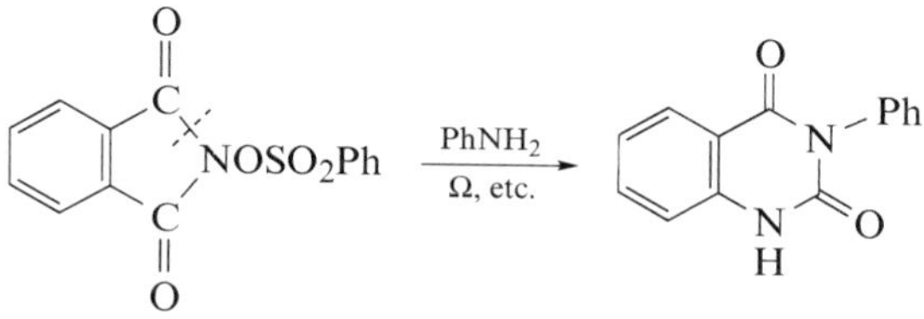

(296)　　　　　　　　　　**(297)**

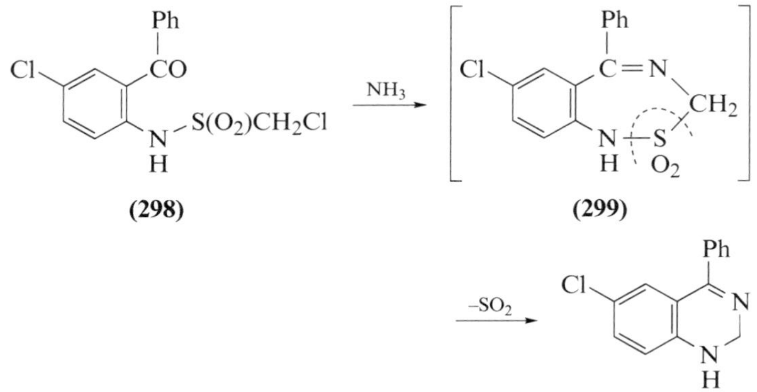

(298)   **(299)**

(300)

1.2.4. Where the Synthon Supplies C4 (*H* 331)

This synthesis has had limited use, mainly because there is little flexibility in the substrate, which must be a ureido, isoureido, aminomethyleneamino, or similar derivative of benzene. However, there is a little more diversity among the C4-synthons, which accordingly form a basis for classification of the examples that follow.

Carbonic acid equivalents as synthons

N-(α-Amino-β,β,β-trichloroethylidene)-*p*-chloroaniline (**301**) with phosgene
   gave 6-chloro-2-trichloromethyl-4(3*H*)-quinazolinone (**302**) (PhMe, 140°,
   1 h: > 90%).[741]

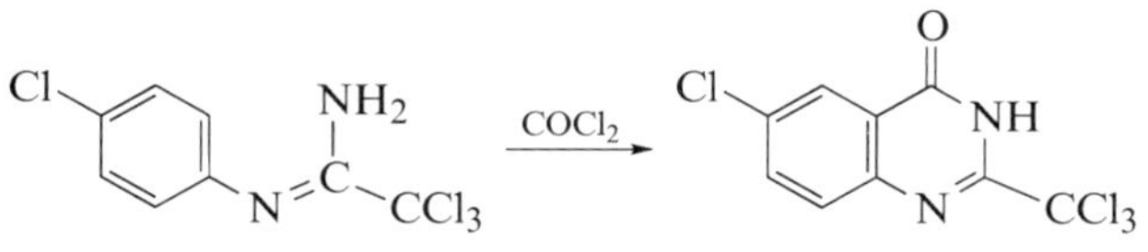

(301)   **(302)**

N-(α-Amino-3,4-dichlorobenzylidene)-*p*-toluidine with ethyl chloroformate
   gave 2-(3,4-dichlorophenyl)-6-methyl-4(3*H*)-quinazolinone (quinoline, reflux).[1500]

A formic acid equivalent (carbon monoxide) as synthon

A cyclic Pd²⁺ complex of *N*-(α-aminobenzylidene)aniline with carbon monoxide gave 2-phenyl-4(3*H*)-quinazolinone (**303**) (MeOH, 20°, 5 atm, 24 h:
   97%).[1787]

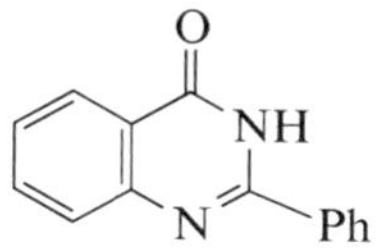

(303)

An attempt to activate such a cyclization by *o*-thallation of the benzene ring failed.[2226]

Aldehydes as synthons

p-(*N*-Cyclopropylmethylureido)anisole (304) with benzaldehyde gave 1-cyclo-propylmethyl-6-methoxy-4-phenyl-3,4-dihydro-2(1*H*)-quinazolinone (305) (PhMe, MeSO$_3$H, reflux, 20 h: 65%) and thence 1-cyclopropylmethyl-6-methoxy-4-phenyl-2(1*H*)-quinazolinone (KMnO$_4$, dioxane–H$_2$O, 20°, 1 h: 85%).[581, cf. 544]

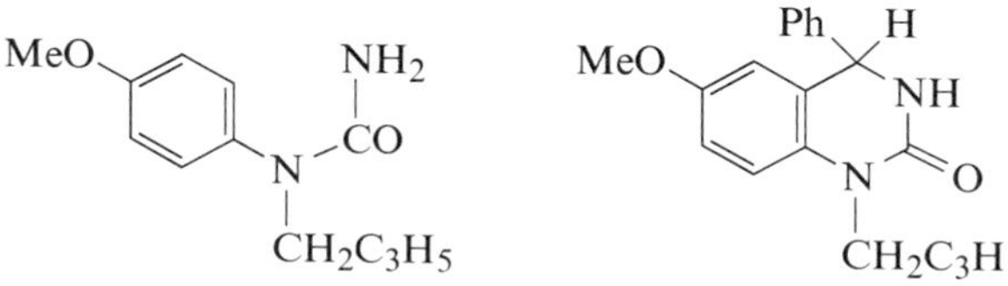

(304)   (305)

N-[*o*-(Triphenylphosphoranylideneamino)benzylidene]-*p*-toluidine (306) with benzaldehyde or *p*-nitrobenzaldehyde gave 6-methyl-2,4-diphenyl-3,4-dihydroquinazoline (307, R = Ph) (xylene, reflux, 48 h: 71%) or 6-methyl-4-*p*-nitrophenyl-2-phenylquinazoline (308, R = C$_6$H$_4$NO$_2$-*p*) (xylene, reflux, 12 h: 70%), respectively; other substituted benzaldehydes gave mixtures of appropriate quinazolines and dihydroquinazolines.[2097,2160]

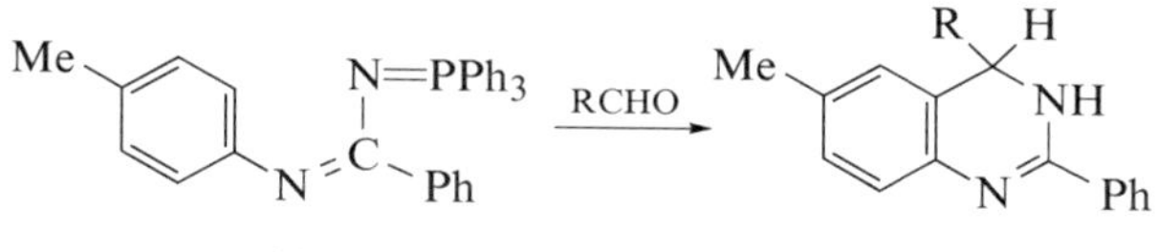

(306)   (307)

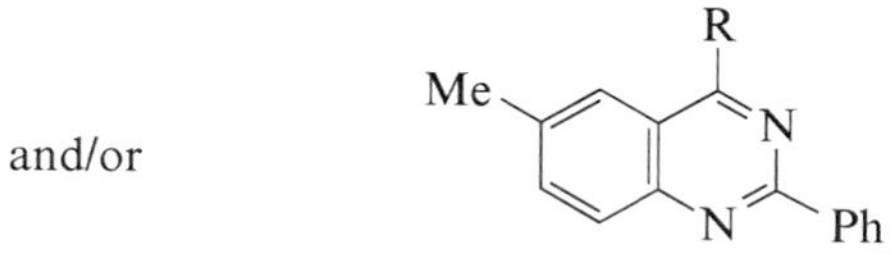

and/or

(308)

Also other examples in the foregoing references and elsewhere.[1431]

Aldehyde equivalents as synthons

N-(α-Aminobenzylidene)aniline (**309**) with chloromethyl methyl sulfide ($\simeq$ HCHO) gave 2-phenylquinazoline (**310**) (decalin, 200°, sealed, 18 h: 5%).[989]

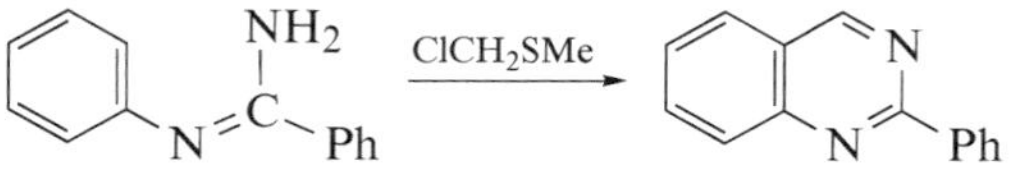

(**309**)   (**310**)

N-(α-Dimethylsulfimidobenzylidene)-p-toluidine (**311**) with N-(isopropy-lidenemethyl)morpholine ($\simeq$ Pr^{i}CHO) gave 4-isopropyl-6-methyl-2-phenylquinazoline (**312**) (tetralin, reflux, 14 h: 43%);[2088] several analogs were made similarly.[2088,2101]

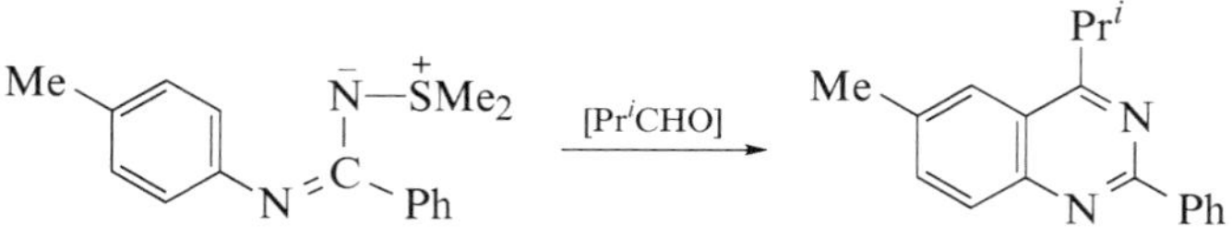

(**311**)   (**312**)

Isothiocyanates as synthons

p-Chloro-N-[(isothiocyanato)(morpholino)methylene]aniline (**313**) with phenyl isothiocyanate gave 4-anilino-6-chloro-2-morpholinoquinazoline (**314**) (Me$_2$NCHO, reflux, 1 h: 60%); appropriate variations gave several analogous products, such as 6-nitro-4-p-nitroanilino-2-piperidinoquinazoline (65%).[1328]

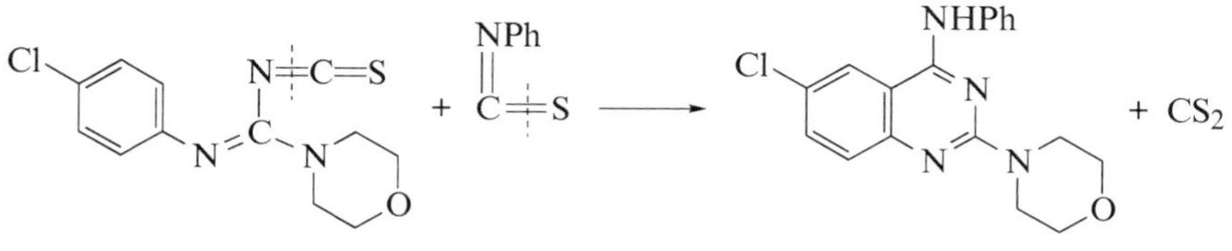

(**313**)   (**314**)

1.2.5.   Where the Synthon(s) Supply N1 + C2

This type of synthesis has been little used, and most of the examples have employed a cyclohexane rather than a benzene substrate simply because most leaving groups are less strongly held to the reduced ring.

A successful procedure leading to aromatic quinazolines is illustrated by the reaction of o-azidomethylbenzenediazonium tetrafluoroborate (**315**) with acetonitrile at 70° for 10 min to afford 2-methylquinazoline (**316**) in 70% yield;[890] appropriate nitriles likewise gave 2-methylthio-, 2-phenyl-, and 2-benzylquinazoline.[890] Another involves o-lithiation of N-methylbenzamide and subsequent treatment with di-t-butyl azodicarboxylate to give 1-butoxycarbonylamino-(44%) and thence, by thermolysis at 190°, 1-amino-3-methyl-2,4(1H,3H)-quinazolinedione (72%).[2574]

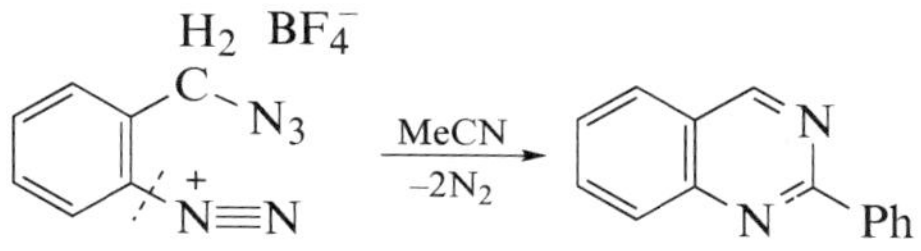

(315)   (316)

Procedures leading to reduced quinazolines are typified by the condensation of 2-(aminomethylene)-6-isopropyl-3-methylcyclohexanone (317) with refluxing formamide (containing piperidine acetate) for 12 h to give 8-isopropyl-5-methyl-5,6,7,8-tetrahydroquinazoline (318, R = H) (78%)[69,563] or with guanidine to give the corresponding 2-quinazolinamine (318, R = NH$_2$) (20%);[563] of 2-ben-zylaminomethylcyclohexanone (319) (preformed in situ) with aqueous potassium isocyanate to give 3-benzyl-1,4,5,6,7,8-hexahydro-2(3H)-quinazolinone (320) (67%);[626] of 2-morpholino-N-phenylcyclohex-1-enecarboxamide with cyanamide in refluxing dimethylformamide for 20 min to give 2-amino-3-phenyl-5,6,7,8-tetrahydro-4(3H)-quinazolinone (50%);[1831] and of 2-oxocyclohexanecarboxamide (321) with benzaldehyde in refluxing aqueous ammonia for 4 h to give 2-phenyl-1,2,5,6,7,8-hexahydro-4(3H)-quinazolinone (322) (66–73%),[757,1939,1980] easily 1,2-dehydrogenated.[1939]

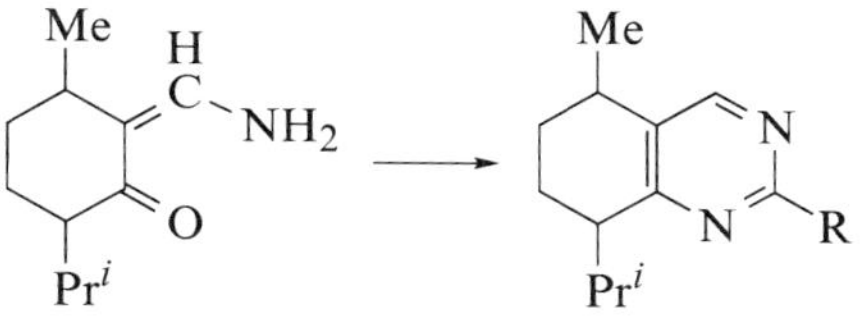

(317)   (318)

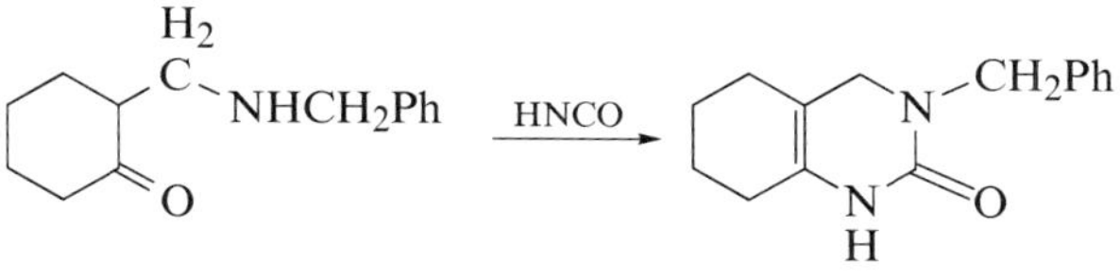

(319)   (320)

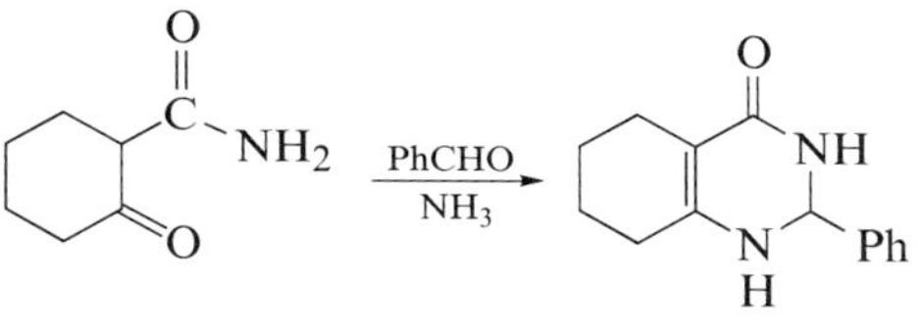

(321)   (322)

1.2.6.   Where the Synthon(s) Supply C2 + N3

Numerous substrates have been used for this type of primary synthesis, and in each case various synthons have been employed to supply the missing carbon and nitrogen atoms. The following treatment is divided according to substrate type, with each such division further subdivided according to the type of synthon.

1.2.6.1.   With o-Aminobenzoic Acids as Substrates
(H 74,81,88,91,116,119,286)

o-Aminobenzoic acid (anthranilic acid) and its N-, 3-, 4-, 5-, or 6-derivatives have proved very popular starting materials for the synthesis of 4-quinazolinones, as indicated in the following examples.

Carboxamides as synthons (Niementowski's synthesis[2450])

o-Aminobenzoic acid (**323**) gave 4(3H)-quinazolinone (**324**, X = O) (neat HCONH$_2$, 120°, 2 h, then 170° 2 h: 69%;[1189] HCONH$_2$, 145°, 4 h: 65%)[1558] or 4(3H)-quinazolinethione (**324**, X = S) (HCONH$_2$, S, 120°, 1 h, then 230°, 1 h: 76%).[1864]

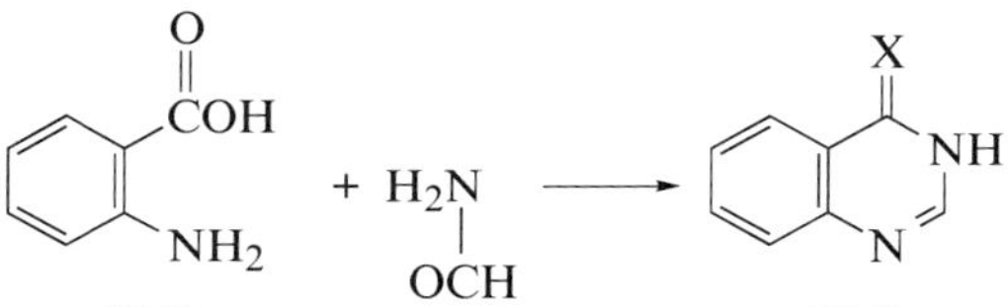

(323)   (324)

2-Amino-3,5-dibromobenzoic acid gave 6,8-dibromo-4(3H)-quinazolinone (HCONH$_2$, 210°, 30 min: 86%).[907]

2-Amino-4-trifluoromethylbenzoic acid gave 7-trifluoromethyl-4(3H)-quinazolinone (HCONH$_2$, 125°, 45 min; then 171°, 2.5 h: ∼ 60%).[2165]

o-Aminobenzoic acid gave 2-cyanomethyl-4(3H)-quinazolinone (**325**) [NCCH$_2$CSNH$_2$, Et$_3$N, reflux, 6 h: 80%; neat NCCH$_2$CSNH$_2$, 135°: 70%; or NCCH$_2$C(SMe)=NH, EtOH, reflux: 85%][2275] or 2-furan-2'-yl-4(3H)-quinazolinone (**326**) (2-furancarboxamide, POCl$_3$, 100°, 1 h: 73%).[2322]

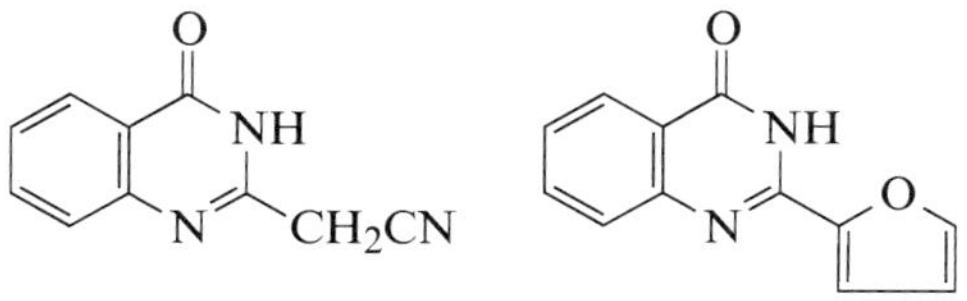

(325)   (326)

o-Methylaminobenzoic acid (**327**) gave 1,3-dimethyl-2-phenacylidene-1,2-dihydro-4(3*H*)-quinazolinone (**329**) [methyl 2-methyl-5-phenylisoxazolium sulfate (**328**) ($\simeq$ *N*-methyl-2-benzoylacetamide), HO$^-$: 72%].[1687]

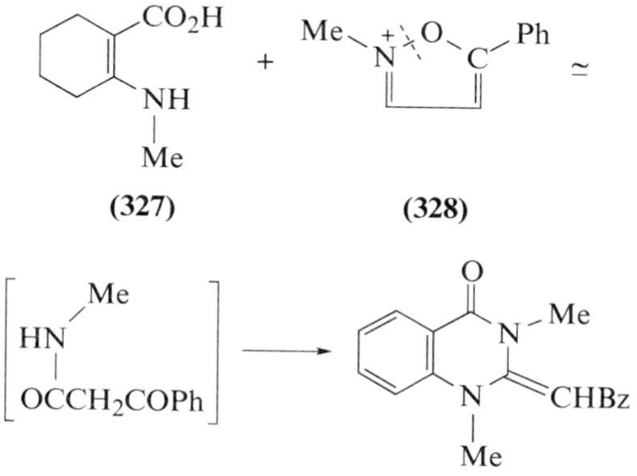

(327)   (328)

(329)

Also other examples.[145,370,384,466,556,833,970,976,1491,1617,1814,2410]

Carboximidates (iminoethers) as synthons

o-Aminobenzoic acid (**330**) gave 2-phenyl-4(3*H*)-quinazolinone (**331**) [PhC(=NH)OEt, Bu^{i}CH$_2$OH, reflux, 3 h: 47%].[39, cf. 13]

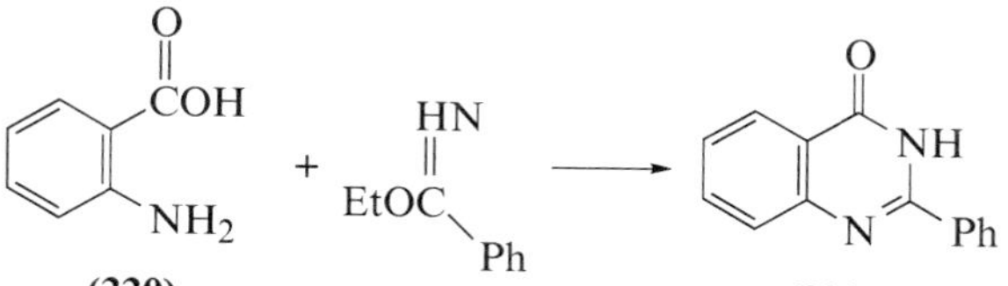

(330)   (331)

2-Amino-5-methylbenzoic acid gave 2-fluoromethyl-6-methyl-4(3*H*)-quinazolinone [FCH$_2$C(=NH)OEt.HCl, EtONa, EtOH, 20°, 60 h: 67%].[2263]

o-Aminobenzoic acid gave 2-ethoxycarbonylmethyl-4(3*H*)-quinazolinone [EtO$_2$CCH$_2$C(=NH)OEt.HCl, EtONa, EtOH, 20°, 24 h: 62%].[531]

2-Aminocyclohex-4-enecarboxylic acid (**332**) gave 2-phenyl-4a,5,8,8a-tetrahydro-4(3*H*)-quinazolinone (**333**), *cis* or *trans* according to the configuration of the substrate [PhC(=NH)OEt, PhCl, reflux, 20 h: 62% or 65%, respectively].[1933]

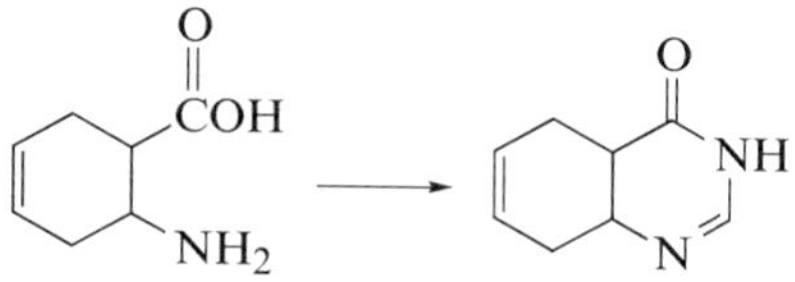

(332)   (333)

2-Amino-6-methylbenzoic acid (**334**) gave 2-(δ-bromo-γ-hydroxybutyl)-6-methyl-4(3*H*)-quinazolinone(**336**) [cyclic imidate (**335**), Et$_3$N, CH$_2$Cl$_2$, 20°, 30 min: 60%].[2389]

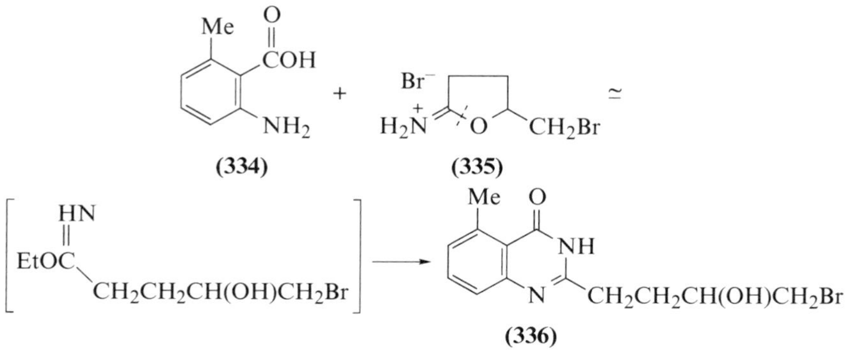

(**334**)   (**335**)

(**336**)

Also other examples.[103,107,293,306,1174,1866]

Urethanes or carbamates as synthons

Note: Unlike regular amides, urethanes normally give 2,4-quinazolinediones.

o-Aminobenzoic acid (**337**) gave 2,4(1*H*,3*H*)-quinazolinedione (**338**) (neat EtO$_2$CNH$_2$, 180°, reflux, 16 h: 60%),[832] 3-phenyl-2-thioxo-1,2-dihydro-4(3*H*)-quinazolinone (PhHNCO$_2$NH$_4$,EtOH, reflux, 5 h: 75%; PhHNCO$_2$NH$_4$, 150°, reflux, 5 h: 60%),[2390] or 3-amino-2,4(1*H*,3*H*)-quinazolinedione (Bu^{t}O$_2$CNHNH$_2$, quinoline, reflux, 8 h: 55%).[1454]

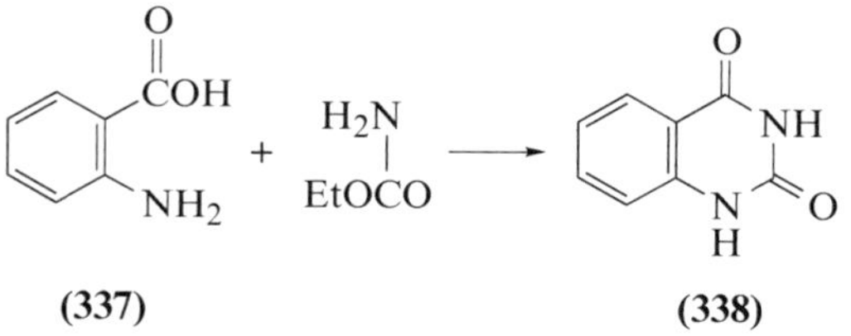

(**337**)   (**338**)

Potassium 2-amino-5-methylbenzoate gave 6-methyl-3-*p*-tolyl-2,4(1*H*,3*H*)-quinazolinedione (*p*-MeC$_6$H$_4$NHCS$_2$Me, red HgO, Me$_2$NCHO, reflux, 13 h: 93%; note the change of 2-thioxo to 2-oxo by HgO).[1312]

Potassium *o*-aminobenzoate gave *p*-bis(2,4-dioxo-1,2,3,4-tetrahydroquinazolin-3-yl)benzene (**339**) [*p*-(MeS$_2$CNH)$_2$C$_6$H$_4$, red HgO, Me$_2$NCHO, reflux, 6 h: 63%].[2212]

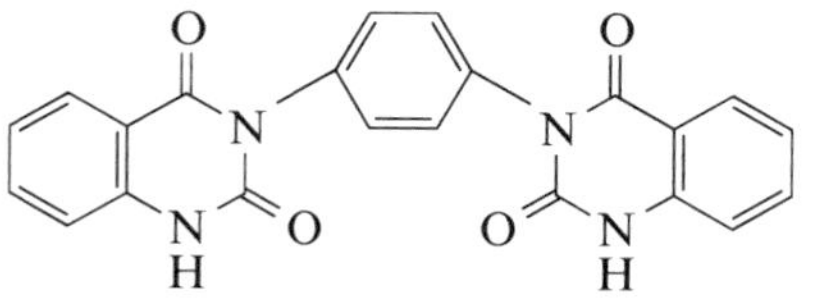

(**339**)

Also other examples in the foregoing references and elsewhere.[1214,1988, 2207,2356]

Amidines as synthons

2-Amino-6-benzyloxy-5-methoxybenzoic acid (340) gave 5-benzyloxy-6-methoxy-4(3H)-quinazolinone (341) [1,3,5-triazine ($\simeq$ HN=CHNH$_2$), HN(CH$_2$)$_5$, EtOH, reflux, N$_2$, 30 h: 80%].[1728]

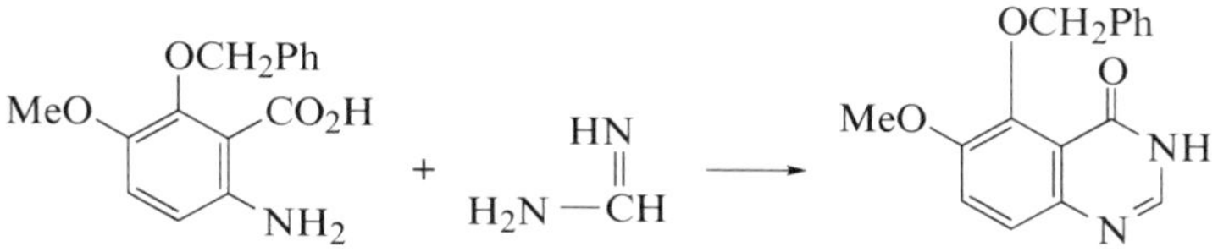

(340)   (341)

o-Aminobenzoic acid gave 2-pyridin-4′-yl-4(3H)-quinazolinone (342) (neat 4-pyridinecarboxamidine. HCl, 180°, 1 h: 76%).[1601]

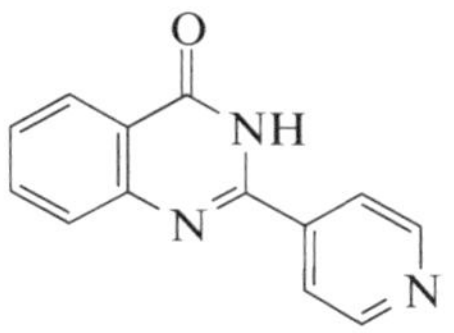

(342)

2-Amino-4-methyl-5-nitrobenzoic acid gave 7-methyl-6-nitro-4(3H)-quinazolinone [H$_2$NCH=NH.AcOH, AcONa, MeCH$_2$CH$_2$OH, 70°, 2 days: 68%].[1157]

Also other examples.[510,628,2363]

Ureas and thioureas as synthons

2-Amino-3,5-diiodobenzoic acid (343) gave 6,8-diiodo-2,4(1H, 3H)-quinazo-linedione (344) (neat H$_2$NCONH$_2$, 175°, 90 min: 73%).[565]

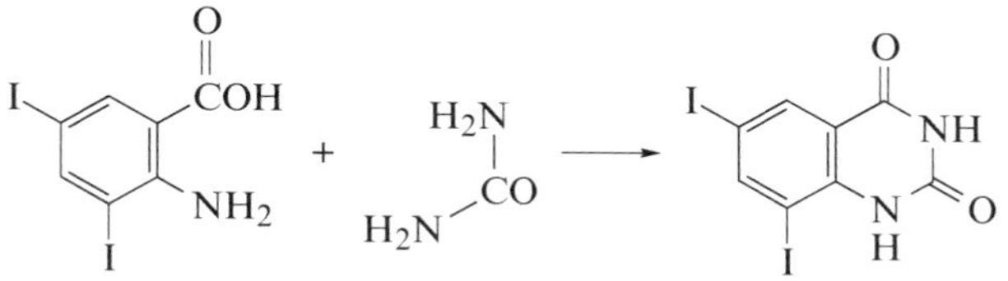

(343)   (344)

o-Ethylaminobenzoic acid gave 1-ethyl-2,4(1H, 3H)-quinazolinedione (345) (neat H$_2$NCONH$_2$, fusion, 6 h: 58%).[718]

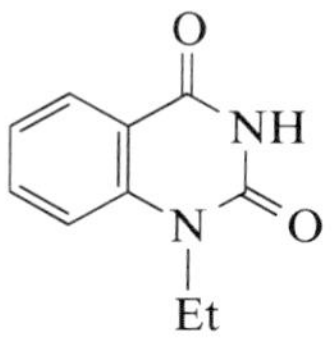

(345)

2-Amino-4-chlorobenzoic acid gave 7-chloro-3-methyl-2,4(1H,3H)-quinazo-linedione (**346**, R = Me, X = O) (neat MeHNCONHMe, 220–240°, 2 h: ?%)[2216] or 7-chloro-2-thioxo-3-p-tolyl-1,2-dihydro-4(3H)-quinazolinone (**346**, R = C$_6$H$_4$Me-p, X = S) (H$_2$NCSNHC$_6$H$_4$Me-p, 160–180°, 2 h: ?%).[566]

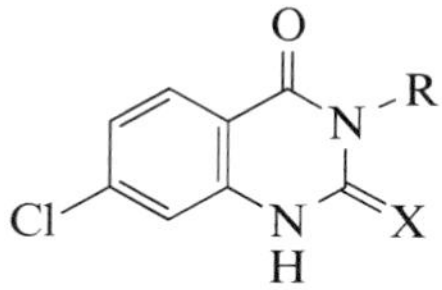

(346)

2-Acetamido-5-nitrobenzoic acid gave 6-nitro-2-thioxo-3-p-tolyl-1,2-dihydro-4(3H)-quinazolinone (neat H$_2$NCSNHC$_6$H$_4$Me-p, 150°, 2 h: 60%; note loss of acetyl group).[2166]

Also other examples in the foregoing references and elsewhere.[476,635,1079, 1490,1513,1539,2295,2448]

Nitriles as synthons

Note: There appear to be no recent data on the use of simple nitriles.

o-Aminobenzoic acid gave 2-(N-phenylcarbamoyl)-4(3H)-quinazolinone (**347**) (NCCONHPh, NEt$_3$, EtOH, reflux, 3 h: 75%),[1780] 2-amino-4(3H)-quinazo-linone (**348**, R = H) (CaNCN, HCl, 90°, 90 min: 91%),[798] or 2-amino-3-benzoyl-4(3H)-quinazolinone (**349**) (BzHNCN, Me$_2$NCHO, 95°, 5 h: 45%).[472]

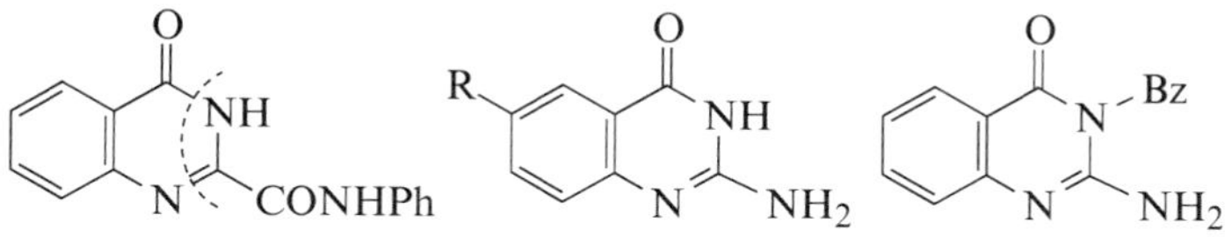

(347)   (348)   (349)

2-Amino-5-bromobenzoic acid gave 2-amino-6-bromo-4(3H)-quinazolinone (**348**, R = Br) (H$_2$NCN, HCl, EtOH, reflux, 8 h: 40%).[2025]

o-Methylaminobenzoic acid gave 1-methyl-2,4(1H,3H)-quinazolinedione (**350**) (BrCN, NaOH, H$_2$O, 100°, 10 min: 93%; hydrolysis included).[41]

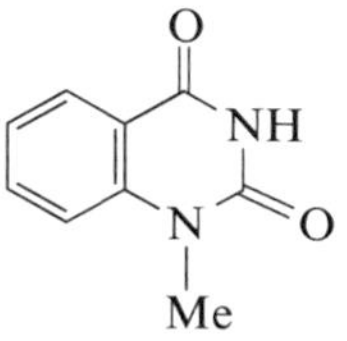

(350)

Also other examples in the foregoing references and elsewhere.[1364,2436]

Isocyanates and isothiocyanates as synthons

2-Amino-3,4-dimethoxybenzoic acid (351, R = OMe) gave 7,8-dimethoxy-2,4(1H, 3H)-quinazolinedione (352, R = OMe) (KNCO, AcOH, H_2O, 20°, 1 h: 38%).[696]

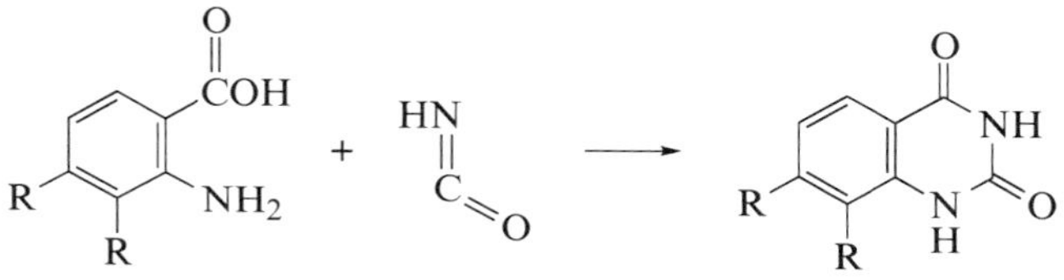

(351)   (352)

o-Aminobenzoic acid (351, R = H) gave 2,4(1H, 3H)-quinazolinedione (352, R = H) (KNCO, AcOH, H_2O, 35°, 1 h: > 85%;[2090] this reaction has been adapted as an analytic method for cyanate ion estimation).[2273]

2-Amino-5-methylbenzoic acid (353) gave 6-methyl-2-thioxo-1,2-dihydro-4(3H)-quinazolinone (354) [H_4NNCS, $(HOCH_2CH_2)_2O$, 150°, 2 h: 49%].[1841]

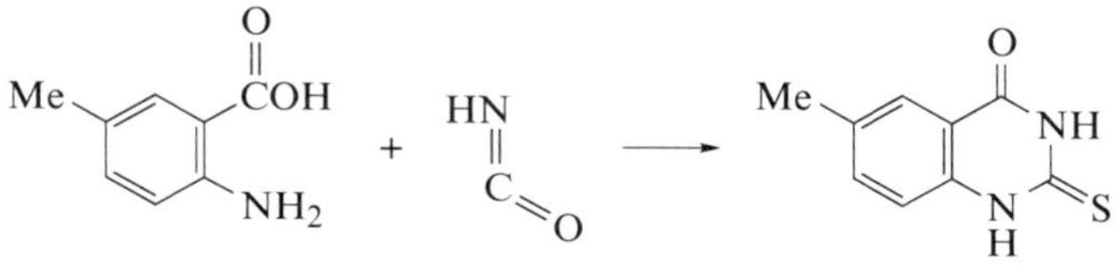

(353)   (354)

o-Methylaminobenzoic acid gave 1,3-dimethyl-2-thioxo-1,2-dihydro-4(3H)-quinazolinone (355) (MeNCS, AcOH, 100°, 2 h: 33%).[1846]

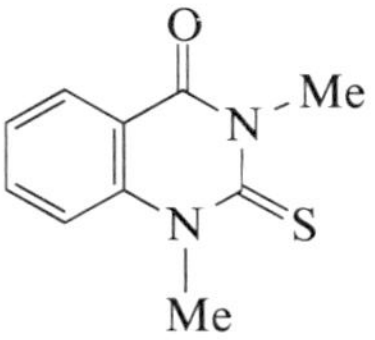

(355)

o-Aminobenzoic acid gave 3-phenacyl- (**356**, R = Bz) (BzCH$_2$NCS, EtOH, reflux, 3 h: 80%)[2179] or 3-β-acetylethyl-2-thioxo-1,2-dihydro-4(3*H*)-quinazolinone (**356**, R = CH$_2$Ac) (AcCH$_2$CH$_2$NCS, EtOH, reflux, 6 h: 70%).[1966]

(356)

Also other examples.[115,131,141,148,160,280,336,377,450,471,529,576,604,615,641, 644,652,665,670,840,864,880,897,978,1009,1046,1070,1087,1175,1632,1636,1674, 1750,1850,1928,2109,2122,2176,2293,2298,2595]

Miscellaneous enimines as synthons

o-Aminobenzoic acid gave 3-p-bromophenyl-2-diphenylmethyl-4(3*H*)-quinazolinone (**357**) (Ph$_2$C=C=NC$_6$H$_4$Br-p, xylene, reflux, 3 h: 75%),[1327] 3-anilino-2-benzoyl-4(3*H*)-quinazolinone (**358**) (BzCBR=NNHPh), NEt$_3$, EtOH, reflux, 4 h: > 80%),[1421] or 4(3*H*)-quinazolinone (**324**, X = O) [Me$_2$NCH=NCH=$\overset{+}{\text{N}}$Me$_2$ Cl$^-$ (Gold's reagent), NaH, dioxane, reflux, 12 h: 48%].[1173]

(357)   (358)

2-Amino-5-bromobenzoic acid gave 6-bromo-4(3*H*)-quinazolinone (Me$_2$NCH=NCH=$\overset{+}{\text{N}}$Me$_2$ Cl$^-$, NaH, dioxane, reflux, 24 h: 82%).[2277]
Also other examples in the foregoing references and elsewhere.[1789,2377,2423]

Sometimes it has proved convenient to treat aminobenzoic acid substrates with *two different synthons* at the same time, one to supply C2 and the other N3. Thus a mixture of α-picoline, aniline, o-aminobenzoic acid, and sulfur (oxidizing agent) was heated at 200° for 8 h to afford 37% of 3-phenyl-2-pyridin-2′-yl-4(3*H*)-quinazolinone (**359**);[370,516] the aniline in this reaction may be replaced by a second molecule of o-aminobenzoic acid to give the same product (**359**), following spontaneous decarboxylation.[1116] The latter point was investigated further by heating γ-picoline, o-aminobenzoic acid, sulfur, and o-toluidine or o-chloroaniline to give some 3-phenyl-2-pyridin-4′-yl-4(3*H*)-quinazolinone (**360**, R = H) accompanied by 2-pyridin-4′-yl-3-o-tolyl-4(3*H*)-quinazolinone (**360**,

R = Me) in one case[789] and by 3-*o*-chlorophenyl-2-pyridin-4′-yl-4(3*H*)-quinazo-linone (**360**, R = Cl) in the other.[375]

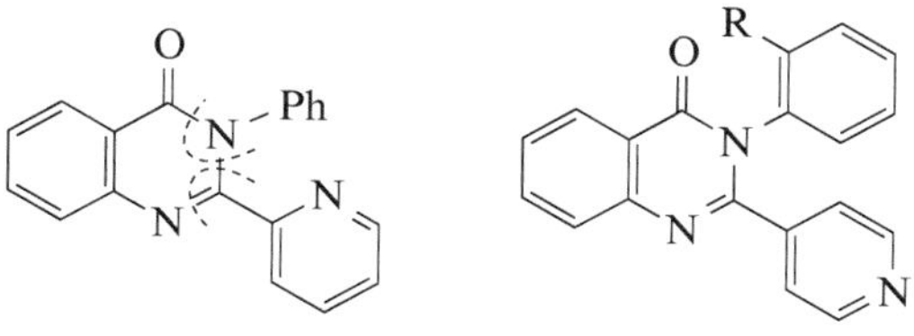

(**359**)   (**360**)

Another advantageous variation has been the conversion of *o*-methyl-aminobenzoic acid into a *bicyclic sulfinamide anhydride intermediate* (**361**) by treatment with thionyl chloride, followed by reaction with acetamide (in benzene at 20° for 12 h) to furnish 1,2-dimethyl-4(1*H*)-quinazolinone (**362**) (33% over-all);[705,1026] for further examples and references, see Section 1.6.13.

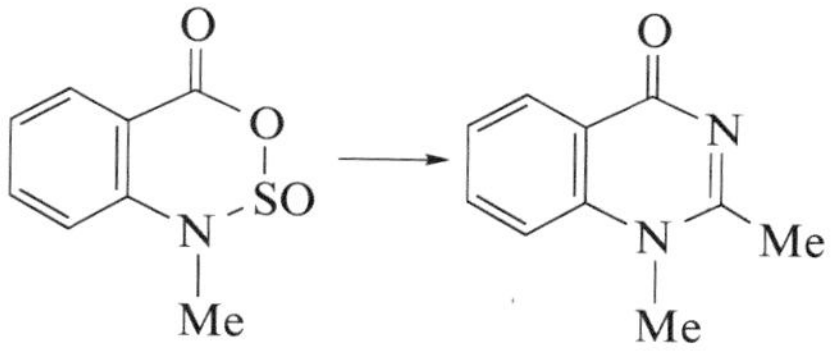

(**361**)   (**362**)

1.2.6.2.   With *o*-Aminobenzoic Esters as Substrates (*H* 127,285)

Although they seem more reactive and often give better yields, *o*-aminoben-zoic esters have been appreciably less used than the corresponding acids (just reviewed) for the synthesis of 4-quinazolinones. The following examples illustrate the use of such esters as substrates with various types of synthon.

Carboxamides as synthons

Methyl 2-amino-3-cyano-4,5-dimethylbenzoate (**363**) gave 6,7-dimethyl-4-oxo-3,4-dihydro-8-quinazolinecarbonitrile (**364**) (HCONH$_2$, HCO$_2$H, 150–170°, sealed?, 24 h: 90%).[1195]

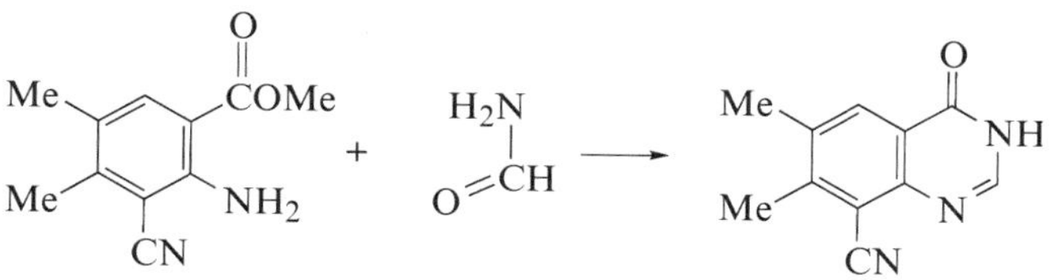

(**363**)   (**364**)

Ethyl *o*-aminobenzoate gave 2,4(1*H*,3*H*)-quinazolinedione [Me$_2$NCMe=
C(CN)CONH$_2$, AcOH, reflux, 5 h: 50%;[1832,1863] Me$_2$NCH=NCONH$_2$,
AcOH, reflux, 5 h: 47%;[2320] rational mechanisms were suggested].

Methyl *o*-aminobenzoate gave 3-benzamido-2-thioxo-1,2-dihydro-4(3*H*)-
quinazolinone [the oxadiazole (365) ($\simeq$ the *N*-acylated thioformohydraz-
ide, HCSNHNHBz), *m*-cresol, 150°, 5 h: 75%].[1457]

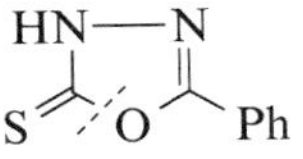

(365)

Also other examples in the foregoing references.

Carboximidates as synthons

Methyl *o*-aminobenzoate gave 2-*β*,*β*-dichloro-*α*,*α*-difluoroethyl-4(3*H*)-quina-
zolinone [CHCl$_2$CF$_2$C(=NH)OMe, AcOH, 24 h: 81%][443] or 2-nitro-
methyl-3-phenyl-4(3*H*)-quinazolinone (366) [O$_2$NCH$_2$C(=NPh)SMe,
AcOH, reflux].[1709, cf. 2416]

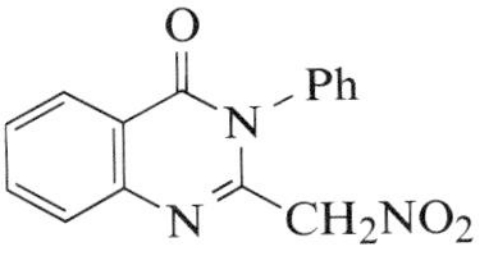

(366)

Also other examples.[85,260,443,1182,1709]

Urethanes as synthons

Methyl *o*-aminobenzoate gave *p*-bis(4-oxo-2-thioxo-1,2,3,4-tetrahydroquina-
zolin-3-yl)benzene (367) [*p*-(MeS$_2$CNH)$_2$C$_6$H$_4$, Me$_2$NCHO, reflux, 48 h:
41%];[2222] analogs with different bridges between the quinazoline nuclei
and bearing additional substituents were made similarly.[2222]

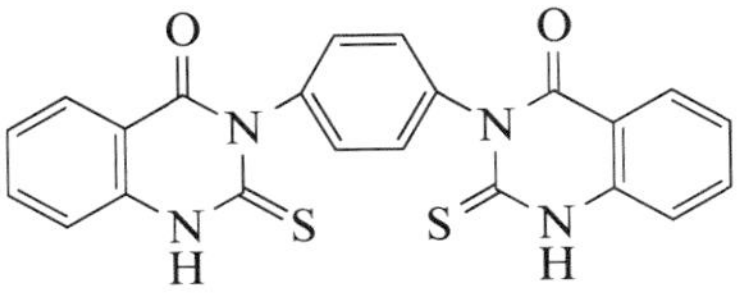

(367)

Also other examples.[2570]

Amidines, guanidine, and imidoyl chlorides as synthons

Methyl *o*-aminobenzoate gave 4(3*H*)-quinazolinone (**368**, R = H) [1,3,5-triazine ($\simeq$ formamidine), trace HN(CH$_2$)$_5$, EtOH, reflux, 8 h: 57%; exceptionally, similar treatment of *o*-aminobenzoic acid gave 77%],[628] 2-*p*-methoxyphenyl-3-phenyl-4(3*H*)-quinazolinone [*p*-MeOC$_6$H$_4$C(=NPh)Cl, Me$_2$CO, 20°, 12 h: ~ 40%],[488] or 3-methyl-2-phenyl-4(3*H*)-quinazolinone [PhC(=NMe)Cl, MeCN, 20°, 1 week: 60%].[2245]

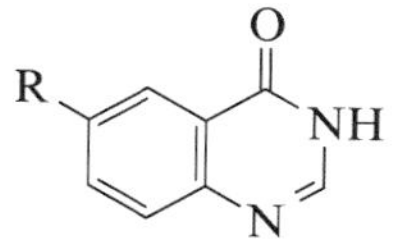

(368)

Ethyl 2-amino-5-methylbenzoate gave 2-amino-6-methyl-4(3*H*)-quinazolinone (**368**, R = Me) [ClC(=NH)NH$_2$.HCl, Me$_2$SO$_2$, 150°, 1 h: > 95%;[2250 cf. 2194] H$_2$NC(=NH)NH$_2$, EtOCH$_2$CH$_2$OH, volatiles removed, then 170° until solid: 80%].[871]

Also other examples in the foregoing references and elsewhere.[94,2198,2411]

Urea as a synthon

Methyl 2-amino-4-chlorobenzoate gave 7-chloro-2,4(1*H*,3*H*)-quinazolinedione (**369**) (neat H$_2$NCONH$_2$, 220°, 1 h: 80%).[1358]

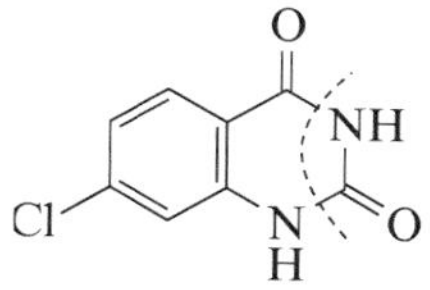

(369)

Also other examples.[1822]

Nitriles and cyanates as synthons

Methyl *o*-aminobenzoate (**370**) gave 2-methyl- (**371**, R = Me) (neat MeCN, HCl gas, 20°, 5 h: 75%),[2151] 2-phenyl- (**371**, R = Ph) (PhCN, dioxane, HCl gas, 20°, 5 h; then reflux, 3 h: 72%),[2151] or 2-trichloromethyl-4(3*H*)-quinazolinone (**371**, R = CCl$_3$) (Cl$_3$CCN, PhMe, NEt$_3$: ?%).[1136]

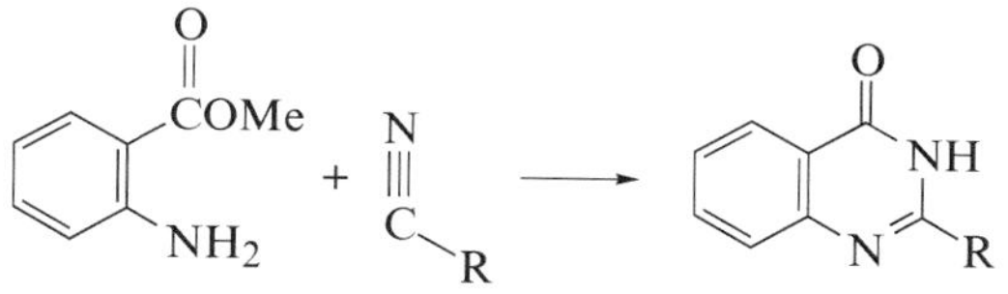

(370)   (371)

Ethyl *o*-aminobenzoate gave ehtyl 4-oxo-3,4-dihdyro-2-quinazolinecarboxylate (EtO$_2$CCN, AcOH, HCl gas, reflux, 2.5 h: 49%).[926]

Ethyl 2-amino-4-chlorobenzoate gave 2-amino-7-chloro-4(3*H*)-quinazolinone (**372**) (H$_2$NCN, EtOH, HCl, reflux, 24 h: 89%).[915]

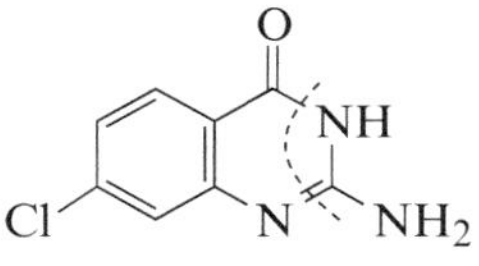

(**372**)

Ethyl *o*-methylaminobenzoate gave 2-*p*-chlorophenoxy-1-methyl-4(1*H*)-quinazolinone (**373**) (*p*-ClC$_6$H$_4$OCN, PhH, PhCO$_2$H as catalyst, reflux, 15 min: 66%).[889]

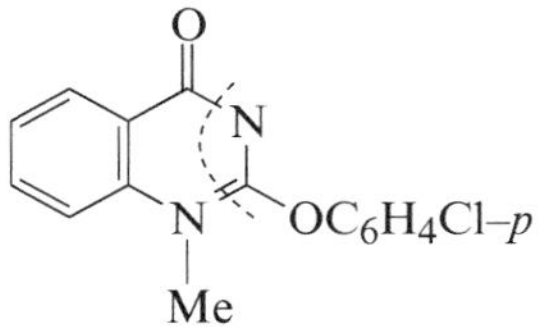

(**373**)

Also other examples in the foregoing references and elsewhere.[364,863,938,2505]

Isocyanates and isothiocyanates as synthons

Dimethyl 3-aminophthalate (**374**), made *in situ* from the nitro analog, gave methyl 2,4-dioxo-1,2,3,4-tetrahydro-5-quinazolinecarboxylate (**375**) (KCNO, AcOH, 95°, 30 min: > 95%).[2203]

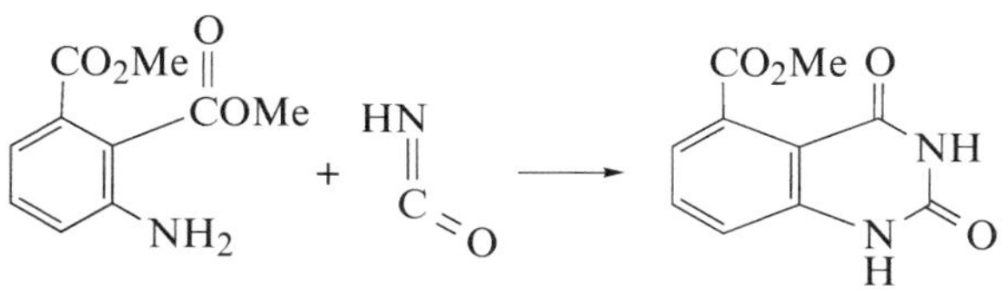

(**374**)   (**375**)

Ethyl 2-aminocyclohexanecarboxylate was converted into its thiocyanate salt (KCNS, AcMe, evaporate), which gave 2-thioxo-1,2,4a,5,6,7,8,8a-octahydro-4(3*H*)-quinazolinone (**376**) (xylene, reflux, 10 h: *cis* + *trans*, 87%).[2214]

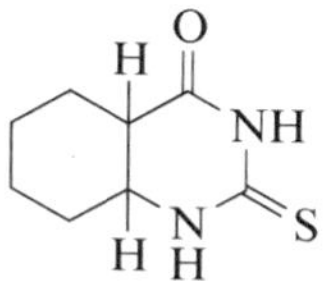

(376)

Methyl *o*-anilinobenzoate gave 3-methyl-1-phenyl-2,4(1*H*, 3*H*)-quinazo-linedione (**377**) (MeNCO, NaH, Me$_2$NCHO, 20°, 2 h: 64%).[1427]

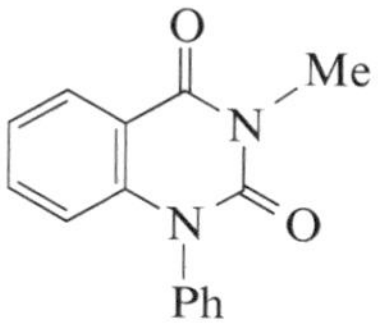

(377)

Methyl 4-nitro-2-phosphoranylideneaminobenzoate (**378**) gave 2-methoxy-7-nitro-3-phenyl-4(3*H*)-quinazolinone (**379**) (PhNCO, MeCN, reflux, 4 h; volatiles removed, MeOH, reflux, 1 h: 60%, probably via a carbo-diimide.[200]

(378)   **(379)**

Methyl *o*-aminobenzoate gave 3-methyl- (**380**, R = Me) (neat MeNCS, 20°, 3 days: >95%),[1335] 3-phenyl- (380, R = Ph) (PhNCS, 20°, 5 h),[1335] or 3-*β*-morpholinoethyl-2-thioxo-1,2-dihydro-4(3*H*)-quinazolinone [380, R = CH$_2$CH$_2$N(CH$_2$CH$_2$)$_2$O] [O(CH$_2$CH$_2$)$_2$NCH$_2$CH$_2$NCS, MeOH, re-flux, 3 h: 85%].[1196]

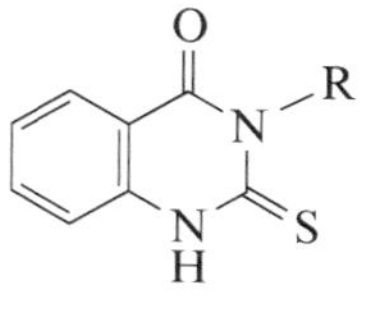

(380)

Also other examples.[112,212,289,322,335,371,434,460,622,630,1106,1122,1199,1483,1735,1822,2222,2261,2448,2508]

Miscellaneous reagents as synthons

Methyl *o*-methylaminobenzoate (**381**) gave 1,3-dimethyl-2,4(1*H*, 3*H*)-quinazo-linedione (**383**) by treatment with *N*-(dichloromethylene)-*N,N*-dimethy-lammonium chloride (**382**).[1531]

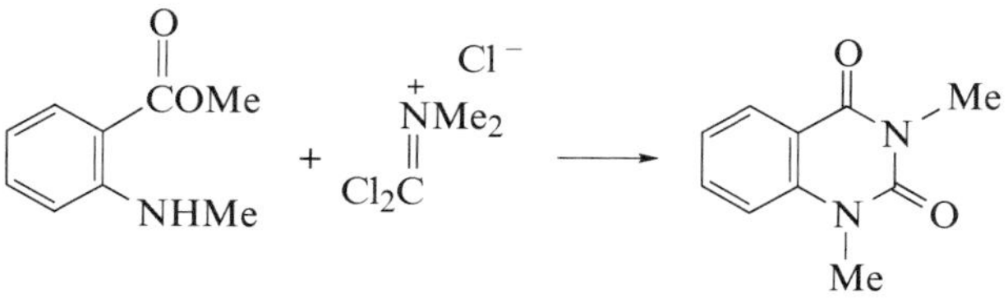

(**381**)   (**382**)   (**383**)

Methyl *o*-aminobenzoate (**384**) gave 3-tetrazol-5′-yl-4(3*H*)-quinazolinone (**385**) on treatment with two synthons, triethyl orthoformate (for C2) and 5-tetrazolamine (for N3) [in PrOH, 70°, 90 min, then NaOH, H$_2$O, reflux, 1 h: 80% as sodium salt];[2217] somewhat similarly, the substrate (**384**) gave 3-amino-4(3*H*)-quinazolinone (**387**) by using dimethylformamide dimethyl acetal (**386**) and hydrazine as synthons.[1497]

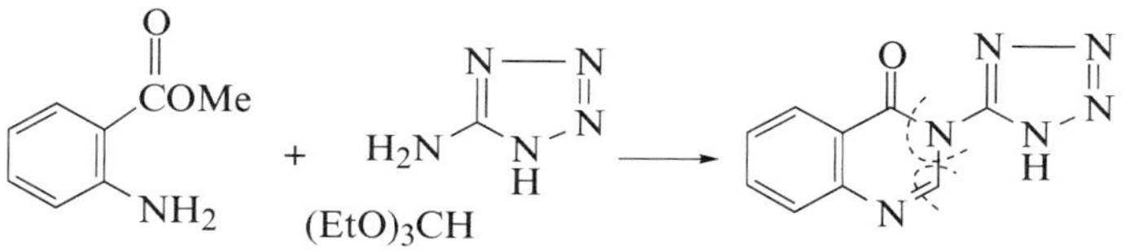

(**384**)   (**385**)

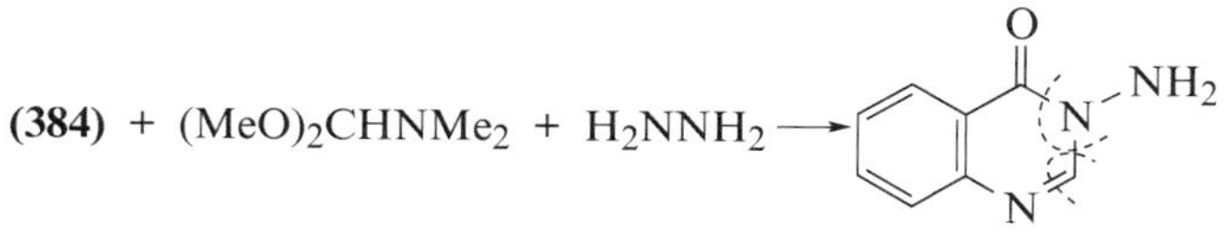

(**384**) + (MeO)$_2$CHNMe$_2$ + H$_2$NNH$_2$ ⟶

(**386**)   (**387**)

1.2.6.3.   With *o*-Aminobenzamides as Substrates (*H* 82)

The reaction of an *o*-aminobenzamide with an alkyl or aryl isothiocyanate appears always to give a 3-substituted 2-thioxo-1,2-dihydro-4(3*H*)-quinazo-linone with loss of ammonia from the original amide. Cyanates behave similarly, but other synthons have been explored inadequately. The following examples illustrate these generalities.

o-Aminobenzamide (**388**) gave 3-phenyl-(**389**, R = Ph) (PhNCS, EtOH, reflux, 8 h: 70%),[62] 3-methyl-(**389**, R = Me) (MeNCS, MeCN, 20°, 24 h: 77%),[1774] or 3-allyl-2-thioxo-1,2-dihydro-4(3H)-quinazolinone (**389**, R = CH$_2$CH:CH$_2$)(CH$_2$:CHCH$_2$NCS, MeCN, 20°, 24 h: 74%).[2227]

o-(β-Dimethylaminoethyl)aminobenzamide gave 1-β-dimethylaminoethyl-3-methyl-2,4(1H,3H)-quinazolinedione (**390**) (MeNCO, CHCl$_3$, reflux, 1 h, volatiles removed, HCl, EtOH, reflux, 2 h: 58%).[668]

o-Aminobenzamide (**388**) gave 3-isopropyl-2-(N-isopropylacetamido)-4(3H)-quinazolinone (**391**)[Pr^{i}N=CClNPr^{i}Ac (made *in situ* from Pr^{i}N=C=NPr^i and AcCl), dioxane, 20°, 24 h: 61%] and thence 3-isopropyl-2-isopropylamino-4(3H)-quinazolinone (**392**) (EtONa, EtOH, dioxane, reflux, 1 h: 86%).[621]

o-Amino-N,N-diethylbenzamide gave 2-phenyl-4(3H)-quinazolinone (**393**) (N-lithiation, then PhCN, THF, 20°, 15 min: 75%).[2085]

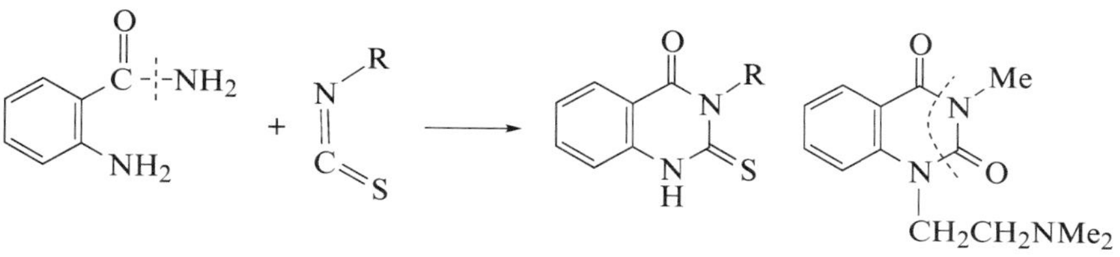

(**388**)   (**389**)   (**390**)

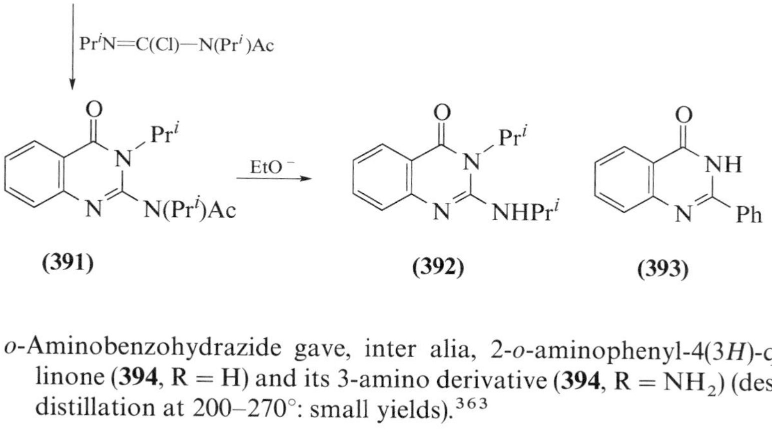

Pr^{i}N=C(Cl)—N(Pr^i)Ac

(**391**)   EtO$^-$ (**392**)   (**393**)

o-Aminobenzohydrazide gave, inter alia, 2-o-aminophenyl-4(3H)-quinazolinone (**394**, R = H) and its 3-amino derivative (**394**, R = NH$_2$) (destructive distillation at 200–270°: small yields).[363]

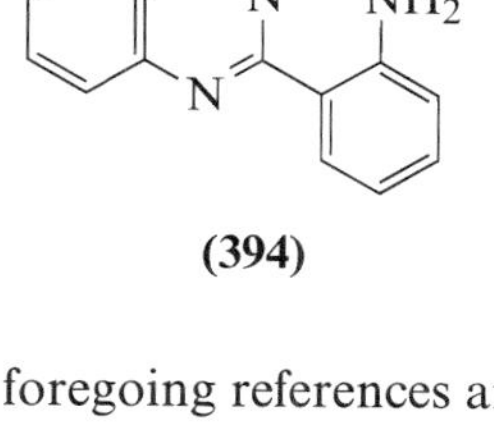

(**394**)

Also other examples in the foregoing references and elsewhere.[351,589]

1.2.6.4.   With *o*-Aminobenzonitriles as Substrates (*H* 328,332)

o-Aminobenzonitriles have been used often to produce 4-quinazolinamines, sometimes with 2- and/or 3-substituents as determined by the synthon employed. The following examples illustrate some of the possibilities.

Amidines or guanidines as synthons

Note: The use of amidines has been confined largely to chloroformamidine, which affords not a 2-chloro-4-quinazolinamine but a 2,4-quinazolinediamine.

2-Amino-6-methylbenzonitrile (**395**) gave 5-methyl-2,4-quinazolinediamine (**396**) [$H_2NC(=NH)Cl.HCl$, $(MeOCH_2CH_2)_2O$, 130–150°, 2 h: 73%].[2246]

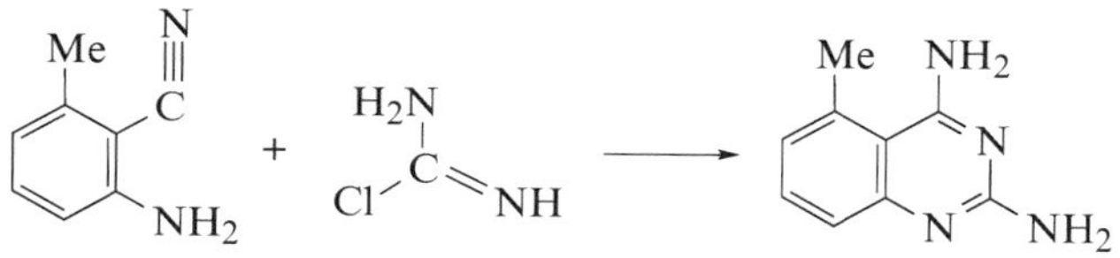

(**395**)   (**396**)

2-Amino-5-cyclohexylthiobenzonitrile gave 6-cyclohexylthio-2,4-quinazo-linediamine [$H_2NC(=NH)Cl.HCl$, $(MeOCH_2CH_2)_2O$, 150°, 30 min: 71%].[811]

2-Amino-5-methylbenzonitrile gave 6-methyl-4-quinazolinamine (**397**) [1,3,5-triazine ($\simeq HN=CHNH_2$), EtOH, reflux, 8 h: 64%].[339]

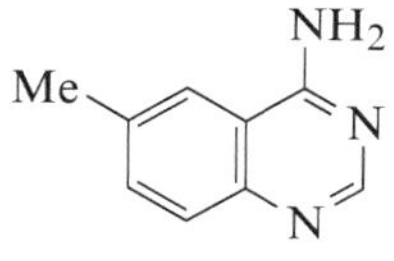

(**397**)

2-Amino-5-nitrobenzonitrile gave 6-nitro-2,4-quinazolinediamine [$H_2NC(=NH)NH_2$, EtOH, reflux, 5 h: 73%].[956]

2-Aminocyclohex-1-enecarbonitrile gave 2-methylamino-5,6,7,8-tetrahydro-4-quinazolinamine (**398**) [$H_2NC(=NH)NHMe$, $EtOCH_2CH_2OH$, reflux, N_2, 20 h: 13%].[534]

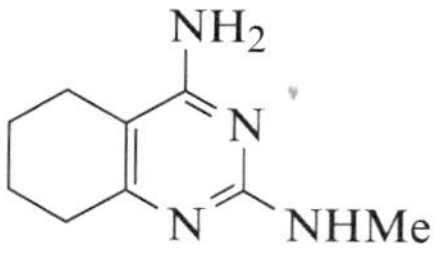

(**398**)

Also other examples.[238,661,680,1429,1551,2041,2077,2225,2326,2466]

Nitriles as synthons

Note: This subsection includes the use of regular nitriles, cyanoamides, and cyanoguanidine.

o-Aminobenzonitrile (**399**) gave 2-phenyl- (**400**, R = Ph) (PhCN, KOH, EtOH, 120°, sealed, 44 h: 59%)[1456] or 2-chloromethyl-4-quinazolinamine (**400**, R = CH$_2$Cl) (ClCH$_2$CN, dioxane, HCl gas, 0°, 6 h, then 20°: 85%).[2190]

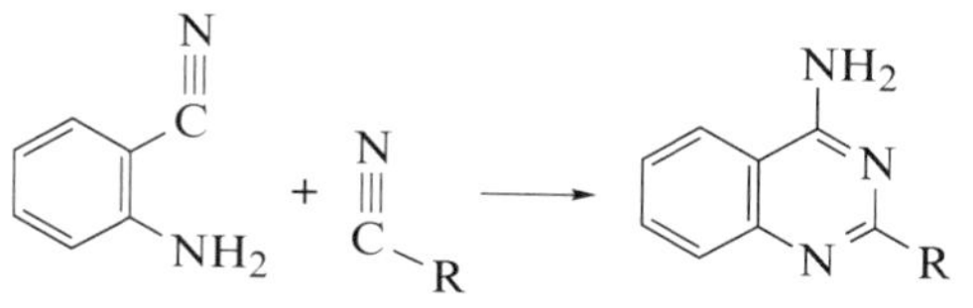

(**399**)   (**400**)

o-Aminobenzonitrile (**401**, R = H) (two molecules) gave 2-*o*-aminophenyl-4-quinazolinamine (**402**, R = H) (NaH, Me$_2$SO, 0°, then 25°, 24 h: ~ 70% as dihydrochloride);[1018] likewise, 2-aminoisophthalonitrile (**401**, R = CN) gave 4-amino-2-(2-amino-3-cyanophenyl)-8-quinazolinecarbonitrile (**402**, R = CN) (KOBu^t, Me$_2$SO, 20°, 7 days: ~ 60%),[2235] accompanied by trimeric material.

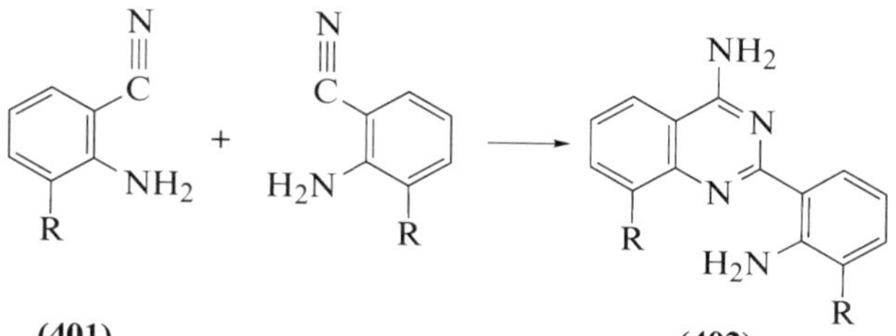

(**401**)   (**402**)

2-Amino-5-chlorobenzonitrile gave 6-chloro-2,4-quinazolinediamine [H$_2$NCN, pyridine.HCl, 150–160°, 10 min: 37%; or H$_2$NC(=NH)NHCN, substrate.HCl, 160°, 10 min: 33%].[661]

o-Aminobenzonitrile hydrochloride gave 2-morpholino-4-quinazolinamine (**403**) [O(CH$_2$CH$_2$)$_2$NCN, AcOEt, to dryness, then 115°, 1 h: 40%].[439]

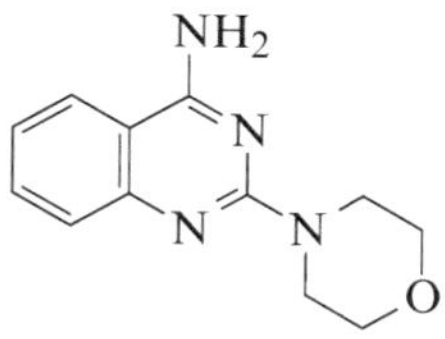

(**403**)

2-Amino-5-methylbenzonitrile (**404**) gave 6-methyl-2,4-quinazolinediamine
(**405**) [$H_2NC(=NH)NHCN$, 1M HCl, reflux, 90 min: 58%].[1912]

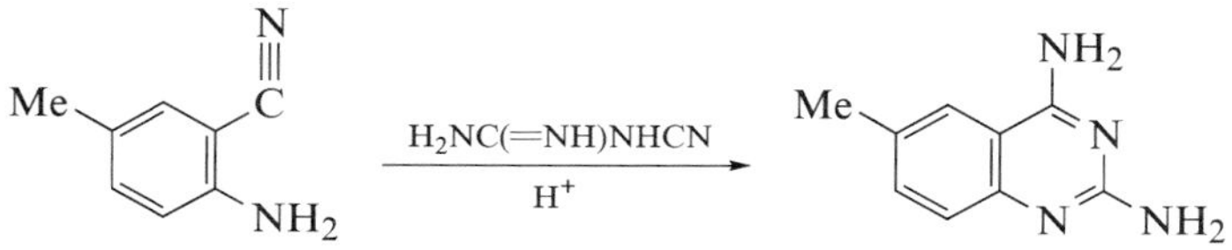

(**404**)   (**405**)

Also other examples in the foregoing references and elsewhere.[201,600,
1368,1602]

Isothiocyanates and isocyanates as synthons

o-Aminobenzonitrile (**406**) gave 4-amino-3-methyl- (**407**, R = Me) (neat
MeNCS, 110°, sealed, 4 h: 80%),[925] 3-allyl-4-amino- (**407**, R = $CH_2CH=$
CH_2) (H_2C:CHCH$_2$NCS, EtOH, reflux, 6 h: 80%),[848] or 4-amino-3-β-
morpholinoethyl-2(3*H*)-quinazolinethione   [**407**,   R = CH_2CH_2N
$(CH_2CH_2)_2O$] [$O(CH_2CH_2)_2NCH_2CH_2NCS$, MeOH, reflux, 3 h: 88%];
[1196] also analogs.[2051]

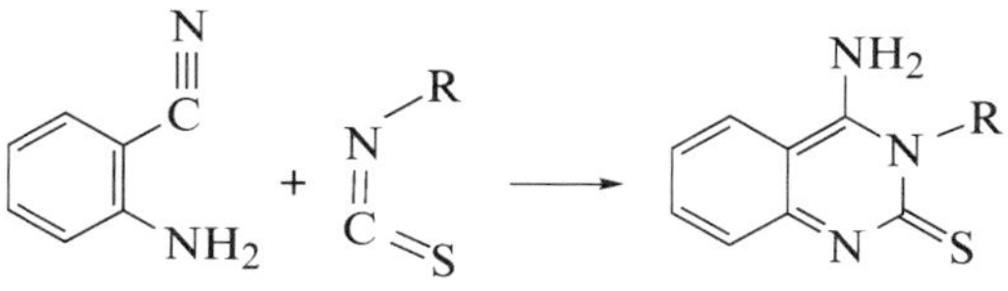

(**406**)   (**407**)

o-Amino- (**408**, R = H), *o*-methylamino- (**408**, R = Me), or *o*-phenethy-
laminobenzonitrile (**408**, R = CH_2CH_2Ph) gave 4-amino- (**409**, R = H),
4-amino-1-methyl-   (**409**,   R = Me),   or   4-amino-1-phenethyl-2(1*H*)-
quinazolinone (**409**, R = CH_2CH_2Ph), respectively (ClSO$_2$NCO, CH_2Cl_2,
20°, 3 h; then H_2O, 20°, 12 h: 88–96%), in each case via a 3-chlorosulfonyl
derivative that was unstable in water.[1925]

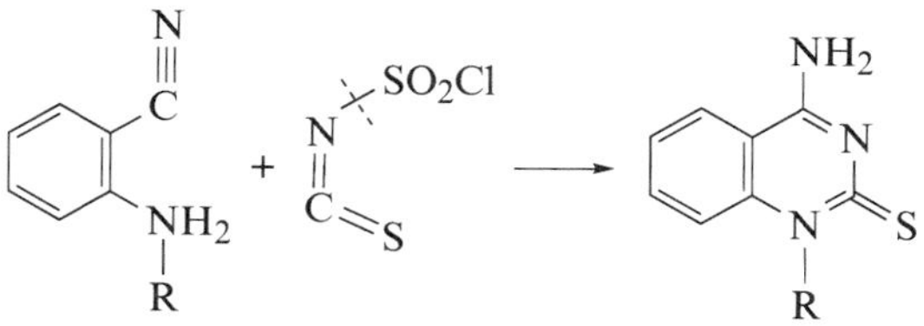

(**408**)   (**409**)

Other reagents as synthons

2-Amino-5-piperidinobenzonitrile (**410**) gave 6-piperidino-2,4(1H,3H)-
quinazolinedione (**412**) (neat H_2NCONH_2, 200°, 2.5 h; then HCl, reflux, 3 h:
79%), presumably via the 4-amino analog (**411**).[1584]

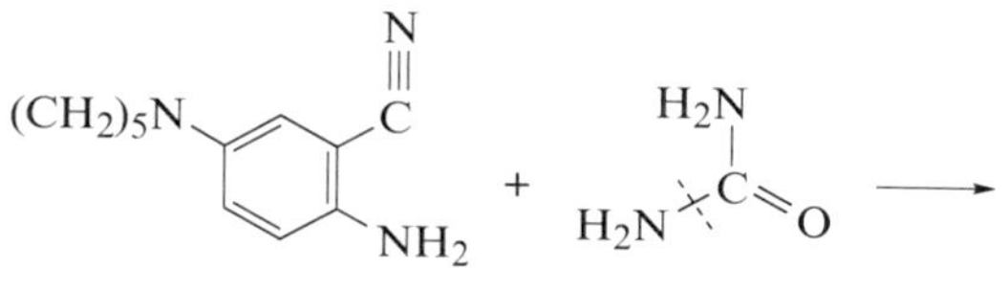

(**410**)

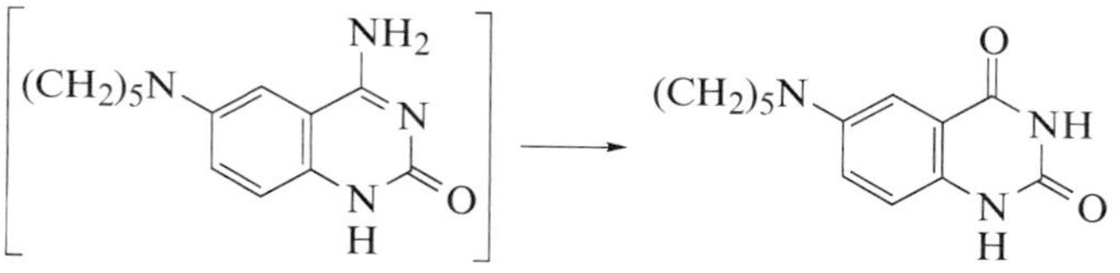

(**411**)   (**412**)

2-Amino-5-naphthalen-2′-ylthiobenzonitrile gave 4-amino-6-naphthalen-2′-
ylthio-2(1H)-quinazolinethione [H_2NCSNH_2, $(CH_2)_4SO_2$, 200°, 90 min:
15%].[673]

o-Aminobenzonitrile gave ethyl 4-amino-2-quinazolinecarboxylate (EtO
(HN=)CCO$_2$Et.HCl, no details: 255).[1182]

2-Amino-4,6,6-trimethylcyclohex-1-enecarbonitrile gave 5,5,7-trimethyl-
5,6,7,8-tetrahydro-4-quinazolinamine (**413**) (HCONH$_2$, HCO$_2$H, 200°,
1 h: ?%).[1158]

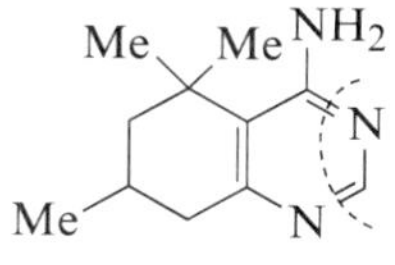

(**413**)

Also other examples in the cited references and elsewhere.[2466]

1.2.6.5. With o-Aminobenzaldehydes or Related Ketones as Substrates (H 41,70,72,272,324)

These aldehydic or ketonic substrates afford 4-unsubstituted-,4-alkyl-, or
4-arylquinazolines, with or without 2- and/or 3-substituents as determined by the

synthone employed for cyclization. The scope of such reactions is indicated by the following examples.

Amides or imidates as synthons

o-Aminoacetophenone (**414**) gave 4-methylquinazoline (**415**) (HCONH$_2$, trace BF$_3$.Et$_2$O, 130°, 2 h: 55%).[2170]

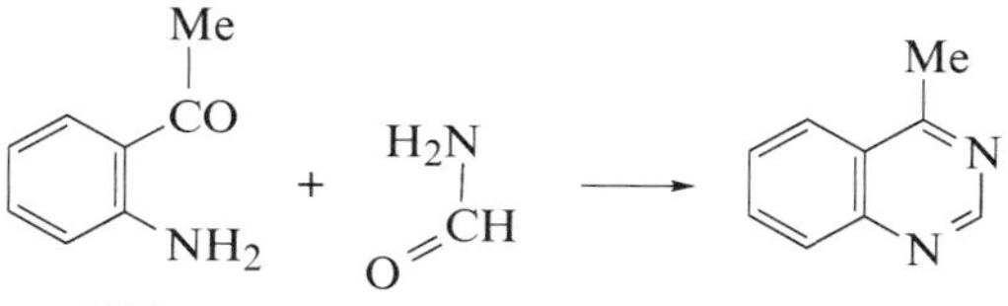

(**414**)   (**415**)

3-Acetyl-2-amino-5,6-dimethylbenzonitrile (**416**) gave 4,6,7-trimethyl-3,4-dihydro-8-quinazolinecarbonitrile (HCONH$_2$, HCO$_2$H, 170°, 4 h: > 74%),[1195] or 4,6,7-trimethyl-8-quinazolinecarbonitrile (HCONH$_2$, AcOH, 170°, 4 h: 73%).[1570]

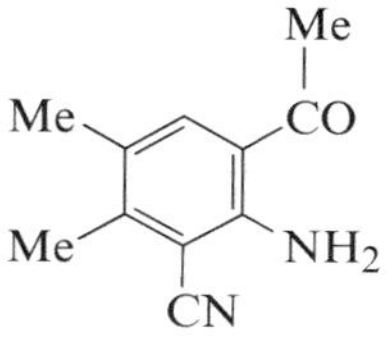

(**416**)

2-Amino-5-chlorobenzophenone hydrazone (**417**) gave 6-chloro-4-phenyl-2-m-trifluoromethylphenylquinazoline [m-F$_3$CC$_6$H$_4$C(=NH)OEt. HCl, Me$_2$NCHO, reflux, 3 h: 36%].[679]

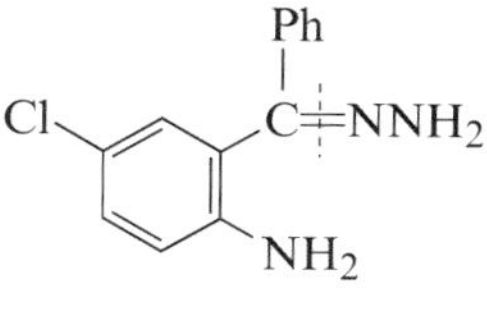

(**417**)

Also other examples in the cited references and elsewhere.[355,959,1110,1643]

Isocyanates or isothiocyanates as synthons

o-Aminoacetophenone (**414**) gave 3,4-dimethyl-2(3H)-quinazolinone (**418**, X = O) (MeNCO, PhH, 20°, 12 h: > 80%)[217] or the corresponding thione (**418**, X = S) (MeNCS, EtOH, reflux, 3 h: 50%);[217,2381] both products were formulated as 4-methylene tautomers.

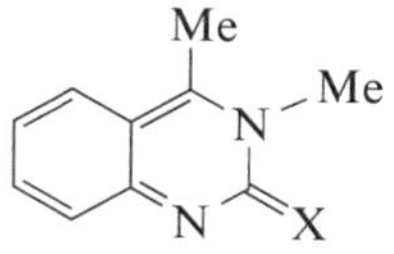

(418)

2-Amino-5,6-dimethoxyacetophenone gave 5,6-dimethoxy-4-methyl-2(1*H*)-quinazolinone (KNCO, AcOH, H_2O, 25°, 12 h: 88%, after purification).[1592,1930,2567]

o-Methylaminobenzophenone **(419)** gave 1-methyl-4-phenyl-2(1*H*)-quinazolinone **(420)** (ClO_2SNCO, PhH, 5°, 1 h, then 28°, 3 h, then H_2O: 78%, after removal of a byproduct).[200]

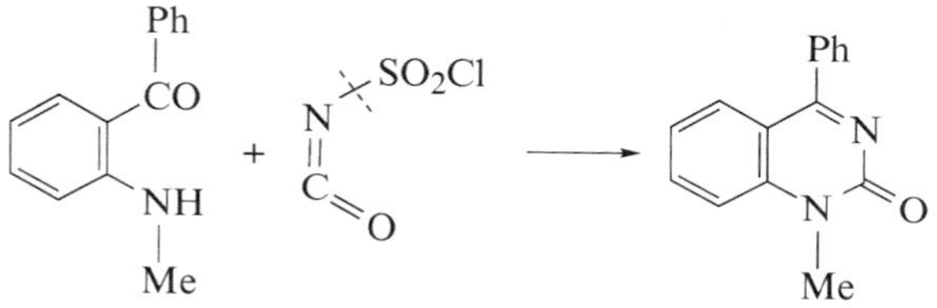

(419)   **(420)**

Also other examples.[5,6,580,955,1078,2575]

Nitriles as synthons

2-Amino-5-chlorobenzophenone **(421)** gave 6-chloro-2-methyl- **(422, R = H)** (MeCN, PBr_3, $CHCl_3$, 20°, 3 h, then reflux, 3 h: 55%),[237] 6-chloro-2-chloromethyl- **(422, R = Cl)**($ClCH_2CN$, likewise: 55%;[237] $ClCH_2CN$, HCl gas, 20°, 5 h: 70%),[1426] or 6-chloro-2-ethyl-4-phenylquinazoline **(422, R = Me)** (EtCN, PBr_3, as before: 43%).[237]

(421)   **(422)**

Urethane, urea, or guanidine as synthons

2-Methylaminobenzophenone **(423)** gave 1-methyl-4-phenyl-2(1*H*)-quinazolinone **(424)** (neat H_2NCO_2Et, $ZnCl_2$, 180°, 90 min: 82%);[247] also other examples.[349,681]

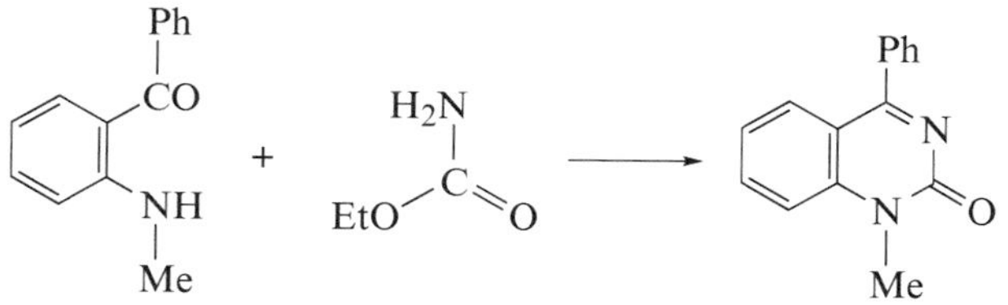

(423)   **(424)**

2-Amino-4-methoxybenzophenone (**425**) gave 7-methoxy-4-phenyl-2(1*H*)-quinazolinone (**426**) (neat H_2NCONH_2, 195°, 15 min: 29%).[79]

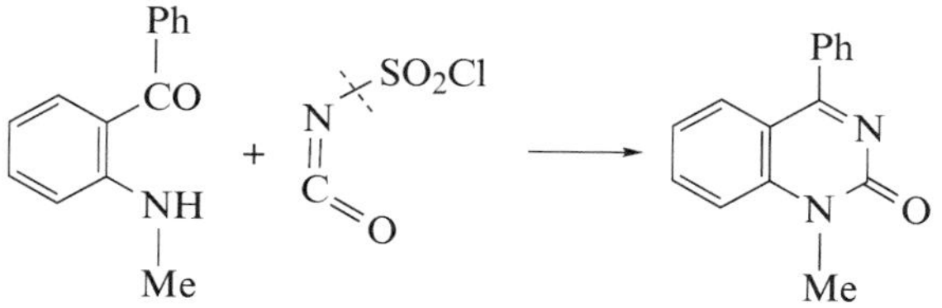

2-Amino-5-fluorobenzaldehyde gave 6-fluoro-2-quinazolinamine (**427**) [$H_2NC(=NH)NH_2$·HCl, Na_2CO_3, decalin, 200°, 1 h: 15%].[647]

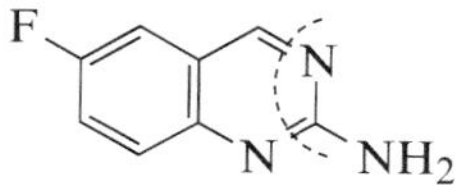

(427)

Separate synthons for C2 and N3

4-*o*-Aminobenzoylpyridazine (**428**) gave 4-pyridazin-4′-ylquinazoline (**429**) [HC(OEt)$_3$, 160°, 6 h; then NH_3, MeOH, 20°, 15 h: 61%].[2294]

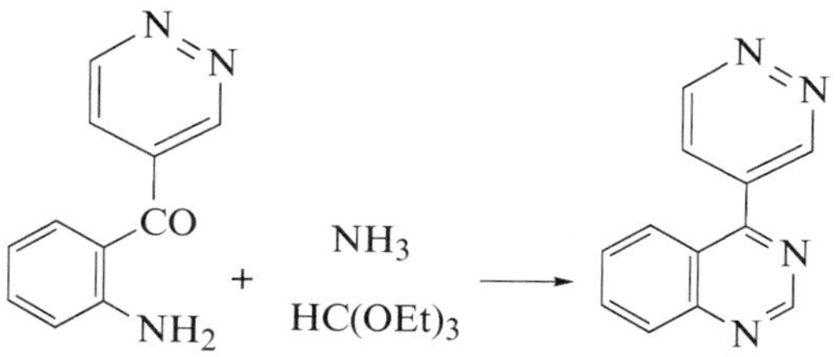

(428)   **(429)**

2-Amino-2′-fluorobenzophenone gave 2-chloromethyl-4-*o*-fluorophenyl-1,2-dihydroquinazoline 3-oxide (**430**) ($ClCH_2CHO$, H_2NOH·H_2SO_4, EtOH, 20°, 18 h: 91%), easily oxidized to its aromatic analog (**431**) (MnO_2, CH_2Cl_2, reflux, 8 h: 82%).[861]

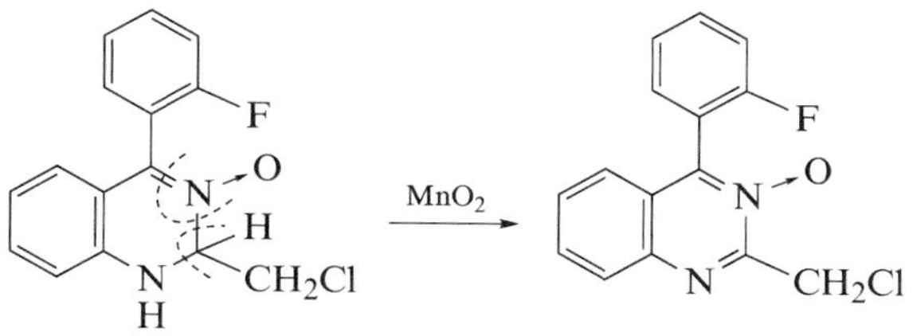

(430)   **(431)**

Also other examples.[2473]

1.2.6.6. With Miscellaneous *o*-Disubstituted Benzenes as Substrates

Several other types of benzene derivative, bearing—NR and —CR substituents situated *ortho* to one another, have been used as substrates occasionally. Their uses are illustrated by the following examples.

2-*o*-Nitrobenzoylthiophene (**432**) gave 4-thien-2′-ylquinazoline (**433**) (neat HCO$_2$NH$_4$, 200°, 50 h: 80%; probably involving a Leuchart reaction on the ketonic group, reduction of the nitro group, and cyclization).[347,710]

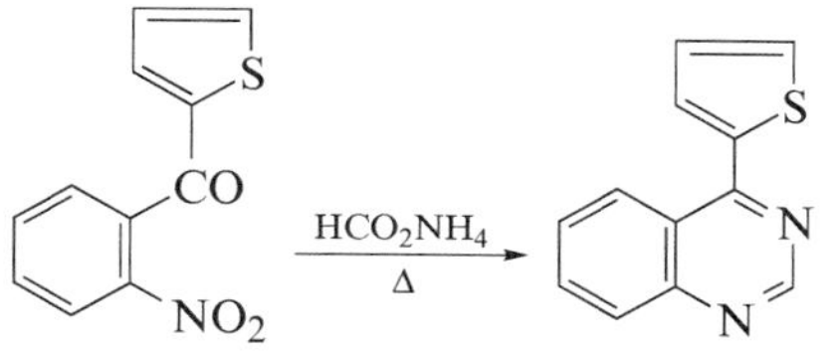

(432)   **(433)**

o-Chlorobenzoyl chloride (**434**) gave 2-phenyl-4(3*H*)-quinazolinone (**435**) (excess PhCONH$_2$, heat: ?%).[550]

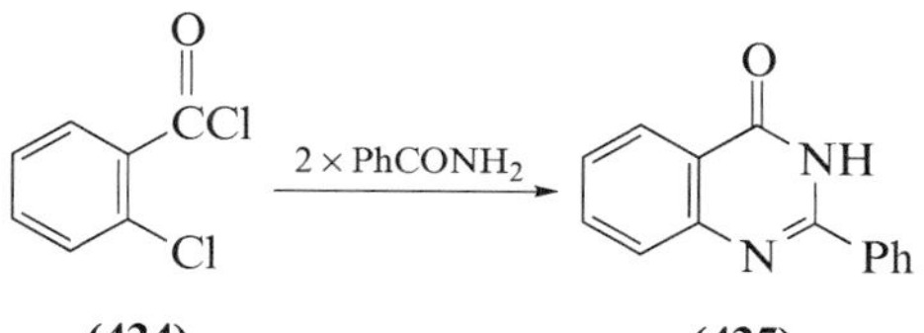

(434)   **(435)**

o-Aminobenzoyl chloride (**436**) gave 2-phenyl-3,4-dihydroquinazoline (**437**) (neat PhCN, POCl$_3$, P$_2$O$_5$, 120°, 3 h: < 34%).[389]

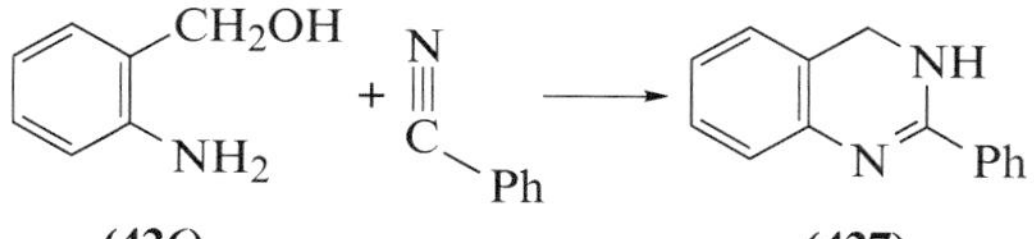

(436)   **(437)**

o-Sulfinylaminobenzyl chloride (**438**) gave 3,4-dihydro-2(1*H*)-quinazolinone (**439**) (KSCN, Me₂NCHO, 140°, 15 min, then H₂O, 20°: 94%).[1837]

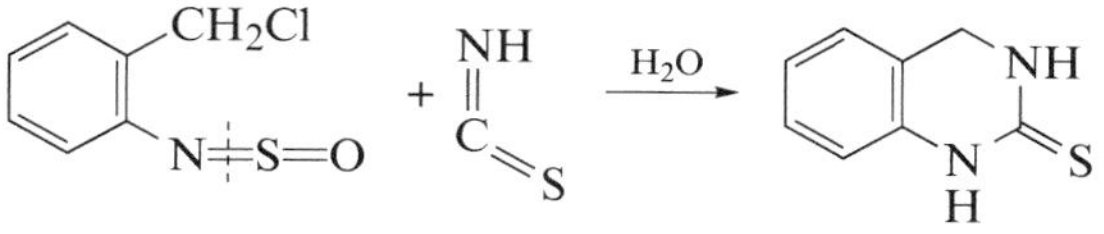

(438)   (439)

Dimethylsulfoxonium *o*-methylaminobenzoylmethylide (**440**) gave 3-benzyl-1-methyl-2-thioxo-1,2-dihydro-4(3*H*)-quinazolinone (**441**) (PhCH₂NCS, PhMe, reflux, 5 h: 36% after separation from another product).[1649]

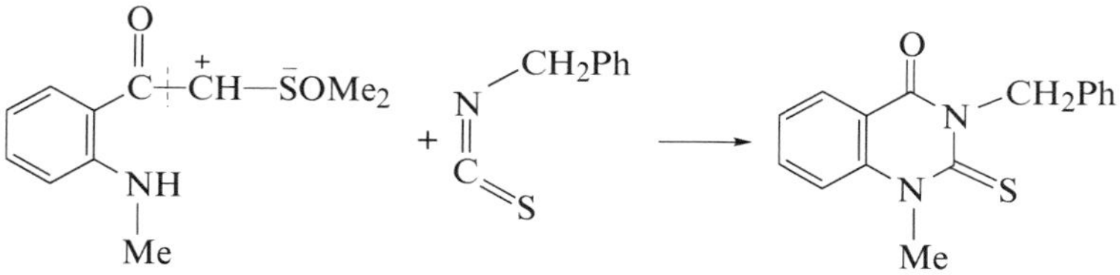

(440)   (441)

2,4,6-Trinitrotoluene gave 4-chloro-5-7-dinitro-2-(2,4,6-trinitrophenyl) quina-zoline (**442**) (NOCl, pyridine, 0 → 25°, 4 h: ∼ 30%; a rational mechanism was proposed).[469]

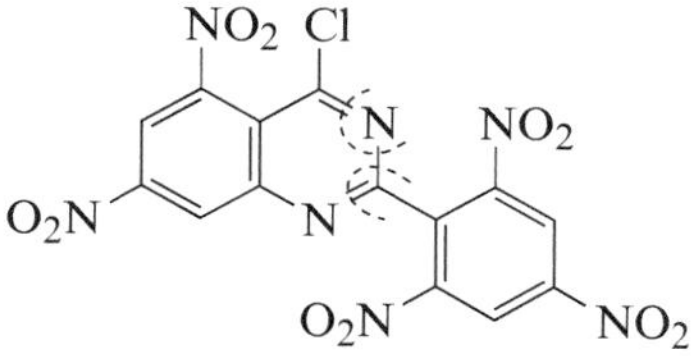

(442)

Also other examples.[389,2350]

1.2.7.   Where the Synthon(s) Supply N3 + C4 (*H* 125)

Such syntheses have been used in moderation to convert several related types of benzene substrate into quinazolines. Unsymmetrical substrates can give products of ambiguous structure.

1.2.7.1.   With *N*-Acylanilines as Substrates (*H* 94)

Although such substrates can undergo self-condensation to quinazolines,[1809] it is more usual to employ a different synthon as in the following examples.

Urethane as a synthon

p-Isobutyramidotoluene (**443**) gave 2-isopropyl-6-methyl-4(3*H*)-quinazolinone (**444**) (H_2NCO_2Et, P_2O_5, xylene, 140°, 5 h: > 45%).[1049] This is an example of the so-called Sen–Ray synthesis (*H* 94).

(**443**) (**444**)

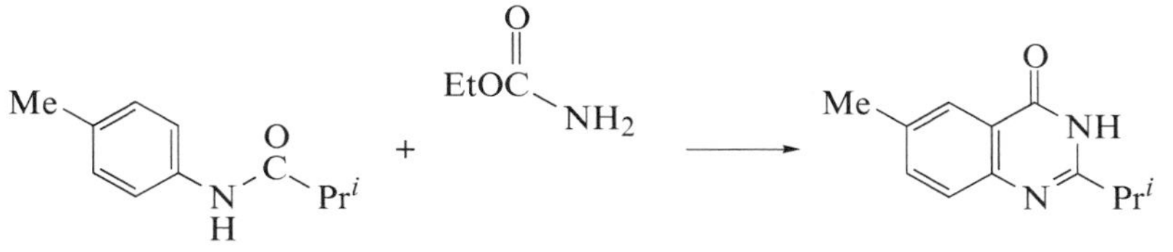

Also other examples.[503,766,782,1807,2264]

Nitriles as synthons

N-*t*-Butoxycarbonylaniline (**445**) gave 4-phenyl-2(1*H*)-quinazolinone (**447**) (Bu^tLi, pentane–THF, − 78°, 15 min, then − 20°, 2.5 h; then PhCN, − 20°, 2 h, then 20°, 1 h; then H_2O; 83%), probably via the dianion (446).[1360]

(**445**) (**446**) (**447**)

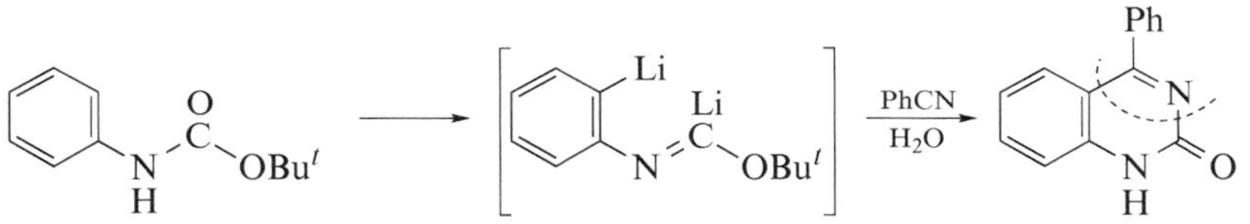

p-Chloropivalamidobenzene gave 2-*t*-butyl-6-chloro-4-*o*-fluorophenylquinazoline (**448**) (dilithiation; then *o*-FC_6H_4CN, THF, 25°, 16 h; then H_2O: 57%).[1101, cf. 2494]

(**448**)

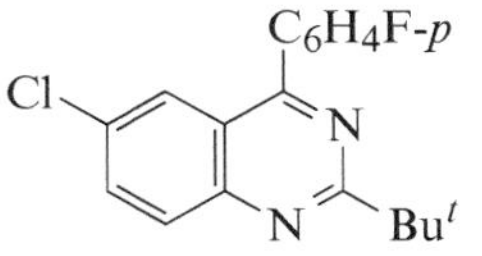

o-Bromoacetanilide (**449**) gave 2-methyl-4(3*H*)-quinazolinone (**450**) [$Pd(Ph_3P)_4$, CO, K_2CO_3, a Ti/NCO complex, $MeN(CH_2)_4$, 100°, 24 h: 54%].[1678]

(**449**) (**450**)

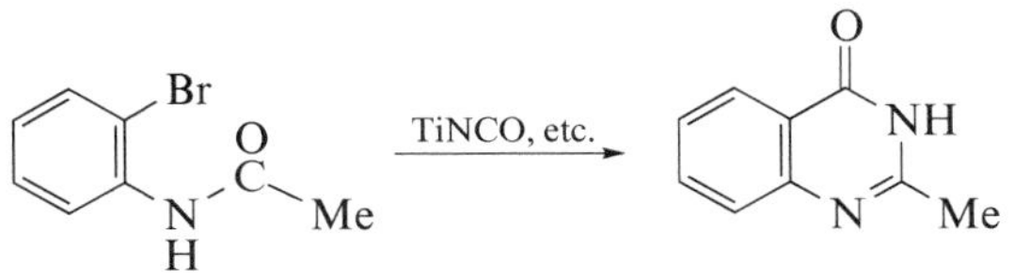

Also other examples,[231,1360,1809,2494] including one employing chlorocarbonyl isocyanate as synthon.[2542]

1.2.7.2. With N-(α-Chloromethylene)anilines or Related
Compounds as Substrates (H 43)

This type of substrate has proved quite useful, as indicated in the following
examples.

p-Chloro-(α-chlorobenzylidene)aniline (**451**) gave 6-chloro-2,4-diphenyl-
quinazoline (**452**) (PhCN, AlBr$_3$, ClC$_6$H$_4$Cl-p. 140°, 1 h: 69%).[1181]

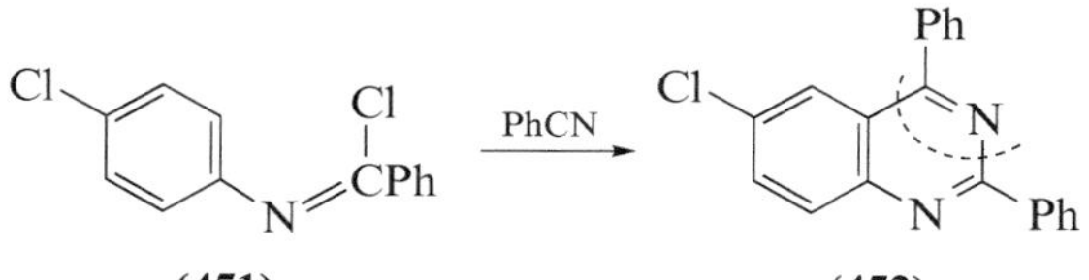

(**451**)   (**452**)

α-Chlorobenzylideneaniline gave 2-phenyl-4(3H)-quinazolinethione (**453**,
X = S) (NaNCS, Me$_2$NCHO, reflux, 5 min: 59%)[210] or 2-phenyl-4(3H)-
quinazolinone (**453**, X = O) (AgNCO, MeCN, reflux, 1 h: 13%).[824]

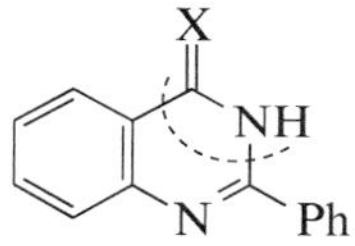

(**453**)

N-(α-chloro-α-morpholinomethylene)aniline (**454**) gave 2,4-dimorpholino-
quinazoline [neat O(CH$_2$CH$_2$)$_2$NCN, 100°, 90 min: 10%][428] or 4-anilino-
2-morpholinoquinazoline (PhHNCN, NEt$_3$, AcNMe$_2$, 20°, 2 h: 10%).[721]

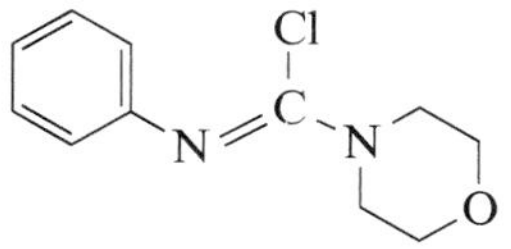

(**454**)

α-Chlorobenzylideneaniline gave 4-phenoxy-2-phenylquinazoline (**455**)
(PhOCN, SnCl$_4$, PhNO$_2$, 150°, 1 h: 68%).[213]

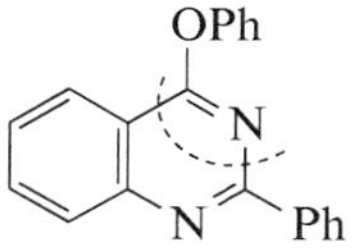

(**455**)

Also other examples in the foregoing references and elsewhere.[24,32,855,1508]

1.2.7.3.   With Other Aniline Derivatives as Substrates

Miscellaneous aniline derivatives, such as symmetrical or unsymmetrical iminomethyleneanilines (carbodiimides) and various vinylideneanilines, have been used for this synthesis, as hereafter exemplified.

Diphenylcarbodiimide (**456**, R = Ph) gave 2-anilino-4(3*H*)-quinazolinone (**458**) [ethyl allophanate (**457**), Me_2NCHO, reflux, 3 h; 64%] and *N*-isobutyl-*N'*-phenylcarbodiimide (**456**, R = Bu^i) likewise gave 2-iso-butylamino-4(3*H*)-quinazolinone (**458**, R = Bu^i) (81%).[1372]

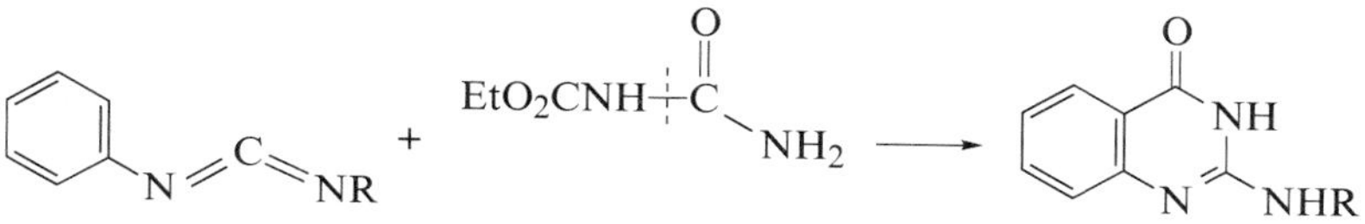

(**456**)   (**457**)   (**458**)

Diphenylcarbodiimide (**456**, R = Ph) self-condensed to 2-anilino-3-phenyl-4-phenylimino-3,4-dihydroquinazoline (**459**) (HBF_4, CH_2Cl_2, 0°, 10 min: 85%).[71,440]

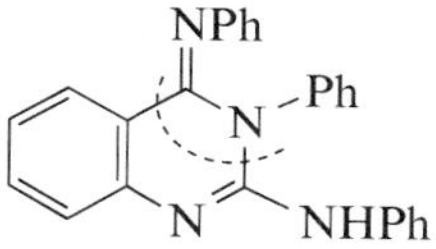

(**459**)

Isopropylidenemethyleneaniline ($Me_2C{=}C{=}NPh$) self-condensed to 2-isopropyl-4-isopropylidene-3-phenyl-3,4-dihydroquinazoline (**460**) (125°, sealed, 1 week: 52% yield after separation from byproducts).[400]

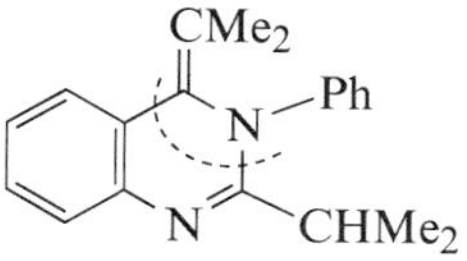

(**460**)

The nitrilium salt (PhN$\overset{+}{\equiv}$CMe.SbCl$_6$) gave 2,3-dimethyl-4(3*H*)-quinazo-linone (**461**) (MeNCO, $ClCH_2CH_2Cl$, reflux, 3 h: 70%)[2361] or 4-dimethy-lamino-2-methylquinazoline (Me_2NCN, CH_2Cl_2, 25°, 10 h: 65%).[2506]

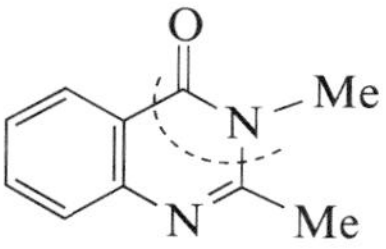

(**461**)

Also other examples in the cited references and elsewhere.[1896]

1.2.8.   Where the Synthon(s) Supply N1 + C2 + N3 (*H* 325)

Although relatively few fully aromatic quinazolines have been so made, this general procedure has been used quite extensively to convert cyclohexane or cyclohexene derivatives into quinazolines that are reduced or partly reduced in the benzene ring. The following treatment is subdivided according to the type of benzene (or reduced benzene) derivative used as substrate.

1.2.8.1.   With *o*-Halogeno or *o*-Alkoxybenzonitriles as Substrates

Despite the foregoing generalization, this type of substrate has been used mainly to produce aromatic quinazolinamines as hereafter exemplified.

o-Fluorobenzonitrile (**462**, R = H) gave 2,4-quinazolinediamine (**463**, R = H) [$H_2NC(=NH)NH_2.H_2CO_3$, AcNMe$_2$, 140°, 3 h: 68%];[1932] 2,5-difluoro- (**462**, R = F) and 2-fluoro-5-trifluoromethylbenzonitrile (**462**, R = CF$_3$) likewise gave 6-fluoro- (**463**, R = F)[1932] and 6-trifluoromethyl-2,4-quina- zolinediamine (**463**, R = CF$_3$),[2246] respectively, each in $\sim$ 80% yield.

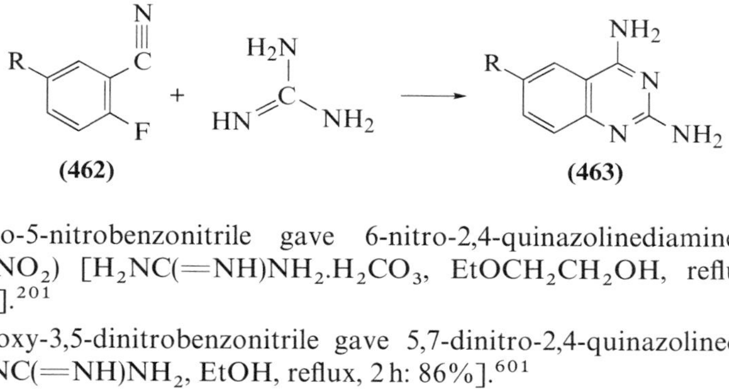

(462)   +   (463)

2-Chloro-5-nitrobenzonitrile gave 6-nitro-2,4-quinazolinediamine (**463**, R = NO$_2$) [$H_2NC(=NH)NH_2.H_2CO_3$, EtOCH$_2$CH$_2$OH, reflux, 3 h: 85%].[201]

2-Methoxy-3,5-dinitrobenzonitrile gave 5,7-dinitro-2,4-quinazolinediamine [$H_2NC(=NH)NH_2$, EtOH, reflux, 2 h: 86%].[601]

2-Chlorooctafluorocyclohex-1-enecarbonitrile (**464**) gave 5,5,6,6,7,7,8,8-octa- fluoro-2-trichloromethyl-5,6,7,8-tetrahydro-4-quinazolinamine (**465**) [$Cl_3C(=NH)NH_2$, Et$_2$O, 20°, 1 h: 84%].[1856]

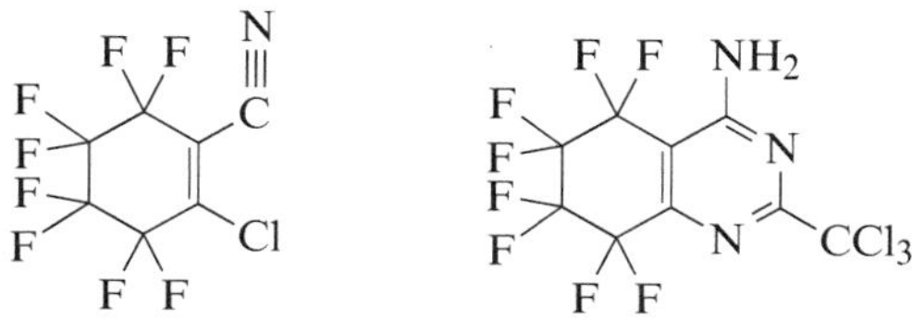

(464)   (465)

Also other examples.[396,2211,2247,2284,2335]

1.2.8.2.   With *o*-Substituted Benzoic Esters or Related Compounds as Substrates

These substrates normally give 4-quinazolinones, as indicated in the following examples.

Methyl 2-chloro-3,5-dinitrobenzoate (466) gave 2-amino-6,8-dinitro-4(3*H*)-quinazolinone (467) [$H_2NC(=NH)NH_2$.HCl, AcONa, EtOH, reflux, ? h: 86%].[117]

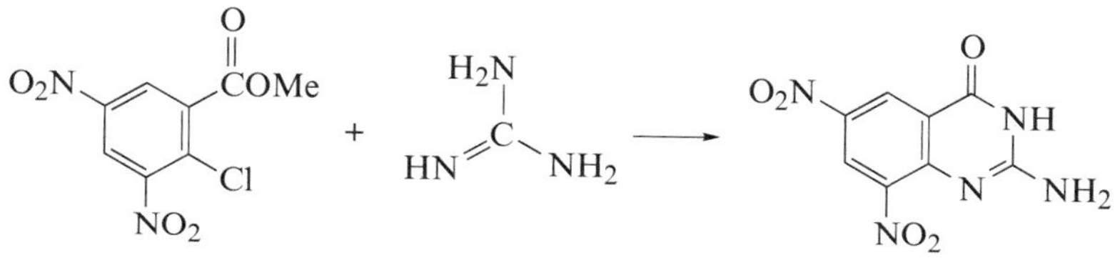

(466)   (467)

Ethyl *o*-hydroxybenzoate gave 2,3-diamino-4(3*H*)-quinazolinone (468) [$H_2NC(=NH)NHNH_2$.H_2CO_3, BuOH, reflux, 5 h: ?%].[1438]

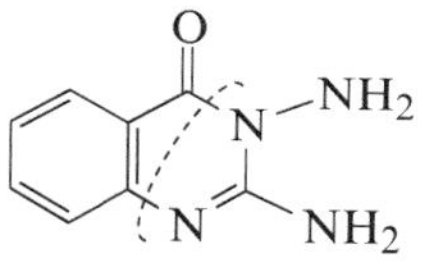

(468)

o-Hydroxybenzoyl chloride gave 2-phenyl-4(3*H*)-quinazolinone [2 equiv (equivalents) BzNH$_2$, reflux, 2 h: 40%]; 2-pyridin-4'-yl-4(3*H*)-quinazolinone (30%) was made similarly using isonicotinamide.[386]

Ethyl 2-oxocyclohexanecarboxylate (469) gave 5,6,7,8-tetrahydro-4(3*H*)-quinazolinone (470, R = H) ($H_2NCH=NH$.AcOH, EtONa, EtOH, reflux, 4 h: 87%);[281, cf. 1635] other amidines have been used in similar reactions.[815,1303,1686]

Ethyl 2-oxocyclohexanecarboxylate (469) gave 2-amino-5,6,7,8-tetrahydro-4(3*H*)-quinazolinone (470, R = NH$_2$)[$H_2NC(=NH)NH_2$.H_2CO_3, trace HCl, EtOH, reflux, 90 min: ~ 75%].[1723] Other such reactions have been reported.[1571,2402,2498]

Ethyl 2-oxocyclohexanecarboxylate (469) gave 3-allyl-2-thioxo-1,2,5,6,7,8-hexahydro-4(3*H*)-quinazolinone (471) ($H_2NCSNHCH_2CH=CH_2$, MeONa, MeOH, reflux, 10 h: ~ 60%).[1336] Ureas and other thioureas have been so used.[19,1264,1499,2006]

In contrast, ethyl 2-oxocyclohexanecarboxylate (469) gave 4-ethoxy-5,6,7,8-tetrahydro-2(1*H*)-quinazolinone (472) (H_2NCN, EtOH, reflux, 30 min: 97%, probably via a benzoxazine intermediate).[1813]

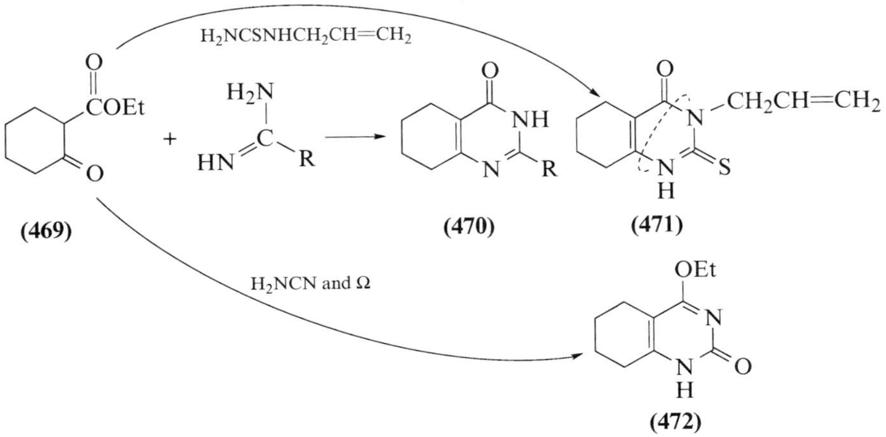

(469) + **(470)** → **(471)** and **(472)**

1.2.8.3. With o-Substituted Acetophenones or Related Compounds as Substrates

These substrates should afford quinazolines bearing an appropriate 4-substituent. However, most of the available examples (sampled hereafter) employ analogous ring-reduced substrates.

2-Benzoylcyclohexanone (**473**) gave 4-phenyl-5,6,7,8-tetrahydro-2(1H)-quina-zolinone (**474**) (H$_2$NCONH$_2$, HCl, EtOH, reflux, 8 h: 73%);[1416] also analogs using urea or thiourea as a synthon.[870,1416,1478,2039]

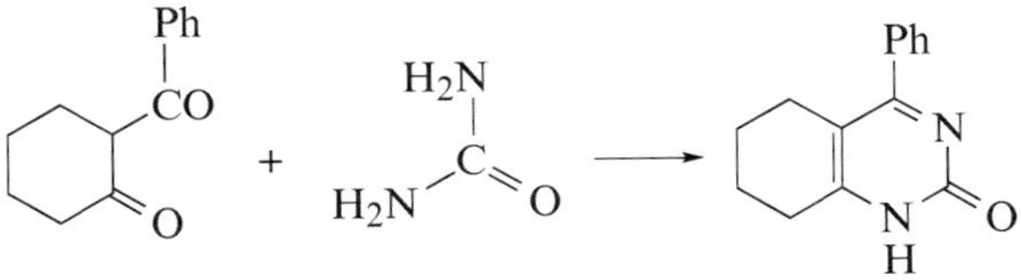

(473)   **(474)**

2-Oxalocyclohexanone gave 2-amino-5,6,7,8-tetrahydro-4-quinazolinecar-boxylic acid (**475**) [H$_2$NC(=NH)NH$_2$.H$_2$CO$_3$, H$_2$O–MeOH, reflux, 2 h: 68%);[418] cyanoguanidine and several amidines have been used rather similarly.[283,533,1520]

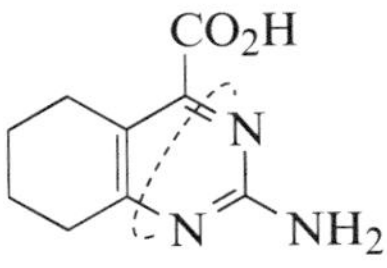

(475)

2-Acetyl-3-dimethylamino-5-hydroxy-5-methylcyclohex-2-enone (**476**) gave 7-hydroxy-4,7-dimethyl-2-phenyl-5,6,7,8-tetrahydro-5-quinazolinone (**477**) [PhC(=NH)NH$_2$, EtOH, reflux, 5 h: 62%].[1633] Several analogous 5,8-quinazolinediones have been made rather similarly.[736]

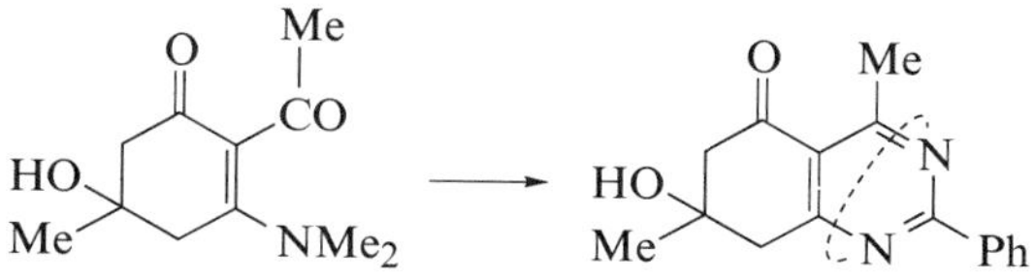

(476)   (477)

2-Acetyl-3,6-dimethoxy-1,4-benzoquinone (**478**) gave 6-methoxy-4-methyl-2-methylthio-5,8-dihydro-5,8-quinazolinedione (**479**) [MeSC(=NH)NH$_2$, MeOH: 65%];[68] amidines gave analogous products.[297]

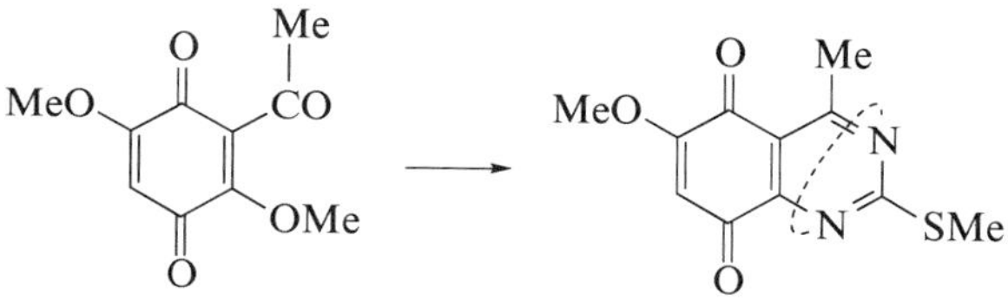

(478)   (479)

Also other examples.[780,2453,2569]

1.2.8.4.   With 2-Methylenecyclohexanones as Substrates

Quinazolines, reduced at least in the benzene ring, have been made often from various 2-(substituted-methylene)cyclohexanones, as illustrated in the following examples.

2-Benzylidenecyclohexanone (**480**) gave 4-phenyl-3,4,5,6,7,8-hexahydro-2(1H)-quinazolinethione (**481**)(H$_2$NCSNH$_2$, KOH, EtOH, reflux, 3 h: 69%)[2348] or 4-phenyl-5,6,7,8-tetrahydro-2-quinazolinamine [H$_2$NC(=NH)NH$_2$.HCl, NaOH, H$_2$O–EtOH, reflux, 4 h: 65%; 3,4-dehydrogenation during workup?].[2153]

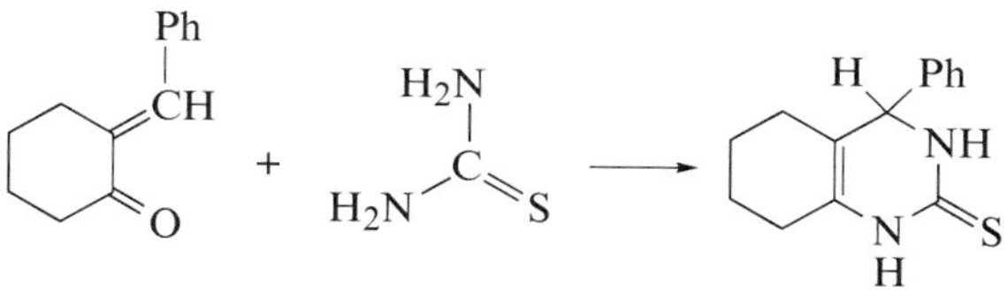

(480)   (481)

2-α-Ethoxyethylidene-1,3-cyclohexanedione (**482**) gave 4-methyl-2-phenyl-5,6,7,8-tetrahydro-5-quinazolinone (**483**) [PhC(=NH)NH$_2$.HCl, K$_2$CO$_3$, MeOH, reflux, 3 h: > 34%].[2105]

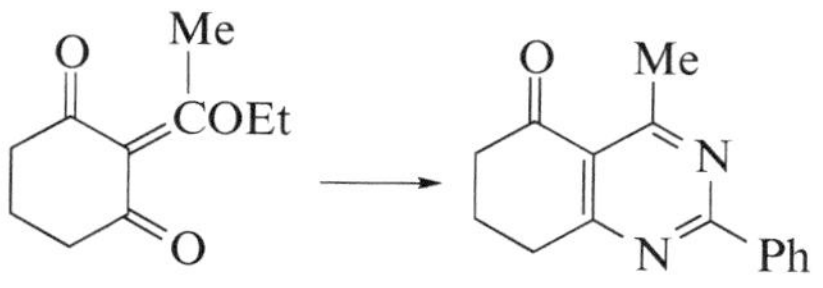

(**482**)   (**483**)

2-(Bismethylthiomethylene)cyclohexanone gave 4-ethoxy-5,6,7,8-tetrahydro-2-quinazolinamine [H$_2$NC(=NH)NH$_2$.HNO$_3$, EtONa, EtOH, reflux, 10 h: 54%; SMe → OEt during reaction?].[704]

Also other examples.[298,302,541,779,813,1068,1472,1701,1740,1883,2586]

1.2.8.5. With Miscellaneous Substrates

The use of less important substrates is illustrated in the following examples.

1-Dimethylamino-2-dimethylimmoniomethylcyclohex-1-ene perchlorate (**484**) gave 5,6,7,8-tetrahydroquinazoline (HN=CHNH$_2$.HCl, MeONa, MeOH, reflux, 2 h: 67%).[203]

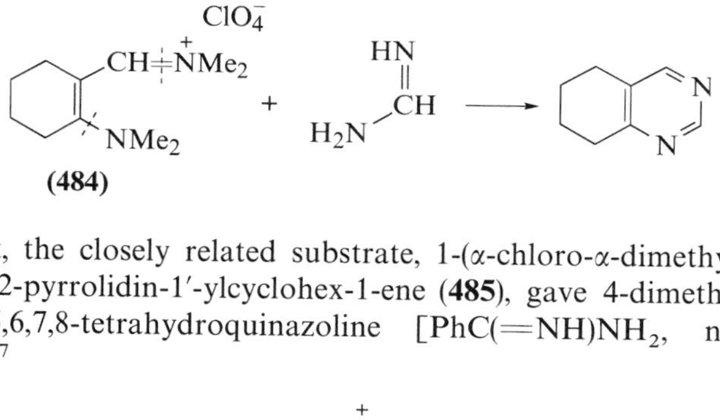

(**484**)

In contrast, the closely related substrate, 1-(α-chloro-α-dimethylimmonio-methyl)-2-pyrrolidin-1′-ylcyclohex-1-ene (**485**), gave 4-dimethylamino-2-phenyl-5,6,7,8-tetrahydroquinazoline [PhC(=NH)NH$_2$, no details: 37%].[337]

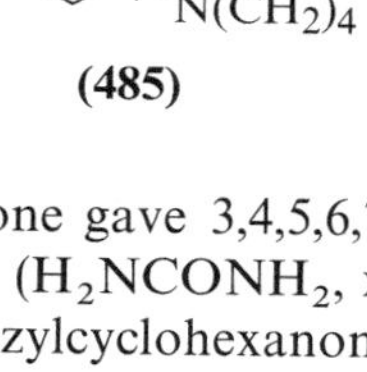

(**485**)

2-Hydroxymethylcyclohexanone gave 3,4,5,6,7,8-hexahydro-2(1*H*)-quinazo-linone (**486**, R = H, X = O) (H$_2$NCONH$_2$, xylene, reflux, 4 h: ~ 60%);[775] similarly, 1-α-hydroxybenzylcyclohexanone gave 4-phenyl-3,4,5,6,7,8-

hexahydro-2(1H)-quinazolinethione (**486**, R = Ph, X = S) (NH₄SCN, PhMe, reflux with water removal, 7 h: 40%).[1299]

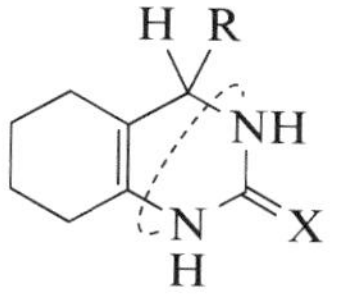

(486)

Also other examples.[512]

1.2.9. Where the Synthon(s) Supply C2 + N3 + C4 (*H* 43)

Although not extensively used, this synthesis is characterized by the ingenuity required to devise a synthon (or combination of synthons) to supply the missing ring atoms to each quite simple aniline derivative normally used as a substrate. For simplicity, the present scope of this synthesis is illustrated by representative examples, without any attempt at systematic treatment.

Using a single synthon

p-Toluidine (**487**) gave ethyl 3,6-dimethyl-3,4-dihydro-4-quinazolinecarboxylate (**489**) [synthon (**488**), CH₂Cl₂, 20°, 2 h; residue in air, 12 h: 37%, as hydrochloride; undoubtedly via the 1,2,3,4-tetrahydro analog].[1030]

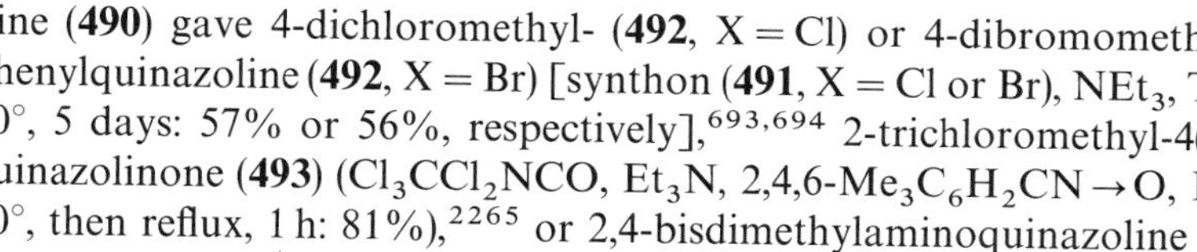

(487)   **(488)**   **(489)**

Aniline (**490**) gave 4-dichloromethyl- (**492**, X = Cl) or 4-dibromomethyl-2-phenylquinazoline (**492**, X = Br) [synthon (**491**, X = Cl or Br), NEt₃, THF, 20°, 5 days: 57% or 56%, respectively],[693,694] 2-trichloromethyl-4(3H)-quinazolinone (**493**) (Cl₃CCl₂NCO, Et₃N, 2,4,6-Me₃C₆H₂CN → O, PhH, 20°, then reflux, 1 h: 81%),[2265] or 2,4-bisdimethylaminoquinazoline (**494**) (Me₂N—CCl=N̶=CCl—NMe₂ Cl⁻, NEt₃, CH₂Cl₂, Et₂O: 50%).[758]

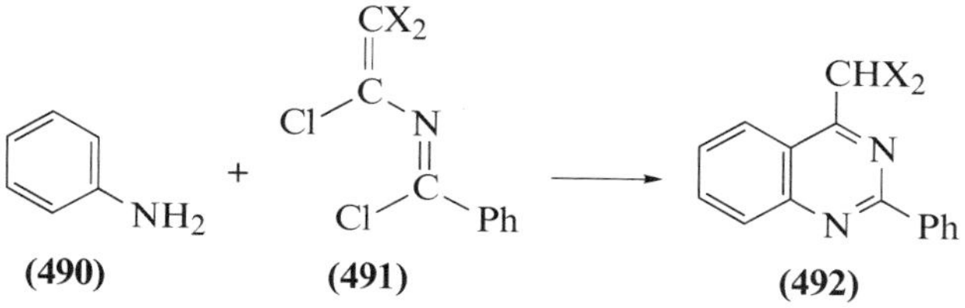

(490)   **(491)**   **(492)**

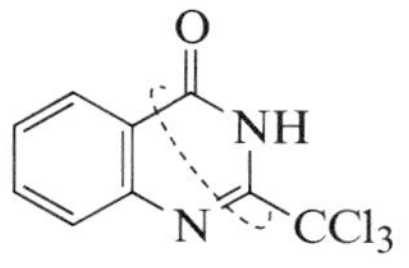

(493) 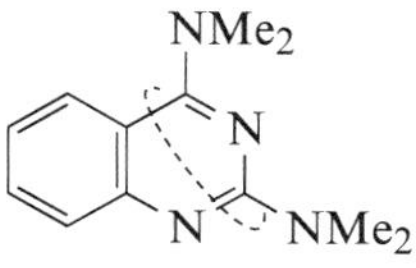   **(494)**

N-Ethylaniline gave 1-ethyl-2,4,4-tristrifluoromethyl-1,4-dihydroquinazoline
(495) $[(F_3C)_2C=NC(O)CF_3$, $CHCl_3$, 20°, 13 weeks: 87%; or 80°, sealed,
2 h: 56%].[1788,2202]

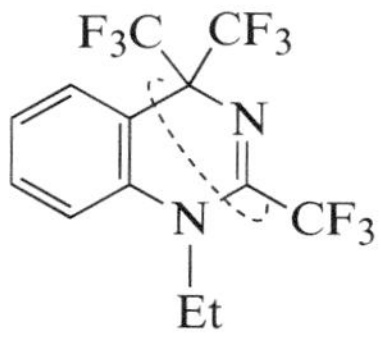

(495)

N-(Cyclohexylidene)cyclohexylamine $(C_6H_{10}=NC_6H_{11})$ gave   1-cyc-
lohexylamino-5,6,7,8-tetrahydro-2,4(1*H*, 3*H*)-quinazolinedithione
$[Et_2N-C(=S)-N=C=S$, reflux, 2 h: 38%].[202]
Also other examples in cited references and elsewhere.[951,1802]

Using more than one synthon

N-Ethylaniline **(496)** gave 1-ethyl-4-phenyl-2(1*H*)-quinazolinone **(497)**
(PhCN, CO_2, Li-assisted: 45%).[1955]

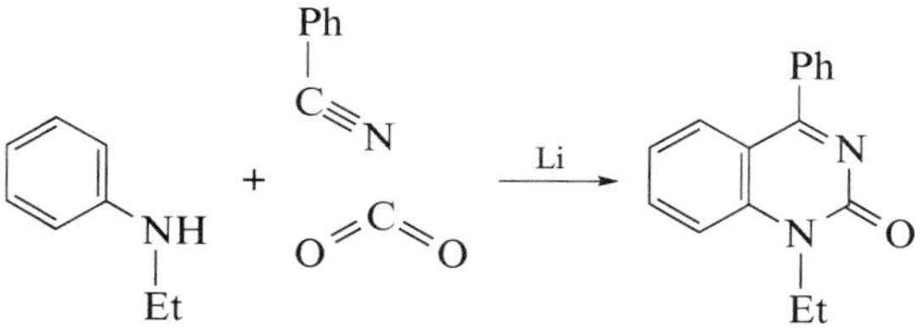

(496)   **(497)**

3-Anilino-5,5-dimethylcyclohex-2-enone gave 7,7-dimethyl-1-phenyl-1,2,3,4,
5,6,7,8-octahydro-5-quinazolinone (**498**, R = H) or its 3-benzyl derivative
(**498**, R = CH_2Ph) (CH_2O, $AcONH_4$ or $PhCH_2NH_2.AcOH$, $EtOH-H_2O$,
reflux, 4 h: 52% or 68%, respectively);[1777] 3-amino-5,5-dimethylcyclohex-
2-enone likewise gave 3-benzylamino-7,7-dimethyl-1,2,3,4,5,6,7,8-octahy-
dro-5-quinazolinone (64% as hydrochloride) and some analogs.[629]

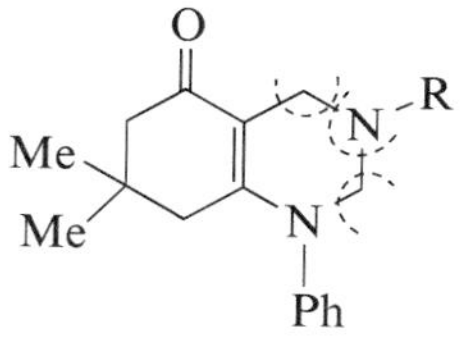

(498)

m-Chloroaniline gave 7-chloro-3-*m*-chlorophenyl-4(3*H*)-quinazolinone **(499)** [CCl$_4$, CO (100 atm), Mo(CO)$_6$, 150°, 16 h: 81%]; also analogs likewise.[47]

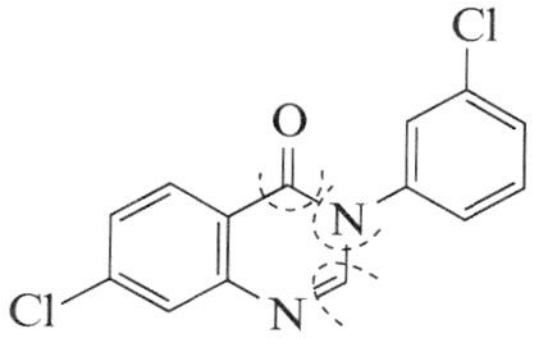

(499)

Also other examples.[1051,1523,1778,1840,1992,2048,2419]

Occasionally, substrates other than simple anilines have been used. Thus *o*-iodoaniline, in which the iodo substituent proved to be a leaving group, reacted with phenethylamine and carbon monoxide (5 atm) in the presence of Ph$_3$PPd(OAc)$_3$ at 130° during 3 days to afford a small yield of 3-phenacyl-2,4(1*H*,3*H*)-quinazolinedione **(500)**.[1767] Nitrosobenzene **(501)** with the cyclic phospholine **(502)** under thermolytic conditions (xylene, 150°, 20 h) gave 2-phenyl-4,4-bistrifluoromethyl-3,4-dihydroquinazoline 1-oxide **(503)** in small yield after separation from other products; analogs were made similarly.[729] Phenyl isothiocyanate (2 mol) with dimethylcyanamide at ~8000 atm/100° for 20 h gave 4-anilino-2-dimethylaminoquinazoline (86%, as hydrochloride).[2576]

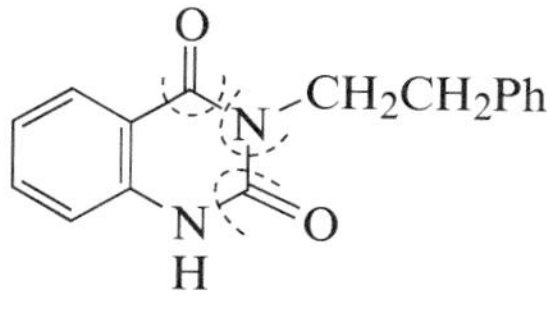

(500)

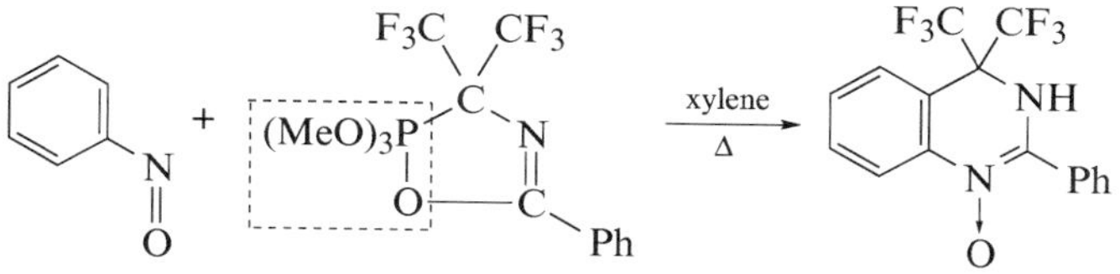

(501)   **(502)**   **(503)**

1.2.10.   Where the Synthon(s) Supply N1 + C2 + N3 + C4

This type of synthesis has been employed to a limited extent to make 5,6,7,8-tetrahydro and other reduced quinazolines, but it has remained almost unused for the direct preparation of fully aromatic quinazolines. Its present scope and utility may be inferred from the following examples.

Using a single synthon

1-Morpholinocyclohexene (**504**) with triethyl 1,3,5-triazine-2,4,6-tricar-boxylate (**505**) gave diethyl 5,6,7,8-tetrahydro-2,4-quinazolinedicar-boxylate (**506**) by an initial cycloaddition and subsequent loss of appropriate fragments (AcOH, CH_2Cl_2, 25°, then 100°, 20 h: 72%);[1971] the same substrate (**504**) with 4,6-diphenyl-1,2,3,5-oxathiadiazine 2,2-dioxide (**507**) likewise gave 2,4-diphenyl-5,6,7,8-tetrahydroquinazoline (**508**) (no solvent, 130°, 1 h: 48%).[111] Similarly, 1-pyridiniocyclohexene with 1,3,5-triazine gave 5,6,7,8-tetrahydroquinazoline (dioxane, 90°, 48 h: 47%).[1378]

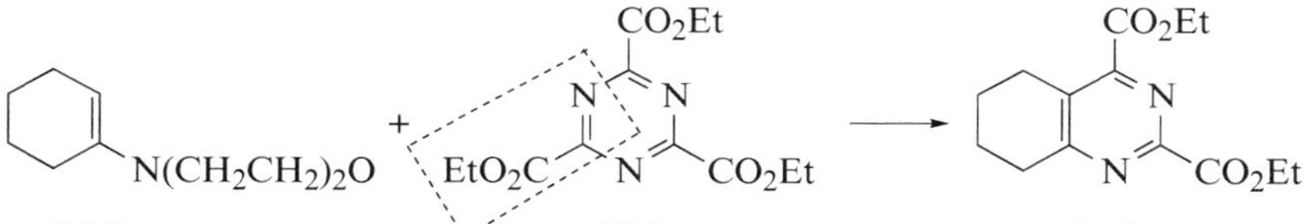

(**504**)   (**505**)   (**506**)

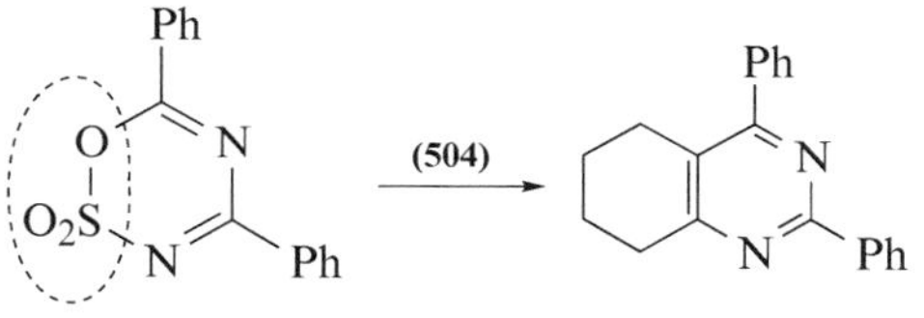

(**507**)   (**508**)

Cyclohexanone (**509**) with cyanoguanidine (**510**, R = NH_2) gave 5,6,7,8-tet-rahydro-2,4,-quinazolinediamine (**511**, R = NH_2) (no solvent, NEt_3, reflux, water separation, 5 h: 67%);[91] the same substrate (**509**) with N-cyano-S-methylisothiourea (**510**, R = SMe) likewise gave 2-methylthio-5,6,7,8-tet-rahydro-4-quinazolinamine (**511**, R = SMe) [no solvent, $HN(CH_2)_4$ as a catalyst, 150°, sealed, 7 h: 67%].[1398]

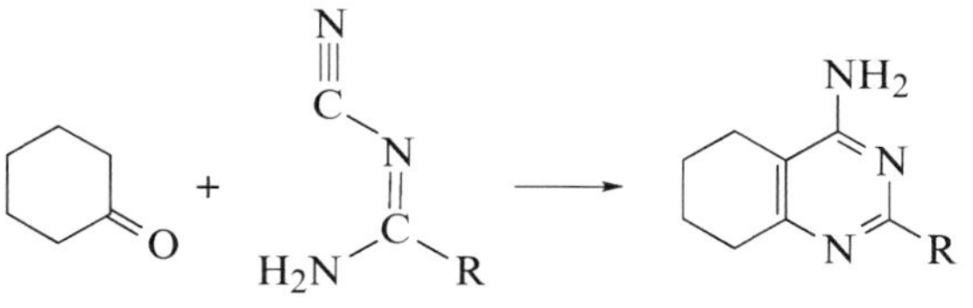

(**509**)   (**510**)   (**511**)

Also other examples.[1338,2126,2545,2566]

Using two similar synthons

2-Phenylcyclohexanone (**512**, R = Ph) gave 8-phenyl-5,6,7,8-tetrahydroquinazoline (**513**, R = Ph) [H_2NCHO, $POCl_3$, 80°, 8 h: 17% after separation from other products;[984] or $HC(NHCHO)_3$, H_2NCHO, TsOH, 160°, 14 h: 25%].[985]

2-Methylcyclohexanone (**512**, R = Me) gave 8-methyl-5,6,7,8-tetrahydro-2,4(1H,3H)-quinazolinedione (**514**, R = Me) (H_2NCONH_2, TsOH, 280°, 30 min: 50%);[803] analogs were made similarly.[803]

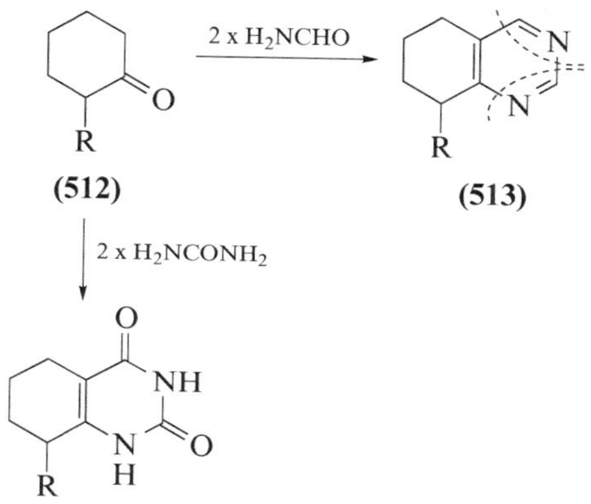

(512)   (513)

(514)

o-Benzylbenzenediazonium tetrafluoroborate (**515**) gave 8-benzyl-2,4-diphenylquinazoline (**516**) (neat reactants, 40°: 53% after purification from byproducts).[204]

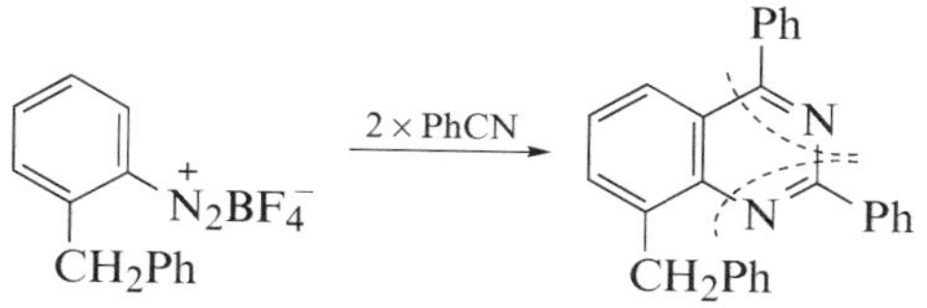

(515)   (516)

Pentachlorophenyllithium (from hexachlorobenzene) gave 5,6,7,8-tetrachloro-2,4-diphenylquinazoline (PhCN, Et_2O, 0°, then reflux, 3 h: 42%).[157,198]

1-Morpholinocyclohexene (**504**) gave 2,4-bistrifluoromethyl-5,6,7,8-tetrahydroquinazoline (F_3CCN, C_6H_{14}, 40°, sealed, 14 days: ~ 80%).[1145,1213]

1-Chlorocyclohexene gave 2,4-diphenyl-5,6,7,8-tetrahydroquinazoline (PhCN, F_3CSO_3H, 95°, 20 min: 63%;[2138] PhCN, $SbCl_5$, Bu^{t}Cl, 100°, 5 min: 53%, initially as hexachloroantimonate).[2477]

Cyclohexanone gave 2,4-dimethyl- (**517**, R = H) [MeCN, $(F_3CSO_2)_2O$, CH_2Cl_2, 20°, 24 h: 80%][2036] and likewise, 2-chlorocyclohexanone gave

8-chloro-2,4-dimethyl-5,6,7,8-tetrahydroquinazoline (**517**, R = Cl) [MeCN, $(F_2CSO_2)_2O$, 20°, 5 days: 50%].[2404]

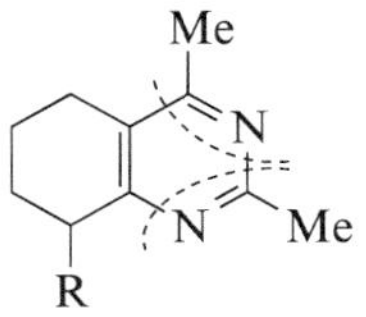

(517)

Cyclohexanone (**518**) gave 1,3-dimethyl-5,6,7,8-tetrahydro-2,4($1H,3H$)-quinazolinedithione (**519**, R = Me) (MeNCS, PhH, Bu^tCH_2ONa, 0°, 3 h, then 20°, 12 h: 50%) or its 1,3-diphenyl analog (**519**, R = Ph) (PhNCS, likewise: 19%);[1304,2300] also 2,4-bismethylthio-5,6,7,8-tetrahydroquinazoline (MeSCN, Tf_2O, CH_2Cl_2, 20°, 20 h: 60%).[2548]

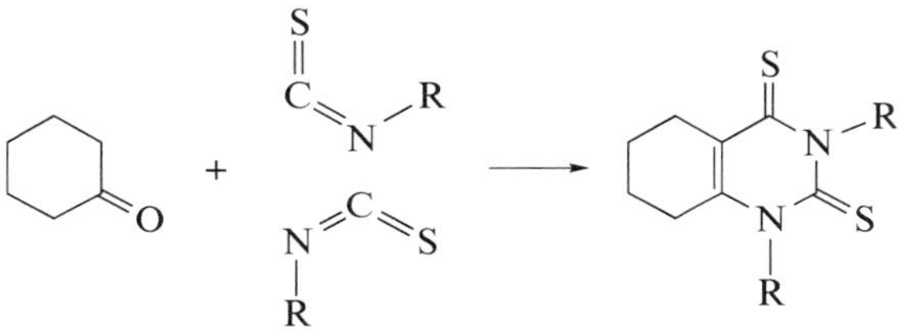

(518)   **(519)**

Also some similar examples.[1334,1349]

Using two dissimilar synthons

1-Pyrrolidinocyclohexene (**520**) gave 2-phenyl-5,6,7,8-tetrahydro-4($3H$)-quinazolinone (**521**) (BzNCO, THF, 0°, then $AcONH_4$, AcOH, reflux, 2 h: 93%).[1913]

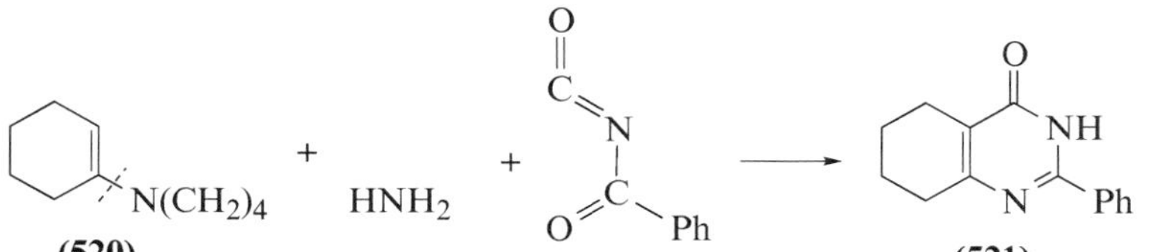

(520)   **(521)**

1-Morpholinocyclohexene rather similarly gave 1-p-methoxyphenyl-2-phenyl-5,6,7,8-tetrahydro-4($1H$)-quinazolinethione (BzNCS, $H_2NC_6H_4$OMe-p, THF, 100°, 4 h: 49%).[839]

5,5-Dimethyl-1,3-cyclohexanedione (**522**) gave 2-amino-7,7-dimethyl- (**523**, R = NH_2) or 2,7,7-trimethyl-4-phenyl-3,4,5,6,7,8-hexahydro-5-quinazolinone (**523**, R = Me) (PhCHO, $H_2NC(=NH)NH_2$ or $MeC(=NH)NH_2$: 55–70%).[2366]

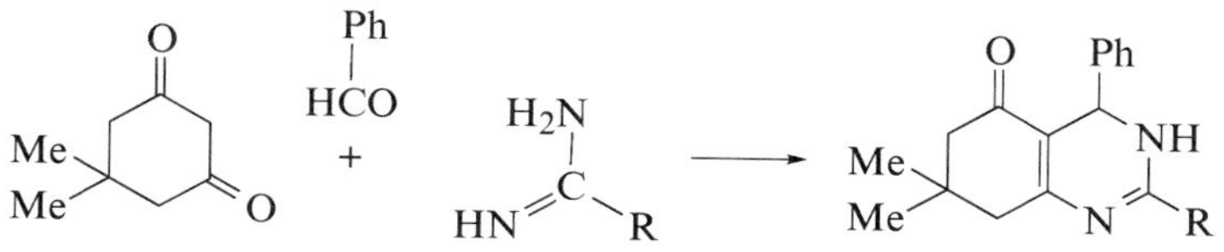

(522) + (523)

Cyclohexanone or 1-morpholinocyclohexene with 3-methyl-5-nitro-4(3H)-pyrimidinone gave 5,6,7,8-tetrahydroquinazoline (NH$_3$ gas, MeCN, 100°, sealed, 3 h: 85% or 78%, respectively; a "plausible path" has been devised for the synthesis, whereby the product's nuclear atoms are derived as follows: N1 from the ammonia, C2 + N3 + C4 from C2 + N1 + C6 of the pyrimidine, and the remaining C-atoms from the cyclohexane or cyclohexene derivative).[2556]

1.3. FROM A SINGLE PYRIMIDINE SUBSTRATE

This theoretically large category of syntheses appears to have been represented in the last 30 years by only one example. 6-δ,δ-Dimethoxybutyl-1-(2,3-O-isopropylidene-β-D-ribofuranosyl)-2,4(1H,3H)-pyrimidinedione (524), easily made from 1-(2,3-O-isopropylidene)uridine by 6-lithiation and subsequent alkylation, was converted into 5-hydroxy-1-(2,3-O-isopropylidene-β-D-ribofuranosyl)-5,6, 7,8-tetrahydro-2,4(1H,3H)-quinazolinedione (525) in 75% yield by stirring in aqueous acetonic trifluoroacetic acid at 25° for 36 h: the two isomeric alcohols (525) underwent dehydration on prolonged refluxing in acetonic p-toluenesulfonic acid to afford 82% of 1-(2,3-O-ispropylidene-β-D-ribofuranosyl)-7,8-dihydro-2,4(1H,3H)-quinazolinedione (526).[2163]

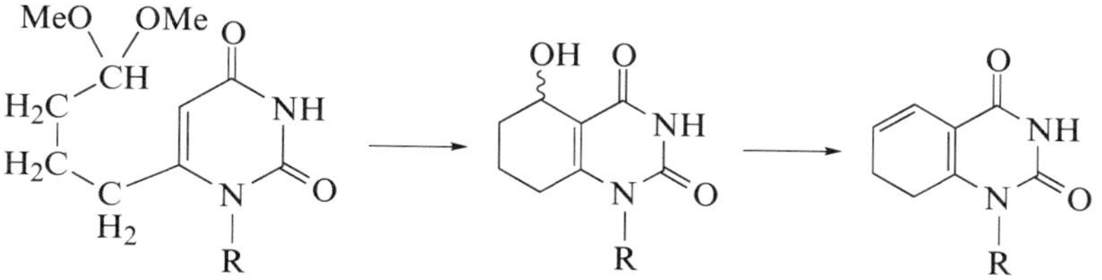

(524)   (525)   (526)

In (524)–(526), R stands for the following:

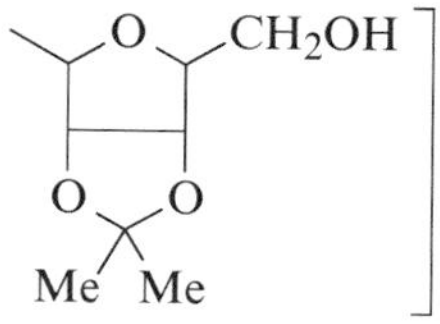

1.4. FROM A PYRIMIDINE SUBSTRATE AND ANCILLARY SYNTHON(S)

The elaboration of pyrimidine derivatives to quinazolines with the aid of synthones, which may introduce one or more of the carbon atoms required for the carbocyclic ring, has never been much used. Indeed, since 1965 only 4 of the 10 possible broad categories of such syntheses have been represented in the literature. The following examples illustrate these available data.

Where the synthon(s) supply C5 + C6

6-β-Dimethylaminovinyl-1,3-dimethyl-2,4(1H, 3H)-pyrimidinedione (**528**) with methyl vinyl ketone (**527**, X = H, Y = Ac) gave 6-acetyl-1,3-dimethyl-2,4(1H,3H)-quinazolinedione (**529**, X = H, Y = Ac)[PhMe, reflux, 16 h: 46%;[2140] MeCN, Pd(OAc)$_4$, reflux, < 5 h: 73%];[2465] the same substrate (**528**) with dimethyl fumarate (**527**, X = Y = CO$_2$Me) gave dimethyl 1,3-dimethyl-2,4-dioxo-1,2,3,4-tetrahydro-5,6-quinazolinedicarboxylate (**529**, X = Y = CO$_2$Me) (PhMe: 20%);[2140] (**528**) with methyl acrylate (**527**, X = H, Y = CO$_2$Me) gave either methyl 1,3-dimethyl-2,4-dioxo-1,2,3,4,6,7-hexahydro-6-quinazolinecarboxylate (**530**) (PhMe: 17%)[2140] or its 1,2,3,4-tetrahydro analog (**529**, X = H, Y = CO$_2$Me) [MeCN, Pd(OAc)$_4$:64%];[2465] and (**528**) with styrene (**527**, X = H, Y = Ph) gave a separable mixture of 1,3-dimethyl-6-phenyl-(**529**, X = H, Y = Ph) and 7-dimethylamino-1,3-dimethyl-6-phenyl-2,4(1H, 3H)-quinazolinedione [MeCN, Pd(OAc)$_4$: 60% and 30%, respectively].[2465]

3,4,4-Trimethyl-6-styryl-3,4-dihydro-2(1H)-pyrimidinone with maleic anhydride gave the Diels–Alder adduct, 3,4,4-trimethyl-2-oxo-7-phenyl-1,2,3,4,4a,5,6,7-octahydro-5,6-quinazolinedicarboxylic anhydride (**531**) (PhCl, reflux, 20 min: 70%).[778]

(527)   (528)   (529)

(530)   (531)

Also other interesting examples involving dimerization of pyrimidine substrates.[66,228]

Where the synthon(s) supply C6 + C7

Methyl 2,4-dimethoxy-6-methyl-5-pyrimidinecarboxylate (533) with dimethyl acetylenedicarboxylate (532) gave dimethyl 5-hydroxy-2,4-dimethoxy-5,6-quinazolinedicarboxylate (534) (lithiation of 6-Me; then $MeO_2CC\equiv CCO_2Me$, Et_2O, $-78°$, then $20°$ briefly: 38%);[1773,2334] Substrate (533) with methyl crotonate likewise gave the reduced product, methyl 2,4-dimethoxy-7-methyl-5-oxo-5,6,7,8-tetrahydro-6-quinazolinecarboxylate (535) (47%).[1773,2334]

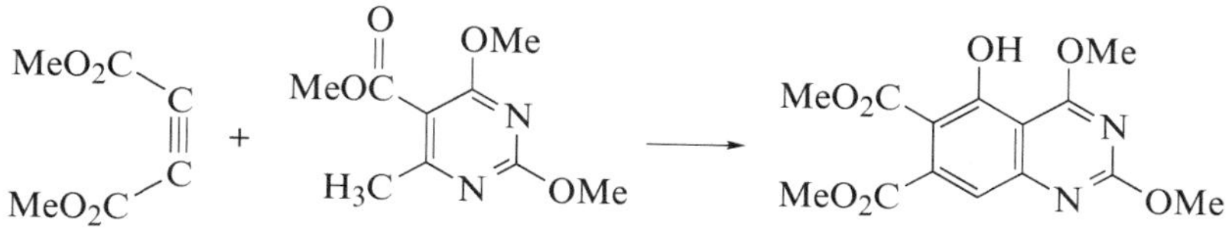

(532)   (533)   (534)

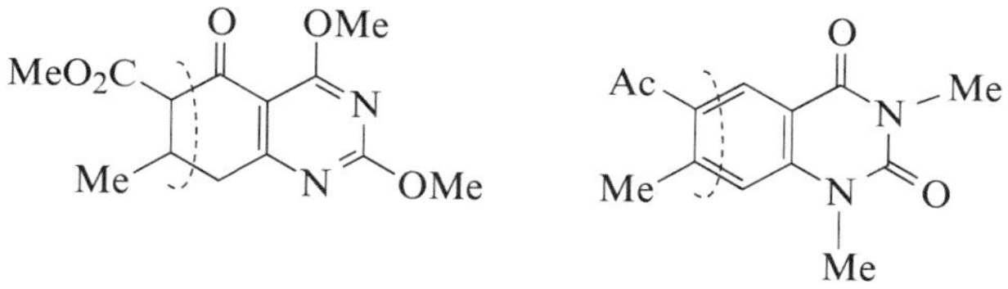

(535)   (536)

1,3,6-Trimethyl-2,4-dioxo-1,2,3,4-tetrahydro-5-pyrimidinecarbaldehyde with acetylacetone gave 6-acetyl-1,3,7-trimethyl-2,4(1H,3H)-quinazolinedione (536) (PhH, trace AcOH, trace $HN(CH_2)_5$, reflux, water separation, 12 h: 46%);[1390] the same substrate with dimethyl acetylenedicarboxylate gave dimethyl 1,3-dimethyl-2,4-dioxo-1,2,3,4-tetrahydro-6,7-quinazolinedicarboxylate (537) (lithiation; then synthon, $-70° \rightarrow 20°$: 83%).[1149, cf. 2177]

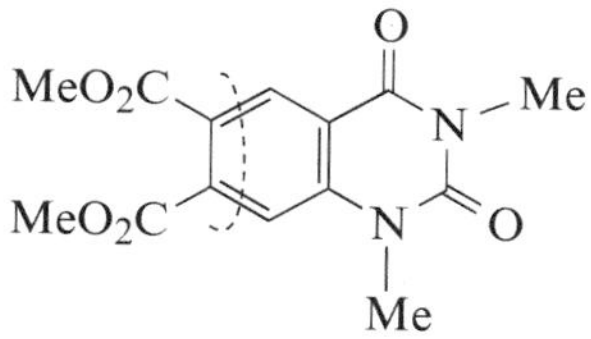

(537)

4-Methyl-2-phenyl-6-thioxo-1,6-dihydro-5-pyrimidinecarbonitrile with α-benzylidenemalononitrile gave 5-amino-2,7-diphenyl-4-thioxo-3,4-dihydro-6-quinazolinecarbonitrile (538, R = Ph) (pyridine, reflux, 4 h: 42%)[2064] or with malononitrile and paraformaldehyde gave 5-amino-

2-phenyl-4-thioxo-3,4-dihydro-6-quinazolinecarbonitrile (**538**, R = H) [Me$_2$NCHO, trace HN(CH$_2$)$_5$, reflux, 4 h: 44%].[2147] 5-Amino-7-phenyl-2-thioxo-1-*p*-tolyl-1,2-dihydro-6-quinazolinecarbonitrile was made somewhat similarly.[2486]

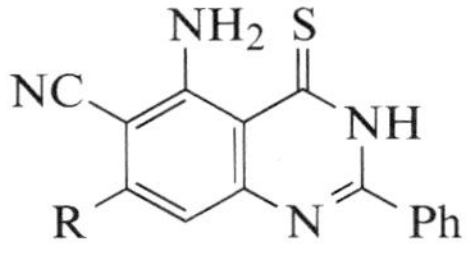

(**538**)

Also other examples.[1510,2005,2208,2403,2479,2531]

Where the synthon(s) supply C5 + C6 + C7

1,3,6-Trimethyl-2,4(1*H*,3*H*)-pyrimidinedione (**540**) with α-benzoyl-β,β-bis-methylthioethylene (**539**) gave 1,3-dimethyl-7-methylthio-5-phenyl-2,4(1*H*,3*H*)-quinazolinedione (**541**) [lithiation of (**540**) in THF; then (**539**), 20°, 12 h: 86%].[2398]

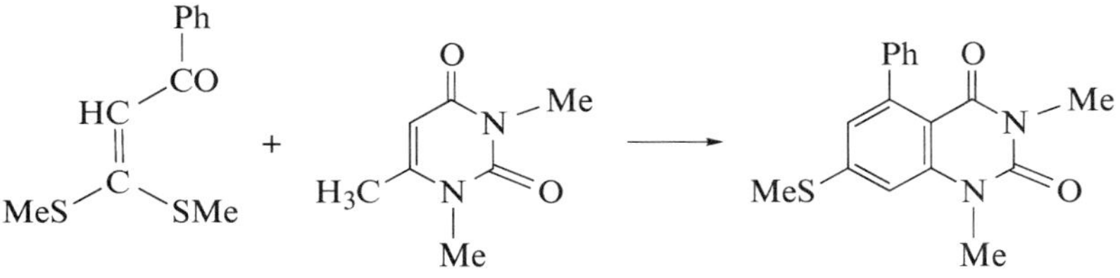

(**539**)   (**540**)   (**541**)

6-Methyl-4-phenyl-3,4-dihydro-2(1*H*)-pyrimidinone with α-benzoyl-β-dimethylaminoethane gave 4,7-diphenyl-3,4,5,6-tetrahydro-2(1*H*)-quinazolinone (**542**) (Bu^{t}OK, Pr^{i}OH, reflux, 8 h: ∼ 40%).[776]

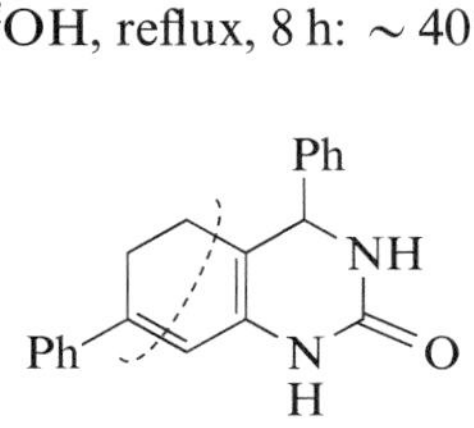

(**542**)

4,4,6-Trimethyl-3,4-dihydro-2(1*H*)-pyrimidinone with bis(2,4,6-trichlorophenyl) α-ethylmalonate gave 6-ethyl-5,7-dihydroxy-4,4-dimethyl-3,4-dihydro-2(1*H*)-quinazolinone (**543**) (no solvent, 200°, 15 min: ?%).[777]

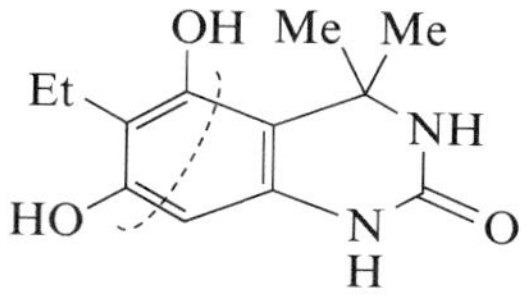

(543)

Also other examples in the cited references and elsewhere.[325]

Where the synthon(s) supply C5 + C6 + C7 + C8

Methyl 2,6-dioxo-1,2,3,6-tetrahydro-4-pyrimidinecarboxylate (**544**) with bu-
tadiene gave methyl 2,4-dioxo-1,2,3,4,4a,5,8,8a-octahydro-8a-quinazo-
linecarboxylate (**545**) (THF, trace HOC_6H_4OH-*p*, 165°, sealed, 48 h:
78%);[252] analogs were made similarly.[252,919]

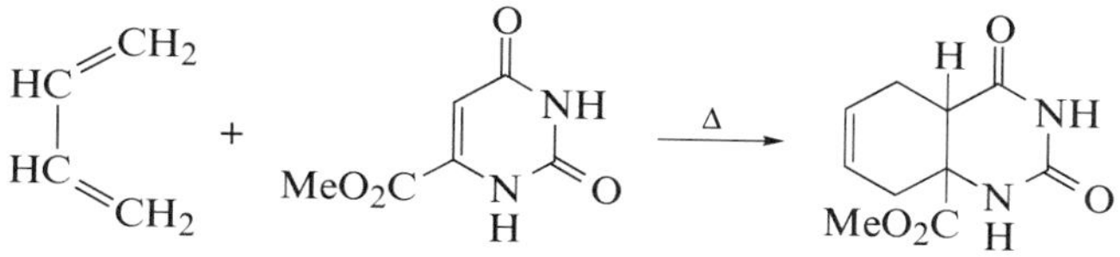

(544)   **(545)**

2-*t*-Butyldidehydropyrimidine (**548**), made *in situ* by oxidation of 5-*t*-butyl-
3*H*-*v*-triazolo[4,5-*d*]pyrimidin-3-amine (**546**), reacted with 2,3,4,5-tet-
raphenylcyclopentadiene (**547**) to give 2-*t*-butyl-5,6,7,8-tetraphenyl-
quinazoline (**549**) [Pb(OAc)$_4$, CH_2Cl_2, 20°: 24%].[2390]

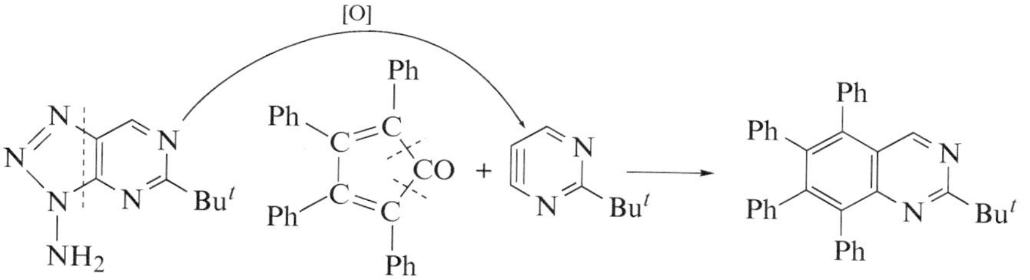

(546)   **(547)**   **(548)**   **(549)**

1.5. FROM A HETEROMONOCYCLIC SUBSTRATE OTHER THAN A PYRIMIDINE

As might be expected, this is only a small group of syntheses. In most cases, the
benzene ring of the resulting quinazoline is supplied as a phenyl substituent on
the heteromonocyclic substrate. For pragmatic reasons, such substrates are
arranged alphabetically to produce the following subsections.

1.5.1. Imidazoles as Substrates

N-Arylimidazoline-4,5-diones have been converted by thermolysis into 4-quinazolinones in good yield. Thus 1,2-diphenylimidazoline-4,5-dione (**550**, R = H) at its melting point (170°) lost carbon monoxide and eventually solidified to give 2-phenyl-4(3*H*)-quinazolinone (**551**, R = H) in 90% yield;[214] 1-*o*-nitrophenyl-2-phenylimidazoline-4,5-dione (**550**, R = NO$_2$) likewise gave 8-nitro-2-phenyl-4(3*H*)-quinazolinone (**551**, R = NO$_2$) in 96% yield;[214] 1-*o*-methoxyphenyl-2-trichloromethylimidazoline-4,5-dione gave 8-methoxy-2-trichloromethyl-4(3*H*)-quinazolinone (**552**) (200°: 47%);[755] and other analogs were made similarly.[214,755,1034]

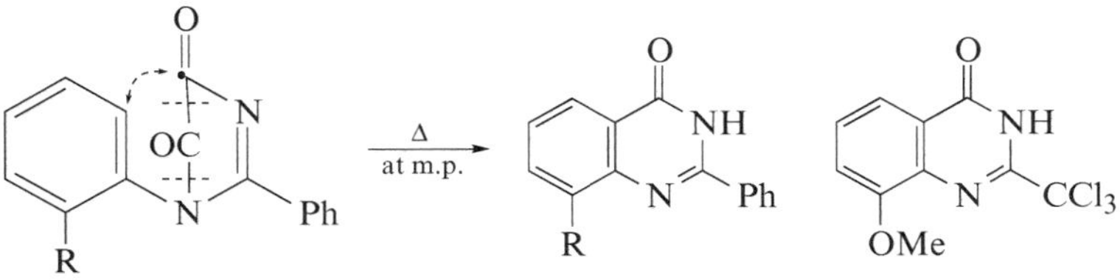

(550)   (551)   (552)

1.5.2. 1,3,5-Oxadiazines as Substrates

Although severely restricted in scope at present, this type of synthesis is exemplified by the conversion of 4,5-diphenyl-2,2,6,6-tetrakis(trifluoromethyl)-5,6-dihydro-2*H*-1,3,5-oxadiazine (**553**) into 2-phenyl-4,4-bistrifluoromethyl-3,4-dihydroquinazoline (**554**) by loss of hexafluoroacetone on heating at 170–180°/15 mmHg until [19]Fnmr (Fluorine-19 nuclear magnetic resonance) indicated completion of the reaction;[1263] several analogs were made rather similarly.[1263]

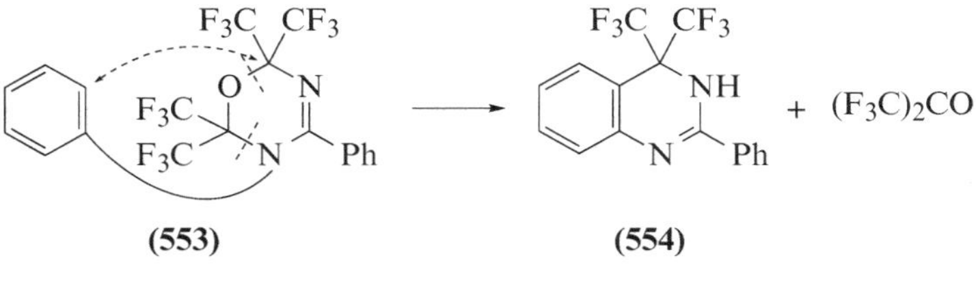

(553)   (554)

1.5.3. 1,2,4-Oxadiazoles as Substrates

Such oxadiazoles have been used as substrates in several ways. Two of these ways have already been mentioned in Sections 1.1.1 (reductive cyclizations) and 1.1.2, respectively. Other ways have been exemplified in the irradiation of 3-methyl-5-phenyl- (**555**, R = Me) or 3,5-diphenyl-1,3,4-oxadiazole (**555**,

R = Ph), in the presence of diphenylacetylene as a sensitizer, to give 2-methyl-
(**556**, R = Me) (46%) or 2-phenyl-4(3*H*)-quinazolinone (**556**, R = Ph) (40%),
respectively;[2186, cf. 1940] and in the scantily described reaction of 3,4-diphenyl-
1,2,4-oxadiazol-5(4*H*)-one (**557**) with benzylideneaniline to give 2,4-diphenyl-
quinazoline (**558**).[27]

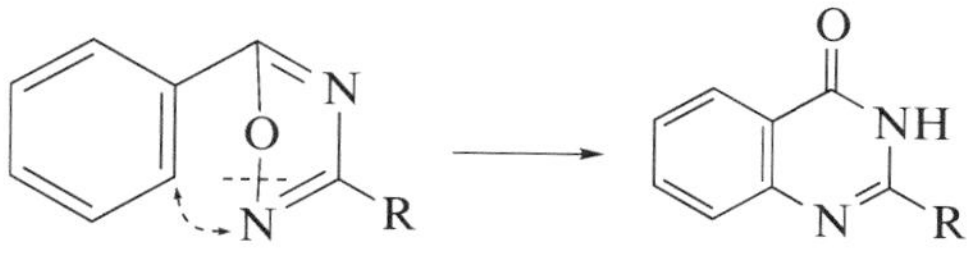

(**555**)   (**556**)

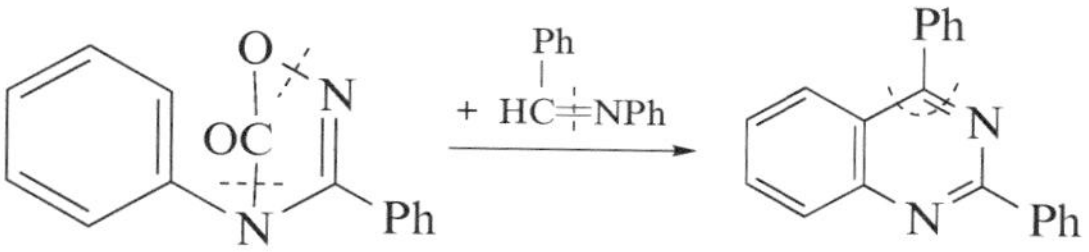

(**557**)   (**558**)

1.5.4. 1,3,4-Oxadiazoles as Substrates

This little-used synthesis is illustrated by the thermal conversion of 5-*o*-
aminophenyl- (**559**, R = H) or 5-*o*-ethylaminophenyl-1,3,4-oxadiazol-2(3*H*)-one
(**559**, R = Et) into 3-amino-2,4(1*H*, 3*H*)-quinazolinedione (**560**, R = H) (200°,
30 min: 86%) or its 1-ethyl derivative (**560**, R = Et) (240°, 1 h: 55%), respectively;[1829] the mechanism has not been elucidated, but several analogs were made
similarly.[1829]

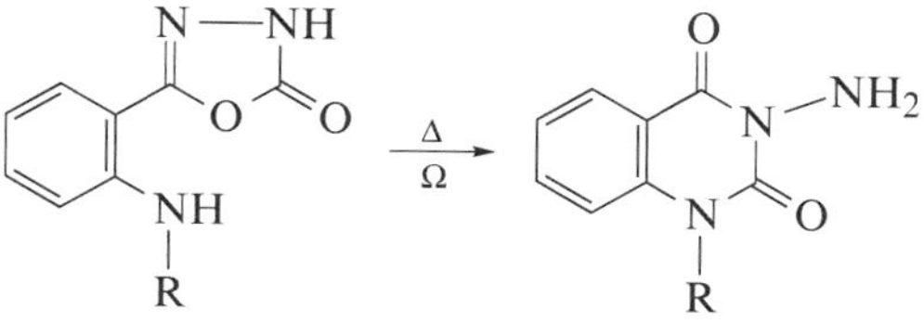

(**559**)   (**560**)

1.5.5. 1,2,3,5-Oxathiadiazines as Substrates

This synthesis is represented presently by a single example: irradiation of
4,6-diphenyl-1,2,3,5-oxathiadiazine 2,2-dioxide (**561**) in dichloromethane containing ethanol under nitrogen gave up to 60% of 2-phenyl-4(3*H*)-quinazolinone
(**562**) plus other separable products; a mechanism has been suggested.[1377]

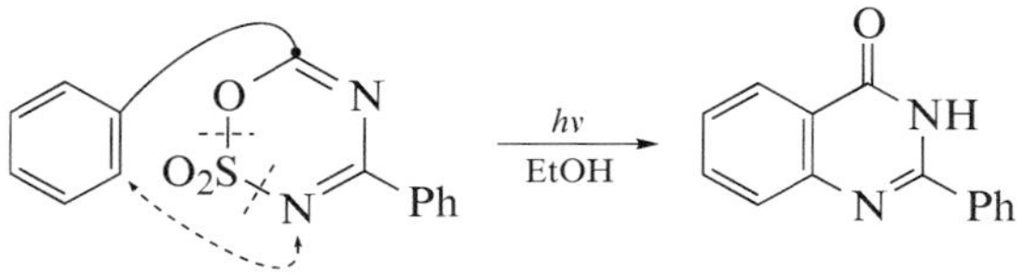

(561)   (562)

1.5.6.   Oxazoles as Substrates

Examples of such a synthesis have been described briefly: irradiation of
4-bromo-2-*o*-nitrophenyl-5-phenyl- (**563**, R = H) or 4-bromo-5-*p*-bromophenyl-
2-*o*-nitrophenyloxazole (**563**, R = Br) in benzene at 70° gave 2-benzoyl- (**564**,
R = H) (70°) or 2-*p*-bromobenzoyl-4(3*H*)-quinazolinone (**564**, R = Br) (80%),
respectively;[888] the transformation did not proceed in the absence of a 4-bromo
substituent, and a necessarily complicated mechanism has been suggested.[888]

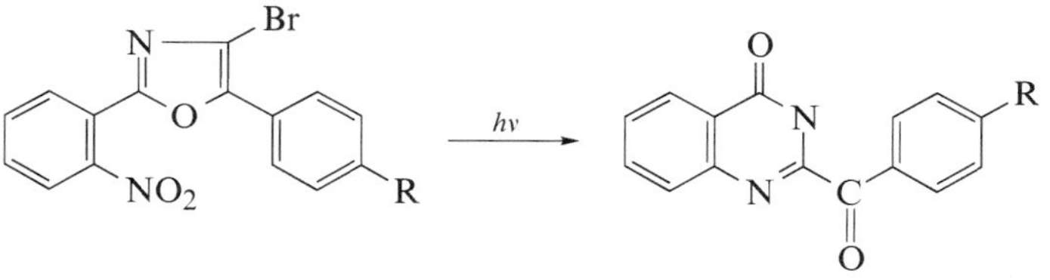

(563)   (564)

1.5.7.   1,2,4,5-Tetrazines as Substrates

The known examples of such transformations are more interesting than
useful. Thus thermolysis of 1,3,6-triphenyl-1,4-dihydro-1,2,4,5-tetrazine (**565**,
R = H) at 200° for 15 min gave three separable products that included 2,4-
diphenylquinazoline (**566**, R = H) (~ 15%);[1942] 3,6-diphenyl-1-*p*-tolyl-1,4-
dihydro-1,2,4,5-tetrazine (**563**, R = Me) likewise gave 6-methyl-2,4-diphenyl-
quinazoline (**566**, R = Me) (~ 22%), and other analogs were so made.[1942]

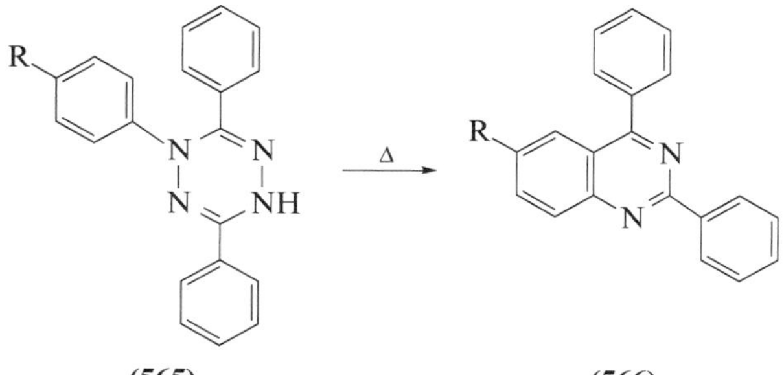

(565)   (566)

1.5.8. 1,2,3,4-Thiatriazoles as Substrates

Only one example of such a reaction has been reported. 5-Diphenylmethyleneaminooxy-1,2,3,4-thiatriazole was converted into 2-phenyl-4(3H)-quinazolinone (36%) by simply heating under reflux in carbon tetrachloride for 1 h. The structure of the product was checked by an independent synthesis and an X-ray analysis.[824]

1.6.   FROM A HETEROBICYCLIC SUBSTRATE

Each heterobicyclic substrate used for the synthesis of quinazolines does, in fact, contain a fused benzene ring that may bear a variety of passenger groups. Some such types of substrate have been used to afford only one or two quinazolines, but others have been widely employed for this purpose. The following treatment is subdivided alphabetically according to the heterobicyclic system used. Please note that the use of 3,1,4-benzoxadiazepines (H 128) and 4,1-benzoxazepines (H 100,340) as substrates for quinazoline syntheses is not covered in this Supplement for want of new data.

1.6.1.   Benzazetines as Substrates

1-t-Butyl-2(1H)-benzazetinone (568), which appears to be in ring–chain tautomeric equilibrium with (567), was easily made and reacted with phenyl isocyanate in refluxing 1,2-dichloroethane during 16 h to afford 1-t-butyl-3-phenyl-2,4(1H, 3H)-quinazolinedione (569) (80%);[1392] when the reactants were heated without solvent to 145°, an exothermic reaction occurred with loss of isobutene to give 3-phenyl-2,4(1H, 3H)-quinazolinedione (570) (87%). [1392]

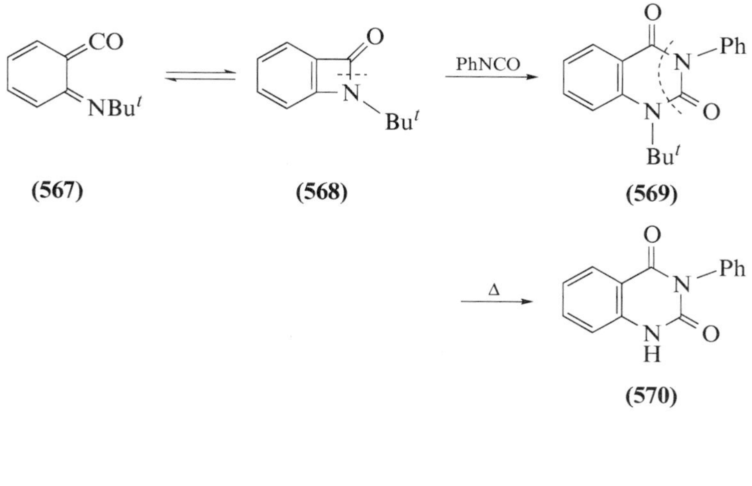

(567)   (568)   (569)

(570)

1.6.2.   2,1-Benzisothiazoles as Substrates

In attempting a quite different reaction, it was observed that two molecules of
2,1-benzisothiazol-3-amine (**571**) reacted in boiling aqueous sodium bisulfite
during 60 h to afford 2-*o*-aminophenyl-4(3*H*)-quinazolinone (**572**) (66%) with the
bisulfite acting as a sulfur scavenger.[1455]

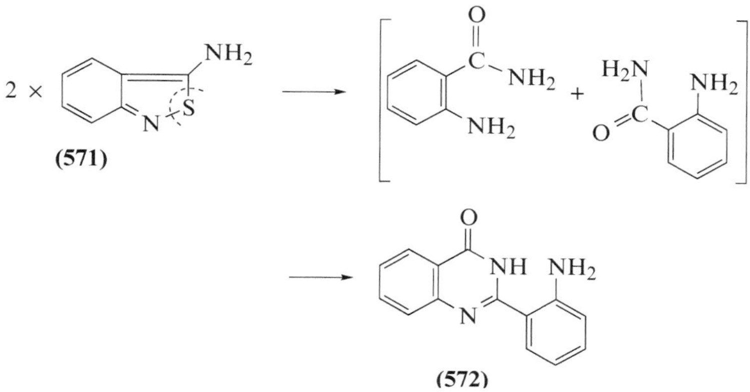

(571)

(572)

1.6.3.   2,1-Benzisoxazoles as Substrates

The use of these substrates is represented in two completely different sequences.
3-Phenyliminomethyl-2,1-benzoxazole (**573**) rearranged in refluxing anisole during
24 h to afford 3-phenyl-4(3*H*)-quinazolinone (**574**) (25%).[1150] 3-Phenyl-2,1-ben-
zoxazole (**575**) with 5-methoxy-3-phenyl-2,4-imidazolidinedione (**576**), in refluxing
dioxane containing TiCl$_4$ as catalyst, gave 4,*N*-diphenyl-2-quinazolinecar-
boxamide (**577**) (82%);[2178] the reaction probably involved cycloaddition and loss
of methanol and carbon dioxide. Many analogs were made similarly.[2178]

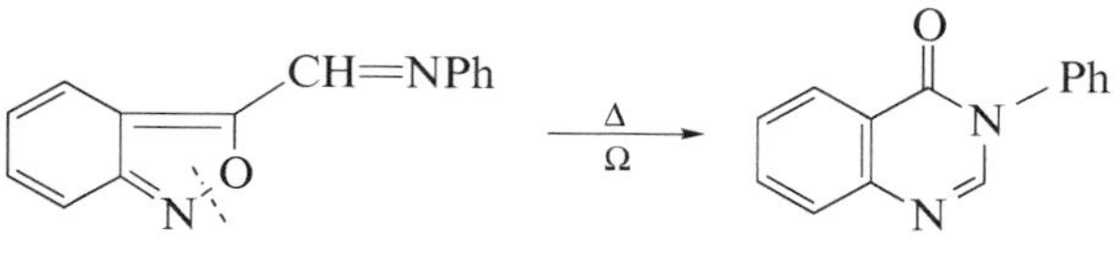

(573)   (574)

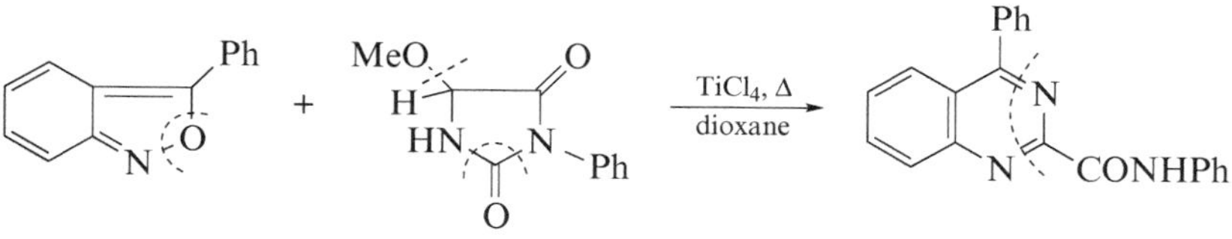

(575)   (576)   (577)

1.6.4.   1,4-Benzodiazepines as Substrates

Because of vast effort devoted to the preparation, reactions, and pharmacology of benzodiazepines as tranquillizers, many of these esoteric compounds have been observed to furnish quinazolines under a variety of conditions. Such transformations can be useful and are illustrated in the following examples, classified according to the processes used for ring contraction.

By hydrolysis, etc.

2-Methoxycarbonylmethyl-3H-1,4-benzodiazepine-3,5(4H)-dione (**578**) gave 2-acetyl-4(3H)-quinazolinone (**579**) (6M HCl, reflux, 2 h: 86%).[246]

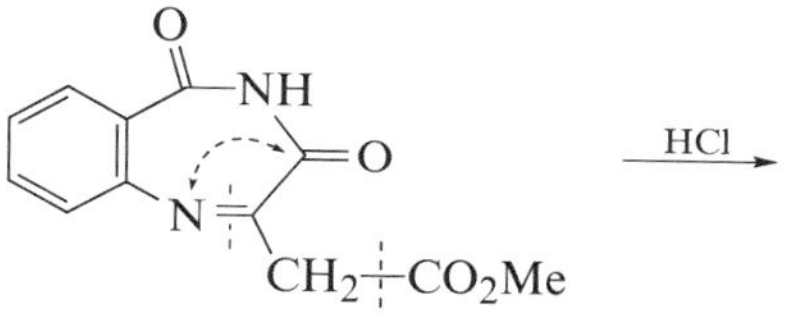

(**578**)   (**579**)

3-Hydroxy-5-phenyl-3H-1,4-benzodiazepin-2(1H)-one (**580**, R = H) gave 4-phenyl-2-quinazolinecarbaldehyde (**581**, R = H) AcOH, reflux, 10 min: 35%);[632] the 7-chloro substrate (oxazepam) (**580**, R = Cl) likewise gave 6-chloro-4-phenyl-2-quinazolinecarbaldehyde (**581**, R = Cl),[1029,1904] which was also made by thermolysis (see next subsection)[1231] or from 7-chloro-3-(2-methylimidazol-1-yl)-5-phenyl-3H-1,4-benzodiazepin-2-amine   (50% H$_2$SO$_4$, 95°, 30 min: ~ 15%).[1395]

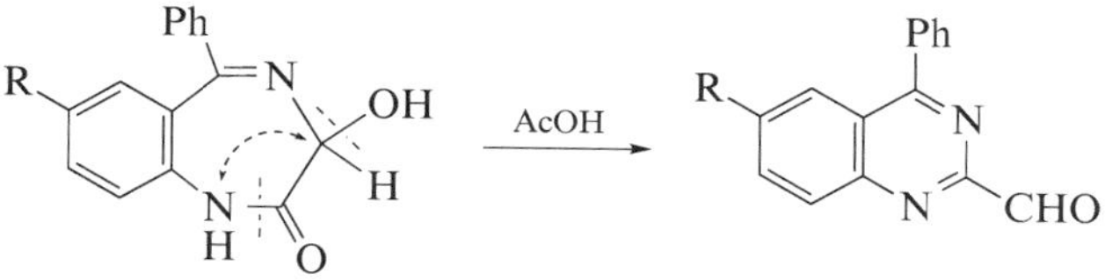

(**580**)   (**581**)

7-Chloro-4,5-epoxy-5-phenyl-4,5-dihydro-3H-1,4-benzodiazepin-2(1H)-one (**582**) gave 6-chloro-4-phenyl-2(1H)-quinazolinone (**583**) (H$_2$O–THF, 20°, 7 days: 43%).[254]

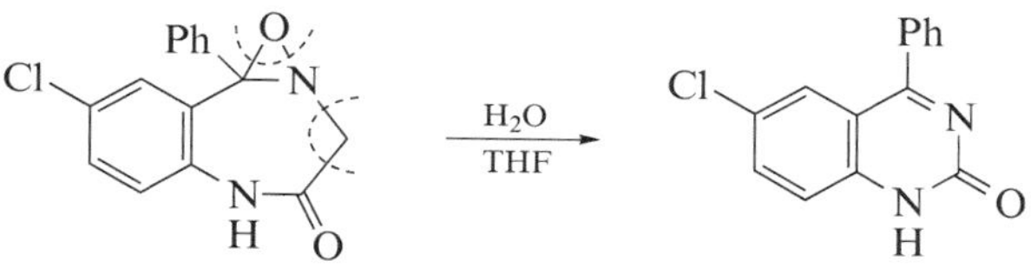

(**582**)   (**583**)

Also other examples.[222,259,402,435,760,1461,2427,2449]

By thermolysis

7-Bromo-3-hydroxy-5-pyridin-2'-yl-3*H*-1,4-benzodiazepin-2(1*H*)-one   (**584**)
   gave 6-bromo-4-pyridin-2'-yl-2-quinazolinecarbaldehyde(**585**)(no solvent,
   220°, 10 min: 84%; hot AcOH gave a lower yield).[1461]

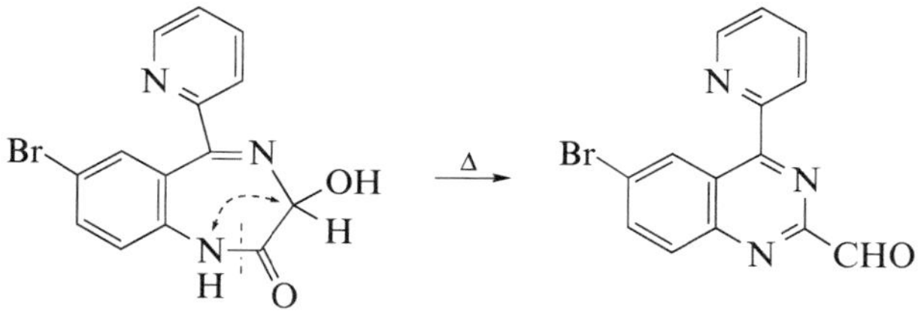

(584)   **(585)**

7-Chloro-3-hydroxy-5-phenyl-3*H*-1,4-benzodiazepin-2(1*H*)-one(**580**, R = Cl)
   gave 6-chloro-4-phenyl-2-quinazolinecarbaldehyde (**581**, R = Cl) (PhH,
   150°, sealed, 1 h, beware CO_2 pressure: 89%;[1231] or neat: ?%).[1119, cf. 822]
Also other examples.[46,2441]

By oxidation

N-Methyl-7-nitro-5-phenyl-1*H*-1,4-benzodiazepine-3-carboxamide   (**586**)
   gave *N*-methyl-6-nitro-4-phenyl-2-quinazolinecarboxamide (**587**) (CrO_3,
   AcOH, 20°, 5 min: 29%).[242]

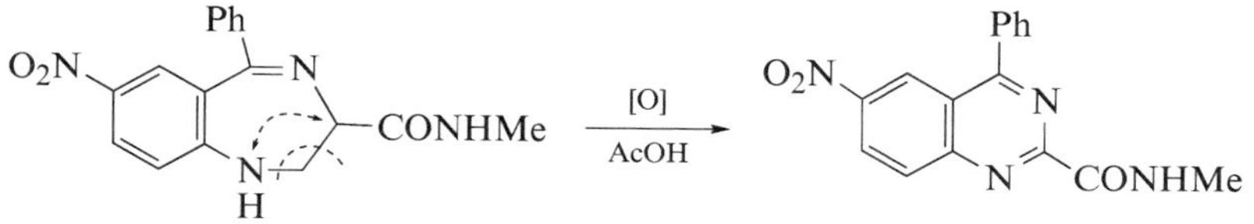

(586)   **(587)**

7-Chloro-1-methyl-5-phenyl-3*H*-1,4-benzodiazepin-2(1*H*)-one (**588**) gave, inter alia, 6-chloro-1-methyl-4-phenyl-2(1*H*)-quinazolinone (**589**) (RuO_4,
   $CHCl_3$, 20°, 30 h: 21%).[222]

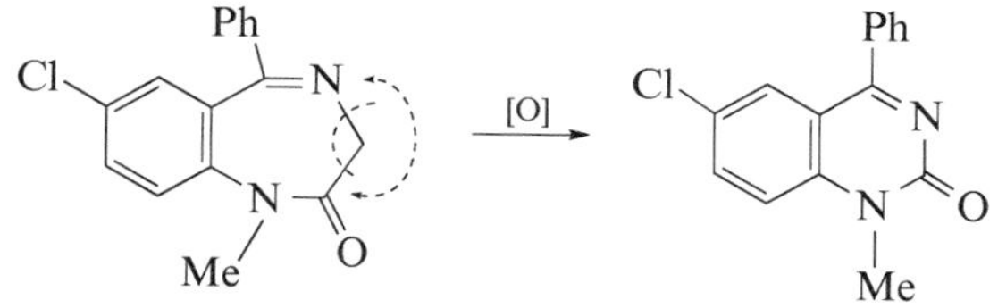

(588)   **(589)**

Also other poor-yielding examples.[1440,2009]

By reduction

7-Chloro-2-(cyclopropylmethyl)amino-4-phenyl-3H-1,4-benzodiazepine   4-
   oxide (590) gave 6-chloro-2-methyl-4-phenyl-3,4-dihydroquinazoline (591)
   (electrolytic, MeOH, AcOH, AcONa, H_2O, N_2, 14 h: 80%).[2376]

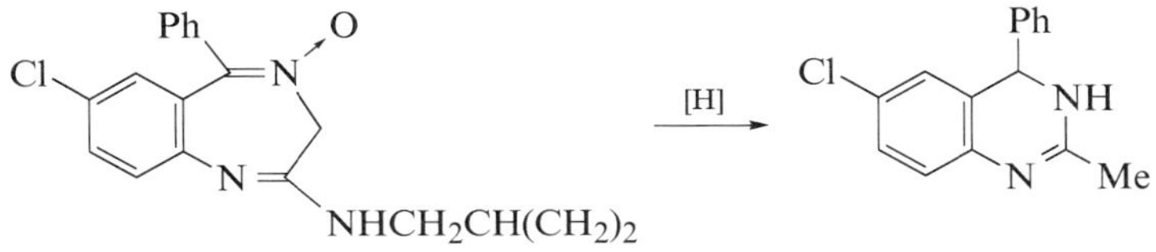

(590)   (591)

By irradiation

7-Chloro-1-methyl-5-phenyl-3H-1,4-benzodiazepin-2(1H)-one (588) has been
   reported to afford 4-phenylquinazolines, 4-phenylquinazolinones, and
   other products on irradiation in methanol.[1011]

By treatment with hydroxylamine

4-Chloro-2-methylamino-5-phenyl-3H-1,4-benzodiazepine (592) gave 2-
   aminomethyl-6-chloro-4-phenylquinazoline 3-oxide (593) ($H_2NOH.HCl$,
   MeOH, reflux, 20 min: 74%), but the 4-oxide of substrate (592), on similar
   treatment for 40 min, gave 6-chloro-4-phenyl-2-quinazolinecarbaldehyde
   3-oxide oxime (594) (20% yield, plus other products);[361] rational mechanisms were proposed.[361]

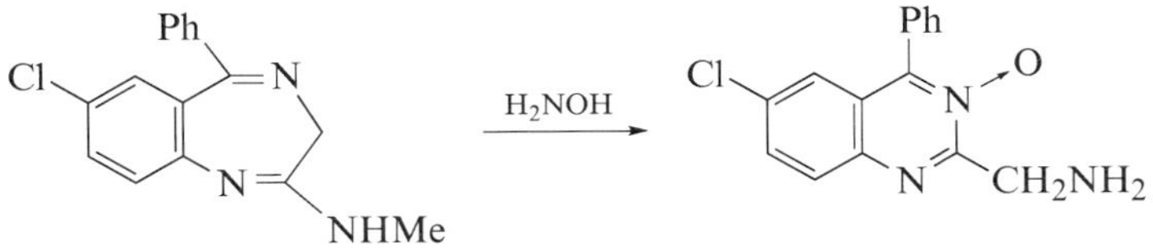

(592)   (593)

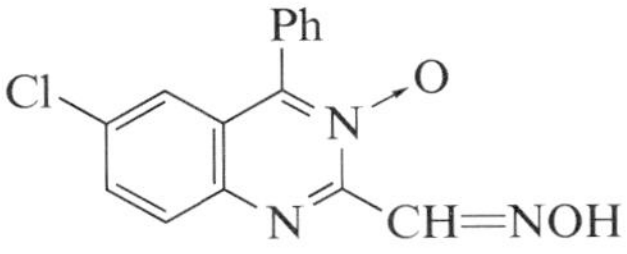

(594)

By treatment with phosphorus-based reagents

5-Methyl-3H-1,4-benzodiazepin-2(1H)-one (595, R = H) gave 2-chloro-
   methyl-4-methylquinazoline (596, R = H) ($POCl_3$, reflux, 30 min:
   52%);[1923] the 7,8-dimethoxy substrate (595, R = OMe) likewise gave 2-
   chloromethyl-6,7-dimethoxy-4-methylquinazoline (596, R = OMe) (1 h:
   52%).[1923]

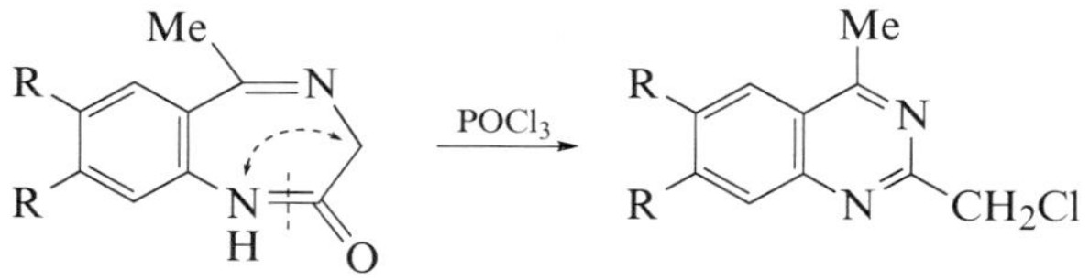

(595)   **(596)**

7-Chloro-3-hydroxy-5-phenyl-3H-1,4-benzodiazepin-2(1H)-one(**580**, R = Cl)
gave 6-chloro-2-diethylaminomethyl-4-phenylquinazoline (**597**) [2-di-
ethylamino-4,5-dihydro-1,3,2-dioxophosphole (**598**), PhMe, reflux, 4 h:
10%);[1279] analogs were made similarly, also in poor yield.[1279]

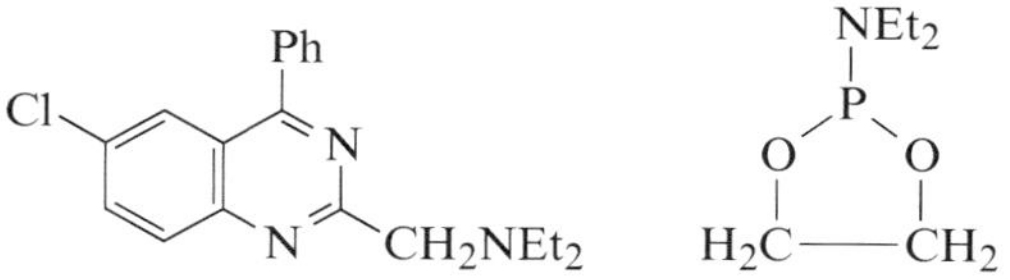

(597)   **(598)**

1.6.5.   1,5-Benzodiazocines as Substrates

A few partially reduced 1,5-benzodiazocines have been reported to undergo
ring contraction to quinazolines, usually by treatment with alcoholic hydrogen
chloride. Thus 8-chloro-6-phenyl-3,4-dihydro-1,5-benzodiazocin-2-amine (**599**,
R = H) in refluxing methanolic hydrogen chloride for 2.5 h gave 2-β-aminoethyl-
6-chloro-4-phenylquinazoline (**600**, R = H) in 53% yield;[1362] the same transformation occurred in refluxing methanolic methylamine hydrochloride during
40 min, but in 91% yield![1362] In much the same way, 8-chloro-10-ethoxycar-
bonylmethyl-6-phenyl-3,4-dihydro-1,5-benzodiazocin-2-amine (**599**, R = CH$_2$
CO$_2$Et) gave 2-β-aminoethyl-6-chloro-8-ethoxycarbonylmethyl-4-phenyl-
quinazoline (**600**, R = CH$_2$CO$_2$Et) in 80% yield by refluxing in ethanolic hydrogen chloride for 1 h.[1396] 1,5-Methano-6-phenyl-3,4,5,6-tetrahydro-1,5-benzo-
diazocin-2(1H)-one (**601**) underwent reductive cleavage by sodium borohydride
in ethanolic methylene chloride to give 3-γ-hydroxypropyl-4-phenyl-1,2,3,4-
tetrahydroquinazoline (28%).[249]

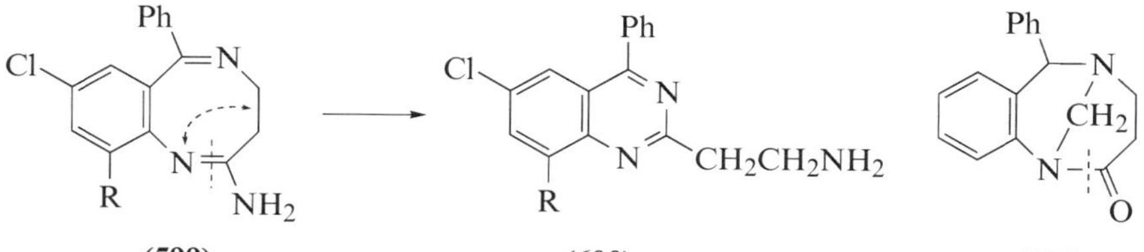

(599)   **(600)**   **(601)**

1.6.6.   2,1,4-Benzothiadiazepines as Substrates

Only one such substrate has been used recently to make quinazolines. 7-Chloro-5-phenyl-1,3-dihydro-2,1,4-benzothiadiazepine 2,2,4-trioxide (603) gave either 6-chloro-4-phenylquinazoline (602) (sublimation at $160°/10^{-3}$ mmHg: ?%) or its 1,2-dihydro derivative (604) (triethyl phosphate, dioxane, reflux, 3 h: 27%).[752]

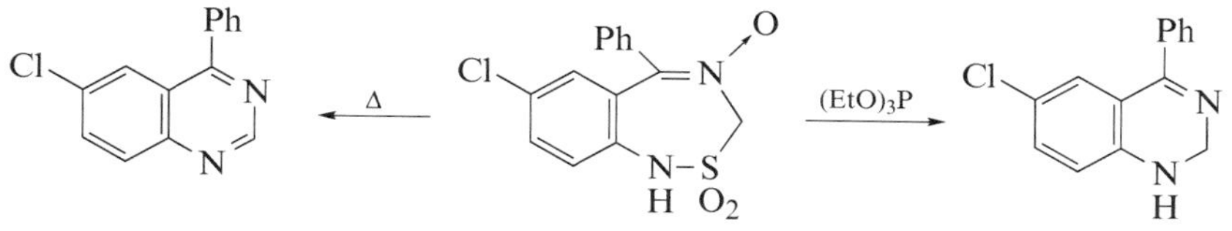

(602)   (603)   (604)

1.6.7.   2,1,3-Benzothiadiazines as Substrates

An oxidized member of this system, the so-called 1,3-didehydro-2,1,3-benzothiadiazin-4-one (606),[2445] reacted with boiling acetone or cyclohexanone to afford 2,2-dimethyl- (605, X = Y = Me) and 2,2-pentamethylene-1,2-dihydro-4(3H)-quinazolinone [605, $X + Y = (CH_2)_5$], respectively, both in excellent yield;[1627] the same substrate (606) in a refluxing mixture of aniline and carbon disulfide gave 3-phenyl-2-thioxo-1,2-dihydro-4(3H)-quinazolinone (607) in 63% yield.[1627]

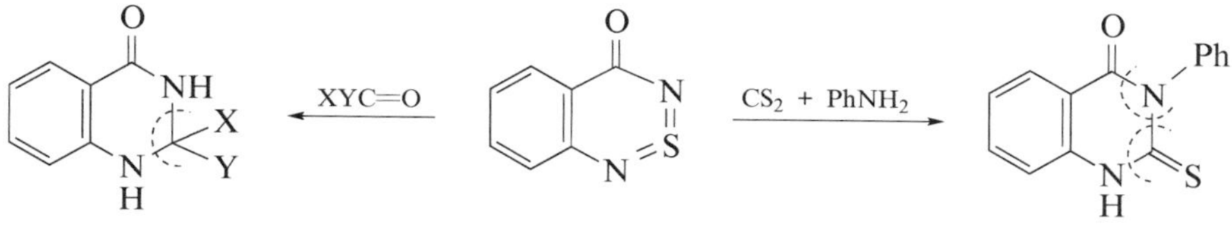

(605)   (606)   (607)

1.6.8.   1,3- and 3,1-Benzothiazines as Substrates (H 339)

The only example of the use of a 1,3-benzothiazine to make quinazolines is given with its oxygen analog in Section 1.6.14. Whatever the mechanism, the conversion of a 3,1-benzothiazine into a quinazoline must involve ring opening and reclosure. These processes become more understandable if each of the three common types of such benzothiazine substrates (608–610) is considered as an anhydride (or lactone) of the diacid (608a), the acylamino acid (609a), or the hydroxymethylamino acid (610a), respectively. The scope and practical aspects of this useful synthesis are illustrated by the following examples.

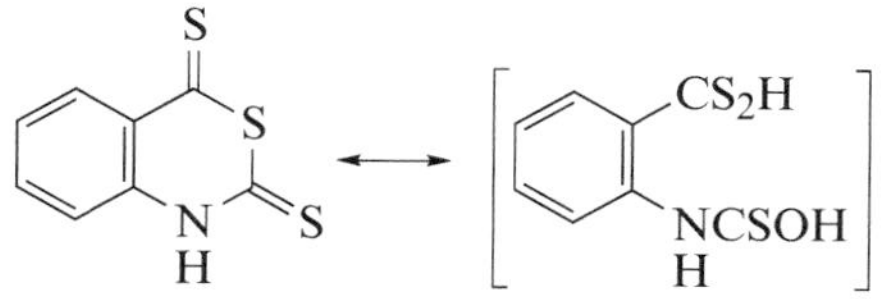

(608)   **(608a)** 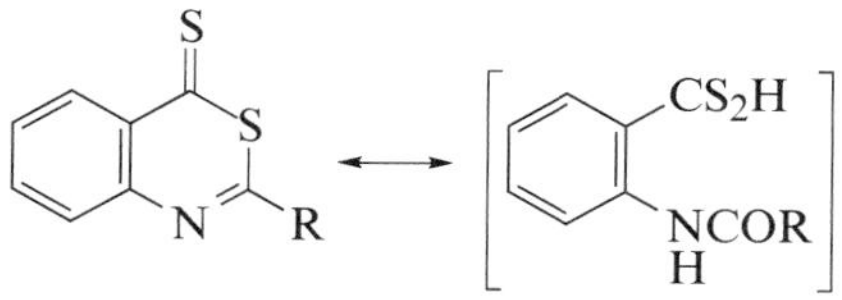   **(609)**   **(609a)**

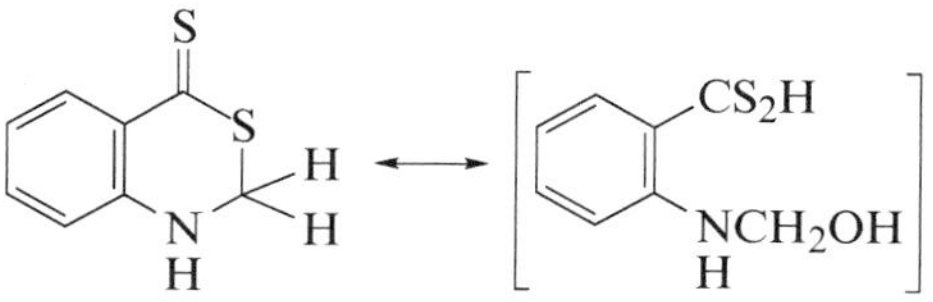

(610)   **(610a)**

From the dithione (608) or a derivative

1-Phenyl-4H-3,1-benzothiazine-2,4(1H)-dithione **(611)** gave 3-isopropyl-1-phenyl-2,4(1H,3H)-quinazolinedithione **(612)** (Pr^{i}NH$_2$, MeOH, 20°, 3 days: 42%).[1271]

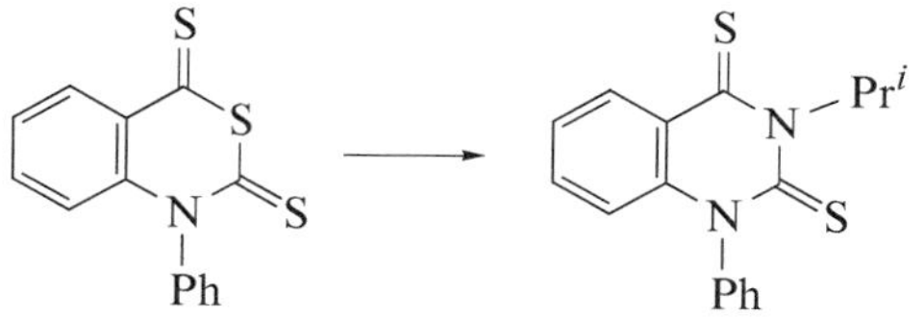

(611)   **(612)**

6-Bromo-4H-3,1-benzothiazine-2,4(1H)-dithione **(613**, X = S) gave 6-bromo-3-β,β-dimethoxyethyl-2,4(1H,3H)-quinazolinedithione [**613**, X = NCH$_2$CH(OMe)$_2$] [H$_2$NCH$_2$CH(OMe)$_2$, reflux: 87%].[1292]

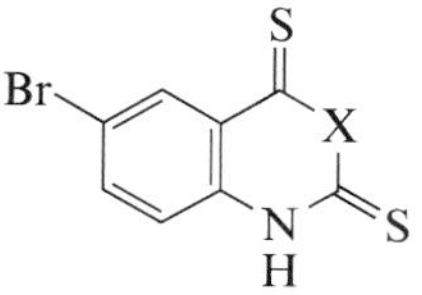

(613)

Also other examples.[81,274,442,547,1010,1294,1481]

From the 4-thione (609) or a derivative

2-Phenyl-4H-3,1-benzothiazine-4-thione (**614**, X = S) gave 3-allyl-2-phenyl-4(3H)-quinazolinethione (**614**, X = NCH$_2$CH:CH$_2$) (H$_2$NCH$_2$CH:CH$_2$, EtOH, reflux 1 h: 81%),[1658] 3-γ-methylaminopropyl-2-phenyl-4(3H)-

quinazolinethione (**614**, X = NCH$_2$CH$_2$CH$_2$NHMe) (H$_2$NCH$_2$CH$_2$CH$_2$NHMe, PhH, 20°, briefly: 85%),[716] 2-phenyl-4(1H)-quinazolinethione 3-oxide (**615**) [H$_2$NOH.HCl, AcONa, H$_2$O–EtOH, reflux (?), 15 min: 33%][2209] or 3-anilino-2-phenyl-4-phenylhydrazono-3,4-dihydroquinazoline (**616**) (PhNHNH$_2$, EtOH, reflux, 3 h: 60%).[455]

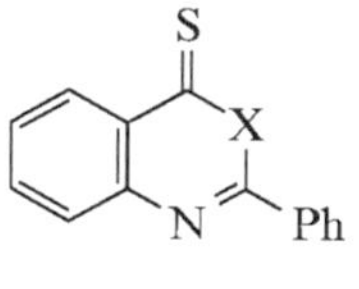

(614) 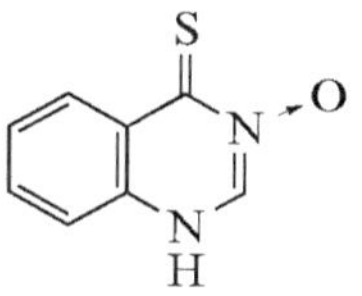   (615)

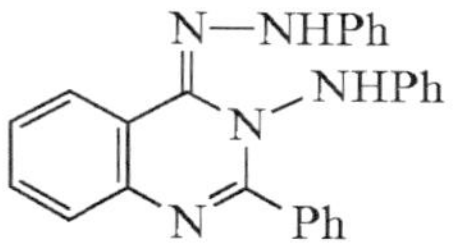

(616)

Also other examples in the foregoing references and elsewhere.[59,105,719,806,1565,1679]

Note: There is significant evidence that this type of reaction may proceed via an initial aminolysis of the 4-thioxo substituent to give a benzothiazinimine, which then undergoes a Dimroth-like rearrangement to the quinazoline, at a rate depending (at least in some part) on steric factors.[953, cf. 1295,2095]

From the 1,2-dihydro-4-thione (610) or a derivative

1-Methyl-1,2-dihydro-4H-3,1-benzothiazine-4-thione (**617**, R = H, X = S) gave 3-ethoxycarbonylamino-1-methyl-1,2-dihydro-4(3H)-quinazoline-thione (**617**, R = H, X = NHMCO$_2$Et) (H$_2$NNHCO$_2$Et, EtOH, reflux, 3 h: 85%);[1278] also many analogs. [1228]

7-Chloro-1-methyl-1,2-dihydro-4H-3,1-benzothiazine-4-thione (**617**, R = Cl, X = S) gave 3-amino-1-methyl-1,2-dihydro-4(3H)-quinazolinethione (**617**, X = NNH$_2$) (H$_2$NNH$_2$.H$_2$O, EtOH, reflux: 86%, plus a separable byproduct);[1230] also analogs.[1230]

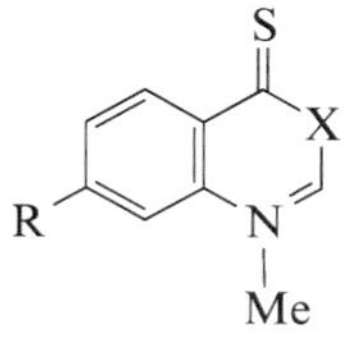

(617)

Also other examples.[1229,1439,1449,1725]

1.6.9. 1,2,4-Benzotriazepines as Substrates

This rather rare system has been observed to furnish quinazolines on acidic treatment. Thus 4-methyl-1H-1,2,4-benzotriazepin-5(4H)-one (**618**) in refluxing aqueous ethanolic hydrochloric acid for 2 h gave 1-amino-3-methyl-4-oxo-3,4-dihydroquinazolinium chloride (**619**, R = H, X = Cl) in 61% yield;[1445] rather similarly, 2,4-dimethyl-5-oxo-4,5-dihydro-1H-1,2,4-triazepinium fluorosulfonate (**620**) in refluxing hydrochloric acid for 30 min gave 3-methyl-1-methylamino-4-oxo-3,4-dihydroquinazolinium fluorosulfonate (**619**, R = Me, X = FO$_3$S$^-$) in 60% yield.[1445]

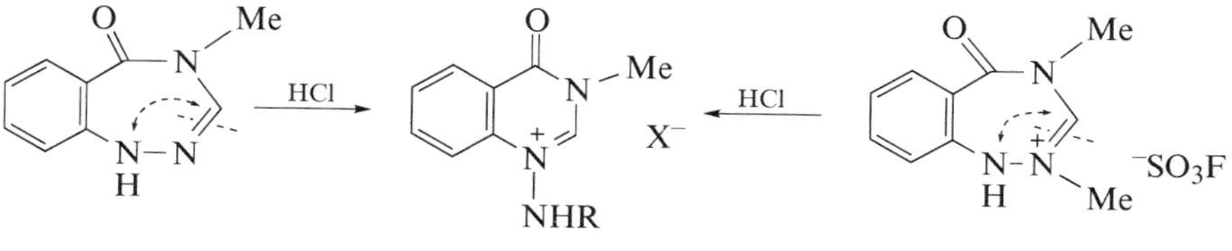

(**618**)   (**619**)   (**620**)

1.6.10. 1,3,4-Benzotriazepines as Substrates

Several quinazolines have been made from these substrates, as indicated in the following examples.

1,4-Dimethyl-1H-1,3,4-benzotriazepin-5(4H)-one (**621**) gave 2-imino-1,3-dimethyl-1,2-dihydro-4(3H)-quinazolinone (**622**) (EtONa, EtOH, reflux, 45 min: 65% as hydrobromide) and thence by Dimroth rearrangement, 1-methyl-2-methylamino-4(1H)-quinazolinone (**623**) (similar treatment but for 15 h: 33%);[947] Dimroth rearrangement alone was involved in other such syntheses.[1976]

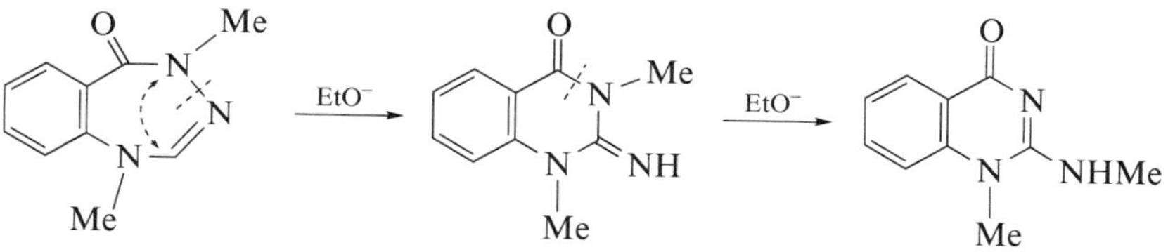

(**621**)   (**622**)   (**623**)

In contrast, 2,4-dimethyl-3H-1,3,4-benzotriazepin-5(4H)-one (**624**) gave 2-methyl-3-methylamino-4(3H)-quinazolinethione (**625**), apparently by direct fission and reclosure (EtONa, EtOH, reflux, 20 h: 96%);[934] other examples also seemed to so proceed.[911]

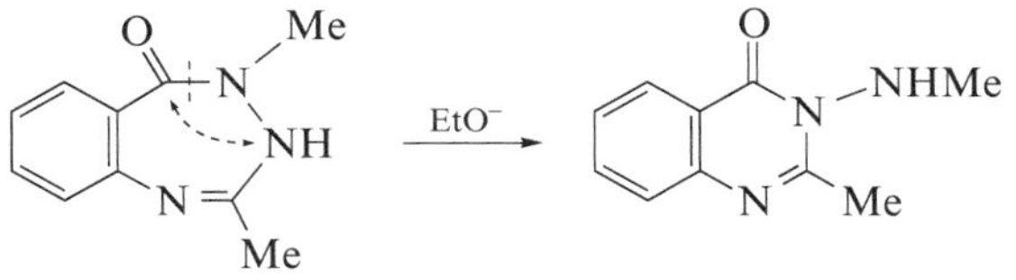

(624)   **(625)**

2-Phenyl-3H-1,3,4-benzotriazepin-5(4H)-one **(626)** gave 2-phenyl-4(3H)-
quinazolinone **(627**, R = H), probably via its 3-amino derivative **(627**,
R = NH$_2$) (KMnO$_4$, AcMe, reflux, 8 h: > 70%).[1704]

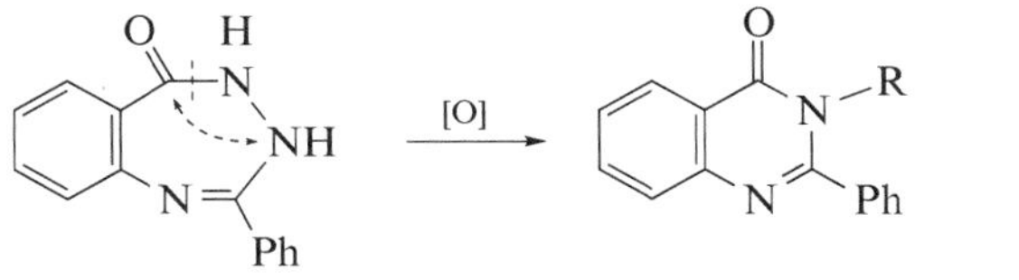

(626)   **(627)**

1.6.11.   1,2,3-Benzotriazines as Substrates

The conversion of these substrates into quinazolines has not been studied
systematically: the present position is indicated by the following typical examples.

1,2,3-Benzotriazine-4(3H)-one **(629)** gave 4(3H)-quinazolinone **(628)**
(HCONH$_2$, 200°, 1 h: ∼ 25%) or 2,4(1H, 3H)-quinazolinedione **(630)**
(H$_2$NCONH$_2$, 200°, 30 min: ∼ 80%), in each case by replacement of N2 and
naturally involving ring fission and reclosure.[174]

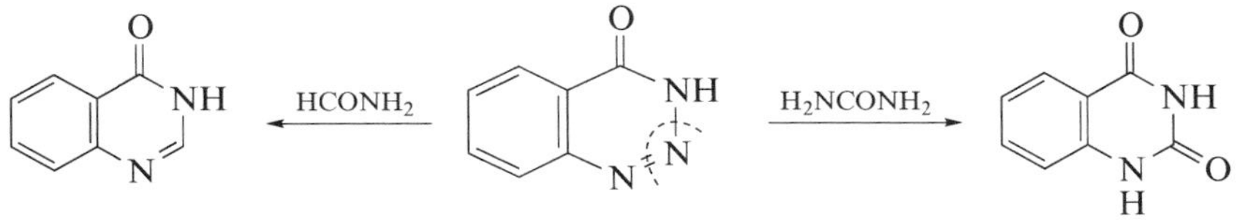

(628)   **(629)**   **(630)**

3-Benzylideneamino-1,2,3-benzotriazin-4(3H)-one **(631)** gave 2-phenyl-4(3H)-
quinazolinone **(632)** (paraffin oil, 300°, ∼ 10 min: 56%; by a complicated
mechanism involving fragmentation and recombination);[542] in contrast,
3-α-(phenylimino)benzyl-1,2,3-benzotriazin-4(3H)-one **(633)** gave 1,2-
diphenyl-4(1H)-quinazolinone **(634)** (paraffin oil, 300°, 5 min: 80%; by
a simple ring fission, loss of nitrogen, and reclosure).[698] The second of
the foregoing reactions also occurred on photolysis but only in small
yield.[1342]

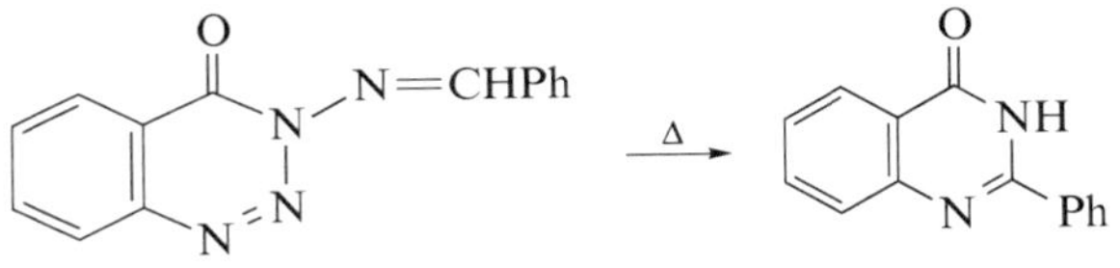

(631)   (632)

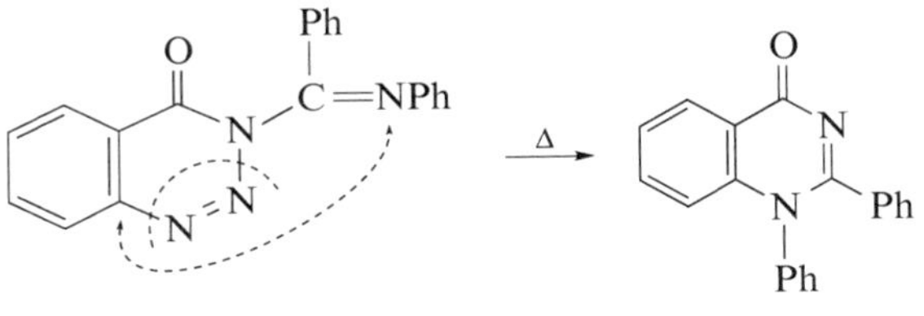

(633)   (634)

4-*p*-Nitroanilino-1,2,3-benzotriazine gave 3-*p*-nitrophenyl-4(3*H*)-quinazo-
linone (HCONH$_2$, reflux, 30 min: ~85%; clearly involving hydrolysis at
some stage).[1352]

Also other examples in the foregoing references and elsewhere.[1353]

1.6.12. 4,1,5-Benzoxadiazocines as Substrates

These unlikely substrates have been converted into quinazolines. Thus 8-chloro-
6-phenyl-3*H*-4,1,5-benzoxadiazocin-2(1*H*)-one oxime (**635**) gave 6-chloro-2-
hydroxymethyl-4-phenylquinazoline (**636**, R = Cl) (H$_2$, Raney-Ni, EtOH: 50%)
or its dechloro analog (**636**, R = H) (H$_2$, Pd/C, EtOH: 76%).[229] In a different way,
8-chloro-3-methyl-6-phenyl-3*H*-4,1,5-benzoxadiazocin-2(1*H*)-one (**637**) gave 2-
acetyl-6-chloro-4-phenylquinazoline (**638**) (MeONa, MeOH, 20°, 20 h: ~50%).[240]

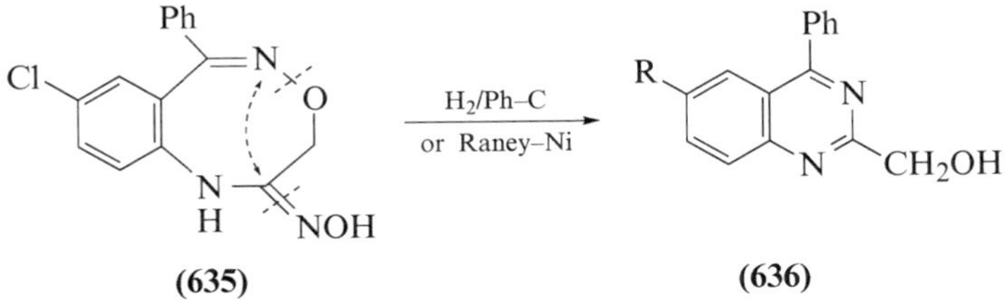

(635)   (636)

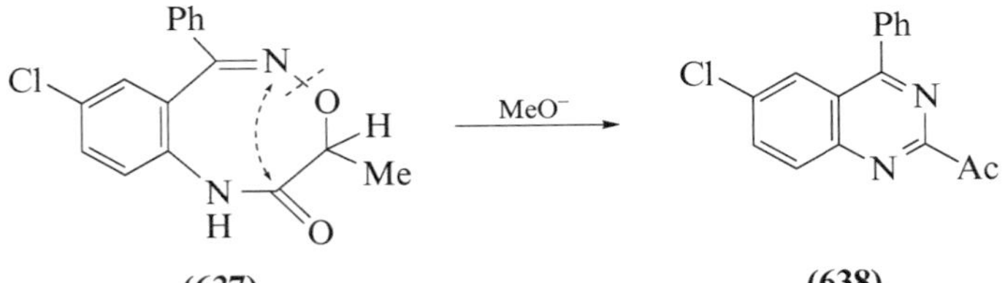

(637)   (638)

1.6.13. 3,2,1-Benzoxathiazines as Substrates

The particular benzoxathiazines involved in this synthesis may be considered as intramolecular anhydrides of *o*-sulfinaminobenzoic acids: accordingly, an example has been given at the end of Section 1.2.6.1. The method is illustrated further by the following examples.

3,2,1-Benzoxathiazin-4(1*H*)-one 2-oxide (**639**) gave 4(3*H*)-quinazolinone (**640**, R = H)   HCSNH$_2$,   PhH,   20°,   12 h: ~40%)   or   2-methyl-4(3*H*)-quinazolinone (**640**, R = Me) (MeCSNH$_2$, PhH, 20°, 12 h: ~70%).[908]

(639)   (640)

1-Methyl-3,2,1-benzoxathiazin-4(1*H*)-one 2-oxide (**641**) gave 2-benzyl-1-methyl-4(1*H*)-quinazolinone (**642**, R = Ph) (PhCH$_2$CONH$_2$, CHCl$_3$, 20°, 12 h: 55%)[705,733,831] or its 2-ethyl analog (**642**, R = Me) (EtCONH$_2$, PhH, 20°, 12 h: low yield).[1026]

(641)   (642)

1.6.14. 1,3-Benzoxazines as Substrates

This is a minor route to quinazolines. 2,4-Diphenyl-5,6,7,8-tetrahydro-1,3-benzoxazinium hexachloroantimonate (**643**) reacted with aqueous ammonia at 20° for 2 h to afford a 90% yield of 2,4-diphenyl-5,6,7,8-tetrahydroquinazoline (**644**).[2364] In a second way, 2*H*-1,3-benzoxazine-2,4(3*H*)-dione (**645**, X = O) and benzamide gave 2-phenyl-4(3*H*)-quinazolinone (**646**);[591] the corresponding 1,3-benzothiazine (**645**, X = S) behaved similarly, and a few analogous quinazolines were made from one or other substrate.[591] In a third manner, 4-anilino-2*H*-1,3-benzoxazin-2-one (**647**, R = H) underwent thermal rearrangement in diphenyl ether at 170° for 1 h to furnish 2-*o*-hydroxyphenyl-4(3*H*)-quinazolinone (**648**, R = H) in 92% yield;[1469] its 6-chloro (**648**, R = Cl) and 6-methyl derivatives (**648**, R = Me) were made similarly.[1469] Other examples have been reported.[898,1128]

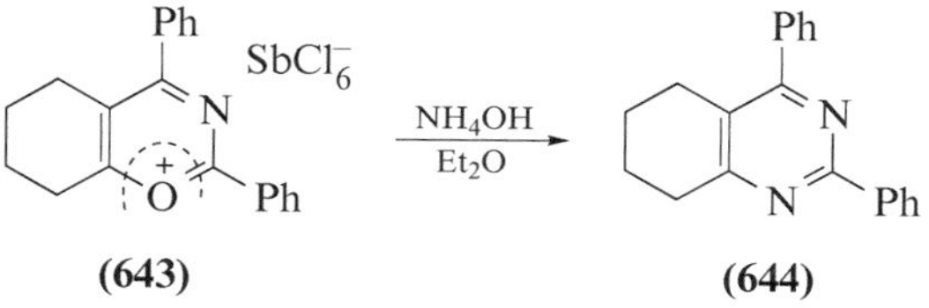

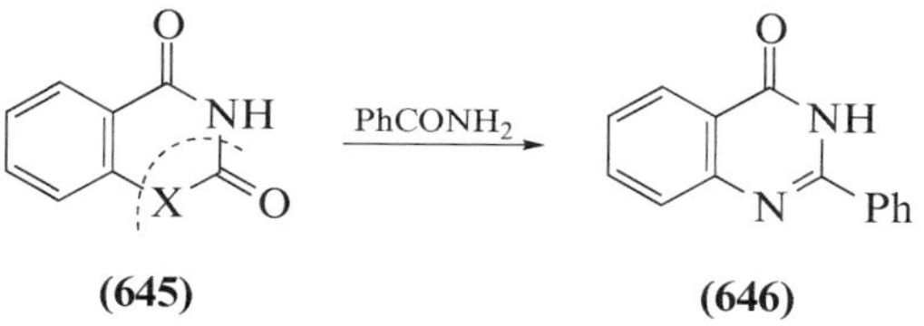

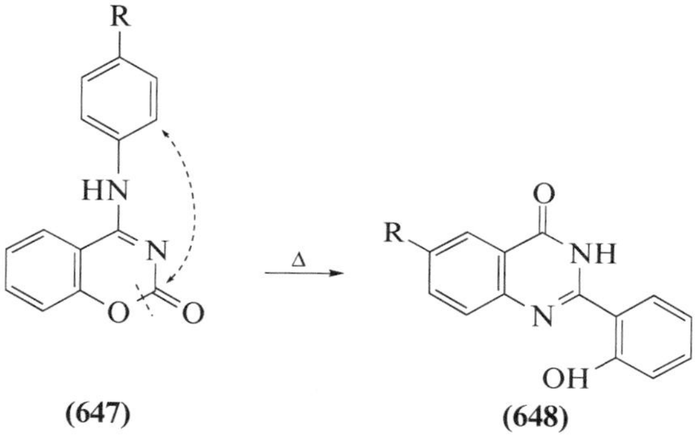

(643)   (644)

(645)   (646)

(647)   (648)

1.6.15. 3,1-Benzoxazines as Substrates (*H* 83–88, 117, 121, 338)

These benzoxazines have been used very extensively, especially by Indian and Egyptian workers, to make a great variety of 4((3*H*)-quinazolinones. Most such reactions have involved either treatment of a 4*H*-3,1-benzoxazin-4-one (**649a**) with almost any primary amine to achieve overall replacement of ring-O by ring-N with the formation of a 4-quinazolinone (**649**), or treatment of a 4*H*-3,1-benzoxazine-2,4(1*H*)-dione (**649b**) (an "isatoic anhydride") with a carboxamide or the like to achieve overall replacement of both C2 and O3, again with the formation of a 4-quinazolinone (**649**). A few other processes have been used occasionally to convert 3,1-benzoxazines into quinazolines.

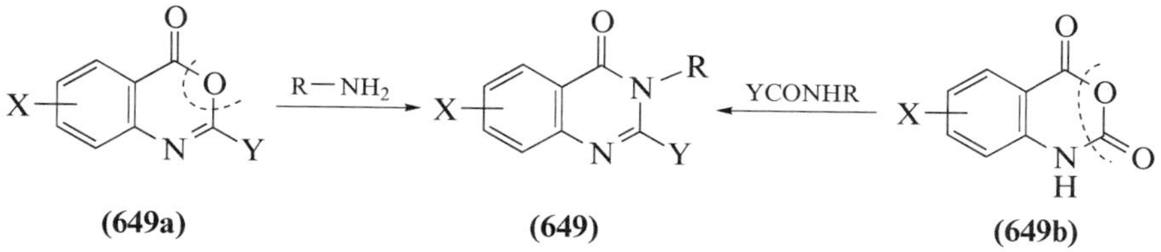

(649a)   (649)   (649b)

The extensive factual literature appears to be best surveyed in the form of a few selected examples from each natural broad category, as follows.

4H-3,1-Benzoxazin-4-ones with ammonia

5-Bromo-2,6-dimethyl-4H-3,1-benzoxazin-4-one (650) gave 5-bromo-2,6-dimethyl-4(3H)-quinazolinone (652) [liquid NH_3, 3 h; evaporation; then 1M NaOH, reflux, 1 h: 83%, probably via (651)].[2411]

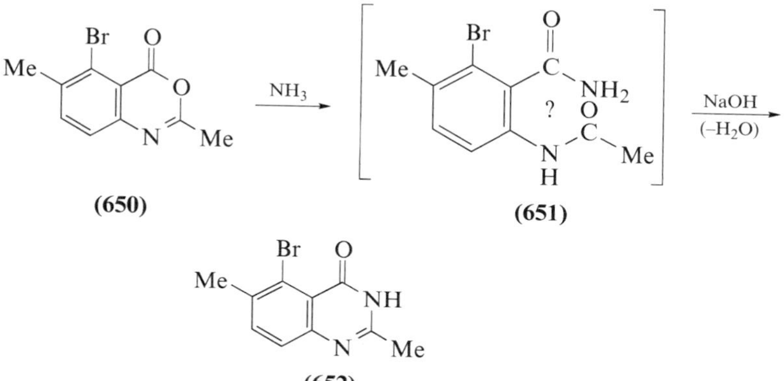

2-α-Chloroethyl-4H-3,1-benzoxazin-4-one gave 2-α-chloroethyl-4(3H)-quinazolinone (AcONH$_4$, ZnCl$_2$, 130°, 2 h: 65%; note the survival of an α-chloro substituent).[2341]

2-Propyl-4H-3,1-benzoxazin-4-one gave 2-propyl-4(3H)-quinazolinone (HCONH$_2$, 190°, 3 h: 65%).[1184]

Also other examples.[97,218,560,784,1057,1098,1161,1386,1506,1514,1556,1660,1765, 2073,2087,2158,2174,2285,2332,2369,2392,2395,2431,2629]

4H-3,1-Benzoxazin-4-ones with aliphatic amines

6,8-Dibromo-2-methyl-4H-3,1-benzoxazin-4-one (653) gave 6,8-dibromo-2,3-dimethyl- (654, R = Me) (MeNH$_2$, EtOH, reflux, 5 h: 70%) or 3-benzyl-6,8-dibromo-2-methyl-4(3H)-quinazolinone (654, R = CH$_2$Ph) (PhCH$_2$NH$_2$, likewise: 79%).[1371]

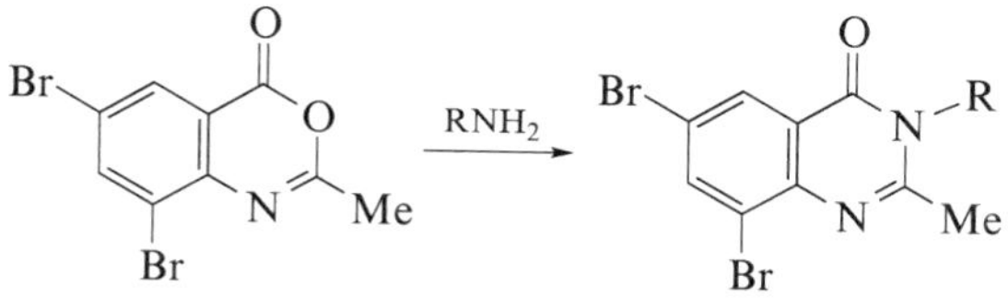

7-Nitro-2-phenyl-4H-3,1-benzoxazin-4-one gave 3-carboxymethyl-7-nitro-2-phenyl-4(3H)-quinazolinone (**655**, R = NO$_2$) (H$_2$NCH$_2$CO$_2$H, pyridine–H$_2$O, reflux, 5 h: 80%).[1588]

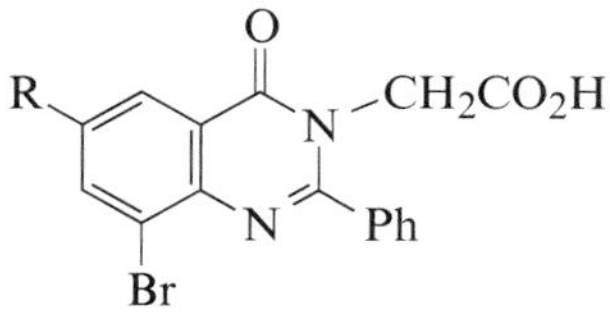

(655)

2-Phenyl-4H-3,1-benzoxazin-4-one gave 3-carboxymethyl-2-phenyl-4(3H)-quinazolinone (**655**, R = H) (neat H$_2$NCH$_2$CO$_2$H, 180°, 1 h: 82%).[2316]
Also other examples.[134,444,486,613,838,924,935,942,963,979,997,1047,1141,1219, 1238,1265,1278,1485,1527,1716,1775,1907,1949,2002,2130,2162,2299,2327,2337,2470]

4H-3,1-Benzoxazin-4-ones with aromatic amines

Note: Despite innumerable examples in this category, no satisfactory work on optimization of conditions has been reported. The following cases appear to represent some of the better procedures.

2-Phenyl-4H-3,1-benzoxazin-4-one (**656**) gave 2,3-diphenyl-4(3H)-quinazolinone (**657**, R = H) (neat PhNH$_2$, reflux, 4 h: 97%).[446]

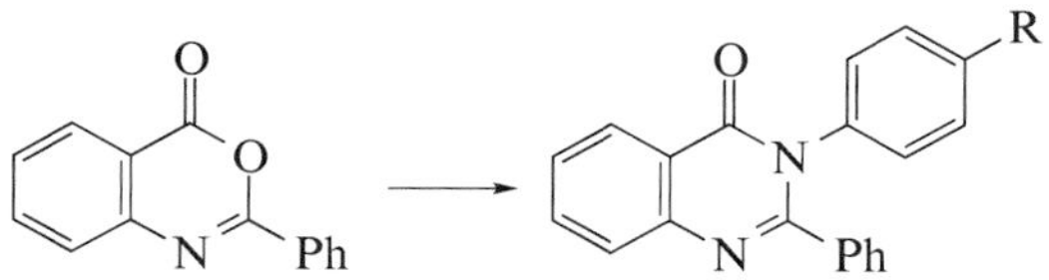

(656)   **(657)**

6,8-Dichloro-2-methyl-4H-3,1-benzoxazin-4-one gave 6,8-dichloro-3-p-methoxyphenyl-2-methyl-4(3H)-quinazolinone (neat H$_2$NC$_6$H$_4$OMe-p, reflux, 10 min: 85%).[573]

2-Methyl-4H-3,1-benzoxazin-4-one gave 2-p-chlorostyryl-3-phenyl-4(3H)quinazolinone (**658**) (neat PhN=CHC$_6$H$_4$Cl-p, 150°, 2 h: >65%).[1236]

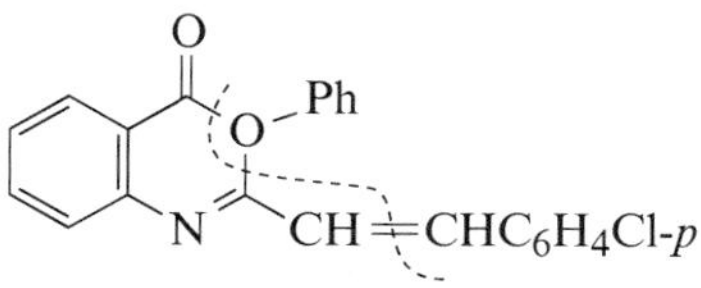

(658)

2-Methyl-6-nitro-4H-3,1-benzoxazin-4-one gave 2-methyl-6-nitro-3-m-nitrophenyl-4(3H)-quinazolinone (**659**) (H$_2$NC$_6$H$_4$NO$_2$-m, ClC$_6$H$_4$Cl-o, reflux with water removal, 3 h: 84%).[2]

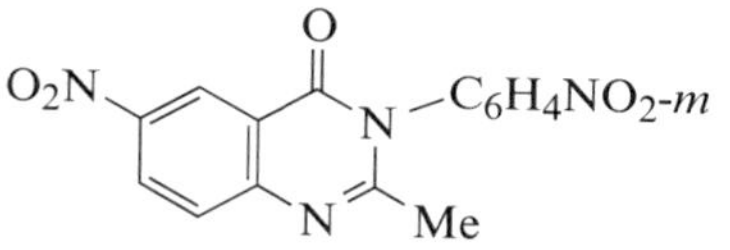

$$O_2N \quad \text{(ring system)} \quad N-C_6H_4NO_2\text{-}m$$

(659)

2-Phenyl-4H-3,1-benzoxazin-4-one **(656)** gave 3-p-hydroxyphenyl-2-phenyl-4(3H)-quinazolinone **(657, R = OH)** ($H_2NC_6H_4OH$-p, pyridine, reflux, 4 h: 80%;[1185] or neat, 180°, 30 min: 80%).[1253]

2-Trifluoromethyl-4H-3,1-benzoxazin-4-one gave 3-phenyl-2-trifluoro-methyl-4(3H)-quinazolinone **(660)** [PhNH$_2$, P(OPh)$_3$, pyridine, reflux, N$_2$, 20 h: 84%].[2351]

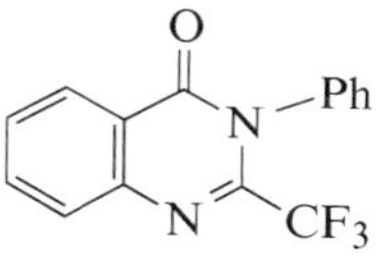

$$\text{(ring system)} \quad N-Ph \quad CF_3$$

(660)

2-Methyl-4H-3,1-benzoxazin-4-one gave 3-p-hydroxyphenyl-2-methyl-4(3H)-quinazolinone ($H_2NH_6H_4OH$-p, Me$_2$NCHO, reflux, 2 h: 71%)[2240] or 2-methyl-3-phenyl-4(3H)-quinazolinone (PhNH$_2$, PhH, 20°, 1 h: then storage of solid at 20° for 48 h to complete recyclization: 75%).[1074]

2-Methyl-6-nitro-4H-3,1-benzoxazin-4-one gave 2-methyl-6-nitro-3-p-tolyl-4(3H)-quinazolinone **(661)** (AcOH, 100°, 45 min; then POCl$_3$ added, 100°, 1 h to recyclize: >90%).[823]

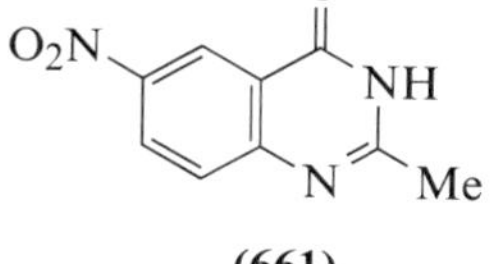

$$O_2N \quad \text{(ring system)} \quad NH \quad Me$$

(661)

Also other examples in the cited references and elsewhere.[11,136,142,311, 356,359,467,485,515,524,579,687,717,781,787,842,845,892,912,983,988,1016,1019, 1020,1045,1091,1151,1185,1202,1217,1234,1237,1252,1266,1503,1543,1544,1586, 1597,1667,1672,1729,1801,1839,1892,1972,1975,2023,2029,2034,2282,2352,2387, 2393,2394,2467,2572,2577,2591] Polymers have been so made.[326]

4H-3,1-Benzoxazin-4-ones with heterocyclic amines

5-Fluoro-2-methyl-4H-3,1-benzoxazin-4-one **(662)** gave 3-(5-t-butyl-1,3,4-thiadiazol-2-yl)-5-fluoro-2-methyl-4(3H)-quinazolinone **(663)** (5-t-butyl-1,3,4-thiadiazol-2-amine, pyridine, reflux, 28 h: 72%).[2331]

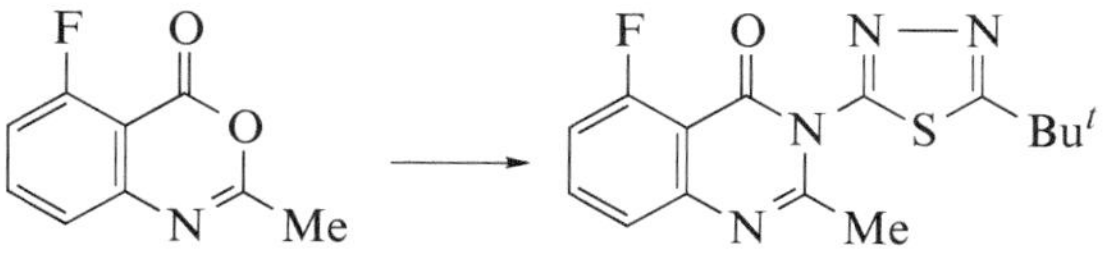

(662)   (663)

2-Methyl-4H-3,1-benzoxazin-4-one gave 2-methyl-3-(4-methylpyridin-2-yl)-4(3H)-quinazolinone (4-methyl-2-pyridinamine, heat over free flame: 60% as hydrochloride),[574] 2-methyl-3-γ-(4-phenylpiperazin-1-yl)propyl-4(3H)-quinazolinone [neat γ-(4-phenylpiperazin-1-yl)propylamine, 175°, 1 h: ~60% as hydrochloride),[671] or 3-(5-methylisoxazol-3-yl)-2-styryl-4(3H)-quinazolinone (664) (3-benzylideneamino-5-methylisoxazole, AcONa, AcOH, reflux, 3 h: 60%).[1745]

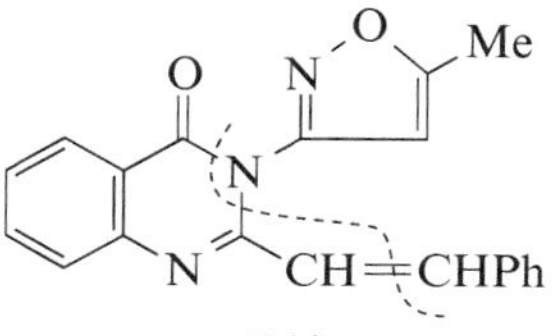

(664)

Also   other   examples.[132,572,623,971,1043,1099,1152,1183,1187,1191,1193,1216,1246,1487,1504,1626,2289,2385]

4H-3,1-Benzoxazin-4-ones with hydroxylamine or hydrazines

2-α-Hydroxybenzyl-4H-3,1-benzoxazin-4-one (665) gave 2-α-hydroxybenzyl-4(1H)-quinazolinone 3-oxide (666) (H$_2$NOH.HCl, pyridine, reflux, 6 h: ~50%).[841]

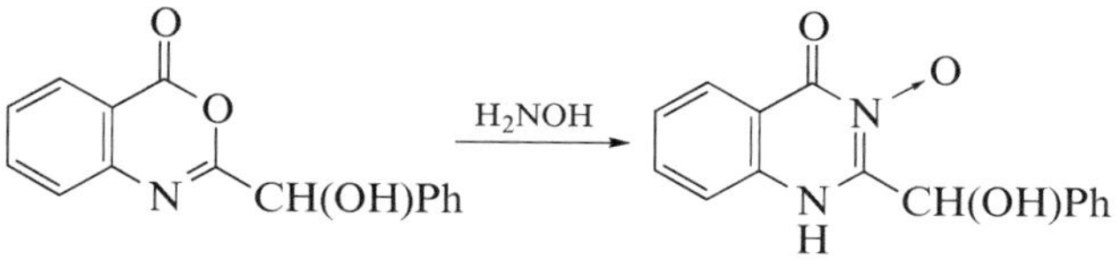

(665)   (666)

2-Phenyl-4H-3,1-benzoxazin-4-one gave 3-amino- (667, R = H) (H$_2$NNH$_2$.H$_2$O, pyridine, reflux, 6 h: 78%)[2340,cf.1655] or 3-anilino-2-phenyl-4(3H)-quina-zolinone (667, R = Ph) (PhNHNH$_2$, pyridine, reflux, 5 h: 60%).[2307]

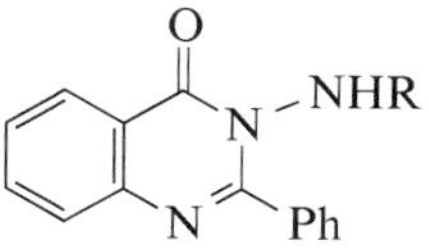

(667)

2-Indol-2′-yl-4H-3,1-benzoxazin-4-one gave 3-amino-2-indol-2′-yl-4(3H)-quinazolinone (neat H$_2$NNH$_2$.H$_2$O, reflux, 1 h: 63%).[1301]

Also other examples.[506,569,825,844,1075,1121,1164,1222,1262,1434,1488,1648,1659, 1675,1703,1717,2012,2042,2183,2310,2318,2323,2349,2354,2360,2370,2549,2581,2582]

4H-3,1-Benzoxazine-2,4(1H)-diones with amides etc.

6-Nitro-4H-3,1-benzoxazine-2,4(1H)-dione (**668**) gave 2-methyl-6-nitro-4(3H)-quinazolinone (**669**) (neat MeCONH$_2$, 160°, 2.5 h: 79%).[2232]

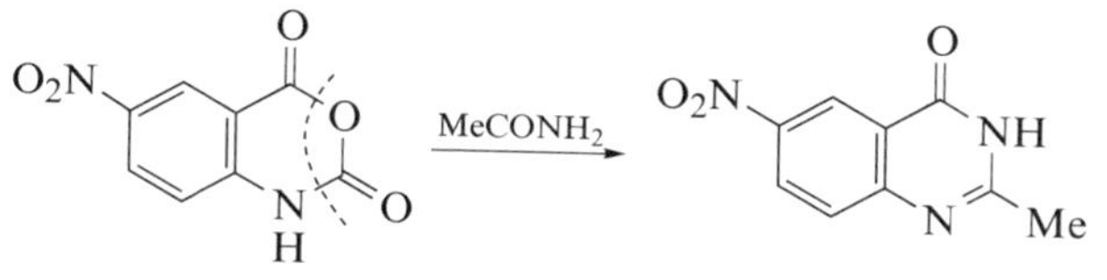

(**668**)   (**669**)

4H-3,1-Benzoxazine-2,4-(1H)-dione (**670**) gave 3-amino-2-phenyl-4(3H)-quinazolinone (**671**) (PhCONHNH$_2$, TsOH, AcOH, reflux, 9 h: 37%).[1385]

1-Methyl-4H-3,1-benzoxazine-2,4(1H)-dione gave 1,2-dimethyl-4(1H)-quinazolinone (neat MeCSNH$_2$, 180°, 3 h: 58%),[754] 1-methyl-2,4(1H,3H)-quinazolinedione (**672**, X = O) (neat H$_2$NCONH$_2$, 180°, 2h: 50%; or H$_2$NCONH$_2$, Me$_2$NCHO, reflux, 8h: 50),[748] or 1-methyl-2-thioxo-1,2-dihydro-4(3H)-quinazolinone (**672**, X = S) (H$_2$NCSNH$_2$, likewise: 55% by either procedure).[748]

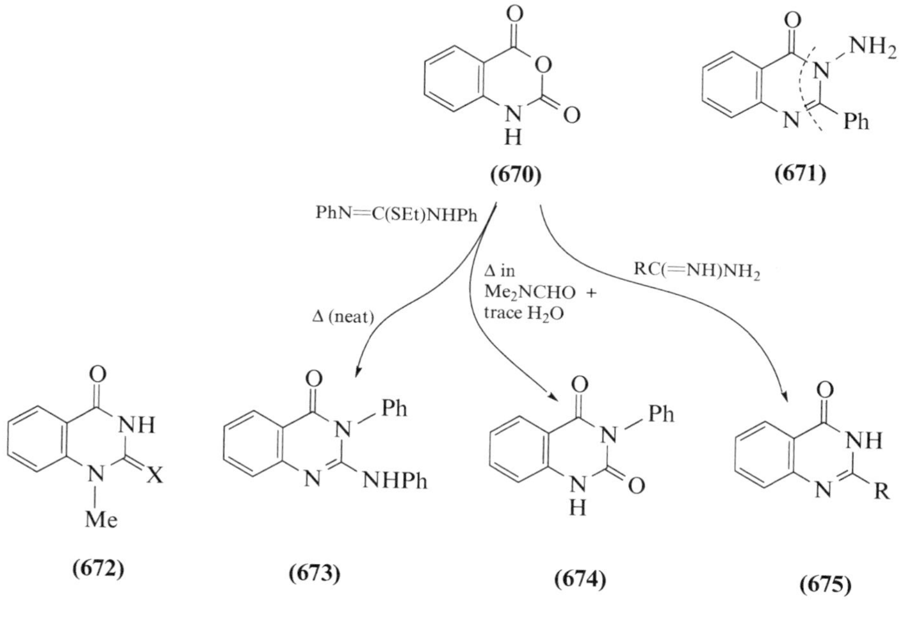

(**670**)   (**671**)

PhN=C(SEt)NHPh

Δ (neat)

Δ in Me$_2$NCHO + trace H$_2$O

RC(=NH)NH$_2$

(**672**)   (**673**)   (**674**)   (**675**)

4*H*-3,1-Benzoxazine-2,4(1*H*)-dione (**670**) gave either 2-anilino-3-phenyl-4(3*H*)-quinazolinone (**673**) [neat PhHNC(SEt)=NPh, $> 150°$, > 2 h: 85%] or 3-phenyl-2,4(1*H*,3*H*)-quinazolinedione (**674**) (same reagent, Me$_2$NCO, trace H$_2$O, reflux, 8 h: 65%, via the ethylthio analog?).[769]

4*H*-3,1-Benzoxazine-2,4(1*H*-dione (**670**) gave 4(3*H*)-quinazolinone (**675**, R = H) (H$_2$NCH=NH.AcOH, EtOH, reflux, 3h: 81%)[1951] or its 2-car-bamoylmethyl derivative (**675**, R = CH$_2$CONH$_2$) [H$_2$NC(=NH) CH$_2$CONH$_2$.HCl, pyridine, reflux, 5 h: 74%];[1293] 5,8-dihydro-4(3*H*)-quinazolinone was made likewise from the correspondingly reduced sub-strate (H$_2$NCH=NH.AcOH, MeOCH$_2$CH$_2$OH, reflux, 90 min: 62%).[1951]

6-Nitro-4*H*-3,1-benzoxazine-2,4(1*H*)-dione gave 2-amino-6-nitro-4(3*H*)-quinazolinone [H$_2$NC(=NH)NH$_2$.H$_2$CO$_3$, Me$_2$NCHO, reflux, 48 h: 75%].[1564]

Also other examples.[233,747,770,1562,1569.1622,1679,1974,2211,2502]

4*H*-3,1-Benzoxazine-2,4(1*H*)-diones with other synthons

4*H*-3,1-Benzoxazine-2,4(1*H*)-dione (**677**) gave 3-phenyl-2,4(1*H*,3*H*)-quinazolinedione (**676**, X = O) (neat PhNCO, 260°, 2 h: > 70%) or 3-phenyl-2-thioxo-1,2-dihydro-4(3*H*)-quinazolinone (**676**, X = S) PhNCS, similarly: > 70%).[1650]

4*H*-3,1-Benzoxazine-2,4(1*H*)-dione (**677**) gave 2,3-diphenyl-1,2-dihydro-4(3*H*)-quinazolinone (**678**) (neat PhCH=NPh, $> 150°$, > 2 h: 57%)[746] or 3-*p*-chlorophenyl-2,4(1*H*.3*H*)-quinazolinedione (**679**) [PhHN—P(OEt)$_2$, NaH, PhH–Me$_2$NCHO, reflux, 5 h: 70%].[1282]

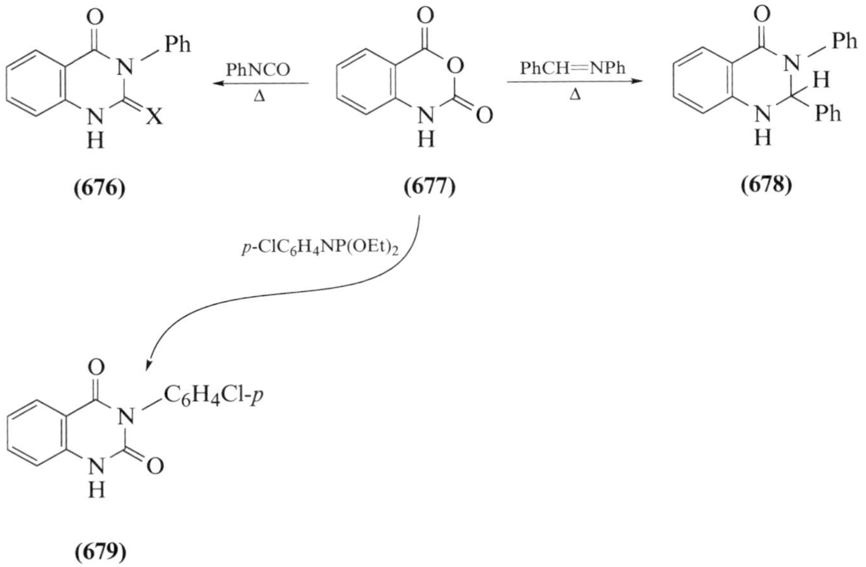

(**676**)   (**677**)   (**678**)

(**679**)

1-β-Bromoethyl-4H-3,1-benzoxazine-2,4(1H)-dione **(680)** gave 1-β-hy-droxyethyl-3-methyl-2,4(1H,3H)-quinazolinedione **(681)** (MeNH$_2$, diox-ane, 20°, 1 h: 39%),[1775,1792] rather than the tricyclic isomer previously postulated.[1566]

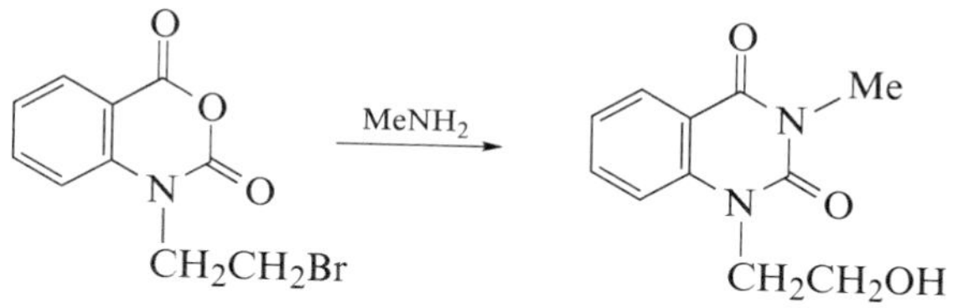

(680)   **(681)**

Also other examples.[308,808,1285,1477,1693]

Rearrangement of amino- or imino-4H-3,1-benzoxazines

2-Methyl-4-methylimino-4H-3,1-benzoxazine **(682)** gave 2,3-dimethyl-4(3H)-quinazolinone **(683)** (HCl gas, ClCH$_2$CH$_2$Cl, 25°, 3 days: 93%; 70°, sealed, 6 h: 89%).[2267]

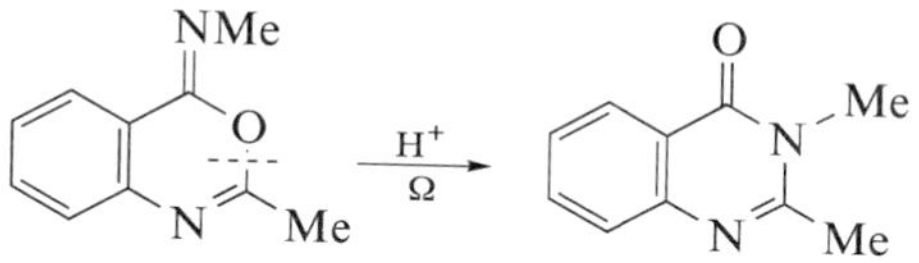

(682)   **(683)**

6-Chloro-2-methylamino-4-phenylimino-4H-3,1-benzoxazine **(684)** gave 4-anilino-6-chloro-3-methyl-2(3H)-quinazolinone **(685)** (BF$_3$.Et$_2$O, PhH, re-flux, 16 h: 63%).[772]

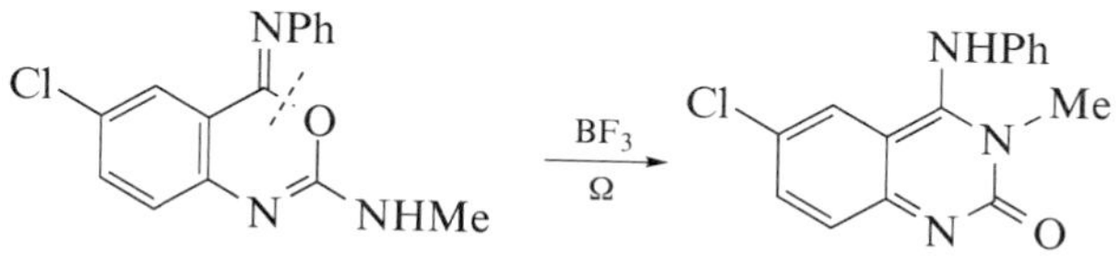

(684)   **(685)**

2-Methylamino-4H-3,1-benzoxazine-4-one **(686)** (made *in situ* by transamina-tion; structure based on analogy with that of its characterized 2-di-

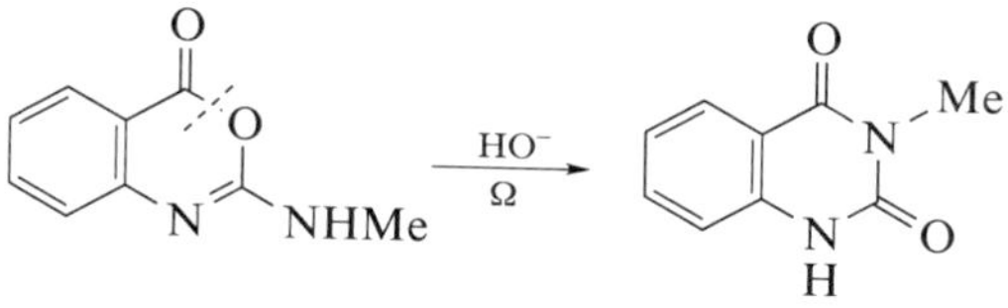

(686)   **(687)**

ethylamino analog) gave 3-methyl-2,4(1*H*,3*H*)-quinazolinedione (**687**) (MeNH$_2$, H$_2$O, 20°, 4 h: 89% overall).[1893]

Also other examples.[227,2474,2493]

An unclassified transformation

1-Methyl-2,4-diphenyl-4*H*-3,1-benzoxazinium fluorosulfonate (**688**) gave 1-methyl-4-phenyl-3,4-dihydro-2(1*H*)-quinazolinethione (**689**) (KNCS, MeCN, 20°, 3 h: 78%).[1249]

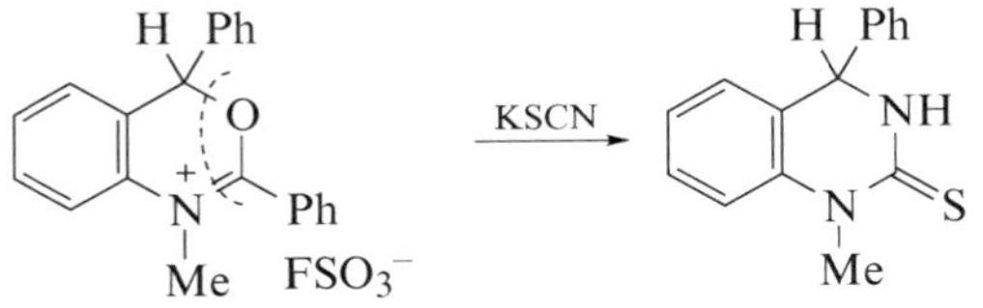

(688)   (689)

1.6.16.   Cinnolines as Substrates

Just as there is a tendency for pyridazines to give pyrimidines under highly energetic conditions,[2446] so have cinnolines been reported to afford quinazolines. Thus uv irradiation of hexafluorocinnoline (**690**) at 100° gave hexafluoroquinazoline (**692**, R = F), possibly via the valence isomer (**691**): the product was isolated (after gentle aminolysis) as pentafluoro-4-quinazolinamine (**692**, R = NH$_2$).[158] More satisfactory was the preparation-scale conversion of 2-methylcinnolinium-4-olate (**693**), by uv irradiation of a reflxuing ethanolic solution during 5 h, into 3-methyl-4(3*H*)-quinazolinone (**694**) in 80% yield; several analogs were made similarly.[990]

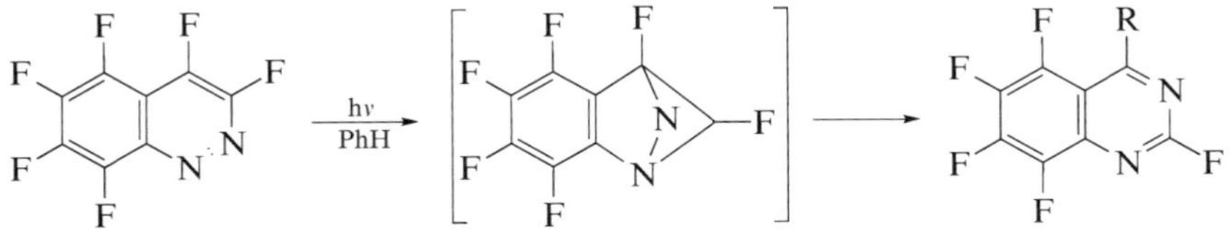

(690)   (691)   (692)

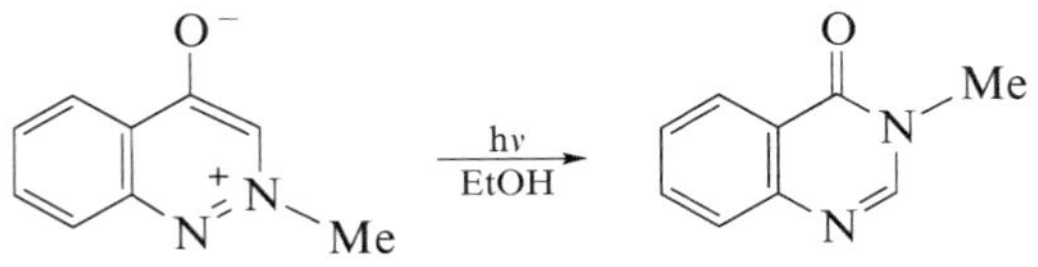

(693)   (694)

1.6.17. Indazoles as Substrates

The ring expansion of N-alkylated indazoles to quinazolines does not appear to be a general reaction, but several diverse examples have been reported.

1-Benzylindazole (**695**) gave 2-phenyl-1,2-dihydroquinazoline (**696**) (PhMgBr, MeOCH$_2$CH$_2$OMe–Et$_2$O–PhMe, reflux, 2.5 h: 63%; the Grignard reagent is involved only in α-metallation, and its phenyl group is not that in the product).[767]

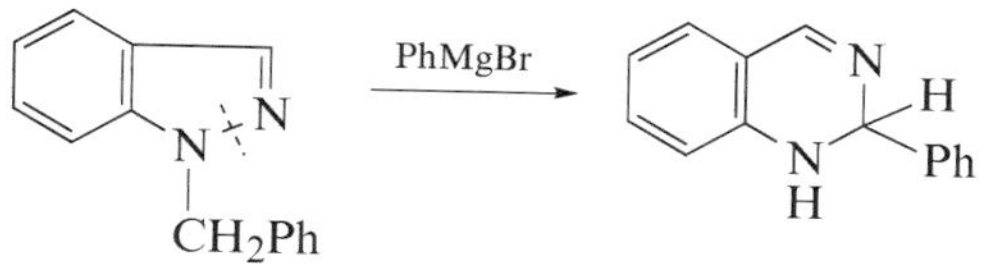

(**695**)   (**696**)

1-Benzyl-2-methyl-3(2H)-indazolone (**697**) gave 3-methyl-2-phenyl-1,2-dihydro-4(3H)-quinazolinone (**698**) (NaH, PhMe, 80°, 15 min: 88%), whereas the isomeric substrate, 2-benzyl-1-methyl-3(2H)-indazolone (**699**), gave 1-methyl-2-phenyl-1,2-dihydro-4(3H)-quinazolinone (**700**) (likewise but 18 h: 59%).[1756]

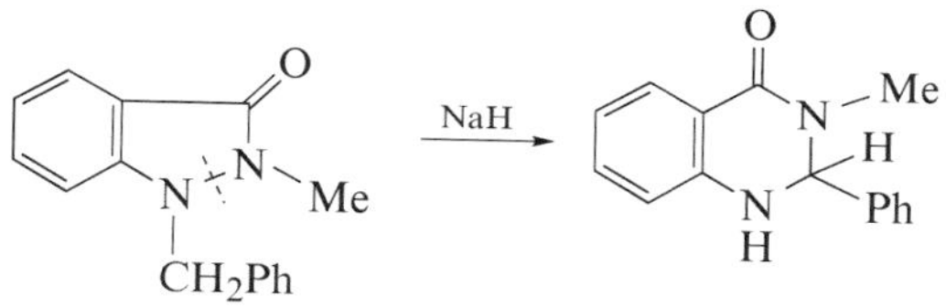

(**697**)   (**698**)

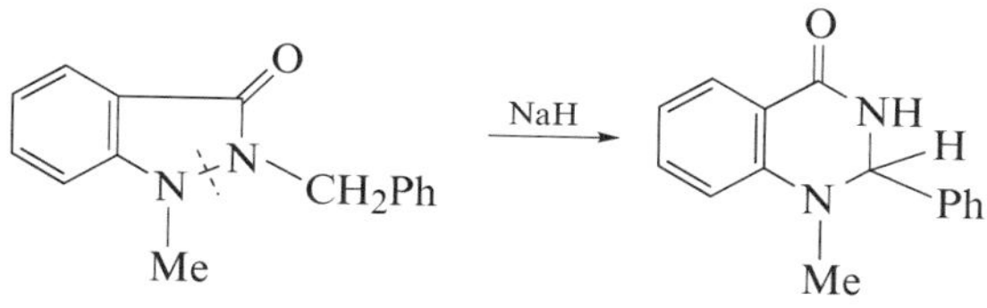

(**699**)   (**700**)

3-(α-Methoxycarbonyl-α-triphenylphosphoranylidenemethyl)-1-methyl-3(2H)-indazolone (**701**) gave methyl 1-methyl-4-oxo-1,2-dihydro-2-quinazolinecarboxylate (**702**) (TsOH, PhMe, 45°, 2 h: >95%).[1915]

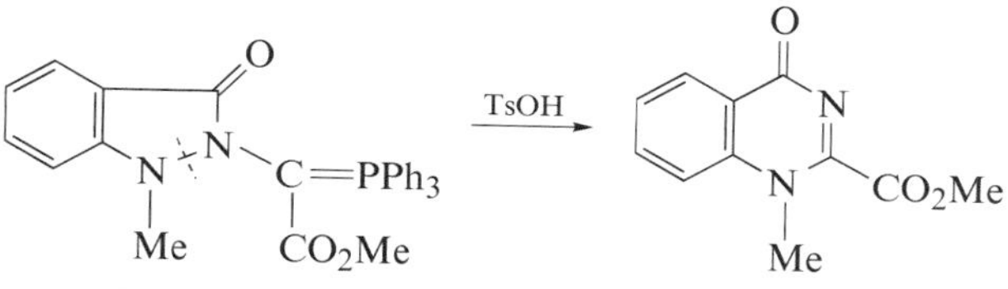

(**701**)   (**702**)

Indazole gave 2-chloroquinazoline ($CHCl_3$, 555°, 30 min: 68%).[1344]
Also other examples in the foregoing references and elsewhere.[258,2355,2471]

1.6.18.   Indoles as Substrates (*H* 96–98, 125)

Most of the indoles used in this way are, in fact, derivatives of isatin (2,3-indolinedione) bearing a nitrogen-containing substituent. The use of these and other indoles is illustrated in the following typical examples.

N-Methyl-2,3-dioxo-1-indolinecarboxamide (**703**) gave methyl 4-hydroxy-3-methyl-2-oxo-1,2,3,4-tetrahydro-4-quinazolinecarboxylate (**704**) (NEt_3, MeOH, 1 h: 95%);[727] 2,3-indoline dione gave ethyl 4-hydroxy-3-phenyl-2-thioxo-1,2,3,4-tetrahydro-4-quinazolinecarboxylate in an essentially similar way (PhNCS, NEt_3, EtOH, 20°, 1 day: 98%).[206]

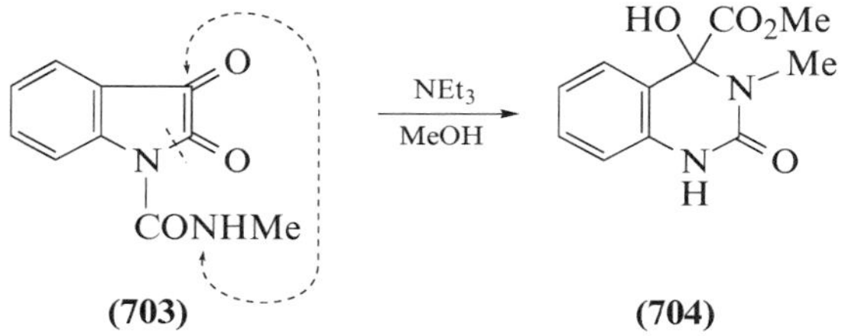

(703)   (704)

5,7-Dibromo-3-hydroxyimino-2-indolinone (**705**) gave 6,8-dibromo-2,4 (1*H*,3*H*)-quinazolinedione (**706**) (Vilsmeier reagent from *N*-methylpyrrolidione + $POCl_3$, 70°, 3 h: NaOH, H_2O workup: 50% by a rational mechanism).[869]

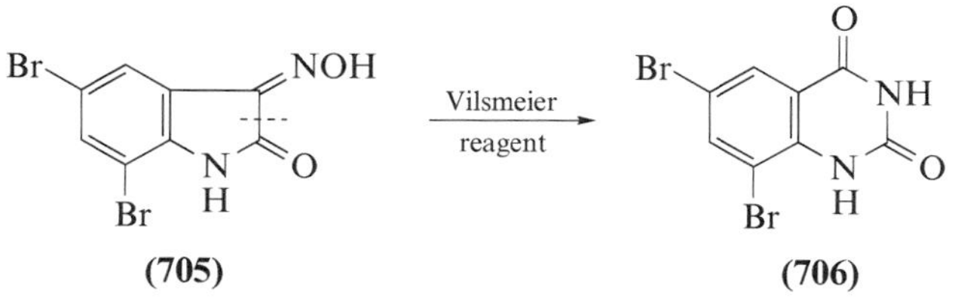

(705)   (706)

3-Nitroso-2-phenylindole (**707**) gave 2-phenyl-4(3*H*)-quinazolinone(**708**) [$POCl_3$, $(CH_2)_4SO_2$, 200°, 1 h; H_2O workup: 90%].[346]

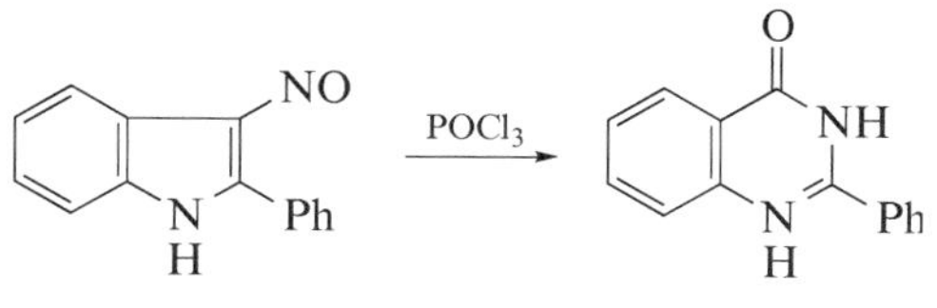

(707)   (708)

3-Azido-3-methyl-2-phenyl-3H-indole (**709**) (from 3-methyl-2-phenylindole with IN_3) gave 4-methyl-2-phenylquinazoline (**710**) (Me_2NCHO, reflux, 16 h: 26%, plus other separable products).[1083,cf.894]

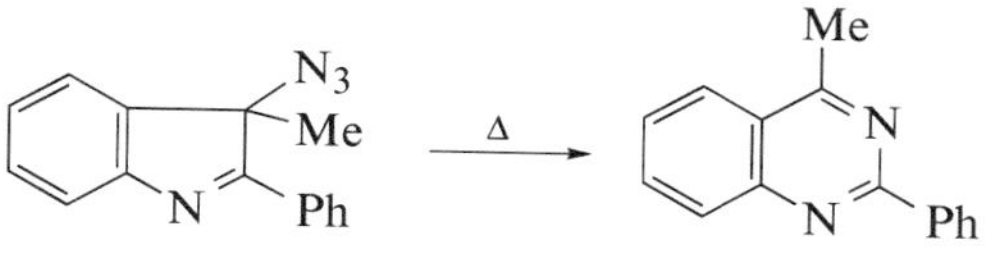

(**709**)   (**710**)

Also other examples.[64,67,93,168,209,226,286,393,722,723,739,1757,2104]

1.6.19.   Isoindoles as Substrates (*H* 123)

Several quinazolines have been made from isoindoles that naturally include phthalimide derivatives. The present limited scope of the reaction is illustrated by the following examples.

2-Phenylsulfonyloxy-1H-isoindole-1,3(2H)-dione (*N*-phenylsulfonyloxyphthalimide) (**711**) gave 2,4(1H,3H)-quinazolinedione (**712**, R = H) (NH_3, EtOH, reflux, 2 h: 74%).[917] or 3-phenyl-2,4-(1H,3H)-quinazolinedione (**712**, R = Ph) (neat $PhNH_2$, reflux, 15 min: 70%).[1254]

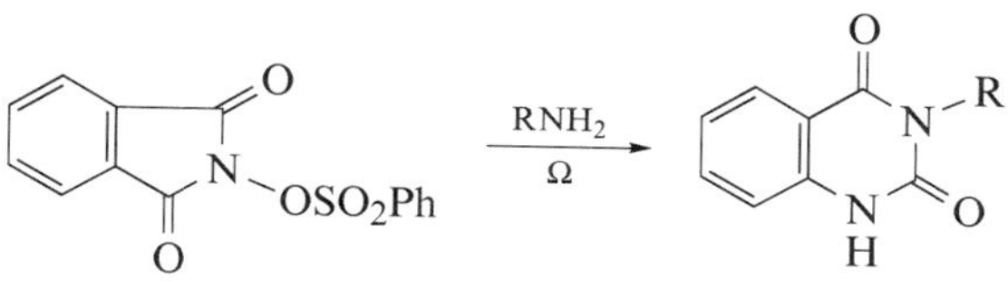

(**711**)   (**712**)

3a-(*N'*-Phenylureido)-2,3,3a,4,5,6,7,7a-octahydro-1H-isoindol-1-one (**713**) gave 8a-aminomethyl-3-phenyl-4a,5,6,7,8a-hexahydro-2,4(1H,3H)-quinazolinedione (**714**) (5M HCl, 100°, 1 h: 92% as hydrochloride).[194]

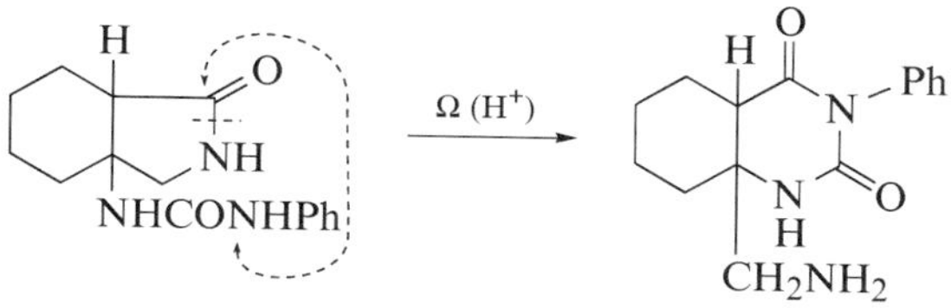

(**713**)   (**714**)

3-Azido-2,3-diphenyl-2,3-dihydro-1H-isoindol-1-one (**715**) appears to have given 2,3-diphenyl-4(3H)-quinazolinone (**716**) (xylene, reflux, N_2, >3 h: 53%).[1918]

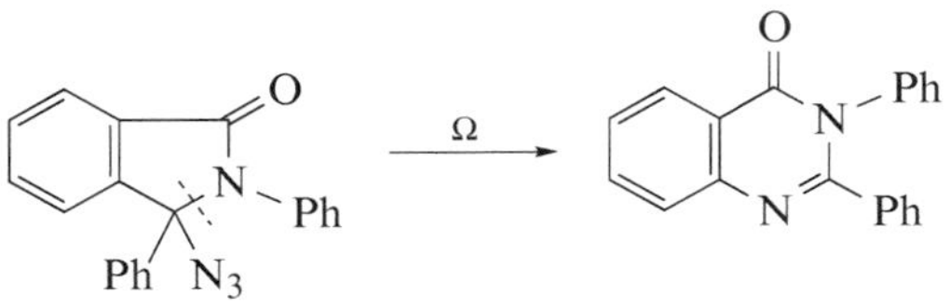

(715)   (716)

Also other examples.[1998]

1.6.20. Phthalazines as Substrates

Thermal rearrangement of neat 1,4,5,8-tetraphenylphthalazine at 360° for 30 min gave 2,4,5,8-tetraphenylquinazoline in 70% yield; the structure was confirmed by X-ray analysis.[2588] Steric crowding in the substrate appears to be an important factor in the rearrangement; for example, 1,4-bis-*p*-fluorophenyl-5,6,7,8-tetraphenylphthalazine gave 2,4-bis-*p*-fluorophenyl-5,6,7,8-tetraphenyl-quinazoline in 75% yield, whereas 1,4-bis-*p*-fluorophenylphthalazine gave 2,4-bis-*p*-fluorophenylquinazoline in only 23% yield under similar conditions.[2585]

1.6.21. Quinolines as Substrates

This synthetic reaction is represented in the conversion of 2-bromoquinoline (717) into 2-methylquinazoline (718) (~20%, plus other separable products) by treatment with sodium amide in liquid ammonia and ether at −33° for about a minute;[20] the thermolysis of 3,3-diazido-6-chloro-3,4-dihydro-2,4(1*H*)-quinolinedione (719) to give 6-chloro-2,4-dioxo-1,2,3,4-tetrahydro-3-quinazolinecarbonitrile (720) in 49% yield by refluxing in toluene for 8 h;[877] and the transformation of 2,2,4-trimethyl-1,2-dihydroquinoline (721) into 2,2,4-trimethyl-1-(*N*-methylcarbamoyl)-1,2-dihydroquinoline (722) (MeNCO, 30°, 20 h) and thence into 3,4-dimethyl-2(3*H*)-quinazolinone (723) (10M HCl, 65–90°, 1 h, with loss of Me$_3$CCl: 95%; several analogs likewise).[110]

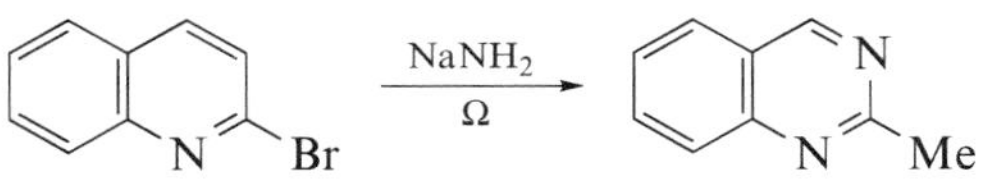

(717)   (718)

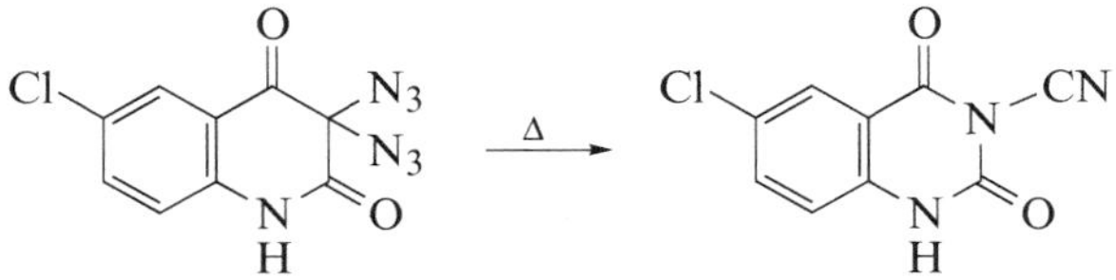

(719)   (720)

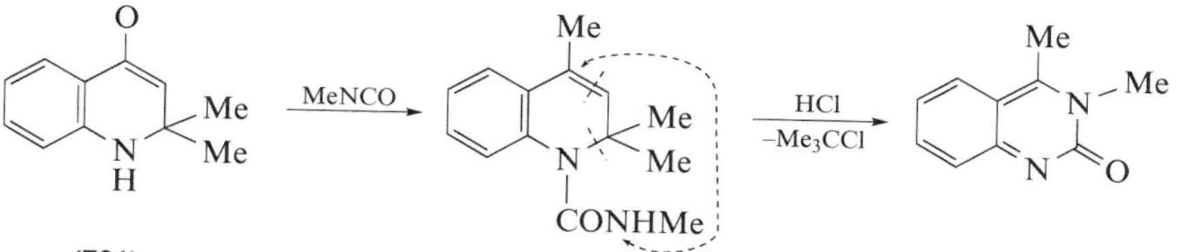

(721)   (722)   (723)

1.7.   FROM A HETEROTRICYCLIC SUBSTRATE

Quinazolines have been made from a variety of heterotricyclic systems. Most such systems are quinazolines with an additional fused ring, and many have themselves been prepared from quinazolines: accordingly, some of these routes to quinazolines are of little practical importance, but most have intrinsic interest. The tricyclic substrate systems are arranged in alphabetical order: some of the resulting subsections necessarily contain very few data.

1.7.1.   Azirino[2,3-b]indoles as Substrates

2-Benzenesulfonyl-1-phthalimido-1,1a,2,6b-tetrahydroazirino[2,3-b]-indole (724, R = H) in methanolic alkali for 1 h gave quinazoline (725, R = H) as a minor product (20%); the 1a-phenyl substrate (724, R = Ph), in methanolic alkali or dimethyl sulfoxide containing sodium hydride, likewise gave 2-phenylquinazoline (725, R = Ph) in 30% or 70% yield, respectively; and the 1a-methyl substrate (724, R = Me) gave 2-methylquinazoline (725, R = Me) (Me$_2$SO, NaH: 30%).[1652,1763]

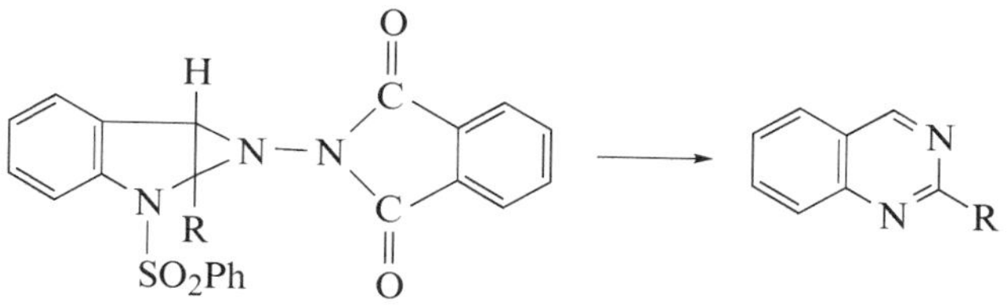

(724)   (725)

1.7.2.   Benzo[c]cinnolines as Substrates

Although the benzocinnoline nucleus was not involved directly, the dipolar compound, benzo[c]cinnoline N-[N-(N-phenylbenzimidoyl)benzimidoyl]imide (727), underwent thermolysis at its melting point (∼ 185°) during 15 min to afford 2,4-diphenylquinazoline (726) in 49% yield;[1025] when the isopropyl analog (727, Pr^i for Ph*) was treated similarly (∼ 235°), the reaction took a different course to furnish 4-isopropylamino-2-phenylquinazoline (**728**) in 27% yield;[1025] in both cases, benzo[*c*]cinnoline was recovered in high yield.

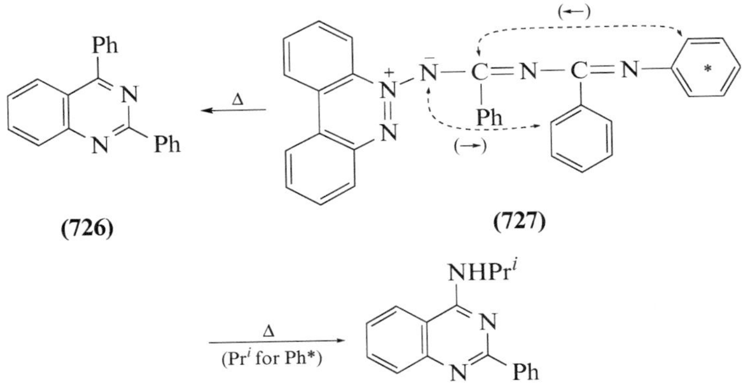

(**726**)   (**727**)

(**728**)

1.7.3. 5,8-Epoxyquinazolines as Substrates

Furo[3,4-*d*]pyrimidines have been converted indirectly into regular quinazolines via the tricyclic 5,8-epoxyquinazolines. Thus 3-[β-(4-*o*-methoxyphenyl-piperazin-1-yl)ethyl]furo[3,4-*d*]pyrimidine-2,4(1*H*,3*H*)-dione (**729**) underwent Diels–Alder addition with ethylene (in Me₂NCHO, 80°, sealed, 3 days) to give 3-[β-(4-*o*-methoxyphenylpiperazin-1-yl)ethyl]-5,6,7,8-tetrahydro-5,8-ep-oxyquinazoline-2,4(1*H*,3*H*)-dione (**730**) in 40% yield: subsequent treatment of the epoxide (**730**) with concentrated sulfuric acid at 20° for 45 min gave 3-[β-(4-*o*-methoxyphenylpiperazin-1-yl)ethyl]-2,4(1*H*,3*H*)-quinazolinedione (**731**) in 63% yield.[2114] Several other dienophiles gave analogous products.[2114]

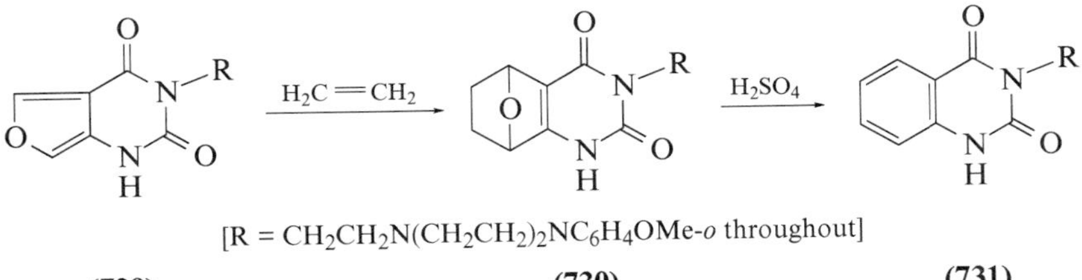

[R = CH₂CH₂N(CH₂CH₂)₂NC₆H₄OMe-*o* throughout]

(**729**)   (**730**)   (**731**)

1.7.4. Imidazo[1,2-*c*]quinazolines as Substrates

3-Methyl-2,3-dihydroimidazo[1,2-*c*]quinazolin-2-one (**732**), itself made from a quinazoline, underwent hydrolytic ring fission in dilute alkali at 100° during 90 min to give 3-α-carboxyethyl-4(3H)-quinazolinone (733) (80%).[783,1085] The reaction seems quite capable of extension.

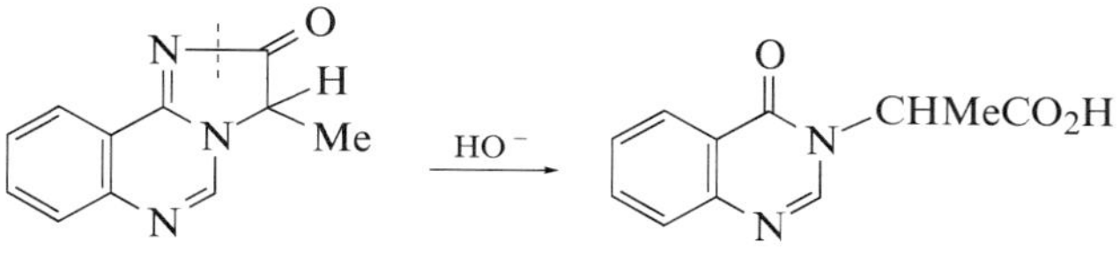

(732)   (733)

1.7.5. [1,2,4]Oxadiazolo[2,3-c]quinazolines as Substrates

Thermolysis of 5-methyl-2H-[1,2,4]oxadiazolo[2,3-c]quinazolin-2-one (734) in a sealed tube at 190° for 6 h gave a separable mixture of 2-methyl-4-quinazolinamine (735) (15% yield), 4-N'-acetylureido-2-methylquinazoline 3-oxide (736) (80%), and other nonquinazoline products; photolysis did not give quinazolines.[1757]

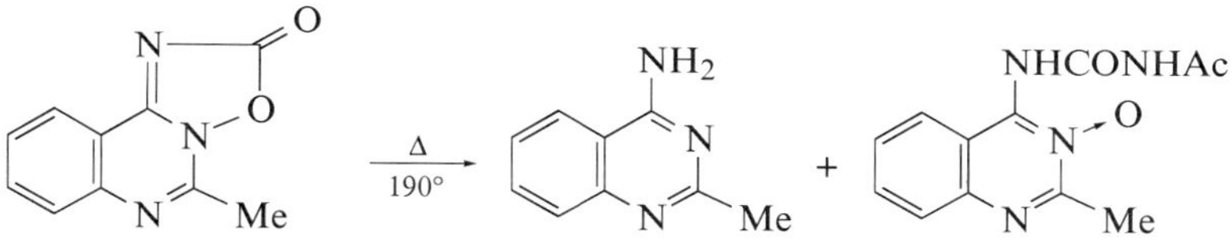

(734)   (735)   (736)

1.7.6. [1,3]Oxazino- and Oxazolo[2,3-b]quinazolines as Substrates

Although not particularly useful, these syntheses are exemplified in the conversion of 2,3-dihydro-5H-oxazolo[2,3-b]quinazolin-5-one (737, n = 2) into 3-β-chloroethyl- (738, n = 2, R = Cl) (10M HCl, 95°, 1 min: 83%), 3-β-bromoethyl- (738, n = 2, R = Br) (HBr likewise: 90%), 3-β-methoxyethyl- (738, n = 2, R = OMe) (MeOH, TsOH, reflux, 1 h: 71%), or 3-β-anilinoethyl-2,4(1H,3H)-quinazolinedione (738, n = 2, R = NHPh) (PhNH₂, TsOH, PhH, reflux, 90 min: 39%);[1428] in the similar conversion of 3,4-dihydro-2H,6H-[1,3]oxazino[2,3-b]quinazolin-6-one (737, n = 3) into 3-γ-bromopropyl-2,4(1H,3H)-quinazolinedione (738, n = 3, R = Br) (HBr, 95°, 5 min: 93%) or analogs with other γ-substituents;[1435] and in the apparent rearrangement of 3,3,10a-trimethyl-2-methylene-2,3,10,10a-tetrahydro-5H-oxazolo[2,3-b]quinazolin-5-one (739) into 3-(α-acetyl-α-methylethyl)-2-methyl-4(3H)-quinazolinone (740) in alkali.[2219] Analogous systems have also been used as substrates.[1476]

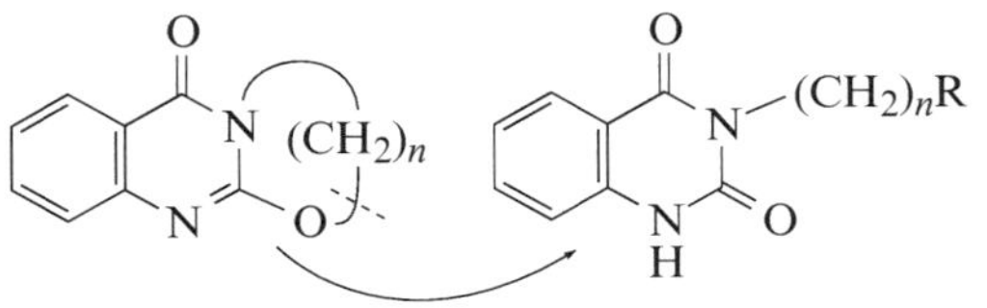

(737)   (738)

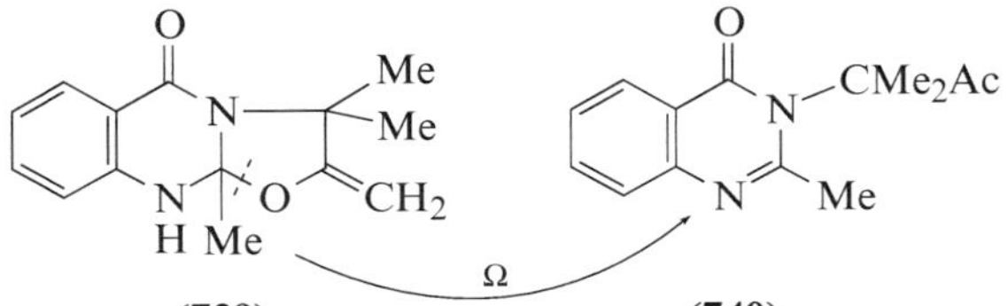

(739)   (740)

1.7.7. Pyrazolo[1,5-c]quinazolines as Substrates

1,10b-Dihydropyrazolo[1,5-c]quinazoline-5(6H)-thione (**741**), easily made from o-isothiocyanatocinnamaldehyde and hydrazine, underwent ring fission and subsequent hydrolysis of an intermediate nitrile on dissolution in refluxing aqueous ethanolic alkali during 4 h, to give 4-carboxymethyl-3,4-dihydro-2(1H)-quinazolinethione (**742**) (~ 80%).[975]

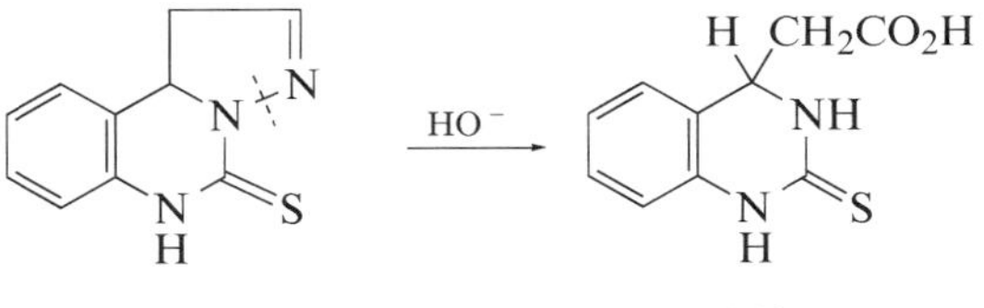

(741)   (742)

1.7.8. Pyrido[1,2-a]quinazolines as Substrates

5-Methyl-6-oxo-5,6-dihydropyrido[1,2-a]quinazolin-11-ium iodide (**743**) underwent facile ring fission by reacting at its activated 1-position with pyrrolidine to afford 3-methyl-2-[4-(pyrrolidin-1-yl)buta-1,3-dienyl]-4(3H)-quinazolinone (**744**) in 81% yield;[2220] 6-methoxypyrido[1,2-a]quinazolin-11-ium tetrafluoroborate reacted somewhat similarly with pyrrolidine to give 68% of 4-(pyrrolidin-1-yl)-2-[4-(pyrrolidin-1-yl)buta-1,3-dienyl]quinazoline (**745**), in which the methoxy group had suffered incidental aminolysis.[2220]

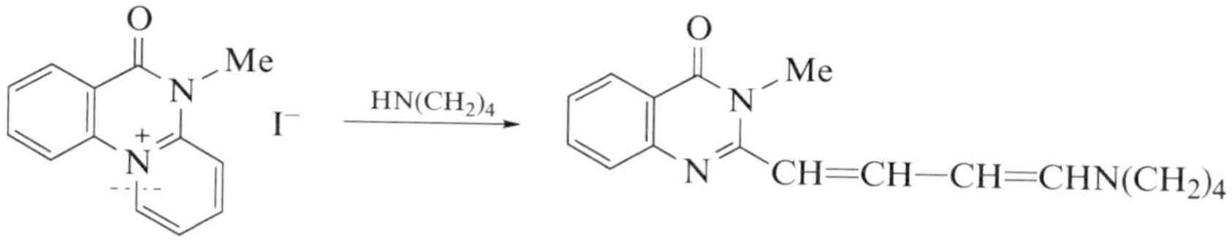

(743)   (744)

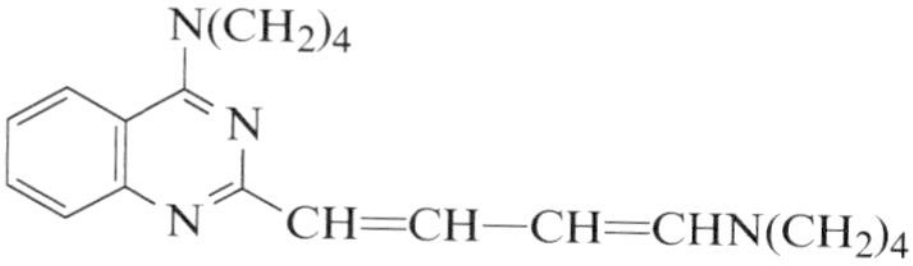

(745)

1.7.9. Pyrido[2,1-*b*]quinazolines as Substrates

6-Benzoyl-5-methyl-7,8-dihydro-9*H*-pyrido[2,1-*b*]quinazolin-11(5*H*)-one (**746**) underwent ring fission by refluxing in aqueous ethanol for 3.5 h to give 3-δ-benzoylbutyl-1-methyl-2,4(1*H*,3*H*)-quinazolinedione (**747**) in 75% yield;[7,29] the corresponding ethoxycarbonylbutyl product was made similarly.[29,cf. 7]

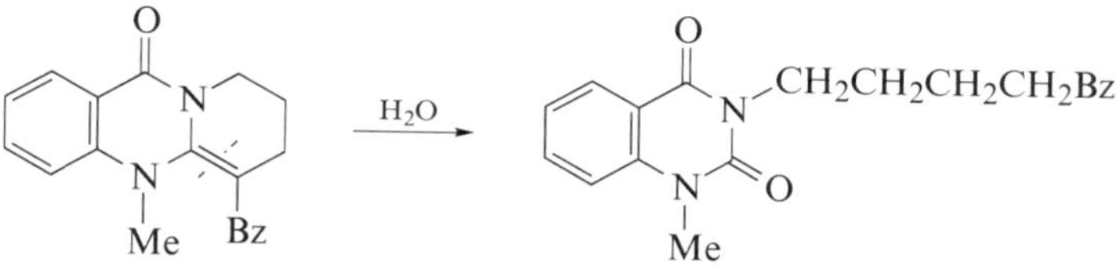

(746)   (747)

1.7.10. Pyrimido[1,6,5-*de*]-1,2,4-benzothiadiazines as Substrates

8-Chloro-3,5-diphenyl-2,3-dihydro-5*H*-pyrimido[1,6,5-*de*]-1,2,4-benzothiadiazin-7(6*H*)-one 1,1-dioxide (**748**) was easily made; it suffered hydrolysis (with loss of benzaldehyde) in boiling aqueous dimethylformamide during 10 min to afford a 50% yield of 5-chloro-4-oxo-2-phenyl-1,2,3,4-tetrahydro-8-quinazolinesulfonamide (**748a**).[1384]

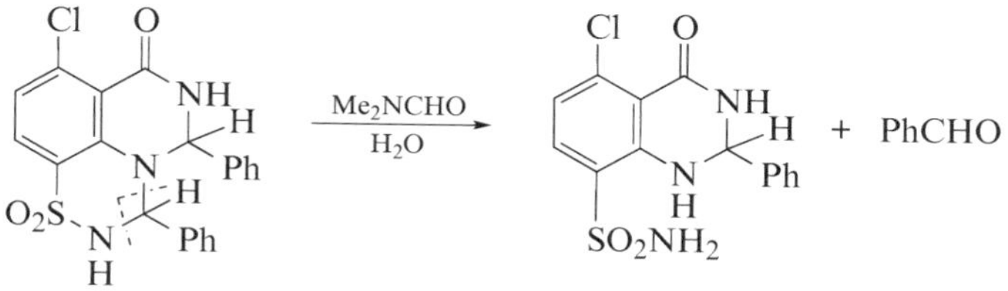

(748)   (748a)

1.7.11. Pyrrolo[1,2-*a*]quinazolines as Substrates

Although of little practical value, 7-chloro-5-hydroxy-5-phenyl-1,2,3,5-tetrahydropyrrolo[1,2-*a*]quinazolin-1-one (**749**) suffered a net isomerization into 2-β-carboxyethyl-6-chloro-4-phenylquinazoline (**749a**) in ~45% yield by brief treatment in warm alkali.[220]

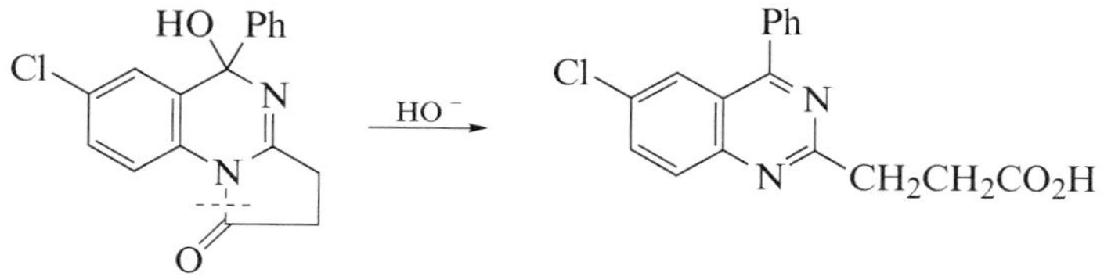

(749)   (749a)

1.7.12. Pyrrolo[2,1-*b*]quinazolines as Substrates

1,2,3,9-Tetrahydropyrrolo[2,1-*b*]quinazoline-1,9-dione(**750**) was easily made, and on refluxing in methanol (for 14 h), ethanol (for 14 h), or ethanolic methylamine (for 1 h), it gave quantitative yields of 2-β-methoxycarbonylethyl- (**751**, R = OMe), 2-β-ethoxycarbonylethyl- (**751**, R = OEt), or 2-β-*N*-methylcarbamoylethyl-4(3*H*)-quinazolinone (**751**, R = NHMe), respectively;[904] rather similarly, the further reduced substrate, 3a-methyl-1,2,3,3a,4,9-hexahydropyrrolo[2,1-*b*]quinazoline-1,9-dione (**752**), was converted by boiling in alkali during 20 min into 2-β-carboxyethyl-2-methyl-1,2-dihydro-4(3*H*)-quinazolinone in 95% yield;[279] and other examples of such fission reactions have been reported.[7,29,1694]

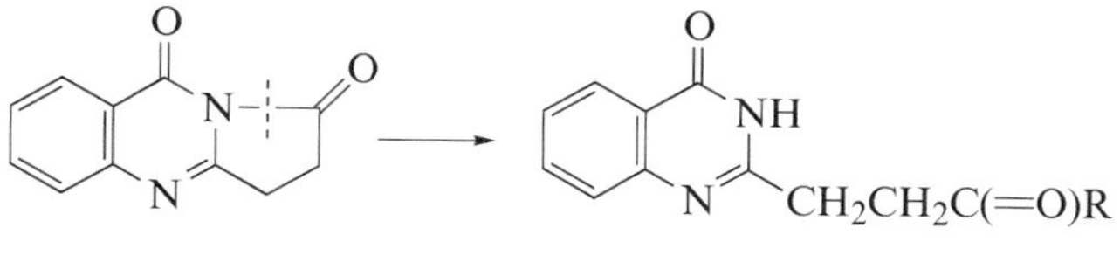

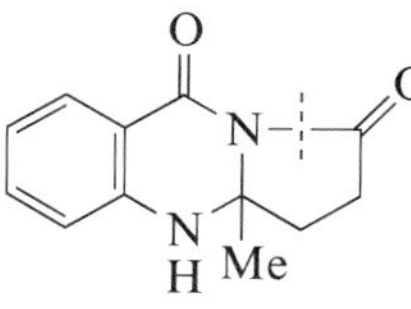

(750)   (751)   (752)

1.7.13. Tetrazolo[1,5-*c*]quinazolines as Substrates

This type of substrate is represented only in the hydrolysis of 9-bromo-5-phenyltetrazolo[1,5-*c*]quinazoline by boiling 5M hydrochloric acid during 3 h to afford 6-bromo-2-phenyl-4(3*H*)-quinazolinone in 5% yield, along with other nonquinazoline products.[749]

1.7.14. [1,3,4]Thiadiazino[3,2-c]quinazolin-5-iums as Substrates

9,10-Dimethoxy-6-methyl-3,4-dihydro-3-oxo-2H-[1,3,4]thiadiazino-[3,2-c]quinazolin-5-ium (754) (with its gegenion, A⁻) was fairly easily made; on alkaline hydrolysis it gave mainly 3-α-mercaptoacetamido-6,7-dimethoxy-2-methyl-4(3H)-quinazolinone (753) and 3-amino-6,7-dimethoxy-2-methyl-4(3H)-quinazolinone (755, X = O), whereas on photolysis the same substrate (754) gave mainly 3-amino-6,7-dimethoxy-2-methyl-4(3H)-quinazolinethione (755, X = S).[1058]

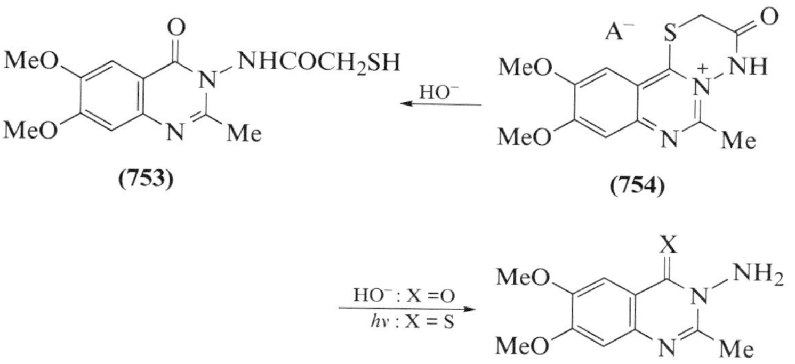

(753)   (754)

(755)

1.7.15. 1,3,4-Thiadiazolo[2,3-b]quinazolines as Substrates

This synthesis is represented only by the fission of 2-phenyl-2,3-dihydro-5H-1,3,4-thiadiazolo[2,3-b]quinazolin-5-one(756) in warm aqueous alkali for 25 min to give 3-amino-2-thioxo-1,2-dihydro-4(3H)-quinazolinone (757) (~40% yield) and benzaldehyde (as its dinitrophenylhydrazone).[129]

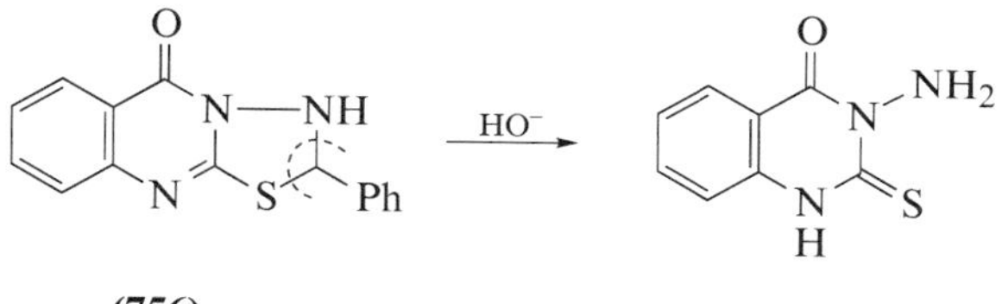

(756)   (757)

1.7.16. 1,3,4-Thiadiazolo[3,2-c]quinazolines as Substrates

The only example of this synthesis appears to be the conversion of 2-(3,4-dimethoxyphenyl)-8,9-dimethoxy-5-methyl-10bH-1,3,4-thiadiazolo[3,2-c]quinazoline (758) into 6,7-dimethoxy-2-methyl-4(3H)-quinazolinethione (759) in ≤51% yield by irradiation of an ethanolic solution under nitrogen for ~2 h.[1354]

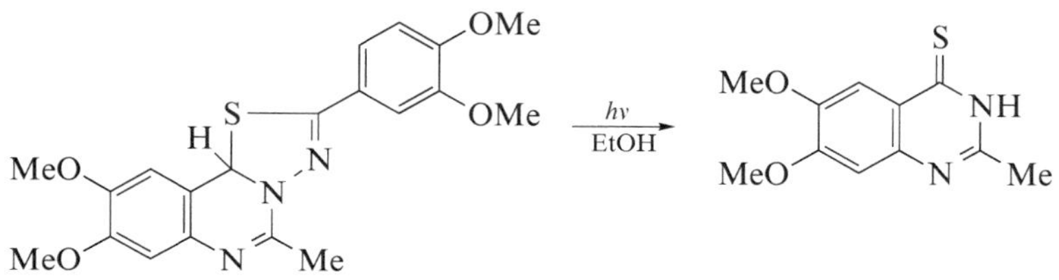

(758)   (759)

1.7.17.   [1,4]Thiazino[3,4-*b*]quinazolines as Substrates

1,3,4,6-Tetrahydro[1,4]thiazino[3,4-*b*]quinazolin-6-one   (**760**)   underwent
oxidative ring fission in peroxyacetic acid at 20° during 18 h to furnish 3-*β*-
sulfoethyl-4(3*H*)-quinazolinone (**761**) in 47% yield (with loss of CO_2?).[1854]

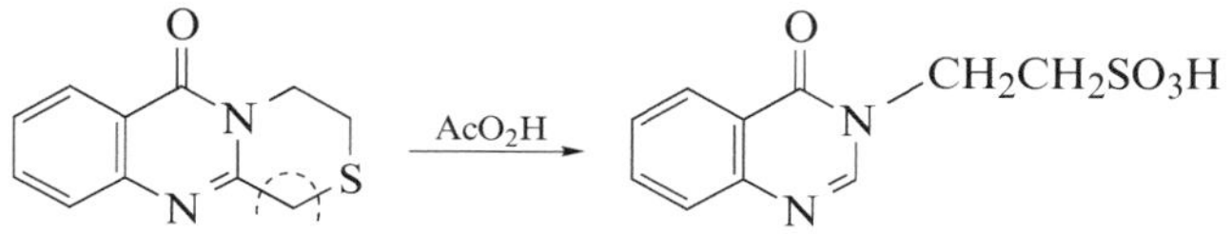

(760)   (761)

1.7.18.   Thiaziolo[4,3-*b*]quinazolines as Substrates

3,9-Dihydro-1*H*-thiazolo[4,3-*b*]quinazolin-9-one (**762**) was converted into
44% of 3-sulfomethyl-4(3*H*)-quinazolinone (**763**) by treatment with peroxyacetic
acid at 20° for 18 h.[1854]

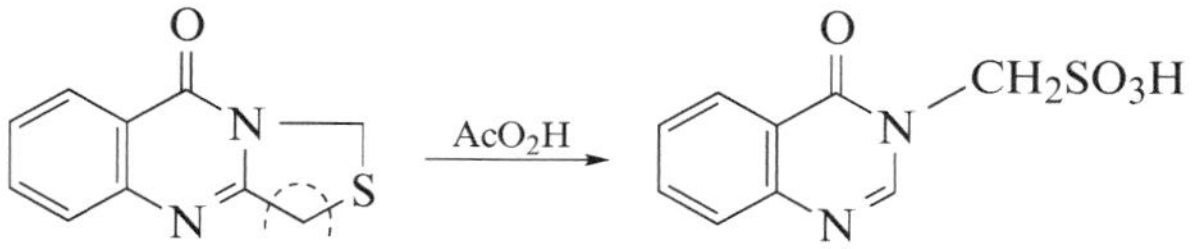

(762)   (763)

1.7.19.   Thioxanthenes as Substrates

Treatment of 1-amino-4-methyl-9*H*-thioxanthen-9-one 10,10-dioxide (**764**)
with formic acid–formamide at 180° for 3 h gave a mixture of products from
which 6-methyl-4-phenylquinazoline (**765**) was isolated in low yield.[633]

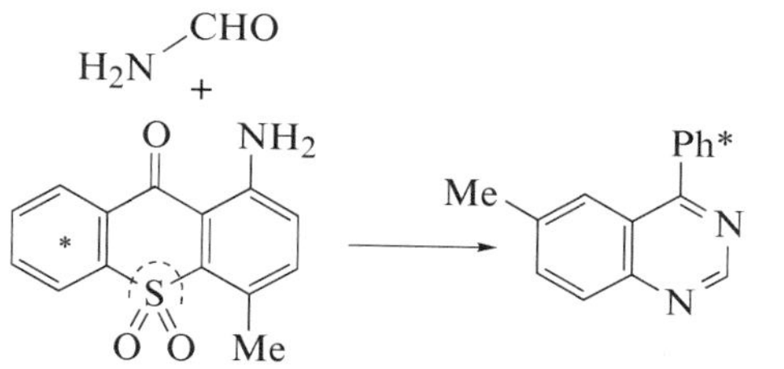

(764)   **(765)**

1.7.20. [1,2,3]Triazolo[1,5-*a*]quinazolines as Substrates

Substrates such as the triazoloquinazolines (**766**, R = H or Ph) are quite easily made and may be converted into useful quinazolines.[154] For example, [1,2,3]triazolo[1,5-*a*]quinazolin-5-amine (**766**, R = H) gave 2-hydroxymethyl-4(3*H*)-quinazolinone (**767**, R = H) (H$_2$SO$_4$, H$_2$O–EtOCH$_2$CH$_2$OH, reflux, 3 h: >60%);[426] the 3-phenyl substrate (**766**, R = Ph) likewise gave 2-α-hydroxybenzyl-4(3*H*)-quinazolinone (**767**, R = Ph) (H$_2$SO$_4$, H$_2$O–EtOH, reflux, 2 h: >70%) but, when the reaction was terminated after 30 min, a separable mixture of the same product (**767**, R = Ph) and 2-α-hydroxybenzyl-4-quinazolinamine (**768**) was obtained;[426] and the simple substrate (**766**, R = H) with acetyl bromide in refluxing acetic acid for 4 h gave 4-acetamido-2-bromomethylquinazoline (**769**) in ~80% yield.[426]

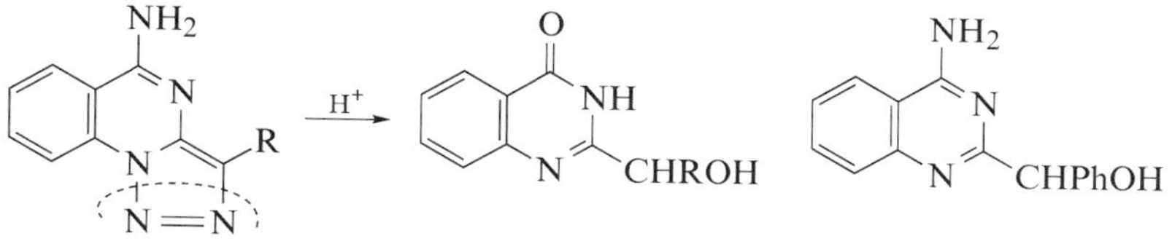

(766)   **(767)**   **(768)**

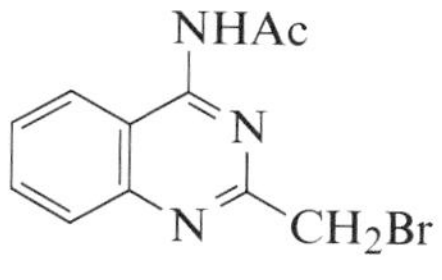

(769)

1.8.   FROM A HETEROTETRACYCLIC SUBSTRATE

A few quinazolines have been made from heterotetracyclic intermediates, but naturally the categories are small in both number and content.

1.8.1.   Benzothiazolo[3,2-a]quinazolines as Substrates

5H-Benzothiazolo[3,2-a]quinazolin-5-one (**770**) underwent desulfurization slowly on stirring with an excess of Raney-nickel catalyst in refluxing ethanol to give 1-phenyl-4(1H)-quinazolinone (**771**) in poor yield.[1467]

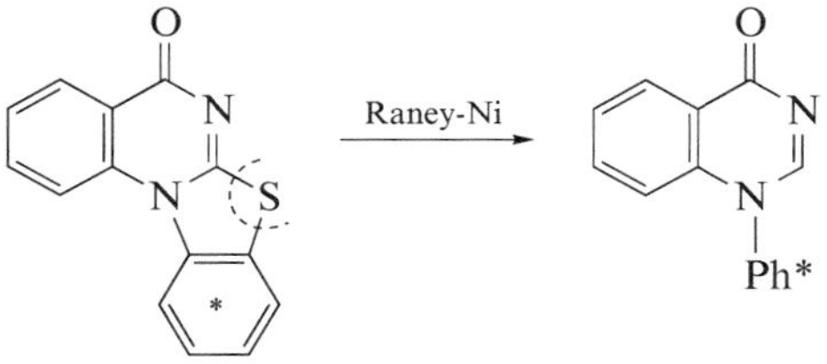

(**770**)   (**771**)

1.8.2.   Benzothiazolo[3,2-c]quinazolin-7-iums as Substrates

8,9,10,11-Tetrahydrobenzothiazolo[3,2-c]quinazolin-7-ium perchlorate (**772**) suffered hydrolysis in 0.5M sodium hydroxide to give two products, one of which proved to be 3-(2-mercaptocyclohex-1-enyl)-4(3H)-quinazolinone (**773**).[730]

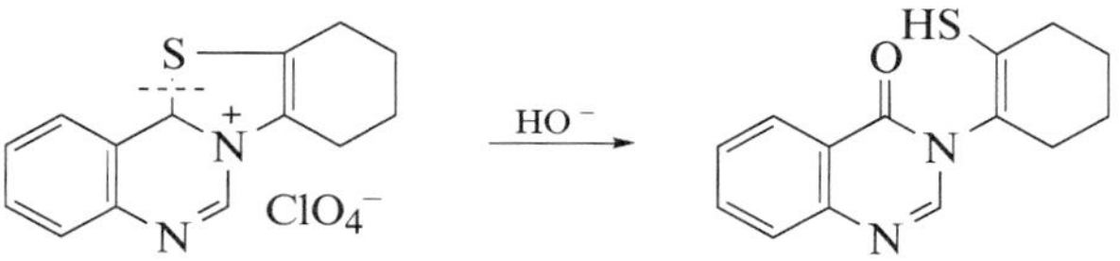

(**772**)   (**773**)

1.8.3.   [1]Benzothiopyrano[4,3,2-de]quinazolines as Substrates

6-Methyl[1]benzothiopyrano[4,3,2-de]quinazoline 7,7-dioxide (**774**) and formamide at 200° for 2h gave a separable mixture of products from which 6-methyl-4-phenylquinazoline (**775**) and di(6-methyl-4-phenylquinazolin-5-yl) disulfide (**776**) (or an isomer) were isolated in low yields.[633]

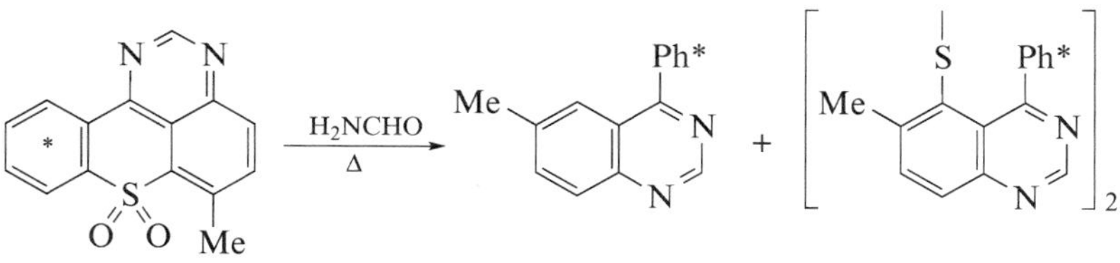

(**774**)   (**775**)   (**776**)

1.8.4. Benzoxazolo[3,2-*a*]quinazolines as Substrates (cf. *H* 128)

In refluxing ethanolic alkali for 1 h, 10-chloro-5*H*-benzoxazolo-[3,2,-*a*]quinazolin-5-one (**778**) underwent alcoholysis to give 55% of 1-(3-chloro-6-hydroxyphenyl)-2-ethoxy-4(1*H*)-quinazolinone (**777**), but, in refluxing aqueous alkali for 2 h, the same substrate (**778**) suffered hydrolysis to afford 1-(3-chloro-6-hydroxyphenyl)-2,4(1*H*,3*H*)-quinazolinedione (**779**) in 77% yield.[1463]

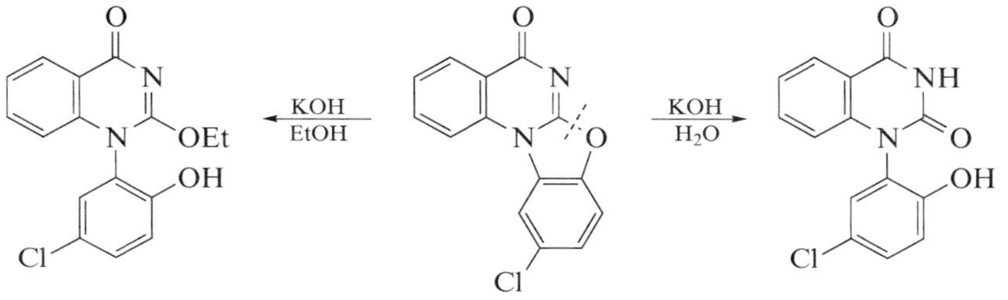

(777)　　　　　　　　(778)　　　　　　　　(779)

1.8.5. Quinazolino[3,2-*c*][1,2,3]benzotriazines as Substrates

Although it could not be described as a general or useful synthetic route, several such quinazolinobenzotriazines have been converted into quinazolines. Thus 8*H*-quinazolino[3,2-*c*][1,2,3]benzotriazin-8-one (**781**) in refluxing aqueous ethanol for 15 h gave 2-phenyl-4(3*H*)-quinazolinone (**780**) in ~80% yield;[174, cf. 76] the same substrate (**781**) in warm ethanolic piperidine for a few hours gave 2-*o*-(piperidinoazo)phenyl-4(3*H*)-quinazolinone (**782**);[950] and 8*H*-quinazolino[3,2-*c*][1,2,3]benzotriazin-8-imine (**783**) in refluxing neat piperidine or morpholine for 1–2h gave 2-*o*-(piperidinoazo)phenyl- (**784**, X = CH₂) or 2-*o*-(morpholinoazo)phenyl-4-quinazolinamine (**784**, X = O), both in ~90% yield.[427]

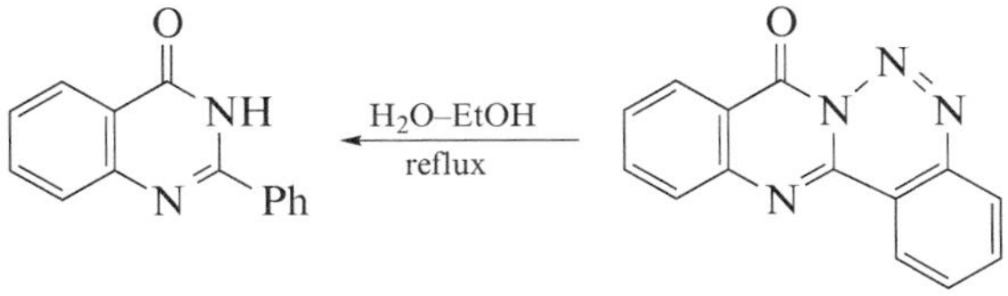

(780)　　　　　　　　　　　(781)

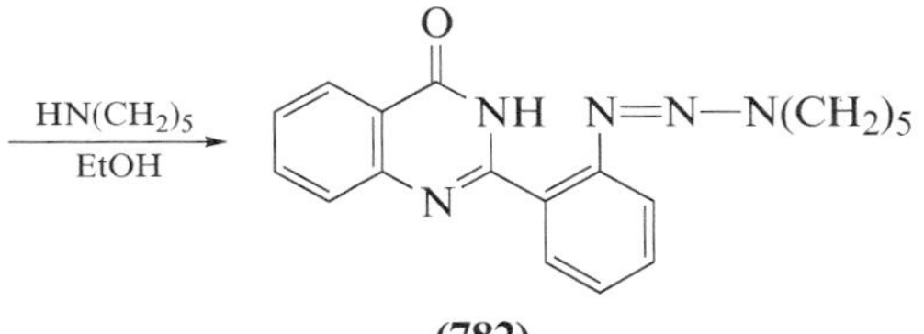

(782)

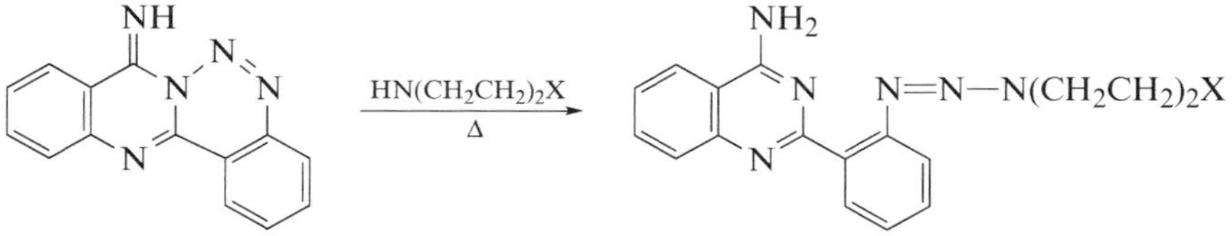

(783)   **(784)**

1.8.6. Quinazolino[3,2-*a*][3,1]benzoxazines as Substrates

5*H*,12*H*-Quinazolino[3,2-*a*][3,1]benzoxazin-5,12-dione (**785**) was easily made and reacted with methylamine in dioxane to give 3-*o*-(*N*-methylcar-bamoyl)phenyl-2,4-(1*H*,3*H*)-quinazolinedione (**786**) in 94% yield; other such amides were made similarly.[270]

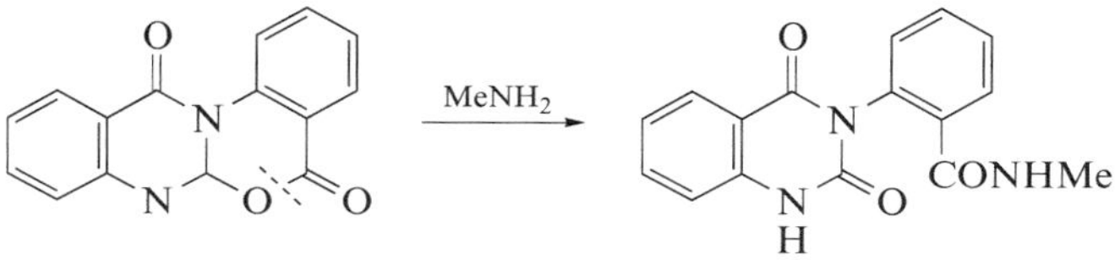

(785)   **(786)**

1.8.7. Quinazolino[3,2-*c*][2,3]benzoxazines as Substrates

5,8-Dihydroquinazolino[3,2-*c*][2,3]benzoxazin-8-one (**787**) was quite easily made, and on reduction with zinc in acetic acid it gave 2-*o*-(hy-droxymethyl)phenyl-4(3*H*)-quinazolinone (**788**).[465]

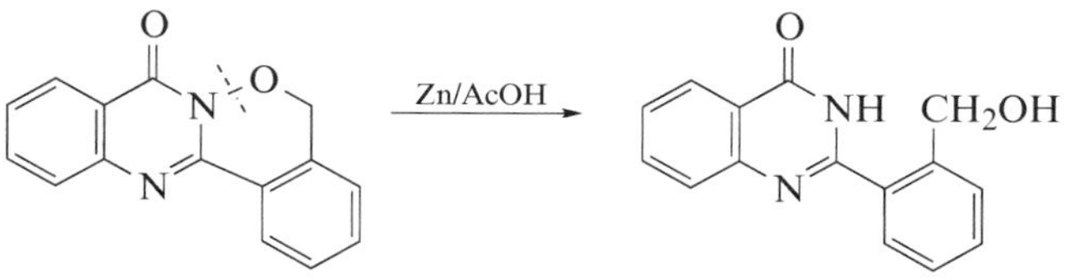

(787)   **(788)**

1.8.8. Quinazolino[1,2-*a*]quinazolines as Substrates

6a-Chloromethyl-6a*H*-quinazolino[1,2-*a*]-quinazoline-5,8(6*H*,7*H*)-dione (**789**) underwent dechlorination and reductive fission on catalytic hydrogenation to afford 1-*o*-carbamoylphenyl-2-methyl-4(1*H*)-quinazolinone (**789a**) (H$_2$, Pd/C, AcONa, EtOH: 68%);[1357] the method appears to have little potential.

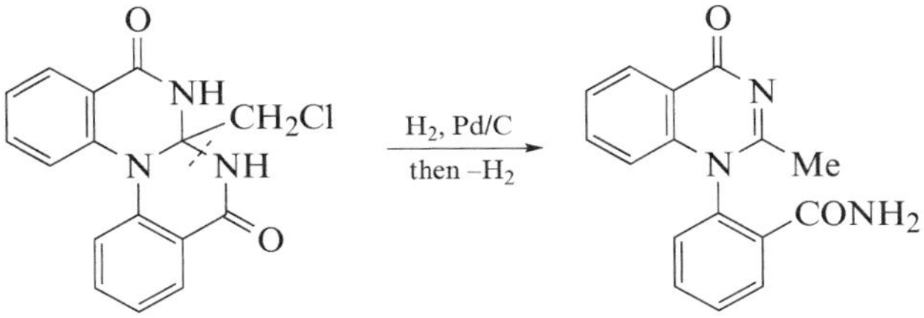

(789)   **(789a)**

1.9. FROM A HETEROPENTACYCLIC SUBSTRATE

Only two heteropentacyclic systems appear to have been used as substrates for the formation of quinazolines.

1.9.1. Benz[4,5]isoquino[2,1-*a*]quinazolines as Substrates

5*H*,13*H*-Benz[4,5]isoquino[2,1-*a*]quinazoline-5,13-dione (**790**) in refluxing aqueous alkali for 2 h gave 2-(8-carboxynaphthalen-1-yl)-4(3*H*)-quinazolinone (**790a**) in > 70% yield;[695] the solid product was reconverted into the starting material (**790**) on prolonged warming.[695]

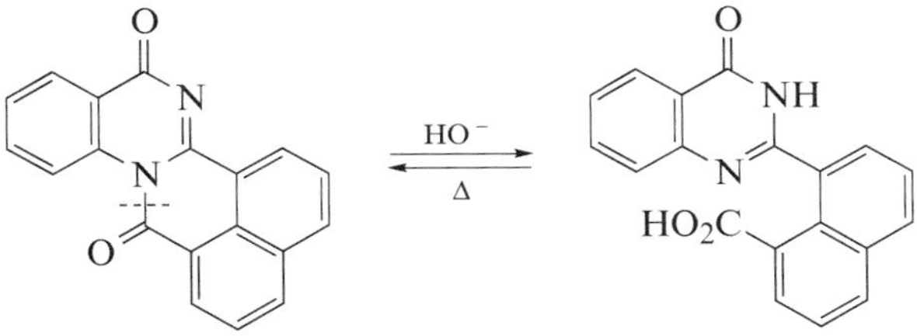

(790)   **(790a)**

1.9.2. Indolo[2',3':3,4]pyrido[2,1-*b*]quinazolines as Substrates

5,7,8,13-Tetrahydroindolo [2',3':3,4]pyrido[2,1-*b*]quinazolin-5-one (**791**) suffered fission of its 6,7-bond to afford 68% of 2-(3-vinylindol-2-yl)-4(3*H*)-quinazolinone (**792**) when treated with sodium hydride in dimethylformamide at 100° for 22 h:[2026] compare with analogous substrates.[445]

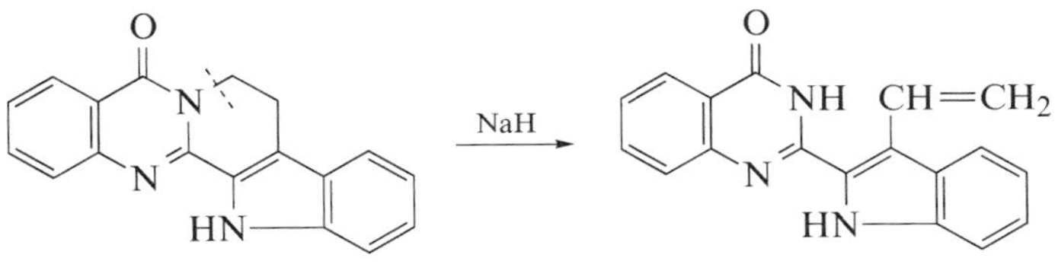

(791)   **(792)**

1.10.   FROM A SPIRO HETEROCYCLIC SUBSTRATE

This uncommon route to quinazolines is exemplified by three diverse reports. 3-Methyl-1,2,3,4-tetrahydroquinazoline-2-spiro-2'-tetrahydrofuran(**793**) was reduced to 2-γ-hydroxypropyl-3-methyl-1,2,3,4-tetrahydroquinazoline (**794**) in 70% yield on treatment in isopropyl alcohol with sodium borohydride for 2h:[594] 2-γ-hydroxypropyl-3-methyl-1,2-dihydro-4(3H)-quinazolinone was made similarly (50%) from the appropriate substrate.[594] 5,5'-Propano-4,4'-spirobi(2-oxo-6-phenylhexahydropyrimidine) (**795**) in a refluxing mixture of hydrochloric acid and acetic acid for 90 min gave 4-phenyl-5,6,7,8-tetrahydro-2,8(1H)-quinazolinedione (**796**) (isolated as its dinitrophenylhydrazone), benzaldehyde (also isolated as its DNPH), and urea (apparently not isolated).[743] 2-Phenyl-4H-3,1-benzothiazine-4-spiro-2'-(4',5',6',7'-tetrachloro-1',3'-benzodioxole) (**797**) reacted with aniline or other aromatic amines to give 2,3-diphenyl-4(3H)-quinazolinethione (**799**, R = Ph) (no details: 77%) or analogs, respectively, possibly via the imines (**798**).[345] Other such specialized syntheses have also been observed.[73,1023,1399,1405,2368]

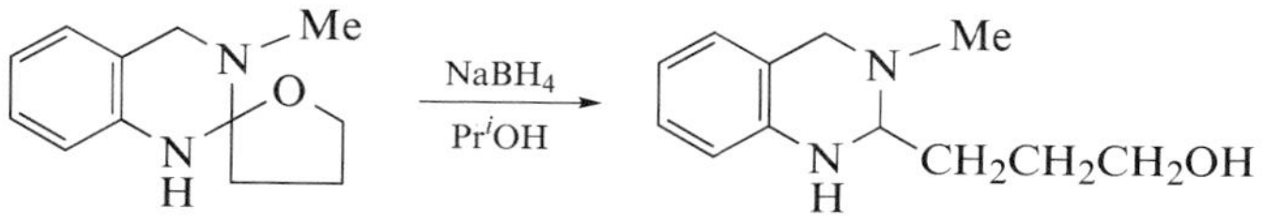

(**793**)   (**794**)

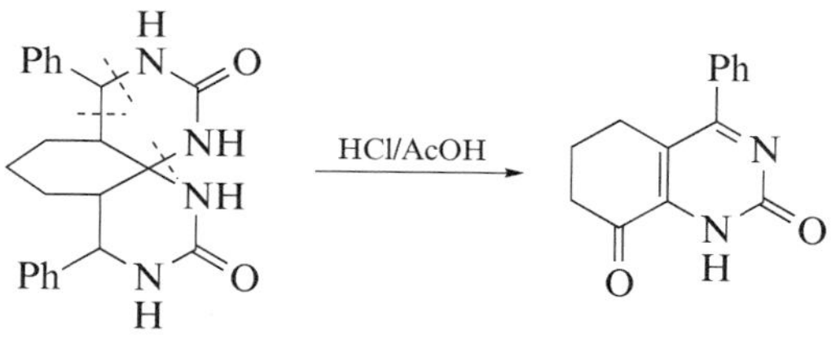

(**795**)   (**796**)

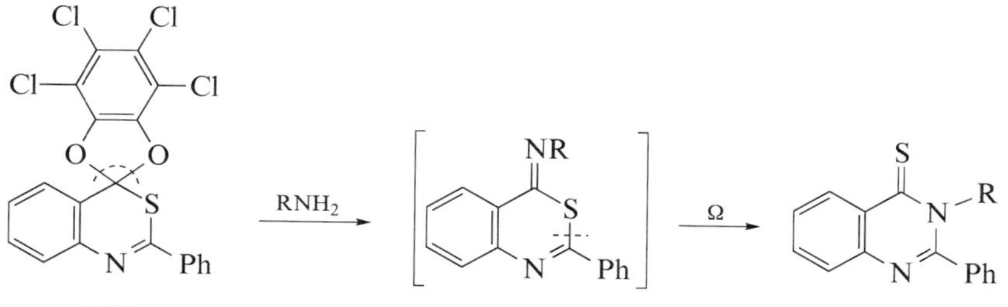

(**797**)   (**798**)   (**799**)

CHAPTER 2

Quinazolines, Alkylquinazolines, and Arylquinazolines

This chapter deals with the preparations, properties, and reactions of quinazoline and its simple *C*-alkyl, *N*-alkyl, and *C*-aryl derivatives. It also includes such data on reduced quinazolines in the foregoing categories, methods for introducing *C*-alkyl or *C*-aryl groups into quinazolines or hydroquinazolines already bearing other substituents (functional or otherwise), and reactions that affect alkyl or aryl groups in such entities. For the obvious pragmatic reason, the term *alkylquinazoline* is deemed to include alkyl, alkenyl, alkynyl, aryl, aralkyl, and heteroaryl quinazolines.

2.1. QUINAZOLINE (*H* 11)

Relatively little has been added to our knowledge of quinazoline and its hydro derivatives during the last 25 years. Such new data are summarized briefly in the following subsections.

2.1.1. Preparation of Quinazoline (*H* 11,34,36)

Apart from primary syntheses (Chapter 1; see also Index), quinazoline (2) has been made by oxidation of 4-hydrazinoquinazoline (1) with a stream of oxygen through an alkaline ethanolic solution at 25° for 2 h (60% yield) or with silver oxide in a refluxing aqueous solution for 1 h (33%);[184] by the reaction of quinazoline 3-oxide (3) with 2-butanone (145°, sealed, 9 h: 8% plus other products);[372] or by alkaline hydrolysis of the Reissert compound, 3-benzoyl- 3,4-dihydro-4-quinazolinecarbonitrile (4) (3% NaOH, MeOH–H$_2$O, 20°, 1 h: 62%).[1419]

Both 1,2- and 3,4-dihydroquinazoline have been obtained by reduction of quinazoline (2), the first (5) using sodium borohydride (F$_3$CCO$_2$H, THF, 20°, 1 h; 85%)[1123,1359] and the second (6) by hydrogenation (Pt, EtOH, 20°, dark, 10 days: 69%).[627] Only *trans*-decahydroquinazoline (see Section 2.1.2) was obtained on reduction of 5,6,7,8-tetrahydroquinazoline with sodium/ethanol (reflux, 5 h: 57%, initially as picrate).[179]

NHNH$_2$ [O]   H CN   Bz

(1)   (2)   (3)   (4)

NaBH$_4$ Pt/H$_2$ dimerization

[O] [H]

(5)   (6)   (7)   (8)

The quinazoline dimer, 4,4′(3H,3′H)-biquinazolinylidene (7), has been made conveniently by dissolution of quinazoline (2) in dimethyl sulfoxide containing potassium cyanide (20°, N$_2$, 12 h: 70%);[1768] subsequent oxidation gave 4,4′-biquinazoline (8) (heating in air, long standing in air, or treatment with alkaline ferricyanide: unstated yields).[1768,cf.933] The oxidized product (8) has also been made by simply heating quinazoline 3-oxide (sealed, 145°, 10 h: 11%),[372] by treatment of quinazoline with phenylacetonitrile (MeONa MeOH, 20°, 3 days: 45%),[162] by somewhat similar treatment of quinazoline with aldehydes (ArCHO, KCN, MeOH, 5°, 15 h: poor yields),[373] or by electrolytic reduction of 4-methoxyquinazoline followed by aerial oxidation;[22] the redox system (7)⇌(8) has been studied polarographically.[22]

2.1.2.   Physical Properties of Quinazoline (H 13)

Revised X-ray diffraction data for quinazoline at 20° indicated a triclinic unit cell with two molecules in the asymmetrical unit: both molecules were identical in shape with a small but significant deviation from planarity.[700] A (−)-trans-decahydroquinazoline (made by reduction) has been shown to have a 4a-R,8a-S absolute configuration by analysis of its ^{1}Hnmr spectrum.[179]

The ^{1}Hnmr spectrum of quinazoline has been analyzed carefully in relation to those of its simple derivatives in order to correlate substituent effects with chemical shifts and to predict chemical shifts from coupling constants.[16,167] All ^{13}Cnmr resonances for 5,6,7,8-tetrahydroquinazoline have been assigned by a combination of techniques.[2251] The ^{14}Nnmr chemical shifts for quinazoline and related parent azaheterocycles have been shown to depend mainly on the π-charge density on the nitrogen atom(s).[48] A new measurement and assignment of ^{15}Nnmr resonances to N1 and N3 in quinazoline[1147] indicated that a previous assignment,[1138] based on calculated screening constants,[408] must be reversed; this has important implications for the ^{15}Nnmr of quinazoline N-oxides.[1147]

The preliminary electron-density diagram and dipole moment (2.4–2.55 D) for quinazoline (H 14) have been confirmed;[10] the complex uv spectrum of quinazoline vapor, due to $n \to \pi^*$ transitions, has been recorded and the principal bands have been assigned; the unpublished ms (mass spectrum) data for quinazoline (H 19) has been fully reported for comparison with those for key derivatives;[172] and pK_a values (3.5 and − 4.5) for quinazoline have been redetermined for comparison with those of related diazaheterocycles.[169]

2.1.3.   Reactions of Quinazoline (H 19,33)

Quinazoline undergoes variety of reactions, or which dimerization, oxidation to 4,4′-biquinazoline, and nuclear reduction have been covered in Section 2.1.2. Many of the remaining reactions occur via the addition of a reagent across the 3,4- or 1,2-bond, but it seems best to avoid mechanistic grouping here in favor of a more discursive treatment.

Covalent hydration.   Additional data on water addition to the 3,4-bond of quinazoline cation to afford the hydrate (**9**) has been confined to physicochemical aspects (kinetics, etc.),[170,484] to the ready detection of such hydration by ^{13}Cnmr spectroscopy,[338] to the occurrence of hydration in the neutral molecule of quinazoline,[168] and to the effects of substitution on hydration in quinazoline and its aza analogs.[70,168,170,338,484] Analogous 3,4-additions have been reviewed briefly.[2382]

(**9**)

Deuteration.   When quinazoline was heated in deuterium oxide containing a platinum catalyst (150°, sealed, 12 h), about 40% proton replacement occurred, predominantly at the 2- and 4-positions, in that order.[543] In contrast, a kinetic study of the deuteration of quinazoline in MeOK/MeOH (50–180°, sealed) indicated that exchange occurred more quickly at the 4- than at the 2-position.[527]

Quaternization.   Contrary to previous reports (H 35), quaternization of quinazoline with methyl iodide (EtOH, reflux, 4 h) produced both 1- (**10**) and 3-methylquinazolinium iodide (11) in a ratio of ∼ 1:5;[74] both underwent hydroxide attack at C4 to afford pseudobases, such as (**12**).[74,75] The addition of primary or secondary amines to 3-methylquinazolinium iodide (**11**) gave, for example, 4-anilino-3-methyl-3,4-dihydroquinazoline (**13**), whereas the addition of tertiary amines initially gave products such as 4-*p*-dimethylaminophenyl-3- methyl-3,4-

dihydroquinazolinium iodide (**14**).[796] Polarographic reduction of the 3-methyl-quinazolinium ion has been studied.[802]

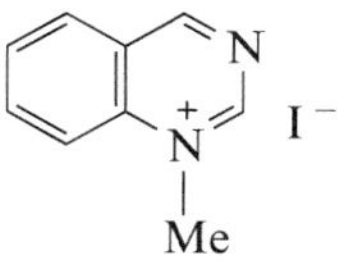

(**10**) 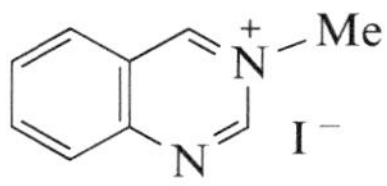   (**11**) 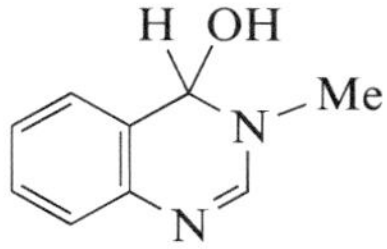   (**12**)

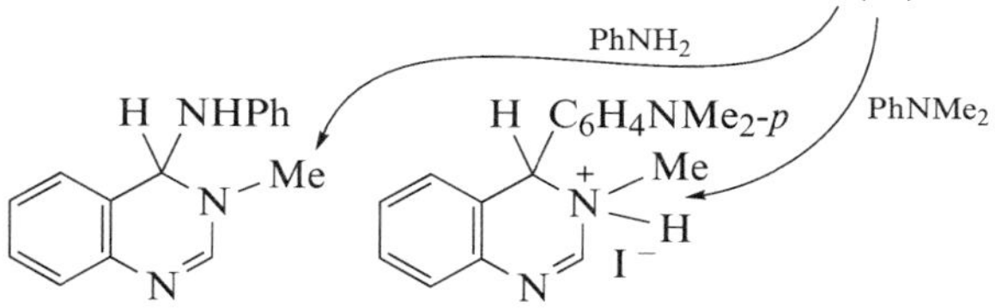

(**13**)   (**14**)

Oxidation.   The oxidation of quinazoline ordinarily yields 4,5-pyrimidinedicarboxylic acid (**15**) (using permanganate) or 4(3*H*)-quinazolinone (**16**) (using peroxyacids) (*H* 33). However, it has now been reported that oxidation in aqueous hydrogen peroxide containing sodium tungstate (40°, 6 h) gave a chromatographically separable mixture of quinazoline 1- (**17**) and 3-oxide (**18**) as well as the quinazolinone (**16**) in 5%, 5%, and 10% yields, respectively.[546] Unlike 4-methyl, which gives its 3-oxide, treatment of quinazoline with hydroxylamine *O*-sulfonic acid (H_2NOSO_3H) gave only the 3,4-adduct (**19**) and thence degradation products.[559] Quinazoline underwent biological oxidation in growing *Pseudomonas putida* to give 4(3*H*)-quinazolinone (**16**) and the more interesting products, *cis*-5,6-dihydro-5,6-quinazolinediol (**20**) and *cis*-5,6,7,8- tetrahydro-5,6-quinazolinediol.[2432]

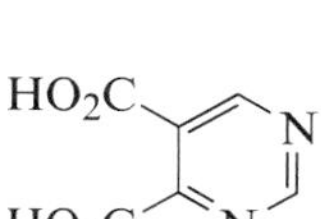

(**15**) 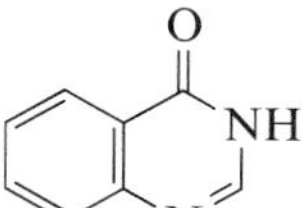   (**16**) 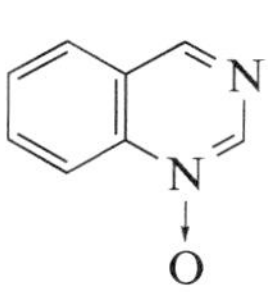   (**17**)

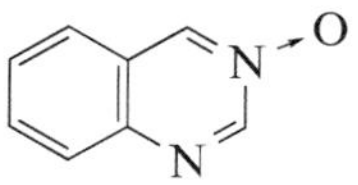

(**18**) 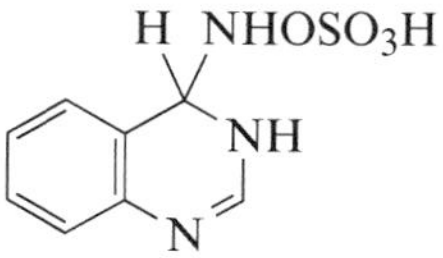   (**19**) 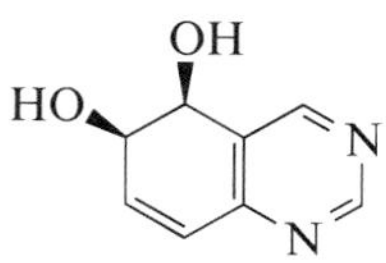   (**20**)

C-Alkylation or arylation.   Most reported alkylations of quinazoline have occurred at the 4-position by addition and subsequent oxidation. Thus treatment of quinazoline with isopropylmagnesium bromide in THF followed by hydrolysis in aqueous ammonium chloride and subsequent oxidation of the 3,4-dihydro product with ferricyanide gave 4-isopropylquinazoline (**21**) in ~ 40% yield;[56] 4-aryl- and 4-(substituted-alkyl)quinazolines were made similarly.[56] The addition of aromatic hydrocarbons to quinazolines has also been used to achieve 4-arylation; for example, stirring quinazoline with anthracene in aqueous alkaline ferricyanide for 5 days gave 4-(anthracen-9-yl)quinazoline (**22**) in 47% yield;[891,1028] the intermediate 3,4-dihydro products have been isolated prior to oxidation[1028,1084] and substituted hydrocarbons have been used quite effectively (in two stages) to give, for example, 4-*p*-methoxyphenylquinazoline (**23**, R = OMe) (89%)[1028] and 4-*p*-aminophenylquinazoline (**23**, R = NH$_2$) (20%).[829] 4-(Substituted-alkyl)quinazolines have also been made from quinazoline with carbanionic reagents[162,562] or by the Reformatsky reaction,[2450b] but yields were generally very poor. Treatment of quinazoline with *t*-butyl radicals (made *in situ* from pivalic acid, silver nitrate, and aqueous ammonium persulfate[159]) gave, not the expected mixture of 2- and 4-*t*-butylated quinazolines, but 2-*t*-butyl-4(3*H*)-quinazolinone (**24**) (> 95%) via the covalent hydrate of quinazoline.[1094] Treatment of quinazoline with ethyl chloroformate and allyltributyltin resulted in addition and *N*-alkylation to afford ethyl 4-allyl-3,4-dihydro-3-quinazolinecarboxylate (73%).[2538]

(21)   (22)   (23)   (24)

Reissert reaction.   Under classic Reissert conditions (KCN, H$_2$O–CHCl$_3$, BzCl, 20°, 30 min), quinazoline gave only a separable mixture of benzene derivatives.[424,1420] However, the typical Reissert product, 3-benzoyl-3,4-dihydro-4-quinazolinecarbonitrile (**27**), has been made by a two-stage process: quinazoline (**25**) was converted into its known hydrogen cyanide adduct, 3,4-dihydro-4-quinazolinecarbonitrile (**26**), and thence into the benzoylated product (**27**) (BzCl, pyridine, 50°, 5 min: 82%).[1419,1738] The same product (**27**) was made more conveniently from quinazoline (**25**) in one stage by using trimethylsilyl cyanide under anhydrous conditions (Me$_3$SiCN, BzCl, CH$_2$Cl$_2$, trace AlCl$_3$, 20°, 48 h: 67%);[1135,1885] an excess of reagents gave 1,3-dibenzoyl-1,2,3,4-tetrahydro-2,4-quinazolinedicarbonitrile (**28**) in ⩽ 69% yield.[1424,2193] Analogous Reissert compounds, such as ethyl 4-cyano-3,4-dihydro-3-quinazolinecarboxylate (57%) and 1,3-diacetyl- (57%),[2193] 1,3-dicrotonoyl- (41%),[2422] or 1,3-dicinnamoyl-1,2,3,4-

tetrahydro-2,4-quinazolinedicarbonitrile(35%),[2422] were made similarly. Such products underwent a variety of useful reactions.[1419,1714,2193]

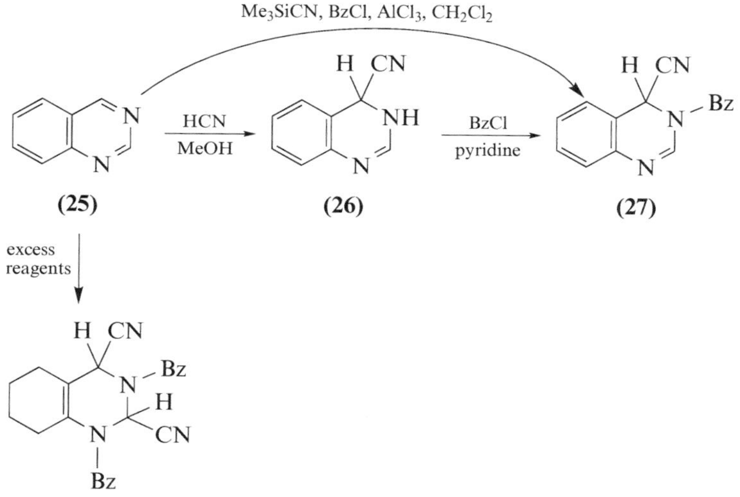

(25)   **(26)**   **(27)**

(28)

Amination. Dissolution of quinazoline in liquid ammonia containing sodium amide for 10 min, followed by the addition of potassium permanganate, gave 62% of 4-quinazolinamine (**30**), undoubtedly via the intermediate adduct (**29**); a small amount of 4,4'-diquinazolinylamine was also isolated.[1474]

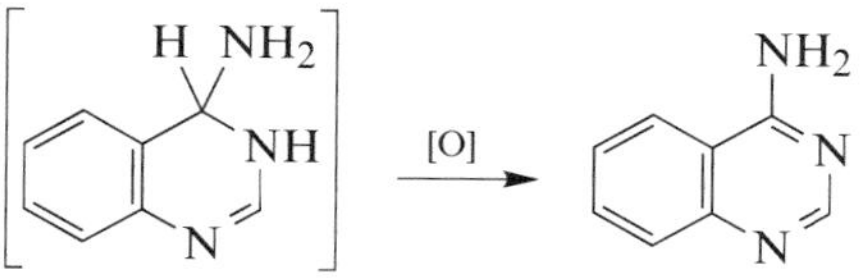

(29)   **(30)**

Acylation. Quinazoline reacted with acetic anhydride (reflux, 2 h) to afford the N-acyl adduct, 3-acetyl-3,4-dihydro-4-quinazolinol (**31**) (34%) accompanied by a separable byproduct, 2(1H)-quinolinone (18%);[1724] in contrast, quinazoline with nitroacetic acid at < 20° for 1 h gave the C-alkylated product, 4-nitromethyl-3,4-dihydroquinazoline as its hydrogen carbonate (**32**) (65%).[178] C-Acylation of quinazoline has been observed, albeit under rather peculiar conditions; thus treatment with benzaldehyde in dimethyl sulfoxide containing cyanide ion gave 4-benzoylquinazoline (**33**) in low yield, accompanied by its 3,4-dihydro precursor (**32b**), which was formulated as its tautomer, 4-α-hydroxybenzyl-quinazoline (**32a**).[790] 5,6,7,8-Tetrahydroquinazoline gave its 4-acetyl derivative (MeCHO, t-BuO₂H, FeSO₄, H₂O, 20°, 15 min: 34%).[828,1393] Reduction of quinazoline in the presence of benzyl chloroformate gave 1,3-dibenzyloxycarbonyl-1,2,3,4-tetrahydroquinazoline [PhCH$_2$O$_2$CCl, NaB(CN)H$_3$, MeOH, N$_2$, 20°, 16 h: 37%].[2066]

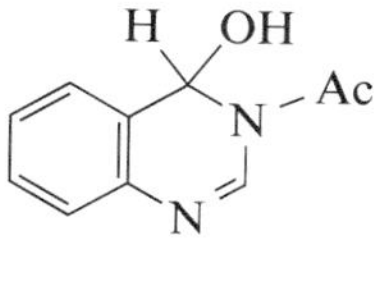

(31) 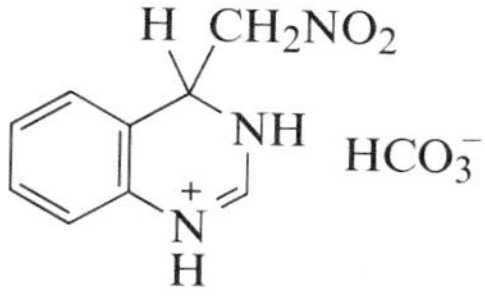   (32)

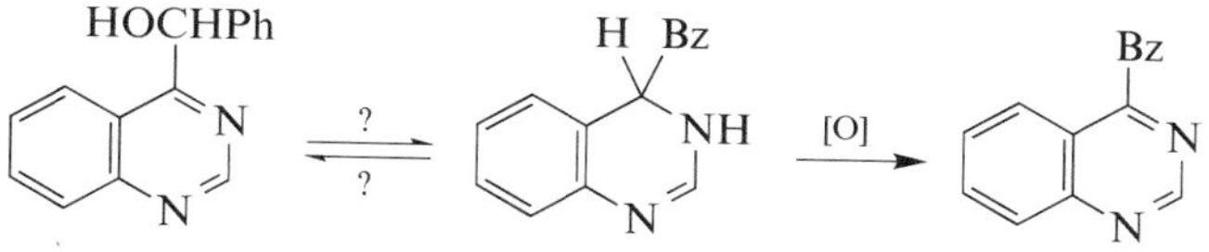

(32a)   (32b)   (33)

Conversion into quinolines. In addition to the instance mentioned in the preceding paragraph, quinazoline has been converted into 2-amino-3-quinolinecarbonitrile (**34**) by simple treatment with malononitrile (MeOH, 20°, 12 h: 52%;[419] neat, 75°, 15 min: 81%).[162]

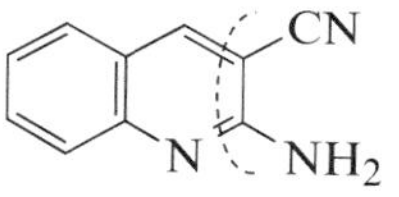

(34)

Conversion into a pyrroloquinazoline. Diphenylcyclopropenone (**35**) converted quinazoline into 2,3-diphenylpyrrolo[1,2-c]quinazolin-1-ol (**36**), simply by prolonged refluxing in methanol or dioxane.[2113]

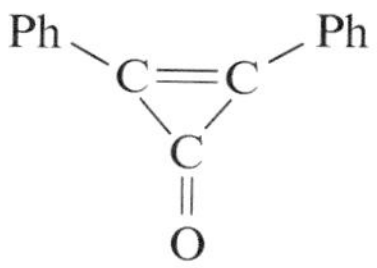

(35) 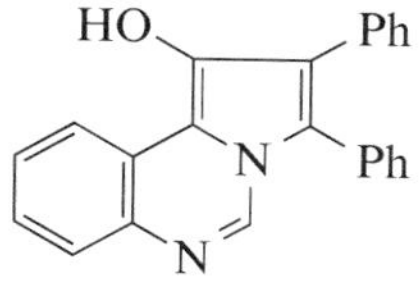   (36)

Degradation. Quinazoline underwent ring fission by mild acidic or alkaline treatment to afford *o*-aminobenzaldehyde and a variety of separable dimeric or polymeric products, some of which contained intact quinazoline entities.[175]

Nitrosation. Treatment of 5,6,7,8-tetrahydroquinazoline with nitrous acid under anhydrous conditions (PrONO, EtOH, HCl gas, < 5°, 2 h) gave 8-nitroso-5,6,7,8-tetrahydroquinazoline, formulated as 8-hydroxyimino-5,6,7,8-tetrahydroquinazoline (**37**, X = NOH), in 63% yield; this was subsequently converted into its parent ketone, 5,6,7,8-tetrahydro-8-quinazolinone (**37**, X = O) by boiling with pyruvic acid in acidified aqueous ethanol.[1414] See also Section 2.2.2.5.

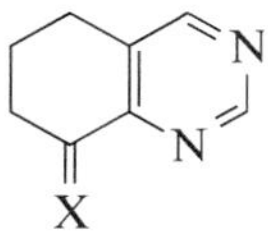

(37)

Vilsmeier reaction.   When 5,6,7,8-tetrahydroquinazoline and the Vilsmeier reagent (prepared *in situ* from formamide and phosphoryl chloride) were heated in formamide at 75° for 2 h, the main product was bis(5,6,7,8-tetrahydro-8-quinazolinylidenemethyl)amine (**38**) in 23% yield after separation from a little 8-formamidomethyl-5,6,7,8-tetrahydroquinazoline.[993]

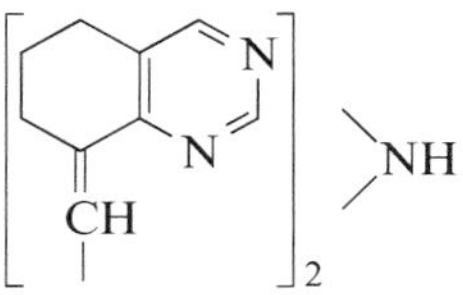

(38)

2.2.   *C*-ALKYL- AND *C*-ARYLQUINAZOLINES (*H* 39)

In earlier days "nonfunctional" alkyl or aryl groups on quinazoline were considered of little importance. It is now realized that such groups not only undergo many useful reactions but also have profound effects on the physical and biological properties of the molecule and on the reactivity of any other attached substituents.

2.2.1.   Preparation of Alkyl- and Arylquinazolines (*H* 39)

The following treatment is not confined entirely to methods for the formation of simple alkyl- or arylquinazolines. It does include representative examples in which the resulting molecule may bear one or more functional passenger groups that have remained unaffected during the procedure(s) involved. For the many primary syntheses of alkyl- or arylquinazolines, see Chapter 1.

2.2.1.1.   By *C*-Alkylation

Alkylations of quinazoline has been discussed in Section 2.1.3. Alkylations of other quinazolines are illustrated in the following examples.

Using organometallic reagents

2,4-Diphenylquinazoline gave 4-methyl-2,4-diphenyl-3,4-dihydroquinazoline
(**39**), naturally stable to oxidation (MeLi, Et$_2$O, 20°, 3 h: 55%).[398]

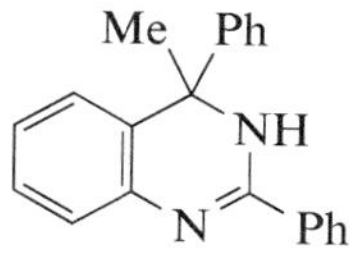

(**39**)

6-Chloroquinazoline gave 6-chloro-4-phenyl-3,4-dihydroquinazoline (PhLi,
Et$_2$O–PhH, 5°, 1 h: 76%).[249]

4-*p*-Tolylaminoquinazoline gave 2-butyl-4-*p*-tolylaminoquinazoline (BuLi,
THF, 20°, N$_2$↓, 3 h, then O$_2$↓, 3 h: 80%) or its 1,2-dihydro derivative
(similarly, N$_2$↓ only, 8 h: 89%).[1784]

5,6-Dimethoxy-2(1*H*)-quinazolinone   gave   5,6-dimethoxy-4-methyl-3,4-
dihydro- 2(1*H*)-quinazolinone (**40**) (MeMgBr, THF–Et$_2$O, N$_2$, 5° then 20°,
16 h: 88%); oxidation (KMnO$_4$, AcMe, 20°, 4 days) gave a little of the
didehydro product but mainly 5,6-dimethoxy-2-oxo-1,2-dihydro-4-quina-
zolinecarboxylic acid.[1930]

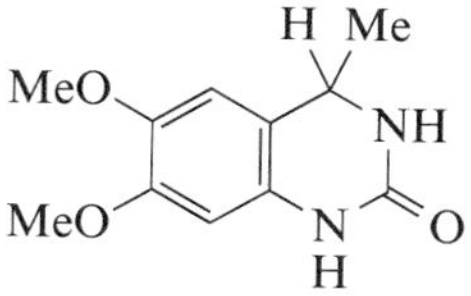

(**40**)

3-Methyl-2-styryl-4(3*H*)-quinazolinone gave 3-methyl-4,4-diphenyl-2-styryl-
3,4-dihydroquinazoline (PhLi, Et$_2$O–PhH, 20°, 12 h: 60%).[520]
Also other examples.[8,163,446,1924,1946,2539,2575,2585]

Using other reagents

2(1*H*)-Quinazolinone gave 4-*p*-dimethylaminophenyl-2(1*H*)-quinazolinone
(**41**) (PhNMe$_2$, S, Me$_2$NCHO, 170°, 2 h: 60%).[342]

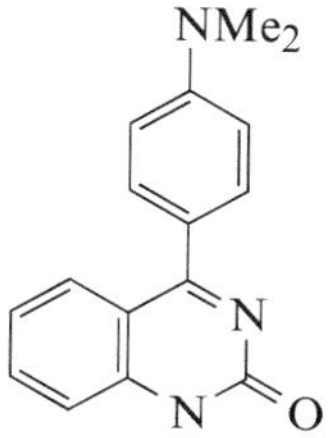

(**41**)

4(3H)-Quinazolinone gave 2-pyridin-2′-yl-4(3H)-quinazolinone and 2,2′-bi-quinazoline-4,4′(3H,3′H)-dione (pyridine or pyridine N-oxide, Pd and Pt/C, 140°, sealed, 3h: ∼ 25% each).[558]

3-Phenyl-4(3H)-quinazolinone methobromide gave 2-(α-ethoxy-α-ethoxy-carbonylmethyl)-1-methyl-3-phenyl-1,2-dihydro-4(3H)-quinazolinone (42) (EtO₂CCHN₂, CuSO₄, EtOH, 70°, 3 h: ∼ 15%).[352] Analogs were made rather similarly.[353]

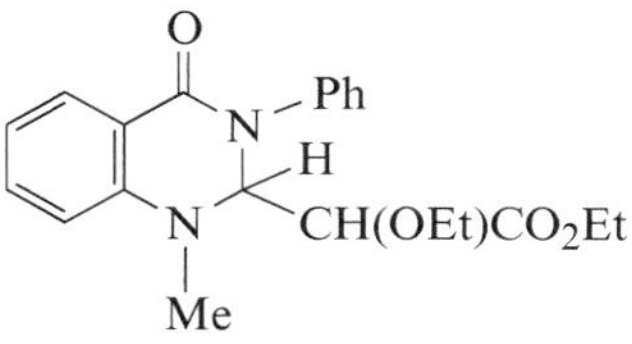

(42)

2-t-Butyl- gave 2,4-di-t-butylquinazoline [Bu^{t}CO₂H, AgNO₃, (NH₄)₂S₂O₈, H₂O, 80°, 30 min: 70%].[1094]

2-Phenyl-5,6,7,8-tetrahydro-4(3H)-quinazolinone gave its 8-benzylidene de-rivative (PhCHO, TsOH, reflux, 30 min: ∼ 75%).[774]

2.2.1.2. By Displacement Reactions (H 44)

Several methods have been used for the displacement of appropriate func-tional groups from quinazoline to afford corresponding alkylquinazolines, as indicated in the following examples.

From halogenoquinazoline substrates

4-Chloroquinazoline gave 4-phenylethynylquinazoline (43) [PhC≡CH, Me₃N, Pd(PPh₃)₂Cl₂, CuI, N₂, 20°, 3 h: 44%].[1093]

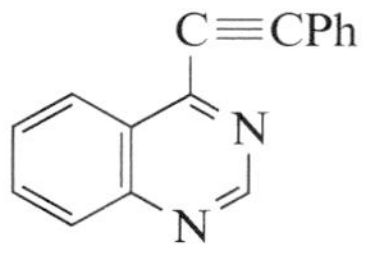

(43)

6-Iodo- gave 6-hept-1′-ynyl- [Me(CH₂)₄C≡CH, Pd(OAc)₂, P(C₆H₄Me-o)₃, CuI, (CH₂)₅NH, Me₂NCHO, A, 20°, 24 h: 55%] and thence 6-heptyl-2,4-quinazolinediamine (Pd/C, H₂, AcOH–Me₂NCHO, 40°: 55%);[2326] also analogous reactions.[2061,2077,2326,2597]

2,4-Dichloroquinazoline gave 2-chloro-4-thien-2′-ylquinazoline (44) (thien-2-yllithium, Et₂O, 0° then 20°, 1 h: 76%).[1963] Adamantyl radicals have been used rather similarly.[2434]

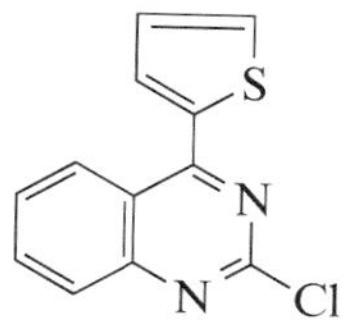

(44)

4-Chloroquinazoline gave 4-(α-ethoxycarbonylacetonylidene)-3,4-dihydro-quinazoline (**45**) as a mixture of *cis*- and *trans*-isomer (AcCHNaCO₂Et, Et₂O, reflux, 30 h: > 95%).[887,1660]

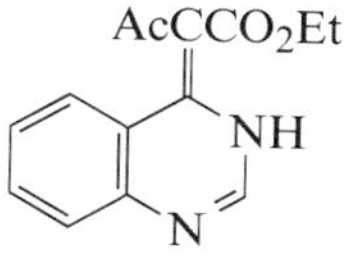

(45)

2-Bromomethyl-3-phenyl-4(3*H*)-quinazolinone or the corresponding thione gave 2-β, β-diacetylethyl-3-phenyl-4(3*H*)-quinazolinone (**46**, X = O) (66%) or the thione (**46**, X = S) (85%), respectively (Ac₂CHNa, MeOH–THF, 20°, 1 h);[137,479] also analogous reactions.[632,1114,2413]

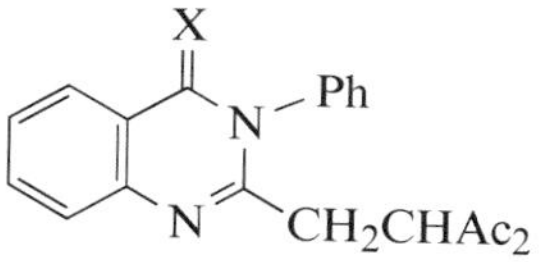

(46)

4-Chloro- gave 4-(2,4-dihydroxyphenyl)quinazoline (HOC₆H₄OH-*m*, AlCl₃, PhH, 80°, 2.5 h: ?%).[226]

4-Chloro- gave 4-ethyl-5,6,7,8-tetrahydroquinazoline (EtMgBr, Ni-complex catalyst, Et₂O, < 5°, 5 h: 84%).[1393]

Also other examples.[1707]

From quinazoline substrates bearing other leaving groups

2- or 4-Methoxyquinazoline gave 2-α-cyanobenzylquinazoline (**47**) (57%) or its 4-isomer (78%), respectively [PhCH₂CN, NaH, C₆H₁₄–THF, reflux, N₂, until substrate used (thin-layer chromatography, tlc)];[2111] 4-(α-cyano-α-methoxycarbonylmethylene-3,4-dihydroquinazoline (**48**) was made somewhat similarly,[555] as were analogs.[43,275]

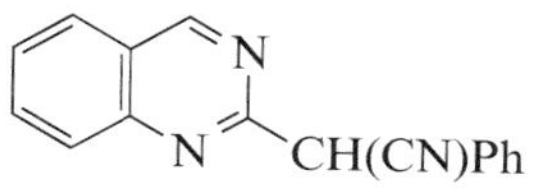

(47)

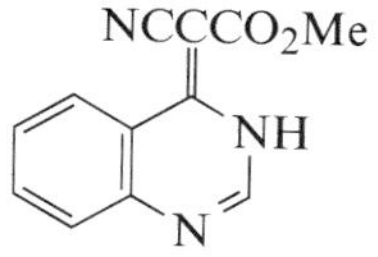

(48)

2-p-Tolylsulfonyl-4(3H)-quinazolinone gave 2-ethyl-4(3H)-quinazolinone (**49**) (EtMgBr, MeOBu^t–THF, reflux, 3 h: $\sim$ 50%).[1915]

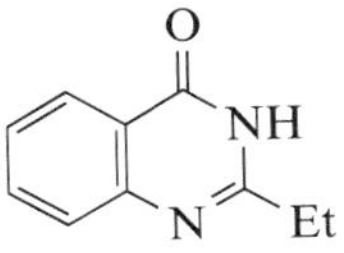

(49)

2-Amino-6-trimethylphosphoniomethyl-4(3H)-quinazolinone bromide gave 2-amino-6-p-[N-(α, γ-diethoxycarbonylpropyl)carbamoyl]styryl-4(3H)-quinazolinone (**50**) [p-OHCC$_6$H$_4$CONHCH(CO$_2$Et)CH$_2$CH$_2$CO$_2$Et, NaOEt, EtOH–Me$_2$NCHO, N$_2$, 20°, 12 h: 81%] and thence to the corresponding 6-(substituted-phenethyl)quinazoline (Pt, H$_2$, MeSO$_3$H, Me$_2$NCHO, $\sim$ 24 h: 84%);[801] also other examples,[416,441]

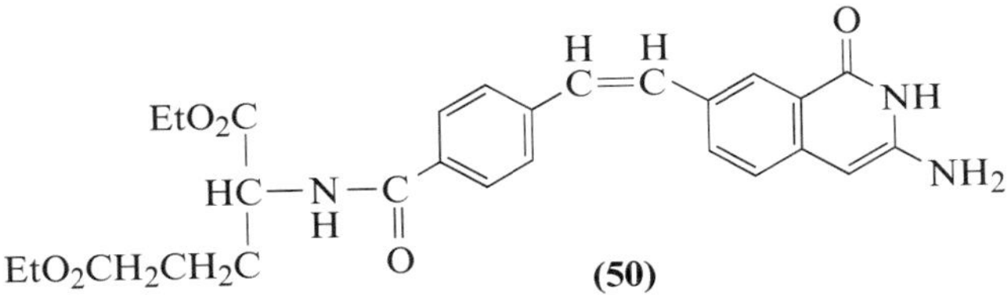

(50)

1-Methyl-4-oxo-1,4-dihydro-2-quinazolinecarbonitrile gave 2-diethoxycarbonyl-methylene-1-methyl-1,2-dihydro-4(3H)-quinazolinone (EtO$_2$CCH$_2$CO$_2$Et, NaH, Me$_2$NCHO, 20°, 30 min; then substrate, 20°, 1h: 72%).[1415]
Also analogous examples.[275]

2.2.1.3. By Eliminative Reactions (H 48)

Alkylquinazolines have been made by eliminating unwanted functional groups from the side chains of appropriate quinazolines in several ways, as illustrated in the following examples.

2-Chloromethyl-6-methyl- (**51**, R = Cl) gave 2,6-dimethyl-4(3H)-quinazolinone (**51**, R = H) (Zn, EtOH, reflux, 8 h: 58%).[510]

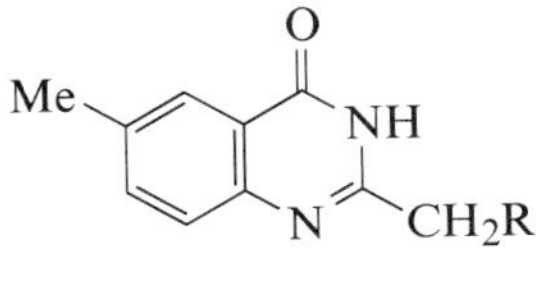

(51)

2-α-Acetoxybenzyl- (**52**, R = OAc) gave 2-benzyl-4-quinazolinamine (**52**, R = H) (H$_2$, Pd/C, EtOH: > 95%).[426]

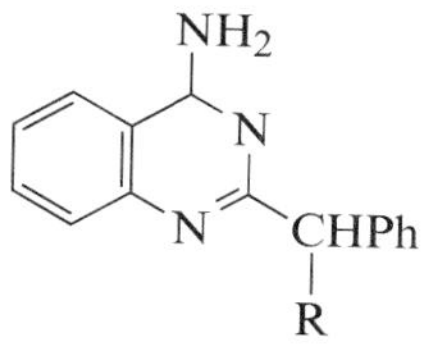

(52)

2-Chloromethyl-6-nitro-3-phenyl-1,2-dihydro-4(3*H*)-quinazolinone (**53**) gave 2-methyl-6-nitro-3-phenyl-4(3*H*)-quinazolinone (**54**) by a nonreductive elimination of hydrogen chloride (NaOEt, EtOH, 0°, 90 min, then 20°, 4 h: 80%).[376]

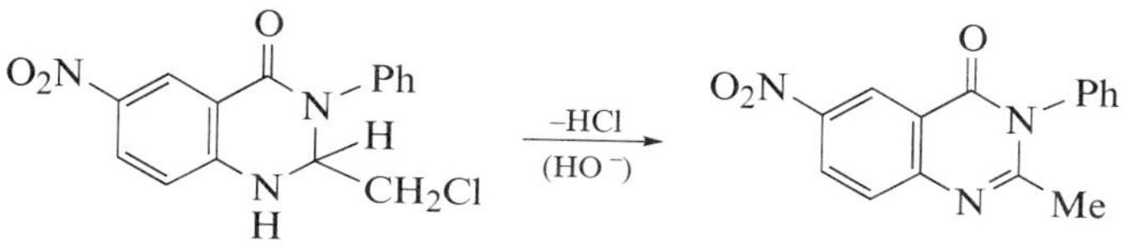

(53)   (54)

4-Phenacyl-2-phenylquinazoline (**55**, R = Bz, Z = Ph) gave 4-methyl-2-phenylquinazoline (**55**, R = H, Z = Ph) by *C*-debenzoylation (90% H$_3$PO$_4$, 150°, reflux, 5 h: 95%).[737]

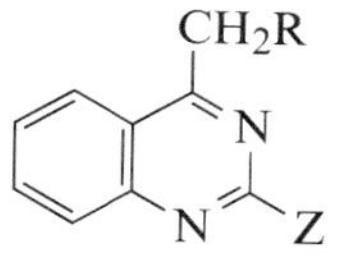

(55)

2-β-Bromoethyl-5,8-dihydroxy- gave 5,8-dihydroxy-2-vinyl-4(3*H*)-quinazolinone (Me$_2$SO, pH 7.4 buffer, N$_2$, 20°, 36 h: 73%).[2430,cf.2205]

4-Hexanoylmethyl-2-methyl- (**55**, R = COC$_5$H$_{11}$, Z = Me) or 2-methyl-4-phenacylquinazoline (**55**, R = Bz, Z = Me) gave 2,4-dimethylquinazoline (**55**, R = H, Z = Me) (NaOMe, MeOH, reflux, 1 h: 86% or > 95%, respectively).[1781]

4(3*H*)-Quinazolinethione gave 4-acetonylthioquinazoline (**56**) (ClCH$_2$Ac, NaOH, EtOH–H$_2$O, 20°, 2 h: 82%), which underwent sulfur extrusion via (?) the intermediate (**57**) to afford 4-acetonylquinazoline, formulated as the tautomeric 4-acetonylidene-3,4-dihydroquinazoline (**58**) on nmr evidence (NaOEt, Me$_2$NCHO, or EtOH, 20°, 10 h: 60%).[895,1869] Analogs were made similarly,[731,1869] some under acidic conditions[1274] and some without isolation of the thioether substrate.[922,1274,1870]

(56)   (57)   (58)

Also other examples.[18,251,353,1914,2304]

2.2.1.4.   By Interconversion of Alkyl Groups (H 51)

The interconversion of C-alkyl groups attached to the quinazoline nucleus has been done directly in three ways. Of these, *alkylation* is represented by the conversion of 2-methyl- (**59**, R = H) into 2-lithiomethyl- (**59**, R = Li) and thence into 2-ethyl-3-o-tolyl-4(3H)-quinazolinone (**59**, R = Me) [Pr^i_2NH, BuLi, THF–C_6H_{14}, 20°, 15 min; then MeI, 20°, 30 min: 53%);[1388] by the analogous preparation of homologs, such as 2-but-3'-enyl-3-o-tolyl-4(3H)-quinazolinone (**59**, R = $CH_2CH_2:CH_2$) (60%);[1388] by the conversion of 2-methyl- into 2-so-diomethyl- and thence into 2-γ-cyanopropyl-3-phenyl-4(3H)-quinazolinone ($NaNH_2$, Me_2NCHO, < 10°, 1 h; then CH_2=CHCN, < 0°, 2 h; then 20°, 16 h: 55%);[1154] and by the conversion of 2-methyl-4(3H)-quinazolinone with its C,O-dianion and thence into 2-phenethyl-4(3H)-quinazolinone (BuLi, THF–C_6H_{14}, 0°, 1 h; then $PhCH_2Cl$, 0°, 2 h: 58%) or related products.[456] Analogous alkylation of 4-methylquinazoline has been reported.[2565] Alkylation or arylation of alkenyl groups has been done by Grignard addition (see Section 2.2.2.6).

(59)

The *reduction* of unsaturated to saturated C-alkyl groups attached to a quinazoline has been exemplified in Section 2.2.1.2; rather similarly, 3-prop-2'-ynyl- (**60**, R = CH_2C:CH) underwent prototropy to 3-allenyl- (**60**, R = CH:C:CH_2) (EtONa, EtOH, reflux, 5 min: 50%) and subsequent hydrogenation to 3-propyl-4(3H)-quinazolinone (**60**, R = Pr) (H_2, Pd/C, AcOEt, 20°).[23,30] Other examples have been reported.[220,632,956,2597]

(60)

The most used interconversions of alkyl groups in the quinazoline series have involved *alkylidenation* of *C*-methyl groups (or the α-methylene portion of higher alkyl groups) to afford styryl or related derivatives. The scope of such reactions and the variety of applicable conditions are illustrated in the following examples (all of which are, quite fortuitously, 4-quinazolinones).

2-Methyl-6-nitro- gave 6-nitro-2-styryl- (61, R = H) (PhCHO, neat, 180°, 1 h: > 90%)[97] or 2-*p*-methylstyryl-6-nitro-4(3*H*)-quinazolinone (61, R = Me) (MeC$_6$H$_4$CHO-*p*, Ac$_2$O, reflux, 2 h: 80%).[2146]

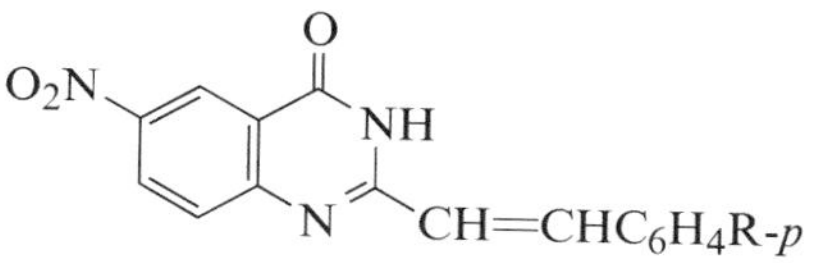

(61)

2-Methyl-3-phenyl- gave 3-phenyl-2-styryl-4(3*H*)-quinazolinone (PhCHO, Ac$_2$O, AcOH, reflux, 2 h: ~ 75%)[146] or 2-phenacylidenemethyl-3-phenyl-4(3*H*)-quinazolinone (PhCOCHO, Ac$_2$O, reflux, 15 min: 90%).[2257]

2-Methyl- gave 2-styryl-3-*o*-tolyl-4(3*H*)-quinazolinone (PhCHO, AcOH, reflux, 15 min: 81% as HClO$_3$ salt;[744] PhCHO, AcOH, trace H$_2$SO$_4$, reflux, 1 h: 52% as base).[1979]

2-Methyl-3-*p*-styrylphenyl- gave 2-*p*-methoxystyryl-3-*p*-styrylphenyl-4(3*H*)-quinazolinone (MeOC$_6$H$_4$CHO-*p*, Pr^{i}OH, trace AcOH, reflux, 6 h: ?%).[1729]

3-Butyl-2-methyl- gave 3-butyl-2-β-(naphthalen-2-yl)vinyl-4(3*H*)-quinazolinone (62) (C$_{10}$H$_7$CHO-2, EtONa, EtOH, 20°, 24 h: > 80%).[515]

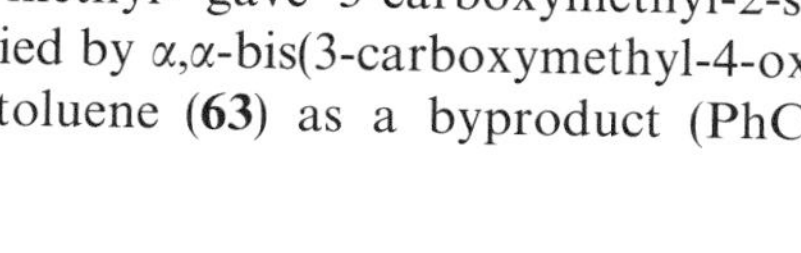

(62)

3-Carboxymethyl-2-methyl- gave 3-carboxymethyl-2-styryl-4(3*H*)-quinazolinone accompanied by α,α-bis(3-carboxymethyl-4-oxo-3,4-dihydroquinazolin-2-ylmethyl)toluene (63) as a byproduct (PhCHO, AcOH, reflux, 2 h).[597]

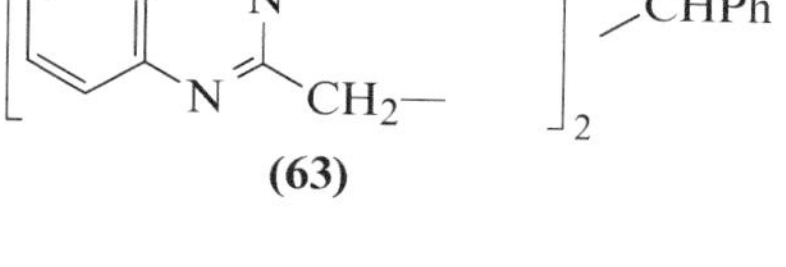

(63)

2-Methyl-3-phenyl- gave 3-phenyl-2-(γ,γ,γ-trichloro-β-hydroxypropyl)-4(3H)-quinazolinone (64) (chloral, pyridine, 80°, 1 h: $\sim$ 70%; as in the pyrimidine series;[2446b] this alkylidenation clearly stalled prior to the usually spontaneous dehydration);[146] 1-methyl-2-(γ,γ,γ-trichloro-β-hydroxypropyl)-4(1H)-quinazolinone ($\sim$ 85%) was made similarly.[244] Other such "aldol" intermediates have been isolated.[1388]

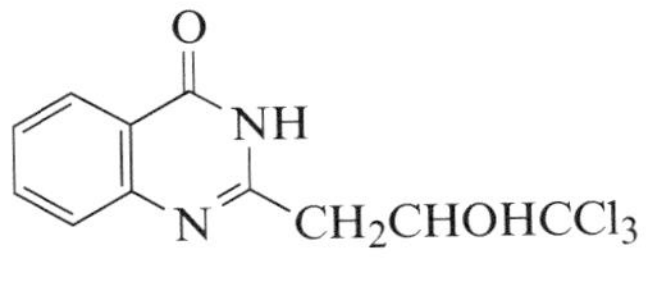

(64)

2-Propyl- gave 2-α-ethylstyryl-4(3H)-quinazolinone (65, R = Et) (PhCHO, Ac$_2$O, AcOH, reflux, 2 h: 65%).[1184]

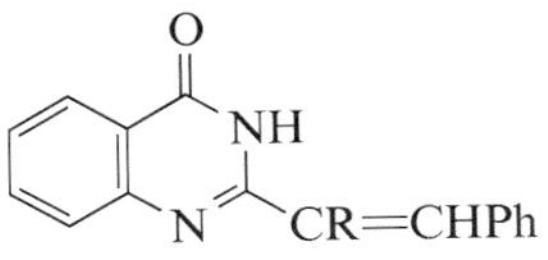

(65)

2-Cyanomethyl- gave 2-α-cyanostyryl-4(3H)-quinazolinone (65, R = CN) (PhCHO, AcONa, AcOH, reflux, 2 h: 85%).[2275]

2-Benzyl-1-methyl- gave 1-methyl-2-α-phenylstyryl-4(1H)-quinazolinone (66) (PhCHO, neat, reflux, 2 h: 65%).[1190]

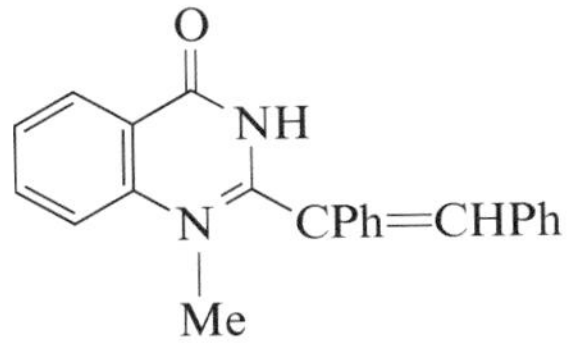

(66)

Also many other examples in the foregoing references and elsewhere.[53,134,378, 388,390,449,458,517,520,590,1042,1055,1098,1113,1118,1152,1202,1482,1501,1505, 1587,1588,1597,1600,1667,1749,1892,1907,1910,1959,1977,2130,2169,2185,2317, 2421]

2.2.1.5. By Modification of Carbaldehydes or Ketones

Although direct reduction of regular quinazolinecarbaldehydes appears to have been neglected, 6-methyl-3-*p*-tolyl-1,2,3,4-tetrahydro-1-quinazolinecarb-aldehyde (**67**, R = CHO) has been reduced by lithium aluminum hydride to afford 1,6-dimethyl-3-*p*-tolyl-1,2,3,4-tetrahydroquinazoline (**67**, R = Me).[153,192] However, the conversion of quinazolinecarbaldehydes into corresponding (sub-stituted-alkyl)quinazolines by condensation with methylene reagents has proved more useful. Thus 6-chloro-4-phenyl-2-quinazolinecarbaldehyde (**68**, X = O) gave 2-*β*-carboxyvinyl-6-chloro-4-phenylquinazoline (**68**, X = CHCO₂H) [CH₂(CO₂H)₂, trace (CH₂)₅NH, pyridine, 95°, 3.5 h: ~60% with loss of CO_2] and thence the corresponding 2-*β*-carboxyethyl derivative (H₂, Pd/C, 1 M NaOH: ~55%);[220] 4-phenyl-2-quinazolinecarbaldehyde likewise gave 2-*β*-carboxyvinyl-4-phenylquinazoline (35%);[632] and 2,4-diamino-6-quinazolinecarbaldehyde re-acted with the appropriate triphenylphosphonium bromide to give 6-*p*-[*N*-(α,γ-dimethoxycarbonylpropyl)carbamoyl]styryl-2,4-quinazolinediamine (**69**) (NaH, Me₂NCHO, 20°, 5 days: 73%).[956] 2-Benzoyl- gave 2-benzyl-4(3*H*)-quin-azolinone (Ni–Al, 2.5 M NaOH, 40°: ?%)[888] 4-Acetyl- gave 4-ethyl-5,6,7,8-tet-rahydroquinazoline (82%) by a Wolff–Kishner[2450f] reduction.[1393]

(**67**)

(**68**)

(**69**)

2.2.1.6. By Rearrangement (*H* 134,245)

The *ortho*-Claisen rearrangement has been used to convert *C*-allyloxy-quinazolines into *C*-allylquinazolinols. Thus 7-allyloxy-6-methoxy- gave 8-allyl-7-hydroxy-6-methoxy-2,3-diphenyl-4(3*H*)-quinazolinone (**70**) (PhNEt₂, reflux, CO_2, 6 h: 90%);[151] in a similar way, the isomeric 6-allyloxyl-7-methoxy- and

7-allyloxy-8-methoxy- gave 5-allyl-6-hydroxy-7-methoxy- (50%)[140] and 6-allyl-7-hydroxy-8-methoxy-2,3-diphenyl-4(3H)-quinazolinone (70%),[149] respectively; and further such examples have been reported.[143] In a different way, 3-p-allyloxyphenyl- gave 3-(3-allyl-4-hydroxyphenyl)-2-methyl-4(3H)-quinazolinone (71) (neat, 200°, 2 h: 75%)[573] and analogs were made similarly.[573,892]

(70)   (71)

An interesting formation of mercaptoalkylquinazolines, which must involve rearrangement at some stage, is illustrated by the reaction of 4(3H)-quinazolinethione with methacrylonitrile under irradiation to give 4-(α-cyano-β-mercapto-α-methylethyl)quinazoline (72) (MeOCH$_2$CH$_2$OMe, A, hv, > 2 h: > 95%):[1953] other electron-poor alkenes gave analogous products, but electron-rich alkenes simply gave alkylthioquinazolines.[1953] 2-Thioxo-1,2-dihydro-4(3H)-quinazolinone reacted rather similarly with methacrylonitrile to afford 2-(α-cyano-β-mercapto-α-methylethyl)-4(3H)-quinazolinone (MeOH, A, hv, 2 h: 60%).[2021]

(72)

2.2.1.7.   By Oxidation of Hydro Derivatives (H 41,49)

Although a variety of individual cases are mentioned elsewhere in passing, the following nuclear oxidations of simple dihydro or tetrahydro phenylquinazolines to their aromatic counterparts will serve as typical examples. 2-Phenyl-1,2-dihydroquinazoline gave 2-phenylquinazoline (MnO$_2$, PhH, reflux, 2 h: 41%);[767] 2-p-tolyl-3,4-dihydroquinazoline gave 2-p-tolylquinazoline (MnO$_2$, PhH, reflux, 88%);[1369] 4-phenyl-3,4-dihydroquinazoline gave 4-phenylquinazoline [KMnO$_4$, H$_2$O–dioxane, 20°, 1 h: 89%;[247] K$_3$Fe(CN)$_6$, KOH, H$_2$O–PhH, shake, 20°, 2 h: ∼ 94%];[56] 2,4-diphenyl-1,2-dihydroquinazoline gave 2,4-diphenylquinazoline

(2,3-dichloro-5,6-dicyano-1,4-benzoquinone, PhMe, 100°, 10 min: > 95%);[1759] 2,4-diphenyl-5,6,7,8-tetrahydroquinazoline gave 2,4-diphenylquinazoline (tetrachlorobenzoquinone, solvent?, reflux, 3.5 h: 66%);[111] 6-chloro-2,4-diphenyl-1,2,3,4-tetrahydroquinazoline gave 6-chloro-2,4- diphenylquinazoline (KMnO$_4$, AcOH–H$_2$O, 20°, briefly: 63%);[1824] and 8-phenyl-5,6,7,8-tetrahydroquinazoline gave 8-phenylquinazoline (S, neat, 230°, 1 h: 19%).[985] Partial oxidation: 6-methyl-3-p-tolyl-1,2,3,4-tetrahydroquinazoline gave its 3,4-dihydro analog [PhCH(NHCONH$_2$)$_2$, BuOH, HCl gas, 135°, 3 h: 21%].[420]

2.2.2. Properties and Reactions of Alkyl- and Arylquinazolines (H 49)

Interesting new material on the physical properties of simple alkyl- or arylquinazolines may be summarized briefly as follows. The conformations of 1-methyl-, 3-methyl-, and 1,3-dimethyl-*trans*-decahydroquinazoline have been reassessed;[701,2204] the X-ray molecular structures of 6-isopropyl-2,4-diphenylquinazoline[1942] and α,β-bis(2-methylquinazolin-4-yl)ethylene[2189] ave been reported; the photoisomerization associated with the benzylidene group in 8-benzylidene-2-methylthio-4-phenyl-5,6,7,8-tetrahydroquinazoline (**73**, R = Me) and homologs (**73**, R = Et, Pr, Bu, etc.) has been studied;[1474] the ms fragmentations of 2-, 4-, 5-, 6-, 7-, and 8-methylquinazoline,[172] of a variety of simple functionally substituted quinazolines,[172] and of several 2-alkyl-3- phenylquinazolines[1286] have been detailed; the uv spectra of simple methyl- and phenylquinazolines have been compared with those of their N-oxides;[575] attention has been drawn to certain characteristics within the ir spectra of 2-styrylquinazolinones;[1088] and the polarography of 4-thien-2′-ylquinazoline has been studied.[1316]

(73)

Alkyl or aryl groups attached to the quinazoline nucleus undergo a variety of reactions, of which reduction and alkylation have already been discussed in Section 2.2.1.4. Other reactions are detailed in the following subsections.

2.2.2.1.   Halogenation at Alkyl Groups (H 226)

Halogenation of saturated or unsaturated alkyl groups may be done by substitution or addition, respectively. The reagents and required conditions are illustrated in the following examples.

By substitution

2-Methyl-3-o-tolyl-4(3H)-quinazolinone (**74**, R = X = Y = Z = H) gave 2-bromomethyl-3-o-tolyl- (**74**, R = Br, X = Y = Z = H) (Br$_2$, AcOH, reflux, 3 h: 50%),[786] 2-dibromomethyl-3-o-tolyl- (**74**, R = X = Br, Y = Z = H) (Br$_2$, Me$_2$SO, 20°, 12 h: 63%,[786] 3-o-(bromomethyl)phenyl-2-methyl- (**74**, R = X = Y = H, Z = Br) (N-bromosuccinimide, Bz$_2$O$_2$, CCl$_4$, reflux, 6 h: ~90%),[619] or 3-o-(dibromomethyl)phenyl-2-methyl-4(3H)-quinazolinone (**74**, R = X = H, Y = Z = Br) [N-bromosuccinimide (2 mol), Bz$_2$O$_2$, CCl$_4$, reflux, 2 h: 44%].[1286]

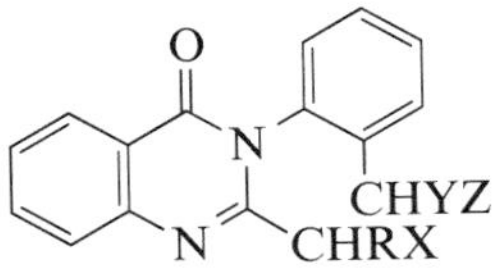

(**74**)

4-Methyl- (**75**, R = H) gave 4-trichloromethyl- (**75**, R = Cl) (Cl$_2$, AcOH, AcONa, 70°, 1 h: 82%)[2001] or 4-tribromomethlquinazoline (**75**, R = Br) (Br$_2$, likewise: 81%).[1651]

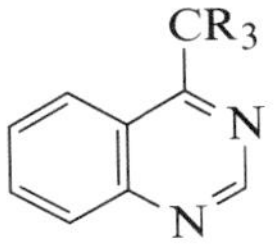

(**75**)

6-Methyl- (**76**, R = H) gave 6-bromomethyl-4-methylthioquinazoline (**76**, R = Br) (N-bromosuccinimide, CCl$_4$, hv, reflux, 30 min: 40%).[2255]

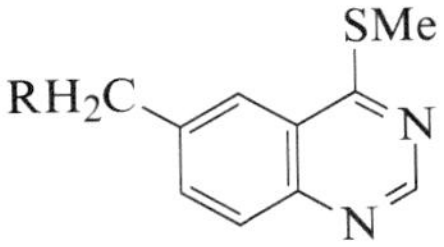

(**76**)

2-Ethyl- or 2-propyl- gave 2-α-bromoethyl- or 2-α-bromopropyl-4-(3H)-quinazolinone, respectively (Br$_2$, AcONa, 60°, 3h: both 79%).[2544]

2-Isopropyl-8-methoxy-4(3H)-quinazolinone (**77**) gave 4-chloro-2-(α-chloro-α-methylethyl)-8-methoxyquinazoline (**78**) (PCl$_5$, POCl$_3$, 115°, 90 min: 25%).[1224]

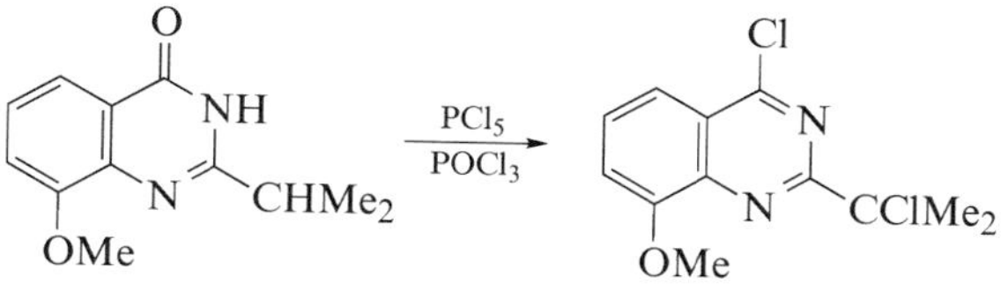

(77)   (78)

2,4-Bisbenzamido-6-methyl- (**79**, R = H) gave 2,4-bisbenzamido-6-bromo-methylquinazoline (**79**, R = Br) (1,3-dibromo-5,5-dimethylhydantoin, Bz$_2$O$_2$, CCl$_4$, *hv*, reflux, 20 min: 70%).[1912, cf. 1608]

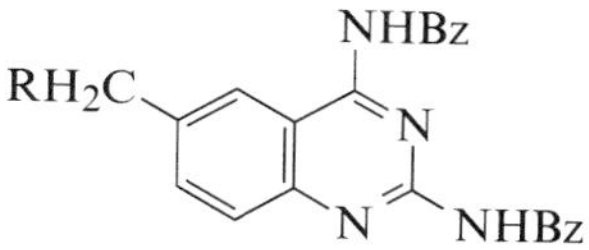

(79)

3-(4-Carboxy-3-hydroxyphenyl)-2-phenyl- (**80**, X = H, Y = CO$_2$H) gave 2-phenyl-3-(2,4,6-tribromo-3-hydroxyphenyl)-4(3*H*)-quinazolinone (**80**, X = Y = Br) with loss of CO$_2$ (Br$_2$, H$_2$O–MeOH, 20°, 30 min: 83%).[1252]

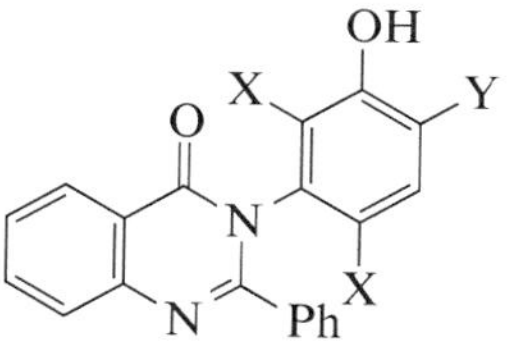

(80)

Also other examples.[146,152,479,871,881,884,1170,1191,1402,1492,1616,1617,1867, 1902,2092,2470]

By addition

3-Allyl- gave 3-β-bromopropyl- (**81**, R = H, X = O) 48% HBr, reflux, 2 h: 70%)[477] or 3-β,γ-dibromopropyl-2,4(1*H*,3*H*)-quinazolinedione (**81**, R = Br, X = O) (pyridinium bromide perbromide, AcOH, 20°, 1 h: 80%;[477] Br$_2$, CCl$_4$, −5° then 0°, 4 h: 92%);[2227] 3-β,γ-dibromopropyl-2-thioxo-1,2-dihydro-4(3*H*)-quinazolinone (**81**, R = Br, X = S) was also made by the second method, in 94% yield.[2227]

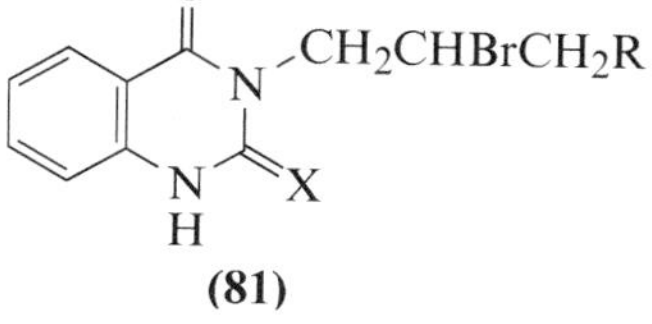

(81)

6,8-Dibromo-3-(2,5-dichlorophenyl)-2-p-dimethylaminostyryl- gave 6,8-di-bromo-2-(α,β-dibromo-p-dimethylaminophenethyl)-3-(2,5-dichlorophenyl)-4(3H)-quinazolinone (**82**) (Br$_2$, CHCl$_3$, 20°, 6 h: 65%);[2150] analogs were made rather similarly.[1258,1600,2150]

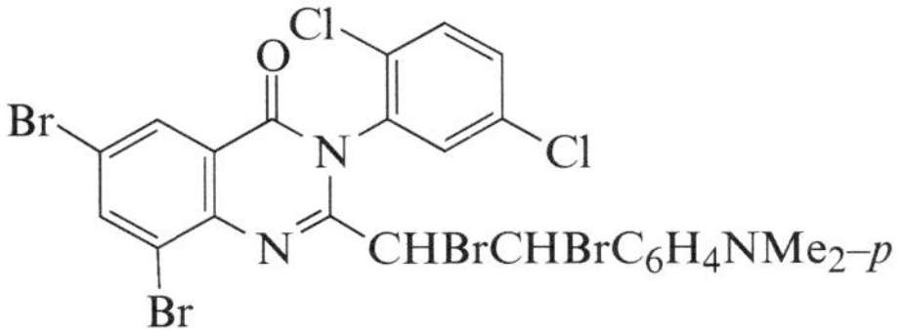

(82)

Also other examples.[143,149,151,1492,1658,2430]

2.2.2.2.   Oxidation at Alkyl Groups (H 54)

Alkyl- but not arylquinazolines undergo oxidation by a variety of reagents to afford a surprising diversity of product types, as indicated in the following examples.

Predictable oxidations

Ethyl 2,5,7-trimethyl- (**83**, R = Me) gave ethyl 2-formyl-5,7-dimethyl-4-oxo-3-phenyl-3,4-dihydro-6-quinazolinecarboxylate (**83**, R = CHO)(SeO$_2$);[2071] 2-methyl-3-phenyl-4(3H)-quinazolinone gave 4-oxo-3-phenyl-3,4-dihydro-2-quinazolinecarbaldehyde (SeO$_2$, EtOH, reflux, 4 h: ~ 50%).[146]

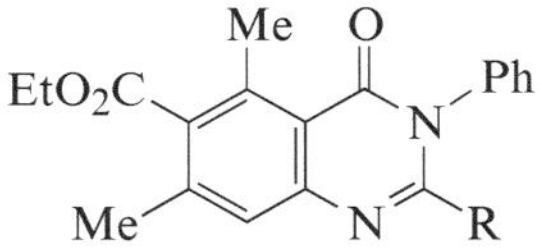

(83)

5,6-Dimethoxy-4-methyl-3,4-dihydro-2(1H)-quinazolinone gave 5,6-dime-thoxy-2-oxo-1,2-dihydro-4-quinazolinecarboxylic acid (**84**) (KMnO$_4$, AcMe, 20°, 4 days: 50% after purification from byproducts).[1930]

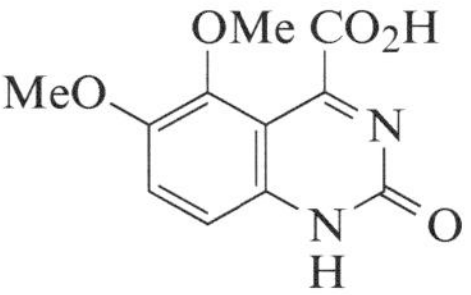

(84)

2-Benzyl- gave 2-benzoyl-4(3*H*)-quinazolinone (**85**) (CrO_3, AcOH–H_2O, 100°, 30 min: ∼ 75%).[182, cf. 555, 918] Such oxidations with permanganate or selenium dioxide appear to have been less effective.[97,134]

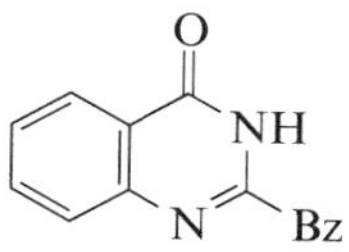

(85)

1-Allyl- gave 1-β,γ-dihydroxypropyl-5,6,7,8-tetrahydro-2,4(1*H*,3*H*)-quinazolinedione (**86**) ($NaClO_3$, OsO_4, H_2O–MeOH, 20°, 12 h: 45%).[1336]

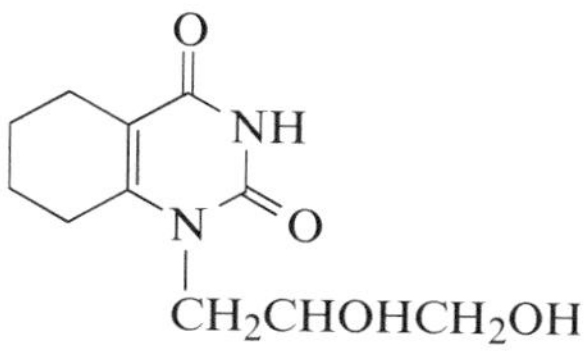

(86)

Also other examples.[97,134,1157,1492]

Less predictable oxidations

2,4-Dimethylquinazoline gave α,β-bis(2-methylquinazolin-4-yl)ethylene (**87**; structure revised on an X-ray analysis[2189]) (SeO_2, AcOEt, reflux, 30 min: 20%).[2268]

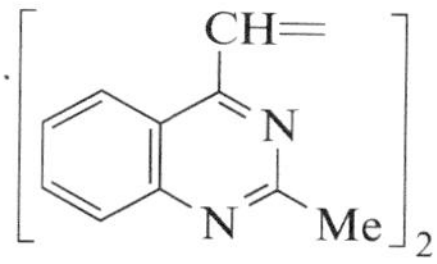

(87)

3-Allyl- gave 3-formylmethyl-4(3*H*)-quinazolinone (OsO_4, dioxane, dark, 20°, 1 h, then $NaIO_4/H_2O$, 20°, 12 h: 40%).[1772]

2-Methylquinazoline gave 2,5-pyrimidinedicarboxylic acid (**88**) ($KMnO_4$, KOH, H_2O, 50°, 3 h: 39%)[2067] or 2,2'-dimethyl-4,4'-biquinazoline (**89**, R = Me) [dimerization in KCN/Me_2SO, then $K_3Fe(CN)_6$, NaOH, PhH, H_2O, 20°, 1 h: cf. quinazoline, Section 2.1.1].[1768] A somewhat analogous process produced 2,2'-diphenyl-4,4'-biquinazoline (**89**, R = Ph).[933]

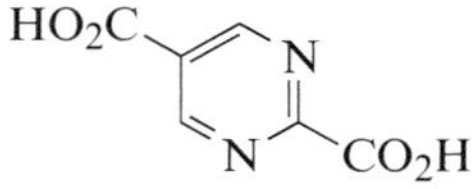

HO_2C [structure] CO_2H

(88)

[structure with 4-methyl] $R]_2$

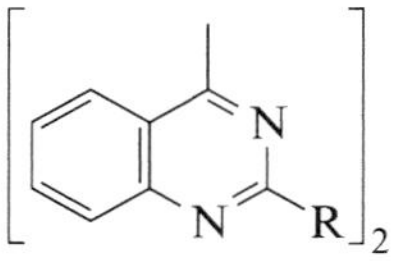

(89)

2-Diphenylmethyl- (**90**, R = H) gave a separable mixture of 2-(α-hydroxy-α,α-diphenylmethyl)- (**90**, R = OH) and 2-benzoyl-4(3H)-quinazolinone (**85**) (CrO_3, AcOH, 95°, 30 min: ~ 40% and ~ 25%, respectively).[481]

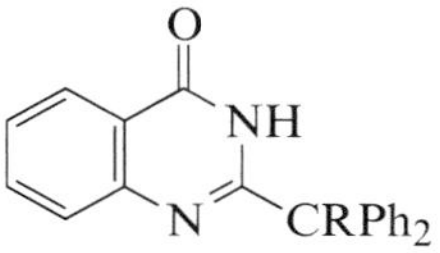

(90)

2-Benzyl-1,2-dihydro-4-quinazolinamine gave 4-quinazolinamine with loss of toluene [NaH, $(MeOCH_2CH_2)_2O$, reflux, N_2, 16 h: 43%; or $K_3Fe(CN)_6$, K_2CO_3, M_2O–EtOH, 20°, 2 h: 45%],[258, cf. 35]

2-Isopropyl- gave 2-(α-acetoxy-α-methylethyl)-6-methyl-4(3H)-quinazolinone (30% H_2O_2, AcOH, reflux, 10 h: ~ 1%).[1049]

2-Benzoylmethylene-1,3-dimethyl-1,2-dihydro-4(3H)-quinazolinone gave 1,3-dimethyl-2,4(1H,3H)-quinazolinedione, probably via a cyclic peroxide (AcMe, air ↓, hv: 76%).[1687]

2.2.2.3. Mannich Reactions at Alkyl or Aryl Groups (H 53,114)

Mannich reactions[2450c] can occur at any activated methyl (or methylene) group, at a ring NH grouping, or at a suitable aryl group (usually phenolic). Although not used extensively in the quinazoline series, the following examples will serve to illustrate some such possibilities.

6,7-Dimethoxy-4-methyl-   gave   4-β-dimethylaminoethyl-6,7-dimethoxy-quinazoline (**91**) (CH_2O, $Me_2NH.HCl$, H_2O–EtOH, warm, then 20°, 22 h: 64% as hydrochloride).[1425]

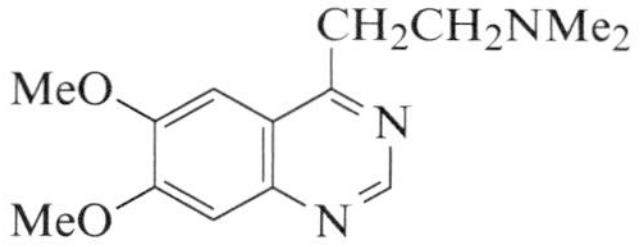

$CH_2CH_2NMe_2$

MeO [structure]

MeO

(91)

2-Methyl- gave (exceptionally) 2-β-ethoxyethyl-3-phenyl-4(3*H*)-quinazo-
linone (paraformaldehyde, HNEt$_2$.HCl, EtOH, reflux, 15 h: 85%; via 2-
vinyl?).[847]

3-*m*-Acetylphenyl-2-methyl- gave 3-*m*-(3-anilinopropionyl)phenyl-2-methyl-
4(3*H*)-quinazolinone (**92**) (paraformaldehyde, PhNH$_2$, EtOH, trace HCl,
reflux, 10 h: ?%).[2286]

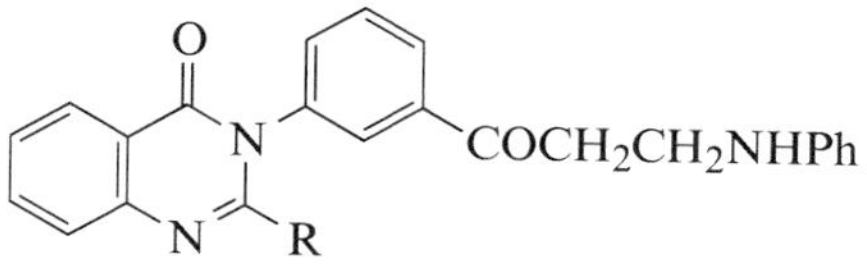

(92)

4(3*H*)-Quinazolinone gave 3-*p*-(carboxymethylthio)anilinomethyl-4(3*H*)-
quinazolinone (CH$_2$O, HO$_2$CCH$_2$SC$_6$H$_4$NH$_2$-*p*, H$_2$O–EtOH, reflux,
10 min: 60%).[1597]

3-*p*-Hydroxyphenyl- gave 3-(3-diethylaminomethyl-4-hydroxyphenyl)-2-
phenyl-4(3*H*)-quinazolinone (CH$_2$O, H$_2$O–EtOH, reflux, 1 h, then Et$_2$NH,
reflux, 6 h: > 60%).[1253]

4-*p*-Hydroxyanilino- gave 4-[4-hydroxy-3,5-bis(pyrrolidinomethyl)anilino]-
quinazoline (**93**) [CH$_2$O, excess HN(CH$_2$)$_5$, H$_2$O–EtOH, reflux, 3 h:
62%].[2291]

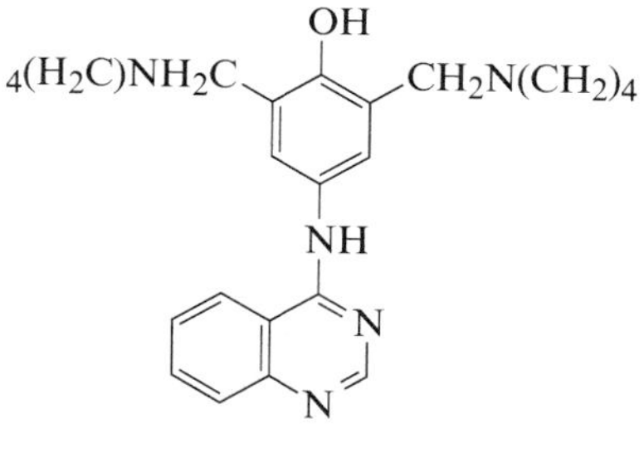

(93)

Also other examples.[503,664,1041,1069,1118,1154,1178,1192,1208,1491,1519,1690,2087,2144,2306,2307]

2.2.2.4.   Acylations at Alkyl Groups (*H* 109)

Activated methyl groups undergo acylation by the Claisen reaction[2450e] or (in
a more complicated way) by the Vilsmeier reaction,[2450d] as indicated in the
following examples.

Claisen reactions

2,4-Dimethyl- gave 2-methyl-4-phenacylquinazoline (**94**) (PhCO$_2$Et, KOEt, PhH, reflux, 24 h: 62%).[1062]

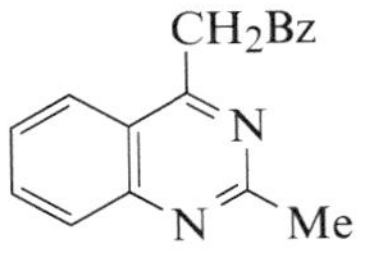

(**94**)

1,2-Dimethyl- gave 2-acetonyl-1-methyl-4(1*H*)-quinazolinone (**95**) (Ac$_2$O, pyridine, 95°, 3 h: 40%).[244]

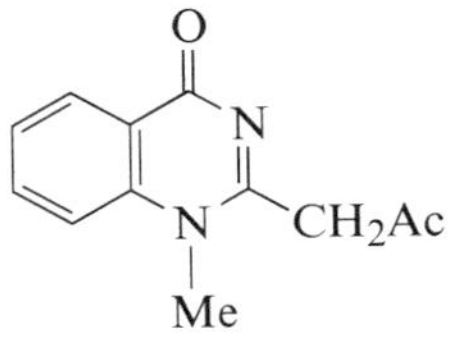

(**95**)

2,3-Dimethyl- gave 2-ethoxalylmethyl-3-methyl-4(3*H*)-quinazolinone (EtO$_2$CCO$_2$Et, NaOEt, Et$_2$O–EtOH, 0°, 12 h: 97%).[2118]

2-Methyl- gave 2-nicotinoylmethyl-3-phenyl-4(3*H*)-quinazolinone (ethyl nicotinate, NaH, MeOCH$_2$CH$_2$OMe, reflux, < 10 h: 72%).[2223]

Also other examples in the foregoing references and elsewhere;[599,1540] see also Section 2.2.2.5.

Vilsmeier reactions

2-Methyl- gave 2-(β-dimethylamino-α-formylvinyl)- (**96**, R = Ph) (Me$_2$NCHO, POCl$_3$, 0°, 10 min, then substrate, Me$_2$NCHO, 75°, 4 h: ~ 80%) and thence 2-diformylmethyl-3-phenyl-4(3*H*)-quinazolinone (**97**, R = Ph) (1.25M NaOH, reflux, 20 min: ~ 80%).[506]

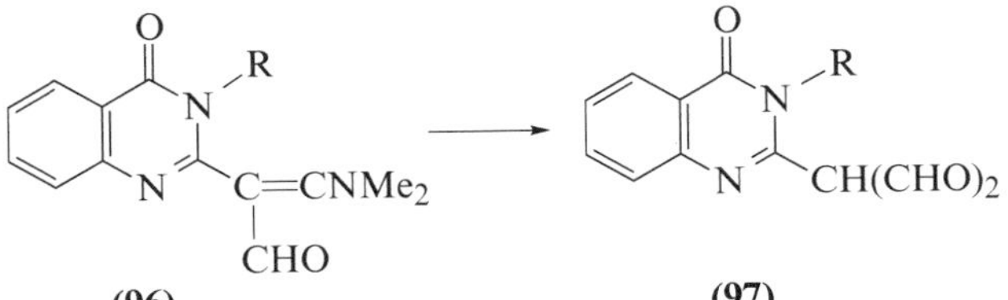

(**96**)   (**97**)

Likewise, 3-benzyloxy-2-methyl- gave 3-benzyloxy-2-(β-dimethylamino-α-formylvinyl)- (**96**, R = OCH$_2$Ph) (83%) and thence 3-benzyloxy-2-diformylmethyl-4(3*H*)-quinazolinone (**97**, R = OCH$_2$Ph) (90%).[1691]

3-Ethyl-4-methyl- gave 3-ethyl-4-formylmethyl-2(3*H*)-quinazolinone (Me$_2$ NCHO, POCl$_3$, CHCl$_3$, 60°, 5 h: ~ 90%);[110] also other examples.[1259]

2.2.2.5. Nitrosation at Alkyl Groups

As in the pyrimidine series,[2446c] activated methyl or methylene groups undergo nitrosation by treatment with a nitrite ester, either by a Claisen procedure under basic conditions or by a classic (nonaqueous) nitrosation procedure under acidic conditions. Thus dissolution of 2,4-dimethylquinazoline in liquid ammonia containing sodium amide and a little ferric chloride, followed by the addition of ethyl nitrite, gave 2-methyl-4-nitrosomethylquinazoline, that is, 2-methyl-4-quinazolinecarbaldehyde oxime (**98**, R = Me), in ~ 80% yield.[830,1063] In contrast, treatment of ethanolic 4-methylquinazoline hydrochloride with ethyl nitrite at 5°, followed by standing at 20° for 5 h, gave 4-quinazolinecarbaldehyde oxime (**98**, R = H) (35%, as hydrochloride);[106] 5,6,7,8-tetrahydroquinazoline underwent a similar nitrosation at its 8-methylene grouping to afford 8-hydroxyimino-5,6,7,8-tetrahydroquinazoline (**99**) in > 95% yield.[106, cf. 1414]

(**98**)   (**99**)

2.2.2.6. Other Reactions at Alkyl Groups

The remaining reactions of alkylquinazolines have not been used frequently of late. Accordingly, each is represented by one or two typical examples in the following paragraphs.

Deuteration. The deuteration of 2- or 4-methylquinazoline in NaOD/D$_2$O–CD$_3$OD was hastened 10–100-fold by *N*-oxidation of each substrate to the 1- or 3-oxide.[1363]

Diazo coupling. 2-Methyl-3-*p*-nitrophenyl-4(3*H*)-quinazolinone (**100**, R = H) underwent coupling with diazotized aniline under the usual conditions (NaOAc buffer, 0°, dark, 72 h) to give 3-*p*-nitrophenyl-2-phenylazomethyl-4(3*H*)-quinazolinone (**100**, R = N:NPh) in ~ 50% yield;[825] analogs were made similarly.[110,794,799,825]

(100)

To a thioamide.   Treatment of 4-methyl-2-phenylquinazoline with thionyl chloride (reflux, 3 h) gave an uncharacterized sulfur-containing intermediate that reacted with diethylamine in ether to afford *N*,*N*-diethyl-2-phenyl-4-quinazolinecarbothioamide **(101)** in 10% yield.[2196]

(101)

Phenylthiation.   Conversion of 2-methyl-3-*o*-tolyl-4(3*H*)-quinazolinone **(102,** $X = Y = H$) into its 2-lithiomethyl analog and subsequent treatment with diphenyl disulfide gave a separable mixture of 2-phenylthiomethyl- **(102,** $X = H$, $Y = SPh$) and 2-bis(phenylthio)methyl-3-*o*-tolyl-4(3*H*)-quinazolinone **(102,** $X = Y = SPh$) in $\leqslant 65\%$ and $\leqslant 85\%$ yields, respectively, according to the conditions used.[1388]

(102)

Addition reactions.   Alkenyl groups undergo several addition reactions, two of which—*hydrogenation* and *halogen-addition*—have been discussed in Sections 2.2.1.4 and 2.2.2.1, respectively. Other types of addition reaction involving alkenyl groups are exemplified in the conversion of 3-methyl-2-*p*-methylstyryl-4(3*H*)-quinazolinone into 3-methyl-2-*β*-*p*-tolylphenethyl-4(3*H*)-quinazolinone **(103)** by a *Grignard reaction* (PhMgBr, Et$_2$O–PhH, 20°, 12 h, then reflux, 3 h: 62%);[520] in similar conversions;[520,537] in the transformation of 5,8-dihydroxy-2-vinyl- into 2-*β*-(*β*-hydroxyethylthio)ethyl-4(3*H*)-quinazolinone (HOCH$_2$ CH$_2$SH, Me$_2$SO, pH 7.4 buffer, A, 20°, 1 h: 59%)[2430] and of 2-styryl-3-*p*-tolyl-4(3*H*)-quinazolinone into 2-*β*-(phenylthio)phenethyl-3-*p*-tolyl-4(3*H*)-quina-zolinone **(104)** by *thiol-addition* (PhSH, no further details);[598] in the conversion of 2-styryl-4(3*H*)-quinazolinone into 1-amino-3-benzyl-9-oxo-1,2,3,9-tetrahyd-ropyrrolo[2,1-*b*]quinazoline-2-carbonitrile **(105)** by an initial *methylene addition* and subsequent cyclization (NCCH$_2$CN, pyridine, reflux, 6 h: 70%);[2256] and in the combination of 3-phenyl-2-styryl-4(3*H*)-quinazolinone with tetrachloro-*o*-benzoquinone by *quinone addition* to furnish 3-phenyl-2-(5,6,7,8-tetrachloro-3-phenyl-2,3-dihydro-1,4-benzodioxin-2-yl)-4(3*H*)-quinazolinone (**106**).[1977]

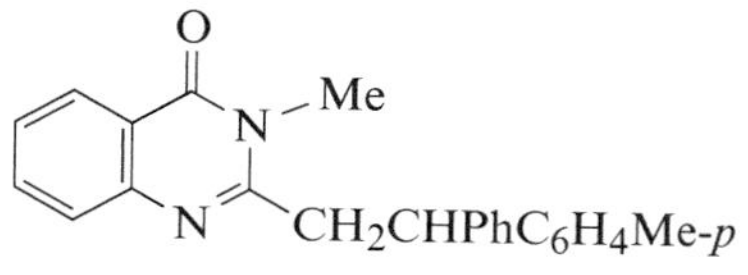

(103) 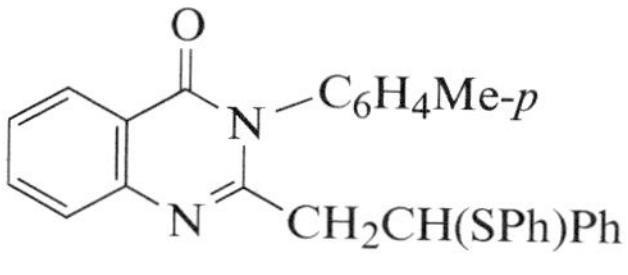   (104)

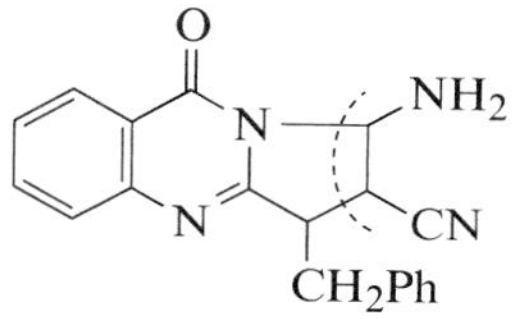

(105) 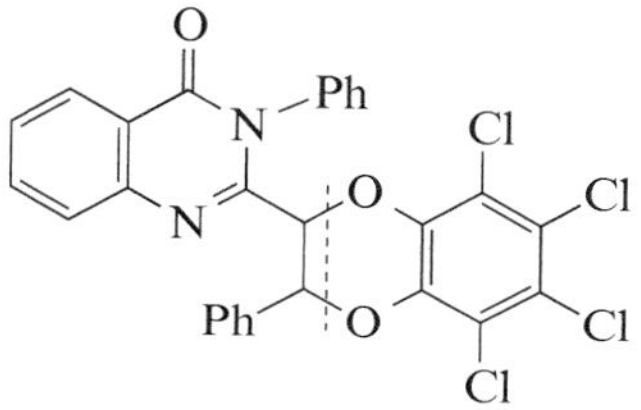   (106)

Intramolecular cyclizations. In attempting to *N*-alkylate 4-methylquinazoline with ethyl bromopyruvate, the methyl group became involved to afford ethyl pyrrolo[1,2-*c*]quinazoline-2-carboxylate (**107**) (excess ester, EtOH, 16 h: 84%);[1552] analogs were made similarly;[1552] and the isolable 5-diazonio-6-methyl-4(3*H*)-quinazolinone tetrafluoroborate (**108**) underwent facile intramolecular diazo-coupling in pyridine to afford 1*H*-pyrazolo[3,4-*f*]quinazolin-9(8*H*)-one (**109**) in 48% yield.[1159]

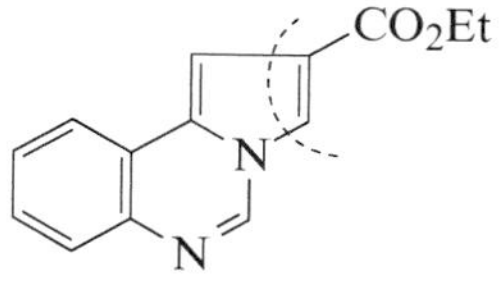

(107) 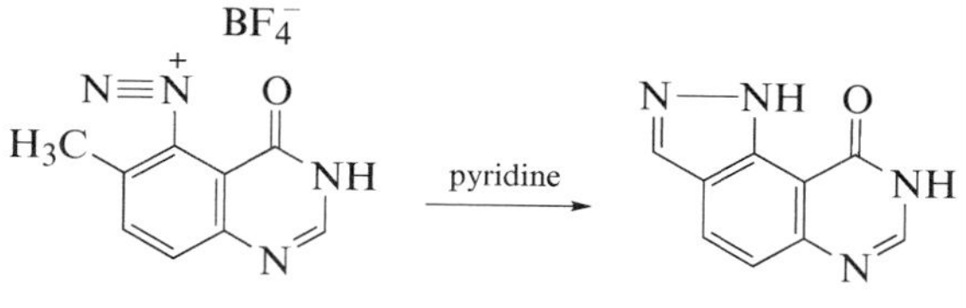   (108)   (109)

Debenzylation. As well as the oxidative debenzylations of 2-benzyl-1,2-dihydro-4-quinazolinamine mentioned in Section 2.2.2.2, an apparently unprecedented reductive removal of a *C*-benzyl group with sodium borohydride has been reported; thus 2-benzyl-3-methyl-4(3*H*)-quinazolinone gave 3-methyl-4(3*H*)-quinazolinone (NaBH₄, MeOH, 0° then 20°, 12 h: 15%).[156]

Reissert reactions. Although the Reissert reaction does not affect alkyl substituents directly, alkylquinazolines do undergo this reaction much as does quinazoline itself (Section 2.1.3). However, some minor variations in the course of the reaction would be expected and indeed are evident.[1420,1722,1738,1819]

Miscellaneous reactions.   Treatment of 4-phenylquinazoline with methyl iodide under reflux for 6 h gave a mixture (~90% yield) of the 1- (**110**) and 3-methiodide in a ratio of 7:1; subsequent reduction with sodium borohydride in methanol–methylene chloride gave a separable mixture of 1- (**111**) and 3-methyl-4-phenyl-1,2,3,4-tetrahydroquinazoline.[247] Reduction of 2,4-diphenylquinazoline with sodium in tetrahydrofuran at $-78°$ during 15 h gave 2,4-diphenyl-3,4-dihydroquinazoline (**112**), initially present as its dianion before addition of water; the dianion proved to be a good substrate for several *N*- and *C*-alkylations.[905] 2-Methyl-3-*o*-tolyl-4(3*H*)-quinazolinone (**74**, R = X = Y = Z = H) reacted with dimethyl acetylenedicarboxylate in refluxing acetonitrile during 3 h to give a product, said to be tetramethyl 5-oxo-6-*o*-tolyl-5,6,10,11-tetrahydroazepino[1,2-*a*]quinazoline-7,8,9,10-tetracarboxylate (**113**) on nmr and ms evidence,[381] although it is not clear how an ester grouping came to occupy the 7-rather than the 11-position. 4-α-Cyanobenzylquinazoline underwent alcoholysis to 4-methoxyquinazoline (MeONa, MeOH, reflux, 5 h: ~10%) or aminolysis to 4-butylaminoquinazoline (BuNH$_2$, 175°, sealed, 5 h: 62%).[555]

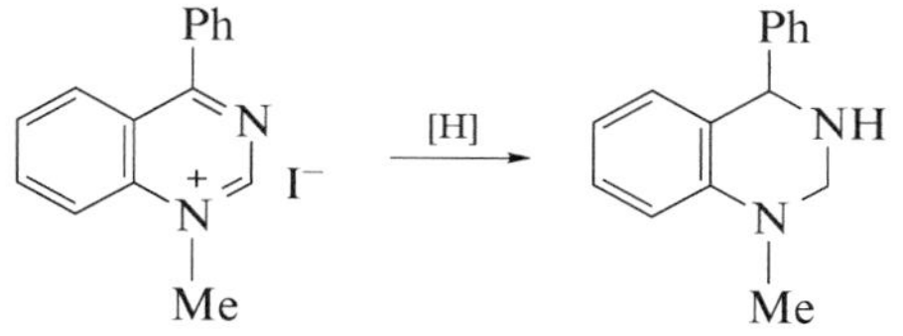

(**110**)   (**111**)

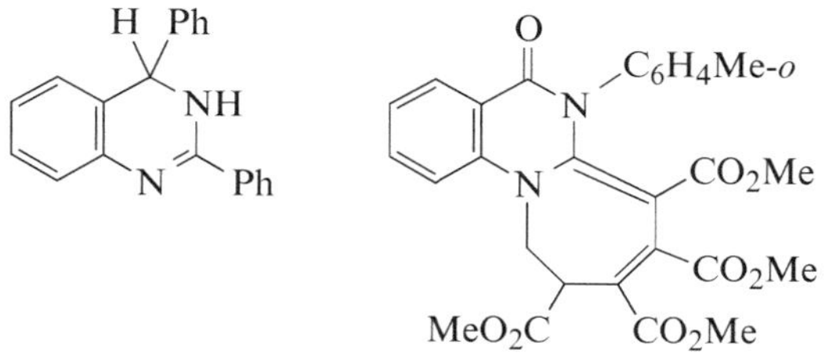

(**112**)   (**113**)

CHAPTER 3

Halogenoquinazolines

Halogeno substituents at the 2- or 4-position of quinazoline are naturally activated as leaving groups, in which respect they resemble the chloro substituent in chloro-2,4-dinitrobenzene. In contrast, halogeno substituents at the 5- to 8-positions are only mildly activated and require quite vigorous conditions for nucleophilic displacement. Extranuclear halogeno substituents are usually little affected by the quinazoline nucleus, and their reactions are reasonably predictable from their molecular environment. Because of their convenient reactivities, halogenoquinazolines (especially 2-, 4-, and some extranuclear halogeno derivatives) have proved to be very useful intermediates for quinazoline metatheses. Since there is little practical difference between corresponding chloro-, bromo-, iodo-, and even fluoroquinazolines, the more easily obtained chloro compounds are the ones usually employed for synthetic procedures; only in molecules designed as candidate drugs are other halogeno substituents used to any extent.

3.1. PREPARATION OF 2- OR 4-CHLOROQUINAZOLINES (*H* 219, 222, 225)

Most 2- or 4-chloroquinazolines have been made from corresponding quinazolinones with phosphoryl chloride, phosphorus pentachloride, or a Vilsmeier reagent.

3.1.1. From Quinazolinones with Phosphoryl Chloride

Phosphoryl chloride may be used alone or with a tertiary amine like *N,N*-diethylaniline, as illustrated in the following examples.

Phosphoryl chloride alone

2(1*H*)-Quinazoline gave 2-chloroquinazoline (**1**) (neat POCl$_3$, reflux, 2h: 71%).[849]

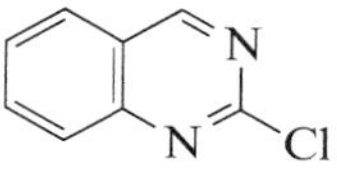

(**1**)

6-Acetamido-7-methoxy-4(3H)-quinazolinone gave 6-acetamido-4-chloro-7-methoxyquinazoline (neat POCl$_3$, reflux, 2h: 74%).[2017]

2-Cyanomethyl-4(3H)-quinazolinone gave 4-chloro-2-cyanomethylquinazoline (2) (neat POCl$_3$, 95°, 2h: 50%).[1161]

(2)

4-Ethyl-5,6,7,8-tetrahydro-2(1H)-quinazolinone   gave   2-chloro-4-ethyl-5,6,7,8-tetrahydroquinazoline (neat POCl$_3$, reflux, 4h: 67%).[1416]

Also other examples.[123,519,707,1075,1097,1113,1501,2016,2020,2564] including a reported use of P$_2$O$_5$/POCl$_3$.[887]

Phosphoryl chloride with a tertiary amine

2-Diethylamino-6,7-dimethoxy-4(3H)-quinazolinone gave 4-chloro-2-diethyl-amino-6,7-dimethoxyquinazoline (3) (POCl$_3$, Me$_2$NPh, reflux, 2h: 85%);[1810] 4-chloro-2-quinazolinamine likewise (1h: 75%).[1723]

(3)

8-Chloro-6,7-dimethoxy-2,4(1H,3H)-quinazolinedione gave 2,4,8-trichloro-6,7-dimethoxyquinazoline (POCl$_3$, Et$_2$NPh, reflux, 4h: >95%).[1146]

6,7-Diethoxy-2,4-(1H,3H)-quinazolinedione gave 2,4-dichloro-6,7-diethoxy-quinazoline (POCl$_3$, Me$_2$NPh, reflux, 3h: >95%).[642]

2-Propyl-4(3H)-quinazolinone gave 4-chloro-2-propylquinazoline (POCl$_3$, Me$_2$NPh, PhH, reflux, 2h: 85%);[188] homologs likewise.[188,1127]

5-Benzyloxy-6-methoxy-4(3H)-quinazolinone gave 5-benzyloxy-4-chloro-6-methoxyquinazoline (POCl$_3$, NEt$_3$, PhH, reflux, N$_2$, 3h: 80%).[1728]

6-Piperidino-2,4(1H,3H)-quinazolinedione gave 2,4-dichloro-6-piperidino-quinazoline (POCl$_3$, EtNPr$^i_2$, reflux, 18h: 66%).[1584]

2,4(1H,3H)-Quinazolinedione   gave   2,4-dichloroquinazoline   [POCl$_3$, Bu^{t}N(CH$_2$)$_4$, 80°, 20 min: 89%] or 4-chloro-2-(N-δ-chlorobutyl-N-methyl-amino)quinazoline (4) [POCl$_3$, MeN(CH$_2$)$_4$, 80°, 20 min: 72% plus a trace of the 2,4-dichloro product].[1380,1412,cf.1374,1410,2068]

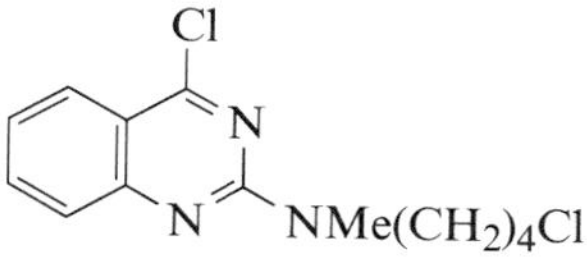

(4)

2,4-(1H,3H)-Quinazolinedione gave 4-chloro-2-(4-methylpiperazin-1-yl)quinazoline [POCl$_3$, NPr$_3$, MeN(CH$_2$CH$_2$)$_2$NMe, dioxane, 100°, 1h: 74%].[2068]

Also other examples.[674,696,756,785,1037,1175,2363,2595]

Note: In view of the foregoing and other data, it would seem best to avoid using the lower (less hindered) trialkylamines and related cyclic entities as adjuvants to chlorinations with phosphoryl chloride.

3.1.2. From Quinazolinones with Phosphorus Pentachloride

The most powerful reagent for converting quinazolinones into chloroquinazolines is phosphorus pentachloride. Today it is employed widely with phosphoryl chloride as a convenient solvent, but, if fact, such a mixture seldom offers any real advantage over the use of phosphoryl chloride alone or with a tertiary amine. The following examples indicate the conditions used and the yields to be expected.

4(3H)-Quinazolinone gave (unstable) 4-chloroquinazoline (PCl$_5$, POCl$_3$, reflux, 2h: 77–79%);[1139,1189, cf. 1266] 4-chloro-5-methylquinazoline likewise (75%).[1814]

6-Chloro-4-phenyl-2(1H)-quinazolinone gave 2,6-dichloro-4-phenylquinazoline **(5)** (PCl$_5$, POCl$_3$, reflux, 4h: 60%);[836] 4-chloro-6-methoxy-2-phenylquinazoline likewise (45 min: 79%).[766]

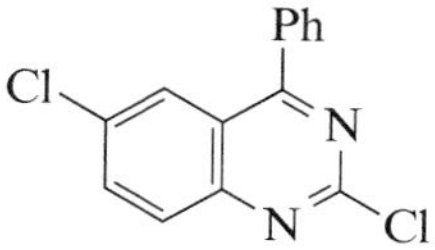

(5)

2-Pyridin-2′-yl-4(3H)-quinazolinone gave 4-chloro-2-pyridin-2′-ylquinazoline (PCl$_5$, POCl$_3$, reflux, 4h: 94%);[1602] 4-chloro-2-ethoxyquinazoline likewise (60%).[2230]

3-Phenyl-2,4(1H,3H)-quinazolinedione gave 2-chloro-3-phenyl-4(3H)-quinazolinone **(6**, R = Ph) (PCl$_5$, POCl$_3$, reflux, 6h: 84%);[1225] 2-chloro-3-ethoxycarbonylmethyl-4(3H)-quinazolinone **(6**, R = CH$_2$CO$_2$Et) likewise (5 days: 40%).[1137]

(6)

7-Trifluoromethyl-4(3H)-quinazolinone gave 4-chloro-7-trifluoromethyl-quinazoline (PCl₅, POCl₃, reflux, 3h: ~60%).[2165]

cis-3,4,4a,5,6,7,8,8a-Octahydro-2(1H)-quinazolinone gave the (unpurified) 2-chloro analog (**7**) (PCl₅, POCl₃, 130°, sealed, 3h), characterized by conversion into cis-3,4,4a,5,6,7,8,8a-octahydro-2-quinazolinamine hydrochloride (**8**) (NaNH₂, liquid NH₃, briefly: 83% overall).[1004]

(7)   **(8)**

2,4(1H,3H)-Quinazolinedione gave 2-4-dichloroquinazoline (PCl₅, POCl₃, reflux, 2h: 90%).[814]

Also other example.[123,128,443,503,553,672,686,1001,1021,1069,1184,1186,1660, 1866,2087,2096]

3.1.3. From Quinazolinones with a Vilsmeier Reagent

The Vilsmeier reagent (ClHC=N̄Me₂Cl⁻) may be made *in situ* from dimethylformamide with phosphoryl chloride, phosgene, thionyl chloride, or the like and then used immediately to convert a quinazolinone into its chloro analog. The method, as illustrated in available examples, appears to offer no particular advantage.

Treatment of 6-acetoxy-2-phenyl-4(3H)-quinazolinone with an equal weight of dimethylformamide and an excess of thionyl chloride under reflux for 30 min gave 6-acetoxy-4-chloro-2-phenylquinazoline (**9**) (84%);[766] 2-p-methoxyphenyl-4(3H)-quinazolinone with dimethylformamide and phosphoryl chloride under reflux for 1h gave 4-chloro-2-p-methoxyphenylquinazoline (80%);[968] passing phosgene into a solution of 2-o-hydroxyphenyl-4(3H)-quinazolinone and dimethylformamide in o-dichlorobenzene at 20° for 2h gave 4-chloro-2-o-hydroxyphenylquinazoline in unstated yield;[226] 2-acetyl-5,6,7,8-tetrahydro-4(3H)-quinazolinone in refluxing chloroform containing dimethylformamide and thionyl chloride for 1h gave 2-acetyl-4-chloro-5,6,7,8-tetrahydroquinazoline (**10**) (72%);[1982] and 6,8-dichloro-2,4(1H,3H)-quinazolinedione with phosphorus pentachloride, phosphoryl chloride, and dimethylformamide under reflux for 16h gave 2,4,6,8-tetrachloroquinazoline (43%).[341] Other examples have been reported.[2547]

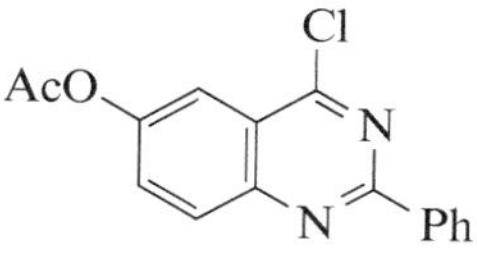

(9) 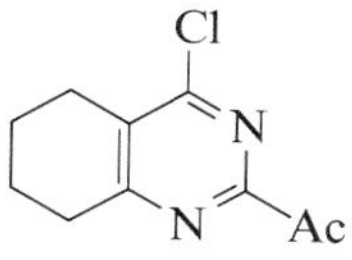   **(10)**

3.2.   PREPARATION OF 2- OR 4-(BROMO, IODO, OR FLUORO)QUINAZOLINES (*H* 222, 225)

Very few such halogenoquinazolines have been made recently, and only two routes are represented.

From a quinazolinethione. Treatment of 4-carboxymethyl-3,4-dihydro-2(1*H*)-quinazolinethione (**11**) in 90% acetic acid at <40° with bromine over 30 min gave 2-bromo-4-carboxymethyl-3,4-dihydroquinazoline (**12**) as hydrobromide (~75%).[975]

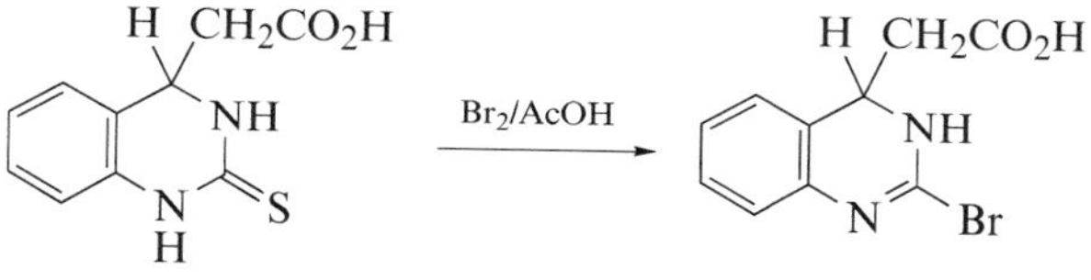

(11)   **(12)**

Transhalogenation. This method has been used to convert 4-chloro- into 4-fluoroquinazoline (**13**, R = H) (48%) by heating with neat potassium fluoride at 250° for 8 h;[699] to make 2,4-difluoroquinazoline (**13**, R = F) similarly (53%);[699] to convert appropriately substituted 4-chloroquinazolines into 4-fluoro-6-nitro- (24%), 4-fluoro-7-nitro- (38%), 4-fluoro-8-nitro- (26%), 5-chloro-4-fluoro- (61%), 6-chloro-4-fluoro- (35%), 7-chloro-4-fluoro- (49%), 8-chloro-4-fluoro- (51%), and 6,8-dichloro-4-fluoroquinazoline (51%), in each case by heating with potassium fluoride in diethylene glycol at 120–150° for 8–20 h;[553] and to transform 2,4,5,6,7,8-hexachloro- into 2,4,5,6,7,8-hexafluoroquinazoline in good yield by heating with neat potassium fluoride at 350°![134]

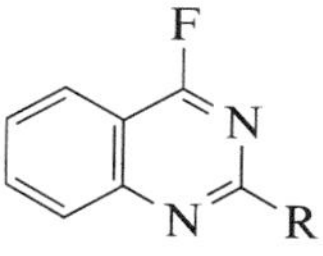

(13)

3.3. PREPARATION OF 5- TO 8-HALOGENOQUINAZOLINES (*H* 226)

Most of these halogenoquinazolines have been made by *primary syntheses* (see Cahpter 1), but there are two other commonly used routes—direct halogenation and via diazotized quinazolinamines—illustrated in the following examples.

Halogenations

5-Ethoxy-8-quinazolinol gave its 6-chloro derivative (**14**) (Cl$_2$, AcOH, 20°, 3 days: ∼ 50%);[595] 6,7-dichloro-5-methoxy-8-quinazolinol was made somewhat similarly.[1095]

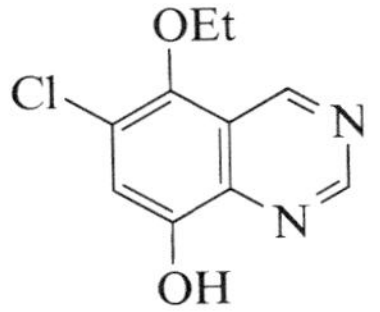

(**14**)

5,8-Dimethoxyquinazoline gave 6,7-dichloro-5,8-dihydro-5,8-quinazolinedione (**15**) (10M HCl, HNO$_3$, 90°, 15 min: 10%; note additional hydrolysis and oxidation).[1604]

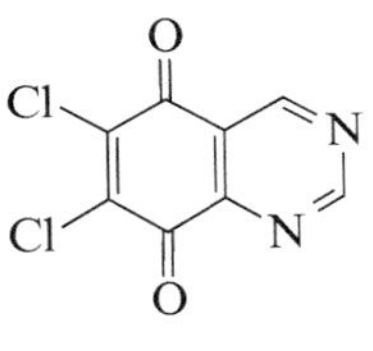

(**15**)

2,4-Dichloro- gave 2,4,5,6,7,8-hexachloroquinazoline (PCl$_5$, sealed, 300°).[34]

1-Hydroxy- gave 6-chloro-2,4(1*H*,3*H*)-quinazolinedione (PCl$_5$, POCl$_3$, reflux, 90 min; then 10M HCl, reflux, 3h: 20%; via a 1-dichlorophosphinyloxy intermediate?).[471]

3-Methyl-4(3*H*)-quinazolinone gave its 6-bromo derivative (**16**) Br$_2$, MeOH–H$_2$O, 20°, 5h: 69% as hydrobromide).[907]

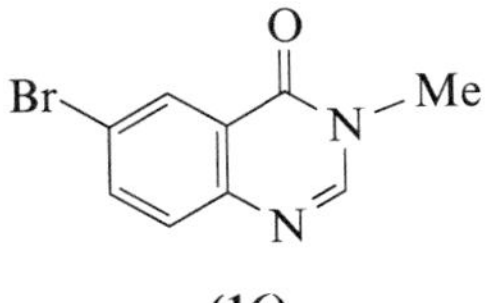

(**16**)

6-Methoxy-5,8-dihydro-5,8-quinazolinedione gave its 7-bromo derivative (Br_2, $CHCl_3$, 0°, 60h, N_2: 80%).[1577]

6-Acetamido-5,8-dihydro- gave 6-amino-7-bromo-5,8-dihydro-2,4,5,8 ($1H,3H$)-quinazolinetetrone (17) (Br_2, AcOH, reflux, 5 min: 87%; note deacylation).[1987]

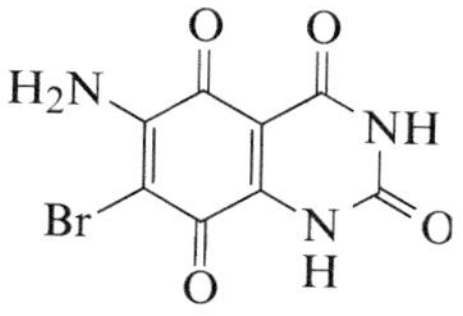

(17)

2,4-Bistrifluoromethyl-5,6,7,8-tetrahydroquinazoline gave its 5,8-dibromo derivative (18) (N-bromosuccinimide, CCl_4, reflux, 1h: 54%) and thence 2,4-bistrifluoromethylquinazoline by a double dehydrobromination (Et_3N, Et_2O, 0°, 12h: 80%).[1145]

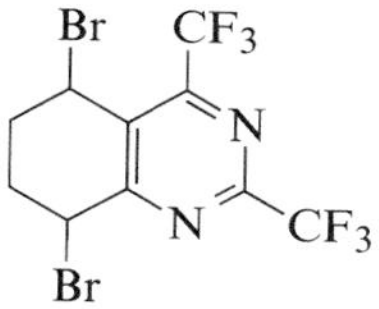

(18)

Also other examples.[1490,1848,1886]

From diazotized quinazolinamines

5-Methyl-2,4,6-quinazolinetriamine (19, R = NH_2) was diazotized ($NaNO_2$, 2M HCl, 0°, 30 min), and the solution was treated with CuCl/HCl (20°, 6h) or KI/H_2O (0° → 70°, 2h) to give 6-chloro- (19, R = Cl) (78%) or 6-iodo-5-methyl-2,4-quinazolinediamine (19, R = I) (70%), respectively; the bromo analog (19, R = Br) was made rather similarly (diazotization in 2M $MeSO_3H$; CuBr/HBr: 75%).[2246]

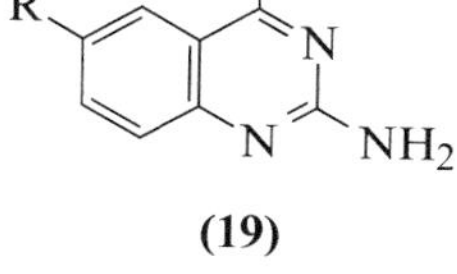

(19)

5-Fluoro-2,4,8-quinazolinetriamine gave 8-chloro-5-fluoro-2,4-quinazolinediamine ($NaNO_2$, 2M HCl, 0°; then CuCl, HCl, 0° → 20°, 5h: 63%).[2335]

5-Methyl-2,4,6-quinazolinetriamine gave 6-bromo-5-methyl-2,4-quinazoline-diamine (diazotization in HBr, 10°; then CuBr, HBr, 75°, 1 h: 70%, as hydrobromide).[2041]

6,8-Diamino- (**20**, R = NH$_2$) gave 6,8-diiodo-2,4(1H,3H)-quinazolinedione (**20**, R = I) (diazotization in HCl, 0°; then KI, trace Cu, H$_2$O, 0° → 100°, 20 min: 50%).[565]

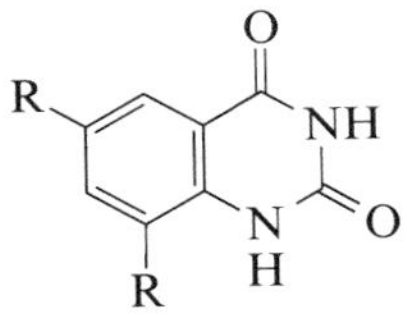

(20)

Also other examples.[242,678,1513]

3.4. PREPARATION OF EXTRANUCLEAR HALOGENOQUINAZOLINES (*H* 226)

Quinazolines, in which attached groups bear halogeno substituents, have been made by *primary syntheses* (see Chapter 1), by *direct halogenation*, (see Section 2.2.2.1) and by several other procedures illustrated in the following examples.

From hydroxyalkylquinazolines

3-β-Hydroxyethyl- (**21**, R = OH) gave 3-β-chloroethyl-2,4-(1H,3H)-quinazo-linedione (**21**, R = Cl) (SOCl$_2$, reflux, 30 min: >95%).[1310,1442]

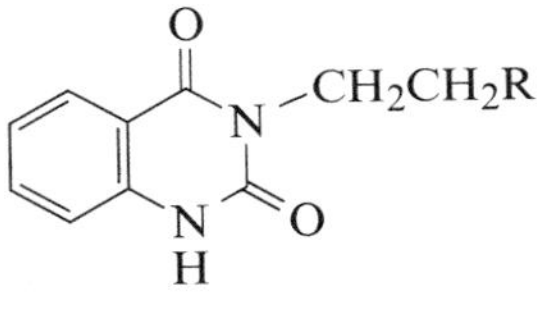

(21)

2-Amino-6-β-hydroxyethyl- (**22**, R = OH) gave 2-amino-6-β-bromoethyl-5,6,7,8-tetrahydro-4(3H)-quinazolinone (**22**, R = Br) (49% HBr, reflux, 4 h: >90%, as hydrobromide).[1591]

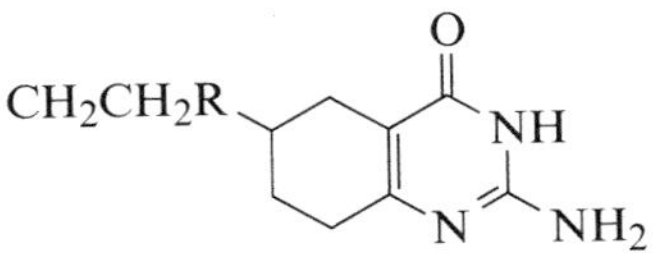

(22)

2-Hydroxymethyl-4(3H)-quinazolinone gave 4-chloro-2-chloromethylquina-
zoline (POCl$_3$, reflux, 2h: ~90%; note that nuclear chlorination by
POCl$_3$).[123]

Also other examples.[405,561,845,1073,1397,1611]

From aryloxyalkylquinazolines

5,8-Dimethoxy-2-phenoxymethyl- gave 2-bromomethyl-5,8-dihydroxy-
4(3H)-quinazolinone (BBr$_3$, PhH, reflux, 90 min: 68%; note the subtle
difference in reactivity of the two types of ether grouping).[1999]

Via diazoalkylquinazolines

8a-Aminomethyl- (**23**, R = NH$_2$) gave 8a-bromomethyl-3-phenyl-4a,5,6,
7,8,8a-hexahydro-2,4(1H,3H)-quinazolinedione (**23**, R = Br) (NaNO$_2$,
KBr, H$_2$SO$_4$/H$_2$O, 0°, 30 min; then 20°, 1h: 80%).[194, cf. 190]

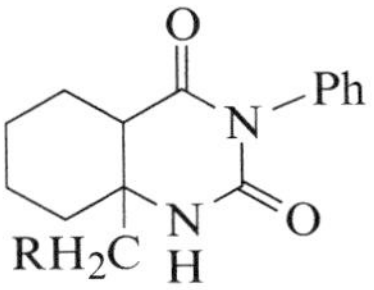

(**23**)

2-o-Piperidinoazophenyl- [**24**, R = N:NN(CH$_2$)$_5$] gave 2-o-iodophenyl-4-
quinazolinamine (**24**, R = I) (NaI, AcOH, reflux, 2h: ?%).[1014]

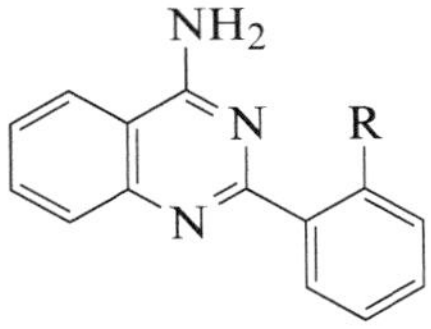

(**24**)

4-Phenylquinazoline 1-methiodide (**25**, X = I) or 1-methobromide (**25**, X = Br)
gave 2-iodomethyl- (**27**, X = I) or 2-bromomethyl-1-methyl-4-phenyl-1,2-
dihydroquinazoline (**27**, X = Br), respectively [CH$_2$N$_2$, THF–Et$_2$O,
15 min: 70% or 28% (the latter as a picrate), respectively, probably via
a diazonio intermediate (**26**)].[353]

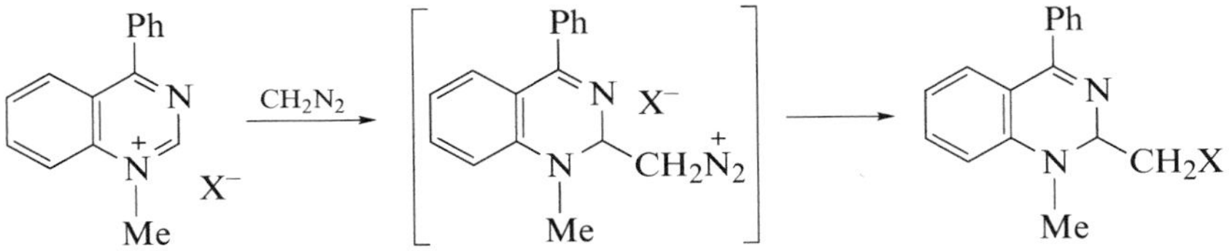

(**25**)   (**26**)   (**27**)

2-p-Chlorophenyl-5-diazoacetyl- (**28**, R = CH$_2$N$_2$) gave 5-bromoacetyl-2-p-chlorophenylquinazoline (**28**, R = CH$_2$Br) (HBr gas, Et$_2$O, 20°, 12 h: 60%, as hydrobromide).[672]

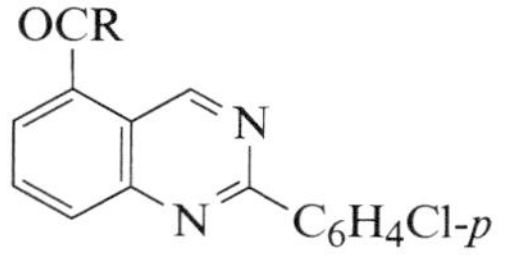

(28)

By transhalogenation

2-Chloromethyl- gave 2-fluoromethyl-3-o-tolyl-4(3H)-quinazolinone (**29**, R = X = H) (KF, HOCH$_2$OH, 160°, 2h; 69%);[1550, cf. 687] likewise, the 6-nitro derivative (**29**, R = NO$_2$, X = H) (CsF, HOCH$_2$CH$_2$OH, 130°, 50 min: 67%).[1550]

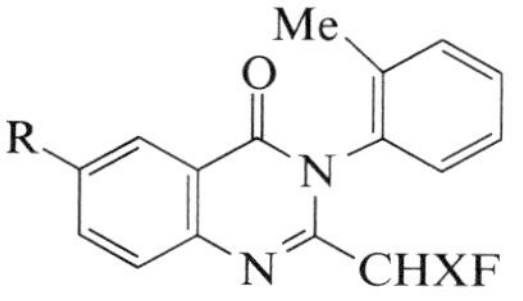

(29)

6-Amino-2-dichloromethyl- gave 6-amino-2-difluoromethyl-3-o-tolyl-4(3H)-quinazolinone (**29**, R = NH$_2$, X = F) (CsF, HOCH$_2$CH$_2$OH, 160°, 6h: 42%).[1397]

2-Chloromethyl- gave 2-iodomethyl-4-phenylquinazoline 3-oxide (**30**) (NaI, AcMe, reflux, 1h: 75%).[632]

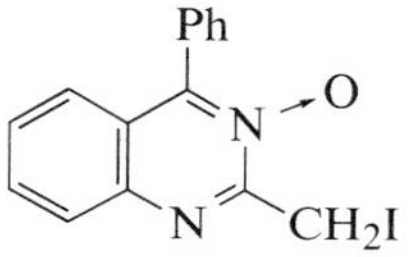

(30)

2-Bromomethyl- gave 2-fluoromethyl- (**29**, R = X = H) [KF, "digol" (?), 130°, 2h: 76%] or 2-iodomethyl-3-o-tolyl-4(3H)-quinazolinone (KI, AcMe, reflux, 1h: 87%).[786]

2-Bromomethyl- gave 2-chloromethyl-5,8-dihydroxy-4(3H)-quinazolinone (**31**) (10M HCl, 20°, 22h: 87%, as a hydrate).[1998]

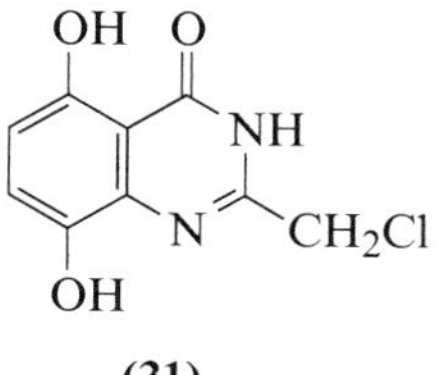

(31)

Also other examples in the foregoing references and elsewhere.[1276, 2430]

In addition, extranuclear halogeno substituents have often been introduced *as passenger groups.* For example, 5-chloro-4(3H)-quinazolinone underwent alkylation by neat 1-chloro-2,3-epoxypropane at 100° during 4h to afford 30% of 5-chloro-3-(γ-chloro-β-hydroxypropyl)-4(3H)-quinazolinone (32),[883] and so on.[2469]

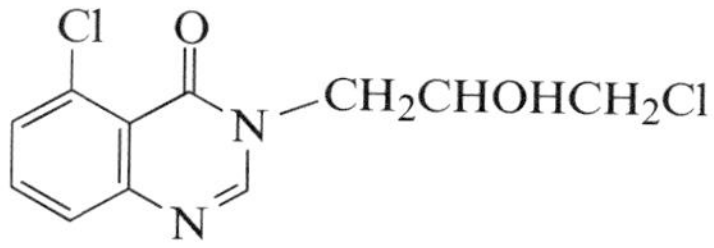

(32)

3.5.   REACTIONS OF 2- OR 4-HALOGENOQUINAZOLINES
(*H* 227)

These active halogenoquinazolines have been used extensively as substrates, mainly for nucleophilic reactions. Their *conversion into alkyl- or (substituted alkyl) quinazolines* has been discussed in Section 2.2.1.2. In the classification of other reactions, substituents at the 6- to 8- or extranuclear positions have been ignored as being almost irrelevant.

3.5.1.   Aminolysis of 2- or 4-Halogenoquinazolines
(*H* 323, 327, 330)

Good correlation has been obtained between an HMO (highest molecular orbital) approach[42] and experimental kinetic data (*H* 228) for the reaction of 2- or 4-chloroquinazoline with piperidine in ethanol; the 4-isomer is appreciably the more reactive. However, a great variety of both 2- and 4-halogenoquinazolines has been used successfully for all manner of aminolyses, as indicated in the following subsections.

3.5.1.1.   With Simple 2-Halogenoquinazolines
as Substrates

The scope of such reactions is indicated in the following examples.

2-Chloroquinazoline gave 2-quinazolinamine (33, R = H) (NH$_3$ gas, dioxane, 20°, 1h; then reflux, 3h: 52%),[849] 2-methylaminoquinazoline (MeNH$_2$, likewise: 89%),[849] 2-dimethylaminoquinazoline (Me$_2$NH, likewise: 71%),[849] or 2-trimethylammonioquinazoline chloride (Me$_3$N, PhH, 5°, 24h: ~80%).[165]

2-Chloro-4-phenylquinazoline   gave   4-phenyl-2-quinazolinamine   (**33**,
R = Ph) (NH$_3$ gas, EtOH, 150°, sealed 24h: >95%)[343] or 4-phenyl-
2-trimethylammonioquinazoline   chloride   (**34**)   (Me$_3$N, AcMe, 20°,
12h: 90%).[2206]

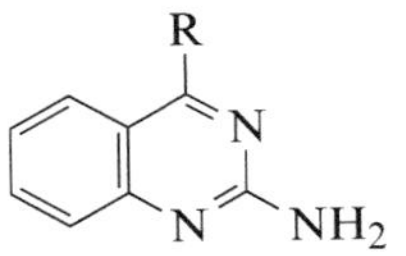

(**33**) 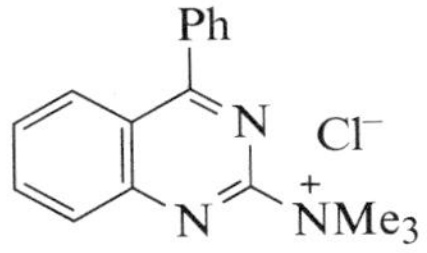   (**34**)

2,6-Dichloro-   gave   6-chloro-2-hydrazino-4-phenylquinazoline
(H$_2$NNH$_2$.H$_2$O, PhH, <20°, 2h: 40%).[836, cf. 1718]
Also other examples in the foregoing references and elsewhere.[349,647,1021, 1142,2530]

3.5.1.2.   With Simple 4-Halogenoquinazolines as Substrates

Although 4-halogenoquinazolines are more active than their 2-isomers, this
has seldom been reflected in the conditions used for aminolyses because of an
almost universal tendency to use excessively vigorous conditions to ensure
completion of the reaction. The following examples illustrate the scope and
conditions actually employed for such preparative aminolyses.

4-Fluoro- (**35**, R = F) or 4-chloro-6-nitroquinazoline (**35**, R = Cl) gave 6-nitro-
4-quinazolinamine (**35**, R = NH$_2$) (NH$_3$, EtOH, 20°, 10 min: 79% or 82%,
respectively).[553]

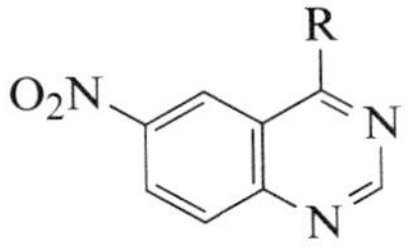

(**35**)

4-Chloro- gave 4-p-hydroxyanilino- (HOC$_6$H$_4$NH$_2$-p, EtOH, 20°, then reflux,
2.5h: 60%),[2291] 4-o-hydroxyanilino-(HOC$_6$H$_4$NH$_2$-o, EtOH, reflux, 2h:
71%),[1559] 4-β,β-diethoxyethylamino-[(EtO)$_2$CH$_2$CH$_2$NH$_2$, PhH, 20°, 1h:
42%][1762]   or   4-β-piperidinoethylaminoquinazoline   [(CH$_2$)$_5$NCH$_2$
CH$_2$NH$_2$, PhOH, 140°, 8h: ?%].[514]
4-Chloro- gave 4-isopropylamino- (Pr^{i}NH$_2$, AcMe, reflux, 2h: 92%),[1025]
4-β-dimethylaminoethylamino- (neat Me$_2$NCH$_2$CH$_2$NH$_2$, reflux, 2h:

89%, as dihydrochloride),[1601] 4-α-methylhydrazino (MeHNNH$_2$, CH$_2$Cl$_2$, <30°, 45 min, then 0°, 12h: 80%).[2230] 4-anilino- (PhNH$_2$, AcMe–H$_2$O, trace HCl, reflux, 40 min: 59%),[728] 4-p-cyanoanilino(H$_2$NC$_6$H$_4$CN-p, AcMe–H$_2$O, trace HCl, reflux, 1 h: 90%),[1353] or 4-p-hydroxyanilino-2-phenylquinazoline (HOC$_6$H$_4$NH$_2$-p, EtOH, reflux, 45 min: ~80%).[503]

4-Chloro- gave 4-(3,5-bisdimethylaminomethyl-4-hydroxyanilino)-7-trifluoromethylquinazoline [3,5-(Me$_2$NCH$_2$)$_2$-4-OHC$_6$H$_2$NH$_2$, MeOH–H$_2$O, trace HCl, reflux, 5 h: ~60%];[2165] analogs likewise.[2119,2165]

4-Chloro- gave 4-piperidino- [neat HN(CH$_2$)$_5$, N$_2$, 20°, 2 h: 70%][1139, cf. 340] or 4-aziridin-1'-ylquinazoline [HN(CH$_2$)$_2$, Et$_3$N, PhH, 20°, 3 h: 63%;[219,473] HN(CH$_2$)$_2$, CH$_2$Cl$_2$, <5°, 90 min: 85%];[474] also 4-trimethylammonioquinazoline chloride (Me$_3$N, PhH, 4°, 8 days: ~60%).[165]

4-Chloro-2-phenylquinazoline gave 2-phenyl-4-trimethylammonioquinazoline chloride (Me$_3$N, AcMe, 20°, 12h: 91%;[2206] Me$_3$N, PhMe, 20°, 7 days: 68%),[923] 4-benzimidazol-1'-yl-2-phenylquinazoline (sodiobenzimidazole, PhMe–MeCN, reflux, 7 h: 70%),[2480] or 4-hydroxyamino-2-phenylquinazoline (H$_2$NOH.HCl, Na$_2$CO$_3$, H$_2$O–EtOH–Me$_2$NCHO, reflux, 6 h: 84%).[1872]

4-Chloro- gave 4-hydrazine-2-methylquinazoline (36) (H$_2$NNH$_2$.H$_2$O, EtOH, <5°, then 20°, 2 h: 81%; see original for details).[1127]

(36)

5-Benzyloxy-4-chloro-6-methoxyquinazoline gave α,β-bis(5-benzyloxy-6-methoxyquinazolin-4-ylamino)ethane (37) (H$_2$NCH$_2$CH$_2$NH$_2$, Et$_3$N, EtOH, 20°, 24 h: 74%)[1728] or analogs with diverse bridges.[1728]

(37)

4-Chloroquinazoline gave 2,3-dihydroimidazo[1,2-c]quinazoline (39) (ClCH$_2$CH$_2$NH$_2$, excess NaHCO$_3$, EtOH–H$_2$O, reflux, 6 h: ~75%), presumably via the unisolated intermediate (38) or (40).[795]

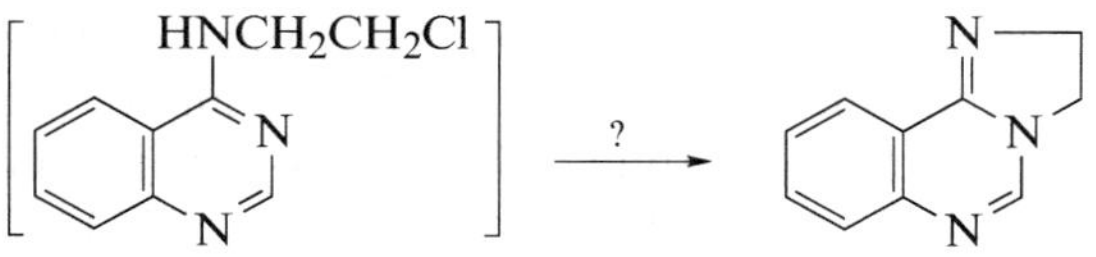

(38)   (39)

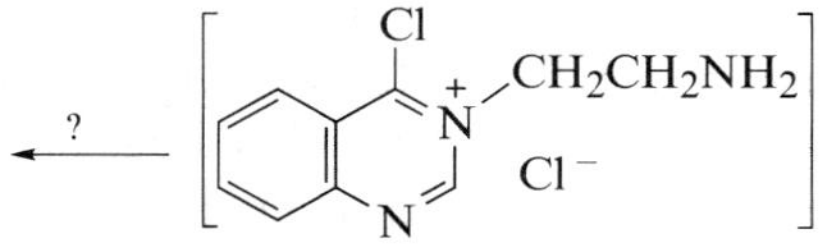

(40)

Also other examples.[188,305,315,324,551,561,593,707,749,756,764,766,1037,1069, 1113,1142,1189,1208,1501,1761,1909,2102,2129,2428,2451,2564]

3.5.1.3.   With 2,4-Dihalogenoquinazolines as Substrates

2,4-Dihalogenoquinazolines undergo aminolysis by *ammonia*, *primary amines*, or *secondary amines* at the 4-position under minimal conditions and subsequently at the 2-position only under more vigorous conditions; under sufficient stimulation, both steps occur in the same reaction mixture. 4-Mono-aminolyses and 2,4-diaminolyses of such dichloroquinazolines are illustrated in the following examples but second-stage (2-aminolyses) are included in Section 3.5.1.4.

2,4-Dichloro-6,7-dimethoxyquinazoline (**41**, X = Y = Cl) gave 2-chloro-6,7-dimethoxy-4-quinazolinamine (**41**, X = NH$_2$, Y = Cl) (NH$_3$/THF, 20°, 44 h: 68%)[785] or 6,7-dimethoxy-2,4-quinazolinediamine (**41**, X = Y = NH$_2$) (NH$_3$/EtOH, 160°, sealed, 65 h: 77%;[785] 180°, 7 h: 71%).[696]

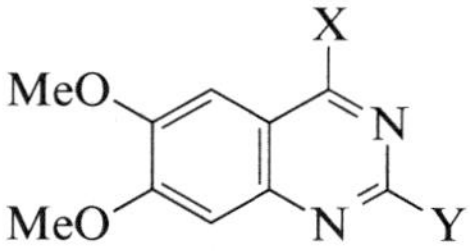

(41)

2,4,6-Trichloroquinazoline gave 6-chloro-2,4-quinazolinediamine (NH$_3$/ EtOH, 150°, sealed, 17 h: 70%).[647]

2,4-Dichloro-7-nitroquinazoline gave 7-nitro-2,4-quinazolinediamine (NH$_3$↓, PhOH, reflux, 2 h: 83%).[600]

2,4-Dichloroquinazoline   gave   2-chloro-4-prop-2′-ynylamino-   [HC≡
CCH$_2$NH$_2$, Et$_3$N, THF, 20°, until substrate gone (tlc): 90%)][643] or 2-
chloro-4-β-pyrrolidinoethylaminoquinazoline   [(CH$_2$)$_4$NCH$_2$CH$_2$NH$_2$,
Et$_2$O, 20°, 90 min: 56%].[1568]

2,4-Dichloro-   gave   4-aziridin-1′-yl-2-chloro-   [42,   X + Y = (CH$_2$)$_2$]
[(CH$_2$)$_2$NH, K$_2$CO$_3$, PhH, 10°, 30 min: 81%],[474] 2-chloro-4-diethy-
laminoquinazoline (42, X = Y = Et) (Et$_2$NH, Et$_3$N, EtOH–H$_2$O, 20°, 1 h,
then reflux, 10 min: 93%),[166] or 4-(N-butyl-N-methylamino)-2-chloro-
quinazoline (BuMeNH$_2$, CH$_2$Cl$_2$, 25°, 30 min: 84%).[1412]

(42)

2,4-Dichloro- gave 2,4-bisdiethylamino-6-nitroquinazoline (Et$_2$NH, EtOH,
reflux, 4 h: 75%).[600]

Also other examples in the foregoing references and elsewhere,[34,610,674,
1146,1179,1266,1810,2230,2336,2551,2595]

2,4-Dihalogenoquinazolines behave quite differently with *tertiary amines*, but
it does not seem possible to rationalize the following examples of available data in
a totally satisfactory way.

2,4-Dichloro- (44) gave 4-chloro-2-piperidinoquinazoline (45) [MeN(CH$_2$)$_5$,
dioxane, 95–100°, 1h: 92%], presumably by loss of methyl chloride from
a thermally unstable quaternary amine (43, n = 5).[2358,cf. 2068]

2,4-Dichloro- (44) gave 2-chloro-4-(N-methylpyrrolidinio)quinazoline chloride (47) [MeN(CH$_2$)$_4$, AcMe, reflux, 2 h: 93%] and thence 2-chloro-4-(N-δ-
chlorobutyl-N-methylamino)quinazoline (48) (100°, 1 h: 67%);[1409] the
same reactants with phosphoryl chloride as solvent (!) gave only the
isomeric 4-chloro-2-(N-δ-chlorobutyl-N-methylamino)quinazoline (46)
(100°, 30 min: 30%).[2068]

2,4-Dichloro- (44) gave 2,4-bis(trimethylammonio)quinazoline dichloride (?)
(Me$_3$N, PhH, 5° → 20°, 2h: 81% crude).[1142]

2,4(1H,3H)-Quinazolinedione gave a separable mixture of 2,4-dich-
loroquinazoline (44) and 2-chloro-4-(N-δ-chlorobutyl-N-methylamino)
quinazoline (48) [POCl$_3$, MeN(CH$_2$)$_4$, 80°, 20 min: 3% and 77%, respect-
ively]; homologous N-alkylpyrrolidines gave increasinly higher proportions of dichloroquinazoline.[1412,cf.1374,1375,1380]

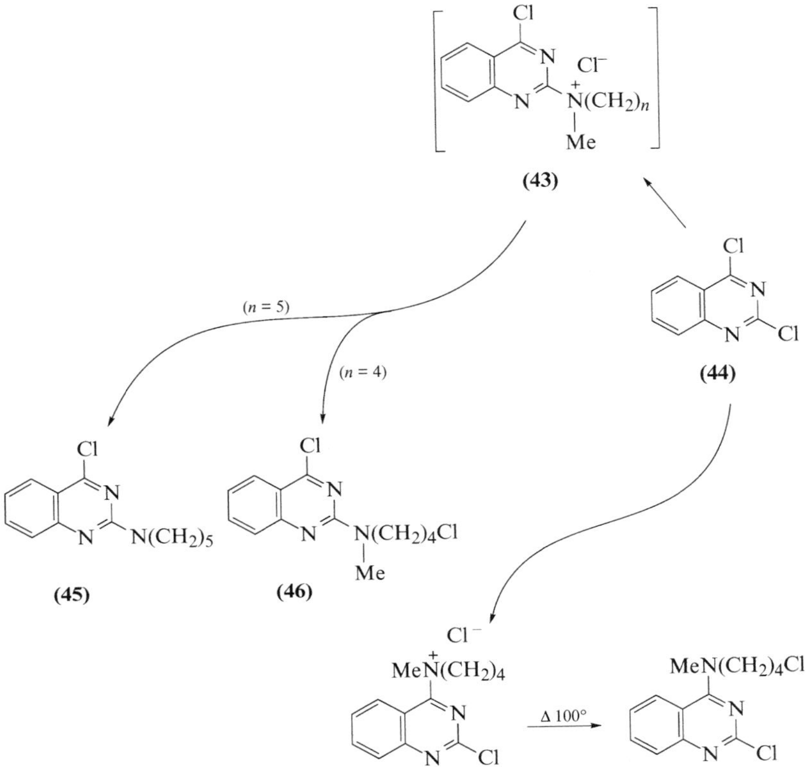

(43)

(n = 5)

(n = 4)

(44)

(45)   (46)

(47) $\xrightarrow{\Delta\ 100^\circ}$ (48)

3.5.1.4. With 2-Halogeno-4- or 4-Halogeno-2-quinazolinamines as Substrates

The amino grouping in such substrates has an appreciable deactivating effect on the chloro substituent. This is evident in the relatively vigorous conditions employed, especially in view of the necessity to avoid overly severe conditions for fear of transamination. The following examples illustrate the important type of reaction that has been used extensively to make analogs of the antihypertensive drug, prazosin, the hydrochloride of [2-[4-(fur-2-oyl)piperazin-1-yl]-6,7-dimethoxy-4-quinazolinamine (49, R = H).

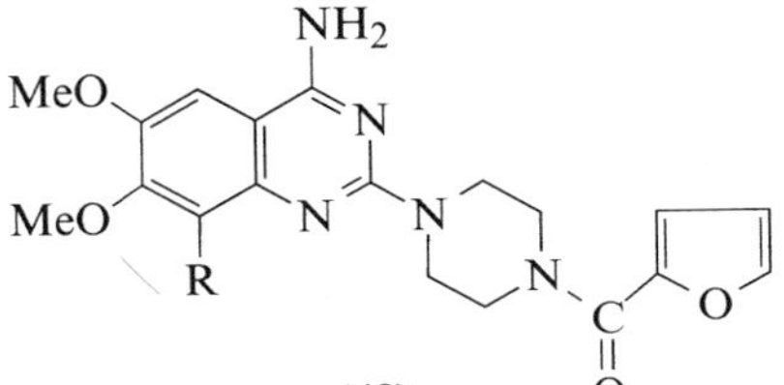

(49)

2-Chloro-6,7-dimethoxy-(**50**, R = Cl) gave 6,7-dimethoxy-2-piperazin-1′-yl-4-quinazolinamine [**50**, R = N(CH$_2$CH$_2$)$_2$NH] [HN(CH$_2$CH$_2$)$_2$NH.HBr, EtOH–H$_2$O, reflux, 5h: 75%, as the dihydrochloride].[785]

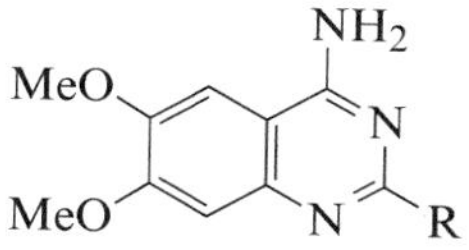

MeO — NH$_2$ — N — MeO — N — R

(50)

2,8-Dichloro- gave 8-chloro-2-[4-(fur-2-oyl)piperazin-1-yl]-6,7-dimethoxy-4-quinazolinamine (**49**, R = Cl) (1-fur-2′-oylpiperazine, Pr^{i}CH$_2$CH$_2$OH, reflux, 2.5 h: 73%);[1146] many prasozin analogs were made rather similarly in isopentyl alcohol[1557,1583,1599,1610,1618,1875,2238,2412] or butanol.[1594,1598,1624]

4-Chloro- gave 4-piperazin-1′-yl-5,6,7,8-tetrahydro-2-quinazolinamine (**51**) (HN(CH$_2$CH$_2$)$_2$NH, EtOH, reflux, 5 h: 87%).[1723]

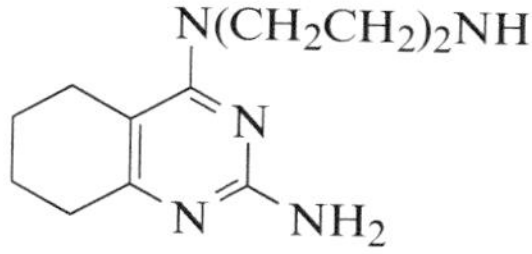

N(CH$_2$CH$_2$)$_2$NH

(51)

4-Chloro-2-p-chloroanilino- gave 2-p-chloroanilino-4-[3-(dimethylamino-methyl)-4-ethoxyanilino]quinazoline (3-Me$_2$NCH$_2$-4-EtOC$_6$H$_3$NH$_2$, AcOH, 95°, 2 h: ∼50%)[1266]

Also other examples in the foregoing references and elsewhere.[128,292,1579,2224,2595]

3.5.1.5. With Other 2-halogeno-4- or 4-Halogeno-2-substituted-quinazolines as Substrates

In these substrates, the accompanying substituents appear to have little activating or deactivating effect on the halogeno substituents, although some are themselves prone to aminolysis, especially if conditions are unnecessarily vigorous. The following examples indicate typical acceptable conditions and the resulting yields of products.

Halogenoquinazolinones

2-Chloro- (**52**, R = Cl) gave 2-benzylamino-4(3H)-quinazolinone (**52**, R = NHCH$_2$Ph) (PhCH$_2$NH$_2$, neat or in Me$_2$NCHO, 100°, <5 h: 83%).[1580]

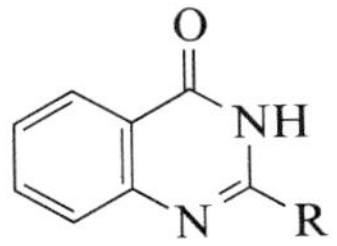

(52)

2-Chloro- gave 2-diethylamino- (Et_2NH, EtOH, 130°, sealed, 5 h: $> 89\%$)[641] or 2-anilino-6,7-dimethoxy-4(3H)-quinazolinone (neat $PhNH_2$, 100°, > 1 h: 86%).[1580]

2-Chloro- gave 2-hydrazino- **(53**, $R = NH_2$) (80% $H_2NNH_2 \cdot H_2O$, EtOH, reflux, 6 h: 92%;[1225] H_2NNH_2, EtOH, reflux, 3 min: 85%)[1894] or 2-butylamino-3-phenyl-4(3H)-quinazolinone **(53**, $R = Bu$) ($BuNH_2$, PhH, reflux, 6 h: 78%).[1225]

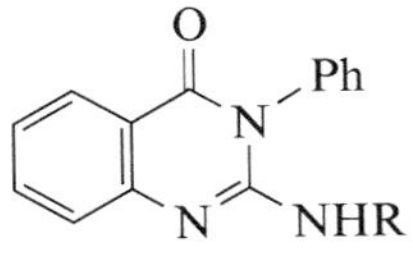

(53)

2-Chloro- gave 2-α-aminobenzylideneamino-4-oxo-3,4-dihydro-3-quinazolinecarbonitrile **(54)** [$PhC(=NH)NH_2$, "glyme," $-25° \rightarrow 20°$, 2 h: 22%] and thence 4-amino-2-phenyl-6H-[1,2,5]triazino[2,1-b]quinazolin-6-one **(55)** (Me_2NCHO, reflux, 3 min: complete; or MeOH, reflux: slowly).[234]

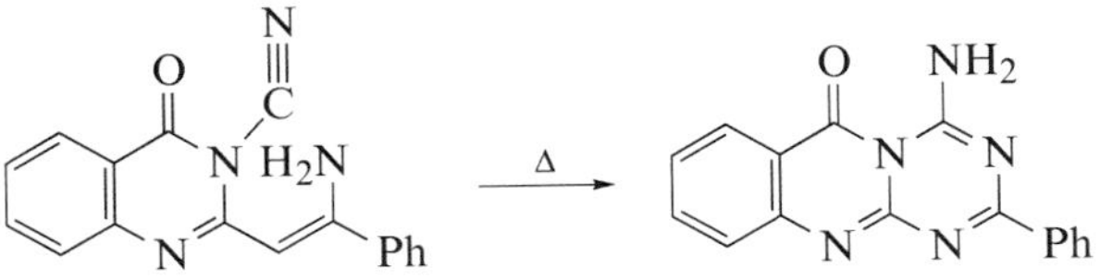

(54)   **(55)**

Also other examples in the foregoing references and elsewhere.[1524]

Halogenoquinazoline esters

Ethyl 4-chloro- gave ethyl 4-(4-methylpiperazin-1-yl)-2-quinazolinecarboxylate **(56)** [$MeN(CH_2CH_2)_2NH$, PhH, reflux, 12 h: $> 70\%$].[1186]

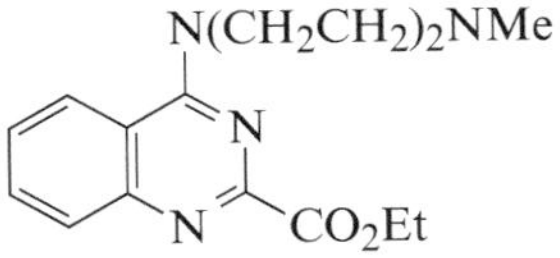

(56)

Dimethyl 2-chloro- gave dimethyl 2-anilino-4-quinazolinephosphonate ($PhNH_2$, trace HCl gas, PhMe, reflux, 2 h: 76%).[571]

Alkylthiohalogenoquinazolines

2-Chloro-4-methylthio- (57, X = Cl, Y = SMe) gave mainly 2-ethylamino-4-methylthio- (57, X = NHEt, Y = SMe) and 2-chloro-4-ethylaminoquinazoline (57, X = Cl, Y = NHEt) ($EtNH_2$, Et_3N, $H_2O–Me_2NCHO$, 80°, 20 min: 38% and 54%, respectively); the same substrate (57, X = Cl, Y = SMe) gave entirely 2-dipropylamino-4-methylthioquinazoline (57, X = NPr_2, Y = SMe) (Pr_2NH, Et_3N, Me_2NCHO, reflux, 3 h: 92%).[1379,1410]

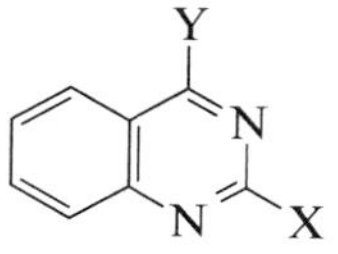

(57)

Note: From the foregoing and other data, it is clear that the bulkiness of the alkylthio group, the nature of the attacking amine, and the conditions all profoundly affect the nature and ratio of products.[1410,2547]

Halogenoquinazolines bearing extranuclear functional groups

2-Bromo-4-carboxymethyl- (58, R = Br) gave 4-carboxymethyl-2-hydrazino-3,4-dihydroquinazoline (58, R = $NHNH_2$) (H_2NNH_2, H_2O, 90°, 30 min: 64%).[975]

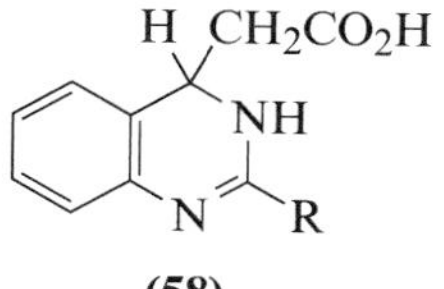

(58)

4-Chloro-2-(β,β-dichloro-α,α-difluoroethyl)quinazoline gave 2-(β,β-dichloro-α,α-difluoroethyl)-4-quinazolinamine ($NH_3\downarrow$, PrOH, reflux, 4 h: 72%).[443]

4-Chloro- gave 4-β-hydroxyethylamino-2-(5-nitrothien-2-yl)quinazoline (59) ($HOCH_2CH_2NH_2$, Me_2NCHO, 100°, 5 h: 83%).[686]

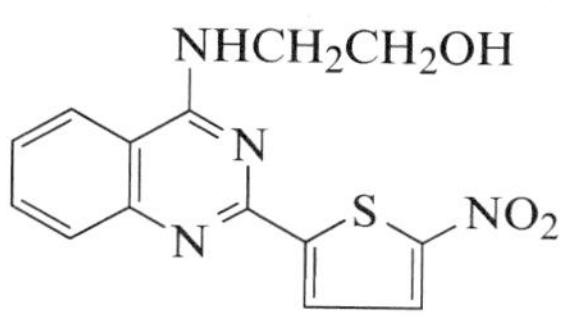

(59)

Also other examples.[123,226,893,1660,1866]

3.5.2. Alcoholysis of 2- or 4-Halogenoquinazolines (*H* 228, 235, 237)

The alkoxyl ion is such an avid nucleophile that any differences in the activity of the halogeno leaving group, occasioned by its position or by the presence of other groups, are seldom noticed under preparative conditions; however, preferential monoalcoholysis of 2,4-dihalogenoquinazolines has been achieved sometimes. The following examples of alcoholysis and phenolysis typify the conditions usually adopted and the yields that may be expected.

2-Halogenoquinazolines as substrates for alcoholysis

2-Chloro- gave 2-methoxy- (**60**, R = H) (MeONa, MeOH, reflux, 1 h: 76%) or 2-ethoxyquinazoline (EtONa, EtOH, likewise: 69%).[849]

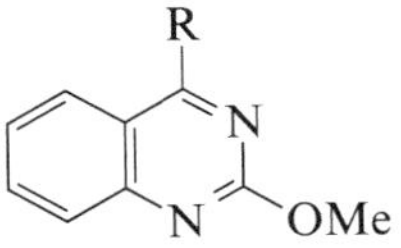

(60)

2-Chloro- gave 2-ethoxy-4(3*H*)-quinazolinone (EtONa, EtOH, reflux, 24 h: 86%).[2230]

Dimethyl 2-chloro-4-quinazolinephosphonate (**61**) gave 2,4-dimethoxy-quinazoline (**60**, R = OMe) (MeONa, MeOH, reflux, 3 h: 60%).[571]

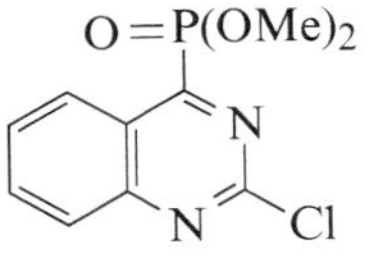

(61)

Also other examples in the foregoing references and elsewhere.[1033]

4-Halogenoquinazolines as substrates for alcoholysis

4-Chloro- gave 4-ethoxy- (**62**, R = Et) (EtONa, EtOH, reflux, 2 h: 90%),[1001] 4-prop-2'-ynyloxy- (HC≡CCH$_2$ONa, HC≡CCH$_2$OH, reflux, 2 h: 57%),[1935] or 4-benzyloxyquinazoline (**62**, R = CH$_2$Ph) (neat PhCH$_2$OH, NaH, reflux, ? h: 92%).[2329]

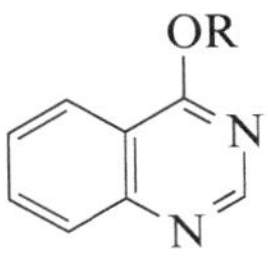

(62)

4-Chloro- gave 4-methoxy-2-phenylquinazoline (MeONa, MeOH, reflux, 1 h: 95%).[2206]

5,6,8-Trimethoxy-7-methyl-4(3H)-quinazolinone gave 4-chloro- (**63**, R = Cl) (POCl$_3$, Et$_3$N, PhH, reflux, 23 h: unisolated) and thence 4,5,6,8-tetramethoxy-7-methylquinazoline (**63**, R = OMe) (foregoing mixture, MeOH, 20°, 2.5 h: 65% overall).[2363]

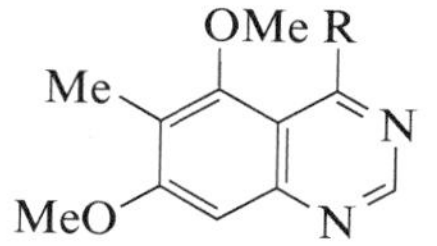

(63)

Also other examples,[42,310(?),315,514,868,968,1635,2451,2547] including the commercial preparation of 4-(p-t-butylphenylethoxy)quinazoline (fenazaquin), apparently involving simple admixture of 4-chloroquinazoline with β-(p-t-butylphenyl)ethanol in a nonprotic solvent.[2546]

4-Halogenoquinazolines as substrates for phenolysis

4-Chloro- gave 4-phenoxyquinazoline (**64**, R = H) (neat PhOH, 40°, 24 h: 72%).[1139]

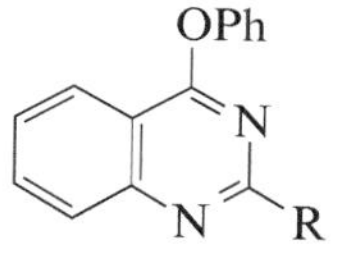

(64)

4-Chloro- gave 4-phenoxy- (**64**, R = Ph) [PhONa, (MeOCH$_2$CH$_2$)$_2$O, 110°, 10 min: 70%],[260] 4-p-nitrophenoxy- (HOC$_6$H$_4$NO$_2$-p, K$_2$CO$_3$, AcMe, reflux, N$_2$, 48 h: 98%),[2032] or 4-(4-nitro-2-pyridin-2′-ylphenoxy)-2-phenylquinazoline (sodium 4-nitro-2-pyridin-2′-ylphenoxide, AcNMe$_2$, reflux, 10 h: 69%).[406]

4-Chloro-2-phenyl- gave 2-phenyl-4-p-(2,2,4-trimethylchroman-4-yl)phenoxyquinazoline (**65**) [sodium p-(2,2,4-trimethylchroman-4-yl)phen-

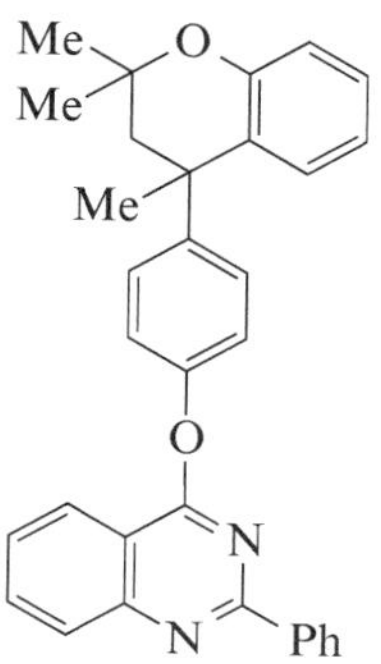

(65)

oxide, $(MeOCH_2CH_2)_2O$, 145°, 40 min: 80%];[903] also analogs rather similarly.[726,809,2164]

Also other examples.[1546,1814,1969,2020]

2,4-Dihalogenoquinazolines as substrates for alcoholysis

2,4-Dichloro-6,7-dimethoxy- (**66**, R = Cl) gave 2-chloro-4,6,7-trimethoxy- (**66**, R = OMe) (NaOH, H_2O–MeOH, 20°, 5 h: 64%)[641] or 2-chloro-4-(O,O-isopropylidene-β,γ-dihydroxypropoxy)-6,7-dimethoxy- (isopropylideneglycerol, Bu_4NBr, NaOH, H_2O–PhCl, 20°, 2 h: 76%) and thence 2-chloro-4-(β,γ-dihydroxypropoxy)-6,7-dimethoxyquinazoline (**66**, R = OCH_2 $CHOHCH_2OH$) (HCl, MeOH–H_2O, 20°, 3 h: 83%).[1432]

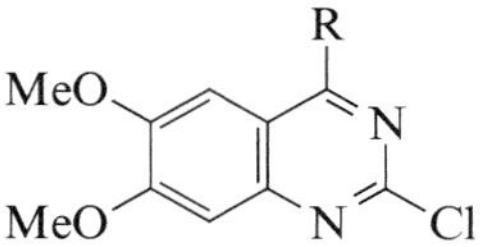

(66)

2,4-Dichloro- gave 2,4-diethoxy-6-methylquinazoline (**67**, R = Me, Y = Et) (EtONa, EtOH, reflux, 2 h: 74%).[1616]

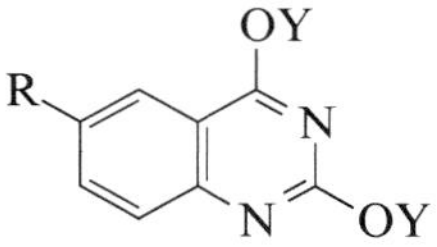

(67)

2,4-Dichloro- gave 2,4-bis(β-diethylaminoethoxy)quinazoline (**67**, R = H, Y = $CH_2CH_2NEt_2$) ($Et_2NCH_2CH_2ONa$, PhH, reflux, 2 h: 54%).[616]

2,4,5,6,7,8-Hexafluoro- gave 5,6,8-trifluoro-2,4,7-trimethoxy- or 6,8-difluoro-2,4,5,7-tetramethoxyquinazoline (MeONa, MeOH, 30° or reflux, respectively; minimal detail);[34] additional relevant ^{19}Fnmr data have been reported.[837]

Also other example.[1635]

2,4-Dihalogenoquinazolines as substrates for phenolysis

2,4-Dichloro- gave 2-chloro-4-o-chlorophenoxy- (**68**, R = Cl) (HOC_6H_4Cl-o, EtONa, EtOH, warm, a few minutes; 80%) or 2,4-bis-o-chlorophenoxyquinazoline (**68**, R = OC_6H_4Cl-o) (neat HOC_6H_6Cl-o, 160–200°, until HCl↑ ceased: 73%);[814] also many analogs and variations of conditions.[814]

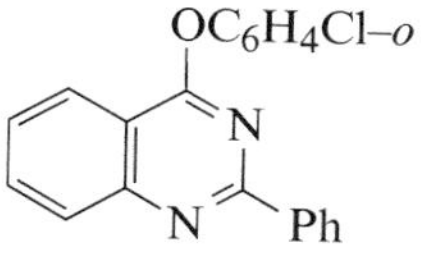

(68)

2,4-Dichloro-6-methyl- gave 6-methyl-2,4-diphenoxyquinazoline (PhONa, PhOH, 180°, 1 h: 93%).[1616]

2,4-Dichloro- gave 2-chloro-4-phenoxy-5,6,7,8-tetrahydroquinazoline (PhOH, 1,8-diazabicyclo(5,4,0)-7-undecene, Me$_2$NCHO, 100°, 3 h: 49%).[1635]

2,4-Dichloro-5,6,7-trimethoxyquinazoline gave 5,6,7-trimethoxy-2-phenoxy-4-quinazolinamine (**69**) (PhOH, NH$_3$↓, 125°, 2 h: 52%).[696]

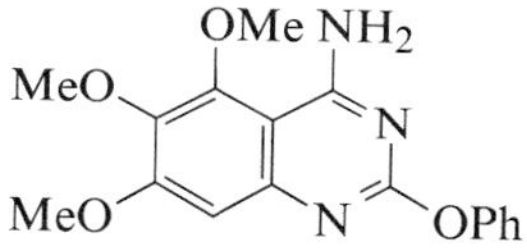

(69)

Also other examples.[640,814]

3.5.3. Hydrolysis of 2- or 4-Halogenoquinazolines (*H* 229)

Since most 2- or 4-halogenoquinazolines have been made from the corresponding quinazolinones, the reverse hydrolytic procedure is normally required only for specialized processes, such as the preferential hydrolysis of a 2,4-dichloroquinazoline. The following examples illustrate two such cases and some unexpected hydrolyses.

2,4-Dichloroquinazoline gave 2-chloro-4(3*H*)-quinazolinone (**70**, R = H, (2M NaOH, 20°, ?h: ?%), identified by aminolysis to 2-*p*-chloroanilino-4(3*H*)-quinazolinone.[1266]

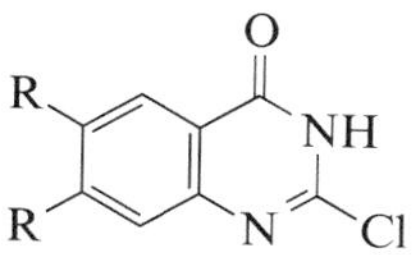

(70)

2,4-Dichloro-6,7-dimethoxyquinazoline gave 2-chloro-6,7-dimethoxy-4(3*H*)-quinazolinone (**70**, R = OMe) (NaOH, H$_2$O–THF, 20°, 8 h: >83%;[1810] likewise, N$_2$, 4 h: >95%).[641]

4-Chloroquinazoline gave 4(3*H*)-quinazolinone [attempted recrystallization from boiling EtOH: initially as hydrochloride;[887] Bu$_4$NOH, MeCN–H$_2$O, 20°, 5 min: 77% plus 4-cyanomethylquinazoline (10%)].[2010]

4-Chloro-2-(α-chloro-α-methylethyl)-6-methylquinazoline gave 2-(α-chloro-α-methylethyl)-6-methyl-4(3H)-quinazolinone (71) (MeOH, reflux, 15 min: 90%; note survival of the α-Cl).[1224]

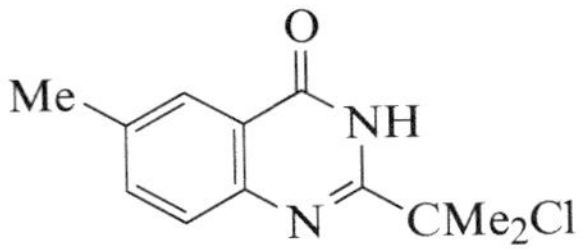

(71)

4,7-Dichloroquinazoline gave 7-chloro-4(3H)-quinazolinone (Me$_2$SO, exothermic, 5 min: >95%).[1076]

3.5.4. Thiolysis or Alkanethiolysis of 2- or 4-Halogenoquinazolines (H 270, 278, 284)

These reactions appear to occur rather less readily than those with the corresponding oxygen nucleophiles, as indicated in the following examples.

2-Chloroquinazoline gave 2(1H)-quinazolinethione (72) (H$_2$NCSNH$_2$, MeOH, reflux, 1 h; then 2.5M NaOH, 95°, 1 h: 94%).[281, cf. 1039]

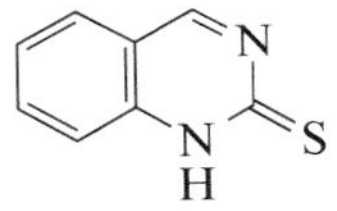

(72)

2,4-Dichloroquinazoline gave 2,4(1H,3H)-quinazolinedithione (H$_2$NCSNH$_2$, EtOH, reflux, 3h: 77%)[38] or 2-chloro-4-methylthioquinazoline (MeSNa, AcMe–H$_2$O, 15°, 1 h: 80%).[1410]

4-Chloro- gave 4-ethoxycarbonylmethylthioquinazoline (73) (EtO$_2$CCH$_2$SH, PhH, reflux, 4h: 60%)[2134] or 4-phenylthioquinazoline [neat PhSH (exothermic), N$_2$, 30 min: 75%;[1139] PhSH, NaOH, dioxane–H$_2$O, 20°, 2 h: 45%].[650]

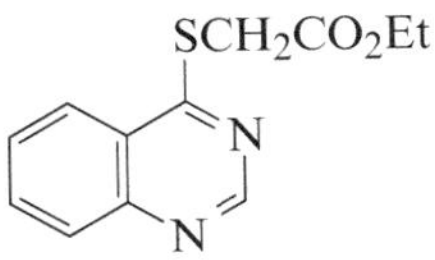

(73)

2-Chloro- gave 2-methylthio-5,6,7,8-tetrahydro-4-quinazolinamine (MeSNa, EtOH–H$_2$O, 180°, sealed, 6h: 56%; note the apparent deactivating effect of the amino group).[1398]

Also other examples.[667,688,1189,1866,1963,2539]

3.5.5. Other Reactions of 2- or 4-Halogenoquinazolines (*H* 227, 331)

Although important, the remaining reactions of these halogenoquinazolines have been used infrequently. Where possible, they are illustrated hereafter by simple examples.

Dehalogenation

2-Chloro-4-methoxy- (**74**, R = Cl) gave 4-methoxy-5,6,7,8-tetrahydroquinazoline (**74**, R = H) (H$_2$, Pd/C, MeOH: ~40%).[1635]

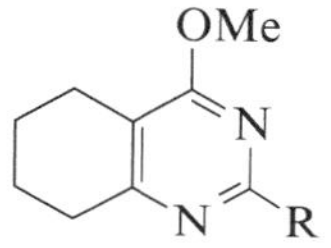

(**74**)

Methyl 4-chloro-2-*p*-chlorophenyl- (**75**, R = Me, Y = Cl) gave methyl 2-*p*-chlorophenyl-4-*β*-tosylhydrazino-5-quinazolinecarboxylate (**75**, R = Me, Y = NHNHTs) (TsNHNH$_2$, CHCl$_3$, 20°, 12 h: 86%, as hydrochloride) and thence 2-*p*-chlorophenyl-5-quinazolinecarboxylic acid (**75**, R = Y = H) (1M NaOH, 95°, 3 h: 75%; a McFadyen–Stevens dechlorination, which induced ester hydrolysis but not ω-dechlorination).[672]

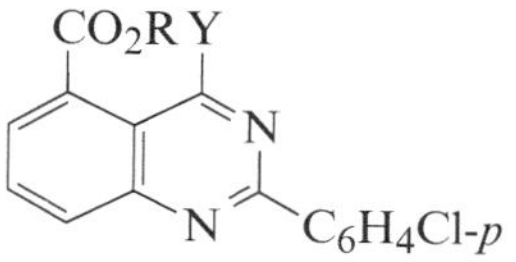

(**75**)

4-Chloroquinazoline underwent addition, dechlorination, and ring fission to *o*-(benzylideneamino)benzonitrile (PhMgBr, Et$_2$O–THF, reflux, 3 h: 75%).[2539]
Also other examples.[1097,1201]

Conversion into ketonic quinazolines

4-Chloroquinazoline gave the dimethyloxosulfonium quinazolinemethylide (**76**, R = H) (H$_2$C̄–S̟O$_2$Me$_2$, THF, reflux, N$_2$, 5 h: crude) and thence the acetyl derivative (**76**, R = Ac) (Ac$_2$O, dioxane, 20°, 1 h: 55%), which underwent desulfurization to 4-acetonylquinazoline (**77**) (Raney-Ni, MeOH, reflux, 30 min: 57%).[1404]

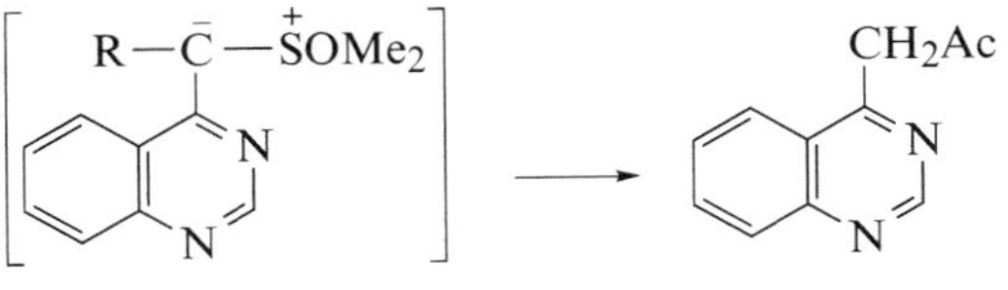

(76)   (77)

4-Chloro- gave 4-propionyl- (36%) or 4-butyrylquinazoline (37%) somewhat similarly (see original for details).[431]

4-Chloro- gave 4-benzoylquinazoline (PhCHO, trace 1,3-dimethylbenzimida-zolium iodide, NaH, THF, N_2, reflux, 1 h: 96%) plus a trace of 4-benzyloxy-quinazoline;[2329] many analogs were made similarly.[2329]

Conversion into arylsulfonylquinazolines

4-Chloro- gave 4-p-tolylsulfonylquinazoline (PhSO$_2$Na, Me$_2$NCHO, 5–10°, 5 min: 70%)[792] and thence 4-quinazolinecarbonitrile (KCN, Me$_2$NCHO, 10°, 30 min: 70%).[792]

Conversion into quinazolinecarbonitriles or thiocyanatoquinazolines

4-Chloro-2-(β,β-dichloro-α,α-difluoroethyl)quinazoline (**78**, R = Cl) gave 2-(β,β-dichloro-α,α-difluoroethyl)-4-quinazolinecarbonitrile (**78**, R = CN) (KCN, MeCN, 20°, 14 h: 50%)[443] or 2-(β,β-dichloro-α,α-difluoroethyl)-4-thiocyanatoquinazoline (**78**, R = SCN) (KSCN, likewise: 51%).[443]

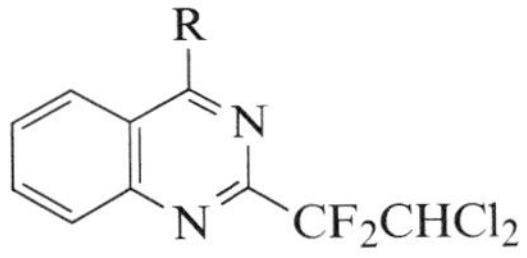

(78)

Note: For an indirect Cl → CN conversion via a sulfone, see preceding subsection.

Conversion into quinazoline phosphorus esters

2,4-Dichloroquinazoline gave dimethyl 2-chloro-4-quinazolinephosphonate (**61**) [(MeO)$_3$P, reflux, 16 h: 22%];[341] in contrast, 2,4,6,8-tetrachloro-quinazoline gave tetramethyl 6,8-dichloro-2,4-quinazolinediphosphonate (**79**) (likewise, 2 h: 11%).[341]

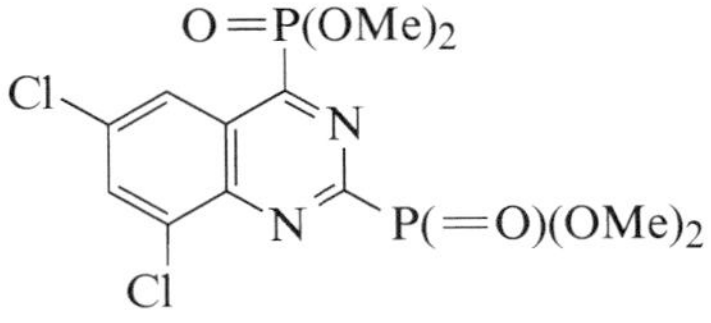

(79)

4-Chloro-2-(β,β-dichloro-α,α-difluoroethyl)quinazoline (**78**, R = Cl) gave 2-(β,β-dichloro-α,α-difluoroethyl)-4-dimethoxyphosphinothioylthioquinazoline [**78**, R = SP($=$S) (OMe)$_2$] [KSP($=$S) (OMe)$_2$, MeCN, 20°, 14h: 29%].[443]

Conversion into azidoquinazolines

4-Chloro- gave 4-azido-2-phenylquinazoline (**80**) (NaN$_3$, EtOH, 55°, sealed, 6h; then 20°, 15h: 97%),[1683] which appears to exist as 5-phenyltetrazolo[1,5-c]quinazoline (**81**).[1080]

(80)   **(81)**

2,4-Dichloro- gave 2,4-diazidoquinazoline (NaN$_3$, Me$_2$NCHO, 20°, 1h: 87%),[2201] present as several tautomers at least in solution.[2201, cf. 1096, 1489] Also other examples.[749, 2087, 2369, 2429]

3.6. REACTIONS OF 5- TO 8-HALOGENOQUINAZOLINES (*H* 227)

Although 5- to 8-halogenoquinazolines are much less reactive than their 2- or 4-isomers, the former do undergo some nucleophilic and other reactions, although under quite severe conditions. Within the 5- to 8-halogeno grouping, a 5- or 7-halogeno substituent would be expected to be more reactive than one at the 6- or 8-position: this postulate has been confirmed by kinetic data on the displacement of halogen ion from simple 5- to 8-chloro- and 5- to 8-fluoroquinazolines, indicating that the order of reactivity toward methoxide ion in both series was $7 > 5 > 6 > 8$ and that each fluoroquinazoline was more reactive than the corresponding chloroquinazoline.[171] The use of such halogenoquinazolines for synthetic purposes is illustrated in the following examples.

Aminolysis

2-Amino-5-chloro- (**82**, R = Cl) gave 2-amino-5-methylamino-6-nitro-4(3*H*)-quinazolinone (**82**, R = NHMe) (MeNH$_2$, BuOH, 130°, sealed, 24h: >95%).[2015]

(82)

5-Chloro-6-nitro-2,4-quinazolinediamine   gave   6-nitro-2,4,5-quinazoline-triamine (NH$_4$OH, MeOH, 100°, sealed, 7 h: 51%).[1443].

7-Chloro- (**83**, R = Cl) gave 7-amino-6-nitro-4(3*H*)-quinazolinone (**83**, R = NH$_2$) (NH$_3$, BuOH, 170°, sealed, 24 h: >95%).[501]

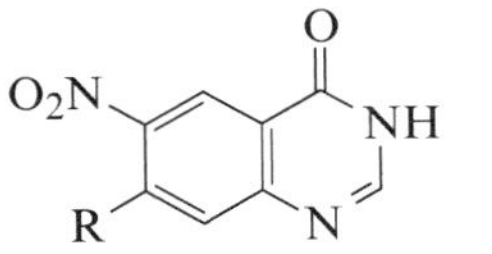

(83)

7-Chloro- gave 7-dodecylamino-3-methyl-2,4(1*H*,3*H*)-quinazolinedione (neat C$_{12}$H$_{25}$NH$_2$, 220°, A, 10 h: 60%).[2216]

7-Chloro-1,3-dimethyl- gave 1,3-dimethyl-7-methylamino-6-nitro-2,4(1*H*, 3*H*)-quinazolinedione (MeNH$_2$, BuOH, 120°, sealed, 24 h: 95%).[1358]

7-Chloro- gave 7-dimethylamino-6-nitro-4(3*H*)-quinazolinone (Me$_2$NH, EtOH, reflux, 30 min: 72%).[900]

Also other examples.[384,915,1165,1466,1581,2324]

Alcoholysis and the like

7-Chloro- (**84**, R = Cl) gave 7-methoxycarbonylmethylthio-1,3-dimethyl-6-nitro-2,4(1*H*,3*H*)-quinazolinedione (**84**, R = SCH$_2$CO$_2$Me) (HSCH$_2$CO$_2$Me, Et$_3$N, MeOH, reflux, 2 h: 89%).[1713]

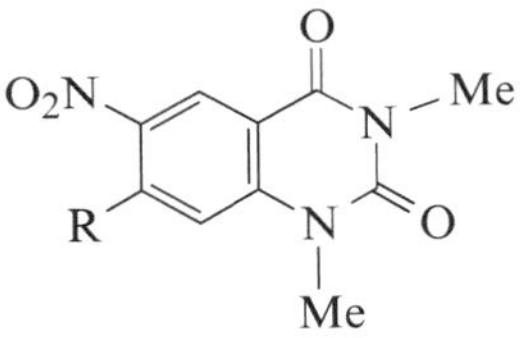

(84)

5-Chloro-2-methyl- gave 2-methyl-5-phenoxy- (PhONa, CuBr, Cu$_2$O, AcNMe$_2$, 130°, 12 h: 52%) or 2-methyl-5-phenylthio-4(3*H*)-quinazolinone (PhSNa, likewise: 43%).[2411]

5-Bromo-2,6-dimethyl- gave 2,6-dimethyl-5-pyridin-4′-ylthio-4(3*H*)-quinazolinone (sodium 4-pyridinethiolate, CuBr, Cu$_2$O, AcNMe$_2$, 90°, 4 h: 48%).[2411]

Conversion into azidoquinazolines

7-Bromo- (**85**, R = Br) gave the unstable 7-azido-6-methoxy-5,8-dihydro-5,8-quinazolinedione (**85**, R = N$_3$) (NaN$_3$, AcMe–H$_2$O, 20°, 2 h) and thence the stable 7-azido-6-methoxy-5,8-quinazolinediol (Na$_2$S$_2$O$_4$, AcMe–H$_2$O, 20°, N$_2$, 30 min: 59% overall).[1577]

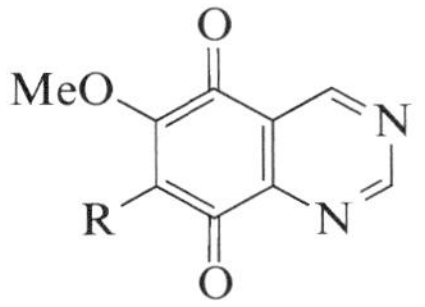

(85)

Dehalogenation

6-Chloro-3-β-hydroxyethyl- (**86**, R = Cl) gave 3-β-hydroxyethyl-8-β-hydroxyethylamino-4(3H)-quinazolinone (**86**, R = H) (H$_2$, Pd/C, EtOH, 20°, 2 h: 74%).[384] Also analogous examples.[220,2584]

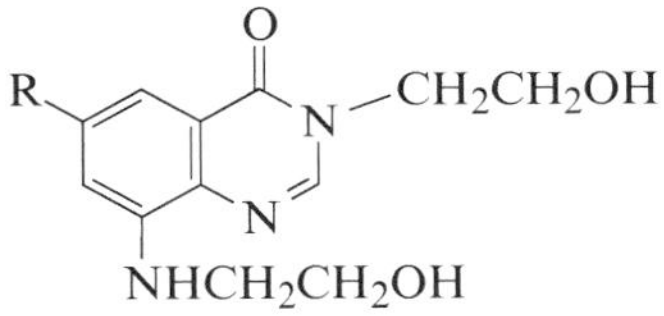

(86)

2,3-Diamino-5,8-dibromo-5,6,7,8-tetrahydro-4(3H)-quinazolinone (**87**) (made *in situ* by bromination) gave 2,3-diamino-4(3H)-quinazolinone by elimination of 2 × HBr (Me$_3$pyridine or Et$_3$N in THF, reflux, 1 h: ?%);[1886] 2,4-bistrifluoromethylquinazoline was made rather similarly.[1213]

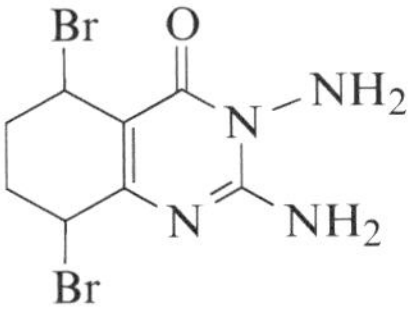

(87)

Conversion into quinazolinecarbaldehydes

6-Bromo-5,8-dimethoxy-1,2-dihydro-4(3H)-quinazolinone (**88**, R = Br) gave 5,8-dimethoxy-4-oxo-1,2,3,4-tetrahydro-6-quinazolinecarbaldehyde (**88**, R = CHO) (BuLi, then Bu^{t}Li, THF, −78°, 3 h; then Me$_2$NCHO, 20°, 12 h: 79%).[2094]

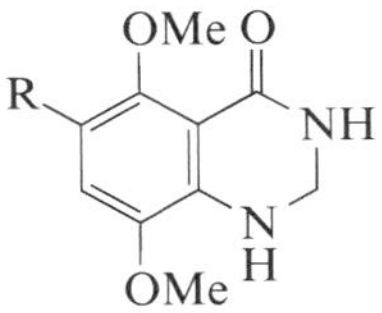

(88)

6-Bromomethyl- gave 6-(*p*-*t*-butoxycarbonyl-*N*-prop-2'-ynylanilino)methyl-
2-methyl-4(3*H*)-quinazolinone (**95**) (HC≡CCH$_2$NH—C$_6$H$_4$CO$_2$Bu^t,
CaCO$_3$, AcNMe$_2$, 50°, dark, 17 h: 48%);[2283] also many other dideazafolate
analogs rather similarly.[871,1582,1591,1606,1614,1616,1617,2024,2249,2250,
2254,2258,2263,2269,2276,2371]

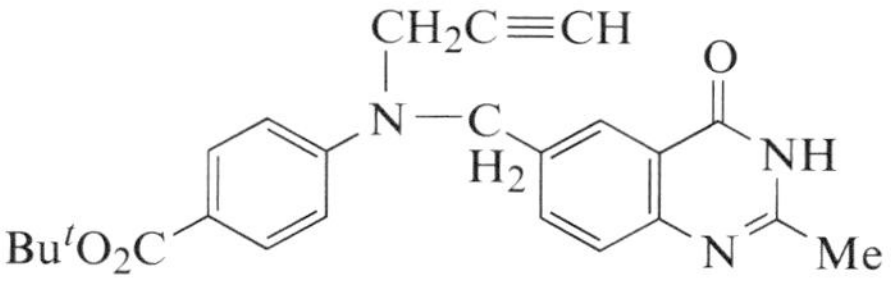

(**95**)

2-Bromomethyl-3-phenyl- gave 3-phenyl-2-phthalimidomethyl- (K phthalimide, Me$_2$NCHO, 100°, 1 h: 90%) and thence 2-aminomethyl-3-phenyl-
4(3*H*)-quinazolinone (80% H$_2$NNH$_2$.H$_2$O, MeOH, reflux, 1 h; then HCl,
reflux, 1 h: 52%);[152] the corresponding thiones were made somewhat
similarly.[505]

2-Dibromomethyl-3-*β*-hydroxyethyl-4(3*H*)-quinazolinone gave 3-*β*-hydro-
xyethyl-4-oxo-3,4-dihydro-2-quinazolinecarbaldehyde phenylhydrazone
(PhNHNH$_2$, EtOH, reflux, 30 min: 62%).[2470]

4-Trichloromethyl- gave 4-propylaminoquinazoline (PrNH$_2$, MeCN, 20°, 3 h:
96%; mechanism?).[2060]

Also other examples.[63,307,387,510,683,787,1104,1156,1251,1272,1330,1397,1401,
1441,1458,1480,1505,1600,1605,1611,1663,1670,1717,1908,1912,1937,2018,2027,
2130,2142,2169,2304,2310,2360,2409,2499,2500,2535,2544]

Alcoholysis or phenolysis

2-Bromomethyl- (**96**, R = Br) gave 2-methoxymethyl- (**96**, R = OMe) (MeONa,
MeOH, reflux, 3 h: ~90%;[146] NaOH, MeOH, 20°, 4 h: 88%).[1154,1156] or
2-phenoxymethyl-3-phenyl-4(3*H*)-quinazolinone (**96**, R = OPh) (PhOH,
K$_2$CO$_3$, Me$_2$NCHO, 20°, 12 h: ~80%).[146]

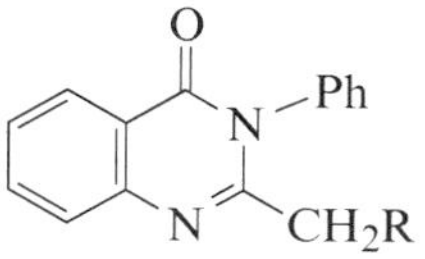

(**96**)

2-*β*-Chloroethyl- gave 2-*β*-ethoxyethyl-1-phenyl-4(1*H*)-quinazolinone (EtOH,
20°, 1 h: 88% as hydrochloride).[1394]

2-(α,β-Dibromo-p-methoxyphenethyl)- (**97**, X = Y = Br) gave 2-(α-bromo-β-ethoxy-p-methoxyphenethyl)-(**97**, X = Br, Y = OEt) (EtOH, reflux, 1 h: ?%) or 2-(p-methoxy-α,β-diphenoxyphenethyl)-3-phenyl-4(3H)-quinazolinone (**97**, X = Y = OPh) (PhOH, PhH, reflux 1 h: ?%).[1223]

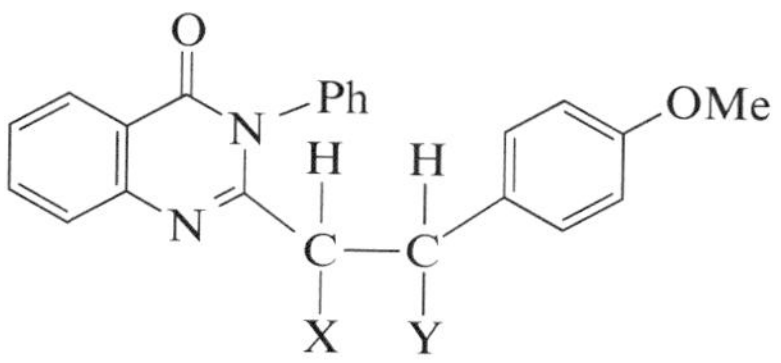

(97)

5-Chloro-3-(γ-chloro-β-hydroxypropyl)- (**98**, X = OH, Y = Cl) gave 5-chloro-3-(β,γ-epoxypropyl)-4(3H)-quinazolinone (**98**, X + Y = O) (EtOH–H$_2$O, pH ~ 12, boil, briefly: 81%).[883]

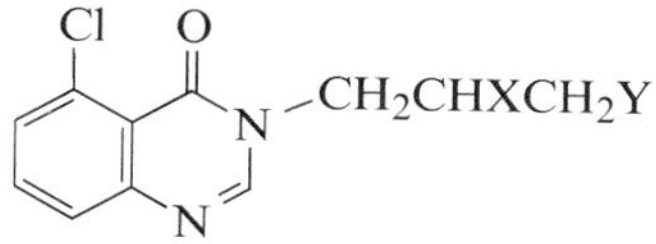

(98)

4-Trichloromethyl- gave 4-methoxyquinazoline (MeONa, MeOH, 25°, 4 h: 97%; by loss of CHCl$_3$ from a covalent 3,4-adduct).[2001]
Also other examples.[786,1067,1258,1397,1492,2373,2388]

Hydrolysis

2-Chloromethyl-5,8-dihydroxy- (**99**, R = Cl) gave 5,8-dihydroxy-2-hydroxymethyl-4(3H)-quinazolinone (**99**, R = OH) (Me$_2$SO–H$_2$O, N$_2$↓, 55°, 10 h: 64%).[1999]

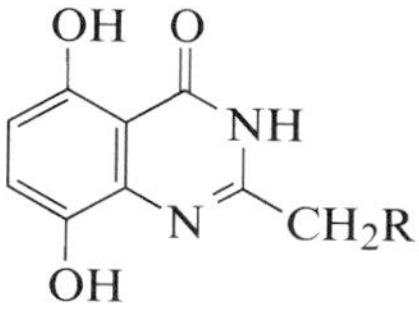

(99)

6-Bromomethyl-2-pivalamido- gave 2-amino-6-hydroxymethyl-4(3H)-quinazolinone (0.5M HCl, reflux, 1 h: 77%; note additional deacylation).[871]

3-o-(Bromomethyl)phenyl- (**100**, R = Br) gave 3-o-(hydroxymethyl)phenyl-2-methyl-4(3H)-quinazolinone (**100**, R = OH) [Na$_2$CO$_3$, H$_2$O–(MeOCH$_2$CH$_2$)$_2$O, reflux, 6 h: 86%][619]

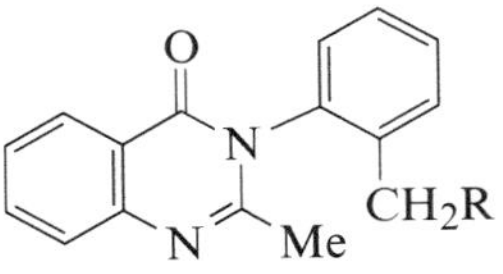

(100)

4-Trichloromethylquinazoline gave 4(3H)-quinazolinone (Bu$_4$NOH, H$_2$O–MeCN, 20°, 5 min: 77%, after separation from 10% of 4-cyanomethyl-quinazoline).[2010]

Also other examples.[477,510,810,1402,1441,1476,1867]

Alkane- or arenethiolysis

2-Bromomethyl- (**101**, X = Br, Y = H) gave 2-propylthiomethyl-3-o-tolyl-4(3H)-quinazolinone (**101**, X = SPr, Y = H) (PrSH, K$_2$CO$_3$, MeOH, 20°, 12 h: 83%).[786]

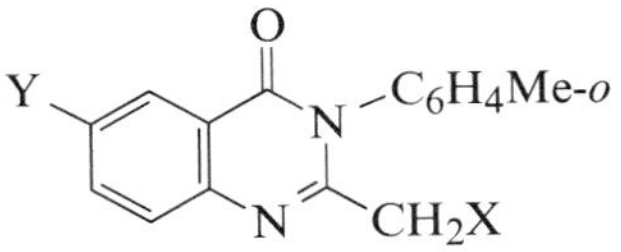

(101)

6-Acetamido-2-chloromethyl- (**101**, X = Cl, Y = NHAc) gave 6-acetamido-2-phenylthiomethyl- (**101**, X = SPh, Y = NHAc) (PhSH, NaH, THF, 20°, 2 h: 80%)[1397] or 6-acetamido-2-methylthiomethyl-3-o-tolyl-4(3H)-quinazolinone (**101**, X = SMe, Y = NHAc) (MeSNa, THF, 20°, 7 h: 93%).[1418]

3-β-Chloroethyl- gave 3-β-(imidazol-2-ylthio)ethyl-2-methyl-4(3H)-quinazo-linone [2(3H)-imidazolethione, NaOH, EtOH, reflux, 4 h: 86%].[2378]

Also other examples.[801,874,1223,1623,1938]

Thiolysis

Note: Indirect thiolysis via an isothiouronio halide has been used, but it is unattractive because of rapid aerial oxidation of the product.

2-Bromomethyl-3-o-tolyl-4(3H)-quinazolinone (**101**, X = Br, Y = H) gave 2-isothiouroniomethyl-3-o-tolyl-4(3H)-quinazolinone bromide (**102**) (H$_2$NCSNH$_2$, AcMe, reflux, 15 min: 76%) and thence 2-mercaptomethyl-3-o-tolyl-4(3H)-quinazolinone (**101**, X = SH, Y = H) (2M NaOH, briefly, then pH 2: ∼80% crude), which was characterized as 2-ethoxycarbonyl-thiomethyl-3-o-tolyl-4(3H)-quinazolinone (**101**, X = SCO$_2$Et, Y = H) (EtO$_2$CCl, Et$_3$N, CHCl$_3$, 20°, 24 h: 60%).[786]

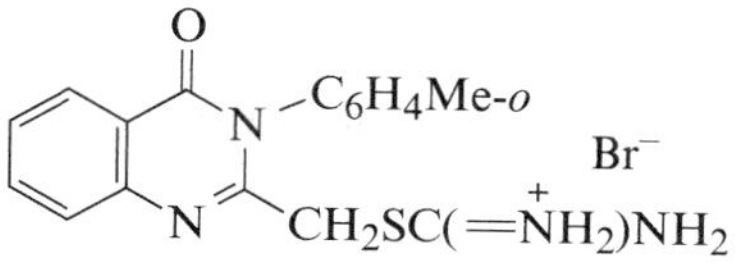

(102)

2-Chloromethyl-4(3*H*)-quinazolinone gave 2-isothiouroniomethyl-4(3*H*)-quinazolinone bromide (H_2NCSNH_2, EtOH, reflux, 3 h: 86%) and thence bis(4-oxo-3,4-dihydroquinazolin-2-ylmethyl) disulfide **(103)** [KOH (?), EtOH, reflux, 1 h: 83%].[683, cf. 605]

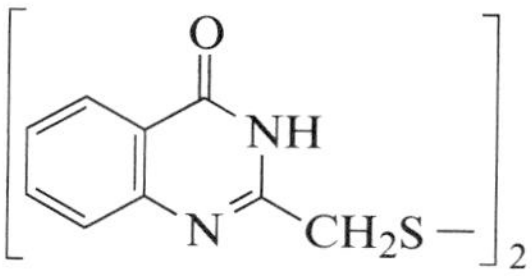

(103)

Also other examples.[510,1276]

Acyloxylation or acylthiolation

2-Chloromethyl- gave 2-acetoxymethyl-1-phenyl-4(1*H*)-quinazolinone **(104)** (AcOK, AcOH, reflux, 90 min: 59%).[1394]

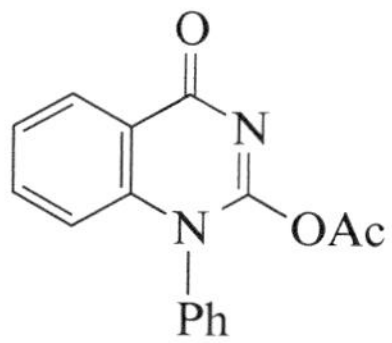

(104)

2-Bromomethyl- gave 2-acetoxymethyl-3-phenyl-4(3*H*)-quinazolinone (AcOK, Me_2NCHO, 95°, 1 h: ~70%).[146]

6,8-Dibromo-2-chloromethyl-3-*o*-chlorophenyl- gave 6,8-dibromo-3-*o*-chlorophenyl-2-[piperidino(thiocarbonyl)thiomethyl]-4(3*H*)-quinazo-linone **(105)** [$(CH_2)_5NCS_2NH_4$, AcMe, reflux, 3 h: 72%].[2308]

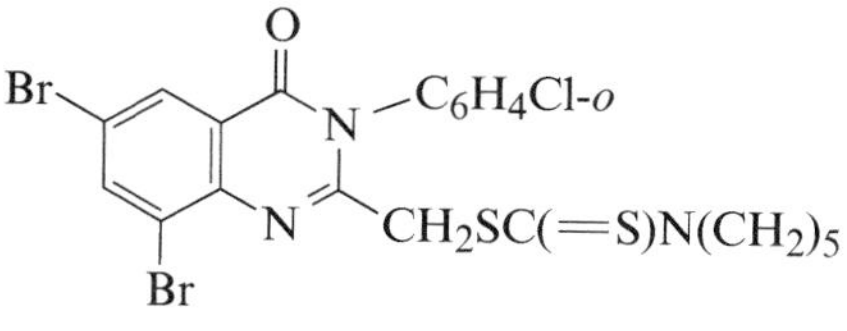

(105)

Also other examples.[1004,1223,1364]

Dehalogenation (see also Section 2.2.1.3)

4-Tribromomethyl- gave 4-dibromomethyl- and 4-bromomethylquinazoline (PhAc, trace F_3CCO_2H, PhH, reflux, $> 80\%$ and $\sim 5\%$, respectively);[1651] related examples also.[2060]

3-Acetamido-6,8-dibenzyloxy-2-p-chlorophenyl- (**106**, R = CH_2Ph, X = Cl) gave 3-acetamido-6,8-dihydroxy-2-phenyl-4(3H)-quinazolinone (**106**, R = X = H) (H_2, ~ 3 atm, Pd/C: 40%; note additional debenzylation).[844]

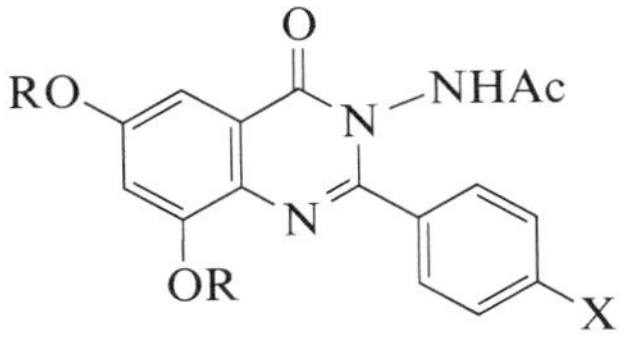

(106)

2-Bromomethyl-3-phenyl-4(3H)-quinazolinone gave a separable mixture of α,β-bis(4-oxo-3-phenyl-3,4-dihydroquinazolin-2-yl)ethane and the corresponding ethylene (**107**) (EtMgBr, Et_2O–PhH, 20°, 12 h: 50% and 32%, respectively).[1239]

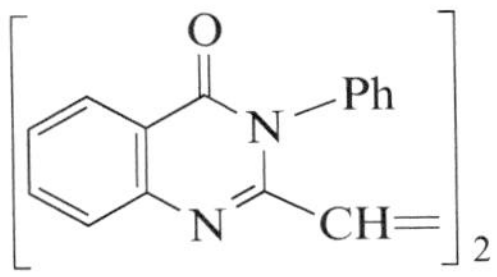

(107)

Conversion into cyano or thiocyanato derivatives

2-Bromomethyl- gave 2-cyanomethyl-3-phenyl-4(3H)-quinazolinone (KCN, Me_2NCHO, 95°, 90 min: $\sim 90\%$)[146] or the 2-thiocyanatomethyl analog (KSCN, MeOH, reflux, 6 h: ?%).[1286]

8a-Bromomethyl- (**108**, R = Br) gave 8a-cyanomethyl-3,4,4a,5,6,7,8,8a-octahydro-2(1H)-quinazolinone (**108**, R = CN) (KCN, EtOH–H_2O, reflux, 6 h: 98%).[1004]

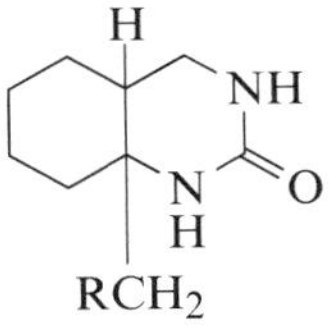

(108)

6-Chloro-2-chloromethyl-4-phenyl- gave 6-chloro-4-phenyl-2-thiocyanatomethylquinazoline 3-oxide (KSCN, Me_2SO, 20°, 12 h: 85%).[861]

Conversion into azido derivatives

2-α-Bromoethyl- (**109**, R = Br) gave 2-α-azidoethyl-4(3H)-quinazolinone (**109**, R = N$_3$) (NaN$_3$, Pr^{i}OH–H$_2$O, reflux, 12h: 72%).[2069]

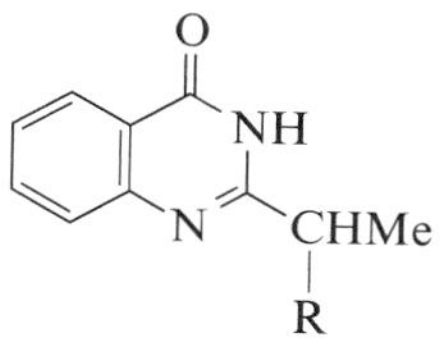

(109)

2-Chloromethyl- gave 2-azidomethyl-3-m-nitrophenyl-4(3H)-quinazolinone (NaN$_3$, Me$_2$NCHO, 0° → 20°, 2h: 87%);[1806] analogs likewise.[1806]

Other displacement reactions

2-Chloromethyl- gave 2-sulfomethyl-4(3H)-quinazolinone (K$_2$SO$_3$, no details: as potassium salt).[605]

2-Amino-6-bromomethyl- gave 2-amino-6-triphenylphosphoniomethyl-4(3H)-quinazolinone bromide (Ph$_3$P, Me$_2$NCHO, 100°, 2h: >95%).[801]

2-Bromomethyl-3-phenyl-4(3H)-quinazolinone gave 4-oxo-3-phenyl-3,4-dihydro-2-quinazolinecarbaldehyde (AgNO$_3$, EtOH, reflux, ?h: 75%; via the hydroxymethyl analog).[1156]

7-Dibromomethyl-4-p-fluorophenyl-1-isopropyl-2(1H)-quinazolinone (**110**, R = CHBr$_2$) gave 4-p-fluorophenyl-1-isopropyl-2-oxo-1,2-dihydro-7-quinazolinecarbaldehyde (**110**, R = CHO) (AgNO$_3$, EtOH–THF–H$_2$O, reflux, 2h: ?%).[1867]

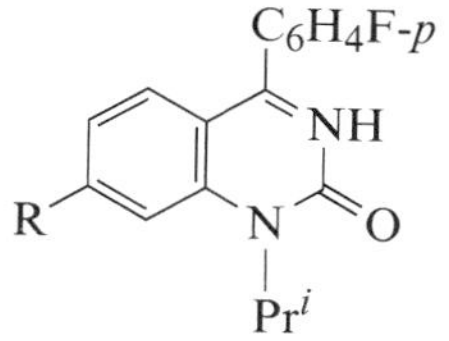

(110)

4-Dichloromethyl-2-phenylquinazoline gave 2-phenylquinazoline (H$_2$SO$_4$, 220°, 7 min: 18%; via the 2-carbaldehyde?).[693]

Cyclizations

2-δ-Bromo-γ-hydroxybutyl-4(3H)-quinazolinone (**111**) gave a separable mixture of 8-hydroxy-6,7,8,9-tetrahydro-11H-pyrido[2,1-b]quinazolin-11-one (**112**) and 2-hydroxy-1,2,3,4-tetrahydro-6H-pyrido[1,2-a]quinazolin-6-one (**113**) [MeCN, reflux, >3h (tlc): 23% and 71%, respectively].[2389]

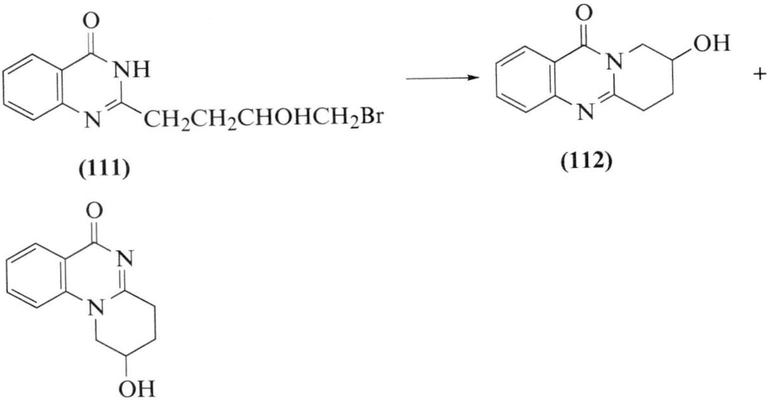

(111)

(112)

+

(113)

6-Chloro-2-chloromethyl-4-phenyl-1,2-dihydroquinazoline 3-oxide (**114**) gave 5-chloro-3-phenyl-1,1a-dihydroazirino[1,2-*a*]-quinazoline 2-oxide (**115**) (Bu^{t}OK, THF, $-78° \rightarrow 20°$, 4.5 h: 57%).[18]

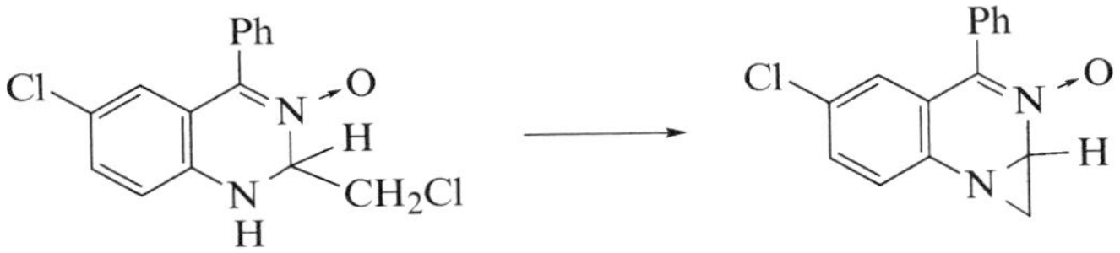

(114)

(115)

CHAPTER 4

Oxyquinazolines

The conveniently imprecise term *oxyquinazolines* is used to include the cyc-loamidic quinazolinones [both tautomeric (**1**) and nontautomeric (**2**)], the phenolic 5- to 8-quinazolinols (**3**), the corresponding quinazoline quinones (**4**), the alcoholic hydroxyalkylquinazolines (**5**), the etherial alkoxyquinazolines (**6**), the ester-like acyloxyquinazolines (**7**), and the quinazoline *N*-oxides (**8**). A high proportion of all known quinazolines (including natural products, drugs, and potential drugs) are oxyquinazolines.[2382]

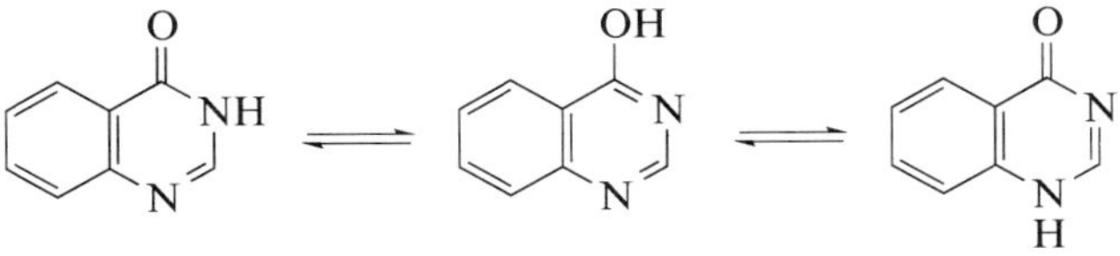

(**1**) 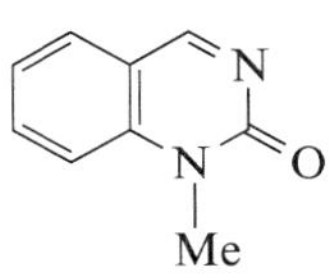   (**2**)

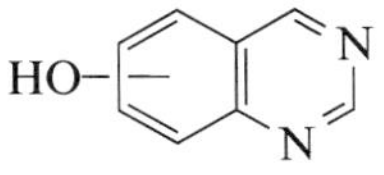

(**3**) 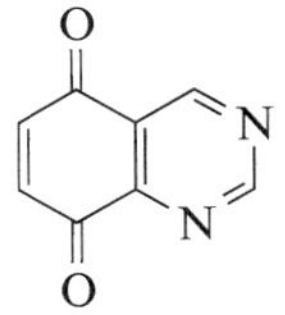   (**4**) 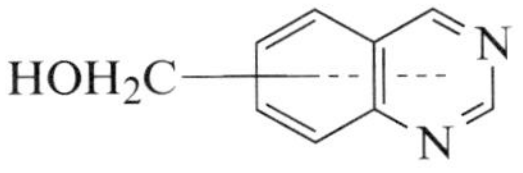   (**5**)

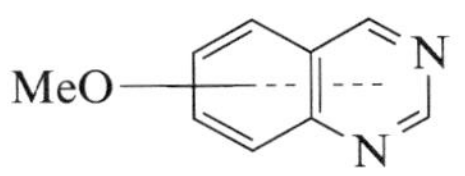

(**6**) 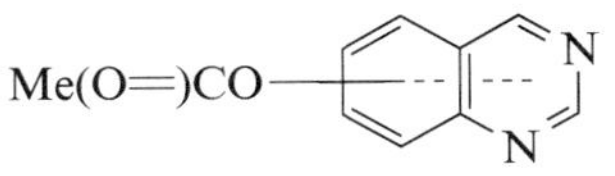   (**7**) 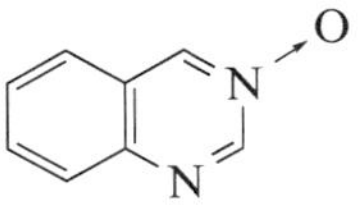   (**8**)

4.1.   TAUTOMERIC QUINAZOLINONES (*H* 69)

Only 2- and 4-quinazolinols are considered to exist predominantly as the corresponding quinazolinones; although transannular tautomerism, for example, 5-quinazolinol⇌5(3H)-quinazolinone, is possible and indeed does occur in some pteridinones,[941] there is no physical or chemical evidence to suggest that 5- to 8-quinazolinols actually undergo any such process.

4.1.1. Preparation of Tautomeric Quinazolinones (H 69,116)

The method of choice for making tautomeric 2- or 4-quinazolinones is usually *primary synthesis* (see Chapter 1). However, several other routes are available as convenient, including *hydrolysis of halogenoquinazolines* (see Section 3.5.3) and those detailed in the following subsections.

4.1.1.1. From Quinazolinamines (H 333)

The conversion of 2- or 4-quinazolinamines into the corresponding quinazolinones has been done by treatment with nitrous acid or by direct hydrolysis in alkaline or acidic media. The nitrous acid method appears to be specific for 2-(primary-amino)quinazolines, whereas both direct hydrolytic methods have been used successfully with any type of amino group in either position. However, preferential hydrolysis at the 4-position of a 2,4-quinazolinediamine may be achieved easily under controlled conditions. The following examples illustrate these procedures.

By nitrous acid treatment

2,4-Quinazolinediamine (**9**) gave 4-amino-2(1H)-quinazolinone (**10**) (NaNO$_2$, 1 M HCl, warm, 15 min: 86%).[2330]

5,6,7,8-Tetrahydro-2,4-quinazolinediamine gave 4-amino-5,6,7,8-tetrahydro-2(1H)-quinazolinone (NaNO$_2$, 55% H$_2$SO$_4$, 20°, 3 days: ∼ 35%).[91]

Also other examples.[50,1398]

By alkaline hydrolysis

4-Quinazolinamine gave 4(3H)-quinazolinone (kinetics in 5 M KOH at 80°: 200 times faster than a similar hydrolysis of 4-pyrimidinamine).[164]

2- or 4-Trimethylammonioquinazoline chloride gave 2(1H)-quinazolinone (0.4 M NaOH, 20°, 60 min: ∼ 60%) or 4(3H)-quinazolinone (1 M NaOH, 20°, 8 min: ∼ 65%), respectively; also kinetics: 4-isomer 30 times faster than the 2-isomer.[165]

5-Trifluoromethyl-2,4-quinazolinediamine gave 2-amino-5-trifluoromethyl-4(3H)-quinazolinone (**11**) (1 M NaOH, N$_2$, reflux, 5 h: 94%).[2211]

Also other examples.[1077]

By acidic hydrolysis

4-Quinazolinamine gave 4(3*H*)-quinazolinone (2 M HCl, reflux, 1 h: 50%;[196] 10 M HCl, 150°, sealed, 15 h: 35%).[340]

2-Pyridin-2'-yl-4-quinazolinamine gave 2-pyridin-2'-yl-4(3*H*)-quinazolinone (**12**) (1 M HCl, reflux, N$_2$, 65 h: 79%).[1602]

2-Anilino-6-chloro-4-phenylquinazoline gave 6-chloro-4-phenyl-2(1*H*)-quinazolinone (50% H$_2$SO$_4$, reflux, 2 h: ~75%).[899]

2,4-Quinazolinediamine (**9**) gave 2-amino-4(3*H*)-quinazolinone (**13**) (6 M HCl, reflux, 30 min: 88%).[2330, cf. 678]

2,4-Diamino- gave 2-amino-4-oxo-3,4-dihydro-6-quinazolinecarbonitrile (1 M HCl, reflux, 6 h: 96%).[201]

4-Benzotriazol-1'-yl-2-phenylquinazoline (**14**) gave 2-phenyl-4(3*H*)-quinazoline (H$_3$PO$_4$, 185°, 15 min: 12%).[404]

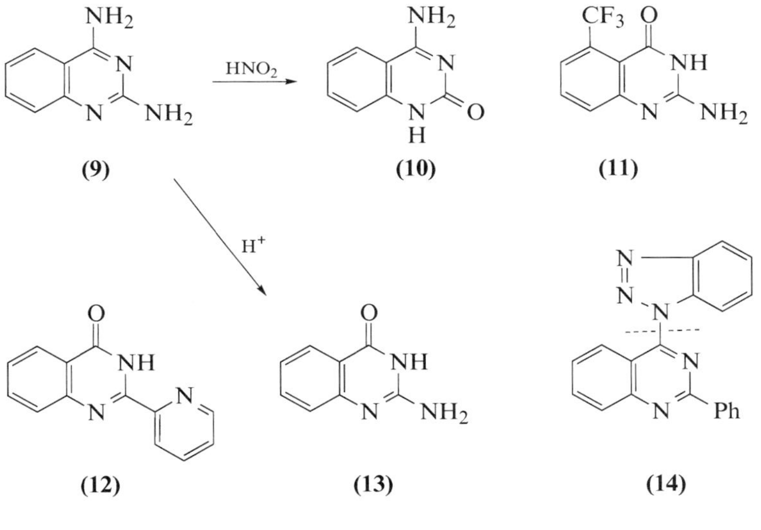

Also other examples.[91,343,600,611,618,1077,1584,1593,1958,1976,2194,2284,2374]

4.1.1.2. From Quinazolinethiones or Derivatives (*H* 283)

Quinazolinones have been made from the corresponding thiones, mainly by oxidative hydrolysis, or from alkylthioquinazolines, acylthioquinazolines, or preoxidized derivatives, usually by regular acidic hydrolysis. These useful processes are illustrated in the following examples.

With quinazolinethiones as substrates

2-t-Butyl-4(3H)-quinazolinethione (**15**, X = S) gave 2-t-butyl-4(3H)-quinazo-linone (**15**, X = O) (H$_2$O$_2$, NaOH, H$_2$O–EtOH, 50°, 1 h: 65%); homologs likewise.[105]

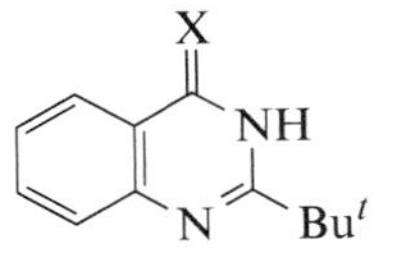

(15)

4(3H)-Quinazolinethione gave 4(3H)-quinazolinone (Bu^{t}OK, Bu^{t}OH, trace Br$_2$, reflux, 20 h: 70%; trace I$_2$, 48 h: 35%; trace Cl$_2$, 16 h: 65%).[1291,1340]

4-Carboxymethyl-3,4-dihydro-2(1H)-quinazolinethione (**16**, X = S) gave the corresponding quinazolinone (**16**, X = O), (H$_2$O$_2$, NaOH, H$_2$O, 20°, 12 h: ∼85%).[975]

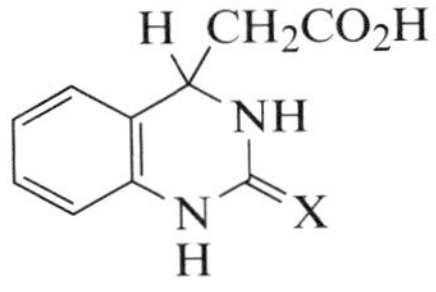

(16)

3-β-Hydroxyethyl-4-phenyl-3,4-dihydro-2(1H)-quinazolinethione gave the corresponding quinazolinone (H$_2$O$_2$, NaOH, H$_2$O–dioxane, 5°, 2 h: 85%).[1073]

2-Thioxo-1,2-dihydro-4(3H)-quinazolinone gave mainly 2,4(1H,3H)-quinazo-linedione (O$_3$, AcOH–H$_2$O, 25°, 30 min: 75%) or entirely 4(3H)-quina-zolinone (likewise in glacial AcOH: 82%).[2420,2558]

Also other examples,[231,603,840,1312,2070]

With alkylthio or acylthioquinazolines as substrates

2-Ethylthio-4(3H)-quinazolinone gave 2,4(1H,3H)-quinazolinedione (H$_2$SO$_4$, AcOH–H$_2$O, reflux, 8 h: ∼70%).[1459]

3-Methyl-2-methylthio-4(3H)-quinazolinethione gave 3-methyl-4-thioxo-3,4-dihydro-2(1H)-quinazolinone(**17**, X = S) (HCl, AcOH–H$_2$O, reflux, 3 h: 97%);[40] likewise, 3-methyl-2-propylthio-4(3H)-quinazolinone gave 3-methyl-2,4(1H, 3H)-quinazolinedione (**17**, X = O) (HCl, EtOH–H$_2$O, re-flux, 4 h: ∼75%).[160]

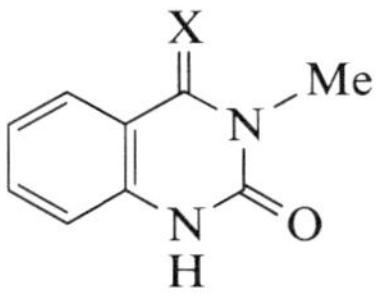

(17)

2-Benzoylthio-3-methyl-4(3H)-quinazolinone   gave   3-methyl-2,4(1H,3H)-qinazolinedione (**17**, X = O) (HCl, EtOH–H$_2$O, reflux, 4 h: ~85%).[1750]

4-Carboxymethyl-2-methylthio-3,4-dihydroquinazoline gave 4-carboxymethyl-3,4-dihydro-2(1H)-quinazolinone (**16**, X = O) (NaOH, H$_2$O, 90°, 2 h: ~65%).[975]

Also other examples.[115,450,477,496,511,547,807,897,1017,2006,2292,2570]

With preoxidized substrates

2-p-Tosyl-4(3H)-quinazolinone (**18**) gave 2,4(1H,3H)-quinazolinedione (**19**, R = H) (pH 8, H$_2$O–EtOH, 20°, 1 h: ~70%).[1915]

4-Oxo-3-phenyl-3,4-dihydroquinazolin-2-yl p-tolyl disulfide (**20**) gave 3-phenyl-2,4(1H,3H)-quinazolinedione (**19**, R = Ph) (HCl: see original for details).[410]

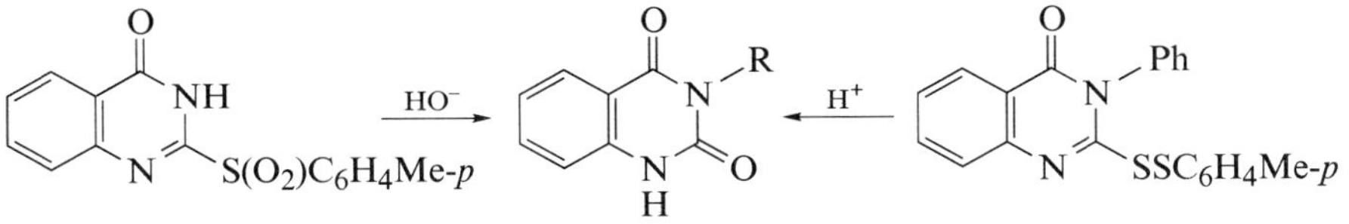

(18)   (19)   (20)

Also other exzmple.[281,1031]

4.1.1.3. From Alkoxy- or Aryloxyquinazolines (*H* 241)

Since alkoxyquinazolines are usually made directly or indirectly from the corresponding quinazolinones, the reverse procedure is seldom used. However, 4-phenoxy-2-phenylquinazoline (**21**) (made by primary synthesis) conveniently afforded 2-phenyl-4(3H)-quinazolinone (**22**) in 86% yield by hydrolysis in refluxing 2.5 M aqueous alcoholic alkali during 3 h;[213] similarly, 2-methoxy-4(3H)-quinazolinone 1-oxide gave >70% of 1-hydroxy-2,4(1H,3H)-quinazolinedione by heating in acid or alkali for 1 h.[182] Other examples are more interesting than intrinsically useful: 6-chloro-4-methoxy-2-phenylquinazoline with neat pyridine hydrochloride at 180° gave 6-chloro-2-phenyl-4(3H)-quinazolinone in 60% yield;[58] 2,4-dimethoxyquinazoline gave 2,4(1H,3H)-quinazolinedione (70%) on attempted aminolysis with neat methylhydrazine in a sealed tube at 80° for 3 h, followed by an aqueous workup;[2230] bis(3-phenyl-2-thioxo-1,2,3,4-tetrahydroquinazolin-4-yl) ether (23), on heating in 2% alkali for 1 h, gave not a single product (24) but its disproportionation products, 3-phenyl-2-thioxo-1,2-dihydro-4(3H)-quinazolinone (25) (in 20% yield) and 3-phenyl-3,4-dihydro-2(1H)-quinazolinethione (26) (in 50% yield);[84] and 6-hydroxy-4-methoxy-2-phenyl-5,8-dihydro-5,8-quinazolinedione gave 6-hydroxy-2-phenyl-5,8-dihydro-4,5,8(3H)-quinazolinetrione (dilute HCl/MeOH–H$_2$O, reflux, 2 h: 88%).[763]

(21)   (22)   (23)

(24)   (25)   (26)

4.1.1.4.   From Other Types of Quinazoline Substrate

Although the routes are of little importance, several other types of quinazoline substrate have been used to make tautomeric quinazolinones. Thus 4-dicyanomethyl-3-methyl-2-methylimino-2,3-dihydroquinazoline (27) in refluxing 10 M HCl for 12 h gave a 73% yield of 3-methyl-2-methylamino-4(3H)-quinazolinone (28);[723] 2-methyl-4-quinazolinecarbonitrile gave 2-methyl-4(3H)-quinazolinone (80%), simply by stirring in 2 M KOH at 20°;[1063] 3-phenyl-2-pyridiniomethyl-4(3H)-quinazolinone bromide (29) gave 3-phenyl-2,4(1H,3H)-quinazolinedione (80%) by refluxing in dilute ethanolic alkali for 2 h;[1284] 3-methyl-3,4-dihydro-4-quinazolinol (30) appears to have undergone a thermal rearrangement, for example, in boiling toluene to afford a 45% yield of 3-methyl-3,4-dihydro-2(1H)-quinazolinone (31);[415] the fixed quinazolinone, 3-benzyl-6,7-dibenzyloxy-2-methyl-4(3H)-quinazolinone, was completely debenzylated by prolonged hydrogenolysis (H$_2$, Pd/C, EtOH) to give 6,7-dihydroxy-2-methyl-4-(3H)-quinazolinone (~40%);[218] and oxidative,[35] hydrolytic,[472,1966,2439,2511] and thermal[901] procedures have been used also to convert fixed into tautomeric quinazolinones. The acidic hydrolysis of 4-azido-6-bromo-2-phenylquinazoline (32) to 6-bromo-2-

phenyl-4(3*H*)-quinazoline[749] has been mentioned in Section 1.7.13. Pyrolysis of 2,2'-di(*o*-chlorophenyl)-1,1',2,2'-tetrahydro-3,3'-biquinazoline-4,4'(3*H*,3'*H*)-dione at 270° gave a separable mixture of 2-*o*-chlorophenyl-4(3*H*)-quinazolinone (30%) and its 1,2-dihydro derivative (40%).[2532]

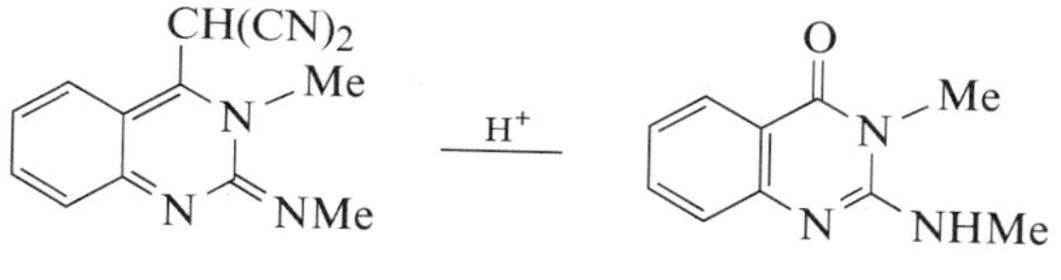

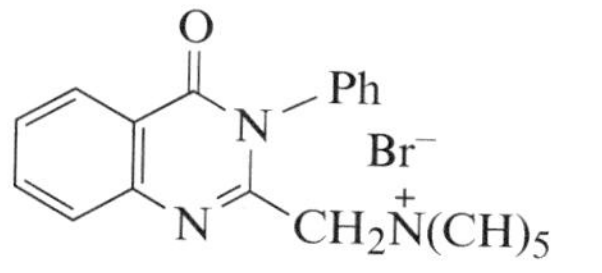

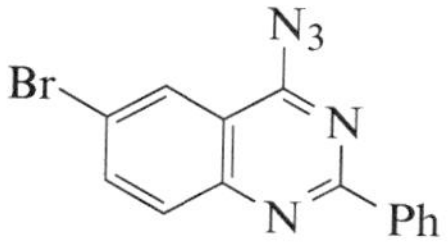

4.1.1.5. By Nuclear Oxidation of Quinazolines or Hydroquinazolines (*H* 96,404)

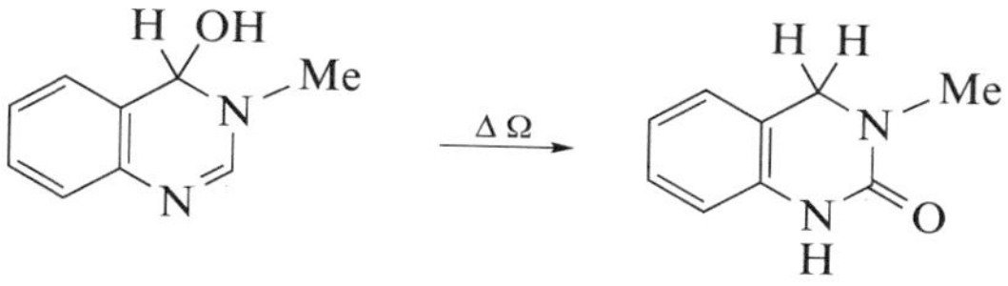

Tautomeric quinazolinones have been made by direct oxidative hydroxylation of quinazolines or by dehydrogenation of dihydroquinazolinones through regular oxidation or an eliminative reaction. These processes are illustrated in the following examples.

By oxidative hydroxylation

3-Methyl-2(3*H*)-quinazolinimine hydriodide (**33**) gave 2-amino-3-methyl-4(3*H*)-quinazolinone (**34**) (1 M H_2SO_4, H_2O_2, 20°, 3 days: 20%).[50]

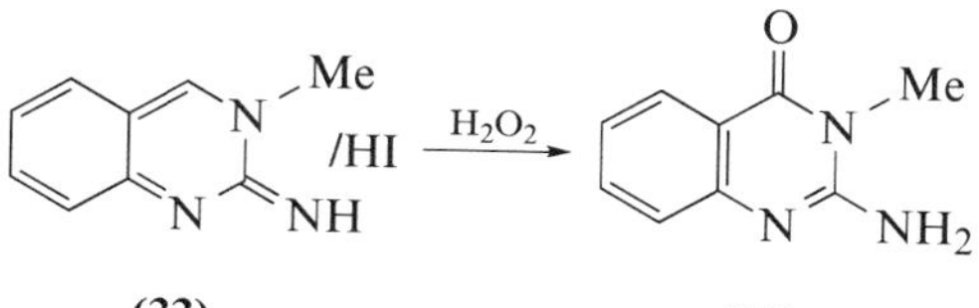

1-Phenyl-1,2-dihydro-4(3*H*)-quinazolinone (**35**) gave 1-phenyl-2,4(1*H*,3*H*)-quinazolinedione (**36**) ($KMnO_4$, H_2O, 20°, 4 h: 64%).[186]

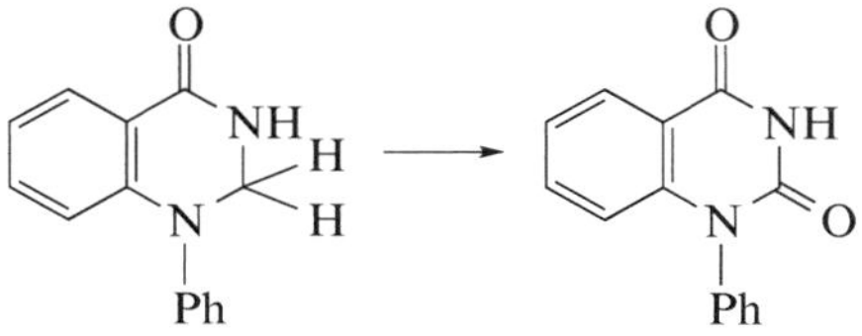

(35)   (36)

Also other examples.[497,1542(?)]

By regular oxidation of a dihydroquinazolinone

1,2-Dihydro-4(3H)-quinazolinone (**37**, R = H) gave 4(3H)-quinazolinone (**38**, R = H) [Hg^{2+}/(HO$_2$CH$_2$C)$_2$NCH$_2$CH$_2$N(CH$_2$CO$_2$H)$_2$ complex, ? solvent, ?°, 90 min: 94%];[818] 2-methyl (**38**, R = Me), 2-phenyl (**38**, R = Ph), and other analogs likewise.[818]

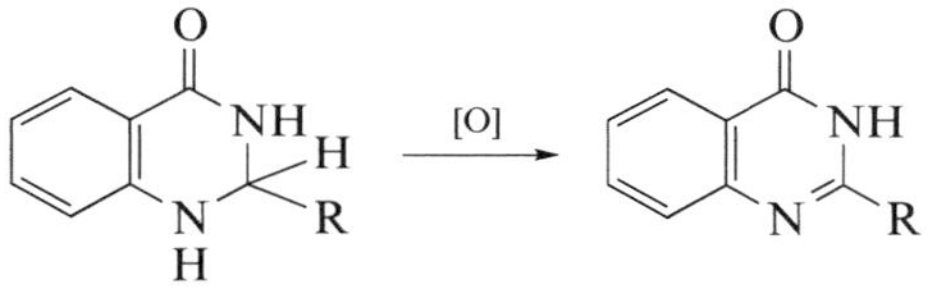

(37)   (38)

2-p-Chlorophenyl-1,2-dihydro- (**37**, R = C$_6$H$_4$Cl-p) gave 2-p-chlorophenyl-4(3H)-quinazolinone (KMnO$_4$, AcMe, reflux, 90 min).[566]
Also other examples with air, HgO, or benzoquinone as the oxidant.[251,577,686]

By eliminative oxidation of a dihydroquinazolinone

2-Methyl-2-(N-phenylcarbamoylmethyl)-1,2-dihydro- (**39**) gave 2-methyl-4(3H)-quinazolinone (**38**, R = Me) (neat, 180°, 20 min: 54%; by loss of PhNHAc).[1370]

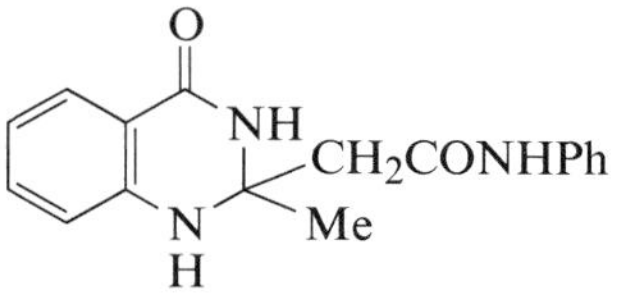

(39)

Methyl 2-benzyl-4-oxo-1,2,3,4-tetrahydro-2-quinazolinecarboxylate (**40**) gave methyl 4-oxo-3,4-dihydro-2-quinazolinecarboxylate (**38**, R = CO$_2$Me) (xylene, reflux, 48 h: 54%; by loss of PhMe).[948]

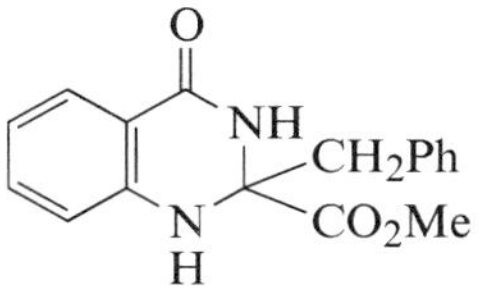

(40)

2-(2-Hydroxynaphthalen-1-yl)-1,2-dihydro-4(3H)-quinazolinone gave 4(3H)-quinazolinone (**38**, R = H) (ClC$_6$H$_4$Cl-o, reflux, N$_2$, 6 h: 56%; by loss of C$_{10}$H$_7$OH-2);[1943] 4(1H)-quinazolinone 3-oxide (41%) similarly.[1943]

2-Phenyl-1,2,5,6,7,8-hexahydro- gave 2-phenyl-5,6,7,8-tetrahydro-4(3H)-quinazolinone [Pd/C, PhMe, H$_2$(!), 20°, 30 min: 92%; other catalysts or EtOH as solvent gave lower yields].[1980]

Also other examples involving Pd-induced loss of H$_2$[1771] or the thermally induced loss of H$_2$O[552] or CHF$_3$.[1727]

4.1.2. Reactions of Tautomeric Quinazolinones (*H* 72,104,130)

Tautomeric quinazolinones undergo few but important reactions. Of these, *conversion into halogenoquinazolines* has been covered in Sections 3.1 and 3.2; the remainder are discussed in subsections hereafter.

4.1.2.1. Conversion into Quinazolinethiones (*H* 278)

Both 2- and 4-quinazolinones (tautomeric or fixed) undergo thiation with phosphorus pentasulfide in pyridine, xylene, or such. However, the 4-thiones are formed more readily than their 2-isomers, so that preferential 4-thiation of 2,4-quinazolinediones is possible. The thiation of tautomeric quinazolinones is illustrated by the following typical examples, in which P$_2$S$_5$/pyridine under reflux has been used unless otherwise indicated; for thiation of fixed quinazolinones, see Section 4.6.2.1.

6-Chloro-4-phenyl-2(1H)-quinazolinone gave the thione (**41**) (45 min: 30%).[836]

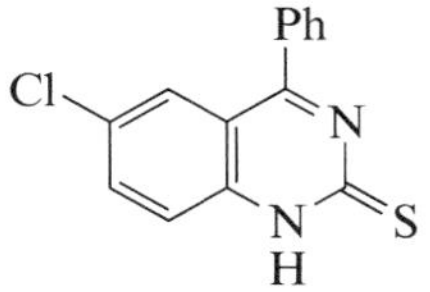

(41)

4(3H)-Quinazolinone gave the thione (**42**, R = H) (1 h: 62%).[105]

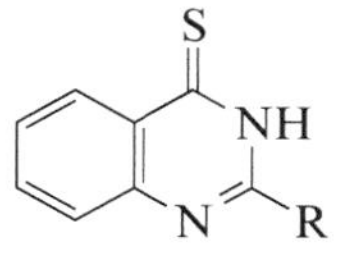

(42)

2-Phenyl-4(3H)-quinazolinone gave the thione (**42**, R = Ph) (P$_2$S$_5$, MgO, PhH, reflux, 14 h: 17%).[210]

2-α-Chloroethyl-4(3H)-quinazolinone gave the thione (**42**, R = CHClMe) (xylene, reflux, 12 h: 58%).[2343]

6-Chloro-2-styryl-4(3H)-quinazolinone gave the corresponding thione (90 min: 48%).[2264]

2-Amino-6-methyl-4(3H)-quinazolinone gave the corresponding thione (3 h: > 90% crude).[2255]

2,4(1H,3H)-Quinazolinedione gave only 4-thioxo-3,4-dihydro-2(1H)-quinazolinone (**43**, R = H, X = O) (2 h: ∼ 70%).[2330]

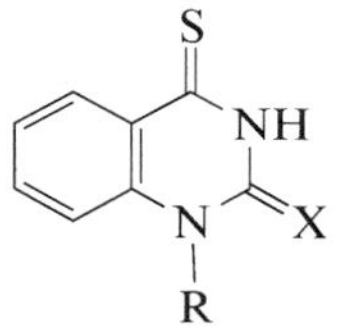

(43)

1-Methyl-4-thioxo-3,4-dihydro-2(1H)-quinazolinone (**43**, R = Me, X = O) gave 1-methyl-2,4(1H,3H)-quinazolinedithione (**43**, R = Me, X = S) (ClC$_6$H$_4$Cl-o, reflux, 90 min: ∼ 80%).[231]

Methyl 2-methoxycarbonylmethyl-4-oxo- (**44**, X = O) gave methyl 2-methoxycarbonylmethyl-4-thioxo-1,2,3,4-tetrahydro-2-quinazolinecarboxylate (**44**, X = S) (PhH, reflux, 2 h: 33%).[232]

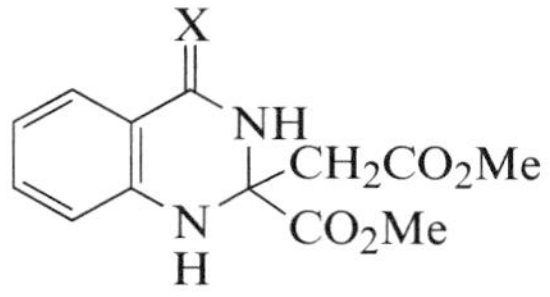

(44)

Also other examples.[40,41,223,673,678,825,968,1184,1280,1354,1595,1903,2086,2527]

4.1.2.2. *O*- or *N*-Acylation and *O*-Silylation Reactions (*H* 104)

Although tautomeric quinazolinones may undergo *O*- or *N*-acylation (according to the reagent and/or molecular environment), they appear to afford only *O*-trimethylsilyl derivatives, at least under the usual silylation conditions. The following examples illustrate the formation of such products, used sometimes to direct subsequent alkylation reactions.

O-Acylation

3-Allyl-2,4(1*H*,3*H*)-quinazolinedione gave 3-allyl-2-benzoyloxy-4(3*H*)-quinazolinone (BzCl, NaOH, H_2O, < 5°, ? h: 33%).[477]

6-Chloro-4-phenyl-2(1*H*)-quinazolinone gave 6-chloro-4-phenyl-2-tosyloxyquinazoline (**45**) (TsCl, pyridine, 20°, 4 days: ~ 30%).[254]

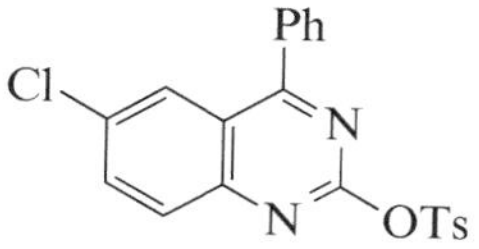

(45)

2-Phenyl-4(3*H*)-quinazolinone gave 4-(dimorpholinophosphinyloxy)-2-phenylquinazoline (**46**) {NaH, THF, 20°, 1 h; then ClP(=O)[N(CH$_2$CH$_2$)$_2$O]$_2$, 20°, 4 h: 77%}.[920]

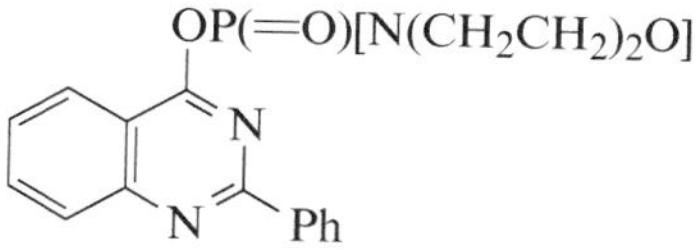

(46)

2-(β,β-Dichloro-α,α-difluoroethyl)-4(1*H*)-quinazolinone 3-oxide gave 2-(β,β-dichloro-α,α-difluoroethyl)-4-(*N*-methylcarbamoyloxy)quinazoline 3-oxide (**47**) (MeNCO, Et$_2$O, 20°, 6 h: 70%).[443]

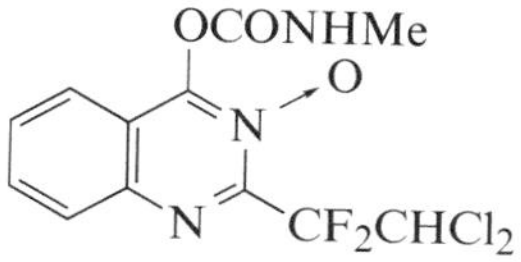

(47)

Also other examples.[183,443]

N-Acylation

2-Methyl-4(3*H*)-quinazolinone gave 3-chloroacetyl-2-methyl-4(3*H*)-quina-zolinone (**48**) (ClCH$_2$COCl, PhH, reflux, 4 h: ?%).[2142]

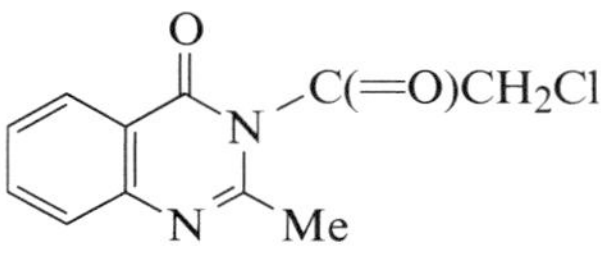

(48)

1-Ethyl-2,4-(1*H*,3*H*)-quinazolinedione gave its 3-benzoyl derivative (BzCl, pyridine, Me$_2$NCHO, heat: ?%).[1493]

4(3*H*)-Quinazolinone gave 3-tosyl-4(3*H*)-quinazolinone (TsCl, pyridine, <10°: >95%).[650]

Also other examples.[1405,1407,1493,1525,1939]

O-Trimethylsilylation

4(3*H*)-Quinazolinone gave 4-trimethylsiloxyquinazoline (**49**) (Me$_3$SiCl, Et$_3$N, PhMe, reflux, 3 h: 81%).[224]

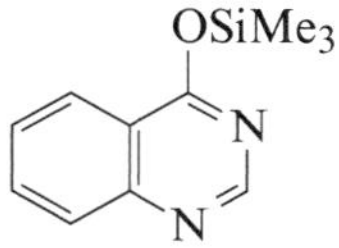

(49)

2,4(1*H*,3*H*)-Quinazolinedione gave 2,4-bis(trimethylsiloxy)quinazoline (**50**), (Me$_3$SiCl, Et$_3$N, PhMe, reflux, 6 h: 83%;[243] Me$_3$SiNHSiMe$_3$, Me$_2$NCHO, sealed, 150°, ? h: 97%).[78]

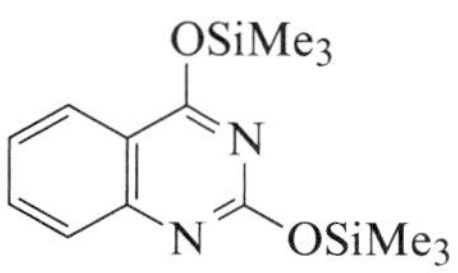

(50)

4.1.2.3. *O*- or *N*-Alkylation Reactions (*H* 73,104,130)

Irrespective of the type of reagent or the conditions used, alkylation of a tautomeric quinazolinone, such as (**51**), usually gives an *N*-alkylquinazolinone, sometimes accompanied by a smaller amount of the isomeric alkoxyquinazoline. Occasionally, the alkoxyquinazoline may predominate, especially when a diazoalkane is used or the steric and/or electronic factors associated with the reagent or substrate are favorable; in this regard, a functional passenger group on the pyrimidine ring may affect the outcome appreciably,[607,653] whereas any on the benzene ring appear to have little effect.

Formation of the following products illustrates the results that may be expected on alkylation of various types of tautomeric quinazolinone using a variety of reagents and conditions. The examples are grouped according to substrate types, ignoring transannular passenger groups as almost irrelevant; any necessary subgrouping follows obvious lines. Individual substrates are not mentioned except where confusion might arise; given percentages represent isolated yields.

From simple 2-quinazolinones: *O*-alkylation

6-Chloro-2-cyclopropylmethoxy-4-phenylquinazoline (**52**) plus 6-chloro-1-acylopropylmethyl-4-phenyl-2(1*H*)-quinazolinone (**53**) [NaH, Me_2NCHO, 100°, 30 min; then $BrCH_2CH(CH_2)_2$, 100°, 6 h: 23% and 51%, respectively].[580]

(51)   (52)   (53)

6-Chloro-2-β-dimethylaminoethoxyquinazoline plus 6-chloro-1-β-dimethylaminoethyl-2(1*H*)-quinazolinone (KOBu^t, Me_2NCHO, 0°, 1 h; then $ClCH_2CH_2NMe_2$/PhH, reflux, 2 h: 24% and 31%, respectively).[676]

From simple 2-quinazolinones: N1-alkylation

6-Chloro-1-methyl-4-phenyl-2(1*H*)-quinazolinone (NaOH, Me_2NCHO, 20°, 30 min; then MeI, 1 h: 40%).[836]

1-*o*-Fluorobenzyl-3-methyl-3,4-dihydro-2(1*H*)-quinazolinone (**54**, R = Me) [from 3-methyl-3,4-dihydro-2(1*H*)-quinazolinone with NaH, *o*-FC_6H_4-CH_2Cl, Me_2SO, 20°, 20 h: 88%];[2339] also many analogs rather similarly.[654]

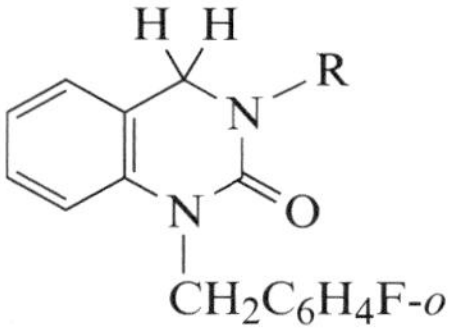

(54)

Also other examples.[2563,2575]

From simple 4-quinazolinones: *O*-alkylation

2-*t*-Butyl-4-pentyloxyquinazoline (**55**, R = Bu^t) (NaH, C$_5$H$_{11}$I, Me$_2$NCHO, 70°, 2 h: 88%);[2482] compare with 2-isopropyl-4-pentyloxyquinazoline (**55**, R = Pr^i) plus 2-isopropyl-3-pentyl-4(3*H*)-quinazolinone (**56**, R = Pr^i) (likewise: 52% and 17%, respectively) or only 3-pentyl-4(3*H*)-quinazolinone (**56**, R = H) (likewise: 60%),[2482] clearly indicating a steric effect of the 2-substituent.

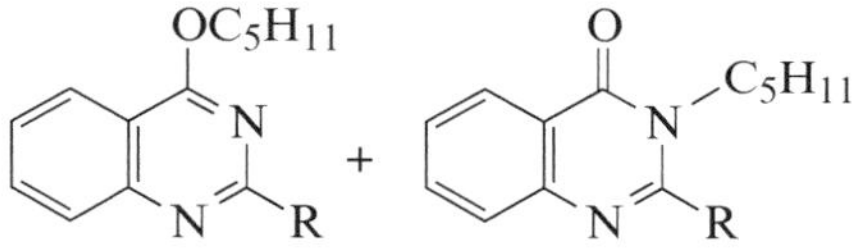

(55)   **(56)**

2-*o*-Hydroxyphenyl-4-methoxyquinazoline only (CH$_2$N$_2$, Me$_2$NCHO–Et$_2$O, 20°, ? h: ?%) but the same product plus 2-*o*-methoxyphenyl-3-methyl-4(3*H*)- quinazolinone (MeI, K$_2$CO$_3$, AcMe, reflux, 48 h: ∼40% each).[226]

4-α-Ethoxycarbonylethoxyquinazoline only (MeCHBrCO$_2$Et, K$_2$CO$_3$, EtCOMe, reflux, 6 h: ∼78%).[1189]

4-Isopropoxy-2-*p*-methoxyphenylquinazoline only (Pr^{i}OH, Ph$_3$P, EtO$_2$CN=NCO$_2$Et, THF, 20°, <4 days: 71%).[433]

Also other examples.[1056,1161,1184,1227,2285]

From simple 4-quinazolinones: *N*1-alkylation

1-Allyl-7-chloro-2-methyl-4(1*H*)-quinazolinone (**57**) (NaH, dioxane, 100°, 2 h; hen BrCH$_2$CH=CH$_2$, reflux, 14 h: 70%; reason for 1-allyl assignment unclear).[863]

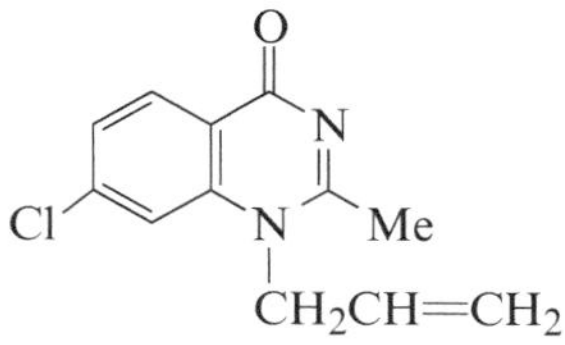

(57)

From simple 4-quinazolinones: N3-alkylation

1. *Using unsubstituted alkyl halides*

3-Benzyl-7-chloro-6-nitro-4(3H)-quinazolinone (**58**, R = Ph) (PhCH$_2$Br, KOH, MeOH, reflux, 1 h: 89%);[501] 7-chloro-3-methyl-6-nitro-4(3H)-quinazolinone (**58**, R = H) (MeI, EtOH, 20°, 11 h: ~55%).[900]

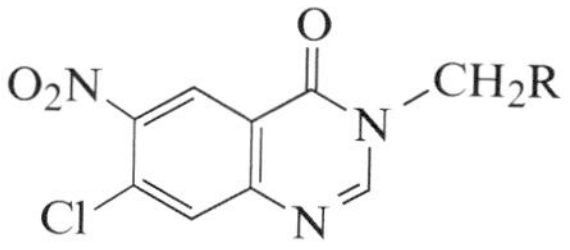

(58)

3-*o*-Chlorobenzyl-4(3H)-quinazolinone (ClCH$_2$C$_6$H$_4$Cl-*o*, NaHCO$_3$, Me$_2$NCHO, 70°, until CO$_2$↑ finished: 46%);[1939] also the 3-prop-2′-ynyl analog (BrCH$_2$C≡CH, MeONa, MeOH, 20°, 12 h: 91%).[1935]

2-Isopropyl-3,6-dimethyl-4(3H)-quinazolinone (MeI, NaOH, MeOH, reflux, 6 h: >40%);[1050] also related compounds.[1050]

Also other examples.[35,498,1165,1525,1727,1772,1835,2341,2392,2584]

2. *Using functionally substituted alkyl halides*

3-β-Hydroxyethyl-4(3H)-quinazolinone (**59**, R = CH$_2$OH) (BrCH$_2$CH$_2$OH, MeONa, MeOH, reflux, 24 h: 62%);[1772] 3-acetonyl-4(3H)-quinazolinone (**59**, R = Ac) (AcCH$_2$Br, KOH, MeOH, 20°, 12 h: 50%);[1956] 3-trimethyl-silylmethyl-4(3H)-quinazolinone (**59**, R = SiMe$_3$) (ClCH$_2$SiMe$_3$, K$_2$CO$_3$, Me$_2$SO, 25°, 3 days: 91%).[1965]

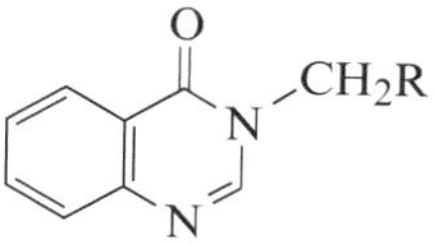

(59)

3-β-Methoxycarbonylvinyl-6-methylthio-4(3H)-quinazolinone (**60**) ClCH=CHCO$_2$Me, K$_2$CO$_3$, AcMe–Me$_2$NCHO, reflux, 18 h: 90%; or NaH, Me$_2$NCHO, 50°, 30 min, then ClCH=CHCO$_2$Me, 50°, 1 h: 68%).[1558]

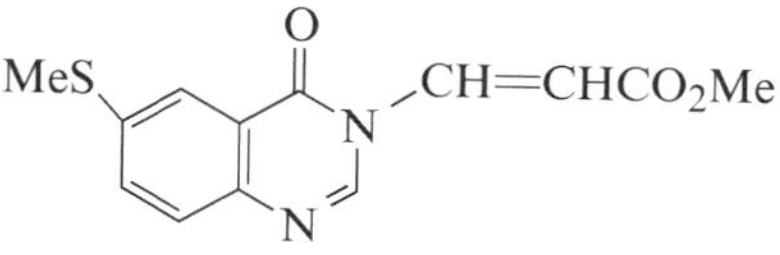

(60)

3-α-Carboxyethyl-4(3H)-quinazolinone (MeCHBrCO$_2$H, NaOH, H$_2$O, reflux, 2 h: 66%).[783,1085]

Also other examples.[330,384,833,1108,1617,1689,1772,1907,2043,2277,2433,2451,2528,2580]

3. *Using (activated) aryl halides*

3-*o*-Cyanophenyl-7-methyl-4(3*H*)-quinazolinone (**61**) (FC$_6$H$_4$CN-*o*, K$_2$CO$_3$, Me$_2$SO, 100°, 16 h: 42%).[2488]

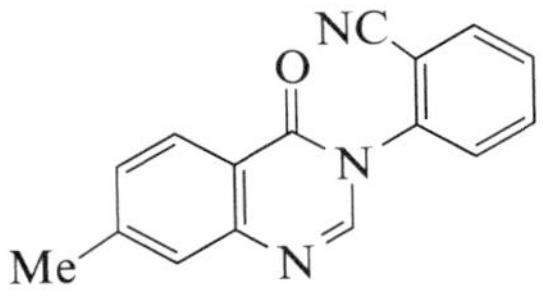

(61)

3-(4-Formyl-2-nitrophenyl)-4(3*H*)-quinazolinone (4-Cl-3-NO$_2$-C$_6$H$_3$CHO, pyridine, EtOH, reflux, 1 h: > 70%);[1218] also analogs.[1218]

4. *Using other types of alkylating agent*

3-Anilinomethyl-4(3*H*)-quinazolinone (**59**, R = NHPh) (CH$_2$O, PhNH$_2$, H$_2$O–EtOH, reflux, 15 min: 60%);[2302] 3-aminomethyl-4(3*H*)-quinazolinone (**59**, R = NH$_2$) (CH$_2$O, NH$_4$Cl, H$_2$O–EtOH, reflux, 3 h: ?%);[982,1192] also analogs.[1192,2302] For other such Mannich alkylations, see Section 2.2.2.3.

2-α-Chloroethyl-3-(2-hydroxynaphthalen-1-yl)methyl-4(3*H*)-quinazolinone (**62**) (CH$_2$O, C$_{10}$H$_7$OH-2, trace HCl, EtOH, reflux, 3 h: 72%; a more traditional Mannich reaction).[2341]

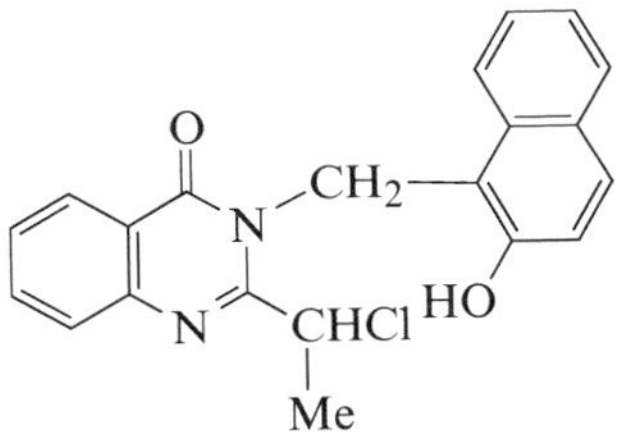

(62)

3-β-Nitroethyl-4(3*H*)-quinazolinone (H$_2$C=CHNO$_2$, PhH, 20°, 12 h: 95%).[1172]

3-δ-Bromo-β-hydroxybutyl-4(3*H*)-quinazolinone (**63**) (1-bromo-3,4-epoxybutane, MeONa, MeOH, 20°, 3 h: 56%);[253] 3-(1-benzyl-4-hydroxypiperidin-4-yl)methyl-4(3*H*)-quinazolinone [epoxide (**64**), MeONa, MeOH, reflux, 12 h: 28%].[221]

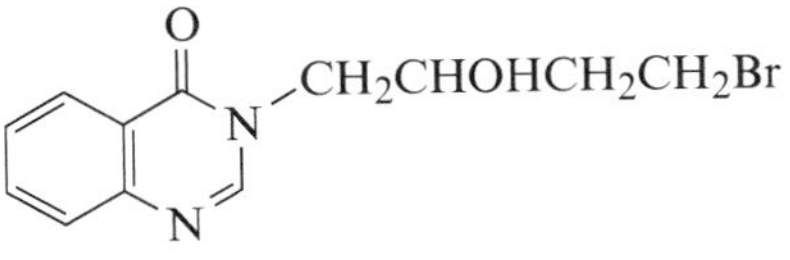

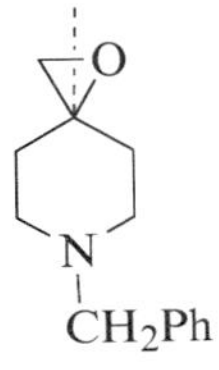

(63)   **(64)**

3-Benzyl-2-methyl-4(3H)-quinazolinone (**65**) ($Cl_3CCO_2CH_2Ph$, 4-Me_2N-pyridine, 120°, 2 h: 20%; with loss of CO_2 and $CHCl_3$).[2120]

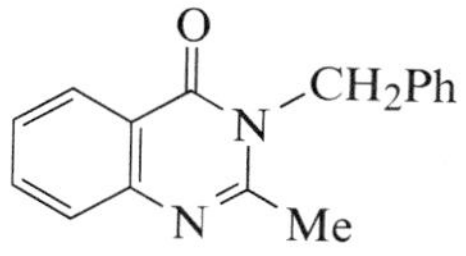

(**65**)

Also the formation of 3-glucosyl derivatives.[1973]

From 2-(functionally substituted)-4-quinazolinones

Ethyl 4-oxo-3,4-dihydro-2-quinazolinecarboxylate (**67**) gave only ethyl 3-ethoxycarbonylmethyl-4-oxo-3,4-dihydro-2-quinazolinecarboxylate (**66**) (NaH, Me_2SO, 1 h: then $BrCH_2CO_2Et$, 20°, 5 h: 48%) or only ethyl 3-α-ethoxycarbonylethoxy-2-quinazolinecarboxylate (**68**) (likewise but $MeCHBrCO_2Et$: 63%); note the clear steric effect.[1639]

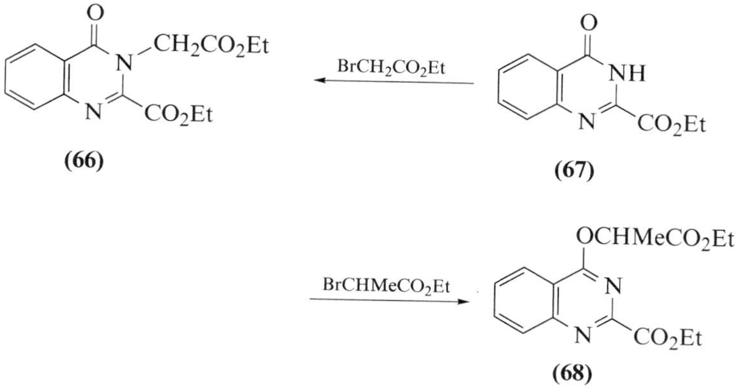

(**66**)   (**67**)

(**68**)

4-But-3′-enyloxy-2-trifluoromethylquinazoline only ($BrCH_2CH_2CH{=}CH_2$, K_2CO_3, AcMe, reflux, 12 h: 46%);[1727] compare with 3-but-3′-enyl-4(3H)-quinazolinone plus 4-but-3′-enyloxyquinazoline (likewise: 92% and 4%, respectively).[1727] Steric effect?

3-Methyl-2-α,β,β,β-tetrafluoroethyl-4(3H)-quinazolinone only (CH_2N_2, Et_2O–EtOH, 20°, 12 h: 84%; difficult to explain!).[712]

3-Methyl-2-methylthio-4(3H)-quinazolinone (MeI, KOH, MeOH, reflux, 4 h: 77%).[1846, cf. 1845]

Also other examples.[1115,2016,2197,2263,2491,2495,2587]

From 2,4-quinazolinediones: first-stage alkylation

3-[β-(4-p-Fluorobenzoylpiperidino)ethyl]-5,6,7,8-tetrahydro-2,4(1H,3H)-
quinazolinedione (**69**) [HOCH$_2$CH$_2$N(CH$_2$CH$_2$)$_2$CHCOC$_6$H$_4$F-p, Ph$_3$P,
EtO$_2$CN=NCO$_2$Et, dioxane, 20°, 40 min: 12%].[2103]

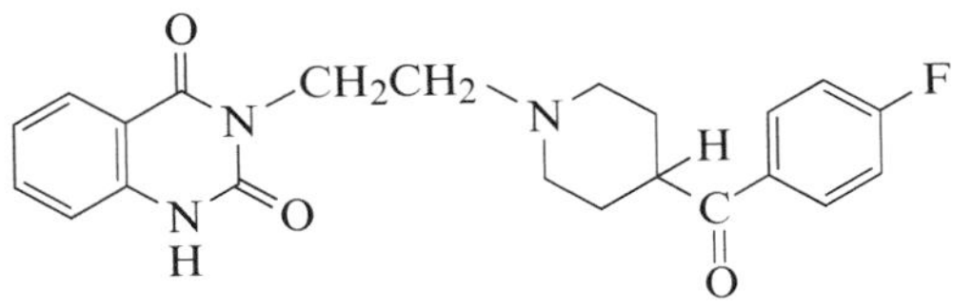

(69)

Also other examples including data on O/N-monoalkylations with various
agents under a variety of conditions.[998,2040,2054]

From 2,4-quinazolinediones: second-stage alkylation

Note that substrates in the following examples were not made by simple
first-stage alkylations.

3-Phenyl-2,4(1H,3H)-quinazolinedione (**71**, R = Ph) gave 2-allyloxy-3-
phenyl-4(3H)-quinazolinone (**70**) (BrCH$_2$CH=CH$_2$, K$_2$CO$_3$, Me$_2$NCHO,
100°, 2 h: 90%) but 3-allyl-2,4(1H,3H)-quinazolinedione (**71**, R = CH$_2$
CH:CH$_2$) gave only 1,3-diallyl-2,4(1H,3H)-quinazolinedione (**72**) (BrCH$_2$
CH=CH$_2$, 2.5 M NaOH, 20°, 12 h: 60%);[1698] the reason must be other than
steric.

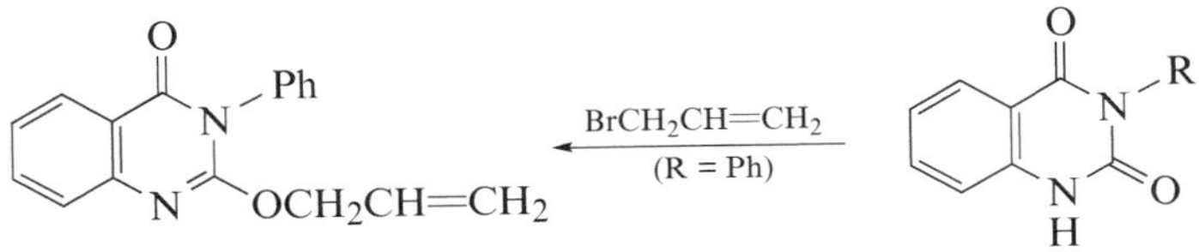

(70)   **(71)**

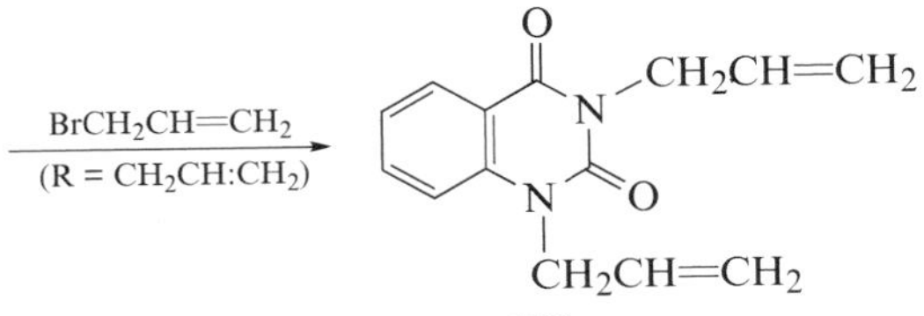

(72)

3-(2,4-Dimethylphenyl)-1-methyl-2,4(1H,3H)-quinazolinedione(MeI, NaOH,
MeOH, reflux, 3 h: 85%).[857]

3-Amino-1-methyl-2,4(1H,3H)-quinazolinedione (MeI, NaOH, MeOH, reflux, 3 h: 85%).[857]

3-Methyl-1-(β-trimethylsilylethoxy)methyl-2,4(1H,3H)-quinazolinedione ($Me_3SiCH_2CH_2OCH_2Cl$, MeOH, 20°, slowly: 36%); also 1-glycosyl and other analogs.[2508]

Ethyl 4-methoxy-3-methyl-2-oxo-1,2,3,4-tetrahydro-4-quinazolinecarboxylate (**73**, R = Me) from the corresponding 4-hydroxy analog (**73**, R = H) (CH_2N_2, MeOH–AcMe–Et_2O, 20°, 5 days: < 50%).[209]

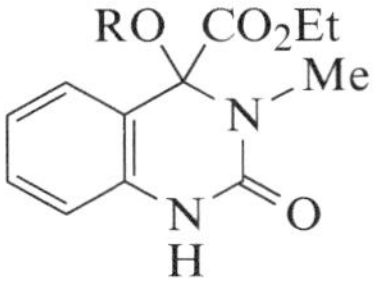

(73)

2-(2,6-Dimethylphenoxy)-3-methyl-4(3H)-quinazolinone (MeI, EtONa, EtOH, reflux, 2 h: 93%).[889]

Also other examples.[447,523,664,977,1480,1521,1538,1539,1547,1735,2109,2262]

From 2,4-quinazolinediones: dialkylation

1,3-Bis(ethoxycarbonylmethyl)-2,4(1H,3H)-quinazolinedione (**74**, R = CO_2Et) (NaH, Me_2SO, 20°, until $H_2\uparrow$ finished; then $BrCH_2CO_2Et$, 5°, 2 h: 57%); 1,3-bis(cyanomethyl)-2,4(1H,3H)-quinazolinedione (**74**, R = CN) (likewise but $ClCH_2CN$: 45%).[1822]

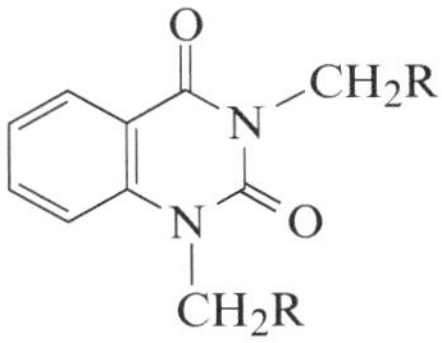

(74)

7-Chloro-1,3-diethyl-6-nitro-2,4(1H,3H)-quinazolinedione (EtI, K_2CO_3, Me_2NCHO, 60°, 6 h: 84%).[2261]

1,3-Bisprop-2′-ynyl-2,4(1H,3H)-quinazolinedione (**74**, R = C≡CH) ($HOCH_2$ C≡CH, Ph_3P, EtO_2CN=NCO_2Et, dioxane, 20°, then reflux, 3 h: 28%).[2481]

1,3-Bis(2,3-epoxypropyl)-2,4(1H,3H)-quinazolinedione [1-bromo-2,3-epoxypropane, Et_4NBr, 85°, 2 h: crude (**75**); with NaOH, $MeOCH_2CH_2OMe$, 20°, 1 h: 36%, by loss of 2 × HBr].[1842]

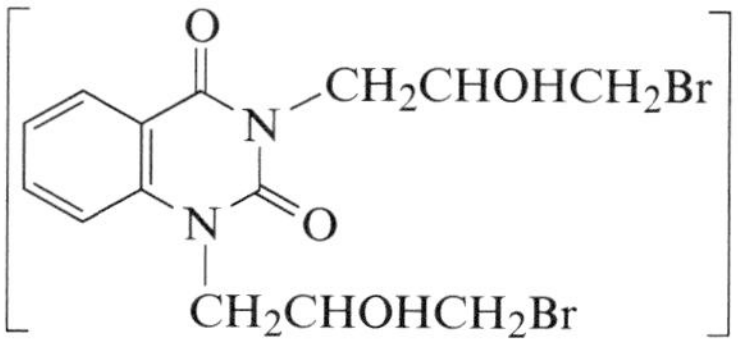

(75)

Also other examples.[1464,1493,2125,2281,2469]

4.1.2.4. Conversion into Quinazolinamines

Tautomeric quinazolinones are usually converted into the corresponding quinazolinamines indirectly via derived chloroquinazolines. However, one or more of several procedures may be used if both the substrate and product can survive rather drastic conditions, as illustrated in the following examples.

4(3H)-Quinazolinone (76) gave 4-quinazolinamine (77, R = H) [neat PhOP(=O) (NH$_2$)$_2$, 230°, open, 30 min: 47%].[382]

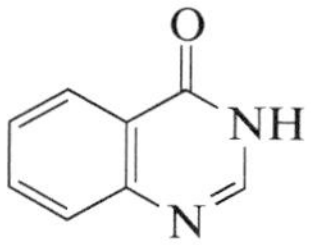

(76) 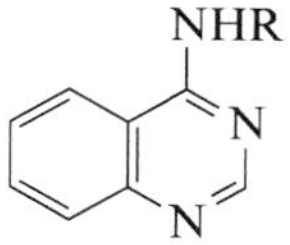   (77)

4-Phenyl-2(1H)-quinazolinone gave 4-phenyl-2-quinazolinamine [PhOP(=O) (NH$_2$)$_2$, Ph$_2$O, 250°, 3 h: 69%].[1764]

4(3H)-Quinazolinone (76) gave 4-benzylamino- (77, R = CH$_2$Ph) (PhCh$_2$NH$_2$, P$_2$O$_5$, Me$_2$NC$_6$H$_{11}$, 5° → 200°; then substrate ↓, 200°, 2 h: 82%), 4-dimethylamino- (Me$_2$NH likewise: 80%), or 4-anilinoquinazoline (77, R = Ph) (PhNH$_2$ likewise: 77%).[1786]

2,4(1H,3H)-Quinazolinedione gave 2-(4-β-hydroxyethylpiperazin-1-yl)-4(3H)-quinazolinone (78) [HNCH$_2$CH$_2$)$_2$NCH$_2$CH$_2$OH, Me$_3$SiNHSiMe$_3$, TsOH, 130°, open, 8 h; then pyridine, 130°, 10 h: 68%] or 2,4-bis(4-β-hydroxyethylpiperazin-1-yl)quinazoline (79) [HN(CH$_2$CH$_2$)$_2$NCH$_2$CH$_2$ OH, octamethylcyclotetrasiloxane, TsOH, 180°, 48 h; then MeOH, reflux: 56% plus a byproduct].[1244]

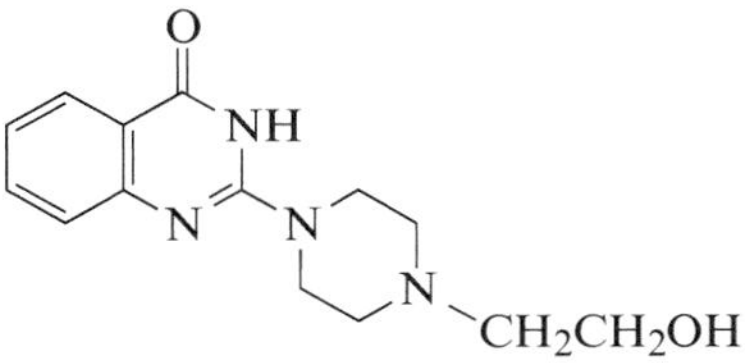

(78) 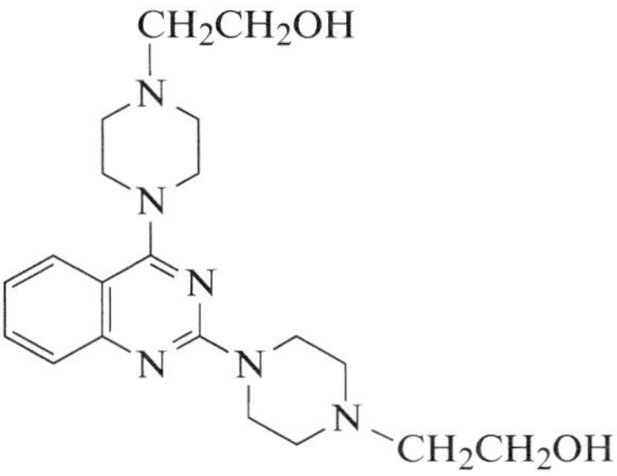   **(79)**

Ethyl 4-hydroxy-3-methyl-2-oxo-1,2,3,4-tetrahydro-4-quinazolinecarboxylate (**80**, R = OH) gave the ethyl 4-methylamino analog (**80**, R = NHMe) (MeNCO, Et$_3$N, CHCl$_3$–EtOH, 20°, 5 days: ~70%; probably via the N-methylcarbamoyloxy intermediate (**80**, R = OCONHMe).[206]

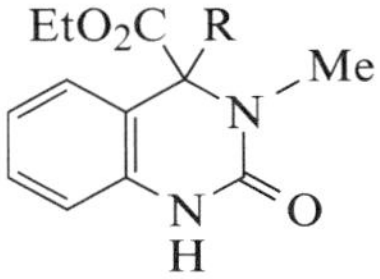

(80)

Quinazolinones can also undergo N-amination: for example, treatment of 4(3H)-quinazolinone (**76**) with hydroxylamine-O-sulfonic acid in aqueous alkali at 20° for 5 h gave a 22% yield of 3-amino-4(3H)-quinazolinone (**81**) by a process akin to N-alkylation.[1895]

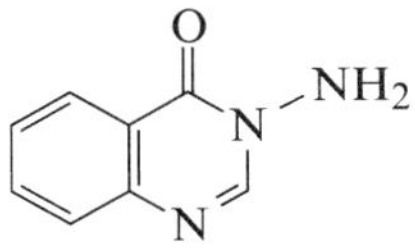

(81)

4.1.2.5. Nuclear Reduction with or without Deoxygenation
(H 110,402)

Tautomeric quinazolinones can undergo nuclear reduction that may or may not involve deoxygenation according to the site(s) attacked. The following examples will clarify the possibilities.

With deoxygenation

2-Amino-5,6-dimethoxy-4 (3*H*)-quinazolinone (**82**) gave 5,6-dimethoxy-3,4-dihydro-2-quinazolinamine (**83**) (BH$_3$, THF, 25°, 24 h: 54%).[1562]

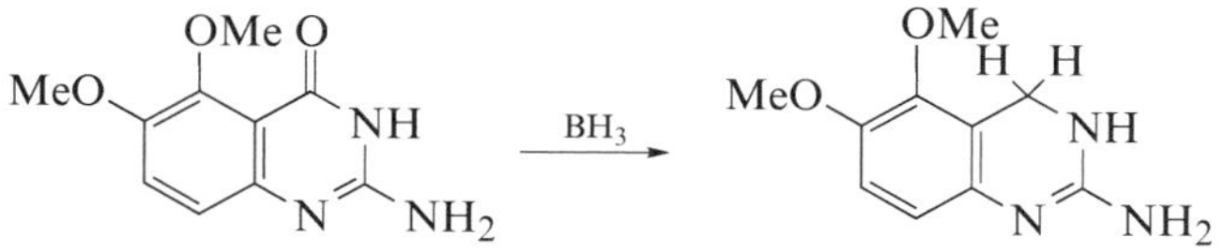

      (82)   **(83)**

2-*o*-Hydroxyphenyl-4(*H*)-quinazolinone gave 2-*o*-hydroxyphenyl-3,4-dihydroquinazoline (LiAlH$_4$).[465]

Without deoxygenation

4-Phenyl-2(1*H*)-quinazolinone (**84**) gave 4-phenyl-3,4-dihydro-2(1*H*)-quinazolinone (**85**) (NaBH$_4$, AcOH, N$_2$, 20°, 22 h: 83%).[1946]

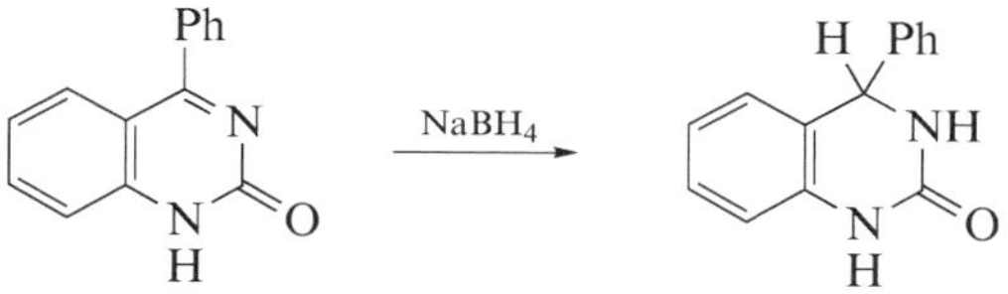

      (84)   **(85)**

6-Chloro-4-phenyl-2(1*H*)-quinazolinone gave its 3,4-dihydro derivative (H$_2$, Ni, EtOH, 20°, 15 h: 91%; electrolytic reduction at pH 4.8: 90%).[836]

2,6-Dimethyl-4(3*H*)-quinazolinone gave 2,6-dimethyl-1,2-dihydro-4(3*H*)-quinazolinone (**86**) (NaBH$_4$, AcOH, 25°, 12 h: 70%).[2254]

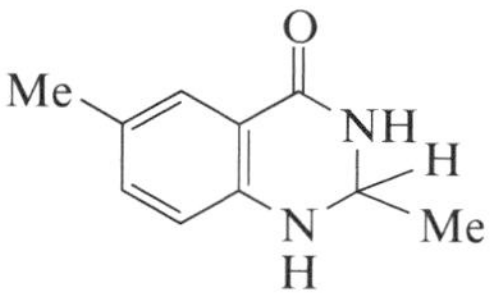

(86)

5-Chloro-2-methyl-4-oxo-3,4-dihydro-6-quinazolinesulfonamide gave its 1,2,3,4-tetrahydro analog (**87**) (NaBH$_4$, MeOCH$_2$CH$_2$OCH$_2$CH$_2$OMe, 85°, 1 h: 50%).[1384]

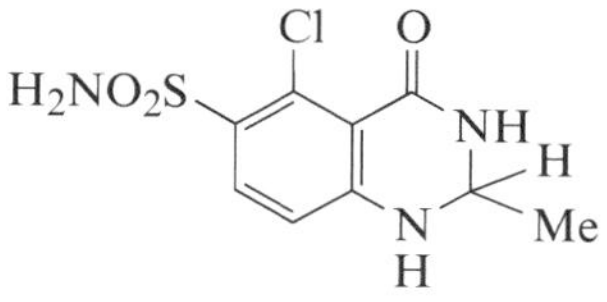

(87)

2-Amino-5,5-dimethyl-7-pyrrolidin-1'-yl-5,6-dihydro-4(3H)-quinazolinone
gave its 5,6,7,8-tetrahydro analog (**88**) (H$_2$, Pt, HCl–MeOH, 20°, 22 h: 87%,
as dihydrochloride).[2402]

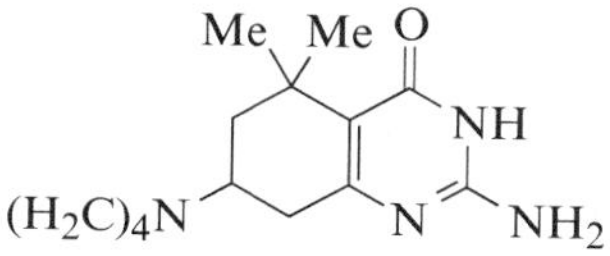

(88)

Also other examples.[2410]

4.1.2.6.   Other Reactions

Tautomeric quinazolinones undergo several other reactions, some of more
interest than importance. Thus irradiation of α,γ-bis(2,4-dioxo-1,2,3,4,5,6,7,8-
octahydroquinazolin-1-yl)propane (**89**) caused slow *photoisomerization* to a molecule in which the two quinazoline moieties lay side-by-side and head-to-head,
still joined by a 1,1'-propano hinge but also joined by 4a,4a'- and 8a,8a'-bonds by
virtue of the original 4a,8a- and 4a',8a'-double bonds being replaced by single
bonds;[1509] the molecule thus rather resembled some of the "cyclobutane dimers,"
so familiar in uracil and thymine chemistry.[2446d]

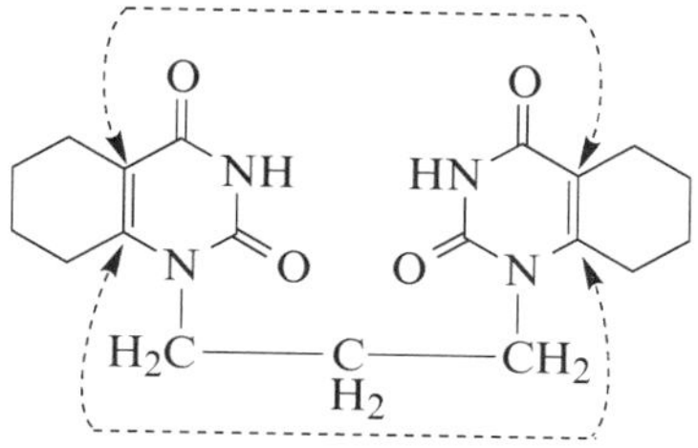

(89)

4(3H)-Quinazolinone (**90**) underwent a *three-point condensation* with 2,2'-
iminodibenzoyl chloride (**91**) to afford 4b,9a,14b-triazatribenzo[b,e,j]phenalene-
9,10,15-trione (**92**) [NEt$_3$, THF, reflux, 4 h: 55%].[1005]

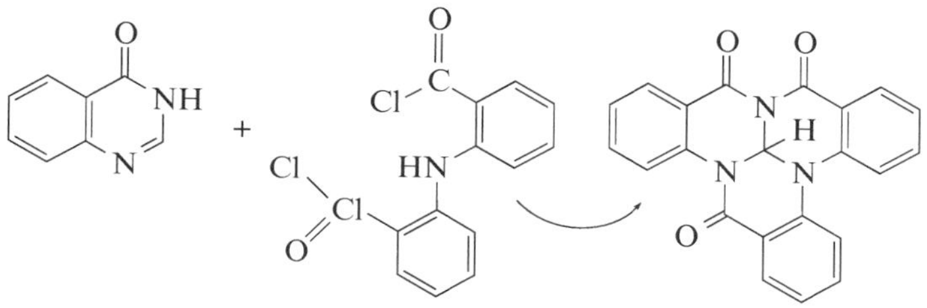

(90)   **(91)**   **(92)**

In the presence of aluminum metal, 2,4(1H,3H)-quinazolinedione suffered *thermolysis* to *o*-aminobenzonitrile in good yield.[2328]

2(1H)-Quinazolinones underwent facile covalent *addition* of trivalent phosphorus reagents to give, for example, 4-methyl-2-oxo-1,2,3,4-tetrahydro-4-quinazolinephosphinic acid (**93**) (H$_3$PO$_2$, H$_2$O, reflux, 12 h: 92%) or dimethyl 4-methyl-2-oxo-1,2,3,4-tetrahydro-4-quinazolinephosphonate (**94**) [substrate. HCl, (MeO)$_3$P, reflux, N$_2$, 5 h: >95%].[1212]

(93)   (94)

Treatment of 2-benzyl-2-ethyl-1-methyl-1,2-dihydro-4(3H)-quinazolinone (**95**) with acetic anhydride in pyridine induced a *rearrangement* with loss of acetamide to give 2-ethyl-1-methyl-3-phenyl-4(1H)-quinolinone (**96**) (100°, 4 h: 55%).[1407]

(95)   (96)

Some *metal complexes* from the following simple quinazolinones have been prepared and studied: 4(3H)-quinazolinone (Cu);[2543] 2-methyl-, 2-ethyl-, 2-propyl-, and 2-pentyl-4(3H)-quinazolinone (Cu, Sb, Sn);[1126,2243] 2-pyridin-2′-yl-1,2-dihydro-4(3H)-quinazolinone (Ni);[545] and 2-methyl- and 2-phenyl-4(1H)-quinazolinone 3-oxide (Cd, Co, Cu, Ni, Pd, Pt, Ru, Zn).[1260,1684,1688,2159] For mention of complexes from fixed quinazolinones, aminoquinazolinones, and thioxodihydroquinazolinones, see Sections 4.6.2.3, 6.2.2.8, and 5.1.2.4, respectively.

For a *Grignard reaction*, see Section 4.6.2.3

4.2. 5- TO 8-QUINAZOLINOLS (*H* 133)

Relative to the amidic quinazolinones, these phenolic quinazolinols are few in number and of limited importance.

4.2.1. Preparation of Quinazolinols (*H* 133)

Although quinazolinols have been made by direct *primary synthesis* (see Chapter 1), it is often more convenient to prepare them from the corresponding

5- to 8-alkoxyquinazolines or other preformed quinazolines. The following examples illustrate some of these possibilities.

From alkoxyquinazolines: acidic hydrolysis

6-Methoxy- (**97**, R = Me) gave 6-hydroxy-2-methyl-4(3*H*)-quinazolinone (**97**, R = H) (48% HBr, reflux, 48 h: 65%).[782]

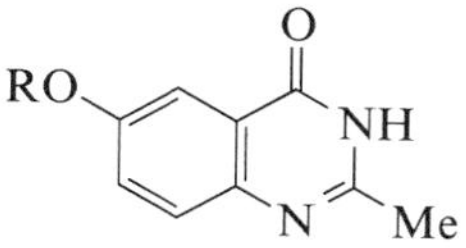

(97)

2-Diethylamino-6,7-dimethoxy- gave 2-diethylamino-6,7-dihydroxy-4(3*H*)-quinazolinone (48% HBr, reflux, 3 h: ~75%, as hydrobromide).[641]

5,8-Dipropoxyquinazoline (**98**, R = Pr) gave 5-propoxy-8-quinazolinol (**98**, R = H) (57%, HI, 100°, 30 min: ~55%);[595] 5,8-dimethoxyquinazoline gave 3,4-dihydro-5,8-quinazolinediol (57% HI–AcOH, 100°, 28 h: ~40%, as hydriodide: note nuclear reduction).[586]

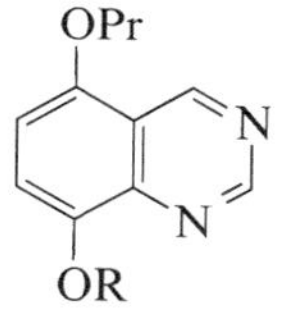

(98)

Also other examples.[120,140,681,762,766,1562,2411]

From alkoxyquinazolines: Lewis acid splitting

5-Chloro-7-methoxy- (**99**, R = Me) gave 5-chloro-7-hydroxy-3-phenyl-4(3*H*)-quinazolinone (**99**, R = H) (AlCl$_3$, PhH, reflux 2 h; then H$_2$O: 61%).[2112]

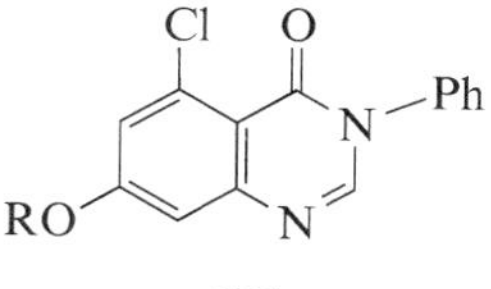

(99)

8-Methoxy-5-β-(trifluoroacetamido)ethyl- gave 5-β-aminoethyl-8-hydroxy-4(3*H*)-quinazolinone (AlCl$_3$, PhMe, reflux, 2 h; then ice: 82%; note additional deacylation).[2502]

5,8-Dimethoxy- (**100**, R = Me) gave 5,8-dihydroxy-4(3H)-quinazolinone (**100**, R = H) (BBr$_3$, CH$_2$Cl$_2$, reflux, 4 h; then H$_2$O: 85%).[2324]

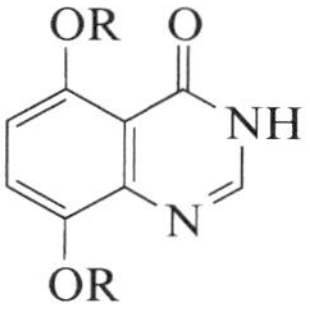

(100)

5-Chloro-6-methoxy-3,4-dihydro-2-quinazolinamine gave 2-amino-5-chloro-3,4-dihydro-6-quinazolinol (BBr$_3$, CH$_2$Cl$_2$, 20°, 12 h; then MeOH, 20°, 30 min: 59%).[1711]

5,8-Dimethoxy-2-vinyl- gave 2-β-bromoethyl-5,8-dihydroxy-4(3H)-quinazolinone (BBr$_3$, PhH, reflux, 2 h; then MeOH: 69%; note incidental addition of HBr to vinyl group).[2430]

Also other examples.[436,595,659,764,1999]

From allyloxyquinazolinones: Claisen rearrangement

This limited method has been discussed in Section 2.2.1.6.

From benzyloxyquinazolines: hydrogenolysis or trifluoroacetic acid

6,7-Dibenzyloxy- gave 6,7-dihydroxy-2-methyl-4(3H)-quinazolinone (H$_2$, Pd/C, EtOH–AcOH, 2 atm, 1 h: 45%).[218]

α,β-Bis(5-benzyloxy-6-methoxyquinazolin-4-ylamino)ethane (**101**, R = CH$_2$Ph, X = CH$_2$CH$_2$) gave the corresponding 5-hydroxy compound (**101**, R = OH, X = CH$_2$CH$_2$) (neat F$_3$CCO$_2$H, reflux, N$_2$, 5 h; then MeOH: 85%).[1728]

N,N-Bis[γ-(5-benzyloxy-6-methoxyquinazolin-4-ylamino)propyl]-N-methyl-amine [**101**, R = CH$_2$Ph, X = CH$_2$CH$_2$CH$_2$NMeCH$_2$CH$_2$CH$_2$] gave the corresponding 5-hydroxy compound [**101**, R = H, X = CH$_2$CH$_2$CH$_2$NMeCH$_2$CH$_2$CH$_2$] (H$_2$, Pd/C, dioxane–MeOH, ? h: >95%).[1728]

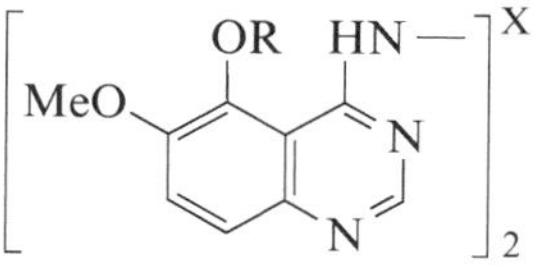

(101)

Also other examples.[142,844]

From other substrates

6-Acetoxy- (**102**, R = Ac) gave 6-hydroxy-7-methoxy-2-methyl-3-phenyl-
4(3*H*)-quinazolinone (**102**, R = H) (MeONa, MeOH, 20°, 12 h: 75%).[143]

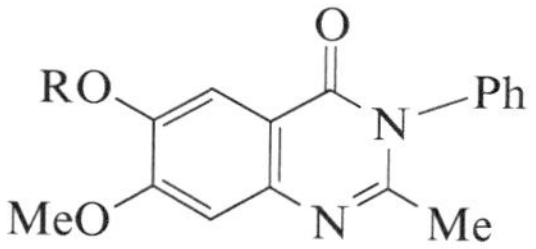

(102)

6-Anilino- (**103**, R = NHPh) gave 6-hydroxy-4-methyl-2-phenyl-5,8-dihydro-
5,8-quinazolinedione (**103**, R = OH) (NaOH, MeOH–H$_2$O, 65°, 1 h:
65%; note the strong activation of the anilino group in the quinone
environment).[736]

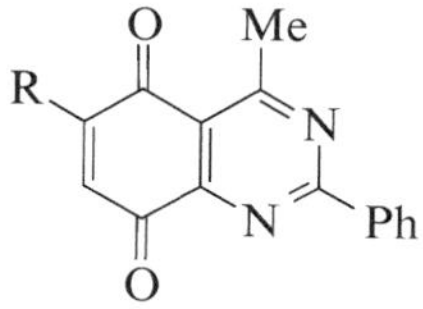

(103)

4-Methyl-2-phenyl-5,6,7,8-tetrahydro-5-quinazolinone (**104**, R = Me) gave
4-methyl-2-phenyl-5,6,7,8-tetrahydro-5-quinazolinol (NaBH$_4$, H$_2$O) and
thence 4-methyl-2-phenyl-7,8-dihydroquinazoline (dehydration in 85%
H$_3$PO$_4$).[309]

2-Phenyl-5,6,7,8-tetrahydro-5-quinazolinone (**104**, R = H) gave 2,5-diphenyl-
5,6,7,8-tetrahydro-5-quinazolinol (PhLi, C$_6$H$_{14}$–Et$_2$O–THF, $-10° \rightarrow 20°$,
30 min, then reflux, 1 h: 80%).[2105]

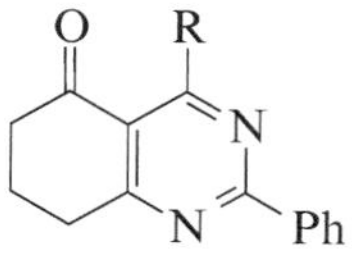

(104)

Also other examples.[140,763]

4.2.2.   Reactions of Quinazolinols (*H* 134)

Quinazolinols have been involved recently in only three major types of
reaction, as illustrated in the following examples.

O-Alkylation

2-Butyl-7-hydroxy-8-methoxy- (**105**, R = H) gave 2-butyl-7,8-dimethoxy-3-phenyl-4(3*H*)-quinazolinone (**105**, R = Me) (CH$_2$N$_2$,Et$_2$O, 20°, 12 h: 66%).[142]

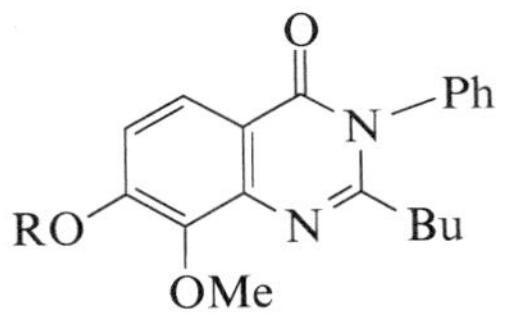

(105)

6-Hydroxy- (**106**, R = H) gave 6-allyloxy-7-methoxy-2-methyl-3-phenyl-4(3*H*)-quinazolinone (**106**, R = CH$_2$CH:CH$_2$) (BrCH$_2$CH=CH$_2$, K$_2$CO$_3$, Me$_2$NCHO, 20°, 12 h: 92%)[143] or 7-methoxy-2-methyl-6-phenacyloxy-3-phenyl-4(3*H*)-quinazolinone (**106**, R = CH$_2$Bz) (BrCH$_2$Bz, K$_2$CO$_3$, AcMe, reflux, 4 h: 90%).[139]

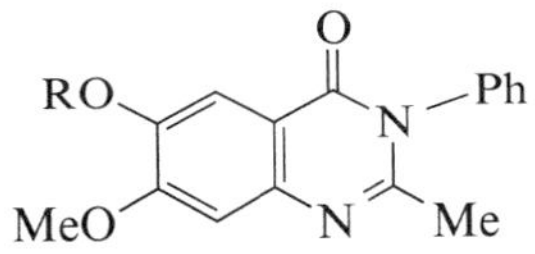

(106)

2-Phenyl-5,6,7,8-tetrahydro-5-quinazolinol (**107**, R = H) gave 2-phenyl-5-prop-2′-ynyloxy-5,6,7,8-tetrahydroquinazoline (**107**, R = CH$_2$C≡CH) (NaH, THF, then BrCH$_2$C≡CH, reflux, 12 h: 77%).[2105]

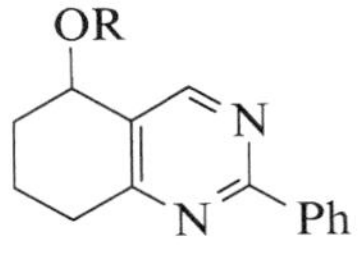

(107)

5-Chloro-7-hydroxy- gave 5-chloro-7-ethoxycarbonylmethoxy-3-phenyl-4(3*H*)-quinazolinone (BrCH$_2$CO$_2$Et, K$_2$CO$_3$, Me$_2$NCHO, 60°, 2 h: unisolated) and thence 7-carboxymethoxy-5-chloro-3-phenyl-4(3*H*)-quinazolinone (NaOH, H$_2$O, 60°, 1 h: 77% overall).[2112]

Also other examples.[120,136,141,681,736]

O-Acylation

8-Allyl-7-hydroxy- (**108**, R = H) gave 7-acetoxy-8-allyl-6-methoxy-2-methyl-3-phenyl-4(3*H*)-quinazolinone (**108**, R = Ac) (Ac$_2$O, pyridine, 20°, 12 h: 90%).[151]

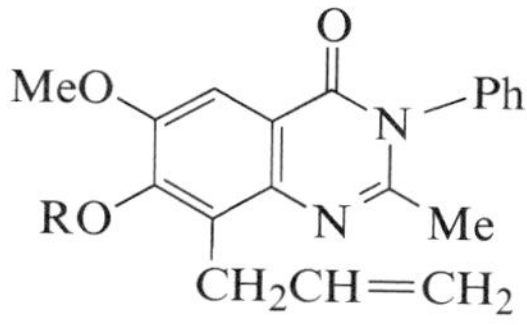

(108)

6-Hydroxy- gave 6-acetoxy-2-phenyl-4(3*H*)-quinazolinone (Ac$_2$O, pyridine: 97%).[766]

Also other examples.[140,149,586]

Oxidation (to quinones)

5,8-Quinazolinediol (**109**) gave 5,8-dihydro-5,8-quinazolinedione (**110**), (K$_2$Cr$_2$O$_7$, 10% H$_2$SO$_4$, <5°, 15 min: 75%).[659]

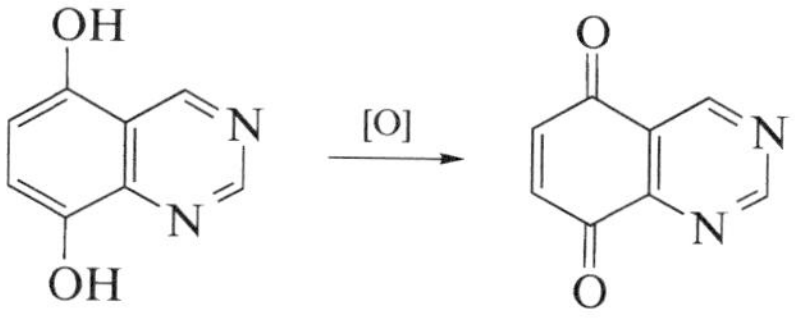

(109)   **(110)**

5,8-Dihydroxy-4(3*H*)-quinazolinone (**111**, R = H) gave 5,8-dihydro-4,5,8(3*H*)-quinazolinetrione (**112**) (dichlorodicyanobenzoquinone, MeOH, 25°, 1 h: 49%);[2324] similar procedures gave the 2-methyl (**112**, R = Me) (74%),[1999] 2-vinyl (**112**, R = CH:CH$_2$) (67%),[2430] 2-bromomethyl (**112**, R = CH$_2$Br) (32%),[1999] and 2-β-bromoethyl (**112**, R = CH$_2$CH$_2$Br) (56%) analog.[2430]

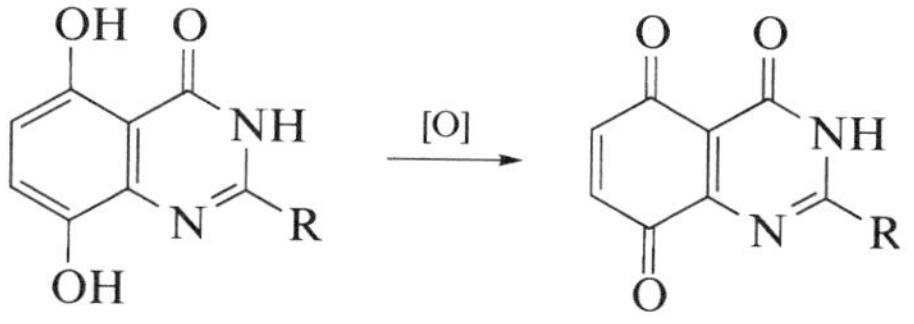

(111)   **(112)**

2,4-Dimethoxy-5-quinazolinol gave 2,4-dimethoxy-5,8-dihydro-5,8-quina-zolinedione [(CF$_3$CO$_2$)$_2$IPh, MeCN, 0°: 52%].[2531]

For less simple examples, see Section 4.3.

4.3. QUINAZOLINE QUINONES

Both *ortho* (5,6-) and *para* (5,8-)quinones are known in the quinazoline series. A few have been made by *primary synthesis* (see Section 1.2.8.3), some by simple *oxidation of quinazolinols* (see preceding paragraph), and some by other oxidative procedures illustrated in the following examples.

By oxidative hydrolysis

4,5,6,8-Tetramethoxy-7-methyl- (**113**, R = Me) or 8-ethoxy-4,5,6-trimethoxy-7-methylquinazoline (**113**, R = Et) gave 4,6-dimethoxy-7-methyl-5,8-dihydro-5,8-quinazolinedione (**114**) [(NH$_4$)$_2$Ce(NO$_3$)$_6$, 2,6-pyridinedicarboxylic acid *N*-oxide, MeCN–H$_2$O, 15°, 15 min: 31% or 20%, respectively; note survival of the 4- and 6-OMe].[2363]

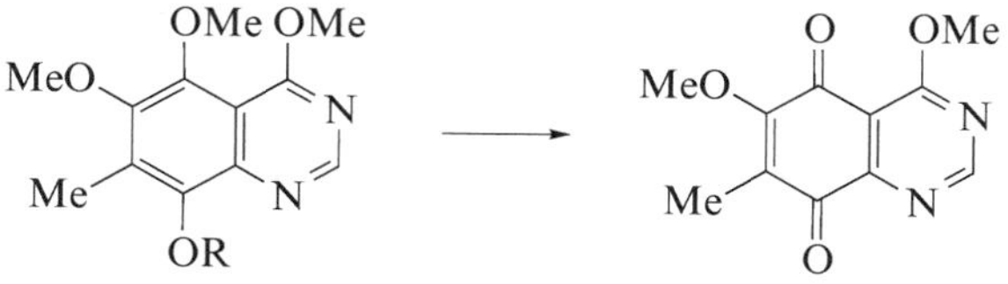

(113)   (114)

5,8-Dimethoxy-6-methyl-4(3*H*)-quinazolinone gave 6-methyl-5,8-dihydro-4,5,8(3*H*)-quinazolinetrione [(NH$_4$)$_2$Ce(NO$_3$)$_6$, H$_2$O, 25°, 30 min: 80%].[2324]

6-Acetamido-5,8-dimethoxy-2,4(1*H*,3*H*)-quinazolinedione gave 6-acetamido-5,8-dihydro-2,4,5,8(1*H*,3*H*)-quinazolinetetrone (**115**, R = Ac, X = O) (fuming HNO$_3$, < 5°, 60 min: 76%).[1987]

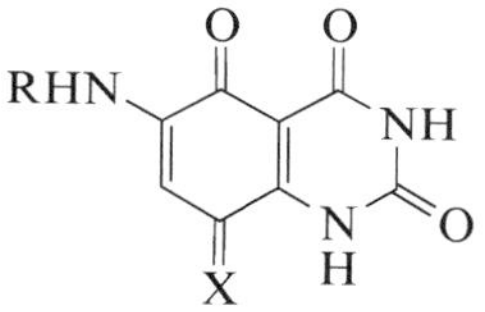

(115)

Also other examples.[897,1604]

By oxidative hydrolysis and nuclear oxidation

6-Methoxy-5-quinazolinamine (**116**) gave 6-methoxy-5,8-dihydro-5,8-quinazolinedione (**117**) [(KSO$_3$)$_2$NO (Fremy's salt), K$_2$HPO$_4$, H$_2$O–AcMe, 20°, N$_2$, ~12 h: 36%].[1577]

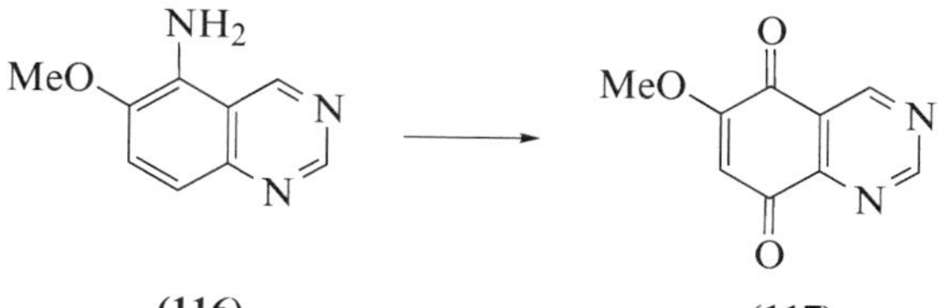

(116)   (117)

By substituent oxidation, nuclear oxidation, and amination

4-Chloro-2-phenyl- (**118**, R = Cl) or 2-phenyl-4-piperidino-6-quinazolinol [**118**, R = N(CH$_2$)$_5$] gave 2-phenyl-4,8-dipiperidino-5,6-dihydro-5,6-quinazolinedione (**119**) [HN(CH$_2$)$_5$, Cu(OAc)$_2$, MeOH, O$_2$↓, 20°, 3 h: ~65%];[592] also analogs.[592,636]

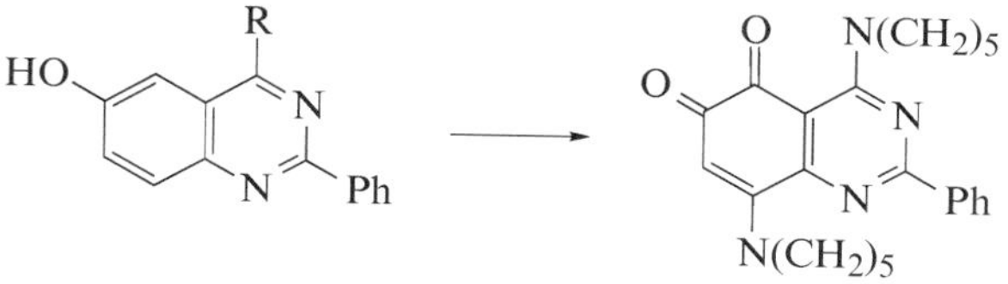

(118)   (119)

6-Quinazolinol (**120**) gave 2,8-dipiperidino-5,6-dihydro-5,6-quinazolinedione (**121**) [HN(CH$_2$)$_5$, Cu(OAc)$_2$, MeOH, O$_2$↓, 20°, 2 h: 88%].[764,765]

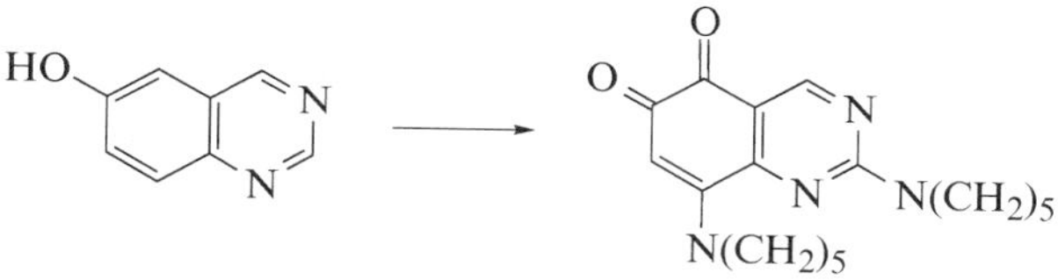

(120)   (121)

6-Hydroxy-4(3H)-quinazolinone (**122**, R = H) gave 2,8-dipiperidino-5,6-dihydro-4,5,6(3H)-quinazolinetrione (**123**) [HN(CH$_2$)$_5$, Cu(OAc)$_2$, MeOH, O$_2$↓, 20°, 3 h: 75%];[762] the 2-methyl substrate (**122**, R = Me) gave the same product (**123**) (likewise: 18%; note displacement of Me).[782]

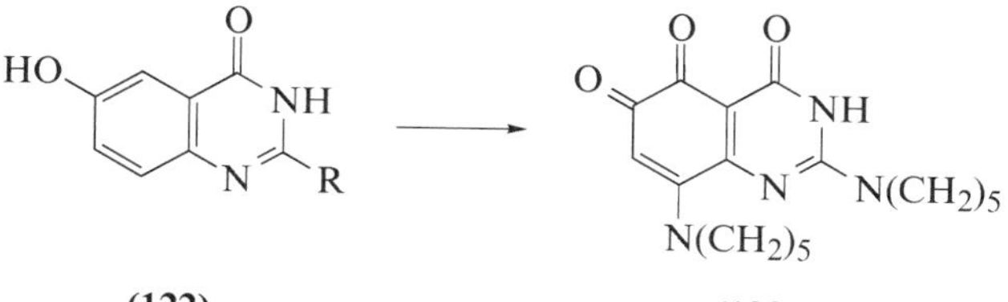

(122)   (123)

Also other examples.[763]

Few *reactions* of quinazoline quinones have been reported. Reductive acylation of 4-methoxy-2-phenyl-8-piperidino- (**124**, R = OMe) or 2-phenyl-4,8-dipiperidino-5,6-dihydro-5,6-quinazolinedione [**124**, R = N(CH$_2$)$_5$] with zinc dust in acetic anhydride–pyridine gave 5,6-diacetoxy-4-methoxy-2-phenyl-8-piperidino- (**125**, R = OMe) (92%)[763] or 5,6-diacetoxy-2-phenyl-4,8-dipiperidinoquinazoline [**125**, R = N(CH$_2$)$_5$],[636] respectively. 6-Acetamido-5,8-dihydro-2,4,5,8(1H,3H)-quinazolinetetrone (**115**, R = Ac, X = O) in refluxing aqueous methylamine underwent Schiff base formation, deacylation, and trans-amination to afford 56% of 6-methylamino-8-methylimino-5,8-dihydro-2,4,5(1H,3H)-quinazolinetrione (**115**, R = Me, X = NMe), but similar treatment with the weaker base, aniline, in refluxing dimethylformamide induced only Schiff base formation to give 20% of 6-acetamido-8-phenylimino-5,8-dihydro-2,4,5(1H,3H)-quinazolinetrione (**115**, R = Ac, X = NPh).[1987] 2-Phenyl-4,8-dipiperidino-5,6-dihydro-5,6-quinazolinedione (**119**) in aqueous alcoholic alkali at 20° for 2 h underwent hydrolysis of its 8-piperidino group and prototropy to furnish 75% of 6-hydroxy-2-phenyl-4-piperidino-5,8-dihydro-5,8-quinazol-inedione (**126**, R = H), whereas the same substrate (**119**) in methanolic sulfuric acid under reflux suffered not only the same reaction but also an additional O-methylation to give 64% of the 6-methoxy analog (**126**, R = Me).[592] A copper complex of 2-phenyl-8-piperidino-5,6-dihydro-4,5,6(3H)-quinazolinetrione has been characterized.[762] The activating effect of a quinone system on appropriately placed leaving groups was illustrated by the aminolysis of 6-methoxy-5,8-dihydro-5,8-quinazolinedione (**117**) in methanolic piperidine at 0° in only 2 h to give 6-piperidino-5,8-dihydro-5,8-quinazolinedione in 88% yield.[1567]

(124)   (125)   (126)

4.4.   EXTRANUCLEAR HYDROXYQUINAZOLINES

Extranuclear hydroxyquinazolines are mainly simple primary or secondary alcohols (or occasionally phenols), in which the nature and reactivity of the hydroxy groups are only mildly affected by their indirect attachment to the quinazoline system.

4.4.1.   Preparation of Extranuclear Hydroxyquinazolines

Of the rather mundane methods available for the synthesis of such hy-droxyquinazolines, *primary synthesis* (Chapter 1), *hydrolysis of extranuclear halogenoquinazolines* (Section 3.7), and *oxidative hydroxylation of alkenyl-quinazolines* (Section 2.2.2.2) have been covered already; in addition, *derivatiza-tion* (involving the direct or indirect attachment of an hydroxyalkyl or hy-droxyaryl grouping to existing quinazolines in various ways, such as aminolysis with ethanolamine), is well exemplified in most chapters.

The remaining methods are illustrated briefly in the following examples.

By reduction of esters, aldehydes, or ketones

Butyl   4-oxo-2-pivalamido-3,4,5,6,7,8-hexahydro-6-quinazolinecarboxylate
(**127**) gave 6-hydroxymethyl-2-pivalamido-5,6,7,8-tetrahydro-4(3*H*)-quina-
zolinone (**128**) (LiBEt$_3$H, THF, N$_2$, 3 h: ~50%).[1591]

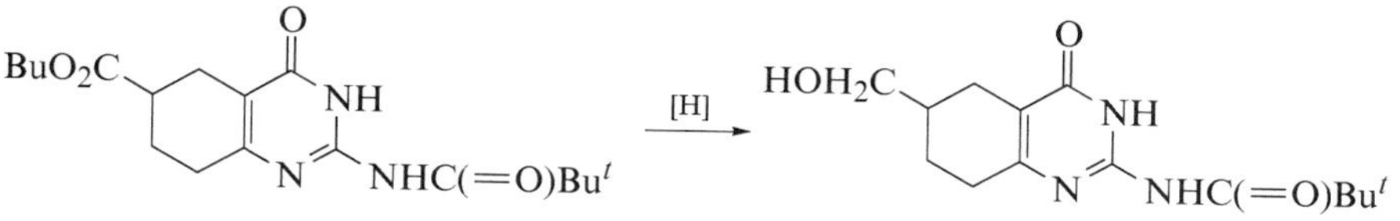

(127)   **(128)**

Ethyl 2-formyl- (**129**, R = CHO) gave ethyl 2-hydroxymethyl-6,8-dimethyl-4-
oxo-3-phenyl-3,4-dihydro-7-quinazolinecarboxylate (**129**, R = CH$_2$OH)
(NaBH$_4$, no details; note preferential reduction of the aldehydic group).[1364]

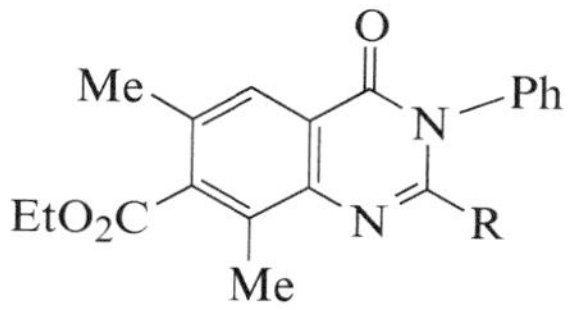

(129)

2-Phenacylthio- (**130**, R = Bz) gave 2-*β*-hydroxyphenethylthio-3-phenyl-
4(3*H*)-quinazolinone (**130**, R = CHOHPh) (NaBH$_4$, MeOCH$_2$CH$_2$-
OCH$_2$CH$_2$OMe–MeOH, trace NaOH, 20°, 12 h: ~75%).[144]

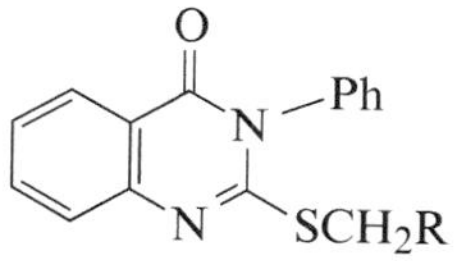

(130)

Also other examples.[138,181,587,618,1221,1364,1772,2071]

By hydrolysis of acyloxy derivatives

2-Acetoxymethyl- (**131**, R = Ac) gave 2-hydroxymethyl-3-phenyl-4(3*H*)-
quinazolinone (**131**, R = H) (MeONa, MeOH, 20°, 12 h: ~80%).[146]

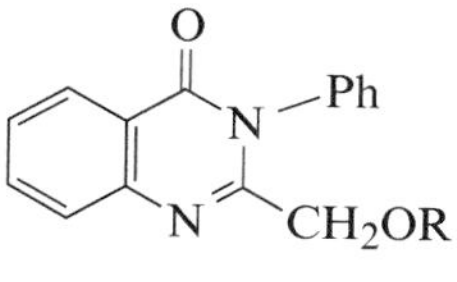

(131)

6-α-Acetoxyacetamido-2-fluoromethyl- gave 2-fluoromethyl-6-α-hydroxya-cetamido-3-o-tolyl-4(3H)-quinazolinone (NaHCO$_3$, MeOH–H$_2$O, reflux, 1 h: 81%).[1418]

2-α-Acetoxybenzyl- gave 2-α-hydroxybenzyl-3-methyl-4(3H)-quinazolinone (H$_2$SO$_4$–AcOH–H$_2$O, reflux, 2 h: 85%).[181]

2-β-Benzoyloxyethyl- gave 2-β-hydroxyethyl-4(3H)-quinazolinone (KOH, EtOH–H$_2$O, 20°, 12 h: 73%).[2069]

Also other examples.[426,510,786,1364]

By hydrolysis of alkoxy derivatives

2-α-Benzyloxyethyl- (132, R = CH$_2$Ph) gave 2-α-hydroxyethyl-4(3H)-quina-zolinone (132, R = H) (HCl, EtOH–H$_2$O, reflux, 28 h: ∼ 50%).[1026]

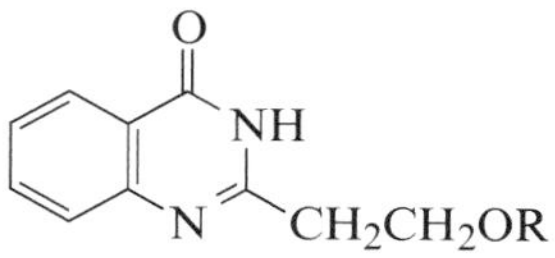

(132)

2-p-Methoxyphenyl- gave 2-p-hydroxyphenyl-3-phenyl-4(3H)-quinazolinone (48% HBr, reflux, 12 h: ∼ 90%).[488]

3-p-Methoxyphenyl- gave 3-p-hydroxyphenyl-2-methyl-4(3H)-quinazolinone (HBr, H$_2$O–AcOH, reflux, ? h: 80%).[573]

3-β-(3,4-Dimethoxyphenyl)ethyl- gave 3-β-(3,4-dihydroxyphenyl)ethyl-2-methyl-4(3H)-quinazolinone (47% HI, reflux, 4 h: 80%, as hydrochloride).[963]

Also other examples in the foregoing references and elsewhere.[447,657,892]

By hydrogenolysis of benzyloxy derivatives

4-p-Benzyloxyanilino- (133, R = CH$_2$Ph) gave 4-p-hydroxyanilinoquinazo-line (133, R = H) (H$_2$, Pd/C, AcOH, 3 atm, 20°, 4 h: 88%).[1208]

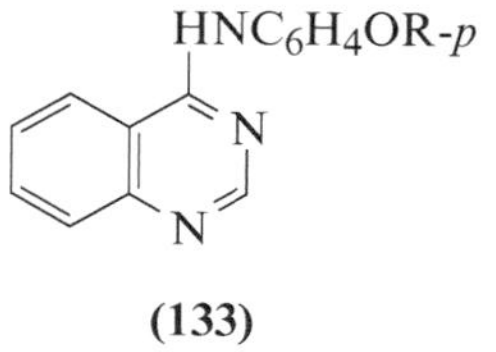

(133)

2-γ-Benzyloxypropyl- gave 2-γ-hydroxypropyl-3-methyl-4(3H)-quinazoli-none (H$_2$, Pd/C, EtOH, 2 atm, 20°, 8 h: 14%, plus a separable byproduct).[594]

By other means

2-*o*-(Piperidinoazo)phenyl-4-quinazolinamine (**134**) gave 2-*o*-hydroxyphenyl-4(3*H*)-quinazolinone (2M H_2SO_4, reflux, 2 h: ∼ 25%; note additional hydrolysis of the amino group).[1018]

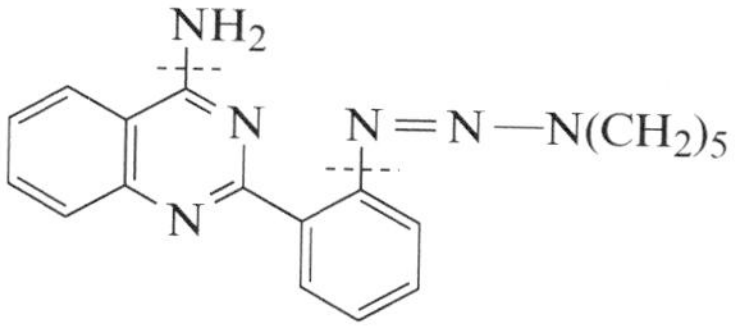

(134)

2-Methyl-3-*o*-tolyl- gave 3-*o*-(hydroxymethyl)phenyl-2-methyl-4(3*H*)-quinazolinone (by biological hydroxylation in human subjects, but only after massive doses of methaqualone over a long period).[1112, cf. 619]

3-Trimethylsilylmethyl- gave 3-(2-*p*-chlorophenyl-2-hydroxypropyl)-4(3*H*)-quinazolinone (AcC_6H_4Cl-*p*, Bu_4NF, THF, 20°, 2 h: 43%).[1965]

4.4.2.   Reactions of Extranuclear Hydroxyquinazolines

The conversion of *hydroxyalkyl-* into *chloroalkylquinazolines* (Section 3.4) and the *Mannich reaction* on (hydroxyaryl)quinazolines (Section 2.2.2.3) have been covered. The remaining reported reactions are illustrated in the following examples.

Acylation and trimethylsilylation

2-α-Hydroxybenzyl- (**135**, R = H) gave 2-α-acetoxybenzyl-4(3*H*)-quinazolinone (**135**, R = Ac) (Ac_2O, AcONa, reflux, 2 h: ∼ 85%).[475]

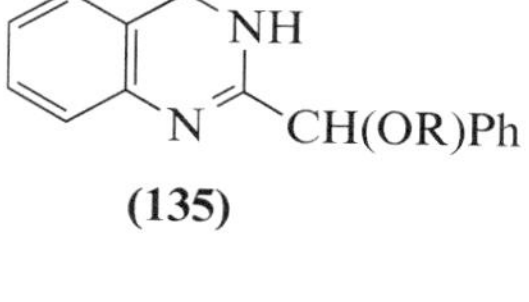

(135)

3-(4-Carboxy-3-hydroxyphenyl)- gave 3-[4-carboxy-3-(*p*-nitrobenzoyloxy)phenyl]-2-phenyl-4(3*H*)-quinazolinone ($O_2NC_6H_4COCl$-*p*, pyridine, 20°, 2 h: 70%).[1252]

2-Ethoxy-3-β-hydroxyethyl-4(3*H*)-quinazolinone gave 3-β-acetoxyethyl-2,4(1*H*,3*H*)-quinazolinedione (Ac_2O, reflux, 4 h: 96%; note unexpected conversion of OEt to OH/=O).[1473]

7-Bromo-6-chloro-3-γ-(3-hydroxypiperidin-2-yl)acetonyl- (**136**,R = H) gave
7-bromo-6-chloro-3-γ-(3-trimethylsiloxypiperidin-2-yl)acetonyl-4(3H)-
quinazolinone (**136**, R = SiMe$_3$) (N-trimethylsilylimidazole, Me$_2$NCHO,
20°, 20 min: 95%).[2476]

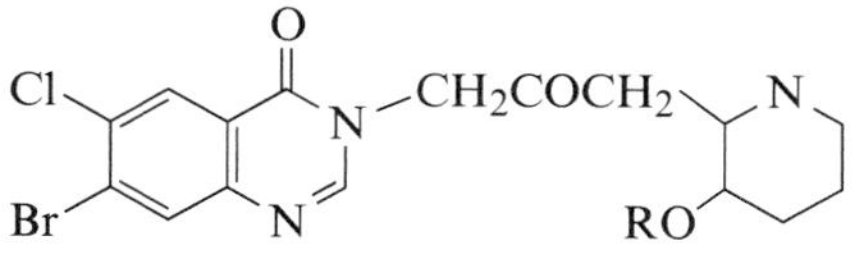

(**136**)

2-Hydroxymethyl- gave 2-sulfooxymethyl-3-o-tolyl-4(3H)-quinazolinone
(ClSO$_3$H, pyridine: 53%).[969]

1-β-D-Ribofuranosyl-2,4(1H,3H)-quinazolinedione gave the tetraisopropyl-
disiloxanediyl derivative (**137**) (1,3-dichloro-1,1,3,3-tetraisopropyldisi-
loxane, pyridine, 20°, 12 h: 85%).[2367]

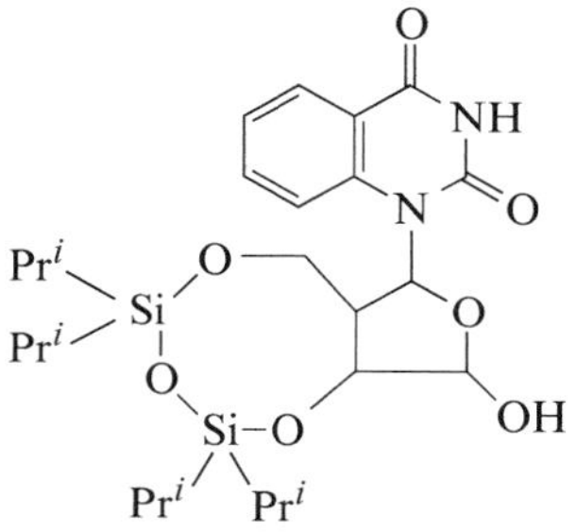

(**137**)

Also other examples.[486,1103,2069,2185,2367]

Alkylation

3-p-Hydroxyphenyl- (**138**, R = H) gave 3-p-allyloxyphenyl-2-methyl-4(3H)
quinazolinone (**138**, R = CH$_2$CH:CH$_2$)(BrCH$_2$CH = CH$_2$, K$_2$CO$_3$, AcMe,
reflux, 12 h: 90%).[573]

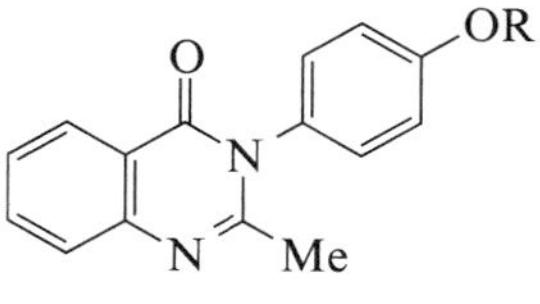

(**138**)

3-β-Hydroxyethyl- gave 3-β-(2,3-epoxypropoxy)ethyl-2-methyl-4(3H)-quina-
zolinone (**139**) (1-chloro-2,3-epoxypropane, NaOH, EtOH, reflux, 8 h:
75%).[2185]

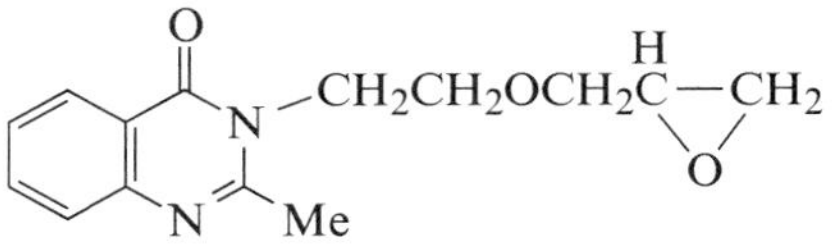

(139)

6-Methyl-2-[γ,γ,γ-trifluoro-β-hydroxy-β-(trifluoromethyl)propyl]-4,4-bistri-
fluoromethyl-3,4-dihydroquinazoline (**140**, R = H) gave its β-methoxy analog (**140**, R = Me) (CH$_2$N$_2$, Et$_2$O, 0°: 98%).[1263]

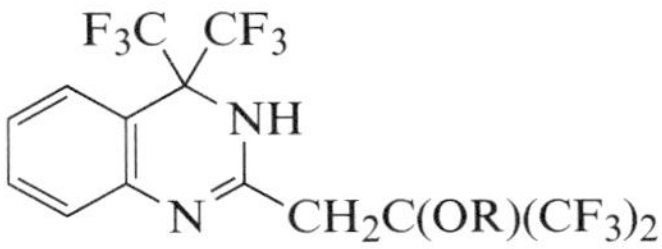

(140)

Also other examples.[488,892,1849,2240]

Oxidation

4-(*m*-Chloro-α-hydroxybenzyl)quinazoline (**141**, R = *m*-Cl) gave 4-*m*-chloro-
benzoylquinazoline (**142**, R = *m*-Cl) (28% HNO$_3$, 20°, 12 h: 56%).[373]

4-(α-Hydroxy-*o*-methoxybenzyl)quinazoline (**141**, R = *o*-OMe) gave 4-*o*-
methoxybenzoylquinazoline (**142**, R = *o*-OMe) (K$_2$CO$_3$, MeOH–H$_2$O,
20°, 2 days: 68%; aerial oxidation?).[373]

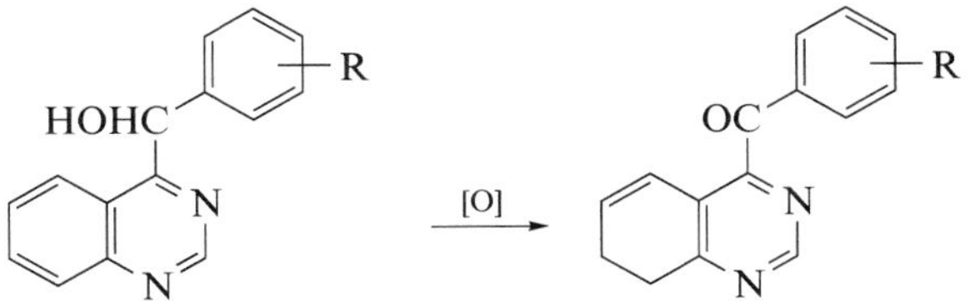

(141)   **(142)**

2-α-Hydroxyethyl- gave 2-acetyl-4(3*H*)-quinazolinone [Me$_2$SO, (CH$_2$COCl)$_2$,
CH$_2$Cl$_2$, − 60°, 10 min; then substrate ↓, − 60°, 2 h; then Et$_3$N, → 20°: 29%
(a Swern[2450g] oxidation);[2069] or CrO$_3$/H$_2$O, AcMe, 20°: ∼ 25%].[185]

3-(γ-Anilino-β-hydroxypropyl)- gave 3-[γ-(*N*-phenylacetamido)acetonyl]-
4(3*H*)-quinazolinone (Me$_2$SO, Ac$_2$O, 20°, 48 h: 59%).[833]

Also examples using MnO$_2$ or CrO$_3$ as oxidant.[181]

Other reactions

3-β-Hydroxyethyl- (**143**) gave 3-β-(2-oxocyclohexyl)ethyl- (**144**) [(CH$_2$)$_5$CO,
conc (concentrated) H$_2$SO$_4$, 5°, 30 min; then 20°, 12 h: 80%][2305] or 3-β-

benzamidoethyl-2-phenyl-4(3*H*)-quinazolinone(**145**)(BzNH₂, HCl, EtOH, reflux, 3 h: ?%).[838]

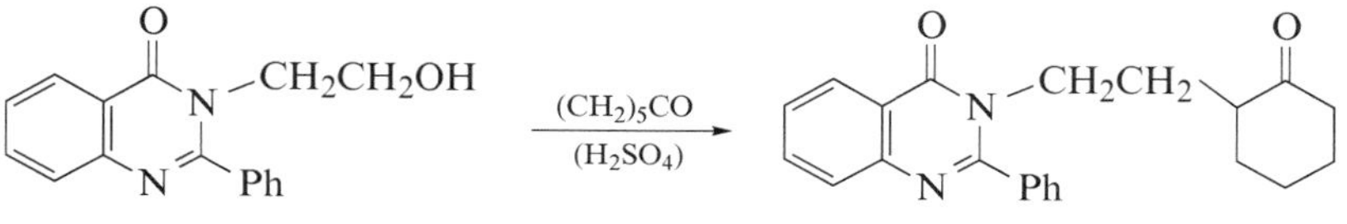

Bis(3-*o*-, *m*-, or *p*-Hydroxyphenyl-2-methyl-4-oxo-3,4-dihydroquinazolin-6-yl)methane (**146**) with 1-chloro-2,3-epoxypropane gave useful polymeric resins.[327]

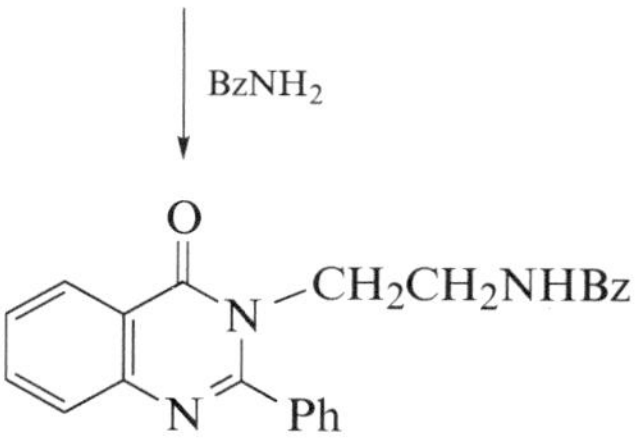

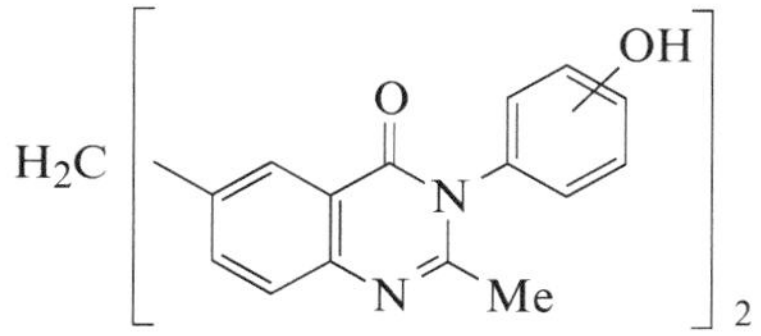

4.5. ALKOXY- AND ARYLOXYQUINAZOLINES (*H* 235)

The 2-, 4-, 5-, 6-, 7-, 8-, and extranuclear alkoxyquinazolines are all covered in this section, despite their inherent differences in preparative methods and in reactivities. However, discussion of *N*1- and *N*3-alkoxyquinazolines is postponed to Section 4.7 with the quinazoline *N*-oxides.

4.5.1. Preparation of Alkoxy- and Aryloxyquinazolines (*H* 236,239)

The preparation of alkoxyquinazolines by *primary synthesis* (Chapter 1); by *O-alkylation* of quinazolinones, quinazolinols, or ω-hydroxyquinazolines (Sections 4.1.2.3, 4.2.2, or 4.4.2); and by *alcoholysis* of halogenoquinazolines (Sections 3.5.2, 3.6, and 3.7) have been discussed. The remaining preparative methods are of minor importance and are illustrated in the following examples.

From quinazolinamines

2-Trimethylammonioquinazoline chloride (**147**) gave 2-methoxyquinazoline (**148**) (MeONa, MeOH, 66°, 10 min: 56%, as picrate);[189] 4-methoxy-quinazoline similarly (20°, 5 min: 98%, as picrate).[189]

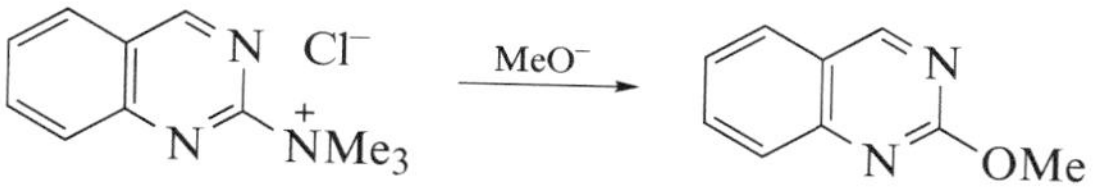

(147)   **(148)**

2-Chloro-4-α-methylhydrazino- (**149**, X = Cl, Y = NMeNH$_2$) gave 2,4-di-ethoxyquinazoline (**149**, X = Y = OEt) (EtONa, EtOH, reflux, 2 h: 72%).[2230]

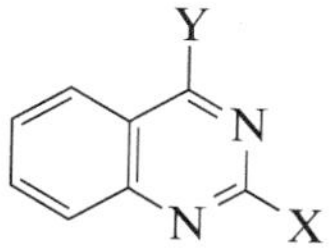

(149)

From quinazolinethiones or derivatives

4-Methylthio- (**150**, R = SMe) gave 4-ethoxy-6-(naphthalen-2-ylsulfonyl)-2-quinazolinamine (**150**, R = OEt) (EtONa, EtOH, 100°, sealed, 5 h: ~45%).[1595]

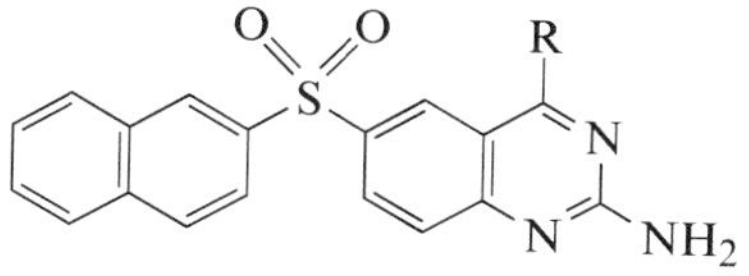

(150)

2-Thioxo-1,2-dihydro-4(3H)-quinazolinone gave 2-methoxy-4(3H)-quinazo-linone (**152**) [EtO$_2$CN=NCO$_2$Et, MeOH, reflux, 3 h: 15%, after separation from several other products: probably via (**151**)].[1460]

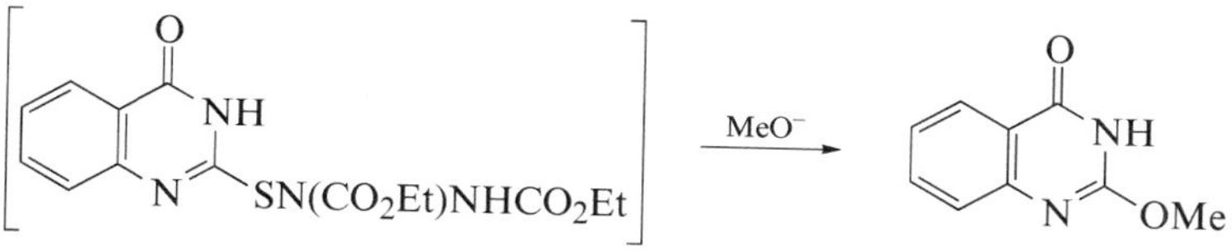

(151)   **(152)**

Also other examples.[1779,1915]

From quinazolinecarbonitriles

1-Methyl-4-oxo-1,4-dihydro-2-quinazolinecarbonitrile gave 2-carboxyme-thoxy-1-methyl-4(1H)-quinazolinone (HOCH$_2$CO$_2$H, Et$_3$N, Me$_2$NCHO, 60°, 2 h: 41%).[1415]

4.5.2. Reactions of Alkoxy- and Aryloxyquinazolines (*H* 241)

Alkoxyquinazolines are useful intermediates for nucleophilic displacement reactions, especially (but not exclusively) when the ether grouping occupies the 2- or 4-position. The reactions of trialkylsiloxyquinazolines are included in this section. The *hydrolysis* of alkoxyquinazolines to quinazolinones, quinazolinols, or extranuclear hydroxyquinazolines (Sections 4.1.1.3, 4.2.1, and 4.4.1), their *oxidative hydrolysis* to quinazoline quinones (Section 4.3), their direct *conversion into extranuclear halogenoquinazolines* (Section 3.4), and the *ortho*-Claisen rearrangement of alkenyloxyquinazolines (Section 2.2.1.6) have been covered already. A ruthenium *complex* of 6,7-dimethoxy-4-phenylquinazoline has been described.[1647] The remaining quite important reactions are discussed in the following subsections.

4.5.2.1. Aminolysis of Alkoxyquinazolines (*H* 243)

Although of considerable potential use, the aminolysis of alkoxy and aryloxyquinazolines has been rather neglected. The following examples illustrate the conditions and yields to be expected.

Regular aminolyses

5-Methyl-4-phenoxy- (**153**, R = OPh) gave 5-methyl-4-piperidinoquinazoline
  [**153**, R = N(CH$_2$)$_5$] [neat HN(CH$_2$)$_5$, 20°, 3 h: 98%; also kinetics in
  2-methylheptane].[1814]

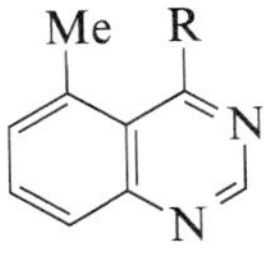

(153)

4-Ethoxy- (**154**, R = OEt) gave 4-anilino- (**154**, R = NHPh) (neat PhNH$_2$,
  reflux, 3 days: 83%) or 4-*p*-chlorobenzylamino-5,6,7,8-tetrahydro-2 (1*H*)-
  quinazolinone (**154**, R = NHCH$_2$C$_6$H$_4$Cl-*p*) (H$_2$NCH$_2$C$_6$H$_4$Cl-*p*,
  Me$_2$NCHO, reflux 21 h: > 95%).[1853]

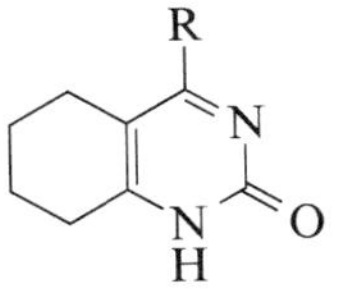

(154)

2-Methoxy- gave 2-butylamino-5,6,7,8-tetrahydro-4(3H)-quinazolinone [BuNHNa (made *in situ*), tetralin, reflux, 8 h: 66%].[1982]
Also other examples.[60,1473,1969,2547]

Related reactions with amines

3-β,γ-Epoxypropyl- (155) gave 3-(γ-cyclohexylamino-β-hydroxypropyl)-4(3H)-quinazolinone (156) (C$_6$H$_{11}$NH$_2$, 100°, 45 min: 49%);[833] also analogs similarly.[833]

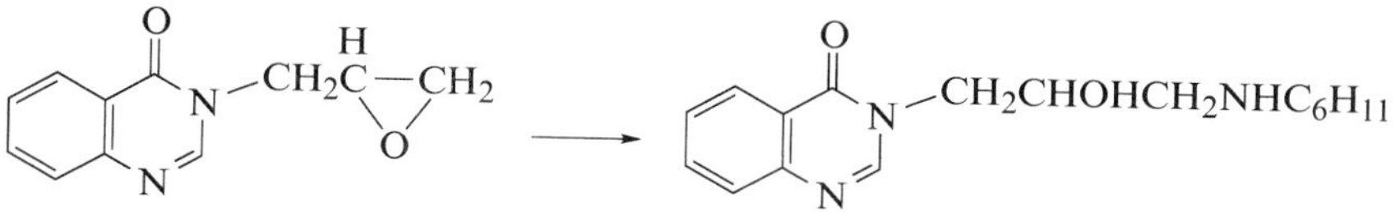

(155) **(156)**

4-γ-Dimethylaminopropylamino-6-methoxy- (158) gave 4-γ-dimethylaminopropylamino-6-piperidino-5,8-dihydro-5,8-quinazolinedione (157) [HN(CH$_2$)$_5$, MeOH, 0°, 5 h: 37%; a regular reaction] but the same substrate (158) gave 6,7-bisaziridin-1'-yl-4-γ-dimethylaminopropylamino-5,8-dihydro-5,8-quinazolinedione (160) [excess neat HN(CH$_2$)$_2$, 0°, N$_2$, 4.5 h: 35%; probably via aerial oxidation of the adduct (159)].[2278]

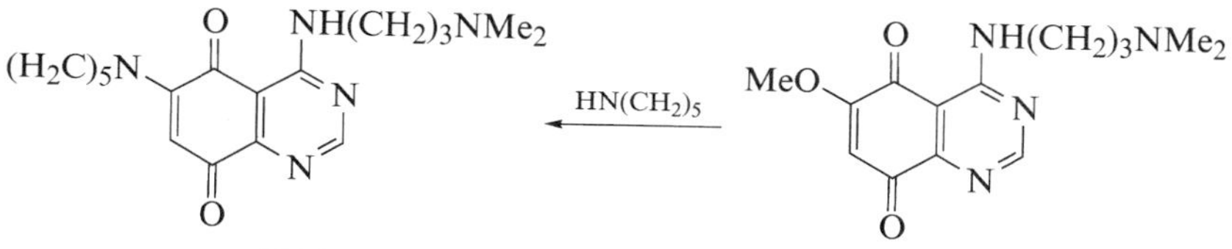

(157) **(158)**

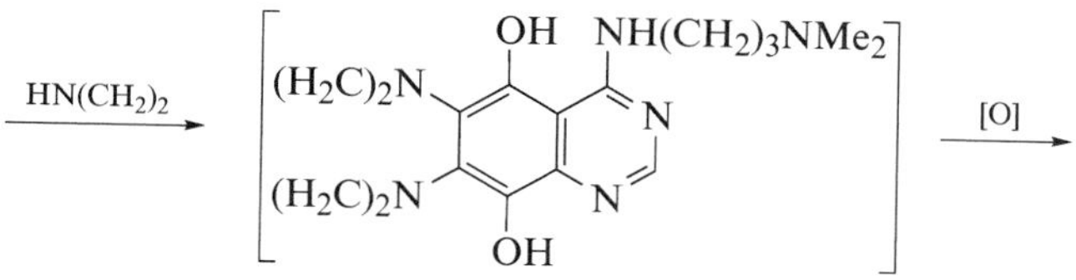

(159)

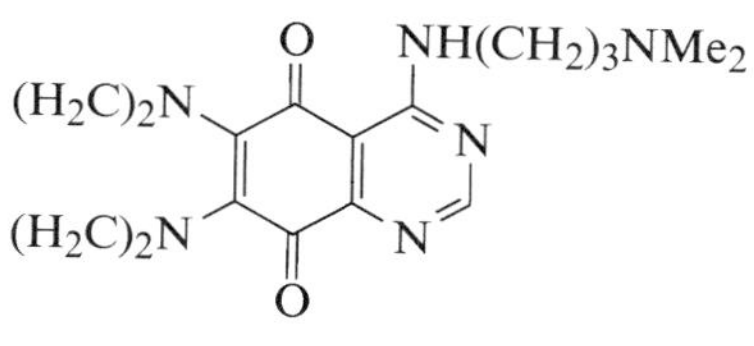

(160)

4.5.2.2.   Thermal Rearrangement of 2- or 4-alkoxyquinazolines (*H* 243)

Both 2- and 4-alkoxyquinazolines undergo thermal rearrangement to the isomeric 1-alkyl-2(1*H*)- or 2-alkyl-4(3*H*)-quinazolinones, respectively. The limited practicability of this attractive reaction is illustrated in the following examples.

4-Prop-2′-ynyloxyquinazoline gave 3-prop-2′-ynyl-4(3*H*)-quinazolinone (**161**) (ClC$_6$H$_4$Cl-*o*, 190°, 12 h: 22%, after purification).[1935]

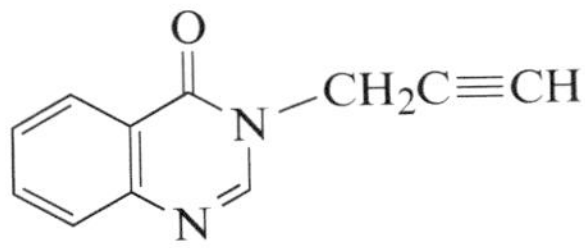

(**161**)

4-Phenoxyquinazoline gave 3-phenyl-4(3*H*)-quinazolinone (heavy mineral oil, 320°, N$_2$, 5 h: 59%).[260]

4-(4-Nitro-2-pyridin-2′-ylphenoxy)-2-phenylquinazoline gave 3-(4-nitro-2-pyridin-2′-ylphenyl)-2-phenyl-4(3*H*)-quinazolinone (**162**) (no solvent, N$_2$, 300°, 30 min: 77%).[406]

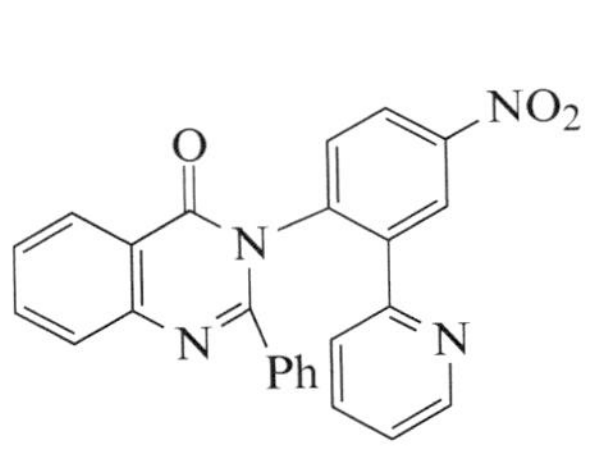

(**162**)

2-Phenyl-4-*p*-(2,2,4-trimethylchroman-4-yl)phenoxyquinazoline gave 2-phenyl-3-*p*-(2,2,4-trimethylchroman-4-yl)phenyl-4(3*H*)-quinazolinone (neat substrate, sealed in a vacuum, 340°, 5.5 h: 80%).[903]

2,4-Bis(2,6-dichloro-3-methylphenoxy)quinazoline gave 1,3-bis(2,6-dichloro-3-methylphenyl)-2,4(1*H*,3*H*)-quinazolinedione (**163**) (mineral oil, 345°, N$_2$, 130 min: > 40%).[640]

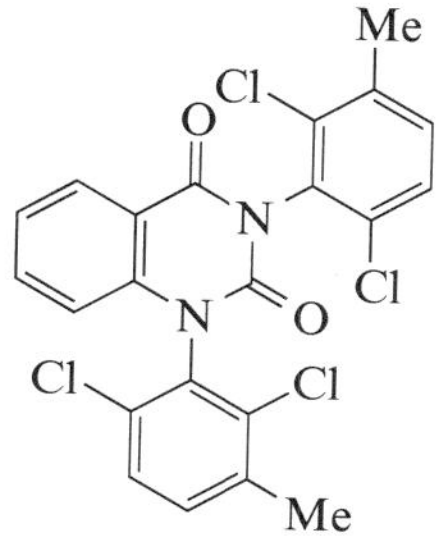

(163)

4-(Tetra-O-acetyl-β-D-glycopyranosyloxy)quinazoline gave 3-(tetra-O-acetyl-β-D-glucopyranosyl)-4(3H)-quinazolinone (HgBr$_2$, PhMe, reflux, 2 h: 80%).[651]

Also other examples.[726,809,1772,2032,2164]

4.5.2.3. Hilbert–Johnson Reactions on Trimethylsiloxyquinazolines (H 242)

As pioneered in the pyrimidine series,[1446d] the Hilbert–Johnson reaction[2450h] involved formation of an alkoxypyrimidine methiodide with subsequent loss of alkyl iodide to afford an N-methylpyrimidinone, a process that became widely used in nucleoside chemistry. It was later found advantageous to replace the alkoxy substrate by a trialkylsiloxypyrimidine, and it is this modified version (sometimes known as a *silyl–Hilbert–Johnson reaction*) that has been used of late in the quinazoline series. Although unfortunately restricted in range, the following examples will illustrate the reaction.

2,4-Bistrimethylsiloxyquinazoline (**164**) gave 1-methyl- (**165**) (MeI, reflux, 2 h: 88%)[78] or 1-(2,3,5-tri-O-benzoyl-β-D-ribofuranosyl)-2,4(1H,3H)-quinazolinedione (2,3,5-tri-O-benzoyl-D-ribofuranosyl bromide, MeCN, 20°, 50 h; 90%).[243]

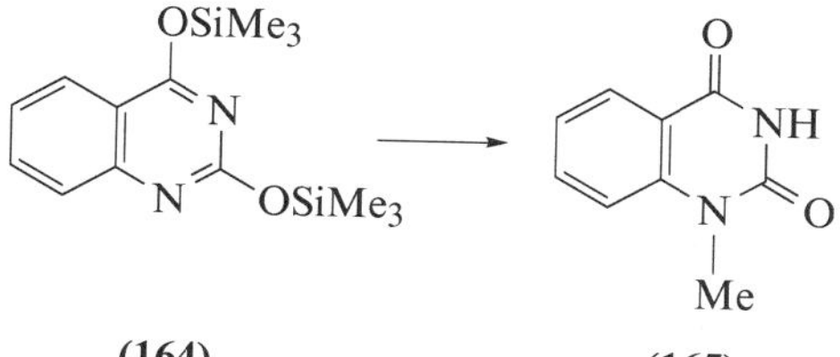

(164)   **(165)**

2,4-Bistrimethylsiloxy-5,6,7,8-tetrahydroquinazoline gave 1-methyl- (**166**, X = Me, Y = H) (MeI, AgClO$_4$, PhH)[1499] or 1,3-bis-γ-bromopropyl-5,6,7,8-tetrahydro-2,4(1H,3H)-quinazolinedione (**166**, X = Y = CH$_2$CH$_2$CH$_2$Br) (using BrCH$_2$CH$_2$CH$_2$Br; also two separable byproducts).[1476]

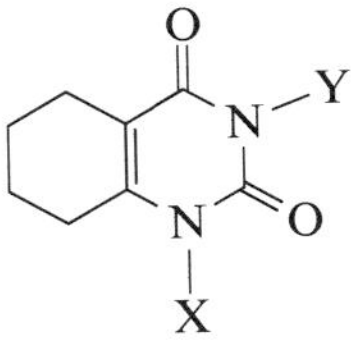

(166)

2-Methoxy-4-trimethylsiloxy-5,6,7,8-tetrahydroquinazoline gave 1-benzoyl-2-methoxy-5,6,7,8-tetrahydro-4(1H)-quinazolinone (**167**) (BzCl, SnCl₄, MeCN, warm, 30 min: 70%; note that this is an *N*-acylation).[1753]

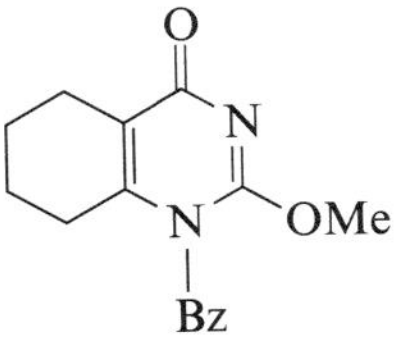

(167)

Also other examples.[78,224,1413,2424,2437,2568]

4.6. NONTAUTOMERIC (FIXED) QUINAZOLINONES

Fixed *N*-substituted 2- or 4-quinazolinones differ from the corresponding tautomeric quinazolinones mainly by their inability to form anions and hence to undergo reactions dependent thereon. The fixed quinazolinones include several established therapeutic agents such as the hypnotic methaqualone,[483] its analog Lonetil,[1090] and other related compounds of diverse potential.[295,1611]

4.6.1. Preparation of Fixed Quinazolinones

The main synthetic routes to fixed quinazolinones, namely by *primary synthesis* (Chapter 1), by *N-alkylation* of tautomeric quinazolinones (Section 4.1.2.3), by *thermal rearrangement* of alkoxyquinazolines (Section 4.5.2.3), and by the *Hilbert–Johnson reaction* (Section 4.5.2.3) have been discussed already. Several less important routes are illustrated by the following examples.

From fixed quinazolinethiones

3-γ-Ethylaminopropyl-2-phenyl-4(3H)-quinazolinethione (**168**, X = S) gave the corresponding quinazolinone (**168**, X = O) (EtOH–H₂O, reflux, 8 h: 92%).[716]

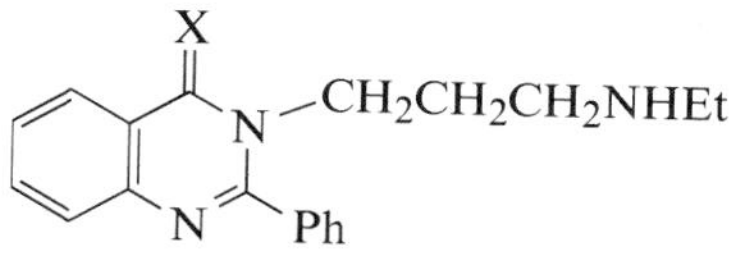

(168)

1-Phenyl-4(1*H*)-quinazolinethione gave the corresponding quinazolinone [PhC(NOH)Cl, Et₃N, reflux, until decolorized: 90%; reagent is PhCN → O];[1439] also analogs likewise.[1228,1439,1725]

1,3-Bis-*p*-chlorophenyl-4-thioxo-3,4,5,6,7,8-hexahydro-2(1*H*)-quinazolinone **(169**, X = S) gave 1,3-bis-*p*-chlorophenyl-5,6,7,8-tetrahydro-2,4(1*H*,3*H*)-quinazolinedione **(169**, X = O) (HgO, AcOH, reflux, 1 h: ∼ 60%).[268]

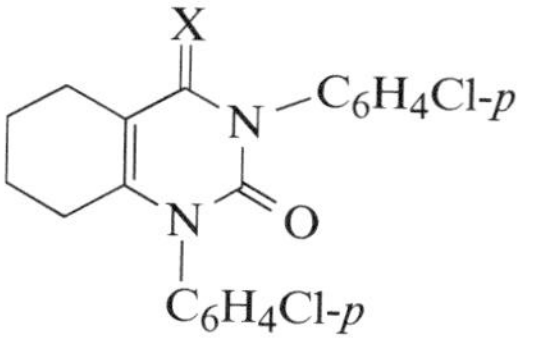

(169)

Also other examples.[994]

From fixed quinazolinimines

6-Chloro-2-hydrazono-1-methyl-4-phenyl-1,2-dihydroquinazoline **(170**, X = NNH₂) gave 6-chloro-1-methyl-4-phenyl-2(1*H*)-quinazolinone **(170**, X = O) (hydrolysis in alkali?).[1976]

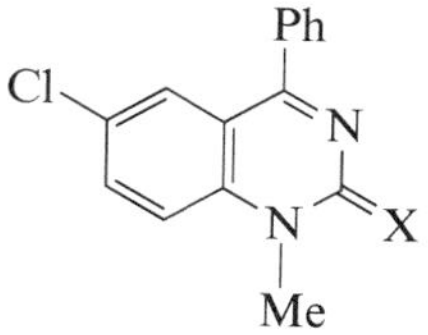

(170)

1-*p*-Bromophenacyl-4(1*H*)-quinazolinimine gave the corresponding quinazolinone (hydrolysis in alkali?).[995]

Also other examples.[50,1077]

From fixed hydroquinazolinones

2-Isopropyl-3-phenyl-1,2-dihydro-4(3*H*)-quinazolinone **(171)** gave 2-isopropyl-3-phenyl-4(3*H*)-quinazolinone **(172)** (KMnO₄, AcMe, reflux, 90 min: 81%).[1590]

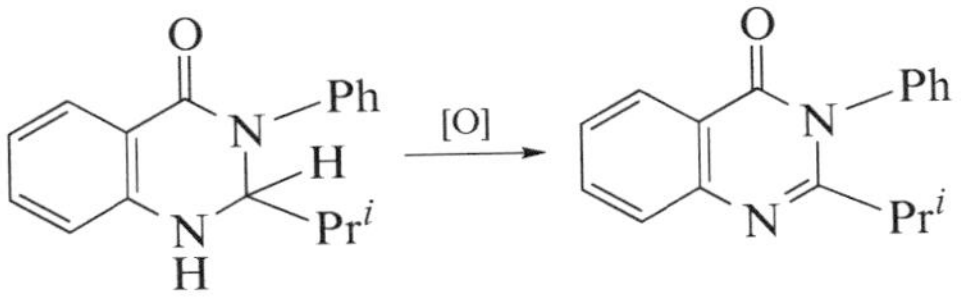

(171)   **(172)**

1-Cyclopropylmethyl-6-methoxy-4-phenyl-3,4-dihydro-2(1H)-quinazolinone
gave its didehydro derivative (**173**, R = OMe) (KMnO$_4$, dioxane–H$_2$O, 20°,
1 h: 85%);[544,581] 6-chloro-1-cyclopropylmethyl-4-phenyl-2(1H)-quinazo-
linone (**173**, R = Cl) was made differently from its 3,4-dihydro derivative
(Me$_2$SO, hv, 50 h: 87%; AcMe, hv, air↓, 28 h: 68%).[986]

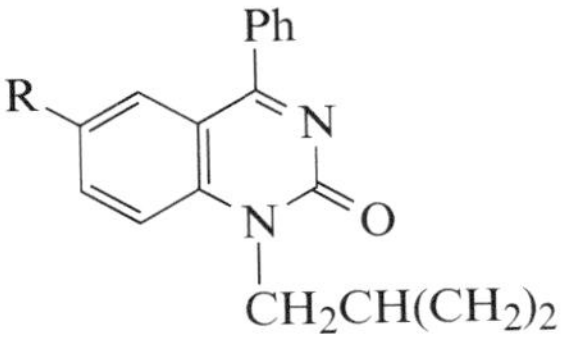

(173)

2-Ethyl-1-methyl-2,3-dihydro-4(1H)-quinazolinone gave 2-ethyl-1-methyl-
4(1H)-quinazolinone (**174**) [complex from HgO + (HO$_2$C)$_2$NCH$_2$CH$_2$N
(CO$_2$Na)$_2$, EtOH–H$_2$O, reflux; then NH$_4$OH: 91%].[821]

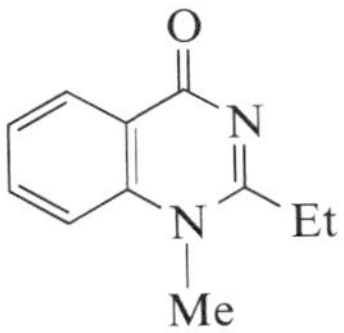

(174)

Also other examples.[186,642,983,1275,1285,2107]

From fixed N-acylquinazolinones

3-Tosyl-4(3H)-quinazolinone gave 3-phenyl-4(3H)-quinazolinone (PhNH$_2$,
PhH, 20°, 10 h: 78%; mechanism unclear).[650]

N3-Alkyl- from N1-alkylquinazolinones

1-Methyl-1-phenyl-4(1H)-quinazolinone (**176**, R = Me, Y = Me, Y = Ph) gave
3-methyl-2-phenyl-4(3H)-quinazolinone (**175**) (neat, 300°, 30 min: 53%;[39]
260°, 4 h: > 40%).[1275]

1-Ethyl-2-o-tolyl-4(1H)-quinazolinone (**176**, R = Et, Y = C$_6$H$_4$Me-o) gave 3-
ethyl-2-o-tolyl-4(3H)-quinazolinone (**177**) and 2-o-tolyl-4(3H)-quinazo-

linone (**178**) (neat, 260°, 4 h: ~30% and ~15%, respectively);[1275] mechanistic studies, inconclusive.[39,1275]

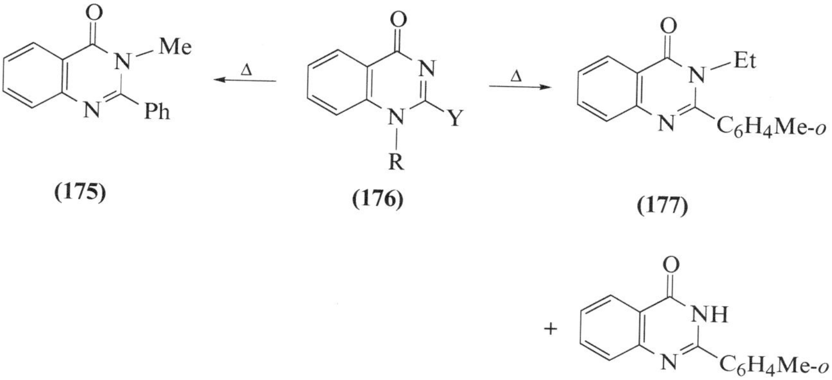

(**175**)   (**176**)   (**177**)

+

(**178**)

4.6.2. Reactions of Fixed Quinazolinones

Unlike tautomeric quinazolinones, fixed quinazolinones undergo only a few general reactions, which are discussed in the following subsections.

4.6.2.1. Conversion into Fixed Quinazolinethiones (*H* 278)

Although it is just as easy to thiate fixed quinazolinones as tautomeric quinazolinones (Section 4.1.2.1) with phosphorus pentasulfide, a much greater variety of media and conditions have been employed to make fixed quinazolinethiones, as indicated in the formation of the following products.

2-Methyl-3-phenyl-4(3*H*)-quinazolinethione (**179**, R = H) (neat P_2S_5, 155°, 30 min: 79%).[479]

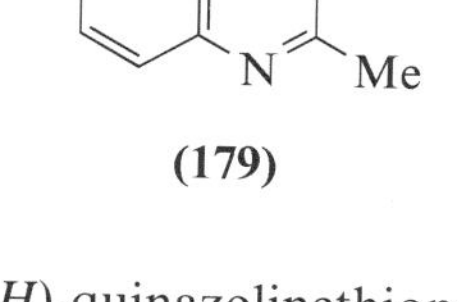

(**179**)

3-Phenyl-2-pyridin-2'-yl-4(3*H*)-quinazolinethione (neat P_2S_5, 180°, 4 h: >60%; xylene, reflux, 4 h: <55%).[994]

1,2-Diphenyl-4(1*H*)-quinazolinethione (**180**) (P_2S_5, xylene, reflux, 2 h: 57%).[937]

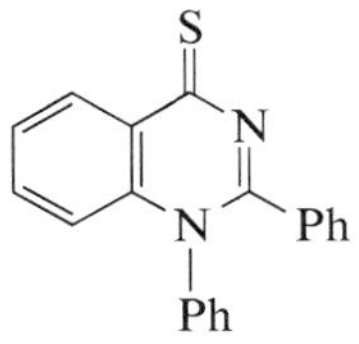

(180)

6-Fluoro-2-methyl-3-phenyl-4(H)-quinazolinethione (**179**, R = F) (P$_2$S$_5$, xylene, reflux, 4 h: 74%).[489]

3-Ethoxycarbonylmethyl-2-methyl-4(3H)-quinazolinethione (**181**) (P$_2$S$_5$, dioxane, reflux, 15 h: 66%).[1209]

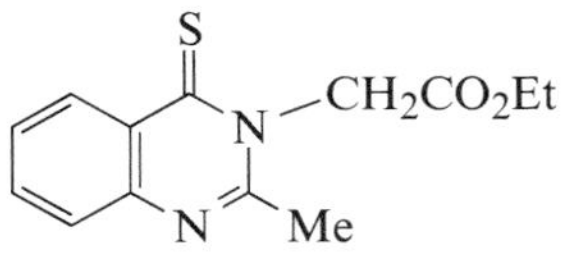

(181)

1-Isopropyl-7-methyl-4-phenyl-2(1H)-quinazolinethione (**182**) (P$_2$S$_5$, pyridine, reflux, 2.5 h: 32%; note poor yield from this 2-thiation).[681]

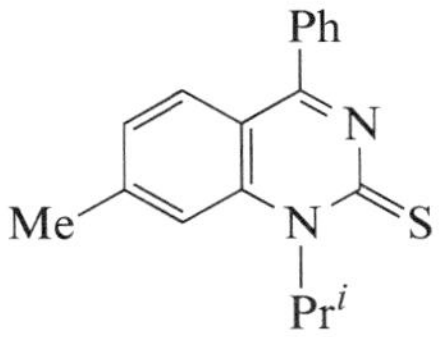

(182)

6,7-Dimethoxy-2,3-dimethyl-4(3H)-quinazolinethione (P$_2$S$_5$, pyridine, sealed, 150°, 8 h, then 170°, 3 h: 56%).[1354]

3-Ethyl-1-methyl-4-thioxo-3,4-dihydro-2(1H)-quinazolinone (**183**) (P$_2$S$_5$, 1,2,4-Me$_3$C$_6$H$_3$, reflux, 30 min: 56%; selective 4-thiation).[86]

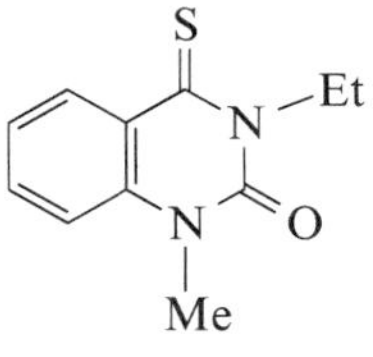

(183)

3-Amino-4(3H)-quinazolinethione (**184**, R = H) (P$_2$S$_5$, pyridine, reflux, 2.5 h: 73%);[882] its 2-napthalen-2′-yl derivative (**184**, R = C$_{10}$H$_7$-β) (P$_2$S$_5$, xylene, reflux, 6 h: 60%).[1731]

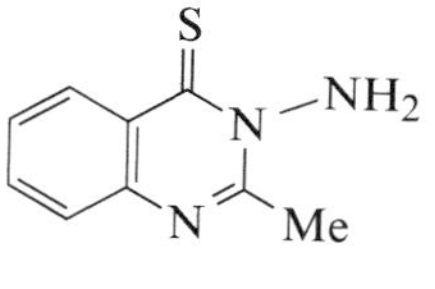

(184)

Also other examples.[105,319,320,460,617,720,1313,1507,1513]

4.6.2.2.   Reductive Reactions (*H* 110)

Gentle reduction of 1-alkyl-2(1*H*)-, 1-alkyl-4(1*H*)-, and 3-alkyl-4(3*H*)-quinazolinones normally gives their respective 3,4-, 2,3-, and 1,2-dihydro derivatives, but more vigorous treatment may cause further reduction, ring fission, and/or deoxygenation; 1,3-dialkyl-2,4(1*H*,3*H*)-quinazolinediones can react only in the second way. These possibilities are illustrated in the following examples.

From 1-alkyl-2(1*H*)-quinazolinones

1-Isopropyl-4-phenyl-2(1*H*)-quinazolinone gave its 3,4-dihydro derivative
  (**185**, R = Pr^i) (NaBH$_4$, EtOH–H$_2$), 20°, 30 min: 96%).[681]

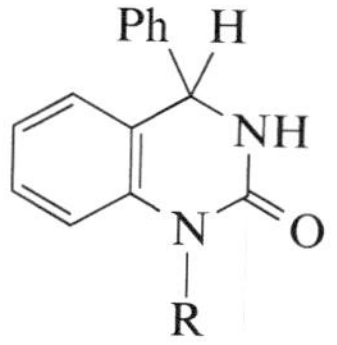

(185)

1-Methyl-4-phenyl-2(1*H*)-quinazolinone gave its 3,4-dihydro derivative (**185**,
  R = Me) (NaBH$_4$, MeOH, 20°, 30 min: 84%; Et$_3$N, PhH, *hv*, 20°, 15h:
  10%).[2037]

1-Benzyl-5,6,7,8-tetrahydro-2(1*H*)-quinazolinone gave its 3,4,5,6,7,8-hexahydro analog (**186**) (H$_2$, PtO$_2$, EtOH, 20°, 1 atm: 81%) and thence the
  3,4,4a,5,6,7,8,8a-octahydro analog (**187**) (H$_2$, PtO$_2$, AcOH, 20°, 5 atm, 5 h:
  56%).[423]

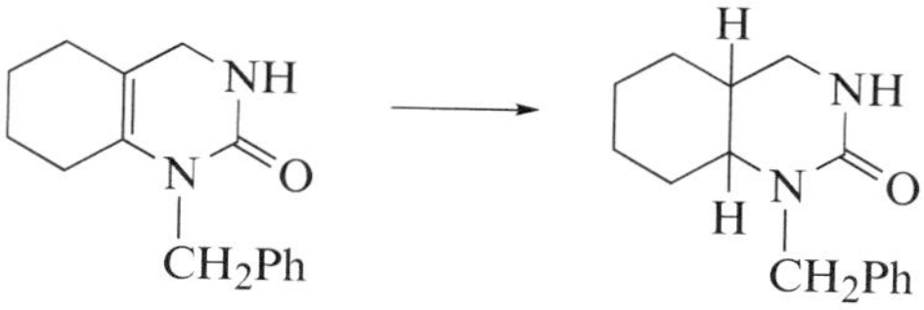

(186)   **(187)**

From 1-alkyl-4(1H)-quinazolinones

1-Benzyl-4(1H)-quinazolinone gave its 2,3-dihydro derivative (188) (NaBH$_4$, Pr^{i}OH, 20°, 18 h: 78%).[186]

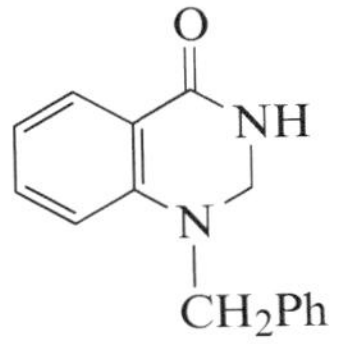

(188)

1-p-Methoxyphenyl-4(1H)-quinazolinone gave its 2,3-dihydro derivative (HCONH$_2$, 175°, 5 h: 92%).[245]

Also other examples.[235,264]

From 3-alkyl-4(3H)-quinazolinones

2-Methyl-3-o-tolyl-4(3H)-quinazolinone gave its 1,2-dihydro derivative (189) (substrate hydrochloride, NaBH$_4$, THF–MeOCH$_2$CH$_2$OCH$_2$CH$_2$OMe, 5°, 4 h: 85%;[642] likewise, LiBH$_4$, ~30 min: ~55%;[637] H$_2$, Pt, MeOH: ?%).[1797]

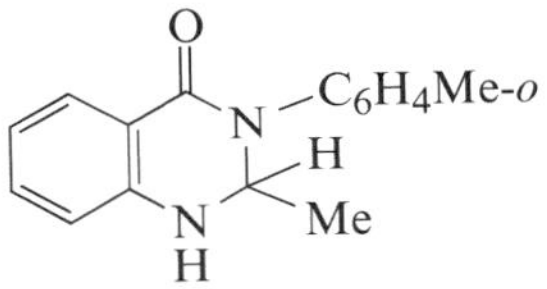

(189)

2-Benzyl-3-methyl-4(3H)-quinazolinone (190) gave its 1,2-dihydro derivative (191) (H$_2$, Pd/C, EtOH, 20°, 15 h: 65%) and thence, or directly from substrate (190), N-methyl-o-phenethylaminobenzamide (192) (LiAlH$_4$, THF, 20°, 3 h: 73% or 55%, respectively).[264]

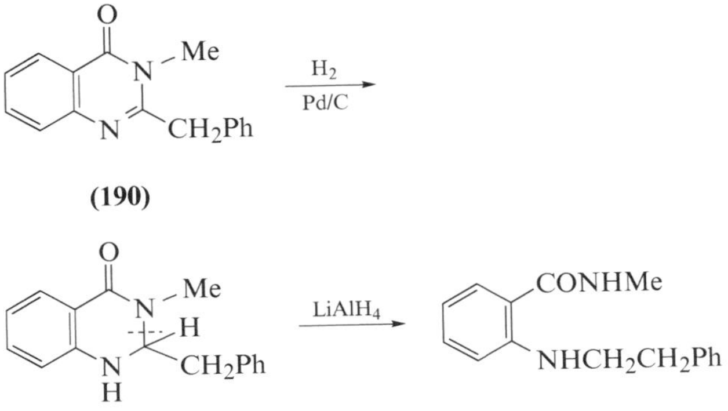

(190)

(191)   **(192)**

3-Benzyl-4(3*H*)-quinazolinone gave 3-benzyl-1,2,3,4-tetrahydroquinazoline (**193**) (LiAlH$_4$, Et$_2$O, reflux, 48 h: 67%; note deoxygenation).[2239]

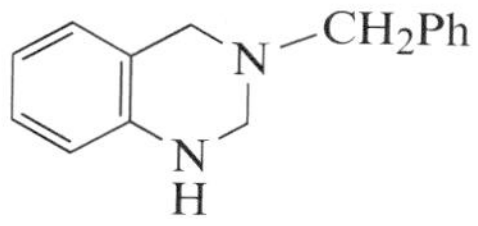

(**193**)

Also other examples in the foregoing references and elsewhere.[273,645,687,2002]

From *N*-alkylated 2,4(1*H*,3*H*)-quinazolinediones

1-Phenyl- (**194**, R = H) or 3-methyl-1-phenyl-2,4(1*H*,3*H*)-quinazolinedione (**194**, R = Me) gave *o*-methylaminomethyl- (**195**, R = H) or *o*-dimethyl-laminomethyldiphenylamine (**195**, R = Me), respectively (LiAlH$_4$, THF, reflux, 12 h: each > 76%, as hydrochloride).[1589]

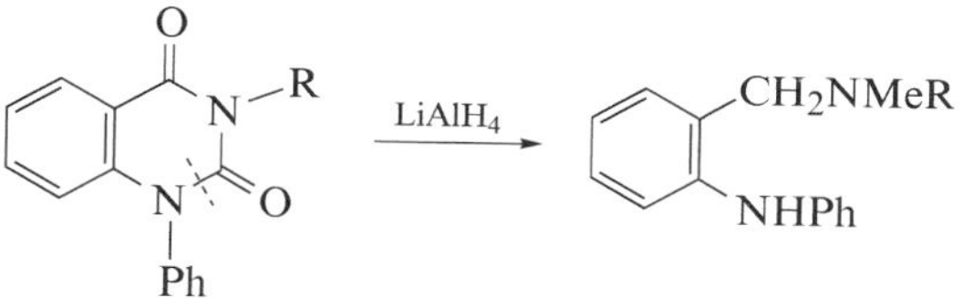

(**194**)   (**195**)

A rather analogous result was achieved by treatment of 2(1*H*)-quinazolinone 3-methiodide (made by primary synthesis) with aqueous alkali to afford 4-hydroxy-3-methyl-1,4-dihydro-2(3*H*)-quinazolinone (**196**) in 80% yield,[1027] a case of nuclear reduction by covalent hydration.

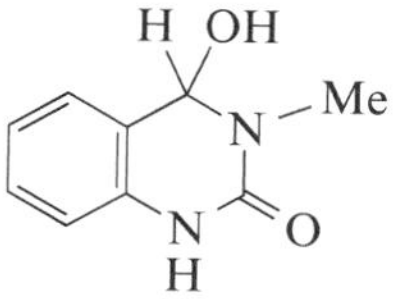

(**196**)

4.6.2.3. Other Reactions (*H* 112,113)

Grignard reactions. 2-Phenyl-3-*p*-tolyl-4(3*H*)-quinazolinone (**197**, R = C$_6$H$_4$Me-*p*) reacted with methylmagnesium iodide to afford 4-methyl-2-phenyl-3-*p*-tolyl-3,4-dihydro-4-quinazolinol (**198**, R = C$_6$H$_4$Me-*p*), whereas the corresponding tautomeric substrate, 2-phenyl-4(3H)-quinazolinone (**197**, R = H), suf-fered additional ring fission on similar treatment to furnish N-o-(α-hydroxy-ethyl)phenylbenzamidine (**199**).[1109]

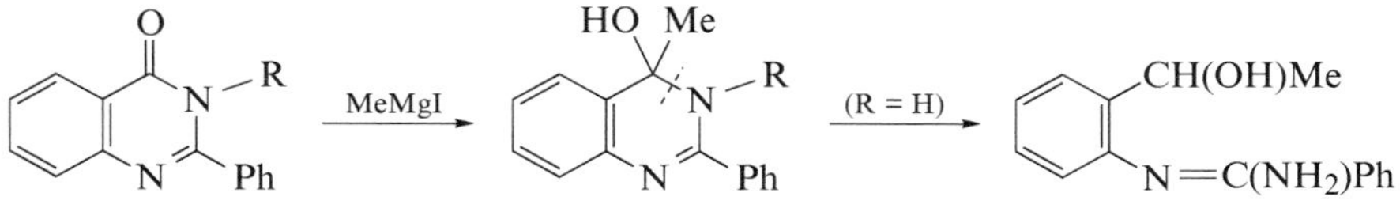

$$(197) \qquad (198) \qquad (199)$$

Nonreductive fissions.   3-Methyl-4(3H)-quinazolinone methiodide (**200**) underwent ring fission at pH 10.2 during 10 min to give N-methyl-o-(N-methyl-formamido)benzamide (**201**, R = CHO) and thence, more slowly, N-methyl-o-methylaminobenzamide (**201**, R = H); kinetics of the several steps were re-ported.[1321] 2-Benzyl-3-phenyl-4(3H)-quinazolinone (**202**) in the presence of RO⁻/ROH has been reported (without detail) to undergo ring fission and reclosure to (**203**, R = NHPh), followed by hydrolysis to 2-hydroxy-3-phenyl-4(1H)-quinolinone (**203**, R = OH).[1804] 2-Methyl-3-o-tolyl-4(3H)-quinazolinone in refluxing dilute alkali gave o-acetamido- and o-amino-N-o-tolylbenzamide.[1985]

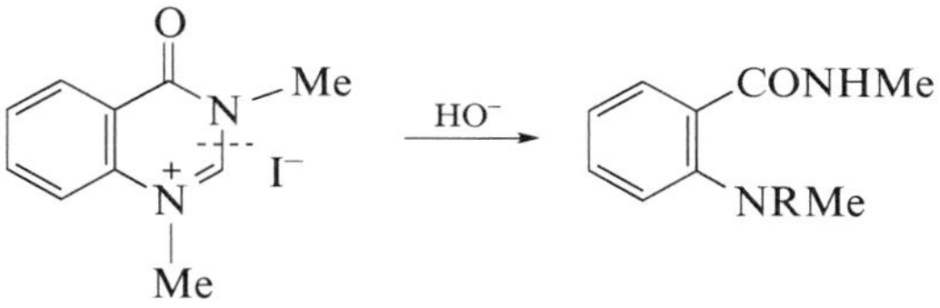

$$(200) \qquad (201)$$

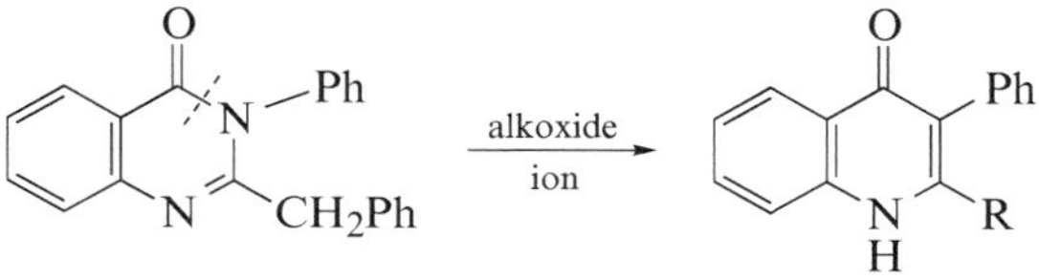

$$(202) \qquad (203)$$

Metal complexes.   Many 2,3-disubstituted 4(3H)-quinazolinones have been converted into metal complexes, some with auxiliary ligands included. The following list embraces most of the recent examples.

3-Acetamido-2-methyl-4(3H)-quinazolinone: Cu,[2082] Pd,[2342] Ru.[1684]

3-Acetamido-2-phenyl-4(3H)-quinazolinone: Cu,[2082] Pd.[2342]

3-Amino-2-methyl-4(3H)-quinazolinone: Cd, Co, Cu, Ni, Zn;[1220] Pd,[1260] Pt,[2159] Ru.[1684]

3-*o*-Aminophenyl-2-methyl-4(3*H*)-quinazolinone: Pt,[2159] Ru.[1684]

3-*o*-Aminophenyl-2-phenyl-4(3*H*)-quinazolinone: Pt.[2159]

3-Amino-2-phenyl-4(3*H*)-quinazolinone: Cd, Co, Cu, Ni, Zn;[1220] Pd,[1260] Pt,[2159] Ru.[1684]

3-Anilino-2-methyl-4(3*H*)-quinazolinone: Cd, Co, Cu, Ni, Zn;[1261] Pd,[1260] Pt,[2159] Ru.[1684]

3-Anilino-2-phenyl-4(3*H*)-quinazolinone: Cd, Co, Cu, Ni, Zn;[1261] Pd,[1260] Pt,[2159] Ru.[1684]

3-Carboxymethyl-2-methyl-4(3*H*)-quinazolinone: Co,[2019] Cu,[2019,2082] Ni,[2019] Pd,[2019,2342] Pt,[2019,2159] Ru.[2019]

3-Carboxymethyl-2-phenyl-4(3*H*)-quinazolinone: Co,[2019] Cu,[2019,2082] Ni,[2019] Pd,[2019,2342] Pt.[2019,2159]

3-Furan-2′-ylmethyleneamino-2-methyl-4(3*H*)-quinazolinone: Pd.[2342]

3-Furan-3′-ylmethyleneamino-2-methyl-4(3*H*)-quinazolinone: Cu.[2082]

3-Furan-2′-ylmethyleneamino-2-phenyl-4(3*H*)-quinazolinone: Pd.[2342]

3-Furan-3′-ylmethyleneamino-2-phenyl-4(3*H*)-quinazolinone: Cu.[2082]

3-*o*-Hydroxybenzylideneamino-2-methyl-4(3*H*)-quinazolinone: Co, Ni, Pt, Ru, Zn;[1676] Cu,[1676,2082] Pd.[1676,2342]

3-*o*-Hydroxybenzylideneamino-2-phenyl-4(3*H*)-quinazolinone: Co, Ni, Pt, Ru, Zn;[1676] Cu,[1676,2082] Pd.[1676,2342]

3-*β*-(*β*-Hydroxyethylamino)ethyl-2-methyl-4(3*H*)-quinazolinone: Ru.[1684]

3-*β*-Hydroxyethyl-2-methyl-4(3*H*)-quinazolinone: Ru.[1684]

3-(4-Hydroxy-3-methoxybenzylidene)amino-2-methyl-4(3*H*)-quinazolinone: Cd, Co, Cr, Cu, Hg, Ni, Zn.[2137]

3-(4-Hydroxy-3-methoxybenzylidene)amino-2-phenyl-4(3*H*)-quinazolinone: Cd, Co, Cr, Cu, Hg, Ni, Zn.[2137]

3-Hydroxymethyl-2-methyl-4(3*H*)-quinazolinone: Pt.[2159]

3-Hydroxymethyl-2-phenyl-4(3*H*)-quinazolinone: Pt.[2159]

3-*o*-Hydroxyphenyl-2-methyl-4(3*H*)-quinazolinone: Pt,[2159] Ru.[1684]

3-*o*-Hydroxyphenyl-2-phenyl-4(3*H*)-quinazolinone: Pt.[2159]

2-Methyl-3-pyridin-2′-yl-4(3*H*)-quinazolinone: Cd, Co, Cu, Ni, Zn;[1688] Pd,[1260] Pt,[2159] Ru.[1684]

2-Methyl-3-thioureido-4(3*H*)-quinazolinone: Cu,[2082] Pd.[2342]

2-Methyl-3-*o*-tolyl-4(3*H*)-quinazolinone: Cd, Co, Zn;[272,278] Cr;[272] Cu.[1126]

2-Methyl-3-ureido-4(3*H*)-quinazolinone: Cu,[2082] Pd.[2342]

2-Phenyl-3-pyridin-2′-yl-4(3*H*)-quinazolinone: Cd, Co, Cu, Ni, Zn;[1688] Pd,[1260] Pt,[2159] Ru.[1684]

2-Phenyl-3-thioureido-4(3*H*)-quinazolinone: Cu,[2082] Pd.[2342]

3-Phenyl-2-thioxo-1,2-dihydro-4(3*H*)-quinazolinone: Co, Ni.[1645]

2-Phenyl-3-ureido-4(3*H*)-quinazolinone: Cu,[2082] Pd.[2342]

2-Pyridin-2'-yl-3-pyridin-2'-ylmethyleneamino-1,2-dihydro-4(3H)-quinazo-linone: Co, Cu, Mn, Ni, Zn;[500] Sn.[1132]

4.7.   QUINAZOLINE N-OXIDES (H 446)

All quinazoline N-oxides are here formulated and named as such, except for rather rare cases in which the N-oxide must be formulated as an N-hydroxy entity to avoid violation of the well-established carbonyl structure of a quinazolinone. Thus, for example, 3-hydroxy-2,4(1H,3H)-quinazolinedione (**204**) is used instead of 4-hydroxy-2(1H)- (**205**) or 2-hydroxy-4(1H)-quinazolinone 3-oxide (**206**).

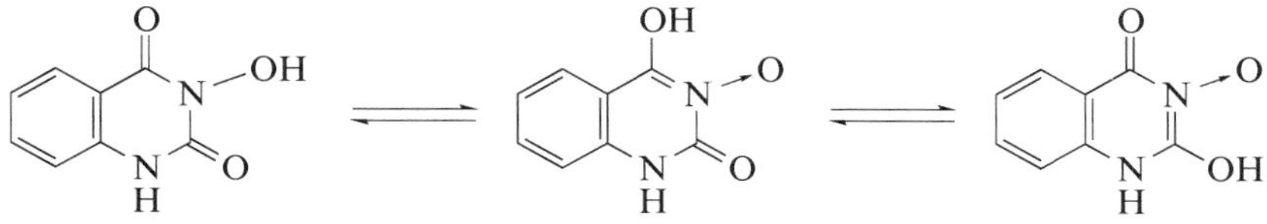

| (204) | (205) | (206) |

4.7.1.   Preparation of Quinazoline N-Oxides (H 446,450,457)

The most frequently used route to quinazoline N-oxides is *primary synthesis* (see Chapter 1) because it usually affords, either directly or indirectly, products of unambiguous structure. The other major route, *N-oxidation* of a quinazoline, has been mentioned in respect to unsubstituted quinazoline (Section 2.1.3), and such oxidations of substituted quinazolines, as well minor approaches to N-oxides, are illustrated in the following recent examples.

By N-oxidation

4-Methylquinazoline gave a separable mixture of its 1- (**207**, R = Me) and 3-oxide (**208**, R = Me) (HO$_2$CC$_6$H$_4$CO$_3$H-o, Et$_2$O, < 20°, 5 h: 6% and 5%, respectively); also the 4-ethyl homologs (**207**, R = Et; **208**, R = Et) (2% each) and the 4-phenyl analog (**207**, R = Ph) (8%).[163]

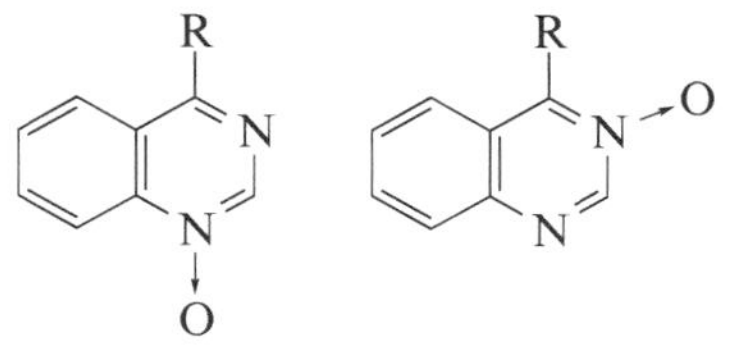

| (207) | (208) |

2,4-Dimethylquinazoline gave its 1-oxide (PhCO$_3$H, CHCl$_3$, reflux, 7 h: 24%).[874]

3-Phenyl-2-pyridin-2′-yl-4(3*H*)-quinazolinone gave its 1′-oxide (**209**) (30%
H_2O_2/AcOH, 75°, 10 h: 38%) or its 1,1′-dioxide (**210**) (30% H_2O_2/
F_3CCO_2H, 75°, 1 h: 40%).[1887, cf. 994]

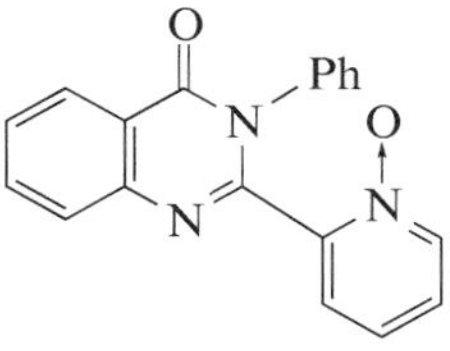

(209) 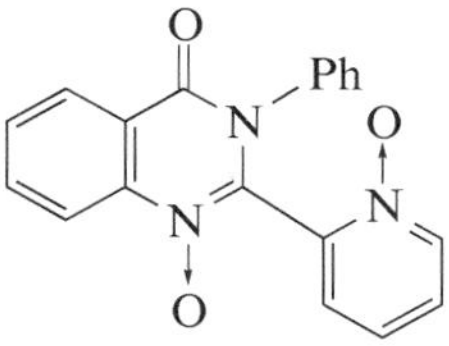   **(210)**

4-Methylquinazoline gave its 3-oxide (H_2NOSO_3H, 60°, 5 min: 25%, by a
complicated but rational mechanism).[559]

4-Ethoxyquinazoline gave its 1-oxide (17%) plus 4-ethoxy-2(3*H*)-quinazo-
linone 1-oxide (51%) [peroxyphthalic acid (2 equiv), Et_2O].[471]

2-Methyl-3-*o*-tolyl-4(3*H*)-quinazolinone gave its 1-oxide as a major meta-
bolite (from human urine: ∼ 9%).[1107]

6-Chloro-1-*β*-dimethylaminoethyl-4-phenyl-2(1*H*)-quinazolinone gave only
its *ω*-*N*-oxide (*m*-ClC$_6$H$_4$CO$_3$H, CH$_2$Cl$_2$, 20°, 20 min: 75%) and thence 6-
chloro-4-phenyl-1-vinyl-2(1*H*)-quinazolinone (PhMe, reflux, 1.5 h: 78%).[676]

Also other examples.[1418]

By nuclear dehydrogenation

2-Chloromethyl-4-*o*-fluorophenyl-1,2-dihydroquinazoline 3-oxide gave 2-
chloromethyl-4-*o*-fluorophenylquinazoline 3-oxide (**211**) (MnO$_2$, CH$_2$Cl$_2$,
reflux, 8 h: 82%);[861] also analogs.[407]

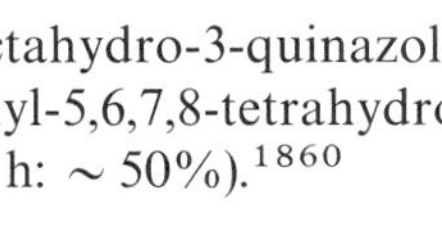

(211)

2-Methyl-3,4,4a,5,6,7,8,8a-octahydro-3-quinazolinol 1-oxide (**212** or a
tautomer) gave 2-methyl-5,6,7,8-tetrahydroquinazolinone 1,3-dioxide
(**213**) (Ni$_2$O$_3$, H$_2$O, 20°, 1 h: ∼ 50%).[1860]

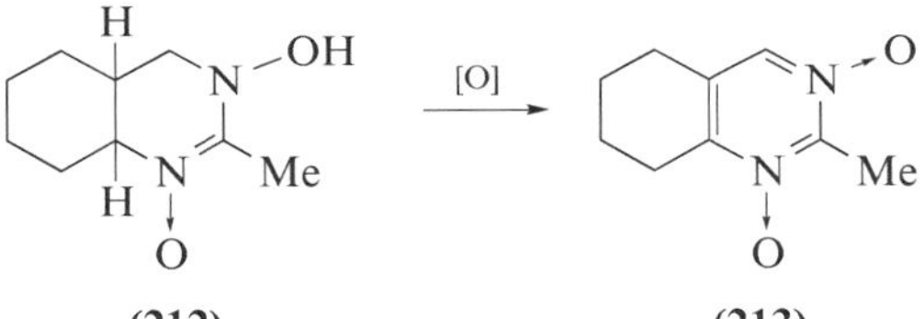

(212)   **(213)**

From *N*-acyloxyquinazolines

3-Methyl-1-(*N*-methylcarbamoyloxy)- **(214**, R = CONHMe, X = O) gave 1-hydroxy-3-methyl-2,4(1*H*,3*H*)-quinazolinedione **(214**, R = H, X = O) (0.25M NaOH, boil, 4 min: > 95%).[212]

3-Methyl-1-(*N*-methylcarbamoyloxy)- **(214**, R = CONHMe, X = S) gave 1-hydroxy-3-methyl-2-thioxo-1,2-dihydro-4(3*H*)-quinazolinone **(214**, R = H, X = S) (0.25M NaOH, boil, 4 min: 80%).[211]

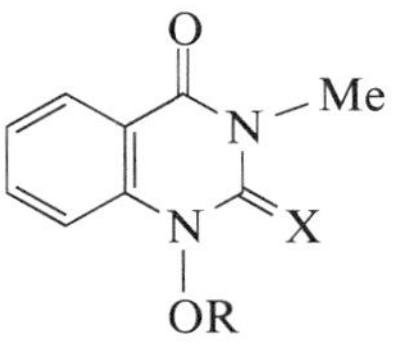

(214)

From alkoxyquinazolines

4-Methoxy- or 4-phenoxyquinazoline gave 4-quinazolinamine 3-oxide (H_2NOH, MeOH–H_2O, reflux?, 3.5 h: 71%, by a rational mechanism involving an initial 3,4-addition of H_2NOH and subsequent Dimroth-like rearrangement).[554]

3-Benzyloxy- gave 3-hyroxy-2,4(1*H*,3*H*)-quinazolinedione (HBr, AcOH, reflux, 2.5 h: ?%).[2553]

4.7.2. Reactions of Quinazoline *N*-Oxides (*H* 447,455)

Quinazoline *N*-oxides display occasional biological activity[973] and undergo a variety of disperate reactions, which are grouped into several categories in the following subsections.

4.7.2.1. Deoxygenation (*H* 447,455,457)

Removal of the *N*-oxide function can occur by elimination of water or the like from a dihydroquinazoline *N*-oxide or by regular reduction or phosphorus ʻrichloride treatment of a quinazoline *N*-oxide. Such procedures are illustrated in the following examples.

By eliminative reactions

4-*o*-Fluorophenyl-1,2-dihydroquinazoline 3-oxide **(215)** gave 4-*o*-fluoro-phenylquinazoline **(216)** (Ac_2O, reflux, 15 min: 84%).[407]

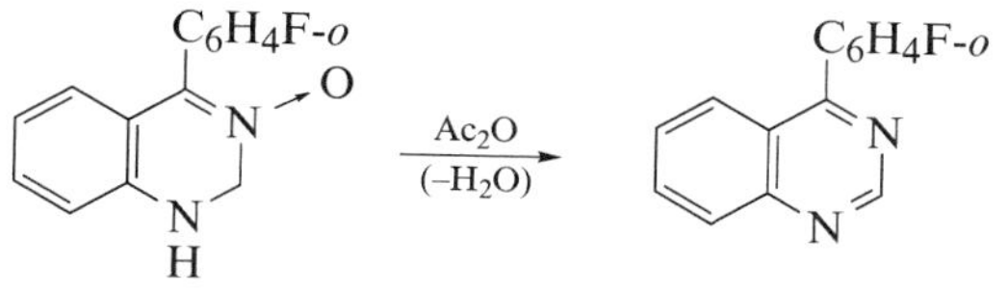

(215)   **(216)**

6-Chloro-2-methyl-4-phenyl-1,2-dihydro-2-quinazolinecarboxylic acid 3-oxide (**217**) gave 6-chloro-2-methyl-4-phenylquinazoline (**218**) (neat, N$_2$, 137° → 160°, 9 min: 49%).[490]

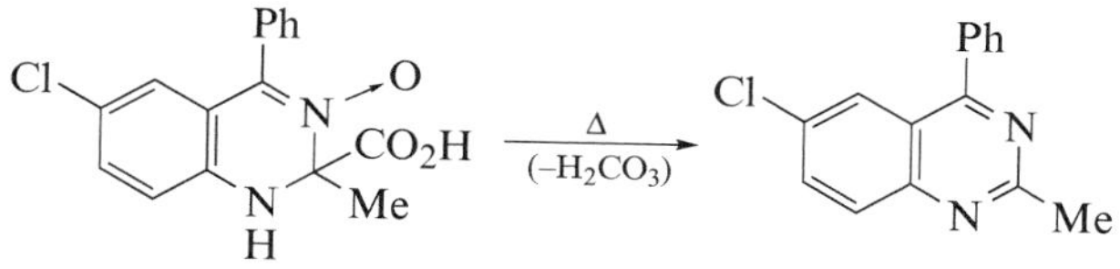

(217)   **(218)**

2-Methylquinazoline 3-oxide gave the unisolated adduct (**219**) and thence spontaneously, 2-methyl-4-quinazolinecarbonitrile (**220**) {Me$_3$SiCN, Et$_3$N (or 1,8-diazobicyclo[5,4,0]-7-undecene), THF, reflux, 2 h (or 30 min): 67% (or 77%)};[2110] also analogs.[2110]

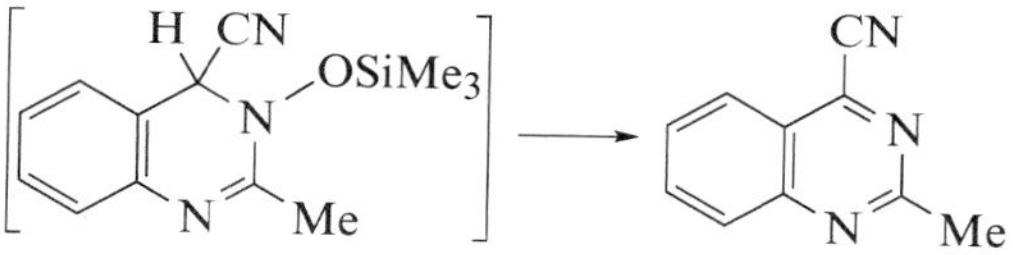

(219)   **(220)**

Also other examples.[454]

By regular reductions

2-Anilino-6-chloro-4-phenylquinazoline 3-oxide gave its parent quinazoline (H$_2$, Pd/C, EtOH?, 30 min: 60%).[399]

4-Methylquinazoline 1-oxide gave its parent quinazoline (H$_2$, Ni, MeOH: 43%, as picrate).[163]

1-Hydroxy-3-methyl- (**221**, R = OH) gave 3-methyl-2,4(1*H*,3*H*)-quinazolinedione (**221**, R = H) (Zn, HCl, EtOH–H$_2$O, reflux, 90 min: ~ 55%).[212]

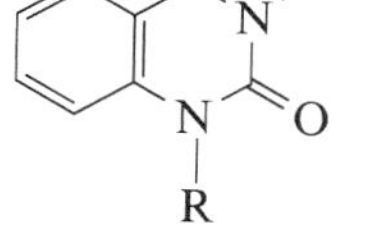

(221)

4-Methyl-2-quinazolinamine 3-oxide gave the parent methylquinazolinamine (TiCl$_3$, MeOH–H$_2$O, 0°, ~ 15 min: 83%).[267]

6-Methoxy-2-quinazolinamine 3-oxide gave its parent quinazolinamine (Fe powder, FeSO$_4$, EtOH–H$_2$O, reflux, 8 h: 93%).[4]

Also other examples using H$_2$S,[899] Ni catalyst without H$_2$,[875] electrolytic reduction,[2399] and triethyl phosphite.[578]

By phosphorus trichloride treatment

2-β,β-Di(ethoxycarbonyl)propyl-4-phenylquinazoline 3-oxide gave its parent quinazoline (PCl$_3$, CHCl$_3$, reflux, 30 min: 60%).[632]

6-Chloro-2-(β,β-dimethylhydrazono)methyl-4-phenylquinazoline 3-oxide gave its parent quinazoline (neat PCl$_3$, 20°, 1 h: 62%).[176]

6-Chloro-4-phenyl-2-quinazolinecarbaldehyde 3-oxide oxime (**222**) gave 6-chloro-4-phenyl-2-quinazolinecarbonitrile [neat PCl$_3$ (exothermic), 30 min: 67%; note additional dehydration of the oxime].[361]

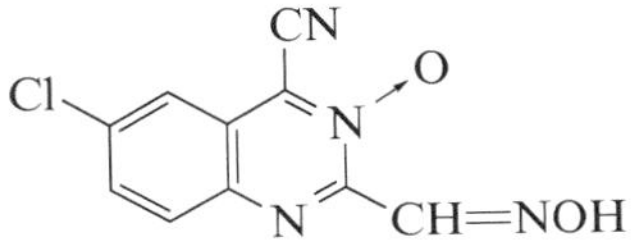

(222)

Also other examples.[82,874]

4.7.2.2.   *O*-Acylation or *O*-Alkylation (*H* 458)

These reactions can occur only when the *N*-oxide function is in equilibrium with its *N*-hydroxy form, by virtue of a mobile proton from elsewhere in the molecule. Examples follow.

Acylations

1-Hydroxy- (**223**, R = H) gave 1-acetoxy-2,4(1*H*,3*H*)-quinazolinedione (**223**, R = Ac) (Ac$_2$O, reflux, 4 h: 55%),[182,471] but the same substrate (**223**, R = H) gave 8-mesyloxy- (**224**, R = Ms) (MsCl, pyridine, 20°, 72 h: 30%) or 8-tosyloxy-2,4(1*H*,3*H*)-quinazolinedione (**224**, R = Ts) (TsCl, pyridine, 20°, 48 h: 39%).[471]

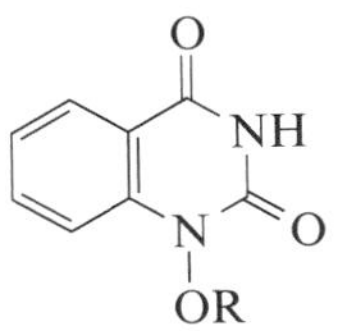

     (223) 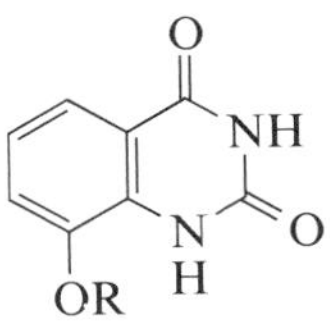   **(224)**

2-Methyl-4(1*H*)-quinazolinone 3-oxide gave 3-acetoxy (**225**, R = Ac) (Ac$_2$O, reflux, 2 h: 80%) or 2-benzoyloxy-2-methyl-4(3*H*)-quinazolinone (**225**, R = Bz) (BzCl, 1.25M NaOH, 20°, 10 min: 75%).[236]

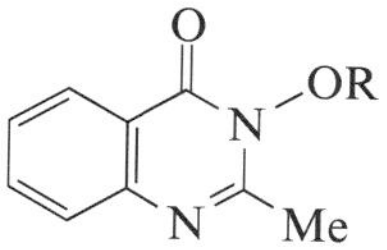

(225)

1-Hydroxy-3-methyl- gave 3-methyl-1-palmitoyloxy-2-thioxo-1,2-dihydro-4(3*H*)-quinazolinone (no details: 96%).[1168]

Also other examples.[187,457,2209] (cf. Section 4.7.2.4).

Alkylations

2-Phenyl-4(1*H*)-quinazolinone 3-oxide, as its Na salt (**226**), gave 3-benzyloxy-2-phenyl-4(3*H*)-quinazolinone (**226**, R = CH$_2$Ph) (PhCH$_2$Br, Me$_2$NCHO, 75°, 20 min: 86%).[851,1343] On photolysis in benzene or thermolysis, the product gave 2-phenyl-4(3*H*)-quinazolinone and benzaldehyde in good yield; analogs were made similarly.[851,1343]

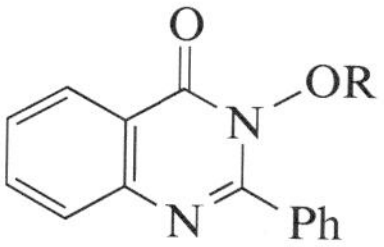

(226)

3-Hydroxy- gave 3-methoxy-2,4(1*H*,3*H*)-quinazolinedione (Me$_2$SO$_4$, NaOH, H$_2$O, 80°, 2 h: 86%).[457]

4-Quinazolinamine 3-oxide (**227**) gave a separable mixture of 3-methoxy-4(3*H*)-quinazolinimine (**228**, R = H) and 3-methoxy-4-methylimino-3,4-dihydroquinazoline (**228**, R = Me) (CH$_2$N$_2$, Et$_2$O–MeOH: 19% and 18%, respectively).[554]

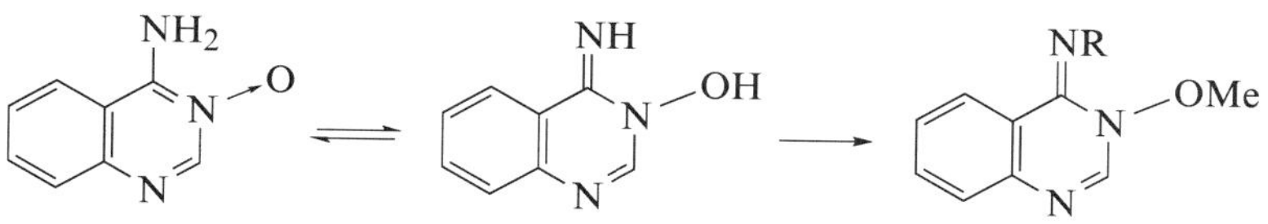

(227)   **(228)**

Also other examples.[236,660,825]

4.7.2.3. Reactions Involving Ring Scission (*H* 451,455,456,460)

Treatment of 2-methyl-4-phenylquinazoline 3-oxide with diethyl acetylene-dicarboxylate resulted in a 1,3-dipolar addition and subsequent *scission* to afford ethyl 2-(*o*-acetamidophenyl)-3-amino-3-phenylacrylate(**229**) as the main product (∼ 20%) with ethyl 2-methyl-4-phenyl-1*H*-1,3-benzodiazepine-5-carboxylate (**230**) as a byproduct (from its cyclization?).[57] Quinazoline 3-oxide (**231**) with ethyl cyanoacetate in refluxing benzene gave three separable products: *o*-for-mamidobenzaldehyde oxime (**232**), ethyl α-cyano-*o*-(β-cyano-β-ethoxycarbonyl-vinylamino)cinnamate (**233**), and ethyl 2-amino-3-quinolinecarboxylate (**234**), each in <5% yield.[365]

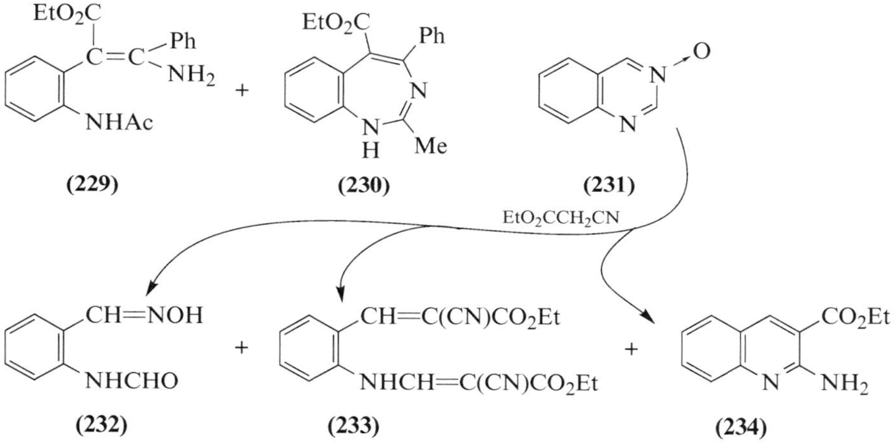

2-Phenyl-1,2-dihydro-4-quinazolinamine 3-oxide (**236**) underwent a themally induced *Dimroth rearrangement* to give 4-hydroxyamino-2-phenyl-1,2-dihyd-roquinazoline (**235**) (neat, 195°, 2 min: 64%; PhMe, 140°, sealed, 4 h: 87%);[1828] the same substrate (**236**) in acetic anhydride underwent a similar rearrangement, replacement of (PhC2) by (MeC2), and concomitant oxidation to give 4-acetoxyamino-2-methylquinazoline (**237**) plus benzaldehyde (Ac₂O, Et₂O, 20°, 4 h: ∼ 50%);[61] and 4-amino-2(1*H*)-quinazolinone 3-oxide gave 4-hydroxyamino-2(1*H*)-quinazolinone (95%) by refluxing in dimethylformamide for 4 h.[1857,1858]

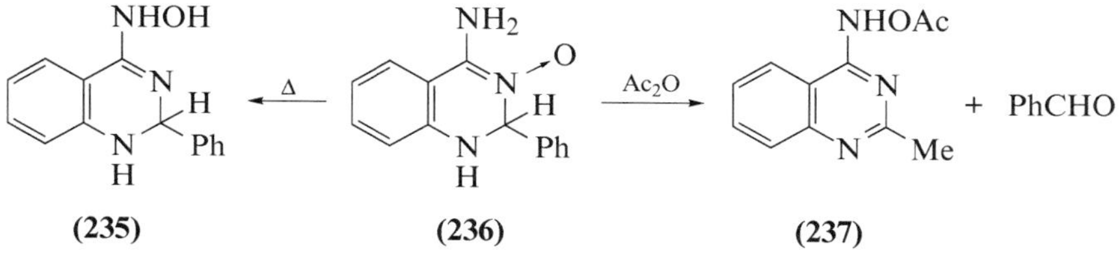

Two types of *ring expansion* have been reported. Irradiation of 6-chloro-2-methyl-4-phenyl- (**238**, R = Me, Y = Cl) or 2,4-diphenylquinazoline 3-oxide (**238**, R = Ph, Y = H) in benzene gave 8-chloro-4-methyl-2-phenyl- (**239**, R = Me, Y = Cl) or 2,4-diphenyl-1,3,5-benzoxadiazepine (**239**, R = Ph, Y = H), respectively, each in >80% yield.[248,1024] However, irradiation of the 2-unsubstituted substrate, 6-chloro-4-phenylquinazoline 3-oxide (**238**, R = H, Y = Cl), gave only 6-chloro-2-phenylbenzoxazole (**240**) in 25% yield, probably by degradation of the initial product (**239**, R = H, Y = Cl).[248, cf. 28] In a quite different way, the treatment of 6-chloro-2-chloromethyl-4-*p*-methoxyphenylquinazoline 3-oxide (**242**, R = OMe) in methanolic ammonia gave a 54% yield of 7-chloro-5-*p*-methoxyphenyl-1*H*-1,4-benzodiazepin-2-amine 4-oxide (**241**);[367] however, the analogous substrate, 6-chloro-2-chloromethyl-4-phenylquinazoline 3-oxide (**242**, R = H), with methanolic hydrazine hydrate, gave 8-chloro-2-hydroxyamino-6-phenyl-3,4-dihydro-1,4,5-benzotriazocine (**243**) in 57% yield.[366]

(238)   (239)   (240)

(241)   (242)   (243)

4.7.2.4. Minor Reactions

Several *rearrangements*, apparently without ring scission, have been reported recently. Thus 2-methyl-4(3*H*)-quinazolinone 1-oxide (**244**) was converted rapidly in hot acetic anhydride into 2-acetoxymethyl-4(3*H*)-quinazolinone (**245**)[187] (cf. Section 4.7.2.2). The sodium salt (**246**?) of 2-phenyl-4(1*H*)-quinazolinethione 3-oxide reacted slowly with *N*-(*p*-tolyl)benzimidoyl chloride (**247**) in dimethylformamide at 20° to give 4-(*N*-benzoyl-*p*-tolylamino)thio-2-phenylquinazoline (**249**) in 87% yield, presumably via an intermediate (**248**).[1986] Photolysis of quinazoline 3-oxide induced isomerization into 4(3*H*)-quinazolinone (no details).[28]

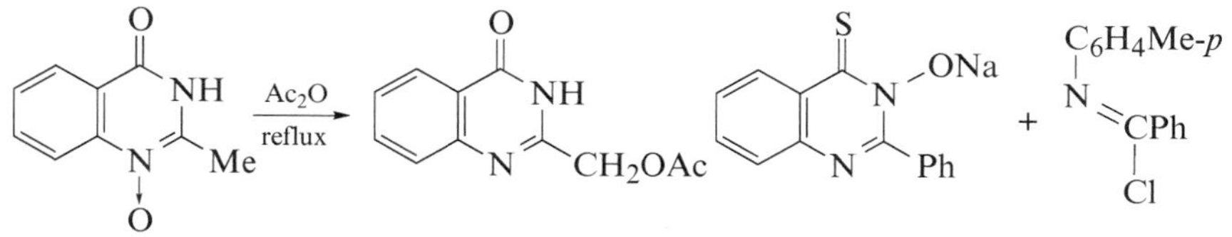

(244)   (245)   (246)   (247)

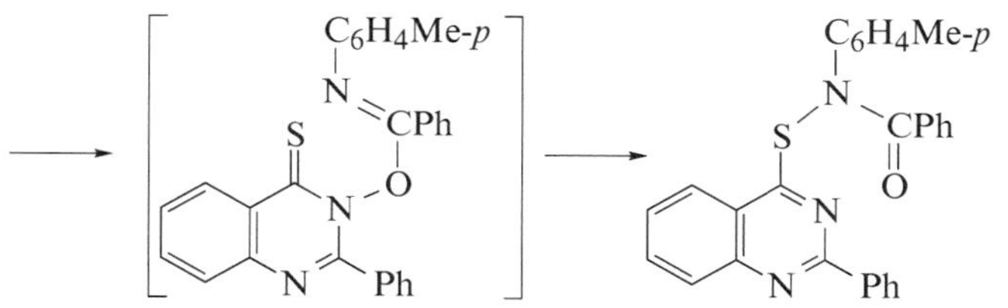

(248)   (249)

A facile *cyclization*, directly involving an *N*-oxide grouping, occurred on treatment of 2-methyl-4-quinazolinamine 3-oxide (**250**) with phosgene to give 5-methyl-2*H*-[1,2,4]oxadiazolo[2,3-*c*]quinazolin-2-one (**251**) in good yield;[876] the structure was checked by X-ray analysis, and the same product was obtained by rearrangement and cyclization of *N*-hydroxy-2-methyl-4-quinazolinecarboxamide 3-oxide in acetic anhydride.[876]

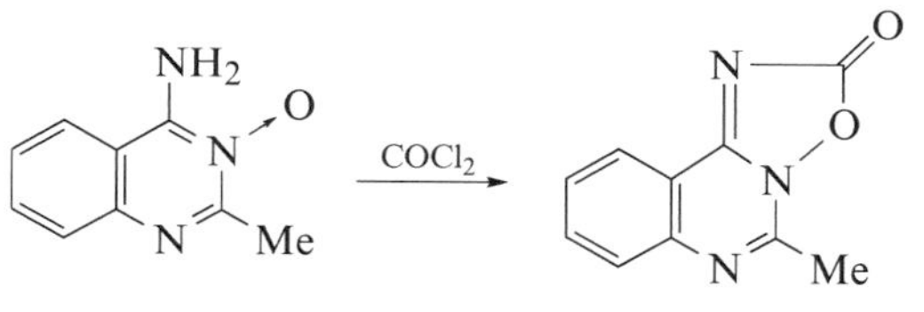

(250)   (251)

The *deuterium exchange* rates for quinazoline 1,3-dioxide and related heterocyclic *N*-oxides have been studied under alkaline conditions.[528]

4.8. ANCILLARY ASPECTS OF OXYQUINAZOLINES (*H* 490)

Because many individual papers on the structure, spectra, or natural occurrence of oxyquinazolines involve various types thereof, it seems appropriate to review the bulk of such recent information briefly in one central location.

4.8.1.   X-Ray Structural Analyses for Oxyquinazolines

The following gross structures of oxyquinazolines have been determined or confirmed by X-ray analyses.

2-Acetyl-4(3H)-quinazolinone.[1268, cf. 684]

3-Amino-6-bromo-1-methyl-2,4(1H,3H)-quinazolinedione.[928]

2-p-Bromophenyl-4a,5,6,7,8,8a-hexahydro-4(3H)-quinazolinone.[1133]

3-o-Chlorophenyl-2-phenacyl-4(3H)-quinazolinone.[2443]

3-o-Chlorophenyl-2-(pyridin-4-ylcarbonyl)methyl-4(3H)-quinazolinone.[2443]

3-p-Chlorophenyl-2-(pyridin-4-ylcarbonyl)methyl-4(3H)-quinazolinone.[2443]

1-Cyclopropylmethyl-6-methoxy-4-phenyl-2(1H)-quinazolinone.[713]

2,2-Dimethyl-3-isopropylideneamino-1,2-dihydro-4(3H)-quinazolinone.[1140]

3-β-(4-o-Methoxyphenylpiperazin-1-yl)ethyl-2,4(1H,3H)-quinazolinedi-
one.[1791]

3-Methyl-2-p-nitrophenyl-1,2,4a,5,6,7,8,8a-octahydro-4(3H)-quinazoli-
none.[1961]

2-p-Nitrophenyl-1,2,4a,5,6,7,8,8a-octahydro-4(3H)-quinazolinone.[1961]

2-Phenyl-4a,5,6,7,8,8a-hexahydro-4(3H)-quinazolinone.[1134]

3-Phenyl-2-(pyridin-4-ylcarbonyl)methyl-4(3H)-quinazolinone.[2443]

3-Phenyl-2,4(1H,3H)-quinazolinedione.[1]

2-Phenylquinazoline 1,3-dioxide.[1100]

2-(Pyridin-4-ylcarbonyl)methyl-3-o-tolyl-4(3H)-quinazolinone.[2443]

Also other structures mentioned specifically elsewhere (see Index).

4.8.2.   Studies of Fine Structure in Oxyquinazolines (H 73,102,128,446)

There has been little recent quantitative evidence to confirm or deny our generally accepted wisdom on the fine structure of quinazolinones or quinazoline N-oxides, much of which appears to have been based on analogy with corresponding pyrimidines and the like. However, contributions have been made to our knowledge of fine structure in specific compounds or small groups thereof, as indicated in the following listing.

The ^{13}Cnmr spectra of 4(3H)-quinazolinone and its 1-alkyl, 2-alkyl, 3-alkyl, 1,2-dialkyl, 2,3-dialkyl, and related derivatives confirmed their essentially oxo structure in dimethyl sulfoxide and (in some cases) suggested additional minor contributors to the equilibria.[1367,1770]

The ir, ^{1}Hnmr, and uv spectra of 2-benzyl-1-methyl-4(1H)-quinazolinone (and related models) have been used to disprove a former quaint belief that it existed as the tautomeric 2-benzylidene-1-methyl-1,2-dihydro-4(3H)-quinazolinone (**252**).[2380]

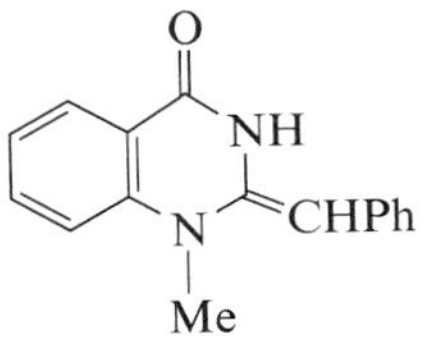

(**252**)

Calculated *aromaticity indices* for possible tautomers confirmed the existence of 4(3H)-quinazolinone as such;[2362] other diverse calculations reached similar conclusions in respect of 2(1H)-quinazolinone and 2,4(1H,3H)-quinazolinedione.[2229,2507]

The *ms fragmentation* pattern from 2-methyl-4(1H)-quinazolinone 3-oxide contained a very abundant m/e 146 ion from an initial loss of nitric oxide; this might suggest that the substrate exists as an N-oxide rather than an N-hydroxy tautomer;[15] *ir evidence* has been used to suggest that 4(1H)-quinazolinone 3-oxide exists as the N-oxide, formulated as (**253**).[532]

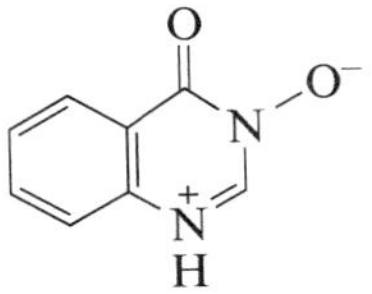

(**253**)

4.8.3.   General Spectral and Related Studies on Oxyquinazolines

Only papers that describe essentially nonroutine applications of spectral or related physical methods to oxyquinazolines are included in the following listings.

Fluorescence spectra

The fluorescence emission yield was far greater from a β,β,β-trifluoroethanolic solution of 4(3H)-quinazolinone than from an ethanolic solution thereof; this proved to be general for azaheterocyclic but not for homocyclic solutes, thereby offering a useful diagnostic test.[2180]

Gas chromatography

Retention times for 4(3H)-quinazolinone and about 35 of its 2-alkyl, 3-alkyl, and 2,3-dialkyl derivatives showed good correlation with the size and location of alkyl groups.[2272]

Ionization

The acidic pK_a values for 4a,5,6,7,8,8a-hexahydro-2,4(1H,3H)-quinazo-linedione (*cis* 11.8; *trans* 12.1) have been compared with those for six 5/6-methylated 5,6-dihdyro-2,4(1H,3H)-pyrimidinediones.[606] The acidic and basic pK_a values for 2-alkyl-4(3H)- and 4-alkyl-2(1H)-quinazolinones have been determined for correlation with polarographic reduction poten-tials of 2-alkyl-4-chloro- and 4-alkyl-2-chloroquinazolines, respectively.[751]

Infrared spectra

The unusual abilities of 3-phenyl-2-pyridin-2′-yl-4(3H)-quinazolinone and seven analogs to form well-defined intermolecular hydrogen bonds with phenol, p-chlorophenol, or benzyl alcohol have been quantified by measur-ing the appreciable shifts in ir frequencies for HO stretching.[161]

Mass spectra

Most of the reported mass spectra have been for fixed quinazolinones, like 2-methyl-3-o-tolyl-4(3H)-quinazolinone (methaqualone) and its hypnotic analogs, usually in connection with metabolic studies or identification of those on the illicit-drug market. The range of reported spectra is indicated hereafter.

3-[4-(p-Acetylphenyl)piperazin-1-ylcarbonyl]methyl-1-methyl-2,4(1H,3H)-quinazolinedione and six analogs.[2157]

1-Allyl-5,6,7,8-tetrahydro-2,4(1H,3H)-quinazolinedione and the 3-allyl isomer.[1803]

3-o-Chlorophenyl-2-methyl-4(3H)-quinazolinone and eight halogenophenyl analogues.[996,1129,1619]

2-α,β-Dibromophenethyl-3-phenyl-4(3H)-quinazolinone and seven analogs.[1296]

2-Diphenylmethyl-3-phenyl-4(3H)-quinazolinone and seven p-substituted phenyl analogs.[1298]

2-Ethyl-3-(3-methylisoxazol-5-yl)-4(3H)-quinazolinone and 12 analogs.[1433]

3-(2-Methoxy-5-nitrophenyl)-2-methyl-4(3H)-quinazolinone and three ana-logs.[996,1621]

3-p-Methoxyphenyl-2-p-tolyl-1,2-dihydro-4(3H)-quinazolinone and seven analogs.[1297]

3-Methyl-6-nitro-7-piperidino-4(3H)-quinazolinone and two analogues.[1669]

2-Methyl-3-pyrazol-3′-yl-4(3H)-quinazolinone and 22 analogs.[2156,2338]

2-Methyl-3-o-tolyl-4(3H)-quinazolinone and 18 analogs,[294,996,1008,1129,1619,2157]

3-Phenyl-2,4(1H,3H)-quinazolinedione and two substituted phenyl ana-logs.[294,2444]

4-Phenylquinazoline 1-oxide and several analogs.[2510]

3-Phenyl-2-styryl-4(3H)-quinazolinone and five analogs.[1739]

2-Thioxo-1,2-dihydro-4(3H)-quinazolinone.[2444]

Molecular-orbital calculations

Calculations on 8-quinazolinol indicated that N1 is more basic than N3,[49] a finding contrary to earlier results based on HO stretching frequencies.[2509]

^{13}C nuclear-magnetic-resonance spectra

The ^{13}Cnmr spectra for the following quinazolines have been reported for reasons other than mere confirmation of structure.

1-Cyclopropylmethyl-6-methoxy-4-phenyl-2(1H)-quinazolinone and some 4-(substituted phenyl) analogs.[1061]

3-(3,4-Dihydroxyphenethyl)-2-methyl-4(3H)-quinazolinone and five derivatives.[967]

3-(3,4-Dimethoxyphenethyl)-2-methyl-4(3H)-quinazolinone and five derivatives.[967]

3-Methyl-2,4(1H,3H)-quinazolinedione and nine 3-alkyl/aryl analogs.[1446]

2-Methyl-3-o-tolyl-4(3H)-quinazolinone and 11 analogs, mostly hydroxylated in order to identify metabolites.[886,962]

1-Phenyl-2-thioxo-1,2-dihydro-4(3H)-quinazolinone and several analogs.[2426]

5,6,7,8-Tetrahydro-2,4(1H,3H)-quinazolinedione.[2343]

^{1}H nuclear-magnetic-resonance spectra

The ^{1}Hnmr spectra for the following quinazolines have been reported for reasons other than characterization or structural confirmation on a routine basis.

2-Benzyl-3-o-bromophenyl-4(3H)-quinazolinone and seven analogs, for study of hindered internal rotation.[702]

3-(2,4-Dimethylphenyl)-2-methyl-4(3H)-quinazolinone, for identification of this methaqualone analog.[2442]

4a,5,6,7,8,8a-Hexahydro-2,4(1H,3H)-quinazolinedione, for conformational analysis.[44]

4,4,5,5,7-Pentamethyl-3,4,4a,5-tetrahydro-2(1H)-quinazolinone, as a proposed structure for "tetraacetone urea."[285]

2,4(1H,3H)-Quinazolinedione and thirteen 1-, 3-, and/or 6-substituted analogs.[362]

5,6,7,8-Tetrahydro-2,4(1H,3H)-quinazolinedione.[2343]

Polarography

Polarographic reduction of 2-methyl-3-o-tolyl-4(3H)-quinazolinone has been studied as a means to determine methaqualone in biofluids.[1797]

Ultraviolet spectra

The uv spectra for 2-amino-4(3H)-quinazolinone, the corresponding pteridinone (pterin), and several other deazapterins have been analyzed in detail.[2487]

The uv spectra of 2,3-dimethyl-4(3H)-quinazolinone and some 25 homologs and analogs have been examined for common characteristics.[133]

4.8.4. Naturally Occurring Oxyquinazolines (H 490)

Most naturally occurring oxyquinazolines have been isolated from plant materials, but a few have come from other sources such as microorganisms. Those from plants are usually called *quinazoline alkaloids*, but this term also includes many products that are not true quinazolines but fused quinazolines, and hence are nominally excluded from this book. In fact, the true oxyquinazolines that occur naturally appear to be of minor chemical or even biological interest. Accordingly, those that have been discussed in recent literature or included in periodic reviews[288,1798,1799,2013,2014,2081,2133,2383,2440] are simply listed under their trivial names, along with leading references; to avoid confusion, some fused quinazolines from natural sources are listed separately.

2-Acetyl-4(3H)-quinazolinone (no trivial name) from the fungi, *Alternaria citri* and *Fusarium culmorum*.

7-Bromo-2,4(1H,3H)-quinazolinedione (no trivial name) from the marine tunicate, *Pyura sacciformis*.[1967]

Arborine, 2-benzyl-1-methyl-4(1H)-quinazolinone, from *Glycosmis arborea*.[35,118,754,821,831,908,1367,1807,2380]

Chrysogine, 2-α-hydroxyethyl-4(3H)-quinazolinone, from *Penicillium chrysogenum*.[831,1026,2069]

Echinozolinone, probably 3-β-hydroxyethyl-4(3H)-quinazolinone, from *Echinops echinatus*.[940,2030,2557]

Febrifugine, *trans*-3-γ-(3-hydroxypiperidin-2-yl)acetonyl-4(3H)-quinazolinone, from the common Easter hydrangea.[451,452]

Glomerine, 1,2-dimethyl-4(1H)-quinazolinone, from the millipede, *Glomeris marginata*.[754,831,908,945,946,1026]

Glycophymine; see "Glycosminine."[961,1227,2511]

Glycophymoline, 2-benzyl-4-methoxyquinazoline, from *Glycosmis pentaphylla*.[1227]

Glycorine, 1-methyl-4(1H)-quinazolinone, from *Glycosmis arborea*.[35,831,918,1367,1807]

Glycosine; see "Arborine."[2380]

Glycosmicine, 1-methyl-2,4(1H,3H)-quinazolinedione, from *Glycosmis arborea*.[35]

Glycosmine; see "Glycosminine."[831]

Glycosminine, 2-benzyl-4(3H)-quinazolinone, from *Glycosmis arborea* and *G. pentaphylla.*[35,118,908,961,1227,2511]

Homoglomerine, 2-ethyl-1-methyl-4(1H)-quinazolinone, from *Glomeris marginata.*[831,945,1026]

7-Hydroxyechinozolinone, 7-hydroxy-3-β-hydroxyethyl-4(3H)-quinazolinone (?), from *Echinops echinatus.*[2401,2557]

3-*p*-Methoxyphenethyl-1-methyl-2,4(1H,3H)-quinazolinedione (no trivial name) from *Zanthoxylum arborescens.*[1243,1361]

1-Methyl-3-phenethyl-2,4(1H,3H)-quinazolinedione (no trivial name) from *Z. arborescens.*[1243,1361]

Pegamine, 2-γ-hydroxypropyl-4(3H)-quinazolinone, from the plant *Peganum harmala.*[300]

Tryptoquivaline-G and several analogs, all 4(3H)-quinazolines with complicated heterocyclic 3-substituents, from the fungus *Aspergillus fumigatus.*[1322]

Some naturally occurring fused oxyquinazolines

Aniflorine.[36]

Anisessine.[36]

Anisotine.[36]

Deoxyaniflorine.[36]

Euxylophoricine-A.[831,913,1026]

Euxylophoricine-C.[831,913,1026]

Evodiamine.[913,1324]

Rutaecarpine.[831,913,1760]

Sessiflorine.[36]

Tetrodotoxin.[1899,2154,2425]

CHAPTER 5

Thioquinazolines

The term *thioquinazoline* covers any quinazoline with a sulfur-containing substituent that is attached directly or indirectly to the quinazoline ring system through its sulfur atom. The thioquinazolines form a closely knit family. Thus a parent tautomeric quinazolinethione, quinazolinethiol, or mercaptoalkylquinazoline may be S-alkylated to an alkylthioquinazoline (thioether, RSR'), which may then be oxidized to an alkylsulfinyl- [sulfoxide, RS(=O)R'] or an alkylsulfonylquinazoline [sulfone, RS(=O)$_2$R']; alternatively, the parent may be oxidized directly to a diquinazolinyl disulfide (RSSR), a quinazolinesulfenic acid (S-oxide, RSOH), quinazolinesulfinic acid (RSO$_2$H), or a quinazolinesulfonic acid (RSO$_3$H). The nontautomeric quinazolinethiones, a few known selenium-containing analogs, and diquinazoline sulfides are included in this chapter, but discussion of thiocyanatoquinazolines (RSC≡N) is deferred to Chapter 7 in order to be included with isothiocyanatoquinazolines (RN=C=S), quinazolinecarbonitriles (RC≡N), and related entities.

5.1. QUINAZOLINETHIONES (*H* 270)

Since nearly all the available information on nontautomeric (fixed) quinazolinethiones has been discussed already, any further material on such compounds is included here with material on the tautomeric 2- and 4-quinazolinethiones.

5.1.1. Preparation of Quinazolinethiones

Most tautomeric and fixed quinazolinethiones have been made by either *primary synthesis* (see Chapter 1) or *phosphorus pentasulfide thiation* of corresponding quinazolinones (see Sections 4.1.2.1 and 4.6.2.1); in addition, some tautomeric quinazolinethiones have been made by *thiolysis* of halogenoquinazolines (see Section 3.5.4). The remaining minor synthetic routes are illustrated in the following examples.

Direct introduction of sulfur

4(3*H*)-quinazolinone (1, R = H), its 3-methyl derivative (1, R = Me), or its 1-methyl derivative (3) gave 2-thioxo- (2, R = H), 3-methyl-2-thioxo- (2,

R = Me), or 1-methyl-2-thioxo-1,2-dihydro-4(3*H*)-quinazolinone (**4**) (neat S, fusion at 225°, 20 min: 91%, 60%, or 81%, respectively; or S, Me$_2$NCHO, reflux, ?h: ?% yields).[1846,1862,cf.1864,2197]

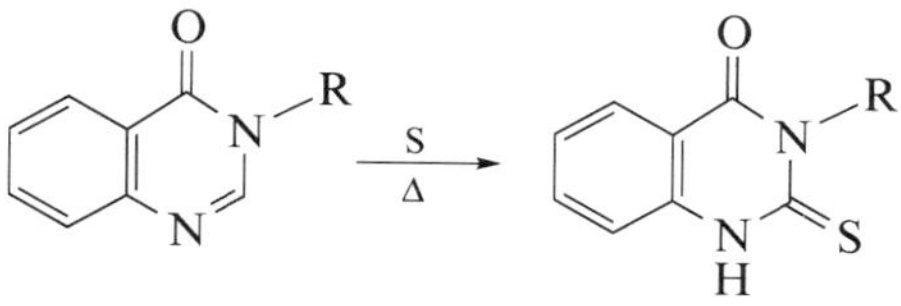

(1)   (2)

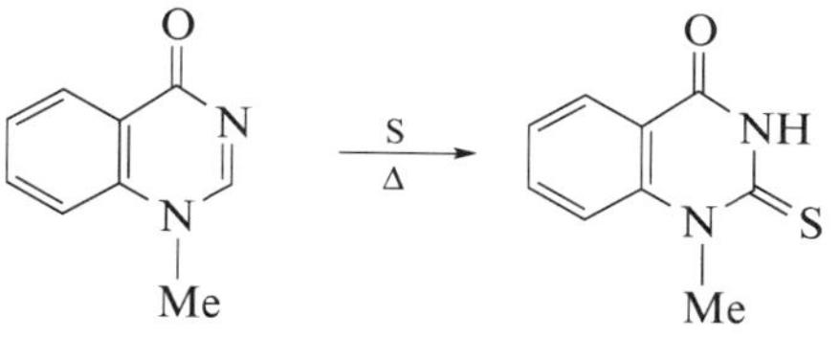

(3)   (4)

1-Ethoxycarbonylmethyl-3-*p*-methoxybenzyl-1,2-dihydro-4(3*H*)-quinazolin-one (**5**) gave 1-ethoxycarbonylmethyl-3-*p*-methoxybenzyl-2-thioxo-1,2-dihydro-4(3*H*)-quinazolinone (**6**) (neat S, 180°, 30 min: 35%; note included oxidation.[2262]

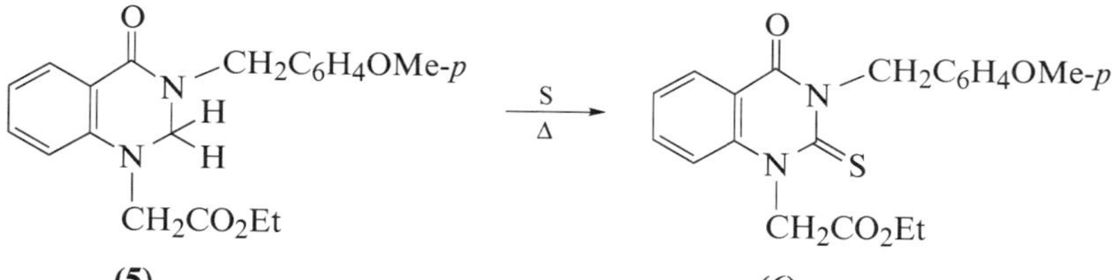

(5)   (6)

Also other examples.[720,1725]

From alkylthioquinazolines or disulfides

4-Benzylthioquinazoline gave 4(3*H*)-quinazolinethione (substrate.HClO$_4$, NaBH$_4$, pH 3–4, 12h: 15% after separation from other products).[539]

4-Cyclohex-2′-enylthioquinazoline gave 4(3*H*)-quinazolinethione (48% HBr, reflux; HClO$_4$–H$_2$SO$_4$, 20°, 30 min; HClO$_4$, MeOH, 20°, 7 days; no yields given).[1221]

2-Benzylthio-4-methylthioquinazoline gave 4-methylthio-2(1*H*)-quinazoline-thione (AlCl$_3$: note preferential splitting of the benzyl thioether group-ing).[1948]

2-Methylthio-4(3H)-quinazolinone gave 2-selenoxo-1,2-dihydro-4(3H)-quin-
azolinone (7) (NaSeH, ? solvent: 59%).[1833]

(7)

Bis(4-oxo-3,4-dihydroquinazolin-2-yl) disulfide gave 2-thioxo-1,2-dihydro-
4(3H)-quinazolinone (glucose, EtOH–H$_2$O, slightly alkaline, 100°, 4h:
good yield); also 3-aryl derivatives.[535]

5.1.2.   Reactions of Quinazolinethiones (H 271, 273, 282)

As expected, X-ray analysis of 2-thioxo-1,2-dihydro-4(3H)-quinazolinone
(2, R = H) indicated that it existed as such in the solid state.[1993] The effects of
4-thiation on the ^{1}Hnmr spectra of 7-bromo-2,3-diphenyl-4(3H)-quinazolinone
and analogs proved to be especially pronounced on signals for H5 and for H2′
and H6′ of the 2-phenyl substituent.[1997, cf. 2573]
   The conversion of tautomeric or fixed quinazolinethiones into the corresponding quinazolinones by *regular or oxidative hydrolysis* has been discussed already
(Sections 4.1.1.2 and 4.6.1). The remaining reactions, some of considerable
importance, are covered in the following subsections.

5.1.2.1.   S-Acylation and S-Alkylation (H 280)

Although these processes are usually applied only to tautomeric quinazolinethiones, it is possible to S-alkylate fixed quinazolinethiones, provided that the
product is acceptable as a quaternary salt.
   In fact, *S-acylation* has not been used much, despite the fact that it appears to
occur readily and in good yield. For example, 3-phenyl-2-thioxo-1,2-dihydro-
4(3H)-quinazolinone (8) with p-chlorophenoxyacetyl chloride in ethanolic alkali
under reflux for 6 h, gave a 65% yield of 2-p-chlorophenoxyacetylthio-3-phenyl-
4(3H)-quinazolinone (9), and analogs were made similarly;[1574] also 2-thioxo-1,2-
dihydro-4(3H)-quinazolinone with benzoyl chloride and triethylamine in
dimethylformamide at 20° for 1h, gave 2-benzoylthio-4(3H)-quinazolinone in
53% yield.[1836] In less simple ways, 2-phenyl-4(3H)-quinazolinethione (10) with
tosyl isothiocyanate at 140° for 3h gave the thioacetylthio intermediate (11),
which lost CS$_2$ to afford 2-phenyl-4-tosylaminoquinazoline (12) in 60% yield;[697]
and 3-amino-2-thioxo-1,2-dihydro-4(3H)-quinazolinone (13) with ethyl car-
bazate gave 62% of 2-hydrazino-5H-1,3,4-thiadiazolo[2,3-b]quinazolin-5-one

(14), probably via an acylthio intermediate.[1816] Other examples have been reported.[1750,2161]

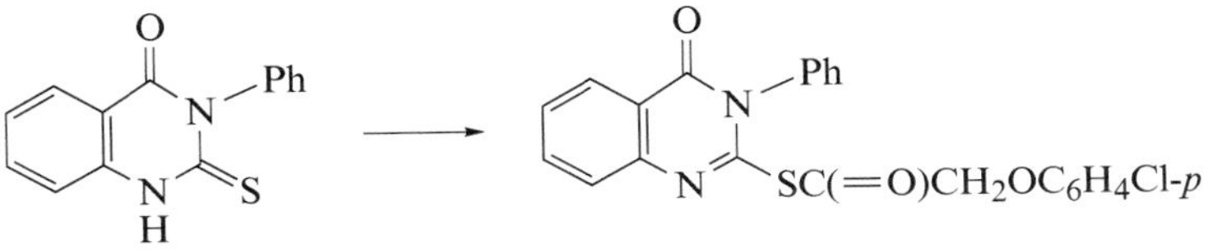

(8)   (9)

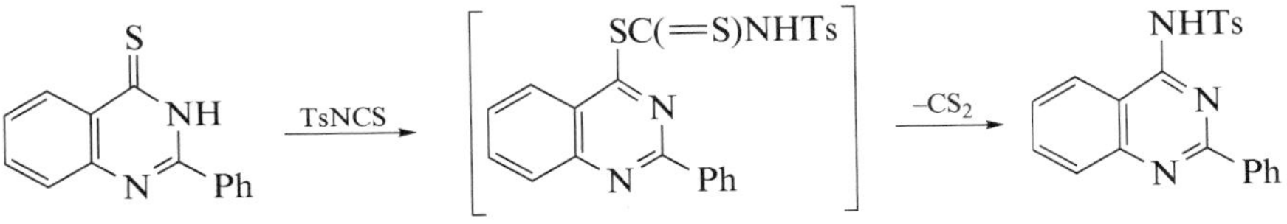

(10)   (11)   (12)

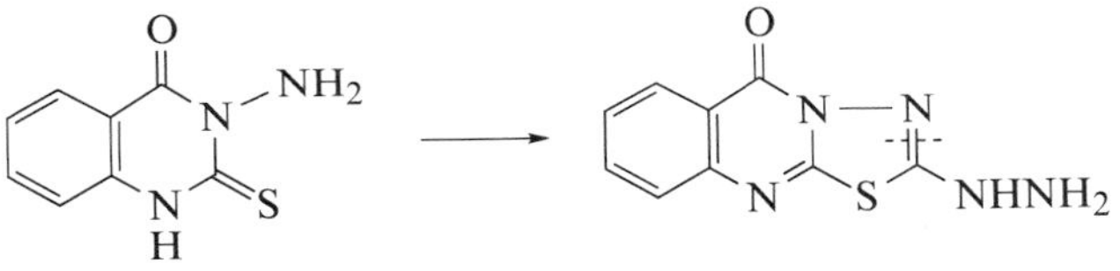

(13)   (14)

In contrast, *S-alkylation* has been used extensively. It is usually done by treatment of the quinazolinethione with an alkyl, substituted-alkyl, or activated aryl halide in the presence of a base under quite gentle conditions; occasionally, other types of alkylating agent have been used. An additional functional group in the pyrimidine ring may affect the ease of alkylation, but such groups attached elsewhere to the substrate molecule have such little effect that they have been ignored in classification of alkylations, leading to the following typical alkyl-thioquinazolines.

Simple 2/4-alkylthioquinazolines

4-Methylthioquinazoline   (**15**,   R = H)   (CH$_2$N$_2$,   Et$_2$O–MeOH–
    MeOCH$_2$CH$_2$OCH$_2$CH$_2$OMe, 20°, 2 h: 25%); also the 2-isopropyl deriva-
    tive (**15**, R = Pr^i) (59%) and others.[104]

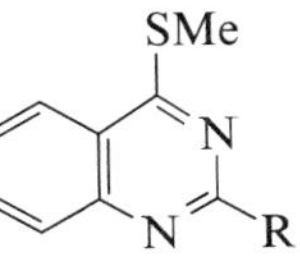

(15)

6,7-Dimethoxy-2-methyl-4-methylthioquinazoline (MeI, MeONa, MeOH, reflux, 3 h: 72%).[1354]

2-*p*-Methoxyphenyl-4-methylthioquinazoline (**15**, R = C$_6$H$_4$OMe-*p*) (MeI, THF, reflux, 3 h: 82%, as hydriodide).[968]

2-Methylthio-3,4-dihydroquinazoline (**16**) (MeI, EtOH, reflux, 45 min: 89%).[1569]

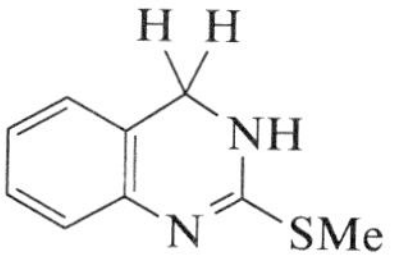

(16)

2,4-Bismethylthioquinazoline (**17**, R = Me) (MeI, NaOH, H$_2$O, 20°, 24 h: 87%);[1143,1948] also the bisethylthio homolog (86%).[1143]

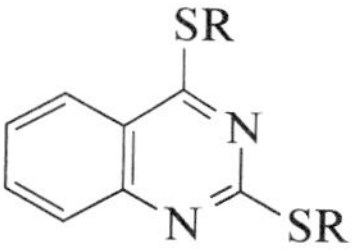

(17)

4-Methylthio-1-phenylquinazolinium iodide (**18**) [from 1-phenyl-4(1*H*)-quinazolinethione, neat MeI, 20°, 1 h: ?%].[1439]

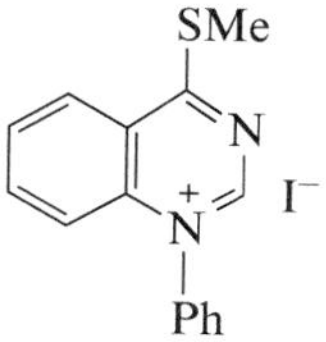

(18)

1,2-Bis(3,4-dihydroquinazolin-2-ylthio)ethane (**19**) [from 3,4-dihydro-2(1*H*)-quinazolinethione, BrCH$_2$CH$_2$Br, EtOH, reflux, 4 h: 57%].[2346]

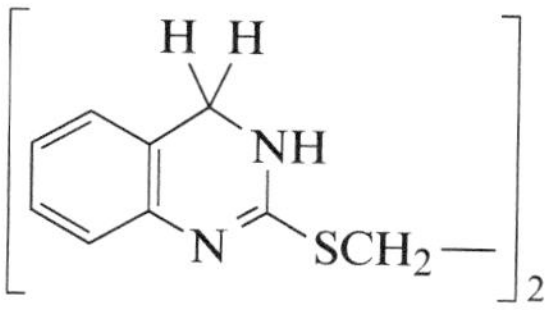

(19)

3-Methyl-2-methylthio-4(3*H*)-quinazolinethione (**21**) [by selective *S*-methylation of 3-methyl-2,4(1*H*,3*H*)-quinazolinedithione (**20**), MeI, NaOH, MeOH–H$_2$O, 20°, 12 h: 58%].[40]

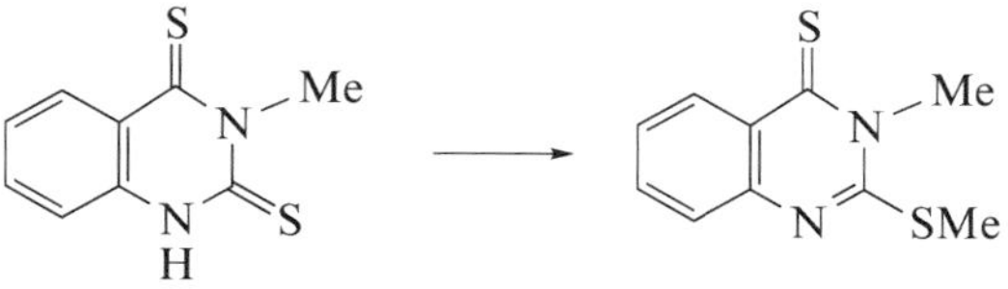

(20)   **(21)**

4-Methylthio-2-phenylquinazoline 3-oxide (MeI, Et$_3$N, MeOH, reflux, 2h: 79%).[1926]

Also other examples.[975,1073,1536,1567,2057,2076,2086,2209,2539]

Simple 2/4-alkylthio-4/2-quinazolinamines

6-Chloro-4-ethylthio-2-morpholinoquinazoline (EtI, EtONa, EtOH, reflux, 10 h: 77%).[413]

2-Methylthio-6-naphthalen-2′-ylthio-4-quinazolinamine (MeI, 1M KOH, 20°, 1 h: 86%).[673]

4-Methylthio-6-naphthalen-2′-ylsulfonyl-2-quinazolinamine (MeI, Me$_2$NCHO, 20°, 45 min: 43%).[1595]

3-Dimethylamino-4-methylthio-1-phenyl-1,2-dihydroquinazolinium iodide (**22**) [from 3-dimethylamino-1-phenyl-1,2-dihydro-4(3H)-quinazoline-thione, MeI, PhH, 20°, 12 h: ?%]; also analogs.[1229]

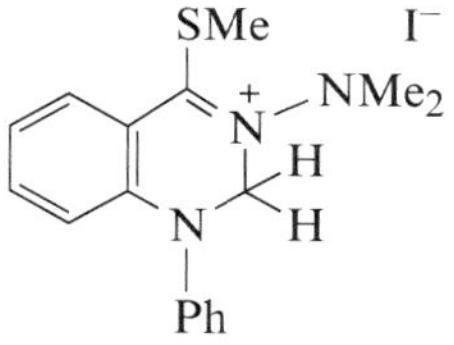

(22)

Also other examples.[2255]

Simple 2/4-alkylthio-4/2-quinazolinamines

2-Ethylthio-4(3H)-quinazolinone (**23**) (EtI, EtONa, EtOH, reflux, 10 min: 72%;[943] EtI, KOH, EtOH, reflux, 3 h: 75%).[1845]

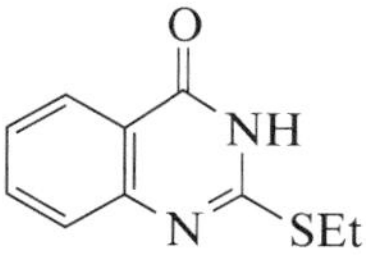

(23)

3-Methyl-2-propylthio-4(3H)-quinazolinone (PrBr, NaOH, EtOH–H$_2$O, 25°, 1 h: 60%);[160] also analogs.[160]

1-Methyl-4-methylthio-2(1H)-quinazolinone (**24**) (MeI, KOH, MeOH–H$_2$O, 20°, 30 min: 56%).[41]

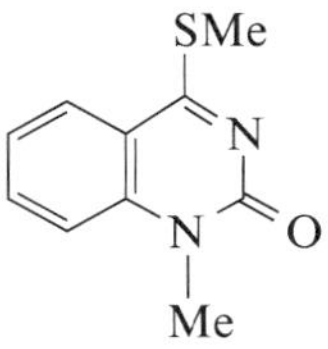

(24)

6-Chloro-2-ethylthio-3-phenethyl-4(3H)-quinazolinone (EtI, NaOH, EtOH–H$_2$O, 20°, 90 min: 74%).[615]

3-Allyl-2-methylthio-4(3H)-quinazolinone (MeI, K$_2$CO$_3$, Me$_2$NCHO, 20°, 12 h: 90%).[477]

2-Methylseleno-4(3H)-quinazolinone (MeI, NaOH, EtOH, ? h: 62%).[1833]

1,2-Bis(4-oxo-3-phenyl-3,4-dihydroquinazolin-2-ylthio)ethane (BrCH$_2$CH$_2$Br, AcONa, EtOH, reflux, 5 h: 80%).[1726]

1,4-Bis(2-ethylthio-4-oxo-3,4-dihydroquinazolin-3-yl)butane (**25**) (EtI, NaOH, EtOH–H$_2$O, 75°, 10 min: ?%).[371, cf. 547]

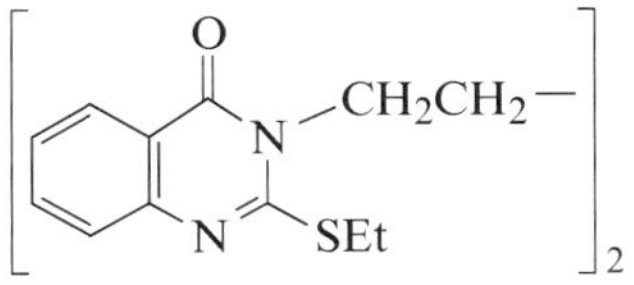

(25)

Also other examples.[115,131,135,296,450,522,529,566,576,588,644,670,845,1017, 1070, 1089,1137,1250,1483,1846,1996,2304,2570]

2/4-Alkenylthioquinazolines

4-Allylthio-2-phenylquinazoline (**26**) (CH$_2$=CHCH$_2$Br, Et$_3$N, PhH, reflux, 3 h: 91%).[1658]

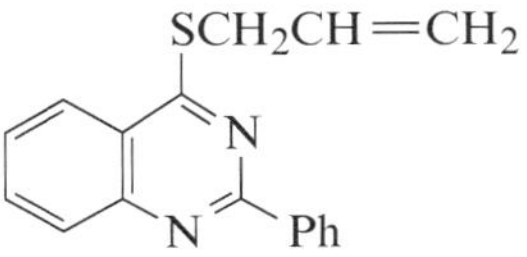

(26)

4-Cyclohex-2′-ethylthioquinazoline (**27**) (3-bromocyclohexene, NaOH, H$_2$O–EtOH, reflux, 3 h: 70%).[2221]

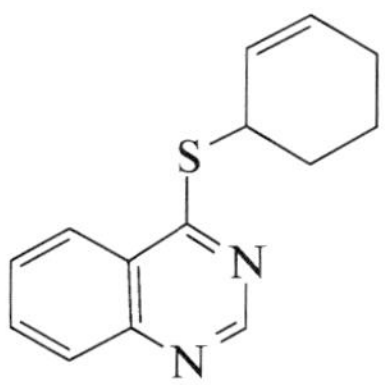

(27)

4-Allylthio-1-methyl-5,6,7,8-tetrahydro-2(1H)-quinazolinone **(28)** (CH_2=
$CHCH_2Br$, MeONa, MeOH–MeOCH$_2$CH$_2$OCH$_2$CH$_2$OMe, reflux, 3 h:
87%).[231]

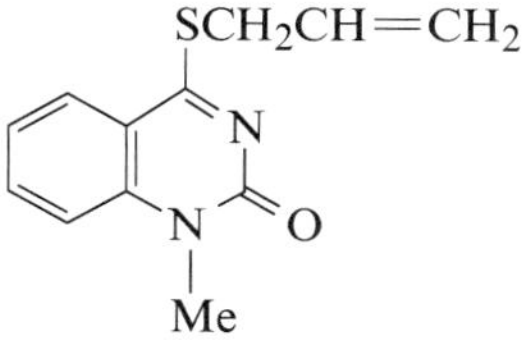

(28)

2-Allylthio-5,6,7,8-tetrafluoro-3-p-fluorophenyl-4(3H)-quinazolinone   **(29)**
(CH_2=$CHCH_2Cl$, NaOH, EtOH–H$_2$O, 20°, 2 h: 95%).[617]

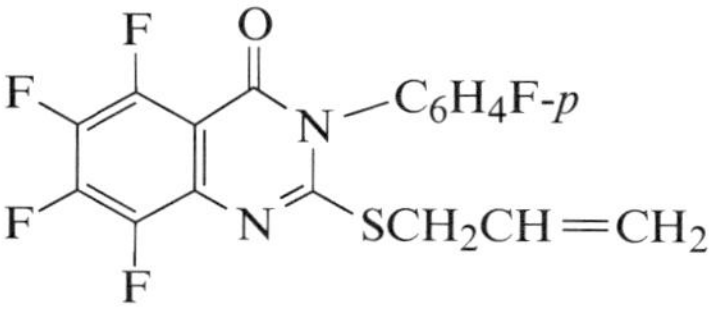

(29)

Also other examples.[313,336,511,980,1820]

Arylthio- and aralkylthioquinazolines

3-Butyl-2-(2,4-dinitrophenyl)thio-4(3H)-quinazolinone **(30)** [2,4-(NO$_2$)$_2$C$_6$H$_3$Cl,
KOH, EtOH–H$_2$O?] and analogs.[603]

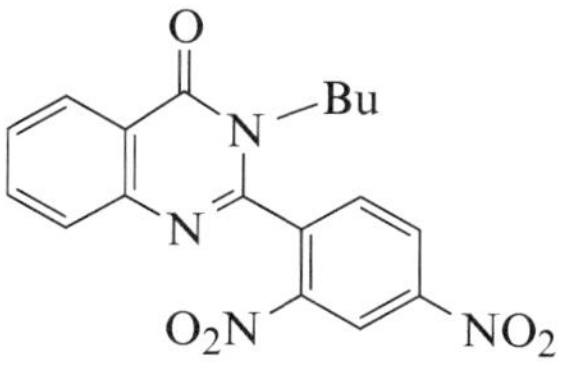

(30)

2-(3-Methyl-5-oxo-2-pyrazolin-4-yl)thio-4(3H)-quinazolinone **(31)** (4-bromo-
3-methyl-2-pyrazolin-5-one, NaOH, EtOH, 20°, 12 h, then reflux, 1 h:
75%).[1180]

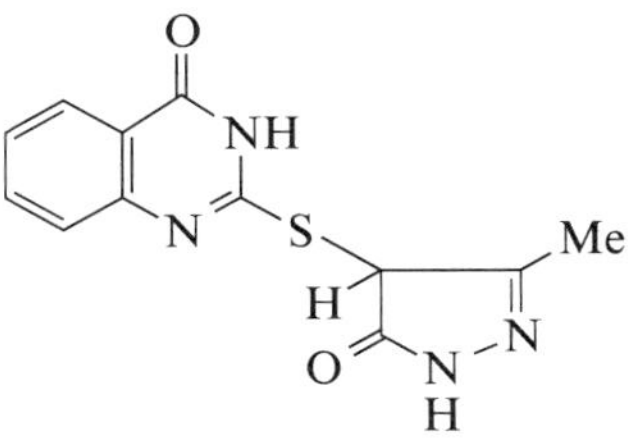

(31)

6,8-Dibromo-2-*o*-nitrobenzylthio-3-phenyl-4(3*H*)-quinazolinone (**32**) (ClCH$_2$-
  C$_6$H$_4$NO$_2$-*o*, NaOH, EtOH, 20°, ?h: 70%).[603]

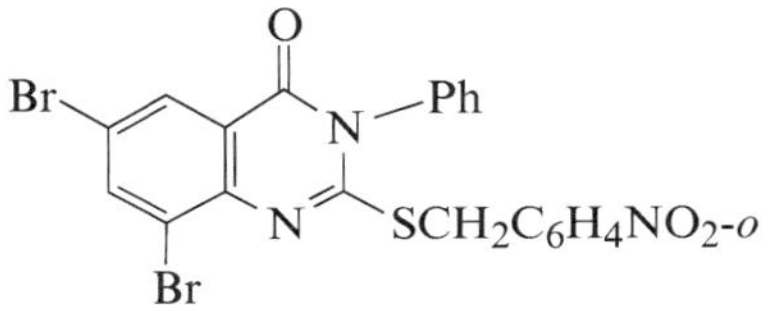

(32)

Also other examples.[354,377,504,840,1235,1957,2062]

(Aminoalkylthio)quinazolines

2-*β*-Dimethylaminoethylthio-4-phenylquinazoline (**33**) (Me$_2$NCH$_2$CH$_2$Cl.
  HCl, 2M NaOH, 25°, 2h: 91%, as hydrochloride); the 4-*β*-dimethylamino-
  ethylthio-2-phenyl isomer likewise (88%).[1601]

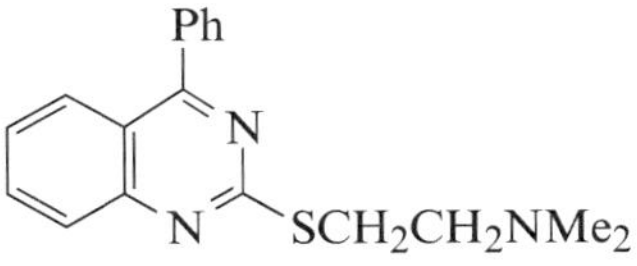

(33)

2-*β*-Diethylaminoethylthio-3-phenyl-4(3*H*)-quinazolinone (Et$_2$NCH$_2$CH$_2$Cl-
  HCl, Na$_2$CO$_3$, EtOH–H$_2$O, 20°, 10h: 71%).[62]
2,4-Bis(*β*-dimethylaminoethylthio)quinazoline (**34**) (Me$_2$NCH$_2$CH$_2$Cl.HCl,
  1M NaOH, 20°, 1h: ~25%, characterized as the dihydrobromide).[1677]

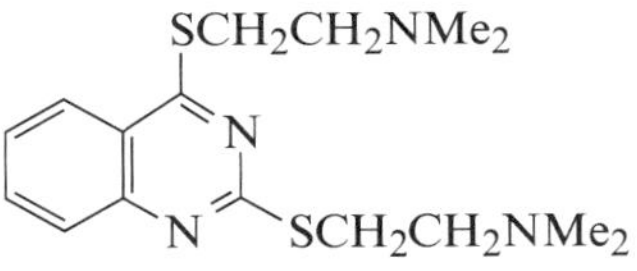

(34)

3-Ethyl-6-iodo-2-β-isopropylaminoethylthio-4(3H)-quinazolinone (**35**) (Pr^i-NHCH$_2$CH$_2$Br.HBr, NaOH, EtOH, 20°, 40 min: 70%).[1750]

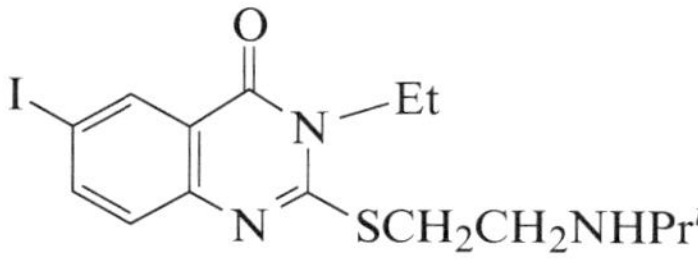

(**35**)

Also other examples.[496,1475,1507,1632,2292]

(Acylalkylthio)quinazolines

4-(β-Acetylethyl)thioquinazoline (**36**) (ClCH$_2$CH$_2$Ac, NaOH, EtOH–H$_2$O, 20°, 2 h: 75%).[1280]

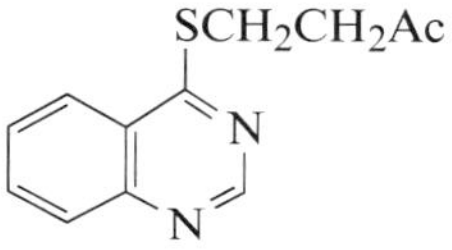

(**36**)

2-Acetonylthio-3-amino-4(3H)-quinazolinone (**37**) (BrCH$_2$Ac, NaOH, EtOH–H$_2$O, 20°, briefly: 85%; extended time gave a cyclized product).[1820]

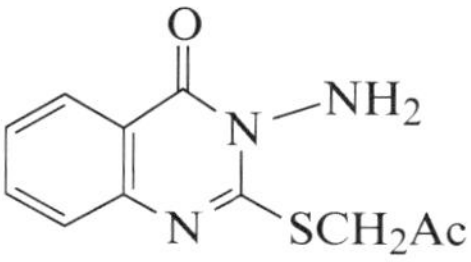

(**37**)

4-Phenacylthio-2-phenylquinazoline (**38**) (BzCH$_2$Br, Et$_3$N, MeCN, reflux, 3 h: 95%).[737]

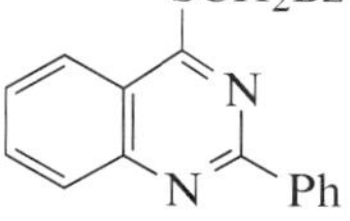

(**38**)

4-p-Bromophenacylthio-2-phenylquinazoline 3-oxide (**39**, R = SCH$_2$COC$_6$-H$_4$Br-p) (BrCH$_2$COC$_6$H$_4$Br-p, MeCN, reflux, 75 min: 61%);[1468] under more vigorous conditions, sulfur extrusion ensued to give 4-p-bromophenacyl-2-phenylquinazoline 3-oxide (**39**, R = CH$_2$COC$_6$H$_4$Br-*p*) (Me$_2$NCHO, reflux, 2 h: 66%; MeONa/MeOH, reflux, 30 min: 92%).[1468]

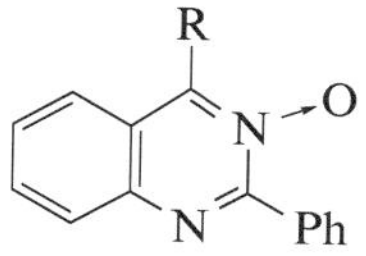

(39)

4-(2-Oxocyclohexyl)thioquinazoline (**40**) [4(3*H*)-quinazolinethione Na salt, 2-chlorocyclohexanone, Me$_2$NCHO, 20°, 12 h: 70%] and thence 4-(2-oxocyclohexyl)quinazoline (**41**) (Me$_2$NCHO, reflux, 4 h: 40%).[846, cf. 538]

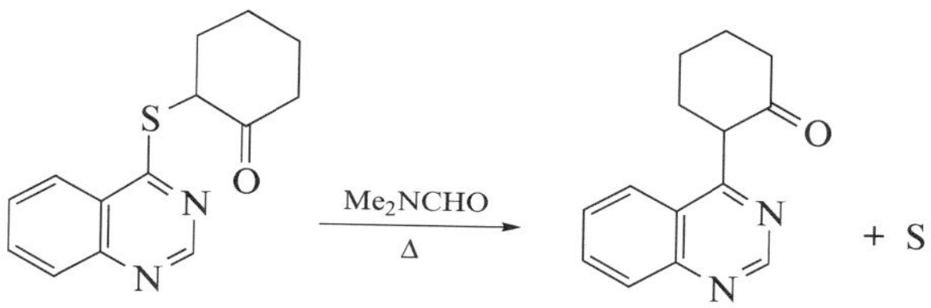

(40)   **(41)**

2-[*β*-Carboxy-*α*-(*p*-chlorobenzoyl)ethyl]thio-3-phenyl-4(3*H*)-quinazolinone (**42**) (*p*-ClC$_6$H$_4$COCH=CHCO$_2$H, MeONa, MeOH, reflux, 6 h: 75%).[2408]

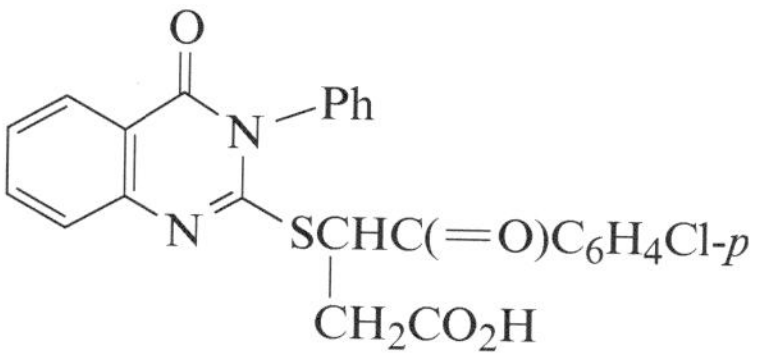

(42)

Also other examples.[138,478,508,521,1166,1281,1554,2003,2260,2396]

(Carboxyalkylthio)quinazolines

2-Carboxymethylthioquinazoline (**43**, R = H) (ClCH$_2$CO$_2$H, 2.5M NaOH, reflux, 1 h: 73%); 2-*α*-carboxybenzylthio-4-phenylquinazoline (**43**, R = Ph) (ClCHPhCO$_2$H, likewise: 71%);[2314] 4-*α*-carboxybenzylthioquinazoline (BrCHPhCO$_2$H, likewise: >95%);[2315] and 4-carboxymethylthioquinazoline (ClCH$_2$CO$_2$Na, NaOH, H$_2$O, 25°, 2 h: 71%).[880]

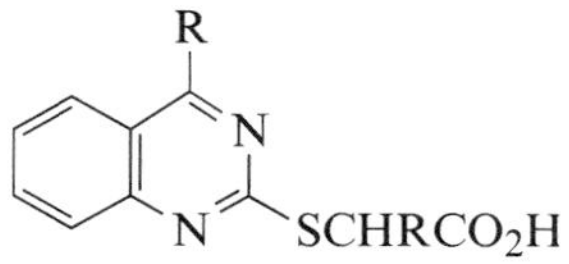

R

(43)

SCHRCO₂H structure

(43)

2-Carboxymethylthio-6-methyl-4(3H)-quinazolinone (**44**) (ClCH₂CO₂Na, NaOH, H₂O–Me₂NCHO, 100°, 1 h: 80%).[2176]

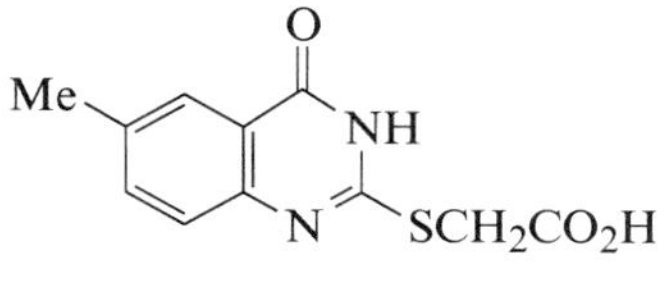

(44)

2-Carboxymethylthio-3-phenyl-4(3H)-quinazolinone (ClCH₂CO₂H, NaOH, H₂O–Me₂NCHO, 100°, 2h: >90%);[144] also the 3-p-tolyl analog (ClCH₂CO₂Na, NaOH, H₂O, 20°, 2.5h: 74%).[1015]
Also other examples.[98,141,144,478,646,1215,1365,1674,1876]

(Alkoxycarbonylalkylthio)quinazolines

7-Chloro-2-ethoxycarbonylmethylthio-3-phenyl-4(3H)-quinazolinone (**45**) (EtO₂CCH₂Cl, K₂CO₃, AcMe, reflux, 18 h: 63%).[1588,2536]

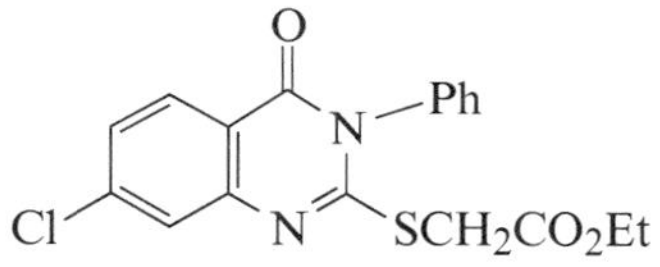

(45)

4-Ethoxycarbonylmethylthioquinazoline (EtO₂CCH₂Cl, K₂CO₃, AcMe, ?°, ?h: ∼90%).[880]

4-α-Ethoxycarbonylpropylthio-2-phenyl-5,6,7,8-tetrahydroquinazoline (substrate, NaH, Me₂SO, 20°, 1h; then EtO₂CCHBrEt, 20°, 5h: 65%).[1639]

3-Amino-2-ethoxycarbonylthio-4(3H)-quinazolinone (**46**) (ClCO₂Et, 1.25M NaOH, 20°, 2h: 60%).[1642]

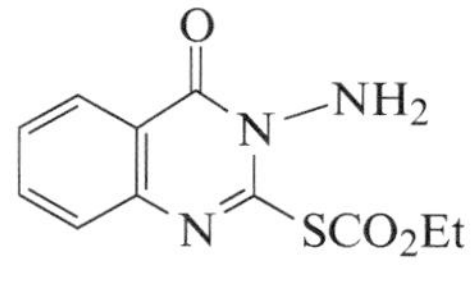

(46)

Also other examples.[478,630,1006,1516,1685,1983,2347]

(Carbamoylalkylthio)quinazolines

2-Carbamoylmethylthioquinazoline (**47**) (ClCH$_2$CONH$_2$, NaHCO$_3$, H$_2$O, reflux, 2.5 h: ~50%).[949]

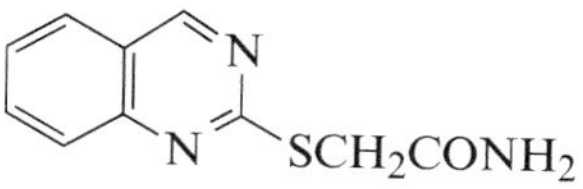

(**47**)

2-N-Phenylcarbamoylmethylthio-4(3H)-quinazolinone (ClCH$_2$CONHPh, KOH, EtOH, reflux, 3 h: >60%).[807]

3-Benzyl-6-bromo-2-[N-(5-cyclohexyl-1,3,4-thiadiazol-2-yl)carbamoyl-methylthio]-4(3H)-quinazolinone (**48**) (2-α-chloroacetamido-5-cyclohexyl-1,3,4-thiadiazole, NaOH, EtOH–H$_2$O, reflux, 5 h: 53%).[1636]

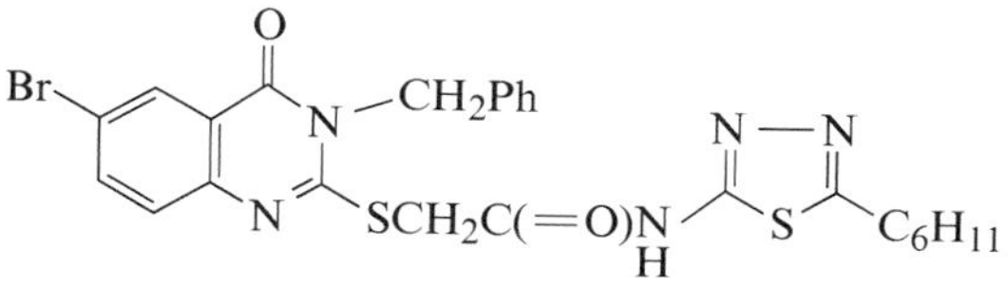

(**48**)

2,4-Bis(carbamoylmethylthio)- (**49**, R = H) (ClCH$_2$CONH$_2$, 1M NaOH, 20°, 24 h: 83%)[1143,1677] and 2,4-bis(α-carbamoylethylthio)quinazoline (**49**, R = Me) (MeCHBrCONH$_2$, likewise: 90%).[1143]

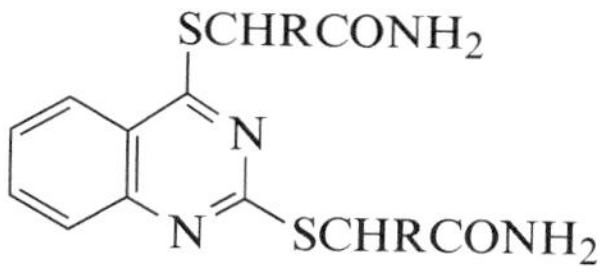

(**49**)

Also other examples.[148,518,646,864,974,987,1022,1048,1092,1496,1843,2293]

Other types of (substituted-alkylthio)quinazoline

2-Chlorocarbonylmethylthio-3-phenyl-4(3H)-quinazolinone (ClCH$_2$COCl, pyridine, ?°, ?h: 70%);[2503] and the 3-p-tolyl analog (likewise).[2503]

2-Cyanomethylthio-4(3H)-quinazolinone (ClCH$_2$CN, K$_2$CO$_3$, Bu$_4$NBr, PhH, ?°, ?h: ?%).[1983]

5.1.2.2. Aminolysis (H 283, 328)

Both tautomeric and nontautomeric quinazolinethiones can undergo aminolysis to afford quinazolinamines and quinazolinimines, respectively. However, apart from hydrazinolysis of tautomeric quinazolinethiones, such reactions have not been used extensively. The following classified examples illustrate the conditions required and the yields to be expected.

Aminolyses of tautomeric quinazolinethiones

4-Thioxo-3,4-dihydro- gave 4-amino-2(1H)-quinazolinone (**50**) (NH$_4$OH, reflux, 1 h: ~75%).[2330]

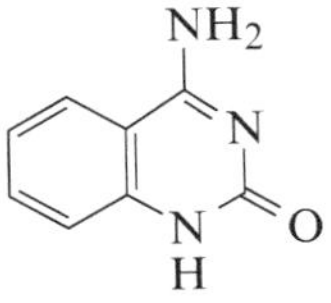

(**50**)

3-Phenyl-2-thioxo-1,2-dihydro-4(3H)-quinazolinone gave 2-anilino-3-phenyl- (**51**, R = Ph) (PhNH$_2$, Δ),[2386] 2-benzylamino-3-phenyl- (**51**, R = CH$_2$Ph) (PhCH$_2$NH$_2$, Δ),[2386] or 3-phenyl-2-thioureido-4(3H)-quinazolinone (**51**, R = CSNH$_2$) (H$_2$NCSNH$_2$, EtOH, reflux, 6 h: 70%).[2408]

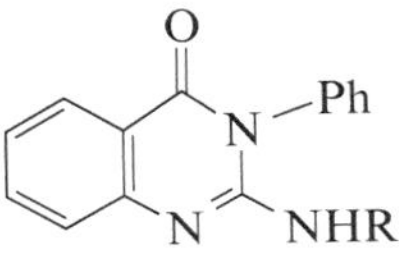

(**51**)

4(3H)-Quinazolinethione (**52**) gave 2,4-quinazolinediamine (**53**) (fresh NaNH$_2$, PhNMe$_2$, 125°, 4.5 h: 71%; note additional Chichibabin[2450i] 2-amination).[383]

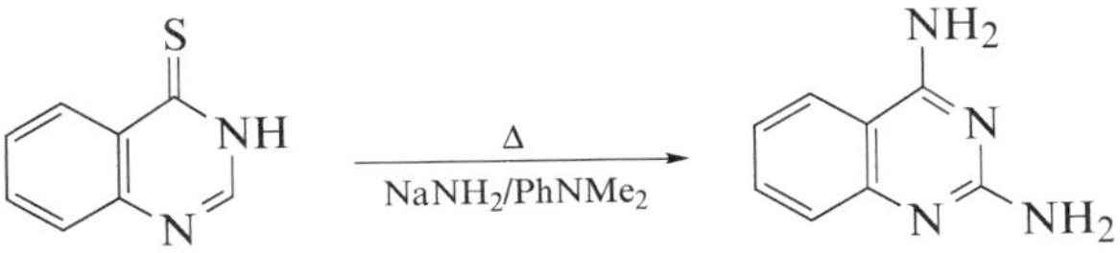

(**52**)   (**53**)

Hydrazinolysis of tautomeric quinazolinethiones

6-Methyl-2-phenyl-4(3H)-quinazolinethione gave 4-hydrazino-6-methyl-2-phenylquinazoline (**54**, X = Ph, Y = Me) (H$_2$NNH$_2$.H$_2$O, EtOH, reflux, 8 h: 75%).[756]

2-Diethylamino-4(3H)-quinazolinethione gave 2-diethylamino-4-hydrazinoquinazoline (**54**, X = NEt$_2$, Y = H) (H$_2$NNH$_2$.H$_2$O, EtOH, reflux, <10 h: 62%).[1337]

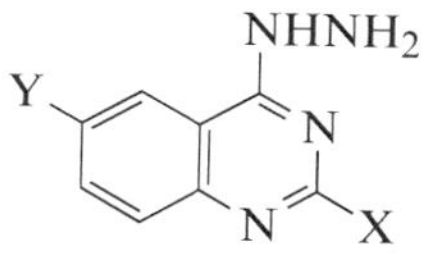

(54)

4-Thioxo-3,4-dihydro- gave 4-(2-β-hydroxyethyl-2-methylhydrazino)-2(1H)-
quinazolinone (H$_2$NNMeCH$_2$CH$_2$OH, Me$_2$NCHO, 20°, 4 days: 74%).[405]

6-Chloro-3-methyl-2-thioxo-1,2-dihydro- gave 6-chloro-2-hydrazino-3-
methyl-4(3H)-quinazolinone (H$_2$NNH$_2$.H$_2$O, EtOH, reflux, 10 h:
75%).[2279]

3-Phenyl-2-thioxo-1,2-dihydro- (55) gave a separable mixture of 2-hydrazino-
3-p-phenyl- (56) and 3-amino-2-anilino-4(3H)-quinazolinone (57), the second probably from Dimroth rearrangement of the first product (H$_2$NNH$_2$,
EtOH, reflux, 10 h: 40% and 25%, respectively).[1894]

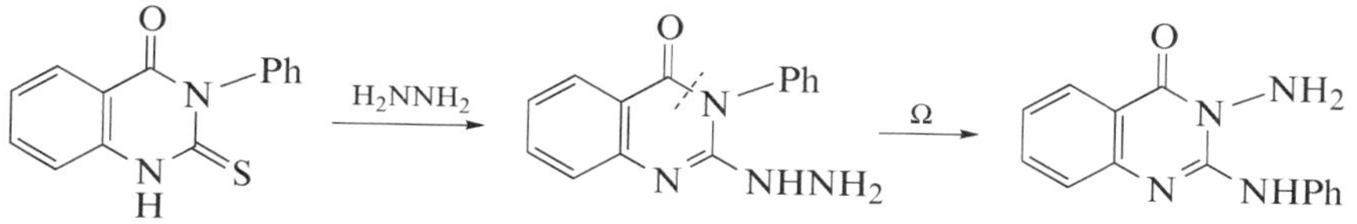

(55)   **(56)**   **(57)**

Also other examples.[1089,1522,1841,1878,1884,1928,1960,2044,2050,2123,2386]

Aminolyses of fixed quinazolinethiones

2,3-Diphenyl-4(3H)-quinazolinethione (58) gave 2,3-diphenyl-4-phenylimino-
3,4-dihydroquinazoline (59) (PhNH$_2$, HgO, PhH, reflux, 6 h: 65%); also
many analogs likewise.[1167]

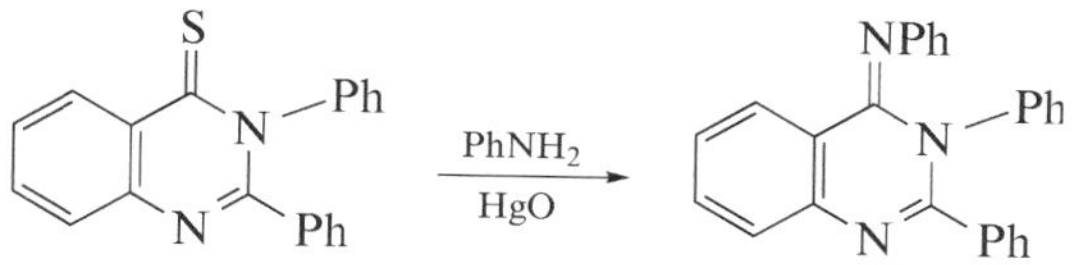

(58)   **(59)**

2-Methyl-3-phenyl-4(3H)-quinazolinethione gave 2-methyl-3-phenyl-4-
tosylimino-3,4-dihydroquinazoline (TsNClNa, MeOH–AcMe–CH$_2$Cl$_2$,
−15°, 1 h, 20°, 15 min: 30%; by a rational mechanism).[1130]

Also other examples.[88,332,1731]

5.1.2.3.   Oxidative Reactions

A typical tautomeric quinazolinethione (**60**) might be oxidized to the disulfide (**61**), the sulfenic acid (**62**), the sulfinic acid (**63**), or the sulfonic acid (**64**); in contrast, a typical fixed quinazolinethione (**65**) might give only the *S*-oxide (**66**), a tautomer of the sulfenic acid (**62**) fixed by *N*-methylation, or the *S,S*-dioxide (**67**), a tautomer of the sulfinic acid (**63**) fixed by *N*-methylation. In fact, apart from the diquinazolinyl disulfides, such as (**61**), very few such oxidation products have been isolated in the quinazoline series. Recently reported examples of these products, each made by oxidation of the corresponding quinazolinethione, are listed below.

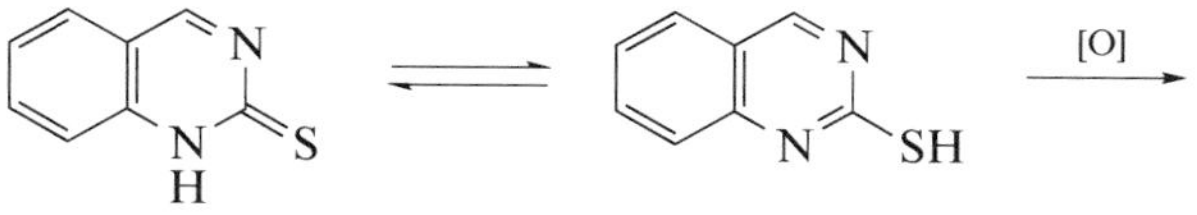

(**60**)

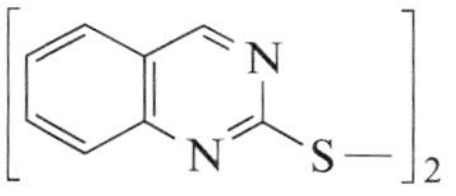

(**61**)   *n* = 1: (**62**)
                                  n = 2: (**63**)
                                  n = 3: (**64**)

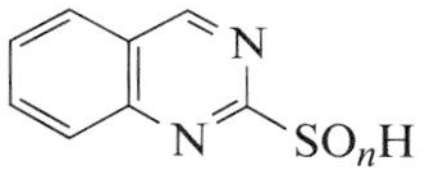

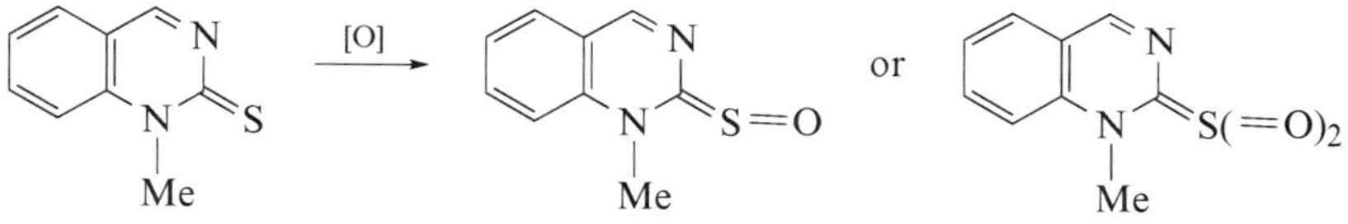

(**65**)   (**66**)   (**67**)

Diquinazolinyl disulfides

Diquinazolin-2-yl disulfide (**61**) (I_2, 1M KI, $NaHCO_3$, 20°, briefly: ?%).[281]
Bis(2-phenylquinazolin-4-yl) disulfide (**68**) (H_2O_2, pyridine, 20°, briefly: 75%;[105]
    TsNClNa, MeOH–AcMe–CH_2Cl_2, −15°, 1 h, 20°, 15 min: 84%).[1130]

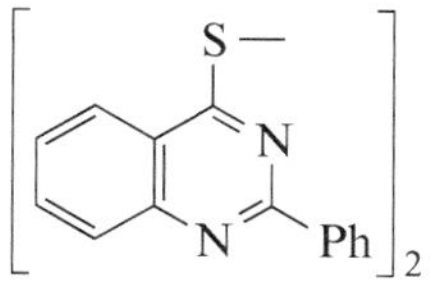

(**68**)

Bis(3-o-chlorophenyl-4-oxo-3,4-dihydroquinazolin-2-yl)disulfide (**69**, R = Cl)
(I$_2$, H$_2$O–EtOH–pyridine, 20°, 30 min: 74%);[2122] bis(4-oxo-3-phenyl-3,4-
dihydroquinazolin-2-yl) disulfide (**69**, R = H) (likewise: 88%; Br$_2$, EtOH,
20°, briefly: 90%).[1726]

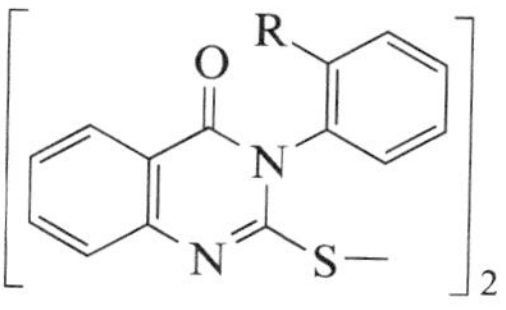

(69)

Also other examples,[241,535,1365,2303,2420,2558] including oxidations during
other reactions.[104,1536]

Quinazolinesulfenic acids

4-Quinazolinesulfenic acid (**70**, R = H) (30% H$_2$O$_2$, Pr^{i}OH–CHCl$_3$, −20°,
20 days: 36%); also its 2-methyl (**70**, R = Me) (H$_2$O$_2$, MeOH–CHCl$_3$,
−20°, 3 days: 6%), 2-isopropyl (**70**, R = Pr^i) (likewise, 2 days: 42%), and
2-t-butyl derivative (**70**, R = Bu^t) (H$_2$O$_2$, MeOH, 40°, 15 min: 56%).[105]

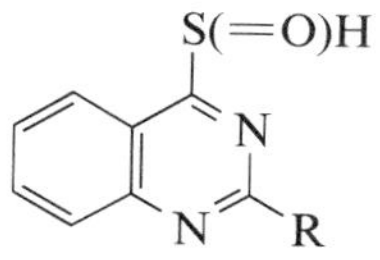

(70)

Quinazolinesulfonic acids

4-Quinazolinesulfonic acid (**71**) (KMnO$_4$, EtOH–H$_2$O, 20°, 10 min: 48%, as
K salt); also its 2-isomer (**64**) (likewise) or 2-quinazolinesulfonyl fluoride (**72**)
(Cl$_2$↓, KHF, H$_2$O–MeOH, −10°, 30 min: 58%).[281]

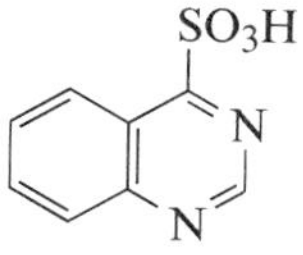

(71) 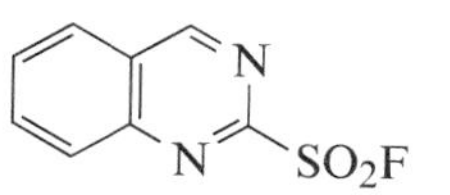   **(72)**

5,6,7,8-Tetrahydro-4-quinazolinesulfonic acid (**73**, R = H) (KMnO$_4$, EtOH–
H$_2$O, 20°, 10 min: 23%, as K salt) and the 2-methyl derivative (**73**, R = Me)
(likewise: 21%, as K salt).[281]

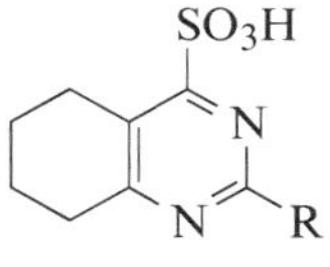

(73)

Fixed quinazolinethione S-oxides

2-Methyl-3-phenyl-4(3H)-quinazolinethione S-oxide (74, R = Me) (m-ClC$_6$H$_4$CO$_3$H, CH$_2$Cl$_2$, −15°, ?h: 22%) and the 2,3-diphenyl analog (74, R = Ph) (likewise: 41%).[1130]

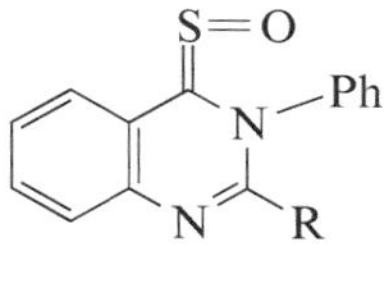

(74) 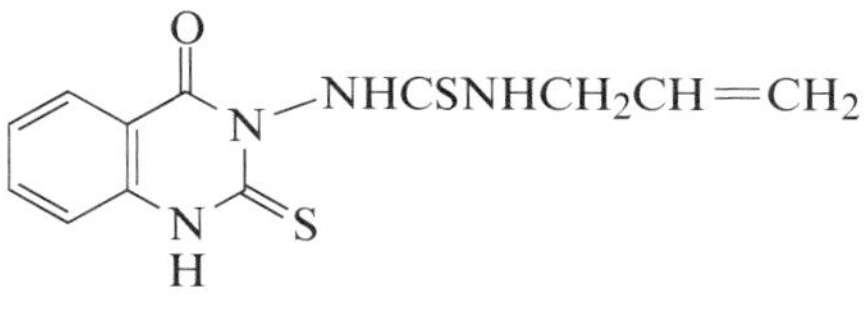   (75)

5.1.2.4. Formation of Metal Complexes (H 274)

Many heavy-metal complexes have been made from both tautomeric and fixed quinazolinethiones (including thioxodihydroquinazolinones), sometimes with auxiliary ligands also involved. Most such complexes are listed here, according to the principal ligand.

3-Allyl-6-bromo-2-thioxo-1,2-dihydro-4(3H)-quinazolinone: Co, Cu, Ru.[1747]

3-[3-Allyl(thioureido)]-2-thioxo-1,2-dihydro-4(3H)-quinazolinone (75): Cd, Co, Cu, Ni.[2313]

3-Allyl-2-thioxo-1,2-dihydro-4(3H)-quinazolinone: Ni.[1646]

3-Amino-2-phenyl-4(3H)-quinazolinethione: Cu, Ni.[1861]

3-Amino-2-thioxo-1,2-dihydro-4(3H)-quinazolinone: Cd, Co, Hg, Zn.[1124]

6-Bromo-3-butyl-2-thioxo-1,2-dihydro-4(3H)-quinazolinone: Co, Cu, Ni.[1747]

2-o-Bromophenyl-4(1H)-quinazolinethione 3-oxide: V.[806]

2-p-Bromophenyl-4(1H)-quinazolinethione 3-oxide: V.[806]

3-p-Bromophenyl-2-thioxo-1,2-dihydro-4(3H)-quinazolinone: Cu,[1668] Ni.[1646]

6-Bromo-3-propyl-2-thioxo-1,2-dihydro-4(3H)-quinazolinone: Co, Cu, Ni.[1747]

6-Bromo-2-thioxo-1,2-dihydro-4(3H)-quinazolinone: Co,[1747] Cu.[1682]

2-t-Butyl-4(1H)-quinazolinethione 3-oxide: V.[806]

3-Butyl-2-thioxo-1,2-dihydro-4(3H)-quinazolinone: Cd, Cu, Hg, Pb, Sn, Zn;[2410] Ni.[1646]

2-*o*-Chlorophenyl-4(1*H*)-quinazolinethione 3-oxide: V.[806]

3-*p*-Chlorophenyl-2-thioxo-1,2-dihydro-4(3*H*)-quinazolinone: Cu,[1668] Ni.[1646]

5,7-Dibromo-2-phenyl-4(1*H*)-quinazolinethione 3-oxide: V.[806]

6,8-Dibromo-3-phenyl-2-thioxo-1,2-dihydro-4(3*H*)-quinazolinone: Cu,[1682] Ni.[1747]

6,8-Dibromo-2-thioxo-1,2-dihydro-4(3*H*)-quinazolinone: Cu.[1682]

5,7-Dibromo-2-*o*-tolyl-4(1*H*)-quinazolinethione 3-oxide: V.[806]

3-*p*-Ethoxyphenyl-2-thioxo-1,2-dihydro-4(3*H*)-quinazolinone: Cu,[1669] Nl.[1646]

3-Ethyl-2-thioxo-1,2-dihydro-4(3*H*)-quinazolinone: Ni.[1646]

6-Iodo-2-thioxo-1,2-dihydro-4(3*H*)-quinazolinone: Cu,[1682], Ni.[1747]

2-Isopropyl-4(1*H*)-quinazolinethione 3-oxide: V.[806]

3-Isopropyl-2-thioxo-1,2-dihydro-4(3*H*)-quinazolinone: Ni.[1646]

2-*p*-Methoxyphenyl-4(1*H*)-quinazolinethione 3-oxide: V.[806]

3-*p*-Methoxyphenyl-2-thioxo-1,2-dihydro-4(3*H*)-quinazolinone:Cu,[1668] Ni.[1646]

2-Methyl-4(1*H*)-quinazolinethione 3-oxide: V.[806]

3-Methyl-2-thioxo-1,2-dihydro-4(3*H*)-quinazolinone: Ni,[1646] Ti,[1125,1205,1769] Zr;[1405] stability constants: Cd, Dy, Gd, Mn, Nd, Ni, Sm, Zn[1273]

2-Phenyl-4(1*H*)-quinazolinethione 3-oxide: Cu, Ni;[1861] V.[806]

3-Phenyl-2-thioxo-1,2-dihydro-4(3*H*)-quinazolinone: Ag,[1484,2459] Bi,[2459] Cd,[1484,2460] Co,[1242,1484] Cu,[1242,1484,1668,2460] Hg,[1484,2459,2460] Ir,[2540] Mo, Nb;[316] Ni,[1205,1242,1484,1646] Pb,[2460] Pd, Pt;[2457,2540] Rh,[2540] Ru,[2459] Sn,[2460] Ta,[316] Ti,[1125,1769] V, W;[316] Zn,[1484,2460] Zr.[316,1205]

3-Propyl-2-thioxo-1,2-dihydro-4(3*H*)-quinazolinone: Ni.[1646]

2,4(1*H*,3*H*)-Quinazolinedithione: Pd, Pt.[9]

2(1*H*)-Quinazolinethione: Cu.[2091]

2-Thioxo-1,2-dihydro-4(3*H*)-quinazolinone: Ag,[2210] Bi,[1163] Cd,[691,2460,2483] Co,[691,1242] Cu,[1242,2139,2213,2460] Fe,[1207] Hg,[691,2405,2460,2483] Ir,[2540] Mo, Nb;[316] Ni,[476,1163,1242] Pb,[2460] Pd, Pt,[1163,1207,2540] Rh,[1207,2540] Sb,[1163] Sn,[2460] Ta,[316] Tl,[2405] V, W;[316] Zn,[691,2460] Zr.[316]

2-Thioxo-3-*m*-tolyl-1,2-dihydro-4(3*H*)-quinazolinone: Ag, Cd, Co, Cu, Hg, Ni;[1484] Ir, Pd, Pt, Rh;[2540] Zn.[1484]

2-Thioxo-3-*o*-tolyl-1,2-dihydro-4(3*H*)-quinazolinone: Ti,[1125,1205,1769] Zr.[1205]

2-Thioxo-3-*p*-tolyl-1,2-dihydro-4(3*H*)-quinazolinone: Cu,[1668] Ni.[1646]

2-*o*-Tolyl-4(1*H*)-quinazolinethione 3-oxide: V.[806]

5.1.2.5.   Other Reactions (*H* 288)

Quinazolinethiones undergo several other rather neglected reactions, illustrated in the following examples.

Conversion into thiocyanatoquinazolines

3-Phenyl-2-thioxo-1,2-dihydro-4(3*H*)-quinazolinone (**76**) gave 3-phenyl-2-thiocyanato-4(3*H*)-quinazolinone (**77**) [BrCN (made *in situ*), MeOH, −5°, briefly: ?%];[620] also many analogs.[450,620,663]

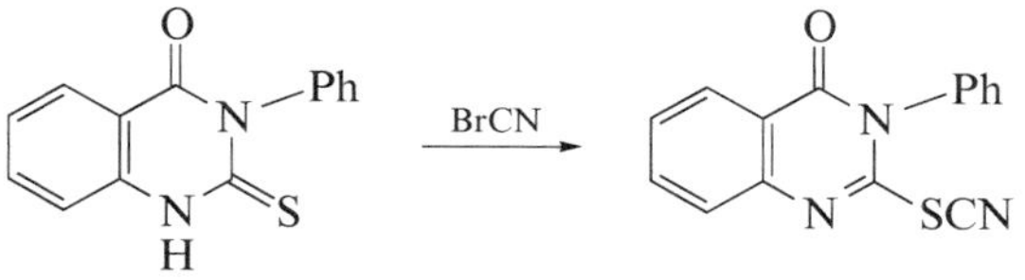

(**76**)   (**77**)

Reductive desulfurization

2-Benzyl-4(3*H*)-quinazolinethione (**78**, R = CH$_2$Ph) gave a separable mixture of 2-benzylquinazoline and 2-benzyl-3,4-dihydroquinazoline (Raney-Ni, EtOH, reflux, ?h: 44% and 10%, respectively).[1373]

4(3*H*)-Quinazolunethione (**78**, R = H) gave 3,4-dihydroquinazoline (electrolytic reduction).[22]

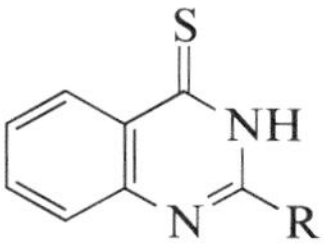

(**78**)

4-Selenoxo-3,4-dihydro-2(1*H*)-quinazolinone (**79**) gave 3,4-dihydro-2(1*H*)-quinazolinone (Raney-Ni, EtOH, reflux, 2 h: 82%).[1897]

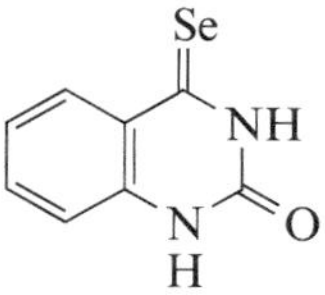

(**79**)

Oxidative desulfurization

2-Thioxo-1,2-dihydro-4(3*H*)-quinazolinone gave 4(3*H*)-quinazolinone (O$_3$, AcOH, 25°, 30 min: 82%; via the sulfinic acid?).[2420]

Conversion into a quinolinethione

A substrate, formulated as 4-methyl-3,5,6,7,8,8a-hexahydro2(1*H*)-quinazolinethione (**80**), gave 4-dimethylamino-4a,5,6,7,8,8a-hexahydro-

2(1*H*)-quinolinethione (**81**) (Me$_2$NCHO, 180°, sealed, 24 h: ~25%; probably via a Dimroth rearrangement and subsequent transamination).[1815]

(**80**)   (**81**)

Conversion into diquinazolinyl sulfides

3-Phenyl-2-thioxo-1,2-dihydro-4(3*H*)-quinazolinone   gave   bis(4-oxo-3-phenyl-3,4-dihydroquinazolin-2-yl) sulfide (SOCl$_2$, 20°, 3 h: 75%); also analogs.[1726, cf. 2579]

Cyclization reactions

3-Phenyl-2-thioxo-1,2-dihydro-4(3*H*)-quinazolinone (**82**) gave 12*H*-benzothiazolo[2,3-*b*]quinazolin-12-one (**83**) (*N*-bromosuccinimide, H$_2$SO$_4$, 70°, 7 h: 96%);[1620] irradiation of the 3-*o*-chlorophenyl substrate in MeOH also gave the product (**83**) in poorer yield, by elimination of HCl.[2122]

(**82**)   (**83**)

2-Thioxo-1,2-dihydro-4(3*H*)-quinazolinone (**84**) gave a separable mixture of dimethyl 5-oxo-1,2-dihydro-5*H*-thiazolo[3,2-*a*]quinazoline-1,2-dicarboxylate (**85**), dimethyl 5-oxo-2,3-dihydro-5*H*-thiazolo[2,3-*b*]quinazoline-2,3-dicarboxylate (**86**), its 2,3-didehydro derivative (**87**), and a fourth (bicyclic) product (MeO$_2$CC≡CCO$_2$Me, MeOH, reflux, 3 h: each ~20%);[1827] also somewhat analogous reactions.[1332, 1625]

3-Amino-2-thioxo-1,2-dihydro-4(3*H*)-quinazolinone (**89**) gave 5*H*-1,3,4-thiadiazolo[2,3-*b*]quinazolin-5-one (**88**) [(EtO)$_3$CH, TsOH, reflux, 6 h: 69%)][1736] or 2-*p*-methoxyphenyl-2,3-dihydro-5*H*-1,3,4-thiadiazolo[2,3-*b*]quinazolin-5-one (**90**) (MeO$_6$H$_4$CHO-*p*, EtOH, reflux, 30 min: 56%).[129]

3-Amino-2-phenyl-4(3*H*)-quinazolinethione gave 5-phenyl-1,3,4-thiadiazolo-[3,2-*c*]quinazolinium-2-yl (dicyanomethanide) (**91**) [(MeS)$_2$C=C(CN)$_2$, Bu^{t}OK, Me$_2$NCHO, 20°, 2.5 h: 30%]; also analogs likewise.[1710]

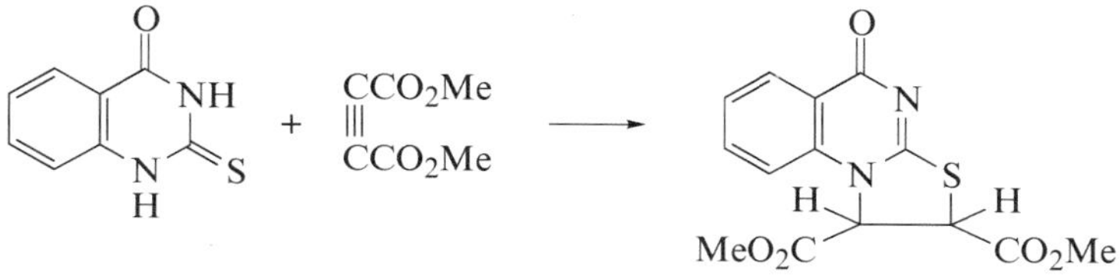

(84)   (85)

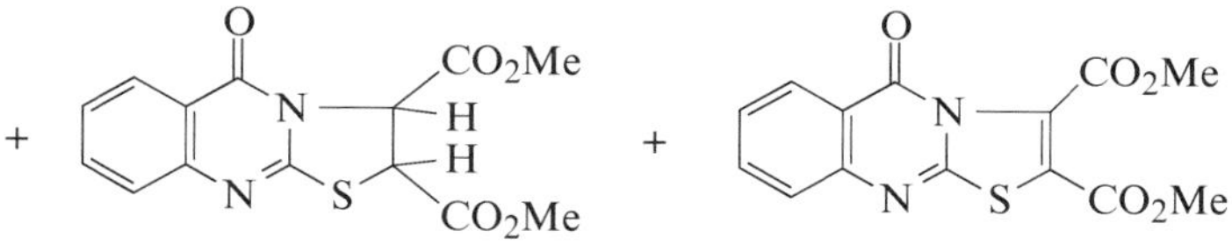

(86)   (87)

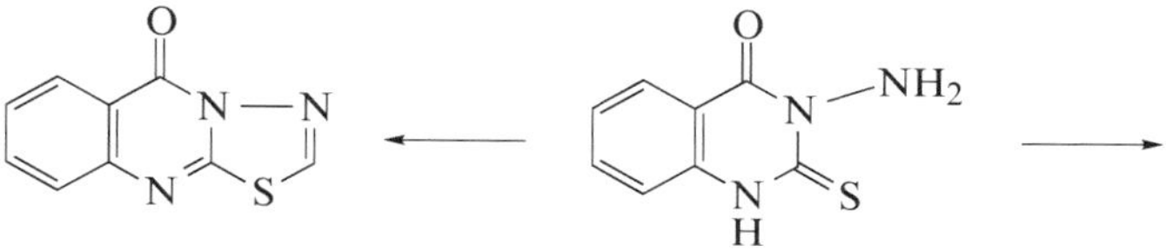

(88)   (89)

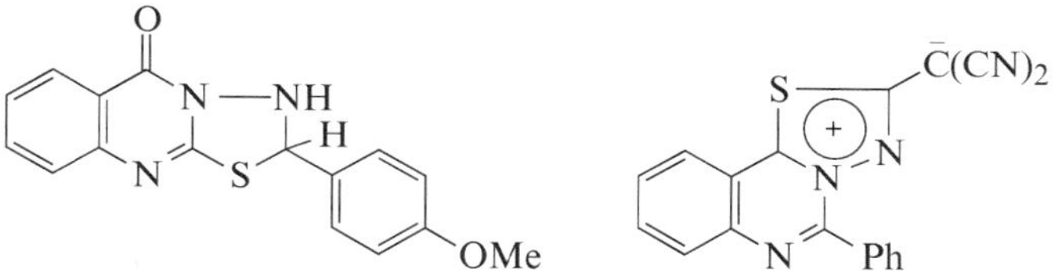

(90)   (91)

Formation of piperazinediium salts

Crystalline salts from piperazine and two molecules of 3-cyclohexyl-2-thioxo-1,2-dihydro-4(3H)-quinazolinone and related thioxoquinazolines have been isolated and characterized.[1194]

5.2. 5- TO 8-QUINAZOLINETHIOLS AND EXTRANUCLEAR MERCAPTOQUINAZOLINES (H 287)

Very little work within these areas has been reported in the recent literature.
The *preparation* of such compounds is represented in a few primary syntheses (see Chapter 1), by thiolyses of corresponding halogenoquinazolines (see Section 3.7), by occasional derivatization reactions, and by the conversion of 3-*p*-thiocyanatophenyl- (92, R = CN) into 3-*p*-mercaptophenyl-2-thioxo-1,2-

dihydro-4(3*H*)-quinazolinone (**92**, R = H) by treating with boiling aqueous alcoholic sodium sulfide.[665]

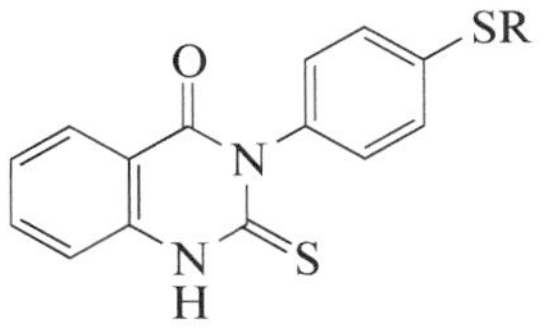

(**92**)

Their *reactions* are exemplified in the *S*-alkylation of 6-bromo-3-(5-mercapto-1,3,4-thiadiazol-2-yl)- (**93**, R = H) to 6-bromo-3-(5-carboxymethylthio-1,3,4-thiadiazol-2-yl)-2-phenyl-4(3*H*)-quinazolinone (**93**, R = CH$_2$CO$_2$H) in 86% yield by boiling with chloroacetic acid in pyridine for 6 h;[1626] in the *S*-arylation of 2-mercaptomethyl- to 2-(2,4-dinitrophenylthio)methyl-6-methyl-4(3*H*)-quinazolinone (**94**) in 56% yield by treatment with 1-chloro-2,4-dinitrobenzene;[510] and by the conversion of 2-methyl-3-*p*-(4-phenyl-5-thioxo-1,3,4-triazolin-3-yl)phenyl-4(3*H*)-quinazolinone (**95**) into bis{5-[*p*-(2-methyl-4-oxo-3,4-dihydroquinazolin-3-yl)phenyl]-4-phenyl-1,2,4-triazol-3-yl} sulfide (**96**, X = S) (50%) by treatment with thionyl chloride at 20° or into the corresponding disulfide (**96**, X = S$_2$) (53%) by oxidation with cold methanolic bromine.[2135]

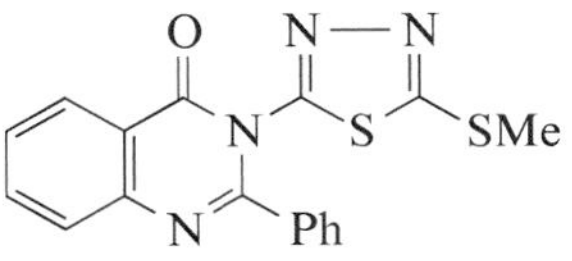

(**93**)

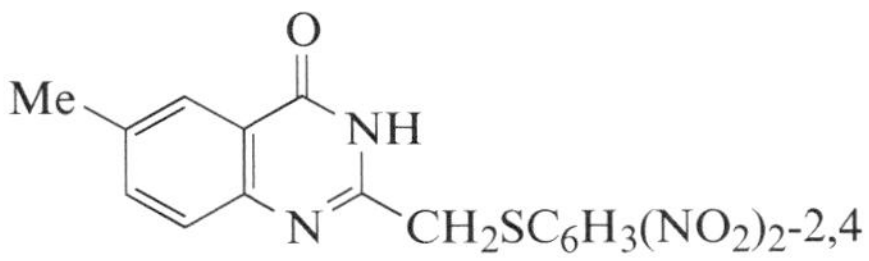

(**94**)

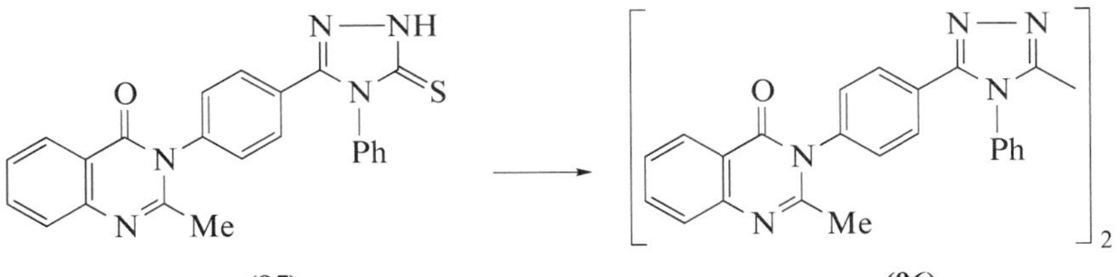

(**95**)

(**96**)

5.3. ALKYLTHIO- AND ARYLTHIOQUINAZOLINES (*H* 280)

All types of such thioethers are discussed in this section except for diquinazolinyl sulfides, which are covered with analogous disulfides in Section 5.4 for pragmatic reasons.

5.3.1. Preparation of Alkylthio- and Arylthioquinazolines (*H* 280, 284)

Most alkylthio- and arylthioquinazolines have been made by *primary syn-thesis* (see Chapter 1), by *alkanethiolysis* of halogenoquinazolines (see Sections 3.5.4, 3.6, and 3.7), or by *S-alkylation* of quinazolinethiones or the like (see Sections 5.1.2.1 and 5.2).

A few additional *minor preparative routes* are illustrated in Section 6.1.2 and in the following examples.

1-Methyl-4-oxo-1,4-dihydro-2-quinazolinecarbonitrile (**97**, R = CN) gave 2-β-carboxyethylthio-1-methyl-4(1*H*)-quinazolinone (**97**, R = SCH$_2$CH$_2$-CO$_2$H) (HSCH$_2$CH$_2$CO$_2$H, Et$_3$N, Me$_2$NCHO, 20°, 23 h: 47%).[1415]

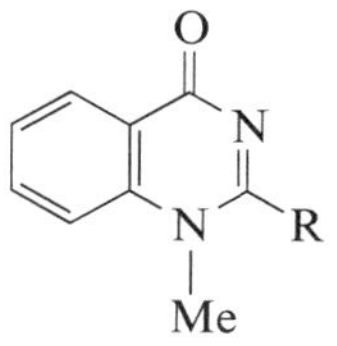

(**97**)

2-*p*-Tolylsulfonyl- (**98**, R = Ts) gave 2-butylthio-4(3*H*)-quinazolinone (**98**, R = SBu) [BuSH, NaOH, (C$_8$H$_{17}$)$_4$NBr, CHCl$_3$, reflux, 24 h: ∼65%].[1915]

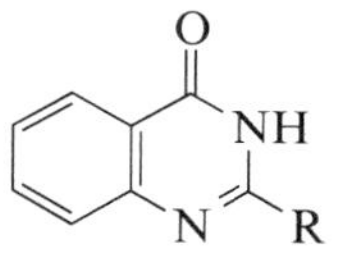

(**98**)

6-Amino-7-methoxy- (**99**, R = NH$_2$) gave 7-methoxy-6-methylthio-4(3*H*)-quinazolinone (**99**, R = SMe) (NaNO$_2$, 30% HBF$_4$, 5°, 45 min; then MeSNa, Cu, 20°, 15 min: 32%).[2017]

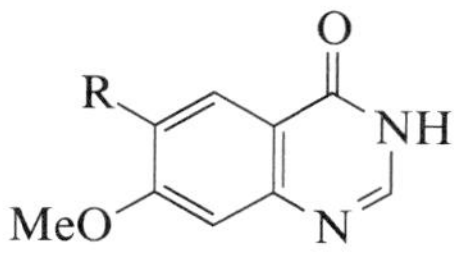

(**99**)

1-Methyl-4-methylthioquinazolinium hydroxide (**101**, X = OH) or nitrate (**101**, X = NO$_3$) with the ylid, Me$_2$S$^+$–C$^-$HCO$_2$Et, gave 2-[α-ethoxycarbonyl-α-(methylthio)methyl]-1-methyl-4-methylthio-1.2-dihydroquinazoline (**102**) (80%) or 4-[α-ethoxycarbonyl-α-(methylthio)methylene]-1-methyl-1,4-dihydroquinazoline (**100**) (90%), respectively (no experimental details).[31]

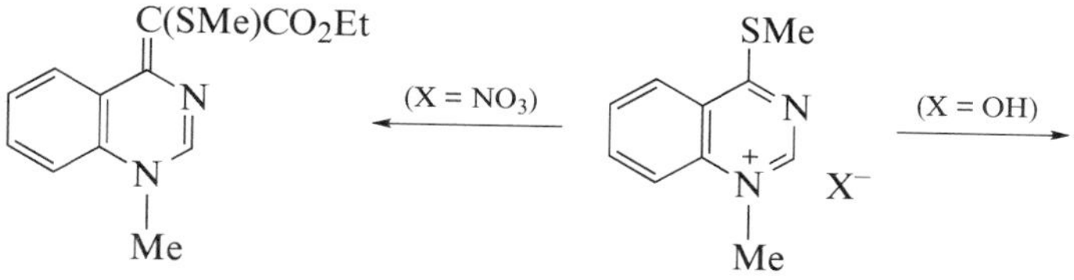

(100)   **(101)**

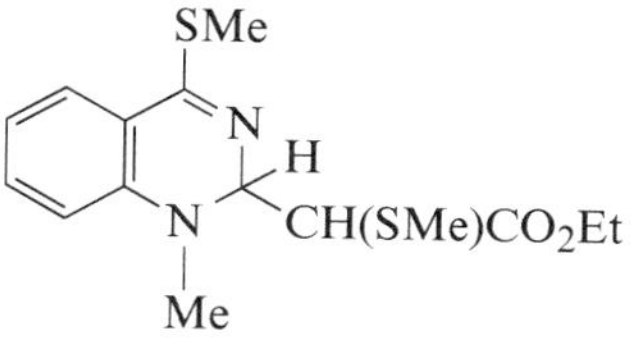

(102)

5.3.2.   Reactions of Alkylthio- and Arylthioquinazolines (*H* 282)

The *hydrolysis of alkylthioquinazolines* to quinazolinones (see Section 4.1.1.2) and the *S-dealkylation of alkylthioquinazolines* to quinazolinethiones (see Section 5.1.1) have been discussed already. The remaining reactions are illustrated in the following examples, classified according to type.

Aminolyses

2-Methylthio-3,4-dihydroquinazoline hydriodide (**103**, R = SMe) gave 3,4-dihydro-2-quinazolinamine hydriodide (**103**, R = NH$_2$) (NH$_3$, H$_2$O–EtOH, reflux, 4 h: 67%) or 2-methylamino-3,4-dihydroquinazoline hydriodide (**103**, R = NHMe) (MeNH$_2$, likewise: 84%).[1569]

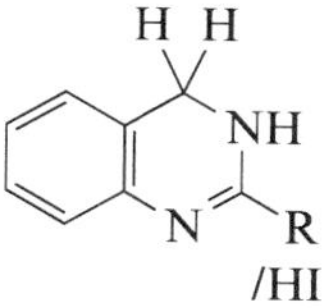

(103)

2-Ethylthio-4(3*H*)-quinazolinone   gave   2-(4-formylpiperazin-1-yl)-4(3*H*)-quinazolinone (**104**, R = CHO) [neat HN(CH$_2$CH$_2$)$_2$NCHO, 160°, 4 h: 49%][2016] or 2-(4-benzylpiperazin-1-yl)-4(3*H*)-quinazolinone (**104**, R = CH$_2$Ph) [neat HN(CH$_2$CH$_2$)$_2$NCH$_2$Ph, 160°, 3 h: 87%].[2020]

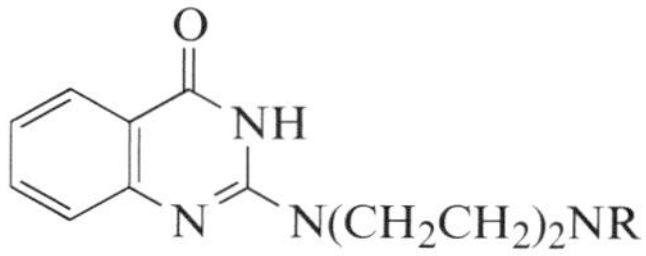

(104)

2-*p*-Methoxyphenyl-4-methylthio- gave 4-*m*-methoxyanilino-2-*p*-methoxy-phenylquinazoline hydriodide (H$_2$NC$_6$H$_4$OMe-*m*, Pr^{i}OH, reflux, 18 h: 84%).[968]

2-Ethylthio- (**105**, R = SEt) gave 2-hydrazino-3-phenyl-4(3*H*)-quinazolinone (**105**, R = NHNH$_2$) (H$_2$NNH$_2$.H$_2$O, Pr^{i}OH, reflux, 10 h: 78%).[128] In contrast, 2-methylthio-3-phenyl- (**105**, R = SMe) gave 3-amino-2-hydrazino-4(3*H*)-quinazolinone (**106**, X = H) (neat H$_2$NNH$_2$.H$_2$O, reflux, 10 h: 87%; note additional transamination of ring-N);[2199] the 6-bromo (**106**, X = Br)[1250] and 6-fluoro (**106**, X = F)[1046] analogs were made similarly.

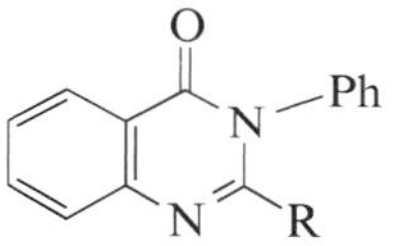

(105) 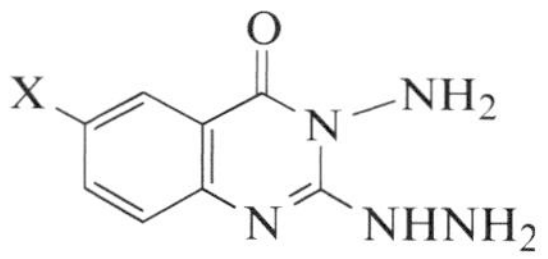   **(106)**

2,4-Bismethylthioquinazoline gave 4-butylamino-2-methylthioquinazoline (BuNH$_2$, Me$_2$NCHO, 80°, 3 h: 77%; note the selectivity).[1410,cf. 1379]

4-Methylthio- gave 4-methoxyamino-6-*m*-trifluoromethylphenyl-2-quinazo-linamine (MeONH$_2$. HCl, pyridine, 45°, 6 h: ∼ 30%).[1595]

Also other examples in the foregoing references and elsewhere.[126,845,943,1572,1894,2057,2076,2149,2547,2595]

Oxidation to alkylsulfinyl- or alkylsulfonylquinazolines

5-Methylthio- (**107**, R = SMe) gave 5-methylsulfonyl-2,4-quinazolinediamine (**107**, R = SO$_2$Me) (KMnO$_4$, H$_2$O–AcOH, 20°, 12 h: 34%).[2225]

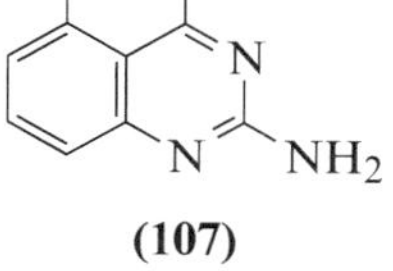

(107)

5-Chloro-6-*p*-chlorobenzylthio- (**108**, *n* = 0) gave 5-chloro-6-*p*-chlorobenzyl-sulfinyl- (**108**, *n* = 1) (30% H$_2$O$_2$, AcOH, 20°, 4 h: 55%) or 5-chloro-6-*p*-chlorobenzylsulfonyl-2,4-quinazolinediamine (**108**, *n* = 2) (30% H$_2$O$_2$, AcOH, 20°, 48 h: 42%).[811]

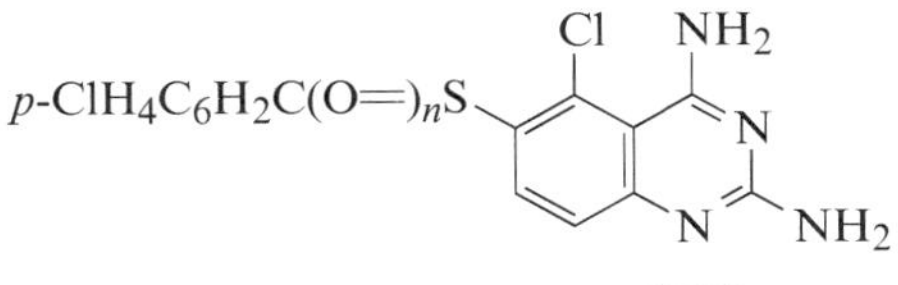

$$p\text{-ClH}_4\text{C}_6\text{H}_2\text{C(O=})_n\text{S}$$

(108)

6-p-Chlorobenzylthio- gave 6-p-chlorobenzylsulfinyl-2,4-quinazolinediamine (diazabicyclo[2,2,2]octane–Br$_2$ complex, AcOH–H$_2$O, 20°, 18 h: 74%);[811] 5-chloro-6-p-fluorophenylsulfinyl-2,4-quinazolinediamine (likewise, 4 days: 80%).[1548]

6-Phenylthio- gave 6-phenylsulfonyl-2,4-quinazolinediamine (30% H$_2$O$_2$, AcOH–H$_2$O, 20°, 20 h: 65%).[1548]

4-Methyl-2-methylthiomethyl- (**109**, R = SMe) gave 4-methyl-2-methylsulfinylmethyl- (**109**, R = SOMe) [PhCO$_3$H (1 mol), CHCl$_3$, 20°, 35 h: 89%] or 4-methyl-2-methylsulfonylmethylquinazoline 3-oxide (**109**, R = SO$_2$Me) [PhCO$_3$H (> 2 mol), CHCl$_3$, 20°, 23 h: 91%].[874]

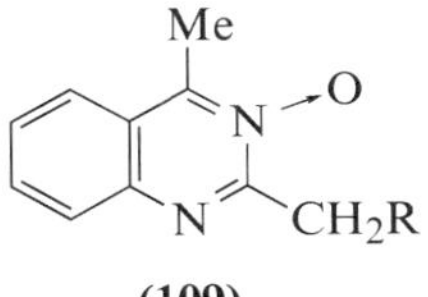

(109)

2-Methyl-3-β-phenylthioethyl- (**110**, R = SPh) gave 2-methyl-3-β-phenylsulfonylethyl-4(3H)-quinazolinone (**110**, R = SO$_2$Ph) (H$_2$O$_2$, AcOH, 70°, 3 h: 87%).[2378]

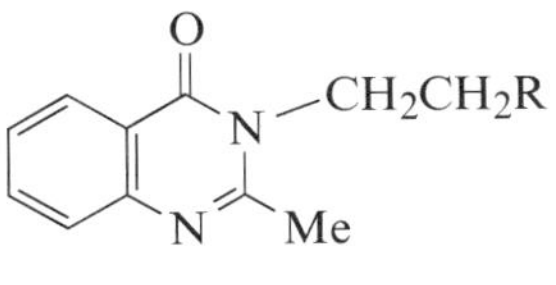

(110)

2-Propylthiomethyl- gave 2-propylsulfonylmethyl–3-o-tolyl-4(3H)-quinazolinone (ClC$_6$H$_4$CO$_3$H-m, CH$_2$Cl$_2$, 20°, 2 days: 58%).[786] Also other examples.[1398,1418,1558,1716,1892,1938,2161,2207]

Desulfurizations and nuclear reductions

2-Ethylamino-4-methylthioquinazoline (**111**, R = SMe) gave 2-ethylaminoquinazoline (**111**, R = H) (Raney-Ni, EtOH, reflux, 3 h: 38%); 4-methylaminoquinazoline likewise (55%).[1410]

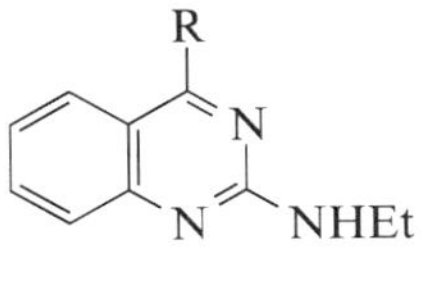

(111)

5,8-Dimethoxy-2-methylthio-3-phenyl- (**112**, R = SMe) gave 5,8-dimethoxy-3-phenyl-4(3H)-quinazolinone (**112**, R = H) (H$_2$, Ni, EtOH, 12 h: 10%).[897]

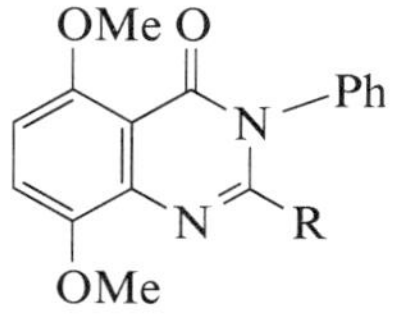

(112)

4-Allylthioquinazoline gave mainly 4-allylthio-1,2,3,4-tetrahydroquinazoline (**113**) (30%) and 4(3H)-quinazolinone (**114**, X = O) (50%), with a little 4(3H)-quinazolinethione (**114**, X = S) (5%) (LiAlH$_4$, THF, 20°, 30 min; aqueous workup).[896]

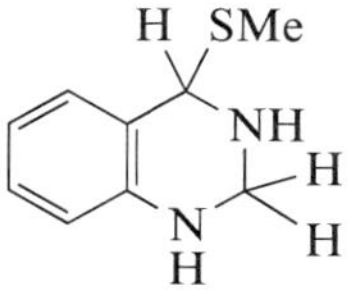

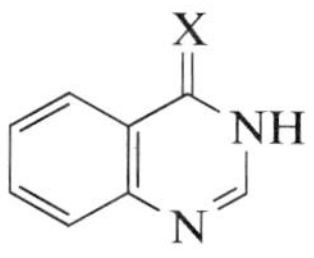

(113)   **(114)**

Also other examples.[511]

Rearrangements

4-β-Acetylethylthioquinazoline (**115**, R = Ac) gave mainly 3-β-acetylethyl-4(3H)-quinazolinone (**116**, R = Ac) (POCl$_3$, reflux, 6 h: 45%);[1203,1280] likewise, 4-β-benzoylethylthioquinazoline (**115**, R = Bz) gave 3-β-benzoylethyl-4(3H)-quinazolinone (**116**, R = Bz) (60%).[1203,1280]

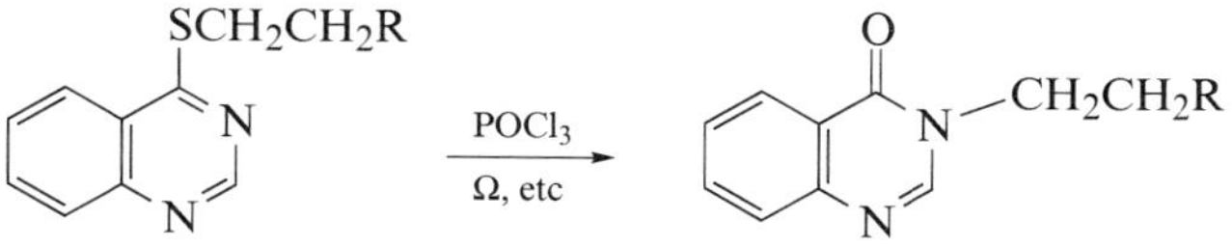

(115)   **(116)**

4-[α-Ethoxycarbonyl-α-(methylthio)methylene]-1-methyl-1,4-dihydroquinazoline (**117**) gave 4-amino-1-methyl-3-methylthio-2(1H)-quinolinone (**118**) (NaOH, EtOH, ?°, 3 h: 70%).[31]

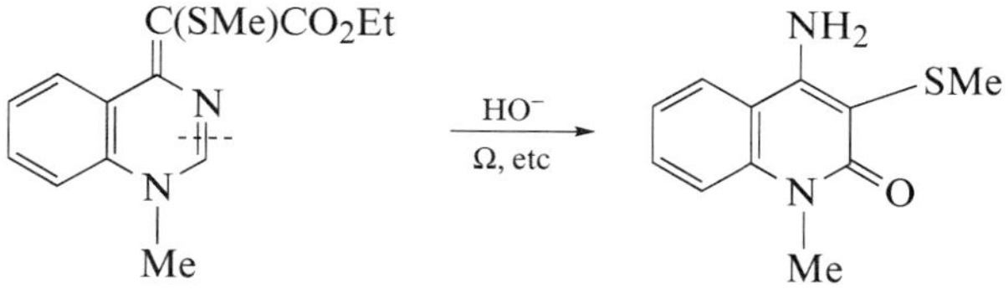

(117)   **(118)**

Sulfur extrusion

2-p-Chlorophenacylthio-3,4-dihydro- **(119)** gave 2-p-chlorophenacyl-3,4-dihydroquinazoline (Ph_3P, Me_2NCHO, hot: 66%).[1233]

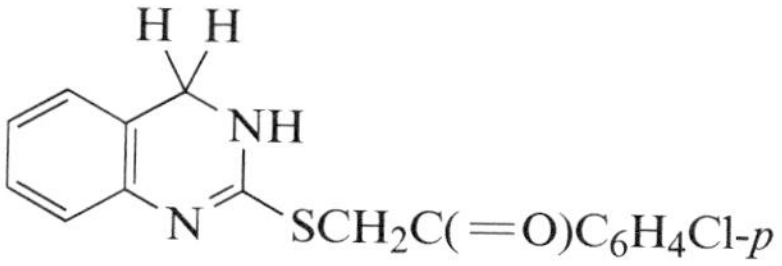

(119)

4-Cyanomethylthio- gave 4-cyanomethylquinazoline (NaOEt, Me_2NCHO, 20°, 20 h: 40%);[1869] analogs likewise.[895,1869]

Displacement by a carbanion

1-Ethyl-2-methylthio- gave 2-diethoxycarbonylmethyl-1-ethyl-4(1H)-quinazolinone **(120)** [$(EtO_2C)_2CH_2$, Na, 165°, 1 h: 93%].[275]

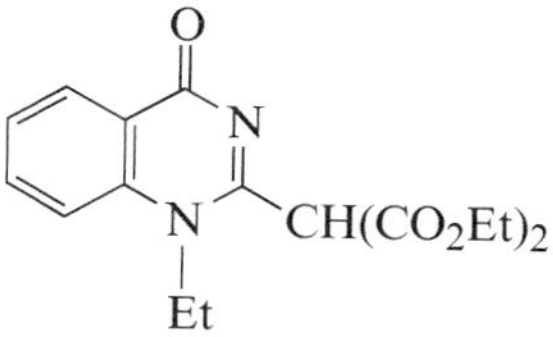

(120)

Other oxidations

8-Benzylidene-2-methylthio-4-phenyl-3,4,5,6,7,8-hexahydroquinazoline hydriodide gave the 5,6,7,8-tetrahydro analog **(121)** [$K_3Fe(CN)_6$, KOH, H_2O: ?%].[1533]

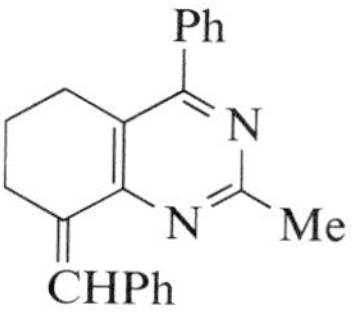

(121)

5-Benzylthio-6-chloro-3-methyl-4(3H)-quinazolinone gave 6-chloro-3-methyl-4-oxo-3,4-dihydro-5-quinazolinesulfonyl chloride ($Cl_2\downarrow$, $AcOH$–H_2O, 5°, 30 min: 75%).[2584]

5.4. DIQUINAZOLINYL DISULFIDES AND SULFIDES

Most of the meager recent information on such disulfides and sulfides has been discussed already.

Thus the *preparation of disulfides* by primary syntheses (see Chapter 1) and by controlled oxidation of quinazolinethiones (Section 5.1.2.3) or extranuclear mercaptoquinazolines (Section 5.2) has been covered. A rarely used route to disulfides/diselenides is illustrated by the reaction of 4-chloroquinazoline (**122**) with "sodium hydroselenide" (?) in ethanol–THF at 20° for 5 h to give diquinazolin-4-yl diselenide (**123**) in unstated yield;[1247] and a route to unsymmetrical disulfides is exemplified in the treatment of 3-phenyl-2-thioxo-1,2-dihydro-4(3H)-quinazolinone with p-toluenesulfenyl chloride to give 4-oxo-3-phenyl-3,4-dihydroquinazoline-2-yl p-tolyl disulfide (**124**).[410]

(122)   (123)   (124)

The *preparation of diquinazolinyl sulfides* has been done usually by treatment of quinazolinethiones (Section 5.1.2.5) or extranuclear mercaptoquinazolines (Section 5.2) with thionyl chloride, but occasionally such sulfides have arisen as byproducts during other reactions; for example, thioxo-1,2-dihydro-4(3H)-quinazolinone with diethyl azodicarboxylate gave not only the expected pentacyclic product but also bis(4-oxo-3,4-dihydroquinazolin-2-yl) sulfide (**125**) in 17% yield.[1460] Naturally, all preparative information on the (unsymmetrical) alkyl quinazolinyl sulfides, such as (**126**), has been collected under alkylthioquinazolines (Section 5.3.1).

(125)   (126)

Of the few *reactions of disulfides*, reduction to quinazolinethiones has been exemplified in Section 5.1.1. Although simple oxidation to quinazolinesulfonic acids or the like must be possible, it does not appear to have been used recently. However, bis(4-oxo-3-phenyl-3,4-dihydroquinazolin-2-yl) disulfide (**127**, R = H) underwent oxidative cyclization by N-bromosuccinimide in sulfuric acid at 70° for 7 h to afford 12H-benzothiazolo[2,3-b]quinazolin-12-one (**129**) in 90% yield; in sulfuric acid alone (under the same conditions), the disulfide (**127**, R = H) suffered disproportionation to a separable mixture of the tetracyclic product (**129**) (72%) and 3-phenyl-2-thioxo-1,2-dihydro-4(3*H*)-quinazolinone (**128**) (60%).[1776] Irradiation of the kindred substrate, bis(3-*o*-chlorophenyl-4-oxo-3,4-dihydroquinazolin-2-yl)disulfide (**127**, R = Cl) (at 300 nm in CHCl$_3$ − MeOH for 32 h), also gave the foregoing product (**129**) in 48% yield, although the eliminative mechanism is not clear.[2122] The nmr spectrum of bis[*o*-(6-methylquinazolin-4-yl)phenyl] disulfide (**130**)[633] has been studied in some detail.[805]

(**127**) (**128**) (**129**)

(**130**)

Although alkyl quinazolinyl sulfides undergo a variety of reactions (Section 5.3.2), the reported *reactions of diquinazolinyl sulfides* appear to be confined to oxidation; for example, bis(-4-oxo-3-phenyl-3,4-dihydroquinazolin-2-yl) sulfide (**130**, *n* = 0) gave a 60% yield of the corresponding sulfone (**130**, *n* = 2) on treatment with hydrogen peroxide in refluxing acetic acid during 1 h.[1726]

5.5. QUINAZOLINESULFONIC ACIDS AND RELATED DERIVATIVES (*H* 480)

Recently reported information in these areas is so scrappy that it is simply collected in the following paragraphs, each covering available data on a single type of derivative.

Quinazolinesulfonic acids. Most such acids have been made by oxidation of quinazolinethiones (see Section 5.1.2.3) and occasionally from halogenoquinazolines with sulfite ion (see toward the end of Section 3.7). An interesting extranuclear sulfothioquinazoline has been made by maintaining an aqueous solution of 4-aziridin-1'-ylquinazoline (**131**) and sodium thiosulfate at pH 5 during 4 h to give 4-*β*-(sulfothio)ethylaminoquinazoline (**132**) in 91% yield.[219]

(131)   (132)

Potassium 4-quinazolinesulfonate has been converted into 4-hydrazino-quinazoline (15%) by boiling for 1 h with aqueous alcoholic hydrazine.[281]

Quinazolinesulfonyl halides. The formation of quinazolinesulfonyl fluorides by oxidation of quinazolinethiones in the presence of fluoride ion has been covered in Section 5.1.2.3. Several 6-quinazolinesulfonyl chlorides have been made by direct chlorosulfonation; for example, 2,4-quinazolinediamine (**133**, R = H) as its sulfate salt reacted with chlorosulfonic acid (containing sodium chloride) at 160° to afford 2,4-diamino-6-quinazolinesulfonyl chloride (**133**, R = SO$_2$Cl) (>95%, as its sulfate salt);[1563] 7-chloro-3-dimethylamino-2,4-di-oxo-1,2,3,4-tetrahydro-6-quinazolinesulfonyl chloride (**134**, R = Cl)[1603] and 7-chloro-2-ethyl-4-oxo-3,4-dihydro-6-quinazolinesulfonyl chloride[970] were made somewhat similarly. Other approaches are illustrated by diazotization of 5-methyl-2,4,6-quinazolinetriamine and subsequent treatment with a solution of sulfur dioxide and cuprous chloride in glacial acetic acid to give 2,4-diamino-5-methyl-6-quinazolinesulfonyl chloride in ~60% yield,[1562] by the conversion of 6-(1-hydroxy-6-methylamino-3-sulfonaphthalen-2-ylazo)- into 6-(3-chlorosul-fonyl-1-hydroxy-6-methylaminonaphthalene-2-ylazo)-2-methyl-4(3*H*)-quinazolinone (substrate Na salt, SOCl$_2$, Me$_2$NCHO, 30°→95°, 2 h: >72%),[2158] and by peculiar chlorination of a benzylthioquinazoline outlined in Section 5.3.2.

(133)   (134)

Aminolysis of quinazolinesulfonyl halides can give either quinazolinesul-fonamides or quinazolinamines, according to the position of the halogenosul-fonyl group. Thus 2,4-diamino-6-quinazolinesulfonyl chloride (**133**, R = SO$_2$Cl) gave 2,4-diamino-*N*-*p*-chlorophenyl-6-quinazolinesulfonamide (**133**, R = SO$_2$NHC$_6$H$_4$Cl-*p*) (H$_2$NC$_6$H$_4$Cl-*p*) (*p*-chloroaniline, EtOH, reflux, 1 h: 37%);[1563] rather similar procedures afforded 7-chloro-3-dimethylamino-2,4-dioxo-1,2,3,4-tetrahydro-6-quinazolinesulfonamide (NH$_4$OH: 63%)[1603] and many other such sulfonamides.[970,1486,1537,1563,2584] In contrast, 2-quinazolinesulfonyl fluoride (**135**, R = SO$_2$F) reacted exothermally with neat dipropylamine to give 2-dipropylaminoquinazoline (**135**, R = NPr$_2$) ($\sim$ 50%);[281] 2-quinazolinamine (**135**, R = NH$_2$) and 2-diethylaminoquinazoline (**135**, R = NEt$_2$) were made rather similarly, using liquid ammonia and diethylamine, respectively.[281]

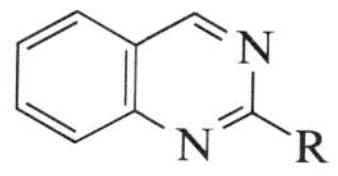

(**135**)

Quinazolinesulfonamides. A well-established synthetic route, at least to 6-quinazolinesulfonamides, has been discussed in the preceding paragraph. Other recently used approaches are of little interest; prolonged treatment of commercially available 7-chloro-2-ethyl-4-oxo-1,2,3,4-tetrahydro-6-quinazolinesulfonamide (**136**) in hot dilute hydrochloric acid gave mainly the fission product, 2-amino-3-chloro-4-sulfamoylbenzamide (**137**), but also a 12% yield of the dehydrogenation product, 7-chloro-2-ethyl-4-oxo-3,4-dihydro-6-quinazolinesulfonamide (**138**);[577] aminolysis of 6,8-dibromo-3-α-chloroacetamido-2-methyl-4(3*H*)-quinazolinone with sulfanilamide in refluxing ethanolic pyridine gave 65% of 6,8-dibromo-2-methyl-3-α-(*p*-sulfamoylanilino)acetamido-4(3*H*)-quinazolinone.[2310]

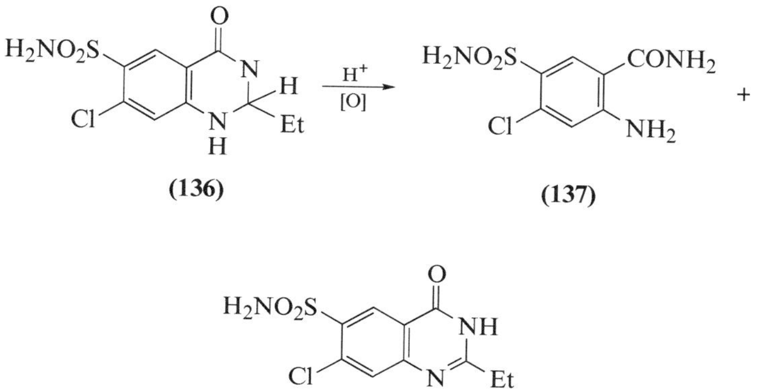

(**136**)   (**137**)

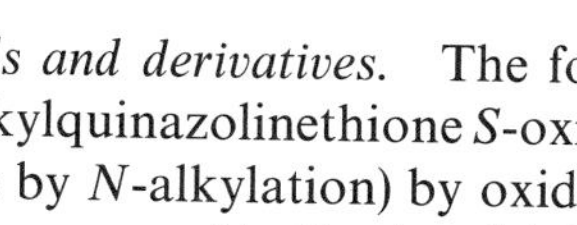

(**138**)

Quinazolinesulfenic acids and derivatives. The formation of both quinazolinesulfenic acids and *N*-alkylquinazolinethione *S*-oxides (i.e., sulfenic acids fixed as their *S*-oxide tautomers by *N*-alkylation) by oxidation of the corresponding quinazolinethiones has been covered in Section 5.1.2.3.

Treatment of 2-*t*-butyl-4-quinazolinesulfenic acid (**139**, R = H) with diazomethane gave methyl 2-*t*-butyl-4-quinazolinesulfenate (36%), formulated as structure (**140**) rather than as the sulfoxide (**139**, R = Me).[104]

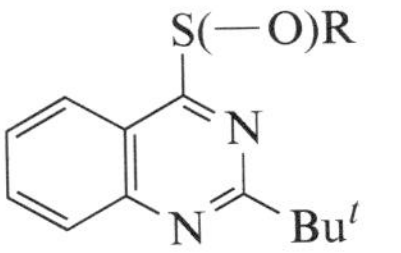

(139) 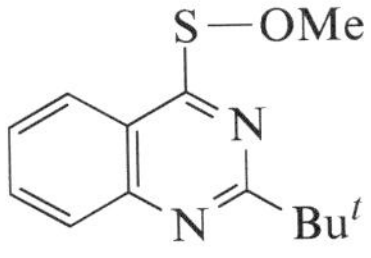   (140)

5.6.   QUINAZOLINE SULFOXIDES AND SULFONES

The *preparation* of sulfoxides and sulfones from alkylthioquinazolines or diquinazolinyl sulfides has been discussed in Sections 5.3.2 and 5.4, respectively. The formation of 4-*p*-tolylsulfonylquinazoline (**141**, R = Ts) from 4-chloroquinazoline (**141**, R = Cl) and its subsequent conversion into 4-quinazolinecarbonitrile (**141**, R = CN) has been covered in Section 3.5.5. 6-Naphthalen-2′-ylsulfonyl-5,6,7,8-tetrahydro-2,4-quinazolinediamine (**142**) has been resolved using 9-(α-carboxyethoxyimino)-2,4,5,7-tetranitrofluorene: neither optical isomer showed better antimalarial activity than the racemic mixture.[812]

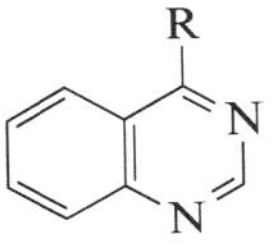

(141) 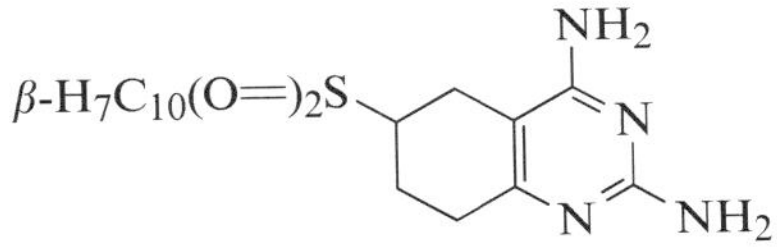   (142)

Potential *reactions* of such quinazolines are poorly represented. However, aminolysis of 2-methylsulfinyl-5,6,7,8-tetrahydro-4-quinazolinamine (**143**, R = SOMe) gave 5,6,7,8-tetrahydro-2,4-quinazolinediamine (**143**, R = NH$_2$) (NH$_4$OH, EtOH, 180°, sealed, 7 h: 36%), 2-methylamino-5,6,7,8-tetrahydro-4-quinazolinamine (**143**, R = NHMe) (MeNH$_2$, EtOH–H$_2$O, likewise: 58%), or other such products.[1398]

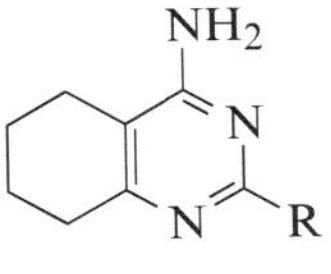

(143)

CHAPTER 6

Nitro, Amino, and Related Quinazolines

This chapter includes quinazolines bearing nitrogenous substituents that are joined directly or indirectly to the nucleus through their nitrogen atom; the only exceptions are isocyanato- and isothiocyanatoquinazolines, which are covered in Chapter 7, to be included with quinazolinecarbonitriles and the like.

6.1. NITROQUINAZOLINES

Although several 2- or 4-nitropyrimidines have been made,[2446f] only 6- to 8- or extranuclear-nitroquinazolines are known at present.

6.1.1. Preparation of Nitroquinazolines (*H* 34,55,340)

The most versatile route to nitroquinazolines is by *primary synthesis*, already covered in Chapter 1.

An alternative approach by *direct nitration* can afford a variety of mono- or dinitroquinazolines. On the basis of theoretical considerations for quinazoline (*H* 34) and subsequent experimental data, the position(s) at which nitration will occur may be forecast with reasonable confidence, although each product still requires structural confirmation by nmr or other means. As a general rule, a 6-nitro derivative is the usual product, but it may be accompanied by its 8-isomer or (under more vigorous conditions) by the 6,8-dinitro derivative; when the 6-position is already occupied in the substrate, a 5-nitro derivative is more often formed than the expected 8-nitro derivative; and extranuclear nitration, such as at a secondary amino or a phenyl substituent, may occur in addition to (or even in lieu of) regular nuclear nitration.

Typical products available by nitration, as well as the conditions used in their formation, are listed hereafter.

6/8-Nitrations

6-Nitro-2,4-quinazolinediamine (**1**, R = H) (substrate.HNO$_3$, fuming HNO$_3$/ conc H$_2$SO$_4$, < 10°, 15 min; then 20°, 1 h: 89%);[201] also its 7-methyl derivative (**1**, R = 7-Me) (free substrate, likewise: 93%).[600]

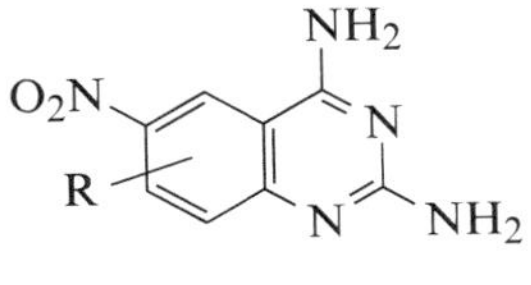

(1)

7-Fluoro-6-nitro-2,4-quinazolinediamine (**1**, R = 7-F) (fuming HNO_3/conc H_2SO_4, 0°, 15 min; then 20°, 5 h: 82%);[2335] its 8-fluoro isomer (**1**, R = 8-F) (likewise);[2335] and a separable mixture of 5-fluoro-6-nitro- (**1**, R = 5-F) plus 5-fluoro-8-nitro-2,4-quinazolinediamine (**2**) (likewise: 40% and 16%, respectively).[2221]

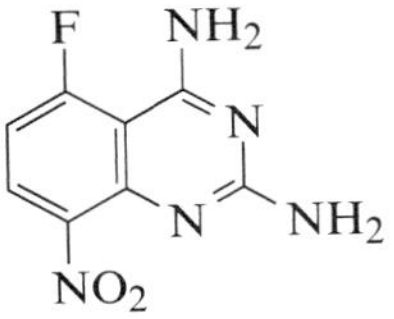

(2)

2-Amino-6-nitro-5-trifluoromethyl-4(3H)-quinazolinone (**3**, R = CF_3) (fuming HNO_3/conc H_2SO_4, 0°, 30 min; then 20°, 24 h: 61%);[2211,2344] an inseparable mixture of 2-amino-5-fluoro-6-nitro-4(3H)-quinazolinone (**3**, R = 5-F) and its 8-nitro isomer (**4**, Q = H, R = 5-F) (likewise: 96% total);[2374, cf. 2344] and 2-amino-7-chloro-6-nitro- (**3**, R = 7-Cl) (fuming HNO_3/conc H_2SO_4, − 10°, then 90°, 10 min: 90%) or 2-amino-7-chloro-6,8-dinitro-4(3H)-quinazolinone (**4**, Q = NO_2, R = 7-Cl) (likewise but more HNO_3: 68%).[915]

2-Amino-6-bromo-8-nitro-4(3H)-quinazolinone (**4**, Q = Br. R = H) (fuming HNO_3, 80°, 2 h: 54%).[2025]

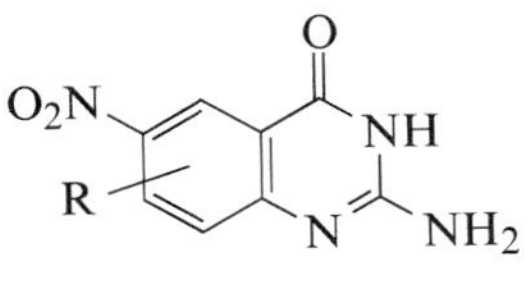

(3) 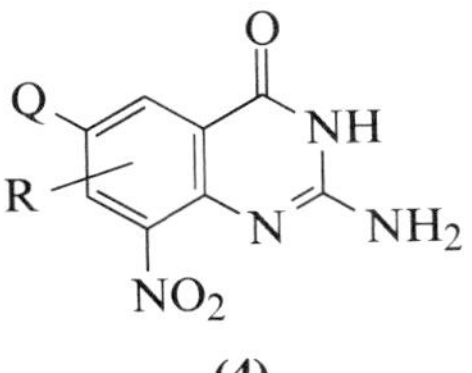   **(4)**

5-Chloro-6-nitro-2,4(1H,3H)-quinazolinedione (**5**, R = 5-Cl) (fuming HNO_3/conc H_2SO_4, − 10°, then 95°, 10 min: 97%);[1466,2261] its 7-chloro isomer (**5**, R = 7-Cl) (likewise: 85%);[1358] and 5-chloro-3-methyl-6-nitro-2,4(1H,3H)-quinazolinedione (likewise: 84%).[2261]

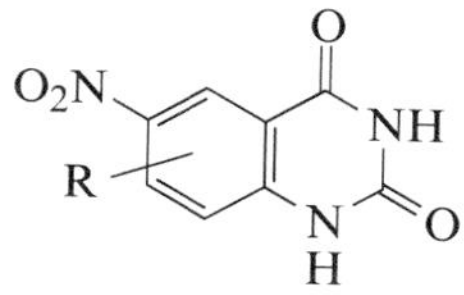

$$\text{(5)}$$

2-Methyl-6-nitro- (**6**, R = Me) (fuming HNO$_3$/conc H$_2$SO$_4$, < 75°: 95%),[1118,2158] 6-nitro-2-*p*-nitrophenyl- (**6**, R = C$_6$H$_4$NO$_2$-*p*) (HNO$_3$/ H$_2$SO$_4$, 20°, 10 h: ?%; note that the *p*-nitro group was present in the substrate),[666] and 7-chloro-6-nitro-4(3*H*)-quinazolinone (HNO$_3$/conc H$_2$SO$_4$, 100°, 2 h: 57%).[909]

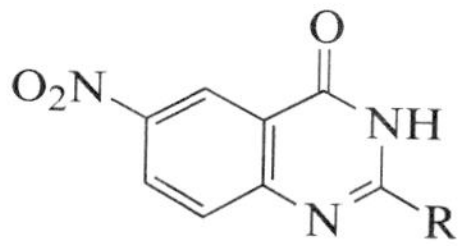

$$\text{(6)}$$

Also other examples in the foregoing references and elsewhere.[123,501,586,681, 1987,2015]

5-Nitrations

6-Methyl- (**7**, R = Me) (fuming HNO$_3$/conc H$_2$SO$_4$, 20°, 30 min; then 95°, 1 h: 67%[1159,1366] KNO$_3$/conc H$_2$SO$_4$, 30°, 24 h: 65%),[1387] 6-chloro- (**7**, R = Cl) (fuming HNO$_3$/conc H$_2$SO$_4$, 95°, 2 h: 26%),[502,2584] and 6-acet-amido-5-nitro-4(3*H*)-quinazolinone (**7**, R = NHAc) (fuming HNO$_3$/ conc H$_2$SO$_4$, < 10°, 30 min: 40%).[502]

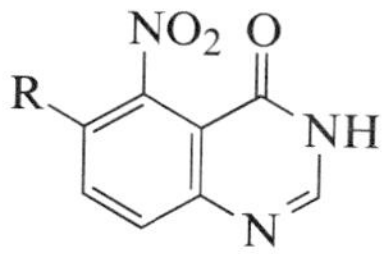

$$\text{(7)}$$

6-Methoxy-5-nitroquinazoline (**8**) (KNO$_3$/conc H$_2$SO$_4$, − 5°, 3 h: 87%).[1577]

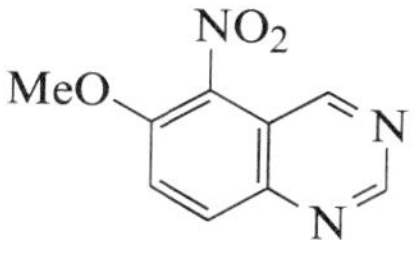

$$\text{(8)}$$

Also other examples.[121]

6,ω-Dinitrations

6-Acetamido-7-(N-methyl-N-nitroamino)-8-nitro-4(3H)-quinazolinone   (**9**)
(fuming HNO_3/conc H_2SO_4–AcOH, $0° \rightarrow 20°$, 90 min: 60%).[1962]

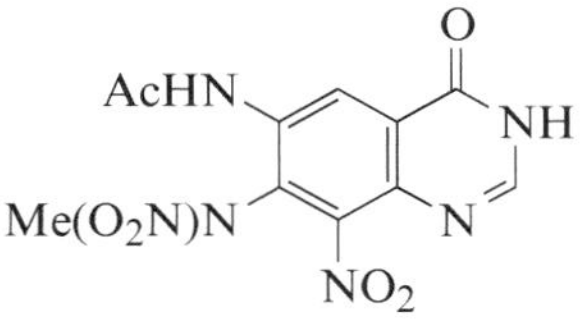

(**9**)

6-Nitro-3-m-nitrophenyl-2-pyridin-2′-yl-4(3H)-quinazolinone (**10**)   (fuming
HNO_3/conc H_2SO_4, 5°; then 25°, 5 h: 52%);[791] also related compounds
similarly.[1882]

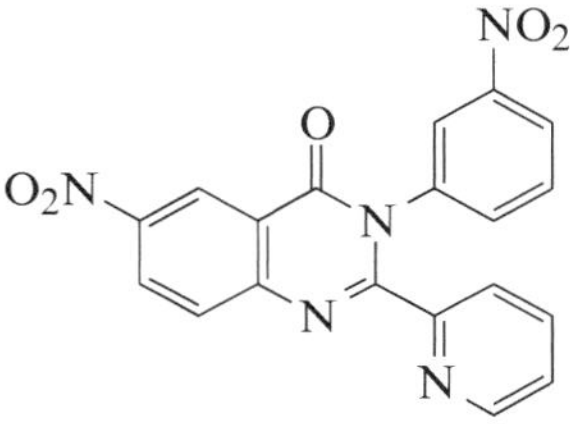

(**10**)

6.1.2.   Reactions of Nitroquinazolines (*H* 340)

Almost every known nitroquinazoline has been made only for subsequent
reduction to the corresponding quinazolinamine. Although catalytic hydrogenation has been favored by many workers, a variety of classic reducing agents has
been used quite effectively; the choice is affected sometimes by the presence of
passenger groups. The following primary quinazolinamines have been made
from the corresponding nitroquinazolines and will serve to exemplify reduction
methods, conditions, and yields.

Hydrogenation over palladium

6-Amino-4(3H)-quinazolinone (**11**) (H_2, Pd/C, $MeOCH_2CH_2OH$, 20°,
92%).[1622, cf. 321]

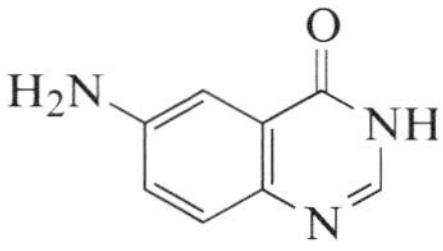

(**11**)

7-Amino-3-*p*-methoxyphenyl-4(3*H*)-quinazolinone   (H$_2$,   2 atm,   Pd/C,
  Me$_2$NCHO, 20°, 30 min: 52%).[2467]

Ethyl 6-amino-4-oxo-3,4-dihydro-2-quinazolinecarboxylate (12) (H$_2$, Pd/C,
  MeOH: ∼ 75%).[123]

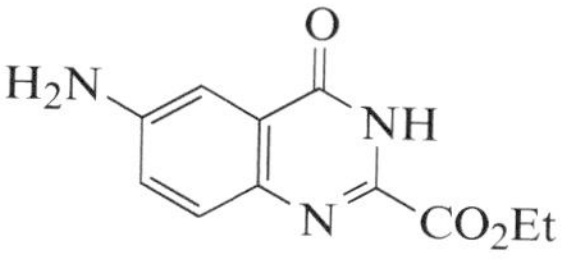

(12)

2-*o*-Aminophenyl-1-methyl-4(1*H*)-quinazolinone (H$_2$, Pd/C, AcOH, 20°,
  30 min: 69%).[40]

2,4,6-Quinazolinetriamine (13, Q = H, R = NH$_2$) [from the 6-nitro analog (13,
  Q = H, R = NO$_2$); H$_2$, 3 atm, Pd/C, 20°, 90 min: 94%];[956] 2,4,5,6-
  quinazolinetetramine (13, Q = R = NH$_2$) [from the 6-nitro analog (13,
  Q = NH$_2$, R = NO$_2$); H$_2$, 3 atm, Pd/C, MeOH–AcOH, 20°, 42 h:
  88%].[1443]

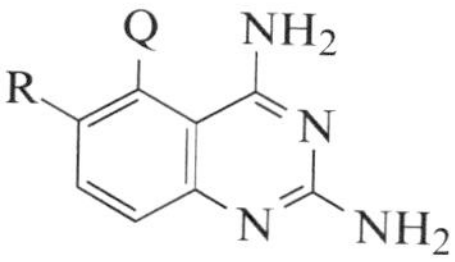

(13)

2,4,5-Quinazolinetriamine (from the 5-nitro analog; H$_2$, Pd/C, HCl, EtOH–
  H$_2$O, 20°, 45 min: 87%).[2225]

2,4,6,8-Quinazolinetetramine (from the 6,8-dinitro analog; H$_2$, Pd/C,
  Me$_2$NCHO, exothermic, 2 h: 93%).[601]

2,8-Diamino-4(3*H*)-quinazolinone (14, Q = H, R = NH$_2$) [from 2-amino-6-
  bromo-8-nitro-4(3*H*)-quinazolinone (14, Q = Br, R = NO$_2$); H$_2$, 3 atm,
  Pd/C, MeOH, 20°, 24 h: 69%, as hydrochloride; note debromination].[2025]

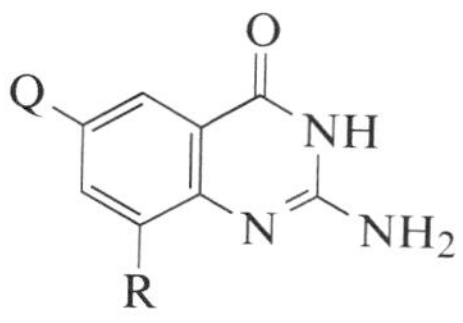

(14)

Also other examples.[201,282,501,638,1159,1513,1564,1987,2211,2221,2232,2335]

Hydrogenation over Raney-nickel

6-Methoxy-5-quinazolinamine (H_2, Ni, MeOH–dioxane, 20°: 90%).[1577]

6-Amino-N-methyl-4-phenyl-2-quinazolinecarboxamide (**15**) (H_2, Ni, EtOH, 20°: 85%).[242]

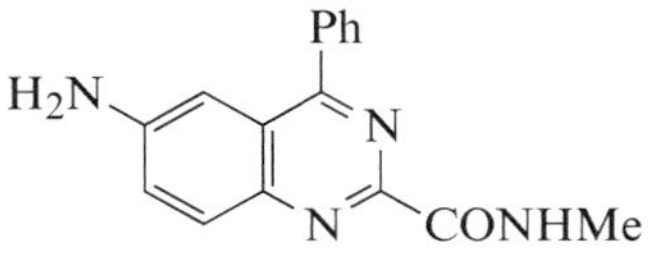

(15)

2,4,7-Quinazolinetriamine [from the 7-nitro analog; H_2, 3 atm, Ni, MeOCH$_2$CH$_2$OH, 25°, 17 h: 38%].[600]

6-Amino-3-p-aminophenyl-1-methyl-2,4($1H,3H$)-quinazolinedione (**16**, R = NH$_2$) [from the dinitro analog (**16**, R = NO$_2$); H_2, 60 atm, Ni, Me$_2$NCHO, 70°: 67%].[3]

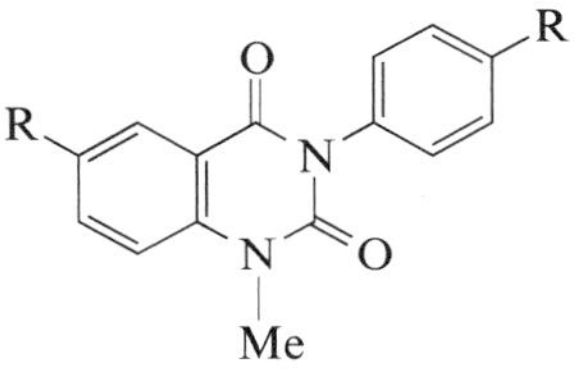

(16)

1-Isopropyl-7-methyl-6-nitro- gave 6-dimethylamino-1-isopropyl-7-methyl-4-phenyl-2($1H$)-quinazolinone (H_2, 3 atm, Ni, HCHO, MeOH–H$_2$O–di-oxane, 20°, 2 h: 87%; a useful reductive alkylation).[681]

Also other examples.[2,113,121,2152,2374]

Hydrazine catalyzed by palladium

5-Amino-6-methyl-4($3H$)-quinazolinone (H_2NNH$_2$.H$_2$O, Pd/C, EtOH, 0° → 20°, 1 h; then reflux, 2 h: 86%;[1387] likewise in EtOH–H$_2$O, reflux, 1 h: ~ 75%).[1366]

6,7-Diamino-4($3H$)-quinazolinone [from the 6-nitro analog; H_2NNH$_2$.H$_2$O, Pd/C, EtOH, N$_2$, 20°, 1 h; then reflux, 1 h: 91%].[501]

Also other examples.[952,1157,1165]

Hydrazine catalyzed by Raney-nickel

2-o-Aminophenyl-7-ethyl-4-o-methoxycarbonylanilinoquinazoline (**17**) (H_2N NH$_2$.H$_2$O, Ni, BuOH, 95°, 1 h: 81%).[195]

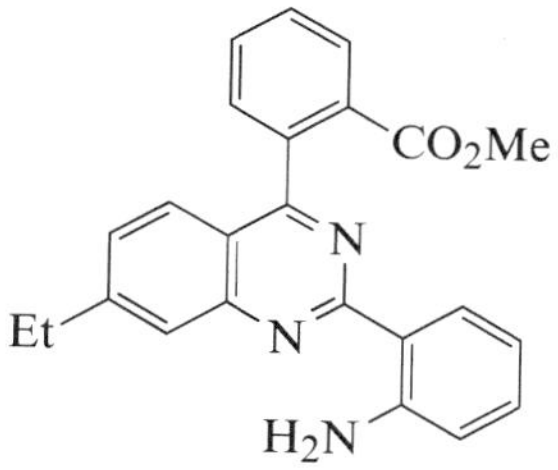

(17)

4-o-Aminoanilino-6-bromo-2-piperidinoquinazoline ($H_2NNH_2.H_2O$, Ni, THF, reflux, < 10 h: 98%).[2480]

Also many other examples in the foregoing references and elsewhere.[1672,1746]

Stannous chloride or tin/acid

2,4,6-Quinazolinetriamine (18, R = H) [from the 6-nitro analog; $SnCl_2.2H_2O$, HCl–AcOH, < 30°; then 4°, 24h: 92%];[1919] 5-chloro-2,4,6-quinazoline-triamine (18, R = Cl) [from the 6-nitro analog; likewise: 64%].[201]

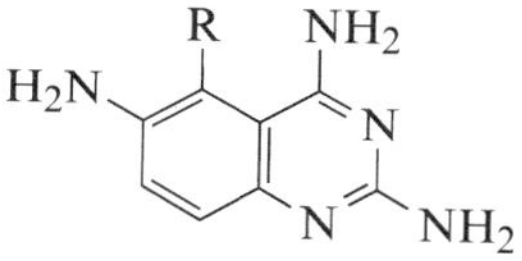

(18)

3-p-Aminophenyl-2,4(1H,3H)-quinazolinedione (19) (Sn, 10M HCl, hot, 2 h: 74%).[2248]

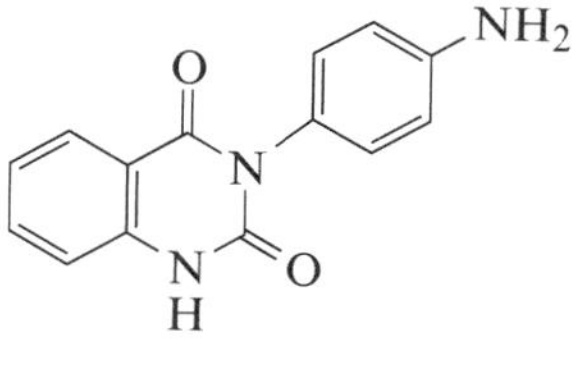

(19)

Also other examples.[356,666]

Iron/acid

6-Amino-1-isopropyl-4-phenyl-2(1H)-quinazolinone (20) (Fe filings, HCl, EtOH–H_2O, reflux, < 3 h: 81%).[681]

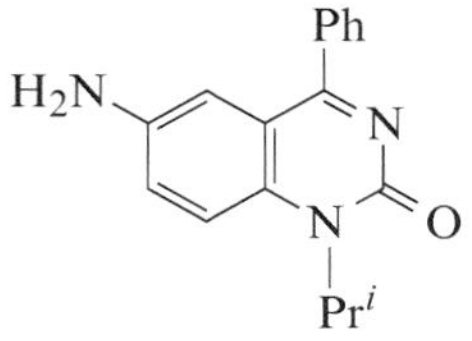

(20)

6-Amino-2-methyl-3-*o*-tolyl-4(3*H*)-quinazolinone (Fe powder, PhH–H$_2$O–
HCl, reflux, 3.5 h: ∼ 70%).[823]

2,4,6-Quinazolinetriamine (**18**, R = H) [from the 6-nitro analog; Fe, NaCl,
1.5 M HCl, reflux, 3 h: 70%].[1541]

Also other examples.[564,2264]

Sodium dithionite or sodium sulfide

3-(4-Amino-2-methoxyphenyl)-2-methyl-4(3*H*)-quinazolinone   (**21**,   R = H)
(Na$_2$S$_2$O$_4$, 0.2 M NaOH, 80°, 10 min: 80%);[1849] also the 2-hydroxymethyl
analog (**21**, R = OH) (likewise: 92%).[1849]

6-Amino-2-methyl-4(3*H*)-quinazolinone (Na$_2$S, H$_2$O, reflux, 2 h:87%).[2158]

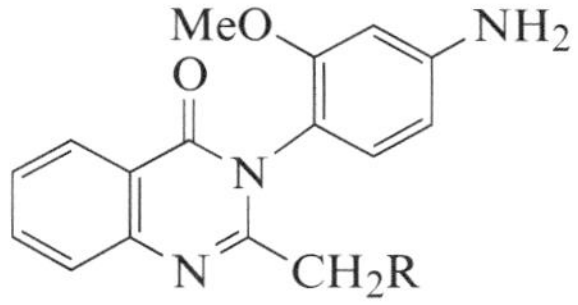

(21)

Titanous chloride

6-Quinazolinamine (TiCl$_3$, AcMe, N$_2$, 20°, 1 h: 83%).[2075]

A rare *aminolysis* of a nitroquinazoline has been reported: treatment of
6-chloro-5-nitro-4(3*H*)-quinazolinone (**22**, R = NO$_2$) and butanolic ammonia in
an autoclave at 175° for 24 h gave—not the 6-amino-5-nitro derivative as might
be expected, but—5-amino-6-chloro-4(3*H*)-quinazolinone (**22**, R = NH$_2$) in
83% yield.[502] Likewise, *alkanethiolysis* of 6-chloro-3-methyl-5-nitro-4(3*H*)-
quinazolinone, with benzyl mercaptan in potassium carbonate/dimethylforma-
mide at 20° for 24 h under nitrogen, gave 5-benzylthio-6-chloro-3-methyl-4(3*H*)-
quinazolinone in 96% yield.[2584]

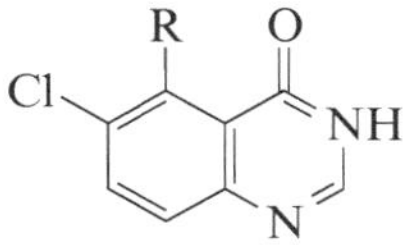

(22)

Several *reductive cyclizations* of appropriately substituted nitroquinazolines have been observed. For example, 5-amino-1,3-dimethyl-6-nitro-2,4(1H,3H)-quinazolinedione (23) gave 6,8-dimethyl-1H-imidazo[4,5-f]quinazoline-7,9(6H,8H)-dione (24) in 98% yield on hydrogenation over Pd/C in formic acid, followed by a period under reflux;[1466] a rather analogous reaction gave 7-amino-1H-imidazo[4,5-f]quinazolin-9(8H)-one (58%);[2015] and 6-acetamido-7-(N-methyl-N-nitroamino)-8-nitro-4(3H)-quinazolinone (9) gave 4-amino 2,3-dimethyl-3H-imidazo[4,5-g]quinazolin-8(7H)-one (25) in 80% yield (as hydrochloride) by hydrogenation over Pd/C in methanolic alkali and subsequent acidification (by elimination of hydroxylamine?).[1962] Other examples have been reported.[1277]

(23)　　　　　　　　(24)　　　　　　　(25)

The *structure* of 6-nitro-4-phenyl-1,2-dihydroquinazoline has been checked by X-ray crystallography.[2215]

6.2. REGULAR AMINOQUINAZOLINES (*H* 322)

This section covers primary, secondary, tertiary, and quaternary amino-quinazolines but not N-(functionally substituted)aminoquinazolines such as hydrazino-, hydroxyamino-, or ureidoquinazolines.

6.2.1. Preparation of Regular Aminoquinazolines (*H* 322,337,340)

The *main routes* (and some minor ones) to regular aminoquinazolines have been discussed already as indicated in the following list, which includes the potential scope of each method.

By *primary synthesis* (primary, secondary, tertiary; all positions): Chapter 1.

By *aminolysis of halogenoquinazolines* (primary, secondary, tertiary, quaternary; all positions): Sections 3.5.1, 3.6, and 3.7.

By *aminolysis of alkoxyquinazolines* (primary, secondary, tertiary; 2/4-position): Section 4.5.2.1.

By *aminolysis of alkylthioquinazolines* (primary, secondary, tertiary; 2/4-position): Section 5.3.2.

By *aminolysis of tautomeric quinazolinethiones* (primary, secondary, tertiary; 2/4-position): Section 5.1.2.2.

By *aminolysis of tautomeric quinazolinones* (primary, secondary, tertiary; 2/4-position): Section 4.1.2.4.

By *aminolysis of alkylsulfinyl- or alkylsulfonylquinazolines* (primary, secondary, tertiary; 2/4-position): Section 5.6.

By *aminolysis of quinazolinesulfonyl fluorides* (primary, secondary, tertiary; 2/4-position): Section 5.5.

By *direct C-amination* (primary only; 4-position): Sections 2.1.3 and 2.2.2.6.

By *reduction of nitroquinazolines* (primary only; 5- to 8- and extranuclear-positions): Section 6.1.2.

By the *Mannich reaction* (secondary, tertiary; extranuclear positions): Section 2.2.2.3.

The remaining *minor synthetic routes* to regular aminoquinazolines are illustrated briefly in the following examples, to which notes are added as required.

By deacylation of acylaminoquinazolines

4-Acetamido- (**26**, R = Ac) or 4-methoxycarbonylamino- (**26**, R = CO$_2$Me) gave 4-amino-2(1H)-quinazolinethione (**26**, R = H) (1.25M NaOH, reflux, 3 min: > 89% each).[2095]

6-Acetamido- gave 6-amino-5-nitro-4(3H)-quinazolinone (1M HCl, reflux, N$_2$, 2 h: 75%).[502]

6-Acetamido- gave 6-amino-3-o-tolyl-2-trifluoromethyl-4(3H)-quinazolinone (HCl/MeOH, reflux, 30 min: 93%).[1397]

3-Acetamido- (**27**, R = Ac) gave 3-amino-2-thioxo-1,2-dihydro-4(3H)-quinazolinone (**27**, R = H) (10M HCl, 95°, 2 h: 70%).[1166]

2-Phenyl-1-β-phthalimidoethyl-1,2-dihydro-4(3H)-quinazolinone (**28**) gave 1-β-aminoethyl-2-phenyl-1,2-dihydro-4(3H)-quinazolinone (H$_2$NNH$_2$. H$_2$O, EtOH, 50°, then reflux, 2 h: 80%).[1634]

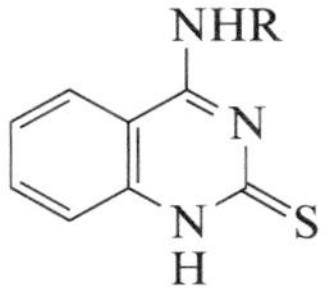

(26) 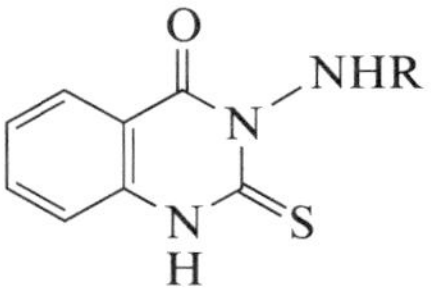   (27)

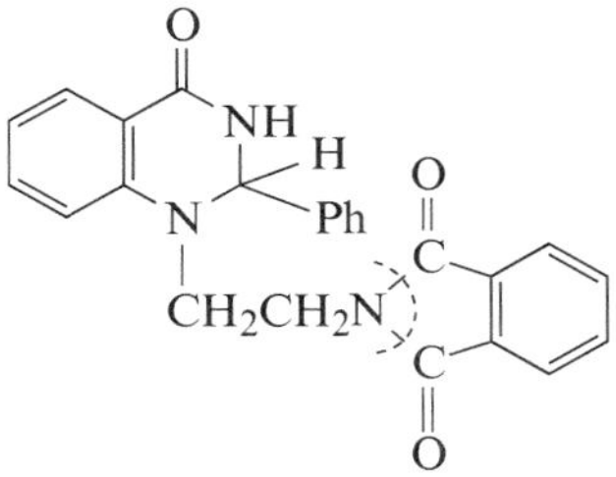

(28)

Also other examples,[152,833,871,911,1418,1557,1706,1721,2008,2017,2502,2586,2593]

By hydrolysis or reduction of Schiff bases

3-Dimethylaminomethyleneamino- (29) gave 3-amino-4-oxo-3,4-dihydro-2-quinazolinecarbonitrile (HCl, MeOH–H_2O, reflux, 1 h: 42%).[869]

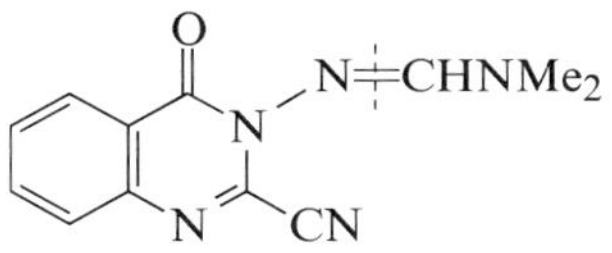

(29)

3-(Indol-3-ylmethylene)amino- gave 3-(indol-3-ylmethyl)amino-2-methyl-4(3H)-quinazolinone ($H_2NNH_2.H_2O$, Pd/C, Me_2NCHO, reflux, 18 h: ?%);[2447] also analogs.[2447]

Also other examples.[519,568,612,1593,2590] See also the reduction of a different type of Schiff base.[2566]

By reduction or splitting of isothiocyanatoquinazolines

4-Isthiocyanato-2-morpholinoquinazoline (30, R = N:C:S) gave 4-methylamino-2-morpholinoquinazoline (30, R = NHMe) ($LiAlH_4$, Et_2O, reflux, 1 h: 35%) or 2-morpholino-4-quinazolinamine (30, R = NH_2) (HCl ↓, Bu_2O, 80°, 2 h: ?%).[439]

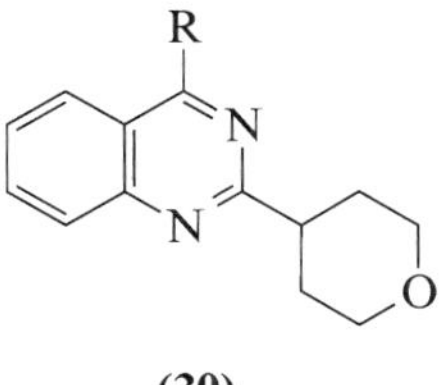

(30)

By aminolysis of (activated-alkyl)quinazolines

4-(α-Benzoyl-α-methylethyl)- (31, R = CMe_2Bz) or 4-α-cyanobenzyl- [31, R = CH(CN)Ph] gave 4-anilinoquinazoline (31, R = NHPh) (neat $PhNH_2$, 175°, > 5 h: 15% or 49%, respectively;[555] cf. Section 2.2.2.6.

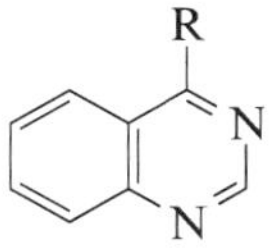

(31)

Ethyl 4-dicyanomethyl-3-methyl-2-oxo-1,2,3,4-tetrahydro-4-quinazolinecar-
boxylate [**32**, R = CH(CN)$_2$] gave the 4-anilino analog (**32**, R = NHPh)
(PhNH$_2$, Me$_2$NCHO, 110°, briefly: 46%).[727]

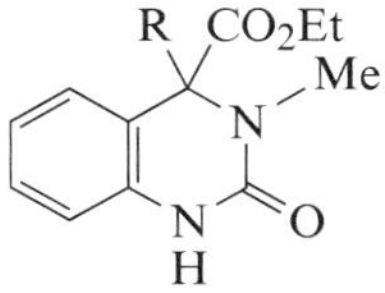

(**32**)

By aminolysis of quinazolinecarbonitriles

4-Oxo-1-phenyl-1,4-dihydro-2-quinazolinecarbonitrile (**33**, R = CN) gave 2-
dimethylamino-1-phenyl-4(1*H*)-quinazolinone (**33**, R = NMe$_2$) (40%
Me$_2$NH, 20°, 2 h: 78%).[1590]

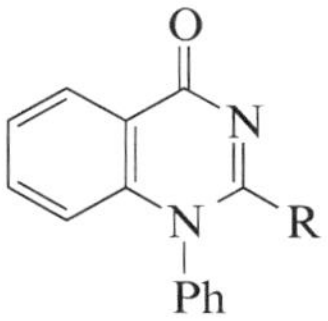

(**33**)

1-Methyl-4-oxo-1,4-dihydro-2-quinazolinecarbonitrile gave 2-amino-1-methyl-
(NH$_3$ ↓, Me$_2$NCHO, 5°, 1 h: 63%) or 1-methyl-2-methylamino- 4(1*H*)-
quinazolinone (MeNH$_2$, H$_2$O, 20°, 3 h: 61%).[1415]
Also other examples.[2031]

By reductive transamination of quinazolinecarbonitriles

Note: This well-developed route (for converting —C≡N into —CH$_2$NHR by
reduction in the presence of an amine) seems to involve two reductive
steps, addition of the amine and loss of ammonia in a yet undetermined
sequence.

2,4-Diamino-6-quinazolinecarbonitrile (**34**, R = CN) gave 6-amino methyl-
(**34**, R = CH$_2$NH$_2$) (H$_2$, Raney-Ni, NH$_3$/Me$_2$NCHO, 70 atm, 75°, 3 h:
80%)[601] or 6-(3,4-dichloroanilino)methyl-2,4-quinazolinediamine (**34**,
R = CH$_2$NC$_6$H$_3$Cl$_2$-3,4) (H$_2$, Ni, H$_2$NC$_6$H$_3$Cl$_2$-3,4, AcOH–H$_2$O, 3 atm,
28°, 22 h: 57%, as acetate salt).[1578]
2-(*N*-β-Cyanoethyl-*N*-methylamino)- (**35**, R = CN) gave 2-(*N*-γ-amino-
propyl-*N*-methylamino)-6,7-dimethoxy-4-quinazolinamine (**35**, R =
CH$_2$NH$_2$) (H$_2$, Ni, NH$_3$/EtOH, 70 atm, 70°, ? h: 52%).[1579]

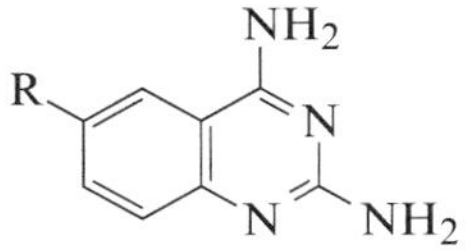

(34) 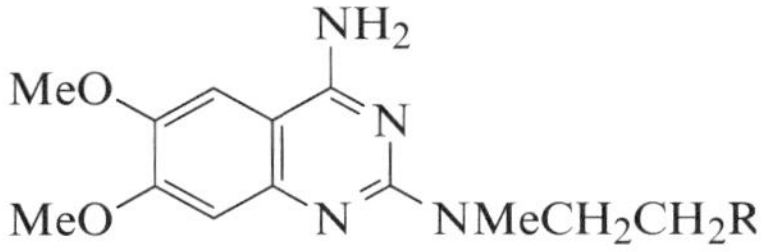   (35)

2,4-Dimino-5-methyl-6-quinazolinecarbonitrile gave 5-methyl-6-(3,4,5-tri-methoxyanilino)methyl-2,4-quinazolinediamine [H_2, Ni, $H_2NC_6H_2$-$(OMe)_3$-3,4,5, AcOH–H_2O, 3 atm, 20°, 3 h: 52%].[2041]
Also other examples in the foregoing references and elsewhere.[1585,2224,2489,2590,2597]

By treatment of a quinazolinecarbaldehyde with secondary amines

2-Methyl-5-oxo-5,6,7,8-tetrahydro-6-quinazolinecarbaldehyde (36, Q = Me) gave 6-dimethylaminomethylene-2-methyl- (37, Q = R = Me) (Me_2NH, PhMe, 20°, 24 h: 84%) or 2-methyl-6-piperidinomethylene-5,6,7,8-tet-rahydro-5-quinazolinone [37, Q = Me, $R_2 = (CH_2)_5$] [$HN(CH_2)_5$, like-wise: 72%];[1916] 5-oxo-2-phenyl-5,6,7,8-tetrahydro-6-quinazolinecarb-aldehyde (36, Q = Ph) gave 6-diphenylaminomethylene -2-phenyl-5,6,7,8-tetrahydro-5-quinazolinone (37, Q = R = Ph) (Ph_2NH, PhMe, reflux, Dean–Stark, 4 h: 80%).[1916]

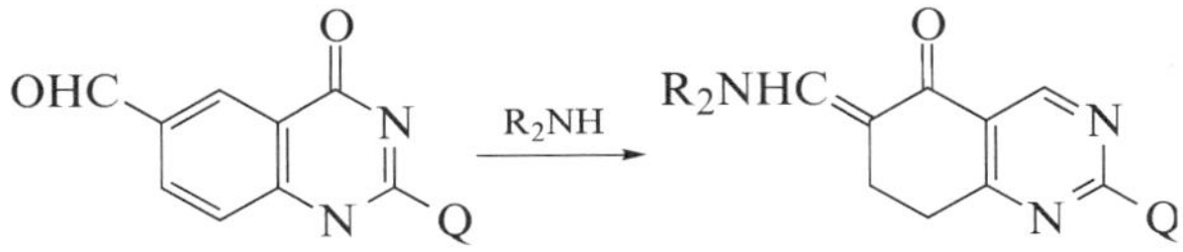

(36)   (37)

By reduction of quinazoline oximes or hydroxyaminoquinazolines

6-Chloro-4-hydroxy-3-methyl-4-phenyl-3,4-dihydro-2-quinazolinecarbal-dehyde oxime (38, R = CH:NOH) gave 2-aminomethyl-6-chloro-3-methyl-4-phenyl-3,4-dihydro-4-quinazolinol (38, R = CH_2NH_2) (H_2, Ni, THF–Pr^iOH, 1 h: 55%).[402]

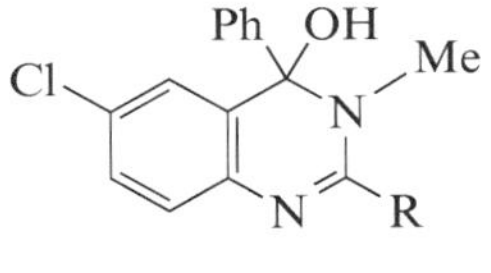

(38)

4,7,7-Trimethyl-2-phenyl-5,6,7,8-tetrahydro-5-quinazolinone oxime (**39**, QR = NOH) gave 4,7,7-trimethyl-2-phenyl-5,6,7,8-tetrahydro-5-quinazolina-mine (**39**, Q = H, R = NH$_2$) (H$_2$, Ni, 20°, reduced pressure?).[334]

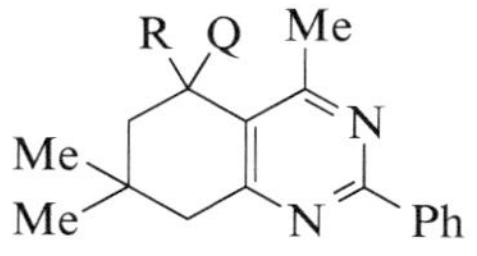

(39)

1-Benzyl-2-phenyl-4(1*H*)-quinazolinone oxime gave 2-phenyl-4-quinazolin-amine (H$_2$, Pd/C, HCl–dioxane: 78%; note additional debenzylation).[2132]

4-Hydroxyamino-2-phenylquinazoline gave 2-phenyl-4-quinazolinamine (H$_2$, Ni, MeOH, 20°: 95%).[1872]

By thermolysis of an *N*-hydroxyquinazolinecarboxamide

N-Hydroxy-1-methyl-4-oxo-1,4-dihydro-2-quinazolinecarboxamide(**40**, R = CONHOH) gave 2-amino-1-methyl-4(1*H*)-quinazolinone (**40**, R = NH$_2$) (C$_{10}$H$_8$, 200°, 3 h: 68%; a Lossen[2450j] rearrangement).[1415]

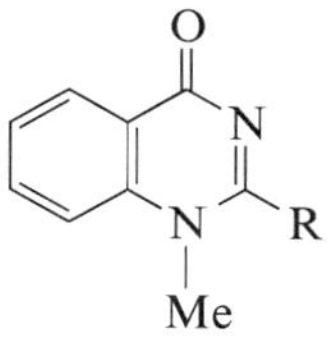

(40)

By indirect aminations

Note: The following aminations appear to occur by an addition–elimination mechanism, usually in rather poor yield.

4(3*H*)-Quinazolinone (**41**, R = H) gave 3-amino-4(3*H*)-quinazolinone (**41**, R = NH$_2$) (H$_2$NOSO$_3$Na, NaOH, H$_2$O, 20°, 5 h: 22%).[1895]

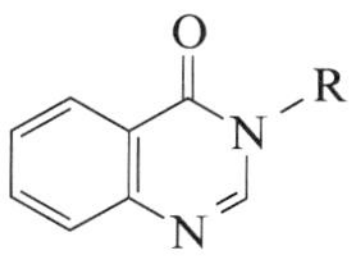

(41)

1-Phenyl-2,4(1*H*,3*H*)-quinazolinedione gave its 3-amino derivative (likewise, 3 days: ∼ 10%).[1830]

4-Phenyl-2(1H)-quinazolinone gave a separable mixture of two isomeric
N-amino-4-phenyl-2(1/3H)-quinazolinones (similarly, 60°, 1 h: 25% and
12%); 4-methyl-2(1H)-quinazolinone likewise gave a single N-amino derivative (85%). Structures were not assigned.[432]

6-Methoxyquinazoline 3-oxide (**42**, R = H) gave 6-methoxy-2-quinazolin-
amine 3-oxide (**42**, R = NH$_2$) (H$_2$NOH, 2.5M NaOH, 140°, sealed, 30 min:
20%).[4]

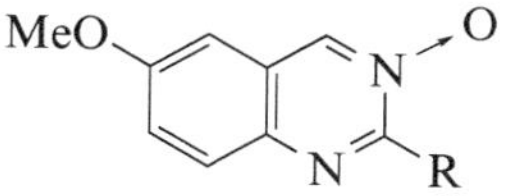

(**42**)

Also other examples.[403]

By transamination of N3 and its substituent

Note: This reaction must involve ring fission, transamination, and recyclization.

2-Hydrazino-4(3H)-quinazolinone (**43**, R = H) or its 3-ethyl derivative (**43**,
R = Et) gave 3-amino-2-hydrazino-4(3H)-quinazolinone (**43**, R = NH$_2$)
(neat H$_2$NNH$_2$.H$_2$O, reflux, 12 h: > 80% in both cases).[1884]

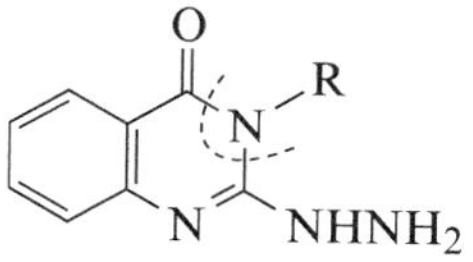

(**43**)

3-*p*-Ethoxycarbonylphenyl- (**44**) gave 3-amino-2-methyl-4(3H)-quinazol-
inone plus *p*-aminobenzanilide (H$_2$NNH$_2$.H$_2$O, EtOH, reflux, 30 min:
isolated in 53% and 35% yield, respectively).[1881]

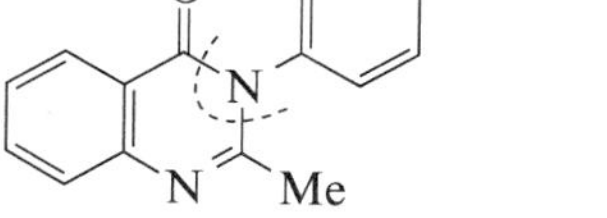

(**44**)

3-Methyl- gave 3-benzyl-4-thioxo-3,4-dihydro-2(1H)-qunazolinone (neat
PhCH$_2$NH$_2$, reflux, 8 h: 20%, after purification; note survival of the thioxo
substituent).[40]

Also other examples.[614,1046,1127,1434,1772]

By regular transamination (*H* 335)

2-Trimethylammonioquinazoline chloride (**45**) gave 2-quinazolinamine (**46**, R = H) (NH$_4$OH, NH$_4$Cl, 50°, 3 h: 62%), 2-propylaminoquinazoline (**46**, R = Pr) (neat PrNH$_2$, 50°, sealed, 3 h: 67%, as picrate), or 2-hydrazinoquinazoline (**46**, R = NH$_2$) (neat H$_2$NNH$_2$.H$_2$O, 20°, 15 min: 69%);[189] their 4-isomers were made similarly (25%, 67% as picrate, and 49%, respectively).[189]

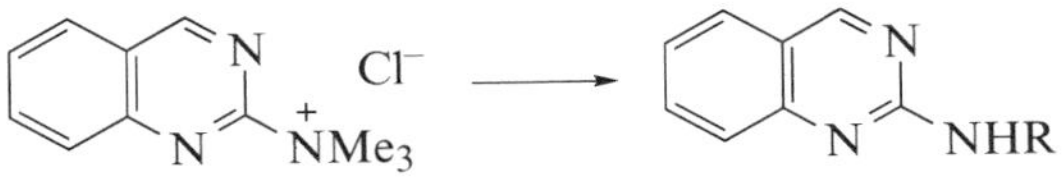

(**45**)   (**46**)

2-(5-Amino-4-methoxycarbonyl-3-methylthiopyrazol-1-yl)- (**47**) gave 2-amino-6-chloro-3-methyl-4(3*H*)-quinazolinone (**48**) (HCONH$_2$, 190°, 4 h: 79%);[2279] analogs were made similarly.[2143,2241]

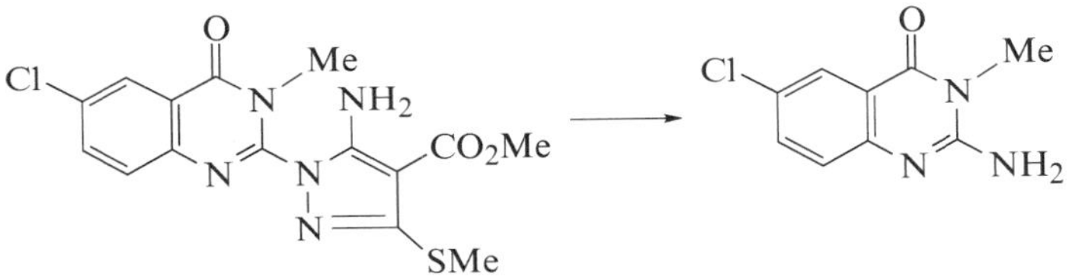

(**47**)   (**48**)

6-Acetamido-5,8-dihydro-2,4,5,8(1*H*,3*H*)-quinazolinetetrone gave 6-methylamino-8-methylimino-5,8-dihydro-2,4,5(1*H*,3*H*)-quinazolinetrione (MeNH$_2$, H$_2$O, reflux, 15 min: 56%; note initial deacetylation and final Schiff base formation at the 8-position).[1987]

Also other examples.[2072]

By *N*-alkylation or *N*-dealkylation (*H* 334)

5-β-Aminoethyl- (**49**, R = H) gave 5-β-(dipropylamino)ethyl-8-hydroxy-2,4(1*H*,3*H*)-quinazolinedione (**49**, R = Pr) (EtCHO, H$_2$, MeOH, 3 atm, 50°, 3 h: 55%, as hydrochloride).[2502]

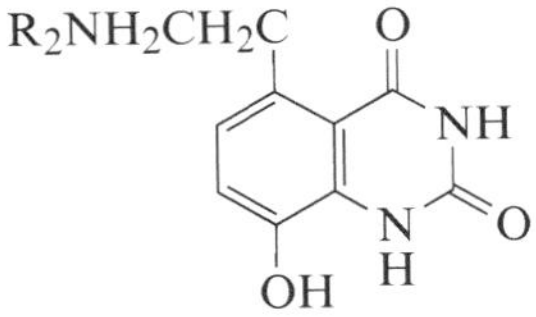

(**49**)

2,4,6-Quinazolinetriamine   gave   6-ethylamino-2,4-quinazolinediamine
(MeCHO, H_2, Pt/C, AcOH–MeOCH$_2$CH$_2$OH, 2 atm, 25°, 23 h: 27%).[602]

2-Amino- gave 2-(β,β-diethoxycarbonylvinyl)amino-4(3H)-quinazolinone
[neat (EtO$_2$C)$_2$C=CHOEt, 180°, 1 h: 90%].[2218]

6-Amino-2-methyl- gave 2-methyl-6-prop-2'-ynylamino-4(3H)-quinazolinone
(BrCH$_2$C≡CH, CaCO$_3$, PhMe—Me$_2$NCHO, A, 70°, 8 h: 84%).[2271]

2-(4-Benzylpiperazin-1-yl)- (**50**, R = CH$_2$Ph) gave 2-piperazin-1'-yl-5,6,7,8-
tetrahydro-4-quinazolinamine (**50**, R = H) (H_2, Pd/C, MeOH, 60°, 7 h:
59%, as hydrochloride).[1398]

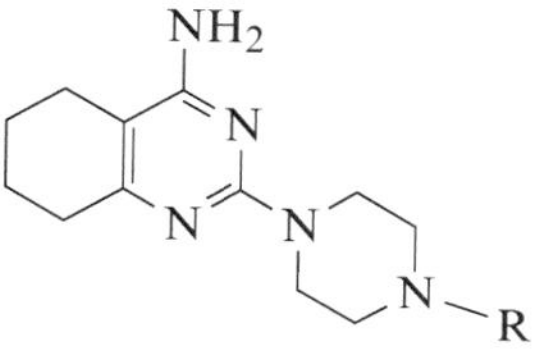

(50)

Also other examples.[392,616,681,845,879,1012,1262,1517,1594,1622,1662,1706,1719,
1721,1812,1891,1900,1905,2035,2058,2080,2221,2232,2374,2497]

From (functionally substituted amino)quinazolines

Note: Deacylation of acylaminoquinazolines was covered at the beginning of
this section.

3-t-Butoxycarbonylamino- (**51**, R = CO$_2$Bu^t) gave 3-amino-2-chloromethyl-
6-nitro-4(3H)-quinazolinone (**51**, R = H) (AcOH, reflux until CO$_2$ ↑ ceased:
78%; or neat, 185°: 57%);[2360] also analogs.[2360]

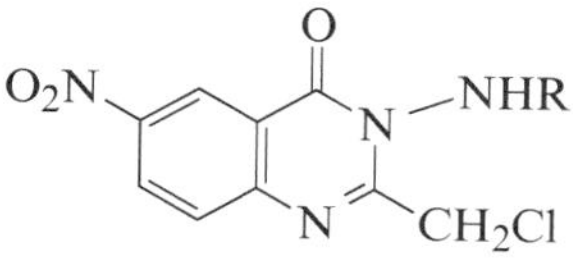

(51)

3-Benzyl-2-cyanoamino- (**52**, R = CN) gave 2-amino-3-benzyl-4(3H)-quin-
azolinone (**52**, R = H) (10M HCl, Me$_2$SO, reflux, 6 h: 81%, as hydrochloride.[2000]

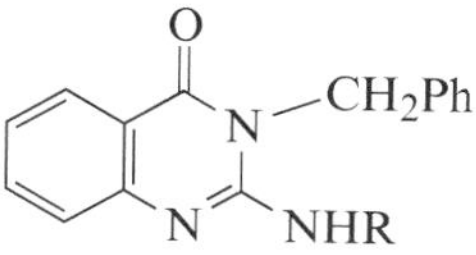

(52)

2-Azidomethyl- (**53**, R = N$_3$) gave 2-aminomethyl-3-phenyl-4(3H)-quinazo-
linone (**53**, R = NH$_2$) (H$_2$S ↓, pyridine–H$_2$O, 20°, 90 min: 83%);[1806] also
analogs.[1806]

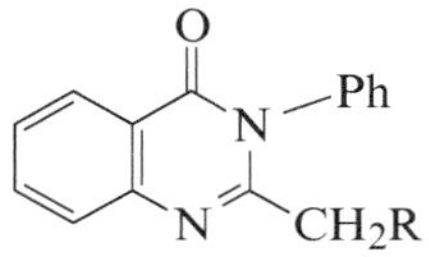

(53)

2-α-Azidoethyl- gave 2-α-aminoethyl-4(3H)-quinazolinone (H$_2$, Pd/C, EtOH:
98%).[2069]

7-Azido-6-methoxy-5,8-quinazolinediol gave 7-amino-6-methoxy-5,8-dihydro-
5,8-quinazolinedione (Cl$_2$HCCH$_2$Cl, reflux, N$_2$, 4 h: 31%).[1577]

2-o-(Piperidinoazo)phenyl- (**54**) gave 2-o-aminophenyl-4-quinazolinamine
(**55**) (KMnO$_4$, KHCO$_3$, H$_2$O–AcMe, 25°, dark, 10 h: small yield after
separation from another product).[1013]

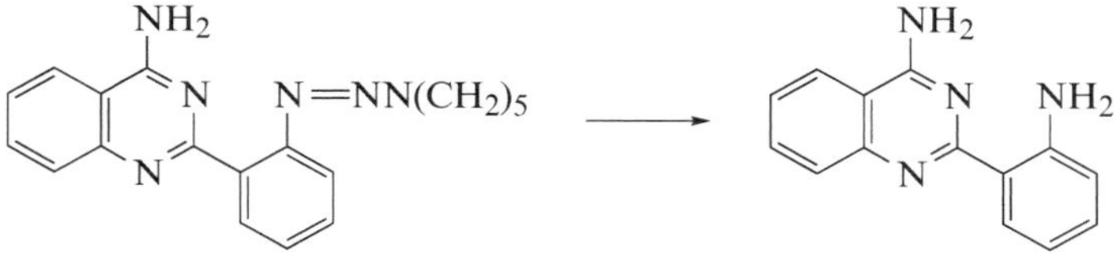

(54)   **(55)**

Also other examples in the foregoing references and elsewhere.[1131,1877,2568,2574]

By Dimroth rearrangement of quinazolinimines (*H* 335)

3-Methyl-7-(4-methyl-6-oxo-1,4,5,6-tetrahydropyridazin-3-yl)-4(3H)-quinazo-
linimine (**56**) gave 4-methylamino-7-(4-methyl-6-oxo-1,4,5,6-tetrahydro-
pyridazin-3-yl)quinazoline (**57**) (2M NaOH, 100°, 5 h, 76%).[2033]

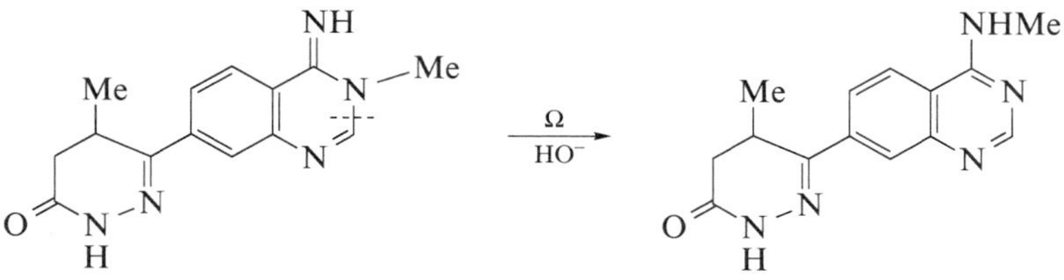

(56)   **(57)**

3-Methyl-4(3H)-quinazolinimine gave 4-methylaminoquinazoline (1M NaOH,
60°, 1 h: 94%); also analogs.[1000]

3-Methyl-2(3*H*)-quinazolinimine hydriodide gave 2-methylaminoquinazoline (2M KOH, 20°, 1 h: 80%).[50]

4-Amino-3-phenyl-2(3*H*)-quinazolinethione, tautomeric with 4-imino-3-phenyl-3,4-dihydro-2(1*H*)-quinazolinethione (**58**), gave 4-anilino-2(1*H*)-quinazolinethione (**59**) (Me$_2$NCHO–H$_2$O, reflux: good yield?).[2051]

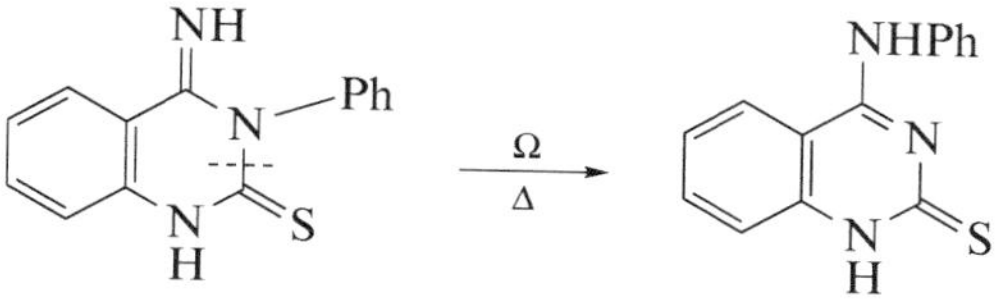

(**58**)   (**59**)

3-Allyl-4-amino-2(3*H*)-quinazolinethione gave 4-allylamino-2(1*H*)-quinazolinethione (no solvent, 187°, 2 min: 60%).[848]
Also other examples.[417,442,947,1857,1858,1958,1991,2083]

By introduction of a passenger amino group

Note: Many diverse instances of such processes will be found throughout this book. Two random examples are given here.

6,8-Dichloro-4(3*H*)- quinazolinone gave 6,8-dichloro-3-β-diethylaminoethyl-4(3*H*)-quinazolinone (NaH, Me$_2$SO, 20°, 1 h; then ClCH$_2$CH$_2$NEt$_2$, 20°, 22 h: 32%, as oxalate salt).[384]

Ethyl 3-amino- gave ethyl 3-β-aminoethylamino-4-oxo-3,4-dihydro-2-quinazolinecarboxylate [neat (CH$_2$)$_2$NH, 20°, 12 h: 25%].[124]

6.2.2.   Reactions of Regular Aminoquinazolines (*H* 333,340)

Although regular aminoquinazolines undergo a great variety of reactions as a class, the potential reactions of any individual aminoquinazoline vary according to the type of amino group (primary, secondary, tertiary, or quaternary) and to its position in the molecule.
The following reactions of aminoquinazolines have been discussed already.

Conversion into *halogenoquinazolines* via diazotization (primary only; 6- to 8- and extranuclear positions): Sections 3.3 and 3.4.

Conversion into *tautomeric quinazolinones* (all types; 2/4-position): Section 4.1.1.1.

Conversion into *alkylthioquinazolines* via diazotization (primary only; 6- to 8-position): Section 5.3.1.

The *interconversion* of aminoquinazolines by transamination, alkylation, dealkylation, or Dimroth rearrangement (all types; all positions): Section 6.2.1.

The remaining reactions of aminoquinazolines are discussed in the subsections that follow.

6.2.2.1.   Acylation Reactions (H 334)

All primary and secondary aminoquinazolines appear to undergo N-acylations quite readily irrespective of position. However, some selectivity has been achieved in the acylation of di- or polyamines. Also included in this subsection are 1/3-acylations of those reduced quinazoline systems in which N1 and/or N3 bear(s) a hydrogen atom. The following classified list of typical acylaminoquinazolines, each made from the corresponding aminoquinazoline, will serve to exemplify the conditions, yields, and other aspects of such acylations.

To 2-acylaminoquinazolines

2-Propionamidoquinazoline (60) (EtCOCl, Et$_3$N, dioxane, 20°, 1 h, then 60°, 2 h: 64%).[849]

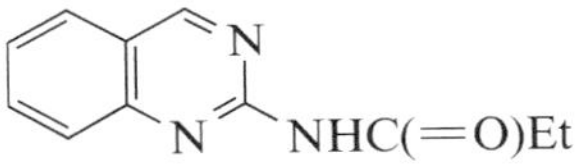

(60)

6-Methyl-4-methylthio-2-pivalamidoquinazoline (Bu^{t}COCl, NEt$_3$, CH$_2$Cl$_2$, reflux, 18 h: 84%).[2255]

2-(α-Acetylacetamido)-3-amino-4(3H)-quinazolinone (AcCH$_2$CO$_2$Et, xylene, reflux, 1 h: 85%; note selectivity) but 2,3-bisacetamido-4(3H)-quinazolinone (61) (neat Ac$_2$O, reflux, 90 min: 90%).[1715]

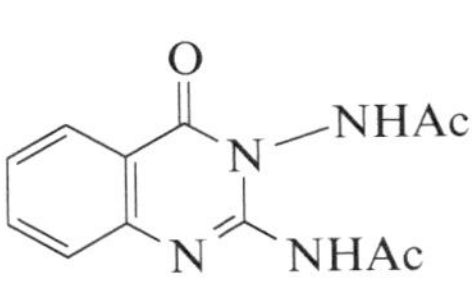

(61)

2,4-Bisacetamido-5,6,7,8-tetrahydro-6-quinazolinecarbaldehyde   dimethyl acetal (Ac$_2$O, 100°, 20 min: 66%).[2566]

Also other examples.[871,1591,1764,1834,2194]

To 3-acylaminoquinazolines

3-Acetamido- (62, R = H) and 3-diacetylamino-2-phenyl-4(3H)-quinazolinone (62, R = Ac) (Ac$_2$O, AcOH, reflux, 2 h: 60% and 12%, respectively, after separation).[1981]

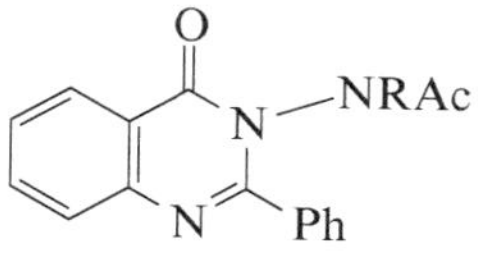

(62)

3-Benzamido-2,4($1H,3H$)-quinazolinedione(PhCOCl, Et$_3$N, THF, 0°, 50 min, then 20°, 3 h: 93%).[1457]

3-N-Methylacetamido-4($3H$)-quinazolinone (neat Ac$_2$O, reflux, 2 h: 78%).[934]

Ethyl 3-α-chloroacetamido-4-oxo-3,4-dihydro-2-quinazolinecarboxylae **(63)** (ClCH$_2$COCl, pyridine–PhMe, reflux, 1 h: 74%).[124]

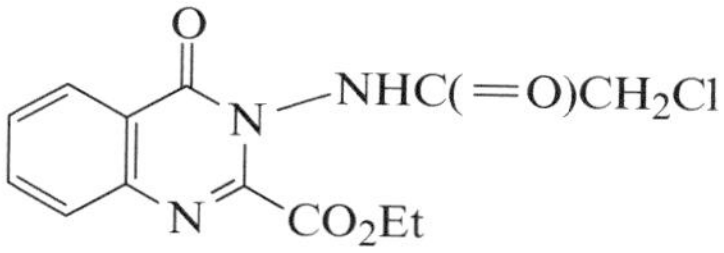

(63)

3-Chlorocarbonylamino-2-naphthalen-2′-yl-4($3H$)-quinazolinone (ClCO$_2$Et, BuOH, reflux, 24 h: 88%).[1731]

Also other examples.[81,1429,1449,1609,1662,1731,1888,1984,1995,2042,2310,2318, 2349,2393,2582,2593] See also a review of N-quinazolinioamidates.[266]

To 4-acylaminoquinazolines

4-Benzamido-2-phenylquinazoline (PhCOCl, Et$_3$N, AcMe, 20°, 24 h: 69%).[1232]

4-(3,5-Dimethylisoxazol-4-ylcarbonylamino)quinazoline **(64)** (3,5-dimethyl-4-isoxazolcarbonyl chloride, pyridine, 20°, 12 h: 68%).[401]

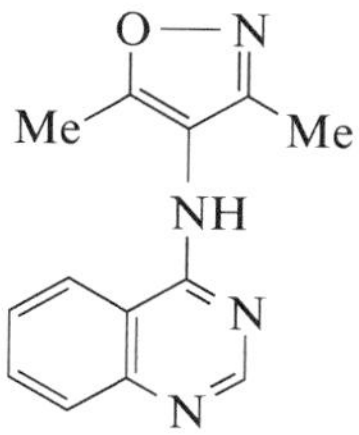

(64)

Also other examples.[605,1602,1912]

To 6-acylaminoquinazolines

Note: There is an extraordinary lack of examples for the acylation of 5-, 7-, or 8-aminoquinazolines, but there is no reason to believe that such substrates will resist acylation.

6-*p*-Bromophenylacetamido-5-chloro-2,4-quinazolinediamine **(65)** (*p*-BrC$_6$H$_4$CH$_2$COCl, Me$_2$NCHO, 20°, 48 h: 56%; note selective acylation); also many analogs.[609]

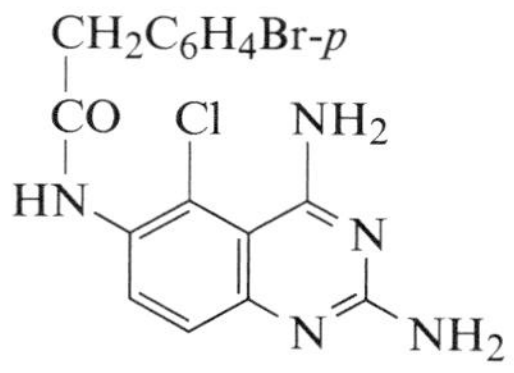

(65)

6-α-Methoxyacetamido-7-methylamino-4(3*H*)-quinazolinone **(66)** (MeOCH$_2$COCl, pyridine–PhH, 20°, 12 h: 64%; note selectivity).[1962]

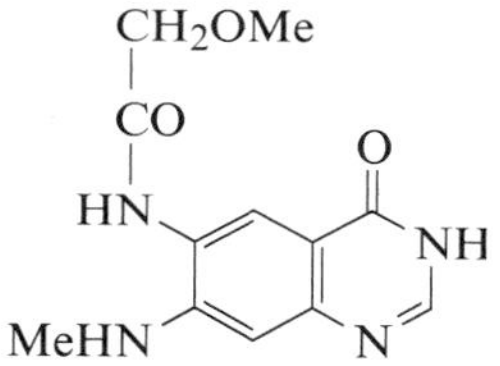

(66)

2,4,6-Trisacetamidoquinazoline (?) (from 2,4,6-quinazolinetriamine, Ac$_2$O, reflux, 30 min; structure unconfirmed).[1919]

Also other examples[121,356,1157,1529,2497,2583] including *in vivo* acetylation.[582]

To extranuclear acylaminoquinazolines

2-Formamidomethyl- **(67, R = H)** (HCO$_2$H/AcOH, 60°, then 20°, 30 min: 90%), 2-acetamidomethyl- **(67, R = Me)** (neat Ac$_2$O, warm, briefly: 86%), or 2-benzamidomethyl-3-phenyl-4(3*H*)-quinazolinone **(67, R = Ph)** (BzCl, 4M NaOH, 20°, 15 min: 70%).[152]

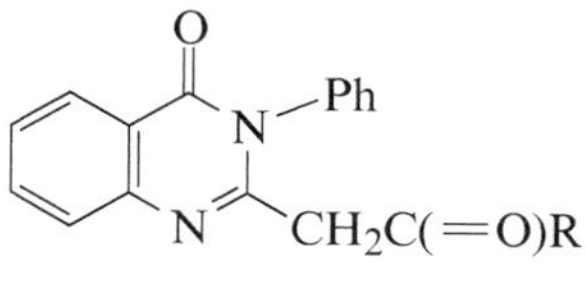

2-(4-Cinnamoylpiperazin-1-yl)-5,6,7,8-tetrahydro-4-quinazolinamine **(68)** (PhCH=CHCOCl, Et$_3$N, THF, < −10°, 12 h: 51%; note selectivity).[1638]

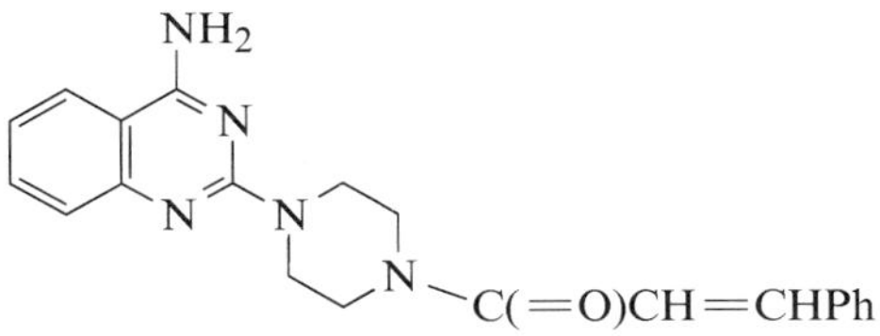

(68)

2-[N-γ-(2-Furoamido)propyl-N-methylamino]-4-pyrrolidin-1′-ylquinazoline
(**69**) (2-furoyl chloride, Et$_3$N, CH$_2$Cl$_2$, N$_2$, 0°, then 20°, 18 h: > 70%).[2224]

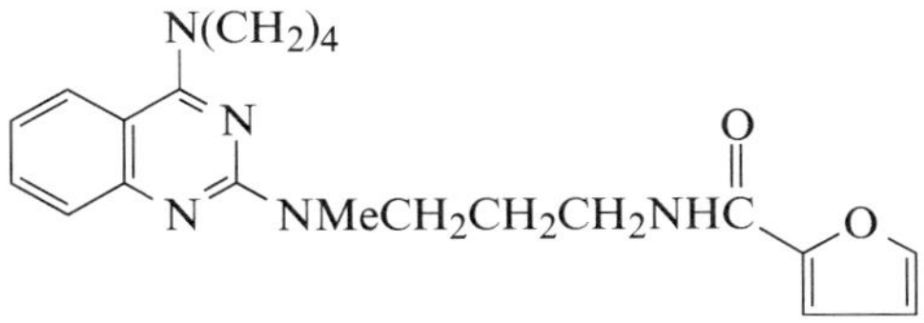

(69)

Also other examples.[45,64,623,1543,1612,1613,1670,2018,2288,2476]

To N-acylamino-1,2/3,4-dihydroquinazolines

2,6-Dimethyl-1-trifluoroacetyl-1,2-dihydro-4(3H)-quinazolinone (**70**) [(F$_3$
CCO)$_2$O, 15°, 12 h: 65%].[2254]

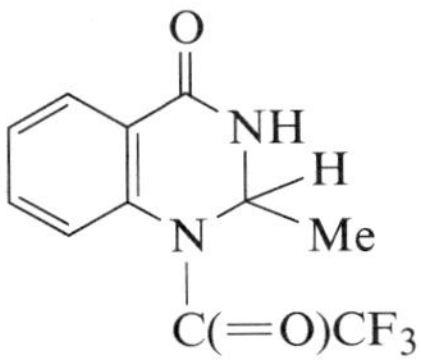

(70)

3-Benzoyl-3,4-dihydro-4-quinazolinecarbonitrile(**71**) (BzCl, pyridine, 0°, then
50°, 5 min: 82%).[1419]

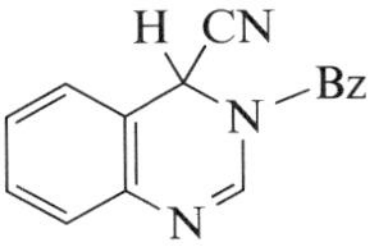

(71)

Also other examples.[153,192,395,1534,2370]

6.2.2.2.   Conversion into Schiff Bases (*H* 340)

Quinazolines with a primary amino group at any position can react with aldehydes or occasionally ketones to afford Schiff base–type derivatives (Q—N=CHR or Q—N=CRR', respectively). Naturally, Schiff base types with reversed orientation in respect of the quinazoline nucleus (Q—CH=NR or Q—CR'=NR) can be made from quinazoline aldehydes or ketones with primary aliphatic or aromatic amines (see Sections 7.6 and 7.7).

Schiff bases from aminoquinazolines are exemplified in the following products.

From 2/4-aminoquinazolines

4-Benzylideneamino-3-phenyl-5,6,7,8-tetrahydro-2(3*H*)-quinazolinone   (**72**, R = H) (neat PhCHO, reflux, 10 min: 38%);[1851] 4-*p*-chlorobenzylideneamino analog (**72**, R = Cl) (*p*-ClC$_6$H$_4$CHO, xylene, trace TsOH, reflux, 16 h: 65%).[1853]

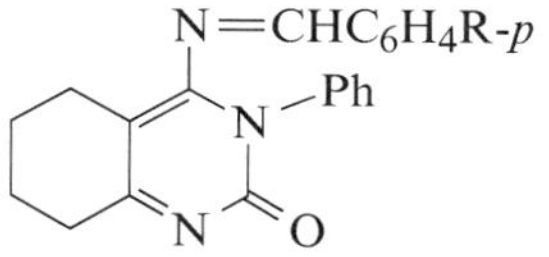

(72)

2-Dimethylaminomethyleneamino-3-methyl-4(3*H*)-quinazolinone (**73**) Me$_2$NCHO, heat: > 37%); also analogs.[1502]

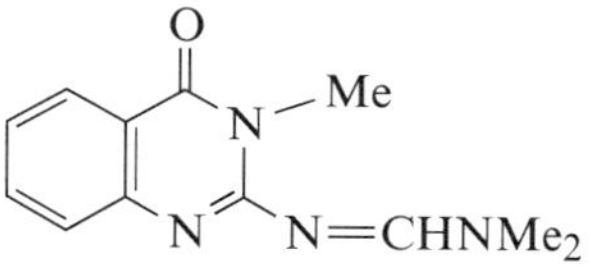

(73)

2,4-Bisdimethylaminomethyleneaminoquinazoline   (**74**)   [neat   Me$_2$NCH(OMe)$_2$, 130°, ? h:80%]; also analogs.[1852]

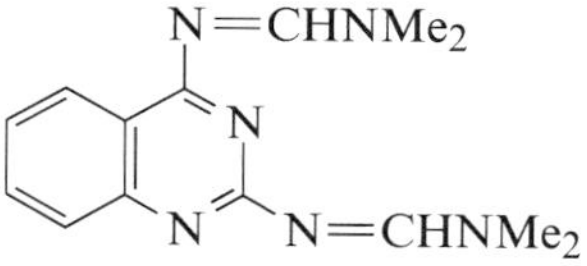

(74)

Also other examples.[2124,2578]

From 3-aminoquinazolines

3-But-2′-enylideneamino- (**75**, R = CH:CHMe) (neat MeCH=CHCHO, reflux, 2 h: 90%),[191] 3-ethoxymethyleneamino- (**75**, R = OEt) [neat (EtO)$_3$CH, reflux, 40 h: 97%],[1453] and 3-benzylideneamino-2-methyl-4(3H)-quinazolinone (**75**, R = Ph) (PhCHO, EtOH, reflux, 20 min: 73%;[1197] PhCHO, EtOH, trace AcOH, reflux, 5 h: 80%).[1653]

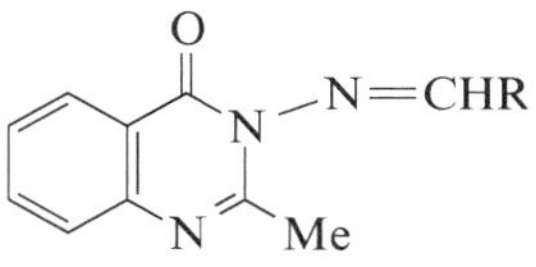

(75)

1-Methyl-3-p-nitrobenzylideneamino-2,4(1H,3H)-quinazolinedione   **(76)**
(O$_2$NC$_6$H$_4$CHO-p, EtOH, reflux, 8 h: 77%).[495]

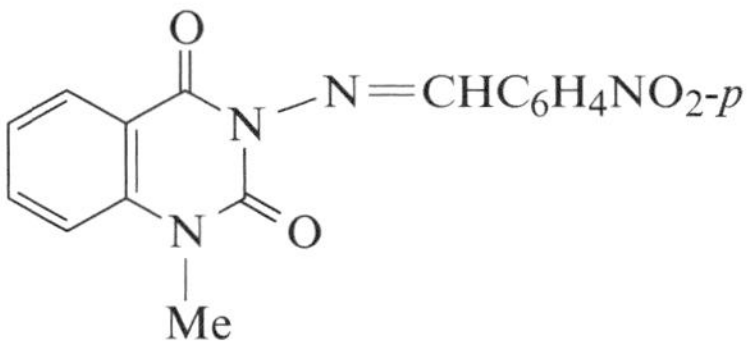

(76)

6,8-Dibromo-3-isopropylideneamino-(**77**, R=Me) (AcMe, EtOH, trace HCl, reflux, 10 min: 45%) and 6,8-dibromo-3-diphenylmethyleneamino-2-methyl-4 (3H)-quinazolinone (**77**, R = Ph) (Ph$_2$CO, likewise: 60%).[2354]

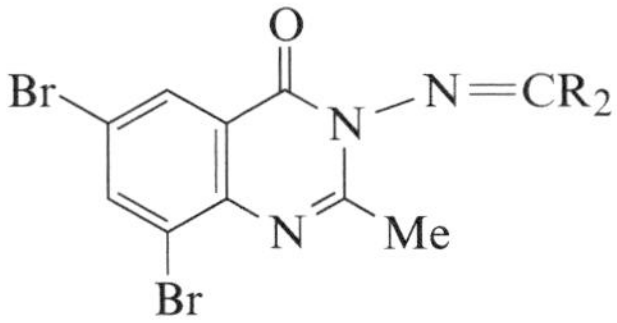

(77)

Also other examples.[113,124,391,468,568,862,1245,1656,1680,1720,1731,2023,2173,2234,2301,2370,2447,2533]

From 6-aminoquinazolines

5-Chloro-6-p-ethoxycarbonylbenzylideneamino-2,4-quinazolinediamine (**78**) (EtO$_2$CC$_6$H$_4$CHO-p, EtOH, reflux, 35 h: 66%; note selectivity).[1593, cf. 612]

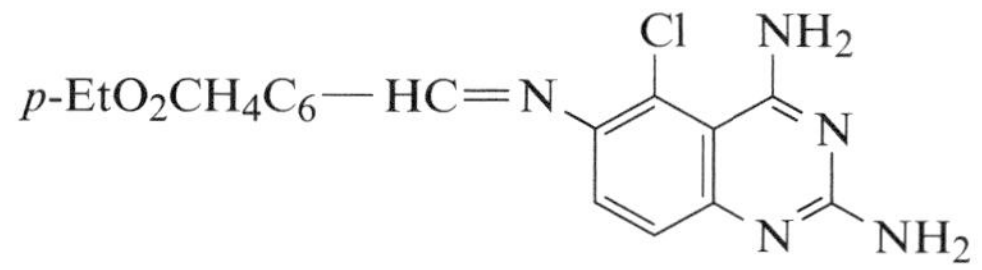

(78)

2-Amino-6-p-[N-(α,γ-diethoxycarbonylpropyl)carbamoyl]benzylideneamino-
5-methyl-4(3H)-quinazolinone **(79)** [EtO$_2$CCH$_2$CH$_2$CH(CO$_2$Et)NHC
(=O)C$_6$H$_4$CHO-p, 4-Å molecular sieves, Me$_2$SO, reflux, 6 h: 63%].[611]

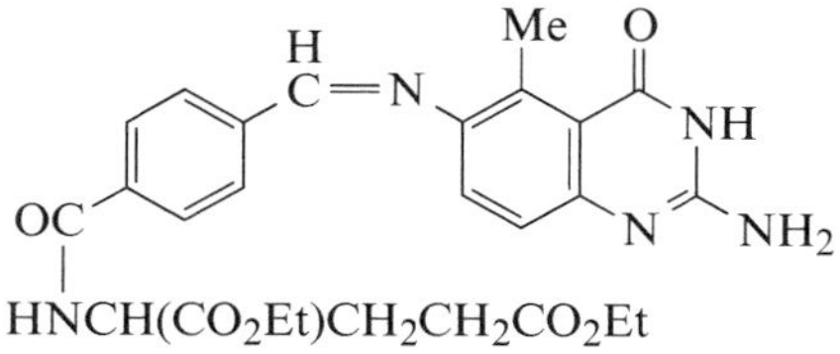

(79)

Also other examples in the foregoing references and elsewhere.[1906,2541]

From extranuclear aminoquinazolines

3-p-(Benzylideneamino)phenyl-2-phenyl-4(3H)-quinazolinone **(80,** R = N:
CHPh) (PhCHO, trace AcOH, EtOH, reflux 4 h: 50%); also analogs.[1217]
3-[4'-(Benzylideneamino)biphenyl-4-yl]-2-phenyl-4(3H)-quinazolinone [**80,**
R = C$_6$H$_4$ (N=CHPh)-p] (PhCHO, trace AcOH, EtOH, reflux, 6 h: ?%);
also analogs.[1673]

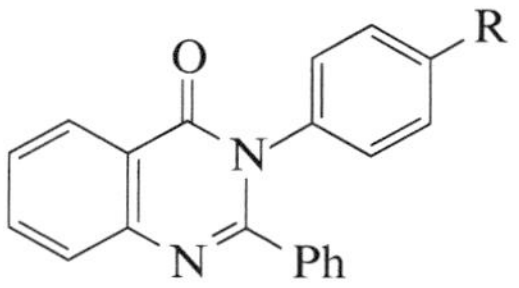

(80)

Also other examples.[360,1479,1543]

6.2.2.3. Conversion into Alkoxycarbonylaminoquinazolines

Treatment of an aminoquinazoline with an alkyl chloroformate (ClCO$_2$R)
may give a chlorocarbonylaminoquinazoline[1731] but usually gives an alkoxycar-
bonylaminoquinazoline, especially in the presence of an organic base; to avoid
any ambiguity, dialkyl carbonates [(RO)$_2$CO] may be used. The following
examples typify the usual course of such reactions.

2-Quinazolinamine gave 2-ethoxycarbonylaminoquinazoline (**81**) [(EtO)$_2$CO, dioxane–pyridine, reflux, 6 h: 52%].[849]

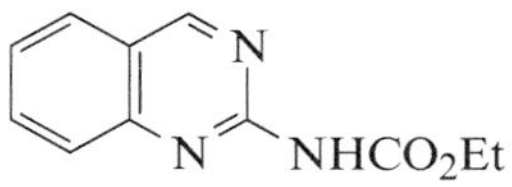

(**81**)

Ethyl 6-amino- (**82**, R = H) gave ethyl 6-ethoxycarbonylamino-4-oxo-3,4-dihydro-2-quinazolinecarboxylate (**82**, R = CO$_2$Et) (EtO$_2$CCl, pyridine–PhH, reflux, 4 h: ∼ 65%).[123]

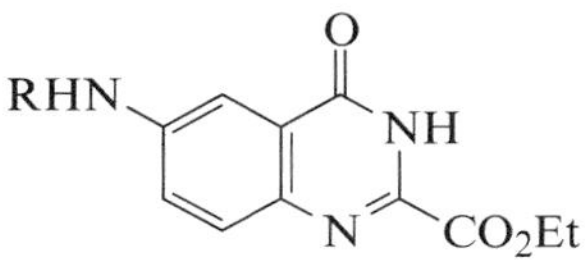

(**82**)

2-*o*-Aminophenyl- gave 2-*o*-(methoxycarbonylamino)phenyl-3,4-dihydroquinazoline (MeO$_2$CCl, Et$_3$N, CHCl$_3$, 20°, 30 min: 68%).[73]

4-[4-(Aminomethyl)piperidino]-6,7-dimethoxyquinazoline (**83**, R = H) gave 6,7-dimethoxy-4-{4-[methylthio(thiocarbonyl)aminomethyl]piperidino}-quinazoline (**83**, R = CS$_2$Me) (CS$_2$, Et$_3$N, EtOH, 20°, 2 h; then MeI, 20°, 1 h: > 40% but characterized only as a derivative).[2108]

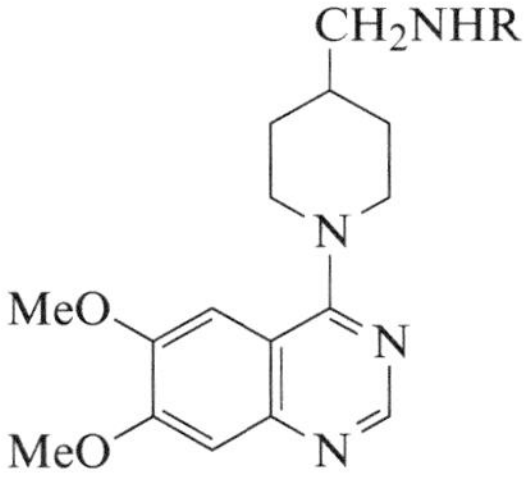

(**83**)

Also other examples.[380,2297]

6.2.2.4. Conversion into Ureidoquinazolines

Primary or secondary aminoquinazolines are converted readily into ureidoquinazolines by treatment with isocyanates or isothiocyanates; such conversions may also be done indirectly by aminolysis of derived alkoxycarbonylaminoquinazolines.

The *direct conversion* is illustrated in the formation of the following ureido- or thioureidoquinazolines.

From 2/4-aminoquinazolines

2-N'-Methylureidoquinazoline **(84)** [MeNCO, trace (EtBuCHCO$_2$)$_2$Sn, dioxane, 50°, 2 h: 59%).[849]

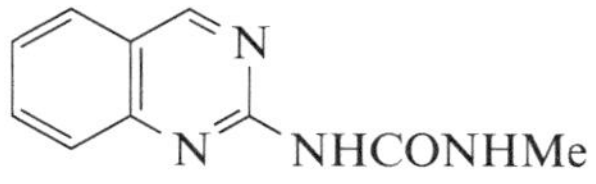

(84)

4-Phenyl-2-N'-phenyl(thioureido)-5,6,7,8-tetrahydroquinazoline   (PhNCS, EtOH, reflux, 3 h: 66%).[2153]

4-N'-Methylureido-2-β,β-dichloro-α,α-difluoroethylquinazoline **(85)** (MeNCO, Et$_3$N, CH$_2$Cl$_2$, 20°, 12 h: 78%).[443]

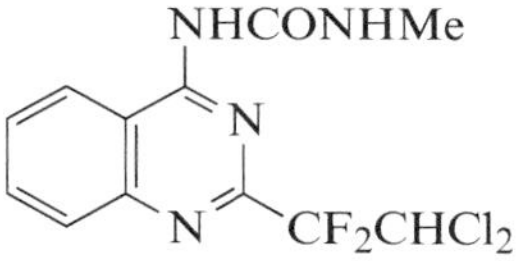

(85)

Also other examples.[725]

From 3-aminoquinazolines

2-Phenyl-3-N'-phenylureido-4(3H)-quinazolinone **(86,** X = O) (PhNCO, PhH, reflux, 3 h: 88%); also analogs.[1679]

2-Methyl-3-N'-phenyl(thioureido)-4(3H)-quinazolinone (PhNCS, PhH, reflux, 2 h: 62%);[1555] likewise in EtOH, > 3 h: 60%).[1631]

2-Phenyl-3-N'-phenyl(thioureido)-4(3H)-quinazolinethione   **(86,**   X = S) (PhNCS, Me$_2$NCHO, reflux, < 7 h: 55%);[1307] when the reaction was attempted in Et$_3$N/MeCN under reflux for 24 h, cyclization ensued to give a 1,2,4-triazolo[3,2-c]quinazoline (59%).[1307]

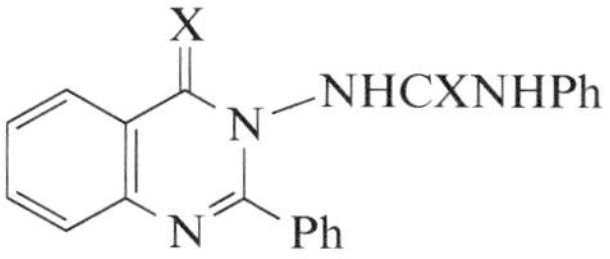

(86)

Also other examples.[124,1204,1675,1921,2188,2313,2317,2552]

From 6-aminoquinazolines

6-N'-Methyl(thioureido)-2,4-quinazolinediamine (**87**) (limited MeCNS, conditions?: ?%; note selectivity);[1528] also analogs.[1528]

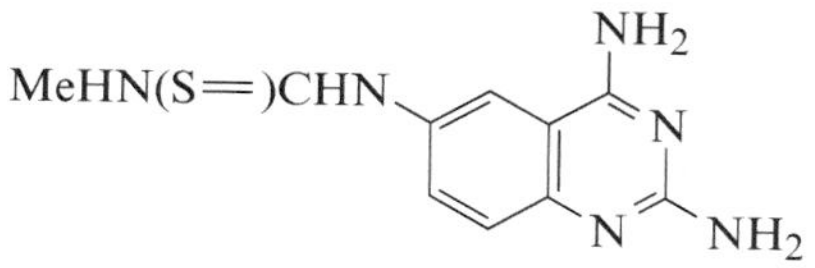

(**87**)

Also other examples.[570]

From extranuclear aminoquinazolines

3-p-[N'-p-chlorophenyl(thioureido)]phenyl-2-phenoxymethyl-4(3H)-quinazolinone (**88**) (ClC$_6$H$_4$NCS-p, PhH, reflux, 4 h: 53%); also analogs.[1746]

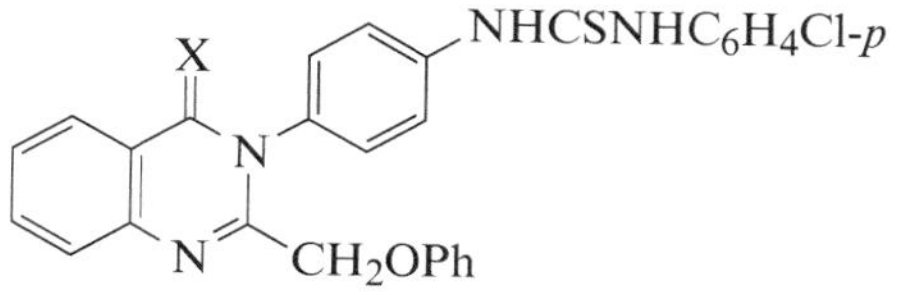

(**88**)

3-β-(N-β-Hydroxyethyl-N'-phenylureido)ethyl-2-methyl-4(3H)-quinazolinone (**89**) from the 3-β-(N-β-hydroxyethylamino)ethyl analog (PhNCO, AcMe, 20°, 12 h: ~80%); also analogs.[114]

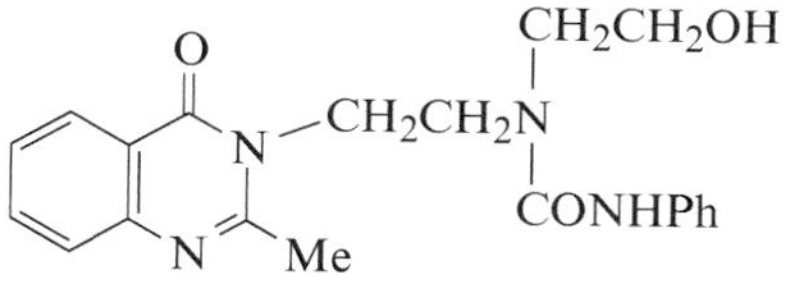

(**89**)

1-γ-[N'-Benzoyl(thioureido)]propyl-2-phenyl-1,2-dihydro-4(3H)-quinazolinone (**90**, R = Bz) (BzNCS, CH$_2$Cl$_2$, reflux, 1 h: 89%);[2237] subsequent

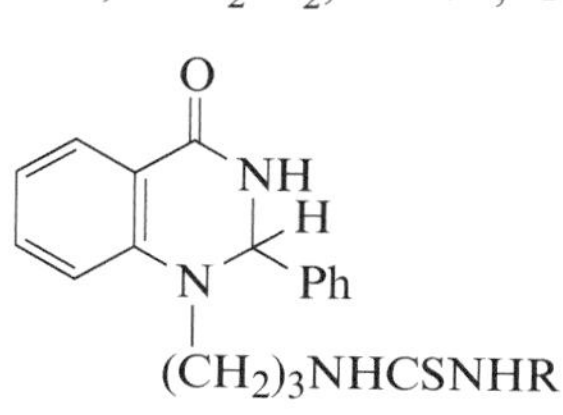

(**90**)

deacylation gave the 1-γ-thioureidopropyl analog (**90**, R = H) (K$_2$CO$_3$, H$_2$O–MeOH, reflux, 30 min: 86%).[2237]

Also other examples.[1044,2581]

The *indirect route* from amino- to ureidoquinazolines is illustrated by the following sequence: 3-amino- (**91**) → 3-dithiocarboxyamino- (**92**, R = H) → 3-carboxymethylthio(thiocarbonyl)amino- (**92**, R = CH$_2$CO$_2$H) → 3-N'-amino (thioureido)-2-methyl-4(3H)-quinazolinone (**93**) (CS$_2$, NH$_4$OH, EtOH, < 30°, 1 h; then ClCH$_2$CO$_2$Na, 20°, briefly; then H$_2$NNH$_2$.H$_2$O, 5°, 12 h: 75% overall).[2243,2297]

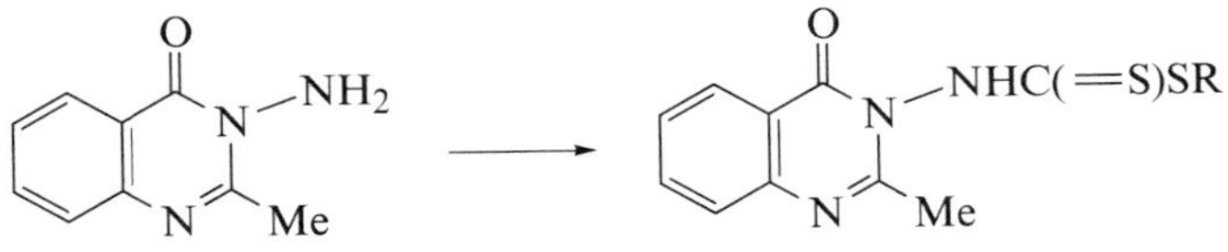

(91)   (92)

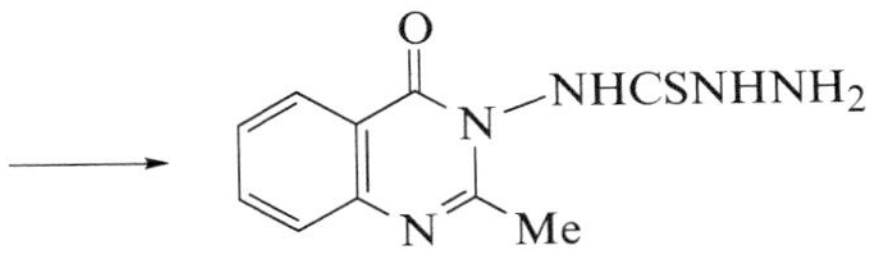

(93)

Thioureidoquinazolines, in particular, have been used as substrates for further interesting cyclizations.[437,1553,2452]

6.2.2.5. Diazotization and Subsequent Reactions (*H* 474)

Most primary aminoquinazolines will undergo diazotization, although the resulting diazonium salts are seldom isolable. The conversions of such diazonium salts into *quinazolinones* (Section 4.1.1.1), *halogenoquinazolines* (Sections 3.3 and 3.4), and *alkylthioquinazolines* (Section 5.3.1) were covered earlier. The remaining reactions occurring via diazonium salts are illustrated in the following classified examples.

To arylazoquinazolines

2-m-Aminophenyl- gave 2-m-(2-hydroxynaphthalen-1-ylazo)phenyl-4-oxo-3,4-dihydro-5-quinazolinecarboxylic acid (**94**) (NaNO$_2$, 2M HCl, 5°; then C$_{10}$H$_7$OH-2, AcONa);[567] also many analogs.[65,567,666]

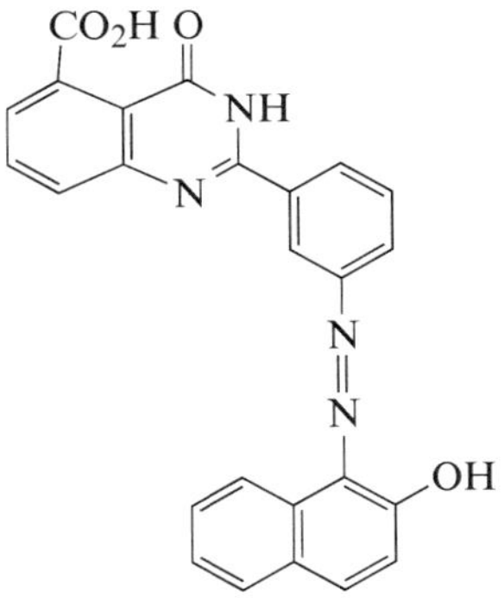

(94)

6-Amino- gave 6-(1-hydroxy-6-methylamino-3-sulfonaphthalen-2-ylazo)-2-methyl-4(3H)-quinazolinone (**95**) (NaNO$_2$, HCl, 5°; then sodium 4-hydroxy-7-methylamino-2-naphthalenesulfonate, H$_2$O maintained at pH 8, 5°, 3 h);[2158] also analogs.[2146,2158]

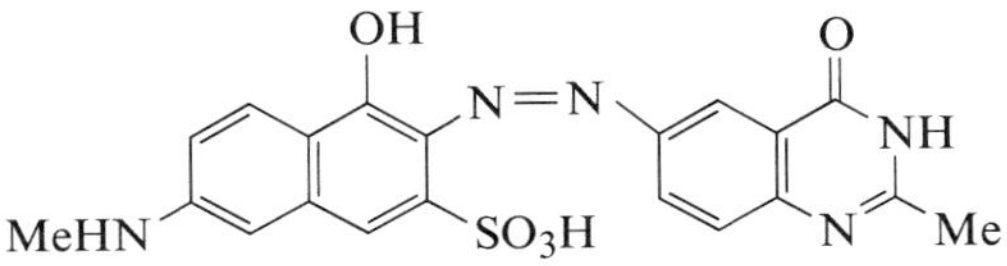

(95)

6-Quinazolinamine gave 6-(2-oxocyclohexyl)azoquinazoline (**96**) (NaNO$_2$, 4M HCl, 0°, 1 h; then 2-oxocyclohexanecarbaldehyde, AcONa, pH6, 24 h: 36%; a peculiar reaction involving loss of CH$_2$O?).[2192]

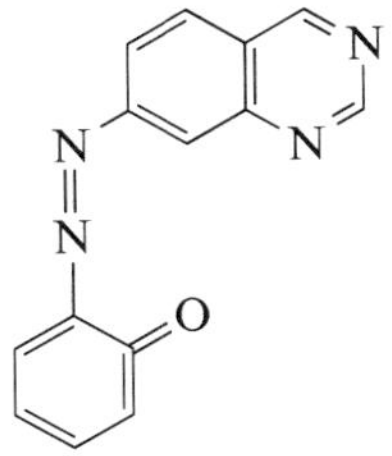

(96)

2,4,6-Quinazolinetriamine gave 6-diazonio-2,4-quinazolinediamine tetrafluoroborate (NaNO$_2$, HBF$_4$, H$_2$O, < 5°, 90 min: 82%) and thence 6-(2-hydroxynaphthalen-1-ylazo)-2,4-quinazolinediamine (C$_{10}$H$_7$OH-2, 0.5 M KOH, 4°, briefly: 25%).[1919]

Also other examples.[284,303,682,2454]

To azidoquinazolines

6-Quinazolinamine gave 6-azidoquinazoline (**97**, R = H) (NaNO$_2$, 4M HCl, < 5°; then NaN$_3$, AcONa, no details: 96%).[2075]

2,4,6-Qunazolinetriamine gave 6-azido-2,4-quinazolinediamine (**97**, R = NH$_2$) (NaNO$_2$, 2M HCl, 0°, 90 min; then NaN$_3$, H$_2$O, 0°, 2 h: 98%).[1919]

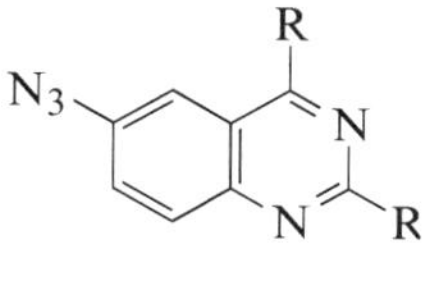

(97)

Also other examples.[2274]

To quinazolinecarbonitriles

2,4,6-Quinazolinetriamine gave 2,4-diamino-6-quinazolinecarbonitrile (NaNO$_2$, 2M HCl, < 15°, 30 min; then CuSO$_4$, KCN, H$_2$O, 50°,   30 min: 40% or 65%).[201,956] Analogs likewise.[201,2590]

2,6-Diamino-5-trifluoromethyl-4(3H)-quinazolinone (**98**, R = NH$_2$) gave 2-amino-4-oxo-5-trifluoromethyl-3,4-dihydro-6-quinazolinecarbonitrile (**98**, R = CN) (NaNO$_2$, 2M HCl, 2°, 30 min; then CuCl, KCN, H$_2$O, 2° → 35°, 60 min: 45%).[2211]

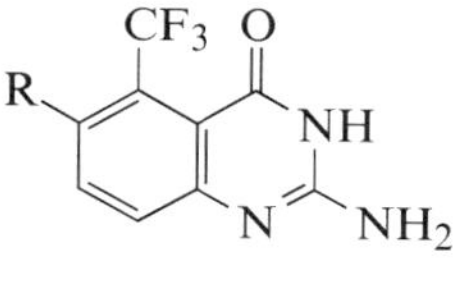

(98)

To hydrazinoquinazolines

2,4,6-Quinazolinetriamine gave 6-hydrazino-2,4-quinazolinediamine (**99**) (NaNO$_2$, 2M HCl, 10°; then SnCl$_2$, 5M HCl, 0°, 90 min, then 20°, 24 h: 42%).[601]

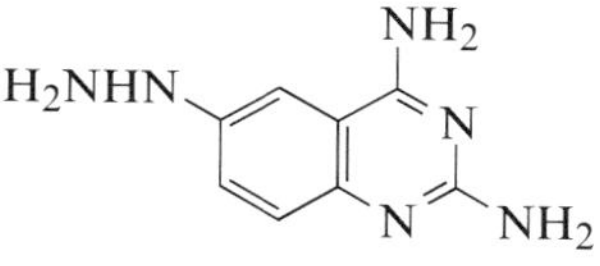

(99)

To arylthioazoquinazolines

2,4,6-Quinazolinetriamine gave 6-phenylthioazo-2,4-quinazolinediamine (**100**) (NaNO$_2$, 2M HCl, 2°, 30 min; then PhSH, KOH, H$_2$O, 40°, 1 h: 95%); likewise for 5-chloro-6-*p*-methoxyphenylthioazo-2,4-quinazolinediamine (98%).[1551]

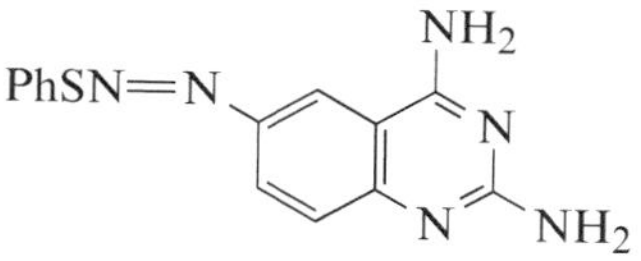

(**100**)

To intramolecular cyclization products

5-Amino-6-methyl-4(3*H*)-quinazolinone (**101**) gave 1*H*-pyrazolo[3,4-*f*] quin-azolin-9(8*H*)-one (**102**) (NaNO$_2$, AcOH, 20°, 3 days: 34%);[1387] alternative-ly, substrate (**101**) gave 5-diazonio-6-methyl-4(3*H*)-quinazolinone tetra-fluoroborate (NaNO$_2$, HBF$_4$, H$_2$)–AcOEt, 5°, 1 h: > 95%) and thence the product (**102**) (Me$_4$NOAc, CHCl$_3$, 20°, 1 h: 50%).[1159,1366]

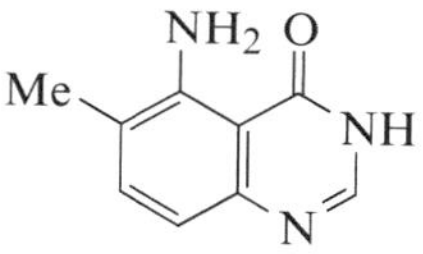

(**101**) 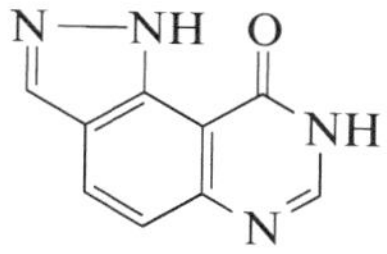   (**102**)

2-*o*-Aminophenyl-4(3*H*)-quinazolinone gave 8*H*-quinazolino[3,2-*c*][1,2,3] benzotriazin-8-one (**103**) (diazotization: no details).[703]

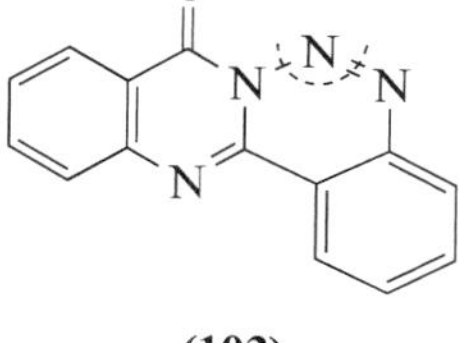

(**103**)

To deaminated products (*H* 340)

Note: Deaminations seem to occur only with 3-amino substrates.

3-Amino-2,4(1*H*,3*H*)-quinazolinedione (**104**, R = NH$_2$) gave 2,4(1*H*,3*H*)-quinazolinedione (**104**, R = H) (NaNO$_2$, AcOH–H$_2$O, 20°, 15 min; then NaOH, H$_2$O, warm: 98%); also analogs.[1454]

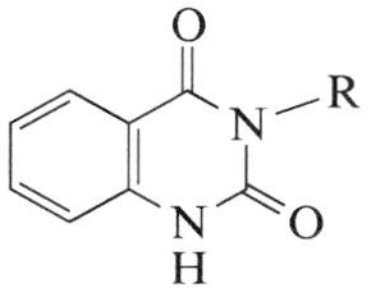

(104)

2,3-Diamino- (**105**, R = NH$_2$) gave 2-amino-4(3H)-quinazolinone (**105**, R = H) (NaNO$_2$, conc H$_2$SO$_4$, < 5°, 45 min; then 20°, 4 h: ~ 80%, as sulfate).[1347]

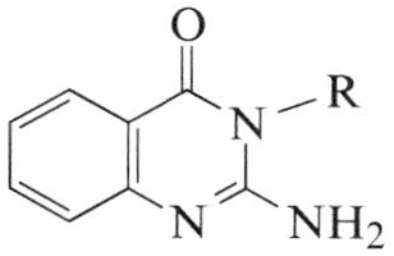

(105)

3-Amino-2-pyrrol-2'-yl- (**106**, R = NH$_2$) gave 2-pyrrol-2'-yl-4(3H)-quinazolinone (**106**, R = H) (NaNO$_2$, 2.5M HCl, 20°, until N$_2$↑ ceased: 89%).[1308]

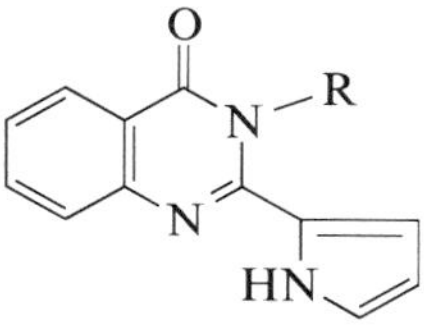

(106)

Also other examples.[1894,2042,2349]

6.2.2.6.   Deamination and Other Displacement Reactions

Displacement of regular amino groups by hydrogen (deamination) or by other groups are covered in this subsection, except for those already discussed: *deamination via diazotization* (Section 6.2.2.5), *hydrolysis* to quinazolinones (Section 4.1.1.1), and *transamination* using regular amines (Section 6.2.1).

The following classified examples illustrate the remaining reactions in this category.

Deaminations

2-Methyl-3-(3-pyrrolin-1-yl)- (**107**) gave 2-methyl-4(3H)-quinazolinone (*hv*, MeCN, 21 h: 90%).[197]

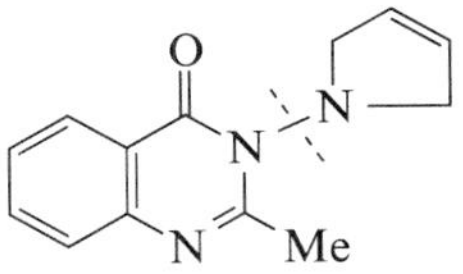

(107)

3-Amino-2-phenyl- (108) gave 2-phenyl-4(3H)-quinazolinone (110) (KMnO₄, AcMe, reflux, 8 h: ~ 70%; analogs likewise);[1655,1704] or via the unisolated intermediate (109) (BzCH₂Br, Me₂SO, 100°, 6 h:?%).[1744]

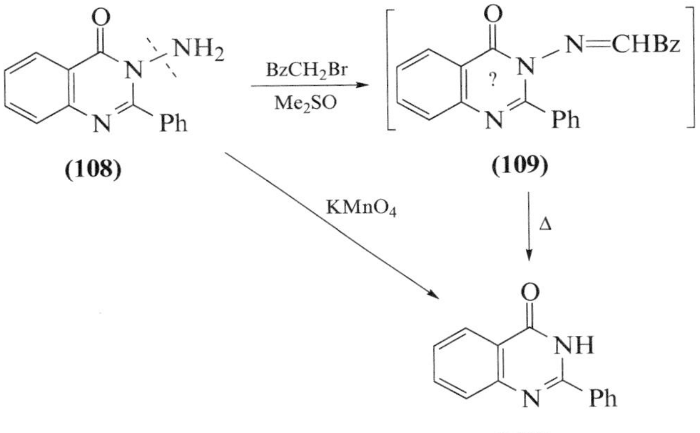

3-Methyl-1-methylamino-4-oxo-3,4-dihydroquinazolin-1-ium   fluorosulfonate (111) gave 3-methyl-4(3H)-quinazolinone (112) (H₂, Pt, EtOH: ~ 80%) or 3-methyl-1,2-dihydro-4(3H)-quinazolinone (113) (H₂, Pd/C, EtOH: ~ 30%).[1445]

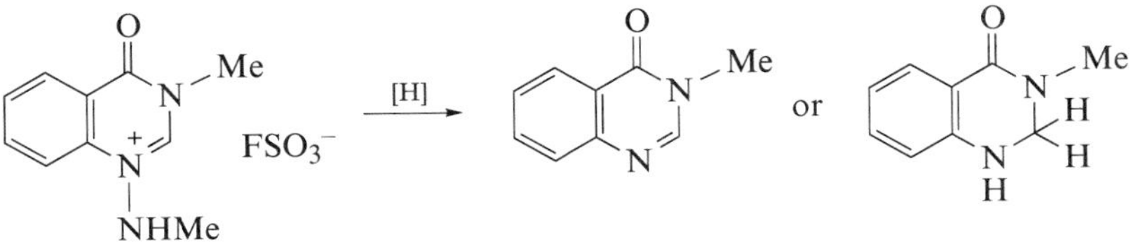

Also other examples.[878,1105,1665,2052]

To quinazolinecarbonitriles

Note: This reaction is limited to trimethylammonioquinazoline salts.

2-Phenyl-4-trimethylammonioquinazoline chloride (114) gave 2-phenyl-4-quinazolinecarbonitrile (115) (KCN, H₂O, 70°, 1 h: 71%;[2206] Et₄NCN, CH₂Cl₂, 20°, 30 min: 86%;[1142] KCN, MeCONH₂, 90°, 2 h: 80%);[923] the isomeric   4-phenyl-2-quinazolinecarbonitrile   was   made   similarly (72%).[2206]

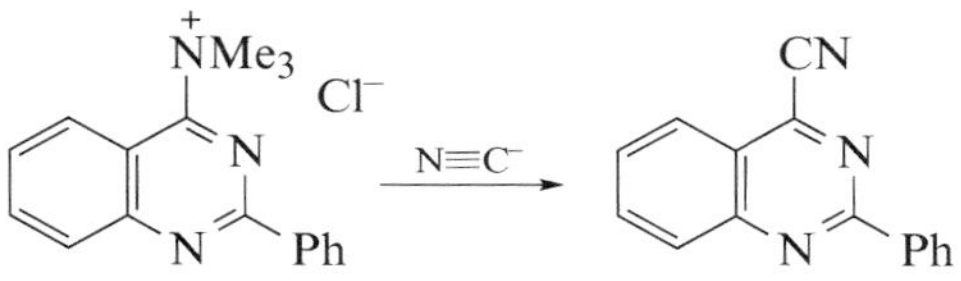

(114)   **(115)**

2-Trimethylammonioquinazoline chloride gave 2-quinazolinecarbonitrile (Et_4NCN, CH_2Cl_2, 20°, 30 min: 67%); also analogs.[1142]
4-Trimethylammonioquinazoline chloride gave 4-quinazolinecarbonitrile (NaCN, Me_2NCHO, 50°, 1 h: $\sim$ 55%).[189]

To hydrazinoquinazolines

2,4-Quinazolinediamine gave 4-hydrazino-2-quinazolinamine (**116**, R = H) (neat $H_2NNH_2.H_2O$, reflux, 90 min: 40%) or 2,4-dihydrazinoquinazoline (**116**, R = NH_2) (likewise but 6 h: 60%).[1919]

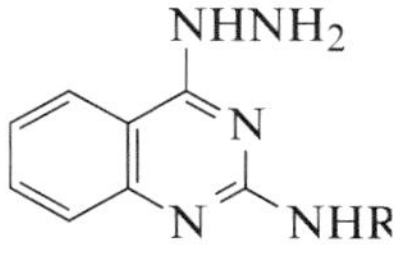

(116)

To hydroxyaminoquinazolines

2-Phenyl-4-quinazolinamine   gave   4-hydroxyamino-2-phenylquinazoline ($NH_2OH.HCl$, Na_2CO_3, MeOH, reflux, 5 h: 74%).[1872]

6.2.2.7.   Nuclear *N*-Alkylation, Oxidation, or Reduction

As well as the nuclear alkylation, oxidation, or reduction of 2/4-amino-quinazolines, this section includes a few 1/3-alkylations of 1,2- or 3,4-dihyd-roquinazolines. These reactions are illustrated by the following examples.

1/3-Alkylation of 2/4-aminoquinazolines

2-Quinazolinamine (**117**, R = H) gave 3-methyl-2(3*H*)-quinazolinimine (**118**, R = H) (MeI, EtOH, 100°, sealed, 3 h: 47%, as hydriodide); the 5-, 6-, or 7-methoxy-3-methyl-2(3*H*)-quinazolinimines (**118**, R = OMe-5, 6, or 7) like-wise.[50]

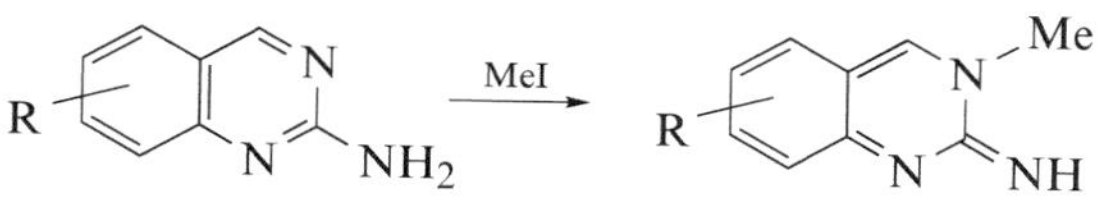

(117)   **(118)**

4-Quinazolinamine gave 1-acetonyl-4 (1H)-quinazolinimine hydrobromide
  (119) (BrCH$_2$Ac, Me$_2$NCHO, ?°, 12 h: > 47%); also analogs.[1077]

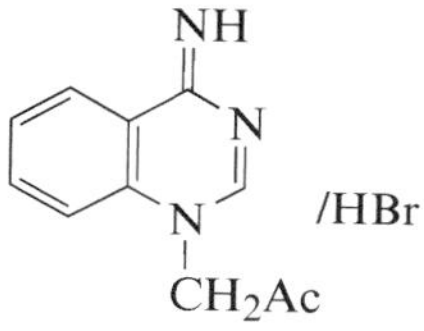

(119)

Also other examples.[965,2231]

N-Alkylation of 1,2/3,4-dihydroquinazolines

3-Lithio-4-methyl-2,4-diphenyl-3,4-dihydroquinazoline **(120)** (made *in situ*)
  gave a separable mixture of 3,4-dimethyl-2,4-diphenyl-3,4-dihydro- **(121)**
  and 1,4-dimethyl-2,4-diphenyl-1,4-dihydroquinazoline **(122)** (MeI, Et$_2$O,
  20°, 12 h: 30% and 60%, respectively).[398]

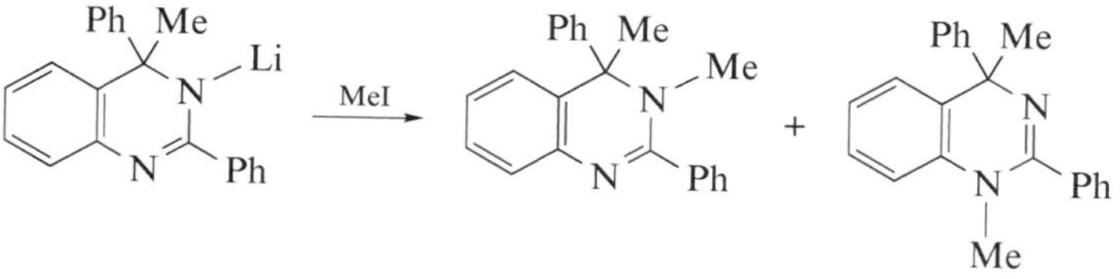

(120)   **(121)**   **(122)**

2-Chloro-5-methyl-3,4-dihydroquinazoline gave 2-chloro-3-ethoxycarbonyl-
  methyl-5-methyl-3,4-dihydroquinazoline (BrCH$_2$CO$_2$Et, K$_2$CO$_3$, AcEt,
  reflux, 3 h: 73%).[1462]
Also other examples.[965,1824,2530]

Nuclear oxidation or reduction

8-Benzylidene-2-ethylamino-4-phenyl-3,4,5,6,7,8-hexahydroquinazoline gave
  8-benzylidene-2-ethylamino-4-phenyl-5,6,7,8-tetrahydroquinazoline (**123**)
  [$K_3Fe(CN)_6$, KOH, H_2O, 20°, 16 h: 39%]; analogs likewise.[1701]

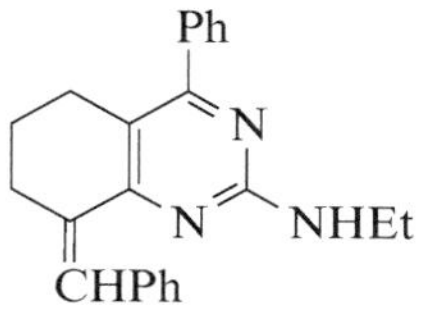

(**123**)

4,7-Dimethyl-5-phenyl-5,6-dihydro-2-quinazolinamine   (**124**)   gave   4,7-
  dimethyl-5-phenyl-2-quinazolinamine (Pd/C, $C_6H_3Me_3$-1,3,5, reflux, CO_2,
  24 h: ∼ 55%).[526]

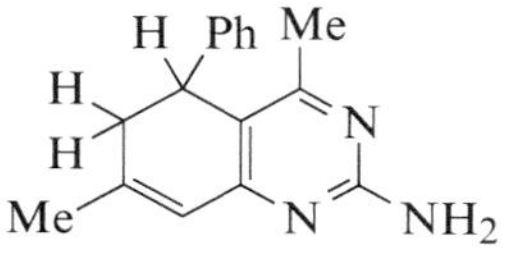

(**124**)

4-Quinazolinamine or 4-diethylaminoquinazoline gave 3,4-dihydroquinazoline
  with loss of ammonia or diethylamine, respectively (controlled-potential
  reduction at pH 2.9).[22]

6.2.2.8.   Miscellaneous Reactions of Minor Application

Apart from the conversion of 3-aminoquinazolinones into 3-aziridin-1-yl-
quinazolinones and related products (reserved for Section 6.2.2.9), the remaining
reactions of aminoquinazolines are illustrated in the following classified examples.

Deuteration

6,7-Diamino-4(3H)-quinazolinone gave mainly its 5,8-D_2 derivative (**125**)
  (D_2O, Me_2NCHO, evaporated; then D_2SO_4, 115°, A, 20 h: 91%).[952]

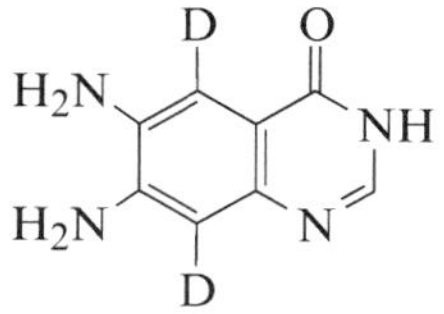

(**125**)

To biguanido derivatives

3-*p*-Aminophenyl- gave 3-*p*-(*N*5-*p*-methoxyphenylsulfonylbiguanido)phenyl-2-methyl-4(3*H*)-quinazolinone (**126**) [*p*-MeOC$_6$H$_4$SO$_2$NHC(=NH)NHCN, HCl, reflux, 10 h: ∼ 65%]; also analogs.[1575]

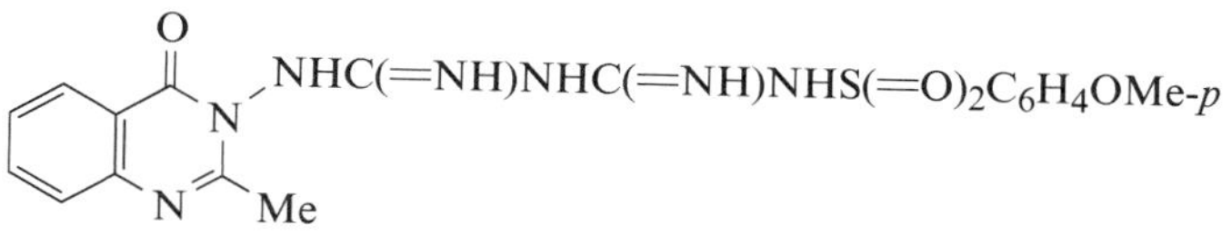

NHC(=NH)NHC(=NH)NHS(=O)$_2$C$_6$H$_4$OMe-*p*

Me

(**126**)

N-Nitrosation

2-*o*-Methylaminophenyl- (**127**, R = H) gave 2-*o*-(*N*-methyl-*N*-nitrosoamino)-phenyl-4(3*H*)-quinazolinone (**127**, R = NO) (NaNO$_2$, 3M HCl, 0°, 5 min, then 100°, briefly: 73%, as hydrochloride).[45]

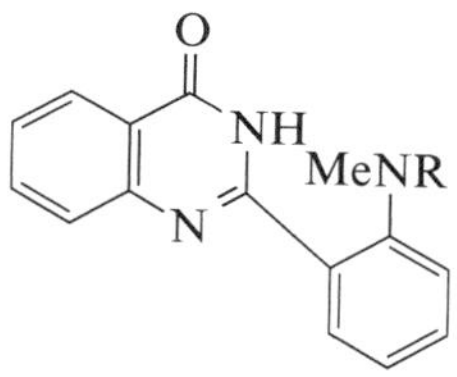

MeNR

(**127**)

8a-Aminomethyl- (**128**, Q = NH$_2$, R = H) gave 8a-bromomethyl-3-nitroso-3,4,4a,5,6,7,8,8a-octahydro-2(1*H*)-quinazolinone (**128**, Q = Br, R = NO) (NaNO$_2$, KBr, HBr–H$_2$SO$_4$–H$_2$O, 0°, 1 h, then 20°, 1 h: 40%; note additional Sandmeyer-type reaction).[190]

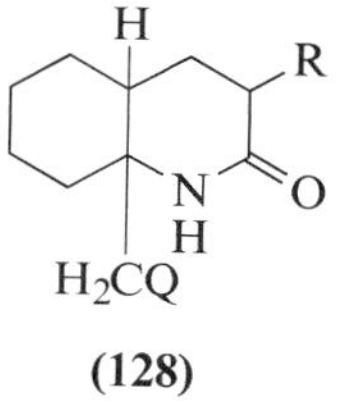

H

R

N

O

H

H$_2$CQ

(**128**)

4(3*H*)-Quinazolinone gave 3-nitroso-4(3*H*)-quinazolinone (NaNO$_2$, AcOH, < 5°, then 20°, 12 h: 82%).[1690]

To isothiocyanato derivatives

3-*p*-Aminophenyl- (**129**, R = NH$_2$) gave 3-*p*-isothiocyanatophenyl-2,4(1*H*,3*H*)-quinazolinedione (**129**, R = N:C:S) (S = CCl$_2$, CaCO$_3$, H$_2$O–CHCl$_3$, < 5°, 2 h: 78%).[2248]

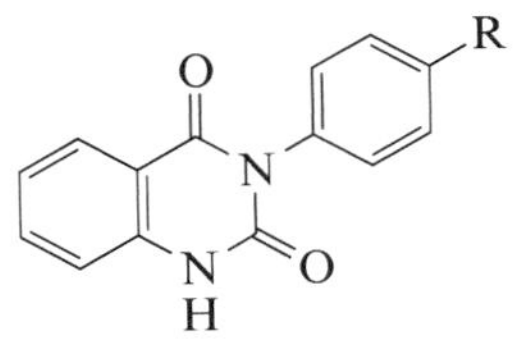

(129)

To metallic complexes

Note: The following complexes are typical of those derived from aminoquinazolines or their derivatives.

4-(2,4-Dinitrophenylhydrazono)-2-methyl-3-quinazolinamine: Cd, Co, Cr, Cu, Fe, Hg, Mn, Ni.[2168]

2-Pyridin-2′-yl-1,2-dihydro-4(3*H*)-quinazolinone: Cu (with dehydrogenation of the ligand),[902] Ni.[545]

2-Methyl-3-*m*-nitrobenzylideneamino-4(3*H*)-quinazolinone: Cd, Co, Cr, Cu, Hg, Ni, Zn.[1720]

Ring fission

3-Amino-2-*p*-chlorophenyl-4(3*H*)-quinazolinethione gave 2-*o*-benzamido-phenyl-5-*p*-chlorophenyl-1,3,4-thiadiazole (**131**), probably by formation of the Schiff base (**130**), ring fission, cyclization, and oxidation as shown (*p*-ClC$_6$H$_4$CHO, HCl, EtOH, reflux, 6 h: 52%); several analogs like-wise.[1210]

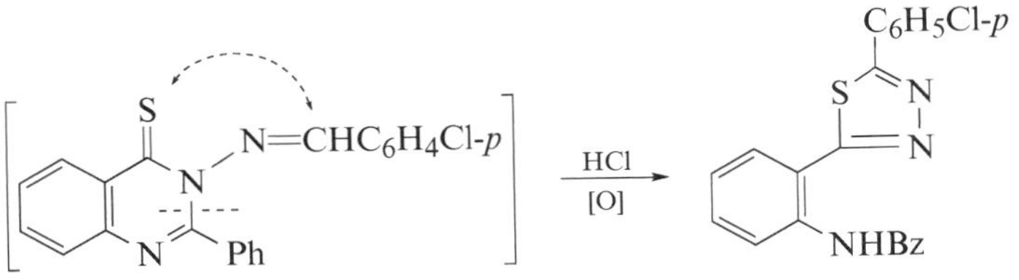

(130)   **(131)**

Cyclization to heterocyclylquinazolines

Note: For the formation of aziridinylquinazolines, see Section 6.2.2.9.

3-Amino-2-chloromethyl- gave 2-chloromethyl-3-(2,5-dimethylpyrrol-1-yl)-4(3H)- quinazolinone (**132**, R = Cl) (neat $AcCH_2CH_2Ac$, 150°, A, 1 h: 67%).[2360]

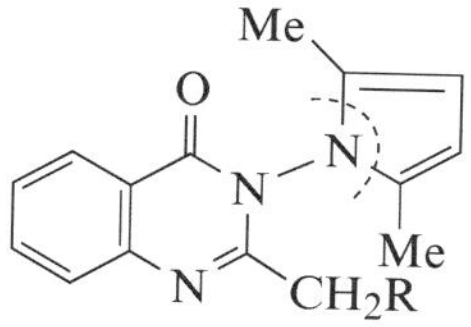

(**132**)

3-Amino- gave 3-(2,5-dimethylpyrrol-1-yl)-4(3H)-quinazolinone (**132**, R = H) ($AcCH_2CH_2Ac$, trace TsOH, EtOH, reflux, 1 h: 80%).[2056]

3-Amino-2-methyl-4(3H)-quinazolinone gave the spiro product, 2-methyl-3-(4-oxo-2,2-tetramethylenethiazolidin-3-yl)-4(3H)-quinazolinone (**133**) (2,2-tetramethylene-1,3-oxathiolan-5-one, EtOH, reflux, 3 h: 82%); also analogs likewise.[2464]

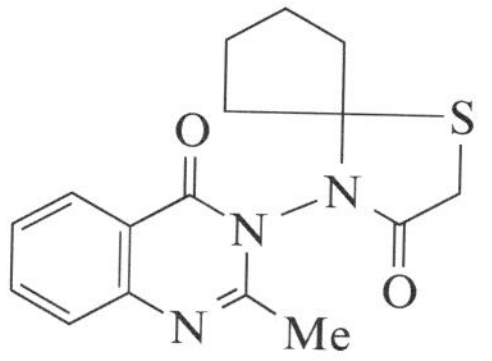

(**133**)

Also other examples.[414,2581]

Cyclization to fused quinazolines

2-*o*-Aminophenyl-4(3H)-quinazolinone (**134**) gave 6-phenyl-8H-quinazolino[4,3-b]quinazolin-8-one (**135**) (neat Bz_2O, 255°, 6 h: ?%).[1153, cf. 2484]

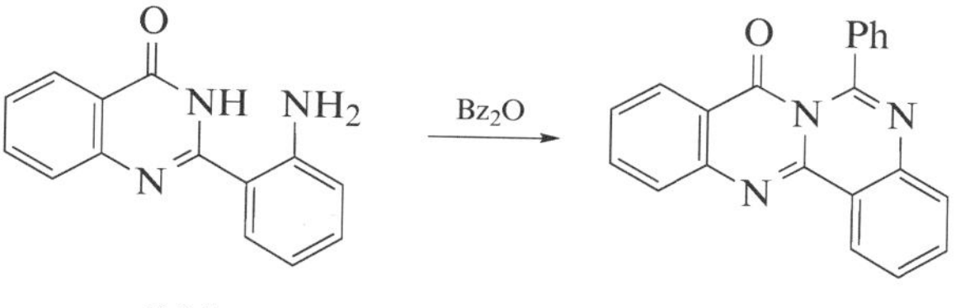

(**134**)   (**135**)

1-Methyl-1,4-dihydro-2-quinazolinamine (**136**) gave 11-methyl-6,11-dihydro-2H-pyrimido[2,1-b]quinazolin-2-one (**137**) (HC≡CCO$_2$Et, EtOH, reflux, 6 h: 45%) or its 3,4,6,11-tetrahydro analog (H$_2$C=CHCO$_2$Et, likewise: 64%);[1165] the rules of nomenclature require the latter product to be named as 11-methyl-6,11-dihydro-4H-pyrimido[2,1-b]quinazolin-2(3H)-one!

(136)   (137)

6,7-Diamino-3-butyl-4(3H)-quinazolinone (**138**) gave 7-butyl-2-methoxy car-bonylamino-1H-imidazo[4,5-g]quinazolin-8(7H)-one (**139**) [MeSC(=NH$_2$, ClCO$_2$Me, EtOH–H$_2$O, pH 5, reflux, 1 h: 80%]; also analogs.[1165]

(138)   (139)

Also other examples.[1877,1917,2124,2533]

6.2.2.9. Conversion of 3-Amino- into 3-Aziridin-1′-yl-4-(3H)-quinazolinones and Related Compounds

*An overview.** The chemistry associated with 3-acetoxyaminoquinazolinones (**141**) should be mentioned here. These unstable compounds are prepared in solution by lead tetraacetate oxidation of the corresponding 3-aminoquinazo-linones (**140**) in excellent yields. They bring about the conversion of a range of alkenes into aziridines (**143**), and the reaction mechanism appears to be analog-ous to that by which peroxy acids convert alkenes into epoxides (**142**).[2063]

The quinazolinone ring plays an important role in this aziridination, not only as a stabilizing appendage for the acetoxyamino group but also in controlling the stereochemistry of the three-membered ring formation (see below). The presence of a chiral 2-substituent on the quinazolinone ring (R = R*) has been found to bring about asymmetrical induction in aziridination of prochiral alkenes, and in some cases single diastereoisomers of the product (**143**, R = R*) are obtained.[1817]

_____

*Dr Robert S. Atkinson, responsible for most of the research in this intricate area has kindly supplied this introductory overview.[2515]

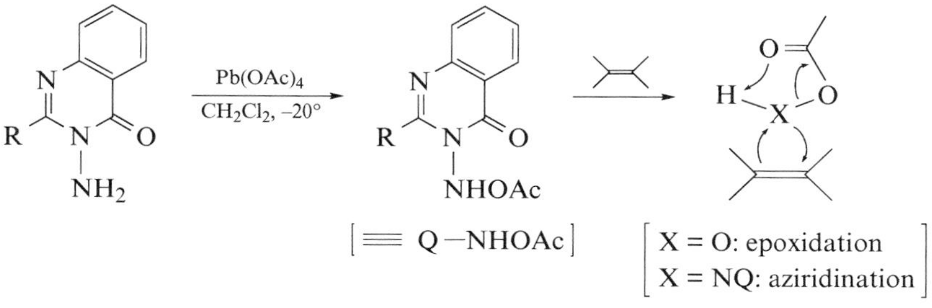

(140)   (141)   (142)

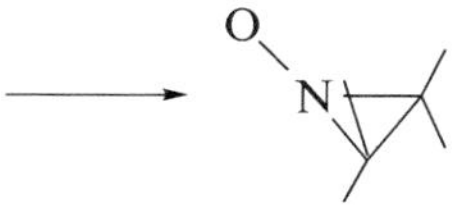

$$\left[\equiv Q-NHOAc \right] \quad \left[\begin{array}{l} X = O: \text{epoxidation} \\ X = NQ: \text{aziridination} \end{array} \right]$$

(143)

The presence and properties of the quinazoline ring (Q) have allowed a detailed description of the transition state for these aziridinations as in (144)[2513] and the means by which chirality is propagated by the existing chiral 2-substituent on Q. A focus of the work in this area is the rational design of the substituents on the chiral center R* in (144) so as to maximize the diastereoselectivity of the aziridination, since the resulting diastereo- and enantiopure aziridines (143) can be converted into enantiopure Q-free products or chirons (e.g., amino acids) by ring cleavage and N—N bond cleavage.[2397]

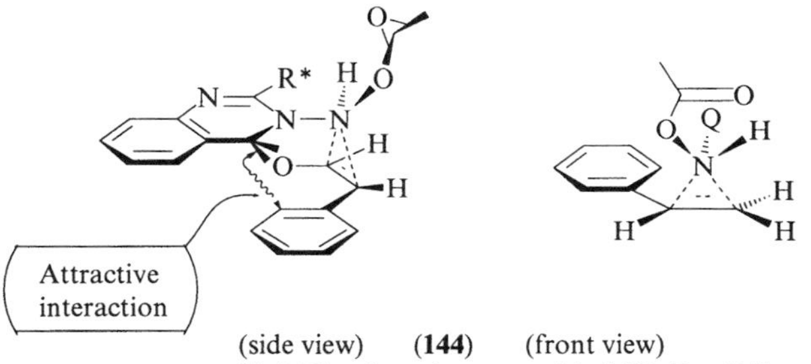

(side view)   (144)   (front view)
[Aziridination of styrene with (141), R = R*]

A further property of the 3-aminoquinazolinones (140), which is being explored, is the chirality of the N,N-diacyl derivatives (145) with $R^1 \neq R^2$. This chirality arises from the absence of rotation around the N—N bond, as a result of which this bond becomes a chiral axis. The presence of an additional chiral center gives rise to diastereoisomers and, in the case of (146a) and (146b), these have been separated and identified by X-ray crystallography. These N-(quinazo-

(145)   **(146)**

(a) R^1 = Ph, R^2 = Me
(b) R^1 = Me, R^2 = Ph

linonyl)imides **(145)** are acylating agents and show some promise as chiral acylating agents when used in enantiopure form.[2512]

Examples of the quinazoline reactions. As mentioned in the preceding overview, the oxidation of appropriate 3-amino-4(3H)-quinazolinones, such as **(147)**, by lead tetraacetate in an inert solvent at $-20°$ furnishes products such as 3-acetoxyamino-2-ethyl-4(3H)-quinazolinone **(148)**, which are stable in cold solution and may be characterized therein by their nmr spectra. Such products can function as intermolecular or intramolecular aziridinating agents for ethylenic double bonds, a role ascribed until recently to the corresponding nitrenes, although their presence could not be detected experimentally.[2063, cf. 2514] Thus oxidation of the substrate **(147)** and subsequent treatment with methyl acrylate or styrene afforded 2-ethyl-3-(2-methoxycarbonylaziridin-1-yl)- **(149**, R = CO_2Me) (80%) or 2-ethyl-3-(2-phenylaziridin-1-yl)4(3H)-quinazolinone **(149**, R = Ph) (79%), respectively.[1699,2063] Similarly, but in an intramolecular fashion, oxidation of 3-amino-2-(4-phenylbut-3-enyl)-4(3H)-quinazolinone **(150**, $n = 1$) gave the tetracyclic product **(151**, $n = 1$) (76%),[1355] and oxidation of 3-amino-2-(6-phenylhex-5-enyl)-4(3H)-quinazolinone **(150**, $n = 3$) gave the analogous product **(151**, $n = 3$) ($\sim$ 30%).[2063] Unlike the intermediate **(148)** and most of its analogs, 3-acetoxyamino-2-trifluoromethyl-4(3H)-quinazolinone proved sufficiently stable for characterization as a crystalline solid, which reacted very satisfactorily with unsaturated entities; for example, with cyclohexene, it gave the product **(152)**

(147)   **(148)**

(149)

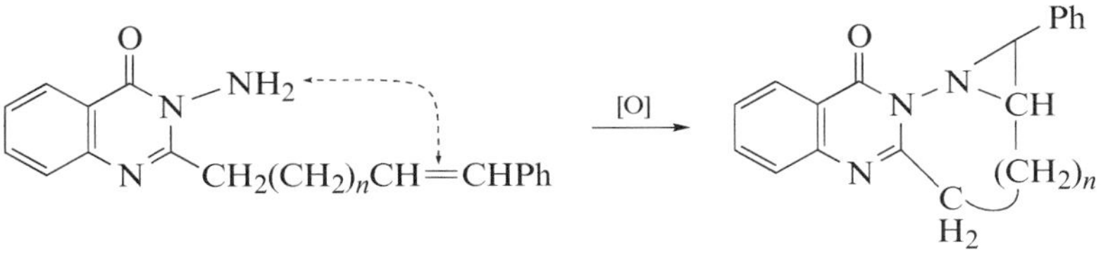

(150)   **(151)**

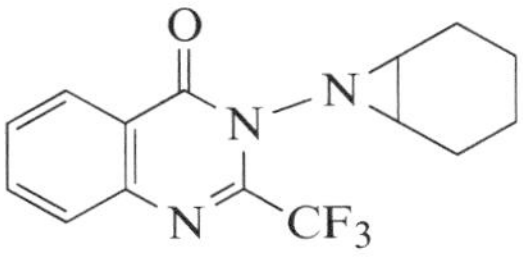

(152)

in 68% yield.[2462] This interesting area of quinazoline chemistry was pioneered[155,191,200] and subsequently developed, with special emphasis on its stereochemical applications, mainly at the University of Leicester.[1326,1355,1696,1697,1699,1754,1817,1818,2049,2052,2055,2063,2106,2127,2128, 2180,2172,2175,2182,2353,2392,2462,2463,2492,2512–2514,2516,2517,2549,2592,2594]

Appropriate 3-aminoquinazolinones may also undergo analogous oxidative reactions of the intramolecular type to afford products other than aziridine derivatives. For example, a lead tetraacetate oxidation of 3-amino-2-pent-4'-ynyl-4(3H)-quinazolinone (153) gave mainly the spiro azirine (154) accompanied

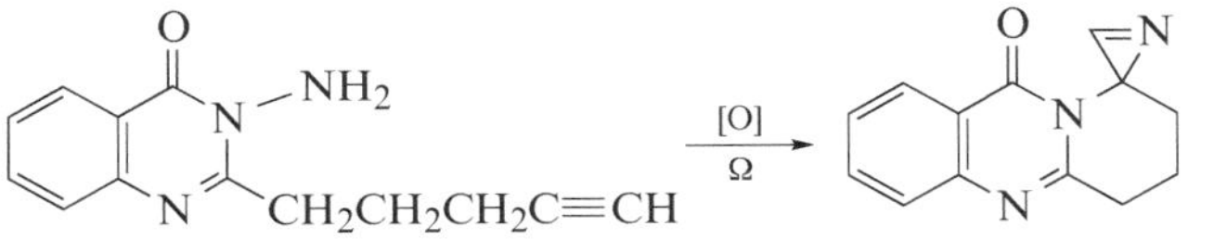

(153)   **(154)**

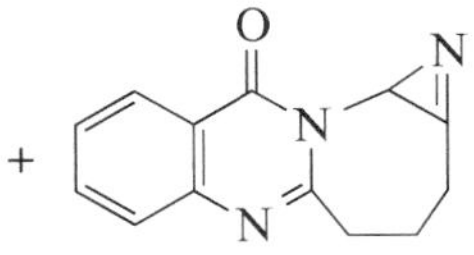

(155) 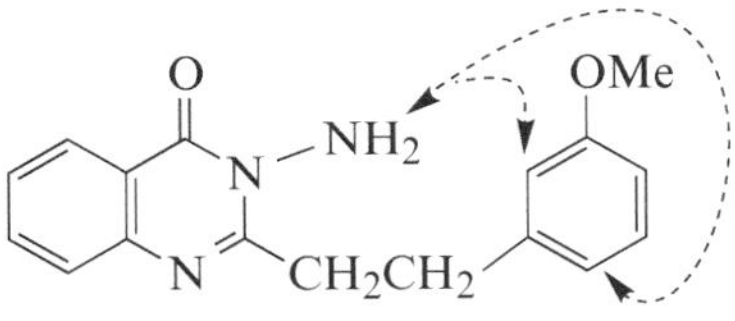   **(156)**

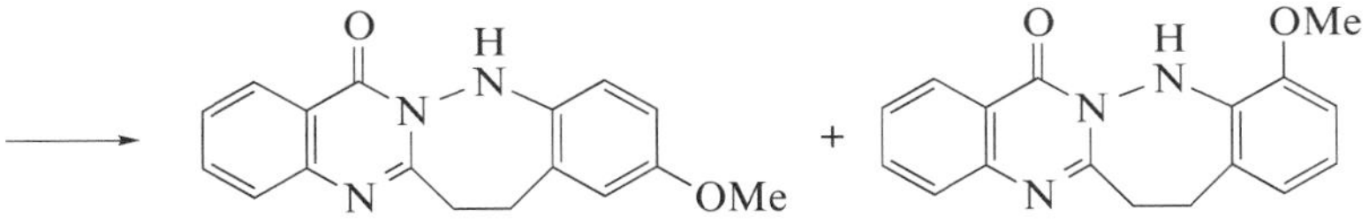

(157)   **(158)**

by the fused-azirine (155);[1733,1752] and 3-amino-2-*m*-methoxyphenethyl-4(3*H*)-quinazolinone (156) gave a separable mixture of the products (157) and (158), each in ~ 30% yield.[1325,1351, cf. 1323,1333]

6.2.2.10.   Ancillary and Biological Aspects (*H* 508)

Most of the recent physicochemical and biological studies on aminoquinazolines have been in connection with the dihydrofolate reductase (DHFR) inhibition exhibited by 2,4-diamino-5,8-dideazafolate analogs and related quinazolines. The following lists include typical examples of compounds so studied.

Structural or other physical studies

2,4-Quinazolinediamine: X-ray analysis;[1270,1795] ionization.[250]

5-Chloro-2,4,6-quinazolinetriamine: X-ray analysis.[1269]

5-Methyl-6-(3,4,5-trimethoxyanilino)methyl-2,4-quinazolinediamine: X-ray analysis.[1968]

5,6,7,8-Tetrahydro-2,4-quinazolinediamine: [15]Nnmr.[1138]

2-*p*-Aminophenyl-4-phenyl-6-quinazolinamine: X-ray analysis.[2152]

1,3/3,5-Diphenyl-5/1-quinazolin-2'-ylformazan: [1]Hnmr and [13]Cnmr.[1641]

6-Chloro-4-*p*-chloroanilino-2-dimethylaminoquinazoline/*o*-chlorobenzoic acid salt: X-ray analysis.[2576]

Biological studies

6-Benzylamino-2,4-quinazolinediamine (also ~ 20 substituted-benzyl analogs, several 2,4-bisalkylamino analogs, ~ 18 heterocyclylmethyl analogs, and ~ 10 other types of analog): DHFR inhibition.[675]

5/6-β-(Naphthalen-2-yl)vinyl-2,4-quinazolinediamine (also ~ 30 analogs with various 5/6-substituents): DHFR inhibition and antimalarial activity.[810]

2-Amino-5/6-naphthalen-2'-ylthio-4(3*H*)-quinazolinone (also ~ 30 assorted analogs): DHFR inhibition and antimalarial activity.[810]

2,4-Quinazolinediamine (also ~ 30 5-alkoxy, amino, alkylthio, halogeno, and other derivatives): DHFR inhibition and antibacterial activity.[2225, cf. 1919]

6-*p*-Carboxyanilinomethyl-5-chloro-2,4-quinazolinediamine (also 17 analogous esters, etc.): DHFR inhibition and leukemia inhibition.[1868]

2-Piperazin-1'-yl-5,6,7,8-tetrahydro-4-quinazolinamine   also   ~ 20   substituted-piperazinyl and several substituted-amino analogs): hypoglycemic activity.[1417]

4-*p*-Hydroxyanilinoquinazoline (also four derivatives): antiarrhythmic activity.[1559]

2-*o*-Piperidinoazophenyl-4-quinazolinamine (and several analogs): antitumor activity.[1873]

See also two general reviews of DHFR inhibition, mainly by 6-substituted-2,4-quinazolinediamines,[2379,2518] a set of papers on the anticancer activities of 2-amino (or methyl)-6-substituted-4(3H)-quinazolinones,[2518–2524] and others.[2597]

6.3.   HYDRAZINO-, HYDROXYAMINO-, AND AZIDOQUINAZOLINES

Although the preparative routes to these (substituted-amino)quinazolines have much in common with those to regular aminoquinazolines, the reactions of the two groups differ markedly in many respects.

6.3.1.   Preparation of Hydrazinoquinazolines

Almost all the recently used routes to hydrazinoquinazolines have been covered already: from *primary syntheses* (Chapter 1), from *hydrazinolysis of halogenoquinazolines* (Sections 3.5.1, 3.6, and 3.7), from *hydrazinolysis of quinazolinethiones* (Section 5.1.2.2), from *hydrazinolysis of alkylthioquinazolines* (Section 5.3.2), from *reduction of diazonioquinazolines* (Section 6.2.2.5), and from *transamination of regular aminoquinazolines* (Section 6.2.2.6).

In addition, 3-(2-hydroxynaphthalen-1-yl)azo- gave 3-β-(2-hydroxynaphthalen-1-yl)hydrazino-2-methyl-4(3H)-quinazolinone (**159**) in 60% yield by reduction in refluxing stannous chloride/hydrochloric acid for 8 h;[2183] 3-β-(p-anilinophenyl)hydrazino-2-butyl-4(3H)-quinazolinone (50%) and several other analogues were made similarly.[2183]

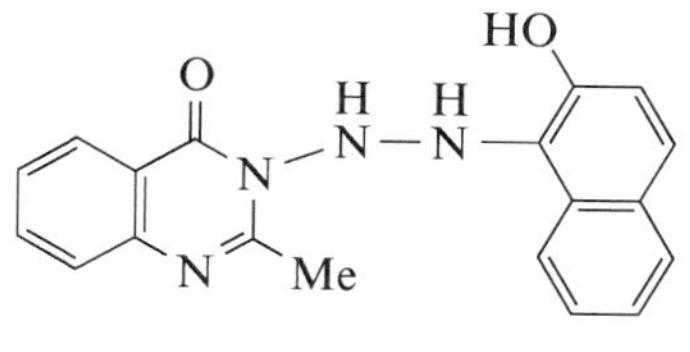

(**159**)

6.3.2.   Reactions of Hydrazinoquinazolines (*H* 336)

The main reactions of hydrazinoquinazolines are illustrated by the following classified examples; some subsequent reactions, not covered elsewhere in this book, are also included here.

Acylations

2-Hydrazino- gave 2-β-acetylhydrazino-3-phenyl-4(3H)-quinazolinone (**160**) (AcCl, K$_2$CO$_3$, CHCl$_3$, reflux, 1 h: 80%) and thence 1-methyl-4-phenyl[1,2,4]triazolo[4,3-a]quinazolin-5(4H)-one (**161**) (neat, 150°, 5 min: 60%).[1319]

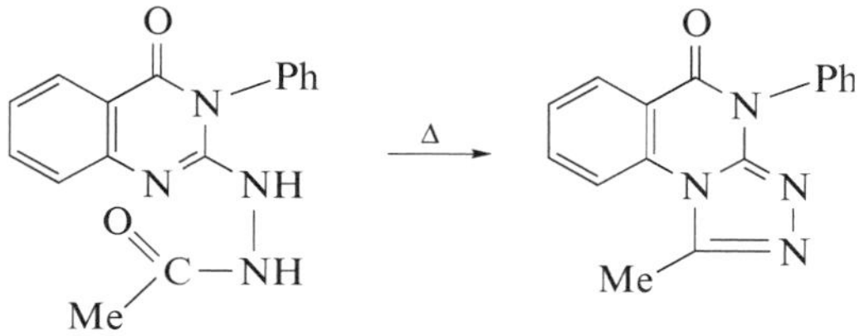

(160)   (161)

3-Amino-2-hydrazino- gave 2-β-acetylhydrazino-3-amino-4(3H)-quinazolinone (Ac$_2$O: 95%; note preferential acetylation) and thence 3-acetamido[1,2,4]triazolo[5,1-b]quinazolin-9(3H)-one (**162**) [HC(OEt)$_3$: ?%].[1990]

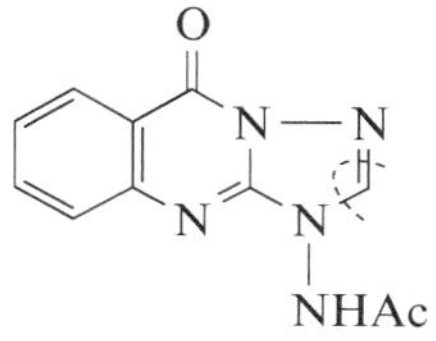

(162)

6-Methoxy-2-phenyl-4-(β-p-tosylhydrazino)quinazoline (made from the corresponding 4-chloroquinazoline) gave 6-methoxy-2-phenylquinazoline (NaOH, Na$_2$CO$_3$, H$_2$O, boiling, 2.5 h: 45%; a type of McFadyan–Stevens reaction,[2450k] rarely used for quinazolines).[766]

Conversion into Schiff bases (quinazolinylhydrazones)

2-Chloro-4-hydrazino- gave 4-benzylidenehydrazino-2-chloroquinazoline (**163**) (PhCHO, EtOH, 20°, 30 min: 69%) and thence 5-chloro-3-phenyl-1,2,4-triazolo[4,3-c]quinazoline (**164**) (Br$_2$, AcONa, AcOH, 20°, 30 min: 62%);[2228] also analogs.[2228]

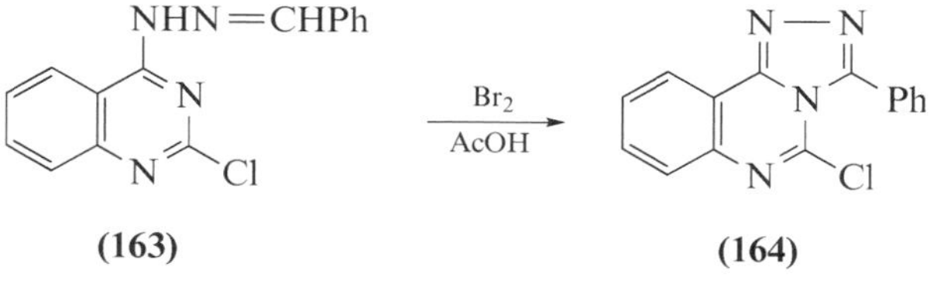

(163)   (164)

3-Amino-2-hydrazino- gave 3-amino-2-(α-carboxybenzylidene)hydrazino-4($3H$)-quinazolinone (BzCO$_2$H, H$_2$O, reflux, 15 min: 95%), 5-amino-2-phenyl-1H-[1,2,4]triazino[4,3-a]quinazoline-1,6($5H$)-dione (**165**, R = NH$_2$) (AcOH, reflux, 5h: 73%), and finally the 5-deamino analog (**165**, R = H) (neat PhCHO, 180°, 2h: 78%, by loss of PhCN and H$_2$O).[2059]

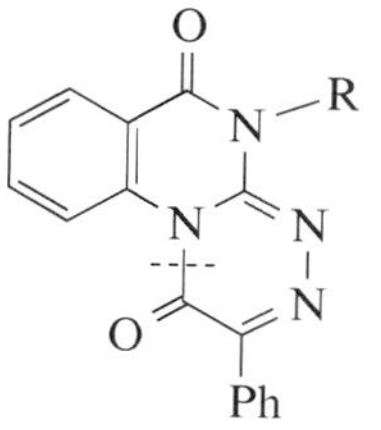

(165)

4-Hydrazinoquinazoline gave 4-p-chlorobenzylidenehydrazinoquinazoline (**166**) (p-ClC$_6$H$_4$CHO, MeOH: ?%; uncharacterized) and thence 3-acetyl-2-p-chlorophenyl-2,3-dihydro[1,2,4]triazolo[1,5-c]quinazoline (**167**) (Ac$_2$O, reflux, 2h: > 50%; note rearrangement), proved in structure by oxidative hydrolysis to the known product, 2-p-chlorophenyl[1,2,4]triazolo[1,5-c]quinazoline (FeCl$_3$, H$_2$O, 100°, 5h).[2461]

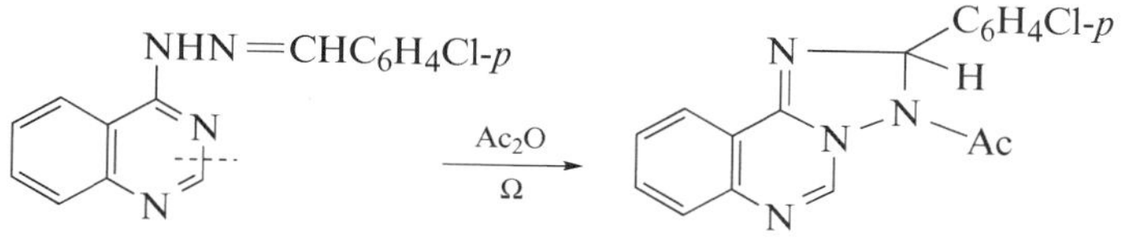

(166)   (167)

Also other examples.[1121,1319,1524,1748,1922,1950,1989,1991,2044, 2050,2084, 2141,2526,2537]

Conversion into semicarbazidoquinazolines

4-Hydrazino-8-methyl-2-phenyl- gave 8-methyl-2-phenyl-4-[4-phenyl(thio-semicarbazido)] quinazoline (**168**) (PhNCS, EtOH, reflux, 1h: 87%) and thence 7-methyl-5-phenyl-1,2,4-triazolo[4,3-c]quinazoline-3($2H$)-thione (**169**) (neat, 220°, until PhNH$_2$ ↑ ceased?: 79%).[2485]

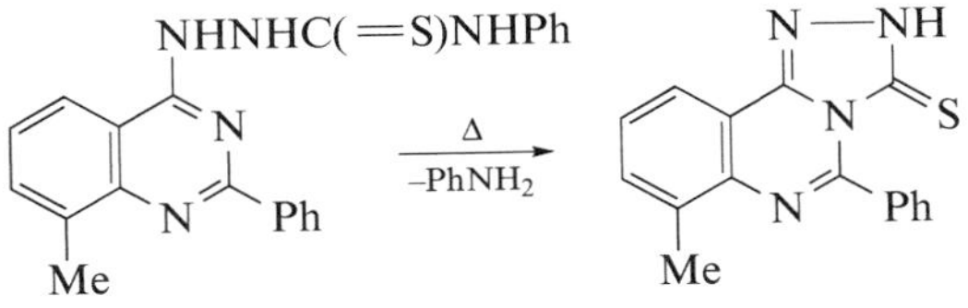

(168)   **(169)**

3-Allyl-2-hydrazino- gave 3-allyl-2-[4-phenyl(thiosemicarbazido)]- (PhNCS, EtOH, reflux, <1 h: 70%) and thence 3-allyl-2-[3,4-diphenyl-2(3H)-thiazolylidene]hydrazino-4(3H)-quinazolinone (170) (BzCH$_2$Br, AcONa, EtOH, reflux, 2 h: 85%); also analogs.[1637]

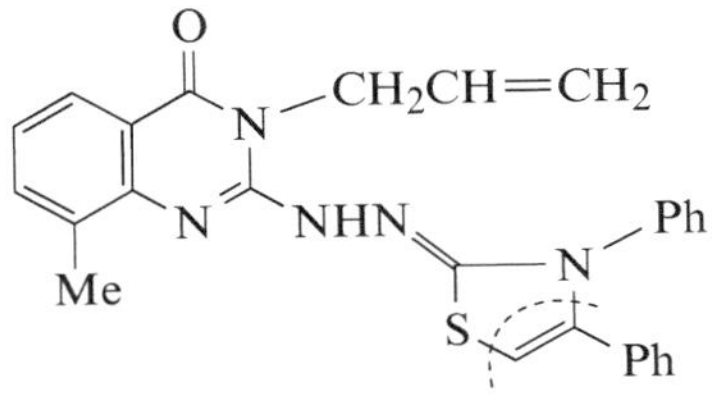

(170)

Conversion into azidoquinazolines

4-Hydrazino-(171) gave 4-azido-2-morpholinoquinazoline(172a), formulated as its cyclic tautomer, 5-morpholinotetrazolo[1,5-c]quinazoline (172b) (NaNO$_2$, dilute HCl, 20°, 1 h: 47%).[2266] (*Note*: In this book, azidoquinazolines are normally formulated as such irrespective of any evidence for the predominant form in each case.)

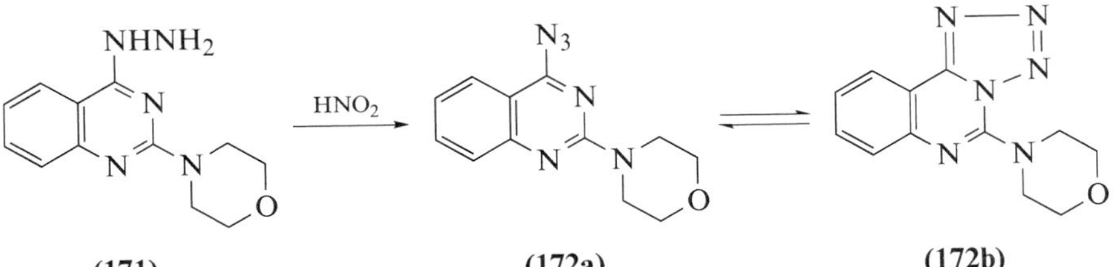

(171)   **(172a)**   **(172b)**

2-Hydrazino- gave 2-azido-3-phenyl-4(3H)-quinazolinone (173) (NaNO$_2$, 6M HCl, 5°, 15 min; then 20°, 30 min: ~55%).[1894]

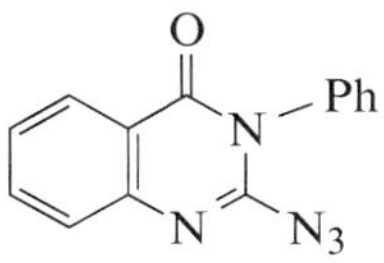

(173)

Also other examples,[740,1021,1518,1522,1760,1841,2022,2053]

Cyclizations (without isolation of intermediates)

6-Chloro-2-hydrazino-3-methyl-4(3H)-quinazolinone gave 7-chloro-4-methyl-1-thioxo-1,2-dihydro[1,2,4]triazolo[4,3-a]quinazolin-5(4H)-one (**174**) (CS$_2$, pyridine, reflux, 5 h: 90%) or 2-(5-amino-4-cyano-3-methyl-thiopyrazol-1-yl)-6-chloro-3-methyl-4(3H)-quinazolinone (**175**) [(MeS)$_2$ C=C(CN)$_2$, MeNCHO–EtOH, reflux, 5 h: 46%).[2279]

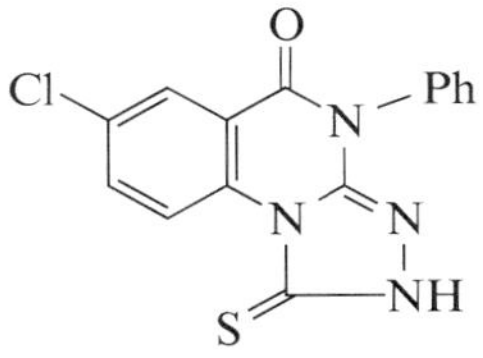

(174)

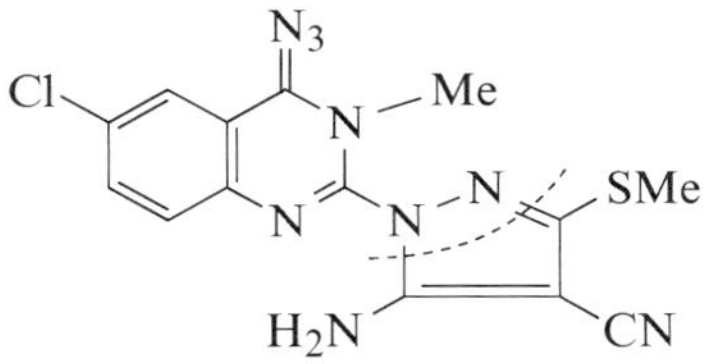

(175)

2-Hydrazino-3-phenyl-4(3H)-quinazolinone gave 4-phenyl[1,2,4]-triazolo[4,3-a]quinazolin-5(4H)-one (**176**) [HCO$_2$H: >58%; HC(OEt)$_3$: >45%].[1064]

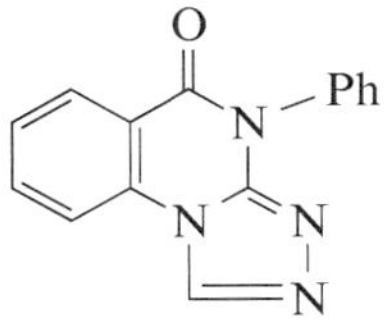

(176)

Also other examples.[1066,1515,2526]

6.3.3. Preparation and Reactions of Hydroxyaminoquinazolines

Under appropriate circumstances, hydroxyaminoquinazolines are sometimes formulated as the tautomeric oximes. For example, 4-hydroxyaminoquinazoline

(177) may be formulated as 4-hydroxyimino-3,4-dihydroquinazoline (178, R = H) and even called "4(3H)-quinazolinone oxime"; naturally, the presence of a 3- or 1-substituent as in (178, R = Me) will fix the compound in the oxime form.[2132]

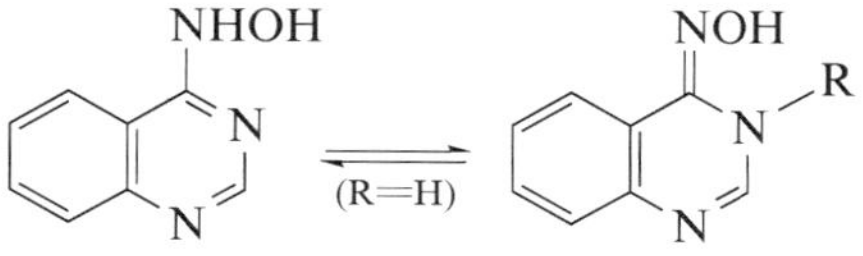

          (177)   (178)

Preparative methods.   Most of the synthetic routes to hydroxyaminoquinazo-lines (for which any recent examples are available) have been covered already: primary syntheses (Chapter 1), *hydroxyaminolysis of halogenoquinazolines* (Sections 3.5.1 and 3.7), *Dimroth rearrangement of quinazolinamine N-oxides* (Section 4.7.2.3), *hydroxyaminolysis of alkylthioquinazolines* (Section 5.3.2), and *trans-amination of regular aminoquinazolines* (Section 6.2.2.6).

In addition, 4-chloroquinazoline reacted with the preformed sodium salt of acetone oxime (in water at 20° for 3 h) to afford a 90% yield of 4-(iso-propylideneaminooxy)quinazoline (179) that underwent thermolysis at 160° during 6 h to furnish 6-hydroxyaminoquinazoline (180) in 20% yield.[1935]

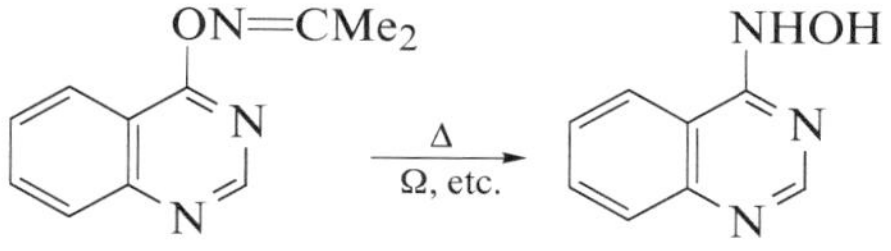

          (179)   (180)

Reactions.   Very few of the possible reactions of hydroxyaminoquinazolines have been used recently. The *reductive deoxygenation* of hydroxyamino- to aminoquinazolines has been exemplified in Section 6.2.1.

In addition, 4-hydroxyamino-2-styryl-1,2-dihydroquinazoline (181) with 1 or 2 equiv of benzoic anhydride in ether at 20° for 2 days gave 4-benzoyloxyamino-(182, R = H) or 1-benzoyl-4-benzoyloxyamino-2-styryl-1,2-dihydroquinazoline (182, R = Bz) (60%), respectively,[2290] and other such acylations have been reported.[2132]

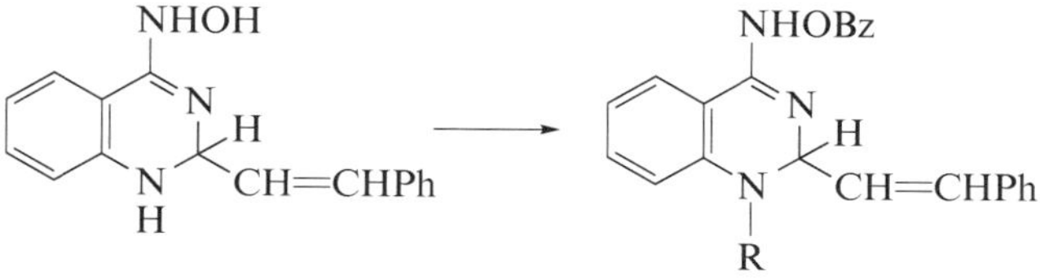

          (181)   (182)

6.3.4. Preparation, Fine Structure, and Reactions of Azidoquinazolines (*H* 337)

Preparative methods.   Most azidoquinazolines have been made by the reaction of *azide ion on halogenoquinazolines* (see Sections 3.5.5, 3.6, and 3.7), *azide ion on preformed diazonioquinazolines* (see Section 6.2.2.5), or *nitrous acid on hydrazinoquinazolines* (see Section 6.3.2). In addition, 2-*o*-(piperidinoazo)phenyl-4-quinazolinamine (**183**) gave 2-azido-4-quinazolinamine (**184**) in 27% yield by boiling with sodium azide in glacial acetic acid for 15 min;[1014] 2-azido-4(3*H*)-quinazolinone (∼ 70%) as made rather similarly.[1014]

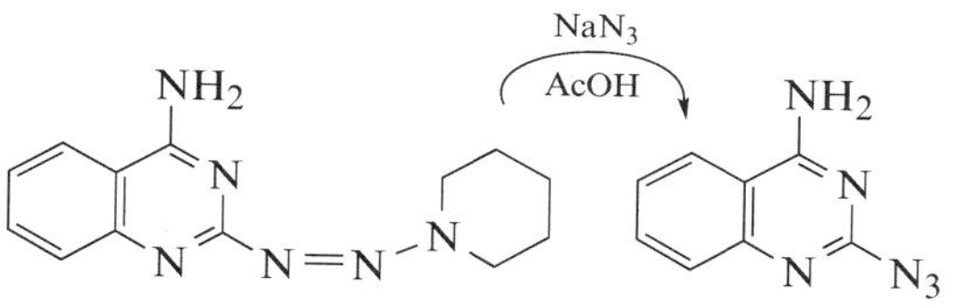

(183)   **(184)**

Fine structure.   The existence of 2/4-azidoqinazolines as tetrazoloquinazoline tautomers has been mentioned previously (Sections 3.5.5 and 6.3.2). A briefly reported ir spectral study of 4-azido-2-phenylquinazoline (**185**, R = H) has indicated that it exists almost entirely as 5-phenyltetrazolo[1,5-*c*]quinazoline (**186**,   R = H)   in   chloroform   solution,   whereas   4-azido-2-*o*-hydroxyphenylquinazoline (**185**, R = OH) exists almost entirely as such in chloroform.[834, cf, 2200]

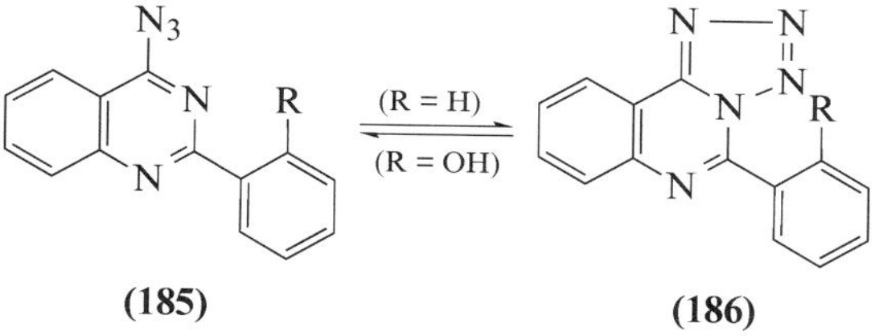

(185)   **(186)**

Reactions. The *reduction* of azido- to aminoquinazolines has been exemplified toward the end of Section 6.2.1. Minor reactions are illustrated in the following examples.

6-Azido-2,4-quinazolinediamine gave 6,6′-azoquinazoline-2,2′,4,4′-tetramine (**187**) (PhNO$_2$, reflux, 2 h: ∼ 80%; *hv*, MeOH, 10 h: ∼ 30%).[1919]

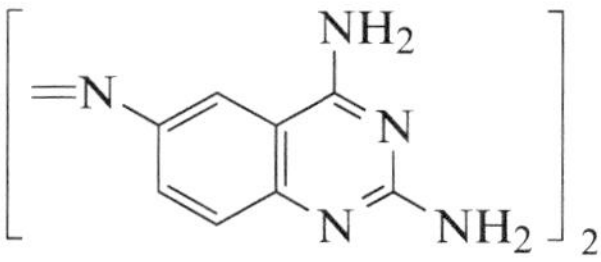

(187)

2-*o*-Azidophenyl-4(3*H*)-quinazolinone   gave   indazolo[2,3-*a*]-quinazolin-
5(6*H*)-one (**188**, R = H) (xylene, reflux, 6 h: 64%); also the 6-methyl (**188**,
R = Me) (70%), 6-phenyl (**188**, R = Ph) (46%), and other derivatives likewise.[1277]

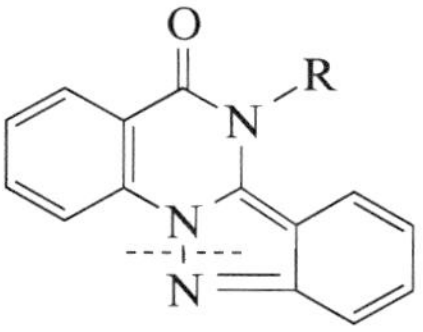

(188)

6-Azidoquinazoline   gave   5,7-dimethoxy-8,9-dihydro-5*H*-pyrimido[5,4-*c*]-
azepine (**189**) (*hv*, MeOH–dioxane, N$_2$, 4 h: 16%).[2075]

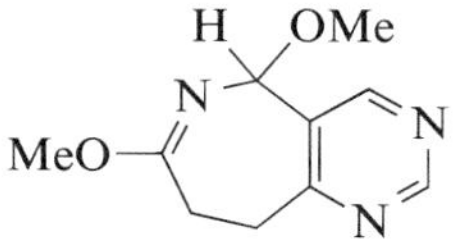

(189)

2-Azidomethyl-3-*p*-chlorophenyl-4(3*H*)-quinazolinone   gave   3-*p*-chloro-
phenyl-4-oxo-3,4-dihydro-2-quinazolinecarbaldehyde (three stages: see
original).[1821]

6.4.   PREPARATION AND REACTIONS OF NONTAUTOMERIC
2/4-QUINAZOLINIMINES (*H* 335)

Nearly all the available information in this area has been discussed already.

Preparative routes.   Such imines have been made by *primary synthesis* (e.g.,
Sections 1.1.4.7 and 1.2.3.7), by *aminolysis of nontautomeric quinazolinethiones*
(Section 5.1.2.2), and by *1/3-alkylation of 2/4-aminoquinazolines* (Section 6.2.2.7).

Reactions. The only two important reactions of these imines are *hydrolysis to fixed quinazolinones* (see Section 4.6.1) and *Dimroth rearrangement*, whereby a quinazolinimine with an alkyl group on an adjacent ring-N undergoes ring fission and recyclization to the isomeric alkylaminoquinazoline (see toward the end of Section 6.2.1). Although these rearrangements, such as (190) → (192), usually occur in alkaline media via intermediates like (191), they can also occur under thermolytic conditions via unidentified intermediates; moreover, it is evident that 3-methyl-4(3H)-quinazolinimine (190, R = Me) isomerizes readily into 4-methylaminoquinazoline (192, R = Me) on electron bombardment during measurement of its mass spectrum.[417]

(190)   +H₂O →   (191)   −H₂O →   (192)

6.5. ARYLAZO- AND NITROSOQUINAZOLINES

Although arylazo and nitroso derivatives are of major importance in the pyrimidine series,[2446g] such derivatives have not been used much in the quinazolines.

Arylazoquinazolines. A few arylazoquinazolines have been made by *primary syntheses* (see Chapter 1), but most have been prepared from aminoquinazolines by *diazotization and subsequent coupling* (see Section 6.2.2.5). The reverse of the latter process—the coupling of diazotized aniline or the like to a quinazoline—has been used sparingly: for example, 3-benzylideneamino-2-methyl- (193) gave 2-methyl-3-α-(phenylazo)benzylideneamino-4(3H)-quinazolinone (194) in 70% yield on coupling with benzenediazonium chloride in aqueous pyridine;[1656] and other such examples have been reported.[1656,1879] A fourth preparative route is exemplified by the condensation of 3-amino-2-methyl-4(3H)-quinazolinone (195) with *p*-nitrosodiphenylamine in refluxing glacial acetic acid for 1 h to give 3-(*p*-anilinophenyl)azo-2-methyl-4(3H)-quinazolinone (196) in 70% yield.[2183]

(193)   PhN₂Cl →   (194)

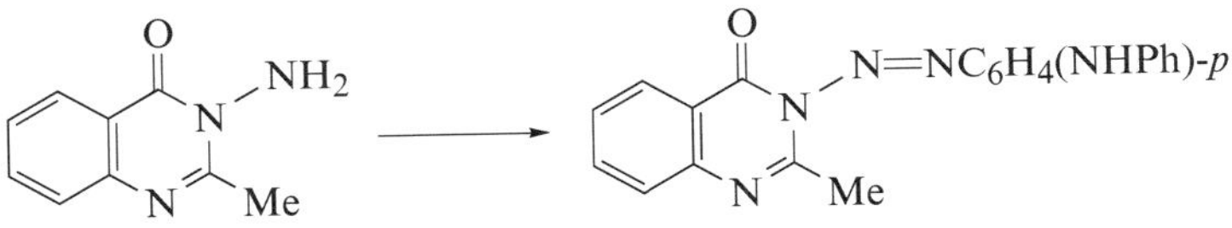

(195)   **(196)**

The reduction of a piperidinoazo- to an aminoquinazoline has been mentioned in Section 6.2.1; and of a phenylazo- to a phenylhydrazinoquinazoline, in Section 6.3.1.

Nitrosoquinazolines. The *N*-nitrosation of secondary amino groupings in quinazolines has been mentioned in Section 2.2.2.8. Although *C*-nitrosations are rare, 6-quinazolinol (**197**, R = H) gave its 5-nitroso derivative (**197**, R = NO) in 80% yield by treatment with aqueous nitrous acid at <5° during 1 h.[2359] In contrast, 5,6,7,8-tetrahydroquinazoline underwent nitrosation at the 8-position (on treatment with propyl nitrite and ethanolic hydrogen chloride at 20° for 2 h) to give a product (63%) formulated as 5,6,7,8-tetrahydro-8-quinazolinone oxime (**198**, R = H);[1414] the 4-methoxy derivative (**198**, R = OMe) was made in 31% yield using isopentyl nitrite in liquid ammonia containing sodium amide as the nitrosating agent (under Claisen conditions).[788]

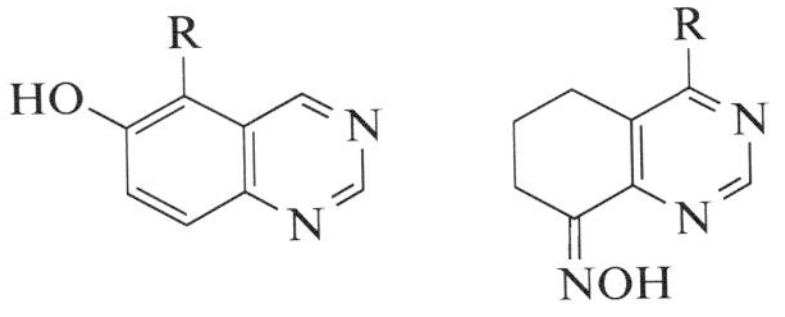

(197)   **(198)**

The *N*-nitroso group of 3-nitroso-4(3*H*)-quinazolinone or its 6,8-dibromo derivative reacted with the methylene grouping in 2-thioxo-4-thiazolidinone (in refluxing Ac$_2$O/AcOH during 6 h) to give 3-(4-oxo-2-thioxothiazolidin-5-ylidene)amino-4(3*H*)-quinazolinone (**199**, R = H) or its 6,8-dibromo derivative (**199**, R = Br), each in >70% yield.[1690]

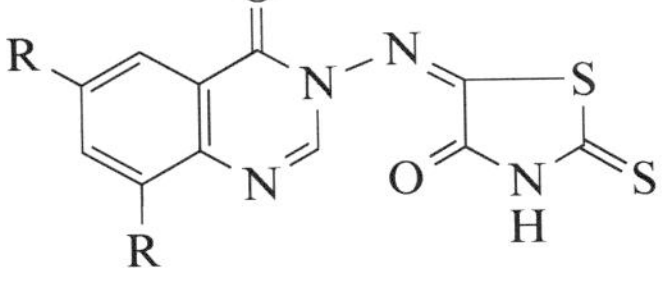

(199)

CHAPTER 7

Quinazolinecarboxylic Acids and Related Derivatives

This chapter embraces not only the quinazolinecarboxylic acids but also an extended family of esters, acyl halides, amides, nitriles, aldehydes, ketones, thiocyanates, isothiocyanates, and derived compounds. Any interconversion is discussed only at the first logical opportunity: for example, the formation of esters from carboxylic acids is covered as a reaction of carboxylic acids rather than as a preparative route to esters, simply because acids are discussed prior to esters. Nevertheless, extensive cross-references are included.

7.1. QUINAZOLINECARBOXYLIC ACIDS (*H* 475)

As well as the regular quinazolinecarboxylic acids, this section includes those acids in which the carboxy group is attached indirectly to the quinazoline nucleus as, for example, in 3-carboxymethyl-4(3*H*)-quinazolinone.

7.1.1. Preparation of Quinazolinecarboxylic Acids (*H* 475)

Some routes to quinazolinecarboxylic acids have been discussed already: by *primary syntheses* (Chapter 1), by *oxidation of alkylquinazolines* (Section 2.2.2.2), or by *carboxylation as a passenger group*, for example, by carboxyalkylation of quinazolinethiones (Section 5.1.2.1) or such like processes.

The *hydrolytic routes* to quinazolinecarboxylic acids are illustrated in the following examples.

From quinazolinecarboxylic esters

Ethyl 6-bromo-4-pyridin-2′-yl-2-quinazolinecarboxylate (**1**, R = Et) gave 6-bromo-4-pyridin-2′-yl-2-quinazolinecarboxylic acid (**1**, R = H) (KOH, MeOH, 20°, 4 h: 65%).[1461]

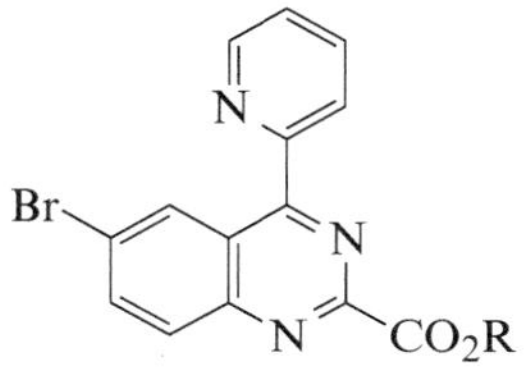

(1)

Ethyl 3-p-hydroxyphenyl-4-oxo-3,4-dihydro-2-quinazolinecarboxylate (**2**, R = CO$_2$Et) gave 3-p-hydroxyphenyl-4-oxo-3,4-dihydro-2-quinazoline-carboxylic acid (**2**, R = CO$_2$H) and thence 3-p-hydroxyphenyl-4(3H)- quinazolinone (**2**, R = H) (HCl to pH 2, 70°, 1 h: 79% overall; thus acidic hydrolysis of such esters should be avoided in order to obviate decarboxylation).[2240]

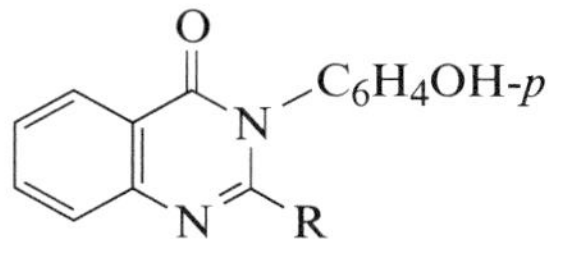

(2)

Methyl 3-methyl-2,4-dioxo-1,2,3,4-tetrahydro-6-quinazolinecarboxylate (**3**, R = Me) gave 3-methyl-2,4-dioxo-1,2,3,4-tetrahydro-6-quinazolinecar-boxylic acid (**3**, R = H) (NaOH, MeOH, reflux, 90 min: 83%).[1737]

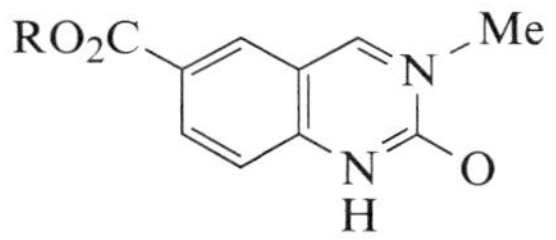

(3)

1-Ethoxycarbonylmethyl- (**4**, R = Et) gave 1-carboxymethyl-4(1H)-quinazolinone (**4**, R = H) (KOH, H$_2$O–MeOH, reflux, 1 h: 89%).[1825]

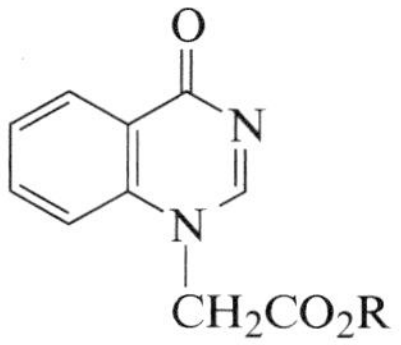

(4)

2-β-Methoxycarbonylvinyl- gave 2-β-carboxyvinyl-6-methylthio-4(3H)-quinazolinone (6M HCl, trace octanol, reflux, 30 min: 81%; note resistance to decarboxylation).[1558]

2-β,β-Diethoxycarbonylethyl- gave 2-β,β-dicarboxyethyl-4-phenylquinazoline (KOH, EtOH–H$_2$O, reflux, 4 h: 75%).[632]

6-Chloro-2-ethoxycarbonylmethylthio- gave 2-carboxymethylthio-6-chloro-4(3H)-quinazolinone (1M NaOH, 95°, 10 min: ∼ 85%).[220]

6-p-Ethoxycarbonylphenoxymethyl-2-pivalamido- gave 2-amino-6-p-carboxyphenoxymethyl-4(3H)-quinazolinone (HCl gas, MeOH–THF–H$_2$O, reflux, 4 h: 90%; note additional deacylation).[801]

Also other examples.[249,531,759,760,1071,1423,1549,1582,1588,1608,1702,1838, 2262, 2277,2547,2597]

Note: Beware of concomitant decarboxylation during hydrolysis of some quinazolinecarboxylic esters.[856]

From quinazolinecarboxamides

N-Methyl-4-oxo-3-phenyl-3,4-dihydro-2-quinazolinecarboxamide (5, R = CONHMe) gave 4-oxo-3-phenyl-3,4-dihydro-2-quinazolinecarboxylic acid (5, R = CO$_2$H) (AcOH, H$_2$O, reflux, 30 min: unisolated) and thence 3-phenyl-4(3H)-quinazolinone (5, R = H) (HCl, hot, until CO$_2$↑ceased; then HCO$_2$H, reflux, 1 h: 75% overall).[768]

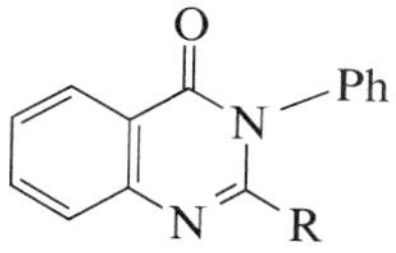

(5)

From quinazolinecarbonitriles

2-Methyl-4-quinazolinecarbonitrile (6, R = CN) gave 2-methyl-4-quinazolinecarboxylic acid (6, R = CO$_2$H) (75% H$_2$SO$_4$, reflux, 1 h: 56%).[1063]

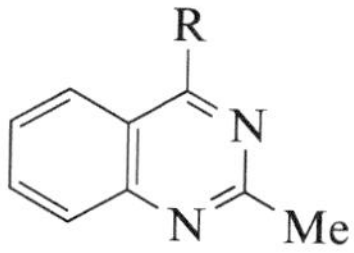

(6)

2,4-Diamino-6-quinazolinecarbonitrile gave 2-amino-4-oxo-3,4-dihydro-6-quinazolinecarboxylic acid (NaOH, HOCH$_2$CH$_2$OH–H$_2$O, reflux, 4 h: 61%; note additional hydrolysis of the 4-NH$_2$).[618]

8a-Cyanomethyl- (7, R = CN) gave 8a-carboxymethyl-3,4,4a,5,6,7,8,8a-oc-
tahydro-2(1H)-quinazolinone (7, R = CO$_2$H) ($\sim$ 70% H$_2$SO$_4$, reflux, 2 h:
87%).[1004]

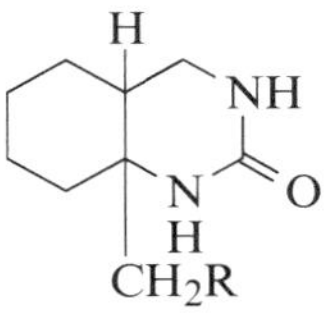

(7)

Also other examples.[1844]

Some *oxidative routes* to quinazolinecarboxylic acids (not yet covered) are
illustrated in the following examples.

From quinazolinecarbaldehydes

4-Oxo-3-phenyl-3,4-dihydro-2-quinazolinecarbaldehyde (5, R = CHO) gave
the corresponding carboxylic acid (5, R = CO$_2$H) [Ag$_2$O (made *in situ*),
H$_2$O, reflux, 4 h: good yield].[146]

7-Dibromomethyl-4-p-fluorophenyl-1-isopropyl-2(1H)-quinazolinone   (8,
R = CHBr$_2$) gave crude 4-p-fluorophenyl-1-isopropyl-2-oxo-1,2-dihydro-
7-quinazolinecarbaldehyde (8, R = CHO) (AgNO$_3$, THF–EtOH, reflux,
2 h) and thence the corresponding carboxylic acid (7, R = CO$_2$H) (KMnO$_4$,
dioxane–H$_2$O, 15°, 1 h: ?%).[1867]

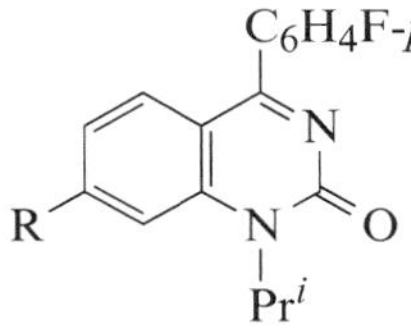

(8)

From quinazoline ketones

2-Acetyl-6-chloro-4-phenylquinazoline gave 6-chloro-4-phenyl-2-quinazo-
linecarboxylic acid (NaOCl, KOH, H$_2$O, 95°, 4 h: $\sim$ 35%).[240]
Also other examples.[2011]

7.1.2. Reactions of Quinazolinecarboxylic Acids (*H* 475)

Except for decarboxylation, the ease with which quinazolinecarboxylic acids
undergo reactions appears to be unaffected by the position of the carboxy group

or by the presence of other groups. The general conditions required for various reactions of quinazolinecarboxylic acids and the yields to be expected therefrom are illustrated in the following classified examples.

Decarboxylation

2-Methyl-4-quinazolinecarboxylic acid (**9**, R $= CO_2H$) gave 2-methyl-quinazoline (**9**, R $=$ H) (PhNO$_2$, reflux, 24 h: 25%, isolated as picrate).[1063]

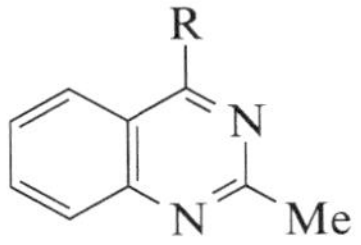

(9)

2-(α-Carboxybenzylidene)hydrazino- (**10**, R $= CO_2H$) gave 2-benzylidenehy-drazino-3-methyl-4(3H)-quinazolinone (**10**, R $=$ H) (Me$_2$NCHO, reflux, 2 h: 72%).[2059]

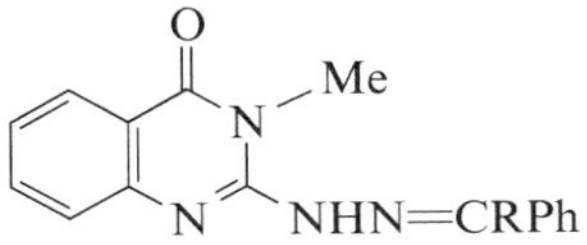

(10)

6-Chloro-4-phenyl-2-quinazolinecarboxylic acid gave 6-chloro-4-phenyl-quinazoline (neat, 230°, 10 min: 73%).[2473]

1-Methyl-4-oxo-1,4-dihydro-2-quinazolinecarboxylic acid gave 1-methyl-4(1H)-quinazolinone (CHCl$_3$, 20°, 3 h: 94%; note the extreme ease of decarboxylation).[1415]

Also other examples.[531,1702,2240]

Conversion into quinazolinecarbonyl chlorides

2-p-Chlorophenyl-5-quinazolinecarboxylic acid (**11**, R $=$ OH) gave 2-p-chlorophenyl-5-quinazolinecarbonyl chloride (**11**, R $=$ Cl) (ClOCCOCl, PhH, reflux, 18 h: 92%).[672]

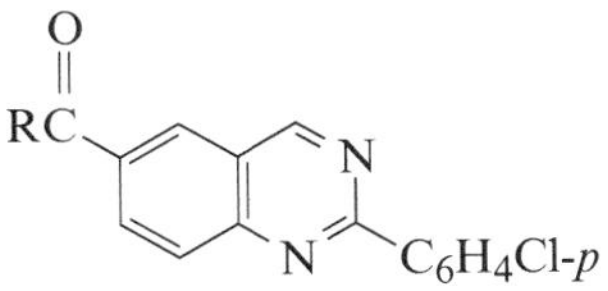

(11)

3-Carboxymethyl-2-phenyl-4(3H)-quinazolinone gave the 3-chlorocarbonyl-methyl analog (neat PCl$_5$, 100°, 30 min; then PhH, reflux, 1 h: ?%, crude).[2316]

6-Chloro-4-phenyl-2-quinazolinecarboxylic acid (**12**, R = OH) gave the corresponding carbonyl chloride (**12**, R = Cl) (SOCl$_2$, PhH, reflux, 2 h: crude) and thence 6-chloro-N-methyl-4-phenyl-2-quinazolinecarboxamide (**12**, R = NHMe) (MeNH$_2$, PhM, 20°, 30 min: 69% overall).[242]

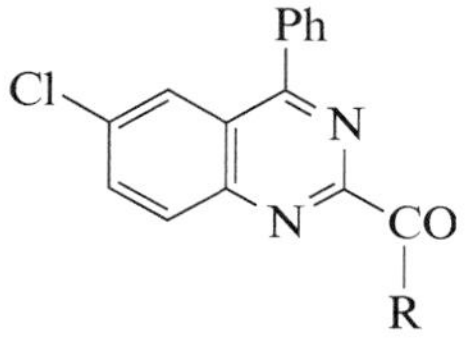

(12)

Note: Crude quinazolinecarbonyl chlorides are often prepared and used to make corresponding esters[2503] or amides.[1040,1047,1141,1265,1278,2327,2336]

Esterification

6-Chloro-4-phenyl-2-quinazolinecarboxylic acid (**12**, R = OH) gave methyl 6-chloro-4-phenyl-2-quinazolinecarboxylate (**12**, R = OMe) (HCl gas, MeOH, reflux, 3 h: 82%).[2473]

2-p-Chlorophenyl-4-oxo-3,4-dihydro-5-quinazolinecarboxylic acid (**13**, R = H) gave methyl 2-p-chlorophenyl-4-oxo-3,4-dihydro-5-quinazolinecarboxylate (MeOH, trace H$_2$SO$_4$, reflux, 24 h: 50%).[672]

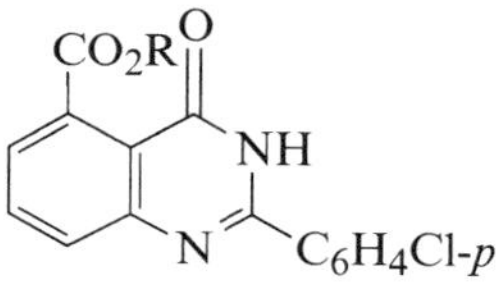

(13)

2-Carboxymethylthio-6,7-dimethoxy- (**14**, Q = Me, R = H) or 2-carboxymethylthio-6-hydroxy-7-methoxy- (**14**, Q = R = H) gave 6,7-dimethoxy-2-methoxycarbonylmethylthio-3-phenyl-4(3H)-quinazolinone (**14**, Q = R = Me) (CH$_2$N$_2$, Et$_2$O, 20°, 12 h: both 90%; note additional O-methylation in the latter case).[141]

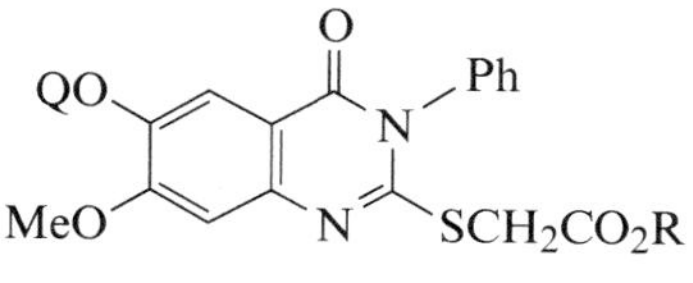

(14)

3-α-Carboxyethyl- gave 3-α-ethoxycarbonylethyl-2-phenyl-4(3H)-quinazo-linone (EtOH, trace H_2SO_4, reflux, 5 h: 79%).[1889]

3-α-Carboxyethyl- gave 3-α-methoxycarbonylethyl-2,4(1H,3H)-quinazo-linedione (MeOH, dicyclohexylcarbodiimide, 4-dimethylaminopyridine, 20°, ? h: 96%).[2435]

3-Carboxymethyl- (15, R = H) gave 3-methoxycarbonylmethyl-2-methyl-4(3H)-quinazolinone (15, R = Me) (SOCl$_2$, MeOH, reflux, 4 h: 65%; per-haps via the acyl chloride).[1681]

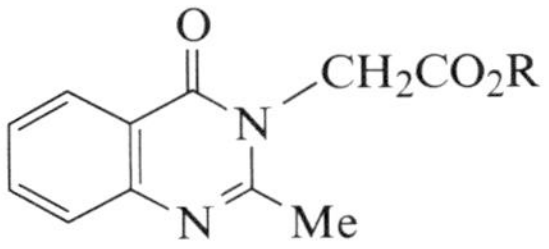

(15)

4-Oxo-3,4-dihydro-2-quinazolinecarboxylic acid gave 2-(isobutoxycar-bonyl)oxycarbonyl-4(3H)-quinazolinone (ClCO$_2$Bu^i, Et$_3$N, THF, 20°, 3 h: 86%).[1186]

Also other examples.[948,975,1252,1283,1565,1659,2503]

Conversion into quinazolinecarboxamides

Note: Quinazolinecarboxylic acids are usually converted into corresponding amides via acyl chlorides or esters, but an occasional direct conversion has been reported.

3-Carboxymethyl- gave 3-(4-p-methoxyphenethylpiperazin-1-yl)carbonyl-methyl-2,4(1H,3H)-quinazolinedione (16) [p-MeOC$_6$H$_4$CH$_2$CH$_2$N-(CH$_2$CH$_2$)$_2$NH, dicyclohexylcarbodiimide, THF–CH$_2$Cl$_2$, < 5°, 15 min, then 20°, 2 h: 53%];[1874] also analogs.[1874]

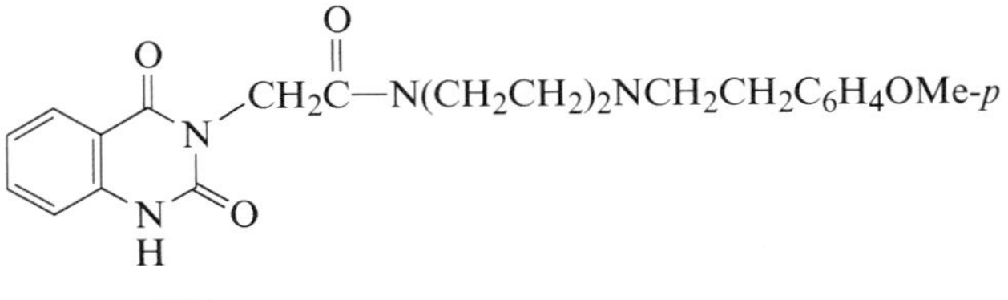

(16)

Cyclization reactions

2-β-Carboxyethyl-6-chloro-4(3H)-quinazolinone gave a separable mixture of 7-chloropyrrolo[1,2-a]quinazoline-1,5(2H,3H)-dione (17) and 7-chloro-pyrrolo[2,1-b]quinazoline-1,9(2H,3H)-dione (18)(Ac$_2$O, Me$_2$NCHO, 95°, 25–45 min: ~ 50% and ~ 20%, respectively).[265]

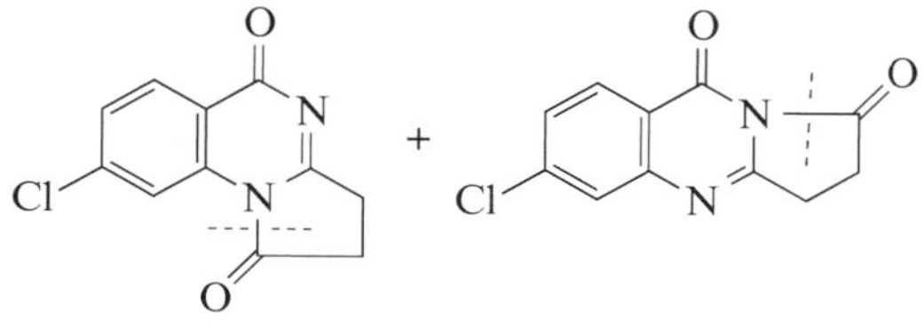

(17)   **(18)**

6-Bromo-3-carboxymethyl- **(19)** gave 3-(benzimidazol-2-yl)methyl-6-bromo-2-phenyl-4(3H)-quinazolinone **(21)** [neat **(20)**, reflux, 30 min: 55%]; also many analogs.[1035]

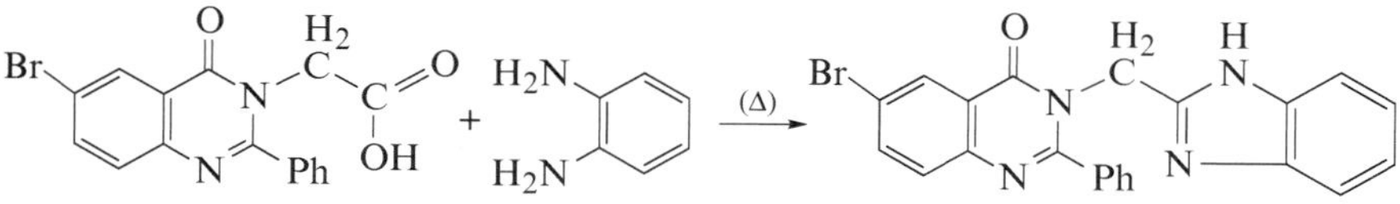

(19)   **(20)**   **(21)**

The *mass spectra* of 3-carboxymethyl-4(3H)-quinazolinone and several derivatives have been compared with those of their 2-aza analogs, including 3-carboxymethyl-1,2,3-benzotriazin-4(3H)-one.[1734]

7.2. QUINAZOLINECARBOXYLIC ESTERS (*H* 475)

This section convers quinazolines with ester groupings attached directly or indirectly to the nucleus at any position; alkoxycarbonylamino groups are included in the latter category.

7.2.1. Preparation of Quinazolinecarboxylic Esters (*H* 475)

The major routes to quinazoline esters have been discussed already: by *primary syntheses* (Chapter 1), by *esterification of carboxylic acids* (Section 7.1.2), y *S*- or *N-alkoxycarbonylation* (Sections 5.1.2.1, and 6.2.2.3), or by *passenger introduction of ester groupings* (Sections 2.2.1.1, 2.2.1.2, 4.1.2.3, 5.1.2.1, and 5.3.2).

The remaining minor synthetic routes are illustrated by the following examples.

From quinazolinecaraldehydes

6-Bromo-4-pyridin-2'-yl-2-quinazolinecarbaldehyde(**22**, R = H) gave ethyl 6-bromo-4-pyridin-2'-yl-2-quinazolinecarboxylate(**22**, R = OEt) [Pb(OAc)$_4$, AcOH–CHCl$_3$–EtOH, 100°, 4 h: 55%].[1461]

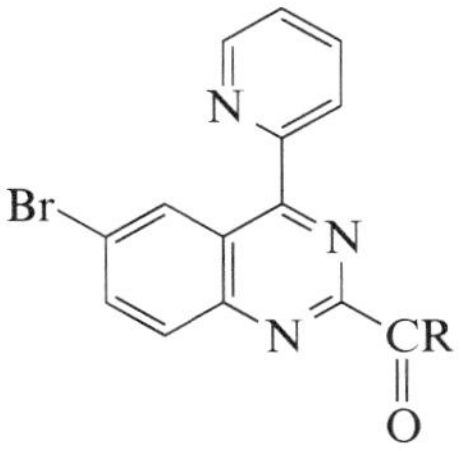

(22)

From quinazolinecarbonyl chlorides

2-Chlorocarbonylmethylthio- gave 2-(β-ethoxyethoxycarbonyl)methylthio-(HOCH$_2$CH$_2$OEt, reflux, 3 h: 68%) or 2-(β-dimethylaminoethoxycarbonyl)methylthio-3-phenyl-4(3H)-quinazolinone (Me$_2$NCH$_2$CH$_2$OH, reflux, 3 h: 62%).[2503]

From isothiocyanatoquinazolines

4-Isothiocyanato- (23) gave 4-methoxy (thiocarbonyl)amino-2-morpholino-quinazoline (24) (MeOH, Et$_2$O, HCl↓, briefly: 85%; the product appears to be remarkably slow to equilibrate with its tautomer (25).[429]

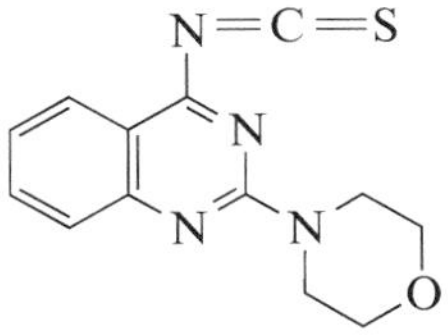

(23)

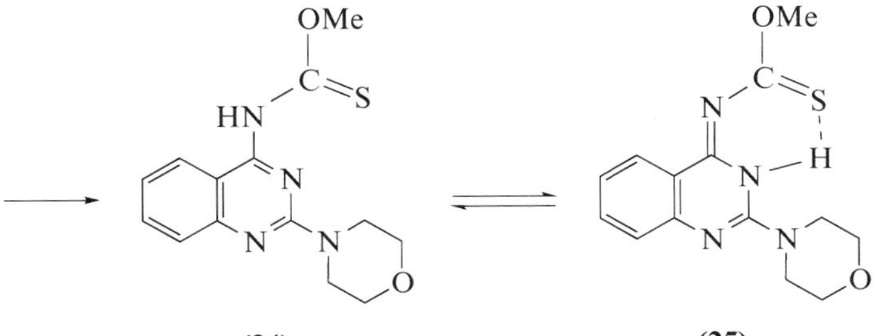

(24)   (25)

From quinazolinecarbonyl azides

3-Azidocarbonylmethyl- (26) gave 3-isocyanatomethyl- (27) (unisolated) and thence 3-allyloxycarbonylaminomethyl-2-methyl-4(3H)-quinazolinone

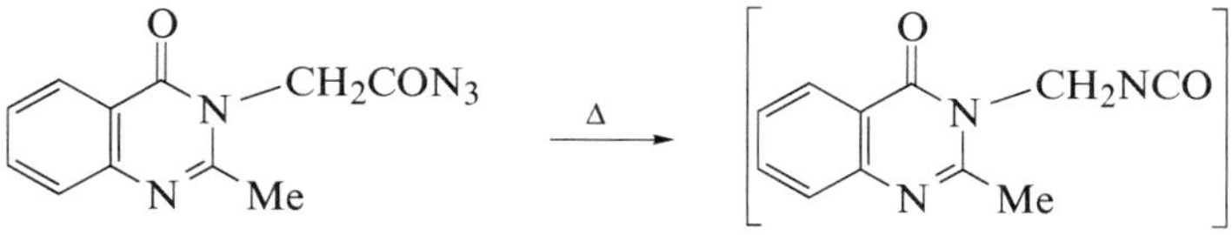

(26)   **(27)**

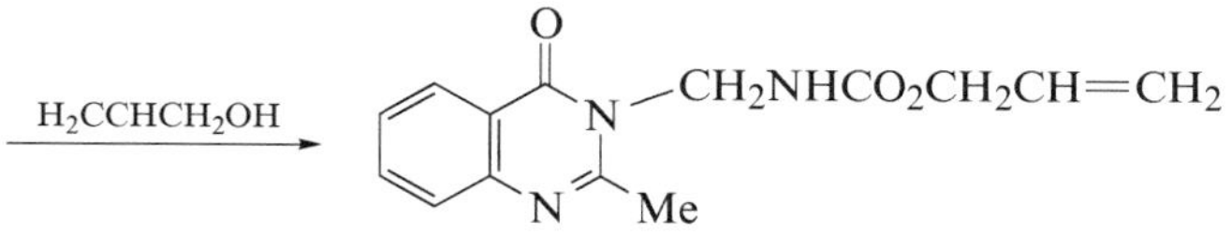

(28)

(28) (PhH, reflux; then $H_2C=CHCH_2OH$, conditions ?: ?%);[290] also analogs similarly.[277,290,291]

7.2.2. Reactions of Quinazolinecarboxylic Esters (*H* 475)

The *hydrolysis of esters* has been discussed in Section 7.1.1. Other reactions are illustrated in the following examples.

Regular aminolyses

Ethyl 4-oxo-1-phenyl-1,4-dihydro-2-quinazolinecarboxylate (**29**, R = OEt) gave 4-oxo-1-phenyl-1,4-dihydro-2-quinazolinecarboxamide (**29**, R = NH_2) (NH_3, EtOH, 80°, sealed, 6 h: 87%).[1743]

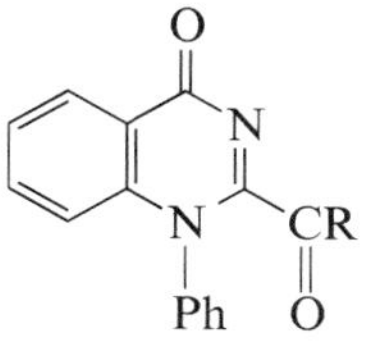

(29)

Ethyl 3-amino-4-oxo-3,4-dihydro-2-quinazolinecarboxylate (**30**, R = OEt) gave 3-amino-*N*-isopropyl-4-oxo-3,4-dihydro-2-quinazolinecarboxamide (**30**, R = NHPr^i) (Pr^{i}NH$_2$, MeOH, reflux, 4 h: 78%);[2121] also analogs likewise.[2121,2173]

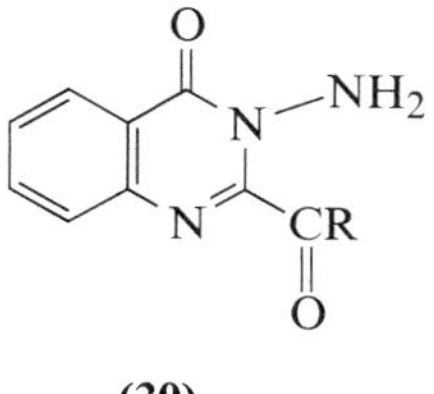

(30)

Ethyl 7-methyl-4-oxo-3,4-dihydro-2-quinazolinecarboxylate gave 7-methyl-4-oxo-3,4-dihydro-2-quinazolinecarboxamide (EtOH, $NH_3\downarrow$, 20°, 1 h: 93%).[1743]

Methyl 4-oxo-3,4-dihydro-2-quinazolinecarbodithioate (**31**) gave *N-o*-carboxyphenyl-4-oxo-3,4-dihydro-2-quinazolinecarbothioamide (**32**) (*o*-$H_2NC_6H_4CO_2H$, Et_3N, reflux, until yellow: 80%)[2325] or 2-piperidino(thiocarbonyl)-4(3*H*)-quinazolinone (**33**) [$HN(CH_2)_5$, EtOH, reflux, until yellow: 75%).[2319]

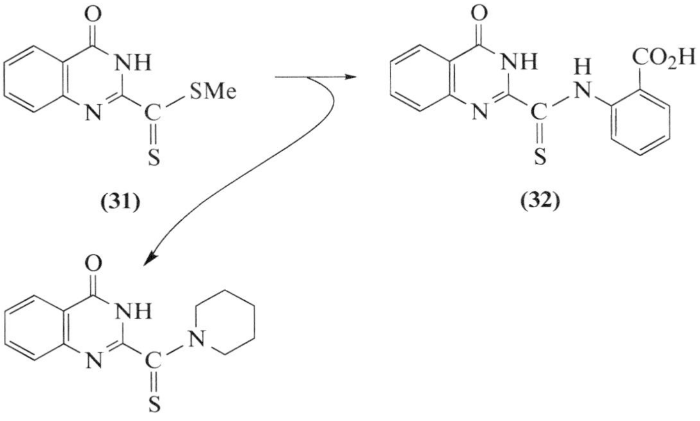

(31)   (32)

(33)

Also other examples, both nuclear and extranuclear.[124,462,531,977, 1186,1516,1695,2073,2582]

Hydrazinolyses

Ethyl 3-*p*-ethoxyphenyl-4-oxo-3,4-dihydro-2-quinazolinecarboxylate (**34**, R = OEt) gave 3-*p*-ethoxyphenyl-4-oxo-3,4-dihydro-2-quinazolinecarbohydrazide (**34**, R = $NHNH_2$) ($H_2NNH_2.H_2O$, EtOH, reflux, 7 h: 80%).[781]

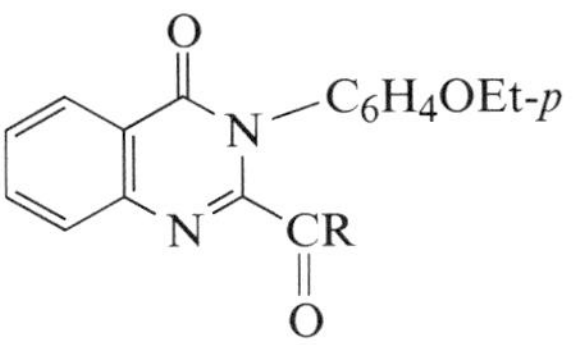

(34)

3-Amino-2-ethoxycarbonylmethyl-(**35**, R = OEt) gave 3-amino-2-hydrazinocarbonylmethyl-4(3*H*)-quinazolinone (**35**, R = $NHNH_2$) ($H_2NNH_2.H_2O$, MeOH, 20°, 12 h: 98%).[2323]

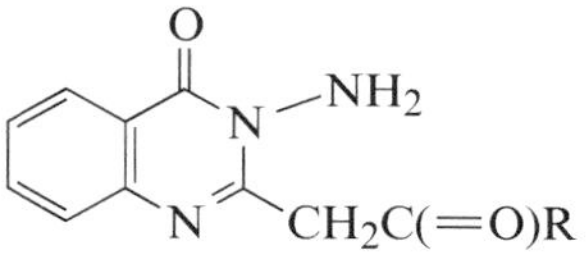

(35)

4-Ethoxycarbonylmethylthio- gave 4-hydrazinocarbonylmethylthioquinazo-
line ($H_2NNH_2.H_2O$, EtOH, reflux, 4 h: 70%).[2134]

3-Methoxycarbonylmethyl- gave 3-hydrazinocarbonylmethyl-2-methyl-
4(3H)-quinazolinone ($H_2NNH_2.H_2O$, MeOH, reflux, 3 h: 75%).[1681]

Also many other examples.[123,277,504,613,699,1036,1054,1071,1078,1255,1659,
1661,1743,1889,2285,2321,2347,2433,2491,2528,2536]

Other reactions

Ethyl 4-oxo-3,4-dihydro-2-quinazolinecarboxylate (**36**, R = OEt) gave 2-α-
(methylsulfinyl)acetyl-4(3H)-quinazolinone [**36**, R = $CH_2S(O)Me$] (Me_2SO,
EtONa, N_2, 20°, 4 h: ~ 50%).[123]

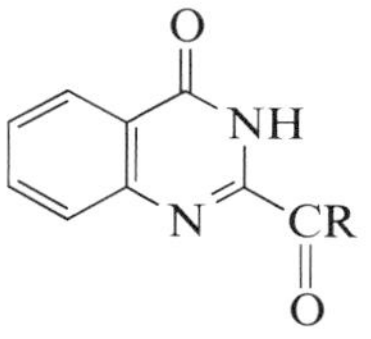

(36)

Methyl 4-oxo-3,4-dihydro-2-quinazolinecarbodithioate (**37**) gave 2-(5-
aminothiazol-2-yl)-4(3H)-quinazolinone (**38**) (H_2NCH_2CN, EtOH, reflux,
< 10 min; then Et_3N, 10 min: 85%).[2309]

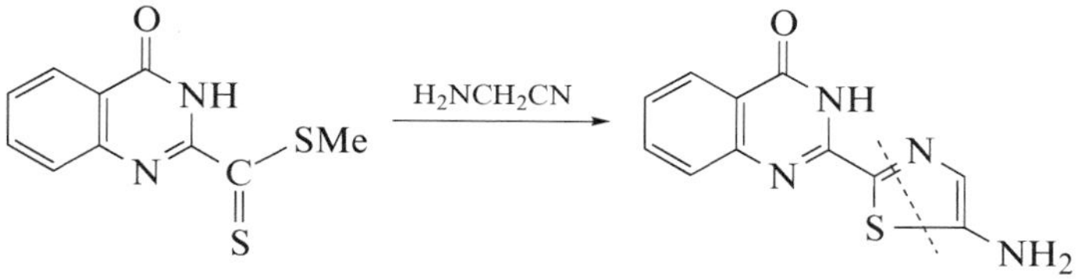

(37)   **(38)**

3-Methoxycarbonylmethyl- gave 3-azidocarbonylmethyl-2-methyl-4(3H)-
quinazolinone (NaN_3, AcOH, H_2O, with cooling: 65%).[277]

7.3. QUINAZOLINECARBONYL CHLORIDES

Preparation. Nearly all known quinazolinecarbonyl chlorides have been made by treatment of *quinazolinecarboxylic acids with thionyl chloride* or the like (see Section 7.1.2).

However, 2-chlorocarbonylmethylthio-3-phenyl-4(3H)-quinazolinone (**40**) was made from 3-phenyl-2-thioxo-1,2-dihydro-4(3H)-quinazolinone (**39**) by treatment with chloroacetyl chloride in pyridine (no further details).[2503]

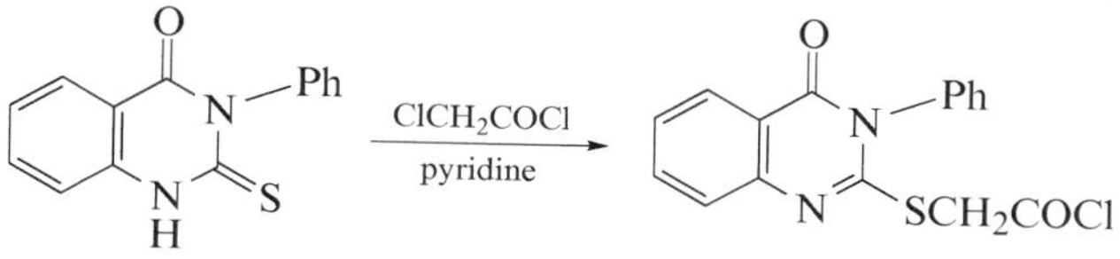

(39)   (40)

Reactions. The main reactions of quinazolinecarbonyl chlorides are their *conversion into corresponding esters* (see Section 7.2.1) or *amides* (see Section 7.1.2).

In addition, 2,4(?)-dichloro-7-quinazolinecarbonyl chloride (**41**, R = Cl) was converted into the corresponding isothiocyanate (**41**, R = NCS) (NaSCN, AcOH–dioxane, 20°, ? h: isolated as a derivative in 65% yield);[215] and 3-chlorocarbonylmethyl-2-phenyl-4(3H)-quinazolinone (**42**) with *o*-(benzylideneamino)phenol (**43**) in benzene containing triethylamine at 20° during 90 min gave 3-(2-*o*-hydroxyphenyl-4-oxo-1-phenylazetidin-3-yl)-2-phenyl-4(3H)-quinazolinone (**44**) in 45% yield.[2315]

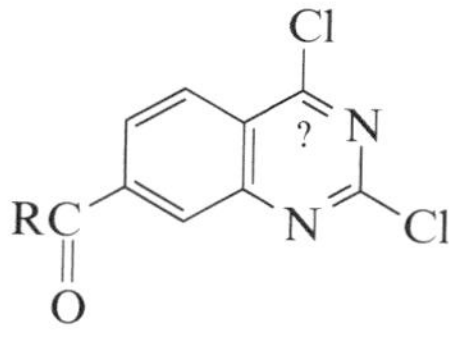

(41)

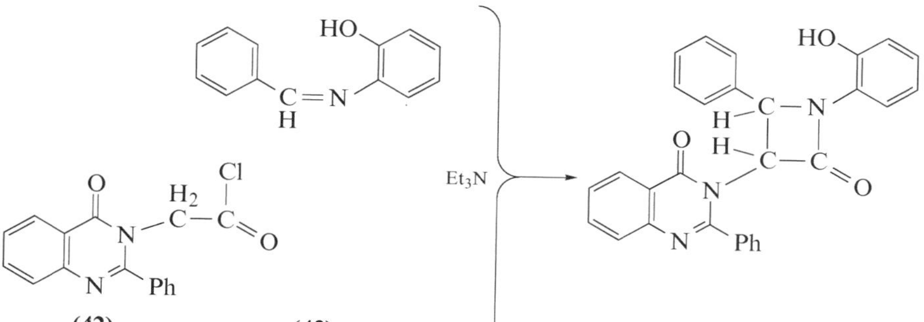

(42)   (43)   (44)

7.4. QUINAZOLINECARBOXAMIDES AND QUINAZOLINECARBOHYDRAZIDES (*H* 473)

Preparation. Most amides and hydrazides are made by *primary syntheses* (see Chapter 1) or by well-known procedures, *from esters* (see Section 7.2.2), *from acyl chlorides* (see Section 7.1.2), or *from carboxylic acids* (see Section 7.1.2). The conversion of a *methyl into a thiocarbamoyl* group has been reported (see Section 2.2.2.6), and the introduction of *passenger carbamoyl groups* has been mentioned elsewhere (e.g., toward the end of Section 5.1.2.1).

The remaining minor synthetic routes to amides are illustrated in the following examples.

Direct *N*-amidation

2-Phenyl-4(3*H*)-quinazolinone (**45**, X = O) or the corresponding thione (**45**, X = S) gave 4-oxo-2,*N*-diphenyl- (**46**, X = Y = O) (PhNCO, PhH, reflux, 3 h: 75%) or 2,*N*-diphenyl-4-thioxo-3,4-dihydro-3-quinazolinecarboxamide (**46**, X = S, Y = O) (PhNCS, likewise: 72%), respectively;[1679] 4-oxo-2,*N*-diphenyl- (**46**, X = O, Y = S) (65%) and 2,*N*-diphenyl-4-thioxo-3,4-dihydro-3-quinazolinecarbothioamide (**46**, X = Y = S) (68%) were made similarly.[1679]

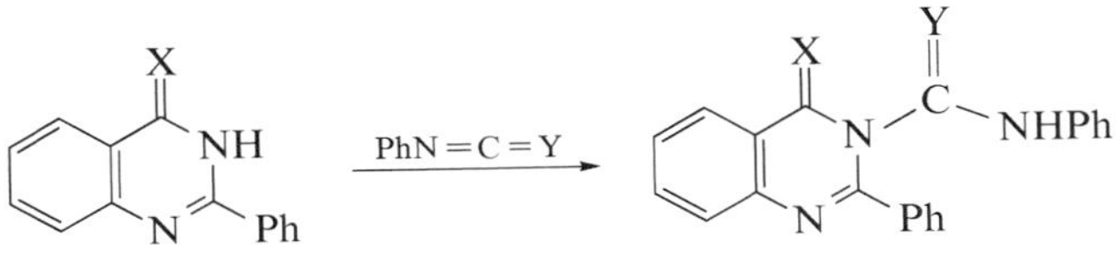

(45)   (46)

In contrast, 4(3*H*)-quinazolinone gave *N*-methyl-4-oxo-1,4-dihydro-1-quinazolinecarboxamide (**47**) [MeNCO, $(Me_2N)_2C{=}NH$, Me_2NCHO, 20°, 2 h: 4%, accompanied by several other separable products].[1381]

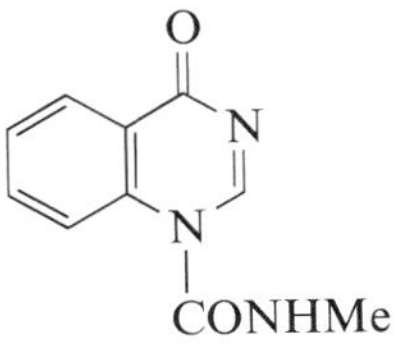

(47)

From tribromomethylquinazolines

4-Tribromomethylquinazoline gave a separable mixture of *N*-propyl-4-quinazolinecarboxamide (**48**) and 4-dibromomethylquinazoline ($PrNH_2$,

MeCN, 20°, 18 h: 68% and 30%, respectively);[2060] 4-piperidinocarbonyl-quinazoline (82%) was made similarly.[2060]

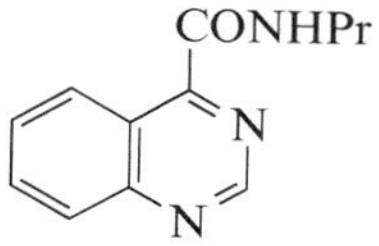

(48)

By hydrolysis of quinazolinecarbonitriles

8a-Cyanomethyl- gave 8a-carbamoylmethyl-3,4,4a,5,6,7,8,8a-octahydro-2(1H)-quinazolinone (49) ($\sim$ 95% H_2SO_4, 95°, 30 min: 42%).[1004]

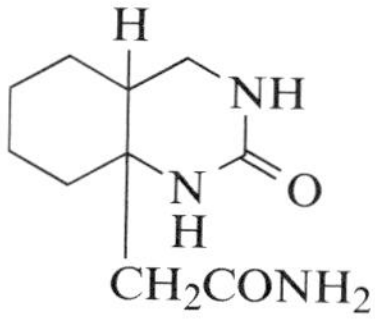

(49)

By transamination within an amide group

3-(3,5-Dimethylpyrazol-1-yl)carbamoylmethyl-2-methyl- (50) gave 2-methyl-3-(N-phenylcarbamoyl)methyl-4(3H)-quinazolinone (51) (PhNH$_2$, AcOH, heat ?: 88%);[1117] analogs similarly.[1002,1117] *Note*: Since the substrate (50) came from 3-hydrazinocarbonylmethyl-2-methyl-4(3H)-quinazolinone, this amounts to a conversion of a hydrazide into a regular amide.

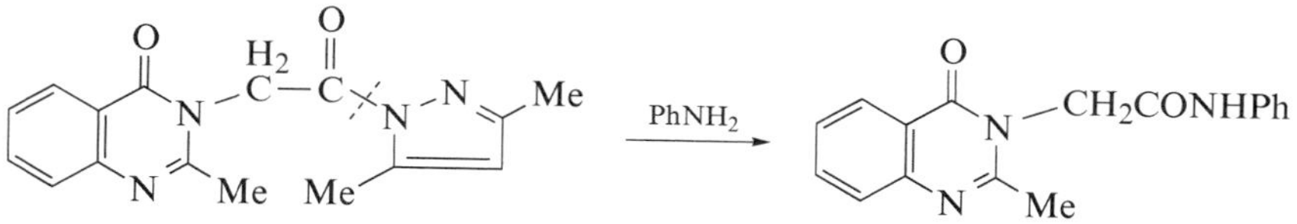

(50)　　　　　　　　　　　　　　　(51)

Reactions. The *hydrolysis of amides* (Section 7.1.1) and their *transamination* (preceding paragraph) have been covered already.

The remaining reactions of amides and hydrazides are illustrated in the following examples.

Dehydration of amides to nitriles

1-Methyl-4-oxo-1,4-dihydro-2-quinazolinecarboxamide (**52**, R = Me) gave the corresponding carbonitrile (**53**, R = Me) [Cl$_2$(O=)P—O—P(=O)-Cl$_2$,[1898] 40°, 3 h: 51%; note that dehydration with POCl$_3$ or SOCl$_2$ was unsuccessful].[1415]

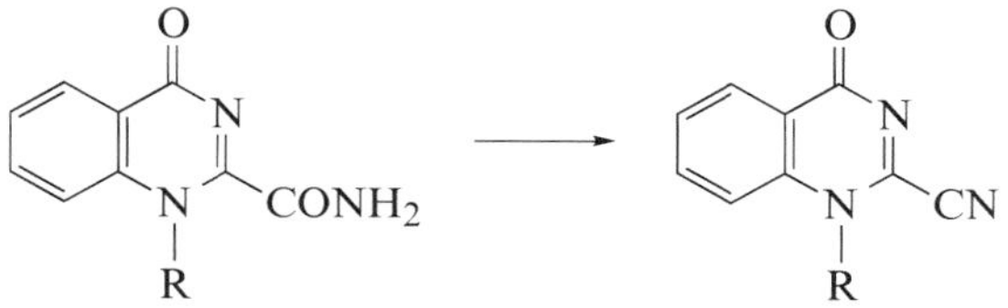

(52)   (53)

4-Oxo-1-phenyl-1,4-dihydro-2-quinazolinecarboxamide (**52**, R = Ph) gave the carbonitrile (**53**, R = Ph) [Cl$_2$(O=)P—O—P(=O)Cl$_2$, 50°, 10 h: 75%].[1590]

4-Oxo-3,4-dihydro-6-quinazolinecarboxamide gave 4-chloro-6-quinazolinecarbonitrile (POCl$_3$ + SOCl$_2$, reflux, 36 h: 35%).[2564]

N-Alkylideneation of hydrazides and subsequent reactions

4-Hydrazinocarbonylmethylthio- (**54**, R = NH$_2$) gave 4-isopropylidenehydrazinocarbonylmethylthio- (**54**, R = N:CMe$_2$) (AcMe, EtOH, reflux, 5 h: 30%) or 4-cinnamylidenehydrazinocarbonylmethylthioquinazoline (**54**, R = N:CHCH:CHPh) (PhCH=CHCHO, EtOH, reflux, 5 h: 75%).[2134]

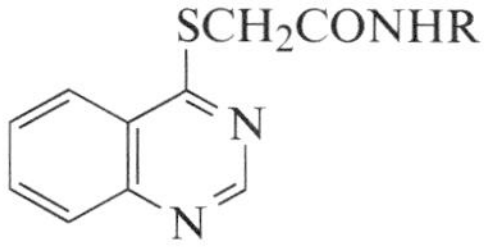

(54)

3-Amino-4-oxo-3,4-dihydro-2-quinazolinecarbohydrazide(**55**) gave 3-amino-2-[α-(ethoxycarbonylmethyl)ethylidene]hydrazinocarbonyl-4(3H)-quinazolinone (**56**) (EtO$_2$CCH$_2$Ac, EtOH, reflux, 6 h: 85%) and thence 2-methyl-6,12-dihydro-4H-pyrazolo[5′,   1′:3,4][1,2,4]triazolo[6,1-b]quinazoline-6,12-dione (**57**) (TsOH, EtOH, reflux, 4 h: 58%).[380]

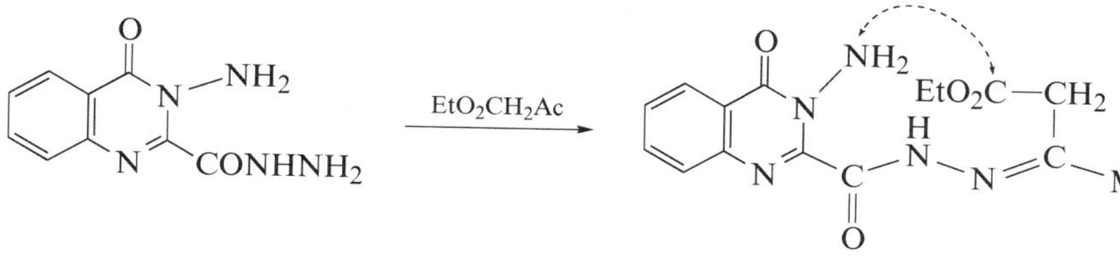

(55)   (56)

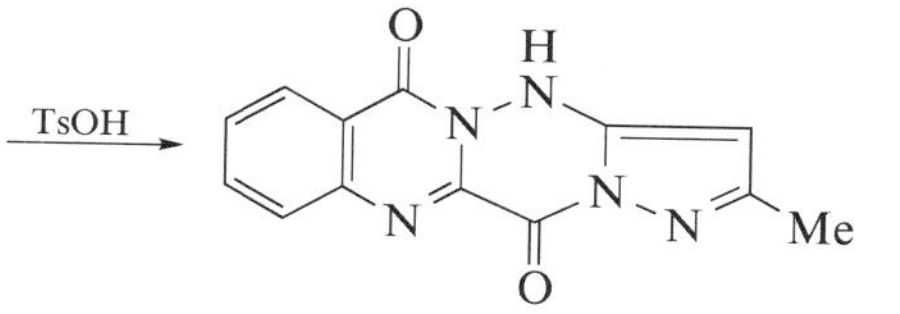

(57)

3-Hydrazinocarbonylmethyl-2-methyl- gave 3-(benzylidenehydrazino)car-
bonylmethyl-2-methyl-(PhCHO, EtOH, trace AcOH, reflux, 3 h: 80%) and
thence 2-methyl-3-N-(4-oxo-2-phenylthiazolidin-3-yl)carbamoylmethyl-
4($3H$)-quinazolinone (58) (HSCH$_2$CO$_2$H, PhH, reflux, 7 h: 80%).[1661]
Also other examples.[504,1054,1071,1179,1512,2148,2285,2528]

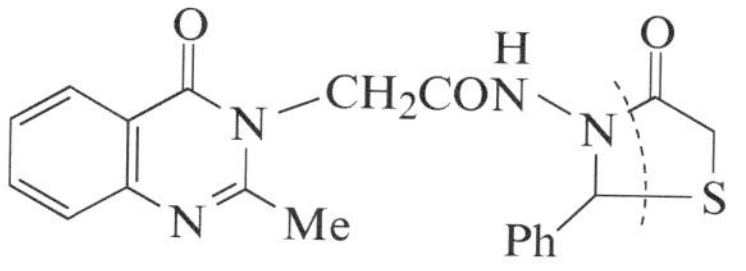

(58)

N'-Acylation of hydrazides

7-Methyl-4-oxo-3,4-dihydro-2-quinazolinecarbohydrazide (59, R = H) gave
7-methyl-4-oxo-N'-p-toluenesulfonyl-2-quinazolinecarbohydrazide (59,
R = Ts) (TsCl, pyridine, 10°, then 20°, 90 min: 49%).[1743]

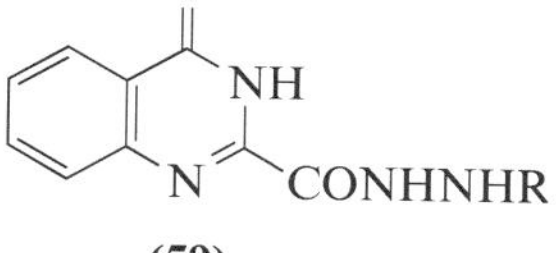

(59)

4-Oxo-3,4-dihydro-2-quinazolinecarbohydrazide gave 4-oxo-N'-trichloro-
acetyl-3,4-dihydro-2-quinazolinecarbohydrazide (Cl$_3$COCl, Et$_3$N, dioxane, reflux, 4 h: 69%).[123]

Reaction of isothiocyanates with hydrazides

3-Hydrazinocarbonylmethyl-2-methyl- (**60**, R = H) gave 2-methyl-3-(thiosemicarbazido)carbonylmethyl- [**60**, R = C(:S)NH$_2$] (KCNS, HCl, H$_2$O, reflux, 4 h: 50%) or 2-methyl-3-[4-p-tolyl(thiosemicarbazido)]carbonylmethyl-4(3H)-quinazolinone [**60**, R = C(:S)NHC$_6$H$_4$Me-p] (p-MeC$_6$H$_4$N=C=S, EtOH, reflux, 4 h: 65%).[613]

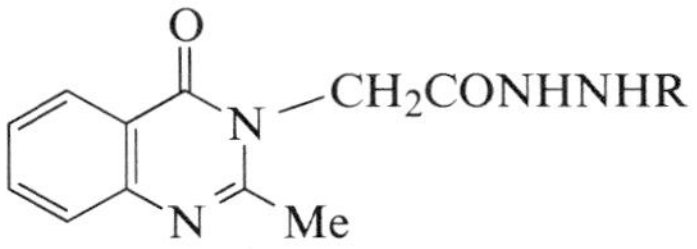

(60)

Cyclization reactions of hydrazides

3-Hydrazinocarbonylmethyl-2-methyl- (**60**, R = H) gave 2-methyl-3-(5-phenyl-1,3,4-oxadiazol-2-yl)methyl-4(3H)-quinazolinone (**61**) (PhCO$_2$H, PCl$_3$, PhH, reflux, 5 h: 60%).[1681]

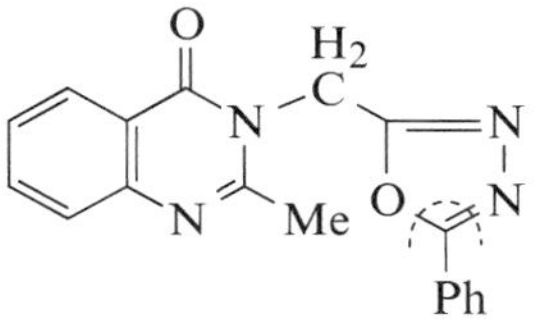

(61)

3-p-Hydrazinocarbonylphenyl-2-methyl- gave 2-methyl-3-p-(4-phenyl-5-thioxo-1,2,4-triazolin-3-yl)phenyl-4(3H)-quinazolinone (**62**) (PhNCS, 2M NaOH, reflux, 5 h: 70%).[2135]

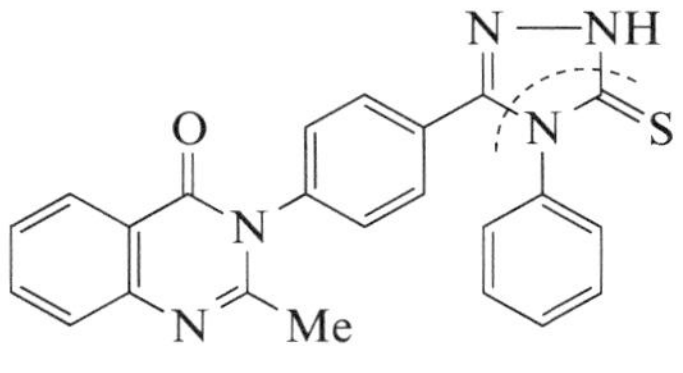

(62)

Also other examples in the foregoing references and elsewhere.[2312,2458]

7.5. QUINAZOLINECARBONITRILES (*H* 473)

Preparation. The following routes to quinazolinecarbinitriles have been discussed already: *by primary synthesis* (Chapter 1), *by Reissert-type reactions* (Sections 2.1.3, 2.2.2.6, and 4.7.2), *from halogenoquinazolines* with cyanide ion (Sections 3.5.5 and 3.7), *from diazonioquinazoline salts* with cyanide ion (Section 6.2.2.5), *from trimethylammonioquinazolines* with cyanide ion (Section 2.2.2.6), *by dehydration of quinazolinecarboxamides* (Section 7.4), and *by introduction of passenger cyano groups* in a number of ways such as the *S*-alkylation of a quinazolinethione with a cyanoalkyl halide (Section 5.1.2.1).

In addition, *dehydration of quinazolinecarbaldehyde oximes* has been used to make nitriles. Thus 6-chloro-4-phenyl-2-quinazolinecarbonitrile (**64**) resulted in > 60% yield by dehydration of 6-chloro-4-phenyl-2-quinazoline-carbaldehyde oxime (**63**) with phosphoryl chloride at 95° for 15 min or by dehydration and deoxygenation of the corresponding substrate 3-oxide (**65**) during its exothermic reaction with phosphorus trichloride.[361] Somewhat similarly, 2-methyl-4-quinazolinecarbaldehyde oxime (**66**) underwent dehydration on heating with neat phosphorus pentoxide at 170° under reduced pressure to give a crystalline sublimate of 2-methyl-4-quinazolinecarbonitrile (~ 45%).[1063]

(**63**)   POCl₃ →   (**64**)   ← PCl₃

(**65**)   (**66**)

Reactions. The *reduction* or *aminolysis* of nitriles to give aminoquinazolines (Section 6.2.1); the *hydrolysis* of nitriles to quinazolinones (Section 4.1.1.4), quinazolinecarboxylic acids (Section 7.1.1), or quinazolinecarboxamides (Section 7.4); and the *alcoholysis* of nitriles to alkoxyquinazolines (Section 4.5.1) have been covered already.

The other recently reported reactions of quinazolinecarbonitriles are illustrated in the following classified examples.

Conversion into aldehydes

2,4-Diamino-5-methyl-6-quinazolinecarbonitrile (**67**, Q = Me, R = CN) gave the corresponding carbaldehyde (**67**, Q = Me, R = CHO) (Raney-Ni, 98% HCO_2H, 70°, N_2, 6 h: 60%).[1871]

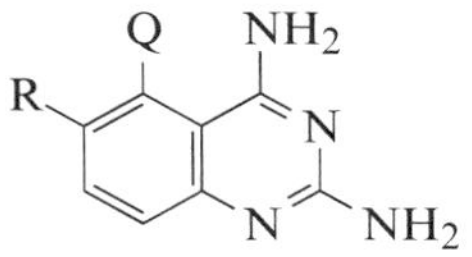

(**67**)

2,4-Diamino-6-quinazolinecarbonitrile (**67**, Q = H, R = CN) gave the carbaldehyde (**67**, Q = H, R = CHO) (H_2, Raney-Ni, $PhNHNH_2$, $AcOH-H_2O$, until 1.2 mol H_2 used: 78%, as the phenylhydrazone; then p-$O_2NC_6H_4CHO$, $AcOH-H_2O$, reflux, 2 h: 60% as free aldehyde);[201] also analogs, rather similarly, each in ~ 45% yield.[956]

Conversion into ketones

2-Methyl-4-quinazolinecarbonitrile (**68**) gave 4-benzoylmethyl-2-methyl-quinazoline (**69**) (BzMe, ~ 10M NaOH, 20°, 2.5 h: 45%; by displacement).[1062]

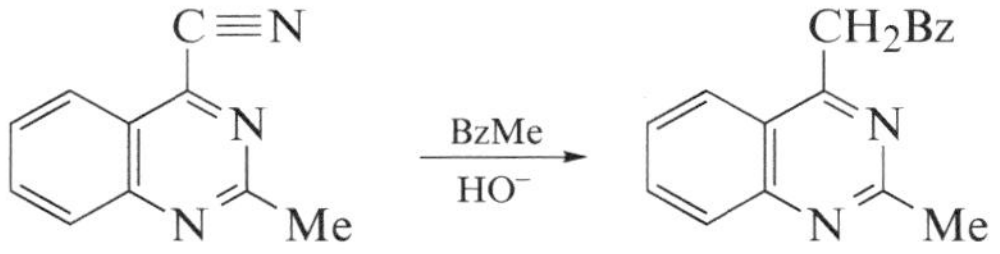

(**68**)   (**69**)

3-o-Cyanophenyl- (**70**, R = CN) gave 3-o-benzoylphenyl-2-methyl-4(3H)-quinazolinone (**70**, R = Bz) (PhMgBr, Et_2O–PhH, reflux, 1 h; then HCl, H_2O, reflux, 2 h: 33%).[884]

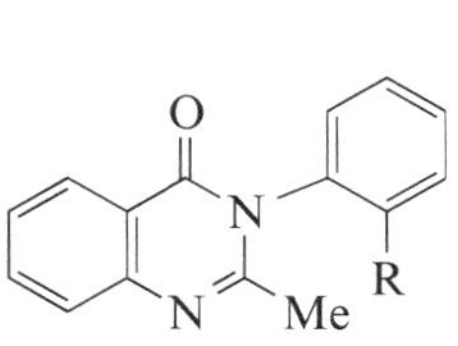

(**70**)

2-α-Cyanobenzyl- gave 2-benzoylquinazoline (NaH, THF, 20°, ~ 5 min; then $O_2\downarrow$ until colorless: 91%); 4-benzoylquinazoline (92%) likewise.[2111]

Conversion into amidines

3-β-Cyanoethyl- (**71**, R = CN) gave 3-β-N-hydroxyamidino-4(3H)-quinazolinone [**71**, R = C(:NOH)NH$_2$] (NH$_2$OH, EtOH: > 60%); also analogs.[317]

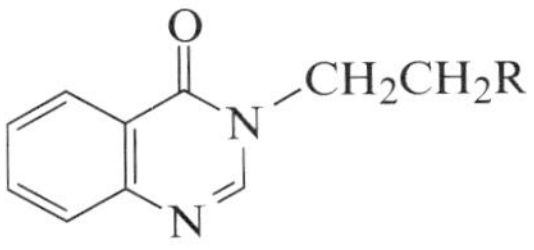

(71)

Cyclization reactions

2-Phenyl-4-quinazolinecarbonitrile (**72**) gave 2-phenyl-4-tetrazol-5'-yl-quinazoline (**73**) (NaN$_3$, NH$_4$Cl, Me$_2$NCHO, 100°, 5 h: 53%) and thence 5-phenyl[1,2,3]triazolo[1,5-c]quinazoline (**74**) (thermolysis in mesitylene, 160°, 78 h: 75%).[923]

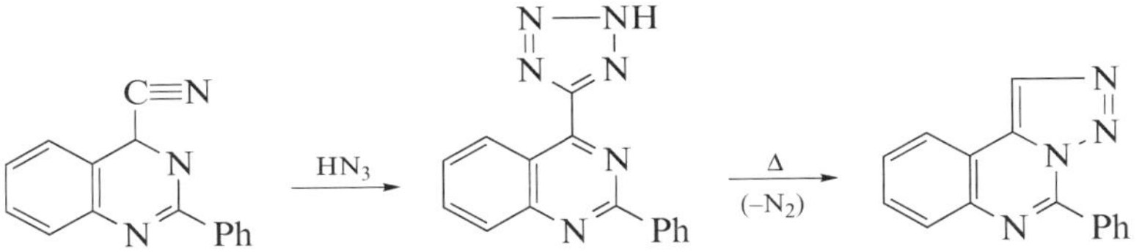

(72)   (73)   (74)

6-Chloro-2-β-cyanoethyl-4(3H)-quinazolinone (**75**) gave 7-chloro-1-imino-1,2,3,3a-tetrahydropyrrolo[1,2-a]quinazolin-5(4H)-one (**76**) (NaOH, EtOH–H$_2$O, 20°, 3 h: 60%).[265]

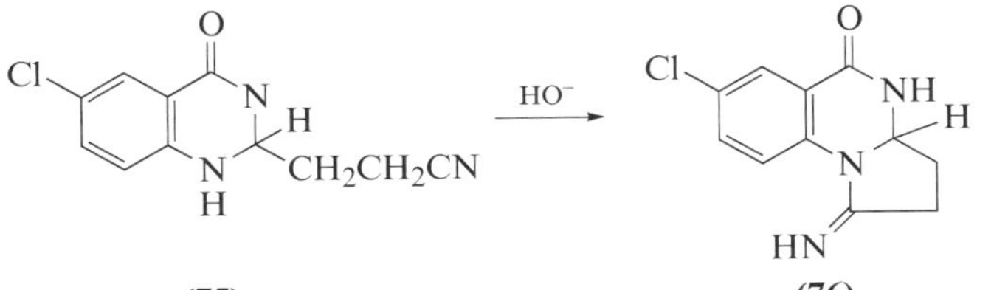

(75)   (76)

The mass-spectral fragmentation of 4-quinazolinecarbonitrile (**77**) has been compared with those of analogous bicyclic nitriles.[1052]

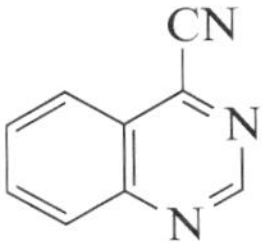

(77)

7.6. QUINAZOLINECARBALDEHYDES (*H* 480)

Preparation. Nearly all the recently used routes to quinazolinecarbaldehydes have been covered already: by *primary syntheses* (usually to acetals initially) (Chapter 1), by *oxidation of alkylquinazolines* (Section 2.2.2.2), by *Vilsmeier formylation of alkylquinazolines* (Section 2.2.2.4), from *halogenoquinazolines by lithium/dimethylformamide* (Section 3.6), from *halogenomethylquinazolines with hydroxylamine* (Section 3.7), from *dihalogenomethylquinazolines with silver nitrate* (Sections 3.7 and 7.1.1), by *reductive hydrolysis of quinazolinecarbonitriles* (Section 7.5), or by the occasional introduction of a *pro-aldehydic function as a passenger group* [e.g., Section 4.1.2.3: alkylation of 4(3H)-quinazolinone to 3-β,β-diethoxyethyl-4(3H)-quinazolinone (63%) by using bromoacetaldehyde diethyl acetal in sodium methoxide/phosphorous triamide at 100° for 24 h].[1772]

In addition, 3,4,5,6,7,8-hexahydro-2(1H)-quinazolinone (**78**, R = H) suffered nuclear *Vilsmeier diformylation* to afford 2-oxo-1,2,3,4,5,6,7,8-octahydro-3,8-quinazolinedicarbaldehyde (**78**, R = CHO) (POCl$_3$, Me$_2$NCHO, 90°, 1 h: ~45%), formulated on spectral grounds as the 1,2,3,4,4a,5,6,7-octahydro isomer.[775]

(**78**)

Reactions. The *oxidation of quinazolinecarbaldehydes* to carboxylic acids (Section 7.1.1) and the *reduction of quinazolinecarbaldehydes* to hydroxymethyl analogs (Section 4.4.1) have been covered already.

The formation of *traditional aldehyde derivatives* is illustrated briefly in the following examples.

3-Formylmethyl- (**79**, X = O) gave the *oxime*, 3-hydroxyiminomethyl-4(3H)-quinazolinone (**79**, X = NOH) (H$_2$NOH.HCl, AcONa, EtOH–H$_2$O, reflux, 30 min: 45%);[1772] analogs likewise.[361] Note that such oximes may be made also by nitrosation of activated methylquinazolines (see Section 2.2.2.5).

(**79**)

6-Chloro-4-phenyl-2-quinazolinecarbaldehyde (**80**, X = O) gave the *hydrazones*, 6-chloro-2-dimethylhydrazonomethyl- (**80**, X = NNMe$_2$) (Me$_2$NNH$_2$, EtOH, 20°, 1 h: 69%) and 6-chloro-2-(2,4-dinitrophenylhydrazono)methyl-4-phenylquinazoline [**80**, X = NNHC$_6$H$_3$(NO$_2$)$_2$-2,4] (usual conditions).[176]

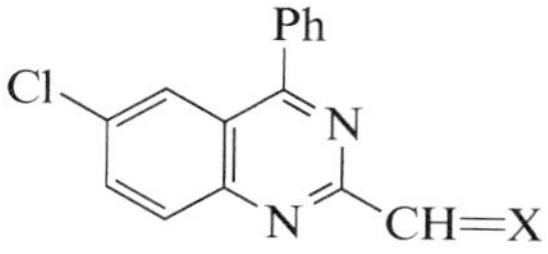

(80)

4-Oxo-3-phenyl-3,4-dihydro-2-quinazolinecarbaldehyde (**81**, R = O) gave the *semicarbazones*, 3-phenyl-2-thiosemicarbazonomethyl- (**81**, X = NNH-CSNH$_2$) (H$_2$NNHCSNH$_2$, EtOH–CHCl$_3$, reflux, 3 h: 65%), 3-phenyl-2-[4-phenyl(thiosemicarbazono)]methyl-4(3H)-quinazolinone (**81**, X = NNHCSNHPh) (H$_2$NNHCSNHPh, likewise: 70%), and the like.[2357]

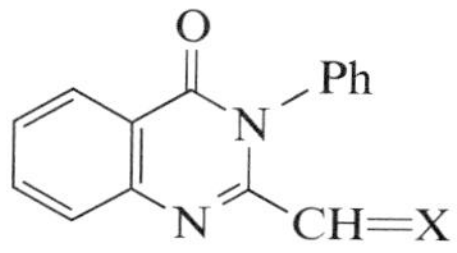

(81)

As well as hydrolytic reversion to quinazolinecarbaldehydes, the aforementioned and other such derivatives do undergo more interesting reactions. For example, the dehydration of oximes affords corresponding nitriles (see Section 7.5); treatment of the oxime, 6-chloro-2-hydroxyiminomethyl-4-phenylquinazoline 3-oxide (**82**, R = H), with acetyl chloride in dimethylformamide at 95° for 10 min, gave 44% of 2-acetoxyiminomethyl-6-chloro-4-phenylquinazoline 3-oxide (**82**, R = Ac);[361] brief thermolysis of the acetal, 6-bromo-3-β,β-dimethoxyethyl-2,4(1H,3H)-quinazolinedithione (**83**) at 180° gave 95% of 7-bromo-2-methoxy-2,3-dihydro-5H-thiazolo[2,3-b]quinazoline-5-thione (**84**);[1292] and the acetal, 6-chloro-3-β,β-dimethoxyethyl-2-methyl-4-phenyl-3,4-dihydroqinazoline (**85**), underwent ring fission and a different reclosure on refluxing in formic acid for 3 h to give 83% of 1-(2-benzyl-4-chlorophenyl)-2-methylimidazole (**86**).[1314]

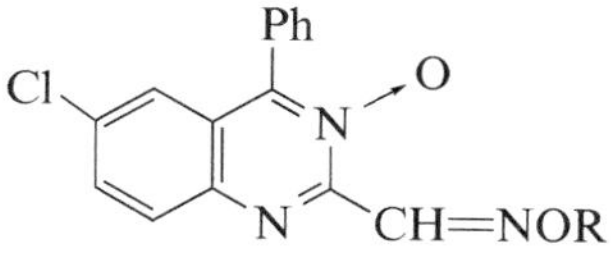

(82)

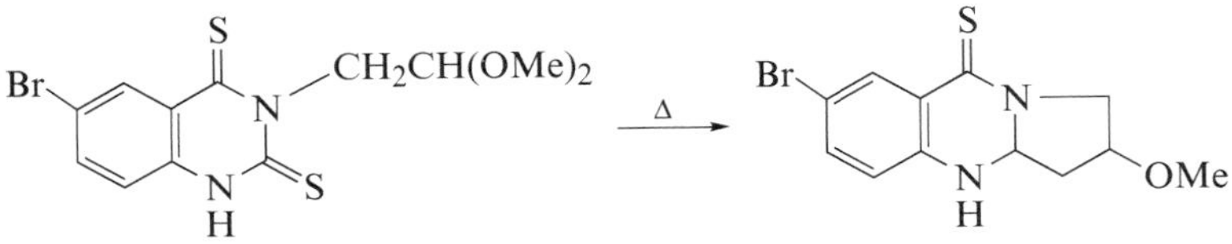

(83)   (84)

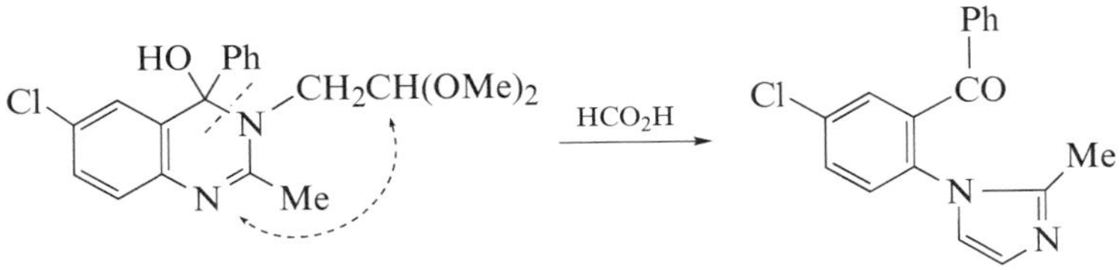

(85)   (86)

7.7.   QUINAZOLINE KETONES (*H* 480)

Preparation.   The major routes to quinazoline ketones have been discussed elsewhere: by *primary synthesis* (Chapter 1), via *hydroxyiminotetrahydroquinazo-lines* (Section 2.1.3), by *oxidation of alkylquinazolines* (Section 2.2.2.2), by *C-acylation of quinazolines* (Sections 2.1.3 and 2.2.2.4), by *oxidation of quinazoline (secondary) alcohols* (Section 4.4.2), *from halogenoquinazolines* (Section 3.5.5), *from quinazolinecarbonitriles* (Section 7.2.2), and by introduction of *ketonic resi-dues as passenger groups*, such as by acylalkylation of quinazolinethiones (Sec-tion 5.1.2.1).

In addition, irradiation of the zwitterion from 3-ethoxy(thiocarbonyl)amino-6,7-dimethoxy-2-methylquinazolinium chloride (**87**) in ethanol gave (among other products) a small yield of 4-acetyl-6,7-dimethoxy-2-methylquinazoline (**88**), in which the acetyl group was derived apparently from the solvent;[1302] alkaline treatment of 4-(α-benzoyloxy-α-cyanobenzyl)quinazoline (**89**, R = CN) in dimethyl sulfoxide at 20° for 30 min gave a 60% yield of 4-benzoylquinazoline (**90**);[1707] and 4-(α-benzoyloxybenzyl)quinazoline (**89**, R = H) also gave 4-ben-zoylquinazoline (**90**) by oxidative hydrolysis (no details).[1419]

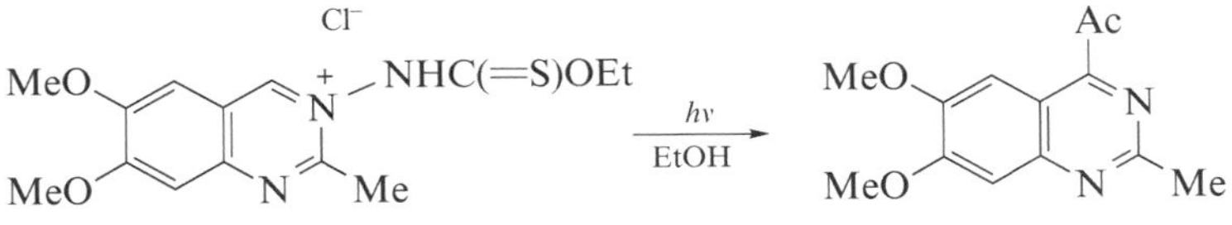

(87)   (88)

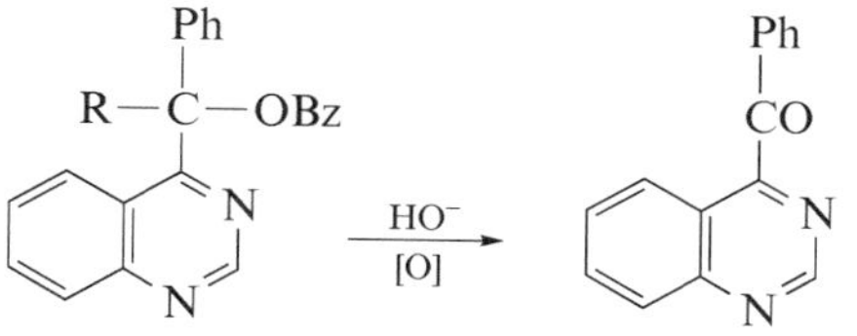

(89) (90)

Reactions. Remarkably few examples of the reactions of quinazoline ketones have been reported. Of such, *oxidation to carboxylic acids* (Section 7.1.1), *reduction to alkylquinazolines* (Section 2.2.1.5), and *reduction to quinazoline (secondary) alcohols* (Section 4.4.1) have been illustrated already.

The remaining reactions are exemplified in the following list.

Deacylation

4-(α-Acetyl-α-ethoxycarbonylmethyl)quinazoline (**91**, R = Ac) gave 4-α-ethoxycarbonylmethylquinazoline (**91**, R = H) (F_3CCO_2H, reflux, 3 h: 36%).[887]

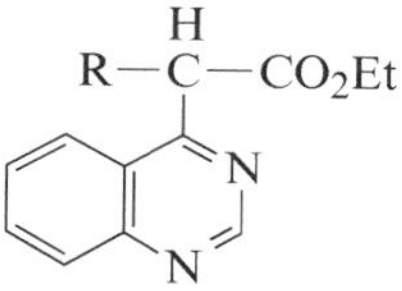

(91)

2-Acetonyl- (**92**, R = Ac) gave 2-methyl-4(3H)-quinazolinone (**92**, R = H) (2.5M KOH, 20°, 5 days: > 45%).[1766]

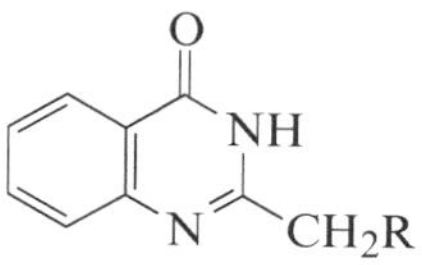

(92)

Also other examples in Section 2.2.1.3 and elsewhere.[251]

Rearrangement

2-Benzoyl-2-phenyl-1,2-dihydro-4(3H)-quinazolinone (**93**) gave 2-(α-hydroxy-α-phenylbenzyl)-4(3H)-quinazolinone (**94**) ($ZnCl_2$, AcOH, reflux, N_2, 10 h: 61% net).[251]

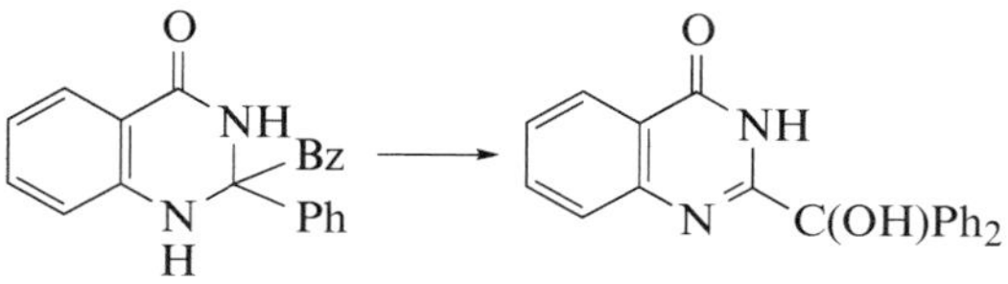

$$\text{(93)} \longrightarrow \text{(94)}$$

(93)   **(94)**

Traditional derivatizations

2-Acetyl-4(3*H*)-quinazolinone (**95**, X = O) gave its *2,4-dinitrophenylhydrazone* [**95**, X = NNHC$_6$H$_3$(NO$_2$)$_2$-2,4] [H$_2$NNHC$_6$H$_3$(NO$_2$)$_2$-2,4, usual conditions).[246]

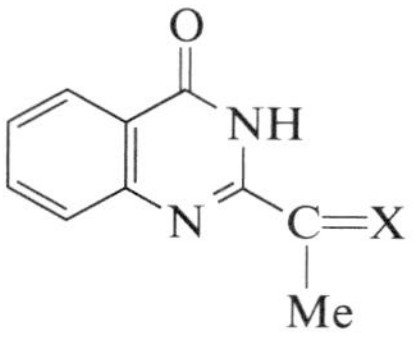

(95)

7,7-Dimethyl-2-phenyl-5,6,7,8-tetrahydro-5-quinazolinone (**96**, X = O) gave its *oxime* (**96**, X = NOH) (usual conditions).[302]

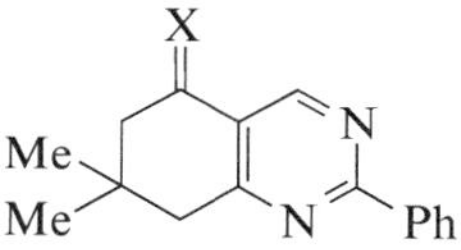

(96)

2-Acetonyl-3-*p*-chlorophenyl-4(3*H*)-quinazolinone (**97**, X = O) gave the *Schiff base*, 3-*p*-chlorophenyl-2-(*β*-*p*-chlorophenylimino)propyl-4(3*H*)-quinazolinone (**97**, X = NC$_6$H$_4$Cl-*p*) (H$_2$NC$_6$H$_4$Cl-*p*, during a primary synthesis); analogs likewise.[2387]

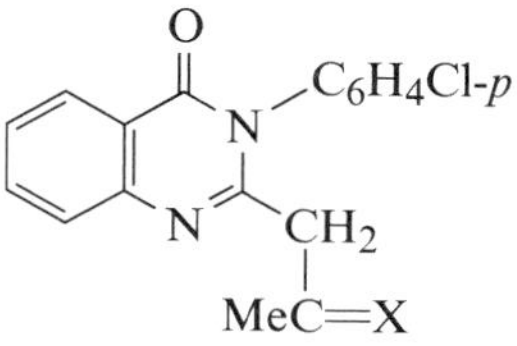

(97)

Also other examples.[2102,2259]

Ring fission

3-Benzoyl-2-methyl-4(3H)-quinazolinone oxime (**98**) gave 5-o-aminophenyl-3-methyl-1,2,4-oxadiazole (**99**) (HCl, H_2O–EtOH, reflux, 8 h: 68%); analogs likewise.[958]

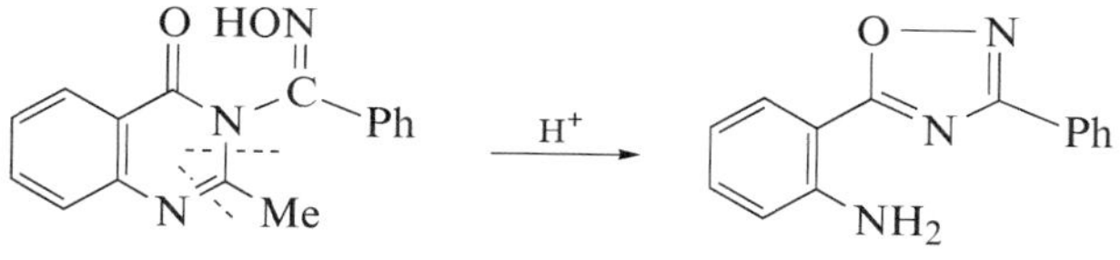

(**98**)　　　　　　(**99**)

Cyclization

2-p-Chlorobenzoylmethylthio-4(3H)-quinazolinone gave 3-p-chlorophenyl-5H-thiazolo[2,3-b]quinazolin-5-one (**100**) (H_2SO_4, 130°, 3 h; or polyphosphoric acid, 180°, 3 h: ?%).[119]

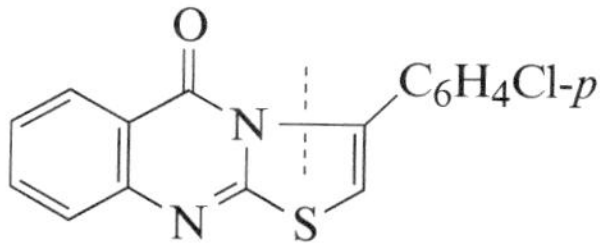

(**100**)

2-β-Benzoylvinyl-3-phenyl- gave 3-phenyl-2-(3-phenyl-Δ^2-1,2-oxazolin-5-yl)-4(3H)-quinazolinone (H_2NOH.HCl, AcONa, AcOH, reflux, 5 h: 72%); also analogs.[2561]

Ring expansion

4,7,7-Trimethyl-2-phenyl-5,6,7,8-tetrahydro-5-quinazolinone (**101**, X = O) gave 4,4,9-trimethyl-7-phenyl-2,3,4,5-tetrahydro-1H-pyrimido[5,4-b]azepin-2-one (**102**) (NaN_3, polyphosphoric acid, heat: 50%);[549] the substrate tosyloxime (**101**, X = NOTs) also gave the product (**102**) (AcOH, heat: 90%).[549]

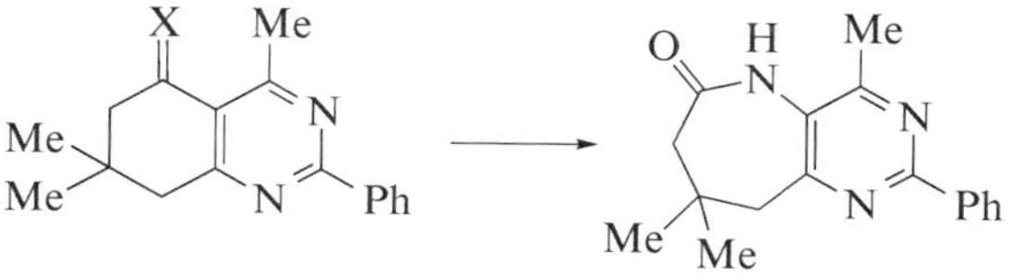

(**101**)　　　　　　(**102**)

Enolization

2-Benzoylmethyl-3-phenyl-4(3*H*)-quinazolinone (**103**) has been shown to exist largely as its enol, 2-(α-hydroxybenzylidene)methyl-3-phenyl-4(3*H*)-quinazolinone (**104**), with a strong intramolecular hydrogen bond.[466]

(103)   (104)

7.8. QUINAZOLINE THIOCYANATES AND ISOTHIOCYANATES

Of the family of derivatives comprising cyanates $(R\!-\!O\!-\!C\!\equiv\!N)$, isocyanates $(R\!-\!N\!=\!C\!=\!O)$, thiocyanates $(R\!-\!S\!-\!C\!\equiv\!N)$, isothiocyanates $(R\!-\!N\!=\!C\!=\!S)$, and nitrile oxides $(R\!-\!C\!\equiv\!N\rightarrow O)$, only the thiocyanates and isothiocyanates are represented to any extent in the quinazoline literature.

Thiocyanatoquinazolines. Such thiocyanates must be made under mild conditions to avoid the risk of rearrangement into their isothiocyanato isomers. Accordingly, they are made usually by *displacement of (active) halogeno substituents* (with thiocyanate ion (see Sections 3.5.5 and 3.7) or, less frequently, by *cyanation of quinazolinethiones* with cyanogen bromide (see Section 5.1.2.5).

Isothiocyanatoquinazolines. The formation of *isothiocyanato- from thiocyanatoquinazolines* appears to have been neglected, although such an isomerization must have occurred during treatment of a quinazolinecarbonyl chloride with thiocyanate ion to give the corresponding uncharacterized isothiocyanatocarbonylquinazoline (see Section 7.3). Another route to isothiocyanatoquinazolines is the treatment of primary *quinazolinamines with thiophosgene* (see Section 6.2.2.8). More *esoteric procedures* are represented in the conversion of 4-[4-(aminomethyl)piperidino]- (**105**, R = NH$_2$) into 66% of 4-[4-(isothiocyanatomethyl)piperidino]-6,7-dimethoxyquinazoline (**105**, R = N:C:S) by initial treatment with carbon disulfide and triethylamine in refluxing ethanol, followed by treatment of the crude product with ethyl chloroformate and triethylamine in chloroform at 20°;[2108] and by a peculiar primary synthesis involving treatment of mercuric *C*-morpholino-*C*-(phenylimino)methanethiolate (**106**) with cyanogen bromide (2.6 mol) in refluxing dioxane for 10 min to afford 4-isothiocyanato-2-morpholinoquinazoline (**107**) in 30% yield.[439]

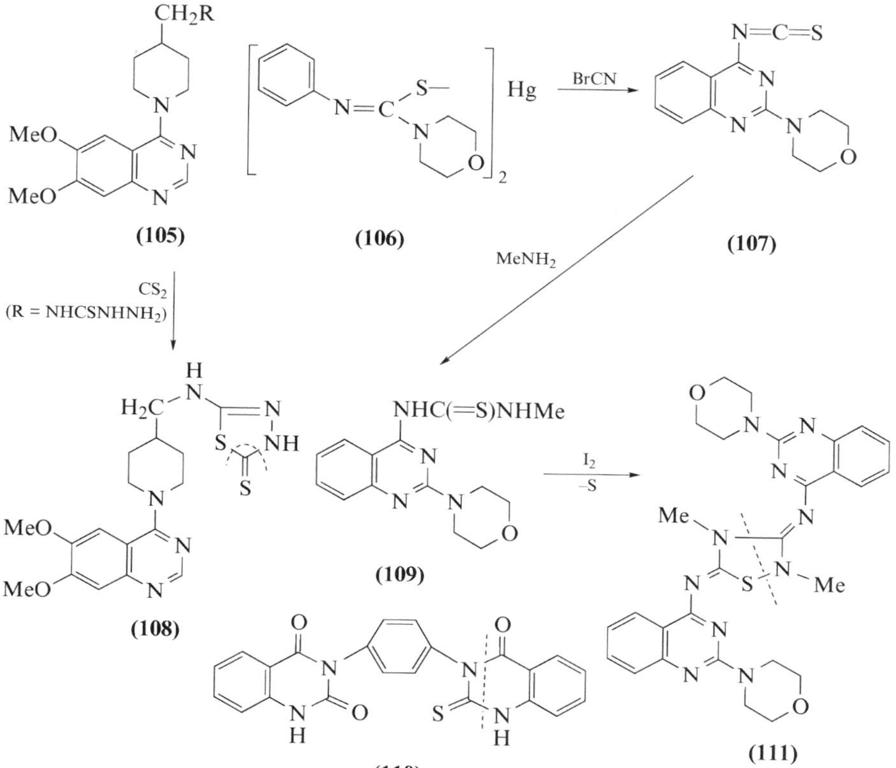

Quinazoline isothiocyanates undergo *reduction or splitting* to primary or secondary aminoquinazolines (see Section 6.2.1). Other reactions are illustrated in the following examples.

4-[4-(Isothiocyanatomethyl)piperidino]-6,7-dimethoxy- (**105**, R = N:C:S) gave 4-{4-[hydrazino(thiocarbonyl)aminomethyl]piperidino}-6,7-di-methoxy- (**105**, R = CH$_2$NHCSNHNH$_2$) (H$_2$NNH$_2$.H$_2$O, MeOH, 20°, 15 min: 89%) and thence 6,7-dimethoxy-4-{4-[(5-thioxo-1,3,4-thiadia-zolin-2-yl)aminomethyl]piperidino}quinazoline (**108**) (CS$_2$, Et$_3$N, Me$_2$NCHO, 20°, 30 min, then 100°, 90 min: 58%).[2028]

4-Isothiocyanato- (**107**) gave 4-N′-methyl(thioureido)-2-morpholinoquinazo-line (**109**) (MeNH$_2$↓, Et$_2$O, 20°, short time: > 95%),[429] which underwent oxidative self-condensation with loss of sulfur to give 2,4-dimethyl-3,5-bis(2-morpholinoquinazolin-4-ylimino)-1,2,4-thiadiazolidine (**111**) (I$_2$, CHCl$_3$, 20°, 15 min: 70%);[735] analogs likewise.[735]

3-*p*-Isothiocyanatophenyl-2,4(1*H*,3*H*)-quinazolinedione gave 3-[*p*-(4-oxo-2-thioxo-1,2,3,4-tetrahydroquinazolin-3-yl)phenyl]-2,4(1*H*,3*H*)-quinazolinedione (**110**) (MeO$_2$CC$_6$H$_4$NH$_2$-*o*, Me$_2$NCHO, 70°, 2 h, then reflux, 24 h: 60%; a primary synthesis!).[2248]

Also other examples.[215]

APPENDIX

Table of Simple Quinazolines

This table aims to be a comprehensive alphabetical list of simple quinazolines described up to mid-1995. For each compound are recorded (1) melting and/or boiling point; (2) an indication of reported spectra or other physical properties; (3) any reported salts or simple derivatives, especially when the parent compound was ill-characterized; (4) an indication of whether any chelate complexes have been described; and (5) direct reference(s) to the original literature from 1967 onward with indirect reference to any previously published data and literature in the form of a page number [e.g., *H* 428 (p. 428 in original *Hauptwerk* volume)] where such material appears in Armarego's original book.[2414]

To keep the table within manageable proportions, the following categories of quinazolines have been *excluded* on the grounds that they are not simple.

Fused or nuclear-reduced quinazolines

Quinazolines with a cyclic substituent other than a cycloalkyl, piperidino, morpholino, or unsubstituted phenyl group

Quinazolines bearing a substituent with > 6 carbon atoms

Quinazolines with two substituents at any one position

Quinazolines with two or more different independent groups on a single substituent

The following conventions and abbreviations have been used in the table.

Melting point. This term covers not only a regular melting point or melting range but also such variations as "decomposing at," "melting with decomposition at," and so on. The use of the symbol > before a melting point indicates that the substance melts or decomposes above that temperature or that it does not melt or decompose below that temperature. When two differing melting points are given in the literature, they appear in the table as, for example, "244 or 263"; when more than two melting points are given, they are recorded in a form such as "218 to 231" in order to distinguish them from the single melting range "218–231."

Boiling point. Boiling points are distinguished from melting points by the presence of a pressure in millimeters of mercury (mmHg) after the temperature: for instance, 108–111/0.01.

Abbreviations for physical data

anal	analytical data (usually assumed)
crude	compound not purified
dipole	dipole moment
el den	electron density
fl sp	fluorescence spectrum
ir	infrared spectrum
ms	mass spectrum
nmr	nuclear magnetic resonance spectrum (any nucleus)
pol	polarographic data
st	fine structure discussed
uv	ultraviolet/visible spectrum
xl st	crystal structure (X-ray data)

Abbreviations for salts or solvates

AcOH	acetate salt
HBr, etc.	appropriate hydrohalide salt
H_2O	hydrate
H_2SO_4	sulfate salt
NH_4	ammonium salt
Na, etc.	appropriate alkali metal salt
pic	picrate
TsOH	*p*-toluenesulfonate salt

Abbreviations for derivatives

Ac	acetyl derivative
acetal	diethyl acetal
dnp	2,4-dinitrophenylhydrazone
$H_2NN=$	hydrazone
oxime	oxime
PhHNN=	phenylhydrazone
sc	semicarbazone
Ts	*p*-toluenesulfonyl derivative
tsc	thiosemicarbazone

Other notes. The use of "cf." before a reference usually indicates suspect, apparently inconsistent, or mildly relevant information therein. A query mark (?) indicates an anomaly, apparent anomaly, or reasonable doubt associated with a datum or reference.

ALPHABETICAL LIST OF SIMPLE QUINAZOLINES, REPORTED TO THE END OF 1994

Quinazoline	m.p. (°C), etc.	Ref.
2-Acetamido-6-acetoxyquinazoline	—	H 342
2-Acetamido-1-acetyl-2,4(1H,3H)-quinazolinedithione	196–199	81
6-Acetamido-5-amino-3-butyl-4(3H)-quinazolinone	195, uv	121
6-Acetamido-3-amino-2-methyl-4(3H)-quinazolinone	—	H 380
7-Acetamido-3-amino-2-methyl-4(3H)-quinazolinone	—	H 380
7-Acetamido-3-anilino-2-methyl-4(3H)-quinazolinone	—	H 380
4-Acetamido-2-benzoylquinazoline	173	426
6-Acetamido-3-benzylideneamino-2-styryl-4(3H)-quinazolinone	—	H 380
7-Acetamido-3-benzylideneamino-2-styryl-4(3H)-quinazolinone	—	H 380
3-Acetamido-7-benzyloxy-6-ethoxy-4(3H)-quinazolinone	165	844
4-Acetamido-2-benzylquinazoline	145, ir	426
4-Acetamido-2-α-bromobenzylquinazoline	178, ir	426
4-Acetamido-2-bromomethylquinazoline	205, ir	426
7-Acetamido-4-butylaminoquinazoline	—	H 363
3-Acetamido-1-butyl-7-chloro-2,4-dioxo-1,2,3,4-tetrahydro-6-quinazolinesulfonamide	239–242	1603
6-Acetamido-3-butyl-5-nitro-4(3H)-quinazolinone	170, ir, nmr, uv	121
6-Acetamido-3-butyl-4(3H)-quinazolinone	203–204, uv	121
4-Acetamido-2-α-chlorobenzylquinazoline	174, ir	426
3-Acetamido-2-α-chloroethyl-4(3H)-quinazolinone	144	2341
6-Acetamido-4-chloro-7-methoxyquinazoline	215, ir, nmr	2017
3-Acetamido-6,8-dibromo-2-methyl-4(3H)-quinazolinone	—	2349
3-Acetamido-6,8-dibromo-2-phenyl-4(3H)-quinazolinone	—	2370
3-Acetamido-6,8-dichloro-2-phenyl-4(3H)-quinazolinone	219	2318
6-Acetamido-5,8-dihydro-2,4,5,8(1H,3H)-quinazolinetetrone	>300, ir, nmr	1987

409

ALPHABETICAL LIST OF SIMPLE QUINAZOLINES (*Continued*)

Quinazoline	m.p. (°C), etc.	Ref.
3-Acetamido-6,8-dihydroxy-2-phenyl-4(3*H*)-quinazolinone	230, ir	844
3-Acetamido-6,7-dimethoxy-2-methyl-4(3*H*)-quinazolinethione	—	1058
6-Acetamido-5,8-dimethoxy-2,4(1*H*,3*H*)-quinazolinedione	285–290, ir, nmr	1987
4-Acetamido-2-diethylaminoquinazoline	174–176, ir, nmr	2418
3-Acetamido-2,7-dimethyl-4-oxo-3,4-dihydro-6-quinazolinecarboxylic acid	—	H 484
7-Acetamido-2,6-dimethyl-3-phenyl-4(3*H*)-quinazolinone	—	H 369
3-Acetamido-2,6-dimethyl-4(3*H*)-quinazolinone	114, ir, nmr	1430
6-Acetamido-2,3-dimethyl-4(3*H*)-quinazolinone	—	H 368
7-Acetamido-2,3-dimethyl-4(3*H*)-quinazolinone	—	H 369
7-Acetamido-2,6-dimethyl-4(3*H*)-quinazolinone	—	H 369
6-Acetamido-3-ethoxycarbonylamino-2-methyl-4(3*H*)-quinazolinone	—	H 380
3-Acetamido-6-ethoxy-7-hydroxy-2-methyl-4(3*H*)-quinazolinone	252, nmr	844
6-Acetamido-3-ethyl-2-methyl-4(3*H*)-quinazolinone	—	H 367
7-Acetamido-3-ethyl-2-methyl-4(3*H*)-quinazolinone	—	H 369
2-Acetamido-3-ethyl-4(3*H*)-quinazolinone	180–181, ir, nmr	2418
3-Acetamido-3-ethyl-4(3*H*)-quinazolinone	135–136, ir, ms, nmr	H 372; 2593
7-Acetamido-3-isopentyl-2-methyl-4(3*H*)-quinazolinone	—	H 369
2-Acetamido-6-isopropyl-8-methyl-5-quinazolinol	—	H 342
7-Acetamido-6-methoxyquinazoline	—	H 260
6-Acetamido-7-methoxy-4(3*H*)-quinazolinone	300, ir, nmr	2017
6-Acetamido-7-methylamino-4(3*H*)-quinazolinone	>350, ms, nmr	1962
7-Acetamido-2-methyl-6,8-dinitro-4(3*H*)-quinazolinone	—	H 369
2-Acetamidomethyl-6-methyl-4(3*H*)-quinazolinone	crude, nmr	2263
6-Acetamido-7-(*N*-methyl-*N*-nitroamino)-8-nitro-4(3*H*)-quinazolinone	234–235, ir, nmr	1962

410

Compound		
3-Acetamido-2-methyl-5-nitro-4-phenylhydrazono-3,4-dihydroquinazoline	—	H 379
3-Acetamido-2-methyl-7-nitro-4-phenylhydrazono-3,4-dihydroquinazoline	—	H 379
3-Acetamido-2-methyl-5-nitro-4(3H)-quinazolinone	—	H 377
3-Acetamido-2-methyl-7-nitro-4(3H)-quinazolinone	—	H 377
2-Acetamidomethyl-3-phenyl-4(3H)-quinazolinone	234–235	152
6-Acetamido-2-methyl-3-phenyl-4(3H)-quinazolinone	—	H 368
7-Acetamido-2-methyl-3-phenyl-4(3H)-quinazolinone	—	H 369
6-Acetamido-2-methyl-3-propyl-4(3H)-quinazolinone	—	H 368
7-Acetamido-2-methyl-3-propyl-4(3H)-quinazolinone	—	H 369
4-Acetamido-2-methylquinazoline	170–172, ir, nmr	1232
3-Acetamido-1-methyl-2,4(3H)-quinazolinedione	240	H 382; 1603
3-Acetamidomethyl-4(3H)-quinazolinone	HCl: 160–161	1192
3-Acetamido-2-methyl-4(3H)-quinazolinone	156 to 176, ir, nmr, complexes	H 372; 1430,1657,1684,2082, 2342,2593
5-Acetamido-2-methyl-4(3H)-quinazolinone	—	H 367
6-Acetamido-2-methyl-4(3H)-quinazolinone	—	H 367
6-Acetamido-3-methyl-4(3H)-quinazolinone	269–271	H 367; 498
6-Acetamido-7-methyl-4(3H)-quinazolinone	>310	1157
7-Acetamido-2-methyl-4(3H)-quinazolinone	352–353	H 369; 103
7-Acetamido-3-methyl-2-styryl-4(3H)-quinazolinone	—	H 369
4-Acetamido-6-nitroquinazoline	—	H 351
4-Acetamido-7-nitroquinazoline	—	H 351
6-Acetamido-5-nitro-4(3H)-quinazolinone	311–312, ms. nmr	502
6-Acetamido-4-oxo-3,4-dihydro-7-quinazolinecarboxylic acid	>300	1157
6-Acetamido-8-phenylimino-5,8-dihydro-2,4,5(1H,3H)-quinazolinetrione	>250, ir, ms, nmr	1987
4-Acetamido-2-phenylquinazoline	202, ir, nmr	1232
3-Acetamido-1-phenyl-2,4(1H,3H)-quinazolinedithione	250	1725
2-Acetamido-3-phenyl-4(3H)-quinazolinimine	142–144, ir, nmr	2418

ALPHABETICAL LIST OF SIMPLE QUINAZOLINES (*Continued*)

Quinazoline	m.p. (°C), etc.	Ref.
3-Acetamido-2-phenyl-4(3*H*)-quinazolinone	118–121, complexes	*H* 371,372; 1981,2082,2342
7-Acetamido-4-quinazolinamine	—	*H* 363
2-Acetamidoquinazoline	—	*H* 342
8-Acetamidoquinazoline	—	*H* 362
3-Acetamido-2,4(1*H*,3*H*)-quinazolinedione	—	*H* 382
3-Acetamido-2,4(1*H*,3*H*)-quinazolinedithione	>236	81
4-Acetamido-2(1*H*)-quinazolinethione	241–242, ir, nmr	2095
2-Acetamido-4(3*H*)-quinazolinone	275–278	678,2186
3-Acetamido-4(3*H*)-quinazolinone	250–206, ir, nmr	*H* 374; 276,1430,1984
5-Acetamido-4(3*H*)-quinazolinone	—	*H* 367
6-Acetamido-4(3*H*)-quinazolinone	333	*H* 367; 498
7-Acetamido-4(3*H*)-quinazolinone	—	*H* 369
3-Acetamido-2-styryl-4(3*H*)-quinazolinone	—	*H* 372
7-Acetamido-2-styryl-4(3*H*)-quinazolinone	—	*H* 370
3-Acetamido-2-thioxo-1,2-dihydro-4(3*H*)-quinazolinone	298, ir	1166
6-Acetamido-2,3,*N*-trimethyl-8-nitro-4-oxo-3,4-dihydro-7-quinazolinecarboxamide	283–286, ir, nmr	1964
2-Acetonyl-3-butyl-4(3*H*)-quinazolinone	—	981
2-Acetonyl-1-cyclohex-1′enyl-4(1*H*)-quinazolinone	161–163, ir, ms	1405
2-Acetonyl-1(1′-ethylprop-1′-enyl)-4(1*H*)-quinazolinone	142–143, nmr	1406
2-Acetonyl-3-ethyl-4(3*H*)-quinazolinone	—	981
2-Acetonyl-3-β-hydroxyethyl-4(3*H*)-quinazolinone	—	981
2-Acetonyl-1-methyl-4(1*H*)-quinazolinone	263–266, st	244
2-Acetonyl-3-methyl-4(3*H*)-quinazolinone	201–202, ir, ms, nmr	1766
2-Acetonyl-1-phenyl-4(1*H*)-quinazolinone	241–243, uv	937

2-Acetonyl-3-phenyl-4(3H)-quinazolinone	—	981
4-Acetonylquinazoline	109–110 or 127–129, ir, ms, nmr	H 61; 1404,1869
2-Acetonyl-4(3H)-quinazolinone	213 or 219, ir	1161,1766
3-Acetonyl-4(3H)-quinazolinone	158 or 166–167, ir, ms, nmr	H 149; 276,1956
4-Acetonyl-2,5,8-trimethylquinazoline	—	H 65
6-Acetoxy-5-allyl-7-methoxy-2,3-diphenyl-4(3H)-quinazolinone	165–166	140
7-Acetoxy-6-allyl-8-methoxy-2,3-diphenyl-4(3H)-quinazolinone	170	149
7-Acetoxy-8-allyl-6-methoxy-2,3-diphenyl-4(3H)-quinazolinone	153–154	151
6-Acetoxy-5-allyl-7-methoxy-2-methyl-3-phenyl-4(3H)-quinazolinone	220–221	143
7-Acetoxy-8-allyl-6-methoxy-2-methyl-3-phenyl-4(3H)-quinazolinone	165–166	151
3-Acetoxyamino-2-ethyl-4(3H)-quinazolinone	solution only, nmr	2052
4-Acetoxyamino-2-methylquinazoline	184, ir, nmr	60,61
4-Acetoxyaminoquinazoline	215–217, ir, nmr	60
1-Acetoxy-3-benzyl-2,4(1H,3H)-quinazolinedione	151	187
3-Acetoxy-2-chloromethyl-4(3H)-quinazolinone	148	236
6-Acetoxy-4-chloro-2-phenylquinazoline	128–130	766
6-Acetoxy-5-β,γ-dibromopropyl-7-methoxy-2,3-diphenyl-4(3H)-quinazolinone	190–191	140
7-Acetoxy-6-β,γ-dibromopropyl-8-methoxy-2,3-diphenyl-4(3H)-quinazolinone	173–174	149
7-Acetoxy-8-β,γ-dibromopropyl-6-methoxy-2,3-diphenyl-4(3H)-quinazolinone	172–173	151
6-Acetoxy-5-β,γ-dibromopropyl-7-methoxy-2-methyl-3-phenyl-4(3H)-quinazolinone	182–183	143
7-Acetoxy-8-β,γ-dibromopropyl-6-methoxy-2-methyl-3-phenyl-4(3H)-quinazolinone	209–210	151
1-Acetoxy-2-ethoxy-4(1H)-quinazolinone	117, ir	182
1-β-Acetoxyethyl-6-nitro-4-phenyl-2(1H)-quinazolinone	155	493
3-β-Acetoxyethyl-2,4(1H,3H)-quinazolinedione	165–167 or 167–168, ir, nmr	1428,1473
3-β-Acetoxyethyl-4(3H)-quinazolinone	81	276
6-Acetoxy-7-methoxy-2-methyl-3-phenyl-4(3H)-quinazolinone	197–198	143

413

ALPHABETICAL LIST OF SIMPLE QUINAZOLINES (*Continued*)

Quinazoline	m.p. (°C), etc.	Ref.
1-Acetoxy-2-methoxy-4(1*H*)-quinazolinone	152, ir	182
2-Acetoxyethyl-6-chloro-4-phenylquinazoline	—	*H* 249
2-Acetoxymethyl-6-chloro-4-phenylquinazoline 3-oxide	236	82
2-(α-Acetoxy-α-methylethyl)-6-methyl-4(3*H*)-quinazolinone	178–179	1049
2-(α-Acetoxy-α-methylethyl)-4(3*H*)-quinazolinone	183–184	475
2-Acetoxymethyl-6-methyl-4(3*H*)-quinazolinone	186–187	510
2-Acetoxymethyl-1-phenyl-4(1*H*)-quinazolinone	150–153, nmr	1394
2-Acetoxymethyl-3-phenyl-4(3*H*)-quinazolinone	99–100 or 101, ir, nmr	146,1276
2-Acetoxymethyl-4-quinazolinamine	203, ir	426
1-Acetoxy-3-methyl-2,4(1*H*,3*H*)-quinazolinedione	132	187
3-Acetoxy-1-methyl-2,4(1*H*,3*H*)-quinazolinedione	155–156, ir	236
2-Acetoxymethyl-4(3*H*)-quinazolinone	196–197 or 198, ir	*H* 139; 187,2069
3-Acetoxy-2-methyl-4(3*H*)-quinazolinone	117	236
7-Acetoxy-4-methyl-2(1*H*)-quinazolinone	245–250	79
1-Acetoxy-3-phenyl-2,4(1*H*,3*H*)-quinazolinedione	202	187
6-Acetoxy-2-phenyl-4(3*H*)-quinazolinone	264–265	766
3-γ-Acetoxypropyl-2,4(1*H*,3*H*)-quinazolinedione	133–134, ir, nmr	1435
1-Acetoxy-2,4(1*H*,3*H*)-quinazolinedione	225 or 232, ir, uv	182,471
3-Acetoxy-2,4(1*H*,3*H*)-quinazolinedione	260, ir	236
6-Acetoxy-2-styryl-4(3*H*)-quinazolinone	293–298, nmr	2264
2-Acetyl-3-anilino-7-nitro-4(3*H*)-quinazolinone	207–209, ir	856
2-Acetyl-3-anilino-4(3*H*)-quinazolinone	151–152, ir, nmr	856
2-Acetyl-6-chloro-4-phenylquinazoline	131–132 or 132–134, ir, nmr	240,259
4-Acetyl-6,7-dimethoxy-2-methylquinazoline	181–182, ir, ms, nmr, uv	1302,1354

414

Compound	Data	Ref
3-(β-Acetyl-α,α-dimethylethyl)-2-thioxo-1,2-dihydro-4(3H)-quinazolinone	299	125
6-Acetyl-2,4-dimethylquinazoline	—	H 486
7-Acetyl-2,4-dimethylquinazoline	—	H 486
8-Acetyl-2,4-dimethylquinazoline	—	H 486
6-Acetyl-1,3-dimethyl-2,4(1H,3H)-quinazolinedione	197 or 207–208, ir, ms, nmr	2140,2465
2-β-Acetylethyl-3-phenyl-4(3H)-quinazolinone	153–155; dnp: 250	137
3-β-Acetylethyl-4(3H)-quinazolinethione	210, ir, ms	1279
3-β-Acetylethyl-4(3H)-quinazolinone	76, ir, ms, nmr	H 148; 1203,1280,1835
3-β-Acetylethyl-2-thioxo-1,2-dihydro-4(3H)-quinazolinone	204–205, ms, nmr	1966
2-β-Acetylhydrazino-3-amino-4(3H)-quinazolinone	—	1990
6-Acetyl-2-isopropyl-1-phenyl-4(1H)-quinazolinone	206–208	1590
2-(γ-Acetyl-β-methylpropyl)-4(3H)-quinazolinone	154–157, nmr	1311
2-Acetyl-4-methylquinazoline	102–103, ir, nmr	2193
2-Acetyl-3-methyl-4(3H)-quinazolinone	78–79, ir	918
7-Acetyl-2-methyl-4(3H)-quinazolinone	242–243, ir, nmr	1712
3-Acetyl-6-nitro-4(3H)-quinazolinone	—	1525
3-Acetyl-2-phenyl-4(3H)-quinazolinone	—	H 170
6-Acetyl-4-phenyl-2(1H)-quinazolinone	266–267	1060
2-γ-Acetylpropyl-4(3H)-quinazolinone	145–147, nmr	1310
3-γ-Acetylpropyl-4(3H)-quinazolinone	—	H 148
2-Acetyl-4(3H)-quinazolinone	197 to 205, ir, ms, nmr, uv, xl st	185,246,684,1268,1915, 2069,2081
3-Acetyl-4(3H)-quinazolinone (?)	84	276,1525
2-Acetylthio-4(3H)-quinazolinone	—	1983
6-Acetyl-1,3,7-trimethyl-2,4(1H,3H)-quinazolinedione	210–211, ms, nmr	1390
3-Allenyl-4(3H)-quinazolinone	130–131, ir, nmr	23,30
2-Allylamino-6-chloro-4-phenylquinazoline 3-oxide	199	399
2-Allylaminomethyl-6-chloro-4-phenylquinazoline 1-oxide	—	H 462

ALPHABETICAL LIST OF SIMPLE QUINAZOLINES (*Continued*)

Quinazoline	m.p. (°C), etc.	Ref.
3-Allyl-4-amino-2(3*H*)-quinazolinethione	nmr	*H* 310; 848
4-Allylamino-2(1*H*)-quinazolinethione	240, ir	848
3-Allyl-4-amino-2(3*H*)-quinazolinone	ir, nmr	2136
3-Allyl-2-benzoyloxy-4(3*H*)-quinazolinone	170–171, ir, uv	477
1-Allyl-6-bromo-3-carboxymethyl-2,4(1*H*,3*H*)-quinazolinedione	160–162	2277
3-Allyl-6-bromo-2-thioxo-1,2-dihydro-4(3*H*)-quinazolinone	complexes	1747
3-Allyl-1-but-2'-enyl-2,4(1*H*,3*H*)-quinazolinedione	52, ir, nmr	718
3-Allyl-2-carboxymethylthio-4(3*H*)-quinazolinone	173–179, ir, uv	478
1-Allyl-7-chloro-3-diethylamino-2,4(1*H*,3*H*)-quinazolinedione	108	1603
1-Allyl-7-chloro-3-dimethylamino-2,4(1*H*,3*H*)-quinazolinedione	100–102	1603
1-Allyl-7-chloro-2-methyl-4(1*H*)-quinazolinone	—	863
3-Allyl-6-chloro-8-methyl-2-thioxo-1,2-dihydro-4(3*H*)-quinazolinone	—	*H* 309
2-Allyl-6-chloro-4-phenylquinazoline	145–147	237
1-Allyl-6-chloro-4-phenyl-2(1*H*)-quinazolinone	184–185	1060
1-Allyl-6-chloro-4(1*H*)-quinazolinone	126–128	655
3-Allyl-6-chloro-2-thioxo-1,2-dihydro-4(3*H*)-quinazolinone	—	*H* 309
3-Allyl-7-chloro-2-thioxo-1,2-dihydro-4(3*H*)-quinazolinone	—	*H* 309
3-Allyl-8-chloro-2-thioxo-1,2-dihydro-4(3*H*)-quinazolinone	—	*H* 309
5-Allyl-6,7-dimethoxy-2,3-diphenyl-4(3*H*)-quinazolinone	140–141	140
8-Allyl-6,7-dimethoxy-2,3-diphenyl-4(3*H*)-quinazolinone	181–182	136
5-Allyl-6,7-dimethoxy-2-methyl-3-phenyl-4(3*H*)-quinazolinone	181–182	143
8-Allyl-6,7-dimethoxy-2-methyl-3-phenyl-4(3*H*)-quinazolinone	224–225	136
1-Allyl-3-dimethylamino-2,4(1*H*,3*H*)-quinazolinedione	95	1603
2-*N*-Allyl-*N*-ethylamino-6,7-dimethoxy-4(3*H*)-quinazolinone	183–185; HCl: 239–240	641

Compound		Reference
6-Allyl-2-ethyl-7,8-dihydroxy-3-phenyl-4(3H)-quinazolinone	anal, ms, nmr	1700
1-Allyl-6-fluoro-4(1H)-quinazolinone	104–105	655
5-Allyl-6-hydroxy-7-methoxy-2,3-diphenyl-4(3H)-quinazolinone	170–171	140
6-Allyl-7-hydroxy-8-methoxy-2,3-diphenyl-4(3H)-quinazolinone	224	149
8-Allyl-7-hydroxy-6-methoxy-2,3-diphenyl-4(3H)-quinazolinone	188–189	151
5-Allyl-6-hydroxy-7-methoxy-2-methyl-3-phenyl-4(3H)-quinazolinone	110–111	143
8-Allyl-7-hydroxy-6-methoxy-2-methyl-3-phenyl-4(3H)-quinazolinone	238–239	151
3-Allyl-1-hydroxy-2,4(1H,3H)-quinazolinedione	147–148 or 148–149	622,750
3-Allyl-1-hydroxy-2-thioxo-1,2-dihydro-4(3H)-quinazolinone	90–92	622
3-Allyl-4-imino-1-methyl-3,4-dihydro-2(1H)-quinazolinone	—	2136
3-Allyl-4-imino-1-phenyl-3,4-dihydro-2(1H)-quinazolinone	—	2136
1-Allyl-6-methoxy-4(1H)-quinazolinone	127–129	655
1-Allyl-7-methoxy-4(1H)-quinazolinone	122–125	655
3-Allyl-2-methyl-5-nitro-4(3H)-quinazolinone	—	H 180
1-Allyl-7-methyl-4-phenyl-2(1H)-quinazolinedione	153–155	681
3-Allyl-1-(2′-methylprop-2′-enyl)-2,4(1H,3H)-quinazolinedione	80, ir, nmr	718
1-Allyl-3-methyl-2,4(1H,3H)-quinazolinedione	93, ir, nmr	718
3-Allyl-2-methyl-4(3H)-quinazolinethione	66, ir, ms, nmr	1658
1-Allyl-6-methyl-4(1H)-quinazolinone	135–137	655
3-Allyl-2-methyl-4(3H)-quinazolinone	78–79	H 156; 935
3-Allyl-2-methylthio-4(3H)-quinazolinone	84–85, ir, uv	477
3-Allyl-1-methyl-2-thioxo-1,2-dihydro-4(3H)-quinazolinone	154–156, ms	1649
3-Allyl-6-methyl-2-thioxo-1,2-dihydro-4(3H)-quinazolinone	—	H 309
3-Allyl-7-methyl-2-thioxo-1,2-dihydro-4(3H)-quinazolinone	—	H 309
3-Allyl-8-methyl-2-thioxo-1,2-dihydro-4(3H)-quinazolinone	—	H 309
3-Allyloxycarbonylaminomethyl-2-methyl-4(3H)-quinazolinone	—	290
7-Allyloxy-2-carboxymethylthio-6-methoxy-3-phenyl-4(3H)-quinazolinone	189–190	141
4-Allyloxy-2-chloroquinazoline	—	H 263

ALPHABETICAL LIST OF SIMPLE QUINAZOLINES   (*Continued*)

Quinazoline	m.p. (°C), etc.	Ref.
6-Allyloxy-2-guanidino-4-methylquinazoline	—	H 343
6-Allyloxy-7-methoxy-2,3-diphenyl-4(3H)-quinazolinone	154–155	140
7-Allyloxy-6-methoxy-2,3-diphenyl-4(3H)-quinazolinone	137–138	151
7-Allyloxy-8-methoxy-2,3-diphenyl-4(3H)-quinazolinone	129 or 132–133	136,149
6-Allyloxy-7-methoxy-2-methyl-3-phenyl-4(3H)-quinazolinone	176–177	143
7-Allyloxy-6-methoxy-2-methyl-3-phenyl-4(3H)-quinazolinone	160–161; HCl: 225–226	136,151
7-Allyloxy-8-methoxy-2-methyl-3-phenyl-4(3H)-quinazolinone	148–149	136
7-Allyloxy-6-methoxy-3-phenyl-2-thioxo-1,2-dihydro-4(3H)-quinazolinone	245–246	141
2-Allyloxy-3-phenyl-4(3H)-quinazolinone	162–163, ir	1698
4-Allyloxyquinazoline	130/0.2, ir, nmr	1935
8-Allyloxyquinazoline	—	H 260
2-Allyl-4-phenylquinazoline	79, ms	1946
3-Allyl-2-phenyl-4(3H)-quinazolinethione	90, ms, nmr	1658
1-Allyl-4-phenyl-2(1H)-quinazolinone	159–160	681
1-Allyl-7-phenyl-4(1H)-quinazolinone	177–178	655
2-Allyl-3-phenyl-4(3H)-quinazolinone	—	H 168
3-Allyl-1-prop-2′-ynyl-2,4(1H,3H)-quinazolinedione	134–136	664
1-Allyl-2,4(1H,3H)-quinazolinedione	218–220	718
3-Allyl-2,4(1H,3H)-quinazolinedione	184 or 188–189, ir, nmr, uv	H 199; 362,718,2227
3-Allyl-2,4(1H,3H)-quinazolinedithione	—	460
2-Allylquinazoline 3-oxide	—	H 462
7-Allyl-8-quinazolinol	—	H 197
1-Allyl-4(1H)-quinazolinone	136–137	655

418

3-Allyl-4(3H)-quinazolinone	65–68, ir, nmr	*H* 144; 276,1772,1935
2-Allylthio-3-amino-4(3H)-quinazolinone	118, ir	1820
2-Allylthio-3-benzyl-6,8-dibromo-4(3H)-quinazolinone	135	644
2-Allylthio-3-benzyl-4(3H)-quinazolinone	—	*H* 304
2-Allylthio-6,8-dibromo-3-butyl-4(3H)-quinazolinone	199	644
2-Allylthio-6,8-dibromo-3-ethyl-4(3H)-quinazolinone	115	644
2-Allylthio-6,8-dibromo-3-methyl-4(3H)-quinazolinone	282	644
2-Allylthio-6,8-dibromo-3-phenyl-4(3H)-quinazolinone	276	644
2-Allylthio-6-methoxy-4(3H)-quinazolinone	184–185	280
4-Allylthio-2-phenylquinazoline	69, ir, ms, nmr	1658
2-Allylthio-3-phenyl-4(3H)-quinazolinethione	—	*H* 302
2-Allylthio-3-phenyl-4(3H)-quinazolinone	153–154	511
1-Allyl-2-thioxo-1,2-dihydro-4(3H)-quinazolinone	—	*H* 303
3-Allyl-2-thioxo-1,2-dihydro-4(3H)-quinazolinone	201 or 205–206, ir, nmr, complexes	*H* 308
		H 306; 460,652,1646,2227
1-Allyl-7-trifluoromethyl-4(1H)quinazolinone	141–142	655
4-Amino-3-anilino-2(3H)-quinazolinethione	210–212, ir, nmr	2417
2-Amino-3-anilino-4(3H)-quinazolinone		1877
3-Amino-2-anilino-4(3H)-quinazolinone	151, ms	1894
2-Amino-3-benzoyl-6-methoxy-4(3H)-quinazolinone	203–204, ir, nmr	472
2-Amino-3-benzoyl-6-methyl-4(3H)-quinazolinone	209–210, ir, nmr	472
2-Amino-3-benzoyl-8-methyl-4(3H)-quinazolinone	226–227, ir, nmr	472
2-Amino-3-benzoyl-4(3H)-quinazolinone	188–190, ir, nmr	472
6-Amino-7-benzylamino-3-methyl-4(3H)-quinazolinone	198	1165
2-Amino-6-benzylamino-4(3H)-quinazolinone	270–271	600
6-Amino-7-benzylamino-4(3H)-quinazolinone	290 or 293–295, ms, nmr	501,1165
3-Amino-2-benzyl-8-hydroxy-4(3H)-quinazolinone	—	*H* 375

ALPHABETICAL LIST OF SIMPLE QUINAZOLINES (*Continued*)

Quinazoline	m.p. (°C), etc.	Ref.
3-Amino-7-benzylideneamino-2-methyl-4(3*H*)-quinazolinone	—	*H* 380
2-α-Aminobenzylideneamino-4-oxo-3,4-dihydro-3-quinazolinecarbonitrile	344–345, ir	234
3-Amino-2-benzyl-8-methoxy-4(3*H*)-quinazolinone	—	*H* 375
7-Amino-3-benzyl-6-nitro-4(3*H*)-quinazolinone	225–227, ms, nmr	501
3-Amino-7-benzyloxy-6-ethoxy-2-methyl-4(3*H*)-quinazolinone	209–210, ir, nmr	844
6-Amino-3-benzyloxy-2-methyl-4(3*H*)-quinazolinone	—	*H* 469
6-Amino-3-benzyloxy-4(3*H*)-quinazolinone	—	*H* 469
3-Amino-2-benzyl-4(3*H*)-quinazolinethione	—	*H* 381
4-Amino-3-benzyl-2(3*H*)-quinazolinethione	209–211, ir, nmr	2417
2-Amino-3-benzyl-4(3*H*)-quinazolinone	195–196 or 202–204, ir, nmr;	938,2000
	HCl: >250, ir, ms, nmr	
3-Amino-2-benzyl-4(3*H*)-quinazolinone	120	1703
3-Amino-2-benzyl-6,7,8-trimethoxy-4(3*H*)-quinazolinone	—	*H* 375
3-Amino-6-bromo-2-β-carboxyethyl-4(3*H*)-quinazolinone	231	1659
2-Amino-5-bromo-6-chloro-4(3*H*)-quinazolinone	HCl: >400	2411
6-Amino-7-bromo-5,8-dihydro-2,4,5,8(1*H*,3*H*)-quinazolinetetrone	>300, ir, ms, nmr, pK_a, uv	1987
3-Amino-6-bromo-2-β-ethoxycarbonylethyl-4(3*H*)-quinazolinone	144	1659
2-Amino-6-bromo-3-ethyl-4(3*H*)-quinazolinone	—	*H* 364
3-Amino-6-bromo-2-ethyl-4(3*H*)-quinazolinone	170	1675
3-Amino-6-bromo-2-β-hydrazinocarbonylethyl-4(3*H*)-quinazolinone	231	1659
3-Amino-6-bromo-2-hydrazino-4(3*H*)-quinazolinone	270, ir	1250
3-Amino-2-bromomethyl-6-nitro-4(3*H*)-quinazolinone	151–152, ir	2360
3-Amino-6-bromo-1-methyl-2,4(1*H*,3*H*)-quinazolinedione	xl st	928
2-Amino-5-bromo-6-methyl-4(3*H*)-quinazolinone	crude, HCl; >390, ir, nmr	2411

420

Compound	Data	Ref
2-Amino-6-bromomethyl-4(3H)-quinazolinone	>400, nmr	871
3-Amino-2-bromomethyl-4(3H)-quinazolinone	185–186, ir	2360
3-Amino-6-bromo-2-methyl-4(3H)-quinazolinone	nmr	1072,1997
3-Amino-8-bromo-2-methyl-4(3H)-quinazolinone	197 or 230	1262,1675
3-Amino-5-bromo-7-nitro-2-phenyl-4(3H)-quinazolinone	—	H 375
2-Amino-6-bromo-8-nitro-4(3H)-quinazolinone	>200, ir, ms, nmr	2025
3-Amino-8-bromo-2-phenyl-4(3H)-quinazolinone	158, nmr	1262,1997
3-Amino-6-bromo-2,4(1H,3H)-quinazolinedione	304–306	1829
2-Amino-6-bromo-4(3H)-quinazolinone	>250, ir, nmr	2025
3-Amino-6-bromo-4(3H)-quinazolinone	224–225	H 374; 1453
3-Amino-2-but-3'-enyl-4(3H)-quinazolinone	105–106, ir, ms, nmr	1355
6-Amino-3-butyl-5-nitro-4(3H)-quinazolinone	145–147, uv	121
7-Amino-3-butyl-6-nitro-4(3H)-quinazolinone	205–208	1165
3-Amino-2-t-butyl-4(3H)-quinazolinethione	—	H 381
3-Amino-2-butyl-4(3H)-quinazolinone	106–108 or 148, ir	1657,2183
6-Amino-3-butyl-4(3H)-quinazolinone	108, uv	121
3-Amino-2-but-3'-ynyl-4(3H)-quinazolinone	157–159, ir, nmr	1826
3-Amino-2-β-carboxyethyl-4(3H)-quinazolinone	198 or 200	1184
3-Amino-2-carboxymethylthio-4(3H)-quinazolinone	247, ir	1215
2-Amino-7-chloro-6,8-dinitro-4(3H)-quinazolinone	>300, ms, nmr	915
2-Amino-6-chloro-3-ethyl-4(3H)-quinazolinone	211–212, ir, ms, nmr	2143,2241,2279
2-Amino-7-chloro-3-ethyl-4(3H)-quinazolinone	>280, ir, nmr	2241
3-Amino-2-α-chloroethyl-4(3H)-quinazolinone	148	2341
3-Amino-2-chloromethyl-6-nitro-4(3H)-quinazolinone	151–152, ir	2360
2-Amino-6-chloro-3-methyl-4(3H)-quinazolinone	218, ir, nmr	2143,2279
2-Amino-7-chloro-3-methyl-4(3H)-quinazolinone	>280, ir, nmr	2241
3-Amino-2-chloromethyl-4(3H)-quinazolinone	158, ir	2360
3-Amino-5-chloro-2-methyl-4(3H)-quinazolinone	—	H 375

ALPHABETICAL LIST OF SIMPLE QUINAZOLINES (*Continued*)

Quinazoline	m.p. (°C), etc.	Ref.
3-Amino-6-chloro-2-methyl-4(3*H*)-quinazolinone	—	H 375
3-Amino-7-chloro-2-methyl-4(3*H*)-quinazolinone	—	H 375
8-Amino-6-chloro-3-methyl-4(3*H*)-quinazolinone	—	H 370
3-Amino-5-chloro-7-nitro-2-phenyl-4(3*H*)-quinazolinone	—	H 375
2-Amino-5-chloro-6-nitro-4(3*H*)-quinazolinone	>355, nmr	2015
2-Amino-7-chloro-5-nitro-4(3*H*)-quinazolinone	>300	915
2-Amino-5-chloro-4-oxo-3,4-dihydro-6-quinazolinecarbonitrile	>350	2284
3-Amino-7-chloro-4-oxo-3,4-dihydro-6-quinazolinesulfonamide	273–274	648
3-Amino-6-chloro-2-phenyl-4(3*H*)-quinazolinethione	—	H 381
3-Amino-7-chloro-2-phenyl-4(3*H*)-quinazolinone	—	H 375
3-Amino-6-chloro-2,4(1*H*,3*H*)-quinazolinedione	311–313	1829
3-Amino-7-chloro-2,4(1*H*,3*H*)-quinazolinedione	285–286 or 306–307	1454,1829
4-Amino-6-chloro-2(1*H*)-quinazolinone	—	H 310
2-Amino-5-chloro-4(3*H*)-quinazolinone	321–325, nmr	678,2015
2-Amino-6-chloro-4(3*H*)-quinazolinone	370–375	618,641
2-Amino-7-chloro-4(3*H*)-quinazolinone	>300 or 397–400, ms, nmr;	641
	HCl: >400	
3-Amino-6-chloro-4(3*H*)-quinazolinone	225, ir, nmr	1453
5-Amino-6-chloro-4(3*H*)-quinazolinone	278–279, ms, nmr	502
3-Amino-2-cyanomethyl-4(3*H*)-quinazolinone	135	1164
6-Amino-7-cyclohexylamino-4(3*H*)-quinazolinone	>320, ms, nmr	501
4-Amino-3-cyclohexyl-2(3*H*)-quinazolinethione	318–320, ir, nmr	2417
3-Amino-6,8-dibromo-2-ethyl-4(3*H*)-quinazolinone	145	1675
3-Amino-6,8-dibromo-2-methyl-4(3*H*)-quinazolinone	134–136 (?), 238 or 241, ir	1262,1675,1717,2042,2310,

422

Compound	mp (°C), properties	Ref.
3-Amino-6,8-dibromo-2-phenyl-4(3H)-quinazolinone	190	2349,2354
5-Amino-1,3-dibutyl-6-nitro-2,4(1H,3H)-quinazolinedione	84–85, nmr	1262
7-Amino-1,3-dibutyl-6-nitro-2,4(1H,3H)-quinazolinedione	158–159, nmr	2261
3-Amino-6,8-dichloro-2-phenyl-4(3H)-quinazolinone	200	2261
3-Amino-2-diethylaminomethyl-4(3H)-quinazolinone	135–137/0.01	2318
5-Amino-1,3-diethyl-6-nitro-2,4(1H,3H)-quinazolinedione	128–130, nmr	816
7-Amino-1,3-diethyl-6-nitro-2,4(1H,3H)-quinazolinedione	201–204, nmr	2261
2-Amino-5,8-dihydro-4,5,8(3H)-quinazolinetrione	190, ir, nmr	2261
3-Amino-6,7-dimethoxy-2-methyl-4(3H)-quinazolinone	—	2430
2-Amino-5,6-dimethoxy-4(3H)-quinazolinone	200–202, ir, nmr	1058
2-Amino-6,7-dimethoxy-4(3H)-quinazolinone	313–316 or 317–319, ir, nmr, uv; HCl: 267–269	1562 / 641,1562
5-Amino-1,3-dimethyl-6-nitro-2,4(1H,3H)-quinazolinedione	261–263, nmr	1466,2261
7-Amino-1,3-dimethyl-6-nitro-2,4(1H,3H)-quinazolinedione	321–323, ir, nmr	1358
3-Amino-N',N'-dimethyl-4-oxo-3,4-dihydro-2-quinazolinecarbohydrazide	230	124
3-Amino-2,7-dimethyl-4-oxo-3,4-dihydro-6-quinazolinecarboxylic acid	—	H 484
4-Amino-5,7-dimethyl-8-quinazolinecarbonitrile	>300, ir, ms, nmr, uv	2466
6-Amino-1,3-dimethyl-2,4(1H,3H)-quinazolinedione	—	1513
7-Amino-1,3-dimethyl-2,4(1H,3H)-quinazolinedione	—	1513
2-Amino-1,6-dimethyl-4(1H)-quinazolinone	309–314, nmr, uv	2194
2-Amino-3,6-dimethyl-4(3H)-quinazolinone	267, ir, nmr, uv	2194
3-Amino-2,6-dimethyl-4(3H)-quinazolinone	170–171, ir, nmr	1451
6-Amino-2,3-dimethyl-4(3H)-quinazolinone	—	H 368
7-Amino-2,3-dimethyl-4(3H)-quinazolinone	—	H 370
7-Amino-2,6-dimethyl-4(3H)-quinazolinone	—	H 370
2-Amino-6,8-dinitro-4(3H)-quinazolinone	227–228, ir, uv	117
3-Amino-2,4-dioxo-1,2,3,4-tetrahydro-6-quinazolinesulfonamide	230–235	1603

ALPHABETICAL LIST OF SIMPLE QUINAZOLINES (*Continued*)

Quinazoline	m.p. (°C), etc.	Ref.
4-Amino-5,7-diphenyl-8-quinazolinecarbonitrile	260–261, ir, ms, nmr, uv	2466
5-Amino-2,7-diphenyl-4-thioxo-3,4-dihydro-6-quinazolinecarbonitrile	245, ir, nmr	2064
6-Amino-3-ethoxycarbonylamino-2-methyl-4(3*H*)-quinazolinone	—	*H* 380
3-Amino-2-β-ethoxycarbonylethyl-4(3*H*)-quinazolinone	119, ir, nmr	1659
3-Amino-2-ethoxycarbonylmethyl-4(3*H*)-quinazolinone	108–110, ir, nmr	2323
3-Amino-2-ethoxy-4(3*H*)-quinazolinone	99–100, ir, nmr	1473
4-β-Aminoethylamino-6-chloroquinazoline	—	*H* 249
4-β-Aminoethylaminoquinazoline	uv	324
3-β-Aminoethyl-6-bromo-2,4(1*H*,3*H*)-quinazolinedione	—	*H* 202
2-β-Aminoethyl-6-chloro-8-ethoxycarbonylmethyl-4-phenylquinazoline	78–79, ir, nmr	1396
2-β-Aminoethyl-6-chloro-4-phenylquinazoline	106–107, ms, nmr, uv	1362
5-β-Aminoethyl-8-hydroxy-2,4(1*H*,3*H*)-quinazolinedione	250, ms	2502
3-Amino-2-ethyl-8-iodo-4(3*H*)-quinazolinone	218	1675
2-Amino-3-ethyl-6-methoxy-4(3*H*)-quinazolinone	230, ir, nmr	2241
3-β-Aminoethyl-2-methyl-4(3*H*)-quinazolinone	170	1216
6-Amino-3-ethyl-2-methyl-4(3*H*)-quinazolinone	—	*H* 368
3-Amino-*N*-ethyl-4-oxo-3,4-dihydro-2-quinazolinecarboxamide	190–191, ir, nmr	2121,2173
3-β-Aminoethyl-2-phenyl-4(3*H*)-quinazolinone	239	1216
3-β-Aminoethyl-2,4(1*H*,3*H*)-quinazolinedione	HCl: 315	2409
3-Amino-1-ethyl-2,4(1*H*,3*H*)-quinazolinedione	153–154	1829
3-Amino-2-ethyl-4(3*H*)-quinazolinethione	—	*H* 381
4-Amino-3-ethyl-2(3*H*)-quinazolinethione	210–211, ir, nmr	2417
2-α-Aminoethyl-4(3*H*)-quinazolinone	192–194, ir, nmr	2069
2-Amino-3-ethyl-4(3*H*)-quinazolinone	150, ir, nmr	2143

Compound		
3-β-Aminoethyl-4(3H)-quinazolinone	—	H 144
3-Amino-2-ethyl-4(3H)-quinazolinone	90 to 153, ir, nmr; dnp: 260, ir	H 371; 449,1245,1657,1675, 1979,1994,2593
4-Amino-1-ethyl-2(1H)-quinazolinone	—	1925
3-Amino-6-fluoro-2-hydrazino-4(3H)-quinazolinone	233–235, ir	220
3-Amino-2-fluoromethyl-4(3H)-quinazolinone	220	1046
6-Amino-5-fluoro-2-methyl-4(3H)-quinazolinone	—	H 371
2-Amino-5-fluoro-6-nitro-4(3H)-quinazolinone	304–306, nmr	2374
2-Amino-5-fluoro-8-nitro-4(3H)-quinazolinone	>400, nmr	2344,2374
6-Amino-5-fluoro-2-pivalamido-4(3H)-quinazolinone	>400, nmr	2344,2374
8-Amino-5-fluoro-2-pivalamido-4(3H)-quinazolinone	205–207, nmr	2374
2-Amino-5-fluoro-4(3H)-quinazolinone	231–235, nmr	2374
3-Amino-2-guanidinocarbonyl-4(3H)-quinazolinone	344–350, nmr	2344,2374
2-Amino-3-guanidino-4(3H)-quinazolinone	250	124
3-Amino-2-β-hydrazinocarbonylethyl-8-methoxy-4(3H)-quinazolinone	—	H 364
3-Amino-2-β-hydrazinocarbonylethyl-4(3H)-quinazolinone	—	H 376
3-Amino-2-hydrazinocarbonylmethyl-4(3H)-quinazolinone	266, ir, nmr	1659
3-Amino-2-hydrazino-6-iodo-4(3H)-quinazolinone	306, nmr	2323
3-Amino-2-hydrazino-4(3H)-quinazolinone	241, ir	1365
3-Amino-2-α-hydroxyethyl-4(3H)-quinazolinone	218 or 219–221, ir, nmr	943,1473,1884,1960,2199
2-Amino-6-hydroxymethyl-8-quinazolinol	—	H 371; 2080
2-Amino-6-hydroxymethyl-4(3H)-quinazolinone	>360, nmr	H 342; 1899,2425
3-Amino-2-hydroxymethyl-4(3H)-quinazolinone	—	618,871
3-Amino-8-hydroxy-2-methyl-4(3H)-quinazolinone	—	H 371
2-Amino-8-hydroxy-2-pentyl-4(3H)-quinazolinone	—	H 376
4-Amino-6-hydroxy-2-phenyl-5,8-dihydro-5,8-quinazolinedione	—	H 376
8-Amino-5-hydroxy-4(3H)-quinazolinone	172–174, ir	592
3-Amino-6-iodo-2-methyl-4(3H)-quinazolinone	crude, nmr; HBr: >250, ir, nmr	2324
	—	2393

425

ALPHABETICAL LIST OF SIMPLE QUINAZOLINES (*Continued*)

Quinazoline	m.p. (°C), etc.	Ref.
3-Amino-8-iodo-2-methyl-4(3*H*)-quinazolinone	203	1675
3-Amino-5-iodo-2-phenyl-4(3*H*)-quinazolinone	—	*H* 376
5-Amino-1-isobutyl-3-isopentyl-6-nitro-2,4(1*H*,3*H*)-quinazolinedione	165–166	1581
7-Amino-1-isobutyl-3-isopentyl-6-nitro-2,4(1*H*,3*H*)-quinazolinedione	144–148, nmr	1581
5-Amino-1-isobutyl-3-methyl-6-nitro-2,4(1*H*,3*H*)-quinazolinedione	184–185, nmr	2261
7-Amino-1-isobutyl-3-methyl-6-nitro-2,4(1*H*,3*H*)-quinazolinedione	226–228, nmr	2261
3-Amino-*N*-isobutyl-4-oxo-3,4-dihydro-2-quinazolinecarboxamide	135–136, ir	2121
3-Amino-2-isobutyl-4(3*H*)-quinazolinone	ir, nmr	2549
3-Amino-*N*-isopropyl-4-oxo-3,4-dihydro-2-quinazolinecarboxamide	147–147, ir, nmr	2121
6-Amino-1-isopropyl-4-phenyl-2(1*H*)-quinazolinone	252	681
7-Amino-2-isopropyl-1-phenyl-4(1*H*)-quinazolinone	>280	1590
3-Amino-2-isopropyl-4(3*H*)-quinazolinethione	—	*H* 381
4-Amino-3-isopropyl-2[3*H*)-quinazolinethione	259–260, ir, nmr	2417
2-Amino-1-methylacryloyl-4(1*H*)-quinazolinone	220, ir, ms, nmr	1834
3-Amino-2-β-methoxycarbonylethyl-4(3*H*)-quinazolinone	179	1659
7-Amino-2-methoxycarbonylmethyl-4(3*H*)-quinazolinone	—	*H* 370
7-Amino-6-methoxy-5,8-dihydro-5,8-quinazolinedione	212, ir, nmr	1577
3-Amino-8-methoxy-2-methyl-4(3*H*)-quinazolinone	—	*H* 376
3-Amino-8-methoxy-2-phenyl-4(3*H*)-quinazolinone	—	*H* 376
3-Amino-8-methoxy-4(3*H*)-quinazolinone	—	*H* 374
6-Amino-7-methoxy-4(3*H*)-quinazolinone	289–291, ir, nmr	2017
2-Amino-5-methylamino-6-nitro-4(3*H*)-quinazolinone	330, ir, nmr	2015
6-Amino-7-methylamino-4(3*H*)-quinazolinone	289–290, ir, nmr	1962
8-Amino-7-methylamino-4(3*H*)-quinazolinone	245–252, ir, nmr	2324

Compound	Properties	Ref.
3-Aminomethyl-6-bromo-4(3H)-quinazolinone	HCl: 180–182	1192
2-Aminomethyl-6-chloro-4-phenylquinazoline 3-oxide	165–167 ir, nmr, uv	361
3-Aminomethyl-6-chloro-4(3H)-quinazolinone	HCl: 193–195	1192
3-Aminomethyl-6,8-dibromo-4(3H)-quinazolinone	HCl: >280	1192
3-Aminomethyl-6,8-dichloro-4(3H)-quinazolinone	HCl: 225–226	1192
3-Aminomethyl-2-methyl-4(3H)-quinazolinone	—	H 156
2-Amino-6-methyl-4-methylthioquinazoline	158, ir, nmr	2255
3-Amino-6-methyl-2-methylthio-4(3H)-quinazolinone	174–175, ir	1841
3-Aminomethyl-6-nitro-4(3H)-quinazolinone	HCl: 226–228	1192
3-Amino-2-methyl-5-nitro-4(3H)-quinazolinone	—	H 376
3-Amino-2-methyl-6-nitro-4(3H)-quinazolinone	278–279, ir, nmr	H 376; 1451
3-Amino-2-methyl-7-nitro-4(3H)-quinazolinone	—	H 376
7-Amino-3-methyl-6-nitro-4(3H)-quinazolinone	270	1165
2-Amino-5-methyl-4-oxo-3,4-dihydro-6-quinazolinecarbonitrile	>350	2284
3-Amino-N-methyl-4-oxo-3,4-dihydro-2-quinazolinecarboxamide	203 or 210–211, ir, nmr	2121,2173
6-Amino-N-methyl-4-phenyl-2-quinazolinecarboxamide	143–145	242
2-Aminomethyl-3-phenyl-4(3H)-quinazolinethione	100–101	505
2-Aminomethyl-3-phenyl-4(3H)-quinazolinone	164–165 or 165–169, nmr; HCl: 250	152,387,1806
6-Amino-2-methyl-3-phenyl-4(3H)-quinazolinone	232–234, uv	12
7-Amino-2-methyl-3-phenyl-4(3H)-quinazolinone	236–238, uv	12
4-Amino-6-methyl-8-quinazolinecarbonitrile	>300, ir, ms, nmr, uv	2466
4-Amino-7-methyl-8-quinazolinecarbonitrile	>300, ir, ms, nmr, uv	2466
6-Aminomethyl-2,4-quinazolinediamine	235–239 or 239–242	601,678
1-Amino-3-methyl-2,4(1H,3H)-quinazolinedione	178–179, ir, nmr	2574
3-Amino-1-methyl-2,4(1H,3H)-quinazolinedione	240–241 or 241–242, ir, nmr	H 382; 495
3-Amino-6-methyl-2,4(1H,3H)-quinazolinedione	264–266 or 266–268	1454,1829
2-Amino-6-methyl-4(3H)-quinazolinethione	261–264, nmr	2255

427

ALPHABETICAL LIST OF SIMPLE QUINAZOLINES (*Continued*)

Quinazoline	m.p. (°C), etc.	Ref.
3-Amino-2-methyl-4(3*H*)-quinazolinethione	140–141	882
4-Amino-3-methyl-2(3*H*)-quinazolinethione	289–291, ir, nmr, uv	925
2-Amino-4-methyl-6-quinazolinol	—	H 342
1/3-Amino-4-methyl2(3/1*H*)-quinazolinone	176–177, ir, ms, nmr, uv	432
2-Amino-1-methyl-4(1*H*)-quinazolinone	>280, ir, ms, nmr	1415
2-Amino-3-methyl-4(3*H*)-quinazolinone	241 or 273–275 (?); pic: 282	H 364; 50,1834
2-Amino-5-methyl-4(3*H*)-quinazolinone	297–300	678
2-Amino-6-methyl-4(3*H*)-quinazolinone	>360 or 439–440, nmr, uv	H 364; 678,871,2194,2250
3-Aminomethyl-4(3*H*)-quinazolinone	HCl: 210–211	H 144; 982,1192
3-Amino-2-methyl-4(3*H*)-quinazolinone	202 (?), 142 to 152, ir, ms, nmr, complexes; dnp: complexes	H 371; 200,391,506,706,878, 1127,1162,1220,1256,1260, 1262,1267,1451,1452,1457, 1675,1684,1703,1881,1901, 1979,1994,2128,2137,2159, 2168,2317,2354
4-Amino-1-methyl-2(1*H*)-quinazolinone	274–275, ir	1925,2115
5-Amino-2-methyl-4(3*H*)-quinazolinone	—	H 368
5-Amino-6-methyl-4(3*H*)-quinazolinone	212 or 260–261, nmr	1159,1366,1387
6-Amino-2-methyl-4(3*H*)-quinazolinone	297 to 315, ms	H 368; 1290,2158,2232,2271
6-Amino-3-methyl-4(3*H*)-quinazolinone	208–211	H 368; 498
6-Amino-7-methyl-4(3*H*)-quinazolinone	>300	1157
7-Amino-2-methyl-4(3*H*)-quinazolinone	316–318	H 370; 103
7-Amino-3-methyl-4(3*H*)-quinazolinone	—	H 370
7-Amino-6-methyl-4(3*H*)-quinazolinone	212, ms, nmr	1356
8-Amino-3-methyl-4(3*H*)-quinazolinone	—	H 370

Compound	Data	Ref.
6-Amino-2-methyl-4(1H)-quinazolinone 3-oxide	—	H 467
7-Amino-3-methyl-2-styryl-4(3H)-quinazolinone	—	H 370
3-Amino-6-methyl-2-thioxo-1,2-dihydro-4(3H)-quinazolinone	295, ir	1841
4-Amino-3-β-morpholinoethyl-2(3H)-quinazolinone	216, ir, nmr	1196
3-Amino-2-morpholinomethyl-4(3H)-quinazolinone	126–128	816
6-Amino-3-morpholino-4(3H)-quinazolinone	—	H 380
5-Amino-6-nitro-1,3-dipropyl-2,4(1H,3H)-quinazolinedione	133–135, nmr	2261
7-Amino-6-nitro-1,3-dipropyl-2,4(1H,3H)-quinazolinedione	155–157, nmr	2261
3-Amino-7-nitro-2-phenyl-4(3H)-quinazolinone	—	H 377
5-Amino-6-nitro-2,4(1H,3H)-quinazolinedione	>340, nmr	1466,2261
2-Amino-6-nitro-4(3H)-quinazolinone	crude, >350, nmr	1564
3-Amino-6-nitro-4(3H)-quinazolinone	210, nmr	H 374; 1453
6-Amino-5-nitro-4(3H)-quinazolinone	310–311, ms, nmr	502
7-Amino-6-nitro-4(3H)-quinazolinone	>300 or >320, ms, nmr	501,1165
7-Amino-8-nitro-4(3H)-quinazolinone	293–295, ms, nmr	502
2-Amino-6-nitro-5-trifluoromethyl-4(3H)-quinazolinone	>300, nmr	2211,2344
3-Amino-7-nitro-2-α,β,β-trimethylpropyl-4(3H)-quinazolinone	187–189, ir, ms, nmr	1954
2-Amino-4-oxo-3,4-dihydro-6-quinazolinecarbaldehyde	>360	201
3-Amino-4-oxo-3,4-dihydro-2-quinazolinecarbohydrazide	200–201, ir	H 481; 124,2121
2-Amino-4-oxo-3,4-dihydro-6-quinazolinecarbonitrile	360	201
3-Amino-4-oxo-3,4-dihydro-2-quinazolinecarbonitrile	209–210	869
3-Amino-4-oxo-3,4-dihydro-2-quinazolinecarboxamide	268–269, ir	124,2121
2-Amino-4-oxo-3,4-dihydro-6-quinazolinecarboxylic acid	>450	618
6-Amino-4-oxo-3,4-dihydro-7-quinazolinecarboxylic acid	>300	1157
3-Amino-4-oxo-N-propyl-3,4-dihydro-2-quinazolinecarboxamide	142–143, ir, nmr	2121
2-Amino-4-oxo-5-trifluoromethyl-3,4-dihydro-6-quinazolinecarbonitrile	>350, nmr	2211
3-Amino-2-pent-4'-enyl-4(3H)-quinazolinone	76–77, ir, nmr	1818
3-Amino-2-pentyl-4(3H)-quinazolinone	156	1657

ALPHABETICAL LIST OF SIMPLE QUINAZOLINES (*Continued*)

Quinazoline	m.p. (°C), etc.	Ref.
3-Amino-2-pent-3′-ynyl-4(3H)-quinazolinone	147–150, ir, nmr	1826
3-Amino-2-pent-4′-ynyl-4(3H)-quinazolinone	106–107, ir, nmr	1752,1826
4-Amino-1-phenethyl-2(1H)-quinazolinone	261–262, ir	1925
4-Amino-2-phenyl-8-piperidino-5,6-dihydro-5,6-quinazolinedione	183–185, ir	592
4-Amino-6-phenyl-8-quinazolinecarbonitrile	> 300, ir, ms, nmr, uv	2466
3-Amino-1-phenyl-2,4(1H,3H)-quinazolinedione	239–240	1829
3-Amino-2-phenyl-4(3H)-quinazolinethione	175–176, ir, uv	H 381; 882
4-Amino-3-phenyl-2(3H)-quinazolinethione	196–198, ir, nmr	H 310,367,2051,2099
1-Amino-4-phenyl-1-quinazolinium (mesitylenesulfonate)	166–167	403
1/3-Amino-4-phenyl-2(1/3H)-quinazolinone (structures not assigned)	100–102, ir; 150–153, ir	432
	—	H 364
2-Amino-3-phenyl-4(3H)-quinazolinone	152 (?); 172 to 185, ir, ms, nmr, complexes	H 371; 1220,1260,1262,1267, 1385,1452,1655,1684,1703, 1704,1981,2137,2159,2340
3-Amino-2-phenyl-4(3H)-quinazolinone		
4-Amino-1-phenyl-2(1H)-quinazolinone	—	2115
4-Amino-3-phenyl-2(3H)-quinazolinone	—	H 366; 1790
5-Amino-2-phenyl-4-thioxo-3,4-dihydro-6-quinazolinecarbonitrile	162, ir, nmr	2147
3-Amino-2-piperidinomethyl-4(3H)-quinazolinone	101–103	816
3-γ-Aminopropyl-4(3H)-quinazolinone	—	H 144
3-Amino-2-propyl-4(3H)-quinazolinone	161 or 195	1184,1657
2-Amino-4-quinazolinecarboxylic acid	—	H 482
3-Amino-2,4(1H,3H)-quinazolinedione	265 (?); 286 to 295	H 382; 227,495,1007,1434,1454, 1829,2004
5-Amino-2,4(1H,3H)-quinazolinedione	—	H 383

430

Compound	Data
6-Amino-2,4(1*H*,3*H*)-quinazolinedione	—	*H* 383
7-Amino-2,4(1*H*,3*H*)-quinazolinedione	—	*H* 383
8-Amino-2,4(1*H*,3*H*)-quinazolinedione	—	*H* 383
3-Amino-2,4(1*H*,3*H*)-quinazolinedithione	246–249	81
6-Amino-2,4(1*H*,3*H*)-quinazolinedithione	—	*H* 302
2-Amino-4(3*H*)-quinazolinethione	>260	678
3-Amino-4(3*H*)-quinazolinethione	175–176, ir, uv	882
4-Amino-2(1*H*)-quinazolinethione	295–297 or 306–308	*H* 310,367; 678,2095,2384,
		2417
2-Amino-6-quinazolinol		*H* 342
1-Amino-4(1*H*)-quinazolinone	3-MeCl: 226–228, ir;	1445
	base therefrom: 261	
2-Amino-4(3*H*)-quinazolinone	>250 to 315; >400 (?), ir, nmr,	*H* 364; 385,641,678,798,1347,
	pK_a, uv; H_2SO_4: 317–318	2186,2330
3-Amino-4(3*H*)-quinazolinone	204 to 215, ir, nmr; pic:	*H* 371; 276,614,862,1452,1453,
	211–212, ir	1497,1628,1657,1895,1994
4-Amino-2(1*H*)-quinazolinone	350 to 360, then 410–412,	*H* 366; 678,1060,1925,2015,2330
	ir, nmr, uv	
5-Amino-4(3*H*)-quinazolinone		*H* 368
6-Amino-4(3*H*)-quinazolinone	302–305, ir, nmr, uv	*H* 368; 321,384,498,1622
7-Amino-4(3*H*)-quinazolinone	—	*H* 370
8-Amino-4(3*H*)-quinazolinone	—	*H* 370
2-Amino-4(3*H*)-quinazolinone 1-oxide	—	*H* 466
4-Amino-2(1*H*)-quinazolinone 3-oxide	H_2O: 272, ir, Raman, st;	1857
	HCl: 262–263	
3-Amino-2-styryl-4(3*H*)-quinazolinone	257–260, nmr	*H* 371
6-Amino-2-styryl-4(3*H*)-quinazolinone	125 (?), 227 to 237, ir	2264
3-Amino-2-thioxo-1,2-dihydro-4(3*H*)-quinazolinone	>250, nmr	52,129,1124,1166,1736
2-Amino-5-trifluoromethyl-4(3*H*)-quinazolinone		2211,2344

ALPHABETICAL LIST OF SIMPLE QUINAZOLINES (*Continued*)

Quinazoline	m.p. (°C), etc.	Ref.
2-Amino-7-trifluoromethyl-4(3*H*)-quinazolinone	339–342, then 364–370, then 420–428!	678
3-Amino-6,7,8-trimethoxy-2-methyl-4(3*H*)-quinazolinone	—	H 376
3-Amino-6,7,8-trimethoxy-2-phenyl-4(3*H*)-quinazolinone	—	H 376
3-Amino-2-α,β,β-trimethylpropyl-4(3*H*)-quinazolinone	118–120, ir, ms, nmr	1954
3-Amino-2,6,7-trimethyl-4(3*H*)-quinazolinone	—	H 376
3-Anilino-2-benzoyl-4(3*H*)-quinazolinone	226	1421
2-Anilino-3-benzylideneamino-4(3*H*)-quinazolinone	153–155, ms	1894
2-Anilino-4-benzyloxyquinazoline	—	H 255
3-Anilino-6-bromo-2-ethyl-4(3*H*)-quinazolinone	145–146, ir, nmr	1994
3-Anilino-6-bromo-2-methyl-4(3*H*)-quinazolinone	nmr	1997
3-Anilino-8-bromo-2-methyl-4(3*H*)-quinazolinone	135, ir	2307
3-Anilino-5-bromo-7-nitro-2-phenyl-4(3*H*)-quinazolinone	—	H 377
3-Anilino-6-bromo-2-phenyl-4(3*H*)-quinazolinone	nmr	1997
3-Anilino-8-bromo-2-phenyl-4(3*H*)-quinazolinone	176–178, ir	2307
2-Anilino-4-butoxyquinazoline	—	H 255
3-Anilino-2-β-carboxyethyl-4(3*H*)-quinazolinone	—	H 272
3-Anilino-2-carboxymethylthio-4(3*H*)-quinazolinone	228, ir	1215
3-Anilino-2-β-carboxyvinyl-4(3*H*)-quinazolinone	—	H 272
3-Anilino-2-α-chloroethyl-4(3*H*)-quinazolinone	132	2341
3-Anilino-6-chloro-2-ethyl-4(3*H*)-quinazolinone	131, ir, nmr	1994
3-Anilino-5-chloro-2-methyl-4(3*H*)-quinazolinone	—	H 377
3-Anilino-6-chloro-2-methyl-4(3*H*)-quinazolinone	195–196, ir, nmr	1994
3-Anilino-7-chloro-2-methyl-4(3*H*)-quinazolinone	—	H 377

Compound	mp (°C), methods	Refs.
4-Anilino-6-chloro-3-methyl-2(3H)-quinazolinone	280–282, ir, uv	771
4-Anilino-6-chloro-2-morpholinoquinazoline	262–264 or 264–265, ir, uv	721,1328
3-Anilino-5-chloro-7-nitro-2-phenyl-4(3H)-quinazolinone	—	H 377
2-Anilino-6-chloro-4-phenylquinazoline	129 or 148–149, ir	399,899
2-Anilino-6-chloro-4-phenylquinazoline 3-oxide	190	399
4-Anilino-6-chloroquinazoline	229–230	916
3-Anilino-6,8-dibromo-2-methyl-4(3H)-quinazolinone	167–170, ir	2307
3-Anilino-6,8-dibromo-2-phenyl-4(3H)-quinazolinone	180–181, ir	2307
2-Anilino-6,7-dimethoxy-4(3H)-quinazolinone	260–265 or 267–270	641,1580
4-Anilino-2-dimethylaminoquinazoline	HCl: 288, ir, ms	2576
6-Anilino-2,4-dimethyl-5,8-dihydro-5,8-quinazolinedione	198, ir, nmr, uv	736
2-Anilino-4-ethoxyquinazoline	—	H 255
2-Anilino-2-ethyl-4-methyl-5,8-dihydro-5,8-quinazolinedione	179, ir, nmr, uv	736
2-Anilino-4-hydrazinoquinazoline	186–187	128,292
3-Anilino-2-hydrazino-4(3H)-quinazolinone	—	1878
2-Anilino-4-hydrazono-3,4-dihydro-3-quinazolinamine	186, ms	1894
3-Anilino-6-iodo-2-methyl-4(3H)-quinazolinone	—	H 377; 2393
3-Anilino-8-iodo-7-methyl-4(3H)-quinazolinone	160–161	2307
3-Anilino-5-iodo-7-nitro-2-phenyl-4(3H)-quinazolinone	—	H 377
3-Anilino-8-iodo-2-phenyl-4(3H)-quinazolinone	170–172, ir	2307
6-Anilino-4-methoxy-2-phenyl-5,8-dihydro-5,8-quinazolinedione	241–243, ir, uv	763
4-Anilino-6-methoxy-2-phenylquinazoline	155–156	766
2-Anilino-4-methoxyquinazoline	—	H 256
4-Anilino-2-methoxyquinazoline	—	H 353
4-Anilino-8-methyl-2-morpholinoquinazoline	158–160	721
3-Anilino-2-methyl-7-nitro-4(3H)-quinazolinone	—	H 377
6-Anilino-4-methyl-2-phenyl-5,8-dihydro-5,8-quinazolinedione	242, ir, nmr, uv	736
2-Anilino-6-methyl-4-quinazolinamine	—	H 361

433

ALPHABETICAL LIST OF SIMPLE QUINAZOLINES (Continued)

Quinazoline	m.p. (°C), etc.	Ref.
4-Anilino-6-methylquinazoline	—	H 351
2-Anilino-4-methylquinazoline 3-oxide	—	1053
4-Anilino-1-methyl-2(1H)-quinazolinethione	—	H 310,367
4-Anilino-3-methyl-2(3H)-quinazolinethione	250–251	88
2-Anilino-3-methyl-4(3H)-quinazolinone	208–210, ir, ms	H 364; 2093
3-Anilino-2-methyl-4(3H)-quinazolinone	165 (?) or 203, ir, complexes	H 372; 1260,1261,1684,1878, 2159,2296,2307
4-Anilino-1-methyl-2(1H)-quinazolinone	—	H 366
4-Anilino-3-methyl-2(3H)-quinazolinone	255–257	88
4-Anilino-8-methyl-2(1H)-quinazolinone	—	H 366
4-Anilino-2-methylthioquinazoline	—	H 310
3-Anilino-2-morpholinomethyl-4(3H)-quinazolinone	—	2012
4-Anilino-2-morpholinoquinazoline	185–186; HCl: 267, ir, ms	721,2576
3-Anilino-7-nitro-2-phenyl-4(3H)-quinazolinone	—	H 377
4-Anilino-6-nitroquinazoline	236–237	H 351; 507
4-Anilino-7-nitroquinazoline	—	H 351
2-Anilino-4-pentyloxyquinazoline	—	H 256
3-Anilino-2-phenylhydrazino-4(3H)-quinazolinone	H₂O: 232–233	943
3-Anilino-2-phenyl-4-phenylhydrazono-3,4-dihydroquinazoline	155	455
2-Anilino-3-phenyl-4-phenylimino-3,4-dihydroquinazoline	171, then 181–185	H 355; 71,439
4-Anilino-2-phenylquinazoline	151–152 or 249 (?)	346,728,1300
2-Anilino-3-phenyl-4(3H)-quinazolinethione	188–189	88
3-Anilino-2-phenyl-4(3H)-quinazolinethione	—	H 381
4-Anilino-3-phenyl-2(3H)-quinazolinethione	283–288	88

434

4-Anilino-2-phenyl-6-quinazolinol	245–247	397,766
2-Anilino-3-phenyl-4(3*H*)-quinazolinone	162–163 or 167–168, ir, nmr	*H* 364; 88,723,769,2386
3-Anilino-2-phenyl-4(3*H*)-quinazolinone	138 (?) or 158–159, ir, complexes	*H* 372; 345,1012,1260,1261, 1684,2159,2307
4-Anilino-3-phenyl-2(3*H*)-quinazolinone	293–294	88
3-Anilino-2-piperidinomethyl-4(3*H*)-quinazolinone	—	2012
4-Anilino-2-piperidinoquinazoline	HCl: 250, ir, ms	2576
2-Anilino-4-quinazolinamine	155–157, ir, nmr	2242
4-Anilino-2-quinazolinamine	261–263, ir, nmr	2418
4-Anilino-6-quinazolinamine	—	*H* 363
4-Anilino-7-quinazolinamine	—	*H* 363
2-Anilinoquinazoline	146	849
4-Anilinoquinazoline	215 to 223, ms	*H* 345; 333,507,555,916,1786
3-Anilino-2,4(1*H*,3*H*)-quinazolinedione	224–225	1007
3-Anilino-2,4(1*H*,3*H*)-quinazolinedione	215–217	81
4-Anilino-2(1*H*)-quinazolinethione	239–240, ir, nmr	*H* 310,367; 2384
2-Anilino-4(3*H*)-quinazolinone	234 (?) or 254 to 262, ir, ms, nmr, uv	*H* 364; 40,292,769,1372,1580, 1894,2242,2431,2487
3-Anilino-4(3*H*)-quinazolinone	166–167 or 170–171, ir, nmr; 1-MeI: 204, ir, nmr	*H* 374; 318,856,1878
4-Anilino-2(1*H*)-quinazolinone	—	*H* 366
3-Anilino-2-styryl-4(3*H*)-quinazolinone	—	*H* 372
3-Anilino-2-thioxo-1,2-dihydro-4(3*H*)-quinazolinone	248–250	1166
4-Anilino-2-trichloromethylquinazoline	—	*H* 348
Arborine, *see* 2-Benzyl-1-methyl-4(1*H*)-quinazolinone		
4-Azido-6-bromo-2-morpholinoquinazoline	126–130, ir	2266
2-Azido-6-bromo-4-phenylquinazoline	211–212	836
3-Azidocarbonylmethyl-2-methyl-4(3*H*)-quinazolinone	110–115	277

ALPHABETICAL LIST OF SIMPLE QUINAZOLINES (*Continued*)

Quinazoline	m.p. (°C), etc.	Ref.
4-Azido-6-chloro-2-diethylaminoquinazoline	116–119, ir	2266
4-Azido-6-chloro-2-morpholinoquinazoline	112–114, ir	2266
2-Azido-6-chloro-4-phenylquinazoline	210	836,1021
4-Azido-6-chloro-2-piperidinoquinazoline	168–172, ir	2266
2-Azido-6-chloro-4(3*H*)-quinazolinone	—	2022
2-α-Azidoethyl-4(3*H*)-quinazolinone	156, ir, nmr	2069
7-Azido-6-methoxy-5,8-quinazolinediol	crude, 208, ir, nmr	1577
2-Azidomethyl-6-chloro-4-phenylquinazoline 3-oxide	152	82
4-Azido-5-methyl-2-morpholinoquinazoline	96–98, ir	2266
4-Azido-6-methyl-2-morpholinoquinazoline	169–170, ir	2266
2-Azido-6-methyl-4-phenylquinazoline	234–236	836
2-Azidomethyl-3-phenyl-4(3*H*)-quinazolinone	98–100, nmr	1806
2-Azido-6-methyl-4(3*H*)-quinazolinone	269–270, ir	1841
4-Azido-2-morpholinoquinazoline	150–152, ir	2266
2-Azido-4-phenylquinazoline	205–207	836
4-Azido-2-phenylquinazoline	162–163, ir, nmr	1080,1131,1683,1761
2-Azido-3-phenyl-4(3*H*)-quinazolinone	192–193, ms	1894
6-Azidoquinazoline	117, ir, nmr	2075
6-Azido-2,4-quinazolinediamine	135; HCl: 135	1919
3-Benzamido-6-chloro-4(3*H*)-quinazolinimine	245	2345
3-Benzamido-6,8-dibromo-2-methyl-4(3*H*)-quinazolinone	—	2349
4-Benzamido-2-diethylaminoquinazoline	186–188, ir, nmr	2418
2-Benzamido-3-ethyl-4(3*H*)-quinazolinimine	213–214, ir, nmr	2418
3-Benzamido-6-iodo-2-methyl-4(3*H*)-quinazolinone	—	2393

Compound		
2-Benzamidomethyl-3-phenyl-4(3*H*)-quinazolinone	209	152
4-Benzamido-2-methylquinazoline	152–154, ir, nmr	1232
2-Benzamido-6-methyl-4(3*H*)-quinazolinone	—	H 365
3-Benzamido-2-methyl-4(3*H*)-quinazolinone	184–185 or 185–186	H 372; 109,1981
4-Benzamido-2-phenylquinazoline	167–169, ir, nmr	1232
3-Benzamido-1-phenyl-2,4(1*H*,3*H*)-quinazolinedithione	305	1725
2-Benzamido-3-phenyl-4(3*H*)-quinazolinimine	181–184, ir, nmr	2418
3-Benzamido-2-phenyl-4(3*H*)-quinazolinone	—	H 371,372
6-Benzamido-2,4-quinazolinediamine	—	1529
3-Benzamido-2,4(1*H*,3*H*)-quinazolinedione	281–282, ir, nmr	1457
4-Benzamido-2(1*H*)-quinazolinethione	258–260, ir, nmr	2095
3-Benzamido-4(3*H*)-quinazolinimine	209–211, nmr	2375
2-Benzamido-4(3*H*)-quinazolinone	185–186	H 374; 276
5-Benzamido-4(3*H*)-quinazolinone	—	H 368
3-Benzamido-2-styryl-4(3*H*)-quinazolinone	—	H 372
3-Benzamido-2-thioxo-1,2-dihydro-4(3*H*)-quinazolinone	264–265, ir, nmr	1457
2-Benzenesulfonamido-7-chloro-4(3*H*)-quinazolinone	—	H 365
2-Benzenesulfonamido-3-cyclohexyl-4(3*H*)-quinazolinone	176–177, ir, nmr	2571
3-Benzenesulfonamido-6,8-dibromo-2-methyl-4(3*H*)-quinazolinone	153	1662
2-Benzenesulfonamido-3-hexyl-4(3*H*)-quinazolinone	117–118, ir, nmr	2571
4-Benzenesulfonamido-2-methoxyquinazoline	—	H 253
2-Benzenesulfonamido-2-methyl-4(3*H*)-quinazolinone	162	1662
3-Benzenesulfonamido-2,4(1*H*,3*H*)-quinazolinedione	—	H 382
3-Benzenesulfonyl-2-carboxymethylthio-4(3*H*)-quinazolinone	195, then 225	138
3-Benzenesulfonyl-5-chloro-2,4(1*H*,3*H*)-quinazolinedione	—	1988
3-Benzenesulfonyloxy-6-bromo-2-methyl-4(3*H*)-quinazolinone	162–163	1313
3-Benzenesulfonyloxy-1-methyl-2,4(1*H*,3*H*)-quinazolinedione	187–189, ir, ms, nmr	457
3-Benzenesulfonyloxy-2,4(1*H*,3*H*)-quinazolinedione	234–235	H 470; 457

ALPHABETICAL LIST OF SIMPLE QUINAZOLINES (*Continued*)

Quinazoline	m.p. (°C), etc.	Ref.
3-Benzenesulfonyl-2,4(1*H*,3*H*)-quinazolinedione	—	1988
3-Benzenesulfonyl-4(3*H*)-quinazolinone (?)	195–197	276
3-Benzenesulfonyl-2-thioxo-1,2-dihydro-4(3*H*)-quinazolinone	327, ir	138
1-Benzoyl-6-bromo-3-carboxymethyl-2,4(1*H*,3*H*)-quinazolinedione	234–236	2277
3-Benzoyl-6-bromo-1-ethyl-2,4(1*H*.3*H*)-quinazolinedione	—	1490,1539
3-Benzoyl-6-chloro-2-ethyl-4(3*H*)-quinazolinone	oxime: 226–228	958
3-Benzoyl-6-chloro-2-methyl-4(3*H*)-quinazolinone	oxime: 218–219	958
2-Benzoyl-6-chloro-4(3*H*)-quinazolinone	225, uv	134
2-Benzoyl-7-chloro-4(3*H*)-quinazolinone	197, uv	134
3-Benzoyl-2-diethylamino-4(3*H*)-quinazolinimine	178–180, ir, nmr	2418
2-Benzoyl-4-ethoxyquinazoline	112, ir	181
3-Benzoyl-1-ethyl-2,4(1*H*,3*H*)-quinazolinedione	—	1493
3-Benzoyl-2-ethyl-4(3*H*)-quinazolinone	oxime: 236–237	958
3-Benzoyl-1-ethyl-2-thioxo-1,2-dihydro-4(3*H*)-quinazolinone	xl st	2455
4-*N*′-Benzoylhydrazinoquinazoline	—	*H* 383
2-Benzoyl-3-isopropylquinazoline	—	*H* 486
2-Benzoyl-4-methoxyquinazoline	128, ir, uv	*H* 486; 181
3-Benzoyl-6-methoxy-2-thioxo-1,2-dihydro-4(3*H*)-quinazolinone	183–184	280
1-Benzoyl-6-methyl-4-phenyl-2(1*H*)-quinazolinone	—	*H* 136
4-Benzoyl-2-methylquinazoline	142–143, ir, nmr	2329
2-Benzoyl-3-methyl-4(3*H*)-quinazolinone	137 or 154, ir, uv	181,918
3-Benzoyl-2-methyl-4(3*H*)-quinazolinone	oxime: 245–247	958
7-Benzoyl-2-methyl-4(3*H*)-quinazolinone	227–228, ir, nmr	1712
1-Benzoyl-3-methyl-2-thioxo-1,2-dihydro-4(3*H*)-quinazolinone	95, ir	1750

Compound		
3-Benzoyl-1-methyl-2-thioxo-1,2-dihydro-4(3*H*)-quinazolinone	229–230	2439
2-Benzoyl-6-nitro-4(3*H*)-quinazolinone	230–233, uv	97
2-Benzoyl-7-nitro-4(3*H*)-quinazolinone	235–237, uv	97
3-Benzoyl-6-nitro-4(3*H*)-quinazolinone	—	1525
3-Benzoyloxy-2-chloromethyl-4(3*H*)-quinazolinone	134–135	236
6-Benzoyloxy-7-methoxy-2,3-diphenyl-4(3*H*)-quinazolinone	175–176	140
4-Benzoyloxy-6-methyl-2-quinazolinamine	—	*H* 342
3-Benzoyloxy-1-methyl-2,4(1*H*,3*H*)-quinazolinedione	198, ir	236
3-Benzoyloxy-2-methyl-4(3*H*)-quinazolinone	160	236
3-Benzoyloxy-2-phenyl-4(3*H*)-quinazolinone	152	236
3-Benzoyloxy-2,4(1*H*,3*H*)-quinazolinedione	275, ir	*H* 470; 236
3-Benzoyloxy-4(3*H*)-quinazolinone	158	236
4-Benzoyl-2-phenylquinazoline	151–152, ir, nmr	2329
2-Benzoyl-3-phenyl-4(3*H*)-quinazolinone	190, ir, nmr	719
2-Benzoyl-4-quinazolinamine	203, ir	426
2-Benzoylquinazoline	100–102, ir, nmr	2111
4-Benzoylquinazoline	96 to 99, ir, nmr	*H* 61; 373,790,1419,1885,2111, 2193,2329
1-Benzoyl-2,4(1*H*,3*H*)-quinazolinedione	—	*H* 199
3-Benzoyl-2,4(1*H*,3*H*)-quinazolinedione	—	*H* 199
2-Benzoyl-4(3*H*)-quinazolinone	175 to 184, ir, ms, nmr	122,181,182,481,888,918,2069
3-Benzoyl-4(3*H*)-quinazolinone (?)	130–132, ir	276,532
2-Benzoyl-4(1*H*)-quinazolinone 3-oxide	76–78, ir	101
2-Benzoylthio-3-methyl-4(3*H*)-quinazolinone	158–160 or 248, ir	1750,1836
2-Benzoylthio-4(3*H*)-quinazolinone	210–212, ir, ms	1836
3-Benzoyl-2-thioxo-1,2-dihydro-4(3*H*)-quinazolinone	139–140, ir, ms	*H* 306; 1836
2-Benzylamino-6-chloro-4-phenylquinazoline 3-oxide	156	399
3-Benzylamino-6-chloro-4(3*H*)-quinazolinone	166–167	1471

439

ALPHABETICAL LIST OF SIMPLE QUINAZOLINES (Continued)

Quinazoline	m.p. (°C), etc.	Ref.
2-Benzylamino-6,7-dimethoxy-4(3H)-quinazolinone	245–247	641,1580
6-Benzylamino-5-fluoro-2,4-quinazolinediamine	—	2541
5-Benzylamino-1-isobutyl-3-methyl-6-nitro-2,4(1H,3H)-quinazolinedione	141–142, nmr	1581
7-Benzylamino-1-isobutyl-3-methyl-6-nitro-2,4(1H,3H)-quinazolinedione	158–160, nmr	1581
7-Benzylamino-3-methyl-6-nitro-4(3H)-quinazolinone	198–200 or 205–206, nmr	900,1165
7-Benzylamino-6-nitro-4(3H)-quinazolinone	172–173, ms, nmr	501,1165
4-Benzylamino-2-phenylquinazoline	118	346
2-Benzylamino-3-phenyl-4(3H)-quinazolinone	—	2386
4-Benzylaminoquinazoline	169, ms	1786
6-Benzylamino-2,4-quinazolinediamine	224–225	2326
4-Benzylamino-2(1H)-quinazolinethione	215–217, ir, nmr	2384
4-Benzylamino-8-quinazolinol	—	H 351
2-Benzylamino-4(3H)-quinazolinone	212 to 217, ir, nmr	938,1473,1580
4-Benzylamino-6,7,8-trimethoxyquinazoline	180–181	2451
3-Benzyl-7-benzylamino-6-nitro-4(3H)-quinazolinone	187–188, ms, nmr	501
3-Benzyl-2-benzylthio-6-chloro-4(3H)-quinazolinone	113	670
3-Benzyl-2-benzylthio-6,8-dibromo-4(3H)-quinazolinone	145	603
3-Benzyl-6-bromo-1-ethyl-2,4(1H,3H)-quinazolinedione	—	1490,1539
3-Benzyl-6-bromo-2-ethylthio-4(3H)-quinazolinone	98	2298
3-Benzyl-6-bromo-2-methylthio-4(3H)-quinazolinone	126 to 138	135,2298
3-Benzyl-6-bromo-2-propylthio-4(3H)-quinazolinone	81	2298
3-Benzyl-6-bromo-2,4(1H,3H)-quinazolinedione	264	98
3-Benzyl-6-bromo-4(3H)-quinazolinone	—	H 177
3-Benzyl-6-bromo-2-thioxo-1,2-dihydro-4(3H)-quinazolinone	231 or 233	135,2298

Compound		
3-Benzyl-1-but-2′-enyl-2,4(1*H*,3*H*)-quinazolinedione	90, ir, nmr	718
3-Benzyl-2-butylthio-6-chloro-4(3*H*)-quinazolinone	71	670
3-Benzyl-2-β-carboxyethyl-4(3*H*)-quinazolinone	207	1184
1-Benzyl-3-carboxymethyl-2,4(1*H*,3*H*)-quinazolinedione	172–175	2262
3-Benzyl-1-carboxymethyl-2,4(1*H*,3*H*)-quinazolinedione	222–224	2262
3-Benzyl-1-carboxymethyl-2-thioxo-1,2-dihydro-4(3*H*)-quinazolinone	234–237	2262
3-Benzyl-1-carboxymethyl-4-thioxo-3,4-dihydro-2(1*H*)-quinazolinone	259–262	2262
1-Benzyl-6-chloro-4-cyclopropyl-2(1*H*)-quinazolinone	anal, nmr	2585
3-Benzyl-2-α-chloroethyl-4(3*H*)-quinazolinone	210, nmr	2341
3-Benzyl-5-chloro-2-ethyl-4(3*H*)-quinazolinone	172	2299
2-Benzyl-6-chloro-1-methyl-8-nitro-4(1*H*)-quinazolinone	292–294	1190
2-Benzyl-7-chloro-1-methyl-6-nitro-4(1*H*)-quinazolinone	312–314, ir, uv	1190
2-Benzyl-6-chloro-1-methyl-4(1*H*)-quinazolinone	174–175, ir, uv	1190
2-Benzyl-6-chloro-3-methyl-4(3*H*)-quinazolinone	104, uv	134
2-Benzyl-7-chloro-1-methyl-4(1*H*)-quinazolinone	154–155, ir, uv	1190
2-Benzyl-7-chloro-3-methyl-4(3*H*)-quinazolinone	95–96, uv	134
3-Benzyl-6-chloro-2-methyl-4(3*H*)-quinazolinone	131–132	942
3-Benzyl-7-chloro-2-methyl-4(3*H*)-quinazolinone	115–116	942
3-Benzyl-6-chloro-2-methylthio-4(3*H*)-quinazolinone	97	670
3-Benzyl-7-chloro-6-nitro-4(3*H*)-quinazolinone	175–177, ms, nmr	501
2-Benzyl-6-chloro-4-phenylquinazoline	145–146	237
4-Benzyl-6-chloro-2-phenylquinazoline	—	*H* 249
1-Benzyl-6-chloro-4-phenyl-2(1*H*)-quinazolinone	179–180	1060
3-Benzyl-6-chloro-2-phenyl-4(3*H*)-quinazolinone	116–118	942
3-Benzyl-6-chloro-4-phenyl-2(3*H*)-quinazolinone	190, uv	552
3-Benzyl-7-chloro-2-phenyl-4(3*H*)-quinazolinone	91–93	942
3-Benzyl-6-chloro-2-propylthio-4(3*H*)-quinazolinone	237	670
2-Benzyl-4-chloroquinazoline	—	*H* 247

ALPHABETICAL LIST OF SIMPLE QUINAZOLINES   (*Continued*)

Quinazoline	m.p. (°C), etc.	Ref.
2-Benzyl-6-chloro-4(3H)-quinazolinone	255 or 256–260, uv	134,2264
2-Benzyl-7-chloro-4(3H)-quinazolinone	260–261 or 263–264, nmr, uv	134,1174
3-Benzyl-6-chloro-4(3H)-quinazolinone	—	H 177
3-Benzyl-6-chloro-2-thioxo-1,2-dihydro-4(3H)-quinazolinone	265	670
3-Benzyl-2-cyanoamino-4(3H)-quinazolinone	>260, ir, nmr	2000
3-Benzyl-6,7-dibenzyloxy-2-methyl-4(3H)-quinazolinone	137–138	218
3-Benzyl-6,8-dibromo-2-butylthio-4(3H)-quinazolinone	265	644
3-Benzyl-6,8-dibromo-2-t-butylthio-4(3H)-quinazolinone	263	840
3-Benzyl-6,8-dibromo-2-carboxymethylthio-4(3H)-quinazolinone	237	646
3-Benzyl-6,8-dibromo-2-cyclohexylthio-4(3H)-quinazolinone	245	840
3-Benzyl-6,8-dibromo-2-ethylthio-4(3H)-quinazolinone	141	603
3-Benzyl-6,8-dibromo-2-isopentylthio-4(3H)-quinazolinone	112	840
3-Benzyl-6,8-dibromo-2-isopropylthio-4(3H)-quinazolinone	275	644
3-Benzyl-6,8-dibromo-2-methyl-4(3H)-quinazolinone	134–135	1371,2098,2184
3-Benzyl-6,8-dibromo-2-methylthio-4(3H)-quinazolinone	165	603
3-Benzyl-6,8-dibromo-2-pentylthio-4(3H)-quinazolinone	262	840
3-Benzyl-6,8-dibromo-2-propylthio-4(3H)-quinazolinone	259	840
3-Benzyl-6,8-dibromo-2,4(1H,3H)-quinazolinedione	225 or 254–255	603,840,1365
3-Benzyl-6,8-dibromo-2-thioxo-1,2-dihydro-4(3H)-quinazolinone	228	644,1365
4-Benzyl-6,8-dichloro-2-phenylquinazoline	—	H 249
2-Benzyl-3-β-diethylaminoethyl-4(3H)-quinazolinone	—	H 168
2-Benzyl-3-N,N-diethylcarbamoylmethyl-4(3H)-quinazolinone	—	H 168
3-Benzyl-6,7-dihydroxy-2-methyl-4(3H)-quinazolinone	263–266	218
3-Benzyl-6,8-diiodo-2-pentylthio-4(3H)-quinazolinone	192	131

442

Compound		
3-Benzyl-6,8-diiodo-2-thioxo-1,2-dihydro-4(3H)-quinazolinone	180	131
4-Benzyl-6,7-dimethoxyquinazoline	133–134	2170
2-Benzyl-6,7-dimethoxy-4(3H)-quinazolinone	—	H 174
4-Benzyl-6,7-dimethoxy-2(1H)-quinazolinone	212–215	79,1592
2-Benzyl-1-β-dimethylaminoethyl-4(1H)-quinazolinone	—	H 137
2-Benzyl-4-dimethylaminoquinazoline	95–97, nmr	2506
1-Benzyl-3-dimethylamino-2,4(1H,3H)-quinazolinedione	130–132	1603
2-Benzyl-3,6-dimethyl-4(3H)-quinazolinone	—	H 188
2-Benzyl-7,8-dimethyl-4(3H)-quinazolinone	—	H 174
8-Benzyl-2,4-diphenylquinazoline	152	204
2-Benzyl-4-ethoxyquinazoline	—	H 256
3-Benzyl-2-ethoxy-4(3H)-quinazolinone	96–97, ir, nmr	1473
2-Benzyl-3-ethyl-4(3H)-quinazolinethione	—	H 297
3-Benzyl-2-ethyl-4(3H)-quinazolinone	—	H 164
2-Benzyl-4-ethylthioquinazoline	—	H 299
3-Benzyl-2-ethylthio-4(3H)-quinazolinone	—	H 305
3-Benzyl-2-α-hydroxybenzyl-4(3H)-quinazolinone	50	842
3-Benzyl-8-hydroxy-2-methyl-4(3H)-quinazolinone	—	H 198
3-Benzyl-1-hydroxy-2,4(1H,3H)-quinazolinedione	233–234 or 237	187,622
3-Benzyl-8-hydroxy-4(3H)-quinazolinone	—	H 177
3-Benzyl-1-hydroxy-2-thioxo-1,2-dihydro-4(3H)-quinazolinone	174–176	622
3-Benzylideneamino-6,8-dibromo-2-methyl-4(3H)-quinazolinone	163–165	2354
3-Benzylideneamino-6,8-dibromo-2-phenyl-4(3H)-quinazolinone	—	2370
3-Benzylideneamino-2,7-dimethyl-4-oxo-3,4-dihydro-6-quinazolinecarboxylicacid	—	H 484
3-Benzylideneamino-1-ethyl-2,4(1H,3H)-quinazolinedione	115–116	1829
3-Benzylideneamino-2-ethyl-4(3H)-quinazolinone	98	1653,1656
6-Benzylideneamino-5-fluoro-2,4-quinazolinediamine	—	2541
3-Benzylideneamino-2-methyl-6-nitro-4(3H)-quinazolinone	—	H 377

ALPHABETICAL LIST OF SIMPLE QUINAZOLINES (Continued)

Quinazoline	m.p. (°C), etc.	Ref.
3-Benzylideneamino-1-methyl-2,4(1H,3H)-quinazolinedione	167–168	H 382; 495
3-Benzylideneamino-2-methyl-4(3H)-quinazolinone	95 (?), 151–152 (?), 180 to 188, ir, ms	H 372; 1197,1259,1653,1656, 2301,2354
3-Benzylideneamino-2-phenyl-4(3H)-quinazolinone	143	1197
6-Benzylideneamino-2,4-quinazolinediamine	224–225	2326
3-Benzylideneamino-2,4(1H,3H)-quinazolinedione	243–245	H 382; 1434
3-Benzylideneamino-4(3H)-quinazolinone	128–129, ir, nmr	H 374; 863
3-Benzylideneamino-2-styryl-4(3H)-quinazolinone	—	H 372
4-Benzylidenehydrazino-6-chloro-2-morpholinoquinazoline	181–183, ir, nmr	2526
4-Benzylidenehydrazino-2-chloroquinazoline	172–174, nmr	2228
2-Benzylidenehydrazino-3-methyl-4(3H)-quinazolinone	158, nmr	2059
4-Benzylidenehydrazino-2-phenylquinazoline	—	2050
2-Benzylidenehydrazino-3-phenyl-4(3H)-quinazolinone	218, ms	1894
2-Benzylidenehydrazinoquinazoline	173, ir	799
4-Benzylidenehydrazinoquinazoline	171–172	H 383; 555
4-Benzylimino-2-chloroethyl-3,4-dihydro-3-quinazolinamine	230	2341
3-Benzyl-6-iodo-2-isopentylthio-4(3H)-quinazolinone	97	522
3-Benzyl-6-iodo-2-isopropylthio-4(3H)-quinazolinone	233	522
3-Benzyl-6-iodo-2-methyl-4(3H)-quinazolinone	—	H 180
3-Benzyl-6-iodo-2-methylthio-4(3H)-quinazolinone	140	135
3-Benzyl-6-iodo-2-propylthio-4(3H)-quinazolinone	103	522
3-Benzyl-6-iodo-2-thioxo-1,2-dihydro-4(3H)-quinazolinone	238	135
3-Benzyl-2-isopentylthio-4(3H)-quinazolinone	89	160
3-Benzyl-2-isopropylthio-4(3H)-quinazolinone	125	160

444

3-Benzyl-2-β-methoxyethyl-4(3H)-quinazolinone	96–97, nmr	584
2-Benzyl-6-methoxy-3-methyl-4(3H)-quinazolinone	—	H 188
2-Benzyl-8-methoxy-3-methyl-4(3H)-quinazolinone	—	H 188
3-Benzyl-8-methoxy-2-methyl-4(3H)-quinazolinone	—	H 180
4-Benzyl-6-methoxy-2-phenylquinazoline	114, nmr	2088
2-Benzyl-4-methoxyquinazoline (glycophymoline ?)	165 (?)	H 256; 1227
2-Benzyl-8-methoxyquinazoline		H 260
2-Benzyl-6-methoxy-4(3H)-quinazolinone		H 174
2-Benzyl-8-methoxy-4(3H)-quinazolinone		H 174
3-Benzyl-8-methoxy-4(3H)-quinazolinone		H 177
4-Benzyl-7-methoxy-2(1H)-quinazolinone	204–207	79,447
2-Benzyl-1-methyl-6-nitro-4(1H)-quinazolinone	288–289, ir, uv	1190
2-Benzyl-1-methyl-7-nitro-4(1H)-quinazolinone	282–283, ir, uv	1190
2-Benzyl-3-methyl-6-nitro-4(3H)-quinazolinone	180–182, uv	97
2-Benzyl-3-methyl-7-nitro-4(3H)-quinazolinone	155–158, uv	97
3-Benzyl-2-methyl-7-nitro-4(3H)-quinazolinone		H 180
4-Benzyl-6-methyl-2-phenylquinazoline	127, nmr	2088
3-Benzyl-1-(2′-methylprop-2′-enyl)-2,4(1H,3H)-quinazolinedione	165, ir, nmr	718
2-Benzyl-4-methylquinazoline	80–81	H 63; 1946
2-Benzyl-3-methyl-4(3H)-quinazolinethione		H 297
3-Benzyl-2-methyl-4(3H)-quinazolinethione		H 297
3-Benzyl-4-methyl-2(3H)-quinazolinethione		2333
1-Benzyl-2-methyl-4(1H)-quinazolinone	186, nmr; oxime: 183–185, ir, nmr	2132
2-Benzyl-1-methyl-4(1H)-quinazolinone (arborine)	154 to 161, ir, ms, nmr, uv; pic: 173–174	35,118,705,754,821,908,961, 1190,1306,1367,1807
2-Benzyl-3-methyl-4(3H)-quinazolinone	88–90, ir, nmr, uv; HCl: 178 or 224; pic: 179	H 168; 35,181,961,990

ALPHABETICAL LIST OF SIMPLE QUINAZOLINES (*Continued*)

Quinazoline	m.p. (°C), etc.	Ref.
2-Benzyl-6-methyl-4(3*H*)-quinazolinone	239–240, nmr	H 174; 1173
2-Benzyl-7-methyl-4(3*H*)-quinazolinone	—	H 174
2-Benzyl-8-methyl-4(3*H*)-quinazolinone	—	H 174
3-Benzyl-2-methyl-4(3*H*)-quinazolinone	228/1.5, ir, ms; HCl: 230–232	H 156; 127,1059,2120,2406
3-Benzyl-4-methyl-2(3*H*)-quinazolinone	211–213, ir, nmr	1400
4-Benzyl-2-methylthiomethylquinazoline 3-oxide	106–107	874
2-Benzyl-4-methylthioquinazoline	—	H 299
3-Benzyl-2-methylthio-4(3*H*)-quinazolinethione	144	40
3-Benzyl-2-methylthio-4(3*H*)-quinazolinone	90	H 305; 135
3-Benzyl-1-methyl-2-thioxo-1,2-dihydro-4(3*H*)-quinazolinone	134, ms	1649
2-Benzyl-6-nitro-3-phenyl-4(3*H*)-quinazolinone	231–233, uv	97
2-Benzyl-7-nitro-3-phenyl-4(3*H*)-quinazolinone	171–173, uv	97
2-Benzyl-6-nitro-4(3*H*)-quinazolinone	258–260, uv	97
2-Benzyl-7-nitro-4(3*H*)-quinazolinone	262–264	H 174; 97
3-Benzyl-4-oxo-3,4-dihydro-2-quinazolinecarbaldehyde	—	H 486
3-Benzyl-6-bromo-2-methyl-4(3*H*)-quinazolinone	174–175	1313
7-Benzyloxy-2-butyl-8-methoxy-3-phenyl-4(3*H*)-quinazolinone	205–206	142
3-Benzyloxy-7-chloro-2-ethyl-4(3*H*)-quinazolinone	—	H 469
6-Benzyloxy-2-chloro-7-methoxy-4-quinazolinamine	225–231	785
7-Benzyloxy-2-chloro-6-methoxy-4-quinazolinamine	199–205	785
5-Benzyloxy-4-chloro-6-methoxyquinazoline	137, nmr	1728
3-Benzyloxy-2-chloromethyl-4(3*H*)-quinazolinone	114	H 469; 441
4-Benzyloxy-2-chloroquinazoline	—	H 263
6-Benzyloxy-2,4-dichloro-7-methoxyquinazoline	180	785

7-Benzyloxy-2,4-dichloro-6-methoxyquinazoline	162–163	785
3-Benzyloxy-6,8-dichloro-2-methyl-4(3H)-quinazolinone	—	H 469
5-Benzyloxy-4-β-diethylaminoethylamino-6-methoxyquinazoline	liquid, anal	2278
3-Benzyloxy-2-diformylmethyl-4(3H)-quinazolinone	149–150	1691
5-Benzyloxy-4-β-dimethylaminoethylamino-6-methoxyquinazoline	liquid, anal	2278
5-Benzyloxy-4-(N-β-dimethylaminoethyl-N-methylamino)-6-methoxyquinazoline	liquid, anal	2278
5-Benzyloxy-4-γ-dimethylaminopropylamino-6-methoxyquinazoline	liquid, ir, nmr	2278
3-Benzyloxy-2-ethyl-4(3H)-quinazoline	—	H 469
6-Benzyloxy-2-guanidino-4-methylquinazoline	—	H 343
7-Benzyloxy-8-methoxy-2,3-diphenyl-4(3H)-quinazolinone	193	149
7-Benzyloxy-8-methoxy-2-methyl-3-phenyl-4(3H)-quinazolinone	157–158	142
6-Benzyloxy-7-methoxy-2,4(1H,3H)-quinazolinedione	254–256	785
7-Benzyloxy-6-methoxy-2,4(1H,3H)-quinazolinedione	285–286	785
5-Benzyloxy-6-methoxy-4(3H)-quinazolinone	200, ir, nmr	1728
3-Benzyloxy-2-methyl-6,8-dinitro-4(3H)-quinazolinone	—	H 469
3-Benzyloxy-2-methyl-6-nitro-4(3H)-quinazolinone	—	H 469
3-Benzyloxy-2-methyl-7-nitro-4(3H)-quinazolinone	—	H 469
4-Benzyloxy-2-methylquinazoline	—	H 256
3-Benzyloxy-2-methyl-4(3H)-quinazolinone	—	H 469
3-Benzyloxy-7-nitro-2-phenyl-4(3H)-quinazolinone	—	H 469
3-Benzyloxy-6-nitro-4(3H)-quinazolinone	—	H 469
3-Benzyloxy-6-nitro-4(3H)-quinazolinone 1-oxide	—	H 466
3-Benzyloxy-2-phenyl-4(3H)-quinazolinone	123–125, nmr	851,1343
4-Benzyloxyquinazoline	liquid; pic.164–165, nmr	H 254; 2329
4-Benzyloxy-2-quinazolinecarbonitrile	—	H 481
4-Benzyloxy-2-quinazolinecarboxamide	—	H 481
5-Benzyloxy-2,4-quinazolinediamine	207–208	2225
6-Benzyloxy-2,4-quinazolinediamine	194–198	238

ALPHABETICAL LIST OF SIMPLE QUINAZOLINES (*Continued*)

Quinazoline	m.p. (°C), etc.	Ref.
3-Benzyloxy-2,4(1*H*,3*H*)-quinazolinedione	—	H 470
4-Benzyloxyquinazoline 1-oxide	—	H 461
3-Benzyloxy-4(3*H*)-quinazolinone	—	H 469
4-Benzyloxy-2(3*H*)-quinazolinone 1-oxide	—	H 466
3-Benzyl-2-pentylthio-4(3*H*)-quinazolinone	68	160
2-Benzyl-4-phenylquinazoline	116–117	1946
4-Benzyl-2-phenylquinazoline	56–59	1781
2-Benzyl-3-phenyl-4(3*H*)-quinazolinethione	—	H 297
3-Benzyl-2-phenyl-4(3*H*)-quinazolinethione	165	H 297; 59
1-Benzyl-2-phenyl-4(1*H*)-quinazolinone	186–187; oxime: 187–189, ir, nmr, xl st; oxime HCl: 207	901,2132
1-Benzyl-4-phenyl-2(1*H*)-quinazolinone	181, ir, ms	1946
2-Benzyl-3-phenyl-4(3*H*)-quinazolinone	109–110 or 112, ir, nmr; HCl: 208–210	H 168; 525,717,2372
3-Benzyl-2-phenyl-4(3*H*)-quinazolinone	137–139	H 170; 525,1865
3-Benzyl-2-propyl-4(3*H*)-quinazolinone	—	H 165
3-Benzyl-2-propylthio-4(3*H*)-quinazolinone	109	160
2-Benzyl-4-quinazolinamine	221 or 233–235, ir, ms, nmr	426,1232
2-Benzylquinazoline	55–56 or 70	H 60; 890,1373
4-Benzylquinazoline	nmr; pic: 152–154	H 61; 1714,1885
1-Benzyl-2,4(1*H*,3*H*)-quinazolinedione	206–208	H 199; 2295
3-Benzyl-2,4(1*H*,3*H*)-quinazolinedione	220 to 236, ir, nmr	H 199; 187,709,769,770, 1473,2337
3-Benzyl-2,4(1*H*,3*H*)-quinazolinedithione	—	460

Compound		
2-Benzyl-4(3H)-quinazolinethione	nmr	H 296, 2573
3-Benzyl-4(3H)-quinazolinethione	—	H 297
2-Benzyl-4(1H)-quinazolinethione 3-oxide	—	H 468
1-Benzyl-4(1H)-quinazolinone	oxime: 157–159	H 137; 1828
2-Benzyl-4(3H)-quinazolinone (glycosminine, glycophymine)	246 to 254, ir, ms, nmr, st, uv; HBr: 284	H 139; 35,36,118,133,181,499, 733,747,888,908,961,1003, 1174,1306,1367,1406,1678, 1766,1940,2133,2151, 2511,2573
3-Benzyl-4(3H)-quinazolinone	112 to 118, ms, nmr	H 144; 36,276,1497,1835
3-Benzyl-2-styryl-4(3H)-quinazolinone	140–141	H 166; 1233
6-Benzylsulfonyl-2,4-quinazolinediamine	277–280	811
2-Benzylthio-3-butyl-6-chloro-4(3H)-quinazolinone	100	670
2-Benzylthio-6-chloro-3-ethyl-4(3H)-quinazolinone	122	670
2-Benzylthio-6-chloro-3-methyl-4(3H)-quinazolinone	123	670
5-Benzylthio-6-chloro-3-methyl-4(3H)-quinazolinone	128–130, ir, ms, nmr	2584
2-Benzylthio-6,8-dibromo-3-butyl-4(3H)-quinazolinone	122	603
2-Benzylthio-6,8-dibromo-3-ethyl-4(3H)-quinazolinone	151	603
2-Benzylthio-6,8-dibromo-3-phenyl-4(3H)-quinazolinone	200	603
2-Benzylthio-6-methoxy-4(3H)-quinazolinone	213–214	280
2-Benzylthio-4-methylthioquinazoline	—	1948
2-Benzylthio-3-phenyl-4(3H)-quinazolinethione	—	H 302
2-Benzylthio-3-phenyl-4(3H)-quinazolinone	—	H 305
4-Benzylthioquinazoline	103–104	667
6-Benzylthio-2,4-quinazolinediamine	192–193	811
2-Benzylthio-4(3H)-quinazolinethione	—	1948
2-Benzylthio-4(3H)-quinazolinone	212–213	1845
3-Benzyl-2-thioxo-1,2-dihydro-4(3H)-quinazolinone	249 to 255	H 306; 87,357,460,685,770,1774
3-Benzyl-4-thioxo-3,4-dihydro-2(1H)-quinazolinone	243 or 246–247	40,86,135

449

ALPHABETICAL LIST OF SIMPLE QUINAZOLINES (*Continued*)

Quinazoline	m.p. (°C), etc.	Ref.
2,4-Bisbenzylamino-6-methylquinazoline	—	H 361
3,7-Bisbenzylideneamino-2-styryl-4(3H)-quinazolinone	—	H 380
2,4-Bisbenzylidenehydrazinoquinazoline	—	H 383
1,3-Bis(4-bromo-3-methylbut-2-enyl)-6,7-dimethoxy-2,4(1H,3H)-quinazolinedione	110–114, ir, ms, nmr, uv	2469
1,3-Bis(4-bromo-3-methylbut-2-enyl)-2,4(1H,3H)-quinazolinedione	91–94, ir, ms, nmr, uv	2469
1,3-Bis(but-2′-enyl)-2,4(1H,3H)-quinazolinedione	50, ir, nmr	718
2,4-Bisbutylaminoquinazoline	n_D^{20} 1.5631, ir, nmr	1410
1,3-Biscarboxymethyl-6-chloro-2,4(1H,3H)-quinazolinedione	357–360, ir, ms, nmr	1822
1,3-Biscarboxymethyl-2,4(1H,3H)-quinazolinedione	343–345, ir, ms, nmr	1822
1,3-Bis-β-chloroethyl-2,4(1H,3H)-quinazolinedione	—	H 204
1,3-Biscyanomethyl-2,4(1H,3H)-quinazolinedione	195, ir, ms, nmr	1822
2,4-Bis(β-diethylaminoethylamino)quinazoline	—	H 356
1,3-Bis-β-diethylaminoethyl-2,4(1H,3H)-quinazolinedione	—	H 205
2,4-Bisdiethylamino-6-nitroquinazoline	91–93	600
2,4-Bisdiethylaminoquinazoline	—	H 356
2,8-Bisdimethylamino-5,6-dihydro-5,6-quinazolinedione	184–185, ir, uv	764
1,3-Bis-β-dimethylaminonethyl-2,4(1H,3H)-quinazolinedione	—	H 205
2,4-Bis(dimethylaminomethyleneamino)-6,7-dimethoxyquinazoline	216	1852
2,4-Bis(dimethylaminomethyleneamino)-6-nitroquinazoline	228	1852
2,4-Bis(dimethylaminomethyleneamino)quinazoline	142	1852
2,4-Bis(dimethylaminomethyleneamino)-6-quinazolinecarbonitrile	270	1852
2,4-Bisdimethylamino-6-methylquinazoline	HClO₄: 259	1331
2,4-Bisdimethylamino-6-nitroquinazoline	199–203	600
4,8-Bisdimethylamino-2-phenyl-5,6-dihydro-5,6-quinazolinedione	182–183, ir	592

450

Compound		
2,4-Bis-γ-dimethylaminopropylquinazoline	HCl: 213–214	56
2,4-Bisdimethylaminoquinazoline	90, ir, nmr	758,1410
3,6-Bisethoxycarbonylamino-2-ethyl-4(3H)-quinazolinone	—	H 380
3,6-Bisethoxycarbonylamino-2-methyl-4(3H)-quinazolinone	—	H 380
3,7-Bisethoxycarbonylamino-2-methyl-4(3H)-quinazolinone	—	H 380
3,6-Bisethoxycarbonylamino-4(3H)-quinazolinone	—	H 380
1,3-Bisethoxycarbonylmethyl-2,4(1H,3H)-quinazolinedione	117, ir, ms, nmr	1822
2,4-Bisethylaminoquinazoline	123–126	1410
2,4-Bisethylthioquinazoline	liquid, anal, nmr	1143
2,4-Bismethylaminoquinazoline	—	H 358
1,3-Bis(2'-methylprop-2'-enyl)-2,4(1H,3H)-quinazolinedione	114, ir, nmr	718
2,4-Bismethylthio-8-phenylthioquinazoline	113	204
2,4-Bismethylthioquinazoline	67–68 or 78	H 302; 1143,1379,1410,1948
1,3-Bisprop-2'-ynyl-2,4(1H,3H)-quinazolinedione	171, ir, ms, nmr	H 205; 2481
2,4-Bistrichloromethylquinazoline	—	H 64
2,4-Bistrifluoromethylquinazoline	46, ir, nmr	1145,1213
2,4-Bistrimethylammonioquinazoline (dichloride)	crude	1142
2,4-Bistrimethylsilyloxyquinazoline	93→98/0.08 or 115–117/0.5, ir, uv	78,243
3-β-Bromoallyl-4(3H)-quinazolinone	—	H 144
3-β-Bromoallyl-2-thioxo-1,2-dihydro-4(3H)-quinazolinone	—	H 306
6-Bromo-2,4-bis(dimethylaminomethyleneamino)quinazoline	204	1852
6-Bromo-3-β-bromoethyl-2,4(1H,3H)-quinazolinedione	—	H 202
6-Bromo-2-bromomethyl-3-phenyl-4(3H)-quinazolinone	—	1492
6-Bromo-3-butoxy-2-methyl-4(3H)-quinazolinone	146–147	1313
6-Bromo-3-butyl-2-ethylthio-4(3H)-quinazolinone	76	2298
6-Bromo-3-butyl-2-methylthio-4(3H)-quinazolinone	89 or 106	135,2298
2-α-Bromobutyl-4-phenylquinazoline 3-oxide	—	H 462
2-α-Bromobutyl-4(3H)-quinazolinone	—	2092

451

ALPHABETICAL LIST OF SIMPLE QUINAZOLINES (*Continued*)

Quinazoline	m.p. (°C), etc.	Ref.
2-δ-Bromobutyl-4(3*H*)-quinazolinone	—	H 139
6-Bromo-3-butyl-2-thioxo-1,2-dihydro-4(3*H*)-quinazolinone	236 or 238, complexes	135,1747,2298
6-Bromo-2-carbamoylmethyl-4(3*H*)-quinazolinone	173–174, ir, ms	1293
6-Bromo-3-α-carboxyethyl-2-phenyl-4(3*H*)-quinazolinone	250	1278
6-Bromo-3-β-carboxyethyl-2-phenyl-4(3*H*)-quinazolinone	243	1278
6-Bromo-3-β-carboxyethyl-4(3*H*)-quinazolinone	218–220, ir, ms	1640
6-Bromo-3-(α-carboxy-γ-methylbutyl)-2-phenyl-4(3*H*)-quinazolinone	254	1278
6-Bromo-3-carboxymethyl-2-ethyl-4(3*H*)-quinazolinone	250	1278
6-Bromo-3-carboxymethyl-2-isopropyl-4(3*H*)-quinazolinone	195–197	2277
6-Bromo-3-carboxymethyl-2-methyl-4(3*H*)-quinazolinethione	245, nmr	2277
6-Bromo-3-carboxymethyl-2-methyl-4(3*H*)-quinazolinone	262–264	2277
6-Bromo-3-carboxymethyl-2-methyl-4(3*H*)-quinazolinone	225–227	2277
6-Bromo-3-carboxymethyl-1-phenethyl-2,4(1*H*,3*H*)-quinazolinedione	255	1278
6-Bromo-3-carboxymethyl-2-phenyl-4(3*H*)-quinazolinone	322–325, ir, ms	1838
6-Bromo-3-carboxymethyl-2,4(1*H*,3*H*)-quinazolinedione	283–285, ir, ms, nmr	1825
6-Bromo-1-carboxymethyl-4(1*H*)-quinazolinone	272–274 or 275–277	1640, 2277
6-Bromo-3-carboxymethyl-4(3*H*)-quinazolinone	190	98
6-Bromo-2-carboxymethylthio-3-phenyl-4(3*H*)-quinazolinone	255, ir	1009
6-Bromo-3-carboxymethylthio-4(3*H*)-quinazolinone	209–210	2277
6-Bromo-3-carboxymethyl-2-trifluoromethyl-4(3*H*)-quinazolinone	194–196, ir, ms	1640
6-Bromo-3-γ-carboxypropyl-4(3*H*)-quinazolinone	245–248, nmr	H 202; 1611
6-Bromo-3-β-chloroethyl-2,4(1*H*,3*H*)-quinazolinedione		H 252
8-Bromo-4-chloro-6-iodoquinazoline	—	441
6-Bromo-2-chloromethyl-3-methoxy-4(3*H*)-quinazolinone	154	874
6-Bromo-2-chloromethyl-4-methylquinazoline 3-oxide	182–185, ir, nmr	

Compound	Properties	Ref.
6-Bromo-2-chloromethyl-4-phenylquinazoline 3-oxide	173–174	H 462
6-Bromo-2-chloro-4-phenylquinazoline	—	836
6-Bromo-4-chloroquinazoline	—	H 252
6-Bromo-5-chloro-2,4-quinazolinediamine	275–277	678
5-Bromo-6-chloro-4(3*H*)-quinazolinone	277–281	833
6-Bromo-4-chloro-2(1*H*)-quinazolinone	>350 (salt?)	72
6-Bromo-8-chloro-4(3*H*)-quinazolinone	—	H 154
6-Bromo-3-crotonamido-2-methyl-4(3*H*)-quinazolinone	—	1995
6-Bromo-2-dibromomethyl-3-phenyl-4(3*H*)-quinazolinone	—	1492
6-Bromo-4,8-dichloroquinazoline	—	H 252
8-Bromo-4,6-dichloroquinazoline	—	H 252
6-Bromo-3-β,β-diethoxyethyl-2,4(1*H*,3*H*)-quinazolinedithione	174–176, ms, uv	1292
6-Bromo-2-diethylamino-4(3*H*)-quinazolinethione	162	413
6-Bromo-2-β-dimethylaminoethyl-4(3*H*)-quinazolinone	—	H 174
6-Bromo-3-dimethylamino-4(3*H*)-quinazolinone	—	H 374
6-Bromo-2,4-dimethylquinazoline	uv	575
6-Bromo-2,4-dimethylquinazoline 3-oxide	—	H 462
5-Bromo-2,6-dimethyl-4(3*H*)-quinazolinone	crude: 288–291, ir, nmr	2411
6-Bromo-2,3-diphenyl-4(3*H*)-quinazolinone	nmr	H 189;1997
6-Bromo-3-ethoxycarbonylmethyl-2-methyl-4(3*H*)-quinazolinone	81	613
2-α-Bromoethyl-6-chloro-4-phenylquinazoline 3-oxide	—	H 462
2-β-Bromoethyl-5,8-dihydro-4,5,8(3*H*)-quinazolinetrione	120, ir, nmr	2430
2-β-Bromoethyl-5,8-dihydroxy-4(3*H*)-quinazolinone	202, ir, nmr	2430
6-Bromo-3-ethyl-2-ethylthio-4(3*H*)-quinazolinone	83	2298
2-α-Bromoethyl-3-β-hydroxyethyl-4(3*H*)-quinazolinone	100–102, nmr	2470
3-β-Bromoethyl-8-hydroxy-4(3*H*)-quinazolinone	—	H 178
2-α-Bromoethyl-3-methoxy-4(3*H*)-quinazolinone	151	441
2-β-Bromoethyl-3-methoxy-4(3*H*)-quinazolinone	141–142	441

ALPHABETICAL LIST OF SIMPLE QUINAZOLINES (*Continued*)

Quinazoline	m.p. (°C), etc.	Ref.
6-Bromo-2-ethyl-3-methylamino-4(3*H*)-quinazolinone	159–160, ir, nmr	1994
2-α-Bromoethyl-4-methylquinazoline 3-oxide	173–174, ir	874
5-Bromo-6-ethyl-2-methyl-4(3*H*)-quinazolinone	230–233, ir, nmr	2411
6-Bromo-6-ethyl-2-methylthio-4(3*H*)-quinazolinone	138	2298
2-(α-Bromo-α-ethylpropyl)-4-phenylquinazoline	92, ir	1946
6-Bromo-3-ethyl-2-propylthio-4(3*H*)-quinazolinone	82	2298
3-β-Bromoethyl-2,4(1*H*,3*H*)-quinazolinedione	206–208, ir, nmr	1428
6-Bromo-1-ethyl-2,4(1*H*,3*H*)-quinazolinedione	—	1490, 1539
6-Bromo-3-ethyl-2,4(1*H*,3*H*)-quinazolinedithione	228–230, ms, uv	1292
2-α-Bromoethyl-4(3*H*)-quinazolinone	234 or 254–255, ir, nmr	2069, 2092, 2544
2-β-Bromoethyl-4(3*H*)-quinazolinone	—	H 139
3-β-Bromoethyl-4(3*H*)-quinazolinone	116, ir, nmr	1772
6-Bromo-2-ethyl-4(3*H*)-quinazolinone	180	H 174; 2285
6-Bromo-2-ethylthio-3-methyl-4(3*H*)-quinazolinone	108	2298
6-Bromo-2-ethylthio-3-phenyl-4(3*H*)-quinazolinone	152	2298
6-Bromo-3-ethyl-2-thioxo-1,2-dihydro-4(3*H*)-quinazolinone	245	2298
6-Bromo-2-guanidino-4-methylquinazoline	—	H 343
6-Bromo-3-hexyl-2-methylthio-4(3*H*)-quinazolinone	60	135
3-ζ-Bromohexyl-4(3*H*)-quinazolinone	—	H 144
6-Bromo-3-hexyl-2-thioxo-1,2-dihydro-4(3*H*)-quinazolinone	195–196	135
6-Bromo-3-hydrazinocarbonylmethyl-2-methyl-4(3*H*)-quinazolinone	248	613
6-Bromo-4-hydrazino-2-morpholinoquinazoline	211–212	1337
6-Bromo-2-hydrazino-4-phenylquinazoline	185–186	836
6-Bromo-4-hydrazino-2-phenylquinazoline	226–228	749

454

6-Bromo-4-hydrazino-2-piperidinoquinazoline	204–206	1337
8-Bromo-3-β-hydroxyethyl-2-phenyl-4(3H)-quinazolinone	194	838
6-Bromo-3-β-hydroxyethyl-2,4(1H,3H)-quinazolinedione	264–265, nmr	H 202; 1611
6-Bromo-8-iodo-2,4(1H,3H)-quinazolinedione	314–316	308
6-Bromo-8-iodo-4(3H)-quinazolinone	—	H 154
8-Bromo-6-iodo-4(3H)-quinazolinone	—	H 154
6-Bromo-8-iodo-2-thioxo-1,2-dihydro-4(3H)-quinazolinone	303–305	308
6-Bromo-2-isobutyl-4(3H)-quinazolinone	—	H 174
6-Bromo-2-isopentyl-4(3H)-quinazolinone	—	H 174
6-Bromo-1-isopropyl-2-phenyl-4(1H)-quinazolinone	>280	1590
6-Bromo-2-isopropyl-4(3H)-quinazolinone	—	H 174
6-Bromo-3-methoxycarbonylmethyl-2-methyl-4(3H)-quinazolinethione	138–140, ir, nmr	2277
6-Bromo-1-methoxycarbonylmethyl-2,4(1H,3H)-quinazolinedione	260–265, ir, ms, nmr	1822
6-Bromo-3-methoxycarbonylmethyl-2,4(1H,3H)-quinazolinedione	263–266, ir, ms, nmr	1708
6-Bromo-1-methoxycarbonylmethyl-4(1H)-quinazolinone	224–226, ir, ms, nmr	1825
6-Bromo-3-methoxycarbonylmethyl-4(3H)-quinazolinone	153–155 or 160–162, ir, nmr	1640,2277
7-Bromo-6-methoxy-5,8-dihydro-5,8-quinazolinedione	202, ir, nmr	1577
5-Bromo-6-methoxy-2-methyl-4(3H)-quinazolinone	273–274, ir, nmr	2411
6-Bromo-2-methylaminomethyl-3-phenyl-4(3H)-quinazoline	—	1492
1-(4-Bromo-3-methylbut-2-enyl)-3-(2-hydroxy-2-methylbut-3-enyl)-6,7-dimethoxy-2,4(1H,3H)-quinazolinedione	102–104, ir, ms, nmr, uv	2469
1-(4-Bromo-3-methylbut-2-enyl)-3-(2-hydroxy-2-methylbut-3-enyl)-2,4(1H,3H)-quinazolinedione	98–100, ir, ms, nmr, uv	2469
1-(4-Bromo-3-methylbut-2-enyl)-3-(2-hydroxy-3-methylbut-3-enyl)-2,4(1H,3H)-quinazolinedione	118–121, ir, ms, nmr, uv	2469
1-(4-Bromo-3-methylbut-2-enyl)-2,4(1H,3H)-quinazolinedione	177–179, ir, ms, nmr uv	2469
3-(4-Bromo-3-methylbut-2-enyl)-2,4(1H,3H)-quinazolinedione	181–184, ir, ms, nmr, uv	2469
2-Bromomethyl-3-t-butoxycarbonylamino-6-nitro-4(3H)-quinazolinone	170–171, ir	2360

ALPHABETICAL LIST OF SIMPLE QUINAZOLINES (*Continued*)

Quinazoline	m.p. (°C), etc.	Ref.
2-Bromomethyl-3-*t*-butoxycarbonylamino-4(3*H*)-quinazolinone	180–181, ir	2360
2-Bromomethyl-6-chloro-4(3*H*)-quinazolinone	—	1902
6-Bromomethyl-2,4-diethoxyquinazoline	crude	1616
2-Bromomethyl-5,8-dihydro-4,5,8(3*H*)-quinazolinetrione	177–181, ir, nmr	1999
2-Bromomethyl-5,8-dihydroxy-4(3*H*)-quinazolinone	211–212, ir, ms, nmr	1999
6-Bromomethyl-2,4-dimethoxyquinazoline	138–143	1616
6-Bromomethyl-2,4-diphenoxyquinazoline	crude	1616
2-Bromomethyl-3-ethoxycarbonylamino-4(3*H*)-quinazolinone	143, ir, nmr	2360
2-(α-Bromo-α-methylethyl)-4-phenylquinazoline	135	1383,1946
6-Bromomethyl-2-fluoromethyl-4(3*H*)-quinazolinone	crude, nmr	2263
6-Bromomethyl-2-methyl-4-methylthioquinazoline	crude	2255
6-Bromomethyl-2-methyl-4(3*H*)-quinazolinone	>300 or >320, ms, nmr	1608,2263
6-Bromo-4-methyl-2-methylsulfinylmethylquinazoline 3-oxide	194–195, ir, nmr, uv	874
6-Bromo-4-methyl-2-methylthiomethylquinazoline 3-oxide	160–161, ir, nmr, uv	874
6-Bromomethyl-4-methylthio-2-pivalamidoquinazoline	crude	2255
6-Bromomethyl-4-methylthioquinazoline	anal, ms, nmr	2255
6-Bromo-3-methyl-2-methylthio-4(3*H*)-quinazolinone	168	2298
2-Bromomethyl-6-nitro-4-phenylquinazoline	—	355
2-Bromomethyl-4-phenylquinazoline	156–157, ir	1946
7-Bromo-4-methyl-2-phenylquinazoline	106–108, uv	1083
6-Bromo-4-methyl-2-phenylquinazoline 3-oxide	—	H 462
2-Bromomethyl-3-phenyl-4(3*H*)-quinazolinethione	300, ir	479
2-Bromomethyl-3-phenyl-4(3*H*)-quinazolinone	176 to 188, ms; MeClO$_4$: 251–253, ir	146,152,786,943,1008
6-Bromo-1-methyl-4-phenyl-2(1*H*)-quinazolinone	238–241	836

456

Compound		
6-Bromo-2-methyl-3-phenyl-4(3H)-quinazolinone	nmr	H 183; 1997
6-Bromomethyl-2-pivalamido-4(3H)-quinazolinone	205–207, nmr	871
6-Bromomethyl-3-pivaloyloxymethyl-4(3H)-quinazolinone	121–122, ir, nmr	1617
6-Bromo-3-methyl-2-propylthio-4(3H)-quinazolinone	84	2298
4-Bromomethylquinazoline	79–80, ms, nmr	1651
6-Bromo-5-methyl-2,4-quinazolinediamine	303–305, nmr; HBr: 350, ms	2041,2246
8-Bromo-5-methyl-2,4-quinazolinediamine	275–276, nmr	2246
6-Bromo-3-methyl-2,4(1H,3H)-quinazolinedione	291	98
6-Bromo-7-methyl-4(3H)-quinazolinethione	274–275	1365
2-Bromomethyl-4(3H)-quinazolinone	>300, ir	943,2069
6-Bromo-2-methyl-4(3H)-quinazolinone	301	H 173; 1365
6-Bromo-3-methyl-4(3H)-quinazolinone	HBr: 338–340, ir, nmr; 1-MeI: 287–288, nmr	907
2-Bromomethyl-4(1H)-quinazolinone 3-oxide	188, ir	236
6-Bromo-2-methyl-4(1H)-quinazolinone 3-oxide	249–250 or 257	207,1313
6-Bromo-2-methylthio-3-pentyl-4(3H)-quinazolinone	75	135
6-Bromo-2-methylthio-3-phenyl-4(3H)-quinazolinone	214	2298
6-Bromo-3-methyl-2-thioxo-1,2-dihydro-4(3H)-quinazolinone	273	2298
6-Bromomethyl-2-trifluoromethyl-4(3H)-quinazolinone	>300, ms, nmr	1608
6-Bromo-2-morpholinomethyl-3-phenyl-4(3H)-quinazolinone	—	1492
6-Bromo-2-morpholino-4(3H)-quinazolinethione	196	413
6-Bromo-3-morpholino-4(3H)-quinazolinone	—	H 374
5-Bromo-7-nitro-2,3-diphenyl-4(3H)-quinazolinone	—	H 189
5-Bromo-7-nitro-2-phenyl-4(3H)-quinazolinone	—	H 176
5-Bromo-7-nitro-2-phenyl-4(1H)-quinazolinone 3-oxide	—	H 467
7-Bromo-4-oxo-3,4-dihydro-2-quinazolinecarbohydrazide	>355	1743
7-Bromo-4-oxo-3-4-dihydro-2-quinazolinecarboxamide	252–253	1743
2-α-Bromopentyl-4(3H)-quinazolinone	—	2092

ALPHABETICAL LIST OF SIMPLE QUINAZOLINES (*Continued*)

Quinazoline	m.p. (°C), etc.	Ref.
6-Bromo-3-pentyl-2-thioxo-1,2-dihydro-4(3*H*)-quinazolinone	198	135
6-Bromo-3-phenyl-2-piperidinomethyl-4(3*H*)-quinazolinone	180–181; HCl: 220–221; pic: 225	745,1492
6-Bromo-3-phenyl-2-propylthio-4(3*H*)-quinazolinone	175	2298
6-Bromo-4-phenyl-2-quinazolinamine	—	*H* 252
4-Bromo-2-phenylquinazoline	—	*H* 252
6-Bromo-3-phenyl-2,4(1*H*,3*H*)-quinazolinedione	314	98
6-Bromo-3-phenyl-2,4(1*H*,3*H*)-quinazolinedithione	267–268, uv	1292
6-Bromo-4-phenyl-2(1*H*)-quinazolinethione	243–244	836
6-Bromo-2-phenyl-4(3*H*)-quinazolinone	303–305	*H*176; 749
6-Bromo-3-phenyl-4(3*H*)-quinazolinone	—	*H* 180
6-Bromo-4-phenyl-2(1*H*)-quinazolinone	305 to 322, ir, uv	836,1060,1946
6-Bromo-3-phenyl-2-thiocyanato-4(3*H*)-quinazolinone	187–189	663
6-Bromo-3-phenyl-2-thioxo-1,2-dihydro-4(3*H*)-quinazolinone	325 or >340	663,864,2298
5-β-Bromopropyl-6-hydroxy-7-methoxy-2-methyl-3-phenyl-4(3*H*)-quinazolinone	265–266	143
3-γ-Bromopropyl-6-methyl-2-thioxo-1,2-dihydro-4(3*H*)-quinazolinone	—	*H* 309
2-α-Bromopropyl-4-phenylquinazoline	110, ir	1946
3-β-Bromopropyl-2,4(1*H*,3*H*)-quinazolinedione	214–215, ir, uv	477
3-γ-Bromopropyl-2,4(1*H*,3*H*)-quinazolinedione	189–191, ir, nmr	1435
2-α-Bromopropyl-4(3*H*)-quinazolinone	215–217, nmr	*H* 139; 2092,2544
6-Bromo-2-propyl-4(3*H*)-quinazolinone	—	*H* 174
6-Bromo-3-propyl-2-thioxo-1,2-dihydro-4(3*H*)-quinazolinone	complexes	1747
6-Bromo-4-quinazolinamine	324–326 or 338–340, ms	180,196,339,507
6-Bromo-4-quinazolinamine 3-oxide	>300	507
6-Bromoquinazoline	—	*H* 252

458

Compound		
7-Bromoquinazoline	—	*H* 252
5-Bromo-2,4-quinazolinediamine	193–198 or 202–204, nmr	1932,2225
6-Bromo-2,4-quinazolinediamine	188–190 (?), 260–265, or 270–271	*H* 254; 678,1852,2326
7-Bromo-2,4-quinazolinediamine	255–256, nmr	2247
6-Bromo-2,4(1*H*,3*H*)-quinazolinedione	—	*H* 202
7-Bromo-2,4(1*H*,3*H*)-quinazolinedione	—	1969
6-Bromo-2,4(1*H*,3*H*)-quinazolinedithione	>360	38
4-Bromo-2(1*H*)-quinazolinone	348–350 (salt?)	72
5-Bromo-4(3*H*)-quinazolinone	—	*H* 153
6-Bromo-4(3*H*)-quinazolinone	>230 (?), 260 to 275, ir, nmr	*H* 153; 196,628,907,2277
7-Bromo-4(3*H*)-quinazolinone	—	*H* 153
6-Bromo-4(1*H*)-quinazolinone 3-oxide	257–260	207
6-Bromo-2-styryl-4(3*H*)-quinazolinone	334–338	2264
6-Bromo-2-styryl-4(1*H*)-quinazolinone 3-oxide	221–222	1313
6-Bromo-2-thioxo-1,2-dihydro-4(3*H*)-quinazolinone	>280 or 323–324, ir, complexes	1009,1682,1747,1862
6-Bromo-2-trichloromethyl-4(3*H*)-quinazolinone	249–250, ir, nmr	2265
4-But-3'-enylamino-2(1*H*)-quinazolinethione	294–296, ir, nmr	2384
3-But-3'-enyl-2-chloro-4(3*H*)-quinazolinone	liquid, ir, nmr	1727
1-But-2'-enyl-3-ethyl-2,4(1*H*,3*H*)-quinazolinedione	70, ir, nmr	718
3-But-2'-enylideneamino-2-methyl-4(3*H*)-quinazolinone	141–142, ir, nmr	191
3-But-2'-enylideneamino-2-phenyl-4(3*H*)-quinazolinone	98–100, ir, nmr	191
3-But-2'-enyleneamino-4(3*H*)-quinazolinone	102–104, ir, nmr	191
3-But-2'-enyl-1-(2-methylprop-2-enyl)-2,4(1*H*,3*H*)-quinazolinedione	72, ir, nmr	718
1-But-2'-enyl-3-(2-methylprop-2-enyl)-2,4(1*H*,3*H*)-quinazolinedione	100–101, ir, nmr	718
1-But-2'-enyl-3-methyl-2,4(1*H*,3*H*)-quinazolinedione	110, ir, nmr	718
4-But-3'-enyloxy-2-chloroquinazoline	49–50, ir, nmr	1727
4-But-3'-enyloxyquinazoline	liquid, ir, ms, nmr	1727

ALPHABETICAL LIST OF SIMPLE QUINAZOLINES (*Continued*)

Quinazoline	m.p. (°C), etc.	Ref.
4-But-3′-enyloxy-2-trifluoromethylquinazoline	49–50, ir, nmr	1727
1-But-2′-enyl-3-prop-2′-ynyl-2,4(1*H*,3*H*)-quinazolinedione	134, ir, nmr	718
1-But-2′-enyl-2,4(1*H*,3*H*)-quinazolinedone	177, ir, nmr	718
1-But-2′-enyl-4(1*H*)-quinazolinone	*trans*: 83–84	655
1-But-3′-enyl-4(1*H*)-quinazolinone	100–101	655
3-But-3′-enyl-4(3*H*)-quinazolinone	58, nmr	1727
2-But-2′-enylthio-3-phenyl-4(3*H*)-quinazolinone	145	1017
3-But-3′-enyl-2-trifluoromethyl-4(3*H*)-quinazolinone	51–53, ir, nmr	1727
3-*t*-Butoxycarbonylamino-2-chloromethyl-6-nitro-4(3*H*)-quinazolinone	178, ir	2360
3-*t*-Butoxycarbonylamino-2-chloromethyl-4(3*H*)-quinazolinone	180, ir	2360
3-Butoxycarbonylaminomethyl-2-methyl-4(3*H*)-quinazolinone	102, ir	277
3-*s*-Butoxycarbonylaminomethyl-2-methyl-4(3*H*)-quinazolinone	24–25, ir	277
1-*t*-Butoxycarbonylamino-3-methyl-2,4(1*H*,3*H*)-quinazolinedione	182–183, ir, nmr	2574
7-Butoxy-3-β-diethylaminoethyl-4(3*H*)-quinazolinone	—	330
3-Butoxy-2-phenyl-4(3*H*)-quinazolinone	93–94	1343
2-Butoxy-4-quinazolinamine	—	*H* 253
4-Butoxyquinazoline	—	*H* 254
4-Butoxy-2-quinazolinecarboxamide	—	*H* 481
5-Butoxy-2,4-quinazolinediamine	187–190	2225
5-Butoxy-8-quinazolinol	116, uv	595
2-Butoxy-4(3*H*)-quinazolinone	151	*H* 143; 126
2-*s*-Butylamino-4-*t*-butylthioquinazoline	85–86, ir, nmr	1410
4-Butylamino-2-chloroquinazoline	115–116, ir, nmr	1410
4-*t*-Butylamino-2-chloroquinazoline	139–141, ir, nmr	1410

460

4-Butylamino-6,7-dimethoxy-2-methylquinazoline	158–159, ms, uv; HCl: 264–265, uv	1354
4-Butylamino-2-ethylquinazoline	—	1070
3-β-Butylaminoethyl-4(3H)-quinazolinone	—	H 144
6-Butylamino-4-methoxy-2-phenyl-5,8-dihydro-5,8-quinazolinedione	184–185, ir, uv	763
4-Butylamino-6-methoxy-2-phenylquinazoline	139–140	766
4-Butylamino-8-methoxyquinazoline	—	H 260
3-t-Butylamino-2-methylamino-4(3H)-quinazolinone	—	H 365
4-s-Butylamino-2-methylquinazoline	164–167 or 166–168, ir, nmr, uv	1267, 1300
2-Butylamino-4-methylthioquinazoline	59, ir, nmr	1410
2-t-Butylamino-4-methylthioquinazoline	87–89, nmr	1410
4-Butylamino-2-methylthioquinazoline	99–100, ir, nmr	1410
4-Butylamino-6-nitroquinazoline	—	551
4-Butylamino-7-nitroquinazoline	—	H 351
4-Butylamino-2-phenyl-8-piperidino-5,6-dihydro-5,6-quinazolinedione	142–143, ir	592
4-Butylamino-2-phenylquinazoline	100–103, ir, nmr	1267
4-t-Butylamino-2-phenylquinazoline	145, ir, nmr	1666
4-Butylamino-2-phenyl-6-quinazolinol	212–213	766
2-Butylamino-3-phenyl-4(3H)-quinazolinone	149–150, ir	1225
4-Butylamino-7-quinazolinamine	—	H 363
4-Butylaminoquinazoline	116	H 346; 333,555
4-t-Butylaminoquinazoline	95–96	1000
4-Butylamino-2-quinazolinecarbonitrile	—	H 481
6-Butylamino-2,4-quinazolinediamine	231–233	602
3-t-Butylamino-2,4(1H,3H)-quinazolinedione	203–204	1437
4-Butylamino-2(1H)-quinazolinethione	249–251, ir, nmr	2384
4-Butylamino-8-quinazolinol	—	H 351
2-Butylamino-4(3H)-quinazolinone	—	H 365

461

ALPHABETICAL LIST OF SIMPLE QUINAZOLINES (*Continued*)

Quinazoline	m.p. (°C), etc.	Ref.
2-*t*-Butylamino-4(3*H*)-quinazolinone	218–220, ir, nmr	1372
3-*t*-Butylamino-4(3*H*)-quinazolinone	104–105	1471
2-Butyl-4-butylaminoquinazoline	93–95, 240/15, nmr	1784
3-Butyl-1-β-carboxyethyl-2,4(1*H*,3*H*)-quinazolinedione	—	1106
3-Butyl-2-carboxymethylthio-4(3*H*)-quinazolinone	140–141, ir, uv	880
1-Butyl-7-chloro-3-dimethylamino-2,4-dioxo-1,2,3,4-tetrahydro-6-quinazolinesulfonamide	160–162	1603
3-Butyl-6-chloro-2-ethylthio-4(3*H*)-quinazolinone	74	670
3-Butyl-5-chloro-2-methyl-4-oxo-3,4-dihydro-6-quinazolinesulfonamide	171–174, ir, nmr	1384
3-Butyl-5-chloro-2-methyl-4(3*H*)-quinazolinone	176–178	2299
3-Butyl-7-chloro-6-nitro-4(3*H*)-quinazolinone	132	1165
3-Butyl-6-chloro-2-propylthio-4(3*H*)-quinazolinone	71	670
2-Butyl-6-chloro-4(3*H*)-quinazolinethione	191, ms, nmr	957
3-Butyl-6-chloro-4(3*H*)-quinazolinone	109–110, ir, nmr	1970
3-Butyl-7-chloro-4(3*H*)-quinazolinone	88–89, ir, nmr	1970
3-Butyl-6-chloro-2-thioxo-1,2-dihydro-4(3*H*)-quinazolinone	220	670
3-Butyl-6,8-diiodo-2-isopentylthio-4(3*H*)-quinazolinone	84	131
3-Butyl-6,8-diiodo-2-pentylthio-4(3*H*)-quinazolinone	298 (?)	131
3-Butyl-6,8-diiodo-2-thioxo-1,2-dihydro-4(3*H*)-quinazolinone	220	131
3-Butyl-6,7-dimethoxy-2-methyl-4(3*H*)-quinazolinethione	132–134, ms, uv	1354
3-Butyl-7,8-dimethoxy-3-phenyl-4(3*H*)-quinazolinone	175–176	142
2-*t*-Butyl-6,7-dimethoxyquinazoline	105–107, ms	1354
2-*t*-Butyl-6,7-dimethoxy-4(3*H*)-quinazolinethione	172–173, ms, uv	1354
2-*t*-Butyl-6,7-dimethoxy-4(3*H*)-quinazolinone	251–252, ir, ms, uv	1354

Compound		
2-t-Butyl-4-β-dimethylaminoethylaminoquinazoline	liquid, ms, nmr; phosphate: 178–179	95
3-Butyl-1-β-dimethylaminoethyl-2,4(1H,3H)-quinazolinedione	187–189	668
3-Butyl-1-γ-dimethylaminopropyl-2,4(1H,3H)-quinazolinedione	162–163	668
2-t-Butyl-6,8-dimethyl-4(3H)-quinazolinone	—	H 175
3-Butyl-2,4-dioxo-1,2,3,4-tetrahydro-6-quinazolinecarboxanilide	—	1122
7-t-Butyl-2,4-diphenylquinazoline	104–106, ms, nmr	905
2-t-Butyl-3-ethoxycarbonylmethyl-4(3H)-quinazolinethione	57, nmr	1565
3-Butyl-2-ethyl-4(3H)-quinazolinone	111	2162
6-Butyl-2-guanidino-4-methylquinazoline	—	H 343
6-t-Butyl-2-guanidino-4-methylquinazoline	—	H 344
3-Butyl-8-hydroxy-7-iodo-4(3H)-quinazolinone	—	H 178
2-Butyl-7-hydroxy-8-methoxy-3-phenyl-4(3H)-quinazolinone	203–204	142
3-Butyl-8-hydroxy-2-methyl-4(3H)-quinazolinone	—	H 198
3-Butyl-1-hydroxy-2,4(1H,3H)-quinazolinedione	120–122	622
3-Butyl-8-hydroxy-4(3H)-quinazolinone	—	H 178
3-Butyl-1-hydroxy-2-thioxo-1,2-dihydro-4(3H)-quinazolinone	59–61	622
3-Butyl-6-iodo-2-methyl-4(3H)-quinazolinone	—	H 180
2-Butyl-4-isopropylaminoquinazoline	87–88, 220/10, nmr	1784
4-Butyl-1-isopropyl-7-methyl-2(1H)-quinazolinone	82–85	681
N-Butyl-4-isopropyl-2-quinazolinecarboxamidrazone	—	H 63
6-t-Butyl-3-methoxycarbonylmethyl-2-methylthio-4(3H)-quinazolinone	122–123	1137
3-Butyl-8-methoxy-2-methyl-4(3H)-quinazolinone	—	H 180
3-Butyl-8-methoxy-4(3H)-quinazolinone	—	H 178
4-(N-Butyl-N-methylamino)-2-chloroquinazoline	59, nmr, uv	1412
3-t-Butyl-2-methylamino-4(3H)-quinazolinone	132	621
3-s-Butyl-2-methyl-5-nitro-4(3H)-quinazolinone	—	H 182
3-Butyl-N-methyl-4-oxo-3,4-dihydro-2-quinazolinecarboxamide	172–173	768

ALPHABETICAL LIST OF SIMPLE QUINAZOLINES (*Continued*)

Quinazoline	m.p. (°C), etc.	Ref.
3-*t*-Butyl-*N*-methyl-4-oxo-3,4-dihydro-2-quinazolinecarboxamide	212–213	768
1-*t*-Butyl-7-methyl-4-phenyl-2(1*H*)-quinazolinone	141–143	681
2-*t*-Butyl-4-methylquinazoline	—	*H* 64
6-Butyl-5-methyl-2,4-quinazolinediamine	284–289	674
1-Butyl-3-methyl-2,4(1*H*,3*H*)-quinazolinedione	64–65, ir, nmr	718
6-Butyl-5-methyl-2,4(1*H*,3*H*)-quinazolinedione	crude	674
2-*t*-Butyl-3-methyl-4(3*H*)-quinazolinethione	79	105,953
3-Butyl-2-methyl-4(3*H*)-quinazolinethione	—	*H* 297
2-Butyl-3-methyl-4(3*H*)-quinazolinone	55–56	990
2-*t*-Butyl-3-methyl-4(3*H*)-quinazolinone	64	105
2-*t*-Butyl-6-methyl-4(3*H*)-quinazolinone	—	*H* 175
3-Butyl-2-methyl-4(3*H*)-quinazolinone	89; HCl: 208 to 222	*H* 156; 1059,1267,2162,2296
3-*s*-Butyl-2-methyl-4(3*H*)-quinazolinone	106/0.1,n_D^{19} 1.5792, ir, nmr	1267
3-*t*-Butyl-2-methyl-4(3*H*)-quinazolinone	110–120/0.1, ir, nmr, uv	412
3-Butyl-5-methyl-4(3*H*)-quinazolinone	98–99, ir, nmr	1970
3-*s*-Butyl-5-methyl-4(3*H*)-quinazolinone	65–68, ir, nmr	1970
3-Butyl-6-methyl-4(3*H*)-quinazolinone	107–108, ir, nmr	1970
3-*s*-Butyl-6-methyl-4(3*H*)-quinazolinone	85–95, ir, nmr	1970
3-*t*-Butyl-6-methyl-4(3*H*)-quinazolinone	121–125, ir, nmr	1970
4-Butyl-2-methylthiomethylquinazoline 3-oxide	92–94	874
2-*t*-Butyl-4-methylthioquinazoline	79–80	104
3-Butyl-6-nitro-4(3*H*)-quinazolinone	121–123, uv	121
2-*t*-Butyl-4-pentyloxyquinazoline	ir, nmr	2482
2-Butyl-4-phenylquinazoline	75–77	1946

464

Compound	Data	Reference
2-t-Butyl-4-phenylquinazoline	96, ir, ms, nmr	2494
1-t-Butyl-3-phenyl-2,4(1H,3H)-quinazolinedione	209, ir, ms, nmr	1392
2-t-Butyl-3-phenyl-4(3H)-quinazolinethione	176	953
3-Butyl-2-phenyl-4(3H)-quinazolinethione	142	H 297; 455
1-Butyl-4-phenyl-2(1H)-quinazolinone	102–103	681
1-t-Butyl-4-phenyl-2(1H)-quinazolinone	128–132	681
2-Butyl-1-phenyl-4(1H)-quinazolinone	177–179	1590
2-t-Butyl-1-phenyl-4(1H)-quinazolinone	221–223	1394,1590
2-Butyl-3-phenyl-4(3H)-quinazolinone	110–112; HCl: 200–207	687
3-Butyl-2-phenyl-4(3H)-quinazolinone	—	H 170
3-s-Butyl-2-phenyl-4(3H)-quinazolinone	155–160/0.1, ir, nmr	1267
2-t-Butyl-4-propylaminoquinazoline	119–121, ms, nmr, uv	1300
3-Butyl-1-prop-2'-ynyl-2,4(1H,3H)-quinazolinedione	140–141	664
2-Butyl-3-prop-2'-ynyl-4(3H)-quinazolinedione	68–69	664
2-t-Butylquinazoline	nmr	H 61; 2478
4-Butylquinazoline	liquid, anal, ms, nmr	1714
4-t-Butylquinazoline	80/0.25	1094
1-Butyl-2,4(1H,3H)-quinazolinedione	127–128, ir, nmr	362,718
3-Butyl-2,4(1H,3H)-quinazolinedione	153 to 157, ir, nmr	H 199; 770,1282,1341,1446,1448
3-s-Butyl-2,4(1H,3H)-quinazolinedione	131–132, ir	227,1341
3-t-Butyl-2,4(1H,3H)-quinazolinedione	192 to 199, ir, ms, nmr	H 201; 227,722
7-t-Butyl-2,4(1H,3H)-quinazolinedione	—	H 202
2-t-Butyl-4-quinazolinesulfenic acid	100	105
2-t-Butyl-3(3H)-quinazolinethione	155–157 or 161–162, ir	H 296; 105,724
3-Butyl-4(3H)-quinazolinethione	—	H 297
2-t-Butyl-4(1H)-quinazolinethione 3-oxide	81	H 468; 105
3-t-Butyl-4(3H)-quinazolinimine	95–96	1000
1-Butyl-4(1H)-quinazolinone	64–66	276

ALPHABETICAL LIST OF SIMPLE QUINAZOLINES (*Continued*)

Quinazoline	m.p. (°C), etc.	Ref.
2-Butyl-4-(3H)-quinazolinone	157 or 158–159, ms, nmr	664,957
2-t-Butyl-4(3H)-quinazolinone	182 to 188, ir, nmr	H 141; 104,105,852
3-Butyl-4(3H)-quinazolinone	69 to 73, ir, nmr, uv	H 144; 121,276,690,1970
3-t-Butyl-4(3H)-quinazolinone	68–74, ir, nmr	1970
4-Butyl-2(1H)-quinazolinone	133, ms	1946
3-Butyl-2-styryl-4(3H)-quinazolinone	105, nmr, uv	515
2-t-Butyl-5,6,7,8-tetraphenylquinazoline	269–271, ms, nmr	2391
2-Butylthio-6-chloro-3-ethyl-4(3H)-quinazolinone	121	670
2-Butylthio-6-chloro-3-methyl-4(3H)-quinazolinone	122	670
2-Butylthio-6-chloro-3-phenyl-4(3H)-quinazolinone	233	670
4-t-Butylthio-2-chloroquinazoline	114–115	1410
2-Butylthio-6-methoxy-4(3H)-quinazolinone	188–189	280
2-Butylthio-3-phenyl-4(3H)-quinazolinethione	—	H 302
2-Butylthio-3-phenyl-4(3H)-quinazolinone	205	H 305; 1017
2-t-Butylthio-3-phenyl-4(3H)-quinazolinone	205	1017
2-Butylthio-4(3H)-quinazolinone	122–124 (?), 141 or 145, ir, nmr	126,1483,1845,1915
3-Butyl-2-thioxo-1,2-dihydro-4(3H)-quinazolinone	171–172 or 173–174, ir, uv	289,460,770,880,1646,1774
2-Butyl-3-valeamido-4(3H)-quinazolinone	114	1657
3-Butyramido-2-propyl-4(3H)-quinazolinone	139	1657
4-Butyrylmethylquinazoline	—	H 62
4-Butyrylquinazoline	—	431
3-β-Carbamoylethyl-6-chloro-2-methyl-4(3H)-quinazolinone	—	H 180
3-β-Carbamoylethyl-7-chloro-2-methyl-4(3H)-quinazolinone	—	H 180
3-β-Carbamoylethyl-7-chloro-4(3H)-quinazolinone	—	H 178

Compound		
3-β-Carbamoylethyl-2-ethyl-4(3H)-quinazolinone	—	H 164
3-β-Carbamoylethyl-2-methyl-4(3H)-quinazolinone	—	H 157
3-α-Carbamoylethyl-2-phenyl-4(3H)-quinazolinone	129	1047
3-β-Carbamoylethyl-2-phenyl-4(3H)-quinazolinone	215	1047
2-β-Carbamoylethyl-4(3H)-quinazolinone	>260, ms	2399
3-β-Carbamoylethyl-4(3H)-quinazolinone	—	H 146
2-β-Carbamoylethyl-4(3H)-quinazolinone 1-oxide	>260, ms	2399
3-(α-Carbamoyl-γ-methylbutyl)-2-phenyl-4(3H)-quinazolinone	140	1047
3-Carbamoylmethyl-6-chloro-2-methyl-4(3H)-quinazolinone	—	H 180
2-Carbamoylmethyl-6-chloro-4(3H)-quinazolinone	261–262 or 272–273, ir, ms	531,1293
3-Carbamoylmethyl-6-chloro-4(3H)-quinazolinone	—	H 178
3-Carbamoylmethyl-6,8-dichloro-2-methyl-4(3H)-quinazolinone	—	H 180
2-Carbamoylmethyl-6,8-dichloro-4(3H)-quinazolinone	274–276, ir, ms	1293
3-Carbamoylmethyl-6,8-dichloro-4(3H)-quinazolinone	—	H 178
4-(α-Carbamoyl-α-methylethyl)-quinazoline	—	H 62
3-Carbamoylmethyl-2-methyl-7-nitro-4(3H)-quinazolinone	—	H 181
3-Carbamoylmethyl-2-methyl-4(3H)-quinazolinone	—	H 157
1-Carbamoylmethyl-2-methylthio-4(1H)-quinazolinone	—	H 308
3-Carbamoylmethyl-2-methylthio-4(3H)-quinazolinone	—	H 305
2-Carbamoylmethyl-6-nitro-4(3H)-quinazolinone	330–333, ir, ms	1293
3-Carbamoylmethyl-2-phenyl-4(3H)-quinazolinone	180	1047
3-(α-Carbamoyl-β-methylpropyl)-4(3H)-quinazolinone	172	1047
2-Carbamoylmethyl-4-quinazolinamine	261–263	1368
2-Carbamoylmethyl-4(3H)-quinazolinone	259–261 or 265–266, ir, ms	531,1293
3-Carbamoylmethyl-4(3H)-quinazolinone	241–243	H 146; 276
1-Carbamoylmethyl-2-thioxo-1,2-dihydro-4(3H)-quinazolinone	—	H 308
3-α-Carbamoylpropyl-4(3H)-quinazolinone	173–175	276
1-δ-Carboxybutyl-6,7-dihydroxy-4-methyl-2(1H)-quinazolinone	> 290	944

ALPHABETICAL LIST OF SIMPLE QUINAZOLINES (*Continued*)

Quinazoline	m.p. (°C), etc.	Ref.
4-δ-Carboxybutylquinazoline	—	H 62
2-δ-Carboxybutyl-4(3H)-quinazolinone	—	H 139
2-β-Carboxyethyl-6-chloro-4-phenylquinazoline	163–165	220
3-β-Carboxyethyl-6-chloro-2-phenyl-4(3H)-quinazolinethione	239, nmr	1565
3-β-Carboxyethyl-6-chloro-2-phenyl-4(3H)-quinazolinone	205	1565
2-β-Carboxyethyl-6-chloro-4(3H)-quinazolinone	254–257, uv	265
3-β-Carboxyethyl-6,8-dichloro-2-methyl-4(3H)-quinazolinethione	210, nmr	1565
3-β-Carboxyethyl-6,8-dichloro-2-phenyl-4(3H)-quinazolinethione	252	1565
3-α-Carboxyethyl-6,8-dichloro-2-phenyl-4(3H)-quinazolinone	283	1565
1-β-Carboxyethyl-6,7-dihydroxy-4-isopropyl-2(1H)-quinazolinone	310–312	944
1-β-Carboxyethyl-5,6-dihydroxy-4-methyl-2(1H)-quinazolinone	H₂O: 250–251, ir, nmr	944
1-β-Carboxyethyl-6,7-dihydroxy-4-methyl-2(1H)-quinazolinone	HBr: 241–243, ir, nmr	944
1-β-Carboxyethyl-7,8-dihydroxy-4-methyl-2(1H)-quinazolinone	H₂O: 218–220	944
1-β-Carboxyethyl-6,7-dihydroxy-4-phenyl-2(1H)-quinazolinone	304–306	944
1-β-Carboxyethyl-6,7-dihydroxy-4-propyl-2(1H)-quinazolinone	K₃: 310–312	944
3-β-Carboxyethyl-6,8-diiodo-2-methyl-4(3H)-quinazolinone	258, ir, nmr	H 181; 444
3-β-Carboxyethyl-6,8-diiodo-2-propyl-4(3H)-quinazolinone	—	H 188
1-β-Carboxyethyl-6,7-dimethoxy-4-methyl-2(1H)-quinazolinone	228–229	944
1-β-Carboxyethyl-4-ethyl-6,7-dihydroxy-2(1H)-quinazolinone	K₃: 320–322	944
3-α-Carboxyethyl-2-ethyl-6,8-diiodo-4(3H)-quinazolinone	232, ir, nmr	444
2-β-Carboxyethyl-3-hydroxy-4(3H)-quinazolinone	210, ir	236
3-α-Carboxyethyl-6-iodo-2-phenyl-4(3H)-quinazolinone	237	1278
3-β-Carboxyethyl-6-iodo-2-phenyl-4(3H)-quinazolinone	246	1278
3-β-Carboxyethyl-1-methyl-2,4(1H,3H)-quinazolinedione	—	1106

468

Compound		
1-β-Carboxyethyl-4-methyl-2(1H)-quinazolinone	255–258, nmr	944
3-β-Carboxyethyl-6-methylthio-4-(3H)-quinazolinone	190–191, nmr	1558
3-α-Carboxyethyl-7-nitro-2-phenyl-4(3H)-quinazolinone	—	2327
3-α-Carboxyethyl-8-nitro-2,4(1H,3H)-quinazolinedione	282–284, ir, ms, nmr	2435
2-β-Carboxyethyl-4-phenylquinazoline	164–167	220,632
3-β-Carboxyethyl-2-phenyl-4(3H)-quinazolinethione	220, nmr	1565
2-β-Carboxyethyl-1-phenyl-4(1H)-quinazolinone	198–200, nmr	1590
2-β-Carboxyethyl-3-phenyl-4(3H)-quinazolinone	217	1184
3-α-Carboxyethyl-2-phenyl-4(3H)-quinazolinone	167	1047,1278
3-β-Carboxyethyl-2-phenyl-4(3H)-quinazolinone	162 or 164	1047,1278
2-β-Carboxyethylquinazoline	—	H 60
1-β-Carboxyethyl-2,4(1H,3H)-quinazolinedione	235–237, ms, nmr	2448
3-α-Carboxyethyl-2,4(1H,3H)-quinazolinedione	268–270, ir, ms, nmr	2435
3-β-Carboxyethyl-4(3H)-quinazolinethione	223, nmr	1565
2-β-Carboxyethyl-4(3H)-quinazolinone	200 or 259, nmr	H 139; 1184,2399
3-α-Carboxyethyl-4(3H)-quinazolinone	205 or 221–223, ir, ms	783,1085,1640,1734
3-β-Carboxyethyl-4(3H)-quinazolinone	155–170 or 187–189, ir, ms	H 146; 690,1640,1734
2-β-Carboxyethyl-4(3H)-quinazolinone 1-oxide	248, nmr	2399
2-β-Carboxyethyl-4(1H)-quinazolinone 3-oxide	195, ir, nmr	800
1-β-Carboxyethyl-2-thioxo-1,2-dihydro-4(3H)-quinazolinone	203–204, nmr	2448
3-α-Carboxyethyl-2-thioxo-1,2-dihydro-4(3H)-quinazolinone	114–115	54
3-(α-Carboxy-γ-methylbutyl)-6,8-diiodo-2-methyl-4(3H)-quinazolinone	200, ir, nmr	444
3-(α-Carboxy-γ-methylbutyl)-6-iodo-2-phenyl-4(3H)-quinazolinone	250	1278
3-(α-Carboxy-γ-methylbutyl)-6-nitro-2-phenyl-4(3H)-quinazolinone	—	2327
3-(α-Carboxy-γ-methylbutyl)-2-phenyl-4(3H)-quinazolinone	168 or 170	1047, 1278
3-(α-Carboxy-γ-methylbutyl)-2-thioxo-1,2-dihydro-4(3H)-quinazolinone	183–184	54
1-Carboxymethyl-7-chloro-3-diethylamino-2,4-dioxo-1,2,3,4-tetrahydro-6-quinazolinesulfonamide	165–170	1603

ALPHABETICAL LIST OF SIMPLE QUINAZOLINES (*Continued*)

Quinazoline	m.p. (°C), etc.	Ref.
3-Carboxymethyl-6-chloro-1-methyl-2,4(1*H*,3*H*)-quinazolinedione	209–211	2277
3-Carboxymethyl-6-chloro-2-methyl-4(3*H*)-quinazolinone	—	*H* 181
3-Carboxymethyl-6-chloro-2-phenyl-4(3*H*)-quinazolinethione	242, nmr	1565
3-Carboxymethyl-6-chloro-2-phenyl-4(3*H*)-quinazolinone	228, nmr	1565
1-Carboxymethyl-6-chloro-2,4(1*H*,3*H*)-quinazolinedione	316–320, ir, ms	1822
3-Carboxymethyl-6-chloro-2,4(1*H*,3*H*)-quinazolinedione	327–329, ir, ms	1838
1-Carboxymethyl-6-chloro-4(1*H*)-quinazolinone	285–288, ir, ms	1825
3-Carboxymethyl-6-chloro-4(3*H*)-quinazolinone	274–276 or 277–279, ir, ms	*H* 178; 1640, 1734, 1838
3-Carboxymethyl-6,8-dichloro-2-methyl-4(3*H*)-quinazolinethione	143, nmr	1565
3-Carboxymethyl-6,8-dichloro-2-methyl-4(3*H*)-quinazolinone	—	*H* 181
3-Carboxymethyl-6,8-dichloro-2-phenyl-4(3*H*)-quinazolinethione	248, then 320, nmr	1565
3-Carboxymethyl-6,8-dichloro-2,4(1*H*,3*H*)-quinazolinedione	328–330, ir, ms	1838
3-Carboxymethyl-6,8-dichloro-4(3*H*)-quinazolinone	284–287, ir, ms	*H* 178; 1640, 1838
1-Carboxymethyl-6,7-dihydroxy-4-methyl-2(1*H*)-quinazolinone	309–310	944
3-Carboxymethyl-6,8-diiodo-2-methyl-4(3*H*)-quinazolinone	260, ir, nmr	444
3-(α-Carboxy-α-methylethyl)-6,8-diiodo-2-methyl-4(3*H*)-quinazolinone	—	596
3-Carboxymethyl-6-iodo-2-phenyl-4(3*H*)-quinazolinone	242	1278
6-Carboxymethyl-2-isopropyl-1-phenyl-4(1*H*)-quinazolinone	254–255	1590
3-Carboxymethyl-1-methyl-2,4(1*H*,3*H*)-quinazolinedione	232–235 or 238–240, ir, ms, nmr	1822,1874
3-Carboxymethyl-2-methyl-4(3*H*)-quinazolinone	258 to 260, ir, nmr, complexes	*H* 157; 1684,1770,1979,2019, 2082,2159,2342,2503
3-Carboxymethyl-2-methylthio-4(3*H*)-quinazolinone	—	*H* 305
3-Carboxymethyl-7-nitro-2-phenyl-4(3*H*)-quinazolinone	250–251	1588,2327
3-Carboxymethyl-8-nitro-2,4(1*H*,3*H*)-quinazolinedione	292–294, ir, nmr	2435

Compound	Data	Value
1-Carboxymethyl-3-phenyl-2,4(1*H*,3*H*)-quinazolinedione	235–236	2262
3-Carboxymethyl-1-phenyl-2,4(1*H*,3*H*)-quinazolinedione	260–262	2277
3-Carboxymethyl-2-phenyl-4(3*H*)-quinazolinethione	242, nmr	1565
3-Carboxymethyl-2-phenyl-4(3*H*)-quinazolinone	120 to 255, nmr, complexes	1047,1278,1565,2019,2082,2159, 2316,2342,2503
3-(α-Carboxy-β-methylpropyl)-6,8-diiodo-2-methyl-4(3*H*)-quinazolinone	249, ir, nmr	444
3-α-Carboxy-β-methylpropyl)-8-nitro-2,4(1*H*,3*H*)-quinazolinedione	160–162, ir, ms, nmr	2435
3-(α-Carboxy-β-methylpropyl)-2-phenyl-4(3*H*)-quinazolinedione	167	1047
3-α-Carboxy-β-methylpropyl-2,4(1*H*,3*H*)-quinazolinedione	202–204, ir, ms, nmr	2435
1-Carboxymethyl-2,4(1*H*,3*H*)-quinazolinedione	295–297, ir, ms, nmr	1822
3-Carboxymethyl-2,4(1*H*,3*H*)-qinazolinedione	290 to 301, ir, ms, nmr	*H* 199; 1465,1838,2435
3-Carboxymethyl-4(3*H*)-quinazolinethione	250, nmr	1565
1-Carboxymethyl-4(1*H*)-quinazolinone	263–265, ir, ms, nmr	1825
3-Carboxymethyl-4(3*H*)-quinazolinone	240–241 or 243–245, ms, nmr	*H* 146; 276,759,1734,1770
3-Carboxymethyl-2-styryl-4(3*H*)-quinazolinone	—	597
2-Carboxymethylthio-7-chloro-3-phenyl-4(3*H*)-quinazolinone	275	1588
2-Carboxymethyltho-6-chloro-4(3*H*)-quinazolinone	225–227	220
2-Carboxymethylthio-6,7-dimethoxy-3-phenyl-4(3*H*)-quinazolinone	205–206	141
2-Carboxymethylthio-6-fluoro-3-phenyl-4(3*H*)-quinazolinone	190	1046
2-Carboxymethylthio-7-fluoro-3-phenyl-4(3*H*)-quinazolinone	196	1046
2-Carboxymethylthio-6-hydroxy-7-methoxy-3-phenyl-4(3*H*)-quinazolinone	148–149	141
2-Carboxymethylthio-6-methoxy-4(3*H*)-quinazolinone	203–204	280
2-Carboxymethylthio-4-methylquinazoline	154–157, nmr	2314
2-Carboxymethylthio-3-methyl-4(3*H*)-quinazolinone	184–185, ir, uv	880
2-Carboxymethylthio-6-methyl-4(3*H*)-quinazolinone	218, ir, nmr	2176
2-Carboxymethylthio-8-methyl-4(3*H*)-quinazolinone	240, ir	1674
2-Carboxymethylthio-3-phenethyl-4(3*H*)-quinazolinone	151–153 or 165–166, uv	615
2-Carboxymethylthio-4-phenylquinazoline	131–132, nmr	2314

471

ALPHABETICAL LIST OF SIMPLE QUINAZOLINES   (*Continued*)

Quinazoline	m.p. (°C), etc.	Ref.
2-Carboxymethylthio-3-phenyl-4(3*H*)-quinazolinone	210–211 or 273–274, ir	144,615,880,2386
2-Carboxymethylthio-3-propyl-4(3*H*)-quinazolinone	169–170, uv	880
2-Carboxymethylthioquinazoline	213–214, nmr	2314
4-Carboxymethylthioquinazoline	205–206, ir, uv	880,2315
2-Carboxymethylthio-4(3*H*)-quinazolinone	211 or 336–338, ir, uv	880, 1845
3-Carboxymethyl-2-thioxo-1,2-dihydro-4(3*H*)-quinazolinone	273–274, nmr, uv	*H* 306; 54
3-(β-Carboxy-α-methylvinyl)-6-methylthio-4(3*H*)-quinazolinone	216–218, nmr	1558
3-α-Carboxypentyl-6,8-diiodo-2-methyl-4(3*H*)-quinazolinone	—	*H* 181
3-ε-Carboxypentyl-6,8-diiodo-2-methyl-4(3*H*)-quinazolinone	186, ir, nmr	*H* 181; 444
4-ε-Carboxypentylquinazoline	93, ir, nmr	*H* 62; 1914
2-ε-Carboxypentyl-4-(3*H*)-quinazolinone	—	*H* 139
3-γ-Carboxypropyl-6,8-dichloro-4(3*H*)-quinazolinone	228, ir, ms	1640
1-γ-Carboxypropyl-6,7-dihydroxy-4-methyl-2(1*H*)-quinazolinone	HBr: 261–263	944
3-γ-Carboxypropyl-6,8-diiodo-2-methyl-4(3*H*)-quinazolinone	225, ir, nmr	444
3-γ-Carboxypropyl-7-nitro-2-phenyl-4(3*H*)-quinazolinone	—	2327
2-γ-Carboxypropyl-4(3*H*)-quinazolinone	—	*H* 139
3-γ-Carboxypropyl-4(3*H*)-quinazolinone	123–126, ir, ms	1640,1734
2-β-Carboxyvinyl-6-chloro-4-phenylquinazoline	270–272	220
3-β-Carboxyvinyl-6-chloro-4(3*H*)-quinazolinone	297–298	1558
3-β-Carboxyvinyl-6-cyclopropyl-2-methyl-4(3*H*)-quinazolinone	213–214	1558
3-β-Carboxyvinyl-6,8-diisopropyl-4(3*H*)-quinazolinone	181–182	1558
3-β-Carboxyvinyl-6,7-dimethyl-4(3*H*)-quinazolinone	303–305	1558
3-β-Carboxyvinyl-6,8-dimethoxy-4(3*H*)-quinazolinone	305–307	1558
3-β-Carboxyvinyl-6,8-dimethyl-4(3*H*)-quinazolinone	284–288	1558

473

Quinazoline	m.p. (°C), etc.	Ref.
2-Chloro-3-β-chloroethyl-4(3*H*)-quinazolinone	—	*H* 172
6-Chloro-3-β-chloroethyl-4(3*H*)-quinazolinone	—	*H* 178
7-Chloro-3-β-chloroethyl-4(3*H*)-quinazolinone	—	*H* 178
4-Chloro-2-(α-chloro-α-methylethyl)-8-methoxyquinazoline	135	1224
4-Chloro-2-(α-chloro-α-methylethyl)-6-methylquinazoline	101	1224
2-Chloro-3-(α-chloro-α-methylethyl)-4(3*H*)-quinazolinone	—	*H* 172
6-Chloro-2-chloromethyl-4-hydrazinoquinazoline	163–165	893
7-Chloro-2-chloromethyl-3-methoxy-4(3*H*)-quinazolinone	157	441
6-Chloro-2-chloromethyl-4-phenylquinazoline	124–125 or 126–128	*H* 249; 237,1426
6-Chloro-2-chloromethyl-4-phenylquinazoline 1-oxide	—	*H* 461
6-Chloro-2-chloromethyl-4-phenylquinazoline 3-oxide	135, ir, ms, nmr	*H* 462; 463(?),2500
6-Chloro-2-chloromethyl-3-phenyl-4(3*H*)-quinazolinone	184–186	376
4-Chloro-2-chloromethylquinazoline	80 (?), 95–96, or 100–102, ir, ms	123,893,1171,2190
6-Chloro-3-β-chloropropyl-4(3*H*)-quinazolinone	—	*H* 178
6-Chloro-3-β-cyanoethyl-2-methyl-4(3*H*)-quinazolinone	—	*H* 181
4-Chloro-2-cyanomethylquinazoline	60, ir	1161
6-Chloro-3-cyclohexyl-2-methyl-4(3*H*)-quinazolinone	—	*H* 181
6-Chloro-3-cyclohexyl-2-propyl-4(3*H*)-quinazolinone	—	*H* 188
6-Chloro-3-cyclohexyl-4(3*H*)-quinazolinone	—	*H* 178
6-Chloro-4-cyclohexyl-2(1*H*)-quinazolinone	269–270	1060
6-Chloro-2-cyclopropylmethoxy-4-phenylquinazoline	120–121, ir, nmr	580
6-Chloro-1-cyclopropylmethyl-4-phenyl-2(1*H*)-quinazolinone	170 to 176, ir, nmr	493,580,986,1060,1408
8-Chloro-1-cyclopropylmethyl-4-phenyl-2(1*H*)-quinazolinone	164–165	1060
6-Chloro-4-cyclopropyl-1-methyl-2(1*H*)-quinazolinone	anal, nmr	2585

Compound	mp (°C), notes	Reference
6-Chloro-4-cyclopropyl-2(1H)-quinazolinone	207–209, nmr	2563,2575
6-Chloro-2-diallylamino-4(3H)-quinazolinone	217–220; HCl: 204–208	641
4-Chloro-2-α,α-dichloroethylquinazoline	—	H 247
6-Chloro-4-dichloromethyl-2-phenylquinazoline	184–185	438
4-Chloro-2-dichloromethylquinazoline	135–137, ir, ms	1171,2190
2-Chloro-6,7-diethoxy-4(3H)-quinazolinone	246–249	641
2-Chloro-4-diethylamino-6,7-dimethoxyquinazoline	145–148	1810
4-Chloro-2-diethylamino-6,7-dimethoxyquinazoline	130–131	1810
7-Chloro-3-diethylamino-2,4-dioxo-1,2,3,4-tetrahydro-6-quinazolinesulfonamide	150	1603
2-Chloro-4-β-diethylaminoethylamino-6,7-dimethoxyquinazoline		H 264
2-Chloro-4-β-diethylaminoethylamino-5-methoxyquinazoline		H 263
2-Chloro-4-β-diethylaminoethylamino-6-methoxyquinazoline		H 263
2-Chloro-4-β-diethylaminoethylamino-7-methoxyquinazoline		H 264
2-Chloro-4-β-diethylaminoethylamino-8-methoxyquinazoline		H 264
2-Chloro-4-β-diethylaminoethylamino-7-methylquinazoline		H 246
6-Chloro-8-β-diethylaminoethylamino-3-methyl-4(3H)-quinazolinone		H 370
2-Chloro-4-β-diethylaminoethylamino-6-nitroquinazoline		H 246
2-Chloro-4-β-diethylaminoethylamino-7-nitroquinazoline	114–116	H 246; 1568
4-Chloro-2-β-diethylaminoethylaminoquinazoline	H_2O: 81–82	H 246; 1568
6-Chloro-4-β-diethylaminoethylaminoquinazoline		H 247
7-Chloro-4-β-diethylaminoethylaminoquinazoline		H 249
7-Chloro-3-diethylamino-1-ethyl-2,4(1H,3H)-quinazolinedione	114	1603
6-Chloro-3-β-diethylaminoethyl-4(3H)-quinazoline		H 178
6-Chloro-2-diethylaminomethyl-4-phenylquinazoline	75–77	1279
6-Chloro-2-diethylaminomethyl-4-phenylquinazoline 3-oxide	83–85	H 463; 1279
7-Chloro-3-diethylamino-1-methyl-2,4(1H,3H)-quinazolinedione	100	1603
2-Chloro-4-diethylaminoquinazoline	75	H 246; 122

ALPHABETICAL LIST OF SIMPLE QUINAZOLINES (Continued)

Quinazoline	m.p. (°C), etc.	Ref.
4-Chloro-2-diethylaminoquinazoline	—	H 247
6-Chloro-2-diethylaminoquinazoline	91	849
7-Chloro-2-diethylaminoquinazoline	36, 138–140/0.1	849
7-Chloro-2-diethylamino-4(3H)-quinazolinone	220–223; HCl: 237–241	641
4-Chloro-1,3-diethyl-6-nitro-2,4(1H,3H)-quinazolinedine	140–141, nmr	2261
6-Chloro-3-β,γ-dihydroxypropyl-2-methyl-4(3H)-quinazolinone	—	H 181
4-Chloro-6,8-diiodoquinazoline	—	H 252
2-Chloro-6,7-diisopropoxy-4(3H)-quinazolinone	201–203	641
2-Chloro-6,7-dimethoxy-3-methyl-4(3H)-quinazolinone	crude, 196–214	641
2-Chloro-6,7-dimethoxy-4-phenethylaminoquinazoline	150–152	1810
2-Chloro-6,7-dimethoxy-4-quinazolinamine	—	1175,2065
4-Chloro-5,6-dimethoxyquinazoline	111–112	2564
4-Chloro-6,8-dimethoxyquinazoline	180–181	2564
6-Chloro-2,4-dimethoxyquinazoline	—	H 264
8-Chloro-6,7-dimethoxy-2,4(1H,3H)-quinazolinedione	240–241, ir, nmr	1146
2-Chloro-6,7-dimethoxy-4(3H)-quinazolinone	269–272	641,1580
7-Chloro-3-dimethylamino-2,4-dioxo-1,2,3,4,-tetrahydro-6-quinazolinesulfonamide	235–238	1603
2-Chloro-4-β-dimethylaminoethylaminoquinazoline	—	H 246
6-Chloro-3-β-dimethylaminoethyl-2-phenyl-4(3H)-quinazolinethione	142	716
6-Chloro-1-β-dimethylaminoethyl-4-phenyl-2(1H)-quinazolinone	165–166	676
6-Chloro-1-β-dimethylaminoethyl-4-phenyl-2(1H)-quinazolinone N(ω)-oxide	170–171, nmr	676
7-Chloro-3-dimethylamino-1-ethyl-2,4(1H,3H)-quinazolinedione	143	1603
7-Chloro-3-dimethylamino-1-isopropyl-2,4-dioxo-1,2,3,4-tetrahydro-6-quinazolinesulfonamide	297	1603

7-Chloro-3-dimethylamino-1-isopropyl-2,4(1H,3H)-quinazolinedione	100	1603
7-Chloro-3-dimethylamino-1-methyl-2,4-dioxo-1,2,3,4-tetrahydro-6-quinazolinesulfonamide	285–292	1603
6-Chloro-2-dimethylaminomethyl-4-phenylquinazoline	73–75	1279
6-Chloro-2-dimethylaminomethyl-4-phenylquinazoline 3-oxide	133–134 or 136, ir, ms, nmr	H 463; 1279,2500
4-Chloro-2-dimethylamino-7-methylquinazoline	ir, nmr	1160
6-Chloro-4-dimethylamino-2-methylquinazoline	130–131, nmr	2506
7-Chloro-3-dimethylamino-1-methyl-2,4(1H,3H)-quinazolinedione	212	1603
5-Chloro-3-dimethylamino-2-methyl-4(3H)-quinazolinone	—	H 378
6-Chloro-3-dimethylamino-2-methyl-4(3H)-quinazolinone	138	H 378; 1629
7-Chloro-3-dimethylamino-2-methyl-4(3H)-quinazolinone	—	H 378
7-Chloro-3-dimethylamino-4-oxo-3,4-dihydro-6-quinazolinesulfonamide	221–222	648
2-Chloro-4-γ-dimethylaminopropylaminoquinazoline	—	H 246
6-Chloro-4-γ-dimethylaminopropylaminoquinazoline	—	H 250
7-Chloro-4-γ-dimethylaminopropylaminoquinazoline	—	H 250
5-Chloro-3-γ-dimethylaminopropyl-2-ethyl-4(3H)-quinazolinone	132–133	2299
5-Chloro-3-γ-dimethylaminopropyl-2-methyl-4(3H)-quinazolinone	136–138	2299
2-Chloro-4-dimethylaminoquinazoline	67 or 104–106 (?)	H 246; 166,1410
4-Chloro-2-dimethylaminoquinazoline	ir, nmr	1160
7-Chloro-2-dimethylaminoquinazoline	132	849
5-Chloro-7-dimethylamino-2,4(1H,3H)-quinazolinedione	>280, nmr	1584
6-Chloro-3-dimethylamino-2,4(1H,3H)-quinazolinedione	260 or 261–263	1437,1603
7-Chloro-3-dimethylamino-2,4(1H,3H)-quinazolinedione	300–302	1603
7-Chloro-6-dimethylamino-2,4(1H,3H)-quinazolinedione	>280, nmr	1584
6-Chloro-3-dimethylamino-4(3H)-quinazolinone	102	1471
2-(β-Chloro-α,α-dimethylethyl)-4-phenylquinazoline	110, ir	1946
5-Chloro-1,3-dimethyl-6-nitro-2,4(1H,3H)-quinazolinone	162–163, nmr	1466,2261
7-Chloro-1,3-dimethyl-6-nitro-2,4(1H,3H)-quinazolinone	181–182, ir, nmr	1358

ALPHABETICAL LIST OF SIMPLE QUINAZOLINES (*Continued*)

Quinazoline	m.p. (°C), etc.	Ref.
7-Chloro-2,3-dimethyl-4-oxo-3,4-dihydro-6-quinazolinesulfonamide	—	H 181,487
2-Chloro-4-(1,1-dimethylprop-2-ynylamino)quinazoline	129–130, ir, ms, nmr	643
2-Chloro-6,7-dimethyl-4(3H)-quinazolinone	239–240	641
6-Chloro-1,2-dimethyl-4(1H)-quinazolinone	150–151, ir, uv	1190
6-Chloro-2,3-dimethyl-4(3H)-quinazolinone	145–147 or 151–152, ir, nmr; SbCl$_5$,HCl: 189–191, ir, nmr	942,2361
6-Chloro-3,4-dimethyl-2(3H)-quinazolinone	239	110
7-Chloro-1,2-dimethyl-4(1H)-quinazolinone	200–202, ir, uv	1190
7-Chloro-2,3-dimethyl-4(3H)-quinazolinone	149–150	942,2406
6-Chloro-2,4-dimorpholinoquinazoline	133–134, ir, nmr	428
6-Chloro-2,4-dioxo-1,2,3,4-tetrahydro-3-quinazolinecarbonitrile	245, ir, nmr	877
7-Chloro-2,4-dioxo-1,2,3,4-tetrahydro-6-quinazolinesulfonamide	—	H 203
6-Chloro-2,4-dipentyloxyquinazoline	—	H 264
6-Chloro-2,4-diphenylquinazoline	192–195, ms	H 251; 773,1181,1802,1824,1946
6-Chloro-2,4-diphenylquinazoline 3-oxide	188	83
6-Chloro-2,3-diphenyl-4(3H)-quinazolinone	175–176	942
7-Chloro-2,3-diphenyl-4(3H)-quinazolinone	180–181	942
6-Chloro-2-dipropylaminoquinazoline	78	849
7-Chloro-2-dipropylaminoquinazoline	32,140–142/0.08	849
5-Chloro-3-ethoxycarbonylamino-2-methyl-4(3H)-quinazolinone	—	H 378
6-Chloro-3-ethoxycarbonylamino-2-methyl-4(3H)-quinazolinone	—	H 378
7-Chloro-3-ethoxycarbonylamino-2-methyl-4(3H)-quinazolinone	—	H 378
5-Chloro-3-ethoxycarbonylamino-4(3H)-quinazolinone	—	H 374
6-Chloro-2-β-ethoxycarbonylethyl-4-phenylquinazoline	72–73	220

Compound		
6-Chloro-2-β-ethoxycarbonylethyl-4(3H)-quinazolinone	227–229	220
6-Chloro-3-ethoxycarbonylmethyl-2-methyl-4(3H)-quinazolinone	80	613
7-Chloro-3-ethoxycarbonylmethyl-8-methyl-2-thioxo-1,2-dihydro-4(3H)-quinazolinone	234	1137
6-Chloro-3-ethoxycarbonylmethyl-2-phenyl-4(3H)-quinazolinethione	170	1565
2-Chloro-3-ethoxycarbonylmethyl-4(3H)-quinazolinone	100–101	1137
6-Chloro-2-ethoxycarbonylmethyl-4(3H)-quinazolinone	198–200	531
6-Chloro-3-ethoxycarbonylmethyl-4(3H)-quinazolinone	146–148, ir, ms, nmr	H 178; 1640,1838
5-Chloro-3-β-ethoxyethyl-2-methyl-4(3H)-quinazolinone	150–151	2299
7-Chloro-6-ethoxy-2-guanidino-4-methylquinazoline	—	H 343
6-Chloro-4-ethoxy-2-methylquinazoline	—	H 264
2-Chloro-4-ethoxyquinazoline	—	H 263
4-Chloro-6-ethoxyquinazoline	120–121	2564
5-Chloro-2-ethoxyquinazoline	crude, 55–57	2230
6-Chloro-2-ethoxyquinazoline	91	849
6-Chloro-4-ethoxyquinazoline	—	H 264
6-Chloro-5-ethoxy-6-quinazolinol	158–159, uv	595
7-Chloro-2-ethoxy-4(3H)-quinazolinone	222–224, nmr	2230
6-Chloro-3-β-ethylaminoethyl-4(3H)-quinazolinone	—	H 178
6-Chloro-2-ethylamino-4-phenylquinazoline 3-oxide	197	399
2-Chloro-4-ethylaminoquinazoline	171	H 246; 1379,1410
2-β-Chloroethyl-5,8-dihydroxy-4(3H)-quinazolinone	197–200, ir, nmr	2430
1-β-Chloroethyl-2-α-ethoxycarbonylethyl-4(1H)-quinazolinone	133–134, ir, nmr	1415
3-β-Chloroethyl-2-ethoxy-4(3H)-quinazolinone	—	H 172
6-Chloro-3-ethyl-2-hydrazino-4(3H)-quinazolinone	218–220, ir, ms	2279
7-Chloro-3-ethyl-2-hydrazino-4(3H)-quinazolinone	230, ms, nmr	2123
3-β-Chloroethyl-6-iodo-4(3H)-quinazolinone	—	H 178
3-β-Chloroethyl-8-methoxy-4(3H)-quinazolinone	—	H 178

ALPHABETICAL LIST OF SIMPLE QUINAZOLINES (Continued)

Quinazoline	m.p. (°C), etc.	Ref.
6-Chloro-2-ethyl-3-methylamino-4(3H)-quinazolinone	154–155	934
3-β-Chloroethyl-N-methyl-4-oxo-3,4-dihydro-2-quinazolinecarboxamide	206–207	768
2-α-Chloroethyl-3-methyl-4(3H)-quinazolinone	238	2341
3-β-Chloroethyl-2-methyl-4(3H)-quinazolinone	163–164	1045
3-β-Chloroethyl-6-methyl-4(3H)-quinazolinone	—	H 178
3-β-Chloroethyl-8-methyl-4(3H)-quinazolinone	—	H 178
6-Chloro-3-ethyl-2-methylthio-4(3H)-quinazolinone	124	670
6-Chloro-2-ethyl-3-morpholino-4(3H)-quinazolinone	176–177	1629
7-Chloro-2-ethyl-4-oxo-3,4-dihydro-6-quinazolinesulfonamide	309–312, ir, ms, nmr, uv	H 174; 577,818,917
5-Chloro-2-ethyl-3-phenethyl-4(3H)-quinazolinone	164–166	2299
6-Chloro-2-ethyl-4-phenylquinazoline	103–104 or 110	237,1946
2-β-Chloroethyl-1-phenyl-4(1H)-quinazolinone	205–210	1394
6-Chloro-1-ethyl-4-phenyl-2(1H)-quinazolinone	167–168 or 173–174	681,1060
6-Chloro-2-ethyl-3-phenyl-4(3H)-quinazolinone	—	H 187
7-Chloro-1-ethyl-4-phenyl-2(1H)-quinazolinone	187–188	681
6-Chloro-3-ethyl-2-propylthio-4-(3H)-quinazolinone	197	670
4-Chloro-2-ethylquinazoline	55–57, nmr	188,1069
6-Chloro-N-ethyl-2-quinazolinecarboxamide	225, nmr	2178
7-Chloro-N-ethyl-2-quinazolinecarboxamide	170–171, nmr	2178
3-β-Chloroethyl-2,4(1H,3H)-quinazolinedione	193–196, ir, nmr	1310,1428,1442,1448,1931
6-Chloro-1-ethyl-2,4(1H,3H)-quinazolinedione	264–266	2295
2-α-Chloroethyl-4(3H)-quinazolinethione	169	2341
2-α-Chloroethyl-4(3H)-quinazolinone	220 or 252, ir, nmr	2069,2341
3-β-Chloroethyl-4(3H)-quinazolinone	120–123	H 144; 276

480

6-Chloro-2-ethyl-4(3H)-quinazolinone	258–259 or 276, ms, nmr	H 174; 566,957
7-Chloro-2-ethyl-4(3H)-quinazolinone	238–240	566,917
7-Chloro-2-ethyl-4(1H)-quinazolinone 3-oxide	—	H 467
6-Chloro-2-ethylthiomethyl-4-phenylquinazoline 3-oxide	162	82
6-Chloro-2-ethylthio-3-methyl-4(3H)-quinazolinone	95	670
6-Chloro-4-ethylthio-2-morpholinoquinazoline	95	413
6-Chloro-2-ethylthio-3-phenethyl-4(3H)-quinazolinone	101–103	615
6-Chloro-2-ethylthio-3-phenyl-4(3H)-quinazolinone	132–135 or 238	566,670
7-Chloro-2-ethylthio-3-phenyl-4(3H)-quinazolinone	135–136	566
6-Chloro-3-ethyl-2-thioxo-1,2-dihydro-4(3H)-quinazolinone	217	670
7-Chloro-6-fluoro-4-methyl-2-quinazolinamine	208–210, nmr	2569
2-Chlorofluoromethyl-4(3H)-quinazolinone	223–225, ir	1188
4-Chloro-5-fluoroquinazoline	141	168
4-Chloro-6-fluoroquinazoline	130	168
4-Chloro-7-fluoroquinazoline	92	168
5-Chloro-4-fluoroquinazoline	124–125, uv	553
6-Chloro-4-fluoroquinazoline	109–110, uv	553
7-Chloro-4-fluoroquinazoline	95–96, uv	553
8-Chloro-4-fluoroquinazoline	114–115	553
6-Chloro-7-fluoro-2,4-quinazolinediamine	268–270, nmr	2335
6-Chloro-8-fluoro-2,4-quinazolinediamine	>295, nmr	2335
8-Chloro-5-fluoro-2,4-quinazolinediamine	265–267, nmr	2335
7-Chloro-6-fluoro-2-styryl-4(3H)-quinazolinone	296–301	2264
7-Chloro-3-formamido-4-oxo-3,4-dihydro-6-quinazolinesulfonamide	284	648
6-Chloro-2-formyloxymethyl-4-phenylquinazoline 3-oxide	118	82
7-Chloro-2-guanidino-4-methylquinazoline	—	H 343
6-Chloro-2-guanidino-4-phenylquinazoline	—	H 343
6-Chloro-3-hydrazinocarbonylmethyl-2-methyl-4(3H)-quinazolinone	260	613

ALPHABETICAL LIST OF SIMPLE QUINAZOLINES (*Continued*)

Quinazoline	m.p. (°C), etc.	Ref.
2-Chloro-4-hydrazino-6-methylquinazoline	—	*H* 246
6-Chloro-2-hydrazino-3-methyl-4(3*H*)-quinazolinone	242–243, ir, ms	2279
7-Chloro-2-hydrazino-3-methyl-4(3*H*)-quinazolinone	242, ms, nmr	2123
6-Chloro-4-hydrazino-2-morpholinoquinazoline	200–202	1337
6-Chloro-2-hydrazino-4-phenylquinazoline	177–178	836,1021,1718,1976
6-Chloro-2-hydrazono-1-methyl-4-phenyl-1,2-dihydroquinazoline	—	1976
6-Chloro-3-β-hydroxyethyl-2-methyl-4(3*H*)-quinazolinone	—	*H* 181
6-Chloro-3-β-hydroxyethyl-4(3*H*)-quinazolinone	169–171	*H* 178; 498
6-Chloro-2-hydroxymethyl-4-phenylquinazoline	129–130, ir, nmr	229
5-Chloro-7-hydroxy-3-phenyl-4(3*H*)-quinazolinone	>280, ir, nmr	2112
6-Chloro-3-γ-hydroxypropyl-2-thioxo-1,2-dihydro-4(3*H*)-quinazolinone	255	125
6-Chloro-2-imino-1,3-dimethyl-1,2-dihydro-4(3*H*)-quinazolinone	157, ir, nmr	947
6-Chloro-2-iodomethyl-4-phenylquinazoline	133–134	632
6-Chloro-2-iodomethyl-4-phenylquinazoline 3-oxide	173	82
4-Chloro-6-iodoquinazoline	—	*H* 252
5-Chloro-6-iodo-2,4-quinazolinediamine	226–228, nmr	2246
6-Chloro-8-iodo-2,4(1*H*,3*H*)-quinazolinedione	310	308
6-Chloro-8-iodo-4(3*H*)-quinazolinone	—	*H* 154
8-Chloro-6-iodo-4(3*H*)-quinazolinone	—	*H* 154
6-Chloro-8-iodo-2-thioxo-1,2-dihydro-4(3*H*)-quinazolinone	320–322	308
5-Chloro-1-isobutyl-3-isopentyl-6-nitro-2,4(1*H*,3*H*)-quinazolinedione	100–101	1581
7-Chloro-1-isobutyl-3-isopentyl-6-nitro-2,4(1*H*,3*H*)-quinazolinedione	110–111, nmr	1581
5-Chloro-1-isobutyl-3-methyl-6-nitro-2,4(1*H*,3*H*)-quinazolinedione	145–146, nmr	2261
7-Chloro-1-isobutyl-3-methyl-6-nitro-2,4(1*H*,3*H*)-quinazolinedione	156–157, nmr	2261

Compound	Data	Ref.
5-Chloro-3-isopentyl-6-nitro-2,4(1*H*,3*H*)-quinazolinedione	218–221, nmr	1581
7-Chloro-3-isopentyl-6-nitro-2,4(1*H*,3*H*)-quinazolinedione	204–206, nmr	1581
5-Chloro-3-isopentyl-2,4(1*H*,3*H*)-quinazolinedione	194–195, nmr	1581
7-Chloro-3-isopentyl-2,4(1*H*,3*H*)-quinazolinedione	262–265, nmr	1581
6-Chloro-2-isopropenyl-4-phenylquinazoline	—	312
6-Chloro-2-isopropenyl-4-phenylquinazoline 3-oxide	—	312
2-Chloro-4-isopropoxyquinazoline	—	H 263
6-Chloro-2-isopropoxyquinazoline	130–134/0.08	849
7-Chloro-1-isopropyl-6-methyl-4-phenyl-2(1*H*)-quinazolinone	190–191	681
6-Chloro-3-isopropyl-2-methyl-4(3*H*)-quinazolinone	83–84, ir, nmr	2361
7-Chloro-2-isopropyl-4-oxo-3,4-dihydro-6-quinazolinesulfonamide	—	H 175, 487
6-Chloro-1-isopropyl-4-phenyl-2(1*H*)-quinazolinethione	216–217, ir, nmr, uv	955
5-Chloro-1-isopropyl-2-phenyl-4(1*H*)-quinazolinone	222–224	1590
6-Chloro-1-isopropyl-2-phenyl-4(1*H*)-quinazolinone	273–274	1590
6-Chloro-1-isopropyl-4-phenyl-2(1*H*)-quinazolinone	138–139 or 149–150	681,1060
7-Chloro-1-isopropyl-2-phenyl-4(1*H*)-quinazolinone	148–150	1590
7-Chloro-1-isopropyl-4-phenyl-2(1*H*)-quinazolinone	165–168	681
7-Chloro-2-isopropyl-1-phenyl-4(1*H*)-quinazolinone	240–243	1590
2-Chloro-4-isopropylquinazoline	—	H 246
4-Chloro-2-isopropylquinazoline	68–70, nmr	188
6-Chloro-2-isopropyl-4(3*H*)-quinazolinethione	222, ms, nmr	957
6-Chloro-2-mercaptomethyl-4-phenylquinazoline 3-oxide	208	82
6-Chloro-2-methanesulfonyloxymethyl-4-phenylquinazoline 3-oxide	—	H 463
6-Chloro-2-β-methoxycarbonylethyl-4-phenylquinazoline	118–120	220
6-Chloro-3-β-methoxycarbonylethyl-2-phenyl-4(3*H*)-quinazolinethione	183	1565
7-Chloro-3-methoxycarbonylmethyl-8-methyl-2-methylthio-4(3*H*)-quinazolinone	164	1137
6-Chloro-3-methoxycarbonylmethyl-2-methylthio-4(3*H*)-quinazolinone	170–171	1137
7-Chloro-3-methoxycarbonylmethyl-2-methylthio-4(3*H*)-quinazolinone	160	1137

ALPHABETICAL LIST OF SIMPLE QUINAZOLINES (*Continued*)

Quinazoline	m.p. (°C), etc.	Ref.
6-Chloro-1-methoxycarbonylmethyl-2-phenyl-4(1*H*)-quinazolinone	198–200, ir, ms, nmr	1825
6-Chloro-1-methoxycarbonylmethyl-2,4(1*H*,3*H*)-quinazolinedione	265–268, ir, ms, nmr	1822
6-Chloro-1-methoxycarbonylmethyl-4(1*H*)-quinazolinone	205, ir, ms, nmr	1825
6-Chloro-3-methoxycarbonylmethyl-4(3*H*)-quinazolinone	162–164, ir, ms	1838
6-Chloro-3-methoxycarbonylmethyl-2-thioxo-1,2-dihydro-4(3*H*)-quinazolinone	247	1137
7-Chloro-3-methoxycarbonylmethyl-2-thioxo-1,2-dihydro-4(3*H*)-quinazolinone	222–223	1137
6-Chloro-3-β-methoxycarbonylvinyl-4(3*H*)-quinazolinone	171–172	1558
6-Chloro-8-methoxy-5-methyl-2-phenyl-4(3*H*)-quinazolinone	277–279	103
4-Chloro-6-methoxy-2-methylquinazoline	—	1113
4-Chloro-7-methoxy-2-methylquinazoline	110	168
6-Chloro-4-methoxy-2-methylquinazoline	—	H 264
4-Chloro-7-methoxy-6-methylthioquinazoline	167–168, ir, nmr	2017
4-Chloro-6-methoxy-7-nitroquinazoline	—	H 264
7-Chloro-3-methoxy-4-oxo-3,4-dihydro-6-quinazolinesulfonamide	249–252	648
4-Chloro-6-methoxy-2-phenylquinazoline	126 to 135	503, 707, 766
6-Chloro-4-methoxy-2-phenylquinazoline	107, nmr, uv	58
5-Chloro-7-methoxy-3-phenyl-4(3*H*)-quinazolinone	175–176, ir, nmr	2112
5-Chloro-3-γ-methoxypropyl-2-methyl-4(3*H*)-quinazolinone	137	2299
6-Chloro-8-methoxy-3-prop-2′-ynyl-4(3*H*)-quinazolinone	—	H 178
2-Chloro-8-methoxy-4-quinazolinamine	228	696
4-Chloro-8-methoxy-2-quinazolinamine	crude: 188–190	2595
2-Chloro-4-methoxyquinazoline	—	H 263
4-Chloro-6-methoxyquinazoline	—	H 264
4-Chloro-7-methoxyquinazoline	—	H 264

484

Compound		
4-Chloro-8-methoxyquinazoline	134–135	2564
6-Chloro-2-methoxyquinazoline	114	849
2-Chloro-6-methoxy-4(3*H*)-quinazolinone	232–235	641
2-Chloro-7-methoxy-4(3*H*)-quinazolinone	231–233	641
2-Chloro-8-methoxy-4(3*H*)-quinazolinone	188–193	641
5-Chloro-6-methoxy-4(3*H*)-quinazolinone	—	*H* 154
5-Chloro-8-methoxy-4(3*H*)-quinazolinone	—	*H* 154
7-Chloro-6-methoxy-4(3*H*)-quinazolinone	—	*H* 154
6-Chloro-2-α-methylaminobutyl-4-phenylquinazoline 3-oxide	—	*H* 463
6-Chloro-3-β-methylaminoethyl-2-phenyl-4(3*H*)-quinazolinone	112	716
6-Chloro-3-β-methylaminoethyl-4(3*H*)-quinazolinone	113	716
6-Chloro-2-methylamino-4-phenylquinazoline	—	*H* 250
6-Chloro-2-methylamino-4-phenylquinazoline 3-oxide	236–238, nmr, uv	771
6-Chloro-2-methylamino-3-phenyl-4(3*H*)-quinazolinone	225–228, ir, uv	771
6-Chloro-3-methylamino-2-phenyl-4(3*H*)-quinazolinone	100–101	934
2-Chloro-4-methylaminoquinazoline	213	1410
6-Chloro-4-methylaminoquinazoline	256–257	916
6-Chloro-3-methylamino-4(3*H*)-quinazolinone	151–152	934
2-Chloromethyl-5,8-dihydro-2,5,8(3*H*)-quinazolinetrione	166–172, ir, nmr	2324
2-Chloromethyl-5,8-dihydroxy-4(3*H*)-quinazolinone	anal, ms, nmr	1999
2-Chloromethyl-6,7-dimethoxy-4-methylquinazoline	154–156, nmr, uv	1923
4-Chloromethyl-5,7-dimethoxy-2(1*H*)-quinazolinone	250–252, ir, nmr, uv	2501
4-Chloromethyl-6,7-dimethoxy-2(1*H*)-quinazolinone	255–257, ir, nmr, uv	2501
2-Chloromethyl-6,7-dimethyl-4-phenylquinazoline 3-oxide	—	*H* 463
2-Chloromethyl-6,8-dimethyl-4-phenylquinazoline 3-oxide	—	*H* 463
2-Chloromethyl-3-ethoxycarbonylamino-6-nitro-4(3*H*)-quinazolinone	146–147, ir	2360
2-Chloromethyl-3-ethoxycarbonylamino-4(3*H*)-quinazolinone	148, ir, nmr	2360
2-Chloromethyl-3-ethoxycarbonyloxy-4(3*H*)-quinazolinone	120	236

485

ALPHABETICAL LIST OF SIMPLE QUINAZOLINES (*Continued*)

Quinazoline	m.p. (°C), etc.	Ref.
2-(α-Chloro-α-methylethyl)-8-methoxy-4(3*H*)-quinazolinone	201	1224
2-(α-Chloro-α-methylethyl)-6-methyl-4(3*H*)-quinazolinone	243	1224
3-(α-Chloro-α-methylethyl)-2,4(1*H*,3*H*)-quinazolinedione	—	*H* 199
6-Chloro-2-*N*-methylhydrazinomethyl-4-phenylquinazoline 3-oxide	—	*H* 463
2-Chloromethyl-4-hydrazinoquinazoline	138–140	893
2-Chloro-4-α-methylhydrazinoquinazoline	153–155, nmr	2230
2-Chloromethyl-3-methoxy-6-nitro-4(3*H*)-quinazolinone	167–168	441
2-Chloromethyl-3-methoxy-4(3*H*)-quinazolinone	127 or 129	236,441
4-Chloromethyl-7-methoxy-2(1*H*)-quinazolinone	220–222, ir, nmr, uv	2501
6-Chloro-1-methyl-2-methylamino-4(1*H*)-quinazolinone	362–363, ir, nmr	947
6-Chloro-2-methyl-3-methylamino-4(3*H*)-quinazolinone	132–133 or 134, ir, nmr	934,1994
4-Chloromethyl-6-methyl-2-phenylquinazoline	126–127	694
2-Chloromethyl-6-methyl-4-phenylquinazoline 3-oxide	—	*H* 463
2-Chloromethyl-4-methylquinazoline	58–60 or 162–164, ir, ms, nmr, uv	874,1923
2-Chloromethyl-4-methylquinazoline 3-oxide	165 or 172–175, ir, ms, nmr	*H* 463; 874,2500
2-Chloromethyl-3-methyl-4(3*H*)-quinazolinone	—	*H* 169
2-Chloromethyl-6-methyl-4(3*H*)-quinazolinone	248–249	510
2-Chloromethyl-6-methylthio-4-phenylquinazoline 3-oxide	—	*H* 463
6-Chloro-3-methyl-2-methylthio-4(3*H*)-quinazolinone	158	670
7-Chloro-1-methyl-3-morpholino-2,4(1*H*,3*H*)-quinazolinedione	272	1603
6-Chloro-2-methyl-3-morpholino-4(3*H*)-quinazolinone	194–195	1629
6-Chloro-2-(β-methyl-γ-nitrobutyl)-4-phenylquinazoline 3-oxide	109–111, ir, ms, nmr	861
2-Chloromethyl-6-nitro-4-phenylquinazoline 3-oxide	—	*H* 463
2-Chloromethyl-6-nitro-3-phenyl-4(3*H*)-quinazolinone	228–230	376

486

4-Chloro-2-methyl-6-nitroquinazoline	—	1501
4-Chloro-7-methyl-6-nitroquinazoline	—	H 247
5-Chloro-3-methyl-6-nitro-2,4(1H,3H)-quinazolinedione	320–322, nmr	2261
6-Chloro-3-methyl-5-nitro-4(3H)-quinazolinone	203–205, ir, nmr	2584
6-Chloro-3-methyl-7-nitro-4(3H)-quinazolinone	235–236, ir, nmr	2584
6-Chloro-3-methyl-8-nitro-4(3H)-quinazolinone	—	H 177
7-Chloro-3-methyl-5-nitro-4(3H)-quinazolinone	200–202, nmr	900,1165
5-Chloro-2-methyl-4-oxo-3,4-dihydro-6-quinazolinesulfonamide	>300, ir, nmr	1384
6-Chloro-3-methyl-4-oxo-3,4-dihydro-5-quinazolinesulfonamide	235–240, ir, ms, nmr	2584
7-Chloro-2-methyl-4-oxo-3,4-dihydro-6-quinazolinesulfonamide	—	H 173,487
7-Chloro-3-methyl-4-oxo-3,4-dihydro-6-quinazolinesulfonamide	—	H 177,487
7-Chloro-2-methyl-4-oxo-1,4-dihydro-6-quinazolinesulfonamide 3-oxide	287–288	648
6-Chloro-3-methyl-4-oxo-3,4-dihydro-5-quinazolinesulfonyl chloride	crude, nmr	2584
5-Chloro-2-methyl-4-oxo-3-phenyl-3,4-dihydro-6-quinazolinesulfonamide	>300, ir	1384
5-Chloro-2-methyl-3-phenethyl-4(3H)-quinazolinone	151–153	2299
6-Chloro-2-methyl-3-phenethyl-4(3H)-quinazolinone	—	H 181
2-Chloro-6-methyl-4-phenylquinazoline	139–140	H 246; 836
4-Chloromethyl-2-phenylquinazoline	89–90	694
4-Chloro-6-methyl-2-phenylquinazoline	111–112	756
4-Chloro-8-methyl-2-phenylquinazoline	99–101	756
6-Chloro-2-methyl-4-phenylquinazoline	106 to 109, ir, nmr	H 250; 257,490,773,1305,1382, 1426,2504
6-Chloro-4-methyl-2-phenylquinazoline	141–143, uv	1083
6-Chloro-N-methyl-4-phenyl-2-quinazolinecarboxamide	194–197	242
2-Chloromethyl-4-phenylquinazoline 3-oxide	159, ir, ms, nmr	H 463; 2500
6-Chloro-2-methyl-4-phenylquinazoline 1-oxide	—	H 461
6-Chloro-2-methyl-4-phenylquinazoline 3-oxide	—	H 463; 2504
6-Chloro-3-methyl-4-phenyl-2-(3H)-quinazolinethione	—	H 292

487

ALPHABETICAL LIST OF SIMPLE QUINAZOLINES (*Continued*)

Quinazoline	m.p. (°C), etc.	Ref.
2-Chloromethyl-1-phenyl-4(1*H*)-quinazolinone	212–213	1394
2-Chloromethyl-3-phenyl-4(3*H*)-quinazolinone	149 to 154, nmr	376,1033,1550,1806
5-Chloro-2-methyl-3-phenyl-4(3*H*)-quinazolinone	144	2299
6-Chloro-1-methyl-2-phenyl-4(1*H*)-quinazolinone	203–205, ir, uv	1190
6-Chloro-1-methyl-4-phenyl-2(1*H*)-quinazolinone	218 to 227	222,492,493,681,836,1060,1200, 1408,1440,1976
6-Chloro-2-methyl-3-phenyl-4(3*H*)-quinazolinone	180–181	*H* 184; 942
6-Chloro-3-methyl-4-phenyl-2(3*H*)-quinazolinone	—	*H* 137
7-Chloro-1-methyl-2-phenyl-4(1*H*)-quinazolinone	152–154, ir, uv	1190
7-Chloro-1-methyl-4-phenyl-2(1*H*)-quinazolinone	188–190	681
7-Chloro-2-methyl-3-phenyl-4(3*H*)-quinazolinone	173–175	942
7-Chloro-8-methyl-2-phenyl-4(3*H*)-quinazolinone	314	103
8-Chloro-2-methyl-3-phenyl-4(3*H*)-quinazolinone	237–238	687
2-Chloromethyl-4-phenyl-6-trifluoromethylquinazoline 3-oxide	—	*H* 463
7-Chloro-1-methyl-3-piperidino-2,4(1*H*,3*H*)-quinazolinedione	231	1603
6-Chloro-2-methyl-3-piperidino-4(3*H*)-quinazolinone	160–162	1629
5-Chloro-2-methyl-3-propyl-4(3*H*)-quinazolinone	194–195	2299
6-Chloro-2-methyl-3-propyl-4(3*H*)-quinazolinone	82–84, ir, nmr; SbCl₅.HCl: 183–185, ir, nmr	2361
6-Chloro-3-methyl-2-propylthio-4(3*H*)-quinazolinone	224	670
2-Chloro-6-methyl-4-quinazolinamine	—	*H* 245
5-Chloro-2-methyl-4-quinazolinamine	199–200, nmr	2247
6-Chloro-4-methyl-2-quinazolinamine	—	*H* 342
2-Chloromethylquinazoline	—	*H* 61

488

2-Chloro-4-methylquinazoline	74–76, ms, uv	*H* 246; 163
4-Chloro-2-methylquinazoline	78–79, ms, nmr	*H* 247; 1127
4-Chloro-5-methylquinazoline	115–116	*H* 247; 1814
4-Chloro-6-methylquinazoline	105–106	*H* 247; 1001
4-Chloro-7-methylquinazoline	—	*H* 247
4-Chloro-8-methylquinazoline	126	*H* 247; 1001
6-Chloro-5-methyl-2,4-quinazolinediamine	281–284, nmr	2246
7-Chloro-8-methyl-2,4-quinazolinediamine	227	1852
8-Chloro-5-methyl-2,4-quinazolinediamine	260–262, nmr	2246
5-Chloro-3-methyl-2,4(1*H*,3*H*)-quinazolinedione	>300, nmr	2261
6-Chloro-1-methyl-2,4(1*H*,3*H*)-quinazolinedione	296–299	2295
6-Chloro-3-methyl-2,4(1*H*,3*H*)-quinazolinedione	271–273, ir, uv	771
7-Chloro-3-methyl-2,4(1*H*,3*H*)-quinazolinedione	280, ir, nmr	2216
6-Chloro-2-methyl-4(3*H*)-quinazolinethione	275, ms, nmr	*H* 296; 957
2-Chloromethyl-4(3*H*)-quinazolinone	246–250, ir, ms, nmr	*H* 139; 2116,2151,2195
3-Chloromethyl-4(3*H*)-quinazolinone	HCl: >280	1192
5-Chloro-2-methyl-4(3*H*)-quinazolinone	292	*H* 173; 147
5-Chloro-6-methyl-4(3*H*)-quinazolinone	—	*H* 154
5-Chloro-8-methyl-4(3*H*)-quinazolinone	—	*H* 154
6-Chloro-2-methyl-4(3*H*)-quinazolinone	282 to 287, >325 (?), ms, nmr	*H* 173; 147,942,957,2264
6-Chloro-3-methyl-4(3*H*)-quinazolinone	111–112	498
6-Chloro-4-methyl-2(1*H*)-quinazolinone	286–288 or >340	1060,1592
6-Chloro-5-methyl-4(3*H*)-quinazolinone	—	*H* 154
6-Chloro-8-methyl-4(3*H*)-quinazolinone	—	*H* 154
7-Chloro-2-methyl-4(3*H*)-quinazolinone	262 to 270, ir, nmr	*H* 179; 103,147,566,863,942
7-Chloro-6-methyl-4(3*H*)-quinazolinone	—	*H* 154
7-Chloro-8-methyl-4(3*H*)-quinazolinone	—	*H* 154
8-Chloro-2-methyl-4(3*H*)-quinazolinone	247	147

489

ALPHABETICAL LIST OF SIMPLE QUINAZOLINES   (*Continued*)

Quinazoline	m.p. (°C), etc.	Ref.
8-Chloro-3-methyl-4(3*H*)-quinazolinone	158–159, nmr	990
8-Chloro-6-methyl-4(3*H*)-quinazolinone	—	*H* 154
2-Chloromethyl-4(1*H*)-quinazolinone 3-oxide	210, ir	236
6-Chloro-1-methyl-2-styryl-4(1*H*)-quinazolinone	194–197, ir, uv	1190
7-Chloro-1-methyl-2-styryl-4(1*H*)-quinazolinone	290–293, ir, uv	1190
6-Chloro-2-methylthiomethyl-4-phenylquinazoline 3-oxide	198	82
6-Chloro-2-methylthio-3-phenyl-4(3*H*)-quinazolinone	186	670
2-Chloro-4-methylthioquinazoline	119–120	*H* 299; 1410
4-Chloro-6-methylthioquinazoline	134–135	2564
6-Chloro-2-methylthio-4(3*H*)-quinazolinone	223–224	126
7-Chloro-2-methylthio-4(3*H*)-quinazolinone	280	126
5-Chloro-2-methyl-3-thioureido-4(3*H*)-quinazolinone	—	*H* 378
7-Chloro-2-methyl-3-thioureido-4(3*H*)-quinazolinone	—	*H* 378
6-Chloro-3-methyl-2-thioxo-1,2-dihydro-4(3*H*)-quinazolinone	214	670
5-Chloro-2-methyl-3-ureido-4(3*H*)-quinazolinone	—	*H* 378
6-Chloro-2-methyl-3-ureido-4(3*H*)-quinazolinone	—	*H* 378
7-Chloro-2-methyl-3-ureido-4(3*H*)-quinazolinone	—	*H* 378
6-Chloro-2-morpholinomethyl-4-phenylquinazoline	116–118	1279
6-Chloro-2-morpholinomethyl-4-phenylquinazoline 3-oxide	189–191	1279
2-Chloro-4-morpholinoquinazoline	—	*H* 246
7-Chloro-4-morpholinoquinazoline	101–102	1189
6-Chloro-3-morpholino-2,4(1*H*,3*H*)-quinazolinedione	324–326	1437
7-Chloro-6-morpholino-2,4(1*H*,3*H*)-quinazolinedione	>280, nmr	1584
6-Chloro-2-morpholino-4(3*H*)-quinazolinethione	190	413

Compound		Reference
6-Chloro-2-morpholino-4(3H)-quinazolinone	280–290 or 304	413,845
6-Chloro-3-morpholino-4(3H)-quinazolinone	187–188	1471
5-Chloro-7-nitro-2,3-diphenyl-4(3H)-quinazolinone	—	H 189
7-Chloro-6-nitro-1,3-dipropyl-2,4(1H,3H)-quinazolinedione	93–94, nmr	2261
4-Chloro-6-nitro-2-phenylquinazoline	—	H 247
4-Chloro-7-nitro-2-phenylquinazoline	—	H 247
5-Chloro-7-nitro-2-phenyl-4(3H)-quinazolinone	—	H 176
5-Chloro-7-nitro-2-phenyl-4(1H)-quinazolinone 3-oxide	—	H 467
6-Chloro-8-nitro-4-quinazolinamine	—	H 344
4-Chloro-5-nitroquinazoline	—	H 247
4-Chloro-6-nitroquinazoline	—	H 247
4-Chloro-7-nitroquinazoline	—	H 247
4-Chloro-8-nitroquinazoline	—	H 247
5-Chloro-6-nitro-2,4-quinazolinediamine	225–227	201
6-Chloro-7-nitro-2,4-quinazolinediamine	—	1852
5-Chloro-6-nitro-2,4(1H,3H)-quinazolinedione	>340, nmr	1466,2261
7-Chloro-6-nitro-2,4(1H,3H)-quinazolinedione	335, ir, nmr	1358
6-Chloro-5-nitro-4(3H)-quinazolinone	308–309, ms, nmr	502,2584
6-Chloro-7-nitro-4(3H)-quinazolinone	crude	2584
6-Chloro-8-nitro-4(3H)-quinazolinone	—	H 154
7-Chloro-6-nitro-4(3H)-quinazolinone	300–303, ms, nmr	501,909
7-Chloro-8-nitro-4(3H)-quinazolinone	>320, ms, nmr	501
4-Chloro-6-nitro-2-trichloromethylquinazoline	—	H 247
7-Chloro-4-ozo-3,4-dihydro-2-quinazolinecarbohydrazide	>355	1743
2-Chloro-4-oxo-3,4-dihydro-3-quinazolinecarbonitrile	174–175, ir	234
7-Chloro-4-oxo-3,4-dihydro-2-quinazolinecarboxamide	257	1743
7-Chloro-4-oxo-3,4-dihydro-6-quinazolinesulfonamide	—	H 154,487
7-Chloro-4-oxo-1,4-dihydro-6-quinazolinesulfonamide 3-oxide	285	648

ALPHABETICAL LIST OF SIMPLE QUINAZOLINES (*Continued*)

Quinazoline	m.p. (°C), etc.	Ref.
7-Chloro-4-oxo-3,4-dihydro-6-quinazolinesulfonic acid	—	H 154
7-Chloro-4-oxo-2-phenyl-3,4-dihydro-6-quinazolinesulfonamide	>320	583
6-Chloro-4-oxo-8-trifluoromethyl-3,4-dihydro-2-quinazolinecarboxylic acid	320	1549
6-Chloro-2-phenethyl-4(3*H*)-quinazolinone	245–250	2264
6-Chloro-3-phenethyl-2-thioxo-1,2-dihydro-4(3*H*)-quinazolinone	242–244	615
2-Chloro-4-phenoxyquinazoline	—	H 263
6-Chloro-4-phenyl-2-piperidinomethylquinazoline	78–80	1279
6-Chloro-4-phenyl-2-piperidinomethylquinazoline 3-oxide	147 or 149–150, ir, ms, nmr	1279,2500
6-Chloro-4-phenyl-2-prop-1′-enylquinazoline	—	312
6-Chloro-4-phenyl-2-prop-1′enylquinazoline 3-oxide	—	312
6-Chloro-3-phenyl-2-propylthio-4(3*H*)-quinazolinone	149	670
6-Chloro-4-phenyl-2-quinazolinamine	—	H 342; 1764
2-Chloro-4-phenylquinazoline	110–112 or 112–114	H 246; 343,836,1963
4-Chloro-2-phenylquinazoline	—	H 248; 260
6-Chloro-4-phenylquinazoline	136 to 139, complexes; 1-MeI: 212–213, ir, nmr; 1-MeBr: 230–231, ir, nmr	H 251; 353,454,772,959,1110, 1647,2170,2415,2473
6-Chloro-4-phenyl-2-quinazolinecarbaldehyde	173 to 178, ir, ms, nmr, uv; Me₂NN= 206–208; dnp: 273–275; oxime: 241–244; Me₂-acetal:—	176,361,578,822,1029,1119, 1231, 1395,1904,2427,2449
6-Chloro-4-phenyl-2-quinazolinecarbonitrile	180–182, ir	361
6-Chloro-4-phenyl-2-quinazolinecarboxamide	265–269	H 481; 222
6-Chloro-4-phenyl-2-quinazolinecarboxylic acid	210 to 216, nmr	H 482; 222,240,760,2009,2473
6-Chloro-3-phenyl-2,4(1*H*,3*H*)-quinazolinedione	303–305 or 308–309	H 203; 227,496,566,1312

492

Compound		
7-Chloro-1-phenyl-2,4(1H,3H)-quinazolinedione	308, ir, nmr	362,718
7-Chloro-3-phenyl-2,4(1H,3H)-quinazolinedione	332–335	566,1079
6-Chloro-2-phenyl-4(3H)-quinazolinethione	240	H 296; 210
6-Chloro-4-phenyl-2(1H)-quinazolinethione	235–236 or 243–244, ir, nmr, uv	836,955
6-Chloro-2-phenyl-4(1H)-quinazolinethione 3-oxide	—	H 468
2-Chloro-3-phenyl-4(3H)-quinazolinone	135–136	H 172; 1224
5-Chloro-2-phenyl-4(3H)-quinazolinone	279	147
6-Chloro-2-phenyl-4(3H)-quinazolinone	294 to 302, uv	58,147,214,942
6-Chloro-4-phenyl-2(1H)-quinazolinone	256–258 (?), 305 to 323, ir, ms, nmr, uv	H 136; 99,254,590,624,836,859, 899,1060,1200,1946,1976
7-Chloro-2-phenyl-4(3H)-quinazolinone	269 (?), 290 or 292	147,863,942
7-Chloro-4-phenyl-2(1H)-quinazolinone	284–285	1060
8-Chloro-2-phenyl-4(3H)-quinazolinone	277	147
8-Chloro-4-phenyl-2(1H)-quinazolinone	262–264	1060
6-Chloro-2-phenyl-4(1H)-quinazolinone 3-oxide	—	H 467
6-Chloro-4-phenyl-2-styrylquinazoline	—	312
6-Chloro-4-phenyl-2-styrylquinazoline 3-oxide	—	312
6-Chloro-3-phenyl-2-styryl-4(3H)-quinazolinone	—	H 188
5-Chloro-6-phenylsulfinyl-2,4-quinazolinediamine	284–285	1548
5-Chloro-6-phenylsulfonyl-2,4-quinazolinediamine	291–293	1548
6-Chloro-4-phenyl-2-thiocyanatomethylquinazoline 3-oxide	141 or 154–156, ir, ms, nmr, uv	82,861
5-Chloro-6-phenylthio-2,4-quinazolinediamine	221–222	1551
6-Chloro-3-phenyl-2-thioxo-1,2-dihydro-4(3H)-quinazolinone	—	H 309
7-Chloro-3-phenyl-2-thioxo-1,2-dihydro-4(3H)-quinazolinone	310–312	566
6-Chloro-4-phenyl-1-β,β,β-trifluoroethyl-2(1H)-quinazolinone	185–186	1060
6-Chloro-4-phenyl-1-vinyl-2(1H)-quinazolinone	186–187, nmr	676
2-Chloro-4-piperidinoquinazoline	—	H 246
4-Chloro-2-piperidinoquinazoline	73–75, ms, nmr, uv	2068,2358

ALPHABETICAL LIST OF SIMPLE QUINAZOLINES (*Continued*)

Quinazoline	m.p. (°C), etc.	Ref.
7-Chloro-2-piperidinoquinazoline	86	849
5-Chloro-6-piperidino-2,4(1*H*,3*H*)-quinazolinedione	>280, nmr	1584
5-Chloro-7-piperidino-2,4(1*H*,3*H*)-quinazolinedione	>280, nmr	1584
6-Chloro-3-piperidino-2,4(1*H*,3*H*)-quinazolinedione	292–294	1437
7-Chloro-6-piperidino-2,4(1*H*,3*H*)-quinazolinedione	>280, nmr	1584
6-Chloro-2-piperidino-4(3*H*)-quinazolinethione	163	413
6-Chloro-2-piperidino-4(3*H*)-quinazolinone	250	845
6-Chloro-3-piperidino-4(3*H*)-quinazolinone	189–190	1471
1-(3-Chloroprop-2-enyl)-4(1*H*)-quinazolinone	*trans*: 121–122	655
6-Chloro-2-prop-1′-enyl-4(3*H*)-quinazolinone	216–222	2264
1-γ-Chloropropyl-2-α-ethoxycarbonylethyl-4(1*H*)-quinazolinone	138–140	1415
3-γ-Chloropropyl-6-methyl-4(3*H*)-quinazolinone	—	*H* 178
2-γ-Chloropropyl-1-phenyl-4(1*H*)-quinazolinone	140	1394
4-Chloro-2-propylquinazoline	80, 100–105/0.3, nmr	188,1184
3-β-Chloropropyl-2,4(1*H*,3*H*)-quinazolinedione	—	*H* 200
3-γ-Chloropropyl-2,4(1*H*,3*H*)-quinazolinedione	167–170 or 173–175, ir, nmr	*H* 200; 1435,1931
3-β-Chloropropyl-4(3*H*)-quinazolinone	—	*H* 144
6-Chloro-2-propyl-4(3*H*)-quinazolinone	246, ms, nmr	957
2-Chloro-4-prop-2′-ynylaminoquinazoline	272–273, ir, ms, nmr	643
2-Chloro-4-quinazolinamine	—	*H* 245
5-Chloro-4-quinazolinamine	211–212 or 217–218, nmr, uv	553,2247
6-Chloro-2-quinazolinamine	258–259, uv	647
6-Chloro-4-quinazolinamine	>310 or 340–345, uv	382,553,916
7-Chloro-4-quinazolinamine	>300 or 305–310, uv	382,553

494

Compound		
8-Chloro-4-quinazolinamine	297–299, uv	553
2-Chloroquinazoline	104–106 or 108, ms	H 245; 42,172,849,1344
4-Chloroquinazoline	95–97	H 247; 42,1001,1139,1189,2010
5-Chloroquinazoline	—	H 249
6-Chloroquinazoline	pol	H 249; 1201
7-Chloroquinazoline	—	H 249
8-Chloroquinazoline	—	H 249
4-Chloro-6-quinazolinecarbonitrile	219–220	2564
6-Chloro-2-quinazolinecarboxamide	—	2178
7-Chloro-2-quinazolinecarboxamide	—	2178
5-Chloro-2,4-quinazolinediamine	180 to 188, nmr	201,678,1932,2225
6-Chloro-2,4-quinazolinediamine	265 to 280	647,661,678,1852,2326
6-Chloro-4,8-quinazolinediamine	—	H 363
7-Chloro-2,4-quinazolinediamine	229–230 or 246–247, nmr	661,2247
5-Chloro-2,4(1H,3H)-quinazolinedione	375–376, nmr	1466,2261
6-Chloro-2,4(1H,3H)-quinazolinedione	344 or 358–360, nmr	H 203; 471,566
7-Chloro-2,4(1H,3H)-quinazolinedione	>330 or 360–362, nmr	H 203; 1079,1358,1454
6-Chloro-2,4(1H,3H)-quinazolinedithione	—	H 302
7-Chloroquinazoline 3-oxide	—	H 462
4-Chloro-2(1H)-quinazolinethione	295–296 (HCl?)	72
5-Chloro-2,4,6-quinazolinetriamine	199–205, xl st	201,678,1269
2-Chloro-4(3H)-quinazolinone	210–213 or 214–218	H 143; 1411,1580
4-Chloro-2(1H)-quinazolinone	213 (HCl?)	72
5-Chloro-4(3H)-quinazolinone		H 154
6-Chloro-2(1H)-quinazolinone	>300	849
6-Chloro-4(3H)-quinazolinone	263 to 269, ir, ms, nmr	H 154; 498,1173,1189,1558
7-Chloro-2(1H)-quinazolinone	>300	849
7-Chloro-4(3H)-quinazolinone	243 to 253, ir, ms, nmr	H 154; 1076,1189,1951

ALPHABETICAL LIST OF SIMPLE QUINAZOLINES (*Continued*)

Quinazoline	m.p. (°C), etc.	Ref.
8-Chloro-2(1*H*)-quinazolinone	>300	849
8-Chloro-4(3*H*)-quinazolinone	—	*H* 154
6-Chloro-4(1*H*)-quinazolinone 3-oxide	—	*H* 467
6-Chloro-4-selenoxo-3,4-dihydro-2(1*H*)-quinazolinone	289–290, ir, nmr	1897
6-Chloro-2-styryl-4(3*H*)-quinazolinethione	284–289, nmr	2264
5-Chloro-2-styryl-4(3*H*)-quinazolinone	288–293	2264
6-Chloro-2-styryl-4(3*H*)-quinazolinone	311 or 315–320, nmr, uv	134,2264
7-Chloro-2-styryl-4(3*H*)-quinazolinone	257–262 or 279, uv	134,2264
8-Chloro-2-styryl-4(3*H*)-quinazolinone	296–301	2264
6-Chloro-3-thioureido-4(3*H*)-quinazolinone	—	*H* 374
6-Chloro-2-thioxo-1,2-dihydro-4(3*H*)-quinazolinone	310 or 312–314, ir, uv	*H* 309; 126,867,2236
7-Chloro-2-thioxo-1,2-dihydro-4(3*H*)-quinazolinone	302	126
4-Chloro-2-trichloromethylquinazoline	—	*H* 248
6-Chloro-2-trichloromethyl-4(3*H*)-quinazolinone	229 to 253, ir, nmr	741,1034,2265
4-Chloro-5-trifluoromethylquinazoline	67	*H* 248; 168
4-Chloro-6-trifluoromethylquinazoline	69	*H* 248; 168
4-Chloro-7-trifluoromethylquinazoline	62–64	*H* 248; 168,2165
4-Chloro-8-trifluoromethylquinazoline	148	*H* 248; 168
2-Chloro-5,6,7-trimethoxy-4-quinazolinamine	232–233	696
2-Chloro-6,7,8-trimethoxy-4-quinazolinamine	241–242	696
2-Chloro-4,6,7-trimethoxyquinazoline	201–204, nmr, uv	641
Chrysogine, *see* 2-α-Hydroxyethyl-4-(3*H*)-quinazolinone		
2-Crotonamido-3-methyl-4(3*H*)-quinazolinone	182–183	1834
3-Crotonamido-2-methyl-4(3*H*)-quinazolinone	—	1995

496

2-Crotonamido-4(3H)-quinazolinone	198–200	1834
Crysogine, *see* 2-α-Hydroxyethyl-4(3H)-quinazolinone		
2-Cyanoamino-4-methylquinazoline	—	H 342
2-Cyanoamino-4(3H)-quinazolinone	307–309, ir, nmr	H 365; 2000
2-δ-Cyanobutyl-4(3H)-quinazolinone	—	H 139
3-β-Cyanoethyl-2-ethyl-4(3H)-quinazolinone	—	H 164
2-β-Cyanoethyl-4(3H)-quinazolinone	—	H 139
3-β-Cyanoethyl-4(3H)-quinazolinone	144–145, ir, nmr	1772
2-Cyanomethyl-4-ethoxyquinazoline	57, ir	1161
2-Cyanomethyl-4-methoxyquinazoline	65, ir	1161
3-Cyanomethyl-2-methyl-7-nitro-4(3H)-quinazolinone	—	H 181
2-Cyanomethyl-3-phenyl-4(3H)-quinazolinone	187 or 248–249, ms	146,1164,1286
2-Cyanomethylquinazoline	—	H 60
4-Cyanomethylquinazoline	127–128 or 175–176, ir, nmr	1869,2010
1-Cyanomethyl-4(1H)-quinazolinone	270–273, ir, ms, nmr	1825
2-Cyanomethyl-4(3H)-quinazolinone	228 or 235, ir, nmr	H 139; 1161,2275,2332
2-ε-Cyanopentyl-4(3H)-quinazolinone	—	H 139
2-γ-Cyanopropyl-3-phenyl-4(3H)-quinazolinone	162–163, ir, nmr	1154
2-γ-Cyanopropyl-4(3H)-quinazolinone	—	H 139
2-Cyclobutyl-1-phenyl-4(1H)-quinazolinone	230–232	1590
1-Cyclohex-1'-enyl-2-methyl-4(3H)-quinazolinone	171–173	1405
1-Cyclohex-2'-enyl-4(1H)-quinazolinone	*cis*: 127–129	655
4-Cyclohex-2'-enylthioquinazoline	55, ms, nmr; HClO₄: anal, nmr	1221
7-Cyclohexylamino-6-nitro-4(3H)-quinazolinone	263–265, ms, nmr	501
4-Cyclohexylaminoquinazoline	148–149, ms	1786
4-Cyclohexylamino-2(1H)-quinazolinethione	282–284, ir, nmr	2384
3-Cyclohexyl-2-cyclohexylamino-4(3H)-quinazolinone	122	H 365; 621
2-Cyclohexyl-4-cyclohexylidene-3-phenyl-3,4-dihydroquinazoline	144–145, ir, ms, nmr	400

497

ALPHABETICAL LIST OF SIMPLE QUINAZOLINES (Continued)

Quinazoline	m.p. (°C), etc.	Ref.
4-Cyclohexyl-6,7-dimethyl-2-oxo-1,2-dihydro-8-quinazolinecarbonitrile	232–234	1570
3-Cyclohexyl-2-ethyl-4(3H)-quinazolinone	—	H 164
3-Cyclohexyl-2-α-hydroxybenzyl-4(3H)-quinazolinone	70	842
3-Cyclohexyl-1-hydroxy-2-thioxo-1,2-dihydro-4(3H)-quinazolinone	140–142	622
3-Cyclohexyl-8-methoxy-2-methyl-4(3H)-quinazolinone	—	H 181
3-Cyclohexyl-N-methyl-4-oxo-3,4-dihydro-2-quinazolinecarboxamide	247–248	768
2-Cyclohexyl-1-methyl-4(1H)-quinazolinone	211–212, ir, nmr	821
3-Cyclohexyl-2-methyl-4(3H)-quinazolinone	—	H 156
1-Cyclohexyl-4-phenyl-2(1H)-quinazolinone	147–148	681
2-Cyclohexyl-1-phenyl-4(1H)-quinazolinone	233–235	1394
3-Cyclohexyl-2-propyl-4(3H)-quinazolinone	—	H 165
3-Cyclohexyl-2,4(1H,3H)-quinazolinedione	265 to 272, ir, nmr	H 200; 385,769,770,1282,1341, 1446,1448,2337
2-Cyclohexyl-4(3H)-quinazolinone	217 or 224–226, ir, ms, nmr	487,2085
3-Cyclohexyl-4(3H)-quinazolinone	—	H 144; 1497
2-Cyclohexyl-4(3H)-quinazolinone 3-oxide	172	487
2-Cyclohexylthio-3-phenyl-4(3H)-quinazolinone	202	1017
6-Cyclohexylthio-2,4-quinazolinediamine	190–192	811
3-Cyclohexyl-2-thioxo-1,2-dihydro-4(3H)-quinazolinone	270–271	770
4-Cyclopentyl-6,7-dimethoxy-2(1H)-quinazolinone	196–197	1592
3-Cyclopentyl-2-ethyl-4(3H)-quinazolinone	—	H 164
3-Cyclopentyl-6-methoxy-2-methyl-4(3H)-quinazolinone	—	H 181
3-Cyclopentyl-2,4(1H,3H)-quinazolinedione	234–235	385
3-N-(Cyclopropylaminomethyl)aminomethyl-4(3H)-quinazolinone	HCl: 198–199	1192

498

Compound	Data	Ref
3-Cyclopropylaminomethyl-4(3H)-quinazolinone	HCl: 170–172	1192
6-Cyclopropyl-3-β-methoxycarbonylvinyl-2-methyl-4(3H)-quinazolinone	127–128	1558
2-Cyclopropylmethoxy-6-nitro-4-phenyl-2(1H)-quinazolinone	142–144	544
1-Cyclopropylmethyl-6-methoxy-4-phenyl-2(1H)-quinazolinone	115–116 or 116–117, ir, nmr, xl st	544,581,713,1060,1061
1-Cyclopropylmethyl-6-nitro-4-phenyl-2(1H)-quinazolinone	172–173	544,1060
1-Cyclopropylmethyl-4-phenyl-2(1H)-quinazolinone	nmr	1061
1-Cyclopropylmethyl-4(1H)-quinazolinone	133–136	655
2-Cyclopropyl-3-methyl-4(3H)-quinazolinone	77–78, ir, nmr	2079,2131
2-Cyclopropyl-1-phenyl-4(1H)-quinazolinone	241–242	1394,1590
1-Cyclopropyl-4(1H)-quinazolinone	126–128	655
4-Deuteroquinazoline	47	168
3,6-Diacetamido-2-methyl-4(3H)-quinazolinone	—	H 380
3,7-Diacetamido-2-methyl-4(3H)-quinazolinone	—	H 380
2,3-Diacetamido-4(3H)-quinazolinone	254–255, ir	1715
3,7-Diacetamido-2-styryl-4(3H)-quinazolinone	—	H 380
5,6-Diacetoxy-4-methoxy-2-phenyl-8-piperidinoquinazoline	187–189, ir	763
3-Diacetylamino-2-methyl-7-nitro-4(3H)-quinazolinone	—	H 378
3-Diacetylamino-2-phenyl-4(3H)-quinazolinone	149–150	H 372; 1981
3-Diacetylamino-2,4(1H,3H)-quinazolinedione	210	H 382; 1603
3-Diacetylamino-4(3H)-quinazolinone	111 or 116, ir, nmr	276,1430
2-β,β-Diacetylethyl-3-phenyl-4(3H)-quinazolinethione	181–182	479
2-β,β-Diacetylethyl-3-phenyl-4(3H)-quinazolinone	125–129 or 151–152, ir, ms, nmr	137,1286
2-Diallylamino-6,7-dimethoxy-4(3H)-quinazolinone	190–191; HCl: 233–235	641
2-Diallylaminoquinazoline	113/0.06	849
2-Diallylamino-4(3H)-quinazolinone	150–151	943
2,4-Diallyloxyquinazoline	—	H 259
5,8-Diallyloxyquinazoline	crude liquid, uv	595
6,8-Diamino-3-benzyloxy-2-methyl-4(3H)-quinazolinone	—	H 469

499

ALPHABETICAL LIST OF SIMPLE QUINAZOLINES (*Continued*)

Quinazoline	m.p. (°C), etc.	Ref.
6,7-Diamino-3-benzyl-4(3*H*)-quinazolinone	192–193, ms, nmr	501
5,6-Diamino-3-butyl-4(3*H*)-quinazolinone	crude	121
6,7-Diamino-3-butyl-4(3*H*)-quinazolinone	152	1165
2,4-Diamino-5-chloro-*N*,*N*-diethyl-6-quinazolinesulfonamide	218–220	1563
2,4-Diamino-5-chloro-*N*,*N*-dimethyl-6-quinazolinesulfonamide	272–274	1563
2,4-Diamino-5-chloro-6-quinazolinecarbonitrile	285 or 287, ir, nmr	201,2590
2,6-Diamino-5-chloro-4(3*H*)-quinazolinone	292–294, nmr	2374
2,4-Diamino-*N*-cyclohexyl-6-quinazolinesulfonamide	—	1537
2,4-Diamino-*N*,*N*-diethyl-6-quinazolinesulfonamide	285–289	1563
2,4-Diamino-*N*,*N*-dimethyl-6-quinazolinesulfonamide	275–277	1563,2326
2,4-Diamino-5,7-diphenyl-8-quinazolinecarbonitrile	267–268, ir, ms, nmr, uv	2466
2,4-Diamino-*N*,*N*-dipropyl-6-quinazolinesulfonamide	>300	1563
2,4-Diamino-*N*-isopropyl-*N*-methyl-6-quinazolinesulfonamide	285–287	1563
2,4-Diamino-5-methyl-6-quinazolinecarbaldehyde	crude, 308–310	1871
2,4-Diamino-5-methyl-6-quinazolinecarbonitrile	260–310 (!), ms, nmr	201,1871,2041
2,6-Diamino-5-methyl-4(3*H*)-quinazolinone	332–334	611
3,6-Diamino-2-methyl-4(3*H*)-quinazolinone	—	*H* 380
3,7-Diamino-2-methyl-4(3*H*)-quinazolinone	—	*H* 380
6,7-Diamino-3-methyl-4(3*H*)-quinazolinone	292	1165
2,5-Diamino-6-nitro-4(3*H*)-quinazolinone	>355, nmr	2015
2,7-Diamino-6-nitro-4(3*H*)-quinazolinone	>300, ms, nmr	915
2,4-Diamino-*N*,*N*-pentamethylene-6-quinazolinesulfonamide	271–274	1486,1563
2,4-Diamino-6-phenyl-8-quinazolinecarbonitrile	>300, ir, ms, nmr, uv	2466
2,4-Diamino-*N*-propyl-5-quinazolinecarboxamide	257–259	2225

500

Compound	Data	References
2,4-Diamino-6-quinazolinecarbaldehyde	>300 to >400; PhNN=: AcOH: 223–226 or 232–234	201,678,956
2,4-Diamino-5-quinazolinecarbonitrile	231–232 or 234–236, nmr	2247,2597
2,4-Diamino-6-quinazolinecarbonitrile	>300 to 383	201,678,956,1852
2,4-Diamino-5-quinazolinecarboxamide	274–276, nmr, pK_a	2225
6,8-Diamino-2,4(1H,3H)-quinazolinedione	—	H 383
2,4-Diamino-6-quinazolinesulfonamide	324	1563
2,4-Diamino-6-quinazolinesulfonyl chloride	H_2SO_4: >300	1563
3,4-Diamino-2(3H)-quinazolinethione	223–225, ir, nmr	2417
2,4-Diamino-5-quinazolinol	HCl: >300, nmr	2225
2,3-Diamino-4(3H)-quinazolinone	175–176, 179–180, or 289–290, ir	1347,1438,1715,1886,2083
2,6-Diamino-4(3H)-quinazolinone	>300 or 330–335, ir, ms, nmr	618,1564
2,8-Diamino-4(3H)-quinazolinone	HCl: >250, ir, nmr	2025
3,4-Diamino-2(3H)-quinazolinone	—	2004
6,7-Diamino-4(3H)-quinazolinone	>300 or >320, ms, nmr	501,952,1165
2,6-Diamino-5-trifluoromethyl-4(3H)-quinazolinone	>235, nmr	2211
2,4-Diamino-5,N,N-trimethyl-6-quinazolinesulfonamide	274–278	1563
2,4-Dianilino-6-methylquinazoline	—	H 361
2,4-Dianilino-6-nitroquinazoline	260–262	600
2,4-Dianilinoquinazoline	—	292
2,4-Diazidoquinazoline	145–146, nmr, st	2201
5-Diazonio-6-methyl-4(3H)-quinazolinone	BF_4^-: 157	1159,1366
6-Diazonio-2,4-quinazolinediamine	BF_4^-: 197–200; Cl^-: 160	1919
2,4-Dibenzamido-6-bromomethylquinazoline	crude, 185–198	H 361; 1912
2,4-Dibenzamido-6-methylquinazoline	187–189	H 361; 1912
6,7-Dibenzoyl-1,3-dimethyl-2,4(1H,3H)-quinazolinedione	—	2177

ALPHABETICAL LIST OF SIMPLE QUINAZOLINES   (*Continued*)

Quinazoline	m.p. (°C), etc.	Ref.
1,3-Dibenzoyl-2,4(1*H*,3*H*)-quinazolinedione	123	1538,2281
2,4-Dibenzyl-5,8-dimethylquinazoline	—	H 65
6,7-Dibenzyloxy-2-methyl-4(3*H*)-quinazolinone	229–231	218
1,3-Dibenzyl-2,4(1*H*,3*H*)-quinazolinedione	123, ir, uv	H 204; 1464
2,3-Dibenzyl-4(3*H*)-quinazolinone	95–95	525
6,8-Dibromo-2-bromomethyl-3-ethyl-4(3*H*)-quinazolinone	115–117	1170
6,8-Dibromo-3-butyl-2-butylthio-4(3*H*)-quinazolinone	255	644
6,8-Dibromo-3-butyl-2-isopropylthio-4(3*H*)-quinazolinone	255	644
6,8-Dibromo-3-butyl-2-methyl-4(3*H*)-quinazolinone	79–80, ir	1371
6,8-Dibromo-3-butyl-2-methylthio-4(3*H*)-quinazolinone	208	603
6,8-Dibromo-3-butyl-2-pentylthio-4(3*H*)-quinazolinone	246	840
6,8-Dibromo-3-butyl-2-propylthio-4(3*H*)-quinazolinone	255	840
6,8-Dibromo-2-butylthio-3-ethyl-4(3*H*)-quinazolinone	225	644
6,8-Dibromo-2-*t*-butylthio-3-ethyl-4(3*H*)-quinazolinone	253	840
6,8-Dibromo-2-butylthio-3-methyl-4(3*H*)-quinazolinone	270	644
6,8-Dibromo-2-*t*-butylthio-3-methyl-4(3*H*)-quinazolinone	255	840
6,8-Dibromo-2-butylthio-3-phenyl-4(3*H*)-quinazolinone	185	644
6,8-Dibromo-3-*t*-butylthio-3-phenyl-4(3*H*)-quinazolinone	275	840
6,8-Dibromo-3-butyl-2-thioxo-1,2-dihydro-4(3*H*)-quinazolinone	234	644
6,8-Dibromo-2-carbamoylmethyl-4(3*H*)-quinazolinone	277–279, ir, ms	1293
6,8-Dibromo-3-α-carboxyethyl-2-phenyl-4(3*H*)-quinazolinone	184	1278
6,8-Dibromo-3-β-carboxyethyl-2-phenyl-4(3*H*)-quinazolinone	190	1278
6,8-Dibromo-3-β-carboxyethyl-4(3*H*)-quinazolinone	230–232, ir, ms	1640
6,8-Dibromo-3-(α-carboxy-γ-methylbutyl)-2-phenyl-4(3*H*)-quinazolinone	186	1278

Compound	mp (°C), methods	Ref.
6,8-Dibromo-3-carboxymethyl-2-methyl-4(3*H*)-quinazolinone	230	1681
6,8-Dibromo-3-carboxymethyl-2-phenyl-4(3*H*)-quinazolinone	230	1278
6,8-Dibromo-3-carboxymethyl-2,4(1*H*,3*H*)-quinazolinedione	310–312, ir, ms	1838
6,8-Dibromo-3-carboxymethyl-4(3*H*)-quinazolinone	283–285, ir, ms	1640,1838
6,8-Dibromo-2-carboxymethylthio-4(3*H*)-quinazolinone	270–274, ir	1009
6,8-Dibromo-3-γ-carboxypropyl-4(3*H*)-quinazolinone	235–238, ir, ms	1640
6,8-Dibromo-2-chloromethyl-3-phenyl-4(3*H*)-quinazolinone	189	1033
6,8-Dibromo-4-chloroquinazoline	—	H 252
6,8-Dibromo-3-cyclohexyl-4(3*H*)-quinazolinone	—	1890
6,8-Dibromo-2-cyclohexylthio-3-methyl-4(3*H*)-quinazolinone	235	840
6,8-Dibromo-2-cyclohexylthio-3-phenyl-4(3*H*)-quinazolinone	162	840
6,8-Dibromo-4-β-diethylaminoethylaminoquinazoline	—	H 252
6,8-Dibromo-2,3-dimethyl-4(3*H*)-quinazolinone	148–149, ir	1371,2098,2184
6,8-Dibromo-2,4-diphenylquinazoline	219, ms	1946
6,8-Dibromo-3-α-ethoxycarbonylethyl-4(3*H*)-quinazolinone	120, ir, ms	1640
6,8-Dibromo-3-ethoxycarbonylmethyl-2-methyl-4(3*H*)-quinazolinone	175	613
6,8-Dibromo-3-ethyl-2-ethylthio-4(3*H*)-quinazolinone	95	603
6,8-Dibromo-3-ethyl-2-isopentylthio-4(3*H*)-quinazolinone	235	840
6,8-Dibromo-3-ethyl-2-isopropylthio-4(3*H*)-quinazolinone	258	644
6,8-Dibromo-3-ethyl-2-methyl-4(3*H*)-quinazolinone	155–157	H 180; 1170
6,8-Dibromo-3-ethyl-2-methylthio-4(3*H*)-quinazolinone	154	603
6,8-Dibromo-3-ethyl-4-oxo-3,4-dihydro-2-quinazolinecarbaldehyde	217–219	1170
6,8-Dibromo-3-ethyl-2-pentylthio-4(3*H*)-quinazolinone	231	840
6,8-Dibromo-1-ethyl-3-phenyl-2-thioxo-1,2-dihydro-4(3*H*)-quinazolinone	242	646
6,8-Dibromo-3-ethyl-2-propylthio-4(3*H*)-quinazolinone	228	840
6,8-Dibromo-3-ethyl-2,4(1*H*,3*H*)-quinazolinedione	—	H 202
6,8-Dibromo-2-ethyl-4(3*H*)-quinazolinone	270	H 174; 2285
6,8-Dibromo-3-ethyl-4(3*H*)-quinazolinone	—	H 177

ALPHABETICAL LIST OF SIMPLE QUINAZOLINES (*Continued*)

Quinazoline	m.p. (°C), etc.	Ref.
6,8-Dibromo-4-ethylthio-2-methylquinazoline	—	H 299
6,8-Dibromo-2-ethylthio-3-methyl-4(3H)-quinazolinone	280	603
6,8-Dibromo-2-ethylthio-3-phenyl-4(3H)-quinazolinone	230	603
6,8-Dibromo-3-ethyl-2-thioxo-1,2-dihydro-4(3H)-quinazolinone	180	644
6,8-Dibromo-3-hydrazinocarbonylmethyl-2-methyl-4(3H)-quinazolinone	285	1681
6,8-Dibromo-3-β-hydroxyethyl-2-phenyl-4(3H)-quinazolinone	180	838
6,8-Dibromo-3-isobutylamino-2-methyl-4(3H)-quinazolinone	137	1662
6,8-Dibromo-2-isobutyl-4(3H)-quinazolinone	—	H 174
6,8-Dibromo-2-isopentylthio-3-methyl-4(3H)-quinazolinone	257	840
6,8-Dibromo-2-isopentylthio-3-phenyl-4(3H)-quinazolinone	268	840
6,8-Dibromo-3-isopropylideneamino-2-methyl-4(3H)-quinazolinone	170–172	2354
6,8-Dibromo-2-isopropyl-4(3H)-quinazolinone	—	H 174
6,8-Dibromo-2-isopropylthio-3-methyl-4(3H)-quinazolinone	264	644
6,8-Dibromo-2-isopropylthio-3-phenyl-4(3H)-quinazolinone	248	644
6,8-Dibromo-3-methoxycarbonylmethyl-2-methyl-4(3H)-quinazolinone	175	1681
6,8-Dibromo-3-methoxycarbonylmethyl-2,4(1H,3H)-quinazolinedione	216–219, ir, ms, nmr	1838
6,8-Dibromo-3-methoxycarbonylmethyl-4(3H)-quinazolinone	215–218, ir, ms	1838
2-Dibromomethyl-3-β-hydroxyethyl-4(3H)-quinazolinone	153–155, nmr	2470
4-Dibromomethyl-6-methyl-2-phenylquinazoline	155–156	694
6-Dibromomethyl-4-methylthio-2-pivalamidoquinazoline	205–209, nmr	2255
6,8-Dibromo-3-methyl-2-methylthio-4(3H)-quinazolinone	150	603
6,8-Dibromo-3-methyl-2-pentylthio-4(3H)-quinazolinone	265	840
4-Dibromomethyl-2-phenylquinazoline	144–145	694
6,8-Dibromo-4-methyl-2-phenylquinazoline 3-oxide	uv	H 462; 575

504

2-Dibromomethyl-3-phenyl-4(3H)-quinazolinone	179–180, ir, ms, nmr	1286
6,8-Dibromo-2-methyl-3-phenyl-4(3H)-quinazolinone	210 or 213–214	1170,2098,2184
6,8-Dibromo-2-methyl-3-propylamino-4(3H)-quinazolinone	143	1612
6,8-Dibromo-3-methyl-2-propylthio-4(3H)-quinazolinone	248	840
4-Dibromomethylquinazoline	134–135, ms, nmr	1651,2060
6,8-Dibromo-3-methyl-2,4(1H,3H)-quinazolinedione	—	H 203
6,8-Dibromo-2-methyl-4(3H)-quinazolinone	—	H 173; 2042,2349
6,8-Dibromo-2-methylthio-3-phenyl-4(3H)-quinazolinone	189	603
6,8-Dibromo-3-methyl-2-thioxo-1,2-dihydro-4(3H)-quinazolinone	229	644
6,8-Dibromo-3-nitroso-4(3H)-quinazolinone	280	1690
6,8-Dibromo-4-oxo-3,4-dihydro-2-quinazolinecarboxylic acid	328	1549
6,8-Dibromo-4-oxo-3-phenyl-3,5-dihydro-2-quinazolinecarbaldehyde	145–146	1170
6,8-Dibromo-2-pentylthio-3-phenyl-4(3H)-quinazolinone	274	840
2-α,β-Dibromophenethyl-3-phenyl-4(3H)-quinazolinone	ms	1296
6,8-Dibromo-3-phenyl-2-piperidinomethyl-4(3H)-quinazolinone	142–143	745
6,8-Dibromo-3-phenyl-2-propylthio-4(3H)-quinazolinone	278	840
6,8-Dibromo-3-phenyl-2,4(1H,3H)-quinazolinedione	238–241	H 203; 93
6,8-Dibromo-2-phenyl-4(1H)-quinazolinethione 3-oxide	231	806
6,8-Dibromo-3-phenyl-2-thioxo-1,2-dihydro-4(3H)-quinazolinone	298, complexes	644,1682,1747
5-β,γ-Dibromopropyl-6-hydroxy-7-methoxy-2-methyl-3-phenyl-4(3H)-quinazolinone	257–258	143
3-β,γ-Dibromopropyl-2,4(1H,3H)-quinazolinedione	217–218 or 219–221, ir, ms, nmr, us	477,2227
6,8-Dibromo-3-propyl-2,4(1H,3H)-quinazolinedione	—	H 203
6,8-Dibromo-2-propyl-4(3H)-quinazolinone	—	H 174
3-β,γ-Dibromopropyl-2-thioxo-1,2-dihydro-4(3H)-quinazolinone	303, nmr	2227
6,8-Dibromo-4-quinazolinamine	>300	507
6,8-Dibromo-4-quinazolinamine 3-oxide	>300	507

ALPHABETICAL LIST OF SIMPLE QUINAZOLINES (*Continued*)

Quinazoline	m.p. (°C), etc.	Ref.
6,8-Dibromo-2,4(1*H*,3*H*)-quinazolinedione	> 300	*H* 202; 869
6,8-Dibromo-4(3*H*)-quinazolinethione	—	*H* 296
5,7-Dibromo-4(3*H*)-quinazolinone	—	*H* 153
6,8-Dibromo-4(3*H*)-quinazolinone	—	*H* 154
6,8-Dibromo-2-thioxo-1,2-dihydro-4(3*H*)-quinazolinone	278, ir, complexes	1009,1682
2,4-Dibutoxy-6-nitroquinazoline	—	*H* 259
2,4-Dibutoxy-6-quinazolinamine	—	*H* 259
2,4-Dibutoxyquinazoline	—	*H* 259
2,4-Di-*s*-butoxyquinazoline	—	*H* 259
2,4-Di-*t*-butoxyquinazoline	—	*H* 259
5,8-Dibutoxyquinazoline	79, uv	595
1,3-Dibutyl-7-chloro-6-nitro-2,4(1*H*,3*H*)-quinazolinedione	118–119, nmr	2261
Dibutyl 2-chloro-4-quinazolinephosphonate	35, 192/8	571
2,4-Di-*t*-butylquinazoline	98/0.3	1094
2-β,β-Dicarboxyethyl-4-phenylquinazoline	101–102	632
3-α,β-Dicarboxyethyl-2,4(1*H*,3*H*)-quinazolinedione	248–250, ir, nmr	2435
6,8-Dichloro-2-chloromethyl-3-methoxy-4(3*H*)-quinazolinone	129	441
6,7-Dichloro-2-chloromethyl-4-phenylquinazoline 3-oxide	—	*H* 462
6,8-Dichloro-2-chloromethyl-4-phenylquinazoline 3-oxide	—	*H* 462
4,6-Dichloro-2-chloromethylquinazoline	92–95	893
2,4-Dichloro-6,7-diethoxyquinazoline	172–174	641
2,4-Dichloro-7,8-diethoxyquinazoline	101–102	696
2,6-Dichloro-4-β-diethylaminoethylaminoquinazoline	HCl: 157–162	*H* 251; 1568
2,7-Dichloro-4-β-diethylaminoethylaminoquinazoline	HCl: 218–222	*H* 251; 1568

506

Compound	mp, etc.	References
6,8-Dichloro-4-β-diethylaminoethylaminoquinazoline	—	H 249
6,8-Dichloro-2-β-diethylaminoethyl-4(3H)-quinazolinone	218–219	H 174; 384
6,7-Dichloro-5,8-dihydro-5,8-quinazolinedione	135, ir, ms, nmr	1604
2,4-Dichloro-6,7-diisopropoxyquinazoline	100–102	641
2,8-Dichloro-6,7-dimethoxy-4-quinazolinamine	258–259, ir, nmr	1146
2,4-Dichloro-6,7-dimethoxyquinazoline	159 or 173–176 (?)	H 248; 696,1175,1580
2,4-Dichloro-7,8-dimethoxyquinazoline	155–156	696
6,8-Dichloro-3-β-dimethylaminoethyl-2-methyl-4(3H)-quinazolinethione	99	716
6,8-Dichloro-3-β-dimethylaminoethyl-2-phenyl-4(3H)-quinazolinethione	144	716
6,8-Dichloro-2-β-dimethylaminoethyl-4(3H)-quinazolinethione	—	H 174
6,8-Dichloro-3-γ-dimethylaminopropyl-2-phenyl-4(3H)-quinazolinethione	181	716
2,4-Dichloro-5-dimethylaminoquinazoline	139–141	1584
4,6-Dichloro-2-dimethylaminoquinazoline	ir, nmr	1160
6,8-Dichloro-5,6-dimethyl-2-phenyl-4(3H)-quinazolinone	289–291	103
2,4-Dichloro-6,7-dimethylquinazoline	138–140	641
2,4-Dichloro-N,N-dimethyl-6-quinazolinesulfonamide	—	H 248
6,8-Dichloro-2,4-diphenylquinazoline	—	H 251
6,8-Dichloro-2,3-diphenyl-4(3H)-quinazolinethione	178	953
6,8-Dichloro-2,3-diphenyl-4(3H)-quinazolinone	158	2318
6,8-Dichloro-3-ethoxycarbonylmethyl-2-methyl-4(3H)-quinazolinethione	160, nmr	1565
6,8-Dichloro-3-ethoxycarbonylmethyl-2-methyl-4(3H)-quinazolinone	160	613
6,8-Dichloro-3-ethoxycarbonylmethyl-2-phenyl-4(3H)-quinazolinethione	188, nmr	1565
6,8-Dichloro-3-ethoxycarbonylmethyl-4(3H)-quinazolinone	—	H 178
2,4-Dichloro-6-ethoxy-7-methoxyquinazoline	162–163	696
6,8-Dichloro-3-β-ethylaminoethyl-2-methyl-4(3H)-quinazolinone	106	716
6,8-Dichloro-3-β-ethylaminoethyl-2-phenyl-4(3H)-quinazolinone	118	716
6,8-Dichloro-2-β-ethylaminoethyl-4(3H)-quinazolinone	—	H 174
6,8-Dichloro-3-γ-ethylaminopropyl-2-phenyl-4(3H)-quinazolinethione	105	716

ALPHABETICAL LIST OF SIMPLE QUINAZOLINES (*Continued*)

Quinazoline	m.p. (°C), etc.	Ref.
6,8-Dichloro-3-ethyl-2,4(1*H*,3*H*)-quinazolinedione	—	*H* 203
6,8-Dichloro-4-fluoroquinazoline	100–101, uv	553
6,8-Dichloro-3-hydrazinocarbonylmethyl-2-methyl-4(3*H*)-quinazolinone	275	613
6,8-Dichloro-3-β-hydroxyethyl-4(3*H*)-quinazolinone	119–121	384
6,7-Dichloro-8-hydroxy-5-methoxy-4(3*H*)-quinazolinone	—	1095
1-Dichloroiodo-3-phenyl-2,4(1*H*,3*H*)-quinazolinedione	—	459
4,6-Dichloro-8-iodoquinazoline	—	*H* 252
4,8-Dichloro-6-iodoquinazoline	—	*H* 252
6,8-Dichloro-3-β-methoxycarbonylethyl-2-methyl-4(3*H*)-quinazolinethione	103	1565
6,8-Dichloro-3-α-methoxycarbonylethyl-2-phenyl-4(3*H*)-quinazolinethione	232	1565
6,8-Dichloro-3-β-methoxycarbonylethyl-2-phenyl-4(3*H*)-quinazolinethione	173	1565
6,8-Dichloro-3-methoxycarbonylmethyl-2,4(1*H*,3*H*)-quinazolinedione	190–194, ir, ms	1838
6,8-Dichloro-3-methoxycarbonylmethyl-4(3*H*)-quinazolinone	224–227, ir, ms	1838
2,4-Dichloro-5-methoxyquinazoline	165–166	*H* 248; 696
2,4-Dichloro-6-methoxyquinazoline	—	*H* 248
2,4-Dichloro-7-methoxyquinazoline	—	*H* 248
2,4-Dichloro-8-methoxyquinazoline	155–157 or 157–159, nmr	*H* 248; 696,2595
6,7-Dichloro-5-methoxy-8-quinazolinol	—	1095
6,8-Dichloro-2-β-methylaminoethyl-4(3*H*)-quinazolinone	—	*H* 175
6,8-Dichloro-3-γ-methylaminopropyl-2-phenyl-4(3*H*)-quinazolinone	120	716
4-Dichloromethyl-6-fluoro-2-phenylquinazoline	134–135	693
2,7-Dichloro-4-α-methylhydrazinoquinazoline	188–190, nmr	2230
4-Dichloromethyl-7-methoxy-2-phenylquinazoline	145–146	438
2-Dichloromethyl-7-methoxyquinazoline	170–171, nmr	951

Compound	mp (°C)	Reference
6,8-Dichloro-2-methyl-3-β-methylaminoethyl-4(3H)-quinazolinone	126	716
4-Dichloromethyl-6-methyl-2-phenylquinazoline	144–145	693
2-Dichloromethyl-4-methylquinazoline	—	H 63
2-Dichloromethyl-4-phenylquinazoline	—	H 63
4-Dichloromethyl-2-phenylquinazoline	115–116 or 117–118	438,693
6,8-Dichloro-2-methyl-3-phenyl-4(3H)-quinazolinone	—	1975
2,4-Dichloro-7-methyl-6-piperidinoquinazoline	113–114	1584
2-Dichloromethylquinazoline	—	H 60
2,4-Dichloro-6-methylquinazoline	—	H 248
2,4-Dichloro-7-methylquinazoline	—	H 248
2,4-Dichloro-8-methylquinazoline	—	H 248
4,7-Dichloro-2-methylquinazoline	94–96, nmr	188
2-Dichloromethyl-4(3H)-quinazolinone	—	H 139
6,7-Dichloro-4-methyl-2(1H)-quinazolinone	—	1592
6,8-Dichloro-2-methyl-4(3H)-quinazolinone	—	H 173
2,4-Dichloro-7-morpholinoquinazoline	220–221	1584
6,8-Dichloro-4-morpholinoquinazoline	169–171	1189
6,8-Dichloro-3-morpholino-2,4(1H,3H)-quinazolinedione	230	1603
2,4-Dichloro-6-nitroquinazoline	—	H 248
2,4-Dichloro-7-nitroquinazoline	—	H 248
4,6-Dichloro-8-nitroquinazoline	—	H 251
2,6-Dichloro-4-phenylquinazoline	158–159 or 160	349,836,1021
4,6-Dichloro-2-phenylquinazoline	147	503
4,7-Dichloro-2-phenylquinazoline	135	503
6,7-Dichloro-4-phenyl-2(1H)-quinazolinone	319–320	1060
6,8-Dichloro-4-phenyl-2(1H)-quinazolinone	330–331	1060
6,8-Dichloro-3-phenyl-2-thioxo-1,2-dihydro-4(3H)-quinazolinone	214	2390
2,4-Dichloro-5-piperidinoquinazoline	liquid, nmr	1584

ALPHABETICAL LIST OF SIMPLE QUINAZOLINES (*Continued*)

Quinazoline	m.p. (°C), etc.	Ref.
2,4-Dichloro-7-piperidinoquinazoline	113–115	1584
6,7-Dichloro-2-quinazolinamine	>300	647
6,8-Dichloro-4-quinazolinamine	296–297, uv	553
2,4-Dichloroquinazoline	115–118	*H* 248; 1412,1580
2,6-Dichloroquinazoline	164	849
2,7-Dichloroquinazoline	165	849
2,8-Dichloroquinazoline	170	849
4,5-Dichloroquinazoline	161–162, uv	*H* 251; 553
4,6-Dichloroquinazoline	155–157	*H* 251; 1189
4,7-Dichloroquinazoline	132–134	*H* 251; 1189
4,8-Dichloroquinazoline	176–177, uv	*H* 251; 553
6,8-Dichloroquinazoline	—	*H* 249
5,6-Dichloro-2,4-quinazolinediamine	251–253	678
5,8-Dichloro-2,4-quinazolinediamine	pK_a, uv	250
6,7-Dichloro-2,4-quinazolinediamine	270	661
6,8-Dichloro-2,4-(1*H*,3*H*)-quinazolinedione	—	*H* 203
2,6-Dichloro-4(3*H*)-quinazolinone	222–225	641
2,7-Dichloro-4(3*H*)-quinazolinone	219–224	641
5,6-Dichloro-2(1*H*)-quinazolinone	>300	849
5,6-Dichloro-4(3*H*)-quinazolinone	—	*H* 154
5,8-Dichloro-2(1*H*)-quinazolinone	>300	849
5,8-Dichloro-4(3*H*)-quinazolinone	—	*H* 154
6,7-Dichloro-4(3*H*)-quinazolinone	—	*H* 154
6,8-Dichloro-4(3*H*)-quinazolinone	>300 or 337–340, ir	*H* 154; 384,628,1189

Compound		Reference
2,4-Dichloro-5,6,7-trimethoxyquinazoline	139–140	696
2,4-Dichloro-6,7,8-trimethoxyquinazoline	149–150	696
4-Dicyanomethyl-3-ethyl-2-ethylimino-2,3-dihydroquinazoline	176, ir, nmr	723
4-Dicyanomethyl-3-methyl-2-methylimino-2,3-dihydroquinazoline	219, ir, nmr	723
4-Dicyanomethyl-3-phenyl-2-phenylimino-2,3-dihydroquinazoline	269, ir, nmr	723
4-Dicyanomethylquinazoline	280 or 281–282, ir, ms, nmr, uv	555,2111
4-Dicyanomethyl-2(1H)-quinazolinone	>300, ir, nmr	1350
1,3-Dicyclohexyl-5,8-dinitro-2,4(1H,3H)-quinazolinedione	193–194	536
2,4-Di(α,β-dimethylpropoxy)quinazoline	—	H 159
4-Diethoxycarbonylmethyl-2(1H)-quinazolinone	226–228, ir, nmr	1350
2,4-Diethoxy-6-methylquinazoline	60–62	1616
2,4-Diethoxyquinazoline	54–56	H 259; 2230
5,8-Diethoxyquinazoline	70–71, uv	595
7,8-Diethoxy-2,4-quinazolinediamine	222–224	696
6,7-Diethoxy-2,4(1H,3H)-quinazolinedione	256–259	641
7,8-Diethoxy-2,4(1H,3H)-quinazolinedione	290–292	696
2-Diethylamino-6,7-diethoxy-4(3H)-quinazolinone	196–199; HCl: 193–197	641
2-Diethylamino-6,7-dihydroxy-4(3H)-quinazolinone	296–300; HCl: 304–306	641
2-Diethylamino-6,7-diisopropoxy-4(3H)-quinazolinone	141–143; HCl: 126–132	641
2-Diethylamino-6,7-dimethoxy-3-methyl-4(3H)-quinazolinone	131–133, nmr, uv; HCl: 218–220	641
2-Diethylamino-6,7-dimethoxy-4-phenethylaminoquinazoline	180–181; HCl: 204–205	1810
4-Diethylamino-6,7-dimethoxy-2-phenethylaminoquinazoline	116–117; HCl: 197–198	1810
2-Diethylamino-6,7-dimethoxy-4(3H)-quinazolinone	216–218, nmr, uv; HCl: 250–251	641,1810
3-Diethylamino-6,7-dimethoxy-4(3H)-quinazolinone	112–114, HCl: 230–232	641
2-Diethylamino-4-dimethylaminoquinazoline	HClO₄: 151	1331
2-Diethylamino-6,7-dimethyl-4(3H)-quinazolinone	253–254; HCl: 237–239	641
2-Diethylamino-4-ethoxycarbonylaminoquinazoline	126–127, ir, nmr	2418
4-β-Diethylaminoethylamino-6-methoxy-5,8-dihydro-5,8-quinazolinedione	174	2278

511

ALPHABETICAL LIST OF SIMPLE QUINAZOLINES (*Continued*)

Quinazoline	m.p. (°C), etc.	Ref.
4-β-Diethylaminoethylamino-6-methoxyquinazoline	—	H 260
4-β-Diethylaminoethylamino-7-methoxyquinazoline	—	H 260
6-β-Diethylaminoethylamino-4-methoxyquinazoline	—	H 258
7-β-Diethylaminoethylamino-4-methoxyquinazoline	—	H 258
8-β-Diethylaminoethylamino-4-methoxyquinazoline	—	H 258
6-β-Diethylaminoethylamino-3-methyl-4(3*H*)-quinazolinone	—	H 368
7-β-Diethylaminoethylamino-3-methyl-4(3*H*)-quinazolinone	—	H 370
8-β-Diethylaminoethylamino-3-methyl-4(3*H*)-quinazolinone	—	H 370
4-β-Diethylaminoethylamino-7-nitroquinazoline	—	H 351
4-β-Diethylaminoethylaminoquinazoline	—	H 346
2-β-Diethylaminoethylamino-4(3*H*)-quinazolinone	—	H 365
2-Diethylamino-4-[β-(*N'*-ethylguanidino)ethyl]aminoquinazoline	HNO₃: 152–154	2224
3-β-Diethylaminoethyl-8-hydroxy-4(3*H*)-quinazolinone	—	H 179
2-β-Diethylaminoethyl-6-iodo-4(3*H*)-quinazolinone	—	H 175
3-β-Diethylaminoethyl-8-methoxy-4(3*H*)-quinazolinone	—	H 179
3-β-Diethylaminoethyl-2-methyl-4(3*H*)-quinazolinethione	liquid	H 297; 716
3-β-Diethylaminoethyl-2-methyl-4(3*H*)-quinazolinone	165/3	H 156; 716
1-β-Diethylaminoethyl-3-phenyl-2,4(1*H*,3*H*)-quinazolinedione	—	1107
3-β-Diethylaminoethyl-2-phenyl-4(1*H*)-quinazolinethione	80	716
1-β-Diethylaminoethyl-2-phenyl-4(3*H*)-quinazolinone	87	H 138; 716
3-β-Diethylaminoethyl-2,4(1*H*,3*H*)-quinazolinedione	146–147	H 200; 2287
3-β-Diethylaminoethyl-4(3*H*)-quinazolinethione	—	H 297
2-β-Diethylaminoethyl-4(3*H*)-quinazolinone	—	H 139
3-β-Diethylaminoethyl-4(3*H*)-quinazolinone	210–215/0.2; HCl: 202–205	H 144; 276,2287

512

3-β-Diethylaminoethyl-2-styryl-4(3H)-quinazolinone	—	H 166
2-Diethylamino-4-(δ-guanidinobutyl)aminoquinazoline	HNO₃: 150–151	2224
2-Diethylamino-4-(β-guanidinobutyl)aminoquinazoline	HNO₃: 161–162	2224
2-Diethylamino-4-(γ-guanidinopropyl)aminoquinazoline	HNO₃: 186–187	2224
2-Diethylamino-4-hydrazinoquinazoline	169–170	1337
2-Diethylamino-7-hydroxy-4(3H)-quinazolinone	HBr: 309–311	641
2-Diethylamino-8-hydroxy-4(3H)-quinazolinone	HBr: 280–283	641
2-Diethylamino-3-isopropyl-7-nitro-4(3H)-quinazolinone	110–111, ir, nmr, uv	2200
4-Diethylamino-6-methoxy-2-methylquinazoline	—	1113
2-Diethylamino-6-methoxy-4(3H)-quinazolinone	193–196; HCl: 209–212	641
2-Diethylamino-7-methoxy-4(3H)-quinazolinone	190–193; HCl: 215–218	641
3-N-(Diethylaminomethyl)aminomethyl-4(3H)-quinazolinone	HCl: 225–226	1192
2-Diethylamino-4-[β-(N'-methylguanidino)ethyl]aminoquinazoline	HCl: 242–245	2224
7-Diethylaminomethyl-1-isopropyl-4-phenyl-2(1H)-quinazolinone	174–176	681
3-Diethylaminomethyl-2-methyl-4(3H)-quinazolinone	—	H 156
2-Diethylaminomethyl-1-phenyl-4(1H)-quinazolinone	101–103, nmr	1394,1590
3-Diethylamino-1-methyl-2,4(1H,3H)-quinazolinedione	178	1603
2-Diethylaminomethyl-4(3H)-quinazolinone	—	H 139
3-Diethylaminomethyl-4(3H)-quinazolinone	HCl: 143–144	1192
2-Diethylamino-7-nitro-3-phenyl-4(3H)-quinazolinone	134–135, ir, nmr, uv	2200
2-Diethylamino-4-quinazolinamine	129–130, ir, nmr	2418
4-Diethylamino-7-quinazolinamine	—	H 363
2-Diethylaminoquinazoline	liquid; pic: 210–211	281
4-Diethylaminoquinazoline	—	H 346
3-Diethylamino-2,4(1H,3H)-quinazolinedione	137	1603
2-Diethylamino-4(3H)-quinazolinone	177–180; HCl: 244–248	641
2,4-Di(β-ethylbutoxy)quinazoline	—	H 259
3-N,N-Diethylcarbamoylmethyl-2-methyl-4(3H)-quinazolinone	—	H 157

ALPHABETICAL LIST OF SIMPLE QUINAZOLINES (*Continued*)

Quinazoline	m.p. (°C), etc.	Ref.
3-*N*,*N*-Diethylcarbamoylmethyl-2-phenyl-4(3*H*)-quinazolinone	—	*H* 170
2-*N*,*N*-Diethylcarbamoylmethyl-4(3*H*)-quinazolinone	236–237	531
Diethyl 2-chloro-4-quinazolinephosphonate	180/8	571
Diethyl 1,3-dimethyl-2,4-dioxo-1,2,3,4-tetrahydro-5,6-quinazolinedicarboxylate	149–150, ir, ms, nmr	2140
2,3-Diethyl-5-nitro-4(3*H*)-quinazolinone	—	*H* 187
N,*N*-Diethyl-2-phenyl-4-quinazolinecarbothioamide	136–138, ir, nmr	2196
2,4-Di(α-ethylpropoxy)quinazoline	—	*H* 259
2,4-Diethylquinazoline	nmr	2193
1,3-Diethyl-2,4(1*H*,3*H*)-quinazolinedione	113, ir, uv	*H* 205; 1464,2281
2,3-Diethyl-4(3*H*)-quinazolinethione	—	*H* 297
2,3-Diethyl-4(3*H*)-quinazolinone	95–96	*H* 164; 1059
2-Difluoromethyl-6-methyl-4(3*H*)-quinazolinone	nmr	2263
2,4-Difluoroquinazoline	69–71, nmr	699
5,6-Difluoro-2,4-quinazolinediamine	256–257, nmr	2335
6,7-Difluoro-2,4-quinazolinediamine	172–173, nmr	2335
7,8-Difluoro-2,4-quinazolinediamine	295–296, nmr	2335
6,8-Difluoro-2,4,5,7-tetramethoxyquinazoline	144–146, nmr, uv	34
2-Diformylmethyl-3-phenyl-4(3*H*)-quinazolinone	204	506
2,4-Dihydrazinoquinazoline	226–229; HCl	*H* 384; 292,1919
5,8-Dihydro-5,8-quinazolinedione	> 350, ir, uv	659
5,8-Dihydro-4,5,8(3*H*)-quinazolinetrione	> 320, ir, nmr	2324
5,8-Dihydroxy-2-hydroxymethyl-4(3*H*)-quinazolinone	crude, nmr	1999
6,7-Dihydroxy-1-β-methoxycarbonylethyl-4-methyl-2(1*H*)-quinazolinone	H_2O: 280–285	944
5,6-Dihydroxy-4-methyl-2(1*H*)-quinazolinone	324–326	1592

Compound		Reference
5,8-Dihydroxy-2-methyl-4(3H)-quinazolinone	HBr: 277–280, ir, nmr	1999
5,8-Dihydroxy-6-methyl-4(3H)-quinazolinone	crude (?), ir, nmr	2324
6,7-Dihydroxy-2-methyl-4(3H)-quinazolinone	>350	218
6,7-Dihydroxy-3-methyl-4(3H)-quinazolinone	340	498
6,8-Dihydroxy-2-phenyl-4(3H)-quinazolinone	—	H 198
3-β,γ-Dihydroxypropyl-6-methoxy-2-methyl-4(3H)-quinazolinone	—	H 181
3-β,γ-Dihydroxypropyl-8-methoxy-2-methyl-4(3H)-quinazolinone	—	H 181
3-β,γ-Dihydroxypropyl-2-methyl-4(3H)-quinazolinone	—	H 158
4-β,γ-Dihydroxypropyl-2-methylthiomethylquinazoline 3-oxide	122–123	874
3-β,γ-Dihydroxypropyl-2-phenyl-4(3H)-quinazolinone	—	H 170
3-β,γ-Dihydroxypropyl-4(3H)-quinazolinone	153–154	H 147; 276
5,8-Dihydroxy-4(3H)-quinazolinone	210–213, ir, nmr	2324
6,7-Dihydroxy-4(3H)-quinazolinone	>400	498
5,8-Dihydroxy-2-vinyl-4(3H)-quinazolinone	201–203, ir, nmr	2205,2430
6,8-Diiodo-2-isopentylthio-3-phenyl-4(3H)-quinazolinone	92	131
6,8-Diiodo-2-β-methylaminoethyl-4(3H)-quinazolinone	—	H 175
6,8-Diiodo-2-methyl-4(3H)-quinazolinone	—	H 173
6,8-Diiodo-2-pentylthio-3-phenyl-4(3H)-quinazolinone	230	131
6,8-Diiodo-3-phenyl-2-thioxo-1,2-dihydro-4(3H)-quinazolinone	197	131
6,8-Diiodo-2,4(1H,3H)-quinazolinedione	316–317, ir	H 203; 565
6,8-Diiodo-4(3H)-quinazolinone	—	H 154
6,8-Diiodo-2-thioxo-1,2-dihydro-4(3H)-quinazolinone	308–309	565
2,4-Diisobutoxyquinazoline	—	H 259
2,4-Diisopentyloxyquinazoline	—	H 259
6,7-Diisopropoxy-2,4(1H,3H)-quinazolinedione	248–250	641
4-Diisopropylamino-2-phenylquinazoline	99–100, nmr	2506
3-Diisopropylamino-2-thioxo-1,2-dihydro-4(3H)-quinazolinone	117–118	25
6,8-Diisopropyl-3-β-methoxycarbonylvinyl-4(3H)-quinazolinone	102–103	1558

ALPHABETICAL LIST OF SIMPLE QUINAZOLINES   (*Continued*)

Quinazoline	m.p. (°C), etc.	Ref.
6,8-Diisopropyl-4(3*H*)-quinazolinone	164–165	1558
2,4-Dimethoxy-5,8-dihydro-5,8-quinazolinedione	—	2531
6,7-Dimethoxy-2,3-dimethyl-4(3*H*)-quinazolinethione	229–231, ms, nmr, uv	1354
5,6-Dimethoxy-1,4-dimethyl-2(1*H*)-quinazolinone	172–174	1592
6,7-Dimethoxy-2,3-dimethyl-4(3*H*)-quinazolinone	218–220 or 223–224, ir, ms, nmr	853,1354
6,8-Dimethoxy-3,4-dimethyl-2(3*H*)-quinazolinone	183	110
6,7-Dimethoxy-2,3-diphenyl-4(3*H*)-quinazolinone	247–248	142
6,7-Dimethoxy-1,3-diprop-2'-ynyl-2,4(1*H*,3*H*)-quinazolinedione	225–227, ir, ms, nmr	2481
2,4-Di(β-methoxyethoxy)quinazoline	—	H 259
5,6-Dimethoxy-1-β-methoxycarbonylethyl-4-methyl-2(1*H*)-quinazolinone	118–119	944
6,7-Dimethoxy-4-β-methoxycarbonylethyl-4-methyl-2(1*H*)-quinazolinone	148–150, ir, nmr	944
6,7-Dimethoxy-1-β-methoxycarbonylethyl-4-phenyl-2(1*H*)-quinazolinone	136–138	944
6,7-Dimethoxy-1-β-methoxycarbonylethyl-4-propyl-2(1*H*)-quinazolinone	crude	944
6,7-Dimethoxy-3-β-methoxycarbonylvinyl-4(3*H*)-quinazolinone	242–243	1558
6,8-Dimethoxy-3-β-methoxycarbonylvinyl-4(3*H*)-quinazolinone	224–226	1558
5,6-Dimethoxy-2-methylamino-4(3*H*)-quinazolinone	199–202, ir, nmr	1562
6,7-Dimethoxy-2-methylamino-4(3*H*)-quinazolinone	289–291 or 294–296, ir, nmr; HCl: 334–336	641,1562
4,6-Dimethoxy-7-methyl-5,8-dihydro-5,8-quinazolinedione	182–183, ir, ms, nmr	2363
6,7-Dimethoxy-2-methyl-4-methylthioquinazoline	191–192	1354
2-Dimethoxymethyl-3-phenyl-4(3*H*)-quinazolinone	122–123, ir, ms, nmr	1154,1286
6,7-Dimethoxy-2-methyl-3-phenyl-4(3*H*)-quinazolinone	247–248	142
7,8-Dimethoxy-2-methyl-3-phenyl-4(3*H*)-quinazolinone	178–179	142
2,4-Dimethoxy-6-methylquinazoline	—	H 259

516

6,7-Dimethoxy-2-methylquinazoline	168–169	H 261; 1354
6,7-Dimethoxy-4-methylquinazoline	150–152	1425,2170
6,7-Dimethoxy-2-methyl-4-quinazolinecarbonitrile	193–194, nmr	2110
6,7-Dimethoxy-3-methyl-2,4(1H,3H)-quinazolinedione	296–298	641
6,7-Dimethoxy-2-methylquinazoline 3-oxide	180–182, nmr	2110
6,7-Dimethoxy-2-methyl-4(3H)-quinazolinethione	—	1058,1354
5,6-Dimethoxy-4-methyl-2(1H)-quinazolinone	230–232 or 237–240, nmr; HCl: 208–210 or 215–217, nmr	1592,1930,2567
5,7-Dimethoxy-4-methyl-2(1H)-quinazolinone	245–248	1592
5,8-Dimethoxy-2-methyl-4(3H)-quinazolinone	195–196	1999
5,8-Dimethoxy-4-methyl-2(1H)-quinazolinone	214–215	1592
5,8-Dimethoxy-6-methyl-4(3H)-quinazolinone	244–255, ir, nmr	2324
6,7-Dimethoxy-2-methyl-4(3H)-quinazolinone	—	H 173; 1354
6,7-Dimethoxy-3-methyl-4(3H)-quinazolinone	207–209 or 214–215, nmr	498, 990
6,7-Dimethoxy-4-methyl-2(1H)-quinazolinone	260–270	79,1592
6,8-Dimethoxy-4-methyl-2(1H)-quinazolinone	254–256	1592
7,8-Dimethoxy-2-methyl-4(3H)-quinazolinone	—	H 173
5,8-Dimethoxy-2-methylthio-3-phenyl-4(3H)-quinazolinone	160, ir, nmr	897
6,7-Dimethoxy-2-methylthio-3-phenyl-4(3H)-quinazolinone	229–230	141
6,7-Dimethoxy-3-morpholino-4(3H)-quinazolinone	239–240	641
4,6-Dimethoxy-7-nitroquinazoline	—	H 258
5,8-Dimethoxy-6-nitro-2,4(1H,3H)-quinazolinedione	>300, ir, nmr	1987
5,6-Dimethoxy-2-oxo-1,2-dihydro-4-quinazolinecarboxylic acid	267–269, nmr	1930
6,7-Dimethoxy-4-oxo-3,4-dihydro-2-quinazolinecarboxylic acid	296–298	1549
6,7-Dimethoxy-2-phenethylamino-4(3H)-quinazolinone	229–231	641
4,6-Dimethoxy-2-phenyl-5,8-dihydro-5,8-quinazolinedione	275–277, ir	763
6,7-Dimethoxy-4-phenylquinazoline	178, complexes	1647, 2170
6,8-Dimethoxy-2-phenylquinazoline	—	H 262

517

ALPHABETICAL LIST OF SIMPLE QUINAZOLINES (*Continued*)

Quinazoline	m.p. (°C), etc.	Ref.
5,8-Dimethoxy-3-phenyl-4(3*H*)-quinazolinone	128, ir, nmr	897
6,7-Dimethoxy-4-phenyl-2(1*H*)-quinazolinone	187–190	79,1592
5,8-Dimethoxy-3-phenyl-2-thioxo-1,2-dihydro-4(3*H*)-quinazolinone	270	897
6,7-Dimethoxy-3-phenyl-2-thioxo-1,2-dihydro-4(3*H*)-quinazolinone	297–298	141
6,7-Dimethoxy-2-piperidino-4(3*H*)-quinazolinone	263–265; HCl: 256–257	641
6,7-Dimethoxy-2-propylamino-4(3*H*)-quinazolinone	216–218; HCl: 299–300	641
6,7-Dimethoxy-4-propyl-2(1*H*)-quinazolinone	—	79
4,6-Dimethoxy-7-quinazolinamine	—	H 258
5,8-Dimethoxy-6-quinazolinamine	—	436
2,4-Dimethoxyquinazoline	74–76	H 259; 2230
2,7-Dimethoxyquinazoline	—	H 253
5,8-Dimethoxyquinazoline	119, ir, ms, nmr, uv; pic: 168	586,1604
6,7-Dimethoxyquinazoline	—	H 261
6,7-Dimethoxy-4-quinazolinecarbonitrile	221–223, nmr	2110
6,7-Dimethoxy-2,4-quinazolinediamine	238 to 248	696,785,1852,2008,2011
7,8-Dimethoxy-2,4-quinazolinediamine	261–262	696
5,8-Dimethoxy-2,4(1*H*,3*H*)-quinazolinedione	>300, ir, nmr	1987
6,7-Dimethoxy-2,4(1*H*,3*H*)-quinazolinedione	275–280 or 323–325	939,1175,1580
7,8-Dimethoxy-2,4(1*H*,3*H*)-quinazolinedione	313–315	696
6,7-Dimethoxyquinazoline 3-oxide	237–238, nmr	2110
2,4-Dimethoxy-5-quinazolinol	—	2531
1,4-Dimethoxy-2(1*H*)-quinazolinone	—	H 466
5,6-Dimethoxy-2(1*H*)-quinazolinone	242–244, nmr	1930
5,8-Dimethoxy-4(3*H*)-quinazolinone	>230 or 256, ir, nmr	897,2324

518

Compound	Data	Ref.
6,7-Dimethoxy-4(3H)-quinazolinone	296–297 or 306–307	498,1558
6,8-Dimethoxy-4(3H)-quinazolinone	280–282	1558
7,8-Dimethoxy-4(3H)-quinazolinone	210–211	498
6,7-Dimethoxy-2-styryl-4(3H)-quinazolinone	296–301	2264
6,7-Dimethoxy-4-trifluoromethyl-2(1H)-quinazolinone	266–268	1592
5,8-Dimethoxy-2-vinyl-4(3H)-quinazolinone	195, ir, nmr	2430
3-(1,1-Dimethylacetonyl)-2-methyl-4(3H)-quinazolinone	108–110, ir, ms, nmr	2219
2-Dimethylamino-6,7-diisopropoxy-4(3H)-quinazolinone	195–196; HCl: 229–231	641
3-Dimethylamino-6,7-dimethoxy-2,4(1H,3H)-quinazolinedione	264	1603
2-Dimethylamino-5,6-dimethoxy-4(3H)-quinazolinone	234–235, ir, nmr	1562
2-Dimethylamino-6,7-dimethoxy-4(3H)-quinazolinone	246–248, ir, nmr, uv; HCl: 278–282	641,1562
3-Dimethylamino-6,7-dimethoxy-4(3H)-quinazolinone	174–176, uv	641
7-Dimethylamino-1,3-dimethyl-6-phenyl-2,4(1H,3H)-quinazolinedione	262–264, nmr, uv	2465
2-Dimethylamino-6,7-dimethyl-4(3H)-quinazolinone	286–288; HCl: 301–303	641
6-Dimethylamino-3,4-dimethyl-2(3H)-quinazolinone	211	110
3-Dimethylamino-2,4-dioxo-1,2,3,4-tetrahydro-6-quinazolinesulfonamide	303–305	1603
4-β-Dimethylaminoethylamino-6-methoxy-5,8-dihydro-5,8-quinazolinedione	158	2278
4-β-Dimethylaminoethylamino-2-methylquinazoline	107–108, ms, nmr	95
2-β-Dimethylaminoethylamino-4-phenylquinazoline	HCl: 234–236	1601
4-β-Dimethylaminoethylamino-2-phenylquinazoline	liquid, ms, nmr; HBr: 282–283; HCl: 273	95,1601
2-β-Dimethylaminoethyl-6,8-diiodo-4(3H)-quinazolinone	—	H 175
4-β-Dimethylaminoethyl-6,7-dimethoxyquinazoline	HCl: 169–171, ir, nmr	1425
4-(N-β-Dimethylaminoethyl-N-ethylamino)quinazoline	—	H 347
1-β-Dimethylaminoethyl-3-ethyl-2,4(1H,3H)-quinazolinedione	204–205	668
2-β-Dimethylaminoethyl-6-iodo-4(3H)-quinazolinone	—	H 175

ALPHABETICAL LIST OF SIMPLE QUINAZOLINES (*Continued*)

Quinazoline	m.p. (°C), etc.	Ref.
4-β-Dimethylaminoethyl-7-methoxyquinazoline	—	*H* 261
4-(*N*-β-Dimethylaminoethyl-*N*-methylamino)quinazoline	—	*H* 347
4-β-Dimethylaminoethyl-2-methylquinazoline	—	*H* 63
1-β-Dimethylaminoethyl-3-methyl-2,4(1*H*,3*H*)-quinazolinedione	203–204	668
3-β-Dimethylaminoethyl-2-methyl-4(3*H*)-quinazolinethione	54	716
3-β-Dimethylaminoethyl-2-methyl-4(3*H*)-quinazolinone	—	*H* 156
1-β-Dimethylaminoethyl-3-phenyl-2,4(1*H*,3*H*)-quinazolinedione	224–226	668
3-β-Dimethylaminoethyl-2-phenyl-4(3*H*)-quinazolinethione	97	716
1-β-Dimethylaminoethyl-2-phenyl-4(1*H*)-quinazolinone	—	*H* 138
1-β-Dimethylaminoethyl-4-phenyl-2(1*H*)-quinazolinone	108–109	681
3-β-Dimethylaminoethyl-2-phenyl-4(3*H*)-quinazolinone	121	716
4-β-Dimethylaminoethylquinazoline	—	*H* 62
3-β-Dimethylaminoethyl-2,4(1*H*,3*H*)-quinazolinedione	HCl (?): 221–222	*H* 200; 2409
3-Dimethylamino-1-ethyl-2,4(1*H*,3*H*)-quinazolinedione	98	1603
2-β-Dimethylaminoethyl-4(3*H*)-quinazolinone	—	*H* 140
3-β-Dimethylaminoethyl-4(3*H*)-quinazolinone	—	*H* 145
3-β-Dimethylaminoethyl-2-thioxo-1,2-dihydro-4(3*H*)-quinazolinone	190–191	52
4-Dimethylamino-6-hydroxy-2-phenyl-5,8-dihydro-5,8-quinazolinedione	181–182	592
6-Dimethylamino-1-isopropyl-7-methyl-4-phenyl-2(1*H*)-quinazolinone	184–186	681
6-Dimethylamino-1-isopropyl-4-phenyl-2(1*H*)-quinazolinone	167–168	681
7-Dimethylamino-1-isopropyl-4-phenyl-2(1*H*)-quinazolinone	181–183	681
2-Dimethylamino-6-methoxy-4(3*H*)-quinazolinone	216–219; HCl: 260–264	641
2-Dimethylamino-7-methoxy-4(3*H*)-quinazolinone	257–261; HCl: 266–268	641
2-Dimethylamino-8-methoxy-4(3*H*)-quinazolinone	254–257; HCl: 223–226	641

Compound		
3-N-(Dimethylaminomethyl)aminomethyl-4(3H)-quinazolinone	HCl: 202–203	1192
2-Dimethylaminomethyl-6,7-dimethyl-4-phenylquinazoline	—	H 65
2-Dimethylaminomethyl-6,7-dimethyl-4-phenylquinazoline 3-oxide	—	H 464
3-Dimethylaminomethyl-1-methyl-2,4-dioxo-1,2,3,4-tetrahydro-6-quinazolinesulfonamide	263–265	1603
4-Dimethylaminomethyleneamino-6-nitroquinazoline	153	1852
4-Dimethylaminomethyleneaminoquinazoline	69	1852
2-Dimethylaminomethyl-4-methylquinazoline 3-oxide	101, ir, ms, nmr	2500
3-Dimethylaminomethyl-2-methyl-4(3H)-quinazolinone	—	H 156
4-Dimethylamino-2-methyl-6-nitroquinazoline	—	H 353
3-Dimethylamino-2-methyl-6-nitro-4(3H)-quinazolinone	196–197	1629
2-Dimethylaminomethyl-4-phenylquinazoline 3-oxide	112, ir, ms, nmr	2500
2-Dimethylamino-4-methylquinazoline	—	H 342
4-Dimethylamino-2-methylquinazoline	41–43, nmr	2506
3-Dimethylamino-1-methyl-2,4(1H,3H)-quinazolinedione	154 or 157–159	1437,1603
3-Dimethylamino-6-methyl-2,4(1H,3H)-quinazolinedione	227–228	1437
3-Dimethylaminomethyl-4(3H)-quinazolinone	HCl: 180–181	1192
3-Dimethylamino-2-methyl-4(3H)-quinazolinone	96–97	H 372; 1629
6-Dimethylamino-4-methyl-2(1H)-quinazolinone	278–281	1592
2-Dimethylamino-4-methylthioquinazoline	65–66, nmr	1410
4-Dimethylamino-2-methylthioquinazoline	100–101, nmr	1410
2-Dimethylamino-6-nitro-4-quinazolinamine	266–268	600
4-Dimethylamino-7-nitroquinazoline	—	H 252
3-Dimethylamino-6-nitro-4(3H)-quinazolinone	—	H 374
7-Dimethylamino-6-nitro-4(3H)-quinazolinone	288–289, nmr	900
4-Dimethylamino-6-nitro-2-trichloromethylquinazoline	—	H 350,353
2-Dimethylamino-4-pentyloxyquinazoline	ir,nmr	2482
2-Dimethylamino-4-phenylquinazoline	125–126 or 127, ir, nmr; HCl: >260	1946,2206

ALPHABETICAL LIST OF SIMPLE QUINAZOLINES (*Continued*)

Quinazoline	m.p. (°C), etc.	Ref.
4-Dimethylamino-2-phenylquinazoline	66–68 or 69–70	2206,2506
6-Dimethylamino-3-phenyl-2,4(1*H*,3*H*)-quinazolinedione	—	*H* 383
4-Dimethylamino-2-phenyl-6-quinazolinol	—	397
2-Dimethylamino-1-phenyl-4(1*H*)-quinazolinone	209–210, nmr	1590
4-γ-Dimethylaminopropylamino-6-methoxy-5,8-dihydro-5,8-quinazolinedione	152, ir, nmr	2278
4-γ-Dimethylaminopropylamino-6-methoxyquinazoline	—	*H* 261
4-γ-Dimethylaminopropylamino-7-methoxyquinazoline	—	*H* 261
4-γ-Dimethylaminopropylamino-6-methoxy-5-quinazolinol	crude, ir, nmr	2278
4-γ-Dimethylaminopropylamino-7-nitroquinazoline	—	*H* 352
4-γ-Dimethylaminopropylamino-6-piperidino-5,8-dihydro-5,8-quinazolinedione	105, ir, nmr	2278
4-γ-Dimethylaminopropylaminoquinazoline	—	*H* 347
1-γ-Dimethylaminopropyl-3-ethyl-2,4(1*H*,3*H*)-quinazolinedione	190–192	668
2-γ-Dimethylaminopropyl-4-isopropylquinazoline	liquid; HCl: 157–159, nmr	56
3-β-Dimethylaminopropyl-2-methyl-4(3*H*)-quinazolinone	—	*H* 156
3-γ-Dimethylaminopropyl-2-methyl-4(3*H*)-quinazolinone	—	*H* 156
2-γ-Dimethylaminopropyl-4-phenylquinazoline	HCl: 174–176	56
1-γ-Dimethylaminopropyl-3-phenyl-2,4(1*H*,3*H*)-quinazolinedione	215–217	668
1-γ-Dimethylaminopropyl-2-phenyl-4(1*H*)-quinazolinone	—	*H* 138
4-γ-Dimethylaminopropyl-2-phenyl-4(3*H*)-quinazolinone	85–87, ir, nmr	1267
4-γ-Dimethylaminopropylquinazoline	HCl: 203–205	56
3-γ-Dimethylaminopropyl-2,4(1*H*,3*H*)-quinazolinedione	—	*H* 200
3-γ-Dimethylaminopropyl-2-thioxo-1,2-dihydro-4(3*H*)-quinazolinone	183–184	52
2-Dimethylaminoquinazoline	86	*H* 342; 849
4-Dimethylaminoquinazoline	118–120/0.1, ms	1786

Compound		
4-Dimethylamino-2-quinazolinecarbonitrile	148–150	1142
2-Dimethylamino-4,6-quinazolinediamine	188–189	600
5-Dimethylamino-2,4-quinazolinediamine	204–205, nmr; HCl: 282–284	1932,2225
6-Dimethylamino-2,4-quinazolinediamine	HCl: > 260, nmr	2326
3-Dimethylamino-2,4(1H,3H)-quinazolinedione	237 or 240–241	1437,1603
5-Dimethylamino-2,4(1H,3H)-quinazolinedione	> 280, nmr	1584
2-Dimethylamino-4(3H)-quinazolinethione	195	439
2-Dimethylamino-4(3H)-quinazolinone	239–241; HCl: 279–282	641
3-Dimethylamino-4(3H)-quinazolinone	68–69	H 374; 1471
7-Dimethylamino-5,6,8-trifluoro-1,3-dimethyl-2,4(1H,3H)-quinazolinedione	186–187, ir, nmr	1317
4-Dimethylamino-2-trimethylammonioquinazoline	Cl⁻: crude	1142
Dimethyl 2-anilino-4-quinazolinephosphonate	273–276	571
2,4-Di(β-methylbutoxy)-6-nitroquinazoline	—	H 259
2,4-Di(α-methylbutoxy)-6-quinazolinamine	—	H 259
2,4-Di(α-methylbutoxy)quinazoline	—	H 259
2,4-Di(β-methylbutoxy)quinazoline	—	H 259
Dimethyl 2-butylamino-4-quinazolinephosphonate	270–274	571
3-N,N-Dimethylcarbamoylmethyl-2-phenyl-4(3H)-quinazolinone	—	H 170
Dimethyl 2-chloro-4-quinazolinephosphonate	64 or 74–76, ir, nmr, uv	341,571
Dimethyl 1,3-dimethyl-2,4-dioxo-1,2,3,4-tetrahydro-5,6-quinazolinedicarboxylate	209–210, ir, ms, nmr	2140
Dimethyl 1,3-dimethyl-2,4-dioxo-1,2,3,4-tetrahydro-6,7-quinazolinedicarboxylate	192–193	1149,2177
1,3-Dimethyl-2,4-dioxo-1,2,3,4-tetrahydro-6-quinazolinecarbonitrile	215–216, nmr, uv	2177,2465
1,3-Dimethyl-2,4-dioxo-1,2,3,4-tetrahydro-5-quinazolinecarbonyl chloride	—	H 485
1,3-Dimethyl-2,4-dioxo-1,2,3,4-tetrahydro-5-quinazolinecarboxylic acid	—	H 485
5,6-Dimethyl-2,4-diphenylquinazoline (?)	—	H 65
5,7-Dimethyl-2,4-diphenylquinazoline	—	H 65
6,7-Dimethyl-2,4-diphenylquinazoline (?)	—	H 65
Dimethyl 5-hydroxy-2,4-dimethoxy-6,7-quinazolinedicarboxylic acid	152–154, ir, nmr	1773,2334

523

ALPHABETICAL LIST OF SIMPLE QUINAZOLINES (*Continued*)

Quinazoline	m.p. (°C), etc.	Ref.
Dimethyl 5-hydroxy-1,3-dimethyl-2,4-dioxo-1,2,3,4-tetrahydro-6,7-quinazolinedicarboxylate	203–204	1149
Dimethyl 8-methoxy-2,4-dioxo-1,2,3,4-tetrahydro-6,7-quinazolinedicarboxylate	298–300, ir, nmr, uv	2005
6,8-Dimethyl-2-methylaminomethyl-4-phenylquinazoline 3-oxide	—	H 464
6,7-Dimethyl-2-methylaminomethylquinazoline 3-oxide	—	H 464
1,3-Dimethyl-5-methylamino-6-nitro-2,4(1*H*,3*H*)-quinazolinedione	200–205, nmr	1466,2261
1,3-Dimethyl-7-methylamino-6-nitro-2,4(1*H*,3*H*)-quinazolinedione	327–329, ir, nmr	1358
1,2-Dimethyl-4-methylimino-1,4-dihydro-6-quinazolinamine	—	H 364
1,3-Dimethyl-7-methylthio-5-phenyl-2,4(1*H*,3*H*)-quinazolinedione	154–155, ir, nmr	2398
2,6-Dimethyl-4-methylthioquinazoline	81–82, nmr	2255
1,3-Dimethyl-5-nitro-2,4(1*H*,3*H*)-quinazolinedione	—	H 206
1,3-Dimethyl-6-nitro-2,4(1*H*,3*H*)-quinazolinedione	—	H 206
1,3-Dimethyl-7-nitro-2,4(1*H*,3*H*)-quinazolinedione	—	H 206
1,3-Dimethyl-8-nitro-2,4(1*H*,3*H*)-quinazolinedione	—	H 206
2,3-Dimethyl-5-nitro-4(3*H*)-quinazolinone	—	H 182
2,3-Dimethyl-6-nitro-4(3*H*)-quinazolinone	—	H 182
2,3-Dimethyl-7-nitro-4(3*H*)-quinazolinone	—	H 182
2,3-Dimethyl-8-nitro-4(3*H*)-quinazolinone	—	H 182
6,7-Dimethyl-4-oxo-3,4-dihydro-8-quinazolinecarbonitrile	289–291, ir, nmr	1195
3,*N*-Dimethyl-4-oxo-3,4-dihydro-2-quinazolinecarboxamide	212–213	H 482; 768
2,3-Dimethyl-4-oxo-3,4-dihydro-6-quinazolinecarboxylic acid	>300	1808
2,3-Dimethyl-4-oxo-3,4-dihydro-7-quinazolinecarboxylic acid	—	H 484
2,7-Dimethyl-4-oxo-3,4-dihydro-6-quinazolinecarboxylic acid	—	H 484
6,7-Dimethyl-2-oxo-4-phenyl-1,2-dihydro-8-quinazolinecarbonitrile	311–312	1570

Compound		
2,7-Dimethyl-4-oxo-3-phenyl-3,4-dihydro-6-quinazolinecarboxylic acid	—	H 484
2,4-Di(α-methylpentyloxy)quinazoline	—	H 259
2,4-Di(β-methylpentyloxy)quinazoline	—	H 259
1,3-Dimethyl-4-phenylimino-3,4-dihydro-2(1H)-quinazolinethione	124–125	88
1,3-Dimethyl-4-phenylimino-3,4-dihydro-2(1H)-quinazolinone	87–88	88
4,7-Dimethyl-5-phenyl-2-quinazolinamine	210–211	526
4,6-Dimethyl-2-phenylquinazoline	94–95, uv	1083
4,7-Dimethyl-2-phenylquinazoline	liquid, uv	1083
6,7-Dimethyl-4-phenyl-8-quinazolinecarbonitrile	185–187, ir, nmr	1568
1,3-Dimethyl-6-phenyl-2,4(1H,3H)-quinazolinedione	190–191, nmr, uv	2465
1,2-Dimethyl-4-phenylquinazolinium iodide	—	H 67
4,7-Dimethyl-2-phenyl-5-quinazolinol	105–108, ir, nmr; HCl: 220, ir	1633
6,8-Dimethyl-2-phenyl-5-quinazolinol	—	H 197
1,6-Dimethyl-4-phenyl-2(1H)-quinazolinone	146–147 or 245–246	681,836
1,7-Dimethyl-4-phenyl-2(1H)-quinazolinone	171–172	681
2,6-Dimethyl-3-phenyl-4(3H)-quinazolinone	—	H 185
2,8-Dimethyl-3-phenyl-4(3H)-quinazolinone	—	H 185
5,7-Dimethyl-2-phenyl-4(3H)-quinazolinone	282–283	103
2,6-Dimethyl-3-piperidino-4(3H)-quinazolinone	136–137	1629
1,6-Dimethyl-2-pivalamido-4(1H)-quinazolinone	221, nmr	2194
3,6-Dimethyl-2-pivalamido-4(3H)-quinazolinone	142–143, nmr	2194
3-(1,1-Dimethylprop-2-enyl)-4(3H)-quinazolinone	97–99, ir, ms, nmr	2219
2,4-Dimethylquinazoline	72, nmr, uv; pic: 169–172	H 63; 575,1305,1946,2193
2,6-Dimethylquinazoline	pKa, uv	H 65; 168
4,5-Dimethylquinazoline	—	H 65
2,6-Dimethyl-4-quinazolinecarboxamide	pKa, uv	H 482; 168,170
2,6-Dimethyl-4-quinazolinecarboxylic acid	—	H 482
5,6-Dimethyl-2,4-quinazolinediamine	260–265	674

ALPHABETICAL LIST OF SIMPLE QUINAZOLINES (Continued)

Quinazoline	m.p. (°C), etc.	Ref.
6,7-Dimethyl-2,4-quinazolinediamine	253–254	H 254; 661
6,8-Dimethyl-2,4-quinazolinediamine	—	H 254
1,3-Dimethyl-2,4(1H,3H)-quinazolinedione	165 to 171	H 205; 216,362,718,722,1427, 1464,1531,2281
1,6-Dimethyl-2,4(1H,3H)-quinazolinedione	254–255, ms, nmr	2194
6,7-Dimethyl-2,4(1H,3H)-quinazolinedione	350–354	641
1,3-Dimethyl-2,4(1H,3H)-quinazolinedithione	267–268	89,274
2,4-Dimethylquinazoline 1-oxide	76–78 or 134–136, ir, uv	163,874
2,4-Dimethylquinazoline 3-oxide	uv	H 464; 575
4,7-Dimethylquinazoline 3-oxide	—	H 464
2,3-Dimethyl-4(3H)-quinazolinethione	—	H 298
2,6-Dimethyl-4(3H)-quinazolinethione	258–259, uv	241,2255
3,4-Dimethyl-2(3H)-quinazolinethione	221–223 or 223–225, ir, nmr	217,2381
1,2-Dimethyl-4(1H)-quinazolinimine	94–98, ir, nmr	1232
2,3-Dimethyl-4(3H)-quinazolinimine	157–158, ms, pK_a	417,1000
1,4-Dimethylquinazolinium iodide	—	H 67
2,4-Dimethyl-8-quinazolinol	—	H 197
6,8-Dimethyl-5-quinazolinol	—	H 197
1,2-Dimethyl-4(1H)-quinazolinone (glomerine)	197 to 208, ir, nmr, uv; HCl: >320; MeI: 255–257	H 138; 118,244,754,821,908,945 946,1026,1190,1232,1367, 2380
2,3-Dimethyl-4(3H)-quinazolinone	104 to 110, ir, nmr, uv; H₂O: 70–72; SbCl₅,HCl: 194–196, nmr	H 156; 133,793,818,853,935, 1059,1267,1367,1388,1644,2079, 2131,2267,2361,2406

526

Compound	Properties	References
2,6-Dimethyl-4(3H)-quinazolinone	248 or 255, ms, nmr	H 173; 510,1608
2,7-Dimethyl-4(3H)-quinazolinone	—	H 173
2,8-Dimethyl-4(3H)-quinazolinone	—	H 173
3,4-Dimethyl-2(3H)-quinazolinone	204–206 or 210–214, ir, nmr	110,217
4,7-Dimethyl-2(1H)-quinazolinone	248–253	79
5,6-Dimethyl-4(3H)-quinazolinone	—	H 155
5,7-Dimethyl-4(3H)-quinazolinone	—	H 155
5,8-Dimethyl-4(3H)-quinazolinone	—	H 155
6,7-Dimethyl-4(3H)-quinazolinone	—	H 155
6,8-Dimethyl-4(3H)-quinazolinone	244–247	H 155; 1558
7,8-Dimethyl-4(3H)-quinazolinone	—	H 155
1,3-Dimethyl-2-thioxo-1,2-dihydro-4(3H)-quinazolinone	173–176 or 182–184, ir, ms, nmr	H 308; 87,1846,1996,2197
1,3-Dimethyl-4-thioxo-3,4-dihydro-2(1H)-quinazolinone	210–211	86,87,685
6,8-Dimethyl-2-trichloromethyl-4(3H)-quinazolinone	235–236	755
Dimethyl 1,3,5-trimethyl-2,4-dioxo-1,2,3,4-tetrahydro-6,7-quinazolinedicarboxylate	156–157	1149
2,8-Dimorpholino-5,6-dihydro-5,6-quinazolinedione	204–206, ir, uv	764
2,8-Dimorpholino-5,6-dihydro-4,5,6(3H)-quinazolinetrione	212–214	762
2,4-Dimorpholino-6-phenylquinazoline	168, ir, nmr	428
2,4-Dimorpholinoquinazoline	175, ir, nmr	H 358; 428
2,4-Dineopentyloxyquinazoline	—	H 259
6,8-Dinitro-2,4-quinazolinediamine	336–337	601
6,8-Dinitro-2,4(1H,3H)-quinazolinedione	—	H 202
6,8-Dinitro-4(3H)-quinazolinone	ir	H 155; 532
2,4-Dioxo-3-phenyl-1,2,3,4-tetrahydro-6-quinazolinecarboxanilide	—	1122
2,4-Dioxo-1,2,3,4-tetrahydro-7-quinazolinarsonic acid	—	H 487
2,4-Dioxo-1,2,3,4-tetrahydro-3-quinazolinecarbonitrile	258–260, ir, nmr	1350,1911
2,4-Dioxo-1,2,3,4-tetrahydro-5-quinazolinecarbonyl chloride	—	H 485
2,4-Dioxo-1,2,3,4-tetrahydro-5-quinazolinecarboxamide	—	H 482

527

Quinazoline	m.p. (°C), etc.	Ref.
2,4-Dioxo-1,2,3,4-tetrahydro-6-quinazolinecarboxanilide	—	1122
2,4-Dioxo-1,2,3,4-tetrahydro-5-quinazolinecarboxylic acid	—	H 485
2,4-Dioxo-1,2,3,4-tetrahydro-7-quinazolinecarboxylic acid	—	H 485
2,4-Dioxo-1,2,3,4-tetrahydro-8-quinazolinecarboxylic acid	—	H 485
2,4-Dioxo-1,2,3,4-tetrahydro-7-quinazolinesulfonamide	—	H 487
2,4-Dipentyloxy-6-quinazolinamine	—	H 259
2,4-Dipentyloxyquinazoline	—	H 259
2,4-Diphenoxyquinazoline	—	H 259
2,3-Diphenyl-4-phenylimino-3,4-dihydroquinazoline	155–156	1167
2,4-Diphenyl-8-phenylthioquinazoline	182	204
2,4-Diphenylquinazoline	116 to 122, ir, uv	H 64; 27,111,398,784,905,1025, 1083,1382,1759,1942
1,3-Diphenyl-2,4(1H,3H)-quinazolinedione	complex	1349
1,3-Diphenyl-2,4(1H,3H)-quinazolinedithione	255–256, uv, complex	1271,1349
2,4-Diphenylquinazoline 3-oxide	—	H 464
1,2-Diphenyl-4(1H)-quinazolinethione	305–307, ir, uv	H 301; 937
2,3-Diphenyl-4(3H)-quinazolinethione	149	H 298; 953
3,4-Diphenyl-2(3H)-quinazolinethione	—	1078
2,3-Diphenyl-4(3H)-quinazolinethione S-oxide	115, ir, ms, nmr	1130
1,2-Diphenyl-4(1H)-quinazolinone	261 (?), 270 or 273–275, ms, uv; Mel: 258–266, ir, uv	H 138; 39,698,937,1342
2,3-Diphenyl-4(3H)-quinazolinone	156–158 or 176–177, ir, nmr, uv	H 170; 39,133,161,260,446,1050, 1167,1285,1422,1918
2,8-Dipiperidino-5,6-dihydro-5,6-quinazolinedione	206–208, ir, nmr, uv	764,765

Compound	Properties	References
4,8-Dipiperidino-5,6-dihydro-5,6-quinazolinedione	142–144, ir, uv	764
2,8-Dipiperidino-5,6-dihydro-4,5,6(3H)-quinazolinetrione	193–194 or 195–196, ir, nmr	762,782
2,4-Dipiperidinoquinazoline	—	H 359
2,4-Dipropoxyquinazoline	—	H 259
5,8-Dipropoxyquinazoline	44–45, uv	595
2-Dipropylamino-5,6-dimethoxy-4(3H)-quinazolinone	137–138, ir, nmr	1562
2-Dipropylamino-6,7-dimethoxy-4(3H)-quinazolinone	197–200 or 201–203, ir, nmr	641,1562
2-Dipropylamino-4-methylthioquinazoline	52–54, nmr	1410
2-Dipropylaminoquinazoline	47	281,849
2-Dipropylamino-4(3H)-quinazolinone	188–190, ir, nmr	1569
1,3-Dipropyl-2,4(1H,3H)-quinazolinedione	liquid, anal, ir, uv	1464,2281
2,4-Dithioxo-1,2,3,4-tetrahydro-6-quinazolinecarbonitrile	328–330	77
Echinopsine, *see* 1-Methyl-4(1H)-quinazolinone		
Echinozolinone [probably 3-β-hydroxyethyl-4(3H)-quinazolinone]	150, ir, ms, nmr	2030,2133; cf. 940, 1799, 2557
2-Ethoxalylmethyl-3-ethoxycarbonylmethyl-4(3H)-quinazolinone	152–153, nmr, st, uv	2118
2-Ethoxalylmethyl-3-β-hydroxyethyl-4(3H)-quinazolinone	193–194, nmr, st, uv	2118
2-Ethoxalylmethyl-1-methyl-4(1H)-quinazolinone	270–273, st	244
2-Ethoxalylmethyl-3-methyl-4(3H)-quinazolinone	173–174, nmr, st, uv	2118
2-Ethoxalyl-3-methyl-4(3H)-quinazolinone		H 486
2-Ethoxycarbonylamino-3-ethyl-4(3H)-quinazolinimine	143–145, ir, nmr	2418
6-Ethoxycarbonylamino-3-isopropoxycarbonylamino-2-methyl-4(3H)-quinazolinone	—	H 380
3-Ethoxycarbonylaminomethyl-2-methyl-4(3H)-quinazolinone	135–138, ir	277
3-Ethoxycarbonylamino-2-methyl-6-nitro-4(3H)-quinazolinone	—	H 378
3-Ethoxycarbonylamino-2-methyl-4(3H)-quinazolinone	126–127	H 373; 109
4-Ethoxycarbonylamino-2-methylthioquinazoline	144–146, ir, nmr	2095
3-Ethoxycarbonylamino-2-methyl-6-ureido-4(3H)-quinazolinone	—	H 380
3-Ethoxycarbonylamino-4-oxo-3,4-dihydro-2-quinazolinecarboxamide	>300	124

ALPHABETICAL LIST OF SIMPLE QUINAZOLINES (*Continued*)

Quinazoline	m.p. (°C), etc.	Ref.
2-Ethoxycarbonylamino-3-phenyl-4(3*H*)-quinazolinimine	161–164, ir, nmr	2418
2-Ethoxycarbonylaminoquinazoline	177	849
3-Ethoxycarbonylamino-2,4(1*H*,3*H*)-quinazolinedione	225–227	*H* 382, 227
4-Ethoxycarbonylamino-2(1*H*)-quinazolinethione	154–156, ir, nmr	2095
2-Ethoxycarbonylamino-4(3*H*)-quinazolinone	—	*H* 365
4-β-Ethoxycarbonylethyl-6,7-dimethoxy-2-methylquinazoline	140, ir, nmr	1941
1-β-Ethoxycarbonylethyl-7,8-dimethoxy-4-methyl-2(1*H*)-quinazolinone	128–130	944
2-α-Ethoxycarbonylethyl-1-β-hydroxyethyl-4(1*H*)-quinazolinone	123–125, ir, nmr	1415
2-α-Ethoxycarbonylethyl-1-γ-hydroxypropyl-4(1*H*)-quinazolinone	122–124, ir, nmr	1415
2-β-Ethoxycarbonylethyl-4-methylquinazoline	40–41, ir, ms, nmr, uv	904
3-β-Ethoxycarbonylethyl-2-methyl-4(3*H*)-quinazolinone	102–103 or 121, ir, nmr	461,1907
2-β-Ethoxycarbonylethyl-4-phenylquinazoline	93–94, ir, ms, nmr, uv	904
2-α-Ethoxycarbonylethyl-1-phenyl-4(1*H*)-quinazolinone	186–188, ir, nmr	1415,1590
3-α-Ethoxycarbonylethyl-2-phenyl-4(3*H*)-quinazolinone	131	1889
3-β-Ethoxycarbonylethyl-2-phenyl-4(3*H*)-quinazolinone	65–70/1	1283
4-α-Ethoxycarbonylethylquinazoline	97, ms, nmr	1869
2-β-Ethoxycarbonylethyl-4(3*H*)-quinazolinone	170 or 175–177, ms, nmr, uv	904,2399
3-α-Ethoxycarbonylethyl-4(3*H*)-quinazolinone	97–99, ir, ms, nmr	1640,2433
2-β-Ethoxycarbonylethyl-4(3*H*)-quinazolinone 1-oxide	150, nmr	2399
3-β-Ethoxycarbonylethyl-2-styryl-4(3*H*)-quinazolinone	109–110, ms, nmr	1907
1-Ethoxycarbonylmethyl-6,7-dimethoxy-4-methyl-2(1*H*)-quinazolinone	168–170, ir, nmr	944
3-Ethoxycarbonylmethyl-6-iodo-2-methyl-4(3*H*)-quinazolinone	108	613
3-Ethoxycarbonylmethyl-2-isopropyl-4(3*H*)-quinazolinethione	100, nmr	1565
2-Ethoxycarbonylmethyl-1-β-methoxycarbonylethyl-4(1*H*)-quinazolinone	160–162	1415

Compound		
2-Ethoxycarbonylmethyl-1-methoxycarbonylmethyl-4(1*H*)-quinazolinone	173–175	1415
3-Ethoxycarbonylmethyl-2-methyl-7-nitro-4(3*H*)-quinazolinone	—	*H* 181
3-Ethoxycarbonylmethyl-*N*-methyl-4-oxo-3,4-dihydro-2-quinazolinecarboxamide	251	768
4-Ethoxycarbonylmethyl-2-methylquinazoline	42–43, ir, nmr; HCl: 208–210	887
1-Ethoxycarbonylmethyl-3-methyl-2,4(1*H*,3*H*)-quinazolinedione	132–133, ir, ms, nmr	911
3-Ethoxycarbonylmethyl-1-methyl-2,4(1*H*,3*H*)-quinazolinedione	130–132, ir, ms, nmr	1822
3-Ethoxycarbonylmethyl-2-methyl-4(3*H*)-quinazolinethione	154–155, ir, nmr	1209
3-Ethoxycarbonylmethyl-2-methyl-4(3*H*)-quinazolinone	130–131	*H* 158; 613
3-Ethoxycarbonylmethyl-5-methyl-4(3*H*)-quinazolinone	—	*H* 179
4-Ethoxycarbonylmethyl-2-phenylquinazoline	120, ir, nmr; HCl: 113–115	887
3-Ethoxycarbonylmethyl-2-phenyl-4(3*H*)-quinazolinethione	143, nmr	1565
1-Ethoxycarbonylmethyl-2-phenyl-4(1*H*)-quinazolinone	153–154, ir, ms, nmr	1825
2-Ethoxycarbonylmethyl-1-phenyl-4(1*H*)-quinazolinone	235–237, ir, nmr	1415
2-Ethoxycarbonylmethyl-3-phenyl-4(3*H*)-quinazolinone	156–158, nmr	2572
3-Ethoxycarbonylmethyl-2-phenyl-4(3*H*)-quinazolinone	146, 120/5	1283,1889
2-Ethoxycarbonylmethyl-4-quinazolinamine	174	1368
4-Ethoxycarbonylmethylquinazoline	91 to 105, ms, nmr, uv; HCl: 181–185; pic: 220–222	*H* 62, 162,562,887,895,1869
1-Ethoxycarbonylmethyl-2,4(1*H*,3*H*)-quinazolinedione	252, ir, ms, nmr	1822
3-Ethoxycarbonylmethyl-2,4(1*H*,3*H*)-quinazolinedione	227 to 242, ir, ms, nmr	738,1137,1448,1465
3-Ethoxycarbonylmethyl-4(3*H*)-quinazolinethione	118, nmr	1565
1-Ethoxycarbonylmethyl-4(1*H*)-quinazolinone	160–162, ir, ms, nmr	1825
2-Ethoxycarbonylmethyl-4(3*H*)-quinazolinone	161–163 or 165–166	*H* 140; 531,1426
3-Ethoxycarbonylmethyl-4(3*H*)-quinazolinone	75–77	*H* 147; 276,1640
3-Ethoxycarbonylmethyl-2-thioxo-1,2-dihydro-4(3*H*)-quinazolinone	213–215	54
3-(β-Ethoxycarbonyl-α-methylvinyl)-6-methylthio-4(3*H*)-quinazolinone	149–150	1558
3-Ethoxycarbonyloxy-1-methyl-2,4(1*H*,3*H*)-quinazolinedione	128	236
3-Ethoxycarbonyloxy-2-methyl-4(3*H*)-quinazolinone	81	236

ALPHABETICAL LIST OF SIMPLE QUINAZOLINES (*Continued*)

Quinazoline	m.p. (°C), etc.	Ref.
3-Ethoxycarbonyloxy-2,4(1*H*,3*H*)-quinazolinedione	185	236
3-Ethoxycarbonyloxy-4(3*H*)-quinazolinone	124–125	236
1-*γ*-Ethoxycarbonylpropyl-6,7-dimethoxy-4-methyl-2(1*H*)-quinazolinone	137–139	944
8-Ethoxy-5,6-dimethoxy-3,7-dimethyl-4(3*H*)-quinazolinone	83–84, ir, ms, nmr	2363
8-Ethoxy-5,6-dimethoxy-7-methyl-4(3*H*)-quinazolinone	171–172, ir, ms, nmr	2363
6-Ethoxy-2,3-dimethyl-4(3*H*)-quinazolinone	—	*H* 181
6-Ethoxy-3,4-dimethyl-2(3*H*)-quinazolinone	172	110
8-Ethoxy-3,4-dimethyl-2(3*H*)-quinazolinone	138	110
2-(*β*-Ethoxyethoxy)methyl-3-phenyl-4(3*H*)-quinazolinone	—	2388
4-Ethoxy-2-ethyl-5-nitroquinazoline	—	*H* 258
2-*β*-Ethoxyethyl-1-phenyl-4(1*H*)-quinazolinone	108–110, nmr	1394
2-*β*-Ethoxyethyl-3-phenyl-4(1*H*)-quinazolinone	165–166, ms, nmr	847
4-Ethoxy-2-ethylquinazoline	—	*H* 256
3-*β*-Ethoxyethyl-2,4(1*H*,3*H*)-quinazolinedione	156–157, ir, nmr	1428
6-Ethoxy-2-guanidino-4-methylquinazoline	—	*H* 343
8-Ethoxy-2-guanidino-4-methylquinazoline	—	*H* 343
2-Ethoxy-3-*β*-hydroxyethyl-4(3*H*)-quinazolinone	84–85, ir, nmr	1473
6-Ethoxy-2-isopropyl-3-methyl-4(3*H*)-quinazolinone	95, ir, nmr	1049
7-Ethoxy-2-isopropyl-3-methyl-4(3*H*)-quinazolinone	141, ir, nmr	1049
8-Ethoxy-2-isopropyl-3-methyl-4(3*H*)-quinazolinone	111, ir, nmr	1049
2-Ethoxy-3-isopropyl-7-nitro-4(3*H*)-quinazolinone	150–151, ir, nmr, uv	2200
7-Ethoxy-1-isopropyl-4-phenyl-2(1*H*)-quinazolinone	117–120	681
6-Ethoxy-2-isopropyl-4(3*H*)-quinazolinone	227, ir, nmr	1049
7-Ethoxy-2-isopropyl-4(3*H*)-quinazolinone	215, ir, nmr	1049

Compound	mp (°C), data	References
8-Ethoxy-2-isopropyl-4(3H)-quinazolinone	225, ir, nmr	1049
2-Ethoxy-4-methoxyquinazoline	—	H 259
4-Ethoxy-2-methoxyquinazoline	—	H 259
6-Ethoxy-7-methoxy-2,4-quinazolinediamine	214–216	696
6-Ethoxy-7-methoxy-2,4(1H,3H)-quinazolinedione	265–273	696
4-Ethoxy-2-methyl-5-nitroquinazoline	—	H 258
4-Ethoxy-2-methyl-7-nitroquinazoline	—	H 258
2-Ethoxymethyl-3-phenyl-4(3H)-quinazolinone	82–83, ir, nmr; HCl: 178–180	376,1154
2-Ethoxy-4-methylquinazoline	45–46, ms, uv	163
4-Ethoxy-2-methylquinazoline	HCl: ir, ms	H 256;1127
4-Ethoxy-6-methylquinazoline	63–64	1001
4-Ethoxy-8-methylquinazoline	36–37	1001
2-Ethoxymethyl-4(3H)-quinazolinone	HCl: 203–204	943
2-Ethoxy-3-methyl-4(3H)-quinazolinone	93	H 172; 889
3-Ethoxy-2-methyl-4(3H)-quinazolinone	—	H 469
6-Ethoxy-2-methyl-4(3H)-quinazolinone	—	H 173
8-Ethoxy-2-methyl-4(3H)-quinazolinone	—	H 173
2-Ethoxy-7-nitro-3-phenyl-4(3H)-quinazolinone	187–188, ir, nmr, uv	2200
4-Ethoxy-2-phenoxyquinazoline	—	H 259
2-Ethoxy-4-phenylquinazoline	—	H 253
4-Ethoxy-2-phenylquinazoline	55–57	H 256; 1059
2-Ethoxy-3-phenyl-4(3H)-quinazolinone	—	H 172
3-γ-Ethoxypropyl-2,4(1H,3H)-quinazolinedione	133–134, ir, nmr	1435
2-Ethoxy-4-quinazolinamine	—	H 253
2-Ethoxyquinazoline	61	H 253; 849
4-Ethoxyquinazoline	46–48 or 48–50, ms	H 254; 172,276,1001
4-Ethoxy-2-quinazolinecarbonitrile	—	H 481
4-Ethoxy-2-quinazolinecarboxamide	—	H 482

ALPHABETICAL LIST OF SIMPLE QUINAZOLINES   (Continued)

Quinazoline	m.p. (°C), etc.	Ref.
5-Ethoxy-2,4-quinazolinediamine	214–216 or 255–258, nmr	1932,2225
1-Ethoxy-2,4(1H,3H)-quinazolinedione	170, uv	471
4-Ethoxyquinazoline 1-oxide	—	H 461; 471
5-Ethoxy-8-quinazolinol	159–160, uv	595
2-Ethoxy-4(3H)-quinazolinone	175 to 182, ir, nmr	H 143; 182,889,1473,2230,2558
4-Ethoxy-2(1H)-quinazolinone	—	H 137
2-Ethoxy-4(3H)-quinazolinone 1-oxide	195, ir	182
4-Ethoxy-2(3H)-quinazolinone 1-oxide	—	H 466; 471
6-Ethoxy-2-styryl-4(3H)-quinazolinone	307–311	2264
3-Ethoxy(thiocarbonyl)amino-6,7-dimethoxy-2-methylquinazolinium chloride	118–119; zwitterionic base: 136–137	1302,1348
4-Ethoxy-2-trichloromethylquinazoline	—	H 256
8-Ethoxy-4,5,6-trimethoxy-7-methylquinazoline	liquid, anal, ms, nmr	2363
Ethyl 2-acetoxymethyl-5,7-dimethyl-4-oxo-3-phenyl-3,4-dihydro-6-quinazolinecarboxylate	98–99	1364
2-Ethylamino-6,7-dimethoxy-4(3H)-quinazolinone	262–264; HCl: 293–294	641
Ethyl 6-amino-2-ethoxyformimidoyl-4-oxo-3,4-dihydro-7-quinazolinecarboxylate	249	85
Ethyl 3-(β-aminoethyl)amino-4-oxo-3,4-dihydro-2-quinazolinecarboxylate	138	124
2-β-Ethylaminoethyl-6-iodo-4(3H)-quinazolinone	—	H 175
3-β-Ethylaminoethyl-4(3H)-quinazolinone	—	H 145
6-Ethylaminomethyl-2-methyl-4(3H)-quinazolinone	HBr: 298, nmr	2269
4-Ethylamino-2-methyl-6-nitroquinazoline	—	H 353
Ethyl 6-amino-2-methyl-4-oxo-3,4-dihydro-7-quinazolinecarboxylate	275–277	85
Ethyl 6-amino-2-methyl-4-oxo-3-phenyl-3,4-dihydro-7-quinazolinecarboxylate	218–219	85
6-Ethylamino-5-methyl-2,4-quinazolinediamine	222–225	602

Compound	mp (°C), data	Reference
4-Ethylamino-3-methyl-2(3*H*)-quinazolinethione	201–203	88
2-Ethylaminomethyl-4(1*H*)-quinazolinone 3-oxide	175, ir	236
2-Ethylamino-4-methylthioquinazoline	79–80, nmr	1379,1410
4-Ethylamino-2-methylthioquinazoline	149–150, nmr	1379, 1410
4-Ethylamino-6-nitroquinazoline	—	551
4-Ethylamino-7-nitroquinazoline	—	H 352
Ethyl 3-amino-4-oxo-3,4-dihydro-2-quinazolinecarboxylate	138, ir, nmr, uv	124
Ethyl 6-amino-4-oxo-3,4-dihydro-2-quinazolinecarboxylate	252, nmr	123
Ethyl 6-amino-4-oxo-2-phenyl-3,4-dihydro-2-quinazolinecarboxylate	270–271	85
3-γ-Ethylaminopropyl-2-phenyl-4(3*H*)-quinazolinethione	104	716
3-γ-Ethylaminopropyl-2-phenyl-4(3*H*)-quinazolinone	76	716
4-Ethylamino-2-quinazolinamine	273–274, ir, nmr	2418
4-Ethylamino-7-quinazolinamine	—	H 363
2-Ethylaminoquinazoline	96–97, nmr	1379, 1410
4-Ethylaminoquinazoline	149–150, nmr	H 347; 1379,1410
Ethyl 4-amino-2-quinazolinecarboxylate	—	1182
6-Ethylamino-2,4-quinazolinediamine	218–220	602
4-Ethylamino-2(1*H*)-quinazolinethione	228–230, ir, nmr	2384
Ethyl 3-anilino-4-oxo-3,4-dihydro-2-quinazolinecarboxylate	153–154, ir, nmr	H 483; 856
Ethyl 3-benzylideneamino-4-oxo-3,4-dihydro-2-quinazolinecarboxylate	165	124
Ethyl 7-benzyloxy-6-methoxy-4-oxo-3,4-dihydro-2-quinazolinecarboxylate	216–217	1549
Ethyl 2-bromomethyl-5,7-dimethyl-4-oxo-3-phenyl-3,4-dihydro-6-quinazolinecarboxylate	129–130	1364
Ethyl 7-bromo-4-oxo-3,4-dihydro-2-quinazolinecarboxylate	187	1743
Ethyl 4-butoxy-2-quinazolinecarboxylate	—	H 482
Ethyl 7-chloro-4-oxo-3,4-dihydro-2-quinazolinecarboxylate	187	1743
Ethyl 6-chloro-4-oxo-8-trifluoromethyl-3,4-dihydro-2-quinazolinecarboxylate	206–207	1549
Ethyl 4-chloro-2-quinazolinecarboxylate	85	123

ALPHABETICAL LIST OF SIMPLE QUINAZOLINES   (*Continued*)

Quinazoline	m.p. (°C), etc.	Ref.
Ethyl 4-chloro-6-quinazolinecarboxylate	107–108	2564
3-Ethyl-6,8-diiodo-2,4(1*H*,3*H*)-quinazolinedione	—	*H* 203
4-Ethyl-6,7-dimethoxy-1-β-methoxyethyl-2(1*H*)-quinazolinone	132–134	944
Ethyl 6,7-dimethoxy-4-oxo-3,4-dihydro-2-quinazolinecarboxylate	281–283	1549
4-Ethyl-6,7-dimethoxy-2-quinazolinamine	190–193	696
4-Ethyl-6,7-dimethoxyquinazoline	149–150	261,2170
4-Ethyl-5,6-dimethoxy-2(1*H*)-quinazolinone	224–226, nmr	1592,1930
4-Ethyl-6,7-dimethoxy-2(1*H*)-quinazolinone	252–255	79,1592
Ethyl 5,7-dimethyl-2,4-dioxo-3-phenyl-1,2,3,4-tetrahydro-6-quinazolinecarboxylate	289–290, ir, ms, nmr	2109
Ethyl 1,3-dimethyl-2,4-dioxo-1,2,3,4-tetrahydro-5-quinazolinecarboxylate	—	*H* 485
3-Ethyl-2,7-dimethyl-4-oxo-3,4-dihydro-6-quinazolinecarboxylic acid	—	*H* 484
4-Ethyl-2,6-dimethylquinazoline	50, 100/0.01, nmr, uv	421
4-Ethyl-2,8-dimethylquinazoline	80/0.1, nmr, uv; pic: 148–149	421
Ethyl 2,6-dimethyl-4-quinazolinecarboxylate	—	*H* 483
2-Ethyl-3,6-dimethyl-4(3*H*)-quinazolinone	—	*H* 187
Ethyl 2,4-dioxo-1,2,3,4-tetrahydro-5-quinazolinecarboxylate	—	*H* 485
Ethyl 6-ethoxycarbonylamino-4-oxo-3,4-dihydro-2-quinazolinecarboxylate	190, nmr	123
Ethyl 3-β-ethoxycarbonylethyl-4-oxo-3,4-dihydro-2-quinazolinecarboxylate	51–53, ms, nmr	1639
Ethyl 3-ethoxycarbonylmethyl-4-oxo-3,4-dihydro-2-quinazolinecarboxylate	85–87 or 92–93, ir, nmr	759,1639
Ethyl 1-γ-ethoxycarbonylpropyl-4-oxo-1,4-dihydro-2-quinazolinecarboxylate	117–119	1415
Ethyl 3-γ-ethoxycarbonylpropyl-4-oxo-3,4-dihydro-2-quinazolinecarboxylate	liquid, ir, ms, nmr	1639
Ethyl 4-ethoxy-2-quinazolinecarboxylate	—	*H* 483
3-Ethyl-2-ethylamino-4(3*H*)-quinazolinone	112, ir, nmr	723

Compound		
2-Ethyl-4-ethylthioquinazoline	HCl: 150–152	*H* 299; 1059
2-Ethyl-6-fluoro-3-phenyl-4(3*H*)-quinazolinone	94	489
4-Ethyl-5-fluoro-2-quinazolinamine	142–144, nmr	2569
3-Ethyl-4-formylmethyl-2(3*H*)-quinazolinone	223–224	110
6-Ethyl-2-guanidino-4-methylquinazoline	—	*H* 343
3-Ethyl-2-hydrazino-6-methoxy-4(3*H*)-quinazolinone	220, ms, nmr	2123
1-Ethyl-2-hydrazino-4(1*H*)-quinazolinone	—	1928
3-Ethyl-2-hydrazino-4(3*H*)-quinazolinone	160–163, ir, nmr	1884
2-Ethyl-3-β-hydroxyethyl-4(3*H*)-quinazolinone	143–145, nmr	2470
3-Ethyl-8-hydroxy-7-iodo-4(3*H*)-quinazolinone	—	*H* 179
Ethyl 2-hydroxymethyl-5,7-dimethyl-4-oxo-3-phenyl-3,4-dihydro-6-quinazolinecarboxylate	107–108	1364,2071
Ethyl 2-hydroxymethyl-6,8-dimethyl-4-oxo-3-phenyl-3,4-dihydro-7-quinazolinecarboxylate	179–180	1364
3-Ethyl-8-hydroxy-2-methyl-4(3*H*)-quinazolinone	—	*H* 198
3-Ethyl-1-hydroxy-2,4(1*H*,3*H*)-quinazolinedione	173–176	622,750
3-Ethyl-8-hydroxy-4(3*H*)-quinazolinone	—	*H* 179, 198
3-Ethyl-8-hydroxy-2-styryl-4(3*H*)-quinazolinone	—	*H* 198
3-Ethyl-1-hydroxy-2-thioxo-1,2-dihydro-4(3*H*)-quinazolinone	121–122	622
3-Ethylideneamino-2-methyl-4(3*H*)-quinazolinone	—	*H* 373
3-Ethylideneamino-2-phenyl-4(3*H*)-quinazolinone	—	*H* 373
3-Ethyl-6-iodo-2-thioxo-1,2-dihydro-4(3*H*)-quinazolinone	280, ir	1750
3-Ethyl-2-isopropyl-4(3*H*)-quinazolinethione	—	*H* 297
2-Ethyl-3-isopropyl-4(3*H*)-quinazolinone	56–60, ir, nmr, uv	412
2-Ethyl-3-methoxycarbonylamino-4(3*H*)-quinazolinone	ir, nmr	2063
Ethyl 1-β-methoxycarbonylethyl-4-oxo-1,4-dihydro-2-quinazolinecarboxylate	110–112	1415
Ethyl 3-methoxycarbonylmethyl-4-oxo-3,4-dihydro-2-quinazolinecarboxylate	68–70, ir, ms, nmr, uv	1639
2-Ethyl-1-methoxycarbonylmethyl-4(1*H*)-quinazolinone	164–166, ir, ms, nmr	1825

ALPHABETICAL LIST OF SIMPLE QUINAZOLINES (*Continued*)

Quinazoline	m.p. (°C), etc.	Ref.
3-Ethyl-8-methoxy-2-methyl-4(3*H*)-quinazolinone	—	*H* 181
Ethyl 7-methoxy-4-oxo-3,4-dihydro-2-quinazolinecarboxylate	210	1743
2-Ethyl-4-methoxyquinazoline	—	*H* 256
N-Ethyl-7-methoxy-2-quinazolinecarboxamide	90–91, nmr	2178
2-Ethyl-3-methoxy-4(3*H*)-quinazolinone	—	*H* 470
3-Ethyl-8-methoxy-4(3*H*)-quinazolinone	—	*H* 179
4-Ethyl-7-methoxy-2(1*H*)-quinazolinone	210–215	79
3-Ethyl-8-methoxy-2-styryl-4(3*H*)-quinazolinone	—	*H* 188
2-Ethyl-3-methylamino-4(3*H*)-quinazolinone	93, ir, nmr	1994
2-Ethyl-3-methyl-5-nitro-4(3*H*)-quinazolinone	—	*H* 187
3-Ethyl-2-methyl-5-nitro-4(3*H*)-quinazolinone	—	*H* 181
3-Ethyl-2-methyl-6-nitro-4(3*H*)-quinazolinone	—	*H* 181
3-Ethyl-2-methyl-7-nitro-4(3*H*)-quinazolinone	—	*H* 181
Ethyl 1-methyl-4-oxo-1,4-dihydro-2-quinazolinecarboxylate	170–172, ir, nmr	1415
Ethyl 3-methyl-4-oxo-3,4-dihydro-2-quinazolinecarboxylate	45, ir	123
Ethyl 7-methyl-4-oxo-3,4-dihydro-2-quinazolinecarboxylate	189, ms, uv	1743
Ethyl 6-methyl-2-phenyl-4-quinazolinecarboxylate	—	*H* 483
1-Ethyl-6-methyl-4-phenyl-2(1*H*)-quinazolinone	178–180	681
1-Ethyl-7-methyl-4-phenyl-2(1*H*)-quinazolinone	157–158	681
2-Ethyl-6-methyl-3-phenyl-4(3*H*)-quinazolinone	—	*H* 187
3-Ethyl-1-(2′-methylprop-2′-enyl)-2,4(1*H*,3*H*)-quinazolinedione	65, ir, nmr	718
2-Ethyl-4-methylquinazoline	liquid, nmr	*H* 63; 2193
2-Ethyl-6-methylquinazoline	—	*H* 65
4-Ethyl-2-methylquinazoline	90/0.05, nmr, uv	421,2193

538

Compound		
2-Ethyl-6-methyl-4-quinazolinecarboxamide	—	H 482
N-Ethyl-4-methyl-2-quinazolinecarboxamide	143, nmr	2178
Ethyl 4-methyl-2-quinazolinecarboxylate 3-oxide	151–153, ir	874
2-Ethyl-6-methyl-4-quinazolinecarboxylic acid	—	H 482
1-Ethyl-3-methyl-2,4(1H,3H)-quinazolinedione	—	H 205
3-Ethyl-1-methyl-2,4(1H,3H)-quinazolinedione	—	H 205
2-Ethyl-4-methylquinazoline 1-oxide	92–94, nmr, uv	163
2-Ethyl-3-methyl-4(3H)-quinazolinethione	—	H 297
3-Ethyl-2-methyl-4(3H)-quinazolinethione	—	H 297
2-Ethyl-1-methyl-4(1H)-quinazolinone (homoglomerine)	145 to 152, ir, nmr, uv	118,821,945,1026,1190, 1367,2380
2-Ethyl-3-methyl-4(3H)-quinazolinone	119–121	H 164; 853,1059
2-Ethyl-6-methyl-4(3H)-quinazolinone	—	H 175
2-Ethyl-7-methyl-4(3H)-quinazolinone	—	H 175
2-Ethyl-8-methyl-4(3H)-quinazolinone	—	H 175
3-Ethyl-2-methyl-4(3H)-quinazolinone	64–65 (?), 78 to 82, ir, nmr	H 156; 935,1059,1267,2162
3-Ethyl-4-methyl-2(3H)-quinazolinone	145	110
2-Ethyl-4-methylthioquinazoline	—	H 299
3-Ethyl-2-methylthio-4(3H)-quinazolinethione	114–115, uv	827
3-Ethyl-2-methylthio-4(3H)-quinazolinone	—	H 306
3-Ethyl-1-methyl-4-thioxo-3,4-dihydro-2(1H)-quinazolinone	153–154	86,87
2-Ethyl-3-morpholinomethyl-4(3H)-quinazolinone	—	1069
2-Ethyl-3-morpholino-4(3H)-quinazolinone	170–171	1629
Ethyl 6-nitro-4-oxo-3,4-dihydro-2-quinazolinecarboxylate	294, nmr	123
2-Ethyl-5-nitro-4(3H)-quinazolinone	—	H 175
3-Ethyl-6-nitro-4(3H)-quinazolinone	—	H 179; 1525
Ethyl 4-oxo-3,4-dihydro-2-quinazolinecarboxylate	180 to 192, ir, ms, nmr, uv	H 483; 123,926,1155,1182,1248, 1639,1743

ALPHABETICAL LIST OF SIMPLE QUINAZOLINES (*Continued*)

Quinazoline	m.p. (°C), etc.	Ref.
Ethyl 4-oxo-3,4-dihydro-3-quinazlinecarboxylate	61–62	276
Ethyl 4-oxo-1-phenyl-1,4-dihydro-2-quinazolinecarboxylate	193–195	1394,1415
Ethyl 4-oxo-3-phenyl-3,4-dihydro-2-quinazolinecarboxylate	108–109	H 483; 781,1544
2-Ethyl-3-pentyl-4(3H)-quinazolinone	114	2162
2-Ethyl-3-phenethylamino-4(3H)-quinazolinone	63–65, 180/0.0004, ir, nmr	2397
Ethyl 3-phenyl-4-phenylimino-3,4-dihydro-2-quinazolinecarboxylate	—	H 483
2-Ethyl-4-phenylquinazoline	85, ir	H 63; 1946
4-Ethyl-2-phenylquinazoline	44–46, uv	H 63; 1083
N-Ethyl-4-phenyl-2-quinazolinecarboxamide	178–179, nmr	2178
Ethyl 4-phenyl-2-quinazolinecarboxylate	131–135, ir, ms	1946
1-Ethyl-3-phenyl-2,4(1H,3H)-quinazolinedione	246, ir, ms, nmr	1650
1-Ethyl-4-phenyl-2(1H)-quinazolinethione	240–241, ir, nmr, uv	955
2-Ethyl-3-phenyl-4(3H)-quinazolinethione	—	H 297
3-Ethyl-2-phenyl-4(3H)-quinazolinethione	116	H 297; 953
1-Ethyl-2-phenyl-4(3H)-quinazolinimine	198–199, ir, nmr; HCl: 266–267, ir	1232
1-Ethyl-2-phenyl-4(1H)-quinazolinone	214	1275
1-Ethyl-4-phenyl-2(1H)-quinazolinone	182–184, ir, nmr	681,1955,2037
2-Ethyl-3-phenyl-4(3H)-quinazolinone	125–127	H 165; 608,912
3-Ethyl-2-phenyl-4(3H)-quinazolinone	128	H 170; 1275
1-Ethyl-3-phenyl-2-thioxo-1,2-dihydro-4(3H)-quinazolinone	258, ir, ms, nmr	1650
2-Ethyl-3-piperidinomethyl-4(3H)-quinazolinone	—	1069
2-Ethyl-3-pivalamido-4(3H)-quinazolinone	194–195, ir, ms, nmr	2593
2-Ethyl-3-propionamido-4(3H)-quinazolinone	125	1657

2-Ethyl-3-propyl-4(3H)-quinazolinone	102	2162
3-Ethyl-1-prop-2'-ynyl-2,4(1H,3H)-quinazolinedione	141–142	664
2-Ethyl-3-prop-2'-ynyl-4(3H)-quinazolinone	93–94	664
2-Ethyl-4-quinazolinamine	225, ir, nmr	1456
4-Ethyl-2-quinazolinamine	103–104, nmr	2569
2-Ethylquinazoline	nmr	H 61; 2478
4-Ethylquinazoline	15–16, ir, ms, nmr; pic: 165–167 or 170–171	H 62; 1885,1914,1946
2-Ethyl-4-quinazolinecarbonitrile	86–88, nmr	2110
N-Ethyl-2-quinazolinecarboxamide	128–129, nmr	2178
5-Ethyl-2,4-quinazolinediamine	162–163	H 354; 2225
6-Ethyl-2,4-quinazolinediamine	—	H 354
8-Ethyl-2,4-quinazolinediamine	pK_a, uv	250
1-Ethyl-2,4(1H,3H)-quinazolinedione	198–200 or 263–266, nmr	H 199; 362,2295
3-Ethyl-2,4(1H,3H)-quinazolinedione	195–197, ir, nmr	H 200; 770,1446,1448
3-Ethyl-2,4(1H,3H)-quinazolinedithione	182–183, nmr	89,460,1446
2-Ethylquinazoline 3-oxide	126–128, nmr	2110
4-Ethylquinazoline 1-oxide	101, ms	163
4-Ethylquinazoline 3-oxide	106, ms	163
1-Ethyl-4(1H)-quinazolinethione	—	H 301
2-Ethyl-4(3H)-quinazolinethione	204, ms, nmr	H 296; 957,1069
3-Ethyl-4(3H)-quinazolinethione	—	H 297
2-Ethyl-4(1H)-quinazolinethione 3-oxide	—	H 468
1-Ethyl-4(1H)-quinazolinone	127–129	276
2-Ethyl-4(3H)-quinazolinone	212 to 237, nmr, uv, complexes	H 140; 23,30,118,133,1003,1059, 1240,1406,1754,1915,2063, 2069,2244,2285,2397
3-Ethyl-4(3H)-quinazolinone	99–101	H 145; 276

ALPHABETICAL LIST OF SIMPLE QUINAZOLINES (*Continued*)

Quinazoline	m.p. (°C), etc.	Ref.
5-Ethyl-4(3*H*)-quinazolinone	—	*H* 154
2-Ethyl-4(1*H*)-quinazolinone 3-oxide	—	*H* 467
3-Ethyl-2-styryl-4(3*H*)-quinazolinone	—	*H* 167
5-Ethylsulfonyl-2,4-quinazolinediamine	203, nmr, pK_a	2225
2-Ethyl-5,6,7,8-tetrafluoro-3-phenyl-4(3*H*)-quinazolinone	98	617
4-Ethylthio-6,7-dimethoxy-2-methylquinazoline	—	1536
2-Ethylthio-6-fluoro-3-phenyl-4(3*H*)-quinazolinone	110	1046
2-Ethylthio-6-methoxy-4(3*H*)-quinazolinone	203–204	280
4-Ethylthio-2-methylquinazoline	HCl: 196–197	*H* 299; 1059
4-Ethylthio-2-morpholinoquinazoline	86	855
2-Ethylthio-3-phenethyl-4(3*H*)-quinazolinone	81–83	615
4-Ethylthio-2-phenylquinazoline	52–54	*H* 299; 1059
2-Ethylthio-3-phenyl-4(3*H*)-quinazolinethione	—	*H* 302
2-Ethylthio-3-phenyl-4(3*H*)-quinazolinone	105–107	*H* 306; 615
4-Ethylthio-2-piperidinoquinazoline	68	855
4-Ethylthioquinazoline	133–135/1.5	*H* 299; 1189
7-Ethylthioquinazoline	76, pK_a, uv	168
5-Ethylthio-2,4-quinazolinediamine	159, pK_a, nmr	2225
2-Ethylthio-4(3*H*)-quinazolinone	152 to 163, ir, nmr	943,1459,1483,1845
2-Ethyl-3-thioureido-4(3*H*)-quinazolinone	280–281	1979
1-Ethyl-2-thioxo-1,2-dihydro-4(3*H*)-quinazolinone	—	1928
3-Ethyl-2-thioxo-1,2-dihydro-4(3*H*)-quinazolinone	244 to 261, complexes	87,460,685,770,1646,2390
3-Ethyl-4-thioxo-3,4-dihydro-2(1*H*)-quinazolinone	239–240	86
4-Ethyl-2-trifluoromethylquinazoline	66, ir, ms, nmr	1946

542

Ethyl 2,5,7-trimethyl-4-oxo-3-phenyl-3,4-dihydro-6-quinazolinecarboxylate	102–104	1364
Ethyl 2,6,8-trimethyl-4-oxo-3-phenyl-3,4-dihydro-7-quinazolinecarboxylate	140–141	1364
3-Ethyl-2,6,8-trimethyl-4(3*H*)-quinazolinethione	—	*H* 181
2-Ethyl-3-ureido-4(3*H*)-quinazolinethione	—	*H* 381
6-Fluoro-1,3-dimethyl-2,4(1*H*,3*H*)-quinazolinedione	—	1513
7-Fluoro-1,3-dimethyl-2,4(1*H*,3*H*)-quinazolinedione	—	1513
7-Fluoro-2,3-diphenyl-4(3*H*)-quinazolinone	153	617
6-Fluoro-1-isopropyl-2-phenyl-4(1*H*)-quinazolinone	232–234	1590
7-Fluoro-1-isopropyl-4-phenyl-2(1*H*)-quinazolinone	142–143	681
2-Fluoromethyl-6-methyl-4(3*H*)-quinazolinone	229–230, nmr	2263
5-Fluoro-2-methyl-6-nitro-4(3*H*)-quinazolinone	crude, 246–248, nmr	2374
5-Fluoro-2-methyl-8-nitro-4(3*H*)-quinazolinone	crude, 246–248, nmr	2374
6-Fluoro-2-methyl-3-phenyl-4(3*H*)-quinazolinethione	142	489
7-Fluoro-2-methyl-3-phenyl-4(3*H*)-quinazolinethione	149	489
2-Fluoromethyl-3-phenyl-4(3*H*)-quinazolinone	101–103; HCl: 195–198	687,1550
6-Fluoro-2-methyl-3-phenyl-4(3*H*)-quinazolinone	174–177	489,687
7-Fluoro-2-methyl-3-phenyl-4(3*H*)-quinazolinone	148	489
5-Fluoro-2-methyl-4-quinazolinamine	162, nmr	2247
5-Fluoro-4-methyl-2-quinazolinamine	186–188, nmr	2569
6-Fluoro-4-methyl-2-quinazolinamine	215–216, nmr	2569
7-Fluoro-4-methyl-2-quinazolinamine	189–190, nmr	2569
6-Fluoro-2-methylquinazoline	137, pK_a	*H* 252; 168
6-Fluoro-2-methyl-4-quinazolinecarboxamide	194, pK_a, uv	*H* 482; 168,170
2-Fluoromethyl-4(3*H*)-quinazolinone	—	*H* 140
5-Fluoro-2-methyl-4(3*H*)-quinazolinone	268–270, nmr	2374
5-Fluoro-6-nitro-2-pivalamido-4(3*H*)-quinazolinone	231–232, nmr	2374
4-Fluoro-6-nitroquinazoline	93–94, uv	553
4-Fluoro-7-nitroquinazoline	131–133, uv	553

ALPHABETICAL LIST OF SIMPLE QUINAZOLINES (*Continued*)

Quinazoline	m.p. (°C), etc.	Ref.
4-Fluoro-8-nitroquinazoline	120–121, uv	553
5-Fluoro-6-nitro-2,4-quinazolinediamine	282–283, nmr	2221
5-Fluoro-8-nitro-2,4-quinazolinediamine	235–238, nmr	2221
7-Fluoro-6-nitro-2,4-quinazolinediamine	350–352, nmr	2335
8-Fluoro-6-nitro-2,4-quinazolinediamine	>350, nmr	2335
5-Fluoro-4-phenyl-2-quinazolinamine	142–143, nmr	2569
2-Fluoro-3-phenyl-4(3*H*)-quinazolinone	—	*H* 172
6-Fluoro-3-phenyl-2-thioxo-1,2-dihydro-4(3*H*)-quinazolinone	328–329	1046
6-Fluoro-4-phenyl-2(1*H*)-quinazolinone	282–283	1060
7-Fluoro-3-phenyl-2-thioxo-1,2-dihydro-4(3*H*)-quinazolinone	345	1046
6-Fluoro-3-phenyl-2-trifluoromethyl-4(3*H*)-quinazolinone	125–126	489
6-Fluoro-7-piperidino-2-styryl-4(3*H*)-quinazolinone	237–239, nmr	2264
5-Fluoro-4-quinazolinamine	204–205, nmr	2247
6-Fluoro-2-quinazolinamine	241–243	647
4-Fluoroquinazoline	84–85, nmr	699
5-Fluoroquinazoline	74, pK_a, uv	*H* 252; 168
6-Fluoroquinazoline	140, pK_a, uv	*H* 252; 168
7-Fluoroquinazoline	131, pK_a, uv	*H* 252; 168
5-Fluoro-2,4-quinazolinediamine	247 to 253, nmr	1932,2225,2335
6-Fluoro-2,4-quinazolinediamine	315–320, nmr	1932
7-Fluoro-2,4-quinazolinediamine	274–277, nmr	1932
8-Fluoro-2,4-quinazolinediamine	285–287, nmr	1932
5-Fluoro-2,4,6-quinazolinetriamine	227–229, nmr	2221
5-Fluoro-2,4,8-quinazolinetriamine	185–186, nmr	2335

Compound	Data	Ref.
7-Fluoro-2,4,6-quinazolinetriamine	274–275, nmr	2335
8-Fluoro-2,4,6-quinazolinetriamine	300–302, nmr	2335
5-Fluoro-4(3H)-quinazolinone	—	H 154
6-Fluoro-4(3H)-quinazolinone	252	H 154; 168
7-Fluoro-4(3H)-quinazolinone	230–233	H 154; 168
6-Fluoro-2-styryl-4(3H)-quinazolinone	274–275	2264
6-Fluoro-2-thioxo-1,2-dihydro-4(3H)-quinazolinone	322	1046
7-Fluoro-2-thioxo-1,2-dihydro-4(3H)-quinazolinone	298–300	1046
2-Formamidomethyl-3-phenyl-4(3H)-quinazolinone	184–185	152
3-Formamido-2-methyl-4(3H)-quinazolinone	—	H 373
7-Formamido-2-methyl-4(3H)-quinazolinone	—	H 370
4-Formamido-2-morpholinoquinazoline	238–239	429
7-β-Formylhydrazino-1,3-dimethyl-6-nitro-2,4(1H,3H)-quinazolinedione	314–316, ir, nmr	1358
3-Formylmethyl-4(3H)-quinazolinone	152, ir; oxime: 163–164, nmr; acetal: 79, nmr	1772
4-Formyloxyaminoquinazoline	192–194, nmr, uv	60
Glomerine, *see* 1,2-Dimethyl-4(1H)-quinazolinone		
Glycophymine, *see* 2-Benzyl-4(3H)-quinazolinone		
Glycophymoline, *see* 2-Benzyl-4-methoxyquinazoline (?)		
Glycorine, *see* 1-Methyl-4(1H)-quinazolinone		
Glycosmicine, *see* 1-Methyl-2,4(1H,3H)-quinazolinedione		
Glycosminine, *see* 2-Benzyl-4(3H)-quinazolinone		
2-Guanidino-4,6-dimethylquinazoline	—	H 344
2-Guanidino-4,7-dimethylquinazoline	—	H 344
2-Guanidino-6-hexyloxy-4-methylquinazoline	—	H 343
2-Guanidino-6-isopropyl-4-methylquinazoline	—	H 343
2-Guanidino-6-methoxy-4-methylquinazoline	—	H 344
2-Guanidino-4-methylquinazoline	—	H 344

ALPHABETICAL LIST OF SIMPLE QUINAZOLINES   (*Continued*)

Quinazoline	m.p. (°C), etc.	Ref.
2-Guanidinoquinazoline	—	H 343
2-Guanidino-4(3H)-quinazolinone	—	H 365
2-Guanidino-4,6,7-trimethylquinazoline	—	H 344
2-Heptafluoropropyl-4(3H)-quinazolinone	160–161, ir, nmr	1287
2,4,5,6,7,8-Hexachloroquinazoline	163–166, uv	34
2,4,5,6,7,8-Hexafluoroquinazoline	37–39, 84–87/50, nmr, uv	34,158,837
3-Hexanoylamino-2-pentyl-4(3H)-quinazolinone	98	1657
2,4,5,6,7,8-Hexaphenylquinazoline	264–265, ms, nmr	2588
4-Hexylaminoquinazoline	106–107, ms	1786
4-Hexylamino-8-quinazolinol	—	H 352
4-Hexyl-6-methyl-2-phenylquinazoline	61–63, nmr	2097,2160
6-Hexyl-5-methyl-2,4-quinazolinediamine	274–283	674
3-Hexyl-2-methyl-4(3H)-quinazolinone	—	H 156
3-Hexyl-5-methyl-4(3H)-quinazolinone	30–32, ir, nmr	1970
3-Hexyl-6-methyl-4(3H)-quinazolinone	71–74, ir, nmr	1970
3-Hexyloxy-2-methyl-4(3H)-quinazolinone	—	H 470
3-Hexyloxy-4(3H)-quinazolinone	—	H 470
2-Hexyl-4(3H)-quinazolinone	149–150, ir	1211
2-Hexylthio-6-methoxy-4(3H)-quinazolinone	159–160	280
2-Hexylthio-4(3H)-quinazolinone	80–81	1483,1845
4-Hex-1′-ynylquinazoline	153/3, ir	1093
Homoglomerine, *see* 2-Ethyl-1-methyl-4(1H)-quinazolinone		
3-α-Hydrazinocarbonylethyl-2-phenyl-4(3H)-quinazolinone	120	1889
3-β-Hydrazinocarbonylethyl-2-phenyl-4(3H)-quinazolinone	102	1283

Compound	Properties	References
3-α-Hydrazinocarbonylethyl-4(3H)-quinazolinone	—	2433
3-Hydrazinocarbonylmethyl-6-iodo-2-methyl-4(3H)-quinazolinone	265	613
3-Hydrazinocarbonylmethyl-2-methyl-4(3H)-quinazolinone	260–261 or 270–273, ir, nmr	277,613,1681
3-Hydrazinocarbonylmethyl-2-phenyl-4(3H)-quinazolinone	105 or 122	1283,1889
3-Hydrazinocarbonylmethyl-4(3H)-quinazolinone	252, ir	999,2528
2-Hydrazino-6,7-dimethoxy-4(3H)-quinazolinone	284–285; HCl: 264–266	641
7-Hydrazino-1,3-dimethyl-6-nitro-2,4(1H,3H)-quinazolinedione	278–279, ir, nmr; EtOCH=: 235–237, ir, nmr	1358
2-Hydrazino-4-hydrazono-3-phenyl-3,4-dihydroquinazoline	184–185	128
4-Hydrazino-6-methoxy-5-nitroquinazoline	—	H 261
2-Hydrazinomethyl-6-methyl-4(3H)-quinazolinone	239–240	510
4-Hydrazino-5-methyl-2-morpholinoquinazoline	193–196	1337
2-Hydrazino-6-methyl-4-phenylquinazoline	156–157	836
4-Hydrazino-6-methyl-2-phenylquinazoline	208–210	756
4-Hydrazino-8-methyl-2-phenylquinazoline	215–217	756
4-Hydrazino-2-methylquinazoline	175–176, ms	1127
4-Hydrazino-6-methylquinazoline	—	H 384
2-Hydrazinoethyl-4(3H)-quinazolinone	185–186; pic: 197–198	683
2-Hydrazino-1-methyl-4(1H)-quinazolinone	—	1928
2-Hydrazino-6-methyl-4(3H)-quinazolinone	335	1841
4-Hydrazino-2-morpholinoquinazoline	239–240	H 384; 1337
4-Hydrazino-6-nitro-2-trichloromethylquinazoline	—	H 384
2-Hydrazino-4-phenylquinazoline	157–158	H 384; 836,1718
4-Hydrazino-2-phenylquinazoline	217–218	H 384; 188,2050
2-Hydrazino-3-phenyl-4(3H)-quinazolinone	151–152 (?), 202–203, or 195, then 215; PhCH=: 218–220, 232 (?); Me₂C=: 185	128,1225,1319,1894,2386
4-Hydrazino-2-piperidinoquinazoline	—	H 384

ALPHABETICAL LIST OF SIMPLE QUINAZOLINES (Continued)

Quinazoline	m.p. (°C), etc.	Ref.
2-Hydrazino-4-quinazolinamine	—	H 383
4-Hydrazino-2-quinazolinamine	206–208, ir, ms	1919
2-Hydrazinoquinazoline	130–131; PhCH=; 173	H 384; 189,799
4-Hydrazinoquinazoline	182 to 189	H 384; 189,281,333,555
6-Hydrazino-2,4-quinazolinediamine	234–235	601
4-Hydrazinoquinazoline 1-oxide	—	H 461
2-Hydrazino-4(3H)-quinazolinone	360, ir, nmr, pK_a	H 383; 1347,1473,1524
4-Hydrazono-2-phenyl-3,4-dihydro-3-quinazolinamine	166–168	455
4-Hydrazono-1-phenyl-1,4-dihydroquinazoline	186, nmr	1439
4-Hydroxyamino-2-methylquinazoline	262	60,61
4-Hydroxyamino-2-phenylquinazoline	223–225, ir, nmr	1872,2078
4-Hydroxyaminoquinazoline	205 or 252, ir, nmr	60,1935
4-Hydroxyamino-2(1H)-quinazolinone	272	1857,1858
2-α-Hydroxybenzyl-4-phenylquinazoline	154–155, ir, nmr	1819
2-α-Hydroxybenzyl-3-phenyl-4(3H)-quinazolinone	96	842
2-α-Hydroxybenzyl-4-quinazolinamine	173, ir	426
4-α-Hydroxybenzylquinazoline	liquid, ir, nmr	790,1885
2-α-Hydroxybenzyl-4(3H)-quinazolinone	208 or 218, ir, nmr	426,2069
3-β-Hydroxybutyl-2-methyl-4(3H)-quinazolinone	—	H 158
2-δ-Hydroxybutyl-4(3H)-quinazolinone	—	H 140
3-β-Hydroxybutyl-4(3H)-quinazolinone	—	H 148
5-Hydroxy-2,4-dimethoxy-6-quinazolinecarbonitrile	199–202, ir, nmr	2334
3-(β-Hydroxy-α,α-dimethylethyl)-2-methyl-4(3H)-quinazolinone	196–197	114
6-Hydroxy-2,3-dimethyl-4(3H)-quinazolinone	304–305	498

548

Compound	Properties	References
8-Hydroxy-2,3-dimethyl-4(3H)-quinazolinone	—	H 198
7-Hydroxyechinozolinone, see 7-Hydroxy-3-β-hydroxyethyl-4(3H)-quinazolinone		
4-α-Hydroxyethyl-6,7-dimethoxy-2-methylquinazoline	67–69, ms, nmr, uv	1302,1354
3-β-Hydroxyethyl-6-iodo-2-methyl-4(3H)-quinazoline	—	H 181
3-β-Hydroxyethyl-6-iodo-4(3H)-quinazolinone	—	H 179
3-β-Hydroxyethyl-8-methoxy-4(3H)-quinazolinone	—	H 179
3-β-Hydroxyethyl-N-methyl-4-oxo-3,4-dihydro-2-quinazolinecarboxamide	138	123
1-β-Hydroxyethyl-3-methyl-2,4(1H,3H)-quinazolinedione	165–168, ir, nmr, xl st	1566,1775,1792
2-α-Hydroxyethyl-1-methyl-4(1H)-quinazolinone	—	H 138
2-α-Hydroxyethyl-3-methyl-4(3H)-quinazolinone	—	H 169
3-β-Hydroxyethyl-2-methyl-4(3H)-quinazolinone	156–157 or 161–163, ir, nmr, complexes; HCl: 195	H 158; 924,1684,1907
4-β-Hydroxyethyl-2-methylsulfinylmethylquinazoline 3-oxide	145–146	874
4-β-Hydroxyethyl-2-methylthiomethylquinazoline 3-oxide	126–127	874
3-β-Hydroxyethyl-6-methyl-2-thioxo-1,2-dihydro-4(3H)-quinazolinone	—	H 309
3-β-Hydroxyethyl-4-oxo-3,4-dihydro-2-quinazolinecarbaldehyde	PhNHN=: 214–219, nmr	2470
3-β-Hydroxyethyl-4-oxo-3,4-dihydro-2-quinazolinecarboxamide	208	123
3-β-Hydroxyethyl-2-phenyl-4(3H)-quinazolinethione	156	455
3-β-Hydroxyethyl-2-phenyl-4(3H)quinazolinone	160	838,979
3-β-Hydroxyethyl-2,4(1H,3H)-quinazolinedione	243 to 252, ir, nmr	H 200; 227,1428,1442,2337
1-β-Hydroxyethyl-4(1H)-quinazolinone	—	2030
2-α-Hydroxyethyl-4(3H)-quinazolinone (chrysogine, crysogine)	178–180 or 190–192, ms, nmr	H 140; 185,940,1026, 2069,2081
3-β-Hydroxyethyl-4(3H)-quinazolinone (probably echinozolinone)	152–153 or 157, ir, ms, nmr, uv	H 148; 276,1772,2030,2557
3-β-Hydroxyethyl-2-styryl-4(3H)-quinazolinone	165 or 180–181, ir, nmr	1907,2185
1-β-Hydroxyethyl-2-thioxo-1,2-dihydro-4(3H)-quinazolinone	—	H 308
3-β-Hydroxyethyl-2-thioxo-1,2-dihydro-4(3H)-quinazolinone	248–250, uv	H 307; 55
7-Hydroxy-3-β-hydroxyethyl-4(3H)-quinazolinone (7-hydroxyechinozolinone)	crude, ir, ms, nmr, uv	2401

ALPHABETICAL LIST OF SIMPLE QUINAZOLINES (*Continued*)

Quinazoline	m.p. (°C), etc.	Ref.
8-Hydroxy-3-β-hydroxyethyl-4(3*H*)-quinazolinone	—	*H* 179
3-[β-Hydroxy-α-(hydroxymethyl)ethyl]-4(3*H*)-quinazolinone	—	*H* 148
3-Hydroxy-2-hydroxymethyl-4(3*H*)-quinazolinone	180, ir	236
8-Hydroxy-7-iodo-3-propyl-4(3*H*)-quinazolinone	—	*H* 179
8-Hydroxy-3-isopentyl-2-methyl-4(3*H*)-quinazolinone	—	*H* 198
6-Hydroxy-1-isopropyl-4-phenyl-2(1*H*)-quinazolinone	285–288	681
7-Hydroxy-1-isopropyl-4-phenyl-2(1*H*)-quinazolinone	266–267	681
8-Hydroxy-3-isopropyl-4(3*H*)-quinazolinone	—	*H* 179
6-Hydroxy-3-β-methoxycarbonylvinyl-4(3*H*)-quinazolinone	255–258	1558
6-Hydroxy-7-methoxy-2,3-diphenyl-4(3*H*)-quinazolinone	231–232	140
7-Hydroxy-8-methoxy-2,3-diphenyl-4(3*H*)-quinazolinone	158	149
6-Hydroxy-7-methoxy-2-methyl-3-phenyl-4(3*H*)-quinazolinone	260–261	143
7-Hydroxy-8-methoxy-2-methyl-3-phenyl-4(3*H*)-quinazolinone	183–184	142
5-Hydroxy-6-methoxy-4-methyl-2(1*H*)-quinazolinone	HCl: 288–293, nmr	1930
6-Hydroxy-4-methoxy-2-phenyl-5,8-dihydro-5,8-quinazolinedione	201–203, ir	763
8-Hydroxy-4-methoxy-2-phenyl-6-phenylimino-5,6-dihydro-5-quinazolinone	243–245, ir, nmr, uv	763
8-Hydroxy-5-methoxy-3-phenyl-2,4(1*H*,3*H*)-quinazolinedione	>280, ir, nmr	897
6-Hydroxy-7-methoxy-3-phenyl-2-thioxo-1,2-dihydro-4(3*H*)-quinazolinone	145–146	141
2-(α-Hydroxy-γ-methylbutyl)-4(3*H*)-quinazolinone	142–144, ir	2069
3-(α-Hydroxy-α-methylethyl)-2,4(1*H*,3*H*)-quinazolinedione	—	*H* 200
2-(α-Hydroxy-α-methylethyl)-4(3*H*)-quinazolinone	148 or 150–151, ir, nmr	918,2069
6-Hydroxymethyl-6-methoxy-2-quinazolinamine	—	*H* 342
2-Hydroxymethyl-1-methyl-4(1*H*)-quinazolinone	—	*H* 138
2-Hydroxymethyl-3-methyl-4(3*H*)-quinazolinone	—	*H* 169

550

551

ALPHABETICAL LIST OF SIMPLE QUINAZOLINES (*Continued*)

Quinazoline	m.p. (°C), etc.	Ref.
1-Hydroxy-3-methyl-2-thioxo-1,2-dihydro-4(3*H*)-quinazolinone	151–152 or 154	211,622
2-ε-Hydroxypentyl-4(3*H*)-quinazolinone	147, ir, nmr	820
2-β-Hydroxyphenethyl-3-methyl-4(3*H*)-quinazolinone	132–133, ir, nmr	1388
6-Hydroxy-2-phenyl-5,8-dihydro-5,8-quinazolinedione	227–229, ir, uv	764
6-Hydroxy-2-phenyl-5,8-dihydro-4,5,8(3*H*)-quinazolinetrione	> 240, ir	762,763
6-Hydroxy-2-phenyl-4-piperidino-5,8-dihydro-5,8-quinazolinedione	176–178, ir	592
1-Hydroxy-3-phenyl-2,4(1*H*,3*H*)-quinazolinedione	175 to 184	187,622,750
6-Hydroxy-2-phenyl-4(3*H*)-quinazolinone	301–302, ir	762,763,766
8-Hydroxy-3-phenyl-2-styryl-4(3*H*)-quinazolinone	—	H 198
1-Hydroxy-3-phenyl-2-thioxo-1,2-dihydro-4(3*H*)-quinazolinone	160	211,622
5-Hydroxy-3-phenyl-2-thioxo-1,2-dihydro-4(3*H*)-quinazolinone	—	H 309
6-Hydroxy-3-phenyl-2-thioxo-1,2-dihydro-4(3*H*)-quinazolinone	—	H 309
8-Hydroxy-3-phenyl-2-thioxo-1,2-dihydro-4(3*H*)-quinazolinone	—	H 309
6-Hydroxy-2-piperidino-5,8-dihydro-5,8-quinazolinedione	174–176, ir, uv	764
6-Hydroxy-2-piperidino-5,8-dihydro-4,5,8(3*H*)-quinazolinetrione	260, ir	762
3-γ-Hydroxypropyl-2-methyl-4(3*H*)-quinazolinone	78–79, ir, nmr	924
3-γ-Hydroxypropyl-6-methyl-2-thioxo-1,2-dihydro-4(3*H*)-quinazolinone	247	125
1-Hydroxy-3-propyl-2,4(1*H*,3*H*)-quinazolinedione	116–118	622
3-β-Hydroxypropyl-2,4(1*H*,3*H*)-quinazolinedione	205–206, ir, uv	477
3-γ-Hydroxypropyl-2,4(1*H*.3*H*)-quinazolinedione	175–177, ir, nmr	1435
2-γ-Hydroxypropyl-4(3*H*)-quinazolinone (pegamine)	ir, ms, nmr, uv	300
3-β-Hydroxypropyl-4(3*H*)-quinazolinone	—	H 148
3-γ-Hydroxypropyl-4(3*H*)-quinazolinone	—	H 148
8-Hydroxy-3-propyl-4(3*H*)-quinazolinone	—	H 179

Compound	mp (°C), etc.	Ref.
1-Hydroxy-3-propyl-2-thioxo-1,2-dihydro-4(3H)-quinazolinone	77–79	622
3-β-Hydroxypropyl-2-thioxo-1,2-dihydro-4(3H)-quinazolinone	196–198	55
3-γ-Hydroxypropyl-2-thioxo-1,2-dihydro-4(3H)-quinazolinone	172–173 or 178–179	55,125
1-Hydroxy-2,4(1H,3H)-quinazolinedione	285 or 286, ir, nmr, uv	H 470; 182,471,622
3-Hydroxy-2,4(1H,3H)-quinazolinedione	320, ir	H 470; 236,457
6-Hydroxy-2,4(1H,3H)-quinazolinedione	—	H 204
8-Hydroxy-2,4(1H,3H)-quinazolinedione	>300	471
5-Hydroxy-4(3H)-quinazolinone	>205, ir, nmr	2324
6-Hydroxy-4(3H)-quinazolinone	>300 or 339–345, ir	498,762,1558
7-Hydroxy-4(3H)-quinazolinone	390–395	498
8-Hydroxy-4(3H)-quinazolinone	304–307	H 198; 498
3-Hydroxy-4(3H)-quinazolinone 1-oxide	—	H 466
4-β-Hydroxystyryl-2-methylquinazoline	126–127, nmr	1781
6-Hydroxy-2-styryl-4(3H)-quinazolinone	323–327, nmr	2264
8-Hydroxy-2-styryl-4(3H)-quinazolinone	—	H 198
6-Hydroxy-2-thioxo-1,2-dihydro-4(3H)-quinazolinone	—	H 309
4-Imino-1,2-dimethyl-1,4-dihydro-6-quinazolinamine	—	H 364
2-Imino-1,3-dimethyl-1,2-dihydro-4(3H)-quinazolinone	131–132, ir, nmr; HBr: 294–295, ir, nmr	947
2-Imino-1,3-dimethyl-6-nitro-1,2-dihydro-4(3H)-quinazolinone	251–252, ir, nmr	947
4-Imino-1-methyl-1,4-dihydro-6-quinazolinamine	—	H 363
4-Imino-1-methyl-3-phenyl-3,4-dihydro-2(1H)-quinazolinethione	215–217	H 310; 88
3-β-Iodoethyl-4(3H)-quinazolinone	—	H 145
6-Iodo-2-isopentylthio-3-methyl-4(3H)-quinazolinone	124	522
6-Iodo-2-isopentylthio-3-phenyl-4(3H)-quinazolinone	121	522
6-Iodo-3-isopropyl-2-methyl-4(3H)-quinazolinone	—	H 181
6-Iodo-3-methyl-2-methylthio-4(3H)-quinazolinone	159	1365
2-Iodomethyl-4-phenylquinazoline	164–165	632

ALPHABETICAL LIST OF SIMPLE QUINAZOLINES (*Continued*)

Quinazoline	m.p. (°C), etc.	Ref.
2-Iodomethyl-4-phenylquinazoline 3-oxide	170–172	632
2-Iodomethyl-3-phenyl-4(3*H*)-quinazolinone	177–178, ir	1276
6-Iodo-2-methyl-3-phenyl-4(3*H*)-quinazolinone	—	*H* 185
6-Iodo-3-methyl-2-propylthio-4(3*H*)-quinazolinone	119	522
6-Iodo-5-methyl-2,4-quinazolinediamine	291–293, nmr	2246
8-Iodo-5-methyl-2,4-quinazolinediamine	226–230, nmr	2246
6-Iodo-2-methyl-4(3*H*)-quinazolinone	345, ir, nmr	*H* 173; 2475
6-Iodo-3-methyl-2-thioxo-1,2-dihydro-4(3*H*)-quinazolinone	270, ir	1365
5-Iodo-7-nitro-2,3-diphenyl-4(3*H*)-quinazolinone	—	*H* 189
5-Iodo-7-nitro-2-phenyl-4(3*H*)-quinazolinone	—	*H* 175
5-Iodo-7-nitro-2-phenyl-4(1*H*)-quinazolinone 3-oxide	—	*H* 467
6-Iodo-3-phenyl-2-propylthio-4(3*H*)-quinazolinone	193	522
6-Iodo-3-phenyl-2-thioxo-1,2-dihydro-4(3*H*)-quinazolinone	295	148,2390
3-γ-Iodopropyl-2,4(1*H*,3*H*)-quinazolinedione	201–202, ir, nmr	1435
3-γ-Iodopropyl-4(3*H*)-quinazolinone	—	*H* 145
5-Iodo-2,4-quinazolinediamine	192–193, nmr	1932, 2225
6-Iodo-2,4-quinazolinediamine	250–255 or 286–288, nmr	2077,2246
7-Iodo-2,4-quinazolinediamine	266–267 nmr	2247
5-Iodo-2,4(1*H*,3*H*)-quinazolinedione	340	308
6-Iodo-2,4(1*H*,3*H*)-quinazolinedione	326–328, ir	565
5-Iodo-4(3*H*)-quinazolinone	—	*H* 154
6-Iodo-4(3*H*)-quinazolinone	268 to 275, ir	498,628,1951
6-Iodo-2-styryl-4(3*H*)-quinazolinone	336–342, nmr	2264
5-Iodo-2-thioxo-1,2-dihydro-4(3*H*)-quinazolinone	324–326	308

554

6-Iodo-2-thioxo-1,2-dihydro-4(3H)-quinazolione	315–317, complexes	565,1682,1747
Isobutoxycarbonyl 3-methyl-4-oxo-3,4-dihydro-2-quinazolinecarboxylate	115, ir, nmr	1186
Isobutoxycarbonyl 4-oxo-3,4-dihydro-2-quinzolinecarboxylate	liquid, crude	1186
3-Isobutylamino-2-methyl-4(3H)-quinazolinone	184	1662
4-Isobutylamino-2-phenylquinazoline	95–97, ir, ms, nmr	1267,1300
2-Isobutylamino-4(3H)-quinazolinone	207–209, ir, nmr	1372
1-Isobutyl-3-methyl-6-nitro-5-pivalamido-2,4(1H,3H)-quinazolinedione	crude, 132–133, ir, nmr	1581
1-Isobutyl-3-methyl-6-nitro-7-pivalamido-2,4(1H,3H)-quinazolinedione	169–170, ir, nmr	1581
3-Isobutyl-2-methyl-5-nitro-4(3H)-quinazolinone	—	H 181
3-Isobutyl-N-methyl-4-oxo-3,4-dihydro-2-quinazolinecarboxamide	180–181	768
2-Isobutyl-3-methyl-4(3H)-quinazolinone	—	H 169
2-Isobutyl-7-methyl-4(3H)-quinazolinone	—	H 175
3-Isobutyl-2-methyl-4(3H)-quinazolinone	71–73	H 156; 935,1267
3-Isobutyl-5-methyl-4(3H)-quinazolinone	57–60, ir, nmr	1970
3-Isobutyl-6-methyl-4(3H)-quinazolinone	52–54, ir, nmr	1970
3-Isobutyl-4-oxo-3,4-dihydro-2-quinazolinecarboxamide	235–236	768
1-Isobutyl-4-phenyl-2(1H)-quinazolinone	121–122	681
2-Isobutyl-1-phenyl-4(1H)-quinazolinone	157–159	1590
3-Isobutyl-2-phenyl-4(3H)-quinazolinone	58–60, ir, nmr	1267
3-Isobutyl-2,4(1H,3H)-quinazolinedione	206–208, ir, nmr	1341
3-Isobutyl-2,4(1H,3H)-quinazolinedithione	204–206	442
2-Isobutylquinazoline 3-oxide	—	H 464
2-Isobutyl-4(3H)-quinazolinone	193, nmr	H 140; 1240
3-Isobutyl-4(3H)-quinazolinone	—	H 145
2-Isobutylthio-6-methoxy-4(3H)-quinazolinone	202–203	280
4-Isopentylamino-8-methoxyquinazoline	—	H 261
4-Isopentylamino-8-quinazolinol	—	H 352
3-Isopentyl-8-methoxy-2-methyl-4(3H)-quinazolinone	—	H 181

ALPHABETICAL LIST OF SIMPLE QUINAZOLINES   (*Continued*)

Quinazoline	m.p. (°C), etc.	Ref.
3-Isopentyl-2-methyl-5-nitro-4(3*H*)-quinazolinone	—	*H* 181
3-Isopentyl-2-methyl-7-nitro-4(3*H*)-quinazolinone	—	*H* 181
6-Isopentyl-5-methyl-2,4-quinazolinediamine	270–273	674
2-Isopentyl-3-methyl-4(3*H*)-quinazolinone	—	*H* 169
4-Isopentyl-2-methylthiomethylquinazoline 3-oxide	76–77	874
4-Isopentyloxy-2-methyl-7-nitroquinazoline	—	*H* 258
1-Isopentyl-2,4(1*H*,3*H*)-quinazolinedione	130	718
2-Isopentyl-4(3*H*)-quinazolinone	—	*H* 140
2-Isopentylthio-6-methoxy-4(3*H*)-quinazolinone	179–180	280
2-Isopentylthio-3-methyl-4(3*H*)-quinazolonone	82	160
2-Isopentylthio-3-phenyl-4(3*H*)-quinazolinone	111 or 160	160,1017
3-Isopropoxycarbonylaminomethyl-2-methyl-4(3*H*)-quinazolinone	110, ir	277
6-Isopropoxy-3-β-methoxycarbonylvinyl-4(3*H*)-quinazolinone	156–157	1558
2-Isopropoxy-7-nitro-3-phenyl-4(3*H*)-quinazolinone	139–141, ir, nmr, uv	2200
3-γ-Isopropoxypropyl-2-methyl-4(3*H*)-quinazolinone	—	*H* 158
2-Isopropoxyquinazoline	—	*H* 253
4-Isopropoxyquinazoline	—	*H* 254
5-Isopropoxy-2,4-quinazolinediamine	185–186	2225
6-Isopropoxy-4(3*H*)-quinazolinone	206–208	1558
2-Isopropylamino-6,7-dimethoxy-4(3*H*)-quinazolinone	244–246; HCl: 283–285	641
4-Isopropylamino-7-nitroquinazoline	—	*H* 352
4-Isopropylamino-2-phenylquinazoline	149–151	1025
4-Isopropylamino-7-quinazolinamine	—	*H* 364
2-Isopropylaminoquinazoline	72	849

556

4-Isopropylaminoquinazoline	175, pK_a	1000
2-Isopropylamino-4(3H)-quinazolinone	212–213, ir, nmr	1372
4-Isopropyl-6,7-dimethoxy-1-β-methoxycarbonylethyl-2(1H)-quinazolinone	112–114	944
1-Isopropyl-6,7-dimethoxy-4-phenyl-2(1H)-quinazolinone	148–150	681
4-Isopropyl-6,7-dimethoxy-2(1H)-quinazolinone	238–240	1592
1-Isopropyl-5,7-dimethyl-4-phenyl-2(1H)-quinazolinone	145–147	681
1-Isopropyl-6,7-dimethyl-4-phenyl-2(1H)-quinazolinone	135–137	681
1-Isopropyl-6,8-dimethyl-4-phenyl-2(1H)-quinazolinone	168–169	681
1-Isopropyl-4,7-dimethyl-2(1H)-quinazolinone	209–212	681
2-Isopropyl-3,6-dimethyl-4(3H)-quinazolinone	77, ir, nmr	1049
2-Isopropyl-3,7-dimethyl-4(3H)-quinazolinone	112, ir, nmr	1049
2-Isopropyl-3,8-dimethyl-4(3H)-quinazolinone	47, ir, nmr	1049
8-Isopropyl-3,4-dimethyl-2(3H)-quinazolinone	178	110
6-Isopropyl-2,4-diphenylquinazoline	78, ir, nmr, xl st	1942
3-Isopropylideneamino-2-methyl-4(3H)-quinazolinone	185–187	2354
3-Isopropylideneamino-2,4(1H,3H)-quinazolinedione	—	H 382
4-Isopropylidenehydrazinoquinazoline	—	H 384
3-Isopropyl-2-N-isopropylacetamido-4(3H)-quinazolinone	140	621
3-Isopropyl-2-isopropylamino-4(3H)-quinazolinone	81	H 365; 621
2-Isopropyl-4-isopropylidene-3-phenyl-3,4-dihydroquinazoline	93–94, ir, ms, nmr	400
6-Isopropyl-3-β-methoxycarbonylvinyl-4(3H)-quinazolinone	142–143	1558
2-Isopropyl-6-methoxy-3-methyl-4(3H)-quinazolinone	93, ir, nmr	1049
2-Isopropyl-7-methoxy-3-methyl-4(3H)-quinazolinone	124, ir, nmr	1049
2-Isopropyl-8-methoxy-3-methyl-4(3H)-quinazolinone	130, ir, nmr	1049
3-Isopropyl-2-methoxy-2-methyl-4(3H)-quinazolinone	118–119, ir, nmr; SbCl$_5$,HCl: 205–207, ir, nmr	2361
4-Isopropyl-6-methoxy-2-phenylquinazoline	109–110, nmr	2088
1-Isopropyl-5-methoxy-2-phenyl-4(1H)-quinazolinone	178–180	1590

ALPHABETICAL LIST OF SIMPLE QUINAZOLINES (*Continued*)

Quinazoline	m.p. (°C), etc.	Ref.
1-Isopropyl-6-methoxy-2-phenyl-4(1*H*)-quinazolinone	190–192	1590
1-Isopropyl-6-methoxy-4-phenyl-2(1*H*)-quinazolinone	140–143	681
1-Isopropyl-7-methoxy-4-phenyl-2(1*H*)-quinazolinone	133–137	681
2-Isopropyl-6-methoxy-1-phenyl-4(1*H*)-quinazolinone	191–192	1590
4-Isopropyl-2-methoxyquinazoline	—	*H* 253
2-Isopropyl-6-methoxy-4(3*H*)-quinazolinone	250, ir, nmr	1049
2-Isopropyl-7-methoxy-4(3*H*)-quinazolinone	217, ir, nmr	1049
2-Isopropyl-8-methoxy-4(3*H*)-quinazolinone	258, ir, nmr	1049
3-Isopropyl-8-methoxy-4(3*H*)-quinazolinone	—	*H* 179
4-Isopropyl-7-methoxy-2(1*H*)-quinazolinone	185–190 or 188–193	79,1148
3-Isopropyl-2-methyl-5-nitro-4(3*H*)-quinazolinone	—	*H* 182
4-Isopropyl-6-methyl-2-phenylquinazoline	120, nmr	2088,2101
1-Isopropyl-7-methyl-4-phenyl-2(1*H*)-quinazolinethione	194–198	681
1-Isopropyl-5-methyl-4-phenyl-2(1*H*)-quinazolinone	153–154	681
1-Isopropyl-6-methyl-4-phenyl-2(1*H*)-quinazolinone	170–171	681
1-Isopropyl-7-methyl-2-phenyl-4(1*H*)-quinazolinone	165–167	1590
1-Isopropyl-7-methyl-4-phenyl-2(1*H*)-quinazolinone	137–138	681
1-Isopropyl-8-methyl-4-phenyl-2(1*H*)-quinazolinone	165–166	681
2-Isopropyl-6-methyl-1-phenyl-4(1*H*)-quinazolinone	251–255	1590
2-Isopropyl-7-methyl-1-phenyl-4(1*H*)-quinazolinone	235–238	1590
2-Isopropyl-4-methylquinazoline	—	*H* 63
5-Isopropyl-8-methyl-2,4-quinazolinediamine	—	*H* 354
2-Isopropyl-3-methyl-4(3*H*)-quinazolinethione	106–107	105
2-Isopropyl-1-methyl-4(1*H*)-quinazolinone	145–146, ir, nmr	821

558

2-Isopropyl-3-methyl-4(3*H*)-quinazolinone	78–79	105
2-Isopropyl-6-methyl-4(3*H*)-quinazolinone	238–239 or 256–257, ir, nmr	1049,1372
2-Isopropyl-7-methyl-4(3*H*)-quinazolinone	235, ir, nmr	H 175; 1049
2-Isopropyl-8-methyl-4(3*H*)-quinazolinone	204–205 or 210, ir, nmr	1049, 1372
3-Isopropyl-2-methyl-4(3*H*)-quinazolinone	83 to 92, ir, nmr, uv; HCl: 263; SbCl$_5$.HCl: 198–199, nmr	412,2162,2296,2361
2-Isopropyl-4-methylthioquinazoline	37–38	104
3-Isopropyl-1-methyl-2-thioxo-1,2-dihydro-4(3*H*)-quinazolinone	123–125	87,685
3-Isopropyl-1-methyl-4-thioxo-3,4-dihydro-2(1*H*)-quinazolinone	149–151	87,685
1-Isopropyl-6-nitro-4-phenyl-2(1*H*)-quinazolinone	190–192	681
2-Isopropyl-7-nitro-1-phenyl-4(1*H*)-quinazolinone	270–278	1590
3-Isopropyl-7-nitro-2-propoxy-4(3*H*)-quinazolinone	97–98, ir, nmr, uv	2200
2-Isopropyl-4-pentyloxyquinazoline	ir, nmr	2482
2-Isopropyl-3-pentyl-4(3*H*)-quinazolinone	ir, nmr	2482
1-Isopropyl-4-phenyl-7-propoxy-2(1*H*)-quinazolinone	142–143	681
2-Isopropyl-4-phenylquinazoline	—	H 64
4-Isopropyl-2-phenylquinazoline	60, ms	1946
4-Isopropyl-2-phenylquinazoline 1-oxide	—	H 461
1-Isopropyl-4-phenyl-2(1*H*)-quinazolinethione	214–216 or 219–221, ir, nmr, uv	681,955
2-Isopropyl-3-phenyl-4(3*H*)-quinazolinethione	173	H 298; 953
3-Isopropyl-2-phenyl-4(3*H*)-quinazolinethione	173	953
1-Isopropyl-2-phenyl-4(1*H*)-quinazolinone	223–226	1590
1-Isopropyl-4-phenyl-2(1*H*)-quinazolinone	119–121	681
2-Isopropyl-1-phenyl-4(1*H*)-quinazolinone	228–230, nmr	1590
2-Isopropyl-3-phenyl-4(3*H*)-quinazolinone	142–143, nmr; HClO$_4$: 241–242	H 166; 761,1590

ALPHABETICAL LIST OF SIMPLE QUINAZOLINES (*Continued*)

Quinazoline	m.p. (°C), etc.	Ref.
3-Isopropyl-2-phenyl-4(3*H*)-quinazolinone	136–137 or 138–140, ir, nmr, uv; AlCl$_3$.HCl: 231–233, ir, nmr; SbCl$_5$.HCl: 237–239, ir, nmr; FeCl$_3$.HCl: 234–239, ir	412,1590,2361
4-Isopropyl-1-phenyl-4(1*H*)-quinazolinone	196–197, ir, nmr	1590
2-Isopropyl-4-quinazolinamine	237–240	2290
2-Isopropylquinazoline	—	*H* 61
4-Isopropylquinazoline	90/0.1	*H* 62; 56,555
4-Isopropyl-2-quinazolinecarbonitrile	—	*H* 481
2-Isopropyl-2-quinazolinecarboxamide	—	*H* 482
4-Isopropyl-2-quinazolinecarboxamidrazone	—	*H* 63
1-Isopropyl-2,4(1*H*,3*H*)-quinazolinedione	175–176	718
3-Isopropyl-2,4(1*H*,3*H*)-quinazolinedione	187 to 195, nmr	227,362,718,2287
3-Isopropyl-2,4(1*H*,3*H*)-quinazolinedithione	165–166	89,274
2-Isopropylquinazoline 3-oxide	—	*H* 464
4-Isopropylquinazoline 1-oxide	—	*H* 461
2-Isopropyl-4-quinazolinesulfenic acid	105	105
2-Isopropyl-4(3*H*)-quinazolinethione	201 or 204, ms, nmr	*H* 296; 105,957
2-Isopropyl-4(1*H*)-quinazolinethione 3-oxide	73–75	*H* 468; 105
3-Isopropyl-4(3*H*)-quinazolinimine	123–125, pK_a	1000
2-Isopropyl-4(3*H*)-quinazolinone	220–222 or 231–233	*H* 140; 105,988,1059,1306
3-Isopropyl-4(3*H*)-quinazolinone	87–89, 210–215/22	*H* 145; 276,2287
4-Isopropyl-2(1*H*)-quinazolinone	188–192, ir, ms, nmr	*H* 137; 1946
6-Isopropyl-4(3*H*)-quinazolinone	152–153	1558

Compound	mp (°C), data	Ref.
6-Isopropylthio-3-β-methoxycarbonylvinyl-4(3H)-quinazolinone	129–131	1558
2-Isopropylthio-3-phenyl-4(3H)-quinazolinone	—	H 306
6-Isopropylthio-4(3H)-quinazolinone	141–143	1558
3-Isopropyl-2-thioxo-1,2-dihydro-4(3H)-quinazolinone	177–178, complexes	87,685,1646
3-Isopropyl-4-thioxo-3,4-dihydro-2(1H)-quinazolinone	205–206	86
4-Isothiocyanato-2-morpholinoquinazoline	112, ir	439
3-Isovaleramido-2-methyl-4(3H)-quinazolinone	—	1984
3-Isovaleramido-4(3H)-quinazolinone	—	1984
3-(2′-Mercaptocyclohex-1′-enyl)-4(3H)-quinazolinone	nmr	730
2-(β-Mercapto-α,α-dimethylpropyl)-4(3H)-quinazolinone	132–135, ir, ms, nmr	2021
6-Mercapto-3-methyl-4(3H)-quinazolinone	143–145	498
2-(β-Mercapto-α,α,β-trimethylpropyl)-4(3H)-quinazolinone	172–174, ir, ms, nmr	2021
8-Mesyloxy-2,4(1H,3H)-quinazolinedione	345, uv	471
3-Mesyl-4(3H)-quinazolinone	145–150	276
2-Methacrylamido-3-methyl-4(3H)-quinazolinone	220–222, ir, ms	1834
2-Methacrylamido-4(3H)-quinazolinone	250–251, ir, ms, uv	1834
6-Methoxy-4,7-bisphenylthioquinazoline	—	H 300
3-δ-Methoxybutyl-1-methyl-2,4(1H,3H)-quinazolinedione	110–111, ir	29
3-Methoxycarbonylaminomethyl-2-methyl-4(3H)-quinazolinone	144–145, ir	277
3-Methoxycarbonylamino-2,4(1H,3H)-quinazolinedione	—	H 382
4-Methoxycarbonylamino-2(1H)-quinazolinethione	163–165, ir, nmr	2095
3-Methoxycarbonylamino-2-(α,β,β-trimethylpropyl)-4(3H)-quinazolinone	nmr	2063
1-β-Methoxycarbonylethyl-4-methyl-2(1H)-quinazolinone	164–167	944
2-β-Methoxycarbonylethyl-1-phenyl-4(1H)-quinazolinone	157–158	1590
2-β-Methoxycarbonylethyl-3-phenyl-4(3H)-quinazolinone	—	467
3-α-Methoxycarbonylethyl-2,4(1H,3H)-quinazolinedione	135, nmr	2435
3-β-Methoxycarbonylethyl-4(3H)-quinazolinethione	52	1565
2-β-Methoxycarbonylethyl-4(3H)-quinazolinone	188–190, ms, nmr, uv	904

Quinazoline	m.p. (°C), etc.	Ref.
3-Methoxycarbonylmethyl-6-methyl-2-methylthio-4(3H)-quinazolinone	145–146	1137
3-Methoxycarbonylmethyl-7-methyl-2-methylthio-4(3H)-quinazolinone	144–146	1137
3-Methoxycarbonylmethyl-8-methyl-2-methylthio-4(3H)-quinazolinone	130–131	1137
1-Methoxycarbonylmethyl-3-methyl-2,4(1H,3H)-quinazolinedione	158–159, ir, ms, nmr	2270
1-Methoxycarbonylmethyl-2-methyl-4(1H)-quinazolinone	185–190, ir, ms, nmr	1825
3-Methoxycarbonylmethyl-2-methyl-4(3H)-quinazolinone	135, ir	H 158; 1681
3-Methoxycarbonylmethyl-2-methylthio-4(3H)-quinazolinone	117	1137
3-Methoxycarbonylmethyl-8-methyl-2-thioxo-1,2-dihydro-4(3H)-quinazolinone	196–197	1137
3-Methoxycarbonylmethyl-2-phenyl-4(3H)-quinazolinethione	172	1565
1-Methoxycarbonylmethyl-2-phenyl-4(1H)-quinazolinone	194–195, ir, ms, nmr	1825
3-Methoxycarbonylmethyl-2-phenyl-4(3H)-quinazolinone	117–118, ir, ms, nmr	1825
2-Methoxycarbonylmethyl-4-quinazolinamine	202–203	1368
1-Methoxycarbonylmethyl-2,4(1H,3H)-quinazolinedione	264–267, ir, ms, nmr	1822
1-Methoxycarbonylmethyl-4(1H)-quinazolinone	201–203, ir, ms, nmr	1825
2-Methoxycarbonylmethyl-4(3H)-quinazolinone	—	H 140
3-Methoxycarbonylmethyl-4(H)-quinazolinone	148, ir, nmr	2528
3-Methoxycarbonylmethyl-2-styryl-4(3H)-quinazolinone	—	597
3-Methoxycarbonylmethyl-2-thioxo-1,2-dihydro-4(3H)-quinazolinone	226–228	1137
3-β-Methoxycarbonylvinyl-6,8-dimethyl-4(3H)-quinazolinone	139–141	1558
3-β-Methoxycarbonylvinyl-6-methylthio-4(3H)-quinazolinone	169–170	1558
3-α-Methoxycarbonylvinyl-4(3H)-quinazolinone	132, ir, ms, nmr	1640
3-β-Methoxycarbonylvinyl-4(3H)-quinazolinone	173–174	1558
6-Methoxy-5,8-dihydro-5,8-quinazolinedione	233, ir, nmr	1577
6-Methoxy-3,7-dimethyl-5,8-dihydro-4,5,8(3H)-quinazolinetrione	217–219, ir, ms, nmr	2363

8-Methoxy-2,4-dimethylquinazoline	—	H 261
6-Methoxy-1,2-dimethyl-4(1H)-quinazolinone	—	H 138
6-Methoxy-2,3-dimethyl-4(3H)-quinazolinone	128–130, ir, nmr	H 182, 498,2361
6-Methoxy-3,4-dimethyl-2(3H)-quinazolinone	173–175	110
7-Methoxy-1,2-dimethyl-4(1H)-quinazolinone	—	H 138
7-Methoxy-2,3-dimethyl-4(3H)-quinazolinone	—	H 182
8-Methoxy-2,3-dimethyl-4(3H)-quinazolinone	—	H 182
8-Methoxy-3,4-dimethyl-2(3H)-quinazolinone	134	110
2-(β-Methoxyethoxy)methyl-3-phenyl-4(3H)-quinazolinone	—	2388
2-β-Methoxyethoxy-4(3H)-quinazolinone	—	H 143
2-β-Methoxyethyl-1-phenyl-4(1H)-quinazolinone	161–163, nmr	1394
2-β-Methoxyethyl-3-phenyl-4(3H)-quinazolinone	139–141, nmr	584,847
3-β-Methoxyethyl-2,4(1H,3H)-quinazolinedione	148–150, ir, nmr	1428
6-Methoxy-3-methoxycarbonylmethyl-2-methylthio-4(3H)-quinazolinone	161	1137
6-Methoxy-3-methoxycarbonylmethyl-2-thioxo-1,2-dihydro-4(3H)-quinazolinone	213	1137
6-Methoxy-3-β-methoxycarbonylvinyl-4(3H)-quinazolinone	184–186	1558
8-Methoxy-3-β-methoxycarbonylvinyl-4(3H)-quinazolinone	203–208	1558
4-Methoxy-6-methylamino-2-phenyl-5,8-dihydro-5,8-quinazolinedione	262–264, ir, uv	763
5-Methoxy-2-methylaminoquinazoline	nmr; pic: 253	50
6-Methoxy-2-methylaminoquinazoline	156, nmr, pK_a, uv; pic: 249	50
7-Methoxy-2-methylaminoquinazoline	nmr; pic: 245	50
6-Methoxy-7-methyl-5,8-dihydro-5,8-quinazolinedione	112–113, ir, ms, nmr	2363
3-Methoxy-4-methylimino-3,4-dihydroquinazoline	120	554
6-Methoxy-4-methyl-2-methylthio-5,8-dihydro-5,8-quinazolinedione	188	68
8-Methoxy-2-methyl-3-N'-methyl(thioureido)-4(3H)-quinazolinone	—	H 379
8-Methoxy-2-methyl-3-pentyl-4(3H)-quinazolinone	—	H 182
8-Methoxy-2-methyl-3-phenethyl-4(3H)-quinazolinone	—	H 182
6-Methoxy-4-methyl-2-phenyl-5,8-dihydro-5,8-quinazolinedione	250, ir, nmr, uv	736

ALPHABETICAL LIST OF SIMPLE QUINAZOLINES (*Continued*)

Quinazoline	m.p. (°C), etc.	Ref.
2-Methoxymethyl-4-phenylquinazoline	116–117	1946
8-Methoxy-4-methyl-2-phenylquinazoline	—	H 261
2-Methoxymethyl-1-phenyl-4(1H)-quinazolinone	172–174, nmr	1394
2-Methoxymethyl-3-phenyl-4(3H)-quinazolinone	109–110 or 180–181, ir, nmr	146,1154,1156
6-Methoxy-2-methyl-3-phenyl-4(3H)-quinazolinone	180–181	687
8-Methoxy-2-methyl-3-phenyl-4(3H)-quinazolinone	—	H 185
6-Methoxy-2-methyl-3-propyl-4(3H)-quinazolinone	92–93, ir, nmr; SbCl₅,HCl:	2361
	190–192, ir, nmr	2361
8-Methoxy-2-methyl-3-propyl-4(3H)-quinazolinone	—	H 182
7-Methoxy-4-methyl-2-quinazolinamine	173–175, nmr	2569
8-Methoxy-5-methyl-2-quinazolinamine	—	H 342
8-Methoxy-6-methyl-2-quinazolinamine	—	H 342
8-Methoxy-7-methyl-2-quinazolinamine	—	H 342
2-Methoxy-4-methylquinazoline	—	H 253
2-Methoxy-6-methylquinazoline	—	H 253
4-Methoxy-2-methylquinazoline	—	H 257
7-Methoxy-2-methylquinazoline	99, pK_a, uv	168
7-Methoxy-4-methylquinazoline	—	H 261
8-Methoxy-2-methylquinazoline	—	H 261
8-Methoxy-4-methylquinazoline	—	H 261
1-Methoxy-3-methyl-2,4(1H,3H)-quinazolinedione	142	212
3-Methoxy-1-methyl-2,4(1H,3H)-quinazolinedione	150–152, ir, ms, nmr	457
7-Methoxy-4-methylquinazoline 3-oxide	—	H 464
8-Methoxy-4-methylquinazoline 3-oxide	—	H 464

564

Compound		
6-Methoxy-2-methyl-4(3H)-quinazolinethione	286–287, uv	241
7-Methoxy-2-methyl-4(3H)-quinazolinethione	274–276, uv	241
5-Methoxy-3-methyl-2(3H)-quinazolinimine	HI: 259–262, nmr, pK_a, uv	50
6-Methoxy-3-methyl-2(3H)-quinazolinimine	HI: 225, nmr, pK_a, uv	50
7-Methoxy-3-methyl-2(3H)-quinazolinimine	HI: 278–280, nmr, pK_a, uv	50
2-Methoxy-3-methyl-4(3H)-quinazolinone	—	H 172
3-Methoxy-2-methyl-4(3H)-quinazolinone	91–92	H 470; 236
6-Methoxy-2-methyl-4(3H)-quinazolinone	264–265	H 173; 782,1113
6-Methoxy-3-methyl-4(3H)-quinazolinone	112–114	498
6-Methoxy-4-methyl-2(1H)-quinazolinone	232–236	1592
7-Methoxy-2-methyl-4(3H)-quinazolinone	270–272	H 173; 241
7-Methoxy-4-methyl-2(1H)-quinazolinone	258–265	79,1148
8-Methoxy-2-methyl-4(3H)-quinazolinone	—	H 173
8-Methoxy-3-methyl-4(3H)-quinazolinone	—	H 177
8-Methoxy-3-methyl-2-styryl-4(3H)-quinazolinone	—	H 188
4-Methoxy-2-methylthioquinazoline	—	H 292
6-Methoxy-2-methylthio-4(3H)-quinazolinone	234–235	280
7-Methoxy-6-methylthio-4(3H)-quinazolinone	295, ir, nmr	2017
1-Methoxy-3-methyl-2-thioxo-1,2-dihydro-4(3H)-quinazolinone	185	211
4-Methoxy-6-morpholino-2-phenyl-5,8-dihydro-5,8-quinazolinedione	210–212, ir, uv	763
6-Methoxy-4-morpholino-2-phenylquinazoline	94–96 or 116–118	707,766
2-Methoxy-7-nitro-3-phenyl-4(3H)-quinazolinone	198–200, ir, nmr, uv	2200
4-Methoxy-6-nitroquinazoline	—	H 258
4-Methoxy-7-nitroquinazoline	—	H 258
4-Methoxy-8-nitroquinazoline	—	H 258
6-Methoxy-5-nitroquinazoline	169, ir, nmr	1577
6-Methoxy-7-nitroquinazoline	—	H 384
5-Methoxy-8-nitro-4(3H)-quinazolinone	212–218, ir, nmr	2324

ALPHABETICAL LIST OF SIMPLE QUINAZOLINES (*Continued*)

Quinazoline	m.p. (°C), etc.	Ref.
6-Methoxy-7-nitro-4(3*H*)-quinazolinone	—	H 154
7-Methoxy-4-oxo-3,4-dihydro-2-quinazolinecarbohydrazide	263–264	1743
7-Methoxy-4-oxo-3,4-dihydro-2-quinazolinecarboxamide	257	1743
8-Methoxy-4-pentylaminoquinazoline	—	H 261
6-Methoxy-2-pentylthio-4(3*H*)-quinazolinone	173–174	280
4-Methoxy-2-phenethylquinazoline	—	H 257
4-Methoxy-2-phenoxyquinazoline	—	H 259
6-Methoxy-2-phenyl-5,8-dihydro-5,8-quinazolinedione	208–209, ir, uv	764
4-Methoxy-2-phenyl-6-piperidino-5,8-dihydro-5,8-quinazolinedione	183–184, ir, uv	763
4-Methoxy-2-phenyl-8-piperidino-5,6-dihydro-5,6-quinazolinedione	192–193, ir	763
6-Methoxy-2-phenyl-4-piperidino-5,8-dihydro-5,8-quinazolinedione	200–201	592
6-Methoxy-2-phenyl-4-piperidinoquinazoline	121–123 or, 125–126	707,766
2-Methoxy-4-phenylquinazoline	86–89, ir, nmr, uv	247,2206
4-Methoxy-2-phenylquinazoline	59–62 or, 66–67, nmr	H 257; 2206,2539
6-Methoxy-2-phenylquinazoline	116–117 or 120–122	766,2088
8-Methoxy-2-phenylquinazoline	—	H 262
4-Methoxy-2-phenylquinazoline 1-oxide	—	H 461
6-Methoxy-2-phenyl-4(3*H*)-quinazolinethione	222 to 228	210,241,707
7-Methoxy-2-phenyl-4(3*H*)-quinazolinethione	196–198, uv	241
4-Methoxy-2-phenyl-6-quinazolinol	219–220	763
2-Methoxy-3-phenyl-4(3*H*)-quinazolinone	—	H 172
6-Methoxy-2-phenyl-4(3*H*)-quinazolinone	245 to 262	214,503,707,766
6-Methoxy-4-phenyl-2(1*H*)-quinazolinone	285–286	1060
7-Methoxy-4-phenyl-2(1*H*)-quinazolinone	272–280	79,1148

566

8-Methoxy-3-phenyl-2-styryl-4(3H)-quinazolinone	—	H 188
6-Methoxy-7-phenylsulfonyl-4(3H)-quinazolinone	—	H 154,310
6-Methoxy-7-phenylthio-4(3H)-quinazolinone	—	H 155,310
1-Methoxy-3-phenyl-2-thioxo-1,2-dihydro-4(3H)-quinazolinone	172	211
6-Methoxy-4-phenyl-1-β,β,β-trifluoroethyl-2(1H)-quinazolinone	157–158	1060
6-Methoxy-4-piperidinoquinazoline	141–142	764
8-Methoxy-4-piperidinoquinazoline	—	H 262
3-γ-Methoxypropyl-2-methyl-4(3H)-quinazolinone	—	H 158
4-Methoxy-2-propylquinazoline	78	1184
8-Methoxy-2-propylquinazoline	—	H 262
8-Methoxy-4-propylquinazoline	—	H 262
3-γ-Methoxypropyl-2,4(1H,3H)-quinazolinedione	145–147	227
7-Methoxy-4-propyl-2(1H)-quinazolinone	225–233	79,1148
8-Methoxy-3-propyl-4(3H)-quinazolinone	—	H 179
6-Methoxy-2-propylthio-4(3H)-quinazolinone	194–195	280
2-Methoxy-2-quinazolinamine	—	H 253
4-Methoxy-6-quinazolinamine	—	H 258
4-Methoxy-7-quinazolinamine	—	H 258
4-Methoxy-8-quinazolinamine	—	H 258
5-Methoxy-2-quinazolinamine	nmr, pK_a, uv	H 342; 4,50
6-Methoxy-2-quinazolinamine	nmr, pK_a, uv	H 342; 4,50
6-Methoxy-5-quinazolinamine	149, ir, nmr	1577
6-Methoxy-7-quinazolinamine	—	H 260
7-Methoxy-2-quinazolinamine	nmr, pK_a, uv	H 342; 4,50
8-Methoxy-2-quinazolinamine	nmr, pK_a, uv	H 342; 50
6-Methoxy-2-quinazolinamine 3-oxide	271, uv	4
2-Methoxyquinazoline	58; pic: 135–136	H 253; 189,849
4-Methoxyquinazoline	34–35 or 36, ms; pic: 174–175	H 254; 172,189,276,417,555,2001

ALPHABETICAL LIST OF SIMPLE QUINAZOLINES (*Continued*)

Quinazoline	m.p. (°C), etc.	Ref.
5-Methoxyquinazoline	—	*H* 261; 4
6-Methoxyquinazoline	69–71, uv	*H* 261; 4
7-Methoxyquinazoline	—	*H* 261; 4
8-Methoxyquinazoline	90–91	*H* 261; 4
4-Methoxy-2-quinazolinecarbonitrile	132–134	*H* 481; 1142
4-Methoxy-2-quinazolinecarboxamide	—	*H* 482
6-Methoxy-2-quinazolinecarboxamide	—	2178
7-Methoxy-2-quinazolinecarboxamide	265–266, nmr	2178
2-Methoxy-4-quinazolinecarboxylic acid	—	*H* 482
5-Methoxy-2,4-quinazolinediamine	200–202 or 208–209, nmr, pK_a; HCl (?): 255–257	661,696,1932,2225
6-Methoxy-2,4-quinazolinediamine	213–214	2326
7-Methoxy-2,4-quinazolinediamine	229–230	661
8-Methoxy-2,4-quinazolinediamine	267–268	696
3-Methoxy-2,4(1*H*,3*H*)-quinazolinedione	224–226 or 244–245, ir, ms, nmr	457,738
5-Methoxy-2,4(1*H*,3*H*)-quinazolinedione	—	*H* 204
6-Methoxy-2,4(1*H*,3*H*)-quinazolinedione	—	*H* 204
7-Methoxy-2,4(1*H*,2*H*)-quinazolinedione	—	*H* 204
8-Methoxy-2,4(1*H*,3*H*)-quinazolinedione	255–257, nmr	*H* 204; 2595
6-Methoxy-2,4(1*H*,3*H*)-quinazolinedithione	350–352	38
4-Methoxyquinazoline 1-oxide	—	*H* 461
5-Methoxyquinazoline 3-oxide	192, uv	*H* 464; 4
6-Methoxyquinazoline 3-oxide	191, uv	*H* 464; 4
7-Methoxyquinazoline 3-oxide	202, uv	*H* 464; 4

Compound		
8-Methoxyquinazoline 3-oxide	186, uv	*H* 464; 4
6-Methoxy-4(3*H*)-quinazolinethione	323–324, uv	241
7-Methoxy-4(3*H*)-quinazolinethione	253–254, uv	241
3-Methoxy-4(3*H*)-quinazolinimine (?)	—	554
5-Methoxy-8-quinazolinol	172, ir, uv	586
2-Methoxy-4(3*H*)-quinazolinone	213 to 235, ir, nmr	*H* 143; 182,1460,1915,1940, 2230
4-Methoxy-2(1*H*)-quinazolinone	—	*H* 137
5-Methoxy-4(3*H*)-quinazolinone	—	*H* 154
6-Methoxy-4(3*H*)-quinazolinone	238–242	*H* 154; 498,1558
7-Methoxy-2(1*H*)-quinazolinone	275–280, ir, nmr	1148
7-Methoxy-4(3*H*)-quinazolinone	262–265	*H* 154; 498
8-Methoxy-2(1*H*)-quinazolinone	—	*H* 137
8-Methoxy-4(3*H*)-quinazolinone	298–300 or 311–312	*H* 154; 498,1558
2-Methoxy-4(3*H*)-quinazolinone 1-oxide	194, ir	182
4-Methoxy-4(3*H*)-quinazolinone 1-oxide	—	*H* 466
6-Methoxy-4-selenoxo-3,4-dihydro-2(1*H*)-quinazolinone	273–274, ir, nmr	1897
6-Methoxy-2-styryl-4(3*H*)-quinazolinone	287–291, ir, nmr	2264
7-Methoxy-2-styryl-4(3*H*)-quinazolinone	275–278	2264
8-Methoxy-2-styryl-4(3*H*)-quinazolinone	—	*H* 175
4-Methoxy(thiocarbonyl)amino-2-morpholinoquinazoline	116	429
3-Methoxy-2-thiocyanatomethyl-4(3*H*)-quinazolinone	190	441
6-Methoxy-2-thioxo-1,2-dihydro-4(3*H*)-quinazolinone	295–296	280
4-Methoxy-2-trichloromethylquinazoline	—	*H* 257
6-Methoxy-2-trichloromethyl-4(3*H*)-quinazolinone	198–200 or 206–207	741,1034
8-Methoxy-2-trichloromethyl-4(3*H*)-quinazolinone	251–252	755
4-Methoxy-2-trimethylammonioquinazoline (chloride)	crude	1142
3-(*N*-Methylacetamido)-2,4(1*H*,3*H*)-quinazolinedione	291–293, ir, nmr	911

569

ALPHABETICAL LIST OF SIMPLE QUINAZOLINES (*Continued*)

Quinazoline	m.p. (°C), etc.	Ref.
3-(N-Methylacetamido)-4(3H)-quinazolinone	120–121 or 125, ir, nmr	934,1430
3-β-Methylaminoethyl-2-phenyl-4(3H)-quinazolinone	87	716
3-N-(Methylaminomethylaminomethyl-4(3H)-quinazolinone	HCl: 205	1192
6-Methylamino-8-methylimino-5,8-dihydro-2,4,5(1H,3H)-quinazolinetrione	220–250, ir, ms, nmr	1987
Methyl 6-amino-2-methyl-4-oxo-3,4-dihydro-7-quiazolinecarboxylate	289–290	85
3-Methylaminomethyl-4(3H)-quinazolinone	HCl: 152–153	1192
2-Methylaminomethyl-4(1H)-quinazolinone 3-oxide	188, ir	236
2-Methylamino-4-methylthioquinazoline	131–132, ir, nmr	1410
4-Methylamino-2-morpholinoquinazoline	204, nmr	439
4-Methylamino-6-nitroquinazoline	—	551
4-Methylamino-7-nitroquinazoline	—	H 352
3-Methylamino-6-nitro-4(3H)-quinazolinone	193–195	934
7-Methylamino-6-nitro-4(3H)-quinazolinone	crude	1962
7-Methylamino-8-nitro-4(3H)-quinazolinone	crude	2324
Methyl 6-amino-4-oxo-2-phenyl-3,4-dihydro-7-quinazolinecarboxylate	295–296	85
4-Methylamino-2-phenylquinazoline	126–127 or 128, ir, nmr	1267,1666
4-Methylamino-2-phenyl-6-quinazolinol	—	397
3-Methylamino-2-phenyl-4(3H)-quinazolinone	—	1012
4-Methylamino-3-phenyl-2(3H)-quinazolinone	180–183, then 254–257	88
3-γ-Methylaminopropyl-2-phenyl-4(3H)-quinazolinethione	150	716
4-Methylamino-7-quinazolinamine	—	H 364
2-Methylaminoquinazoline	89 or 92, nmr, pKₐ, uv	H 343; 849
4-Methylaminoquinazoline	196–197 or 201, pKₐ	554,916,1000
Methyl 2-amino-4-quinazolinecarboxylate	—	H 482

570

Compound	Data	Ref
3-Methylamino-2,4(1H,3H)-quinazolinedione	195–196, ir, nmr	H 382: 911
4-Methylamino-2(1H)-quinazolinethione	223–225, ir, nmr	2384
1-Methylamino-4(1H)-quinazolinone	3-MeSO$_3$F: 206–208, ir nmr; zwitterion: 290–296	1445
2-Methylamino-4(3H)-quinazolinone	238–240	1834
3-Methylamino-4(3H)-quinazolinone	109–110, ir, nmr	934
4-Methylamino-2-trichloromethylquinazoline	—	H 350
6-Methyl-2,4-bismethylthioquinazoline	88–93, nmr	H 302
2-Methyl-5,7-bistrifluoromethyl-4(3H)-quinazolinone	221, ir, nmr	1712
Methyl 6-bromo-1-methyl-4-oxo-1,4-dihydro-2-quinazolinecarboxylate	187, ir, nmr	1915
Methyl 6-bromo-4-oxo-3,4-dihydro-2-quinazolinecarboxylate	230–232, ir, nmr	1915,1945
3-(1-Methylbut-2-enylidene)amino-2-methyl-4(3H)-quinazolinone	135–139, ir, nmr	191
1-(3-Methylbut-2-enyl)-4(1H)-quinazolinone	133–134	655
Methyl 2-t-butyl-4-quinazolinesulfenate	89	104
2-β-(N-Methylcarbamoyl)ethyl-4(3H)-quinazolinone	246–248, ir, ms, nmr, uv	904
Methyl 2-chloromethyl-4-phenyl-6-quinazolinecarboxylate 3-oxide	—	H 463
Methyl 6-chloro-4-phenyl-2-quinazolinecarboxylate	132–134 (?) or 194–195, nmr	760,2473
Methyl 2,4-diamino-5-quinazolinecarboxylate	HCl (?): >300	2225
2-Methyl-5,8-dihydro-4,5,8(3H)-quinazolinetrione	165–168, nmr	1999
6-Methyl-5,8-dihydro-4,5,8(3H)-quinazolinetrione	>215, ir, nmr	2324
Methyl 2,4-dimethoxy-5-quinazolinecarboxylate	—	H 485
Methyl 1,3-dimethyl-2,4-dioxo-1,2,3,4-tetrahydro-5-quinazolinecarboxylate	—	H 485
Methyl 1,3-dimethyl-2,4-dioxo-1,2,3,4-tetrahydro-6-quinazolinecarboxylate	222–223, nmr, uv	2465
Methyl 2,6-dimethyl-4-quinazolinecarboxylate	—	H 483
8-Methyl-2,4-dimorpholinoquinazoline	159–160, ir, nmr	428
3-Methyl-6,8-dinitro-4(3H)-quinazolinone	—	H 177
5-Methyl-6,8-dinitro-4(3H)-quinazolinone	234–236	1159
2-Methyl-6,8-dinitro-4(1H)-quinazolinone 3-oxide	—	H 467

ALPHABETICAL LIST OF SIMPLE QUINAZOLINES (*Continued*)

Quinazoline	m.p. (°C), etc.	Ref.
Methyl 2,4-dioxo-1,2,3,4-tetrahydro-5-quinazolinecarboxylate	>300, nmr	H 485; 2203
1-Methyl-2,4-dioxo-1,2,3,4-tetrahydro-5-quinazolinecarboxylic acid	—	H 485
3-Methyl-2,4-dioxo-1,2,3,4-tetrahydro-5-quinazolinecarboxylic acid	—	H 485
3-Methyl-2,4-dioxo-1,2,3,4-tetrahydro-6-quinazolinecarboxylic acid	>300, ir, nmr	1737
6-Methyl-2,4-diphenoxyquinazoline	184–185	1616
6-Methyl-2,4-diphenylquinazoline	178–179 or 180–181	H 65; 27,53,1802,1942
7-Methyl-2,4-diphenylquinazoline	168–169	53
8-Methyl-2,4-diphenylquinazoline	123, ir, nmr	1942
Methyl 2-ethyl-6-methyl-4-quinazolinecarboxylate	—	H 483
4-α-Methylhydrazinoquinazoline	HCl: 126–128, nmr	2230
4-β-Methylhydrazinoquinazoline	157–159, ms, nmr, uv	1880
Methyl 5-hydroxy-2,4-dimethoxy-6-quinazolinecarboxylate	157–159, ir, nmr	1773,2334
4-Methylimino-2-methylthio-3-phenyl-3,4-dihydroquinazoline	—	H 310,367
4-Methylimino-1-phenyl-1,4-dihydroquinazoline	151, nmr	1439
Methyl 3-methoxycarbonylmethyl-2-methylthio-4-oxo-3,4-dihydro-6-quinazolinecarboxylate	197–198	1137
Methyl 3-methoxycarbonylmethyl-4-oxo-2-thioxo-1,2,3,4-tetrahydro-6-quinazolinecarboxylate	257	1137
Methyl 2-methoxy-4-quinazolinecarboxylate	—	H 483
2-Methyl-3-(*N*-methylacetamido)-4(3*H*)-quinazolinone	148–149, ir, nmr	921
4-Methyl-2-methylaminomethylquinazoline 3-oxide	96, ir, ms, nmr	874
2-Methyl-4-methylamino-6-nitroquinazoline	—	H 353
1-Methyl-2-methylamino-6-nitro-4(1*H*)-quinazolinone	>370, ir, nmr	947
6-Methyl-2-methylamino-4-phenylquinazoline 3-oxide	—	H 464

Compound	Data	Ref.
2-Methyl-4-methylamino-6-quinazolinamine	—	*H* 364
2-Methyl-4-methylaminoquinazoline	134 or 154–157, ms, pK_a	1000,1267,1300
4-Methyl-2-methylaminoquinazoline	—	*H* 343
5-Methyl-4-methylaminoquinazoline	178	1000
1-Methyl-3-methylamino-2,4(1*H*,3*H*)-quinazolinedione	—	*H* 382
3-Methyl-4-methylamino-2(3*H*)-quinazolinethione	269–271	88
1-Methyl-2-methylamino-4(1*H*)-quinazolinone	>280, ir, nmr; HBr (?): 323–324, ir, nmr	947,1415
2-Methyl-3-methylamino-4(3*H*)-quinazolinone	108 to 112, nmr	921,934,1994
3-Methyl-2-methylamino-4(3*H*)-quinazolinone	197, ir, nmr	723
3-Methyl-1-*N*-methylcarbamoyloxy-2,4(1*H*,3*H*)-quinazolinedione	245	211,212
3-Methyl-1-*N*-methylcarbamoyloxy-2-thioxo-1,2-dihydro-4(3*H*)-quinazolinone	161	211
Methyl 3-methyl-2,4-dioxo-1,2,3,4-tetrahydro-5-quinazolinecarboxylate	—	*H* 485
Methyl 3-methyl-2,4-dioxo-1,2,3,4-tetrahydro-6-quinazolinecarboxylate	271–273, ir, nmr	1737
1-Methyl-4-methylimino-3-phenyl-3,4-dihydro-2(1*H*)-quinazolinethione	111–114	88
Methyl 1-methyl-4-oxo-1,4-dihydro-2-quinazolinecarboxylate	153–154, ir, nmr	1915
Methyl 2-methyl-4-oxo-3,4-dihydro-5-quinazolinecarboxylate	—	*H* 485
3-Methyl-1-(2-methylprop-2-enyl)-2,4(1*H*,3*H*)-quinazolinedione	125, ir, nmr	718
3-Methyl-2-methylseleno-4(3*H*)-quinazolinone	78, ms, nmr, uv	1833
4-Methyl-2-methylsulfinylmethylquinazoline 3-oxide	168–169, ir, nmr	874
4-Methyl-2-methylsulfonylmethylquinazoline 3-oxide	213–214, ir, nmr	874
4-Methyl-2-methylthiomethylquinazoline	65–67, ir	874
4-Methyl-2-methylthiomethylquinazoline 3-oxide	110–111	874
2-methyl-4-methylthio-6-nitroquinazoline	—	*H* 300
6-Methyl-4-methylthio-2-pivalamidoquinazoline	197–199, ir, nmr	2255
2-Methyl-4-methylthioquinazoline	—	*H* 300
4-Methyl-2-methylthioquinazoline	—	*H* 292
6-Methyl-4-methylthioquinazoline	96–98, nmr	2255

ALPHABETICAL LIST OF SIMPLE QUINAZOLINES (*Continued*)

Quinazoline	m.p. (°C), etc.	Ref.
3-Methyl-2-methylthio-4(3*H*)-quinazolinethione	107	40
1-Methyl-2-methylthio-4(1*H*)-quinazolinone	159–160, ir, ms, nmr	1846,1996,2197
1-Methyl-4-methylthio-2(1*H*)-quinazolinone	146	41
3-Methyl-2-methylthio-4(3*H*)-quinazolinethione	78–80, ir, nmr	1845,1846,2054
3-Methyl-2-morpholinomethyleneamino-6-nitro-4(3*H*)-quinazolinone	—	1502
3-Methyl-2-morpholinomethyleneamino-4(3*H*)-quinazolinone	—	1502
3-Methyl-7-morpholino-6-nitro-4(3*H*)-quinazolinone	204–206, ms, nmr	900,1669
4-Methyl-2-morpholinoquinazoline	n_D^{19} 1.6205; pic: 206	855
1-Methyl-3-morpholino-2,4(1*H*,3*H*)-quinazolinedione	156	1603
8-Methyl-2-morpholino-4(3*H*)-quinazolinethione	201	439
2-Methyl-3-morpholino-4(3*H*)-quinazolinone	144	*H* 373; 1629
4-Methyl-6-morpholino-2(1*H*)-quinazolinone	H_2O: 233	1592
3-Methyl-2-neopentyl-4(3*H*)-quinazolinethione	88	105
3-Methyl-2-neopentyl-4(3*H*)-quinazolinone	73	105
2-Methyl-7-nitro-4-oxo-3,4-dihydro-6-quinazolinecaroxylic acid	—	*H* 484
2-Methyl-7-nitro-4-oxo-3-phenyl-3,4-dihydro-6-quinazolinecarboxylic acid	—	*H* 484
N-Methyl-6-nitro-4-phenyl-2-quinazolinecarboxamide	276–281	242
1-Methyl-6-nitro-4-phenyl-2(1*H*)-quinazolinone	260–262 or 269–271	493,836,1060
2-Methyl-5-nitro-3-phenyl-4(3*H*)-quinazolinone	—	*H* 186
2-Methyl-6-nitro-3-phenyl-4(3*H*)-quinazolinone	213–215	*H* 186; 376
2-Methyl-7-nitro-3-phenyl-4(3*H*)-quinazolinone	203–205	*H* 186; 687
3-Methyl-6-nitro-7-piperidino-4(3*H*)-quinazolinone	200–202, ms, nmr	900,1669
2-Methyl-5-nitro-3-propyl-4(3*H*)-quinazolinone	—	*H* 182
2-Methyl-7-nitro-3-propyl-4(3*H*)-quinazolinone	—	*H* 182

5-Methyl-6-nitro-2,4-quinazolinediamine	290 or 293; AcOH: 288	201,2284
5-Methyl-8(?)-nitro-2,4-quinazolinediamine	255	201
7-Methyl-6-nitro-2,4-quinazolinediamine	>360	600
7-Methyl-6-nitro-2,4(1*H*,3*H*)-quinazolinedione	—	*H* 202
2-Methyl-6-nitroquinazoline 3-oxide	—	*H* 464
2-Methyl-6-nitro-4(3*H*)-quinazolinethione	255–256	*H* 296; 241
1-Methyl-6-nitro-4(1*H*)-quinazolinone	—	*H* 138
2-Methyl-5-nitro-4(3*H*)-quinazolinone	—	*H* 173
2-Methyl-6-nitro-4(3*H*)-quinazolinone	261–264 or 295–300, nmr	*H* 173; 145,241,1118,1290,2158, 2232
2-Methyl-7-nitro-4(3*H*)-quinazolinone	287, uv	*H* 173; 145
2-Methyl-8-nitro-4(3*H*)-quinazolinone	—	*H* 173
3-Methyl-6-nitro-4(3*H*)-quinazolinone	195–196, ms	*H* 177; 498,1525,1669
3-Methyl-7-nitro-4(3*H*)-quinazolinone	—	*H* 177
3-Methyl-8-nitro-4(3*H*)-quinazolinone	—	*H* 177
6-Methyl-5-nitro-4(3*H*)-quinazolinone	265 or 304–305, ms, nmr	1159,1366,1387
7-Methyl-6-nitro-4(3*H*)-quinazolinone	287 or 303	1157,1159
8-Methyl-6-nitro-4(3*H*)-quinazolinone	269–271	1159
2-Methyl-7-nitro-4(1*H*)-quinazolinone 3-oxide	—	*H* 467
2-Methyl-6-nitro-3-thioureido-4(3*H*)-quinazolinone	—	*H* 379
2-Methyl-5-nitro-3-ureido-4(3*H*)-quinazolinone	—	*H* 379
2-Methyl-6-nitro-3-ureido-4(3*H*)-quinazolinone	—	*H* 379
2-Methyl-7-nitro-3-ureido-4(3*H*)-quinazolinone	—	*H* 379
Methyl 4-oxo-3,4-dihydro-2-quinazolinecarbodithioate	174–175, uv	2319
3-Methyl-4-oxo-3,4-dihydro-2-quinazolinecarbohydrazide	198	123
7-Methyl-4-oxo-3,4-dihydro-2-quinazolinecarbohydrazide	254, uv	1743
1-Methyl-4-oxo-1,4-dihydro-2-quinazolinecarbonitrile	242–245, ir, ms, nmr	1415
2-Methyl-4-oxo-3,4-dihydro-5-quinazolinecarbonitrile	—	*H* 481

Quinazoline	m.p. (°C), etc.	Ref.
2-Methyl-4-oxo-3,4-dihydro-7-quinazolinecarbonitrile	—	H 481
1-Methyl-4-oxo-1,4-dihydro-2-quinazolinecarboxamide	247–249, ir, nmr	1415
7-Methyl-4-oxo-3,4-dihydro-2-quinazolinecarboxamide	243	1743
N-Methyl-4-oxo-1,4-dihydro-1-quinazolinecarboxamide	194–196, ir, ms, nmr	1381
N-Methyl-4-oxo-3,4-dihydro-2-quinazolinecarboxamide	192–194	768
Methyl 2-oxo-1,2-dihydro-4-quinazolinecarboxylate	—	H 483
Methyl 4-oxo-3,4-dihydro-2-quinazolinecarboxylate	209–210, ir, ms, nmr	H 483; 1329,1915
Methyl 4-oxo-3,4-dihydro-3-quinazolinecarboxylate	90–94	276
Methyl 4-oxo-3,4-dihydro-6-quinazolinecarboxylate	—	H 485
Methyl 4-oxo-3,4-dihydro-7-quinazolinecarboxylate	—	H 485
1-Methyl-4-oxo-1,4-dihydro-2-quinazolinecarboxylic acid	139–140, ir	1415
2-Methyl-4-oxo-3,4-dihydro-5-quinazolinecarboxylic acid	—	H 484
2-Methyl-4-oxo-3,4-dihydro-6-quinazolinecarboxylic acid	—	H 484
2-Methyl-4-oxo-3,4-dihydro-7-quinazolinecarboxylic acid	—	H 484
3-Methyl-4-oxo-2,3-dihydro-4-quinazolinecarboxylic acid	170–188, ir	727,739
3-Methyl-4-oxo-3,4-dihydro-8-quinazolinecarboxylic acid	218–220	498
2-Methyl-4-oxo-3,4-dihydro-6-quinazolinesulfonamide	—	H 173
2-Methyl-4-oxo-3,4-dihydro-7-quinazolinesulfonamide	—	H 487
3-Methyl-4-oxo-3,4-diydro-5-quinazolinesulfonamide	180–185, ir, ms, nmr	2584
3-Methyl-4-oxo-3,4-dihydro-6-quinazolinesulfonamide	272–273	498
3-Methyl-4-oxo-3,4-dihydro-7-quinazolinesulfonamide	238	498
3-Methyl-4-oxo-3,4-dihydro-6-quinazolinesulfonic acid	> 300	498
N-Methyl-4-oxo-3-pentyl-3,4-dihydro-2-quinazolinecarboxamide	180–181	768
N-Methyl-4-oxo-3-phenyl-3,4-dihydro-2-quinazolinecarboxamide	249–251	768

Methyl 2-oxo-4-phenyl-1,2-dihydro-6-quinazolinecarboxylate	285–287	1060
Methyl 4-oxo-3-phenyl-3,4-dihydro-2-quinazolinecarboxylate	—	*H* 483
2-Methyl-4-oxo-3-phenyl-3,4-dihydro-5-quinazolinecarboxylic acid	—	*H* 484
2-Methyl-4-oxo-3-phenyl-3,4-dihydro-6-quinazolinecarboxylic acid	—	*H* 484
2-Methyl-4-oxo-3-phenyl-3,4-dihydro-7-quinazolinecarboxylic acid	—	*H* 484
3-Methyl-4-oxo-2-phenyl-3,4-dihydro-2-quinazolinecarboxylic acid	257–259	1556
N-Methyl-4-oxo-3-propyl-3,4-dihydro-2-quinazolinecarboxamide	176–178	768
3-Methyl-2-penta-1',3'-dienyl-4(3*H*)-quinazolinone	140–143, ir, nmr	2079, 2131
6-Methyl-4-pentyl-2-phenylquinazoline	90–92	2160
5-Methyl-6-pentyl-2,4-quinazolinediamine	282–288	674
2-Methyl-3-pentyl-4(3*H*)-quinazolinone	—	*H* 157
3-Methyl-2-pentylthio-4(3*H*)-quinazolinone	55	160
6-Methyl-4-phenethyl-2-phenylquinazoline	120–125, nmr	2097, 2160
1-Methyl-3-phenethyl-2,4(1*H*,3*H*)-quinazolinedione	96–99 or 101–102, ir, ms, nmr	1243, 1361
1-Methyl-2-phenethyl-4(1*H*)-quinazolinone	152–153, ir, ms, nmr	821
2-Methyl-4-phenoxyquinazoline	69–70, ir, nmr, uv	*H* 257; 260
5-Methyl-4-phenoxyquinazoline	90–91, nmr	1814
2-Methyl-5-phenoxy-4(3*H*)-quinazolinone	257–258, ir, nmr	2411
1-Methyl-4-phenylimino-1,4-dihydro-6-quinazolinamine	—	*H* 363
1-Methyl-4-phenylimino-1,4-dihydro-7-quinazolinamine	—	*H* 363
1-Methyl-3-phenyl-4-phenylimino-3,4-dihydro-2(1*H*)-quinazolinethione	206–210	88
1-Methyl-3-phenyl-2-phenylimino-1,2-dihydro-4(3*H*)-quinazolinone	—	748
2-Methyl-4-phenylquinazoline	48	*H* 64; 28,1382,1946
4-Methyl-2-phenylquinazoline	85 to 90, nmr, uv	*H* 64; 163,575,737,894, 1083,1305
6-Methyl-2-phenylquinazoline	130–132, ms, nmr	*H* 65; 2088,2101
6-Methyl-4-phenylquinazoline	93–94, ms, nmr	633
6-Methyl-2-phenyl-4-quinazolinecarboxamide	—	*H* 482

ALPHABETICAL LIST OF SIMPLE QUINAZOLINES (*Continued*)

Quinazoline	m.p. (°C), etc.	Ref.
Methyl 4-phenyl-2-quinazolinecarboxylate	154–157, ir, ms	1946
6-Methyl-2-phenyl-4-quinazolinecarboxylic acid	—	H 482; 393
7-Methyl-2-phenyl-4-quinazolinecarboxylic acid	—	393
8-Methyl-2-phenyl-4-quinazolinecarboxylic acid	—	393
1-Methyl-3-phenyl-2,4(1*H*,3*H*)-quinazolinedione	223–225 or 252, ir, ms, nmr	H 205; 748,1650
3-Methyl-1-phenyl-2,4(1*H*,3*H*)-quinazolinedione	234–236, ir, ms, nmr	H 205; 1427
5-Methyl-3-phenyl-2,4(1*H*,3*H*)-quinazolinedione	—	H 206
6-Methyl-3-phenyl-2,4(1*H*,3*H*)-quinazolinedione	>300, ir, ms, nmr	H 206; 1312
8-Methyl-3-phenyl-2,4(1*H*,3*H*)-quinazolinedione	—	H 206
1-Methyl-3-phenyl-2,4(1*H*,3*H*)-quinazolinedithione	223–224	89,274
2-Methyl-4-phenylquinazoline 1-oxide	165–167, ir	874
2-Methyl-4-phenylquinazoline 3-oxide	183–185	H 464; 57
4-Methyl-2-phenylquinazoline 1-oxide	112–114, ms, nmr	163
4-Methyl-2-phenylquinazoline 3-oxide	uv	H 464; 575
2-Methyl-3-phenyl-4(3*H*)-quinazolinethione	184–185 or 186–187	H 298; 105,479
3-Methyl-2-phenyl-4(3*H*)-quinazolinethione	148 or 150	H 298; 455,953
6-Methyl-2-phenyl-4(3*H*)-quinazolinethione	234–235 or 240–241, uv	241,756
6-Methyl-4-phenyl-2(1*H*)-quinazolinethione	225–226	836
8-Methyl-2-phenyl-4(3*H*)-quinazolinethione	241–243	756
2-Methyl-3-phenyl-4(3*H*)-quinazolinethione S-oxide	107, ir, ms, nmr	1130
4-Methyl-2-phenyl-8-quinazolinol	—	H 197
1-Methyl-2-phenyl-4(3*H*)-quinazolinone	150 to 179, ir, ms, uv	39,821,933,991,1190,1275,2555
1-Methyl-4-phenyl-2(1*H*)-quinazolinone	142–143 or 242–243, ir, nmr	247,681,836
2-Methyl-1-phenyl-4(1*H*)-quinazolinone	226–228 or 231–233, ir, nmr, uv	937,1394,1590

2-Methyl-3-phenyl-4(3H)-quinazolinone	142 to 149, ir, ms, nmr, uv; HCl: 240 to 279; MeI: 241–242; EtI: 236–239, ir; SbCl₅.HCl: 212–213, nmr	H 162; 12,133,146,260,376, 466,530,642,645,886,931,932, 935,1008,1059,1074,1129,1289, 1388,1422,1619,1782,2155, 2223,2267,2361
3-Methyl-2-phenyl-4(3H)-quinazolinone	128 to 137, ir, ms, nmr, uv; AlCl₃.HCl: 223–225, nmr; SbCl₅.HCl: 281–282, nmr	H 170; 39,133,853,933,1050, 1226,1267,1275,1918,2245,2361, 2554,2555
4-Methyl-3-phenyl-2(3H)-quinazolinone	223	110
6-Methyl-2-phenyl-4(3H)-quinazolinone	255–258	H 176; 756,1034
6-Methyl-4-phenyl-2(1H)-quinazolinone	286–287	H 137; 1060
7-Methyl-2-phenyl-4(3H)-quinazolinone	—	H 176
7-Methyl-4-phenyl-2(1H)-quinazolinone	282–284	681
8-Methyl-2-phenyl-4(3H)-quinazolinone	252–254	756
2-Methyl-3-phenyl-4(3H)-quinazolinone 1-oxide	—	H 466
2-Methyl-5-phenylthio-4(3H)-quinazolinone	280, ir, nmr	2411
1-Methyl-3-phenyl-2-thioxo-1,2-dihydro-4(3H)-quinazolinone	287 to 310, ir, ms, nmr	H 308; 87,685,748,1649,1650
1-Methyl-3-phenyl-4-thioxo-3,4-dihydro-2(1H)-quinazolinone	191–192	86
6-Methyl-3-phenyl-2-thioxo-1,2-dihydro-4(3H)-quinazolinone	—	1776
3-Methyl-2-piperidinomethyleneamino-4(3H)-quinazolinone	—	1502
4-Methyl-2-piperidinomethylquinazoline 3-oxide	122, ir, ms, nmr	2500
5-Methyl-4-piperidinoquinazoline	liquid, anal, nmr	1814
1-Methyl-3-piperidino-2,4(1H,3H)-quinazolinedione	163–164	1603
6-Methyl-3-piperidino-2,4(1H,3H)-quinazolinedione	280–281	1437
7-Methyl-6-piperidino-2,4(1H,3H)-quinazolinedione	> 280, nmr	1584
2-Methyl-3-piperidino-4(3H)-quinazolinone	92–93 or 200	114,1629
6-Methyl-3-piperidino-4(3H)-quinazolinone	110–112	1471
4-Methyl-6-piperidino-2(1H)-quinazolinone	H₂O: 130	1592
2-Methyl-3-pivalamido-4(3H)-quinazolinone	161–163, ir, ms, nmr	2593

ALPHABETICAL LIST OF SIMPLE QUINAZOLINES (*Continued*)

Quinazoline	m.p. (°C), etc.	Ref.
6-Methyl-2-pivalamido-4(3*H*)-quinazolinone	215–218, nmr	871
6-Methyl-3-pivaloyloxymethyl-4(3*H*)-quinazolinone	108–109, ir, nmr	1617
6-Methyl-3-pivaloyloxymethyl-2-trifluoromethyl-4(3*H*)-quinazolinone	liquid, nmr	2263
1-(2-Methylprop-2-enyl)-3-propyl-2,4(1*H*,3*H*)-quinazolinedione	73, ir, nmr	718
1-(2-Methylprop-2-enyl)-3-prop-2′-ynyl-2,4(1*H*,3*H*)-quinazolinedione	140, ir, nmr	718
1-(2-Methylprop-2-enyl)-2,4(1*H*,3*H*)-quinazolinedione	205, ir, nmr	718
2-(2-Methylprop-1-enyl)-4(3*H*)-quinazolinone	192–194, ir, nmr	1727
2-Methyl-7-propionamido-4(3*H*)-quinazolinone	—	H 370
2-Methyl-3-propoxycarbonylaminomethyl-4(3*H*)-quinazolinone	100, ir	277
2-Methyl-4-propylaminoquinazoline	168–170 or 171–173, ir, nmr	1267,1300
5-Methyl-6-propylamino-2,4-quinazolinediamine	194–197	602
2-Methyl-3-propylamino-4(3*H*)-quinazolinone	202	1662
4-Methyl-2-propylquinazoline	—	H 64
5-Methyl-6-propyl-2,4-quinazolinediamine	210–214	674
3-Methyl-1-propyl-2,4(1*H*,3*H*)-quinazolinedione	121–122, ir, nmr	362,718
1-Methyl-2-propyl-4(1*H*)-quinazolinone	154–156, ir, nmr	821
2-Methyl-3-propyl-4(3*H*)-quinazolinone	81–83 or 98, nmr; HCl: nmr; SbCl₅,HCl: 203–206, nmr	H 157; 886,935,967,1059,1267, 2162,2361
3-Methyl-2-propyl-4(3*H*)-quinazolinone	76–78	H 165; 1059,1267
4-Methyl-3-propyl-2(3*H*)-quinazolinone	132–134, ir, ms, nmr	2474
1-Methyl-2-propylthio-4(1*H*)-quinazolinone	112–114, ir, nmr	2542
3-Methyl-2-propylthio-4(3*H*)-quinazolinone	84	160
2-Methyl-6-prop-2′-ynylamino-4(3*H*)-quinazolinone	liquid, nmr	2271
1-Methyl-3-prop-2′-ynyl-2,4(1*H*,3*H*)-quinazolinedione	—	H 205

580

Compound	Properties	References
3-Methyl-1-prop-2′-ynyl-2,4(1H,3H)-quinazolinedione		H 205
2-Methyl-3-prop-2′-ynyl-4(3H)-quinazolinone		664,2219
2-Methyl-4-quinazolinamine	224 to 235, ir, ms, nmr	H 344; 60,559,1232,1267, 1456,1757,1828,2132,2190
4-Methyl-2-quinazolinamine	156–158	H 342; 267,2569
6-Methyl-2-quinazolinamine		H 342
6-Methyl-4-quinazolinamine	274–277, ir, uv	H 344; 180,196,339,826
7-Methyl-4-quinazolinamine	286–289, ir, uv	826
2-Methyl-4-quinazolinamine 3-oxide	272–275 or 282; HCl: 239 or 242	60,1828,2078
2-Methylquinazoline	34–35 or 41, 120–125/20, 255/760, ir, ms, nmr, pK_a, uv; pic: 192	H 61; 20,170,890,1063,1373, 1420,1763,1766
4-Methylquinazoline	36–37, 85/1, ir, ms, nmr; pic: 178–180 or 183–186	H 62; 163,172,1885,1914, 2067,2170,2193
5-Methylquinazoline		H 64
6-Methylquinazoline		H 64
7-Methylquinazoline		H 64
8-Methylquinazoline		H 64
2-Methyl-4-quinazolinecarbaldehyde	oxime: 210–212, ir, nmr	830,1063
2-Methyl-4-quinazolinecarbonitrile	127–130, ir, nmr	1063,2110
4-Methyl-2-quinazolinecarbonitrile	145–147 or 148–150, ir, nmr	163,2110
2-Methyl-4-quinazolinecarboxamide	pK_a, uv	H 482; 168,170,875
4-Methyl-2-quinazoliecarboxamide		H 481; 2178
2-Methyl-4-quinazolinecarboxamide 3-oxide		875
2-Methyl-4-quinazolinecarboxylic acid	240–243	H 482; 1063
2-Methyl-4,6-quinazolinediamine		H 364
5-Methyl-2,4-quinazolinediamine	204 to 214, nmr, pK_a	H 254; 661,674,678,2246
6-Methyl-2,4-quinazolinediamine	248 to 259	H 254; 661,678,1912,2326
7-Methyl-2,4-quinazolinediamine	226 to 232, nmr	600,661,2247

ALPHABETICAL LIST OF SIMPLE QUINAZOLINES *(Continued)*

Quinazoline	m.p. (°C), etc.	Ref.
8-Methyl-2,4-quinazolinediamine	—	H 254
1-Methyl-2,4(1H,3H)-quinazolinedione (glycosmicine)	147 (?), 241–242 (?), 263 to 272	H 199; 35,41,362,718,889,2197, 2295,2440
3-Methyl-2,4(1H,3H)-quinazolinedione	230 to 250, ir, ms, nmr	H 200; 46,50,177,187,212,216, 722,770,889,911,1750,1758, 1893,2045,2093
6-Methyl-2,4(1H,3H)-quinazolinedione	314–316	H 202; 1454
7-Methyl-2,4(1H,3H)-quinazolinedione	—	H 202
8-Methyl-2,4(1H,3H)-quinazolinedione	—	H 202
3-Methyl-2,4(1H,3H)-quinazolinedithione	288–291	40,274
6-Methyl-2,4(1H,3H)-quinazolinedithione	350–354	38
2-Methylquinazoline 3-oxide	167 or 168–169, nmr	H 464; 1363,1757,2110,2472
4-Methylquinazoline 1-oxide	158–161, nmr	163,1363
4-Methylquinazoline 3-oxide	166–168 or 178–179, nmr	H 464; 163,559,1363
7-Methylquinazoline 3-oxide	—	H 464
2-Methyl-4(3H)-quinazolineselone	—	H 301
2-Methyl-4-quinazolinesulfenic acid	crude	105
1-Methyl-4(1H)-quinazolinethione	196, nmr	H 301; 1439
2-Methyl-4(3H)-quinazolinethione	214 or 220–222, ir, ms, nmr	H 296; 854,957,1952,2573
3-Methyl-4(3H)-quinazolinethione	139–143, nmr	H 298; 104,2573
4-Methyl-2(1H)-quinazolinethione	210–215 or 215–219, ir, nmr, uv	110,955
6-Methyl-4(3H)-quinazolinethione	>300 or 322–323, nmr, uv	241,2255
2-Methyl-4(1H)-quinazolinethione	—	H 468
5-Methyl-2,4,6-quinazolinetriamine	220–224, uv	201

5-Methyl-2,4,8(?)-quinazolinetriamine	215–217, uv	201
7-Methyl-2,4,6-quinazolinetriamine	268–270, uv	600
1-Methyl-4(1H)-quinazolinimine	HCl: 212; HI: 252; ir, ms, nmr, pK_a, uv	50
3-Methyl-2(3H)-quinazolinimine	HI: 259, nmr, pK_a, uv	50
3-Methyl-4(3H)-quinazolinimine	178, ms, pK_a	417,1000
4-Methyl-8-quinazolinol	—	H 197
1-Methyl-4(1H)-quinazolinone (echinopsine, glycorine)	134 to 152, ir, ms, nmr; HCl: 242; pic: 249	H 138; 35,50,276,821,918, 940,1026,1367,1415,1439,1807
2-Methyl-4(3H)-quinazolinone	233 to 244, ir, ms, nmr, pK_a, uv, xl st, complexes; HCl: 278 (?) or 320	H 141; 60,109,118,133,197,464, 513,531,560,706,747,751,808, 818,854,878,887,908,935, 976,1003,1008,1059,1065, 1126,1197,1267,1290,1306, 1345,1346,1367,1370,1376, 1386,1407,1452,1535,1678, 1765,1766,1770,1901,2085, 2151,2158,2186,2399,2440, 2573
3-Methyl-2(3H)-quinazolinone	nmr	H 137; 50
3-Methyl-4(3H)-quinazolinone	104 to 108, ir, ms, nmr; H_2O: 72; MeI: 277; pic: 214 or 215	H 145; 35,46,50,156,276,818, 853,960,990,1186,1321, 1367,1445,1497,1525,2573
4-Methyl-2(1H)-quinazolinone	230–232	H 137; 1060
5-Methyl-4(3H)-quinazolinone	218–220 or 225–227, ir, ms, nmr	H 155; 1173,1814
6-Methyl-4(3H)-quinazolinone	242 to 265	H 155; 196,498,1001,1951
7-Methyl-4(3H)-quinazolinone	—	H 155
8-Methyl-4(3H)-quinazolinone	250–251 or 264–267, ir, ms, nmr	H 155; 498,1001,1173
2-Methyl-4(1H)-quinazolinone 3-oxide	217–219, ms, complexes	H 467; 15,60,1260,1684, 1688,2159

ALPHABETICAL LIST OF SIMPLE QUINAZOLINES (*Continued*)

Quinazoline	m.p. (°C), etc.	Ref.
2-Methyl-4(3*H*)-quinazolinone 1-oxide	241 or 247	187,2399
4-Methyl-2(1*H*)-quinazolinone 3-oxide	—	*H* 466
2-Methylseleno-4(3*H*)-quinazolinone	208–210, ms, nmr, uv	1833
4-Methyl-2-styrylquinazoline	—	*H* 64
4-Methyl-2-styrylquinazoline 3-oxide	uv	*H* 464; 575
1-Methyl-2-styryl-4(1*H*)-quinazolinone	228–230 or 233, ir, nmr, uv	*H* 138; 821,1190
3-Methyl-2-styryl-4(3*H*)-quinazolinone	171 or 173, ir, nmr, uv	*H* 167; 515,2079,2131
5-Methyl-2-styryl-4(3*H*)-quinazolinone	275–279	2264
6-Methyl-2-styryl-4(3*H*)-quinazolinone	274–276	2264
2-Methylsulfinylmethyl-4-phenylquinazoline 3-oxide	168–169	874
2-Methylsulfonylmethyl-4-phenylquinazoline 3-oxide	182–184	874
6-Methylsulfonyl-4-phenyl-2(1*H*)-quinazolinone	>360	1060
5-Methylsulfonyl-2,4-quinazolinediamine	252–254, nmr, pK_a	2225
5-Methylsulfonyl-4(3*H*)-quinazolinone	—	*H* 155,310
3-Methyl-2-α,β,β,β-tetrafluoroethyl-4(3*H*)-quinazolinone	132–133, ir, nmr	712
2-Methylthiomethyl-4-phenylquinazoline 3-oxide	119–121	874
2-Methylthiomethyl-4-quinazolinecarbonitrile 3-oxide	169–171	874
4-Methylthio-2-morpholino-5-nitroquinazoline	137	439
4-Methylthio-2-morpholinoquinazoline	103	439
4-Methylthio-2-neopentylquinazoline	44–45	104
4-Methylthio-5-nitroquinazoline	—	*H* 300
4-Methylthio-6-nitroquinazoline	—	*H* 300
2-Methylthio-3-phenyl-5,8-dihydro-4,5,8(3*H*)-quinazolinetrione	237, ir, nmr	897
4-Methylthio-2-phenylquinazoline	89–91	*H* 300; 104,2539

4-Methylthio-2-phenylquinazoline 3-oxide	156, ir, ms, nmr	1926
2-Methylthio-3-phenyl-4(3H)-quinazolinethione	—	H 302
2-Methylthio-3-phenyl-4(3H)-quinazolinone	—	H 306
2-Methylthio-4-quinazolinamine	234–236, ir, nmr	H 310,345;826,2095
2-Methylthioquinazoline	58–59, nmr	H 292; 890
4-Methylthioquinazoline	63–64; 1-PhI: 193	H 300; 104,1439
5-Methylthio-2,4-quinazolinediamine	170–172, nmr	1932, 2225
2-Methylthio-4(3H)-quinazolinethione	—	1948
4-Methylthio-2(1H)-quinazolinethione	—	1948
2-Methylthio-4(3H)-quinazolinethione	210–211, ir, nmr	H 304; 1483,1845,1846, 2054,2197
5-Methylthio-4(3H)-quinazolinone	—	H 155,310
6-Methylthio-4(3H)-quinazolinone	203–204	1558
4-Methylthio-2-trichloromethylquinazoline	—	H 300
2-Methyl-3-thioureido-4(3H)-quinazolinone	183–185, complexes	H 373; 1979,2082,2342
1-Methyl-2-thioxo-1,2-dihydro-4(3H)-quinazolinone	228 to 261, ir, nmr, uv; K: 322–324, ir; Li: 342–344; Na: 352–356 Ag: 316–318, ir	748,925,1846,1862,1928, 2197,2439
1-Methyl-4-thioxo-3,4-dihydro-2(1H)-quinazolinone	231	41
3-Methyl-2-thioxo-1,2-dihydro-4(3H)-quinazolinone	247 to 275, ir, nmr, uv, complexes	H 307; 25,87,160,211,216,322, 335,685,880,1125,1273,1335, 1646,1758,1769,1774,1862,2053, 2093
3-Methyl-4-thioxo-3,4-dihydro-2(1H)-quinazolinone	285–286 or 289–290, complexes	40,86,1205
6-Methyl-2-thioxo-1,2-dihydro-4(3H)-quinazolinone	288 to 318, ir, nmr	H 309; 1841,1864,2176,2236
8-Methyl-2-thioxo-1,2-dihydro-4(3H)-quinazolinone	262, ir	H 309; 1674
2-Methyl-4-tribromomethylquinazoline	—	H 64
5/7-Methyl-2-trichloromethyl-4(3H)-quinazolinone	224–226	755

585

ALPHABETICAL LIST OF SIMPLE QUINAZOLINES   (*Continued*)

Quinazoline	m.p. (°C), etc.	Ref.
2-Methyl-5-trifluoromethyl-4-quinazolinamine	196–197, nmr	2247
4-Methyl-2-trifluoromethylquinazoline	—	H 64
2-Methyl-7-trifluoromethyl-4(3H)-quinazolinone	234–236, ir, nmr	1712
6-Methyl-2-trifluoromethyl-4(3H)-quinazolinone	> 300, ms, nmr	1608
Methyl 2,6,7-trimethyl-4-oxo-3-phenyl-3,4-dihydro-8-quinazolinecarboxylate	191–192	1576
2-Methyl-3-ureido-4(3H)-quinazolinethione	—	H 381
2-Methyl-3-ureido-4(3H)-quinazolinone	231–232 or 249–250, complexes	H 373; 109,1979,2082,2342
1-β-Morpholinoethyl-3-phenyl-2,4(1H,3H)-quinazolinedione	—	1106
2-β-Morpholinoethyl-1-phenyl-4(1H)-quinazolinone	162–164, nmr	1394
3-β-Morpholinoethyl-2-thioxo-1,2-dihydro-4(3H)-quinazolinone	230–233, ir, nmr	1196
3-Morpholinomethyl-6-nitro-4(3H)-quinazolinone	—	1118
2-Morpholinomethyl-1-phenyl-4(1H)-quinazolinone	184–186, nmr	1394
2-Morpholinomethyl-3-phenyl-4(3H)-quinazolinone	145–147 or 151–152, ir, ms, nmr; HCl: 248; pic: 206–207	745,943,1276,1286
2-Morpholino-5-nitroquinazoline	crude	439
2-Morpholino-6-nitro-4(3H)-quinazolinethione	188	413
3-Morpholino-6-nitro-4(3H)-quinazolinone	—	H 374
7-Morpholino-6-nitro-4(3H)-quinazolinone	300–302, nmr	900
4-Morpholino-6-nitro-2-trichloromethylquinazoline	—	H 350, 353
2-Morpholino-4-pentyloxyquinazoline	HCl: 137–139	2016
4-Morpholino-2-phenyl-8-piperidino-5,6-dihydro-5,6-quinazolinedione	189–190, ir	592
2-Morpholino-4-phenylquinazoline	112	855
4-Morpholino-2-phenyl-6-quinazolinol	278–279	766
2-Morpholino-3-phenyl-4(3H)-quinazolinone	166–168	H 365; 1225

Compound		
2-Morpholino-4-piperidinoquinazoline	—	H 358
4-Morpholino-2-piperidinoquinazoline	—	H 358
2-Morpholino-4-propylquinazoline	56	855
2-Morpholino-4-quinazolinamine	218–219, ir, nmr	439
4-Morpholinoquinazoline	92–94, ms	H 347; 1786
5-Morpholino-2,4-quinazolinediamine	269–271	2225
3-Morpholino-2,4(1H,3H)-quinazolinedione	265–270 or 289–291	1437,1603
7-Morpholino-2,4(1H,3H)-quinazolinedione	>280, nmr	1584
2-Morpholino-4(3H)-quinazolinethione	222; Hg: 273	439
4-Morpholino-8-quinazolinol	—	H 352
2-Morpholino-4(3H)-quinazolinone	237–239 or 245–248, ir, nmr	H 365; 439,1473,2016
3-Morpholino-4(3H)-quinazolinone	124–125	H 374; 1471
4-Morpholino-2(1H)-quinazolinone	—	H 366
2-Morpholino-4-(thiobenzamido)quinazoline	116–117	429
2-Morpholino-4-(thioformamido)quinazoline	223	429
2-Neopentyl-4-quinazolinesulfenic acid	100	105
2-Neopentyl-4(3H)-quinazolinethione	190–191	105
2-Neopentyl-4(1H)-quinazolinethione 3-oxide	85	105
2-Neopentyl-4(3H)-quinazolinone	201	105
6-Nitro-2,4-dipentyloxyquinazoline	—	H 259
6-Nitro-2,4-diphenylquinazoline	—	1802
7-Nitro-2,3-diphenyl-4(3H)-quinazolinone	—	H 189
4-α-Nitroethylquinazoline	—	H 62
3-β-Nitroethyl-4(3H)-quinazolinone	118–119, ir, nmr	1172,1772
2-Nitromethyl-3-phenyl-4(3H)-quinazolinone	208–209, nmr	1709
4-Nitromethylquinazoline	225–227, ms, uv	H 62; 162
2-Nitromethyl-4(3H)-quinazolinone	—	H 141
6-Nitro-4-oxo-3,4-dihydro-2-quinazolinecarboxylic acid	420	1549

587

ALPHABETICAL LIST OF SIMPLE QUINAZOLINES (*Continued*)

Quinazoline	m.p. (°C), etc.	Ref.
7-Nitro-2-phenoxy-3-phenyl-4(3H)-quinazolinone	211–214, ir, nmr, uv	2200
6-Nitro-4-phenoxyquinazoline	—	H 258
7-Nitro-4-phenoxyquinazoline	—	H 258
6-Nitro-4-phenylquinazoline	135–137	2170
6-Nitro-3-phenyl-2,4(1H,3H)-quinazolinedione	296–298	227
7-Nitro-3-phenyl-2,4(1H,3H)-quinazolinedione	283–286, ir, nmr	2200
6-Nitro-2-phenylquinazoline 3-oxide	—	H 464
6-Nitro-2-phenyl-4(3H)-quinazolinethione	247–248, uv	241
5/7-Nitro-2-phenyl-4(3H)-quinazolinone	310	214
5-Nitro-4-phenyl-2(1H)-quinazolinone	312–313	1060
6-Nitro-2-phenyl-4(3H)-quinazolinone	303–304 or 306	145,214
6-Nitro-4-phenyl-2(1H)-quinazolinone	300 or 314–315, ir	470,544,836,1060
7-Nitro-2-phenyl-4(3H)-quinazolinone	332–334	H 176; 145
7-Nitro-4-phenyl-2(1H)-quinazolinone	300–301	1060
8-Nitro-2-phenyl-4(3H)-quinazolinone	260	H 176; 214
7-Nitro-2-phenyl-4(1H)-quinazolinone 3-oxide	—	H 467
7-Nitro-3-phenyl-2-thioxo-1,2-dihydro-4(3H)-quinazolinone	>300, ir, ms, nmr	1214
6-Nitro-3-piperidinomethyl-4(3H)-quinazolinone	—	1118
6-Nitro-7-piperidino-4(3H)-quinazolinone	268–270, nmr	900
6-Nitro-4-propylaminoquinazoline	—	551
6-Nitro-4-quinazolinamine	315–317 or 319–320	H 345; 507,553,1852
7-Nitro-4-quinazolinamine	>300, uv	H 345; 553
8-Nitro-4-quinazolinamine	284–286, uv	H 345; 553
6-Nitro-4-quinazolinamine 3-oxide	>300	507

588

Name		
5-Nitroquinazoline	—	*H* 487
6-Nitroquinazoline	—	*H* 487
7-Nitroquinazoline	—	*H* 487
8-Nitroquinazoline	ms	*H* 487; 172
5-Nitro-2,4-quinazolinediamine	236–237 or 237–239, nmr, pK_a	2225,2335
6-Nitro-2,4-quinazolinediamine	>300 to >360; EtSO$_3$H: >300	201,956,1852,2231,2326
7-Nitro-2,4-quinazolinediamine	AcOH: 330–332	600
5-Nitro-2,4(1*H*,3*H*)-quinazolinedione		*H* 202
6-Nitro-2,4(1*H*,3*H*)-quinazolinedione		*H* 202
7-Nitro-2,4(1*H*,3*H*)-quinazolinedione		*H* 202
8-Nitro-2,4(1*H*,3*H*)-quinazolinedione		*H* 202
6-Nitro-2,4(1*H*,3*H*)-quinazolinedithione	230	77
6-Nitro-4(3*H*)-quinazolinethione	255–256, uv	*H* 296; 241
7-Nitro-4(3*H*)-quinazolinethione		*H* 296
8-Nitro-4(3*H*)-quinazolinethione		*H* 296
6-Nitro-2,4,5-quinazolinetriamine	AcOH: anal	1443
5-Nitro-4(3*H*)-quinazolinone		*H* 155
6-Nitro-4(3*H*)-quinazolinone	274 to 288	*H* 155; 241,498,628,1118,1622
7-Nitro-4(3*H*)-quinazolinone	265 or 273–276, ir, uv	*H* 155; 145,532,1189,1951
8-Nitro-4(3*H*)-quinazolinone		*H* 155
5-Nitroso-6-quinazolinol	172, ir, nmr	2359
3-Nitroso-4(3*H*)-quinazolinone	263, ir	1690
6-Nitro-2-styryl-4(3*H*)-quinazolinone		*H* 175; 1482
6-Nitro-2-trichloromethyl-4(3*H*)-quinazolinone		*H* 175
N,*N*-(3-Oxapentamethylene)-4-oxo-3,4-dihydro-2-quinazolinecarbothioamide	279–280	2319
4-(2-Oxocyclohexyl)quinazoline	124–125, ir, nmr, uv	731,846
2-(2-Oxocyclohexyl)-4(3*H*)-quinazolinone	270–273, ir, ms	1766

ALPHABETICAL LIST OF SIMPLE QUINAZOLINES (*Continued*)

Quinazoline	m.p. (°C), etc.	Ref.
4-(2-Oxocyclopentyl)quinazoline	—	H 62
4-Oxo-3,4-dihydro-5-quinazolinarsonic acid	—	H 487
4-Oxo-3,4-dihydro-6-quinazolinarsonic acid	—	H 487
4-Oxo-3,4-dihydro-7-quinazolinarsonic acid	—	H 487
4-Oxo-3,4-dihydro-2-quinazolinecarbaldehyde	—	H 486
4-Oxo-3,4-dihydro-2-quinazolinecarbohydrazide	239 or 250; PhCH=: 257–258, nmr	123,1743,2148,2491
4-Oxo-3,4-dihydro-2-quinazolinecarbothiohydrazide	240–245	2321
4-Oxo-3,4-dihydro-2-quinazolinecarboxamide	226	H 482; 1514,1743
4-Oxo-3,4-dihydro-6-quinazolinecarboxamide	>260, nmr	2564
4-Oxo-3,4-dihydro-2-quinazolinecarboxanilide	178, ir, nmr	1780
2-Oxo-1,2-dihydro-4-quinazolinecarboxylic acid	—	H 483
4-Oxo-3,4-dihydro-2-quinazolinecarboxylic acid	212–214	H 483; 1549
4-Oxo-3,4-dihydro-6-quinazolinecarboxylic acid	—	H 484
4-Oxo-3,4-dihydro-7-quinazolinecarboxylic acid	—	H 484
4-Oxo-3,4-dihydro-8-quinazolinecarboxylic acid	310	H 484; 498
4-Oxo-3,4-dihydro-6-quinazolinesulfonamide	325–327	H 155;487; 498
4-Oxo-3,4-dihydro-7-quinazolinesulfonamide	194–195	498
4-Oxo-3,4-dihydro-6-quinazolinesulfonic acid	>300 or >330, ir	498,628
4-Oxo-3,4-dihydro-8-quinazolinesulfonic acid	—	H 487
4-Oxo-N,N-pentamethylene-3,4-dihydro-2-quinazolinecarbothioamide	200–201	2319
4-Oxo-3-phenyl-3,4-dihydro-2-quinazolinecarbaldehyde	208, ms; dnp: 280; oxime: 201–202 or 205, ir, ms; PhNHN=: 247–248, ms; tsc: 221, pol	146,1008,1156,1276,1286,2357
4-Oxo-3-phenyl-3,4-dihydro-2-quinazolinecarbohydrazide	—	1054

2-Oxo-4-phenyl-1,2-dihydro-6-quinazolinecarbonitrile	339–341	1060
4-Oxo-1-phenyl-1,4-dihydro-2-quinazolinecarbonitrile	>280, nmr	1590
4-Oxo-3-phenyl-3,4-dihydro-2-quinazolinecarbonitrile	—	H 481
4-Oxo-2-phenyl-3,4-dihydro-3-quinazolinecarbothioamide	177–178, ir, nmr	1679
4-Oxo-1-phenyl-1,4-dihydro-2-quinazolinecarboxamide	231–233, nmr	1590
4-Oxo-3-phenyl-3,4-dihydro-2-quinazolinecarboxamide	262–264	768
4-Oxo-2-phenyl-3,4-dihydro-3-quinazolinecarboxanilide	201–202, ir, nmr	1679
4-Oxo-3-phenyl-3,4-dihydro-2-quinazolinecarboxanilide	290	781
4-Oxo-2-phenyl-3,4-dihydro-5-quinazolinecarboxylic acid	—	1998
4-Oxo-2-phenyl-3,4-dihydro-6-quinazolinecarboxylic acid	400	1556
4-Oxo-3-phenyl-3,4-dihydro-2-quinazolinecarboxylic acid	191	146
4-Oxo-3-phenyl-2-thioxo-1,2,3,4-tetrahydro-6-quinazolinecarboxanilide	—	1122
2-Oxo-N-propyl-1,2-dihydro-4-quinazolinecarboxamide	>270	739
Pegamine, *see* 2-γ-Hydroxypropyl-4(3H)-quinazolinone		
2-Pentafluoroethyl-4(3H)-quinazolinone	205, ir, nmr	914
2,5,6,7,8-Pentafluoro-4-quinazolinamine	238–240, uv	34,158
4-Pent-4'-enylamino-2(1H)-quinazolinethione	217–218, ir, nmr	2384
3-Pent-4'-enyl-2-trifluoromethyl-4(3H)-quinazolinone	39–41, ir, nmr	1727
4-Pentylamino-8-quinazolinol	—	H 352
5-Pentyloxy-2,4-quinazolinediamine	155–156	2225
4-Pentyloxy-2-trifluoromethylquinazoline	ir, nmr	2482
3-Pentyl-2-pentyloxy-4(3H)-quinazolinone	ir, nmr	2482
1-Pentyl-4-phenyl-2(1H)-quinazolinone (?)	121–122	681
3-Pentyl-1-prop-2'-ynyl-2,4(1H,3H)-quinazolinedione	137–139	614
3-Pentyl-2,4(H,3H)-quinazolinedione	165–166	664
2-Pentyl-4(3H)-quinazolinone	153, ir, nmr, complexes	1240,2244
3-Pentyl-4(3H)-quinazolinone	60–61, ir, nmr	276,2482

Quinazoline	m.p. (°C), etc.	Ref.
2-Pentylthio-3-phenyl-4(3H)-quinazolinethione	—	H 303
2-Pentylthio-3-phenyl-4(3H)-quinazolinone	93 or 96	160,1017
2-Pentylthio-4(3H)-quinazolinone	88–90	1483,1845
4-Phenethylaminoquinazoline	—	H 347
2-Phenethyl-3-phenyl-4(3H)-quinazolinone	174–175	2372
5-Phenethyl-2,4-quinazolinediamine	199–200, ir, nmr;	2225,2597
	HCl (?): > 282, nmr	
6-Phenethyl-2,4-quinazolinediamine	210, nmr	2326
3-Phenethyl-2,4(1H,3H)-quinazolinedione	—	H 201; 1767
3-Phenethyl-2,4(1H,3H)-quinazolinedithione	—	460
2-Phenethyl-4(3H)-quinazolinone	208–210	H 141; 456
3-Phenethyl-2,4(1H,3H)-quinazolinedione	204–206, ir, nmr	2117
3-Phenethyl-2-styryl-4(3H)-quinazolinone	—	H 167
2-Phenethylthio-3-phenyl-4(3H)-quinazolinethione	—	H 303
6-Phenethylthio-2,4-quinazolinediamine	156–159	811
3-Phenethyl-2-thioxo-1,2-dihydro-4(3H)-quinazolinone	239–240 or 244–245, ir, uv	460,615,880
4-Phenoxy-2-phenylquinazoline	119–120 or 122, ir, uv	213,260
2-Phenoxyquinazoline	—	H 253
4-Phenoxyquinazoline	71 to 74, ir, nmr, uv	H 255; 260,276,1139
4-Phenoxy-2-quinazolinecarbonitrile	—	H 481
5-Phenoxy-2,4-quinazolinediamine	244–245, nmr	2225
6-Phenoxy-2,4-quinazolinediamine	205–207 or 212–213	238,2326
2-Phenoxy-4(3H)-quinazolinone	—	H 143
4-Phenoxy-2(3H)-quinazolinone 1-oxide	—	H 466

592

Compound	Properties	Ref.
2-Phenyl-5,8-dihydro-4,5,8(3*H*)-quinazolinetrione	260, ir, nmr	2430
2-Phenyl-4,8-dipiperidino-5,6-dihydro-5,6-quinazolinedione	177–179, ir	592
3-Phenyl-2-phenylhydrazino-4(3*H*)-quinazolinone	—	2386
2-Phenyl-8-piperidino-5,6-dihydro-5,6-quinazolinedione	179–181, ir, nmr, uv	764
2-Phenyl-8-piperidino-5,6-dihydro-4,5,6(3*H*)-quinazolinetrione	261, complex	762
4-Phenyl-2-piperidinomethylquinazoline 3-oxide	115, ir, ms, nmr	2500
1-Phenyl-2-piperidinomethyl-4(1*H*)-quinazolinone	152–153, nmr	1394
3-Phenyl-2-piperidinomethyl-4(3*H*)-quinazolinone	121–122 or 124–125, ir, nmr; HCl: 238; pic: 193–194	745, 1276
4-Phenyl-2-piperidinoquinazoline	104	855
2-Phenyl-4-piperidino-6-quinazolinol	250–253	766
3-Phenyl-2-piperidino-4(3*H*)-quinazolinone	—	*H* 365
3-Phenyl-2-propoxymethyl-4(3*H*)-quinazolinone	89–90, ir, nmr	1154
2-Phenyl-4-propoxyquinazoline	ms	607
2-Phenyl-4-propylaminoquinazoline	104–106	1267,1300
2-Phenyl-4-propylquinazoline	79, uv	1083
4-Phenyl-2-propylquinazoline	100, ir	*H* 64; 1946
3-Phenyl-1-propyl-2,4(1*H*,3*H*)-quinazolinedione	182	112
2-Phenyl-3-propyl-4(3*H*)-quinazolinone	88–91 or 97–98, ir, ms, nmr, uv; SbCl$_5$.HCl: 230–232, ir, nmr; FeCl$_3$.HCl: 205–208, ir	607,1267,2361
3-Phenyl-2-propyl-4(3*H*)-quinazolinone	125	*H* 165; 1184
4-Phenyl-1-propyl-2(1*H*)-quinazolinone	130–131	681
3-Phenyl-2-propylthio-4(3*H*)-quinazolinethione	—	*H* 303
3-Phenyl-2-propylthio-4(3*H*)-quinazolinone	—	*H* 306
2-Phenyl-4-prop-2'-ynyloxyquinazoline	ms	607
3-Phenyl-1-prop-2'-ynyl-2,4(1*H*,3*H*)-quinazolinedione	174–177	523
2-Phenyl-3-prop-2'-ynyl-4(3*H*)-quinazolinone	ms	607

ALPHABETICAL LIST OF SIMPLE QUINAZOLINES (*Continued*)

Quinazoline	m.p. (°C), etc.	Ref.
4-Phenyl-1-prop-2′-ynyl-2(1*H*)-quinazolinone	181	681
2-Phenyl-4-prop-2′-ynythioquinazoline	110, ir, ms, nmr	1658
2-Phenyl-4-quinazolinamine	144 to 148, ir, nmr, uv	*H* 345; 258,1018,1131,1232,1267, 1300,1456,1872,2132
4-Phenyl-2-quinazolinamine	165–166 or 169–170	343,1764,2569
2-Phenyl-4-quinazolinamine 3-oxide	223–224 or 225, ir, nmr; HCl: 270	60,1872,2078
2-Phenylquinazoline	93 to 103, nmr; pic: 162–163	*H* 61; 225,398,693,767,890,989, 1206,1652,1763,2478
4-Phenylquinazoline	99–101, ir; 1-MeI: 196–198, ir, nmr; 3-MeI: crude; 1-MeBr: 243–244, ir, nmr	*H* 63; 56,247,353,1946
8-Phenylquinazoline	107–108	985
4-Phenyl-2-quinazolinecarbaldehyde	130–131	632
2-Phenyl-4-quinazolinecarbonitrile	166 to 170, nmr	*H* 481; 923,1142,2206
4-Phenyl-2-quinazolinecarbonitrile	120–121 or 127–129, ir, nmr	2110,2206
4-Phenyl-2-quinazolinecarboxamide	249–250, nmr	2178
2-Phenyl-4-quinazolinecarboxylic acid	—	*H* 482
4-Phenyl-2-quinazolinecarboxylic acid	100–102, nmr	2468,2473
1-Phenyl-2,4(1*H*,3*H*)-quinazolinedione	296–297 or 299, nmr	*H* 199; 186,362,718
3-Phenyl-2,4(1*H*,3*H*)-quinazolinedione	210–212 (?), 268 to 291, ir, ms, nmr, uv, xl st	*H* 201:1,84,90,92,93,100,112,128, 187,216,434,511,635,709,746, 769,770,832,1007,1156,1254, 1282,1312,1360,1392,1434,1446, 1448,1650,2117,2444

594

Name		References
2-Phenylquinazoline 1,3-dioxide	xl st	1100
3-Phenyl-2,4(1H,3H)-quinazolinedithione	252–254 or 268–269	H 302; 89,274,460
2-Phenylquinazoline 3-oxide	—	H 464
4-Phenylquinazoline 1-oxide	166–167, ms, nmr	163
4-Phenylquinazoline 3-oxide	170–171, ms, nmr	H 464; 163
1-Phenyl-4(1H)-quinazolinethione	178, nmr	1439
2-Phenyl-4(3H)-quinazolinethione	218 to 222, nmr, uv	H 296; 133,210,241,2050, 2539,2573
3-Phenyl-2(3H)-quinazolinethione	—	H 292
3-Phenyl-4(3H)-quinazolinethione	nmr	H 298; 2573
4-Phenyl-2(1H)-quinazolinethione	205–206 (?), 220 to 235, ir, ms, nmr, uv	836, 955,1946,1953
2-Phenyl-4(1H)-quinazolinethione 3-oxide	133 or 143–144	H 468; 455,2209
N-(4-Phenyl-1-quinazolinio)benzamidate	223–224, nmr, uv	403
2-Phenyl-6-quinazolinol	235–236	766
1-Phenyl-4(1H)-quinazolinone	180 to 184, ir, ms, nmr; MeI: 300, ir, nmr	186,235,1439,1467
2-Phenyl-4(3H)-quinazolinone	229–252, 267 (?), ir, ms, nmr, pK_a, uv, xl st	H 143; 13,39,76,174,213,214,233, 251,346,351,386,404,499,542, 550,591,747,751,818,824,852, 872,901,936,976,991,1003, 1034,1050,1082,1197,1211,1226, 1240,1257,1267,1290,1315, 1318,1377,1447,1452,1655, 1704,1787,2085,2151,2186, 2555,2573
3-Phenyl-4(3H)-quinazolinone	125–127 (?), 135 to 140, ir, nmr, uv; MeBr: 222–223, ir	H 146,152; 30,47,260,276,448, 650,690,768,992,1150,1352, 1497,2573

Quinazoline	m.p. (°C), etc.	Ref.
4-Phenyl-2(1*H*)-quinazolinone	243 to 263, ir, ms, nmr, pK_a, uv	H 137; 99,342,681,751,836, 1060,1200,1360,1946,2494
6-Phenyl-4(3*H*)-quinazolinone	—	H 155
7-Phenyl-4(3*H*)-quinazolinone	—	H 155
2-Phenyl-4(1*H*)-quinazolinone 3-oxide	177, ir, complexes	H 467; 236,851,1260,1343, 1684,1688,2159
2-Phenyl-4-styrylquinazoline	uv	575
4-Phenyl-2-styrylquinazoline	154–156	1946
3-Phenyl-2-styryl-4(3*H*)-quinazolinone	193 to 201, ms, nmr	H 167; 1234,1296,1739,1749, 1977
6-Phenylsulfinyl-2,4-quinazolinediamine	283–285	1548
5-Phenylsulfonyl-2,4-quinazolinediamine	290–291	2225
6-Phenylsulfonyl-2,4-quinazolinediamine	297–299	1548,2326
2-Phenylsulfonyl-4(3*H*)-quinazolinone	210, ir, nmr	1915
3-Phenyl-2-thiocyanatomethyl-4(3*H*)-quinazolinone	154–155, ir, nmr	1286
3-Phenyl-2-thiocyanato-4(3*H*)-quinazolinone	138–141 or 149–152	620,663
4-Phenylthioquinazoline	112–113 or 115–116, 135–137/1.5 (?)	650,1139,1189
5-Phenylthio-2,4-quinazolinediamine	241–243	2225
6-Phenylthio-2,4-quinazolinediamine	187	2326
7-Phenylthio-2,4-quinazolinediamine	189–191	1551
5-Phenylthio-4(3*H*)-quinazolinone	—	H 155,310
2-Phenyl-3-thioureido-4(3*H*)-quinazolinethione	156–157, ir, nmr	1679
2-Phenyl-3-thioureido-4(3*H*)-quinazolinone	178–179, complexes	1679,2082,2342

Compound	Properties	References
3-Phenyl-2-thioureido-4(3H)-quinazolinone	260, ir	2408
2-Phenyl-4-thioxo-3,4-dihydro-3-quinazolinecarbothioanilide	162–163, ir, nmr	1679
2-Phenyl-4-thioxo-3,4-dihydro-3-quinazolinecarboxanilide	182–183, ir, nmr	1679
3-Phenyl-2-thioxo-1,2-dihydro-4(3H)-quinazolinone	233 (?), 300 to 310, ir, ms, nmr, uv, complexes	H 307; 62,84,87,90,211,216, 316,357,460,511,615,652,663, 770,880,925,1125,1180,1205, 1242,1335,1484,1627,1645, 1646,1650,1668,1769,1776,2386, 2390,2396,2408,2456,2540
3-Phenyl-4-thioxo-3,4-dihydro-2(1H)-quinazolinone	277–278	86,635,685
2-Phenyl-4-trichloromethylquinazoline	—	H 64
4-Phenyl-2-trichloromethylquinazoline	109–110	H 64; 1946
3-Phenyl-2-triethylammoniomethyl-4(3H)-quinazolinone (bromide)	208–209	1276
4-Phenyl-2-trifluoromethylquinazoline	92, ir, ms	1946
1-Phenyl-2-trifluoromethyl-4(1H)-quinazolinone	235–237, nmr	1590
3-Phenyl-2-trifluoromethyl-4(3H)-quinazolinone	80, ms, nmr	2351
4-Phenyl-6-trifluoromethyl-2(1H)-quinazolinone	226–227	1060
2-Phenyl-4-trimethylammonioquinazoline (chloride)	155	923,1142,2206
4-Phenyl-2-trimethylammonioquinazoline (chloride)	anal	2206
2-Phenyl-3-ureido-4(3H)-quinazolinethione	224–225, ir, nmr, complexes	H 381; 1679,2082
2-Phenyl-3-ureido-4(3H)-quinazolinone	298–299, ir, nmr, complexes	H 373; 1679,2342
1-Phenyl-2-vinyl-4(1H)-quinazolinone	226–230, nmr	1394
6-Piperidino-5,8-dihydro-5,8-quinazolinedione	159, ir, nmr	1577
2-Piperidinoquinazoline	148–150/0.1	849
4-Piperidinoquinazoline	173/25, ms; pic: 193 to 196	H 348; 340,555,1139,1786,2031
5-Piperidino-2,4-quinazolinediamine	248–249	2225
3-Piperidino-2,4(1H,3H)-quinazolinedione	252 or 261–262	1437,1603
5-Piperidino-2,4(1H,3H)-quinazolinedione	>280, nmr	1584
6-Piperidino-2,4(1H,3H)-quinazolinedione	crude, >280	1584

ALPHABETICAL LIST OF SIMPLE QUINAZOLINES *(Continued)*

Quinazoline	m.p. (°C), etc.	Ref.
7-Piperidino-2,4(1*H*,3*H*)-quinazolinedione	>280, nmr	1584
6-Piperidino-2,4(1*H*,3*H*)-quinazolinedithione	>360	38
4-Piperidino-6-quinazolinol	196–197	764
2-Piperidino-4(3*H*)-quinazolinone	—	*H* 365
3-Piperidino-4(3*H*)-quinazolinone	71–72	1471
4-Piperidino-2(1*H*)-quinazolinone	—	*H* 366
2-Prop-1′-enyl-4(3*H*)-quinazolinone	195–198	2264
2-Propionamidoquinazoline	150	849
3-Propionamido-4(3*H*)-quinazolinone	—	*H* 374
3-γ-Propionylacetonyl-4(3*H*)-quinazolone	—	*H* 150
3-2′-Propionylacetyl-4(3*H*)-quinazolinone	—	*H* 150
4-Propionylmethylquinazoline	—	*H* 62
4-Propionylquinazoline	—	431
2-Propoxy-4-quinazolinamine	—	*H* 253
4-Propoxyquinazoline	—	*H* 255
4-Propoxy-2-quinazolinecarbonitrile	—	*H* 481
5-Propoxy-2,4-quinazolinediamine	222–223	2225
4-Propoxyquinazoline 1-oxide	—	*H* 461
5-Propoxy-8-quinazolinol	65, uv	595
2-Propoxy-4(3*H*)-quinazolinone	—	*H* 143
4-Propoxy-2(3*H*)-quinazolinone 1-oxide	—	*H* 466
2-Propylaminoquinazoline	pic: 187	189
4-Propylaminoquinazoline	131–133, nmr, uv; pic: 205–206	189,1300,2060
3-Propyl-1-prop-2′-ynyl-2,4(1*H*,3*H*)-quinazolinedione	140–141	664

2-Propyl-3-prop-2'-ynyl-4(3H)-quinazolinone	95–96	664
2-Propyl-4-quinazolinamine	212–213 or 214–216, ir, nmr	1232,1828
2-Propyl-4-quinazolinamine 3-oxide	192	1828
2-Propylquinazoline		H 61
4-Propylquinazoline	liquid, nmr	H 63; 1914
N-Propyl-4-quinazolinecarboxamide	54–56, ms, nmr	2060
6-Propyl-2,4-quinazolinediamine		H 354
1-Propyl-2,4(1H,3H)-quinazolinedione	177–178, ir, nmr	362,718
3-Propyl-2,4(1H,3H)-quinazolinedione		H 201
2-Propyl-4(3H)-quinazolinethione	160 or 179, ms, nmr	H 296; 957,1184
4-Propyl-8-quinazolinol		H 197
1-Propyl-4(1H)-quinazolinone	94–96	276
2-Propyl-4(3H)-quinazolinone	190 to 200, ms, nmr, uv, complexes	H 141; 133,456,957,1003,1059, 1184,1240,2244
3-Propyl-4(3H)-quinazolinone	95–96	H 146; 23,276
5-Propyl-4(3H)-quinazolinone		H 155
5-Propylthio-2,4-quinazolinediamine	126	2225
2-Propylthio-4(3H)-quinazolinone	130–132	1483,1845
3-Propyl-2-thioxo-1,2-dihydro-4(3H)-quinazolinone	191, ir, uv, complexes	880,1646
4-Prop-2'-ynyloxyquinazoline	127, ir, nmr	1935
3-Prop-2'-ynyl-2,4(1H,3H)-quinazolinedione	239–240	H 201; 227
1-Prop-2'-ynyl-4(1H)-quinazolinone	225–226	655
3-Prop-2'-ynyl-4(3H)-quinazolinone	115–117, ir, ms, nmr	H 146; 276,1935,2219
2-Quinazolinamine	202–204, ms, nmr, pK_a, uv	H 342; 50,165,172,189,281,849
4-Quinazolinamine	257 to 280, ir, nmr, pK_a, uv; pic: 218	H 344; 60,164,165,180,189,196, 258,276,333,339,340,382,507, 554,678,826,916,959,1474, 1852,2058
5-Quinazolinamine		H 362

599

ALPHABETICAL LIST OF SIMPLE QUINAZOLINES (*Continued*)

Quinazoline	m.p. (°C), etc.	Ref.
6-Quinazolinamine	crude, ir, nmr	H 362; 2075
7-Quinazolinamine	—	H 362
8-Quinazolinamine	—	H 362
2-Quinazolinamine 3-oxide	—	H 462
4-Quinazolinamine 3-oxide	227 or 231–233, 272 (?)	H 462; 60,507,554
Quinazoline	45–47 or 48–49, dipole, ms, nmr, pKₐ, uv, xl st; 1- and 3-MeI: nmr; uv	H 13; 10,16,28,48,74,75,96, 165,169,170,172,175,184, 338,559,678,700,933, 1138,1147,1652,1768,2193, 2362
4-Quinazolinecarbaldehyde	oxime: 194	106
4-Quinazolinecarbohydrazide	—	H 482
2-Quinazolinecarbonitrile	162–164, ms	H 481; 1052,1142
4-Quinazolinecarbonitrile	117–119, ir, ms	H 481; 189,792,1052,1142, 1419,1714,1885,2110
2-Quinazolinecarboxamide	—	H 481
4-Quinazolinecarboxamide	169 or 171–173, pKₐ, uv	H 481; 168,170,373,1885,2178
4-Quinazolinecarboxylic acid	NH₄: 205–207, pKₐ, uv	168
2,4-Quinazolinediamine	248 to 260, nmr, pKₐ, uv, xl st; H₂SO₄: 346–348	H 354; 201,250,383,678,1270, 1563,1795,1852,1932,2225, 2326
4,6-Quinazolinediamine	—	H 363
4,7-Quinazlinediamine	—	H 363
4,8-Quinazolinediamine	—	H 363
5,8-Quinazolinediol	253, ir, uv	659

600

Compound	Data	References
2,4(1H,3H)-Quinazolinedione	328–331 (?), 350 to 362, ir, ms, nmr, pK_a	H 199; 26,64,174,177,182,362, 494, 635,678,718,770,807,832, 866,889,917,976,1038,1060, 1076,1434,1446,1448,1454, 1459,1473,1708,1794,1832,1863, 1915,1920,2100,2167,2230,2273, 2320,2420,2507
2,4(1H,3H)-Quinazolinedithione	319–322 or 335–338	H 302; 38,89,274
Quinazoline 1-oxide	68, ir, nmr	546
Quinazoline 3-oxide	153, nmr	H 462; 546, 1137
4-Quinazolinesulfenic acid	130	105
2-Quinazolinesulfonic acid	K: crude, ir, nmr, pK_a, uv	281
4-Quinazolinesulfonic acid	K: 326, ir, nmr, pK_a, uv	H 487; 281
2-Quinazolinesulfonyl fluoride	103–104, ir, nmr, uv	281
2,4,5,6-Quinazolinetetramine	HCl: 280	1442
2,4,6,8-Quinazolinetetramine	240	601
2(1H)-Quinazolinethione	220–221, complexes	H 292; 281,2091
4(3H)-Quinazolinethione	280 to 290 (?), 317 to 330, ir, nmr, uv	H 296; 37,105,276,281,539, 678,896,960,1221,1953, 2573
4(1H)-Quinazolinethione 3-oxide	—	H 468
2,4,5-Quinazolinetriamine	HCl: >292	2225
2,4,6-Quinazolinetriamine	251 to 260, uv	201,678,956,1541,1852, 1919,2326
2,4,7-Quinazolinetriamine	141–143	600
5-Quinazolinol	—	H 197
6-Quinazolinol	239	H 197; 2359
7-Quinazolinol	—	H 197
8-Quinazolinol	el den	H 197; 49

ALPHABETICAL LIST OF SIMPLE QUINAZOLINES   (*Continued*)

Quinazoline	m.p. (°C), etc.	Ref.
2(1*H*)-Quinazolinone	>250 or 280–282, ir, pK_a, uv; HCl: 243–245; MeCl: 245; EtCl: 250; Pr^{i}Cl: 275	*H* 136; 165,343,751,849, 1027,1350,2507
4(3*H*)-Quinazolinone	208 to 220, fl, ir, ms, nmr, pK_a, st, uv; HCl: 194, then 222–224 or 272–273	*H* 153; 28, 35,60,133,165,172, 174,182,196,251,276,340,343, 373,539,546,628,678,690,751, 818,873,896,908,943,959,960, 966,992,1001,1008,1173,1186, 1189,1291,1340,1367,1497,1558, 1770,1785,1895,1927,1943, 1951,2010,2180,2229,2362, 2420,2432,2507,2543,2558,2573
2(1*H*)-Quinazolinone 3-oxide	—	*H* 466
4(1*H*)-Quinazolinone 3-oxide	238 or 247–248	*H* 467; 60,1497,1943
4(3*H*)-Quinazolinone 1-oxide	225–230	*H* 466; 182
2-Selenoxo-1,2-dihydro-4(3*H*)-quinazolinone	245–247, ms, uv	1833
4-Selenoxo-3,4-dihydro-2(1*H*)-quinazolinone	233–234, ir, nmr	1897
2-Styryl-4-quinazolinamine	—	1969
2-Styrylquinazoline	—	*H* 61
4-Styrylquinazoline	94–95, nmr	416
5-Styryl-2,4-quinazolinediamine	246–248, ir, nmr	2597
6-Styryl-2,4-quinazolinediamine	(*E*): 270; (*Z*): 219–224	2326
2-Styrylquinazoline 3-oxide	—	*H* 464
2-Styryl-4(3*H*)-quinazolinone	220–228 or 243, ir, ms, nmr	*H* 141; 542,2085,2264
3-β-Sulfoethyl-4(3*H*)-quinazolinone	341–345, ir, ms, nmr	1854

2-Sulfomethyl-4(3H)-quinazolinone	K: 210	605
3-Sulfomethyl-4(3H)-quinazolinone	350, ir, ms, nmr	1854
5,6,7,8-Tetrachloro-2,4-diphenylquinazoline	251–252, nmr	157,198
2,4,6,8-Tetrachloroquinazoline	150	H 251; 341
5,6,7,8-Tetrafluoro-1,3-dimethyl-2,4(1H,3H)-quinazolinedione	102–104, ir, nmr, uv	1317
2-α,β,β,β-Tetrafluoroethyl-4(3H)-quinazolinone	210–212, ms, nmr, uv	712,1793
5,6,7,8-Tetrafluoro-2,4-quinazolinediamine	215–220, nmr	1932
5,6,7,8-Tetrafluoro-2,4(1H,3H)-quinazolinedione	320–323, uv	34
4,5,6,8-Tetramethoxy-7-methylquinazoline	66–67, ms, nmr	2363
Tetramethyl 6,8-dichloro-2,4-quinazolinebisphosphonate	142, nmr	341
N,N-Tetramethylene-4-quinazolinecarboxamide	90–91, ms, nmr	2060
2,4,5,8-Tetramethylquinazoline	—	H 65
2,3,6,7-Tetramethyl-4(3H)-quinazolinone	—	H 182
2,3,6,8-Tetramethyl-4(3H)-quinazolinone	—	H 182
3,4,5,7-Tetramethyl-2(3H)-quinazolinone	191	110
2,4,5,8-Tetraphenylquinazoline	201–203, ms, nmr, X-ray	2588
2-Thioxo-1,2-dihydro-4(3H)-quinazolinone	278 to 306, ir, nmr, xl st, complexes	H 306, 476,521,630,635,663,685, 691,770,807,866,880,1163, 1207,1242,1335,1483,1774, 1846,1862,1864,1966,2095, 2139(?),2210(?),2213(?),2236, 2252,2405,2483,2540
4-Thioxo-3,4-dihydro-2(1H)-quinazolinone	270–272 or 278–279, uv	86,405,2330
2-Thioxo-3-β-trimethylammonioethyl-1,2-dihydro-4(3H)-quinazolinone (iodide)	203–204	52
2-Thioxo-3-γ-trimethylammoniopropyl-1,2-dihydro-4(3H)-quinazolinone (iodide)	245–248	52
2,4,5-Triacetamidoquinazoline	278–280, ir, ms, nmr	1919
4-Tribromomethylquinazoline	166–168, ms, nmr	1651
2-Tribromomethyl-4(3H)-quinazolinone	—	2092
2,4,8-Trichloro-6,7-dimethoxyquinazoline	172–173, ir, nmr	1146

ALPHABETICAL LIST OF SIMPLE QUINAZOLINES (*Continued*)

Quinazoline	m.p. (°C), etc.	Ref.
2,4,5-Trichloro-7-dimethylaminoquinazoline	194–195	1584
2,4,7-Trichloro-6-dimethylaminoquinazoline	113–114	1584
2-Trichloromethyl-4-quinazolinamine	—	H 345
2-Trichloromethylquinazoline	91, nmr, uv; MeOH adduct: 110–111, nmr	2001,2010
2-Trichloromethyl-4(3H)-quinazolinone	210–212 or 237, ir, nmr	H 141; 741,1136,2265
2,4,7-Trichloro-6-morpholinoquinazoline	146–148	1584
2,4,5-Trichloro-6-piperidinoquinazoline	114–116	1584
2,4,5-Trichloro-7-piperidinoquinazoline	82–86	1584
2,4,7-Trichloro-6-piperidinoquinazoline	136–138	1584
2-(3,3,3-Trichloroprop-1-enyl)-4(3H)-quinazolinone	—	H 141
2,4,6-Trichloroquinazoline	—	H 251
2,5,6-Trichloroquinazoline	145	849
2,5,8-Trichloroquinazoline	126	549
2,6,7-Trichloroquinazoline	—	H 251
4,6,8-Trichloroquinazoline	138–141	H 251; 1189
2-Trifluoromethylquinazoline	—	H 61
5-Trifluoromethylquinazoline	64/0.5, pK_a, uv	H 64; 168
6-Trifluoromethylquinazoline	75, pK_a, uv	H 64; 168
7-Trifluoromethylquinazoline	52, pK_a, uv	H 64; 168
8-Trifluoromethylquinazoline	130, pK_a, uv	H 64; 168
5-Trifluoromethyl-2,4-quinazolinediamine	176–178, nmr	2211
6-Trifluoromethyl-2,4-quinazolinediamine	234–236, nmr	2246
7-Trifluoromethyl-2,4-quinazolinediamine	244–246, nmr	2246

Compound	Properties	References
8-Trifluoromethyl-2,4-quinazolinediamine	195–197, nmr	2246
7-Trifluoromethyl-2,4(1H,3H)-quinazolinedione	—	H 202
2-Trifluoromethyl-4(3H)-quinazolinone	238 to 254, ir, ms, nmr	835,1287,1290,1391,2145
5-Trifluoromethyl-4(3H)-quinazoline	—	H 155
6-Trifluoromethyl-4(3H)-quinazoline	210	H 155; 168
7-Trifluoromethyl-4(3H)-quinazoline	225–227, ir, nmr	H 155; 168,2165
8-Trifluoromethyl-4(3H)-quinazoline	239	H 155; 168
5,6,7-Trihydroxy-2,4(1H,3H)-quinazolinedione	—	H 204
6,7,8-Trihydroxy-4(3H)-quinazolinone	>400	498
5,6,8-Trimethoxy-3,7-dimethyl-4(3H)-quinazolinone	136–137, ir, ms, nmr	2363
5,6,8-Trimethoxy-7-methylquinazoline	84–85, ms, nmr	2363
5,6,7-Trimethoxy-4-methyl-2(1H)-quinazolinone	229–230	1592
5,6,8-Trimethoxy-7-methyl-4(3H)-quinazolinone	232–233, ir, ms, nmr	2363
6,7,8-Trimethoxy-4-methyl-2(1H)-quinazolinone	208–210	1592
5,6,7-Trimethoxy-2-phenoxyl-4-quinazolinamine	154–155	696
2,4,7-Trimethoxyquinazoline	—	H 259
6,7,8-Trimethoxy-2,4-quinazolinediamine	240–241	696
5,6,7-Trimethoxy-2,4(1H,3H)-quinazolinedione	—	H 204
6,7,8-Trimethoxy-2,4(1H,3H)-quinazolinedione	271	696
6,7,8-Trimethoxy-4(3H)-quinazolinethione	158, ir, nmr	2017
6,7,8-Trimethoxy-4(3H)-quinazolinone	226–227	498
2-Trimethylammonioquinazoline	Cl⁻: 160–161, nmr, uv; pic: 179–180	165,1142
4-Trimethylammonioquinazoline	Cl⁻: 98–99, nmr, uv; HgCl₃⁻: 170–172	165,1142
1,3,5-Trimethyl-7-methylthio-2,4(1H,3H)-quinazolinedione	163–164, ir, nmr	2398
4,6,7-Trimethyl-2-oxo-1,2-dihydro-8-quinazolinecarbonitrile	253–254, ir, nmr	1570
2,3,7-Trimethyl-4-oxo-3,4-dihydro-6-quinazolinecarboxylic acid	—	H 484

ALPHABETICAL LIST OF SIMPLE QUINAZOLINES (Continued)

Quinazoline	m.p. (°C), etc.	Ref.
2,6,7-Trimethyl-4-oxo-3-phenyl-3,4-dihydro-8-quinazolinecarboxylic acid	262–264	1576
2,5,7-Trimethyl-4-phenyl-8-quinazolinol	84	742
2,4,5-Trimethylquinazoline	—	H 65
4,6,7-Trimethyl-8-quinazolinecarbonitrile	178–179, ir, nmr	1568
5,6,8-Trimethyl-2,4-quinazolinediamine	—	H 354
2,3,6-Trimethyl-4(3H)-quinazolinone	109–112, nmr	853,2406
2,3,8-Trimethyl-4(3H)-quinazolinone	—	H 182
2,6,8-Trimethyl-4(3H)-quinazolinone	—	H 173
3,4,6-Trimethyl-2(3H)-quinazolinone	199–202	110
4-Trimethylsiloxyquinazoline	crude, 90/0.07	224
3-Trimethylsilylmethyl-4(3H)-quinazolinone	147–149/1,5, ir, nmr	1965
2,4,6-Tris(dimethylaminomethyleneamino)quinazoline	173	1852
2-Undecafluoropentyl-4(3H)-quinazolinone	177–178, ir, nmr	1287
3-Ureidomethyl-4(3H)-quinazolinone	—	982
3-Ureido-4(3H)-quinazolinethione	—	H 381
3-Ureido-4(3H)-quinazolinone	—	H 374
2-Vinyl-5,8-dihydro-4,5,8(3H)-quinazolinetrione	210, ir, nmr	2430

606

References

In each case, information was collected from the original publication except where an additional reference to *Chemical Abstracts, Chemisches Zentralblatt*, or some other secondary source is included below. Each citation of a Russian journal or of *Angewandte Chemie* refers to the original Russian or German version, not to the respective English translation. The abbreviations for journal titles are those recommended in the *Chemical Abstracts Service Source Index* (American Chemical Society, Washington, DC, 1989) and its quarterly supplements.

1. Y. Kitano, M. Kashiwagi, and Y. Kinoshita, *Acta Crystallogr., Sect. B*, 1972, **28**, 1223.
2. H. E. Künzel, G. D. Wolf, F. Bentz, G. Blankenstein, and G. E. Nischk, *Makromol. Chem.*, 1969, **130**, 103.
3. H. E. Künzel, F. Bentz, G. D. Wolf, D. Blankenstein, and G. E. Nischk, *Makromol. Chem.*, 1970, **138**, 223.
4. D. J. Brown and B. T. England, *Isr. J. Chem.*, 1968, **6**, 569.
5. T. Ishiwaki, N. Ojima, K. Isagawa, and Y. Fushizaki, *Nippon Kagaku Zasshi*, 1969, **90**, 917, A50.
6. T. Ishiwaki, N. Ojima, K. Isagawa, and Y. Fushizaki, *Nippon Kagaku Zasshi*, 1970, **91**, 994.
7. A. M. Shkrob, Y. I. Krylova, V. K. Antonov, and M. M. Shemyakin, *Zh. Obshch. Khim.*, 1968, **38**, 2030.
8. A. V. Kazantsev and L. E. Litovchenko, *Zh. Obshch. Khim.*, 1971, **41**, 1057.
9. U. Agarwala and L. Agarwala, *J. Inorg. Nucl. Chem.*, 1972, **34**, 251.
10. R. D. Brown and B. A. W. Coller, *Theor. Chim. Acta*, 1967, **7**, 259.
11. A. Fukami, *Kogyo Kagaku Zasshi*, 1970, **73**, 1239.
12. M. Matsuoka, H. Tanii, T. Kitao, and K. Konishi, *Kogyo Kagaku Zasshi*, 1970, **73**, 2195.
13. M. Suga, T. Shono, and K. Shinra, *Kogyo Kagaku Zasshi*, 1966, **69**, 1529.
14. E. J. Merrill, *J. Labelled Compd.*, 1969, **5**, 346.
15. R. T. Coutts and K. W. Hindmarsh, *Org. Mass Spectrom.*, 1969, **2**, 681.
16. A. R. Katritzky, Y. Takeuchi, B. Ternai, and G. J. T. Tiddy, *Org. Magn. Reson.*, 1970, **2**, 357.
17. H. W. Smith and H. Rapoport, *J. Am. Chem. Soc.*, 1969, **91**, 6083.
18. G. F. Ford, W. J. Zally, and L. H. Sternbach, *J. Am. Chem. Soc.*, 1967, **89**, 332.
19. A. L. Cossey and J. N. Phillips, *Chem. Ind. (London)*, 1970, 58.
20. H. J. den Hertog and D. J. Buurman, *Recl. Trav. Chim. Pays-Bas*, 1967, **86**, 187.
21. A. Aviram and S. Vromen, *Chem. Ind. (London)*, 1967, 1452.
22. S. Kwee and H. Lund, *Acta Chem. Scand.*, 1971, **25**, 1813.
23. C. Bogentoft, Ö. Ericsson, and B. Danielsson, *Acta Chem. Scand.*, 1971, **25**, 551.

24. J. Lykkeberg and N. A. Klitgaard, *Acta Chem. Scand.*, 1970, **24**, 2268.

25. U. Anthoni, C. Larsen, and P. H. Nielsen, *Acta Chem. Scand.*, 1968, **22**, 1898.

26. A. F. Hegarty and T. C. Bruice, *J. Am. Chem. Soc.*, 1969, **91**, 4924.

27. J. Sauer and K. K. Mayer, *Tetrahedron Lett.*, 1968, 325.

28. C. Kaneko and S. Yamada, *Tetrahedron Lett.*, 1967, 5233.

29. A. M. Shkrob, Y. I. Krylova, V. K. Antonov, and M. M. Shemyakin, *Tetrahedron Lett.*, 1967, 2701.

30. C. Bogentoft, Ö. Ericsson, P. Sternberg, and B. Danielsson, *Tetrahedron Lett.*, 1969, 4745.

31. G. G. de Angelis and H.-J. Hess, *Tetrahedron Lett.*, 1969, 1451.

32. R. R. Schmidt, *Tetrahedron Lett.*, 1968, 3443.

33. M. E. Derieg, J. F. Blount, R. I. Fryer, and S. S. Hillery, *Tetrahedron Lett.*, 1970, 3869.

34. C. G. Allison, R. D. Chambers, J. A. H. MacBride, and W. K. R. Musgrave, *Tetrahedron Lett.*, 1970, 1979.

35. S. C. Pakrashi and J. Bhattacharyya, *Tetrahedron*, 1968, **24**, 1.

36. R. R. Arndt, S. H. Eggers, and A. Jordaan, *Tetrahedron*, 1967, **23**, 3521.

37. E. C. Taylor, A. McKillop, and S. Vromen, *Tetrahedron*, 1967, **23**, 885.

38. E. C. Taylor, A. McKillop, and R. N. Warrener, *Tetrahedron*, 1967, **23**, 891.

39. Y. Hagiwara, M. Kurihara, and N. Yoda, *Tetrahedron*, 1969, **25**, 783.

40. G. Doleschall and K. Lempert, *Tetrahedron*, 1968, **24**, 5529.

41. G. Doleschall and K. Lampert, *Tetrahedron*, 1968, **24**, 5547.

42. P. Beltrame, P. L. Beltrame, and M. Simonetta, *Tetrahedron*, 1968, **24**, 3043.

43. K. Lempert and P. Gyulai, *Tetrahedron*, 1970, **26**, 3443.

44. A. R. Katritzky, M. R. Nesbit, B. I. Kurtev, M. Lyapova, and I. G. Pojarlieff, *Tetrahedron*, 1969, **25**, 3807.

45. G. Doleschall and K. Lempert, *Tetrahedron*, 1969, **25**, 2539.

46. M. Luckner, K. Winter, L. Nover, and J. Reisch, *Tetrahedron*, 1969, **25**, 2575.

47. Y. Mori and J. Tsuji, *Tetrahedron*, 1971, **27**, 3811.

48. M. Witanowski, L. Stefaniak, H. Januszewski, and G. A. Webb, *Tetrahedron*, 1971, **27**, 3129.

49. M. R. Chakrabarty, E. S. Hanrahan, and A. R. Lepley, *Tetrahedron*, 1967, **23**, 2879.

50. D. J. Brown and B. T. England, *Aust. J. Chem.*, 1968, **21**, 2813.

51. E. Cherbuliez, O. Espejo, H. Jindra, and J. Rabinowitz, *Helv. Chim. Acta*, 1967, **50**, 2019.

52. E. Cherbuliez, B. Willhalm, S. Jaccard, and J. Rabinowitz, *Helv. Chim. Acta*, 1967, **50**, 2563.

53. A. E. Siegrist, *Helv. Chim. Acta*, 1967, **50**, 906.

54. E. Cherbuliez, O. Espejo, B. Willhalm, and J. Rabinowitz, *Helv. Chim. Acta*, 1968, **51**, 241.

55. E. Cherbuliez, B. Willhalm, O. Espejo, S. Jaccard, and J. Rabinowitz, *Helv. Chim. Acta*, 1967, **50**, 1440.

56. A. Marxer, U. Salzmann, and F. Hofer, *Helv. Chim. Acta*, 1969, **52**, 2351.

57. U. Stauss, H. P. Härter, M. Neuenschwander, and O. Schindler, *Helv. Chim. Acta*, 1972, **55**, 771.

58. J. Schmitt, M. Langlois, G. Callet, and C. Perrin, *Bull. Soc. Chim. Fr.*, 1969, 2008.

59. C. Denis-Garez, L. Legrand, and N. Lozac'h, *Bull. Soc. Chim. Fr.*, 1969, 3727.

60. H. Gonçalves, F. Mathis, and C. Foulcher, *Bull. Soc. Chim. Fr.*, 1970, 2599.

61. H. Gonçalves, C. Foulcher, and F. Mathis, *Bull. Soc. Chim. Fr.*, 1970, 2615.

62. D. Kiffer, *Bull. Soc. Chim. Fr.*, 1970, 2377.

63. A. Arcoria, *Gazz. Chim. Ital.*, 1968, **98**, 729.

64. G. Caronna and S. Palazzo, *Gazz. Chim. Ital.*, 1968, **98**, 911.

65. A. Arcoria and G. Scarlata, *Gazz. Chim. Ital.*, 1966, **96**, 279.

66. H. H. Hatt, *Aust. J. Chem.*, 1970, **23**, 577.

67. D. E. Rivett and J. F. K. Wilshire, *Aust. J. Chem.*, 1971, **24**, 2717.

68. W. Schäfer, A. Aguado, and U. Sezer, *Angew. Chem.*, 1971, **83**, 442.

69. E. Breitmaier, *Angew. Chem.*, 1971, **83**, 287.

70. A. Albert, *Angew. Chem.*, 1967, **79**, 913.

71. K. Hartke and F. Rossbach, *Angew. Chem.*, 1968, **80**, 83.

72. G. Simchen, G. Entenmann, and R. Zondler, *Angew. Chem.*, 1970, **82**, 548.

73. P. de Mayo and J. J. Ryan, *Can. J. Chem.*, 1967, **45**, 2177.

74. J. W. Bunting and W. G. Meathrel, *Can. J. Chem.*, 1970, **48**, 3449.

75. J. W. Bunting and W. G. Meathrel, *Can. J. Chem.*, 1972, **50**, 917.

76. K. Vaughan, *Can. J. Chem.*, 1972, **50**, 1775.

77. H.-J. Kabbe, *Synthesis*, 1972, 268.

78. E. Wittenburg, *Collect. Czech. Chem. Commun.*, 1971, **36**, 246.

79. Z. Budĕšinský and P. Lederer, *Collect. Czech. Chem. Commun.*, 1972, **37**, 2779.

80. G. Kempter, H.-J. Ziegner, and G. Moser, *Z. Chem.*, 1971, **11**, 12.

81. G. Wagner and S. Leistner, *Z. Chem.*, 1971, **11**, 65.

82. K.-H. Wünsch, I. Krumpholz, J. Perez-Zayas, R. Tápanes-Peraza, and G. Schulze, *Z. Chem.*, 1970, **10**, 113.

83. K.-H. Wünsch and H. Bajdala, *Z. Chem.*, 1970, **10**, 144.

84. K. Lempert and P. Gyulai, *Z. Chem.*, 1970, **10**, 384.

85. W. Ried and K. Johne, *Z. Chem.*, 1970, **10**, 397.

86. G. Wagner and L. Rothe, *Z. Chem.*, 1968, **8**, 22.

87. G. Wagner and L. Rothe, *Z. Chem.*, 1968, **8**, 377.

88. G. Wagner and L. Rothe, *Z. Chem.*, 1969 **9**, 106.

89. G. Wagner and L. Rothe, *Z. Chem.*, 1969, **9**, 446.

90. N. S. Kozlov, V. D. Pak, and N. A. Ivanov, *Zh. Org. Khim.*, 1970, **6**, 1867.

91. V. G. Vodop'yanov, V. G. Golov, and Y. I. Mushkin, *Zh. Org. Khim.*, 1972, **8**, 1000.

92. M. Kurihara and N. Yoda, *Bull. Chem. Soc. Jpn.*, 1966, **39**, 1942.

93. T. Sasaki and M. Takahashi, *Bull. Chem. Soc. Jpn.*, 1968, **41**, 1967.

94. T. Sasaki and T. Yoshioka, *Bull. Chem. Soc. Jpn.*, 1968, **41**, 2206.

95. S. E. Patterson, L. Janda, and L. Strekowski, *J. Heterocycl. Chem.*, 1992, **29**, 703.

96. Y. Hasegawa, Y. Amako, and H. Azumi, *Bull. Chem. Soc. Jpn.*, 1968, **41**, 2608.

97. R. K. Thakkar and S. R. Patel, *Bull. Chem. Soc. Jpn.*, 1969, **42**, 3198.

98. P. N. Bhargava and R. Lakhan, *Bull. Chem. Soc. Jpn.*, 1969, **42**, 1444.

99. T. Ishiwaki, M. Sano, K. Isagawa, and Y. Fushizaki, *Bull. Chem. Soc. Jpn.*, 1970, **43**, 135.

100. S. Tohyama, M. Kurihara, and N. Yoda, *Bull. Chem. Soc. Jpn.*, 1970, **43**, 1246.

101. T. Sasaki, T. Yoshioka, and Y. Suzuki, *Bull. Chem. Soc. Jpn.*, 1971, **44**, 185.

102. J. Keck, *Liebigs Ann. Chem.*, 1967, **707**, 107.

103. W. Ried and J. Valentin, *Liebigs Ann. Chem.*, 1967, **707**, 250.

104. W. Walter and J. Voss, *Liebigs Ann. Chem.*, 1966, **698**, 113.

105. W. Walter and J. Voss, *Liebigs Ann. Chem.*, 1966, **695**, 87.

106. H. Bredereck, G. Simchen, and P. Speh, *Liebigs Ann. Chem.*, 1970, **737**, 39.

107. W. Ried, D. Piechaczek, and E. Vollberg, *Liebigs Ann. Chem.*, 1970, **734**, 13.

108. G. Satzinger, *Liebigs Ann. Chem.*, 1969, **728**, 64.

109. W. Ried and B. Peters, *Liebigs Ann. Chem.*, 1969, **729**, 124.

110. A. Brack, *Liebigs Ann. Chem.*, 1969, **730**, 166.

111. H. Weidinger and H. J. Sturm, *Liebigs Ann. Chem.*, 1968, **716**, 143.

112. L. Capuano and M. Zander, *Liebigs Ann. Chem.*, 1968, **712**, 73.

113. A. P. Bhaduri and N. M. Khanna, *Indian J. Chem.*, 1968, **6**, 174.

114. A. P. Bhaduri and N. M. Khanna, *Indian J. Chem.*, 1966, **4**, 447.

115. P. N. Bhargava and K. S. L. Srivastava, *Indian J. Chem.*, 1968, **6**, 281.

116. A. N. Kaushal, A. P. Taneja, and K. S. Narang, *Indian J. Chem.*, 1969, **7**, 444.

117. C. M. Gupta, A. P. Bhaduri, and N. M. Khanna, *Indian J. Chem.*, 1968, **6**, 758.

118. S. C. Pakrashi, A. De, and S. Chattopadhyay, *Indian J. Chem.*, 1968, **6**, 472.

119. H. Singh, S. L. Jain, V. K. Sharma, and K. S. Narang, *Indian J. Chem.*, 1969, **7**, 765.

120. C. M. Gupta, A. P. Bhaduri, and N. M. Khanna, *Indian J. Chem.*, 1969, **7**, 866.

121. C. M. Gupta, A. P. Bhaduri, and N. M. Khanna, *Indian J. Chem.*, 1969, **7**, 1166.

122. S. C. Pakrashi, J. Bhattacharyya, and A. K. Chakravarty, *Indian J. Chem.*, 1971, **9**, 1220.

123. T. George, R. Tahilramani, and D. V. Mehta, *Indian J. Chem.*, 1971, **9**, 1077.

124. T. George, D. V. Mehta, and R. Tahilramani, *Indian J. Chem.*, 1971, **9**, 755.

125. S. Singh, A. N. Kaushal, A. S. Narang, and K. S. Narang, *Indian J. Chem.*, 1971, **9**, 647.

126. C. M. Gupta, A. P. Bhaduri, N. M. Khanna, and S. K. Mukerjee, *Indian J. Chem.*, 1971, **9**, 201.

127. P. B. Sattur and N. Bhanumati, *Indian J. Chem.*, 1971, **9**, 185.

128. C. M. Gupta, A. P. Bhaduri, and N. M. Khanna, *Indian J. Chem.*, 1970, **8**, 1055.

129. S. K. Modi, H. K. Gakhar, and K. S. Narang, *Indian J. Chem.*, 1970, **8**, 389.

130. A. S. Narang, A. N. Kaushal, H. Singh, and K. S. Narang, *Indian J. Chem.*, 1969, **7**, 1191.

131. P. N. Bhargava and J. Singh, *J. Indian Chem. Soc.*, 1972, **49**, 633.

132. S. Giri and H. Singh, *J. Indian Chem. Soc.*, 1972, **49**, 175.

133. V. S. Patel and S. R. Patel, *J. Indian Chem. Soc.*, 1972, **49**, 59.

134. V. S. Patel and S. R. Patel, *J. Indian Chem. Soc.*, 1971, **48**, 1083.

135. P. S. Satpanthi and J. P. Trivedi, *J. Indian Chem. Soc.*, 1971, **48**, 1021.

136. S. K. P. Sinha, *J. Indian Chem. Soc.*, 1971, **48**, 985.

137. B. D. Singh and S. K. P. Sinha, *J. Indian Chem. Soc.*, 1971, **48**, 743.

138. B. Singh and D. N. Chaudhury, *J. Indian Chem. Soc.*, 1971, **48**, 443.

139. S. K. P. Sinha, *J. Indian Chem. Soc.*, 1971, **48**, 439.

140. S. K. P. Sinha, *J. Indian Chem. Soc.*, 1971, **48**, 432.

141. S. K. P. Sinha and D. N. Chaudhury, *J. Indian Chem. Soc.*, 1970, **47**, 1095.

142. S. K. P. Sinha, P. K. Banerjee, and D. N. Chaudhury, *J. Indian Chem. Soc.*, 1970, **47**, 1089.

143. S. K. P. Sinha and D. N. Chaudhury, *J. Indian Chem. Soc.*, 1970, **47**, 925.

144. B. D. Singh and D. N. Chaudhury, *J. Indian Chem. Soc.*, 1970, **47**, 759.

145. H. J. Mehta, V. S. Patel, and S. R. Patel, *J. Indian Chem. Soc.*, 1970, **47**, 125.

146. B. D. Singh and D. N. Chaudhury, *J. Indian Chem. Soc.*, 1968, **45**, 311.

147. V. S. Patel and S. R. Patel, *J. Indian Chem. Soc.*, 1968, **45**, 167.

148. P. N. Bhargava and G. C. Singh, *J. Indian Chem. Soc.*, 1968, **45**, 70.

149. V. Srivastava and D. N. Chaudhury, *J. Indian Chem. Soc.*, 1969, **46**, 651.

150. A. Chatterjee and R. Raychaudhuri, *J. Indian Chem. Soc.*, 1969, **46**, 103.

151. S. K. P. Sinha and D. N. Chaudhury, *J. Indian Chem. Soc.*, 1969, **46**, 31.

152. B. D. Singh and D. N. Chaudhury, *J. Indian Chem. Soc.*, 1969, **46**, 21.

153. G. A. Swan, *J. Chem. Soc., Chem. Commun.*, 1968, 1376.

154. D. R. Sutherland and G. Tennant, *J. Chem. Soc., Chem. Commun.*, 1969, 423.

155. D. J. Anderson, T. L. Gilchrist, D. C. Horwell, and C. W. Rees, *J. Chem. Soc., Chem. Commun.*, 1969, 146.

156. S. C. Pakrashi and A. K. Chakravarti, *J. Chem. Soc., Chem. Commun.*, 1969, 1443.

157. D. J. Berry, J. D. Cook, and B. J. Wakefield, *J. Chem. Soc., Chem. Commun.*, 1969, 1273.

158. R. D. Chambers, J. A. H. MacBride, and W. K. R. Musgrave, *J. Chem. Soc., Chem. Commun.*, 1970, 739.

159. H. Yamamoto, S. Inaba, M. Nakao, and I. Maruyama, *Chem. Pharm. Bull.*, 1969, **17**, 400.

160. R. Lakhan, *Chem. Pharm. Bull.*, 1969, **17**, 2357.

161. T. Hisano and M. Ichikawa, *Chem. Pharm. Bull.*, 1972, **20**, 163.

162. T. Higashino, H. Ito, and E. Hayashi, *Chem. Pharm. Bull.*, 1972, **20**, 1544.

163. T. Higashino, T. Amano, Y. Tamura, N. Katsumata, Y. Washizu, T. Ono, and E. Hayashi, *Chem. Pharm. Bull.*, 1972, **20**, 1874.

164. E. Kalatzis, *J. Chem. Soc.* (*B*), 1969, 96.

165. G. B. Barlin and A. C. Young, *J. Chem. Soc.* (*B*), 1971, 2323.

166. J. Almog, A. Y. Meyer, and H. Shanin-Atidi, *J. Chem. Soc., Perkin Trans. 2*, 1972, 451.

167. A. R. Katritzky and Y. Takeuchi, *J. Chem. Soc., Perkin Trans. 2*, 1972, 1682.

168. W. L. F. Armarego and J. I. C. Smith, *J. Chem. Soc.* (*B*), 1967, 449.

169. T. J. Brignell, C. D. Johnson, A. R. Katritzky, N. Shakir, H. O. Tarhan, and G. Walker, *J. Chem. Soc.* (*B*), 1967, 1233.

170. J. W. Bunting and D. D. Perrin, *J. Chem. Soc.* (*B*) 1967, 950.

171. W. L. F. Armarego and J. I. C. Smith, *J. Chem. Soc.* (*B*), 1968, 407.

172. T. J. Batterham, A. C. K. Triffett, and J. A. Wunderlich, *J. Chem. Soc.* (*B*), 1967, 892.

173. D. A. Cox and P. E. Cross, *J. Chem. Soc.* (*C*), 1970, 2134.

174. A. W. Murray and K. Vaughan, *J. Chem. Soc.* (*C*), 1970, 2070.

175. A. Albert and H. Yamamoto, *J. Chem. Soc.* (*C*), 1968, 1944.

176. M. E. Derieg, R. I. Fryer, and L. H. Sternbach, *J. Chem. Soc.* (*C*), 1968, 1103.

177. A. L. G. Beckwith and R. J. Hickman, *J. Chem. Soc.* (*C*), 1968, 2756.

178. W. L. F. Armarego, *J. Chem. Soc.* (*C*), 1969, 986.

179. W. L. F. Armarego and T. Kobayashi, *J. Chem. Soc.* (*C*), 1969, 1635.

180. A. Kreutzberger and M. F. G. Stevens, *J. Chem. Soc.* (*C*), 1969, 1282.

181. G. Tennant, *J. Chem. Soc.* (*C*), 1966, 2290.

182. G. Tennant and K. Vaughan, *J. Chem. Soc.* (*C*), 1966, 2287.

183. M. J. Kort and M. Lamchen, *J. Chem. Soc.* (*C*), 1966, 2190.

184. A. Albert and G. Catterall, *J. Chem. Soc.* (*C*), 1967, 1533.

185. P. J. Suter and W. B. Turner, *J. Chem. Soc.* (*C*), 1967, 2240.

186. W. J. Irwin, *J. Chem. Soc., Perkin Trans. 1*, 1972, 353.

187. T. W. M. Spence and G. Tennant, *J. Chem. Soc., Perkin Trans. 1*, 1972, 97.

188. R. A. Bowie and D. A. Thomason, *J. Chem. Soc., Perkin Trans. 1*, 1972, 1842.

189. G. B. Barlin and A. C. Young, *J. Chem. Soc., Perkin Trans. 1*, 1972, 1269.

190. W. L. F. Armarego, *J. Chem. Soc.* (*C*), 1971, 1812.

191. C. W. Rees, T. L. Gilchrist, and E. Stanton, *J. Chem. Soc.* (*C*), 1971, 3036.

192. G. A. Swan, *J. Chem. Soc.* (*C*), 1971, 2880.

193. W. L. F. Armarego and T. Kobayashi, *J. Chem. Soc.* (*C*), 1971, 2502.

194. W. L. F. Armarego and T. Kobayashi, *J. Chem. Soc.* (*C*), 1971, 3222.
195. D. J. Brunswick, M. W. Partridge, and H. J. Vipond, *J. Chem. Soc.* (*C*), 1970, 2641.
196. S. M. Mackenzie and M. F. G. Stevens, *J. Chem. Soc.* (*C*), 1970, 2298.
197. T. L. Gilchrist, C. W. Rees, and E. Stanton, *J. Chem. Soc.* (*C*), 1971, 988.
198. D. J. Berry and B. J. Wakefield, *J. Chem. Soc.* (*C*), 1971, 642.
199. W. L. F. Armarego and T. Kobayashi, *J. Chem. Soc.* (*C*), 1971, 238.
200. D. J. Anderson, T. L. Gilchrist, D. C. Horwell, and C. W. Rees, *J. Chem. Soc.* (*C*), 1970, 576.
201. J. J. Davoll and A. M. Johnson, *J. Chem. Soc.* (*C*), 1970, 997.
202. J. Goerdeler and H. Lüdke, *Chem. Ber.*, 1970, **103**, 3393.
203. R. M. Wagner and C. Jutz, *Chem. Ber.*, 1971, **104**, 2975.
204. R. R. Schmidt, W. Schneider, J. Karg, and U. Burkert, *Chem. Ber.*, 1972, **105**, 1634.
205. H. Lackner, *Chem. Ber.*, 1970, **103**, 2476.
206. L. Capuano, M. Welter, and R. Zander, *Chem. Ber.*, 1970, **103**, 2394.
207. W. König and R. Geiger, *Chem. Ber.*, 1970, **103**, 2024.
208. G. Bonola and E. Sianesi, *Chem. Ber.*, 1969, **102**, 3735.
209. L. Capuano and M. Welter, *Chem. Ber.*, 1968, **101**, 3671.
210. J. Goerdeler and D. Weber, *Chem. Ber.*, 1968, **101**, 3475.
211. L. Capuano, W. Ebner, and J. Schrepfer, *Chem. Ber.*, 1970, **103**, 82.
212. L. Capuano and W. Ebner, *Chem. Ber.*, 1969, **102**, 1480.
213. D. Martin and A. Weise, *Chem. Ber.*, 1967, **100**, 3736.
214. J. Goerdeler and R. Sappelt, *Chem. Ber.*, 1967, **100**, 2064.
215. J. Goerdeler and D. Wieland, *Chem. Ber.*, 1967, **100**, 47.
216. L. Capuano and M. Zander, *Chem. Ber.*, 1966, **99**, 3085.
217. R. F. Smith, *J. Heterocycl. Chem.*, 1966, **3**, 535.
218. S. M. Gadekar and A. M. Kotsen, *J. Heterocycl. Chem.*, 1968, **5**, 129.
219. R. N. Castle and S. Takano, *J. Heterocycl. Chem.*, 1968, **5**, 113.
220. S. C. Bell and P. H. L. Wei, *J. Heterocycl. Chem.*, 1968, **5**, 185.
221. M. Fishman and P. A. Cruickshank, *J. Heterocycl. Chem.*, 1968, **5**, 467.
222. A. M. Felix, J. V. Earley, R. I. Fryer, and L. H. Sternbach, *J. Heterocycl. Chem.*, 1968, **5**, 731.
223. R. W. Lamon, *J. Heterocycl. Chem.*, 1968, **5**, 837.
224. M. G. Stout and R. K. Robins, *J. Heterocycl. Chem.*, 1969, **6**, 89.
225. J. H. Boyer and P. J. A. Frints, *J. Heterocycl. Chem.*, 1970, **7**, 59.
226. R. Pater, *J. Heterocycl. Chem.*, 1970, **7**, 1113.
227. R. L. Jacobs, *J. Heterocycl. Chem.*, 1970, **7**, 1337.
228. G. Zigeuner, C. Knopp, and A. Fuchsgruber, *Monatsh. Chem.*, 1970, **101**, 1827.
229. P. N. Giraldi, A. Fojanesi, G. P. Tosolini, E. Dradi, and W. Logemann, *J. Heterocycl. Chem.*, 1970, **7**, 1429.
230. H. L. Yale, *J. Heterocycl. Chem.*, 1971, **8**, 193.
231. J. P. Chupp, *J. Heterocycl. Chem.*, 1971, **8**, 565.
232. N. D. Heindel and M. C. Chun, *J. Heterocycl. Chem.*, 1971, **8**, 685.
233. R. Pater, *J. Heterocycl. Chem.*, 1971, **8**, 699.
234. J. T. Shaw, D. M. Taylor, F. J. Corbett, and J. D. Ballentine, *J. Heterocycl. Chem.*, 1972, **9**, 125.
235. C. Bogentoft and B. Danielsson, *J. Heterocycl. Chem.*, 1972, **9**, 193.
236. C. B. Schapira and S. Lamdan, *J. Heterocycl. Chem.*, 1972, **9**, 569.
237. M. Oklobdžija, M. Japelj, and T. Fajdiga, *J. Heterocycl. Chem.*, 1972, **9**, 161.

238. E. F. Elslager, J. Clarke, J. Johnson, L. M. Werbel, and J. Davoll, *J. Heterocycl. Chem.*, 1972, **9**, 759.

239. J. L. Lafferty and F. H. Case, *J. Org. Chem.*, 1967, **32**, 1591.

240. A. Stempel, I. Douvan, E. Reeder, and L. H. Sternbach, *J. Org. Chem.*, 1967, **32**, 2417.

241. J. A. Zoltewicz and T. W. Sharpless, *J. Org. Chem.*, 1967, **32**, 2681.

242. R. I. Fryer, J. V. Earley, and L. H. Sternbach, *J. Org. Chem.*, 1967, **32**, 3798.

243. M. G. Stout and R. K. Robins, *J. Org. Chem.*, 1968, **33**, 1219.

244. E. C. Taylor and Y. Shvo, *J. Org. Chem.*, 1968, **33**, 1719.

245. A. Chatterjee and R. Raychaudhuri, *J. Org. Chem.*, 1968, **33**, 2546.

246. N. D. Heindel, V. B. Fish, and T. F. Lemke, *J. Org. Chem.*, 1968, **33**, 3997.

247. H. Ott and M. Denzer, *J. Org. Chem.*, 1968, **33**, 4263.

248. G. F. Field and L. H. Sternbach, *J. Org. Chem.*, 1968, **33**, 4438.

249. M. Denzer and H. Ott, *J. Org. Chem.*, 1969, **34**, 183.

250. B. Roth and J. Z. Strelitz, *J. Org. Chem.*, 1969, **34**, 821.

251. J. A. Moore, G. J. Sutherland, R. Sowerby, E. G. Kelly, S. Palermo, and W. Webster, *J. Org. Chem.*, 1969, **34**, 887.

252. J. F. W. Keana, F. P. Mason, and J. S. Bland, *J. Org. Chem.*, 1969, **34**, 3705.

253. P. A. Cruickshank and M. Fishman, *J. Org. Chem.*, 1969, **34**, 4060.

254. R. Y. Ning, I. Dounan, and L. H. Sternbach, *J. Org. Chem.*, 1970, **35**, 2243.

255. S. C. Pakrashi, *J. Org. Chem.*, 1971, **36**, 642.

256. G. F. Field, W. J. Zally, and L. H. Sternbach, *J. Org. Chem.*, 1971, **36**, 777.

257. M. E. Derieg, R. I. Fryer, S. S. Hillery, W. Metlesics, and G. Silverman, *J. Org. Chem.*, 1971, **36**, 782.

258. N. Finch and H. W. Gschwend, *J. Org. Chem.*, 1971, **36**, 1463.

259. A. Walser, G. Silverman, J. Blount, R. I. Fryer, and L. H. Sternbach, *J. Org. Chem.*, 1971, **36**, 1465.

260. R. A. Scherrer and H. R. Beatty, *J. Org. Chem.*, 1972, **37**, 1681.

261. G. D. Madding, *J. Org. Chem.*, 1972, **37**, 1853.

262. M. Ohoka, S. Yanagida, and S. Komori, *J. Org. Chem.*, 1972, **37**, 3030.

263. T. F. Lemke, H. W. Snady, and N. D. Heindel, *J. Org. Chem.*, 1972, **37**, 2337.

264. S. C. Pakrashi and A. K. Chakravarty, *J. Org. Chem.*, 1972, **37**, 3143.

265. S. C. Bell and G. Conklin, *J. Heterocycl. Chem.*, 1968, **5**, 179.

266. K. Lempert, *Lect. Heterocycl. Chem.*, 1982, **6**, 25.

267. J. M. McCall and R. E. Ten-Brink, *Synthesis*, 1975, 335.

268. J. Schoen and K. Bogdanowicz-Szwed, *Rocz. Chem.*, 1969, **43**, 65; *Chem. Abstr.*, 1969, **70**, 115111.

269. V. Zota, A. Berechet, J. Soare, V. Isbasoiu, E. Grigorescu, and E. Constantinescu, *Farmacia (Bucharest)*, 1966, **14**, 529; *Chem. Abstr.*, 1967, **66**, 115669.

270. G. Doleschall and K. Lempert, *Acta Chim. Acad. Sci. Hung.*, 1966, **48**, 77; *Chem. Abstr.*, 1967, **66**, 104982.

271. V. M. Dziomko, I. S. Markovich, and S. L. Zelichenko, *Tr. Vses. Nauch. Issled. Inst. Khim. Reaktivov Osobo Chist. Khim. Veshchestv.*, 1964 (No. 26), 89; *Chem. Abstr.*, 1967, **67**, 90764.

272. R. I. Moskalenko and G. I. Savel'eva, *Farm.-Zh.*, 1967, **22**, 24; *Chem. Abstr.*, 1967, **67**, 57285.

273. P. Pflegel and G. Wagner, *Pharmazie*, 1967, **22**, 60; *Chem. Abstr.*, 1967, **67**, 7499.

274. G. Wagner and L. Rothe, *Pharm. Inst. Z. Chem.*, 1967., **7**, 339; *Chem. Abstr.*, 1967, **67**, 108623.

275. K. Lempert and K. Zauer, *Acta Chim. Acad. Sci. Hung.*, 1966, **50**, 303; *Chem. Abstr.*, 1968, **68**, 12902.

276. J. Maillard, M. Benard, M. Vincent, V. van Tri, R. Jolly, R. Morin, C. Menillet, and Madame Benharkate, *Chim. Ther.*, 1967, **2**, 202; *Chem. Abstr.*, 1968, **68**, 12949.

277. A. Berechet, *Farmacia (Bucharest)*, 1967, **15**, 287; *Chem. Abstr.*, 1968, **68**, 12950.

278. P. I. Savel'eva and R. I. Moskalenko, *Farm. Zh.*, 1967, **22**, 65; *Chem. Abstr.*, 1968, **68**, 49550.

279. G. Westphal and H. H. Stroh, *Z. Chem.*, 1967, **7**, 456; *Chem. Abstr.*, 1968, **68**, 49547.

280. K. M. Murav'eva, N. V. Arkhangel'skaya, M. N. Shchukina, T. N. Zykova, and G. N. Pershin, *Khim.-Farm. Zh.*, 1967, **1** (No. 8), 29; *Chem. Abstr.*, 1968, **68**, 114543.

281. D. J. Brown and J. A. Hoskins, *Aust. J. Chem.*, 1972, **25**, 2641.

282. G. Scarlata and S. Spitaleri, *Boll. Sedute Accad. Gioenia Sci. Natur. Catania*, 1969, **10**, 43; *Chem. Abstr.*, 1971, **73**, 45447.

283. D. Brutane and A. Y. Strakov, *Latv. PSR Zinat. Akad. Vestis, Kim. Ser.*, 1970, 202; *Chem. Abstr.*, 1970, **73**, 56056.

284. A. Arcoria, S. Fisichella, and D. Sciotto, *Boll. Sedute Accad. Gioenia Sci. Natur. Catania*, 1969, **10**, 68; *Chem. Abstr.*, 1970, **73**, 57125.

285. B. R. Stratham and I. M. Downie, *Loughborough Univ. Technol., Dep. Chem., Sum. Final Year Stud. Proj. Theses*, 1969, **10**, 161; *Chem. Abstr.*, 1970, **73**, 25396.

286. P. Gyulai and K. Lempert, *Period. Polytech. Chem. Eng.*, 1970, **14**, 13; *Chem. Abstr.*, 1970, **73**, 35310.

287. A. Y. Strakov, D. Brutane, and V. D. Deich, *Latv. PSR Zinat. Akad. Vestis, Kim. Ser.*, 1970, 248; *Chem. Abstr.*, 1970, **73**, 35316.

288. S. Johne and D. Gröger, *Pharmazie*, 1970, **25**, 22; *Chem. Abstr.*, 1970, **73**, 4059.

289. G. Vasilev, L. Iliev, and R. Vasilev, *Dokl. Akad. Sel'skokhoz. Nauk Bolg.*, 1969, **2**, 157; *Chem. Abstr.*, 1970, **73**, 3642.

290. A. Berechet, V. Kenyeres-Ursu, and I. Kenyeres-Ursu, *Farmacia (Bucharest)*, 1970, **18**, 285; *Chem. Abstr.*, 1970, **73**, 130962.

291. A. Berechet, I. Kenyeres-Ursu, and V. Kenyeres-Ursu, *Farmacia (Bucharest)*, 1970, **18**, 297; *Chem. Abstr.*, 1970, **73**, 130964.

292. T. Brzozowski and W. Dymek, *Diss. Pharm. Pharmacol.*, 1970, **22**, 117; *Chem. Abstr.*, 1970, **73**, 98888.

293. L. Berezowski and W. Dymek, *Acta Pol. Pharm.*, 1970, **27**, 11; *Chem. Abstr.*, 1970, **73**, 98894.

294. C. Bogentoft and B. Danielsson, *Acta Pharm. Suec.*, 1970, **7**, 257; *Chem. Abstr.*, 1970, **73**, 65611.

295. L. Kronberg, C. Bogentoft, D. Westerlund, B. Danielsson, S. Ljungberg, and L. Palzow, *Acta Pharm. Suec.*, 1970, **7**, 37; *Chem. Abstr.*, 1970, **73**, 66536.

296. K. S. L. Srivastava, *Indian J. Pharm.*, 1970, **32**, 97; *Chem. Abstr.*, 1970, **73**, 77185.

297. D. Brutane, A. Y. Strakov, A. M. Moiseenkov, and A. A. Akhrem, *Latv. PSR Zinat. Akad. Vestis, Kim. Ser.*, 1970, 610; *Chem. Abstr.*, 1971, **74**, 53703.

298. A.-M. A. Samour, A. Marei, and M. H. M. Hussein, *J. Chem. UAR*, 1969, **12**, 451; *Chem. Abstr.*, 1971, **74**, 53702.

299. G. Wagner and L. Rothe, *Pharmazie*, 1970, **25**, 595; *Chem. Abstr.*, 1971, **74**, 22785.

300. J. N. Kashimov, M. V. Telezhenetskaya, Y. V. Rashkes, and S. Y. Yunusov, *Khim. Prir. Soedin.*, 1970, **6**, 453; *Chem. Abstr.*, 1971, **74**, 10342.

301. Y. V. Kozhevnikov, *Izv. Vyssh. Ucheb. Zaved., Khim. Khim. Tekhnol.*, 1970, **13**, 989; *Chem. Abstr.*, 1971, **74**, 13085.

302. D. Brutane, A. Y. Strakov, and I. A. Strakova, *Latv. PSR Zinat. Akad. Vestis, Kim. Ser.*, 1970, 485; *Chem. Abstr.*, 1971, **74**, 13089.

303. S. Fisichella, *Boll. Sedute Accad. Gioenia Sci. Natur. Catania*, 1970, **10**, 318; *Chem. Abstr.*, 1971, **74**, 143291.

304. Y. V. Kozhevnikov, P. A. Petyunin, N. E. Kharchenko, and V. M. Grishina, *Khim.-Farm. Zh.*, 1970, **4**, 25; *Chem. Abstr.*, 1971, **74**, 125606.

305. S. M. Deshpande and A. K. Singh, *Agr. Biol. Chem.*, 1971, **35**, 119; *Chem. Abstr.*, 1971, **74**, 125623.

306. W. Dymek and A. Cygankiewicz, *Diss. Pharm. Pharmacol.*, 1970, **22**, 411; *Chem. Abstr.*, 1971, **74**, 99986.

307. Y. V. Kozhevnikov, P. A. Petyunin, and N. E. Kharchenko, *Khim.-Farm. Zh.*, 1970, **4**, 22; *Chem. Abstr.*, 1971, **74**, 99978.

308. M. Cavello, A. Dini, and F. de Simone, *Rend. Accad. Sci. Fis. Mat., Naples*, 1969, **36**, 61; *Chem. Abstr.*, 1971, **75**, 49011.

309. D. Brutane and A. Y. Strakov, *Latv. PSR Zinat. Akad. Vestis, Kim. Ser.*, 1971, 182; *Chem. Abstr.*, 1971, **75**, 49021.

310. D. N. Rout, P. K. Jesthi, and M. K. Rout, *J. Inst. Chem., Calcutta*, 1971, **43**, 20; *Chem. Abstr.*, 1971, **75**, 63725.

311. Y. V. Kozhevnikov and P. A. Petyunin, *Tr. Perm. Farm. Inst.*, 1969 (No. 3), 25; *Chem. Abstr.*, 1971, **75**, 35933.

312. S. Lamdan and C. H. Gaozza, *An. Assoc. Quim. Argent.*, 1968, **56**, 79; *Chem. Abstr.*, 1971, **75**, 5836.

313. Narang, A. N. Kaushal, B. K. Bahl, and K. S. Narang, *Indian J. Appl. Chem.*, 1970, **33**, 228; *Chem. Abstr.*, 1971, **75**, 20334.

314. A. K. D. Bhavani and P. S. N. Reddy, *Indian J. Chem., Sect. B*, 1992, **31**, 736.

315. S. M. Deshpande and K. R. Reddy, *Indian J. Appl. Chem.*, 1970, **33**, 238; *Chem. Abstr.*, 1971, **75**, 20341.

316. R. N. Pandey, R. N. Sharma, L. M. R. Chaudhary, and P. Sharma, *J. Indian Chem. Soc.*, 1992, **69**, 719; *Chem. Abstr.*, 1993, **119**, 216203.

317. S. Somasekhara, V. S. Dighe, and S. L. Mukherjee, *J. Inst. Chem., Calcutta*, 1971, **43**, 106; *Chem. Abstr.*, 1971, **75**, 140783.

318. Z. Csuros, R. Soos, I. Bitter, and J. Palinkas, *Acta Chim. (Budapest)*, 1971, **69**, 361; *Chem. Abstr.*, 1971, **75**, 129751.

319. G. Wagner and I. Rothe, *Pharmazie*, 1971, **26**, 459; *Chem. Abstr.*, 1971, **75**, 110262.

320. G. Wagner and L. Rothe, *Pharmazie*, 1971, **26**, 456; *Chem. Abstr.*, 1971, **75**, 110267.

321. K. M. Murav'eva, N. V. Arkhangel'skaya, M. N. Shchukina, T. N. Zykova, and G. N. Pershin, *Khim.-Farm. Zh.*, 1971, **5** (No. 6), 25; *Chem. Abstr.*, 1971, **75**, 76721.

322. G. Vasilev, *Khim. Ind. (Sofia)*, 1971, **43**, 259; *Chem. Abstr.*, 1972, **76**, 45891.

323. M. Covello, F. de Simone, N. Sacco, and A. Dini, *Rend. Accad. Sci. Fis. Mat., Naples*, 1970, **37**, 170; *Chem. Abstr.*, 1972, **76**, 46161.

324. L. I. Bigar and V. A. Grin, *Khim. Issled. Farm.*, 1970, 6; *Chem. Abstr.*, 1972, **76**, 34199.

325. R. M. Khachatryan, S. K. Pirenyan, and S. A. Vartanyan, *Arm. Khim. Zh.*, 1971, **24**, 610; *Chem. Abstr.*, 1972, **76**, 3790.

326. A. Fukami and S. Nishizaki, *Mitsubishi Denki Lab. Rep.*, 1971, **12**, 37; *Chem. Abstr.*, 1972, **76**, 127515.

327. S. Nishigaki and T. Moriwaki, *Mitsubishi Denki Lab. Rep.*, 1971, **12**, 75; *Chem. Abstr.*, 1972, **76**, 127890.

328. P. S. N. Reddy and A. K. Bhavani, *Indian J. Chem., Sect. B*, 1992, **31**, 740.

329. C. Bogentoft, Ö. Ericsson, M. Kvist, and B. Danielsson, *Acta Pharm. Suec.*, 1971, **8**, 667; *Chem. Abstr.*, 1972, **76**, 113159.

330. C. Bogentoft, Ö. Ericsson, M. Kvist, and B. Danielsson, *Acta Pharm. Suec.*, 1971, **8**, 639; *Chem. Abstr.*, 1972, **76**, 113165.

331. S. Nagar and S. S. Parmar, *Indian J. Pharm.*, 1971, **33**, 61; *Chem. Abstr.*, 1972, **76**, 94566.

332. G. Wagner and L. Rothe, *Pharmazie*, 1971, **26**, 725; *Chem. Abstr.*, 1972, **76**, 99593.

333. Z. Csuros, R. Soos, I. Bitter, and J. Palinkas, *Acta Chim. (Budapest)*, 1972, **72**, 59; *Chem. Abstr.*, 1972, **76**, 153706.

334. D. Brutane and A. Y. Strakov, *Latv. PSR Zinat. Akad. Vestis, Kim. Ser.*, 1971, 616; *Chem. Abstr.*, 1972, **76**, 14465.

335. E. N. Karanov and G. N. Vasilev, *Dokl. Akad. Sel'skokhoz Nauk Bolg.*, 1971, **4**, 129; *Chem. Abstr.*, 1972, **77**, 74993.

336. K. S. L. Srivastava, *Indian J. Appl. Chem.*, 1971, **34**, 145; *Chem. Abstr.*, 1972, **77**, 48387.

337. H. G. Viehe, T. van Vyve, and Z. Janousek, *Angew. Chem.*, 1972, **84**, 991.

338. U. Ewers, H. Günther, and L. Jaenicke, *Angew. Chem.*, 1975, **87**, 356.

339. M. F. G. Stevens and A. Kreutzberger, *Angew. Chem.*, 1969, **81**, 84.

340. J. de Valk, H. C. van der Plas, F. Jansen, and A. Koudijs, *Recl. Trav. Chim. Pays-Bas*, 1973, **92**, 460.

341. J. Almog and E. D. Bergmann, *Isr. J. Chem.*, 1973, **11**, 723.

342. I. Y. Postovskii, O. N. Chupakhin, T. L. Pilicheva, and Y. Y. Popelis, *Dokl. Akad. Nauk SSSR*, 1973, **212**, 1125.

343. A. P. Kroon and H. C. van der Plas, *Recl. Trav. Chim. Pays-Bas*, 1971, **93**, 227.

344. J. S. Walia, S. N. Bannore, A. S. Walia, and L. Guillot, *Chem. Lett.*, 1974, 1005.

345. N. Latif, I. F. Zeid, N. Mishriky, and F. M. Assad, *Tetrahedron Lett.*, 1974, 1355.

346. F. Yoneda, M. Higuchi, and R. Nonaka, *Tetrahedron Lett.*, 1973, 359.

347. M. Srinivasan and J. B. Rampal, *Tetrahedron Lett.*, 1974, 2883.

348. T. P. Forrest, G. A. Dauphinee, and F. M. F. Chen, *Can. J. Chem.*, 1974, **52**, 2725.

349. S. Kwon, F. Ikeda, and K. Isagawa, *Nippon Kagaku Kaishi*, 1973, 1944.

350. N. Ishikawa and T. Muramatsu, *Nippon Kagaku Kaishi*, 1973, 563.

351. M. Takahashi, S. Onizawa, and R. Shioda, Nippon Kagaku Kaishi, 1972, 1259.

352. Y. Yamada, T. Oine, and I. Inoue, *Bull. Chem. Soc. Jpn.*, 1974, **47**, 339.

353. Y. Yamada, T. Oine, and I. Inoue, *Bull. Chem. Soc. Jpn.*, 1974, **47**, 343.

354. P. N. Bhargava and V. N. Choubey, *Indian J. Appl. Chem.*, 1971, **34**, 113; *Chem. Abstr.*, 1972, **77**, 48388.

355. I. Mikami, Y. Hara, and Y. Usui, *Takeda Kenkyusho Ho*, 1972, **31**, 11; *Chem. Abstr.*, 1972, **77**, 48419.

356. Y. V. Kozhevnikov, *Izv. Vyssh. Ucheb. Zaved., Khim. Khim. Tekhnol.*, 1971, **14**, 1685; *Chem. Abstr.*, 1972, **77**, 19599.

357. Y. V. Kozhevnikov, P. A. Petyunin, and I. S. Berdinskii, *Uch. Zap. Perm. Univ.*, 1970 (No. 229), 270; *Chem. Abstr.*, 1972, **77**, 139995.

358. A. K. Chaturvedi and S. S. Parmar, *Indian J. Pharm.*, 1972, **34**, 72; *Chem. Abstr.*, 1972, **77**, 139963.

359. A. Sammour, M. I. B. Selim, and M. A. Abdo, *UAR J. Chem.*, 1971, **14**, 197; *Chem. Abstr.*, 1972, **77**, 114352.

360. R. Kumar, T. K. Gupta, and S. S. Parmar, *Indian J. Pharm.*, 1971, **33**, 108; *Chem. Abstr.*, 1972, **77**, 126550.

361. K. Meguro, H. Natsugari, H. Tawada, and Y. Kuwada, *Chem. Pharm. Bull.*, 1973, **21**, 2366.

362. M. Khalife-El-Saleh, G. Pastor, C. Montginoul, E. Torreilles, L. Giral, and A. Texier, *Bull. Soc. Chim. Fr.*, 1974, 1667.

363. J. Gilbert, *Bull. Soc. Chim. Fr.*, 1974, 2261.

364. M. Hedayatullah, J. Pailler, and L. Denivelle, *Bull. Soc. Chim. Fr.*, 1974, 2495.

365. T. Higashino, Y. Nagano, and E. Hayashi, *Chem. Pharm. Bull.*, 1973, **21**, 1943.

366. K. Meguro and Y. Kuwada, *Chem. Pharm. Bull.*, 1973, **21**, 2375.

367. K. Meguro, H. Towada, H. Miyano, Y. Sato, and Y. Kuwada, *Chem. Pharm. Bull.*, 1973, **21**, 2382.

368. N. Hirose, S. Kuriyama, S. Sohda, K. Sakaguchi, and H. Yamamoto, *Chem. Pharm. Bull.*, 1973, **21**, 1005.

369. K. Meguro, H. Tawada, and Y. Kuwada, *Chem. Pharm. Bull.*, 1973, **21**, 1619.

370. T. Hisano, M. Ichikawa, G. Kito, and T. Nishi, *Chem. Pharm. Bull.*, 1972, **20**, 2575.

371. T. Yabuuchi, M. Hisake, M. Matuda, and R. Kimura, *Chem. Pharm. Bull.*, 1975, **23**, 663.

372. T. Higashino, K. Suzuki, and E. Hayashi, *Chem. Pharm. Bull.*, 1975, **23**, 746.

373. T. Higashino, M. Goi, and E. Hayashi, *Chem. Pharm. Bull.*, 1974, **22**, 2493.

374. F. Takami, K. Tokuyama, S. Wakahara, Y. Fukui, and T. Maeda, *Chem. Pharm. Bull.*, 1974, **22**, 267.

375. T. Hisano and Y. Yabuta, *Chem. Pharm. Bull.*, 1974, **22**, 316.

376. Y. Yamada, T. Oine, and I. Inoue, *Chem. Pharm. Bull.*, 1974, **22**, 601.

377. V. N. Choubey, *UAR J. Chem.*, 1971, **14**, 407; *Chem. Abstr.*, 1972, **77**, 88423.

378. A. Y. Strakov, D. Brutane, and M. Sulca, *Latv. PSR Zinat. Akad. Vestis, Kim. Ser.*, 1972, 267; *Chem. Abstr.*, 1972, **77**, 88457.

379. E. Budeanu and E. Tanase, *An. Stiint. Univ. "Al. I. Cuza" Iasi, Sect. 1c*, 1972, **18**, 73; *Chem. Abstr.*, 1973, **78**, 29704.

380. V. G. Reddy and P. S. N. Reddy, *Indian J. Chem., Sect. B*, 1992, **31**, 764.

381. J. B. Taylor, D. R. Harrison, and F. Fried, *J. Heterocycl. Chem.*, 1972, **9**, 1227.

382. A. Rosowsky and N. Papathanasopoulos, *J. Heterocycl. Chem.*, 1972, **9**, 1235.

383. A. Rosowsky, N. Papathanasopoulos, and E. J. Modest, *J. Heterocycl. Chem.*, 1972, **9**, 1449.

384. L. J. Chinn, *J. Heterocycl. Chem.*, 1973, **10**, 403.

385. R. M. Picirilli and F. D. Popp, *J. Heterocycl. Chem.*, 1973, **10**, 671.

386. S. Palazzo and L. I. Giannola, *J. Heterocycl. Chem.*, 1973, **10**, 675.

387. Y. V. Kozhevnikov and N. V. Pilat, *Tr. Perm. Sel.-Khoz. Inst.*, 1971 (No. 79), 66; *Chem. Abstr.*, 1973, **78**, 16128.

388. M. H. Nosseir, N. N. Messiha, and G. G. Gabra, *UAR J. Chem.*, 1970, **13**, 379; *Chem. Abstr.*, 1972, **77**, 152103.

389. R. Mazurkiewicz, *Rocz. Chem.*, 1972, **46**, 971; *Chem. Abstr.*, 1972, **77**, 164627.

390. L. Zhelyazkov, R. Kolchagova, D. Stefanova, and L. Daleva, *Tr. Nauchnoizsled. Khim.-Farm. Inst.*, 1972, **8**, 29; *Chem. Abstr.*, 1973, **78**, 147893.

391. R. Kolchagova and L. Zhelyazkov, *Tr. Nauchnoizsled. Khim.-Farm. Inst.*, 1972, **8**, 121; *Chem. Abstr.*, 1973, **78**, 147902.

392. L. D. Joshi, R. R. Arora, and S. S. Parmar, *Indian J. Pharm.*, 1972, **34**, 112; *Chem. Abstr.*, 1973, **78**, 92406.

393. G. G. Glukhovets and B. I. Ardashev, *Tr. Novocherkassk. Politekh. Inst.*, 1972 (No. 266), 159; *Chem. Abstr.*, 1973, **78**, 97588.

394. S. Somasekhara, V. S. Dighe, and S. V. Gokhale, *Indian J. Pharm.*, 1972, **34**, 121; *Chem. Abstr.*, 1973, **78**, 72047.

395. F. Gatta, R. Landi-Vittory, M. Tomassetti, and G. Nunez-Barrios, *Chim. Ther.*, 1972, **7**, 480; *Chem. Abstr.*, 1973, **78**, 72086.

396. A. Y. Strakov, M. B. Andaburskaya, A. M. Moiseenkov, and A. A. Akhrem, *Latv. PSR Zinat. Akad. Vestis, Kim. Ser.*, 1973, 330; *Chem. Abstr.*, 1973, **79**, 92147.

397. Y. S. Tsizin, N. B. Karpova, A. V. Luk'yanov, E. A. Rudzit, D. Kulikova, and T. P. Radkevich, *Khim.-Farm. Zh.*, 1973, **7** (No. 7), 16; *Chem. Abstr.*, 1973, **79**, 92149.

398. J. G. Smith and J. M. Sheepy, *J. Heterocycl. Chem.*, 1975, **12**, 231.

399. A. Metallidis, A. Sotiriadis, and D. Theodoropoulos, *J. Heterocycl. Chem.*, 1975, **12**, 359.

400. M. W. Barker and J. D. Rosamond, *J. Heterocycl. Chem.*, 1974, **11**, 241.

401. S. Plescia, E. Ajello, V. Sprio, and M. L. Marino, *J. Heterocycl. Chem.*, 1974, **11**, 603.

402. A. Walser, R. I. Fryer, L. H. Sternbach, and M. C. Archer, *J. Heterocycl. Chem.*, 1974, **11**, 619.

403. Y. Tamura, Y. Miki, J. Minamikawa, and M. Ikeda, *J. Heterocycl. Chem.*, 1974, **11**, 675.

404. A. J. Hubert, *J. Heterocycl. Chem.*, 1974, **11**, 737.

405. D. L. Trepanier, S. Sunder, and W. H. Braun, *J. Heterocycl. Chem.*, 1974, **11**, 747.

406. R. A. Abramovitch and J. Kalinowski, *J. Heterocycl. Chem.*, 1974, **11**, 857.

407. A. Walser, T. Flynn, and R. I. Fryer, *J. Heterocycl. Chem.*, 1974, **11**, 885.

408. K. A. K. Ebraheem, G. A. Webb, and M. Witanowski, *Org. Magn. Reson.*, 1978, **11**, 27.

409. W. L. F. Armarego, *Int. Rev. Sci., Org. Chem.*, Ser. 1, 1973, **4**, 153; Ser. 2, 1975, **4**, 160.

410. G. Stajer, J. Pintye, A. E. Szabo, and F. Klivenyi, *Acta Chim. (Budapest)*, 1973, **77**, 81; *Chem. Abstr.*, 1973, **79**, 66296.

411. G. Subrahmanyam and T. D. Roy, *Tetrahedron*, 1973, **29**, 3173.

412. R. Fuks, *Tetrahedron*, 1973, **29**, 2153.

413. W. Abraham and G. Barnikow, *Tetrahedron*, 1973, **29**, 691.

414. A. R. Katritzky and J. W. Suwinski, *Tetrahedron*, 1975, **31**, 1549.

415. T. L. Pilicheva, I. Y. Postovskii, O. N. Chupakhin, N. A. Klyuev, and V. I. Chernyi, *Dokl. Akad. Nauk SSSR*, 1974, **218**, 1375.

416. E. C. Taylor and S. F. Martin, *J. Am. Chem. Soc.*, 1974, **96**, 8095.

417. D. J. Brown and K. Ienaga, *Heterocycles*, 1975, **3**, 283.

418. W. L. F. Armarego and B. A. Milloy, *J. Chem. Soc., Perkin Trans. 1*, 1973, 2814.

419. A. Albert and W. Pendergast, *J. Chem. Soc., Perkin Trans. 1*, 1973, 1794.

420. M. A. Mikhaleva, S. A. Romanovskaya, K. M. Belova, V. F. Sedova, and V. P. Mamaev, *Zh. Org. Khim.*, 1974, **10**, 859.

421. M. W. Partridge and A. Smith, *J. Chem. Soc., Perkin Trans. 1*, 1973, 453.

422. D. J. Brown and K. Ienaga, *J. Chem. Soc., Perkin Trans. 1*, 1974, 372.

423. W. L. F. Armarego and P. A. Reece, *J. Chem. Soc., Perkin Trans. 1*, 1974, 2313.

424. B. C. Uff, J. R. Kershaw, and S. R. Chhabra, *J. Chem. Soc., Perkin Trans. 1*, 1974, 1146.

425. R. W. Baldock, P. Hudson, A. R. Katritzky, and F. Soti, *J. Chem. Soc., Perkin Trans. 1*, 1974, 1422.

426. D. R. Sutherland and G. Tennant, *J. Chem. Soc., Perkin Trans. 1*, 1974, 534.

427. M. F. G. Stevens, *J. Chem. Soc., Perkin Trans. 1*, 1974, 615.

428. W. Ried, N. Kothe, and W. Merkel, *Chem. Ber.*, 1975, **108**, 181.

429. W. Merkel and W. Ried, *Chem. Ber.*, 1973, **106**, 471.

430. J. Goerdeler and M. Bischoff, *Chem. Ber.*, 1972, **105**, 3566.

431. E. C. Taylor, M. L. Chittenden, and S. F. Martin, *Heterocycles*, 1973, **1**, 59.

432. B. M. Adger, S. Bradbury, M. Keating, C. W. Rees, R. C. Storr, and M. T. Williams, *J. Chem. Soc., Perkin Trans. 1*, 1975, 31.

433. M. S. Manhas, W. A. Hoffman, B. Lal, and A. K. Bose, *J. Chem. Soc., Perkin Trans. 1*, 1975, 461.

434. A. Butt and R. Parveen, *Pak. J. Sci. Ind. Res.*, 1972, **15**, 243; *Chem. Abstr.*, 1973, **79**, 66299.

435. G. Caille, J. Braun, D. Gravel, and R. Plourde, *Can. J. Pharm. Sci.*, 1973, **8**, 42; *Chem. Abstr.*, 1973, **79**, 70285.

436. G. Malesani, F. Marcolin, and G. Rodighiero, *Proc. 7th Int. Congr. Chemother.* Vol. 1, Part 2, (M. Hejzlar, Ed.), University Park Press, Baltimore, 1972, p. 1081: *Chem. Abstr.*, 1973, **79**, 53257.

437. W. Ried, W. Merkel, and O. Mösinger, *Liebigs Ann. Chem.*, 1973, 1362.

438. K. D. Augart, G. Kresze, and N. Schönberger, *Liebigs Ann. Chem.*, 1973, 1457.

439. W. Ried and W. Merkel, *Liebigs Ann. Chem.*, 1973, 122.

440. K. Hartke, F. Rossbach, and M. Radau, *Liebigs Ann. Chem.*, 1972, **762**, 167.

441. H. Kohl and E. Wolf, *Liebigs Ann. Chem.*, 1972, **766**, 106.

442. W. Walter, T. Fleck, J. Voss, and M. Gerwin, *Liebigs Ann. Chem.*, 1975, 275.

443. H. Röchling and G. Hörlein, *Liebigs Ann. Chem.*, 1974, 504.

444. B. Gaux and P. Le Henaff, *Bull. Soc. Chim. Fr.*, 1974, 900.

445. B. Danieli and G. Palmisano, *Gazz. Chim. Ital.*, 1975, **105**, 99.

446. M. A. F. Elkaschef, F. M. E. Abdel-Megeid, and A. Abdel-Kader, *Collect. Czech. Chem.Commun.*, 1974, **39**, 287.

447. C. Bogentoft, U. Bondesson, Ö. Ericsson, and B. Danielsson, *Acta Chem. Scand., Ser. B*, 1974, **28**, 479.

448. E. Denis, P. van Brandt, and A. Bruylants, *Bull. Cl. Sci., Acad. R. Belg.*, 1972, **58**, 1076; *Chem. Abstr.*, 1973, **79**, 136136.

449. A. Sammour, M. I. Selim, M. Elkasaby, and M. Abdalla, *Acta Chim.* (*Budapest*), 1973, **78**, 293; *Chem. Abstr.*, 1973, **79**, 137061.

450. G. Stajer, K. Kottke, and R. Pohloudek-Fabini, *Pharmazie*, 1973, **28**, 433; *Chem. Abstr.*, 1973, **79**, 105178.

451. D. F. Barringer, G. Berkelhammer, S. D. Carter, L. Goldman, and A. E. Lanzilotti, *J. Org. Chem.*, 1973, **38**, 1933.

452. D. F. Barringer, G. Berkelhammer, and R. S. Wayne, *J. Org. Chem.*, 1973, **38**, 1937.

453. E. P. Papadopoulos, *J. Org. Chem.*, 1973, **38**, 667.

454. G. N. Walker, A. R. Engle, and R. J. Kempton, *J. Org. Chem.*, 1972, **37**, 3755.

455. A. Sammour, M. I. Selim, A. F. M. Fahmy, and K. Elewa, *Indian J. Chem.*, 1973, **11**, 437.

456. T. P. Murray, J. V. Hay, D. E. Portlock, and J. F. Wolfe, *J. Org. Chem.*, 1974, **39**, 595.

457. K.-Y. Tserng and L. Bauer, *J. Org. Chem.*, 1973, **38**, 3498.

458. L. Avram, I. Ursu-Kenyeres, V. Ursu-Kenyeres, and M. Neacsu, *Farmacia* (*Bucharest*), 1974, **22**, 93; *Chem. Abstr.*, 1974, **81**, 152144.

459. G. F. Dregval, R. I. Rozvaga, and V. A. Golub, *Ukr. Khim. Zh.* (*Russ. Ed.*), 1974, **40**, 512; *Chem. Abstr.*, 1974, **81**, 120568.

460. A. Lespagnol, C. Lespagnol, and J. L. Bernier, *Ann. Pharm. Fr.*, 1974, **32**, 125; *Chem. Abstr.*, 1974, **81**, 105430.

461. V. S. Misra and S. Prakash, *Indian J. Pharm.*, 1974, **36**, 142; *Chem. Abstr.*, 1975, **82**, 140060.

462. Y. V. Kozhevnikov, *Izv. Vyssh. Uchebn. Zaved., Khim. Khim. Tekhnol.*, 1974, **17**, 1877; *Chem. Abstr.*, 1975, **82**, 140064.

463. S. Iacobescu-Cilianu, D. Beiu, M. Lazarescu, G. Neubauer, C. Ilie, and A. Ciuceanu, *Rev. Chim.* (*Bucharest*), 1974, **25**, 869; *Chem. Abstr.*, 1975, **82**, 125374.

464. O. Gaspar, *Farmacia* (*Bucharest*), 1974, **22**, 241; *Chem. Abstr.*, 1975, **82**, 30514.

465. G. Pifferi, L. F. Zerilli, G. Tuan, and P. Consonni, *Chim. Ind.* (*Milan*), 1974, **56**, 492; *Chem. Abstr.*, 1975, **82**, 31302.

466. K. Nagraba, *Zesz. Nauk. Univ. Jagiellon, Pr. Chem.*, 1974, **19**, 81; *Chem. Abstr.*, 1975, **82**, 43308.

467. Y. V. Kozhevnikov, *Izv. Vyssh. Uchebn. Zaved., Khim. Khim. Tekhnol.*, 1974, **17**, 1541; *Chem. Abstr.*, 1975, **82**, 43312.

468. S. M. Roshdy, K. M. Ghoneim, and M. Khalifa, *Pharmazie*, 1975, **30**, 210; *Chem. Abstr.*, 1975, **83**, 58742.

469. M. E. Sitzmann and J. C. Dacons, *J. Org. Chem.*, 1973, **38**, 4363.

470. K. Ishizumi, S. Inaba, and H. Yamamoto, *J. Org. Chem.*, 1973, **38**, 2617.

471. T.-C. Lee, G. Salemnick, and G. B. Brown, *J. Org. Chem.*, 1973, **38**, 3102.

472. J. H. Maguire and R. L. McKee, *J. Org. Chem.*, 1974, **39**, 3434.

473. F. Claudi, P. Franchetti, M. Grifantini, and S. Martelli, *J. Org. Chem.*, 1974, **39**, 3508.

474. G. E. Hardtmann and H. Ott, *J. Org. Chem.*, 1974, **39**, 3599.

475. S. C. Pakrashi and A. K. Chakravarty, *J. Org. Chem.*, 1974, **39**, 3828.

476. L. D. Dave and M. U. Cyriac, *J. Indian Chem. Soc.*, 1974, **51**, 383.

477. S. K. P. Sinha and M. P. Thakur, *J. Indian Chem. Soc.*, 1974, **51**, 453.

478. S. K. P. Sinha and M. P. Thakur, *J. Indian Chem. Soc.*, 1974, **51**, 457.

479. M. P. Thakur and S. K. P. Sinha, *J. Indian Chem. Soc.*, 1972, **49**, 1185.

480. M. D. Nair, *Indian J. Chem.*, 1973, **11**, 109.

481. S. C. Pakrashi and A. K. Chakravarty, *Indian J. Chem.*, 1973, **11**, 122.

482. G. Kissling-Guerin, D. C. Luu, and A. Boucherle, *Bull. Trav. Soc. Pharm. Lyon*, 1973, **17**, 143; *Chem. Abstr.*, 1975, **82**, 170296.

483. D. M. Patel, A. L. Visalli, J. J. Zalipsky, and N. H. Reavey-Cantwell, *Anal. Profiles Drug Subst.*, 1975, **4**, 245; *Chem. Abstr.*, 1975, **83**, 168342.

484. M. J. Cho and I. H. Pitman, *J. Am. Chem. Soc.*, 1974, **96**, 1843.

485. A. R. E. Ossman and M. Khalifa, *Pharmazie*, 1975, **30**, 256; *Chem. Abstr.*, 1975, **83**, 97185.

486. M. N. Aboul-Enein and A. I. Eid, *Pharm. Acta Helv.*, 1974, **49**, 293; *Chem. Abstr.*, 1975, **83**, 97192.

487. A. Sammour, A. F. M. Fahmy, and M. Mahmoud, *Indian J. Chem.*, 1973, **11**, 222.

488. R. N. Iyer and R. Gopalchari, *Indian J. Chem.*, 1973, **11**, 234.

489. K. C. Joshi and V. K. Singh, *Indian J. Chem.*, 1973, **11**, 430.

490. J. B. Hester, *J. Org. Chem.*, 1974, **39**, 2137.

491. E. P. Papadopoulos, *J. Org. Chem.*, 1974, **39**, 2540.

492. K. Ishizumi, S. Inaba, and H. Yamamoto, *J. Org. Chem.*, 1974, **39**, 2581.

493. K. Ishizumi, S. Inaba, and H. Yamamoto, *J. Org. Chem.*, 1974, **39**, 2587.

494. A. F. Hegarty, L. N. Frost, and J. H. Coy, *J. Org. Chem.*, 1974, **39**, 1089.

495. N. P. Peet and S. Sunder, *J. Org. Chem.*, 1975, **40**, 1909.

496. M. Ghelardoni and F. Russo, *Boll. Chim. Farm.*, 1967, **106**, 688; *Chem. Abstr.*, 1968, **68**, 105148.

497. J. Maillard, M. Vincent, M. Benard, V. van Tri, R. Jolly, R. Morin, C. Menillet, and Madame Benharkate, *Chim. Ther.*, 1968, **3**, 100; *Chem. Abstr.*, 1968, **69**, 34529.

498. J. Maillard, M. Benard, M. Vincent, V. van Tri, R. Jolly, R. Morin, Madame Benharkate, and C. Menillet, *Chim. Ther.*, 1967, **2**, 231; *Chem. Abstr.*, 1968, **69**, 36066.

499. S. Tanimoto, S. Shimojo, and R. Oda, *Yuki Gosei Kagaku Shi*, 1968, **26**, 151; *Chem. Abstr.*, 1968, **69**, 36089.

500. C. Pelizzi and G. Pelizzi, *Gazz. Chim. Ital.*, 1975, **105**, 7.

501. N. J. Leonard, A. G. Morrice, and M. A. Sprecker, *J. Org. Chem.*, 1975, **40**, 356.

502. A. G. Morrice, M. A. Sprecker, and N. J. Leonard, *J. Org. Chem.*, 1975, **40**, 363.

503. J. G. Patel, B. H. Bhide, and S. R. Patel, *J. Indian Chem. Soc.*, 1974, **51**, 674.

504. M. I. Husain and S. K. Agarwal, *J. Indian Chem. Soc.*, 1974, **51**, 1015.

505. M. P. Thakur and S. K. P. Sinha, *Indian J. Chem.*, 1973, **11**, 500.

506. R. S. Pandit and S. Seshadri, *Indian J. Chem.*, 1973, **11**, 532.

507. M. N. Deshpande and S. Seshadri, *Indian J. Chem.*, 1973, **11**, 538.

508. S. K. Modi, S. Singh, and K. S. Narang, *Indian J. Chem.*, 1972, **10**, 605.

509. R. Hurmer and J. Vernin, *Therapie*, 1967, **22**, 1325; *Chem. Abstr.*, 1968, **69**, 42690.

510. W. Dymek, B. Lubimowski, and S. Karwat, *Diss. Pharm. Pharmacol.*, 1968, **20**, 29; *Chem. Abstr.*, 1968, **69** 27376.

511. W. Dymek and B. Lucka-Sobstel, *Diss. Pharm. Pharmacol.*, 1968, **20**, 43; *Chem. Abstr.*, 1968, **69**, 27372.

512. J. Schoen and K. Bogdanowicz-Szwed, *Rocz. Chem.*, 1967, **41**, 1903; *Chem. Abstr.*, 1968, **69**, 10415.

513. E. E. Schweizer and S. V. de Voe, *J. Org. Chem.*, 1975, **40**, 144.

514. S. M. Deshpande and K. R. Reddy, *J. Indian Chem. Soc.*, 1975, **52**, 55.

515. M. Z. A. Badr, H. A. H. El-Sherief, and M. M. Aly, *Indian J. Chem.*, 1975, **13**, 245.

516. G. C. Amin and V. T. Soni, *Indian J. Chem.*, 1975, **13**, 303.

517. N. N. Messiha, A. M. M. Abdel-Kader, and M. H. Nosseir, *Indian J. Chem.*, 1975, **13**, 326.

518. P. N. Bhargava and R. Shyam, *Indian J. Chem.*, 1974, **12**, 779.

519. H. A. Zaher, H. Jahine, Y. Akhnookh, and Z. El-Gendy, *Indian J. Chem.*, 1974, **12**, 1212.

520. N. N. Messiha, N. L. Doss, and M. H. Nosseir, *Indian J. Chem.*, 1973, **11**, 738.

521. H. Singh, K. Lal, and S. Singh, *Indian J. Chem.*, 1973, **11**, 750.

522. M. R. Chaurasia, *Agr. Biol. Chem. (Tokyo)*, 1968, **32**, 711; *Chem. Abstr.*, 1968, **69**, 77225.

523. B. Danielsson, L. Kronberg, and F. Ljungner, *Acta Pharm. Suec.*, 1968, **5**, 77; *Chem. Abstr.*, 1968, **69**, 77223.

524. Y. V. Kozhevnikov and P. A. Petyunin, *Nauch. Tr. Perm. Farma. Inst.*, 1967 (No. 2), 37; *Chem. Abstr.*, 1968, **69**, 77222.

525. P. A. Petyunin, Y. V. Kozhevnikov, and I. S. Berdinskii, *Uch. Zap., Perm. Gos. Univ.*, 1966 (No. 141), 309; *Chem. Abstr.*, 1968, **69**, 77226.

526. W. T. Smith and J. T. Sellas, *Chim. Ther.*, 1967, **2**, 148; *Chem. Abstr.*, 1968, **68**, 39584.

527. I. F. Tupitsyn, N. N. Zatsepina, A. V. Kirova, and Y. M. Kapustin, *Reakts. Sposobnost Org. Soedin.*, 1968, **5**, 601; *Chem. Abstr.*, 1969, **70**, 76940.

528. I. F. Tupitsyn, N. N. Zatsepina, and A. V. Kirova, *Reakts. Sposobnost Org. Soedin.*, 1968, **5**, 626; *Chem. Abstr.*, 1969, **70**, 76942.

529. P. N. Bhargava and K. S. L. Srivastava, *Allg. Prakt. Chem.*, 1968, **19**, 424; *Chem. Abstr.*, 1969, **70**, 87738.

530. S. Gronowitz, J. Fortea-Laguna, S. Ross, B. Sjoberg, and N. E. Stjernstrom, *Acta Pharm. Suec.*, 1968, **5**, 563; *Chem. Abstr.*, 1969, **70**, 87745.

531. L. Berezowski and B. Lubimowski, *Acta Pol. Pharm.*, 1968, **25**, 473; *Chem. Abstr.*, 1969, **70**, 68304.

532. P. Sohar and I. Kosa, *Acta Chim. (Budapest)*, 1968, **57**, 411; *Chem. Abstr.*, 1969, **70**, 4024.

533. D. Brutane and A. Y. Strakov, *Latv. PSR Zinat. Akad. Vestis, Kim. Ser.*, 1969, 248; *Chem. Abstr.*, 1969, **71**, 61332.

534. T. E. Snider and K. D. Berlin, *Phosphorus*, 1972, **2**, 43.

535. C. H. Budeanu, *An. Stiint. Univ. "Al. I. Cuza" Iasi, Sect. 1c*, 1968, **14**, 173; *Chem. Abstr.*, 1969, **71**, 61329.

536. M. Moreno and J. Marcial, *An. Quim.*, 1969, **55**, 175; *Chem. Abstr.*, 1969, **71**, 3072.

537. M. H. Nosseir and N. N. Messiha, *J. Chem. UAR*, 1969, **12**, 57; *Chem. Abstr.*, 1969, **71**, 3354.

538. H. Singh and K. B. Lal, *Indian J. Chem.*, 1973, **11**, 959.

539. H. Singh, V. K. Vij, and K. B. Lal, *Indian J. Chem.*, 1973, **11**, 966.

540. I. Y. Postovskii, O. N. Chupakhin, T. L. Pilicheva, N. A. Klyuev, and S. L. Mertsalov, *Zh. Org. Khim.*, 1975, **11**, 875.

541. R. Neidlein and H. G. Hege, *Synthesis*, 1975, 50.

542. T. M. Paterson, R. K. Smalley, and H. Suschitzky, *Synthesis*, 1975, 187.

543. G. Fischer and M. Puza, *Synthesis*, 1973, 218.

544. A. Yoshitake, Y. Makari, K. Kawahara, and M. Endo, *J. Labelled Compd.*, 1973, **9**, 537.

545. B. A. Corradi, C. G. Palmieri, M. Nardelli, and C. Pelizzi, *J. Chem. Soc., Dalton Trans.*, 1974, 150.

546. Y. Kobayashi, I. Kumadaki, H. Wato, Y. Sekine, and T. Hara, *Chem. Pharm. Bull.*, 1974, **22**, 2097.

547. S. Leistner and G. Wagner, *Z. Chem.*, 1973, **13**, 428.

548. Ö. Ericsson, C. Bogentoft, C. Lindberg, and B. Danielsson, *Acta Pharm. Suec.*, 1973, **10**, 257; *Chem. Abstr.*, 1973, **79**, 137080.

549. A. Y. Strakov and D. Brutane, *Latv. PSR Zinat. Akad. Vestis, Kim. Ser.*, 1973, 225; *Chem. Abstr.*, 1973, **79**, 126438.

550. S. Palazzo and K. I. Giannola, *Atti Accad. Sci., Lett. Arti Palermo, Parte 1*, 1972, **31**, 77; *Chem. Abstr.*, 1973, **79**, 105181.

551. S. Botros, K. M. Ghoneim, and M. Khalifa, *Egypt. J. Pharm. Sci.*, 1972, **13**, 11; *Chem. Abstr.*, 1974, **80**, 70768.

552. Y. Sato, T. Tanaka, and T. Nagasaki, *Yakugaku Zasshi*, 1970, **90**, 629.

553. T. Okano and A. Takadate, *Yakugaku Zasshi*, 1968, **88**, 428.

554. E. Hayashi, T. Higashino, and S. Tomisaka, *Yakugaku Zasshi*, 1967, **87**, 578.

555. T. Higashino, H. Ito, M. Watanabi, and E. Hayashi, *Yakugaku Zasshi*, 1973, **93**, 94.

556. T. Hisano, T. Nishi, and M. Ichikawa, *Yakugaku Zasshi*, 1972, **92**, 582.

557. T. Kato, H. Yamanaka, and S. Konno, *Yakugaku Zasshi*, 1971, **91**, 1004.

558. J. Haginiwa, Y. Higuchi, and Y. Yamamoto, *Yakugaku Zasshi*, 1975, **95**, 8.

559. K. Kasuga, M. Hirobe, and T. Okamoto, *Yakugaku Zasshi*, 1974, **94**, 945.

560. I. Ishiguro, J. Naito, R. Shinohara, and A. Ishikura, *Yakugaku Zasshi*, 1974, **94**, 1232.

561. T. Yoshikawa and K. Shitago, *Yakugaku Zasshi*, 1974, **94**, 417.

562. T. Higashino, W. Washizu, and E. Hayashi, *Yakugaku Zasshi*, 1973, **93**, 1234.

563. K.-H. Spohn and E. Breitmaier, *Chimia*, 1971, **25**, 365.

564. A. Arcoria, S. Fisichella, and G. Scarlata, *Ann. Chim. (Rome)*, 1971, **61**, 864.

565. M. Covello, F. de Simone, and E. Abignente, *Ann. Chim. (Rome)*, 1967, **57**, 595.

566. F. Russo and M. Ghelardoni, *Ann. Chim. (Rome)*, 1966, **56**, 839.

567. A. Arcoria and G. Scarlata, *Ann. Chim. (Rome)*, 1966, **56**, 1394.

568. L. Zhelyazkov, R. Kolchagova, L. Daleva, and D. Stefanova, *Tr. Nauchnoizsled. Khim.-Farm. Inst.*, 1974, **9**, 43; *Chem. Abstr.*, 1975, **83**, 58745.

569. Y. V. Kozhevnikov, *Izv. Vyssh. Uchebn. Zaved., Khim. Khim. Tekhnol.*, 1975, **18**, 235; *Chem. Abstr.*, 1975, **83**, 28178.

570. K. M. Murav'eva, N. V. Arkhangel'skaya, and M. N. Shchukina, *Khim.-Farm. Zh.*, 1968, **2**, 3; *Chem. Abstr.*, 1968, **68**, 103651.

571. K. Issleib and H. P. Abicht, *J. Prakt. Chem.*, 1973, **315**, 649.

572. S. S. Parmar, A. K. Chaturvedi, and B. Ali, *J. Prakt. Chem.*, 1971, **312**, 950.

573. S. S. Parmar, V. K. Rastogi, and R. C. Arora, *J. Prakt. Chem.*, 1971, **312**, 958.

574. K. Kishor, R. C. Arora, and S. S. Parmar, *J. Prakt. Chem.*, 1968, 4th *Ser.*, **38**, 273.

575. L. Fey, A. Kövendi, and J. Rusu, *J. Prakt. Chem.*, 1967, 4th *Ser.*, **35**, 225.

576. K. M. Murav'eva, N. V. Arkhangel'skaya, M. N. Shchukina, T. N. Zykova, and G. N. Pershin, *Khim.-Farm. Zh.*, 1968, **2**, (No. 7), 35; *Chem. Abstr.*, 1968, **69**, 106658.

577. K. A. Kovar, G. P. Wojtovicz, and H. Auterhoff, *Arch. Pharm. (Weinheim, Ger.)*, 1974, **307**, 657.

578. J. Dusemond, *Arch. Pharm (Weinheim, Ger.)*, 1974, **307**, 883.

579. P. Daenens and M. van Boven, *Arzneim.-Forsch.*, 1974, **24**, 195.

580. H. Yamamoto, C. Saito, S. Inaba, H. Awata, M. Yamamoto, Y. Sakai, and T. Komatsu, *Arzneim.-Forsch.*, 1973, **23**, 1266.

581. T. Komatsu, H. Awata, Y. Sakai, T. Inukai, M. Yamamoto, S. Inaba, and H. Yamamoto, *Arzneim.-Forsch.*, 1972, **22**, 1958.

582. D. Hoffter, *Arzneim.-Forsch.*, 1971, **21**, 505.

583. G. Cantarelli, *Farmaco, Ed. Sci.*, 1970, **25**, 770.

584. F. Gatta and R. Landi-Vittory, *Farmaco, Ed. Sci.*, 1970, **25**, 991.

585. M. G. Biressi, G. Cantarelli, M. Carissimi, A. Cattaneo, and F. Ravenna, *Farmaco, Ed. Sci.*, 1969, **24**, 199.

586. G. Malesani, A. Pietrgrande, and G. Rodighiero, *Farmaco, Ed. Sci.*, 1968, **23**, 765.

587. c. Bogentoft, Ö. Ericsson, and B. Danielsson, *Acta Pharm. Suec.*, 1974, **11**, 59; *Chem. Abstr.*, 1974, **81**, 3874.

588. P. N. Bhargava and S. N. Singh, *Egypt. J. Chem.*, 1972, **15**, 495; *Chem. Abstr.*, 1974, **80**, 146101.

589. E. Budeanu and E. Tanase, *An. Stiint. Univ. "Al. I. Cuza" Iasi, Sect. 1c*, 1973, **19**, 183; *Chem. Abstr.*, 1974, **80**, 132978.

590. L. Avram, I. Ursu-Kenyeres, and V. Ursu-Kenyeres, *Farmacia (Bucharest)*, 1973, **21**, 729; *Chem. Abstr.*, 1974, **80**, 133375.

591. S. Palazzo and L. I. Giannola, *Atti Accad. Sci., Lett. Arti Palermo, Parte 1*, 1971–1973, **32**, 29; *Chem. Abstr.*, 1974, **81**, 49651.

592. N. B. Karpova and Y. S. Tsizin, *Khim. Geterotsikl. Soedin.*, 1973, 1403.

593. M. Cardellini, P. Franchetti, M. Grifantini, S. Martelli, and F. Petrelli, *Farmaco, Ed. Sci.*, 1975, **30**, 536.

594. R. Landi-Vittory and G. Gatta, *Farmaco, ed. Sci.*, 1972, **27**, 208.

595. G. Malesani, F. Marcolin, G. Chiarelotto, G. Rodighiero, and F. Ghezzo, *Farmaco, Ed. Sci.*, 1972, **27**, 731.

596. B. Carnmalm, J. Gyllander, N. A. Jonson, and L. Mikiver, *Acta Pharm. Suec.*, 1974, **11**, 167; *Chem. Abstr.*, 1974, **81**, 37360.

597. L. Avram, V. Ursu-Kenyeres, and I. Ursu-Kenyeres, *Farmacia (Bucharest)*, 1973, **21**, 365; *Chem. Abstr.*, 1974, **80**, 14888.

598. A. Sammour, T. Zimaity, and M. A. Abdo, *Egypt. J. Chem.*, 1973, **16**, 215; *Chem. Abstr.*, 1974, **81**, 152159.

599. L. Zhelyazkov, R. Kolchagova, L. Daleva, and D. Stefanova, *Tr. Nauchnoizsled. Khim.-Farm. Inst.*, 1974, **9**, 33; *Chem. Abstr.*, 1975, **83**, 58744.

600. J. Davoll, A. M. Johnson, H. J. Davies, O. D. Bird, J. Clarke, and E. F. Elslager, *J. Med. Chem.*, 1972, **15**, 812.

601. E. F. Elslager, J. Clarke, L. M. Werbel, D. F. Worth, and J. Davoll, *J. Med. Chem.*, 1972, **15**, 827.

602. E. F. Elslager, O. D. Bird, J. Clarke, S. C. Perricone, D. F. Worth, and J. Davoll, *J. Med. Chem.*, 1972, **15**, 1138.

603. P. N. Bhargava and M. R. Chaurasia, *J. Chem. UAR*, 1969, **12**, 289; *Chem. Abstr.*, 1970, **72**, 31737.

604. P. N. Bhargava and H. Singh, *Indian J. Pharm.*, 1969, **31**, 111; *Chem. Abstr.*, 1970, **72**, 121475.

605. J. Tulecki and J. Kalinowska-Torz, *Diss. Pharm. Pharmacol.*, 1970, **22**, 21; *Chem. Abstr.*, 1970, **72**, 111410.

606. I. C. Pozharliev and K. Zakharieva, *Izv. Otd. Khim. Nauki, Bulg. Akad. Nauk*, 1969, **2**, 341; *Chem. Abstr.*, 1970, **72**, 78121.

607. C. Bogentoft and B. Danielsson, *Acta Pharm. Suec.*, 1969, **6**, 589; *Chem. Abstr.*, 1970, **72**, 78137.

608. Y. V. Kozhevnikov, V. N. Aleshina, and S. E. Beketova, *Khim.-Farm. Zh.*, 1969, **3**, 38; *Chem. Abstr.*, 1970, **72**, 78975.

609. J. B. Hynes and W. T. Ashton, *J. Med. Chem.*, 1975, **18**, 263.

610. G. E. Hardtmann, G. Koletar, O. R. Pfister, J. H. Gogerty, and L. C. Iorio, *J. Med. Chem.*, 1975, **18**, 447.

611. J. B. Hynes and C. M. Garrett, *J. Med. Chem.*, 1975, **18**, 632.

612. J. B. Hynes, J. M. Buck, L. d'Souza, and J. H. Freisheim, *J. Med. Chem.*, 1975, **18**, 1191.

613. J. P. Barthwal, S. K. Tandon, V. K. Agarwal, K. S. Dixit, and S. S. Parmar, *J. Pharm. Sci.*, 1973, **62**, 613.

614. M. J. Kornet, *J. Pharm. Sci.*, 1973, **62**, 834.

615. A. C. Glasser, L. Diamond, and G. Combs, *J. Pharm. Sci.*, 1971, **60**, 127.

616. R. E. Orth, *J. Pharm. Sci.*, 1967, **56**, 925.

617. K. C. Joshi, V. K. Singh, D. S. Mehta, R. C. Sharma, and L. Gupta, *J. Pharm. Sci.*, 1975, **64**, 1428.

618. D. G. Priest, J. B. Hynes, C. W. Jones, and W. T. Ashton, *J. Pharm. Sci.*, 1974, **63**, 1158.

619. J. A. Cella, *J. Pharm. Sci.*, 1974, **63**, 1627.

620. K. Kottke, F. Friedrich, and R. Pohloudek-Fabini, *Arch. Pharm. (Weinheim, Ger.)*, 1967, **300**, 583.

621. K. Hartke, R. Alarćon, D. Ramirez, and J. Bartulin, *Arch. Pharm. (Weinheim, Ger.)*, 1966, **299**, 914.

622. R. Stoffel and H.-J. Bresse, *Arch. Pharm. (Weinheim, Ger.)*, 1973, **306**, 579.

623. S. S. Parmar, A. K. Chaturvedi, A. Chaudhary, and S. J. Brumleve, *J. Pharm. Sci.*, 1974, **63**, 356.

624. H. Kohl, P. D. Desai, A. N. Dohadwalla, and N. J. de Souza, *J. Pharm. Sci.*, 1974, **63**, 838.

625. R. Pohloudek-Fabini and M. Selchau, *Arch. Pharm. (Weinheim, Ger.)*, 1969, **302**, 527.

626. H. J. Roth and G. Langer, *Arch. Pharm. (Weinheim, Ger.)*, 1968, **301**, 736.

627. H. Möhrle, D. Schittenhelm, and P. Gundlach, *Arch. Pharm. (Weinheim, Ger.)*, 1972, **305**, 108.

628. A. Kreutzberger and M. U. Uzbek, *Arch. Pharm. (Weinheim, Ger.)*, 1972, **305**, 171.

629. H. J. Roth and H.-E. Hagen, *Arch. Pharm. (Weinheim, Ger.)*, 1971, **304**, 331.

630. A. Urbonas, G. Valentukeviciene, and L. Jasinskas, *Liet. TSR Aukst. Mokyklu Mokslo Darb., Chem. Chem. Technol.*, 1972 (No. 14), 135; *Chem. Abstr.*, 1974, **80**, 59907.

631. M. Uchida, T. Higashino, and E. Hayashi, *Shitsuryo Bunseki*, 1973, **21**, 245; *Chem. Abstr.*, 1974, **80**, 36408.

632. F. Eiden and J. Dusemund, *Arch. Pharm. (Weinheim, Ger.)*, 1971, **304**, 729.

633. J. Dusemund, *Arch. Pharm. (Weinheim, Ger.)*, 1975, **308**, 230.

634. H. J. Roth and H. Mensel, *Arch. Pharm. (Weinheim, Ger.)*, 1975, **308**, 557.

635. A. R. E. Ossman, M. Khalifa, and Y. M. Abou-Zeid, *Egypt. J. Pharm. Sci.*, 1972, **13**, 1; *Chem. Abstr.*, 1974, **80**, 70769.

636. N. B. Karpova, Y. S. Tsizin, E. A. Rudzit, T. P. Radkevich, D. A. Kulikova, and A. V. Luk'yanov, *Khim.-Farm. Zh.*, 1974, **8** (No. 2), 21; *Chem. Abstr.*, 1974, **81**, 3870.

637. S. S. Brown and G. A. Smart, *J. Pharm. Pharmacol.*, 1969, **21**, 466.

638. H. L. Yale and M. Kalkstein, *J. Med. Chem.*, 1967, **10**, 334.

639. J. G. Topliss, E. P. Shapiro, and R. I. Taber, *J. Med. Chem.*, 1967, **10**, 642.

640. P. F. Juby, T. W. Hudyma, and M. Brown, *J. Med. Chem.*, 1968, **11**, 111.

641. H.-J. Hess, T. H. Cronin, and A. Scriabine, *J. Med. Chem.*, 1968, **11**, 130.

642. K. Okumura, T. Oine, Y. Yamada, G. Hayashi, and M. Nakama, *J. Med. Chem.*, 1968, **11**, 348.

643. J. Reisch and C. Usifoh, *J. Heterocycl. Chem.*, 1993, **30**, 659.

644. P. N. Bhargava and M. R. Chaurasia, *J. Med. Chem.*, 1968, **11**, 404.

645. K. Okumura, T. Oine, Y. Yamada, G. Hayashi, M. Nakama, and T. Nose, *J. Med. Chem.*, 1968, **11**, 788.

646. P. N. Bhargava and M. R. Chaurasia, *J. Med. Chem.*, 1968, **11**, 908.

647. J. K. Horner and D. W. Henry, *J. Med. Chem.*, 1968, **11**, 946.

648. M. L. Hoefle and A. Holmes, *J. Med. Chem.*, 1968, **11**, 974.

649. G. Bonola, P. Da-Re, M. J. Magistretti, E. Massarani, and I. Setnikar, *J. Med. Chem.*, 1968, **11**, 1136.

650. J. F. Bunnett and J. Y. Bassett, *J. Org. Chem.*, 1963, **27**, 3714.

651. G. Wagner and F. Suess, *Pharmazie*, 1969, **24**, 35; *Chem. Abstr.*, 1969, **71**, 13319.

652. C. H. Budeanu, *An. Stiint. Univ. "Al. I. Cuza" Iasi, Sect. 1c*, 1968, **14**, 65; *Chem. Abstr.*, 1969, **70**, 106470.

653. C. Bogentoft, L. Kronberg, and B. Danielsson, *Acta Pharm. Suec.*, 1969, **6**, 489; *Chem. Abstr.*, 1970, **72**, 11916.

654. W. E. Coyne and J. W. Cusic, *J. Med. Chem.*, 1968, **11**, 1208.

655. M. Vincent, J. C. Poignant, and G. Remond, *J. Med. Chem.*, 1971, **14**, 714.

656. M. R. Boots, S. C. Boots, and D. E. Moreland, *J. Med. Chem.*, 1970, **13**, 144.

657. M. Fishman and P. A. Cruickshank, *J. Med. Chem.*, 1970, **13**, 155.

658. M. Likar, P. Schauer, M. Japeij, M. Globokar, M. Oklobdžija, A. Povše, and V. Šunjić, *J. Med. Chem.*, 1970, **13**, 159.

659. G. Malesani, F. Marcolin, and G. Rodighiero, *J. Med. Chem.*, 1970, **13**, 161.

660. G. Bonola and E. Sianesi, *J. Med. Chem.*, 1970, **13**, 329.

661. A. Rosowsky, J. L. Marini, M. E. Nadel, and E. J. Modest, *J. Med. Chem.*, 1970, **13**, 882.

662. B. V. Shetty, L. A. Campanella, T. L. Thomas, S. E. Fedorchuk, T. A. Davidson, L. Michelson, H. Volz, S. E. Zimmerman, E. J. Belair, and A. P. Truant, *J. Med. Chem.*, 1970, **13**, 886.

663. R. Pohloudek-Fabini, K. Kottke, and F. Friedrich, *Pharmazie*, 1969, **24**, 833; *Chem. Abstr.*, 1969, **71**, 101811.

664. B. Danielsson, L. Kronberg, and B. Akerman, *Acta Pharm. Suec.*, 1969, **6**, 379; *Chem. Abstr.*, 1969, **71**, 101812.

665. K. Kottke, F. Friedrich, and R. Pohloudek-Fabini, *Pharmazie*, 1969, **24**, 438; *Chem. Abstr.*, 1969, **71**, 101813.

666. A. Arcoria, S. Fisichella, and G. Scarlata, *Bol. Sci. Fac. Chim. Ind. Bologna*, 1969, **27**, 57; *Chem. Abstr.*, 1969, **71**, 103150.

667. J. Tulecki and J. Kalinowska-Torz, *Ann. Pharm. (Poznan)*, 1967, **6**, 69; *Chem. Abstr.*, 1969, **71**, 112881.

668. B. Danielsson and L. Kronberg, *Acta Pharm. Suec.*, 1969, **6**, 389; *Chem. Abstr.*, 1969, **71**, 91426.

669. S. S. Parmar, K. Kishor, P. K. Seth, and R. C. Arora, *J. Med. Chem.*, 1969, **12**, 138.

670. P. N. Bhargava and V. N. Choubey, *J. Med. Chem.*, 1969, **12**, 553.

671. S. Hayao, H. J. Havera, W. G. Styrcker, and E. Hong, *J. Med. Chem.*, 1969, **12**, 936.

672. P. A. Cruickshank and W. E. Hymans, *J. Med. Chem.*, 1974, **17**, 468.

673. J. B. Hynes, W. T. Ashton, H. G. Merriman, and F. C. Walker, *J. Med. Chem.*, 1974, **17**, 682.

674. J. I. de Graw, V. H. Brown, W. T. Colwell, and N. E. Morrison, *J. Med. Chem.*, 1974, **17**, 762.

675. W. E. Richter and J. J. McCormack, *J. Med. Chem.*, 1974, **17**, 943.

676. A. Walser and R. I. Fryer, *J. Med. Chem.*, 1974, **17**, 1228.

677. T. Jen, B. Dienel, F. Dowalo, H. van Hoeven, P. Bender, and B. Loev, *J. Med. Chem.*, 1973, **16**, 633.

678. W. T. Ashton, F. C. Walker, and J. B. Hynes, *J. Med. Chem.*, 1973, **16**, 694.

679. H. Kohl, N. J. de Souza, and P. D. Desai, *J. Med. Chem.*, 1973, **16**, 1045.

680. W. T. Ashton and J. B. Hynes, *J. Med. Chem.*, 1973, **16**, 1233.

681. R. V. Coombs, R. P. Danna, M. Denzer, G. E. Hardtmann, B. Huegi, G. Koletar, J. Koletar, H. Ott, E. Jukniewicz, J. W. Perrine, E. I. Takesue, and J. H. Trapold, *J. Med. Chem.*, 1973, **16**, 1237.

682. A. Arcoria and S. Fisichella, *Boll. Sci. Fac. Chim. Ind. Bologna*, 1968, **26**, 239; *Chem. Abstr.*, 1969, **71**, 92616.

683. B. Lubimowski and L. Berezowski, *Acta Pol. Pharm.*, 1969, **26**, 293; *Chem. Abstr.*, 1970, **72**, 66886.

684. M. M. Blight and J. F. Grove, *J. Chem. Soc., Perkin Trans. 1*, 1974, 1691.

685. G. Wagner and L. Rothe, *Pharmazie*, 1969, **24**, 513; *Chem. Abstr.*, 1970, **72**, 55392.

686. R. J. Alaimo and C. J. Hatton, *J. Med. Chem.*, 1972, **15**, 108.

687. K. Okumura, Y. Yamada, T. Oine, J. Tani, T. Ochiai, and I. Inoue, *J. Med. Chem.*, 1972, **15**, 518.

688. J. Tulecki and J. Karlinowska-Torz, *Ann. Pharm. (Poznan)*, 1969, **7**, 17; *Chem. Abstr.*, 1970, **72**, 55402.

689. H. C. van der Plas and A. Koudijs, *Recl. Trav. Chim. Pays-Bas*, 1978, **97**, 159.

690. Z. Csuros, R. Soos, and J. Palinkas, *Acta Chem. (Budapest)*, 1970, **63**, 215; *Chem. Abstr.*, 1970, **72**, 90396.

691. B. Singh, M. M. P. Rukhaiyar, and R. J. Sinha, *J. Inorg. Nucl. Chem.*, 1977, **39**, 29.

692. A. F. Fentiman and R. I. Foltz, *J. Labelled Compd. Radiopharm.*, 1976, **12**, 69.

693. B. S. Drach, V. A. Kovalev, and A. V. Kirsanov, *Zh. Org. Khim.*, 1976, **12**, 673.

694. B. S. Drach and V. A. Kovalev, *Zh. Org. Khim.*, 1976, **12**, 2319.

695. V. L. Plakidin and E. S. Kosheleva, *Zh. Org. Khim.*, 1976, **12**, 2481.

696. Z. Buděšinský, P. Lederer, F. Roubinek, A. Šváb, and J. Vavřina, *Collect. Czech, Chem. Commun.*, 1976, **41**, 3405.

697. W. Ried, B. Heine, W. Merkel, and N. Kothe, *Synthesis*, 1976, 534.

698. T. M. Paterson, R. K. Smalley, and H. Suschitzky, *Synthesis*, 1975, 709.

699. D. M. W. van den Ham, G. F. S. Harrison, A. Spaans, and D. van der Meer, *Recl. Trav. Chim. Pays-Bas*, 1975, **94**, 168.

700. C. Huiszoon, *Acta Crystallogr., Sect. B*, 1976, **32**, 998.

701. W. L. F. Armarego, R. A. Y. Jones, A. R. Katritzky, D. M. Read, and R. Scattergood, *Aust. J. Chem.*, 1975, **28**, 2323.

702. L. D. Colebrook and H. G. Giles, *Can. J. Chem.*, 1975, **53**, 3431.

703. T. P. Ahern, H. Fong, and K. Vaughan, *Can. J. Chem.*, 1976, **54**, 290.

704. S. M. S. Chauhan and H. Junjappa, *Tetrahedron*, 1976, **32**, 1779.

705. T. Kametani, Chu Van Loc, T. Higa, M. Koizumi, M. Ihara, and K. Fukumoto, *Heterocycles*, 1976, **4**, 1487.

706. I. Ninomiya and O. Yamamoto, *Heterocycles*, 1976, **4**, 475.
707. B. V. Golomolzin, L. D. Shcherak, and I. Y. Postovskii, *Khim. Geterotsikl. Soedin.*, 1969, 1131.
708. K. Uno, K. Niiume, and T. Nakayama, *Nippon Kagaku Kaishi*, 1975, 1584.
709. Y. Imai, M. Ueda, and M. Ishimori, *Nippon Kagaku Kaishi*, 1975, 2154.
710. M. Srinivasan and J. B. Rampal, *Phosphorus Sulfur*, 1976, **2**, 105.
711. N. N. Barashkov, L. A. Zimina, E. N. Teleshov, and A. N. Pravednikov, *Dokl. Akad. Nauk SSSR*, 1978, **240**, 847.
712. T. Nakai, N. M. Hassan, and N. Ishikawa, *Bull. Chem. Soc. Jpn.*, 1977, **50**, 3014.
713. M. Kimura, T. Hirohashi, and H. Yamamoto, *Bull. Chem. Soc. Jpn.*, 1976, **49**, 2696.
714. K. Yamaguchi, *Bull. Chem. Soc. Jpn.*, 1976, **49**, 1366.
715. L. Legrand, *Bull. Soc. Chim. Fr.*, 1976, 1857.
716. L. Legrand and N. Lozac'h, *Bull. Soc. Chem. Fr.*, 1976, 1853.
717. G. Rabilloud and B. Sillion, *Bull. Soc. Chim. Fr.*, 1975, 2682.
718. G. Pastor, C. Blanchard, C. Montginoul, E. Torreilles, L. Giral, and A. Texier, *Bull. Soc. Chim. Fr.*, 1975, 1331.
719. L. Legrand and N. Lozac'h, *Bull. Soc. Chim. Fr.*, 1975, 1411.
720. L. Legrand and N. Lozac'h, *Bull. Soc. Chim. Fr.*, 1975, 1415.
721. W. Ried, N. Kothe, R. Schweitzer, and A. Höhle, *Chem. Ber.*, 1976, **109**, 2921.
722. H. G. Aurich and U. Grigo, *Chem. Ber.*, 1976, **109**, 200.
723. L. Capuano and V. Diehl, *Chem. Ber.*, 1976, **109**, 723.
724. J. Goerdeler, F. M. Panshiri, and W. Vollrath, *Chem. Ber.*, 1975, **108**, 3071.
725. W. Ried and O. Mösinger, *Chem. Ber.*, 1978, **111**, 143.
726. S. Blechert, K. -E. Fichter, and E. Winterfeldt, *Chem. Ber.*, 1978, **111**, 439.
727. L. Capuano and K. Benz, *Chem. Ber.*, 1977, **110**, 3849.
728. J. Goerdeler and H. Lohmann, *Chem. Ber.*, 1977, **110**, 2996.
729. K. Burger, K. Einhelling, W. -D. Roth, and E. Daltrozzo, *Chem. Ber.*, 1977, **110**, 605.
730. H. Singh, S. C. Gandhi, and K. B. Lal, *Chem. Ind. (London)*, 1976, 650.
731. H. Singh, K. S. Kumar, and K. B. Lal, *Chem. Ind. (London)*, 1975, 649.
732. B. A. Burdick, P. A. Benkovic, and S. J. Benkovic, *J. Am. Chem. Soc.*, 1977, **99**, 5716.
733. T. Kametani, Chu Van Loc, T. Higa, M. Koizumi, M. Ihara, and K. Fukumoto, *J. Am. Chem. Soc.*, 1977, **99**, 2306.
734. D. Linke, *Liebigs Ann. Chem.*, 1977, 1787.
735. W. Ried and O. Mösinger, *Liebigs Ann. Chem.*, 1977, 1817.
736. W. Schäfer and C. Falkner, *Liebigs Ann. Chem.*, 1976, 1809.
737. W. Ried and L. Kaiser, *Liebigs Ann. Chem.*, 1976, 2007.
738. E. Wolf and H. Kohl, *Liebigs Ann. Chem.*, 1975, 1245.
739. S. Petersen, H. Heitzer, and L. Born, *Liebigs Ann. Chem.*, 1974, 2003.
740. I. Y. Postovskii and N. N. Vereshchagina, *Khim. Geterotsikl. Soedin.*, 1967, 944.
741. L. I. Samarai, V. A. Bondar, and G. I. Derkach, *Khim. Geterotsikl. Soedin.*, 1968, 182.
742. V. M. Dziomko, I. S. Markovich, and N. V. Kruglova, *Khim. Geterotsikl. Soedin.*, 1968, 536.
743. V. F. Sedova and V. P. Mamaev, *Khim. Geterotsikl. Soedin.*, 1968, 921.
744. R. I. Moskalenko and G. I. Savel'eva, *Khim. Geterotsikl. Soedin.*, 1969, 348.
745. Y. V. Kozhevnikov and P. A. Petyunin, *Khim. Geterotsikl. Soedin.*, 1969, 747.
746. W. Steiger, T. Kappe, and E. Ziegler, *Monatsh. Chem.*, 1969, **100**, 146.
747. E. Ziegler, W. Steiger, and T. Kappe, *Monatsh. Chem.*, 1969, **100**, 150.

748. W. Steiger, T. Kappe, and E. Ziegler, *Monatsh. Chem.*, 1969, **100**, 528.

749. I. Y. Postovskii and B. V. Golomolzin, *Khim. Geterotsikl. Soedin.*, 1970, 100.

750. G. S. Shvindlerman and Y. A. Bashakov, *Khim. Geterotsikl. Soedin.*, 1970, 427.

751. S. L. Mertsalov, L. G. Egorova, and I. Y. Postovskii, *Khim. Geterotsikl. Soedin.*, 1970, 687.

752. O. Hromatka, N. Knollmüller, and D. Binder, *Monatsh. Chem.*, 1969, **100**, 872.

753. O. Hromatka, D. Binder, and M. Knollmüller, *Monatsh. Chem.*, 1969, **100**, 879.

754. E. Ziegler, W. Steiger, and T. Kappe, *Monatsh. Chem.*, 1969, **100**, 948.

755. L. I. Samarai, V. A. Bondar, and G. I. Derkach, *Khim. Geterotsikl. Soedin.*, 1970, 814.

756. B. V. Golomolzin and I. Y. Postovskii, *Khim. Geterotsikl. Soedin.*, 1970, 855.

757. C. Bischoff, H. Herma, and E. Schröder, *J. Prakt. Chem.*, 1976, **318**, 895.

758. H. G. Viehe, G. J. de Voghel, and F. Smets, *Chimia*, 1976, **30**, 189.

759. M. Bonamomi and G. Palazzo, *Farmaco, Ed. Sci.*, 1977, **32**, 490.

760. V. Šunjić, A. Lisini, T. Kovač, B. Belin, F. Kajež and L. Klasinc, *Croat. Chem. Acta*, 1977, **49**, 505.

761. Y. V. Kozhevnikov, *Khim. Geterotsikl. Soedin.*, 1972, 1000.

762. Y. S. Tsizin, N. B. Karpova, and I. E. Shumakovich, *Khim. Geterotsikl. Soedin.*, 1972, 836.

763. Y. S. Tsizin and N. B. Karpova, *Khim. Geterotsikl. Soedin.*, 1972, 841.

764. Y. S. Tsizin and N. B. Karpova, *Khim. Geterotsikl. Soedin.*, 1971, 1698.

765. Y. S. Tsizin and N. B. Karpova, *Khim. Geterotsikl Soedin.*, 1971, 283.

766. Y. S. Tsizin, N. B. Karpova, and O. V. Efimova, *Khim. Geterotsikl. Soedin.*, 1971, 418.

767. B. A. Tertov, P. P. Onishchenko, and V. V. Bessonov, *Khim. Geterotsikl. Soedin.*, 1974, 1410.

768. P. A. Prtyunin, V. A. Bulgakov, and G. P. Petyunin, *Khim. Geterotsikl. Soedin.*, 1974, 609.

769. E. Ziegler, W. Steiger, and T. Kappe, *Monatsh. Chem.*, 1968, **99**, 1499.

770. T. Kappe, W. Steiger, and E. Ziegler, *Monatsh. Chem.*, 1967, **98**, 214.

771. W. Metlesics, G. Silverman, and L. H. Sternbach, *Monatsh. Chem.*, 1967, **98**, 633.

772. M. Knollmüller, *Monatsh. Chem.*, 1971, **102**, 1055.

773. M. Knollmüller, *Monatsh. Chem.*, 1970, **101**, 1443.

774. G. Zigeuner and G. Gübitz, *Monatsh. Chem.*, 1970, **101**, 1547.

775. G. Zigeuner, V. Eisenreich, and W. Immel, *Monatsh. Chem.*, 1970, **101**, 1745.

776. G. Zigeuner, H. Brunetti, H. Ziegler, and M. Bayer, *Monatsh. Chem.*, 1970, **101**, 1767.

777. G. Zigeuner, H. Schmidt, and D. Volpe, *Monatsh. Chem.*, 1970, **101**, 1824.

778. G. Zigeuner, R. Hopmann, C. Knopp, and A. Fuchsgruber, *Monatsh. Chem.*, 1970, **101**, 1829.

779. W. Wendelin, A. Harler, and A. Fuchsgruber, *Monatsh. Chem.*, 1976, **107**, 141.

780. W. Wendelin and A. Harler, *Monatsh. Chem.*, 1975, **106**, 1479.

781. P. A. Petyunin, V. P. Chernykh, G. P. Petyunyn, and Y. V. Kozhevnikov, *Khim. Geterotsikl. Soedin.*, 1970, 1575.

782. N. B. Karpova and Y. S. Tsizin, *Khim. Geterotsikl. Soedin.*, 1973, 1697.

783. I. A. Mazur, R. S. Sinyak, R. I. Katkevich, and P. M. Kochergin, *Khim. Geterotsikl. Soedin.*, 1976, 1268.

784. V. I. Dulenko, N. N. Alekseev, V. M. Golyak, and Y. A. Nikolyukin, *Khim. Geterotsikl. Soedin.*, 1976, 1286.

785. T. H. Althuis and H.-J. Hess, *J. Med. Chem.*, 1976, **20**, 146.

786. I. R. Ager, D. R. Harrison, P. D. Kennewell, and J. B. Taylor, *J. Med. Chem.*, 1977, **20**, 379.

787. S. S. Parmar and R. C. Arora, *J. Med. Chem.*, 1970, **13**, 135.

788. T. Kato, H. Yamanaka, and H. Hiranuma, *Yakugaku Zasshi*, 1970, **90**, 877.

789. T. Hisano, K. Muraoka, and M. Ichikawa, *Yakugaku Zasshi*, 1977, **97**, 808.

790. T. Higashino, M. Goi, and E. Hayashi, *Yakugaku Zasshi*, 1976, **96**, 397.

791. T. Hisano, K. Shoji, and M. Ichikawa, *Yakugaku Zasshi*, 1976, **96**, 886.

792. E. Hayashi, N. Shimada, and A. Miyashita, *Yakugaku Zasshi*, 1976, **96**, 1370.

793. N. B. Marchenko, V. G. Granik, T. F. Vlasova, O. S. Anisimova, and R. G. Glushkov, *Khim. Geterotsikl. Soedin.*, 1976, 665.

794. Y. A. Sedov, N. I. Shumova, and N. I. Gogoleva, *Khim. Geterotsikl. Soedin.*, 1976, 709.

795. A. F. Vlasenko, B. E. Mandrichenko, G. K. Rogul'chenko, R. S. Sinvak, I. A. Mazur, and P. M. Kochergin, *Khim. Geterotsikl. Soedin.*, 1976, 834.

796. T. L. Pilicheva, O. N. Chupakhin, and I. Y. Postovskii, *Khim. Geterotsikl. Soedin.*, 1975, 561.

797. O. N. Chupakhin, T. L. Pilicheva, I. Y. Postovskii, and N. I. Klyuev, *Khim. Geterotsikl. Soedin.*, 1975, 708.

798. K. M. Shakhidoyatov, L. M. Yun, and C. S. Kadyrov, *Khim. Geterotsikl. Soedin.*, 1978, 105.

799. N. P. Bednyagina, E. S. Karavaeva, G. N. Lipunova, L. I. Medvedeva, and B. I. Buzykin, *Khim. Geterotsikl. Soedin.*, 1977, 1268.

800. M. Ghelardoni and V. Pestellini, *Ann. Chim. (Rome)*, 1974, **64**, 445.

801. J. E. Otis and J. B. Hynes, *J. Med. Chem.*, 1977, **20**, 1393.

802. O. N. Chupakhin, V. N. Charushin, I. M. Sosonkin, E. G. Kovalev, G. L. Kalb, and I. Y. Postovskii, *Khim. Geterotsikl. Soedin.*, 1977, 690.

803. C. Bischoff, H. Herma, and E. Schröder, *J. Prakt. Chem.*, 1977, **319**, 230.

804. G. Kempter, D. Rehbaum, and J. Schirmer, *J. Prakt. Chem.*, 1977, **319**, 589.

805. J. Dusemund and K. Roth, *Z. Naturforsch., Teil B*, 1976, **31**, 509.

806. D. Chaigne, J.-F. Hémidy, L. Legrand, and D. Cornet, *J. Chem. Res.*, 1977, *Synop.* 198, *Minipr.* 2210.

807. M. I. Ali, H. A. Hammouda, and A.-E. M. Abd.-Elfattah, *Z. Naturforsch., Teil B*, 1977, **32**, 94.

808. C. Mayer and T. Kappe, *Z. Naturforsch., Teil B*, 1977, **32**, 1214.

809. K. Matsumoto, P. Stark, and R. G. Meister, *J. Med. Chem.*, 1977, **20**, 17.

810. C. Hansch, J. Y. Fukunaga, P. Y. C. Jow, and J. B. Hynes, *J. Med. Chem.*, 1977, **20**, 96.

811. E. F. Elslager, J. Davoll, P. Jacob, A. M. Johnson, J. Johnson, and L. M. Werbel, *J. Med. Chem.*, 1978, **21**, 639.

812. F. I. Carroll, B. Berrang, and C. P Linn, *J. Med. Chem.*, 1978, **21**, 326.

813. G. B. Bennett, R. B. Mason, L. J. Alden, and J. B. Roach, *J. Med. Chem.*, 1978, **21**, 623.

814. B. Serafin, M. Modzelewski, A. Kurnatowska, and R. Kadlubowski, *Eur. J. Med. Chem.*, 1977, **12**, 325.

815. R. Albrecht and K. Schumann, *Eur. J. Med. Chem.*, 1976, **11**, 155.

816. F. Sauter, P. Stanetty, and U. Jordis, *Arch. Pharm. (Weinheim, Ger.)*, 1977, **310**, 680.

817. V. von Weissenborn, H. H. Brandt, and K. E. Schulte, *Arch. Pharm. (Weinheim, Ger.)*, 1977, **310**, 1018.

818. H. Möhrle and C.-M. Seidel, *Arch. Pharm. (Weinheim, Ger.)*, 1976, **309**, 471.

819. H. Möhrle and C.-M. Seidel, *Arch. Pharm. (Weinheim, Ger.)*, 1976, **309**, 503.

820. H. Möhrle and C.-M. Seidel, *Arch. Pharm. (Weinheim, Ger.)*, 1976, **309**, 542.

821. H. Möhrle and C.-M. Seidel, *Arch. Pharm. (Weinheim, Ger.)*, 1976, **309**, 572.

822. W. Sadée and E. van der Kleijn *J. Pharm. Sci.*, 1971, **60**, 135.

823. H. Breuer and A. Roesch, *Arzneim.-Forsch.*, 1971, **21**, 238.

824. A. Holm, C. Christophersen, T. Ottersen, H. Hope, and A. Christensen, *Acta Chem. Scand., Ser. B*, 1977, **31**, 687.

825. T. Zimaity, M. Anwar, and M. F. Abdel-Megeed, *Indian J. Chem., Sect. B*, 1977, **15**, 750.

826. K. Gewald, H. Schäfer, and K. Mauersberger, *Z. Chem.*, 1977, **17**, 223.

827. S. Leistner, A. P. Giro, and G. Wagner, *Z. Chem.*, 1977, **17**, 444.

828. T. Sakamoto, T. Sakasai, and H. Yamanaka, *Heterocycles*, 1978, **9**, 481.

829. T. Higashino, Y. Kuwada, and E. Hayashi, *Heterocycles*, 1977, **8**, 159.

830. T. Sakamoto, S. Konno, and H. Yamanaka, *Heterocycles*, 1977, **6**, 1616.

831. T. Kametani, K. Fukumoto, M. Ihara, and Chu Van Loc, *Heterocycles*, 1977, **6**, 1741.

832. M. Michman, S. Patai, and Y. Wiesel, *Org. Prep. Proced. Int.*, 1978, **10**, 13.

833. Z. Buděšinský, P. Lederer, and J. Daněk, *Collect. Czech. Chem. Commun.*, 1977, **42**, 3473.

834. V. P. Krivopalov and V. P. Mamaev, *Izv. Akad. Nauk SSSR, Ser. Khim.*, 1977, 966.

835. V. I. Gorbatenko, V. N. Fetyukhin, N. V. Mel'nichenko, and L. I. Samarai, *Zh. Org. Khim.*, 1977, **13**, 2320.

836. A. V. Bogatskii, S. A. Andronati, Z. I. Zhilina, and N. I. Danilina, *Zh. Org. Khim.*, 1977, **13**, 1773.

837. R. S. Matthews, *Org. Magn. Reson.*, 1977, **9**, 318.

838. V. K. Pandey, *J. Indian Chem. Soc.*, 1977, **54**, 1084.

839. J. P. Trivedi and R. J. Shah, *J. Indian Chem. Soc.*, 1976, **53**, 952.

840. P. N. Bhargava and M. R. Chaurasia, *J. Indian Chem. Soc.*, 1976, **53**, 46.

841. S. S. Tiwari and V. K. Pandey, *J. Indian Chem. Soc.*, 1975, **52**, 736.

842. S. S. Tiwari and S. B. Misra, *J. Indian Chem. Soc.*, 1975, **52**, 1073.

843. A. Singh and B. M. Bhandari, *Indian J. Chem., Sect. B*, 1976, **14**, 67.

844. G. Kumar, B. Lal, P. Singh, and A. P. Bhaduri, *Indian J. Chem., Sect. B*, 1976, **14**, 133.

845. M. Seth and N. M. Khanna, *Indian J. Chem., Sect. B*, 1976, **14**, 536.

846. H. Singh and K. B. Lal, *Indian J. Chem., Sect. B*, 1976, **14**, 685.

847. M. Z. Kirmani and S. R. Ahmed, *Indian J. Chem., Sect. B*, 1977, **15**, 748.

848. A. Singh, *Indian J. Chem., Sect. B*, 1977, **15**, 751.

849. K. Sasse, *Synthesis*, 1978, 379.

850. K. Burger and S. Penninger, *Synthesis*, 1978, 524.

851. A. R. Katritzky and S. B. Brown, *Synthesis*, 1978, 619.

852. J. Goerdeler and R. Richter, *Synthesis*, 1978, 760.

853. E. B. Pedersen, *Synthesis*, 1977, 180.

854. M. S. Manhas, S. G. Amin, and V. V. Rao, *Synthesis*, 1977, 309.

855. V. Gómez-Parra, R. Madroñero, and S. Vega, *Synthesis*, 1977, 345.

856. P. Trimarco and C. Lastrucci, *J. Heterocycl. Chem.*, 1976, **13**, 913.

857. C. Vigne, M. Buti, C. Montginoul, E. Torreilles, and L. Giral, *J. Heterocycl. Chem.*, 1976, **13**, 921.

858. N. P. Peet and S. Sunder, *J. Heterocycl. Chem.*, 1976, **13**, 967.

859. S. C. Bell, G. Conklin, and R. J. McCaully, *J. Heterocycl. Chem.*, 1976, **13**, 51.

860. C. F. Beam, N. D. Heindel, M. Chun, and A. Stefanski, *J. Heterocycl. Chem.*, 1976, **13**, 421.

861. R. I. Fryer, J. V. Earley, N. W. Gilman, and W. Zally, *J. Heterocycl. Chem.*, 1976, **13**, 433.

862. N. P. Peet, S. Sunder, and D. L. Trepanier, *Indian J. Chem., Sect. B*, 1976, **14**, 701.

863. V. P. Arya, K. G. Dave, V. G. Khadse, and S. J. Shenoy, *Indian J. Chem., Sect. B*, 1976, **14**, 879.

864. A. K. S. Gupta, K. C. Agarwal, and P. K. Seth, *Indian J. Chem., Sect. B*, 1976, **14**, 1000.

865. A. Walser, T. Flynn, and R. I. Fryer, *J. Heterocycl. Chem.*, 1975, **12**, 717.

866. S. Palazzo, L. I. Giannola, and M. Neri, *J. Heterocycl. Chem.*, 1975, **12**, 1077.

867. S. C. Bell, C. Gochman, and P. H. L. Wei, *J. Heterocycl. Chem.*, 1975, **12**, 1207.

868. S. M. Deshpande and K. R. Reddy, *Indian J. Chem., Sect. B*, 1977, **15**, 198.

869. V. Purnaprajna and S. Seshadri, *Indian J. Chem., Sect. B*, 1977, **15**, 335.

870. M. A. Elkasaby, *Indian J. Chem., Sect. B*, 1977, **15**, 690.

871. S. P. Acharya and J. B. Hynes, *J. Heterocycl. Chem.*, 1975, **12**, 1283.

872. A. Mendel, *J. Heterocycl. Chem.*, 1977, **14**, 153.

873. M. S. Manhas and S. G. Amin, *J. Heterocycl. Chem.*, 1977, **14**, 161.

874. H. S. Broadbent, R. C. Anderson, and M. C. J. Kuchar, *J. Heterocycl. Chem.*, 1977, **14**, 289.

875. J. Bergman, R. Carlsson, and J.-O. Lindström, *Tetrahedron Lett.*, 1976, 3611.

876. J. Bergman, J.-O. Lindström, J. Abrahamsson, and E. Hadler, *Tetrahedron Lett.*, 1976, 3615.

877. G. Landen and H. W. Moore, *Tetrahedron Lett.*, 1976, 2513.

878. G. R. Meyer and N. A. Rao, *J. Heterocycl. Chem.*, 1977, **14**, 335.

879. C. F. Beam, C. A. Park, N. D. Heindel, and W. P. Fives, *J. Heterocycl. Chem.*, 1977, **14**, 703.

880. P. B. Talukdar, S. K. Sengupta, A. K. Datta, and T. K. Roy, *Indian J. Chem., Sect. B*, 1977, **15**, 41.

881. M. Z. Kirmani and S. R. Ahmed, *Indian J. Chem., Sect. B*, 1977, **15**, 892.

882. P. B. Talukdar, S. K. Sengupta, and A. K. Datta, *Indian J. Chem., Sect. B*, 1977, **15**, 1110.

883. S. Plescia, G. Daidone, G. Dattolo, and E. Aiello, *J. Heterocycl. Chem.*, 1977, **14**, 1075.

884. D. R. Harrison, P. D. Kennewell, and J. B. Taylor, *J. Heterocycl. Chem.*, 1977, **14**, 1191.

885. H. L. Yale, *J. Heterocycl. Chem.*, 1977, **14**, 1357.

886. S. P. Singh, S. S. Parmar, V. I. Stenberg, and T. K. Akers, *J. Heterocycl. Chem.*, 1978, **15**, 53.

887. T. Yamazaki, K. Matoba, A. Shirokawa, and R. N. Castle, *J. Heterocycl. Chem.*, 1978, **15**, 467.

888. I. A. Silberg, R. Macarovici, and N. Palibroda, *Tetrahedron Lett.*, 1976, 1321.

889. M. Hedayatullah and J. Pailler, *J. Heterocycl. Chem.*, 1978, **15**, 1033.

890. R. Kreher and U. Bergmann, *Tetrahedron Lett.*, 1976, 4259.

891. W. Girke, *Tetrahedron Lett.*, 1976, 3537.

892. V. K. Rastogi, S. S. Parmar, S. P. Sing, and T. K. Akers, *J. Heterocycl. Chem.*, 1978, **15**, 497.

893. H. Breuer, *Tetrahedron Lett.*, 1976, 1935.

894. M. Ikeda, F. Tabusa, Y. Nishimura, S. Kwon, and Y. Tamura, *Tetrahedron Lett.*, 1976, 2347.

895. H. Singh, S. S. Narula, and C. S. Gandhi, *Tetrahedron Lett.*, 1977, 3747.

896. H. Singh, A. Kumar, K. S. Kumar, and K. B. Lal, *Indian J. Chem.*, 1975, **13**, 983.

897. G. Kumar and A. P. Bhaduri, *Indian J. Chem.*, 1975, **13**, 1009.

898. G. Subrahmanyam and T. D. Roy, *Indian J. Chem.*, 1975, **13**, 1119.

899. R. J. Lahoti, J. B. Chattopadhyaya, and A. V. R. Rao, *Indian J. Chem.*, 1975, **13**, 458.

900. J. Roy, M. Seth, and A. P. Bhaduri, *Indian J. Chem., Sect. B*, 1978, **16**, 41.

901. V. B. Rao and C. V. Ratnam, *Indian J. Chem., Sect. B*, 1978, **16**, 144.

902. L. P. Battaglia, A. B. Corradi, M. Nardelli, C. Pelizzi, and M. E. V. Tani, *J. Chem. Soc., Dalton Trans.*, 1976, 1076.

903. C. J. Gilmore, A. D. U. Hardy, D. D. MacNicol, and D. R. Wilson, *J. Chem. Soc., Perkin Trans. 2*, 1977, 1427.

904. T. Nagasaka, F. Hamaguchi, N. Ozawa, and S. Ohki, *Heterocycles*, 1978, **9**, 1375.

905. J. G. Smith, J. M. Sheepy, and E. M. Levi, *J. Org. Chem.*, 1976, **41**, 497.

906. G. M. Coppola, G. E. Hardtmann, and O. R. Pfister, *J. Org. Chem.*, 1976, **41**, 825.

907. O. S. Tee and G. V. Patil, *J. Org. Chem.*, 1976, **41**, 838.

908. T. Kametani, C. V. Loc, M. Ihara, and K. Fukumoto, *Heterocycles*, 1978, **9**, 1585.

909. D. F. Wiemer, D. I. C. Scopes, and N. J. Leonard, *J. Org. Chem.*, 1976, **41**, 3051.

910. E. P. Papadopoulos and B. George, *J. Org. Chem.*, 1977, **42**, 2530.

911. S. Sunder and N. P. Peet, *J. Org. Chem.*, 1977, **42**, 2551.

912. L. A. Errede, H. T. Oien, and D. R. Yarian, *J. Org. Chem.*, 1977, **42**, 12.

913. B. Danieli and G. Palmisano, *Heterocycles*, 1978, **9**, 803.

914. H. A. Hammouda and N. Ishikawa, *Bull. Chem. Soc. Jpn.*, 1978, **51**, 3091.

915. G. E. Keyser and N. J. Leonard, *J. Org. Chem.*, 1976, **41**, 3529.

916. C. H. Foster and E. U. Elam, *J. Org. Chem.*, 1976, **41**, 2646.

917. A. F. M. Fahmy, N. F. Aly, and M. O. Orabi, *Bull. Chem. Soc. Jpn.*, 1978, **51**, 2148.

918. S. C. Pakrashi, S. Chattopadhyay, and A. K. Chakravarty, *J. Org. Chem.*, 1976, **41**, 2108.

919. J. F. W. Keana, J. S. Bland, P. E. Eckler, V. Nelson, and J. Z. Gougoutas, *J. Org. Chem.*, 1976, **41**, 2124.

920. R. Y. Ning, R. I. Fryer, P. B. Madan, and B. C. Sluboski, *J. Org. Chem.*, 1976, **41**, 2720.

921. N. P. Peet, S. Sunder, and R. J. Cregge, *J. Org. Chem.*, 1976, **41**, 2733.

922. H. Singh and C. S. Gandhi, *Synth. Commun.*, 1978, **8**, 469.

923. C. Wentrup, *Helv. Chim. Acta*, 1978, **61**, 1755.

924. L. A. Errede and J. J. McBrady, *J. Org. Chem.*, 1977, **42**, 3863.

925. J. P. Ferris, S. Singh, and T. A. Newton, *J. Org. Chem.*, 1979, **44**, 173.

926. Y. Sugiyama, T. Sasaki, and N. Nagato, *J. Org. Chem.*, 1978, **43**, 4485.

927. T. Hara, Y. Kayama, and T. Sunami, *J. Org. Chem.*, 1978, **43**, 4865.

928. M. H. P. Ardebili and J. G. White, *Acta Crystallogr., Sect. B*, 1978, **34**, 2890.

929. H.-J. Kabbe, *Liebigs Ann. Chem.*, 1978, 398.

930. E. G. Ter-Gabrielyan, N. P. Gambaryan, and Y. V. Zeifman, *Izv. Akad. Nauk SSSR, Ser. Khim.*, 1978, 1888.

931. L. A. Errede, *J. Org. Chem.*, 1976, **41**, 1763.

932. L. A. Errede, J. J. McBrady, and H. T. Oien, *J. Org. Chem.*, 1976, **41**, 1765.

933. J. G. Smith and J. M. Sheepy, *J. Org. Chem.*, 1977, **42**, 78.

934. R. W. Leiby and N. D. Heindel, *J. Org. Chem.*, 1977, **42**, 161.

935. L. A. Errede, J. J. McBrady, and H. T. Oien, *J. Org. Chem.*, 1977, **42**, 656.

936. E. E. Schweizer and S. D. Goff, *J. Org. Chem.*, 1978, **43**, 2972.

937. H. M. Blatter, H. Lukaszewski, and G. de Stevens, *J. Org. Chem.*, 1965, **30**, 1020.

938. R. J. Grout and M. W. Partridge, *J. Chem. Soc.*, 1960, 3540.

939. F. H. S. Curd, J. K. Landquist, and F. L. Rose, *J. Chem. Soc.*, 1948, 1759.

940. P. K. Chaudhury, *Phytochemistry*, 1987, **26**, 587.

941. D. J. Brown, *Fused Pyrimidines: Pteridines*, Wiley, New York, 1988, p. 301.

942. D. R. Desai, V. S. Patel, and S. R. Patel, *J. Indian Chem. Soc.*, 1966, **43**, 351.

943. C. M. Gupta, A. P. Bhaduri, and N. M. Khanna, *J. Med. Chem.*, 1968, **11**, 392.

944. R. K. Russell, M. A. Appollina, V. Bandurco, D. W. Combs, R. M. Kanojia, R. Mallory, E. Malloy, J. J. McNally, D. M. Mulvey, Y. Gray-Nunez, M. S. Rampulla, R. A. Rampulla, S. A. Sisk, L. Williams, S. D. Levine, S. C. Bell, E. C. Giardino, R. Falotico, and A. J. Tobia, *Eur. J. Med. Chem.*, 1992, **27**, 277.

945. Y. C. Meinwald, J. Meinwald, and T. Eisner, *Science*, 1966, **154**, 390.

946. H. Schildknecht and W. F. Wenneis, *Z. Naturforsch., Teil B*, 1966, **21**, 552.

947. R. W. Leiby, E. G. Corley, and N. D. Heindel, *J. Org. Chem.*, 1978, **43**, 3427.

948. R. P. Rhee and J. D. White, *J. Org. Chem.*, 1977, **42**, 3650.

949. D. J. Brown, L. Danckwerts, G. W. Grigg, and Y. Iwai, *Aust. J. Chem.*, 1979, **32**, 453.

950. F. D. Eddy, K. Vaughan, and M. F. G. Stevens, *Can. J. Chem.*, 1978, **56**, 1616.

951. D. van Broeck, Z. Janousek, R. Merényi, and H. G. Viehe, *Angew. Chem.*, 1979, **91**, 355.

952. J. R. Barrio, F.-I. Liu, G. E. Keyser, P. van der Lijn, and N. J. Leonard, *J. Am. Chem. Soc.*, 1979, **101**, 1564.

953. L. Legrand and N. Lozac'h, *Phosphorus Sulfur*, 1978, **5**, 209.

954. S. K. Dubey, S. Sharma, and R. N. Iyer, *Z. Naturforsch., Teil B*, 1979, **34**, 99.

955. Y. Tamura, T. Kawaski, M. Tanio, and Y. Kita, *Synthesis*, 1979, 120.

956. S. J. Yan, L. T. Weinstock, and C. C. Cheng, *J. Heterocycl. Chem.*, 1979, **16**, 541.

957. V. I. Cohen, *J. Heterocycl. Chem.*, 1978, **15**, 1415.

958. K. Nagahara and A. Takada, *Heterocycles*, 1979, **12**, 239.

959. M. Ogata and H. Matsumoto, *Heterocycles*, 1978, **11**, 139.

960. T. Kurihara and Y. Sakamoto, *Heterocycles*, 1978, **9**, 1729.

961. J. Bhattacharyya and S. C. Pakrashi, *Heterocycles*, 1979, **12**, 929.

962. G. A. Brine, M. L. Coleman, and F. I. Carroll, *J. Heterocycl. Chem.*, 1979, **16**, 25.

963. S. S. Parmar and S. P. Singh, *J. Heterocycl. Chem.*, 1979, **16**, 449.

964. S. Plescia, G. Daidone, V. Sprio, E. Aiello, G. Datollo, and G. Cirrincione, *J. Heterocycl. Chem.*, 1978, **15**, 1339.

965. J. A. Bristol, *J. Heterocycl. Chem.*, 1978, **15**, 1409.

966. A. Mendel and G. J. Lillquist, *J. Heterocycl. Chem.*, 1979, **16**, 617.

967. S. P. Singh, S. S. Parmar, and S. A. Farnum, *J. Heterocycl. Chem.*, 1979, **16**, 649.

968. M. S. Manhas, W. A. Hoffman, and A. K. Bose, *J. Heterocycl. Chem.*, 1979, **16**, 711.

969. Ö. Ericsson, *Acta Pharm. Suec.*, 1978, **15**, 81; *Chem. Abstr.*, 1978, **89**, 180294.

970. L. Falco da Fonseca, *Rev. Port. Farm.*, 1978, **28**, 175; *Chem. Abstr.*, 1979, **90**, 152116.

971. S. El-Zanfally, *Egypt. J. Pharm. Sci.*, 1976, **17**, 29; *Chem. Abstr.*, 1979, **90**, 121516.

972. S. Johne , M. Süsse, and B. Jung, *Pharmazie*, 1978, **33**, 821; *Chem. Abstr.*, 1979, **90**, 121531.

973. L. Nováček, J. Belusa, V. Hruskova, and D. Vavrinova, *Cesk. Farm.*, 1978, **27**, 173; *Chem. Abstr.*, 1979, **90**, 81450.

974. P. N. Bhargave and F. Bahadur, *J. Indian Chem. Soc.*, 1978, **55**, 293.

975. R. Hull and M. L. Swain, *J. Chem. Soc., Perkin Trans. 1*, 1976, 653.

976. N. C. Gupta, *Acta Cienc. Indica*, 1978, **4**, 147; *Chem. Abstr.*, 1979, **90**, 87392.

977. S. Botros, *Pharmazie*, 1979, **34**, 113; *Chem. Abstr.*, 1979, **91**, 56948.

978. R. M. Shafik, A. A. B. Hazzaa, and N. S. Habib, *Pharmazie*, 1979, **34**, 148; *Chem. Abstr.*, 1979, **91**, 56950.

979. V. K. Panday, *Acta Cienc. Indica*, 1978, **4**, 230; *Chem. Abstr.*, 1979, **91**, 39421.

980. R. P. Rao, B. Sharma, and N. Zaidi, *Acta Cienc. Indica*, 1978, **4**, 254; *Chem. Abstr.*, 1979, **91**, 39422.

981. M. A. El-Hashash, M. A. Hassan, and M. A. Sayed, *Pak. J. Sci. Ind. Res.*, 1977, **20**, 336; *Chem. Abstr.*, 1979, **91**, 39424.

982. D. D. Mukerji, S. R. Nautiyal, and B. N. Dhawan, *Indian J. Pharm. Sci.*, 1979, **41**, 33; *Chem. Abstr.*, 1979, **91**, 5182.

983. T. Hisano, M. Ichikawa, A. Nakagawa, and M. Tsuji, *Chem. Pharm. Bull.*, 1975, **23**, 1910.

984. T. Koyama, T. Hirota, Y. Shinohara, S. Matsumoto, S. Ohmori, and M. Yamato, *Chem. Pharm. Bull.*, 1975, **23**, 2029.

985. T. Koyama, T. Hirota, C. Bashou, Y. Satoh, Y. Watanabe, S. Matsumoto, Y. Shinohara, S. Ohmori, and M. Yamato, *Chem. Pharm. Bull.*, 1975, **23**, 2158.

986.   K. Ishizumi, K. Mori, S. Inaba, and H. Yamamoto, *Chem. Pharm. Bull.*, 1975, **23**, 2169.

987.   M. Sharma, K. Shanker, K. P. Bhargava, and K. Kishor, *Indian J. Pharm. Sci.*, 1979, **41**, 44; *Chem. Abstr.*, 1979, **91**, 5195.

988.   A. F. M. Fahmy, M. A. El-Hashash, M. M. Habashy, and S. A. El-Wannise, *Rev. Roum. Chim.*, 1978, **23**, 1567; *Chem. Abstr.*, 1979, **91**, 20422.

989.   T. L. Gilchrist, C. J. Moody, and C. W. Rees, *J. Chem. Soc., Perkin Trans. 1*, 1975, 1964.

990.   D. E. Ames, S. Chandrasekhar, and R. Simpson, *J. Chem. Soc., Perkin Trans. 1*, 1975, 2035.

991.   K. Nagahara, K. Takagi, and T. Ueda, *Chem. Pharm. Bull.*, 1976, **24**, 1197.

992.   K. Nagahara, K. Takagi, and T. Ueda, *Chem. Pharm. Bull.*, 1976, **24**, 1310.

993.   T. Koyama, T. Hirota, C. Basho, Y. Watanabe, Y. Kitauchi, Y. Satoh, S. Ohmori, and M. Yamato, *Chem. Pharm. Bull.*, 1976, **24**, 1459.

994.   T. Hisano, M. Ichikawa, K. Muraoka, Y. Yabuta, Y. Kido, and M. Shibata, *Chem. Pharm. Bull.*, 1976, **24**, 2244.

995.   I. A. Mazur, *Farm. Zh. (Kiev)*, 1977 (No. 6), 37; *Chem. Abstr.*, 1978, **88**, 104849.

996.   P. Daenens and M. von Boven, *J. Forensic Sci.*, 1976, **21**, 552; *Chem. Abstr.*, 1978, **88**, 69920.

997.   V. S. Misra, R. N. Pandey, S. Dhar, and K. N. Dhawan, *Indian J. Med. Res.*, 1978, **67**, 310; *Chem. Abstr.*, 1978, **89**, 90178.

998.   G. Wagner and I. Wunderlich, *Pharmazie*, 1978, **33**, 15; *Chem. Abstr.*, 1978, **89**, 43166.

999.   I. A. Mazur, A. A. Martynovskii, L. I. Borodin, R. I. Katkevich, and E. G. Knysh, *Farm. Zh. (Kiev)*, 1978 (No. 1), 56; *Chem. Abstr.*, 1978, **89**, 43167.

1000.   D. J. Brown and K. Ienaga, *J. Chem. Soc., Perkin Trans. 1*, 1975, 2182.

1001.   P. J. Abbott, R. M. Acheson, M. Y. Kornilov, and J. K. Stubbs, *J. Chem. Soc., Perkin Trans. 1*, 1975, 2322.

1002.   V. Zotta, A. Berechet-Paun, I. Chirita, A. Missir, and M. Neacsu, *Farmacia (Bucharest)*, 1977, **25**, 207; *Chem. Abstr.*, 1978, **89**, 43308.

1003.   S. Palazzo, L. I. Giannola, and S. Caronna, *Atti Accad. Sci., Lett. Arti Palermo, Parte 1*, 1976, **34**, 339, *Chem. Abstr.*, 1978, **89**, 43309.

1004.   W. L. F. Armarego and P. A. Reece, *J. Chem. Soc., Perkin Trans. 1*, 1975, 1470.

1005.   A. Banerji, J. C. Cass, and A. R. Katritzky, *J. Chem. Soc., Perkin Trans. 1*, 1977, 1162.

1006.   H. Singh and C. S. Gandhi, *Indian J. Chem., Sect. B*, 1978, **16**, 331.

1007.   A. F M. Fahmy, N. F. Aly, A. Nada, and N. Y. Aly, *Indian J. Chem., Sect. B*, 1978, **16**, 484.

1008.   M. Z. Kirmani and S. R. Ahmed, *Indian J. Chem., Sect. B*, 1978, **16**, 526.

1009.   R. P. Gupta, M. L. Sachdeva, R. N. Handa, and H. K. Pujari, *Indian J. Chem., Sect. B*, 1978, **16**, 537.

1010.   S. Leistner, A. P. Giro, and G. Wangner, *Pharmazie*, 1978, **33**, 185.

1011.   P. J. G. Cornelissen, G. M. J. Beijersbergen-van-Henegouwen, and K. W. Gerritsma, *Int. J. Pharm.*, 1978, **1**, 173; *Chem. Abstr.*, 1978, **89**, 117635.

1012.   M. Anwar, F. I. Abdel–Hay, and M. Fahmy, *Rev. Roum. Chim.*, 1978, **23**, 1085; *Chem. Aabstr.*, 1979, **90**, 54899.

1013.   A. Gescher, C. P. Turrbull, and M. F. G. Stevens, *J. Chem. Soc., Perkin Trans. 1*, 1977, 2078.

1014.   A. Gescher, M. F. G. Stevens, and C. P. Turnbull, *J. Chem. Soc., Perkin Trans. 1*, 1977, 103.

1015.   T. B. Talukdar, S. K. Sengupta, and A. K. Datta, *Indian J. Chem., Sect. B*, 1978, **16**, 678.

1016.   E. A. Soliman and G. Hosni, *Indian J. Chem., Sect. B.*, 1978, **16**, 884.

1017.   R. P. Rao, B. Sharma, and N. Ziadi, *Indian J. Chem., Sect. B*, 1978, **16**, 1023.

1018.   A. Gescher, M. F. G. Stevens, and C. P. Turnbull, *J. Chem. Soc., Perkin Trans. 1*, 1977, 107.

1019.   A. Sammour, A. A. Afify, M. Abdallah, and E. A. Soliman, *Egypt. J. Chem.*, 1976, **19**, 1109; *Chem. Abstr.*, 1979, **91**, 175295.

1020. P. A. Bezuglyi and V. P. Chernykh, *Farm Zh.* (*Kiev*), 1979 (No. 4), 70; *Chem. Abstr.*, 1979, **91**, 193254.

1021. A. V. Bogatskii, Z. I. Zhilina, S. A. Andronati, and N. I. Danilina, *Khim. Promst.*, *Ser. Reakt. Osobo Chist Veshchestva*, 1979, 33; *Chem. Abstr.*, 1979, **91**, 123703.

1022. P. N. Bharvava and H. D. Singh, *Indian J. Chem.*, *Sect. B*, 1977, **15**, 659.

1023. J. W. Cornforth, *J. Chem. Soc.*, *Perkin Trans. 1*, 1976, 2004.

1024. C. W. Rees, R. Somanathan, R. C. Storr, and A. D. Woolhouse, *J. Chem. Soc.*, *Chem. Commun.*, 1975, 740.

1025. J. J. Barr and R. C. Storr, *J. Chem. Soc.*, *Perkin Trans. 1*, 1979, 185.

1026. T. Kametani, Chu Van Loc, T. Higa, M. Ihara, and K. Fukumoto, *J. Chem. Soc.*, *Perkin Trans. 1*, 1977, 2347.

1027. A. Parg and G. Hamprecht, *Liebigs Ann. Chem.*, 1979, 1130.

1028. W. P. K. Girke, *Chem. Ber.*, 1979, **112**, 1348.

1029. J. Gasparič, J. Zimák, P. Sedmera, Z. Breberová, and J. Volke, *Collect. Czech. Chem. Commun.*, 1979, **44**, 2243.

1030. F. Huys, R. Merényi, Z. Janovsek, L. Stella, and H. G. Viehe, *Angew. Chem.*, 1979, **91**, 650.

1031. K. M. Murav'eva, N. V. Arkhangel' skaya, M. N. Shchukina, T. N. Zykova, and G. N. Pershin, *Khim. Geterotsikl. Soedin.*, *Sb. 1*, 1967, 411; *Chem. Abstr.*, 1969, **70**, 96748.

1032. V. F. Sedova and V. P. Mamaev, *Khim. Geterotsikl. Soedin.*, *Sb. 1*, 1967, 349; *Chem. Abstr.*, 1969, **70**, 87722.

1033. P. A. Petyunin and Y. V. Kozhevnikov, *Khim. Geterotsikl. Soedin.*, *Sb. 1*, 1967, 415.

1034. L. I. Samarai, V. A. Bondar, and G. I. Derkach, *Khim. Geterotsikl. Soedin.*, 1968, 1099.

1035. V. S. Misra and S. Dhar, *J. Indian Chem. Soc.*, 1978, **55**, 172.

1036. P. C. Joshi and P. C. Joshi, *J. Indian Chem. Soc.*, 1978, **55**, 465.

1037. S. Botros and M. Shaban, *Pharmazie*, 1978, **33**, 646; *Chem. Abstr.*, 1979, **90**, 54902.

1038. G. A. Brinkman, I. Hass-Lisewska, J. T. Veenboer, and L. Lindner, *Int. J. Appl. Radiat. Isot.*, 1978, **29**, 701; *Chem. Abstr.*, 1979, **90**, 68538.

1039. A. Albert and G. B. Barlin, *J. Chem. Soc.*, 1962, 3129.

1040. V. S. Misra, R. N. Pandey, and P. R. Dua, *Pol. J. Pharmacol. Pharm.*, 1978, **30**, 573; *Chem. Abstr.*, 1979, **91**, 20437.

1041. M. Dubey, V. K. Verma, K. Shanker, J. N. Sinha, K. P. Bhargava, and K. Kishor, *Pharmazie*, 1979, **34**, 18; *Chem. Abstr.*, 1979, **91**, 20440.

1042. M. Z. A. Badr and H. A. H. El-Sherif, *Egypt. J. Chem.*, 1976, **19**, 341; *Chem. Abstr.*, 1979, **91**, 211313.

1043. S. S. Tiwari and R. K. Satsangi, *J. Indian Chem. Soc.*, 1978, **55**, 477.

1044. D. D. Mukerji and S. R. Nautiyal, *J. Indian Chem. Soc.*, 1978, **55**, 709.

1045. P. Singh, *J. Indian Chem. Soc.*, 1978, **55**, 801.

1046. V. K. Singh and K. C. Joshi, *J. Indian Chem. Soc.*, 1978, **55**, 928.

1047. V. S. Misra, R. N. Pandey, and K. N. Dhawan, *J. Indian Chem. Soc.*, 1978, **55**, 1046.

1048. P. N. Bhargava and S. Prakash, *J. Indian Chem. Soc.*, 1977, **54**, 881.

1049. B. P. Joshi and B. D. Hosangadi, *Indian J. Chem.*, *Sect. B*, 1978, **16**, 1067.

1050. G. S. Reddy and K. K. Reddy, *Indian J. Chem.*, *Sect. B*, 1978, **16**, 1109.

1051. J. Fischer, G. Toth, and P. Vago, *Acta Chim. Acad. Sci. Hung.*, 1977, **93**, 95; *Chem. Abstr.*, 1978, **88**, 50787.

1052. M. Uchida, T. Higashino, C. Iijima, N. Shimada, and E. Hayashi, *Shitsurgo Bunseki*, 1977, **25**, 169; *Chem. Abstr.*, 1978, **88**, 21568.

1053. J. Sykulski and J. Czyzewska, *Rocz. Chem.*, 1977, **51**, 1216; *Chem. Abstr.*, 1978, **88**, 22817.

1054. Y. V. Kozhevnikov and P. A. Petyunin, *Nauchn. Tr. Permsk. Gos. Farm. Inst.*, 1975, **8**, 32; *Chem. Abstr.*, 1978, **88**, 22823.

1055. G. P. Zhikhareva, E. A. Berlyand, S. S. Liberman, and L. N. Yakhontov, *Khim.-Farm. Zh.*, 1977, **11** (No. 10) 58; *Chem. Abstr.*, 1978, **88**, 22828.

1056. S. Botros and M. Khalifa, *Egypt. J. Pharm. Sci.*, 1975, **16**, 57; *Chem. Abstr.*, 1977, **87**, 167837.

1057. U. G. Joshi, S. R. Patel, and K. A. Thaker, *J. Inst. Chem. (India)*, 1977, **49**, 153; *Chem. Abstr.*, 1977, **87**, 167972.

1058. M. Lempert-Sreter, K. Lempert, P. Bruck, and G. Toth, *Acta Chim. Acad. Sci. Hung.*, 1977, **94**, 391; *Chem. Abstr.*, 1978, **88**, 170106.

1059. T. Kato, A. Takada, and T. Ueda, *Chem. Pharm. Bull.*, 1976, **24**, 431.

1060. M. Yamamoto, S. Inaba, and H. Yamamoto, *Chem. Pharm. Bull.*, 1978, **26**, 1633.

1061. N. Masai, M. Kimura, M. Yamamoto, T. Hirohashi, and H. Yamamoto, *Chem. Pharm. Bull.*, 1977, **25**, 3018.

1062. H. Yamanaka, H. Abe, and T. Sakamoto, *Chem. Pharm. Bull.*, 1977, **25**, 3334.

1063. H. Yamanaka, H. Abe, T. Sakamoto, H. Hiranuma, and A. Kamata, *Chem. Pharm. Bull.*, 1977, **25**, 1821.

1064. K. Kottke and H. Kühmstedt, *Pharmazie*, 1978, **33**, 462; *Chem. Abstr.*, 1978, **89**, 163534.

1065. K. Kottke and H. Kühmstedt, *Pharmazie*, 1978, **33**, 124; *Chem. Abstr.*, 1978, **89**, 129469.

1066. K. Kottke and H. Kühmstedt, *Pharmazie*, 1978, **33**, 125; *Chem. Abstr.*, 1978, **89**, 129470.

1067. K. Golankiewicz and L. Celewicz, *Pol. J. Chem.*, 1978, **52**, 1035; *Chem. Abstr.*, 1978, **89**, 129475.

1068. M. Ali and A.-E. G. Hammam, *J. Chem. Eng. Data*, 1978, **23**, 351; *Chem. Abstr.*, 1978, **89**, 146865.

1069. M. Selim, A. Sammour, M. Abdalla, and M. Elkasaby, *Pak. J. Sci. Res.*, 1975, **27**, 67; *Chem. Abstr.*, 1978, **89**, 109338.

1070. P. N. Bhargava and S. N. Singh, *J. Sci. Res. Banares Hindu Univ.*, 1976, **26**, 27; *Chem. Abstr.*, 1978, **89**, 109345.

1071. S. Johne and B. Jung, *Pharmazie*, 1978, **33**, 299; *Chem. Abstr.*, 1978, **89**, 109352.

1072. A. Sammour, A. Rabie, M. El-Hashash, and M. Sayed, *Egypt. J. Chem.*, 1976, **19**, 571; *Chem. Abstr.*, 1979, **91**, 211360.

1073. A. Kosasayama, K. Higashi, and F. Ishikawa, *Chem. Pharm. Bull.*, 1979, **27**, 880.

1074. H. Asakawa, M. Matano, and Y. Kawamatsu, *Chem. Pharm. Bull.*, 1979, **27**, 287.

1075. M. El-Kerdawy, A. M. Samour, and A. G. El-Agamey, *Acta Pharm. Jugosl.*, 1976, **26**, 135; *Chem. Abstr.*, 1976, **85**, 46571.

1076. D. Twomey, *Proc. R. Ir. Acad., Sect. B*, 1976, **76**, 79; *Chem. Abstr.*, 1976, **85**, 46579.

1077. R. S. Sinyak, I. A. Mazur, and P. M. Kochergin, *Khim.-Farm. Zh.*, 1976, **10**, (No. 3), 67; *Chem. Abstr.*, 1976, **85**, 63022.

1078. B. Arventiev and H. Wexler, *Bul. Inst. Politeh. Iasi, Sect. 2*, 1976, **22**, 67; *Chem. Abstr.*, 1977, **86**, 106515.

1079. M. I. El-Ashmawi, M. M. Badran, and M. Khalifa, *Pharmazie*, 1976, **31**, 601; *Chem. Abstr.*, 1977, **86**, 72562.

1080. C. Thétaz, F. W. Wehrli, and C. Wentrup, *Helv. Chim. Acta*, 1976, **59**, 259.

1081. S. K. V. Seshavataram and N. V. S. Rao, *Proc. Indian Acad. Sci., Sect. A*, 1977, **85**, 81; *Chem. Abstr.*, 1977, **86**, 183805.

1082. P. Hanumanthu, S. K. V. Seshavatharam, C. V. Ratnam, and N. V. S. Rao, *Proc. Indian Acad. Sci., Sect. A*, 1976, **84**, 57; *Chem. Abstr.*, 1977, **86**, 16634.

1083. Y. Tamura, M. W. Chun, H. Nishida, S. Kwon, and M. Ikeda, *Chem. Pharm. Bull.*, 1978, **26**, 2866.

1084. T. Higashino, K. Suzuki, and E. Hayashi, *Chem. Pharm. Bull.*, 1978, **26**, 3485.

1085. R. S. Sinyak, I. A. Mazur, P. M. Steblyuk, and P. M. Kochergin, *Farm. Zh. (Kiev)*, 1977 (No. 3), 84; *Chem. Abstr.*, 1977, **87**, 102264.

1086. H. Bräuniger, B. Fokken, H. Kristen, and K. Peseke, *Pharmazie*, 1977, **32**, 150; *Chem. Abstr.*, 1977, **87**, 84929.

1087. K. Kottke, F. Friedrich, D. Knoke, and H. Kühmstedt, *Pharmazie*, 1977, **32**, 540; *Chem. Abstr.*, 1978, **88**, 37743.

1088. M. Anwar and S. H. Etaiw, *Rev. Roum. Chim.*, 1977, **22**, 1217; *Chem. Abstr.*, 1978, **88**, 49837.

1089. K. Kottke, and H. Kühmstedt, *Pharmazie*, 1977, **33**, 19; *Chem. Abstr.*, 1978, **89**, 24242.

1090 M. Nikolova, *Med. Actual.*, 1978, **14**, 210; *Chem. Abstr.*, 1978, **89**, 152604.

1091. A. M. El-Abbady, M. Anwar, F. I. Abdel-Hay, and M. F. Abdel-Megeed, *Egypt. J. Chem.*, 1975, **18**, 1063; *Chem. Abstr.*, 1978, **89**, 163528.

1092. P. N. Bhargava and R. Shyam, *Egypt. J. Chem.*, 1975, **18**, 393; *Chem. Abstr.*, 1978, **89**, 163530.

1093. K. Edo, T. Sakamoto, and H. Yamanaka, *Chem. Pharm. Bull.*, 1978, **26**, 3843.

1094. D. A. de Bie, A. Nagel, H. C. van der Plas, G. Geurtsen, and A. Koudijs, *Tetrahedron Lett.*, 1979, 649.

1095. G. Malesani and G. Chiarelotto, *Atti Ist. Veneto Sci. Lett. Arti, Cl. Sci. Mat. Nat.*, 1972, **131**, 9; *Chem. Abstr.*, 1975, **83**, 178983.

1096. V. P. Krivopalov, A. Y. Denisov, Y. V. Catilov, V. I. Mamatyuk, and V. P. Mamaev, *Dokl. Akad. Nauk SSSR, Ser. Khim.*, 1988, **300**, 115.

1097. P. Lederer, V. Trcka, S. Hynie, and Z. Buděšinský, *Cesk. Farm.*, 1975, **24**, 201; *Chem. Abstr.*, 1976, **84**, 105533.

1098. L. Y. Yakhontov, G. P. Zhikhareva, E. V. Pronina, G. N. Pershin, S. S. Liberman, E. N. Padeiskaya, T. N. Zykova, T. A. Gus'kova, and E. A. Berlyand, *Khim.-Farm. Zh.*, 1975, **9** (No. 11), 12; *Chem. Abstr.*, 1976, **84**, 43974.

1099. A. R. E. Ossman and M. Khalifa, *Pharmazie*, 1975, **30**, 540; *Chem. Abstr.*, 1976, **84**, 4886.

1100. K. L. Brown and G. L. Gainsford, *Acta Crystallogr., Sect. B*, 1979, **35**, 2276.

1101. W. Fuhrer and H. W. Gschwend, *J. Org. Chem.*, 1979, **44**, 1133.

1102. Y. Imai, M. Ueda, and M. Ishimori, *J. Polym. Sci., Polym. Chem. Ed.*, 1975, **13**, 1969; *Chem. Abstr.*, 1976, **84**, 5418.

1103. M. Ishikawa and Y. Eguchi, *Iyo Kizai Kenkyusho Hokoku, Tokyo Ika Shika Daigaku*, 1974, **8**, 9; *Chem. Abstr.*, 1976, **84**, 17259.

1104. S. C. Bennur, V. B. Jigajinni, and V. V. Badiger, *Rev. Roum. Chim.*, 1975, **20**, 1995; *Chem. Abstr.*, 1976, **84**, 17268.

1105. J. Bergman, B. Egestad, and N. Eklund, *Tetrahedron Lett.*, 1978, 3147.

1106. A. Lespagnol, J. L. Bernier, C. Lespagnol, J. Cazin, and M. Cazin, *Eur. J. Med. Chem.*, 1974, **9**, 263; *Chem. Abstr.*, 1982, **82**, 16784.

1107. C. N. Reynolds, K. Wilson, and D. Burnett, *Xenobiotica*, 1976, **6**, 113; *Chem. Abstr.*, 1976, **82**, 173542.

1108. I. A. Mazur, R. I. Katkevich, L. I. Borodin, A. A. Martynovskii, and E. G. Knish, *Farm. Zh. (Kiev)*, 1976, **31** (No. 1), 63; *Chem. Abstr.*, 1976, **84**, 164588.

1109. T. Zamaity, M. Anwar, F. I. Abdel-Hay, and M. F. Abdel-Megeed, *Acta Chim. Acad. Sci. Hung.*, 1975, **87**, 251; *Chem. Abstr.*, 1976, **84**, 164710.

1110. N. Blazevic, M. Oklobdzija, V. Sunjic, F. Kajfez, and D. Kolbah, *Acta Farm. Jugosl.*, 1975, **25**, 223; *Chem. Abstr.*, 1976, **84**, 121762.

1111. T. Hisano, K. Shoji, and M. Ichikawa, *Org. Prep. Proced. Int.*, 1975, **7**, 271; *Chem. Abstr.*, 1976, **84**, 121768.

1112. A. F. Delong, R. D. Smyth, A. Polk, R. K. Nayak, and R. H. Reavey-Cantwell, *Arch. Int. Pharmacodyn. Ther.*, 1976, **222**, 322; *Chem. Abstr.*, 1976, **85**, 137141.

1113. G. P. Zhikhareva, E. V. Pronina, E. A. Golovanova, G. N. Pershin, N. A. Novitskaya, T. N. Zykova, T. A. Gus'kova, and L. N. Yakhontov, *Khim.-Farm. Zh.*, 1976, **10** (No. 4), 62; *Chem. Abstr.*, 1976, **85**, 123848.

1114. S. Botros and M. Khalifa, *Pharmazie*, 1976, **31**, 155; *Chem. Abstr.*, 1976, **85**, 46557.

1115. I. A. Mazur, A. F. Vlasenko, B. A. Samura, V. I. Lineko, and P. M. Kocherin, *Khim.-Farm. Zh.*, 1976, **10**, (No. 11), 60; *Chem. Abstr.*, 1977, **86**, 155597.

1116. T. Hisano, K. Muraoka, and M. Ichikawa, *Org. Prep. Proced. Int.*, 1977, **9**, 41; *Chem. Abstr.*, 1977, **87**, 53197.

1117. V. Zotta, L. Avram., A. Berechet-Paun, I. Chirita, and M. Neacsu, *Farmacia (Bucharest)*, 1977, **25**, 13; *Chem. Abstr.*, 1977, **87**, 53207.

1118. S. Botros, K. M. Ghoneim, and M. Khalifa, *Egypt. J. Pharm. Sci.*, 1975, **16**, 43; *Chem. Abstr.*, 1977, **87**, 53208.

1119. M. E. Esteve, S. Lamdan, and C. A. Gaozza, *Rev. Farm. (Buenos Aires)*, 1976, **118**, 118; *Chem. Abstr.*, 1977, **87**, 68307.

1120. P. N. Bhargava and S. Prakash, *Indian J. Pharm.*, 1977, **39**, 18; *Chem. Abstr.*, 1977, **87**, 39401.

1121. A. M. Abbady, M. Anwar, and M. F. Abdel-Megeed, *Acta Chim. Acad. Sci. Hung.*, 1976, **91**, 341; *Chem. Abstr.*, 1977, **87**, 23200.

1122. A. Lespagnol, D. Bar, M. Debaert, M. Polveche, and M. Defossez, *Bull. Soc. Pharm. Lille*, 1977, **33**, 67; *Chem. Abstr.*, 1977, **87**, 152121.

1123. R. C. Bugle and R. A. Osteryoung, *J. Org. Chem.*, 1979, **44**, 1719.

1124. A. K. Singh and R. P. Singh *J. Inorg. Nucl. Chem.*, 1980, **42**, 286.

1125. O. P. Pandey, S. K. Sengupta, and S. C. Tripathi, *Inorg. Chim. Acta*, 1984, **90**, 91.

1126. W. M. Coleman, *Inorg. Chim. Acta*, 1983, **70**, 211.

1127. E. S. Hand and D. C. Baker, *Can. J. Chem.*, 1984, **62**, 2570.

1128. I. V. Vasil'eva, L. N. Kurkovskaya, E. N. Teleshov, and A. N. Pravednikov, *Izv. Akad. Nauk SSSR, Ser. Khim.*, 1980, 647.

1129. T. A. Dal-Cason, S. A. Angelos, and O. Washington, *J. Forensic Sci.*, 1981, **26**, 793.

1130. T. Saito, A. Koide, Y. Kikuchi, and S. Motoki, *Phosphorus Sulfur*, 1983, **17**, 153.

1131. C. Wentrup, C. Thétaz, E. Tagliaferri, H. J. Lindner, B. Kitschke, H. W. Winter, and H. P. Reisenauer, *Angew. Chem.*, 1980, **92**, 556.

1132. C. Pelizzi, G. Pelizzi, and P. Tarasconi, *Polyhedron*, 1983, **2**, 145.

1133. A. Petrović, D. Petrović, and G. Bernáth, *Z. Krystallogr.*, 1980, **151**, 171.

1134. A. Kapor, B. Ribár, G. Argay, A. Kálmán, and G. Bernáth, *Cryst. Struct. Commun.*, 1980, **9**, 347.

1135. J. Kant, F. D. Popp, B. L. Joshi, and B. C. Uff, *Chem. Ind. (London)*, 1984, 415.

1136. F. M. Abdelrazek, Z. E.-S. Kandeel, K. M. H. Hilmy, and M. H. Elnagdi, *Chem. Ind. (London)*, 1983, 439.

1137. F. Kienzle, A. Kaiser, and R. E. Minder, *Helv. Chim. Acta*, 1983, **66**, 148.

1138. W. Städeli, W. von Philipsborn, A. Wick, and I. Kompiš, *Helv. Chim. Acta*, 1980, **63**, 504.

1139. D. Corvi, A. Mitidieri-Costanza, and G. Sleiter, *Gazz. Chim. Ital.*, 1982, **112**, 167.

1140. C. Pelizzi, G. Pelizzi, and G. Predieri, *Gazz. Chim. Ital.*, 1982, **112**, 343.

1141. A. M. El-Khawaga, M. A. Abd-Alla, and A. A. Khalaf, *Gazz. Chim. Ital.*, 1981, **111**, 441.

1142. K. Hermann and G. Simchen, *Liebigs Ann. Chem.*, 1981, 333.

1143. D. J. Brown, G. W. Grigg, Y. Iwai, K. N. McAndrew, T. Nagamatsu, and R. van Heeswyck, *Aust. J. Chem.*, 1979, **32**, 2713.

1144. A. Franke, *Liebigs Ann. Chem.*, 1982, 794.

1145. K. Burger, U. Wassmuth, F. Hein, and S. Rottegger, *Liebigs Ann. Chem.*, 1984, 991.

1146. W. L. F. Armarego and P. A. Reece, *Aust. J. Chem.*, 1981, **34**, 1561.

1147. W. B. Cowden and P. Waring, *Aust. J. Chem.*, 1981, **34**, 1539.

1148. G. Heinicke, T. V. Hung, R. H. Prager, and A. D. Ward, *Aust. J. Chem.*, 1984, **37**, 831.

1149. S. Senda, T. Asao, I. Sugiyama, and K. Hirota, *Tetrahedron Lett.*, 1980, **21**, 531.

1150. R. K. Smalley, R. H. Smith, and H. Suschitsky, *Tetrahedron Lett.*, 1979, 4687.

1151. R. Prasad and A. P. Bhaduri, *Indian J. Chem., Sect. B*, 1979, **18**, 443.

1152. M. Z. A. Badr, H. A. H. El-Sherief, G. M. Al-Naggar, and A. M. Mahmoud, *Indian J. Chem.*, Sect. B, 1979, **18**, 560.

1153. P. Hanumanthu and C. V. Ratnam, *Indian J. Chem., Sect. B*, 1979, **17**, 349.

1154. M. Z. Kirmani and S. R. Ahmed, *Indian J. Chem., Sect. B*, 1979, **17**, 445.

1155. D. Bartholomew and I. T. Kay, *Tetrahedron Lett.*, 1979, 2827.

1156. M. Z. Kirmani and K. Sethi, *Tetrahedron Lett.*, 1979, 2917.

1157. F. W. Lichtenthaler and A. Moser, *Tetrahedron Lett.*, 1981, **22**, 4397.

1158. H. Tucker, G. Golding, and S. R. Purvis, *Tetrahedron Lett.*, 1981, **22**, 1373.

1159. E. Cuny, F. W. Lichtenthaler, and A. Moser, *Tetrahedron Lett.*, 1980, **21**, 3029.

1160. B. Kokel, G. Menichi, and M. Hubert-Habart, *Tetrahedron Lett.*, 1984, **25**, 1557.

1161. A. Essawy, M. A. El-Hashash, and M. M. Mohamed, *Indian J. Chem., Sect. B*, 1980, **19**, 663.

1162. S. K. Phadtare, S. K. Kamat, and G. T. Panse, *Indian J. Chem., Sect. B*, 1980, **19**, 212.

1163. B. Singh, R. N. Pandey, D. K. Sharma, U. S. P. Sharma, and U. Bhanu, *Indian J. Chem., Sect. A*, 1981, **20**, 1097.

1164. M. M. Mohamed, M. A. El-Hashash, A. Esswy, and M. E. Shaban, *Indian J. Chem., Sect. B*, 1981, **20**, 718.

1165. S. Kumar, V. K. Kansel, and A. P. Bhaduri, *Indian J. Chem., Sect. B*, 1981, **20**, 1068.

1166. H. K. Gakhar, S. C. Gupta, and N. Kumar, *Indian J. Chem., Sect. B*, 1981, **20**, 14.

1167. N. Latif, N. Mishriky, and F. M. Assad, *Indian J. Chem., Sect. B*, 1981, **20**, 118.

1168. D. H. R. Barton and G. Kretzschmar, *Tetrahedron Lett.*, 1983, **24**, 5889.

1169. M. A. Elkasaby and N. A. Noureldin, *Indian J. Chem., Sect. B*, 1981, **20**, 290.

1170. M. F. Ismail, N. A. Shams, and M. I. Naguib, *Indian J. Chem., Sect. B*, 1981, **20**, 394.

1171. C. J. Shishoo, M. B. Devani, V. S. Bhadti, S. Ananthan, and G. V. Ullas, *Tetrahedron Lett.*, 1983, **24**, 4611.

1172. D. Ranganathan, S. Ranganathan, and S. Bamezai, *Tetrahedron Lett.*, 1982, **23**, 2789.

1173. J. T. Gupton, K. F. Correia, and G. R. Hertel, *Synth. Commun.*, 1984, **14**, 1013.

1174. I. Ganjian and I. Lalezari, *Synth. Commun.*, 1984, **14**, 33.

1175. J. Allen, *J. Labelled Compd. Radiopharm.*, 1983, **20**, 1283.

1176. M. Berridge, D. Comar, C. Crouzel, and J.-C. Baron, *J. Labelled Compd. Radiopharm.*, 1983, **20**, 73.

1177. P. Singh and I. S. Gupta, *J. Indian Chem. Soc.*, 1979, **56**, 77.

1178. R. S. Varma and A. K. Agnihotra, *J. Indian Chem. Soc.*, 1979, **56**, 162.

1179. A. K Sengupta and U. Chandra, *J. Indian Chem. Soc.*, 1979, **56**, 645.

1180. N. P. Das and A. S. Mittra, *J. Indian Chem. Soc.*, 1979, **56**, 398.

1181. H. H. Kaegi, W. Burger, and G. J. Bader, *J. Labelled Compd. Radiopharm.*, 1982, **19**, 289.

1182. A. McKillop, A. Henderson, P. S. Ray, C. Avendano, and E. G. Molinero, *Tetrahedron Lett.*, 1982, **23**, 3357.

1183. J. S. Skukla and I. Ahmed, *Indian J. Chem., Sect. B*, 1979, **17**, 651.

1184. A. Essawy, M. A. El-Hashash, A. M. El-Gendy, and M. M M. Hamad, *Indian J. Chem. Sect. B*, 1982, **21**, 593.

1185. A. K. Sengupta, M. M. Gupta, and A. A. Gupta, *Indian J. Chem., Sect. B.*, 1982, **21**, 600.

1186. R. Rastogi and S. Sharma, *Indian J. Chem., Sect. B.*, 1982, **21**, 744.

1187. V. K. Pandey and A. K. Agarwal, *J. Indian Chem., Soc.*, 1979, **56**, 706.

1188. A. Takaoka, K. Iwamoto, T. Kitazume, and N. Ishikawa, *J. Fluorine Chem.*, 1979, **14**, 421.

1189. H. Schönowsky and B. Sachse, *Z. Naturforsch., Teil B*, 1982, **37**, 907.

1190. N. R. Naik, A. F. Amin, and S. R. Patel, *J. Indian Chem. Soc.*, 1979, **56**, 708.

1191. V. K. Pandey and H. C. Lohani, *J. Indian Chem. Soc.*, 1979, **56**, 915.

1192. D. D. Mukerji and S. R. Nautiyal, *J. Indian Chem. Soc.*, 1979, **56**, 1226.

1193. J. S. Shukla and S. Saxena, *J. Indian Chem. Soc.*, 1979, **56**, 1237.

1194. S. S. Tiwari, S. M. M. Zaidi, and R. K. Satsangi, *J. Indian Chem. Soc.*, 1979, **56**, 1263.

1195. W. J. Nixon, J. T. Garland, and C. D. Blanton, *Synthesis*, 1980, 56.

1196. J. P. Henichart, J. L. Bernier, and R. Houssin, *Synthesis*, 1980, 311.

1197. A. Arques, P. Molina, and A. Soler, *Synthesis*, 1980, 702.

1198. M. R. Chaurasia and S. K. Sharma, *J. Indian Chem. Soc.*, 1982, **59**, 370.

1199. H. K. Gakhar, S. Kiran, and S. B. Gupta, *J. Indian Chem. Soc.*, 1982, **59**, 666.

1200. A. Kamal, K. R. Rao, and P. B. Sattur, *Synth. Commun.*, 1980, **10**, 799

1201. P. Fuchs, U. Hess, H. H. Holst, and H. Lund, *Acta Chem. Scand., Ser. B*, 1981, **35**, 185.

1202. M. A. Abbady, M. M. Ali, and M. M. Kandeel, *J. Indian Chem. Soc.*, 1981, **58**, 59.

1203. M. S. Bal, K. Deep, and H. Singh, *J. Indian Chem. Soc.,* 1982, **21**, 805.

1204. R. Agarwal, C. Chaudhary, and V. S. Misra, *Indian J. Chem., Sect. B*, 1982, **21**, 1110.

1205. S. K. Sengupta and X. Nizamuddin, *Indian J. Chem., Sect. A*, 1982, **21**, 426.

1206. S. K. Robev, *Tetrahedron Lett.*, 1983, **24**, 4351.

1207. B. Singh, M. M. P. Rukhaiyar, R. K. Mehra, and R. J. Sinha, *Indian J. Chem., Sect. A*, 1979, **17**, 520.

1208. V. K. Agarwal, S. Sharma, R. N. Iyer, R. K. Chatterjee, and A. B. Sen, *Indian J. Chem., Sect. B*, 1980, **19**, 1084.

1209. K. D. Deodhar, A. D. d'Sa, S. R. Pednekar, and D. S. Kanekar, *Synthesis*, 1982, 853.

1210. P. Molina, A. Arques, and M. V. Valcarcel, *Synthesis*, 1982, 944.

1211. Y. Imai, S. Sato, R. Takasawa, and M. Ueda, *Synthesis*, 1981, 35.

1212. D. M. Mulvey and Y. Gray-Nuñez, *Synthesis*, 1981, 533.

1213. K. Burger, F. Hein, U. Wassmuth, H. Krist, *Synthesis*, 1981, 904.

1214. J. Mayoral, E. Melendez, F. Marchán, and J. Sanchez, *Synthesis*, 1981, 962.

1215. H. K. Gakhar, A. Jain, and N. Kumar, *J. Indian Chem. Soc.*, 1981, **58**, 166.

1216. B. Dash, E. K. Dora, and C. S. Panda, *J. Indian Chem. Soc.*,1980, **57**, 835.

1217. I. Husain, S. N. Misra, and S. R. Yadav, *J. Indian Chem. Soc.*,1980, **57**, 924.

1218. S. S. Tiwari, S. M. M. Zaidi, R. Agarwal, and R. K. Satsangi, *J. Indian Chem. Soc.*, 1980, **57**, 1039.

1219. Y. D. Kulkarni, B. Kumar, and S. H. R. Abdi, *J. Indian Chem. Soc.*, 1983, **60**, 906.

1220. K. L. Reddy, S. Srihari, and P. Lingaiah, *J. Indian Chem. Soc.*, 1983, **60**, 1020.

1221. H. Singh, C. S. Gandhi, K. B. Lal, and M. S. Bal, *Indian J. Chem. Sect. B*, 1980, **19**, 844.

1222. M. I. Husain and G. C. Srivastav, *Indian J. Chem. Sect. B*, 1980, **19**, 916.

1223. M. Z. A. Badr, G. M. El-Naggar, and H. A. H. El-Sherief, *Indian J. Chem., Sect. B*, 1980, **19**, 925.

1224. B. P. Joshi and B. D. Hosangardi, *Indian J. Chem., Sect. B*, 1980, **19**, 775.

1225. P. B. Talukdar, S. K. Sengupta, and A. K. Datta, *Indian J. Chem., Sect. B*, 1980, **19**, 638.

1226. C. V. C. Rao, K. K. Reddy, and N. V. S. Rao, *Indian J. Chem., Sect. B*, 1980, **19**, 655.

1227. M. Sarkar and D. P. Chakraborty, *Phytochemistry*, 1979, **18**, 694.

1228. L. Legrand and N. Lozac'h, *Bull. Soc. Chim. Fr.*, 1983, Part II, 226.

1229. L. Legrand and N. Lozac'h, *Bull. Soc. Chim. Fr.*, 1983, Part II, 217.

1230. L. Legrand and N. Lozac'h, *Bull. Soc. Chim. Fr.*, 1982, Part II, 139.

1231. G. Kaupp and B. Knichala, *Chem. Ber.*, 1985, **118**, 462.

1232. D. Korbonits, P. Kiss, K. Simon, and P. Kolonits, *Chem. Ber.*, 1984, **117**, 3183.

1233. M. D. Nair and J. A. Desai, *Indian J. Chem., Sect. B*, 1982, **21**, 4.

1234. P. Kumar and A. K. Mukerjee, *Indian J. Chem., Sect. B*, 1982, **21**, 24.

1235. H. K. Gakhar, P. Baveja, and N. Kumar, *Indian J. Chem., Sect. B*, 1982, **21**, 64.

1236. K. D. Deodhar, S. D. Samant, S. R. Pednekar, D. S. Kanekar, A. A. Inamdar, and P. Y. Patkar, *Indian J. Chem., Sect. B*, 1982, **21**, 67.

1237. J. S. Shukla and R. Rastogi, *Indian J. Chem., Sect. B*, 1982, **21**, 375.

1238. M. I. Husain, G. C. Srivastava, and P. R. Dua, *Indian J. Chem., Sect. B*, 1982, **21**, 381.

1239. M. F. Ismail and F. S. Sayed, *Indian J. Chem., Sect. B*, 1982, **21**, 461.

1240. V. P. Reddy, P. L. Prasunamba, P. S. N. Reddy, and C. V. Ratnam, *Indian J. Chem., Sect. B*, 1983, **22**, 917.

1241. B. K. Misra, Y. R. Rao, and S. N. Mahapatra, *Indian J. Chem., Sect. B*, 1983, **22**, 1132.

1242. L. D. Dave, C. Mathew, and V. Oommen, *Indian J. Chem., Sect. A*, 1983, **22**, 420.

1243. D. L. Dreyer and R. C. Brenner, *Phytochemistry*, 1980, **19**, 935.

1244. H. Vorbrüggen and K. Krolikiewicz, *Chem. Ber.*, 1984, **117**, 1523.

1245. M. M. Mohamed, M. A. El-Hashash, and A. Essawy, *Indian J. Chem., Sect. B*, 1983, **22**, 85.

1246. J. S. Shukla, M. Singh, and R. Rastogi, *Indian J. Chem., Sect. B*, 1983, **22**, 306.

1247. H. Singh and N. Malhitra, *Indian J. Chem., Sect. B*, 1983, **22**, 328.

1248. S. Nakanishi and S. S. Massett, *Org. Prep. Proced. Int.*, 1980, **12**, 219.

1249. R. R. Schmidt and B. Beitzke, *Chem. Ber.*, 1983, **116**, 2115.

1250. M. R. Chaurasia and A. K. Sharma, *J. Indian Chem. Soc.*, 1983, **60**, 1071.

1251. C. V. Shankar, A. D. Rao, B. J. Reddy, and V. M. Reddy, *J. Indian Chem. Soc.*, 1983, **60**, 61.

1252. A. K. Sengupta and T. Bhattacharya, *J. Indian Chem. Soc.*, 1983, **60**, 373.

1253. Y. D. Kulkarni, S. H. R. Abdi, and B. Kumar, *J. Indian Chem. Soc.*, 1983, **60**, 504.

1254. N. F. Aly, M. El-Komy, N. Y. Aly, and M. O. A. Orabi, *Indian J. Chem., Sect. B*, 1983, **22**, 471.

1255. A. M. Mahmoud, H. A. H. El-Sherief, G. M. El-Naggar, and A. E. Abdel-Rahman, *Indian J. Chem., Sect. B*, 1983, **22**, 491.

1256. S. K. Phadtare, S. K. Kamat, and G. T. Panse, *Indian J. Chem., Sect. B*, 1983, **22**, 499.

1257. V. P. Reddy, V. B. Rao, and C. V. Ratnam, *Indian J. Chem., Sect. B*, 1984, **23**, 560.

1258. I. P. Singh, A. K. Saxena, J. N. Sinha, K. P. Bhargava, and K. Shanker, *Indian J. Chem., Sect. B*, 1984, **23**, 592.

1259. S. B. Barnela and S. Seshadri, *Indian J. Chem., Sect. B*, 1984, **23**, 161.

1260. K. L. Reddy, S. Srihari, and P. Lingaiah, *Indian J. Chem., Sect. A*, 1984, **23**, 780.

1261. K. L. Reddy, S. Srihari, and P. Lingaiah, *Indian J. Chem., Sect. A*, 1984, **23**, 172.

1262. V. S. Misra, V. K. Saxena, and R. Srivastava, *J. Indian Chem. Soc.*, 1983, **60**, 610.

1263. K. Burger, S. Penninger, M. Greisel, and E. Daltrozzo, *J. Fluorine Chem.*, 1980, **15**, 1.

1264. M. A. Metwally, E.-S. Afsah, and F. A. Amer, *Z. Naturforsch., Teil B*, 1981, **36**, 1147.

1265. J. S. Shukla, S. Saxena, and R. Misra, *J. Indian Chem. Soc.*, 1982, **59**, 1196.

1266. S. Singhal, I. S. Gupta, and P. C. Bansal, *J. Indian Chem. Soc.*, 1984, **61**, 690.

1267. K. E. Nielsen and E. B. Pedersen, *Acta Chem. Scand.*, Ser. B, 1980, **34**, 637.

1268. D. J. Chadwick and I. W. Easton, *Acta Crystallogr.*, Sect. C, 1983, **39**, 454.

1269. P. K. Rogan and G. J. B. Williams, *Acta Crystallogr.*, Sect. B, 1980, **36**, 2358.

1270. W. E. Hunt, C. H. Schwalbe, K. Brid, and P. D. Mallinson, *Biochem. J.*, 1980, **187**, 533.

1271. S. Leistner, K. Hentschel, and G. Wagner, *Z. Chem.*, 1984, **24**, 328.

1272. Y. D. Kulkarni, S. H. R. Abdi, and V. L. Sharma, *J. Indian Chem. Soc.*, 1984, **61**, 720.

1273. D. H. Raheja and N. B. Laxmeshwar, *J. Indian Chem. Soc.*, 1984, **61**, 507.

1274. H. Singh, C. S. Gandhi, and M. S. Bal, *Indian J. Chem.*, Sect. B, 1981, **20**, 17.

1275. V. B. Rao, P. Hanumanthu, and C. V. Ratnam, *Indian J. Chem.*, Sect. B, 1979, **18**, 493.

1276. M. Z. Kirmani and S. R. Ahmed, *Indian J. Chem.*, Sect. B, 1979, **18**, 22.

1277. Y. P. Reddy, G. S. Reddy, and K. K. Reddy, *Indian J. Chem.*, Sect. B, 1979, **18**, 125.

1278. S. S. Tiwari, S. Misra, and R. K. Satsangi, *Indian J. Chem.*, Sect. B, 1979, **18**, 283.

1279. F. Gatta, M. R. del Giudice, and G. Settimj, *Synthesis*, 1979, 718.

1280. H. Singh and K. Deep, *Tetrahedron*, 1984, **40**, 4937.

1281. H. Singh, S. K. Aggarwal, and N. Malhotra, *Tetrahedron*, 1984, **40**, 4941.

1282. T. Minami, M. Ogata, I. Hirao, M. Tanaka, and T. Agawa, *Synthesis*, 1982, 231.

1283. A. K. Sengupta and U. Chandra, *Indian J. Chem.*, Sect. B, 1979, **18**, 382.

1284. M. Z. Kirmani, *Indian J. Chem.*, Sect. B, 1979, **18**, 387.

1285. V. B. Rao and C. V. Ratnam, *Indian J. Chem.*, Sect. B, 1979, **18**, 409.

1286. M. Z. Kirmani, K. Sethi, and S. R. Ahmed, *Indian J. Chem.*, Sect. B, 1979, **18**, 432.

1287. J. Greiner, R. Pastor, and A. Cambon, *J. Fluorine Chem.*, 1981, **18**, 185.

1288. S. Abuzar and S. Sharma, *Z. Naturforsch.*, Teil B, 1981, **36**, 108.

1289. F. S. G. Soliman, W. Stadbauer, and T. Kappe, *Z. Naturforsch.*, Teil B, 1981, **36**, 252.

1290. G. A. Showell, *Synth. Commun.*, 1980, **10**, 241.

1291. H. Singh, P. Singh, and N. Malhotra, *Synth. Commun.*, 1980, **10**, 591.

1292. S. Leistner, K. Hentschel, and G. Wagner, *Z. Chem.*, 1983, **23**, 215.

1293. M. Süsse and S. Johne, *Z. Chem.*, 1983, **23**, 406.

1294. S. Leistner, G. Wagner, and A. P. Antoni, *Z. Chem.*, 1982, **22**, 259.

1295. M. Mühlstädt, B. Schulze, and I. Schubert, *Z. Chem.*, 1981, **21**, 326.

1296. G. M. El-Naggar, H. A. El-Sherief, A. E. Abdel-Rahman, and A. M. Mahmoud, *Org. Mass Spectrom.*, 1983, **18**, 364.

1297. V. B. Rao, P. S. Reddy, and C. V. Ratnam, *Org. Mass Spectrom.*, 1983, **18**, 317.

1298. F. Tureček and J. Světlik, *Org. Mass Spectrom.*, 1981, **16**, 285.

1299. J. Mirek and T. A. Holak, *Chem. Scr.*, 1983, **22**, 133.

1300. K. E. Nielsen and E. B. Pedersen, *Chem. Scr.*, 1981, **18**, 242.

1301. J. Bergman and N. Eklund, *Chem. Scr.*, 1982, **19**, 193.

1302. M. Lempert-Sréter, K. Lempert, and J. Møller, *Chem. Scr.*, 1979, **13**, 195.

1303. M. Botta, M. Cavalieri, D. Ceci, F. de Angelis, G. Finizia, and R. Nicoletti, *Tetrahedron*, 1984, **40**, 3313.

1304. A. M. Lamazouere and J. Sotiropoulos, *Tetrahedron*, 1981, **37**, 2451.

1305. R. K. Anderson, S. D. Carter, and G. W. H. Cheeseman, *Tetrahderon*, 1979, **35**, 2463.

1306. D. P. Chakraborty, A. K. Mandal, and S. K. Roy, *Synthesis*, 1981, 977.

1307. P. Molina, A. Arques, I. Cartagena, and M. V. Valcarcel, *Synthesis*, 1984, 881.

1308. V. Looney-Dean, B. S. Lindamood, and E. P. Papadopoulos, *Synthesis*, 1984, 68.

1309. J. Garin, E. Melendez, F. L. Merchán, and T. Tejero, *Synthesis*, 1984, 520.

1310. R. J. Grout and M. W. Partridge, *J. Chem. Soc.*, 1960, 3546.

1311. S. Miyano, N. Abe, N. Mibu, K. Takeda, and K. Sumoto, *Synthesis*, 1983, 401.

1312. J. Garin, E. Melendez, F. L. Merchán, T. Tejero, and E. Villarroya, *Synthesis*, 1983, 406.

1313. M. Z. A. Badr, H. A. H. El-Sherief, and A. M. Mahmoud, *Bull. Chem. Soc. Jpn.*, 1980, **53**, 2389.

1314. Y. Kayama, T. Hara, and T. Sunami, *Bull. Chem. Soc. Jpn.*, 1980, **53**, 295.

1315. I. Shibuya, *Nippon Kagaku Kaishi*, 1982, 1518.

1316. T. P. Devi, C. Kalidas, and C. S. Venkatachalam, *Bull. Chem. Soc. Jpn.*, 1982, **55**, 286.

1317. Y. Inukai, Y. Oono, T. Sonoda, and H. Kobayashi, *Bull. Chem. Soc. Jpn.*, 1981, **54**, 3447.

1318. I. Shibuya, *Bull. Chem. Soc. Jpn.*, 1981, **54**, 2387.

1319. H. A. El-Sherief, A. E. Abdel-Rahman, G. M. El-Naggar, and M. A. Mahmoud, *Bull. Chem. Soc. Jpn.*, 1983, **56**, 1227.

1320. M. Nakagawa, M. Taniguchi, M. Sodeoka, M. Ito, K. Yamaguchi, and T. Hino, *J. Am. Chem. Soc.*, 1983, **105**, 3709.

1321. O. S. Tee, M. Trani, R. A. McClelland, and N. E. Seaman, *J. Am. Chem. Soc.*, 1982, **104**, 7219.

1322. G. Büchi, P. R. de Shong, S. Katsumura, and Y. Sugimura, *J. Am. Chem. Soc.*, 1979, **101**, 5084.

1323. R. S. Atkinson, N. A. Gawad, D. R. Russell, and L. J. S. Sherry, *J. Chem. Soc., Chem. Commun.*, 1983, 568.

1324. B. Danieli, G. Lesma, and G. Palmisano, *J. Chem. Soc., Chem. Commun.*, 1982, 1092.

1325. R. S. Atkinson, J. R. Malpass, and K. L. Woodthorpe, *J. Chem. Soc., Chem. Commun.*, 1981, 160.

1326. R. S. Atkinson, J. R. Malpass, K. L. Skinner, and K. L. Woodthorpe, *J. Chem. Soc., Chem. Commun.*, 1981, 549.

1327. J. Světlik and A. Martvoň, *Collect. Czech. Chem. Commun.*, 1981, **46**, 428.

1328. Š. Stankovský and A. Martvoň, *Collect. Czech. Chem. Commun.*, 1980, **45**, 1079.

1329. A. Alemagna, P. del Buttero, E. Licandro, and S. Maiorana, *J. Chem. Soc., Chem. Commun.*, 1981, 894.

1330. Z. Yejdělek, M. Rajšner, E. Svátek, J. Holubek, and M. Protiva, *Collect. Czech. Chem. Commun.*, 1979, **44**, 3604.

1331. G. V. Boyd, P. F. Lindley, and G. A. Nicolaou, *J. Chem. Soc., Chem. Commun.*, 1984, 1105.

1332. K. T. Potts and P. Murphy, *J. Chem. Soc., Chem. Commun.*, 1984, 1348.

1333. R. S. Atkinson and N. M. Gawad, *J. Chem. Soc., Chem. Commun.*, 1984, 557.

1334. M. Uher, J. Foltin, and L. Floch, *Collect. Czech. Chem. Commun.*, 1981, **46**, 2696.

1335. J. Kaválek, M. Kotyk, S. El-Bahaie, and V. Štěrba, *Collect. Czech. Chem. Commun.*, 1981, **46**, 246.

1336. M. Draminski, E. Frass, J. Greger, and K. Fabianowska-Majewska, *Collect. Czech. Chem. Commun.*, 1985, **50**, 280.

1337. Š. Stankovský and M. Sokyrová, *Collect. Czech. Chem. Commun.*, 1984, **49**, 1795.

1338. Š. Stankovský, *Collect. Czech. Chem. Commun.*, 1983, **48**, 3575.

1339. F. Roubinek, J. Vavřina, and Z. Buděšinský, *Collect. Czech. Chem. Commun.*, 1982, **47**, 630.

1340. H. Singh, P. Singh, and N. Malhotra, *J. Chem. Soc., Perkin Trans. 1*, 1981, 2647.

1341. T. K. Au, A. E. Baydar, and G. V. Boyd, *J. Chem. Soc., Perkin Trans. 1*, 1981, 2884.

1342. T. M. Paterson, R. K. Smalley, H. Suschitzky, and A. J. Barker, *J. Chem. Soc., Perkin Trans. 1*, 1980, 633.

1343. A. R. Katritzky, M. J. Cook, S. B. Brown, R. Cruz, G. H. Millet, and A. Anani, *J. Chem. Soc., Perkin Trans. 1*, 1979, 2493.

1344. R. E. Busby, J. Parrick, S. M. H. Rizvi, and C. J. G. Shaw, *J. Chem. Soc., Perkin Trans. 1*, 1979, 2786.

1345. M. C. Etter, *J. Chem. Soc., Perkin Trans. 2*, 1983, 115.

1346. L. A. Errede, M. C. Etter, R. C. Williams, and S. M. Darnauer, *J. Chem. Soc., Perkin Trans. 2*, 1981, 233.

1347. R. A. Bowie, P. N. Edwards, S. Nicholson, P. J. Taylor, and D. A. Thompsom, *J. Chem. Soc., Perkin Trans. 2*, 1979, 1708.

1348. M. Lempert-Sréter, K. Lempert, and J. Møller, *J. Chem. Soc., Perkin Trans. 1*, 1983, 2011.

1349. M. Ghavshou and D. A. Widdowson, *J. Chem. Soc., Perkin Trans. 1*, 1983, 3065.

1350. D. J. Le Count, *J. Chem. Soc., Perkin Trans. 1*, 1983, 813.

1351. R. S. Atkinson, J. R. Malpass, and K. L. Woodthorpe, *J. Chem. Soc., Perkin Trans. 1*, 1982, 2407.

1352. G. U. Baig and M. F. G. Stevens, *J. Chem. Soc., Perkin Trans. 1*, 1984, 2765.

1353. G. U. Baig, M. F. G. Stevens, and K. Vaughan, *J. Chem. Soc., Perkin Trans. 1*, 1984, 999.

1354. M. Lempert-Sréter, K. Lempert, and J. Møller, *J. Chem. Soc., Perkin Trans. 1*, 1984, 1143.

1355. R. S. Atkinson, J. R. Malpass, K. L. Skinner, and K. L. Woodthorpe, *J. Chem. Soc., Perkin Trans. 1*, 1984, 1905.

1356. R. H. Foster and N. J. Leonard, *J. Org. Chem.*, 1979, **44**, 4609.

1357. K. Ozaki, Y. Yamada, and T. Oine, *J. Org. Chem.*, 1981, **46**, 1571.

1358. S. W. Schneller and W. J. Christ, *J. Org. Chem.*, 1981, **46**, 1699.

1359. B. E. Maryanoff, D. F. McComsey, and S. O. Nortey, *J. Org. Chem.*, 1981, **46**, 355.

1360. J. M. Muchowski and M. C. Venuti, *J. Org. Chem.*, 1980, **45**, 4798.

1361. J. A. Grina, M. R. Tatcliff, and F. R. Stermitz, *J. Org. Chem.*, 1982, **47**, 2648.

1362. H. Natsugari, K. Meguro, and Y. Kuwada, *Chem. Pharm. Bull.*, 1979, **27**, 2589.

1363. T. Sakamoto, H. Yoshizawa, H. Yamanaka, Y. Goto, T. Niiya, and N. Honjo, *Heterocycles*, 1982, **17**, 73.

1364. M. Ishikawa and Y. Eguchi, *Heterocycles*, 1981, **16**, 31.

1365. M. R. Chaurasia and S. K. Sharma, *Heterocycles*, 1981, **16**, 621.

1366. F. W. Lichtenthaler and E. Cuny, *Heterocycles*, 1981, **15**, 1053.

1367. J. Bhattacharyya and S. C. Pakrashi, *Heterocycles*, 1980, **14**, 1469.

1368. Y. Tomioka and M. Yamazaki, *Heterocycles*, 1981, **16**, 2115.

1369. R. Kreher and U. Bergmann, *Heterocycles*, 1981, **16**, 1693.

1370. S. C. Shim, D. W. Kim, S. S. Moon, and Y. B. Chae, *J. Org. Chem.*, 1984, **49**, 1449.

1371. M. F. Ismail, N. A. Shams, M. R. Salem, and S. A. Emara, *J. Org. Chem.*, 1983, **48**, 4172.

1372. J. Světlik, *Heterocycles*, 1981, **16**, 1281.

1373. M. Chakrabarty, A. K. Chakravarty, and S. C. Pakrashi, *Heterocycles*, 1983, **20**, 445.

1374. H. Miki and F. Kasahara, *Heterocycles*, 1982, **19**, 1879.

1375. H. Miki, *Heterocycles*, 1982, **19**, 7.

1376. H. Singh and R. Sarin, *Heterocycles*, 1984, **22**, 1101.

1377. D. F. Eaton and B. E. Smart, *J. Org. Chem.*, 1979, **44**, 4435.

1378. D. L. Boger, J. Schumacher, M. D. Mullican, M. Patel, and J. S. Panek, *J. Org. Chem.*, 1982, **47**, 2673.

1379. H. Miki and J. Yamada, *Heterocycles*, 1982, **19**, 11.

1380. H. Miki, *Heterocycles*, 1982, **19**, 15.

1381. M. Sawada, Y. Furukawa, Y. Takai, and T. Hanafusa, *Heterocycles*, 1984, **22**, 501.

1382. J. Bergman, A. Brynolf, and B. Elman, *Heterocycles*, 1983, **20**, 2141.

1383. J. Bergman, A. Brynolf, K.-W. Törnroos, B. Karlsson, and P.-E. Werner, *Heterocycles*, 1983, **20**, 2145.

1384. H. Asakawa and M. Matano, *Chem. Pharm. Bull.*, 1979, **27**, 1287.

1385. M. Gál, E. Tihanyi, and P. Dvortsák, *Heterocycles*, 1984, **22**, 1985.

1386. L. A. Errede, P. D. Martinucci, and J. J. McBrady, *J. Org. Chem.*, 1980, **45**, 3009.

1387. R. H. Foster and N. J. Leonard, *J. Org. Chem.*, 1980, **45**, 3072.

1388. T. L. Rathman, M. C. Sleevi, M. E. Krafft, and J. F. Wolfe, *J. Org. Chem.*, 1980, **45**, 2169.

1389. N. A. Vaidya and C. D. Blanton, *J. Org. Chem.*, 1982, **47**, 1777.

1390. K. Hirota, Y. Kitade, and S. Senda, *J. Org. Chem.*, 1981, **46**, 3949.

1391. J. Bergman and S. Bergman, *J. Org. Chem.*, 1985, **50**, 1246.

1392. R. A. Olofson and R. K. Vander Meer, *J. Org. Chem.*, 1984, **49**, 3377.

1393. T. Sakamoto, T. Sakasai, and H. Yamanaka, *Chem. Pharm. Bull.*, 1980, **28**, 571.

1394. K. Ozaki, Y. Yamada, and T. Oine, *Chem. Pharm. Bull.*, 1980, **28**, 702.

1395. H. Natsugari, K. Meguro, and Y. Kuwada, *Chem. Pharm. Bull.*, 1979, **27**, 2608.

1396. H. Natsugari and Y. Kuwada, *Chem. Pharm. Bull.*, 1979, **27**, 2618.

1397. J. Tani, Y. Yamada, T. Ochiai, R. Ishida, I. Inoue, and T. Oine, *Chem. Pharm. Bull.*, 1979, **27**, 2675.

1398. T. Sekiya, H. Hiranuma, M. Uchida, S. Hata, and S.-I. Yamada, *Chem. Pharm. Bull.*, 1981, **29**, 948.

1399. M. Yamato, J. Horiuchi, and Y. Takeuchi, *Chem. Pharm. Bull.*, 1980, **28**, 2623.

1400. F. Ishikawa, Y. Watanabe, and J. Saegusa, *Chem. Pharm. Bull.*, 1980, **28**, 1357.

1401. F. Ishikawa, A. Kosasayama, and K. Higashi, *Chem. Pharm. Bull.*, 1980, **28**, 2024.

1402. M. Ishikawa, H. Azuma, Y. Eguchi, A. Sugimoto, S. Ito, Y. Takashima, H. Ebisawa, S. Moriguchi, I. Kotoku, and H. Suzuki, *Chem. Pharm. Bull.*, 1982, **30**, 744.

1403. M. Yamato and Y. Takeuchi, *Chem. Pharm. Bull.*, 1982, **30**, 1036.

1404. H. Yamanaka, S. Konno, T. Sakamoto, S. Niitsuma, and S. Noji, *Chem. Pharm. Bull.*, 1981, **29**, 2837.

1405. M. Yamato, J. Horiuchi, and Y. Takeuchi, *Chem. Pharm. Bull.*, 1981, **29**, 3055.

1406. M. Yamato, J. Horiuchi, and Y. Takeuchi, *Chem. Pharm. Bull.*, 1981, **29**, 3124.

1407. J. Horiuchi, Y. Takeuchi, and M. Yamato, *Chem. Pharm. Bull.*, 1981, **29**, 3130.

1408. M. Yamamoto and H. Yamamoto, *Chem. Pharm. Bull.*, 1981, **29**, 2135.

1409. H. Miki and F. Kasahara, *Chem. Pharm. Bull.*, 1982, **30**, 3471.

1410. H. Miki and J. Yamada, *Chem. Pharm. Bull.*, 1982, **30**, 2313.

1411. H. Miki, *Chem. Pharm. Bull.*, 1982, **30**, 3121.

1412. H. Miki, *Chem. Pharm. Bull.*, 1982, **30**, 1947.

1413. T. Sasaki, A. Nakanishi, and M. Ohno, *Chem. Pharm. Bull.*, 1982, **30**, 2051.

1414. T. Sakamoto, T. Sakasai, H. Yoshizawa, K.-I. Tanji, S. Nishimura, and H. Yamanaka, *Chem. Pharm. Bull.*, 1983, **31**, 4554.

1415. K.-I. Ozaki, Y. Yamada, and T. Oine, *Chem. Pharm. Bull.*, 1983, **31**, 2234.

1416. T. Sekiya, H. Hiranuma, T. Kanayama, S. Hata, and S.-I. Yamada, *Chem. Pharm. Bull.*, 1983, **31**, 2254.

1417. T. Sekiya, S. Hata, and S.-I. Yamada, *Chem. Pharm. Bull.*, 1983, **31**, 2432.

1418. Y. Yamada, M. Otsuka, J. Tani, and T. Oine, *Chem. Pharm. Bull.*, 1983, **31**, 1158.

1419. T. Higashino, H. Kokubo, and E. Hayashi, *Chem. Pharm. Bull.*, 1984, **32**, 3900.

1420. T. Higashino, H. Kokubo, A. Goto, M. Takemoto, and E. Hayashi, *Chem. Pharm. Bull.*, 1984, **32**, 3690.

1421. A. S. Shawali, M. Sami, S. M. Sherif, and C. Párkányi, *J. Heterocycl. Chem.*, 1980, **17**, 877.

1422. G. Rabilloud and B. Sillion, *J. Heterocycl. Chem.*, 1980, **17**, 1065.

1423. G. M. Coppola and M. J. Shapiro, *J. Heterocycl. Chem.*, 1980, **17**, 1163.

1424. D. Bhattacharjee and F. D. Popp, *J. Heterocycl. Chem.*, 1980, **17**, 1211.

1425. S. D. Phillips and R. N. Castle, *J. Heterocycl. Chem.*, 1980, **17**, 1489.

1426. K. G. Dave, C. J. Shishoo, M. B. Devani, R. Kalyanaraman, S. Ananthan, G. V. Ullas, and V. S. Bhadti, *J. Heterocycl. Chem.*, 1980, **17**, 1497.

1427. N. P. Peet, S. Sunder, and R. J. Barbuch, *J. Heterocycl. Chem.*,1980, **17**, 1513.

1428. E. P. Papadopoulos, *J. Heterocycl. Chem*, 1980, **17**, 1553.

1429. E. F. Elslager, P. Jacob, J. Johnson, and L. M. Werbel, *J. Heterocycl. Chem.*, 1980, **17**, 129.

1430. J. Y. Mérour, *J. Heterocycl. Chem.*, 1982, **19**, 1425.

1431. W. J. Houlihan, G. Cooke, M. Denzer, and J. Nicoletti, *J. Heterocycl. Chem.*, 1982, **19**, 1453.

1432. M. C. Gómez-Gil, V. Gómez-Parra, F. Sánchez, and T. Torres, *J. Heterocycl. Chem.*, 1984, **21**, 1189.

1433. L. Ceraulo, S. Plescia, G. Daidone, and M. L. Bajardi, *J. Heterocycl. Chem.*, 1984, **21**, 1209.

1434. M. A. Badawy, Y. A. Ibrahim, and A. M. Kadry, *J. Heterocycl. Chem.*, 1984, **21**, 1403.

1435. E. P. Papadopoulos, *J. Heterocycl. Chem.*, 1984, **21**, 1411.

1436. C. Montignoul, M.-J. Richard, C. Vigne, and L. Giral, *J. Heterocycl. Chem.*, 1984, **21**, 1509.

1437. M. J. Kornet, T. Varia, and W. Beaven, *J. Heterocycl. Chem.*, 1984, **21**, 1533.

1438. J. J. Hlavka, P. Bitha, Y.-I. Lin, and T. Strohmeyer, *J. Heterocycl. Chem.*, 1984, **21**, 1537.

1439. L. Legrand and N. Lozac'h, *J. Heterocycl. Chem.*, 1984, **21**, 1615.

1440. W. Milkowski, R. Hüschens, and H. Kuchenbecker, *J. Heterocycl. Chem.*, 1980, **17**, 373.

1441. B. M. Fernandez, C. B. Schapira, and S. Lamdan, *J. Heterocycl. Chem.*, 1980, **17**, 667.

1442. N. P. Peet and P. B. Anzeveno, *J. Heterocycl. Chem.*, 1979, **16**, 877.

1443. J. Johnson, E. F. Elslager, and L. M. Werbel, *J. Heterocycl. Chem.*, 1979, **16**, 1101.

1444. S. Sunder and N. P. Peet, *J. Heterocycl. Chem.*, 1979, **16**, 1339.

1445. M. Bianchi, E. Häusermann, and S. Rossi, *J. Heterocycl. Chem.*, 1979, **16**, 1411.

1446. J. Petridou-Fischer and E. P. Papadopoulos, *J. Heterocycl. Chem.*, 1982, **19**, 123.

1447. W. D. Dean and E. P. Papadopoulos, *J. Heterocycl. Chem.*, 1982, **19**, 171.

1448. E. P. Papadopoulos and C. D. Torres, *J. Heterocycl. Chem.*, 1982, **19**, 269.

1449. L. Legrand and N. Lozac'h, *J. Heterocycl. Chem.*, 1984, **21**, 1625.

1450. M. J. Kornet, T. Varia, and W. Beaven, *J. Heterocycl. Chem.*, 1984, **21**, 1709.

1451. N. P. Peet and S. Sunder, *J. Heterocycl. Chem.*, 1984, **21**, 1807.

1452. P. Scheiner, L. Frank, I. Giusti, S. Arwin, S. A. Pearson, F. Excellent, and A. P. Harper, *J. Heterocycl. Chem.*, 1984, **21**, 1817.

1453. R. W. Leiby, *J. Heterocycl. Chem.*, 1984, **21**, 1825.

1454. I. Lalezari and C. A. Stein, *J. Heterocycl. Chem.*, 1984, **21**, 5.

1455. M. Davis, R. J. Hook, and W. Y. Wu, *J. Heterocycl. Chem.*, 1984, **21**, 369.

1456. N. R. Smyrl and R. W. Smithwick, *J. Heterocycl. Chem.*, 1982, **19**, 493.

1457. N. Chau, Y. Saegusa, and Y. Iwakura, *J. Heterocycl. Chem.*, 1982, **19**, 541.

1458. M. E. Esteve and C. H. Gaozza, *J. Heterocycl. Chem.*, 1981, **18**, 1061.

1459. W. Haede, *J. Heterocycl. Chem.*, 1981, **18**, 1417.

1460. L. I. Giannola, G. Giammona, B. Carlisi, and S. Palazzo, *J. Heterocycl. Chem.*, 1981, **18**, 1557.

1461. T. Kovač, B. Belin, T. Fajdiga, and V. Šunjić, *J. Heterocycl. Chem.*, 1981, **18**, 59.

1462. H. Yamaguchi and F. Ishikawa, *J. Heterocycl. Chem.*, 1981, **18**, 67.

1463. D. H. Kim, *J. Heterocycl. Chem.*, 1981, **18**, 287.

1464. M. Hedayatullah, *J. Heterocycl. Chem.*, 1981, **18**, 339.

1465. E. P. Papadopoulos, *J. Heterocycl. Chem.*, 1981, **18**, 515.

1466. S. W. Schneller and W. J. Christ, *J. Heterocycl. Chem.*, 1981, **18**, 653.

1467. D. H. Kim, *J. Heterocycl. Chem.*, 1981, **18**, 801.

1468. P. Molina, A. Arques, I. Cartagena, J. A. Noguera, and M. V. Valcarcel, *J. Heterocycl. Chem.*, 1983, **20**, 983.

1469. J. Petridou-Fischer and E. P. Papadopoulos, *J. Heterocycl. Chem.*, 1983, **20**, 1159.

1470. R. Granados, M. Alvarez, N. Valls, and M. Salas, *J. Heterocycl. Chem.*, 1983, **20**, 1271.

1471. M. J. Kornet, T. Varia, and W. Beaven, *J. Heterocycl. Chem.*, 1983, **20**, 1553.

1472. L. Mosti, G. Menozzi, and P. Schenone, *J. Heterocycl. Chem.*, 1983, **20**, 649.

1473. W. D. Dean and E. P. Papadopoulos, *J. Heterocycl. Chem.*, 1982, **19**, 1117.

1474. H. Hara and H. C. van der Plas, *J. Heterocycl. Chem.*, 1982, **19**, 1285.

1475. I. P. Tregubenko, B. V. Golomolzin, I. Y. Postovskii, V. I. Filyakova, and E. A. Tarakhtii, *Tr. Inst. Khim., Ural. Nauchn. Tsentr., Akad. Nauk SSSR*, 1978, **37**, 3; *Chem. Abstr.*, 1980, **92**, 110956.

1476. K. Golankiewicz and L. Celewicz, *Pol. J. Chem.*, 1979, **53**, 1367; *Chem. Abstr.*, 1980, **92**, 58714.

1477. S. Leistner, A. P. Giro, and G. Wagner, *Pharmazie*, 1979, **34**, 390; *Chem. Abstr.*, 1980, **92**, 58719.

1478. A. Sammour, M. Elzawahry, M. Elhashash, and A. Nagy, *Egypt. J. Chem.*, 1976, **19**, 779; *Chem. Abstr.*, 1980, **92**, 6196.

1479. M. Anwar, M. Omara, F. I. Abdel-Hay, and M. Fahmy, *Egypt. J. Chem.*, 1977, **20**, 289; *Chem. Abstr.*, 1980, **93**, 8122.

1480. S. Botros, *Pharmazie*, 1979, **34**, 746; *Chem. Abstr.*, 1980, **93**, 26381.

1481. S. Leistner and G. Wagner, *Pharmazie*, 1980, **35**, 124; *Chem. Abstr.*, 1980, **93**, 26382.

1482. H. A. Aboushady, M. M. El-Kerdawy, and M. M. Tayel, *Pharmazie*, 1979, **34**, 805; *Chem. Abstr.*, 1980, **92**, 215384.

1483. K. M. Shakhidoyatov, S. Yangibaev, L. M. Yun, and C. S. Kadyrov, *Khim. Prir. Soedin.*, 1982, 112; *Chem. Abstr.*, 1982, **97**, 6247.

1484. S. D. Khattri, P. Shukla, and M. R. Chaurasia, *J. Sci. Res. Banaras Hindu Univ.*, 1979, **30**, 193; *Chem. Abstr.*, 1982, **96**, 114902.

1485. R. C. Gupta, A. K. Saxena, M. B. Gupta, K. Shanker, K. P. Bhargava, and K. Kishor, *Proc. Natl. Acad. Sci., India, Sect. A*, 1980, **50**, 189; *Chem. Abstr.*, 1982, **96**, 122718.

1486. X. Zhang, D. Shen, X. Zhang, L. Chen, Z. Dai, and K. Shu, *Yaoxue Xuebao*, 1981, **16**, 877; *Chem. Abstr.*, 1982, **96**, 122726.

1487. S. Ahmad, A. K. Saxena, K. Kishor, K. P. Bhargava, and K. Shanker, *Natl. Acad. Sci. Lett. (India)*, 1981, **4**, 125; *Chem. Abstr.*, 1982, **96**, 122745.

1488. N. S. Habib and A. A. B. Hazzaa, *Sci. Pharm.*, 1981, **49**, 246; *Chem. Abstr.*, 1982, **96**, 52261.

1489. A. Y. Denisov, V. P. Krivopalov, V. I. Mamatyuk, and V. P. Mamaev. *Magn. Reson. Chem.*, 1988, **26**, 42.

1490. M. Khalifa, A. N. Osman, M. G. Ibrahim, A. R. E. Ossman, and M. A. Ismail, *Pharmazie*, 1982, **37**, 115; *Chem. Abstr.*, 1982, **97**, 92227.

1491. S. S. Tiwari and M. P. Pandey, *Acta Cienc. Indica, Ser. Chem.*, 1981, **7**, 7; *Chem. Abstr.*, 1982, **97**, 38914.

1492. H. A. H. El-Sherief, G. M. El-Naggar, A. M. Mahmoud, H. Aref, and K. Abdel-Hamied, *Bull. Fac. Sci., Assiut Univ.*, 1982, **11**, 48; *Chem. Abstr.*, 1983, **98**, 72038.

1493. A. N. Osman, M. Khalifa, M. A. Ismail, A. E. Ossman, and M. G. Ibrahim, *Rev. Roum. Chim.*, 1982, **27**, 859; *Chem. Abstr.*, 1983, **98**, 72047.

1494. T. Lorand, D. Szabo, A. Foldesi, and A. Neszmelyi, *Acta Chim. Acad. Sci. Hung.*, 1982, **110**, 231; *Chem. Abstr.*, 1983, **98**, 34018.

1495. P. Richter and O. Morgenstern, *Pharmazie*, 1982, **37**, 379; *Chem. Abstr.*, 1982, **97**, 162928.

1496. H. K. Misra and A. K. S. Gupta, *Pharmazie*, 1982, **37**, 254; *Chem. Abstr.*, 1982, **97**, 182336.

1497. A. Arques and P. Molina, *An. Quim., Ser. C.* 1982, **78**, 156; *Chem. Abstr.*, 1982, **97**, 182346.

1498. S. Plescia, M. L. Bajardi, and G. Daidone, *Boll. Chim. Farm.*, 1982, **121**, 563; *Chem. Abstr.*, 1983, **99**, 22415.

1499. M. Draminski and E. Frass, *Pol. J. Chem.*, 1981, **55**, 1547; *Chem. Abstr.*, 1983, **98**, 179812.

1500. A. Robeva and S. Robev, *Dokl. Bolg. Akad. Nauk*, 1984, **37**, 337; *Chem. Abstr.*, 1984, **101**, 171199.

1501. G. P. Zhikhareva, L. I. Mastafanova, M. I. Evstratova, G. Y. Shvarts, R. D. Syubaev, M. D. Mashkovskii, and L. N. Yakhontov, *Khim.-Farm. Zh.*, 1980, **14** (No. 2), 45; *Chem. Abstr.*, 1980, **93**, 114434.

1502. L. M. Yun, S. A. Makhmuodov, K. M. Shakhidoyatov, C. S. Kadyrov, and S. S. Kasymova, *Khim. Prir. Soedin.*, 1980, 680; *Chem. Abstr.*, 1981, **94**, 175039.

1503. M. A. El-Hashash and M. A. Sayed, *Egypt. J. Chem.*, 1978, **21**, 115; *Chem. Abstr.*, 1981, **94**, 139722.

1504. S. S. Tiwari, S. S. M. Zaidi, and R. K. Satsangi, *Pharmazie*, 1980, **35**, 73; *Chem. Abstr.*, 1980, **93**, 71703.

1505. A. K. El-Ansary, S. Botros, and M. Khalifa, *Pharmazie*, 1979, **34**, 753; *Chem. Abstr.*, 1980, **93**, 46577.

1506. M. A. El-Hashash, M. El-Kady, and M. M. Ammer, *Pak. J. Sci. Ind. Res.*, 1982, **25**, 104; *Chem. Abstr.*, 1983, **98**, 143349.

1507. B. V. Golomolzin, I. P. Tregubenko, E. A. Tarakhtii, L. V. Rasina, and O. N. Tikhonova, *Tr. Inst. Khim., Ural. Nauchn. Tsentr., Adad. Nauk SSSR*, 1978, **37**, 14; *Chem. Abstr.*, 1980, **92**, 146713.

1508. M. D. Nair and J. A. Desai, *Indian J. Chem. Sect. B*, 1980, **19**, 20; *Chem. Abstr.*, 1980, **93**, 95232.

1509. K. Golankiewicz and L. Celewicz, *Pol. J. Chem.*, 1979, **53**, 2075; *Chem. Abstr.*, 1980, **93**, 71684.

1510. S. Senda, K. Hirato, T. Asao, and I. Sugiyama, *Fukusokan Kagaku Toronkai Koen Yoshishu, 12th*, 1979, 261; *Chem. Abstr.*, 1980, **93**, 71698.

1511. F. Bahr and G. Dietz, *Pharmazie*, 1980, **35**, 256; *Chem. Abstr.*, 1981, **94**, 47254.

1512. I. A. Mazur, R. S. Sinyak, R. A. Katkevich, and P. N. Steblyuk, *Farm. Zh. (Kiev)*, 1980 (No. 4), 34; *Chem. Abstr.*, 1981, **94**, 47258.

1513. F. M. E. Abdel-Megeid, M. A. F. El-Kaschef, and A. A. G. Ghattas, *Egypt. J. Chem.*, 1977, **20**, 427; *Chem. Abstr.*, 1981, **94**, 30703.

1514. A. E.-G. M. Osman and S. Botros, *Pharmazie*, 1980, **35**, 439; *Chem. Abstr.*, 1981, **94**, 84050.

1515. M. El-Enany an S. Botros, *Pharmazie*, 1981, **36**, 62; *Chem. Abstr.*, 1981, **95**, 43030.

1516. A. A. Morozova and I. A. Majur, *Farm. Zh. (Kiev)*, 1980 (No. 5), 66; *Chem. Abstr.*, 1981, **95**, 7194.

1517. L. M. Yun, S. B. Babadzhanova, C. S. Kadyrov, and K. M. Shakhidoyatov, *Dokl. Akad. Nauk UzSSR*, 1983 (No. 11), 34; *Chem. Abstr.*, 1984, **101**, 55041.

1518. K. Kottke and H. Hühmstedt, *Pharmazie*, 1984, **39**, 867; *Chem. Abstr.*, 1985, **102**, 220816.

1519. R. R. Mohan, R. Agrawal, and V. S. Misra, *Indian J. Chem., Sect. B*, 1985, **24**, 78; *Chem. Abstr.*, 1985, **102**, 220817.

1520. A. Kreutzberger and M. Sellheim, *Chem.-Ztg.*, 1984, **108**, 326; *Chem. Abstr.*, 1985, **102**, 113409.

1521. O. E. Sattarova, Y. V. Kozhevnikov, V. S. Zalesov, S. N. Nikulina, and N. V. Semyakina, *Khim.-Farm. Zh.*, 1984, **18**, 820; *Chem. Abstr.*, 1985, **102**, 6380.

1522. K. Kottke and H. Kühmstedt, *Pharmazie*, 1980, **35**, 800; *Chem. Abstr.*, 1981, **95**, 7198.

1523. B. Novak, D. Babic, M. Japelj, V. Tišler, E. Kardelja, and D. Krka, *Hem. Ind.*, 1980, **34**, 330; *Chem. Abstr.*, 1981, **94**, 191845.

1524. A. A. Morozova, I. A. Mazur, and P. M. Steblyuk, *Farm. Zh. (Kiev)*, 1980 (No. 6), 53: *Chem. Abstr.*, 1981, **94**, 192255.

1525. M. Anwar, *Rev. Roum. Chim.*, 1981, **26**, 639; *Chem. Abstr.*, 1981, **95**, 115442.

1526. Y. V. Kozhevnikov, N. N. Smirnova, V. S. Zalesov, and I. I. Gradel, *Khim.-Farm. Zh.*, 1981, **15** (No. 6), 55: *Chem. Abstr.*, 1981, **95**, 115450.

1527. G. P. Petyunin, T. A.Kolesnikova, and P. A. Petyunin, *Ukr. Khim. Zh. (Russ. Ed.)*, 1981, **47**, 536; *Chem. Abstr.*, 1981, **95**, 132343.

1528. B. G. Khadse, S. R. Lokhande, M. R. Patel, and V. L. Khairnar, *Bull. Haffkine Inst.*, 1979, **7**, 29; *Chem. Abstr.*, 1981, **95**, 97707.

1529. B. G. Khadse, S. R. Lokhande, M. R. Patel, and V. L. Khairnar, *Bull. Haffkine Inst.*, 1979, **7**, 6; *Chem. Abstr.*, 1981, **95**, 62122.

1530. I. Bitter, L. Szocs, and L. Toke, *Acta Chim. Acad. Sci. Hung.*, 1981, **107**, 57; *Chem. Abstr.*, 1982, **96**, 6663.

1531. I. Bitter, L. Szocs, and L. Toke, *Acta Chim. Acad. Sci. Hung.*, 1981, **107**, 171; *Chem. Abstr.*, 1981, **95**, 220056.

1532. T. A. Voronina, G. N. Gordiichuk, S. A. Andronati, T. L. Garibova, and N. I. Zhilina, *Khim.-Farm. Zh.*, 1981, 15 (No. 7), 55; *Chem. Abstr.*, 1981, **95**, 203865.

1533. T. Lorand, D. Szabo, A. Foldesi, and A. Neszmelyi, *Acta Chim. Acad. Sci. Hung.*, 1981, **108**, 91; *Chem. Abstr.*, 1982, **97**, 6243.

1534. Y. V. Kozhevnikov, O. E. Sattarova, V. S. Zalesov, I. I. Gradel, and A. N. Plaksina, *Khim.-Farm. Zh.*, 1982, **16**, 1349: *Chem. Abstr.*, 1983, **98**, 143361.

1535. M. C. Etter, *Mol. Cryst. Liq. Cryst.*, 1983, **93**, 95; *Chem. Abstr.*, 1983, **99**, 104536.

1536. M. Lemptet-Sréter, K. Lempert, and J. Møller, *Acta Chim. Hung.*, 1983, **112**, 83; *Chem. Abstr.*, 1983, **99**, 88153.

1537. J. Zhong, M. Zhang, Y. Wang, Z. Dong, and R. Deng, *Yaoxue Xuebao*, 1983, **18**, 231; *Chem. Abstr.*, 1983, **99**, 70674.

1538. A. N. Osman, M. Khalifa, M. A. Ismail, A. E. Ossman, and M. G. Ibrahim, *Egypt. J. Chem.*, 1982, **25**, 159; *Chem. Abstr.*, 1983, **99**, 212486.

1539. M. Khalifa, A. N. Osman, M. G. Ibrahim, A. E. Ossman, and M. A. Ismail, *Egypt. J. Chem.*, 1982, **25**, 285; *Chem. Abstr.*, 1983, **99**, 158369.

1540. E. H. Sund, C. E. Beall, and P. A. Borgfeld, *J. Chem. Eng. Data*, 1983, **28**, 428; *Chem. Abstr.*, 1983, **99**, 139885.

1541. D. Wang and G. Li, *Yiyao Gongye*, 1983 (No. 10), 24; *Chem. Abstr.*, 1984, **100**, 191817.

1542. E. I. Georgescu, F. Georgescu, F. Ghiraleu, and M. Petrovanu, *Rev. Roum. Chim.*, 1983, **28**, 841; *Chem. Abstr.*, 1984, **100**, 191824.

1543. G. M. El-Naggar, H. A. H. El-Sherief, H. Aref, and K. Abdelhamid, *Pharmazie*, 1983, **38**, 821; *Chem. Abstr.*, 1984, **100**, 174774.

1544. S. Botros and A. N. Osman, *Egypt. J. Chem.*, 1982, **25**, 579; *Chem. Abstr.*, 1984, **100**, 6440.

1545. Š. Stankovský and D. Mrazova, *Chem. Zvesti*, 1984, **38**, 549; *Chem. Abstr.*, 1985, **102**, 24579.

1546. E. H. Erickson, C. F. Hainline, L. S. Lenon, C. J. Matson, T. K. Rice, K. F. Swingle, and N. van Winkle, *J. Med. Chem.*, 1979, **22**, 816.

1547. H. J. Havera and H. Vidrio, *J. Med. Chem.*, 1979, **22**, 1548.

1548. E. F. Elslager, M. P. Hutt, P. Jacob, J. Jhonson, B. Temporelli, L. M. Werbel, D. F. Worth, and L. Rane, *J. Med. Chem.*, 1979, **22**, 1247.

1549. T. H. Althuis, P. F. Moore, and H.-J. Hess, *J. Med. Chem.*, 1979, **22**, 44.

1550. J. Tani, Y. Yamada, T. Oine, T. Ochiai, R. Ishida, and I. Inoue, *J. Med. Chem.*, 1979, **22**, 95.

1551. E. F. Elslager, P. Jacob, J. Johnson, L. M. Werbel, D. F. Worth, and L. Rane, *J. Med. Chem.*, 1978, **21**, 1059.

1552. V. T. Bandurco, E. M. Wong, S. D. Levine, and Z. G. Hajos, *J. Med. Chem.*, 1981, **24**, 1455.

1553. S. Büyüktimkin, *Arch. Pharm.* (*Weinheim, Ger.*), 1985, **318**, 496.

1554. K.-C. Liu and L.-Y. Hsu, *Arch. Pharm.* (*Weinheim, Ger.*), 1985, **318**, 502.

1555. S Büyüktimkin, S. Elz, M. Dräger, and W. Schunack, *Arch. Pharm.* (*Weinheim, Ger.*), 1984, **317**, 797.

1556. G. Doria, C. Romeo, P. Sberze, M. Tibolla, M.-L. Corno, and G. Cadelli, *Eur. J. Med. Chem.*, 1979, **14**, 247.

1557. T. Sekiya, H. Hiranuma, S. Hata, S. Mizogami, M. Hanazuka, and S.-I. Yamada, *J. Med. Chem.*, 1983, **26**, 411.

1558. R. A. LeMahieu, M. Carson, W. C. Nason, D. R. Parrish, A. F. Welton, H. W. Baruth, and B. Yaremko, *J. Med. Chem.*, 1983, **26**, 420.

1559. D. M. Stout, W. L. Matier, C. Barcelon-Yang, R. D. Reynolds, and B. S. Brown, *J. Med. Chem.*, 1983, **26**, 808.

1560. K. Hirai, T. Fujishita, T. Ishiba, H. Sugimoto, S. Matsutani, Y. Tsukinoki, and K. Hirose, *J. Med. Chem.*, 1982, **25**, 1466.

1561. H. A. Parish, R. D. Gilliom, W. P. Purcell, R. K. Browne, R. F. Spirk, and H. D. White, *J. Med. Chem.*, 1982, **25**, 98.

1562. J. A. Grosso, D. E. Nichols, J. D. Kohli, and D. Glock, *J. Med. Chem.*, 1982, **25**, 703.

1563. E. F. Elslager, N. L. Colbry, J. Davoll, M. P. Hutt, J. L. Johnson, and L. M. Werbel, *J. Med. Chem.*, 1984, **27**, 1740.

1564. J. B. Hynes, Y. C. S. Yang, J. E. McGill, S. J. Harmon, and W. L. Washtien, *J. Med. Chem.*, 1984, **27**, 232.

1565. L. Legrand, R. Baronnet, J. Maugard, O. Foussard-Blanpin, G. Uchida-Ernouf, and M. Ray, *Eur. J. Med. Chem.*, 1979, **14**, 357.

1566. S. Barcza, G. M. Coppola, and M. J. Shapiro, *J. Heterocycl. Chem.*, 1979, **16**, 439.

1567. K.-C. Liu and L.-Y. Hsu, *Arch. Pharm.* (*Weinheim, Ger.*), 1987, **320**, 569.

1568. E. F. Elslager, C. Hess, J. Johnson, D. Ortwine, V. Chu, and L. M. Werbel, *J. Med. Chem.*, 1981, **24**, 127.

1569. J. A. Grosso, D. E. Nichols, M. B. Nichols, and G. K. W. Yim, *J. Med. Chem.*, 1980, **23**, 1261.

1570. A. A. Fatmi, N. A. Vaidya, W. B. Iturrian, and C. D. Blanton, *J. Med. Chem.*, 1984, **27**, 772.

1571. M. M. Angelo, D. Ortwine, D. F. Worth, and L. M. Werbel, *J. Med. Chem.*, 1983, **26**, 1311.

1572. J.-P. Spengler and W. Schunack, *Arch. Pharm.* (*Weinheim, Ger.*), 1983, **316**, 82.

1573. K.-C. Liu, M.-H. Yen, J.-W. Chern, and Y.-O Lin, *Arch. Pharm.* (*Weinheim, Ger.*), 1983, **316**, 379.

1574. K. K. Hajela, A. K. Sengupta, K. Shanker, and D. C. Doval, *Arch. Pharm.* (*Weinheim, Ger.*), 1983, **316**, 431.

1575. J. S. Shukla, K. Agarwal, and R. Rastogi, *Arch. Pharm.* (*Weinheim, Ger.*), 1983, **316**, 525.

1576. N. A. Vaidya, C. H. Panos, A. Kite, W. B. Iturrian, and C. D. Blanton, *J. Med. Chem.*, 1983, **26**, 1422.

1577. J. Renault, S. Giorgi-Renault, M. Baron, P. Mailliet, C. Paoletti, S. Cros, and E. Voisin, *J. Med. Chem.*, 1983, **26**, 1715.

1578. E. F. Elslager, J. L. Johnson, and L. M. Werbel, *J. Med. Chem.*, 1983, **26**, 1753.

1579. P. M. Manoury, J. L. Binet, A. P. Dumas, F. Lefèvre-Borg, and I. Cavero, *J. Med. Chem.*, 1986, **29**, 19.

1580. J. de Ruiter, A. N. Brubaker, J. Millen, and T. N. Riley, *J. Med. Chem.*, 1986, **29**, 627.

1581. S. W. Schneller, A. C. Ibay, E. A. Martinson, and J. N. Wells, *J. Med. Chem.*, 1986, **29**, 972.

1582. T. R. Jones, M. J. Smithers, R. F. Betteridge, M. A. Taylor, A. L. Jackman, A. H. Calvert, L. C. Davies, and K. R. Harrap, *J. Med. Chem.*, 1986, **29**, 1114.

1583. D. Giardinà, R. Bertini, E. Brancia, L. Brasili, and C. Melchiorre, *J. Med. Chem.*, 1985, **28**, 1354.

1584. F. Ishikawa, J. Saegusa, K. Inamura, K. Sakuma, and S.-I. Ashida, *J. Med. Chem.*, 1985, **28**, 1387.

1585. J. B. Hynes, S. J. Harmon, G. G. Floyd, M. Farrington, L. D. Hart, G. R. Gale, W. L. Washtien, S. S. Susten, and J. H. Freisheim, *J. Med. Chem.*, 1985, **28**, 209.

1586. S. Bahadur and M. Saxena, *Arch. Pharm. (Weinheim, Ger.)*, 1983, **316**, 964.

1587. M. R. Chaurasia and S. K. Sharma, *Arch. Pharm. (Weinheim, Ger.)*, 1982, **315**, 377.

1588. A. G. Agnihotra and S. K. Shukla, *Arch. Pharm. (Weinheim, Ger.)*, 1982, **315**, 701.

1589. J. Lehmann and G. Kraft, *Arch. Pharm.(Weinheim, Ger.)*, 1982, **315**, 967.

1590. K. Ozaki, Y. Yamada, T. Oine, T. Ishizuka, and Y. Iwasawa, *J. Med. Chem.*, 1985, **28**, 568.

1591. M. G. Nair, R. Dhawan, M. Glazala, T. I. Kalman, R. Ferone, Y. Gaumont, and R. L. Kisliuk, *J. Med. Chem.*, 1987, **30**, 1256.

1592. V. T. Bandurco, C. F. Schwender, S. C. Bell, D. W. Combs, R. M. Kanojia, S. D. Levine, D. M. Mulvey, M. A. Appollina, M. S. Reed, E. A. Malloy, R. Falotico, J. B. Moore, and A. J. Tobia, *J. Med. Chem.*, 1987, **30**, 1421.

1593. J. B. Hynes, A. Kumar, A. Tomažič, and W. L. Washtien, *J. Med. Chem.*, 1987, **30**, 1515.

1594. S. F. Campbell and R. M. Plews, *J. Med. Chem.*, 1987, **30**, 1794.

1595. l. M. Werbel and M. J. Degnan, *J. Med. Chem.*, 1987, **30**, 2151.

1596. W. B. Wright, A. S. Tomcufcik, P. S. Chan, J. W. Marsico, and J. B. Press, *J. Med. Chem.*, 1987, **30**, 2277.

1597. R. S. Verma, S. Bahadur, and A. K. Agnihotri, *Arch. Pharm. (Weinheim, Ger.)*, 1981, **314**, 97.

1598. S. F. Campbell, M. J. Davey, J. D. Hardstone, B. N. Lewis, and M. J. Palmer, *J. Med. Chem.*, 1987, **30**, 49.

1599. V. A. Alabaster, S. F. Campbell, J. C. Danilewicz, C. W. Greengrass, and R. M. Plews, *J. Med. Chem.*, 1987, **30**, 999.

1600. A. Kumar, S. Gurtu, J. N. Sinha, K. P. Bhargava, and K. Shanker, *Eur. J. Med. Chem.*, 1985, **20**, 95.

1601. D. J. Brown and K. Mori, *Aust. J. Chem.*, 1985, **38**, 467.

1602. M. R. Linschoten, H. D. Gaisser, H. van der Goot, and H. Timmerman, *Eur. J. Med. Chem.*, 1984, **19**, 137.

1603. R. Baronnet, R. Callendret, L. Blanchard, O. Foussard-Blanpin, and J. Bretaudeau, *Eur. J. Med. Chem.*, 1983, **18**, 241.

1604. I. A. Shaikh, F. Johnson, and A. P. Grollman, *J. Med. Chem.*, 1986, **29**, 1329.

1605. W. Wouters, C. G. M. Janssen, J. van Dun, J. B. A. Thijssen, and P. M. Laduron, *J. Med. Chem.*, 1986, **29**, 1663.

1606. M. G. Nair, N. T. Nanavati, I. G. Nair, R. L. Kisliuk, Y. Gaumont, M. C. Hsiao, and T. I. Kalman, *J. Med. Chem.*, 1986, **29**, 1754.

1607. N. P. Peet, L. E. Baugh, S. Sunder, J. E. Lewis, E. H. Matthews, E. L. Olberding, and D. N. Shah, *J. Med. Chem.*, 1986, **29**, 2403.

1608.   S. D. Patil, C. Jones, M. G. Nair, J. Galivan, F. Maley, R. S. Kisliuk, Y. Gaumont, D. Duch, and R. Ferone, *J. Med. Chem.*, 1989, **32**, 1284.

1609.   S. Büyüktimkin, *Arch. Pharm.* (*Weinheim, Ger.*), 1986, **319**, 933.

1610.   D. Giardinà, L. Brasili, M. Gregori, M. Massi, M. T. Picchio, W. Quaglia, and C. Melchiorre, *J. Med. Chem.*, 1989, **32**, 50.

1611.   C. G. M. Janssen, H. A. C. Lenoir, J. B. A. Thijssen, A. G. Knaeps, W. L. M. Verluyten, and J. J. P. Heykants, *J. Labelled Compd. Radiopharm.*, 1988, **25**, 783.

1612.   E. Ehrin, S. K. Luthra, C. Crouzel, and V. W. Pike, *J. Labelled Compd. Radiopharm.*, 1988, **25**, 177.

1613.   J. Pitha, L. Szabo, Z. Szurmai, W. Buchowiecki, and J. W. Kusiak, *J. Med. Chem.*, 1989, **32**, 96.

1614.   K. Pawelczak, T. R. Jones, M. Kempny, A. L. Jackman, D. R. Newell, L. Krzyzanowski, and B. Rzeszotarska, *J. Med. Chem.*, 1989, **32**, 160.

1615.   G. Fenton, C. G. Newton, B. M. Wyman, P. Bagge, D. I. Dron, R. Riddell, and G. D. Jones, *J. Med. Chem.*, 1989 **32**, 265.

1616.   P. R. Marsham, P. Chambers, A. J. Hayter, L. R. Hughes, A. L. Jackman, B. M. O'Connor, J. A. M. Bishop, and A. H. Calvert, *J. Med. Chem.*, 1989, **32**, 569.

1617.   T. R. Jones, T. J. Thornton, A. Flinn, A. L. Jackman, D. R. Newell, and A. H. Calvert, *J. Med. Chem.*, 1989, **32**, 847.

1618.   D. Wu, X. Qi, C. Hou, and Y. Sun, *J. Labelled Compd. Radiopharm.*, 1987, **24**, 1307.

1619.   R. H. Liu and M. G. Legendre, *J. Forensic Sci.*, 1986, **31**, 401.

1620.   J. Garin, C. Guillén, E. Meléndez, F. L. Merchán, J. Orduna, and T. Tejero, *Bull. Soc. Chim. Belg.*, 1987, **96**, 797.

1621.   C. C. Clark, *J. Forensic Sci.*, 1988, **33**, 1035.

1622.   J. B. Hynes, S. A. Patil, R. L. Hagan, A. Cole, W. Kohler, and J. H. Freisheim, *J. Med. Chem.*, 1989, **32**, 852.

1623.   J. B. Hynes, S. A. Patil, A. Tomažič, A. Kumar, A. Pathak, X. Tan, L. Xianqiang, M. Ratnam, T. J. Delcamp, and J. H. Freisheim, *J. Med. Chem.*, 1988, **31**, 449.

1624.   S. F. Campbell, J. C. Danilewicz, C. W. Greengrass, and R. M. Plews, *J. Med. Chem.*, 1988, **31**, 516.

1625.   K.-C. Liu, J.-W. Chern, M.-H. Yen, and Y.-O Lin, *Arch. Pharm.* (*Weinheim, Ger.*), 1983, **316**, 569.

1626.   P. Kumar, K. N. Dhawan, S. Vrat, K. P. Bhargava, and K. Kishore, *Arch. Pharm.* (*Weinheim, Ger.*), 1983, **316**, 759.

1627.   K. Eger, *Arch. Pharm.* (*Weinheim, Ger.*), 1981, **314**, 176.

1628.   S. Plescia, M. L. Bajardi, D. Raffa, G. Daidone, M. Matera, A. Caruso, and M. Amico-Roxas, *Eur. J. Med. Chem.*, 1986, **21**, 291.

1629.   M. J. Kornet, *Eur. J. Med. Chem.*, 1986, **21**, 529.

1630.   Y. S. Sadanandam, K. M. M. Reddy, and A. B. Rao, *Eur. J. Med. Chem.*, 1987, **22**, 169.

1631.   A. M. Farghaly, A. M. M. E. Omar, M. A. Khalil, M. A. Gaber, and H. Abou-Shleib, *Eur. J. Med. Chem.*, 1987, **22**, 369.

1632.   R. Lakhan and O. P. Singh, *Arch. Pharm.* (*Weinheim, Ger.*), 1985, **318**, 228.

1633.   F. Eiden and G. Patzelt, *Arch. Pharm.* (*Weinheim, Ger.*), 1985, **318**, 328.

1634.   S. Büyüktimkin, A. Buschauer, and W. Schunack, *Arch. Pharm.* (*Weinheim, Ger.*), 1988, **321**, 833.

1635.   T. Sekiya, H. Hiranuma, T. Kanayama, S. Hata, and S.-I. Yamada, *Eur. J. Med. Chem.*, 1982, **17**, 75.

1636.   H. K. Misra and A. K. S. Gupta, *Eur. J. Med. Chem.*, 1982, **17**, 216.

1637. A. M. M. E. Omar, S. A. S. El-Dine, A. A. Ghobashy, and M. A. Khalil, *Eur. J. Med. Chem.*, 1981, **16**, 77.

1638. T. Sekiya, H. Hiranuma, T. Kanayama, and S. Hata, *Eur. J. Med. Chem.*, 1980, **15**, 317.

1639. M. Süsse, F. Adler, and S. Johne, *Helv. Chim. Acta*, 1986, **69**, 1017.

1640. M. Süsse and S. Johne, *Helv. Chim. Acta*, 1985, **68**, 892.

1641. L. V. Shmelev, G. N. Lipunova, E. S. Karavaeva, G. M. Adamova, and Y. S. Ryabokobylko, *Zh. Org. Khim.*, 1987, **23**, 2458.

1642. H. K. Gakhar, R. Gupta, and J. K. Gill, *Indian J. Chem., Sect. B*, 1986, **25**, 957.

1643. H. V. Kamath and S. N. Kulkarni, *Indian J. Chem., Sect. B*, 1986, **25**, 967.

1644. R. Ashare and A. K. Mukerjee, *Indian J. Chem., Sect. B*, 1986, **25**, 1180.

1645. H. K. Gupta and S. K. Dikshit, *Indian J. Chem., Sect. B*, 1986, **25**, 842.

1646. L. D. Dave, C. Mathew, and V. Oommen, *Indian J. Chem., Sect. A*, 1986, **25**, 914.

1647. S. Gopinathan, I. R. Unny, S. S. Deshpande, and C. Gopinathan, *Indian J. Chem., Sect. A*, 1986, **25**, 1015.

1648. P. S. N. Reddy and C. N. Raju, *Indian J. Chem., Sect. B*, 1987, **26**, 983.

1649. T. George, M. K. Rao, and R. Tahilramani, *Indian J. Chem., Sect. B*, 1987, **26**, 1127.

1650. C. K. Reddy, P. S. N. Reddy, and C. V. Ratnam, *Indian J. Chem., Sect. B*, 1987, **26**, 882.

1651. P. Mencarelli and F. Stegel, *Gazz. Chim. Ital.*, 1987, **117**, 83.

1652. P. R. Kumar, *Indian J. Chem., Sect. B*, 1987, **26**, 725.

1653. V. K. Srivastava, S. Singh, A. Gulati, and K. Shanker, *Indian J. Chem., Sect. B*, 1987, **26**, 652.

1654. A. A. Dvorkin, S. A. Andronati, A. S. Yavorskii, T. S. Gifeisman, Y. A. Simonov, V. I. Pavlovskii, and T. I. Malinovskii, *Dokl. Akad. Nauk. SSSR*, 1987, **296**, 605.

1655. R. K. Vaid, B. V. Rao, B. Kaushik, and S. P. Singh, *Indian J. Chem., Sect. B*, 1987, **26**, 376.

1656. S. Singh, M. Shukla, V. K. Srivastava, and K. Shanker, *Indian J. Chem., Sect. B*, 1987, **26**, 368.

1657. P. S. N. Reddy and P. P. Reddy, *Indian J. Chem., Sect. B*, 1988, **27**, 763.

1658. P. Molina, A. Arques, and M. V. Vinader, *Liebigs Ann. Chem.*, 1987, 103.

1659. V. Balasubramaniyan and N. P. Argade, *Indian J. Chem., Sect. B*, 1988, **27**, 906.

1660. A. A. Afify, S. El-Nagdy, M. A. Sayad, and I. Mohey, *Indian J. Chem., Sect. B*, 1988, **27**, 920.

1661. A. M. Reddy, G. Achaiah, A. D. Rao, and V. M. Reddy, *Indian J. Chem., Sect. B*, 1988, **27**, 1054.

1662. D. P. Gupta, S. Ahmad, A. Kumar, and K. Shanker, *Indian J. Chem., Sect. B*, 1988, **27**, 1060.

1663. M. I. Husain and M. R. Jamali, *Indian J. Chem., Sect. B*, 1988, **27**, 43.

1664. H. Singh, R. Sarin, and K. Singh, *Indian J. Chem., Sect. B*, 1988, **27**, 132.

1665. P. S. N. Reddy and P. P. Reddy, *Indian J. Chem., Sect. B*, 1988, **27**, 135.

1666. J. Goerdeler and W. Eggers, *Chem. Ber.*, 1986, **119**, 3737.

1667. R. S. Varma, S. Chauhan, and C. R. Prasad, *Indian J. Chem., Sect. B*, 1988, **27**, 438.

1668. L. D. Dave, C. Mathew, and V. Oommen, *J. Indian Chem. Soc.*, 1986, **63**, 273.

1669. J. Roy, B. Ram, and A. P. Bhaduri, *J. Indian Chem. Soc.*, 1986, **63**, 343.

1670. A. D. Rao, C. R. Shankar, P. B. Reddy, and V. M. Reddy, *J. Indian Chem. Soc.*, 1985, **62**, 234.

1671. M. R. Chaurasia and A. K. Sharma, *J. Indian Chem. Soc.*, 1985, **62**, 308.

1672. M. I. Husain and M. Amir, *J. Indian Chem. Soc.*, 1985, **62**, 468.

1673. A. K. Saksena and P. Kant, *J. Indian Chem. Soc.*, 1984, **61**, 722.

1674. G. D. Gupta and H. K. Pujari, *J. Indian Chem. Soc.*, 1984, **61**, 1050.

1675. A. Kumar, S. Singh, A. K. Saxena, and K. Shanker, *Indian J. Chem., Sect. B*, 1988, **27**, 443.

1676. B. Prabhakar, K. L. Reddy, and P. Lingaiah, *Indian J. Chem., Sect. A*, 1988, **27**, 217.

1677. G. B. Barlin and S. J. Ireland, *Aust. J. Chem.*, 1985, **38**, 1685.

1678. Y. Uozumi, N. Kawasaki, E. Mori, M. Mori, and M. Shibasaki, *J. Am. Chem. Soc.*, 1989, **111**, 3725.

1679. M. F. Abdel-Megeed and A. Teniou, *Collect. Czech. Chem. Commun.*, 1988, **53**, 329.

1680. M. I. Husain and S. Shukla, *Indian J. Chem., Sect. B*, 1986, **25**, 545.

1681. A. D. Rao, C. R. Shankar, A. B. Rao, and V. M. Reddy, *Indian J. Chem., Sect. B*, 1986, **25**, 665.

1682. L. D. Dave, C. Mathew, and V. Oommen, *Indian J. Chem., Sect. A*, 1985, **24**, 755.

1683. C. Wentrup and C. Thétaz, *Helv. Chim. Acta*, 1976, **59**, 256.

1684. K. L. Reddy, P. Lingaiah, and K. V. Reddy, *Polyhedron*, 1986, **5**, 1519.

1685. J. Spindler and G. Kempter, *Z. Chem.*, 1987, **27**, 36.

1686. M. Süsse and S. Johne, *Z. Chem.*, 1987, **27**, 69.

1687. H. Haber and H.-G. Henning, *Z. Chem.*, 1987, **27**, 336.

1688. K. L. Reddy, S. Srihari, and P. Lingaiah, *Indian J. Chem., Sect. A*, 1985, **24**, 318.

1689. R. K. Saksena and R. Yasmeen, *Indian J. Chem., Sect. B*, 1986, **25**, 438.

1690. M. I. Husain and S. Shukla, *Indian J. Chem., Sect. B*, 1986, **25**, 552.

1691. S. B. Barnela and S. Seshadri, *Indian J. Chem., Sect. B*, 1986, **25**, 709.

1692. M. Süsse and S. Johne, *Z. Chem.*, 1986, **26**, 134.

1693. M. Süsse and S. Johne, *Z. Chem.*, 1985, **25**, 286.

1694. R. Varadarajan and P. K. Dhar, *Indian J. Chem., Sect. B*, 1986, **25**, 746

1695. J. S. Shukla and R. Rastogi, *Indian J. Chem., Sect. B*, 1986, **25**, 774.

1696. R. S. Atkinson and G. Tughan, *J. Chem., Soc., Chem. Commun.*, 1986, 834.

1697. R. S. Atkinson and G. Tughan, *J. Chem., Soc., Chem. Commun.*, 1987, 456.

1698. S. K. P. Sinha, M. P. Singh, Y. N. Singh, C. S. P. Singh, M. P. Shahi, B. D. Singh, and P. Kumar, *Indian J. Chem., Sect. B*, 1985, **24**, 1035.

1699. R. S. Atkinson and B. J. Kelly, *J. Chem., Soc., Chem. Commun.*, 1987, 1362.

1700. S. K. Sinha and P. Kumar, *Indian J. Chem., Sect. B*, 1985, **24**, 1182.

1701. J. Deli, T. Lóránd, D. Szabó, A. Földesi, and A. Zschunke, *Collect. Czech. Chem. Commun.*, 1985, **50**, 1602.

1702. L. Capuano and K. Djokar, *Liebigs Ann. Chem.*, 1985, 2305.

1703. C. K. Reddy, P. S. N. Reddy, and C. V. Ratnam, *Indian J. Chem., Sect. B*, 1985, **24**, 902.

1704. C. K. Reddy, P. S. N. Reddy, and C. V. Ratnam, *Indian J. Chem., Sect. B*, 1985, **24**, 695.

1705. M. Süsse and S. Johne, *Z. Chem.*, 1985, **25**, 367.

1706. H. Takai, H. Obase, M. Teranishi, A. Karasawa, K. Kubo, K. Shuto, Y. Kasuya, and K. Shigenobu, *Chem. Pharm. Bull.*, 1986, **34**, 1907.

1707. T. Higashino, M. Takemoto, and E. Hayashi, *Chem. Pharm. Bull.*, 1985, **33**, 1351.

1708. M. Süsse and S. Johne, *Z. Chem.*, 1985, **25**, 403.

1709. S. Rajappa and R. Sreenivasan, *Indian J. Chem., Sect. B*, 1985, **24**, 795.

1710. P. Molina, A. Arques, I. Cartagena, M. Asunción-Alias, M. C. Foces-Foces, and F. Hernandez-Cano, *Liebigs Ann. Chem.*, 1988, 133.

1711. H. Stalder, *Helv. Chim. Acta*, 1986, **69**, 1887.

1712. Y. Tomioka, K. Ohkubo, and M. Yamazaki, *Chem. Pharm. Bull.*, 1985, **33**, 1360.

1713. H. Kanazawa, K. Senga, and Z. Tamura, *Chem. Pharm. Bull.*, 1985, **33**, 618.

1714. T. Higashino, H. Kokubo, and E. Hayashi, *Chem. Pharm. Bull.*, 1985, **33**, 950.

1715. S. B. Barnela, S. Padmanabhan, and S. Seshadri, *Indian J. Chem., Sect. B*, 1985, **24**, 873.

1716. J. S. Shukla and V. K. Agarwal, *Indian J. Chem., Sect. B*, 1985, **24**, 886.

1717. C. R. Shankar, A. D. Rao, V. M. Reddy, and P. B. Sattur, *Indian J. Chem., Sect. B*, 1985, **24**, 580.

1718. A. Kamal, A. V. N. Reddy, and P. B. Sattur, *Indian J. Chem., Sect. B*, 1985, **24**, 414.

1719. K. R. Desai and P. H. Desai, *J. Indian Chem. Sec.*, 1988, **65**, 804.

1720. M. Kumar, P. Kumar, and T. Sharma, *J. Indian Chem. Soc.*, 1988, **65**, 869.

1721. H. Takai, H. Obase, N. Nakamizo, M. Teranishi, K. Kubo, K. Shuto, Y. Kasuya, K. Shigenobu, M. Hashikami, and N. Karashima, *Chem. Pharm. Bull.*, 1985, **33**, 1116.

1722. T. Higashino, S. Sato, A. Miyashita, and T. Katori, *Chem. Pharm. Bull.*, 1987, **35**, 4803.

1723. S. Ohno, K. Mizukoshi, O. Komatsu, Y. Kunoh, Y. Nakamura, E. Katoh, and M. Nagasaka, *Chem. Pharm. Bull.*, 1986, **34**, 4150.

1724. T. Higashino, A. Goto, A. Miyashita, and E. Hayashi, *Chem. Pharm. Bull.*, 1986, 1986, **34**, 4352.

1725. L. Legrand and N. Lozac'h, *Bull. Soc. Chim. Fr.*, 1985, 859.

1726. R. K. Saksena and A. Khan, *Indian J. Chem., Sect. B*, 1988, **27**, 295.

1727. C. Kaneko, K. Kasai, N. Katagiri, and T. Chiba, *Chem. Pharm. Bull.*, 1986, **34**, 3672.

1728. S. Giorgi-Renault, J. Renault, M. Baron, P. Gebel-Servolles, J. Delic, S. Cros, and C. Paoletti, *Chem. Pharm. Bull.*, 1988, **36**, 3933.

1729. R. S. Varma and D. Chatterjee, *Indian J. Chem., Sect. B*, 1985, **24**, 1039.

1730. H. Takai, H. Obase, N. Nakamizo, M. Teranishi, K. Kubo, K. Shuto, and T. Hashimoto, *Chem. Pharm. Bull.*, 1985, **33**, 1104.

1731. M. M. Mohamed, A. A. E1-Khamary, S. E1-Nagdy, and S. W. Shoshaa, *Indian J. Chem., Sect. B*, 1986, **25**, 207.

1732. A. O. Abdelhamid, F. A. Khalifa, and S. S. Ghabrial, *Phosphorus Sulfur*, 1988, **40**, 41.

1733. R. S. Atkinson, J. Fawcett, M. J. Grimshire, and D. R. Russell, *J. Chem. Soc., Chem. Commun.*, 1985, 544.

1734. J. Schmidt, M. Süsse, and S. Johne, *Org. Mass Spectrom.*, 1985, **20**, 184.

1735. H. Ogawa, S. Tamada, T. Fujioka, S. Teramoto, K. Kondo, S. Yamashita, Y. Yabuuchi, M. Tominaga, and K. Nakagawa, *Chem. Pharm. Bull.*, 1988, **36**, 2401.

1736. U. S. Pathak, M. B. Devani, C. J. Shishoo, R. R. Kulkarni, V. M. Rakholia, V. S. Bhadti, S. Ananthan, M. G. Dave, and V. A. Shah, *Indian J. Chem., Sect. B*, 1986, **25**, 489.

1737. H. Ogawa, S. Tamada, T. Fujioaka, S. Teramoto, K. Kondo, S. Yamashita, Y. Yabuuchi, M. Tominaga, and K. Nakagawa, *Chem. Pharm. Bull.*, 1988, **36**, 2253.

1738. T. Higashino, S. Sato, H. Suge, K.-I. Tanji, A. Miyashita, and T. Katori, *Chem. Pharm. Bull.*, 1988, **36**, 930.

1739. M. S. Reddy and C. V. Ratnam, *Org. Mass Spectrom.*, 1985, **20**, 698.

1740. A. Bajnati, B. Kokel, and M. Hubert-Habart, *Bull. Soc. Chim. Fr.*, 1987, 318.

1741. N. Ogawa, T. Yoshida, T. Aratani, E. Koshinaka, H. Kato, and Y. Ito, *Chem. Pharm., Bull.*, 1988, **36**, 2955.

1742. B. P. Acharya and Y. R. Rao, *Indian J. Chem., Sect. B*, 1987, **26**, 1133.

1743. V. Joshi and R. P. Chaudhari, *Indian J. Chem., Sect. B*, 1987, **26**, 602.

1744. B. Dash, E. K. Dora, and C. S. Panda, *J. Indian Chem. Soc.*, 1988, **65**, 136.

1745. S. Sailaja, E. T. Rao, E. Rajanarendar, and A. Krishnamurthy, *J. Indian Chem. Soc.*, 1988, **65**, 200.

1746. I. Ahmad, *J. Indian Chem. Soc.*, 1988, **65**, 362.

1747. L. D. Dave, C. Mathew, and V. Oommen, *J. Indian Chem. Soc.*, 1988, **65**, 377.

1748. A. K. Sengupta, S. Anand, and A. K. Pandey, *J. Indian Chem. Soc.*, 1987, **64**, 643.

1749. A. Jain and A. K. Mukerjee, *J. Indian Chem. Soc.*, 1987, **64**, 645.

1750. R. Lakhan, O. P. Singh, and R. L. Singh, *J. Indian Chem. Soc.*, 1987, **64**, 316.

1751. T. Yoshida, N. Kambe, S. Murai, and N. Sonoda, *Tetrahedron Lett.*, 1986, **27**, 3037.

1752. R. S. Atkinson and M. J. Grimshire, *Tetrahedron Lett.*, 1985, **26**, 4399.

1753. M. Botta, F. de Angelis, G. Finizia, R. Nicoletti, and M. Delfini, *Tetrahedron Lett.*, 1985, **26**, 3345.

1754. R. S. Atkinson, C. M. Darrah, and B. J. Kelly, *Tetrahedron Lett.*, 1987, **28**, 1711.

1755. F. Fülöp, K. Pihlaja, J. Mattinen, and G. Bernáth, *Tetrahedron Lett.*, 1987, **28**, 115.

1756. L. Baiocchi and G. Picconi, *Tetrahedron Lett.*, 1986, **27**, 5255.

1757. D. Ranganathan, S. Bamezai, and P. V. Ramachandran, *Heterocycles*, 1985, **23**, 623.

1758. P. Molina, M. Alajarin, and A. Vidal, *Tetrahedron Lett.*, 1988, **29**, 3849.

1759. L. Strekowski, M. T. Cegla, D. B. Harden, J. L. Mokrosz, and M. J. Mokrosz, *Tetrahedron Lett.*, 1988, **29**, 4265.

1760. C. Kaneko, T. Chiba, K. Kasai, and C. Miwa, *Heterocycles*, 1985, **23**, 1385.

1761. L. Giammanco and F. P. Invidiata, *Heterocycles*, 1985, **23**, 1459.

1762. P. D. Woodgate, J. M. Herbert, and W. A. Denny, *Heterocycles*, 1987, **26**, 1029.

1763. P. R. Kumar, *Heterocycles*, 1987, **26**, 1257.

1764. A. Kamal, M. V. Rao, and P. B. Sattur, *Heterocycles*, 1986, **24**, 3075.

1765. M. Mori, Y. Oozumi, and M. Shibasaki, *Tetrahedron Lett.*, 1987, **28**, 6187.

1766. M. L. Bajardi, G. Daidone, D. Raffa, and S. Plescia, *Heterocycles*, 1986, **24**, 1367.

1767. M. Mori, H. Kobayashi, M. Kimura, and Y. Ban, *Heterocycles*, 1985, **23**, 2803.

1768. A. Goto and T. Higashino, *Bull. Chem. Soc. Jpn.*, 1986, **59**, 395.

1769. O. P. Pandey, S. K. Sengupta, and S. C. Tripathi, *Bull. Chem. Soc. Jpn.*, 1985, **58**, 2395.

1770. S. L. Spassov, I. A. Atanassova, and M. Haimova, *Magn. Reson. Chem.*, 1985, **23**, 795.

1771. F. Fülöp and G. Bernáth, *Heterocycles*, 1985, **23**, 3095.

1772. D. Ranganathan, F. Farooqui, S. Mehnotra, and K. Kesavan, *Heterocycles*, 1986, **24**, 2493.

1773. A. Wada, H. Yamamoto, and S. Kanetomo, *Heterocycles*, 1988, **27**, 1345.

1774. C.-H. Chan, F.-J. Shish, K.-C. Liu, and J.-W. Chern, *Heterocycles*, 1987, **26**, 3193.

1775. F. Pinnen, G. Lucente, F. Mazza, and G. Pachetti, *Heterocycles*, 1987, **26**, 2053.

1776. J. Garin, C. Guillén, E. Meléndez, F. L. Merchán, J. Orduna, and T. Tejero, *Heterocycles*, 1987, **26**, 2371.

1777. W. S. Hamama, M. Hammouda, and E. M. Afsah, *Z. Naturforsch., Teil B*, 1988, **43**, 483.

1778. H. Möhrle and P. Arz, *Z. Naturforsch., Teil B*, 1987, **42**, 1035.

1779. T. Kametani, K. Kawamura, T. Akagi, C. Fujita, and T. Henda, *Heterocycles*, 1988, **27**, 2531.

1780. S. M. Sherif, R. M. Mohareb, G. E. H. Elgemeie, and R. P. Singh, *Heterocycles*, 1988, **27**, 1579.

1781. I. Thomsen and K. B. G. Torssell, *Acta Chem. Scand., Ser. B*, 1988, **42**, 309.

1782. K. M. H. Hilmy, J. Morgensen, and E. B. Pedersen, *Acta Chem. Scand., Ser. B*, 1987, **41**, 467.

1783. K. Pihlaja, F. Fülöp, J. Mattinen, and G. Bernáth, *Acta Chem. Scand., Ser. B*, 1987, **41**, 228.

1784. F. Johannsen and E. B. Pedersen, *Chem. Scr.*, 1987, **27**, 277.

1785. F. Johannsen, A. Jørgensen, and E. B. Pedersen, *Chem. Scr.*, 1986, **26**, 347.

1786. N. S. Girgis, J. Møller, and E. B. Pedersen, *Chem. Scr.*, 1986, **26**, 617.

1787. V. A. Dorokhov, K. L. Cherkasova, and I. A. Lutsenko, *Izv. Akad. Nauk SSSR, Ser. Khim.*, 1987, 2351.

1788. N. D. Chkanikov, V. L. Vershinin, A. F. Kolomiets, and A. V. Fokin, *Izv. Akad. Nauk SSSR, Ser. Khim.*, 1986, 952.

1789. N. S. Ibrahim, H. Z. Shams, M. H. Mohamed, and M. H. Elnagdi, *Chem. Ind (London)*, 1988, 563.

1790. I. V. Vasil'eva, E. N. Teleshov, and A. N. Pravednikov, *Izv. Akad. Nauk SSSR, Ser. Khim.*, 1985, 1440.

1791. Y. Nawata, H. Nagano, and K. Ochi, *Acta Crystallogr., Sect. C*, 1988, **44**, 183.

1792. F. Mazza, G. Pachetti, F. Pinnen, and G. Lucente, *Acta Crystallogr., Sect. C*, 1988, **44**, 1014.

1793. A. Takaoka, H. Iwakiri, N. Fujiwara, and Ishikawa, *Nippon Kagaku Kaishi*, 1985, 2161.

1794. T. Yoshida, N. Kambe, S. Murai, and N. Sonoda, *Bull. Chem. Soc. Jpn.*, 1987, **60**, 1793.

1795. C. H. Schwalbe and G. J. B. Williams, *Acta Crystallogr., Sect. C*, 1986, **42**, 1257.

1796. P. P. Reddy, C. K. Reddy, and P. S. N. Reddy, *Bull. Chem. Soc. Jpn.*, 1986, **59**, 1575.

1797. L. G. Chatten, R. E. Moskalyk, A. Chin, and P. Zuman, *Anal. Chim. Acta*, 1987, **200**, 281.

1798. M. F. Grundon, *Nat. Prod. Rep.*, 1987, **4**, 225.

1799. M. F. Grundon, *Nat. Prod. Rep.*, 1988, **5**, 293.

1800. S. Plescia, G. Daidone, L. Ceraulo, M. L. Bajardi, and R. A. Reina, *Farmaco, Ed. Sci.*, 1984, **39**, 120; *Chem. Abstr.*, 1984, **100**, 191825.

1801. E.-S. H. El-Ashry, N. Rashed, and A. Mousaad, *J. Carbohydr. Chem.*, 1987, **6**, 599; *Chem. Abstr.*, 1988, **109**, 231445.

1802. B. M. Burlaka, *Khim. Geterotsikl. Soedin.*, 1980, 708; *Chem. Abstr.*, 1980, **93**, 168221.

1803. M. Draminski, *Khim. Geterotsikl. Soedin.*, 1980, 705; *Chem. Abstr.*, 1980, **93**, 185255.

1804. A. N. Kost, Y. V. Kozhevnikov, and M. E. Konshin, *Khim. Geterotsikl. Soedin.*, 1980, 1286; *Chem. Abstr.*, 1981, **94**, 47264.

1805. E. Honkanen, A. Pippuri, P. Kairisalo, P. Nore, H. Karppanen, and I. Paakkari, *J. Med. Chem.*, 1983, **26**, 1433.

1806. R. Domanig, *Monatsh. Chem.*, 1981, **112**, 1195.

1807. P. Bhattacharyya, M. Sarkar, T. Roychowdhury, and D. P. Chakraborty, *Chem. Ind. (London)*, 1978, 532; *Chem. Abstr.*, 1979, **90**, 39095.

1808. M. T. Bogert, J. D. Wiggin, and J. E. Sinclair, *J. Am. Chem. Soc.*, 1907, **29**, 82.

1809. B. M. Burlaka, *Khim. Geterotsikl. Soedin.*, 1980, 705; *Chem. Abstr.*, 1980, **93**, 168220.

1810. J. Millen, T. N. Riley, I. W. Waters, and M. E. Hamrick, *J. Med. Chem.*, 1985, **28**, 12.

1811. H. Möhrle and J. Herbke, *Monatsh. Chem.*, 1980, **111**, 627.

1812. D. D. Mukerji, S. R. Nautiyal, and C. R. Prasad, *J. Prakt. Chem.*, 1980, **322**, 855.

1813. C. Bischoff and E. Schröder, *J. Prakt. Chem.*, 1985, **327**, 129.

1814. A. Mitidieri-Costanza and G. Sleiter, *J. Prakt. Chem.*, 1985, **327**, 865.

1815. K. Schweiger and G. Zigeuner, *Monatsh. Chem.*, 1978, **109**, 543.

1816. H. K. Gakhar, A. Jain, J. K. Gill, and S. B. Gupta, *Monatsh. Chem.*, 1983, **114**, 339.

1817. R. S. Atkinson and G. Tughan, *J. Chem. Soc., Perkin Trans. 1*, 1987, 2803

1818. R. S. Atkinson and M. J. Grimshire, *J. Chem. Soc., Perkin Trans. 1*, 1987, 1135.

1819. B. C. Uff, B. L. Joshi, and F. D. Popp, *J. Chem. Soc., Perkin Trans. 1*, 1986, 2295.

1820. H. K. Gakhar, S. Kiran, and S. B. Gupta, *Monatsh. Chem.*, 1982, **113**, 1145.

1821. R. Domanig, *Monatsh. Chem.*, 1982, **113**, 213.

1822. M. Süsse and S. Johne, *Monatsh. Chem.*, 1987, **118**, 71.

1823. G. Kempter, W. Ehrlichmann, M. Plesse, and H.-U. Lehm, *J. Prakt. Chem.*, 1982, **324**, 832.

1824. G. Kempter, H.-U. Lehm, M. Plesse, and A. Barth, *J. Perkt. Chem.*, 1982, **324**, 841.

1825. M. Süsse and S. Johne, *Monatsh. Chem.*, 1986, **117**, 499.

1826. R. S. Atkinson and M. J. Grimshire, *J. Chem. Soc., Perkin Trans. 1*, 1986, 1215.

1827. L. I. Giannola, S. Palazzo, L. Lamartina, L. Riva di Sanseverino, and P. Sabatino, *J. Chem. Soc., Perkin Trans. 1*, 1986, 2095.

1828. D. Korbonits and P. Kolonits, *J. Chem. Soc., Perkin Trans. 1*, 1986, 2163.

1829. J. S. Davidson, *Monatsh. Chem.*, 1984, **115**, 565.

1830. C. Bischoff, E. Schröder, and E. Gründemann, *J. Prakt. Chem.*, 1982, **324**, 519.

1831. K. Gewald, H. Schäfer, and P. Bellmann, *J. Prakt. Chem.*, 1982, **324**, 933.

1832. N. Z. Yalysheva, V. V. Chistyakov, and V. G. Granik, *Khim. Geterotsikl. Soedin.*, 1986, 84.

1833. L. M. Yun and K. M. Shakhidoyatov, *Khim. Geterotsikl. Soedin.*, 1986, 417.

1834. L. M. Yun, K. O. Nazhimov, S. Masharipov, R. A. Samiev, S. A. Makhmudov, S. S. Kasymova, S. N. Vergizov, K. A. V'yunov, and K. M. Shakhidoyatov, *Khim. Geterotsikl. Soedin.*, 1987, 1527.

1835. P. Singh, K. Deep, and H. Singh, *J. Chem. Res.*, 1984, *Synop.* 71, *Minipr.* 636.

1836. L. M. Yun, S. Yangibaev, V. Y. Alekseeva, K. A. V'yunov, and K. M. Shakhidoyatov, *Khim. Geterotsikl. Soedin.*, 1987, 1095.

1837. R. Beckert and R. Mayer, *J. Prakt. Chem.*, 1981, **323**, 511.

1838. M. Süsse and S. Johne, *J. Prakt. Chem.*, 1984, **326**, 342.

1839. B. P. Acharya and Y. R. Rao, *J. Chem. Res.*, 1987, *Synop.* 96, *Minipr.* 1001.

1840. V. Peesapati, P. L. Pauson, and R. A. Pethrick, *J. Chem. Res.*, 1987, *Synop.* 194.

1841. P. D. Kennewell, R. M. Scrowston, I. G. Shenouda, R. W. Tully, and R. Westwood, *J. Chem. Res.*, 1986, *Synop.* 232, *Minipr.* 2001.

1842. H. Fischer, H. Möller, M. Budnowski, G. Atassi, P. Dumont, J. Venditti, and O. C. Yoder, *Arzneim.-Forsch.*, 1984, **34**, 663.

1843. A. Lata, R. K. Satsangi, V. K. Srivastava, and K. Kishor, *Arzneim.-Forsch.*, 1982, **32**, 24.

1844. C. Bischoff, E. Schröder, and E. Gründemann, *J. Prakt. Chem.*, 1984, **326**, 849.

1845. L. M. Yun, S. Yangibaev, K. M. Shakhidoyatov, V. Y. Alekseeva, and K. A. V'yunov, *Khim. Geterotsikl. Soedin.*, 1987, 254.

1846. L. M. Yun, S. Yangibaev, K. M. Shakhidoyatov, V. Y. Alekseeva, and K. A. V'yunov, *Khim. Geterotsikl. Soedin.*, 1986, 1236.

1847. M. Süsse and S. Johne, *J. Prakt. Chem.*, 1984, **326**, 1027.

1848. W. Wouters, J. van Dun, and P. M. Laduron, *Biochem. Pharmacol.*, 1986, **35**, 3199.

1849. M. van Boven and P. Daenens, *J. Pharm. Sci.*, 1982, **71**, 1152.

1850. A. K. S. Gupta and H. K. Misra, *J. Pharm. Sci.*, 1980, **69**, 1313.

1851. C. Bischoff and E. Schröder, *J. Prakt. Chem.*, 1983, **325**, 88.

1852. J. D. Warren, S. A. Lang, P. S. Chan, and J. W. Marsico, *J. Pharm. Sci.*, 1978, **67**, 1479.

1853. C. Bischoff and E. Schröder, *J. Prakt. Chem.*, 1988, **330**, 289.

1854. A. D. Dunn and R. Norrie, *J. Prakt. Chem.*, 1987, **329**, 321.

1855. C. Bischoff and E. Schröder, *J. Prakt. Chem.*, 1987, **329**, 177.

1856. A. Y. Il'chenko, V. I. Krokhtyak, and L. M. Yagupol'skii, *Khim. Geterotsikl. Soedin.*, 1982, 1407.

1857. N. V. Abbakumova, A. F. Vasil'ev, and E. B. Nazarova, *Khim. Geterotsikl. Soedin.*, 1981, 968.

1858. N. V. Abbakumova, Y. G. Putsykin, Y. A. Baskakov, and Y. A. Kondrat'ev, *Khim. Geterotsikl. Soedin.*, 1981, 1264.

1859. M. Süsse and S. Johne, *J. Prakt. Chem.*, 1986, **328**, 635.

1860. A. Y. Tikhonov, L. B. Volodarskii, and O. M. Sokhatskaya, *Khim. Geterotsikl. Soedin.*, 1979, 1265.

1861. D. Chaigne, J. F. Hémidy, L. Legrand, and D. Cornet, *J. Chem. Res.*, 1978, *Synop.* 160, *Minipr.* 2066.

1862. L. N. Yun, S. Yangibaev, K. M. Shakhidoyatov, and C. S. Kadyrov, *Khim. Geterotsikl. Soedin.*, 1983, 268.

1863. N. Z. Yalysheva and V. G. Granik, *Khim. Geterotsikl. Soedin.*, 1984, 1430.

1864. S. Yangibaev, L. M. Yun, and K. M. Shakhidoyatov, *Khim. Geterotsikl. Soedin.*, 1985, 853.

1865. A. R. Katritzky, N. E. Grzeskowiak, T. Siddiqui, C. Jayaram, and S. N. Vassilatos, *J. Chem. Res.*, 1982, *Synop.* 26, *Minipr.* 528.

1866. J. G. Meingassner, H. Nesvadba, and H. Mieth, *Arzneim.-Forsch.*, 1981, **31**, 6.

1867. B. A. Orwig, H. A. Dugger, S. I. Bhuta, K. C. Talbot, and H. J. Schwarz, *Arzneim.-Forsch.*, 1981, **31**, 904.

1868. S. S. Susten, J. B. Hynes, A. Kumar, and J. H. Freisheim, *Biochem. Pharmacol.*, 1985, **34**, 2163.

1869. H. Singh, S. S. Narula, and C. S. Gandhi, *J. Chem. Res.*, 1978, *Synop.* 324, *Minipr.* 3957.

1870. H. Singh and C. S. Gandhi, *J. Chem. Res.*, 1978, *Synop.* 407, *Minipr.* 4930.

1871. J. H. Schornagel, P. K. Chang, L. J. Schiarini, B. A. Moroson, E. Mini, A. R. Cashmore, and J. R. Bertino, *Biochem. Pharmacol.*, 1984, **33**, 3251.

1872. D. Korbonits and P. Kolonits, *J. Chem. Res.*, 1988, *Synop.* 209, *Monipr.* 1652; 1989, *Synop.* 328.

1873. M. F. G. Stevens, A. Gescher, and C. P. Turnbull, *Biochem. Pharmacol.*, 1979, **28**, 769.

1874. H. Akgün, U. Hollstein, and L. Hurwitz, *J. Pharm. Sci.*, 1988, **77**, 735.

1875. W. L. F. Armarego, J. G. Altin, R. C. Weir, and F. L. Bygrave, *Biochem. Pharmacol.*, 1987, **36**, 1583.

1876. G. A. Bennett, L. A. Radov, L. A. Trusso, and V. St. Georgiev, *J. Pharm. Sci.*, 1987, **76**, 633.

1877. K.-C. Liu and M.-K. Hu, *T'ai-wan Yao Hsueh Tsa Chih*, 1986, **38**, 125; *Chem. Abstr.*, 1988, **108**, 56049.

1878. K.-C. Liu and M.-K. Hu, *T'ai-wan Yao Hsueh Tsa Chih*, 1986, **38**, 85; *Chem. Abstr.*, 1988, **108**, 56048.

1879. S. Büyüktimkin, *Acta Pharm. Turc.*, 1987, **29**, 94; *Chem. Abstr.*, 1988, **108**, 5964.

1880. R. S. Hosmane, B. B. Lim, M. F. Summers, U. Siriwardane, N. S. Hosmane, and S. S. C. Chu, *J. Org. Chem.*, 1988, **53**, 5309.

1881. N. S. Habib and M. A. Khalil, *J. Pharm. Sci.*, 1984, **73**, 982.

1882. K. Muraoka, M. Ichikawa, and T. Hisano, *Yakugaku Zasshi*, 1980, **100**, 375.

1883. A. L. Weis, F. Frolow, and R. Vishkautsan, *J. Heterocycl. Chem.*, 1986, **23**, 705.

1884. R. Murdoch, W. R. Tully, and R. Westwood, *J. Heterocycl. Chem.*, 1986, **23**, 833.

1885. J. Kant, F. D. Popp, and B. C. Uff, *J. Heterocycl. Chem.*, 1985, **22**, 1313.

1886. J. J. Hlavka, P. Bitha, Y.-I Lin, and T. Strohmeyer, *J. Heterocycl. Chem.*, 1985, **22**, 1317.

1887. T. Hisano, K. Muraoka, and M. Ichikawa, *Yakugaku Zasshi*, 1978, **98**, 1173.

1888. E. Hayashi, T. Higashino, E, Oishi, C. Iijima, M. Yamagishi, C. Ota, Y. Miwa, A. Nakajima, S. Iwata, M. Ohmine, S. Tsuda, Y. Ohhira, A. Hoshi, and K. Kuretani, *Yakugaku Zasshi*, 1978, **98**, 1560.

1889. A. E.-H. N. Ahmed, M. A. Abd-Alla, and M. F. El-Zohry, *J. Chem. Technol. Biotechnol.*, 1988, **43**, 63.

1890. B. Gober, H. Lisowski, and P. Franke, *Pharmazie*, 1988, **43**, 23; *Chem. Abstr.*, 1988, **108**, 192653.

1891. J. Kalinowska-Torz, *Acta Pol. Pharm.*, 1986, **43**, 218; *Chem. Abstr.*, 1988, **108**, 131745.

1892. M. A. Abbady, M. M. Ali, and M. M. Kandeel, *J. Chem. Technol. Biotechnol.*, 1981, **31**, 111.

1893. I. Butula, V. Vale, and B. Zorc, *Croat. Chem. Acta*, 1981, **54**, 105.

1894. M. A. Badawy, S. A. L. Abdel-Hady, Y. A. Ibrahim, and A. M. Kadry, *J. Heterocycl. Chem.*, 1985, **22**, 1535.

1895. P. K. Bridson, R. A. Davis, and L. S. Renner, *J. Heterocycl. Chem.*, 1985, **22**, 753.

1896. G. Barbaro, A. Battaglia, and P. Giorgianni, *J. Org. Chem.*, 1987, **52**, 3289.

1897. T. Yoshida, N. Kambe, S. Murai, and N. Sonoda, *J. Org. Chem.*, 1987, **52**, 1611.

1898. P. C. Crofts, I. M. Downie, and R. B. Heslop, *J. Chem. Soc.*, 1960, 3673.

1899. M. Yotsu, T. Yamazaki, Y. Meguro, A. Endo, M. Murata, H. Naoki, and T. Yasumoto, *Toxicon*, 1987, **25**, 225; *Chem. Abstr.*, 1987, **107**, 3948.

1900. N. P. Peet, S. Sunder, R. J. Barbuch, E. W. Huber, and E. M. Bargar, *J. Heterocycl. Chem.*, 1987, **24**, 1531.

1901. Y. Abe, K. Nishino, S. Nakajima, T. Kamura, K. Imagawa, and Y. Ikutani, *Osaka Kyoiku Daigaku Kiyo, Dai-3-bumon*, 1984, **33**, 113; *Chem. Abstr.*, 1985, **103**, 178222.

1902. A. M. Abdelal, *Orient. J. Chem.*, 1992, **8**, 286; *Chem. Abstr.*, 1993, **118**, 233987.

1903. A. Kalman, G. Argay, J. Lazar, T. Rudisch, and G. Bernáth, *Acta Chem. Hung.*, 1985, **118**, 49; *Chem. Abstr.*, 1985, **103**, 122606.

1904. N. Büyüktimkin and S. Büyüktimkin, *Sci. Pharm.*, 1985, **53**, 147; *Chem. Abstr.*, 1986, **105**, 115035.

1905. J. Kalinowska-Torz, *Acta Pol. Pharm.*, 1985, **42**, 112; *Chem. Abstr.*, 1986, **105**, 133838.

1906. W. Zhou, Z. Dai, Y. Ding, and X. Zhang, *Yaoxue Xuebao*, 1985, **20**, 536; *Chem. Abstr.*, 1986, **104**, 61550.

1907. A. D. Dunn, K. I. Kinnear, R. Norrie, N. Ringan, and D. Martin, *J. Heterocycl. Chem.*, 1987, **24**, 175; 1991, **28**, 2071.

1908. P. Kumar, C. Nath, and K. Shankar, *Pharmazie*, 1985, **40**, 267; *Chem. Abstr.*, 1985, **103**, 215250.

1909. I. A. Mazur, R. S. Sinyak, B. Y. Mandrichenko, S. S. Stoyanovich, V. R. Stets, P. N. Steblyuk, and S. I. Kovalenko, *Farm. Zh. (Kiev)*, 1987 (No. 1), 58; *Chem. Abstr.*, 1987, **106**, 196388.

1910. N. Y. Moskalenko, L. N. Yakhontov, G. P. Zhikhareva, G. N. Pershin, V. V. Peters, M. I. Maksakovskaya, and I. M. Kulikovskaya, *Khim.-Farm. Zh.*, 1986, **20**, 437; *Chem. Abstr.*, 1987, **106**, 32975.

1911. N. P. Peet, *J. Heterocycl. Chem.*, 1987, **24**, 223.

1912. L. M. Werbel, E. F. Elslager, and L. S. Newton, *J. Heterocycl. Chem.*, 1987, **24**, 345.

1913. S. Kawamura and Y. Sanemitsu, *J. Org. Chem.*, 1993, **58**, 414.

1914. H. Singh, S. K. Aggarwal, and N. Malhotra, *Tetrahedron*, 1986, **42**, 1139.

1915. A. Alemagna, P. D. Buttero, E. Licandro, S. Maiorana, and A. Papagni, *Tetrahedron*, 1985, **41**, 3321.

1916. L. Mosti, G. Menozzi, and P. Schenone, *J. Heterocycl. Chem.*, 1987, **24**, 603.

1917. G. Menozzi, A. Bargagna, L. Mosti, and P. Schenone, *J. Heterocycl. Chem.*, 1987, **24**, 633.

1918. H. Takeuchi and S. Eguchi, *J. Chem. Soc., Perkin Trans. 1*, 1988, 2149.

1919. E. A. Bliss, R. J Griffin, and M. F. G. Stevens, *J. Chem. Soc., Perkin Trans. 1*, 1987, 2217.

1920. P. A. Brownsort and R. M. Paton, *J. Chem. Soc., Perkin Trans. 1*, 1987, 2339.

1921. S. Büyüktimkin, *Sci. Pharm.*, 1984, **52**, 296; *Chem. Abstr.*, 1985, **103**, 6299.

1922. K. Kottke and H. Kühmstedt, *Pharmazie*, 1984, **39**, 717; *Chem. Abstr.*, 1985, **103**, 6309.

1923. D. W. Combs, J. B. Press, D. Mulvey, Y. Gray-Nuñez, and S. C. Bell, *J. Hetrocycl. Chem.*, 1986, **23**, 1263.

1924. A. Walser, T. Flynn, C. Mason, and R. I. Fryer, *J. Hetrocycl. Chem.*, 1986, **23**, 1303.

1925. A. V. N. Reddy, A. Kamal, and P. B. Sattur, *Synth. Commun.*, 1988, **18**, 525.

1926. P. Molina, A. Arques, M. L. Garcia, and M. V. Vinader, *Synth. Commun.*, 1987, **17**, 1449.

1927. J. T. Gupton, K. F. Correia, and B. S. Foster, *Synth. Commun.*, 1986, **16**, 365.

1928. K. Kottke and H. Kühmstedt, *Pharmazie*, 1984, **39**, 868; *Chem. Abstr.*, 1985, **103**, 22540.

1929. V. St Georgiev, G. A. Bennett, L. A. Radov, D. K. Kamp, and L. A. Trusso, *J. Heterocycl. Chem.*, 1986, **23**, 1359.

1930. J. B. Press, V. T. Bandurco, E. M. Wong, Z. G. Hajos, R. M. Kanojia, R. A. Mallory, E. G. Deegan, J. J. McNally, J. R. Roberts, M. L. Cotter, D. W. Graden, and J. R. Lloyd, *J. Hetrocycl. Chem.*, 1986, **23**, 1821.

1931. J.-W. Chern, F.-J. Shish, C.-D. Chang, C.-H. Chan, and K.-C. Liu, *J. Heterocycl. Chem.*, 1988, **25**, 1103.

1932. J. B. Hynes, A. Pathak, C. H. Panos, and C. C. Okeke, *J. Heterocycl. Chem.*, 1988, **25**, 1173.

1933. T. T. Tita and M. J. Kornet, *J. Heterocycl. Chem.*, 1988, **25**, 265.

1934. G. Bernáth, G. Stájer, A. E. Szabó, F. Fülöp, and P. Sohár, *Tetrahedron*, 1985, **41**, 1353.

1935. D. Ranganathan, R. Rathi, K. Kesavan, and W. P. Sing, *Tetrahedron*, 1986, **42**, 4873.

1936. F. A. Golec and L. W. Reilly, *J. Heterocycl. Chem.*, 1988, **25**, 789.

1937. S. Büyüktimkin and N. Büyüktimkin, *Acta Pharm. Turc.*, 1986, **28**, 41; *Chem. Abstr.*, 1987, **107**, 7148.

1938. N. Despande and Y. V. Rao, *Indian J. Pharm. Sci.*, 1986, **48**, 13; *Chem. Abstr.*, 1987, **107**, 7149.

1939. F. Fülöp and G. Bernáth *Magy. Kem. Foly.*, 1986, **92**, 24; *Chem. Abstr.*, 1986, **105**, 226479.

1940. S. Buscemi, M. G. Cicero, N. Vivona, and T. Caronna, *J. Heterocycl. Chem.*, 1988, **25**, 931.

1041. K. Lempert, J. Fetter, J. Nyitrai, F. Bertha, and J. Møller, *J. Chem. Soc., Perkin Trans. 1*, 1986, 269.

1942. D. Hunter, D. G. Neilson, and T. J. R. Weakley, *J. Chem. Soc., Perkin Trans. 1*, 1985, 2709.

1943. R. M. Christie and S. Moss, *J. Chem. Soc., Perkin Trans. 1*, 1985, 2779.

1944. R. S. Atkinson and N. A. Gawad, *J. Chem. Soc., Perkin Trans. 1*, 1985, 335.

1945. A. Alemagna, C. Baldoli, E. Licandro, S. Maiorana, and A. Papagni, *Tetrahedron*, 1986, **42**, 5397.

1946. J. Bergman, A. Brynolf, B. Elman, and E. Vuorinen, *Tetrahedron*, 1986, **42**, 3697.

1947. R. S. Atkinson and N. A. Gawad, *J. Chem. Soc., Perkin Trans. 1*, 1985, 341.

1948. M. Lempert-Sréter and K. Lempert, *Acta Chim. Hung.*, 1984, **117**, 121; *Chem. Abstr.*, 1985, **103**, 37440.

1949. M. I. Hussain and M. Amir, *J. Chem. Soc. Pak.*, 1984, **6**, 211; *Chem. Abstr.*, 1985, **103**, 37446.

1950. K. Kottke and H. Kühmstedt, *Pharmazie*, 1984, **39**, 773; *Chem. Abstr.*, 1985, **103**, 22541.

1951. T. M. Stevenson, F. Kaźmierczak, and N. J. Leonard, *J. Org. Chem.*, 1986, **51**, 616.

1952. R. S. Atkinson and N. A. Gawad, *J. Chem. Soc., Perkin Trans. 1*, 1985, 825.

1953. T. Nishio, M. Fujisawa, and Y. Omote, *J. Chem. Soc., Perkin Trans. 1*, 1987, 2523.

1954. R. S. Atkinson and G. Tughan, *J. Chem. Soc., Perkin Trans. 1*, 1987, 2797.

1955. A. R. Katritzky, W.-Q. Fan, and K. Akutagawa, *Tetrahedron*, 1986, **42**, 4027.

1956. D. Ranganathan, F. Farooqui, D. Bhattacharyya, S. Mehrotra, and K. Kesavan, *Tetrahedron*, 1986, **42**, 4481.

1957. R. Beugelmans and M. Bois-Choussy, *Tetrahedron*, 1986, **42**, 1381.

1958. M. Schleuder, K. Kottke, H. Kühmstedt, and I. Szkorupa, *Pharmazie*, 1987, **42**, 412; *Chem. Abstr.*, 1988, **108**, 186685

1959. P. M. Parasharya and A. R. Parikh, *Acta Cienc. Indica, Chem.*, 1985, **11**, 71; *Chem. Abstr.*, 1988, **108**, 94486.

1960. K.-C. Liu, S.-W. Hsu, and M.-K. Hu, *T'ai-wan Yao Hsueh Tsa Chih*, 1986, **38**, 242; *Chem. Abstr.*, 1988, **108**, 94490.

1961. F. Fülöp, G. Bernáth, K. Pihlaja, J. Mattinen, G. Argay, and A. Kálmán, *Tetrahedron*, 1987, **43**, 4731.

1962. C.-H. Lee, J. H. Gilchrist, and E. B. Skibo, *J. Org. Chem.*, 1986, **51**, 4784.

1963. D. B. Harden, M. J. Mokrosz, and L. Strekowski, *J. Org. Chem.*, 1988, **53**, 4137.

1964. E. B. Skibo and J. H. Gilchrist, *J. Org. Chem.*, 1988, **53**, 4209.

1965. S. Shimizu and M. Ogata, *J. Org. Chem.*, 1988, **53**, 5160.

1966. H. Singh and S. Kumar, *Tetrahedron*, 1987, **43**, 2177.

1967. H. Niwa, Y. Yoshida, and K. Yamada, *J. Nat. Prod.*, 1988, **51**, 343; *Chem. Abstr.*, 1988, **109**, 4108.

1968. A. Hempel, N. Camerman, and A. Camerman, *Cancer Biochem. Biophys.*, 1988, **10**, 25; *Chem. Abstr.*, 1988, **109**, 197024.

1969. S. Botros, M. M. El-Enany, F. M. Amine, and L. N. Soliman, *Bull. Fac. Pharm. (Cairo Univ.)*, 1986, **25**, 41; *Chem. Abstr.*, 1988, **109**, 211005.

1970. J. T. Gupton, J. F. Miller, R. D. Bryant, P. R. Maloney, and B. S. Foster, *Tetrahedron*, 1987, **43**, 1747.

1971. D. L. Boger and Q. Dang, *Tetrahedron*, 1988, **44**, 3379.

1972. M. F. Ismail, A. M. A. El-Khamry, H. A. A. Hamid, and S. A. Emara, *Tetrahedron*, 1988, **44**, 3757.

1973. M. F. Abdel-Megeed and M. A. H. Saleh, *Sulfur Lett.*, 1987, **6**, 115; *Chem. Abstr.*, 1988, **109**, 38151.

1974. M. A. Hussain, A. T. Chiu, W. A. Price, P. B. Timmermans, and E. Shefter, *Pharm. Res.*, 1988, **5**, 242; *Chem. Abstr.*, 1988, **109**, 222.

1975. Y. A. Ammar, A. M. S. El-Scharief, Y. A. Mohamed, and H. A. Ahmed, *J. Serb. Chem. Soc.*, 1987, **52**, 633; *Chem. Abstr.*, 1988, **109**, 170357.

1976. P. Richter, M. Schleuder, and P. Stiebert, *Pharmazie*, 1988, **43**, 362; *Chem. Abstr.*, 1988, **109**, 170400.

1977. K. A. M. El-Bayouki, H. H. Moharram, A. S. El-Sayed, H. M. Hosney, and M. Abdel-Hamid, *Orient. J. Chem.*, 1988, **4**, 84; *Chem. Abstr.*, 1988, **109**, 170358.

1978. H. Booth, K. A. Khedhair, and H. A. R. Y. Al-Shirayda, *Tetrahedron*, 1988, **44**, 1465.

1979. R. Soliman and F. S. G. Soliman, *Synthesis*, 1979, 803.

1980. F. Fülöp and G. Bernáth, *Synthesis*, 1985, 1148.

1981. N. P. Peet, *Synthesis*, 1984, 1065.

1982. M. Botta, F. de Angelis, G. Finizia, A. Gambacorta, and R. Nicoletti, *Synth. Commun.*, 1985, **15**, 27.

1983. A. K. El-Shafei and A. M. El-Sayed, *Rev. Roum. Chim.*, 1988, **33**, 291; *Chem. Abstr.*, 1988, **109**, 190359.

1984. N. P. Abdullaev and K. M. Shakhidoyatov, *Uzb. Khim. Zh.*, 1985, 81; *Chem. Abstr.*, 1985, **103**, 87840.

1985. H. Pöhlmann, F. P. Theil, and S. Pfeifer, *Pharmazie*, 1985, **40**, 269; *Chem. Abstr.*, 1985, **103**, 92714.

1986. P. Molina, A. Arques, I. Cartagena, and M. V. Valcarel, *Synth. Commun.*, 1985, **15**, 643.

1987. E. B. Skibo, *J. Org. Chem.*, 1985, **50**, 4861.

1988. J. Garcia-Tercero, A. Lopez-Aliaga, E. Meléndez-Andreu, F. L. Merchán-Alvarez, and T. Tejero-Lopez, *An. Quim., Ser. C*, 1987, **83**, 247; *Chem. Abstr.*, 1988, **109**, 128941.

1989. D. P. Gupta and K. Shanker, *Indian J. Chem., Sect B*, 1987, **26**, 1197; *Chem Abstr.*, 1988, **109**, 230930.

1990. K.-C. Liu and S.-W. Hsu, *T'ai-wan Yao Hsueh Tsa Chih*, 1987, **39**, 54; *Chem. Abstr.*, 1988, **109**, 230936.

1991. K. Kottke, H. Kümstedt, and D. Knoke, *Pharmazie*, 1985, **40**, 54; *Chem. Abstr.*, 1985, **103**, 22545.

1992. S. Brownstein, E. Gabe, F. Lee, and B. Louie, *J. Org. Chem.*, 1988, **53**, 951.

1993. B. Tashkhodazhaev, S. Yangibaev, and K. M. Shakhidayatov, *Zh. Strukt. Khim.*, 1985, **26** (No. 1), 155.

1994. R. W. Leiby, *J. Org. Chem.*, 1985, **50**, 2926.

1995. A. R. E.-N. Ossman, H. M. Safwat, and M. A. Aziza, *Indian J. Chem., Sect. B*, 1985, **24**, 333; *Chem. Abstr.*, 1985, **103**, 196047.

1996. S. Yangibaev, L. M. Yun, N. D. Abdullaev, and K. M. Shakhidoyatov, *Dokl. Akad. Nauk UzSSR*, 1984 (No. 11), 37; *Chem. Abstr.*, 1985, **103**, 160465.

1997. M. F. Abdel-Megeed and A. Teniou, *Spectrosc. Lett.*, 1987, **20**, 583; *Chem. Abstr.*, 1988, **109**, 92169.

1998. L. R. Caswell and A. H. M. Chao, *J. Chem. Eng. Data*, 1987, **32**, 389.

1999. R. H. Lemus and E. B. Skibo, *J. Org. Chem.*, 1988, **53**, 6099.

2000. P. J. Garratt, C. J. Hobbs, and R. Wrigglesworth, *J. Org. Chem.*, 1989, **54**, 1062.

2001. P. Mencarelli and F. Stegel, *J. Org. Chem.*, 1985, **50**, 5415.

2002. L. G. Mesropyan, F. R. Shiroyan, and I. S. Sarkisyan, *Arm. Khim. Zh.*, 1986, **39**, 704; *Chem. Abstr.*, 1988, **108**, 37765.

2003. K.-C. Liu, M.-K. Hu, and Y.O. Lin, *T'ai-wan Yao Hsueh Tsa Chih*, 1986, **38**, 154; *Chem. Abstr.*, 1988, **108**, 75337.

2004. G. Zinner, H. Klein, and H. Kahnert, *Chem.-Ztg.*, 1987, **111**, 341; *Chem. Abstr.*, 1988, **109**, 128939.

2005. J. S. Swenton and J. G. Jurcak, *J. Org. Chem.*, 1988, **53**, 1530.

2006. G. Peczak and M. Draminski, *Pol. J. Chem.*, 1985, **59**, 317; *Chem. Abstr.*, 1986, **105**, 172393.

2007. M. E. Suh, *Yakhak Hoechi*, 1986, **30**, 203; *Chem. Abstr.*, 1987, **107**, 198236.

2008. M. Uchide, Y. Takamatsu, K. Ohnuma, T. Sekiya, and A. Iwamoto, *Yakuri To Chiryo*, 1987, **15**, 1513; *Chem. Abstr.*, 1987, **107**, 146748.

2009. S. F. Sisenwine, C. O. Tio, A. L. Liu, and J. F. Politowski, *Drug Metab. Dispos.*, 1987, **15**, 579; *Chem. Abstr.*, 1987, **107**, 146782.

2010. M. Mattioli, P. Mencarelli, and F. Stegel, *J. Org. Chem.*, 1988, **53**, 1087.

2011. K. Shibata, R. Igusa, K. Inoue, H. Mukouyama, J. Nakajima, A, Fujino, T. Sekiya, and M. Uchide, *Oyo Yakuri*, 1987, **33**, 765; *Chem. Abstr.*, 1987, **107**, 190264.

2012. V. K. Srivastava, G. Palit, A. K. Agarwal, and K. Shanker, *Pharmacol. Res. Commun.*, 1987, **19**, 617; *Chem. Abstr.*, 1988, **108**, 68315.

2013. S. Johne, *Alkaloids (Academic Press)*, 1986, **29**, 99; *Chem. Abstr.*, 1987, **107**, 154555.

2014. A. Numata and T. Ibuka, *Alkaloids (Academic Press)*, 1987, **30**, 193; *Chem. Abstr.*, 1988, **109**, 23150.

2015. S. W. Schneller and A. C. Ibay, *J. Org. Chem.*, 1986, **51**, 4067.

2016. M. Hori, R. Iemura, H. Hara, A. Ozaki, T. Sukamoto, and H. Ohtaka, *Chem. Pharm. Bull.*, 1990, **38**, 1286.

2017. Y. Nomoto, H. Obase, H. Takai, T. Hirata, M. Teranishi, J. Nakamura, and K. Kubo, *Chem. Pharm. Bull.*, 1990, **38**, 1591.

2018. U. A. El-Sabagh, H. H. Hassanein, and M. I. Al-Ashmawi, *Egypt. J. Pharm. Sci.*, 1988, **29**, 595; *Chem. Abstr.*, 1989, **111**, 134083.

2019. B. Prabhakar, K. L. Reddy, and P. Lingaiah, *Proc. Indian Acad. Sci., Chem. Sci.*, 1989, **101**, 121; *Chem. Abstr.*, 1989, **111**, 145718.

2020. M. Hori, R. Iemura, H. Hara, A. Ozaki, T. Sukamoto, and H. Ohtaka, *Chem. Pharm. Bull.*, 1990, **38**, 681.

2021. E. Sato, Y. Ikeda, M. Hasebe, T. Nishio, T. Miura, and Y. Kanaoka, *Chem. Pharm. Bull.*, **36**, 4749.

2022. K.-C. Liu and M.-K. Hu, *Chung-hua Yao Hsueh Tsa Chin*, 1988, **40**, 117; *Chem. Abstr.*, 1989, **111**, 57667.

2023. U. I. El-Sabagh, H. H. Hassanein, and M. I. Al-Ashmawi, *Egypt. J. Pharm. Sci.*, 1988, **29**, 587; *Chem. Abstr.*, 1989, **111**, 7342.

2024. T. R. Jones, R. F. Betteridge, S, Neidle, A. L. Jackman, and A. H. Calvert, *Anti-Cancer Drug Des.*, 1989, **3**, 243; *Chem. Abstr.*, 1989, **111**, 17131.

2025. R. O. Dempcy and E. B. Skibo, *J. Org. Chem.*, 1991, **56**, 776.

2026. B. Lal, R. M. Gidwani, and N. J. de Souza, *J. Org. Chem.*, 1990, **55**, 5117.

2027. J. Mertens, M. Gysemans, C. Bossuyt-Piron, and M. Thomas, *J. Labelled Compd. Radiopharm.*, 1990, **28**, 731.

2028. Y. Nomoto, H. Takai, T. Hirata, M. Teranishi, T. Ohno, and K. Kubo, *Chem. Pharm. Bull.*, 1991, **39**, 86.

2029. B. E. Bauomy, M. I. Al-Ashmawi, and M. El-Sadek, *Egypt. J. Pharm. Sci.*, 1988, **29**, 269; *Chem. Abstr.*, 1989, **111**, 23470.

2030. J. Reisch and G. M. K. B. Gunaherath, *J. Nat. Prod.*, 1989, **52**, 404; *Chem. Abstr.*, 1990, **112**, 98938.

2031. A. Miyashita, N. Taido, S. Sato, K.-I. Yamamoto, H. Ishida, and T. Higashino, *Chem. Pharm. Bull.*, 1991, **39**, 282.

2032. I. Yilmaz and H. J. Shine, *J. Labelled Compd. Radiopharm.*, 1989, **27**, 763.

2033. Y. Nomoto, H. Takai, T. Ohno, and K. Kubo, *Chem. Pharm. Bull.*, 1991, **39**, 352.

2034. A. L. El-Gendy, M. El-Safty, A. Essawy, and H. Y. Moustafa, *Arab. Gulf J. Sci. Res.*, *A*, 1988, **6**, 337; *Chem. Abstr.*, 1989, **111**, 232716.

2035. M. Hori, R. Iemura, H. Hara, T. Sukamoto, K. Ito, and H. Ohtaka, *Chem. Pharm. Bull.*, 1991, **39**, 367.

2036. A. Garcia-Martinez, A. Herrara-Fernández, F. Moreno-Jiménez, A. Garcia-Fraile, L. R. Subramanian, and M. Hanack, *J. Org. Chem.*, 1992, **57**, 1627.

2037. T. Nishino, S. Kameyama, Y. Omote, and C. Kashima, *Heterocycles*, 1990, **30**, 493.

2038. M. Anzini, A. Garofalo, and S. Vomero, *Heterocycles*, 1989, **29**, 1477.

2039. F. Cornea, A. Ciobanu, I. Baciu, O. Maiora, and A. A. El-Bahnasawy, *Rev. Roum. Chim.*, 1989, **34**, 103; *Chem. Abstr.*, 1989, **111**, 232714.

2040. B. A. Urkov, L. M. Yuh, N. D. Abdullaev, and K. M. Shakhidoyatov, *Dokl. Akad. Nauk UzSSR*, 1989 (No. 1), 37; *Chem. Abstr.*, 1989, **111**, 232720.

2041. J. L. Hicks, J. D. Hartman, R. W. Skeean, C. C. Huang, J. A. Keppler, and C.-H. Liang, *J. Labelled Compd. Radiopharm.*, 1991, **29**, 415.

2042. A. E. Ossmann, M. M. El -Zahabi, A. E. El-Hakim, and A. N. Osman, *Pharmazie*, 1989, **44**, 113; *Chem. Abstr.*, 1989, **111**, 194703.

2043. L. Fišnerová, B. Brůnová, Z. Kocfeldová, J. Tikalová, E. Maturová, and J. Grimová, *Collect. Czech. Chem. Commun.*, 1991, **56**, 2373.

2044. K. Špirková and Š. Stankovský, *Collect. Czech. Chem. Commun.*, 1991, **56**, 1719.

2045. H.-W. Yoo, J.-W. Lee, and M.-E. Suh, *Yakhak Hoechi*, 1989, **33**, 246; *Chem. Abstr.*, 1990, **112**, 216885.

2046. T. F. Kozlovskaya, A. Y. Strakov, M. V. Ablovatskaya, and M. V. Petrova, *Latv. PSR Zinat. Akad. Vestis, Kim. Ser.*, 1989, 499; *Chem. Abstr.*, 1990, **112**, 138995.

2047. H. Hulinská, Z. Polivka, J. Jilek, K. Šindelář, J. Holubek, E. Svátek, O. Matoušová, M. Buděšinský, H. Frycová, and M. Protiva, *Collect. Czech. Chem. Commun.*, 1988, **53**, 1820.

2048. J. V. Greenhill, J. Hanaee, and P. J. Steel, *J. Chem. Soc., Perkin Trans. 1*, 1990, 1869.

2049. R. S. Atkinson and B. J. Kelly, *J. Chem. Soc., Perkin Trans. 1*, 1989, 1515.

2050. Š. Stankovský and A, Boulmokh, *Chem. Pap.*, 1989, **43**, 433; *Chem. Abstr.*, 1990, **112**, 118749.

2051. P. Pazdera, E. Nováček, and D. Ondráček, *Chem. Pap.*, 1989, **43**, 465.

2052. R. S. Atkinson and B. J. Kelly, *J. Chem. Soc., Perkin Trans. 1*, 1989, 1627.

2053. F. Örtel, *Pharmazie*, 1990, **45**, 370; *Chem. Abstr.*, 1990, **113**, 231303.

2054. O. G. Rubleva, V. A. Dymshits, R. A. Samiev, B. A. Urakov, L. Yun, and K. M. Shakhidoyatov, *Uzb. Khim. Zh.*, 1990 (No. 3), 38; *Chem. Abstr.*, 1990, **113**, 211940.

2055. R. S. Atkinson and B. J. Kelly, *J. Chem. Soc., Perkin Trans. 1*, 1989, 1657.

2056. P. S. N. Reddy and P. P. Reddy, *Indian J. Chem., Sect. B*, 1988, **27**, 342.

2057. Y. Nomoto, H. Takai, T. Ohno, and K. Kudo, *Chem. Pharm. Bull.*, 1991, **39**, 900.

2058. D. Ranganathan and R. Rathi, *J. Org. Chem.*, 1990, **55**, 2351.

2059. M. A. Badawy, S. A. Abdel-Hady, A. M. Mahmoud, and Y. A. Ibrahim, *J. Org. Chem.*, 1990, **55**, 344.

2060. M. Mattioli and P. Mencarelli, *J. Org. Chem.*, 1990, **55**, 776.

2061. K. Minn, *Synlett*, 1991, 115.

2062. C. Amatore, R. Beugelmans, M. Bois-Choussy, C. Combellas, and A. Thiebault, *J. Org. Chem.*, 1989, **54**, 5688.

2063. R. S. Atkinson, M. J. Grimshire, and B. J. Kelly, *Tetrahedron*, 1989, **45**, 2875.

2064. M. H. Elnagdi, F. M. Abdelrazek, N. S. Ibrahim, and A. W. Erian, *Tetrahedron*, 1989, **45**, 3597.

2065. V. S. Dermugin, I. A. Trushkina, and B. K. Sadybakasov, *Izv. Akad. Nauk Kirg. SSR, Khim.-Tekhnol. Biol. Nauki*, 1989 (No. 2), 16; *Chem. Abstr.*, 1990, **113**, 59082.

2066. J. R. Russell, C. D. Garner, and J. A. Joule, *J. Chem. Soc., Perkin Trans. 1*, 1992, 409.

2067. M. Radojković-Veličković and M. Mišić-Vuković, *J. Serb. Chem. Soc.*, 1989, **54**, 563.

2068. K. Yoshida, T. Tanaka, and H. Ohtaka, *J. Chem. Soc., Perkin Trans. 1*, 1991, 1279.

2069. J. Bergman and A. Brynolf, *Tetrahedron*, 1990, **46**, 1295.

2070. I. Gräfe, K. Kottke, H. Kühmstedt, and D. Knoke, *Pharmazie*, 1990, **45**, 530; *Chem. Abstr.*, 1991, **114**, 81743.

2071. Y. Eguchi, A. Sugimoto, and M. Ishikawa, *Iyo Kizai Kenkyusho Hokoku* (*Tokyo Ika Shika Daigaku*), 1989, **23**, 65; *Chem. Abstr.*, 1991, **114**, 6427.

2072. A. Strakov, V. Trapkov, M. Lukasheva, T. Kozlovskaya, K. Yerzinkyan, J. Kacens, M. Petrova, and N. Tonkih, *Latv. Kim. Z.*, 1992, 98; *Chem. Abstr.*, 1993, **118**, 191680.

2073. A. F. El-Farargy, M. M. Hamad, S. A. Said, and A. Haikal, *An. Quim.*, 1990, **86**, 782; *Chem. Abstr.*, 1991, **114**, 207187.

2074. G. Stájer, Z. Szöke-Molnár, G. Bernáth, and P. Sohár, *Tetrahedron*, 1990, **46**, 1943.

2075. R. Hayes, J. M. Schofield, R. K. Smalley, and D. I. C. Scopes, *Tetrahedron*, 1990, **46**, 2089.

2076. P. Richter and F. Örtel, *Pharmazie*, 1990, **45**, 721; *Chem. Abstr.*, 1991, **114**, 143330.

2077. N. V. Harris, C. Smith, and K. Bowden, *Synlett*, 1990, 577.

2078. D. Korbonits and P. Kolonits, *Acta Chim. Hung.*, 1990, **127**, 793; *Chem. Abstr.*, 1991, **115**, 71523.

2079. H. Takeuchi, S. Hagiwara, and S. Eguchi, *Tetrahedron*, 1989, **45**, 6375.

2080. R. S. Atkinson, B. J. Kelly, and J. Williams, *J. Chem. Soc., Chem. Commun.*, 1992, 373.

2081. J. P. Michael, *Nat. Prod. Rep.*, 1992, **9**, 25.

2082. B. Prabhakar, P. Lingaiah, and K. L. Reddy, *Polyhedron*, 1990, **9**, 805.

2083. M.-K. Hu, L.-Y. Hsu, and K.-C. Liu, *Chung-hua Yao Hsueh Tsa Chih*, 1991, **43**, 151; *Chem. Abstr.*, 1991, **115**, 71526.

2084. M. A. Shaaban, H. I. Alashmawl, R. Abdel-Moati, S. El-Meligie, and F. Amin, *Bull. Fac. Pharm. (Cairo Univ.)*, 1990, **28**, 5; *Chem. Abstr.*, 1991, **115**, 49593.

2085. A. Couture, H. Cornet, and P. Grandclaudon, *Synthesis*, 1991, 1009.

2086. S. D. Sharma and V. Kaur, *Synthesis*, 1989, 677.

2087. A. A. Afify, S. El-Nagdy, M. A. Sayed, and I. Mohey, *Rev. Roum. Chim.*, 1990, **35**, 567; *Chem. Abstr.*, 1991, **115**, 8707.

2088. E. Rossi and R. Stradi, *Synthesis*, 1989, 214.

2089. J. Splinder, *Chem.-Ztg.*, 1991, **115**, 224; *Chem. Abstr.*, 1991, **115**, 232171.

2090. M. Dunkel and W. Pfleiderer, *Nucleosides Nucleotides*, 1991, **10**, 799; *Chem. Abstr.*, 1991, **115**, 208437.

2091. C. Lecomte, S. Skoulika, P. Aslanidis, P. Karagiannidis, and S. Papastefanou, *Polyhedron*, 1989, **8**, 1103.

2092. J. Kokosi, L. Orfi, M. Szabo, K. Takacs-Novak, G. Szasz, and I. Szilagyi, *Magy. Kem. Foly.*, 1991, **97**, 185; *Chem. Abstr.*, 1991, **115**, 159501.

2093. P. Molina, M. Alajarin, A. Vidal, M. C. Foces–Foces, and F. Hernández–Cano, *Tetrahedron*, 1989, **45**, 4263.

2094. S. J. Gould, R. L. Eisenberg, and L. R. Hills, *Tetrahedron*, 1991, **47**, 7209.

2095. P. Pazdera, V. Potůček, E. Nováček, I. Kalviňš, P. Trapencieris, and O. Pugovics, *Chem. Pap.*, 1991, **45**, 527.

2096. S. El-Bahaie, B. E. Bayoumy, M. G. Assy, and S. Yousif, *Pol. J. Chem.*, 1991, **65**, 1059; *Chem. Abstr.*, 1992, **116**, 151703.

2097. E. Rossi, D. Calabrese, and F. Farma, *Tetrahedron*, 1991, **47**, 5819.

2098. M. F. Ismail, S. A. Emara, E. I. Enayat, and O. E. A. Mustafa, *Pol. J. Chem.*, 1991, **65**, 1259; *Chem. Abstr.*, 1992, **116**, 151715.

2099. P. Pazdera and J. Pichler, *Chem. Pap.*, 1991, **45**, 517.

2100. T. Miyata, T. Mizuno, Y. Nagahama, I. Nishiguchi, T. Hirashima, and N. Sonoda, *Heteroat. Chem.*, 1991, **2**, 473; *Chem. Abstr.*, 1991, **115**, 256104.

2101. E. Rossi, R. Stradi, and P. Visentin, *Tetrahedron*, 1990, **46**, 3581.

2102. Y. Nomoto, H. Obase, H. Takai, T. Hirata, M. Teranishi, J. Nakamura, T. Ohno, and K. Kubo, *Chem. Pharm. Bull.*, 1990, **38**, 2467.

2103. Y. Watanabe, H. Usai, T. Shibano, T. Tanaka, and M. Kanao, *Chem. Pharm. Bull.*, 1990, **38**, 2726.

2104. M. Yamagishi, K. Ozaki, H. Ohmizu, Y. Yamada, and M. Suzuki, *Chem. Pharm. Bull.*, 1990, **38**, 2926.

2105. W. A. W. Stolle, J. M. Veurink, A. T. M. Marcelis, and H. C. van der Plas, *Tetrahedron*, 1992, **48**, 1643.

2106. J. M. Dickinson and J. A. Murphy, *Tetrahedron*, 1992, **48**, 1317.

2107. F. Fülöp, M. Simeonov, and K. Pihlaja, *Tetrahedron*, 1992, **48**, 531.

2108. Y. Nomoto, H. Takai, T. Hirata, M. Teranishi, T. Ohno, and K. Kubo, *Chem. Pharm. Bull.*, 1990, **38**, 3014.

2109. Y. Eguchi, F. Sasaki, A. Sugimoto, H. Ebisawa, and M. Ishikawa, *Chem. Pharm. Bull.*, 1991, **39**, 1753.

2110. A. Miyashita, T. Kawashima, C. Iijima, and T. Higashino, *Heterocycles*, 1992, **33**, 211.

2111. H. Yamanaka and S. Ohba, *Heterocycles*, 1990, **31**, 895.

2112. M. Kitagawa, T. Mimura, and M. Tanaka, *Chem. Pharm. Bull.*, 1991, **39**, 2400.

2113. C. H. Weidner, F. M. Michaels, D. J. Beltman, C. J. Montgomery, D. H. Wadsworth, B. T. Briggs, and M. L Picone, *J. Org. Chem.*, 1991, **56**, 5594.

2114. J. J. McNally and J. B. Press, *J. Org. Chem.*, 1991, **56**, 245.

2115. A. Kamal, M. V. Rao, and A. B. Rao, *Heterocycles*, 1990, **31**, 577.

2116. P. S. N. Reddy and C. Nagaraju, *Synth. Commun.*, 1991, **21**, 173.

2117. R. Cortez, I. A. Rivero, R. Somanathan, G. Aquirre, F. Ramirez, and E. Hong. *Synth. Commun.*, 1991, **21**, 285.

2118. K. Horváth, J. Kökösi, and I. Hermecz, *J. Chem. Soc., Perkin Trans. 2*, 1989, 1613.

2119. G. B. Barlin and C. Jiravinyu, *Aust. J. Chem.*, 1990, **43**, 1367.

2120. I. A. Atanassova, J. S. Petrov, V. H. Ognjanova, and N. M. Mollov, *Synth. Commun.*, 1990, **20**, 2083.

2121. P. S. N. Reddy and V. G. Reddy, *Synth. Commun.*, 1990, **20**, 23.

2122. S. Muthusamy and V. T. Ramakrishnan, *Synth. Commun.*, 1992, **22**, 519.

2123. V. J. Ram and M. Verma, *Indian J. Chem., Sect. B*, 1992, **31**, 195.

2124. K. C. Joshi, A. Dandia, and S. Khanna, *Indian J. Chem., Sect. B*, 1992, **31**, 105.

2125. M. Botta, V. Summa, R. Saladino, and R. Nicoletti, *Synth. Commun.*, 1991, **21**, 2181.

2126. J. Barluenga, M. Tomás, A. Ballesteros, and L. A. López, *Tetrahedron Lett.*, 1989, **30**, 4573.

2127. R. S. Atkinson and B. J. Kelly, *Tetrahedron Lett.*, 1989, **30**, 2703.

2128. M. B. Ahmad, M. Rauf, and S. M. Osman, *Indian J. Chem., Sect. B*, 1988, **27**, 1140.

2129. J. S. Shukla and R. Shukla, *J. Indian Chem. Soc.*, 1989, **66**, 209.

2130. S. Saxena, M. Bhalla, M. Verma, A. K. Saxena, and K. Shanker, *J. Indian Chem. Soc.*, 1991, **68**, 142.

2131. H. Takeuchi and S. Eguchi, *Tetrahedron Lett.*, 1989, **30**, 3313.

2132. D. Korbonits, I. Kanzel-Szvoboda, C. Gönczi, K. Simson, and P. Kolonits, *Chem. Ber.*, 1989, **122**, 1107.

2133. J. P. Michael, *Nat. Prod. Rep.*, 1991, **8**, 53.

2134. D. Prakash, S. M. Prasad, S. Kumar, and A. K. Gupta, *J. Indian Chem. Soc.*, 1991, **68**, 301.

2135. P. C. Joshi, M. M. Sah, and C. K. Pant, *J. Indian Chem. Soc.*, 1991, **68**, 416.

2136. A. Kamal and P. B. Sattur, *Tetrahedron Lett.*, 1989, **30**, 1133.

2137. M. Kumar and T. Sharma, *J. Indian Chem. Soc.*, 1991, **68**, 539.

2138. S. V. Borodaev, O. V. Zubkoba, and S. M. Luk'yanov, *Zh. Org. Khim.*, 1988, **24**, 2330.

2139. P. Karagiannidis, P. Aslanidis, S. Papastefanou, D. Mentzafos, A. Hountas, and A. Terzis, *Polyhedron*, 1990, **9**, 2833.

2140. E. B. Walsh and H. Wamhoff, *Chem. Ber.*, 1989, **122**, 1673.

2141. M. A. Likhate and P. S. Fernandes, *J. Indian Chem. Soc.*, 1990, **67**, 862.

2142. Y. D. Kulkarni and A. Rowhani, *J. Indian Chem. Soc.*, 1990, **67**, 46.

2143. V. J. Ram and D. S. Kushwara, *Liebigs Ann. Chem.*, 1990, 697.

2144. K. Das, D. Panda, and B. Dash, *J. Indian Chem. Soc.*, 1990, **67**, 58.

2145. K. Uneyama, F. Yamashita, K. Sugimoto, and O. Morimoto, *Tetrahedron Lett.*, 1990, **31**, 2717.

2146. N. M. Naik and K. R. Desai, *J. Indian Chem. Soc.*, 1990, **67**, 84.

2147. M. H. Elnagdi and A. W. Erian, *Liebigs Ann. Chem.*, 1990, 1215.

2148. P. S. N. Reddy and V. G. Reddy, *Indian J. Chem., Sect. B*, 1990, **29**, 564.

2149. V. J. Ram and D. S. Kushwaha, *Liebigs Ann. Chem.*, 1990, 701.

2150. P. K. Naithani, G. Palit, V. K. Srivastava, and K. Shanker, *Indian J. Chem., Sect. B*, 1989, **28**, 745.

2151. C. J. Shishoo. M. P. Devani, S. Ananthan, K. S. Jain, V. S. Bhadti, S. Mohan, and L. J. Patel, *Indian J. Chem., Sect. B*, 1989, **28**, 1039.

2152. S. V. Lindeman, I. I. Ponomarev, Y. T. Struchkov, and S. V. Vinogradova, *Izv. Akad. Nauk SSSR*, Ser. Khim., 1990, 412.

2153. N. Misra, M. Sen, and A. Nayak, *J. Indian Chem. Soc.*, 1990, **67**, 353.

2154. M. Isobe, Y. Fukuda, T. Nishikawa, P. Chabert, T. Kawai, and T. Goto, *Tetrahedron Lett.*, 1990, **31**, 3327.

2155. J. A. Jensen and E. B. Pedersen, *Chem. Scr.*, 1988, **28**, 435.

2156. L. Ceraulo, P. Agozzino, M. Ferrugia, and S. Plescia, *Rapid Commun. Mass Spectrum.*, 1988, **2**, 223.

2157. H. Akgün and U. Hollstein, *Org. Mass Sepctrum.*, 1990, **25**, 289.

2158. N. M. Naik and K. R. Desai, *J. Indian Chem. Soc., 1989*, **66**, 495.

2159. K. L. Reddy, A. R. Chandraiah, K. A. Reddy, and P. Lingaiah, *Indian J. Chem., Sect. A*, 1989, **28**, 622.

2160. E. Rossi, G. Celentano, R. Stradi, and A. Strada, *Tetrahedron Lett.*, 1990, **31**, 903.

2161. M. K. A. Ibrahim, M. S. El-Gharib, A. M. Farag, and A. H. H. Elghandour, *Indian J. Chem., Sect. B*, 1988, **27**, 836.

2162. S. Malhotra, S. K. Koul, R. L. Sharma, K. K. Anand, O. P. Gupta, and K. L. Dhar, *Indian J. Chem., Sect. B*, 1988, **27**, 937.

2163. B. Wang, J. R. Kagel, T. S. Rao, and M. P. Mertes, *Tetrahedron Lett.*, 1989, **30**, 7005.

2164. S. P. Samant, S. K. Dhande, and B. D. Hosagnadi, *Indian J. Chem., Sect. B*, 1988, **27**, 1134.

2165. G. B. Barlin and C. Jiravinyu, *Aust. J. Chem.*, 1990, **43**, 311.

2166. N. M. Naik and K. R. Desai, *J. Indian Chem. Soc.*, 1989, **66**, 35.

2167. A. L. G. Beckwith and L. K. Dyall, *Aust. J. Chem.*, 1990, **43**, 451.

2168. M. Kumar, H. R. Mahato, V. Sharma, and T. Sharma, *J. Indian Chem. Soc.*, 1989, **66**, 73.

2169. A. Kumar, R. S. Verma, B. P. Jaju, and J. N. Sinha, *J. Indian Chem. Soc.*, 1990, **67**, 920.

2170. A. Byford, P. Goadby, M. Hooper, H. V. Kamath, and S. N. Kulkarni, *Indian J. Chem., Sect. B*, 1988, **27**, 396.

2171. J. M. Dickinson and J. A. Murphy, *J. Chem. Soc., Chem. Commun.*, 1990, 434.

2172. R. S. Atkinson, B. J. Kelly, and C. McNicholas, *J. Chem. Soc., Chem. Commun.*, 1989, 562.

2173. H. K. Gakhar, M. Sangeeta, R. Gupta, and S. B. Gupta, *Indian J. Chem., Sect. B*, 1990, **29**, 174.

2174. A. Y. Soliman, N. B. El-Assy, F. El-Shahed, M. El-Kady, and I. M. El-Deen, *Indian J. Chem., Sect. B*, 1990, **29**, 326.

2175. R. S. Atkinson and B. J. Kelly, *J. Chem. Soc., Chem. Commun.*, 1989, 836.

2176. K. K. Jain, S. P. Rout, and H. K. Pujari, *Indian J. Chem., Sect. B*, 1990, **29**, 379.

2177. M. Noguchi, K. Doi, Y. Kiriki, and S. Kajigaeshi, *Chem. Lett.*, 1989 2115.

2178. K. Ohta, Y. Nakamura, J. Iwaoka, and Y. Nomura, *Nippon Kagaku Kaishi*, 1990, 72.

2179. S. K. P. Sinha and P. Kumar, *Indian J. Chem., Sect. B*, 1989, **28**, 274.

2180. P. Bortolus, G. Galiazzo, and G. Gennari, *Annal. Chim. Acta*, 1990, **234**, 353.

2181. V. Joshi, R. P. Chaudhari, and S. R. Mogera, *Indian J. Chem. Sect. B*, 1989, **28**, 431.

2182. R. S. Atkinson and B. J. Kelly, *J. Chem. Soc., Chem. Commun.*, 1988, 624.

2183. R. K. Saksena and A. M. Khan, *Indian J. Chem., Sect. B.* 1989, **28**, 443.

2184. M. F. Ismail, S. A. Emara, E. I. Enayat, and O. E. A. Mustafa, *Indian J. Chem., Sect. B*, 1990, **29**, 811.

2185. R. Khanna, A. K. Saksena, V. K. Srivastava, and K. Shanker, *Indian J. Chem., Sect. B*, 1990, **29**, 1056.

2186. S. Buscemi and N. Vivona, *J. Chem. Soc., Perkin Trans. 2*, 1991, 187.

2187. P. Kumar, S. K. Agarwal, and D. S. Bhakuni, *Indian J. Chem., Sect. B*, 1992, **31**, 177.

2188. P. S. N. Reddy and V. G. Reddy, *Indian J. Chem., Sect. B*, 1992, **31**, 193.

2189. W. Hiller and M. Akkurt, *Acta Crystallogr., Sect. C*, 1990, **46**, 1157.

2190. C. J. Shishoo, M. B. Devani, V. S. Bhadti, K. S. Jain, and S. Ananthan, *J. Heterocycle. Chem.*, 1990, **27**, 119.

2191. L. Strekowski, M. T. Cegla, S.-B. Kong, and D. B. Harden, *J. Heterocycl. Chem.*, 1989, **26**, 923.

2192. J.-M. Gazengel, J.-C. Lancelot, S. Rault, and M. Robba, *J. Heterocycl. Chem.*, 1989, **26**, 1135.

2193. J.-T. Hahn and P. D. Popp, *J. Heterocycl. Chem.*, 1989, **26**, 1357.

2194. T. R. Jones, R. F. Betteridge, D. R. Newell, and A. L. Jackman, *J. Heterocycl. Chem.*, 1989, **26**, 1501.

2195. N. S. Cho, K. Y. Song, and C. Párkányi, *J. Heterocycl. Chem.*, 1989, **26**, 1807.

2196. A. H. M. Al-Shaar, D. W. Gilmour, D. J. Lythogoe, I. McClenaghen, and C. A. Ramsden, *J. Heterocycl. Chem.*, 1989, **26**, 1819.

2197. L. M. Yun, S. Yangibaev, K. M. Shakhidoyatov, V. Y. Alekseeva, and K. A. V'yunov, *Zh. Org. Khim.*, 1989, **25**, 2438.

2198. V. A. Chervonyi, V. S. Zyabrev, A. V. Kharchenko, and B. S. Drach, *Zh. Org. Khim.*, 1989, **25**, 2597.

2199. M. A. Badawy, S. A. Abdel-Hady, A. H. Mahmoud, and Y. A. Ibrahim, *Liebigs Ann. Chem.*, 1990, 815.

2200. H. Wamhoff, H. Wintersohl, S. Stölben, J. Paasch, N.-J. Zhu, and F. Guo, *Liebigs Ann. Chem.*, 1990, 901.

2201. V. P. Krivopalov, S. G. Baram, A. Y. Denisov, and V. I. Mamatyuk, *Izv. Akad. Nauk SSSR, Ser. Khim.*, 1989, 2002.

2202. N. D. Chkanikov, V. L. Vershinin, M. V. Galakov, A. F. Kolomiets, and A. V. Fokin, *Izv. Akad. Nauk SSSR, Ser. Khim.*, 1989, 126.

2203. D. W. Combs and M. S. Rampulla, *J. Heterocycl. Chem.*, 1989, **26**, 1885.

2204. P. Aped, L. Schleifer, B. Fuchs, and S. Wolfe, *J. Comput. Chem.*, 1989, **10**, 265.

2205. R. O. Dempcy and E. B. Skibo, *Bioorg. Med. Chem.*, 1993, **1**, 39.

2206. S. G. Baram, O. P. Shkurko, and V. P. Mamaev, *Izv. Akad. Nauk SSSR, Ser. Khim.*, 1991, 686.

2207. M. M. Kandeel. *Phosphorus Sulfur Silicon Related Elem.*, 1990, **48**, 149.

2208. M. Noguchi, Y. Kiriki, T. Ushijima, and S. Kajigaeshi, *Bull. Chem. Soc. Jpn.*, 1990, **63**, 2938.

2209. D. H. R. Barton, P. Blundel, and J. C. Jaszberenyi, *J. Am. Chem. Soc.*, 1991, **113**, 6937.

2210. P. Karagiannidis, P. Aslanidis, S. Kokkou, and C. J. Cheer, *Inorg. Chim. Acta*, 1990, **172**, 247.

2211. S. K. Singh, M. Govindan, and J. B. Hynes, *J. Heterocycl. Chem.*, 1990, **27**, 2101.

2212. J. Garin, E. Meléndez, F. L. Merchán, P. Merino, J. Orduna, and T. Tejero, *J. Heterocycl. Chem.*, 1990, **27**, 1341.

2213. P. Karagiannidis, P. Aslanidis, S. Papastefanou, D. Mentzafos, A. Hountas, and A. Terzis, *Inorg. Chim. Acta*, 1989, **156**, 265.

2214. P. Sohár, Z. Szöke-Molnár, G. Stájer, and G. Bernáth, *Magn. Reson. Chem.*, 1989, **27**, 959.

2215. A. T. Johnson, D. A. Keszler, K. Sakuma, and J. D. White, *Acta Crystallogr., Sect. C*, 1989, **45**, 1114.

2216. F. Yoneda and M. Koga, *J. Heterocycl. Chem.*, 1989, **26**, 49.

2217. A. P. Vinogradoff and N. P. Peet, *J. Heterocycl. Chem.*, 1989, **26**, 97.

2218. L. W. Deady, M. F. Mackay, and D. M. Werden, *J. Heterocycl. Chem.*, 1989, **26**, 161.

2219. J. Reisch, C. O. Usifoh, and J. O. Oluwadiya, *J. Heterocycl. Chem.*, 1990, **27**, 1953.

2220. G. Timári, G. Hajós, and A. Messmer, *J. Heterocycl. Chem.*, 1990, **27**, 2005.

2221. A. Tomažič, J. B. Hynes, G. R. Gale, and J. H. Freisheim, *J. Heterocycl. Chem.*, 1990, **27**, 2081.

2222. J. Garin, E. Melénzez, F. L. Merchán, P. Merino, J. Orduna, and T. Tejero, *J. Heterocycl. Chem.*, 1990, **27**, 1345.

2223. J. F. Wolfe, T. L. Rathman, M. C. Sleevi, J. A. Campbell, and T. D. Greenwood, *J. Med. Chem.*, 1990, **33**, 161.

2224. M. J. Yu, J. R. McCowan, M. I. Steinberg, S. A. Wiest, V. L. Wyss, and J. S. Horng, *J. Med. Chem.*, 1990, **33**, 348.

2225. N. V. Harris, C. Smith, and K. Bowden, *J. Med. Chem.*, 1990, **33**, 434.

2226. A. Krantz, R. W. Spencer, T. F. Tam, T. J. Liak, L. J. Copp, E. M. Thomas, and S. P. Rafferty, *J. Med. Chem.*, 1990, **33**, 464.

2227. C.-Y. Shiau, J.-W. Chern, K.-C. Liu, C.-H. Chan, M.-H. Yen, M.-C. Cheng, and Y. Wang, *J. Heterocycl. Chem.*, 1990, **27**, 1467.

2228. M. M. El-Kerdawy, A. E.-K. M. Ismaiel, M. M. Gineinah, and R. A. Glennon, *J. Heterocycl. Chem.*, 1990, **27**, 497.

2229. W. M. F. Fabian, *J. Comput. Chem.*, 1991, **12**, 17.

2230. M. M. Gineinah, A. E.-K. M. Ismaiel, M. M. El-Kerdawy, and R. A. Glennon, *J. Heterocycl. Chem.*, 1990, **27**, 723.

2231. R. J. Hodgkiss, A. C. Begg, R. W. Middleton, J. Parrick, M. R. L. Stratford, P. Wardman, and G. D. Wilson, *Biochem. Pharmacol.*, 1991, **41**, 533.

2232. R. L. Hagan, D. S. Duch, G. K. Smith, M. H. Hanlon, B. Shane, J. H. Freisheim, and J. B. Hynes, *Biochem. Pharmacol.*, 1991, **41**, 781.

2233. H. Singh and R. Sarin, *J. Chem. Res.*, 1988, *Synop.* 322, *Minipr.* 2623.

2234. D. Vlaović, B. L. Milić, and K. Mackenzie, *J. Chem. Res.*, 1989, *Synop.* 156, *Minipr.* 1201.

2235. J. H. Gorvin, *J. Chem. Res.*, 1989, *Synop.* 294.

2236. S. Leistner, M. Gütschow, and J. Stach, *Arch. Pharm.* (*Weinheim, Ger.*), 1990, **323**, 857.

2237. S. Büyüktimkin, A. Buschauer, and W. Schunack, *Arch. Pharm.* (*Weinheim, Ger.*), 1989, **322**, 115.

2238. A. Rampa, P. Valenti, P. Da-Re, M. Carrara, S. Zampiron, L. Cima, and P. Giusti, *Arch. Pharm.* (*Weinheim, Ger.*), 1989, **322**, 359.

2239. D. Gravier, J.-P. Dupin, F. Casadebaig, G. Hou, M. Boisseau, and H. Bernard, *Eur. J. Med. Chem.*, 1989, **24**, 531.

2240. S. Botros and S. F. Saad, *Eur. J. Med. Chem.*, 1989, **24**, 585.

2241. V. J. Ram and M. Verma, *J. Chem. Res.*, 1990, *Synop.* 398.

2242. E. C. Taylor and M. Patel, *J. Heterocycl. Chem.*, 1991, **28**, 1857.

2243. S. A. H. El-Feky and Z. K. A. El-Samii, *J. Chem. Technol Biotechnol.*, 1991, **51**, 61.

2244. R. A. Slavinskaya, N. A. Dolgova, T. A. Kovaleva, V. Z. Gabdrakipov, and G. A. Yuldasheva, *Zh. Obshch. Khim.*, 1990, **60**, 2364.

2245. O. B. Smolii, V. S. Brovarets, V. V. Pirozhenko, and B. S. Drach, *Zh. Obshch, Khim.*, 1988, **58**, 2465.

2246. J. B. Hynes, A. Tomažič, A. Kumar, V. Kumar, and J. H. Freisheim, *J. Heterocycl. Chem.*, 1991, **28**, 1981.

2247. J. B. Hynes, A. Tomažič, C. A. Parrish, and O. S. Fetzer, *J. Heterocycl. Chem.*, 1991, **28**, 1357.

2248. J. Garin, E. Meléndez, F. L. Merchán, P. Merino, J. Orduna, and T. Tejero, *J. Heterocycl. Chem.*, 1991, **28**, 359.

2249. P. R. Marsham, A. L. Jackman, J. Oldfield, L. R. Hughes, T. J. Thornton, G. M. F. Bisset, B. M. O'Connor, J. A. M. Bishop, and A. H. Calvert, *J. Med. Chem.*, 1990, **33**, 3072.

2250. D. J. McNamara, E. M. Berman, D. W. Fry, and L. M. Werbel, *J. Med. Chem.*, 1990, **33**, 2045.

2251. D. Gravier, J.-P. Dupin, F. Casadebaig, G. Hou, M. Petraud, J. Moulines, and B. Barbe, *J. Heterocycl. Chem.*, 1991, **28**, 391.

2252. A. H. H. Elghandour, M. M. M. Ramiz, M. K. A. Ibrahim, and M. R. H. Elmpghayar, *Org. Prep. Proced. Int.*, 1989, **21**, 479.

2253. A. K. D. Bhavani and P. S. N. Reddy, *Org. Prep. Proced. Int.*, 1992, **24**, 1.

2254. S. W. Li, M. G. Nair, D. M. Edwards, R. L. Kisliuk, Y. Gaumont, I. K. Dev, D. S. Duch, J. Humphreys, G. K. Smith, and R. Ferone, *J. Med. Chem.*, 1991, **34**, 2746.

2255. T. J. Thornton, T. R. Jones, A. L. Jackman, A. Flinn, B. M. O'Connor, P. Warner, and A. H. Calvert, *J. Med. Chem.*, 1991, **34**, 978.

2256. O. H. Hishmat, A. A. M. El-Din, and N. A. Ismail, *Org. Prep. Proced. Int.*, 1992, **24**, 33.

2257. A. M. Farghaly, I. Chaaban, M. A. Khalil, and A. A. Behkit, *Arch. Pharm. (Weinheim, Ger.)*, 1990, **323**, 311.

2258. A. L. Jackman, P. R. Marsham, T. J. Thornton, J. A. M. Bishop, B. M. O'Connor, L. R. Hughes, A. H. Calvert, and T. R. Jones, *J. Med. Chem.*, 1990, **33**, 3067.

2259. M. T. M. El-Wassimy, M. T. Omar, M. M. Kamal, E. M. Kassem, and N. M. Khalifa, *Sohag Pure Appl. Sci. Bull.*, 1991, **7**, 41; *Chem. Abstr.*, 1993, **118**, 213014.

2260. A. Gürsoy, S. Büyüktimkin, S. Demirayak, and A. C. Ekinci, *Arch. Pharm. (Weinheim, Ger.)*, 1990, **323**, 623.

2261. S. W. Schneller, A. C. Ibay, W. J. Christ, and R. F. Bruns, *J. Med. Chem.*, 1989, **32**, 2247.

2262. F. Billon, C. Delchambre, A. Cloarec, E. Satori, and J. M. Teulon, *Eur. J. Med. Chem.*, 1990, **25**, 121.

2263. L. R. Hughes, A. L. Jackman, J. Oldfield, R. C. Smith, K. D. Burrows, P. R. Marsham, J. A. M. Bishop, T. R. Jones, B. M. O'Connor, and A. H. Calvert, *J. Med. Chem.*, 1990, **33**, 3060.

2264. J. B. Jiang, D. P. Hesson, B. A. Dusak, D. L. Dexter, G. J. Kong, and E. Hamel, *J. Med. Chem.*, 1990, **33**, 1721.

2265. M. V. Vovk and L. I. Samarai, *Khim. Geterotsikl. Soedin.*, 1991, 698.

2266. Š. Stankovský and K. Špirkova, *Monatsh. Chem.*, 1991, **122**, 849.

2267. R. Mazurkiewicz, *Monatsh. Chem.*, 1989, **120**, 973.

2268. M. Kepez, *Monatsh. Chem.*, 1989, **120**, 127.

2269. P. R. Marsham, L. R. Hughes, A. L. Jackman, A. J. Hayter, J. Oldfield, J. M. Wardleworth, J. A. M. Bishop, B. M. O'Connor, and A. H. Calvert, *J. Med. Chem.*, 1991, **34**, 1594.

2270. H.-G. Henning and H. Haber, *Monatsh. Chem.*, 1988, **119**, 1405.

2271. P. R. Marsham, A. L. Jackman, A. L. Hayter, M. R. Daw, J. L. Snowden, B. M. O'Connor, J. A. M. Bishop, A. H. Calvert, and L. R. Hughes, *J. Med. Chem.*, 1991, **34**, 2209.

2272. O. Papp, G. Szász, L. Örfi, and I. Hermecz, *J. Chromatogr.*, 1991, **537**, 371.

2273. M. Guilloton and F. Karst, *Anal. Biochem.*, 1985, **149**, 291.

2274. W. Wouters, J. van Dun, J. E. Leysen, and P. M. Laduron, *J. Biol. Chem.*, 1985, **260**, 8423.

2275. M. A. Abdel-Aziz, H. A. Daboun, and S. M. Abdel-Gawad, *J. Prakt. Chem.*, 1990, **332**, 610.

2276. A. Rosowsky, R. A. Forsch, H. Bader, and J. H. Freisheim, *J. Med. Chem.*, 1991, **34**, 1447.

2277. M. S. Malamas and J. Millen, *J. Med. Chem.*, 1991, **34**, 1492.

2278. S. Giorgi-Renault, J. Renault, P. Gabel-Servolles, M. Baron, C. Paoletti, S. Cros, M.-C. Bissery, F. Lavelle, and G. Atassi, *J. Med. Chem.*, 1991, **34**, 38.

2279. V. J. Ram, R. C. Smiral, D. S. Kushwaha, and L. Mishra, *J. Prakt. Chem.*, 1990, **322**, 629.

2280. C. Foulcher, J. Barrans, and H. Gonçalves, *C. R. Acad. Sci., Ser. C*, 1967, **265**, 407.

2281. M. Hedayatullah, *C. R. Acad. Sci., Ser. C*, 1979, **289**, 365.

2282. S. Somasekhara, V. S. Dighe, N. M. Patel, and S. L. Mukherjee, *Curr. Sci.*, 1970, **39**, 253.

2283. G. M. F. Bisset, K. Pawalczak, A. L. Jackman, A. H. Calvert, and L. R. Hughes, *J. Med. Chem.*, 1992, **35**, 859.

2284. T. R. Jones, *Eur. J. Cancer*, 1980, **16**, 707.

2285. S. Bahadur, N. Srivastava, and S. Mukerji, *Curr. Sci.*, 1983, **52**, 910.

2286. S. Singh, M. Sharma, C. Nath, K. P. Bhargava, and K. Shanker, *Curr. Sci.*, 1983, **52**, 585.

2287. J. Gilbert and D. Rousselle, *C. R. Acad. Sci., Ser. C*, 1974, **279**, 159.

2288. J. W. Kusiak, Z. Szurmai, and J. Pitha, *Eur. J. Pharmacol.*, 1986, **123**, 19.

2289. M. Shrimali, R. Kalsi, K. S. Dixit, and J. P. Barthwal, *Arzneim.-Forsch.*, 1991, **41**, 514.

2290. H. Gonçalves, M. Bon, J. Barrans, and C. Foulcher, *C. R. Acad. Sci., Ser. C*, 1972, **274**, 1750.

2291. L. Li, Z. Qu, Z. Wang, Y. Zeng, G. Ding, G. Hu, and X. Yang, *Sci. Sin.* (*Engl. Ed.*), 1979, **22**, 1220.

2292. R. Shyam and I. C. Tiwari, *Curr. Sci.*, 1975, **44**, 572.

2293. K. S. L. Srivastava, *Curr. Sci.*, 1968, **37**, 136.

2294. N. Haider, G. Heinisch, and J. Moshuber, *Arch. Pharm.* (*Weinheim, Ger.*), 1992, **325**, 119.

2295. S. Somasekhara, V. S. Dighe, and S. L. Mukherjee, *Curr. Sci.*, 1968, **37**, 529.

2296. P. R. Dua, R. P. Kohli, and K. P. Bhargava, *Curr. Sci.*, 1967, **36**, 72.

2297. S. A. H. E.-Feky and Z. K. A. El-Samii, *Arch. Pharm.* (*Weinheim, Ger.*), 1991, **324**, 381.

2298. P. N. Bhargava and R. Lakhan, *Curr. Sci.*, 1967, **36**, 575.

2299. S. Somasekhara, V. S. Dighe, and S. L. Mukherjee, *Curr. Sci.*, 1966, **35**, 594.

2300. J. Sotiropoulos and A.-M. Lamazouère, *C. R. Acad. Sci., Ser. C*, 1970, **271**, 1592.

2301. R. Kalsi, K. Pande, T. N. Bhalla, J. P. Barthwal, G. P. Gupta, and S. S. Parmar, *J. Pharm. Sci.*, 1990, **79**, 317.

2302. R. S. Varma, *Curr. Sci.*, 1978, **47**, 416.

2303. H. Singh, *Curr. Sci.*, 1970, **39**, 234.

2304. M. Pesson and D. Richer, *C. R. Acad. Sci., Ser. C*, 1968, **266**, 1787.

2305. V. K. Pandey, *Curr. Sci.*, 1981, **50**, 678.

2306. M. R. Chaurasia, S. K. Sharma, and S. Kumar, *Curr. Sci.*, 1981, **50**, 841.

2307. S. S. Tiwari, R. K. Satsangi, and R. Agarwal, *Curr. Sci.*, 1979, **48**, 568.

2308. A. D. Rao, C. R. Shanker, and V. M. Reddy, *Curr. Sci.*, 1985, **54**, 720.

2309. W. Thiel, *J. Prakt. Chem.*, 1990, **322**, 845.

2310. C. R. Shanker, A. D. Rao, A. B. Rao, V. M. Reddy, and P. B. Sattur, *Curr. Sci.*, 1984, **53**, 1069.

2311. C. Bischoff and E. Schröder, *J. Prakt. Chem.*, 1989, **331**, 537.

2312. N. Srivastava, S. Bahadur, H. N. Verma, and M. M. A. A. Khan, *Curr. Sci.*, 1984, **53**, 235.

2313. J. K. Verma, *Curr. Sci.*, 1987, **56**, 1168.

2314. K. V. Fedotov and N. N. Romanov, *Khim. Geterotsiki. Soedin.*, 1989, 817.

2315. K. V. Fedotov and N. N. Romanov, *Khim. Geterotsikl. Soedin.*, 1989, 408.

2316. V. K. Pandey and N. Raj, *Curr. Sci.*, 1986, **55**, 785.

2317. V. K. Pandey, *Curr. Sci.*, 1986, **55**, 243.

2318. Y. A. Ammar, Y. A. Mohamed, N. E. Amin, and M. M. Ghorab, *Curr. Sci.*, 1989, **58**, 1231.

2319. W. Thiel and R. Mayer, *J. Prakt. Chem.*, 1989, **331**, 243.

2320. E. N. Dozorova, S. I. Grizik, and V. G. Granik, *Khim. Geterotsikl. Soedin.*, 1990, 81.

2321. W. Thiel and R. Mayer, *J. Prakt. Chem.*, 1989, **331**, 649.

2322. I. N. Kozlovskaya, L. A. Badovskaya, V. E. Zavodnik, and Z. I. Tyukhteneva, *Khim. Geterotsikl. Soedin.*, 1989, 1463.

2323. I. V. Ukrainets, P. A. Bezuglyi, V. I. Treskaya, A. V. Turov, S. V. Slobodzyan, and O. V. Gorokhova, *Khim. Geterotsikl. Soedin.*, 1991, 1128.

2324. R. O. Dempcy and E. B. Skibo, *Biochemistry*, 1991, **30**, 8480.

2325. W. Thiel, *J. Prakt. Chem.*, 1992, **334**, 92.

2326. N. V. Harris, C. Smith, and K. Bowden, *Eur. J. Med. Chem.*, 1992, **27**, 7.

2327. D. D. Mukerji, S. K. Skukla, A. K. Agnihotra, and S. R. Nautiyal, *Curr. Sci.*, 1982, **51**, 1060; *Chem. Abstr.*, 1983, **98**, 107245.

2328. Z. M. Tadic and M. Radojkovic-Velickovic, *J. Serb. Chem. Soc.*, 1993, **58**, 599; *Chem. Abstr.*, 1994, **120**, 244279.

2329. A. Miyashita, H. Matsuda, C. Iijima, and T. Higashino, *Chem. Pharm. Bull.*, 1992, **40**, 43.

2330. R. B. Trattner, G. B. Elion, G. H. Hitchings, and D. M. Sharefkin, *J. Org. Chem.*, 1964, **29**, 2674.

2331. C. Párkányi, H. L. Yuan, B. H. E. Strömberg, and A. Evenzahav, *J. Heterocycl. Chem.*, 1992, **29**, 749.

2332. A. E. Ossman, M. El-Zahabi, A. E. El-Hakim, and A. N. Osman, *Egypt. J. Chem.*, 1989, **32**, 319; *Chem. Abstr.*, 1992, **117**, 171371.

2333. O. Morgenstern, P. H. Richter, and P. Vainiotalo, *Pharmazie*, 1992, **47**, 297; *Chem. Abstr.*, 1992, **117**, 111569.

2334. A. Wada, H. Yamamoto, K. Ohki, S. Nagai, and S. Kanatomo, *J. Heterocycl. Chem.*, 1992, **29**, 911.

2335. A. Tomažič and J. B. Hynes, *J. Heterocycl. Chem.*, 1992, **29**, 915.

2336. K. A. M. Abou-Zeid, K. M. Youssef, F. M. Amine, and S. Botros, *Egypt. J. Pharm. Sci.*, 1991, **32**, 165; *Chem. Abstr.*, 1991, **117**, 233965.

2337. A. E. Ossman, A. N. Osman, M. El-Zahaby, and A. E. El-Hakim, *Egypt. J. Chem.*, 1991, **32**, 717; *Chem. Abstr.*, 1992, **117**, 233961.

2338. M. Ferrugia, M. L. Bajardi, L. Ceraulo, and S. Plescia, *J. Heterocycl. Chem.*, 1992, **29**, 565.

2339. M. J. Kornet, *J. Heterocycl. Chem.*, 1992, **29**, 103.

2340. M. B. Hogale, A. R. Shelar, and P. B. Chavan, *Indian J. Chem., Sect. B*, 1992, **31**, 456.

2341. M. A. Sayed, A. F. El-Kafrawy, A. Y. Soliman, and F. A. El-Bassiouny, *Indian J. Chem., Sect. B*, 1991, **30**, 980.

2342. B. Prabhakar, P. Lingaiah, and K. L. Reddy, *Indian J. Chem., Sect. A*, 1991, **30**, 904.

2343. S. Frimpong-Manso, K. Nagy, G. Stájer, G. Bernáth, and P. Sohár, *J. Heterocycl. Chem.*, 1992, **29**, 221.

2344. S. K. Singh, O. S. Fetzer, J. B. Hynes, and T. C. Williams, *J. Heterocycl. Chem.*, 1991, **28**, 1459.

2345. F. Gatta, M. R. del Giudice, and A. Borioni, *J. Heterocycl. Chem.*, 1993, **30**, 11.

2346. R. Dahiya, S. Kumar, and H. K. Pujari, *Indian J. Chem., Sect. B*, 1991, **30**, 256.

2347. B. Srivastava, J. S. Shukla, Y. S. Prabhakar, and A. K. Saxena, *Indian J. Chem., Sect. B*, 1991, **30**, 332.

2348. R. Sharma, S. Kumar, and H. K. Pujari, *Indian J. Chem., Sect. B*, 1991, **30**, 425.

2349. A. E. Ossman, M. M. El-Zahabi, A. E. El-Hakim, and A. N. Osman, *Egypt. J. Chem.*, 1989, **32**, 327; *Chem. Abstr.*, 1992, **117**, 212428.

2350. J. Y. Becker, E. Shakkour, and J. A. P. R. Sarma, *J. Org. Chem.*, 1992, **57**, 3716.

2351. B. Boutevin, L. R. Rasoloarijai, A. Rousseau, J. Garapon, and B. Sillion, *J. Fluorine Chem.*, 1992, **58**, 29.

2352. S. Saxena, M. Verma, A. K. Saxena, G. P. Gupta, and K. Shanker, *Indian J. Chem., Sect. B*, 1991, **30**, 453.

2353. R. S. Atkinson, P. J. Edwards, and G. A. Thomson, *J. Chem. Soc., Chem. Commun.*, 1992, 1256.

2354. P. Misra, P. N. Gupta, and A. K. Shakya, *J. Indian Chem. Soc.*, 1991, **68**, 618.

2355. J. Elguero, M. Garcia-Rodriguez, E. Gutiérrez-Puebla, A. de la Hoz, M. A. Monge, C. Pardo, and M. del M. Ramos, *J. Org. Chem.*, 1992, **57**, 4151.

2356. R. H. Khan, and R. C. Rastogi, *J. Chem. Res.*, 1992, *Synop.* 342.

2357. M. I. Ismail, *J. Chem. Soc., Perkin Trans. 2*, 1992, 585.

2358. K. Yoshida and M. Taguchi, *J. Chem. Soc., Perkin Trans. 1*, 1992, 919.

2359. P. Tardieu, R. Dubest, J. Aubard, A. Kellmann, F. Tfibel, A. Samat, and R. Guglielmetti, *Helv. Chim. Acta*, 1992, **75**, 1185.

2360. J. Fetter, T. Czuppon, G. Hornyák, and A. Feller, *Tetrahedron*, 1991, **47**, 9393.

2361. M. Al-Talib, J. C. Jochims, A. Hamed, Q. Wang, and A. El-H. Ismail, *Synthesis*, 1992, 697.

674 References

2362. C. W. Bird, *Tetrahedron*, 1992, **48**, 7857.

2363. Y. Kitahara, S. Nakahara, Y. Tanaka, and A. Kubo, *Heterocycles*, 1992, **34**, 1623.

2364. S. V. Borodaev, O. V. Zubkova, N. V. Shibaeva, A. P. Knyazev, A. O. Pyshchev, and S. M. Luk'yanov, *Zh. Org. Khim.*, 1991, **27**, 1986.

2365. B. Staskun and J. F. Wolfe, *S. Afr. J. Chem.*, 1992, **45**, 5.

2366. V. K. Ahluwalia, C. Gupta, and C. H. Khanduri, *Indian J. Chem., Sect. B*, 1992, **31**, 355; *Chem. Abstr.*, 1992, **117**, 90233.

2367. M. Dunkel and W. Pfleiderer, *Nucleosides Nucleotides*, 1992, **11**, 787; *Chem. Abstr.*, 1992, **117**, 90675.

2368. M. Yamagishi, M. Suzuki, and K. Matsumoto, *Chem. Express*, 1992, **7**, 381; *Chem. Abstr.*, 1992, **117**, 69335.

2369. A. F. El-Farargy, *An. Quim.*, 1991, **87**, 903; *Chem. Abstr.*, 1992, **117**, 26491.

2370. M. F. Ismail, A. M. A. El-Khamry, F. S. Sayed, and S. A. Emara, *Egypt. J. Chem.*, 1989, **32**, 433; *Chem. Abstr.*, 1992, **117**, 191799.

2371. T. J. Thornton, A. L. Jackman, P. R. Marsham, B. M. O'Connor, J. A. M. Bishop, and A. H. Calvert, *J. Med. Chem.*, 1992, **35**, 2321.

2372. M. J. Yu, J. R. McCowan, N. R. Mason, J. B. Deeter, and L. G. Mendelsohn, *J. Med. Chem.*, 1992, **35**, 2534.

2373. G. C. Crawley, R. L. Dowell, P. N. Edwards, S. J. Foster, R. M. McMillan, E. R. H. Walker, D. Waterson, T. G. C. Bird, P. Bruneau, and J.-M. Girodeau, *J. Med. Chem.*, 1992, **35**, 2600.

2374. J. B. Hynes, S. K. Singh, O. Fetzer, and B. Shane, *J. Med. Chem.*, 1992, **35**, 4078.

2375. H. Y. Hassan, A. A. Ismaiel, and H. A. H. El-Sherief, *Eur. J. Med. Chem.*, 1991, **26**, 743.

2376. H. Oelschläger, D. Martienssen, and F. Belal, *Arch. Pharm.* (*Weinheim, Ger.*), 1992, **325**, 503.

2377. A. O. Abdelhamid, S. S. Ghabrial, M. Y. Zaki, and N. A. Ramadan, *Arch. Pharm.* (*Weinheim, Ger.*), 1992, **325**, 205.

2378. M. F. El-Zohry, A. El-H. N. Ahmed, F. A. Omar, and M. A. Abd-Alla, *J. Chem. Technol. Biotechnol.*, 1992, **53**, 329.

2379. P. Mamalis and L. M. Werbel, *Handb. Exp. Pharmacol.*, 1984, **68**, Part 2, 417.

2380. D. Chakravarti, R. N. Chakravarti, L. A. Cohen, B. Dasgupta, S. Datta, and H. K. Miller, *Tetrahedron*, 1961, **16**, 224.

2381. L. Doub, L. M. Richardson, D. R. Herbst, M. L. Black, O. L. Stevenson, L. L. Bambas, G. P. Youmans, and A. S. Youmans, *J. Am. Chem. Soc.*, 1958, **80**, 2205.

2382. W. L. F. Armarego, *Adv. Heterocycl. Chem.*, 1979, **24**, 1.

2383. D. J. Brown, in *Comprehensive Heterocyclic Chemistry*, Vol. 3 A. J. Boulton and A. McKillop, (Eds.), Pergamon Press, Oxford 1984, p. 57.

2384. P. Pazdera, J. Meindl, and E. Nováček, *Chem. Pap.*, 1992, **46**, 322.

2385. B. Pramella, E. Rajanarender, and A. K. Murthy, *Indian J. Heterocycl. Chem.*, 1992, **2**, 115; *Chem. Abstr.*, 1993, **118**, 233996.

2386. I. M. El-Deen and S. El-Desuky, *J. Serb. Chem. Soc.*, 1992, **57**, 719; *Chem. Abstr.*, 1993, **118**, 254854.

2387. H. M. Hassan, Y. M. Darwish, M. M. Yousif, and O. M. O. Habib, *Rev. Roum. Chim.*, 1992, **37**, 473; *Chem. Abstr.*, 1993, **118**, 124482.

2388. A. R. R. Rao and V. M. Reddy, *Pharmazie*, 1992, **47**, 794; *Chem. Abstr.*, 1993, **118**, 124484.

2389. G. Fantin, M. Fogagnolo, A. Medici, and P. Pedrini, *J. Org. Chem.*, 1993, **58**, 741.

2390. R. Lakhan and M. Srivastava, *Proc. Indian Acad. Sci., Chem. Sci.*, 1993, **105**, 11.

2391. M. Tielemans, V. Areschka, J. Colomer, R. Promel, W. Langenaeker, and P. Geerlings, *Tetrahedron*, 1992, **48**, 10575.

2392. A. F. El-Farargy, M. M. Hamad, S. A. Said, and A. Haikal, *Egypt. J. Chem.*, 1990, **33**, 283; *Chem. Abstr.*, 1993, **118**, 124487.

2393. Y. A. Mohamed, M. A. E. Aziza, F. M. Salama, and A. M. Alafify, *J. Serb. Chem. Soc.*, 1992, **57**, 629; *Chem. Abstr.*, 1993, **118**, 101905.

2394. H. M. Hassan, Y. M. Darwish, M. M. Yousif, and O. M. O. Habib, *Rev. Roum. Chim.*, 1992, **37**, 903; *Chem. Abstr.*, 1993, **118**, 101912.

2395. A. F. El-Farargy, *Egypt. J. Pharm.*, 1991, **32**, 565; *Chem. Abstr.*, 1993, **118**, 59665.

2396. S. El-Desuky, I. M. El-Deen, and M. Abdel-Megid, *J. Serb. Chem. Soc.*, 1992, **57**, 513; *Chem. Abstr.*, 1993, **118**, 38866.

2397. R. S. Atkinson, B. J. Kelly, and J. Williams, *Tetrahedron*, 1992, **48**, 7713.

2398. J. Satyanarayana, K. R. Reddy, H. Ila, and H. Junjappa, *Tetrahedron Lett.*, 1992, **33**, 6173.

2399. A. Chibani, R. Hazard, and A. Tallec, *Bull. Soc. Chim. Fr.*, 1991, **128**, 814.

2400. H. A. Abdel-Hamid, S. A. Shiba, A. M. A. El-Khamry, and A. S. A. Youssef, *Phosphorus, Sulfur, Silicon, Relat. Elem.*, 1992, **72**, 237.

2401. P. K. Chaudhuri, *J. Nat. Prod.*, 1992, **55**, 249.

2402. J. V. Greenhill, I. Chaaban, and P. J. Steel, *J. Heterocycl. Chem.*, 1992, **29**, 1375.

2403. N. Yasue, S. Ishikawa, and M. Noguchi, *Bull. Chem. Soc. Jpn.*, 1992, **65**, 2845.

2404. A. Garcia-Martinez, A. Herrara-Fernández, D. Molero-Vilchez, M. Hanack, and I. R. Subramanian, *Synthesis*, 1992, 1053.

2405. M. S. C. Tasende, A. Sanchez, J. S. Casas, and J. Sordo, *Inorg. chim. Acta*, 1992, **201**, 35.

2406. M. Akazome, T. Kondo, and Y. Watanabe, *J. Org. Chem.*, 1993, **58**, 310.

2407. C. F. Beam, B. Kadhodyan, R. A. Taylor, and N. D. Heindel, *Synth. Commun.*, 1993, **23**, 237.

2408. S. El-Desuky, L. M. El-Deen, and M. Abdel-Megid, *J. Indian Chem. Soc.*, 1992, **69**, 340.

2409. J. L. Herndon, A. Ismaiel, S. P. Ingher, M. Teitler, and R. A. Glennon, *J. Med. Chem.*, 1992, **35**, 4903.

2410. H. Sugimoto, Y. Tsuchiya, H. Sugumi, K. Higurashi, N. Karibe, Y. Iimura, A. Sasaki, S. Araki, Y. Yamanishi, and K. Yamatsu, *J. Med. Chem.*, 1992, **35**, 4542.

2411. S. E. Webber, T. M. Bleckman, J. Attard, J. G. Deal, V. Kathardekar, K. M. Welsh, S. Webber, C. A. Janson, D. A. Mathews, W. W. Smith, S. T. Freer, S. R. Jordan, R. J. Bacquet, E. F. Howland, C. L. J. Booth, R. W. Ward, S. M. Hermann, J. White, C. A. Morse, J. A. Hilliard, and C. A. Bartlett, *J. Med. Chem.*, 1993, **36**, 733.

2412. D. Giardinà, U. Gulini, M. Massi, M. G. Piloni, P. Pompei, G. Rafaiani, and C. Melchiorre, *J. Med. Chem.*, 1993, **36**, 690.

2413. M. F. El-Zohry and M. A. Abd-Alla, *J. Chem. Technol. Biotechnol.*, 1992, **55**, 209.

2414. W. L. F. Armarego, *Quinazolines*, Wiley-Interscience, New York, 1967.

2415. C. Podesva, G. Kohan, and K. Vagi, *Can. J. Chem.*, 1969, **47**, 489.

2416. H. Schäfer, K. Gewald, and M. Seifert, *J. Prakt. Chem.*, 1976, **318**, 39.

2417. P. Pazdera, D. Ondràček, and E. Novàček, *Chem. Pap.*, 1989, **43**, 771.

2418. P. Pazdera and V. Potůček, *Chem. Pap.*, 1991, **45**, 77.

2419. D. P. Becker, P. M. Finnegan, and P. W. Collins, *Tetrahedron Lett.*, 1993, **34**, 1889.

2420. C. Crestini, R. Saladino, and R. Nicoletti, *Tetrahedron Lett.*, 1993, **34**, 1631.

2421. V. B. Gaur, V. H. Shah, and A. R. Parikh, *J. Inst. Chem. (India)*, 1991, **63**, 219; *Chem. Abstr.*, 1993, **119**, 49324.

2422. J. T. Hahn, J. Kant, F. D. Popp, S. R. Chhabra, and B. C. Uff, *J. Heterocycl. Chem.*, 1992, **29**, 1165.

2423. A. O. Abdelhamid, F. A. Khalifa, A. A. Fawz, and F. H. El-Shiaty, *Phosphorus Sulfur Silicon Relat. Elem.*, 1992, **72**, 135.

2424. S. Goto, H. Tsuboi, and K. Kagara, *Chem. Express*, 1993, **8**, 761; *Chem. Abstr.*, 1993, **119**, 228489.

2425. Y. Nagashima, T. Nagai, K. Shiomi, M. Tanaka, T. Taguchi, and Y. Shida, *Nippon Suisan Gakkaishi*, 1993, **59**, 1177; *Chem. Abstr.*, 1993, **119**, 243323.

2426. J. Imrich, T. Busova, D. Koscik, and T. Liptaj, *Chem. Pap.*, 1993, **47**, 102; *Chem. Abstr.*, 1993, **119**, 270502.

2427. A. R. Escobal, C. Iriondo, L. A. Berrueta, B. Gallo, and F. Vicente, *J. Chem. Res.*, 1993, *Synop.* 304, *Minipr.* 1937.

2428. D. Ranganathan, B. K. Patel, and R. K. Mishra, *J. Chem. Soc., Chem. Commun.*, 1993, 337.

2429. M. A. E. Shaban, M. A. M. Taha, and E. E. M. Sharshira, *Alexandria J. Pharm. Sci.*, 1992, **6**, 219; *Chem. Abstr.*, 1993, **119**, 95498.

2430. R. O. Dempcy and E. W. Skibo, *Bioorg. Med. Chem. Lett.*, 1992, **2**, 1427; *Chem. Abstr.*, 1993, **119**, 225900.

2431. E. A. Mohamed, L. M. El-Deen, and M. M. Ismail, *Pak. J. Ind. Res.*, 1992, **35**, 226; *Chem. Abstr.*, 1993, **119**, 8778.

2432. D. R. Boyd, N. D. Sharma, M. R. J. Dorrity, M. V. Hand, R. A. S. McMordie, J. F. Malone, H. P. Porter, H. Dalton, J. Chima, and G. N. Sheldrake. *J. Chem. Soc., Perkin Trans. 1*, 1993, 1065.

2433. S. I. Kovalenko, R. S. Siniak, I. A. Mazur, I. F. Belenichev, and P. N. Stebliuk, *Farm. Zh.* (*Kiev*), 1992 (No's. 5–6), 38; *Chem. Abstr.*, 1993, **119**, 8762.

2434. N. Kanomata, M. Igarshi, and M. Tada, *Heterocycles*, 1993, **36**, 1127.

2435. P. Canonne, M. Akssira, A. Dahbouh, H. Kasmi, and M. Boumzebra, *Heterocycles*, 1993, **36**, 1305.

2436. A. Kreutzberger and S. Balbach, *Pharmazie*, 1993, **48**, 17; *Chem. Abstr.*, 1993, **119**, 8769.

2437. M. Dunkel and W. Pfleiderer, *Nucleosides Nucleotides*, 1993, **12**, 125.

2438. N. P. Peet and E. W. Huber, *Heterocycles*, 1993, **35**, 315.

2439. J. Kaválek, V. Macháček, M. Sedlák, and V. Štérba, *Collect. Czech. Chem. Commun.*, 1993, **58**, 1122.

2440. J. P. Michael, *Nat. Prod. Res.*, 1993, **10**, 99.

2441. J. Kadoura, A. Chauvet, E. Torreilles, and J. Masse, *Thermochim. Acta*, 1993, **216**, 177; *Chem. Abstr.*, 1993, **119**, 116641.

2442. S. A. Angelos, D. C. Lankin, J. A. Meyers, and J. K. Raney, *J. Forensic Sci.*, 1993, **38**, 455.

2443. N. E. C. Duke and P. W. Codding, *Acta Crystallogr., Sect. B: Struct. Sci.*, 1993, **49**, 719.

2444. M. Himottu, K. Pihlaja, G. Stájer, and P. Vainiotola, *Rapid Commun. Mass Spectrom.*, 1993, **7**, 374.

2445. H. Grill and G. Kresze, *Liebigs Ann. Chem.*, 1971, **749**, 171.

2446. D. J. Brown, *The Pyrimidines*, Wiley, New York, 1994: (a) p. 117; (b) p. 266; (c) p. 270; (d) p. 474; (e) p. 531; (f) p. 292; (g) p. 303, 319.

2447. M. Bhalla, V. K. Srivastava, T. N. Bhalla, and K. Shanker, *Arzneim.-Forsch.*, 1993, **43**, 595.

2448. V. Y. Mitskyavichyus and R. S. Baltrushis, *Khim. Geterotsikl. Soedin.*, 1992, 1391.

2449. A. R. Escobal, C. Iriondo, L. A. Berrueta, B. Gallo, and F. Vicente, *J. Chem. Res.*, 1993, *Synop.* 304, *Minipr.* 1937.

2450. A. Hassner and C. Stumer, *Oramic Syntheses Based on Name Reactions and Unnamed Reactions*, Pergamon Press, Oxford, 1994; (a) p. 275; (b) p. 312; (c) p. 241; (d) p. 399; (e) p. 65; (f) p. 426; (g) p. 378; (h) p. 168; (i) p. 62; (j) p. 236; (k) p. 248.

2451. Y. Takase, T. Saeki, M. Fuji, and I. Saito, *J. Med. Chem.*, 1993, **36**, 3765.

2452. S. H. El-Feky, *Pharmazie*, 1993, **48**, 894; *Chem. Abstr.*, 1994, **120**, 270299.

2453. M. Hillebrand, E. Volanschi, F. Dimitru, L. Jitariu, I. Baciu, A. Ciobanu, and M. Straut, *An. Univ. Bucuresti, Chim.*, 1992, **1**, 43; *Chem. Abstr*, 1994, **120**, 270836.

2454. B. R. Modi, N. R. Desai, B. D. Mistry, and K. R. Desai, *Dyes Pigm.*, 1993, **23**, 25; *Chem. Abstr.*, 1994, **120**, 32840.

2455. M. Pink, J. Sieler, and M. Gutschow, *Z. Kristallogr.*, 1993, **207**, 316; *Chem. Abstr.*, 1994, **120**, 275744.

2456. I. M. El-Deen and S. M. Mohamed, *Indian J. Heterocycl. Chem.*, 1993, **2**, 233; *Chem. Abstr.*, 1994, **120**, 298578.

2457. R. N. Pandey, R. N. Sharma, A. N. Sahay, S. Kumar, S. Kumar, and S. Sing, *Asian J. Chem.*, 1993, **5**, 813; *Chem. Abstr.*, 1994, **120**, 314195.

2458. I. M. El-Deen, S. M. Mohamed, M. M. Ismail, and M. Abdel-Megid, *An. Quim.*, 1993, **89**, 621; *Chem. Abstr.*, 1994, **120**, 323471.

2459. R. N. Pandey, A. K. Sinha, R. N. Sharma, and R. K. Ranjan, *Asian J. Chem.*, 1994, **6**, 246; *Chem. Abstr.*, 1994, **120**, 337723.

2460. R. N. Pandey, R. S. P. Singh, A. N. Sahay, R. N. Sharma, and A. K. Sinha, *Asian J. Chem.*, 1994, **6**, 197; *Chem. Abstr.*, 1994, **120**, 337722.

2461. M. A. Likhate and P. S. Fernandes, *J. Indian Chem. Soc., Sect. B*, 1992, **69**, 667.

2462. R. S. Atkinson, P. M. Coogan, and C. L. Cornell, *J. Chem. Soc., Chem. Commun.*, 1993, 1215.

2463. M. B. Hogale and P. B. Chavan, *Collect. Czech. Chem. commun.*, 1993, **58**, 1705.

2464. Z. A. Hozien, A. L. H. N. Ahmed, and M. F. El-Zohry, *Collect. Czech. Chem. Commun.*, 1993, **58**, 1944.

2465. K. Hirota, H. Kuki, and Y. Maki, *Heterocycles*, 1994, **37**, 563.

2466. P. Victory, J. A. Borrell, A. Vidal-Ferran, E. Montenegro, and M. L. Jimeno, *Heterocycles*, 1963, **36**, 2273.

2467. C. C. Cheng, D. F. Liu, and T. C. Chou, *Heterocycles*, 1993, **35**, 775.

2468. A. Bischler and D. Barad, *Ber. Dtsch. Chem. Ges.*, 1892, **25**, 3080.

2469. J. Reisch, M. Iding, and C. O. Usifoh, *J. Heterocycl. Chem.*, 1993, **30**, 1117.

2470. I. Hermecz, I. Szilágyi, L. Örfi, J. Kökösi, and G. Szász, *J. Heterocycl. Chem.*, 1993, **30**, 1413.

2471. J. Elguero, E. Gutierrez-Puebla, A. Monge, C. Pardo, and M. Ramos, *Tetrahedron*, 1993, **49**, 11305.

2472. C. W. Bird, *Tetrahedron*, 1993, **49**, 8441.

2473. A. J. Hoefnagel, H. van Koningsveld, F. van Meurs, J. A. Peters, A. Sinnema, and H. van Bekkum, *Tetrahedron*, 1993, **49**, 6899.

2474. P. Molina, c. Conesa, A. Alias, A. Argues, M. D. Valesco, A. L. Llamas-Saiz, and C. Foces-Foces, *Tetrahedron*, 1993, **49**, 7599.

2475. B. Baudoin, Y. Ribeill, and N. Vicker, *Synth. Commun.*, 1993, **23**, 2833.

2476. L. D. Rowe, R. C. Beier, M. H. Elissalde, L. H. Stanker, and R. D. Stipanovic, *Synth. Commun.*, 1993, **23**, 2191.

2477. S. M. Luk'yanov, S. V. Borodaev, and O. V. Zubkova, *Zh. Org. Khim.*, 1992, **28**, 2577.

2478. E. Vanden, J. Jean, J. Godin, A. Mayence, A. Maquestiau, and E. Anders, *Synthesis*, 1993, 867.

2479. A. C. Tomé, P. M. O'Neill, R. C. Storr, and J. A. S. Cavaleiro, *Synlett*, 1993, 347, 397.

2480. Š. Stankovský, T. Dérer, and K. Špirková, *Monatsh. Chem.*, 1993, **124**, 733.

2481. J. Reisch, A. R. R. Rao, and C. O. Usifoh, *Monatsh. Chem.*, 1993, **124**, 1217.

2482. M. Hori and H. Ohtaka, *Chem. Pharm. Bull.*, 1993, **41**, 1114.

2483. F. J. Martinez-Martinez, A. Ariza-Castolo, V. Ramos-Nava, N. Barba-Behrens, and R. Contreras, *Magn. Reson. Chem.*, 1993, **31**, 832.

2484. K. Butler, M. W. Partridge, and J. A. Waite, *J. Chem. Soc.*, 1960, 4970.

678 References

2485. K. Špirkova, Š. Stankovský, and M. Dandárova, *Collect. Czech. Chem. Commun.*, 1994, **59**, 222.

2486. A. W. Erian, S. M. Sherif, A. Z. A. Alassar, and Y. M. Elkholy, *Tetrahedron*, 1994, **50**, 1877.

2487. P. Wormall and J. E. Gready, *Chem. Phys.*, 1993, **179**, 55.

2488. S. R. Stabler and X. Jahangir, *Synth. Commun.*, 1994, **24**, 123.

2489. J. R. Piper, C. A. Johnosn, J. A. Maddry, N. D. Malik, J. J. McGuire, G. M. Otter, and F. M. Sirotnak, *J. Med. Chem.*, 1993, **36**, 4161.

2490. L. Zacharia and M. S. Reddy, *Indian J. Chem., Sect. B*, 1993, **32**, 826; *Chem. Abstr.*, 1994, **120**, 270293.

2491. M. S. Amine, M. A. El-Hashash, and I. A. Attia, *Indian J. Chem., Sect. B*, 1993, **32**, 577; *Chem. Abstr.*, 1994, **120**, 217516.

2492. M. B. Hogale and P. B. Chavan, *Indian J. Chem., Sect. B*, 1993, **32**, 581; *Chem. Abstr.*, 1994, **120**, 217517.

2493. R. H. Kahn and R. C. Rastogi, *Indian J. Chem., Sect. B*, 1993, **32**, 596; *Chem. Abstr.*, 1994, **120**, 217518.

2494. A. Guijarro, D. J. Ramón, and M. Yus, *Tetrahedron*, 1993, **49**, 469.

2495. E. A. Mohamed, I. M. El-Deen, M. M. Ismail, and S. M. Mohamed, *Indian J. Chem., Sect. B*, 1993, **32**, 933; *Chem. Abstr.*, 1994, **120**, 244927.

2496. K. A. Suri, N. K. Satti, B. Mahajan, O. P. Suri, and K. L. Dhar, *Indian J. Chem., Sect. B*, 1993, **32**, 1171; *Chem. Abstr.*, 1994, **120**, 244948.

2497. S. E. de Laszlo, C. S. Quagliato, W. J. Greenlee, A. A. Patchett, R. S. L. Chang, V. J. Lotti, T. B. Chen, S. A. Scheck, K. A. Faust, S. S. Kiulign, T. S. Schorn, G. J. Zingaro, and P. K. S. Sieg, *J. Med. Chem.*, 1993, **36**, 3207.

2498. I. O. Edafiogho, J. A. Moore, V. A. Farrar, J. M. Nicholson, and K. R. Scott, *J. Pharm. Sci.*, 1994, **83**, 79.

2499. G. Bejeuhr, G. Blaschke, U. Holzgrabe, K. Mohr, U. Sürig, and G. Terfloth, *J. Pharm. Pharmacol.*, 1994, **46**, 108.

2500. J. Lessel, *Arch. Pharm. (Weinheim, Ger.)*, 1994, **327**, 77.

2501. M. I. Jaeda, Z. K. A. El-Samii, and A. Z. Britten, *J. Chem. Technol. Biotechnol.*, 1993, **58**, 391.

2502. G. Norcini, L. Allievi, G. Bertolini, C. Casagrande, G. Miragoli, F. Santangelo, and C. Semeraro, *Eur. J. Med. Chem.*, 1993, **28**, 505.

2503. A. R. R. Rao and V. M. Reddy, *Arzneim.-Forsch.*, 1993, **43**, 663.

2504. C. Anghel and M. Rosca, *Stud. Univ. Babes-Bolyai, Chem.*, 1990, **35**, 41; *Chem. Abstr.*, 1994, **120**, 30734.

2505. I. V. Ukrainets, S. G. Taran, P. A. Bezuglyi, S. N. Kovalenko, A. V. Turov, and N. A. Marusenko, *Khim. Geterotsikl. Soedin.*, 1993, 1223.

2506. M. A. Al-Talib, *J. Prakt. Chem.*, 1993, **335**, 711.

2507. M. A. Ashirmatov and K. M. Shkhidoyatov, *Khim. Geterotsikl. Soedin.*, 1993, 1246.

2508. H. Wamhoff, W. Wambach, S. Herrmann, M. Jansen, and B. Brühne, *J. Prakt. Chem.*, 1994, **336**, 129.

2509. F. J. C. Rossotti and H. S. Rossotti, *J. Chem. Soc.*, 1958, 1304.

2510. M. Uchida, T. Higashino, and E. Hayashi, *Shitsuryo Bunseki*, 1973, **21**, 245; *Chem. Abstr.*, 1974, **80**, 36408.

2511. M. Sarkar and D. P. Chakraborty, *Phytocheemistry*, 1977, **16**, 2007.

2512. R. S. Atkinson, E. Barker, C. J. Price, and D. R. Russell, *J. Chem. Soc., Chem. Commun.*, 1994, 1159.

2513. R. S. Atkinson, J. Fawsett, D. R. Russell, and P. J. Williams, *J. Chem. Soc., Chem. Commu.*, 1994, 2031.

2514. R. S. Atkinson and E. Barker, *J. Chem. Soc., Chem. Commun.*, 1995, 819.

2515. R. S. Atkinson (University of Leicester, England), personal communication, 1995.

2516. R. S. Atkinson, in *Azides and Nitrenes* (E. F. V. Scriven, Ed.), Academic Press, Orlando, FL, 1984, Chapter 5.

2517. J. E. G. Kemp, in *Comprehensive Organic Synthesis*, Vol. 7 (B. M. Trost, I. Fleming, and S. V. Ley, Eds.), Pergamon Press, Oxford, 1991, p. 480.

2518. P. R. Marsham, *J. Heterocycl. Chem.*, 1994, **31**, 603.

2519. H. T. Spenser, J. E. Villafranca, and J. E. Appleman, in *Chemistry and Biology of Pteridines and Folates* (J. E. Ayling, M. G. Nair, and C. M. Baugh, Eds.), Plenum, New York, 1993, p. 575.

2520. A. L. Jackman, G. M. F. Bisset, D. I. Jodrell, W. Gibson, R. Kimbell, V. Bavetsias, A. H. Calvert, H. R. Harrap, T. C. Stephens, M. N. Smith, and F. T. Boyle, in *Chemistry and Biology of Pteridines and Folates* (J. E. Ayling, M. G. Nair, and C. M. Baugh, Eds.), Plenum, New York, 1993, p. 579.

2521. F. T. Boyle, Z. S. Matusiak, L. R. Hughes, A. M. Slater, T. C. Stephens, M. N. Smith, M. Brown, R. Kimbell, and A. L. Jackman, in *Chemistry and Biology of Pteridines and Folates* (J. E. Ayling, M. G. Nair, and C. M. Baugh, Eds.), Plenum, New York, 1993, p. 585.

2522. T. C. Stephens, M. N. Smith, S. E. Waterman, M. L. McCloskey, A. L. Jackman, and F. T. Boyle, in *Chemistry and Biology of Pteridines and Folates* (J. E. Ayling, M. G. Nair, and C. M. Baugh, Eds.), Plenum, New York, 1993, p. 589.

2523. V. Bavetsias, A. L. Jackman, T. J. Thornton, K. Pawelczak, F. T. Boyle, and G. M. F. Bisset, in *Chemistry and Biology of Pteridines and Folates* (J. E. Ayling, M. G. Nair, and C. M. Baugh, Eds.), Plenum, New York, 1993, p. 593.

2524. R. Kimbell, A. L. Jackman, F. T. Boyle, A. Hardcastle, and W. Aherne, in *Chemistry and Biology of Pteridines and Folates* (J. E. Ayling, M. G. Nair, and C. M. Baugh, Eds.), Plenum, New York, 1993, p. 597.

2525. S. J. Clarke, A. L. Jackson, and I. R. Judson, in *Chemistry and Biology of Pteridines and Folates* (J. E. Ayling, M. G. Nair, and C. M. Baugh, Eds.), Plenum, New York, 1993, p. 601.

2526. K. Špirková, J. Horňaček, and Š. Stankovský, *Chem. Pap.*, 1993, **47**, 382.

2527. M. M. Hamad, S. A. Said, A. F. El-Farargy, and G. M. El-Gendy, *Pak. J. Sci. Ind. Res.*, 1993, **36**, 228; *Chem. Abstr.*, 1994, **121**, 35525.

2528. M. B. Deshmukh and D. S. Deshmukh, *Indian J. Heterocycl. Chem.*, 1994, **3**, 207.

2529. M. M. Hamad, S. A. Said, A. F. El-Farargy, and G. M. El-Gendy, *J. Bangladesh Chem. Soc.*, 1993, **6**, 73; *Chem. Abstr.*, 1994, **121**, 9341.

2530. M. Hori, K. Suzuki, T. Yamamoto, F. Nakajima, A. Ozaki, and H. Ohtaka, *Chem. Pharm. Bull.*, 1993, **41**, 1832.

2531. A. Wada, S. Hirai, and M. Hanaoka, *Chem. Pharm. Bull.*, 1994, **42**, 416.

2532. P. S. N. Reddy and A. K. D. Bhavani, *Indian J. Chem., Sect. B*, 1994, **33**, 683.

2533. G. Shailaja and P. S. N. Reddy, *Indian J. Chem., Sect. B*, 1994, **33**, 474.

2534. B. Narsaiah, A. Sivaprasad, and R. V. Venkataratnam, *J. Fluorine Chem.*, 1994, **66**, 47.

2535. P. M. Parasharya and A. R. Parikh, *J. Inst. Chem.* (*India*), 1992, **64**, 184; *Chem. Abstr.*, 1994, **121**, 108675.

2536. M. I. Al-Ashmawi, M. A. Shabaan, and R. H. Omah, *Zagazig J. Pharm. Sci.*, 1993, **2**, 158; *Chem. Abstr.*, 1994, **121**, 83267.

2537. Y. A. Ibrahim and A. H. M. Elwahy, *Heteroat. Chem.*, 1994, **5**, 97; *Chem. Abstr.*, 1994, **121**, 300861.

2538. T. Itoh, H. Hasegawa, K. Nagata, Y. Matsuya, and A. Ohsawa, *Heterocycles*, 1994, **37**, 709.

2539. A. Miyashita, T. Sasaki, E. Oishi, and T. Hagashino, *Heterocycles*, 1994, **37**, 823.

2540. R. N. Pandey and J. N. Das, *J. Indian Chem. Soc.*, 1994, **71**, 187; *Chem. Abstr.*, 1994, **121**, 214126.

2541. X. Y. Meng, J. Q. Liu, X. P. Zhang, X. H. Chen, A. Z. Yu, and Z. R. Dai, *Yaoxue Xuebuo*, 1994, **29**, 261; *Chem. Abstr.*, 1994, **121**, 230735.

2542. T. Iwakawa, T. Sato, and A. Murabayashi, *Heterocycles*, 1994, **38**, 1015.

2543. Z. M. Musaev, E. S. Yakubov, N. A. Parpiev, and H. M. Shakhidoyatov, *Dokl. Akad. Nauk Resp. Uzb.*, 1993 (No. 6), 32; *Chem. Abstr.*, 1994, **121**, 287500.

2544. I. Hermecz, J. Kökösi, B. Podányi, and G. Szász, *Heterocycles*, 1994, **37**, 903.

2545. J. Barluenga, M. Tomás, A. Ballesteros, and L. A. López, *Heterocycles*, 1994, **37**, 1109.

2546. R. E. Hackler, R. G. Suhr, J. J. Sheets, C. J. Hatton, P. L. Johnson, L. N. Davis, R. G. Edie, S. V. Kaster, G. P. Jourdan, J. L. Jackson, and E. V. Krumkalns, *Spec. Publ.—R. Soc. Chem.*, 1994, **147**, 70.

2547. M. M. Burbuliene and P. Vainilavicius, *Chemija*, 1993 (No. 2), 61; *Chem. Abstr.*, 1994, **121**, 300844.

2548. A. Garcia-Martinez, A. Herrara-Fernández, F. Moreno-Jiménez, M. J. Luengo-Fraile, and L. R. Subramanian, *Synlett*, 1944, 559.

2549. M. Deshmukh, P. Chavan, and D. Kharade, *Monatsh. Chem.*, 1994, **125**, 743.

2550. J. L. Chicharro, P. Prados, and J. Mendoza, *J. Chem. Soc., Chem. Commun.*, 1994, 1193.

2551. M. Bodajla, Š. Stankovský, and K. Špirková, *Collect. Czech. Chem. Commun.*, 1994, **59**, 1463.

2552. M. Mali, *Proc. Indian Natl. Sci. Acad., Part A*, 1994, **60**, 497; *Chem. Abstr.*, 1994, **121**, 300848.

2553. J. L. Romine, S. W. Martin, N. A. Meanwell, and J. R. Epperson, *Synthesis*, 1994, 846.

2554. G. Shailaja and P. S. N. Reddy, *Indian J. Chem., Sect. B*, 1994, **33**, 321.

2555. B. Gardner, A. J. S. Kanagasooriam, R. M. Smyth, and A. Williams, *J. Org. Chem.*, 1994, **59**, 6245.

2556. N. Nishiwaki, T. Matsunaga, Y. Tohda, and M. Ariga, *Heterocycles*, 1994, **38**, 249.

2557. J. P. Michael, *Nat. Prod. Rep.*, 1994, **11**, 163.

2558. C. Claudia, E. Micione, R. Saladino, and R. Nicoletti, *Tetrahedron*, 1994, **50**, 3259.

2559. J. Lessel, *Arch. Pharm. (Weinheim, Ger.)*, 1994, **327**, 329.

2560. J. Lessel, *Arch. Pharm. (Weinheim, Ger.)*, 1994, **327**, 571.

2561. M. A. Khalil, R. Soliman, A. M. Farghaly, and A. A. Bekhit, *Arch. Pharm. (Weinheim, Ger.)*, 1994, **327**, 27.

2562. M. H. Norman, G. C. Rigdon, F. Navas, and B. R. Cooper, *J. Med. Chem.*, 1994, **37**, 2552.

2563. T. J. Tucker, T. A. Lyle, C. M. Wiscount, S. F. Britcher, S. D. Young, W. M. Sanders, W. C. Lumma, M. E. Goldman, J. A. O'Brien, R. G. Ball, C. F. Homnick, W. A. Schleif, E. A. Emini, J. R. Huff, and P. S. Anderson, *J. Med. Chem.*, 1994, **37**, 2437.

2564. Y. Takase, T. Saeki, N. Watanabe, H. Adachi, S. Souda, and I. Saito, *J. Med. Chem.*, 1994, **37**, 2106.

2565. A. Villalobos, J. F. Blake, C. K. Biggers, T. W. Butler, D. S. Chapin, Y. L. Chen, J. L. Ives, S. B. Jones, D. R. Liston, A. N. Arthur, D. M. Nason, J. A. Nielsen, I. A. Shalaby, and W. F. White, *J. Med. Chem.*, 1994, **37**, 2721.

2566. A. Gangjee, N. Zaveri, S. F. Queener, and R. L. Kisliuk, *J. Heterocycl. Chem.*, 1995, **32**, 243.

2567. R. A. Conley, D. L. Barton, S. M. Stefanik, M. M. Lam, G. C. Lindaberg, C. F. Kasulanis, S. Cesco-Cancian, S. Currey, A. C. Fabian, and S. D. Levine, *J. Heterocycl. Chem.*, 1995, **32**, 761.

568. C. Nielsen, *J. Heterocycl. Chem.*, 1995, **32**, 719.

2569. J. B. Hynes, J. P. Campbell, and J. D. Hynes, *J. Heterocycl. Chem.*, 1995, **32**, 1185.

2570. S. Franco, E. Meléndez, and F. L. Merchán, *J. Heterocycl. Chem.*, 1995, **32**, 1181.

2571. R. Bossio, S. Marcaccini, and R. Pepino, *J. Heterocycl. Chem.*, 1995, **32**, 1115.

2572. I. V. Ukrainets, S. G. Taran, O. V. Gorohova, P. A. Bezuglyi, and A. V. Turov, *Khim. Geterotsikl. Soedin.*, 1994, 225.

2573. M. Chakrabarty, A. Batabyal, M. S. Morales-Rios, and P. Joseph-Nathan, *Monatsh. Chem.*, 1995, **126**, 789.

2574. J. P. Demers, *Tetrahedron Lett.*, 1994, **35**, 6425.

2575. I. N. Houpis, A. Molina, A. W. Douglas, L. Xavier, J. Lynch, R. P. Volante, and P. J. Reider, *Tetrahedron Lett.*, 1994, **35**, 6811.

2576. A. Oishi, M. Yasumoto, M. Goto, T. Tsuchiya, I. Shibuya, and Y. Taguchi, *Heterocycles*, 1994, **38**, 2073.

2577. D. V. Ramana and E. Kantharaj, *Indian J. Heterocycl. Chem.*, 1994, **3**, 215; *Chem. Abstr.*, 1995, **122**, 81279.

2578. A. Strakovs, A. Krasnova, V. Aleksandrov, and M. Petrova, *Latv. Kim. Z.*, 1994, 106; *Chem. Abstr.*, 1995, **122**, 9983.

2579. K. M. Shakhidoyatov and B. A. Urakov, *Dokl. Akad. Nauk Resp. Uzb.*, 1993 (No. 3), 31; *Chem. Abstr.*, 1995, **122**, 31449.

2580. J. Kalinowska-Torz, *Acta Pol. Pharm.*, 1994, **51**, 249; *Chem Abstr.*, 1995, **122**, 187525.

2581. B. R. Shah, J. J. Bhatt, H. H. Patel, N. K. Undavia, P. B. Trivedi, and N. C. Desai, *Indian J. Chem., Sect. B*, 1995, **34**, 201.

2582. I. V. Ukrainets, O. V. Gorohova, S. G. Taran, P. A. Bexuglyi, and A. V. Turov, *Khim. Geterotsikl. Soedin.* 1994, 229.

2583. S. Singh, U. Dave, and A. R. Parikh, *J. Inst. Chem. (India)*, 1993, 65, 97; *Chem. Abstr.*, 1995, **122**, 31447.

2584. S. Bell, A. Bonadio, and K. G. Watson, *Aust. J. Chem.*, 1995, **48**, 227.

2585. M. A. Huffman, N. Yasuda, A. E. de Camp, and E. J. J. Grabowski, *J. Org. Chem.*, 1995, **60**, 1590.

2586. V. A. Dorokhov and M. A. Present, *Izv. Akad. Nauk, Ser. Khim.*, 1994, 888.

2587. M. E. Abd El-Fattah, *Indian J. Heterocycl. Chem.*, 1995, **4**, 199; *Chem. Abstr.*, 1995, **123**, 83306.

2588. K. P. Chan and A. S. Hay, *J. Org. Chem.*, 1995, **60**, 3131.

2589. M. Langlois, J. L. Soulier, V. Rampillon, C. Gallais, B. Brémont, S. Shen, D. Yang, A. Giudice, and F. Sureau, *Eur. J. Med. Chem.*, 1994, **29**, 925.

2590. A. Rosowsky, C. E. Mota, J. E. Wright, and S. F. Queener, *J. Med. Chem.*, 1994, **37**, 4522.

2591. A. R. El-Naser Ossman and S. El-Sayed Barakat, *Arzneim.-Forsch.*, 1994, **44**, 915.

2592. R. S. Atkinson, E. Barker, P. J. Edwards, and G. A. Thomson, *J. Chem. Soc., Chem. Commun.*, 1995, 727.

2593. K. Smith, G. A. El-Hiti, M. A. Abdo, and M. F. Abdel-Megeed, *J. Chem. Soc., Perkin Trans. 1*, 1995, 1029.

2594. R. S. Atkinson, J. Fawcett, D. R. Russell, and P. J. Williams, *Tetrahedron Lett.*, 1995, **36**, 3241.

2595. R. J. Ife, T. H. Brown, P. Blurton, D. J. Keeling, C. A. Leach, M. L. Meeson, M. E. Parsons, and C. J. Theobald, *J. Med. Chem.*, 1995, **38**, 2763.

2596. J. Lessel, *Arch. Pharm. (Weinheim, Ger.)*, 1995, **328**, 397.

2597. A. Rosowsky, C. E. Mota, S. F. Queener, M. Waltham, E. Ereikan-Abali, and J. R. Bertino, *J. Med. Chem.*, 1995, **38**, 745.

Index

This index covers the text but not the Appendix (List of Simple Quinazolines), which is alphabetically self indexing.

The page number(s) immediately following each primary entry refer to synthesis or general information. Although each number indicates that the subject is treated on that and possibly subsequent pages, the actual word(s) of the primary entry may appear there only in part, as a synonym, as a formula number, or even simply by inference.

The few authors who are mentioned by name in the text (usually in connection with some historical contribution or as pioneers of named reactions) are included in this index.